MATHEMATICS

FOR ENGINEERS AND SCIENTISTS

SIXTH EDITION

MATHEMATICS
FOR ENGINEERS AND SCIENTISTS

SIXTH EDITION

Alan Jeffrey

CHAPMAN & HALL/CRC

A CRC Press Company
Boca Raton London New York Washington, D.C.

Library of Congress Cataloging-in-Publication Data

Jeffrey, Alan.
Mathematics for engineers and scientists / A. Jeffrey.—6th ed.
p. cm.
Includes index.
ISBN 1-58488-488-6 (alk. paper)
1. Mathematics. 2. Mathematical analysis. 3. Engineering mathematics. 4. Science—Mathematics. I. Title.
QA37.3.J44 2004
510—dc22

2004050154

Visit the CRC Press Web site at www.crcpress.com

International Standard Book Number 1-58488-488-6
Library of Congress Card Number 2004050154
Printed in the United States of America 2 3 4 5 6 7 8 9 0
Printed on acid-free paper

Contents

Preface to the sixth edition

A sixth edition of this book has become necessary because of the many changes to syllabuses that have occurred since the fifth edition was produced, and also because of the way the computer now influences both the application of mathematics and its teaching. Material in some chapters has been simplified for ease of understanding, and supplementary background material has been added wherever it seems likely to be helpful. In addition, many sections have been revised, the original chapter on functions of a complex variable has been removed, with a little of the material having been relocated, the chapter on Fourier series has also been simplified, and much new material has been added.

An important difference between this sixth edition and the previous one is the provision of the new Chapter 19 dealing with symbolic algebraic manipulation by computer software. This refers to readily available software that enables a computer to act in some ways like a person with pencil and paper, by making it possible for a computer to perform symbolic operations like differentiation, integration and finding the general solution of a differential equation, and then to present the results in symbolic form. Symbolic software is also capable of acting like an ordinary computer, because numerical values can be assigned to symbolic parameters, after which the computer can display any symbolic results it has produced in graphical form. The two examples of such software considered in Chapter 19 are called MAPLE and MATLAB, where each of these names is the registered trademark of a software company quoted in that chapter.

It is hoped that whenever this software is available, the new chapter will encourage readers to take advantage of it, to experiment with it, and to learn more about mathematical functions and the operations that can be performed upon them. MAPLE software was used to prepare some of the new material in this book. A reader should apply the software to some of the problems at the ends of chapters, though only to check the hand calculations that should first be carried out since they are essential when first learning about mathematics. Once the use of the software has been mastered, its power is such that it can then be applied to solve realistic problems encountered in other courses which, if attempted by hand, could result in prohibitively long calculations.

As in previous editions, answers to odd numbered problems are given at the end of the book, some being given in more detail than others if a problem might cause difficulty.

Alan Jeffrey
Newcastle upon Tyne

Preface to the first edition

This book has evolved from an introductory course in mathematics given to engineering students at the University of Newcastle-upon-Tyne during the last few years. It represents the author's attempt to offer the engineering student, and the science student who is not majoring in a mathematical aspect of this subject, a broad and modern account of those parts of mathematics that are finding increasingly important application in the everyday development of his subject.

Although this book does not seek to teach any of the many physical disciplines to which its results and methods may be applied, it nevertheless makes free use of them for purposes of illustration whenever this seems to be helpful. Every effort has been made to integrate the various chapters into a description of mathematics as a single subject, and not as a collection of seemingly unrelated topics. Thus, for example, matrices are not only introduced in an algebraic context, but they are also related in other chapters to change of variables in partial differentiation and to the study of simultaneous differential equations.

Modern notation and terminology have been used freely but, it is hoped, never to the point of becoming pedantic when a simple word or phrase seems more natural. Of necessity, much of the material in this book is standard, though the emphasis and manner of introduction and presentation frequently differs from that found elsewhere. This is deliberate, and is a reflection of the changing importance of mathematical topics in engineering and science today.

In many introductory mathematics texts for engineering and science students no serious attempt is made to offer reasonable proofs of main results and, instead, attention is largely confined to their manipulation. Important though this aspect undoubtedly is, it is the author's belief that knowledge of the proof of a result is often as essential as its subsequent application, and that the modern student needs and merits both. With this thought in mind proofs of results have always been included, and, though they have been kept as simple as possible, no attempt has been made to conceal difficulty where it exists. Only very occasionally, when the proof of a result is lengthy, and its details are largely irrelevant to the subsequent development of the argument, has the treatment been shortened to a summary of the logical steps involved. Even then the interested reader can often find more relevant information amongst the specially selected problems at the end of each chapter.

As implied by the previous remark, the many problems not only comprise those offering manipulative exercise, but also those shedding further light on topics only touched upon in the main text. No serious student can progress in his knowledge of this subject without a proper investment of time and effort spent working at a selection of these problems. The main text is provided with numerous illustrative examples designed to be helpful both when working through the text and when attempting the classified problems. It is hoped that their inclusion also makes the book suitable for private study.

The wide range of material covered in this book represents rather more than would normally be contained in an introductory course of lectures. Whilst allowing for changing approaches in teaching, this fact also permits some flexibility in use of the material and at the same time offers further relevant reading to the ambitious student. In addition to the author's own experience of the application of mathematics in engineering and science, the choice and style of presentation of material has been influenced by two recently published documents: the Council of Engineering Institutions syllabuses in mathematics in Britain and the CUPM recommendations made by the Mathematical Association of America. It is the author's hope that this book complies fully with the former document and with the spirit of the latter insofar as its recommendations are applicable to engineering and science students.

The material has all been class-tested and, as a result, has undergone considerable modification from its first appearance as lecture notes to the form of presentation adopted here. It is a pleasure to acknowledge the help of the publishers who have given me continued encouragement and every possible form of assistance throughout the entire period of preparation of the book.

A.J.

Supplementary computer problems

The sixth edition retains the sections entitled '*Supplementary computer problems*' that were added to the fifth addition. The purpose of problems in these sections is to reinforce understanding of the mathematical content of chapters, to encourage the use of computers in mathematics where this is likely to be helpful, and in some cases to allow the reader to explore the mathematical capabilities of methods that would be impossible without the aid of a computer. In particular, the ease with which computers can display results in graphical form is likely to be of considerable help when seeking to understand the significance of results, and this aspect has been emphasized in the computer problems.

Where problems are not of a routine nature designed to illustrate fundamental mathematical ideas, or in the case of numerical methods to illustrate the difficulties that can arise if they are applied without a proper understanding of what is involved, they are taken from the application of mathematics to typical real-life situations. Examples of this nature are the use of quadratic spline functions to fit a smooth interpolation curve to data known only at discrete points, and the effect of matrix transformations on geometrical shapes, both of which topics have come about largely as a result of the demands made by computer-aided design (CAD). The introduction to the numerical solution of the Laplace equation, when coupled with the use of a computer, enables readers to solve a variety of realistic engineering and scientific problems that are likely to arise in other parts of their courses.

In order to solve the supplementary computer problems it is not necessary for the reader to be able to write programs in a high-level programming language, because many simple and easy-to-use software packages are now readily available in computer laboratories. These packages either provide ready-made routines that carry out the necessary standard mathematical operations, or they can be programmed to do so using a notation that is almost equivalent to the mathematical expression that needs to be implemented. The computer problems can also be solved by using either of the software packages MAPLE or METLAB described in Chapter 19.

Although the more sophisticated programs allow operations such as the determination of a root to be carried out by almost a single key stroke, readers will be likely to learn more by using a simpler program in which it is necessary to piece together by themselves the mathematical operations needed to achieve the required result. When working on some of the more open-ended problems the reader is encouraged to experiment, both to find ways in which to widen the applicability of the method and to determine its limitations. A typical example here might be to use the bisection method to locate an approximate root of a function, and then switch to the Newton–Raphson method in order to accelerate the rate of convergence.

The use of a computer cannot replace a proper understanding of mathematical analysis, but when used wisely it can do much to aid the learning process and to enable readers to be able to solve many real-life problems of a mathematical nature at a very early stage of their studies.

1 Numbers, trigonometric functions and coordinate geometry

The topics reviewed in this chapter represent the essential mathematical prerequisites necessary for a proper understanding of the remainder of the book. Applications of these underlying concepts arise repeatedly, and in many different contexts. The material covered is necessarily of a somewhat diverse nature and not all of it is closely related.

The chapter starts with an account of the real number system, together with its arithmetic laws and their consequences, because these represent the basis for the development of the theory of functions of one or more real variables. Inequalities are introduced because they serve both to define intervals on a line and regions in space when working with functions and, also in connection with functions, to make precise the meaning of analytical concepts like limit and continuity.

The technique of mathematical induction is included because it finds applications throughout mathematics. It enables a conjecture concerning the form taken by a general mathematical proposition depending on an integer n to be tested and found to be either true or false. Thus, for example, mathematical induction can be used to verify or reject an expression for the form taken by the general term in a series that has been found by observing the pattern of a few successive terms, to check the correctness of a closed-form expression for the sum of a finite numerical series that has been arrived at intuitively, or to prove the binomial theorem.

Trigonometric functions and their associated identities are reviewed because they belong to the group of elementary functions on which most of mathematics is based, and because of the need to use their properties in most chapters of the book.

A knowledge of the elements of rectangular Cartesian coordinate geometry and of polar coordinates is essential for an understanding of the geometrical implications of many properties of functions, and also for applications of mathematics to physical problems. Discussions of the effect of a rotation or a shift of origin on a coordinate system, and of a transformation from Cartesian to polar coordinates, are included because changes of this type often simplify problems.

The elementary but very useful algebraic operation of completing the square is described because of its importance when integrating rational functions and inverting Laplace transforms. In conclusion, to familiarize the reader with the Greek characters that are used extensively throughout mathematics, the final section lists the Greek alphabet and names each character.

1.1 Sets and numbers

In applications of mathematics to engineering and science, we use the properties of real numbers. Many of these properties are intuitively obvious, but others are more subtle and depend for their correct use on a proper understanding of the mathematical basis for what is called the *real number system*. This section describes the elements of the real number system in a straightforward manner for subsequent use throughout the book.

The reader will know how to work with finite combinations of numbers, but what is less likely is that he or she will be familiar with the interpretation and use of limiting processes. For example, what is the meaning and what, if any, is the value to be associated with the limit

$$\lim_{n\to\infty}\left[\left(1+\frac{1}{n}\right)^n\right],$$

which is to be interpreted as the value approached by the expression in square brackets as the integer n becomes arbitrarily large?

It was questions such as these and, indeed, far simpler ones that first led to the introduction of the real number system as it is now, and to the study of real numbers. Many properties of numbers, nowadays accepted by all as self-evident, were once regarded as questionable. This is still readily apparent from much of the notation that is in current use.

Thus, for example, although fractions m/n, with m and n positive integers, were accepted long ago and were called *rational* numbers, the fact that $\sqrt{2}$ cannot be expressed in this form led to its being termed an *irrational* number. Even more extreme is the term *imaginary* number that is given to the number $i=\sqrt{-1}$, which first arose when seeking the solution of the simple quadratic equation $x^2+1=0$. However, as we shall see later, although this number does not belong to the real number system and so merits special consideration, it is in fact no more imaginary than the integer 2.

Experience suggests that in any systematic development of the properties of the real number system, the operations of addition and multiplication must play a fundamental role. These conjectures are of course true, but underlying the concepts of real numbers and their algebraic manipulation are the even more fundamental concepts of sets and their associated algebra. Later we will consider sets and their algebra in a little detail, but for the moment we will confine our discussion to sets of numbers.

First, though, we must define the term *set*, for which the alternative terms *aggregate*, *class* and *collection* are also often used. The very general nature of the concept of a set will be seen from the definition of a set and from the examples that follow. Our approach will be direct and pragmatic, and we shall agree that a set comprises a collection of *objects* or *elements*, each of which is chosen for membership of the set because it possesses some required property. Membership of the set is determined entirely by this property; an element only belongs to the set if it possesses the required property, otherwise it does not belong to the set. The properties of membership and non-membership of a set are mutually exclusive, though the order in which elements appear in a set is unimportant.

The elements of a set can be perfectly arbitrary provided they satisfy the membership criterion of the set, as may be seen from the following examples of different types of set:

(a) all people with brown eyes,
(b) all rectangles that can be inscribed in a circle of radius 2,
(c) all rational numbers of the form m/n with $m = 1$, 2 or 3 and $n = 17$, 19 or 31,
(d) all quadratic equations,
(e) all integers strictly between 3 and 27,
(f) all entrants to a tennis tournament aged 30 years or less.

The number of elements in each of sets (a), (c), (e) and (f) is finite, whereas in sets (b) and (d) the number of elements is infinite.

An important numerical set which we shall often have occasion to use is the set of positive integers 1, 2, 3, . . ., used in counting. These are called the *natural* numbers, and they are denoted by the symbol **N**. Notice that there can be no greatest member m of this set, because however large m may be, $m+1$ is larger and yet it is also a member of set **N**. Accordingly, when we use a number m that is allowed to increase without restriction, it will be convenient to indicate this by saying that 'm tends to *infinity*', and to write this statement in the form $m \to \infty$. It is important to recognize that infinity is not a number in the usual sense, but just the outcome of the mathematical process in which m is allowed to increase indefinitely (without *bound*). It is always necessary to relate the symbol ∞ to some mathematical limiting expression, because by itself it has little or no meaning.

When we wish to enumerate (display) the elements of set **N** of natural numbers we write

$$\mathbf{N} = \{1, 2, 3, \ldots\},$$

with the understanding that the dots following the integer 3 signify that the sequence of natural numbers increases indefinitely. This set of three dots, called an ellipsis, is used throughout mathematics to indicate the omission of terms, that in this case are integers.

N is only one important type of numerical set, however, and we have already seen examples of sets whose elements are not numerical. Other examples of non-numerical sets are provided by statistical situations which can give rise to sets of *events*, such as a sequence of heads or tails when tossing a coin, and in the analysis of logic circuits used in computers which involve sets of *decisions*.

To simplify the manipulation of these ideas we need to introduce notations for sets, for the elements of a set and for membership of a set. It is customary to denote sets themselves by upper-case letters $A, B, \ldots, S, \ldots$, and the general elements of these sets by the corresponding lower-case letters $a, b, \ldots, s, \ldots$. If a is an element (member) of set A, we use the symbol $\in$ as an abbreviation to denote membership and write

$$a \in A,$$

which is usually read 'a is an element of A'. Conversely, if a is *not* an element of A we denote this by negating the membership symbol $\in$ by writing instead $\notin$. Thus if a is not a member of set A we write

$$a \notin A.$$

In this notation we have $3 \in \mathbf{N}$, but $\pi \notin \mathrm{N}$ where $\pi = 3.14159...$, while if m and n are positive integers $m/n \in \mathrm{N}$ only if m is exactly divisible by n.

If a set only contains a small number of elements it is often simplest to define it explicitly by enumerating them. Hence, for a set S comprising the four integer elements 3, 4, 5 and 6 we would write $S = \{3, 4, 5, 6\}$. Whereas this set is a *finite* set, in the sense that it only contains a finite number of elements, the set **N** is seen to be an *infinite* set.

It is useful to have a concise notation which indicates the membership criterion that is to be used for a set. Thus, if we were interested in the set **B** of positive integers n whose squares lie strictly between 15 and 39, we would write

$$\mathbf{B} = \{n | n \in \mathbf{N},\ 15 < n^2 < 39\},$$

where we have used the convention that the symbol to the left of the vertical rule signifies the general element of the set in question, while the expressions to the right of the rule express the membership criteria for the set. When expressing the membership criterion $15 < n^2 < 39$, the symbol $<$ used in conjunction with the numbers a and b in the form $a < b$ is to be read as 'a is less than b' or, equivalently, $b - a$ is positive. This last statement can also be written as $b - a > 0$, where the symbol $>$ when used in conjunction with the numbers c and d in the form $c > d$ is to be read as 'c is greater than d'.

Another important set that is frequently used is the set of *ordered pairs*. An element of a set of ordered pairs of numbers a and b will be written (a, b). The ordered pair (a, b) is different from the ordered pair (b, a) unless a and b are equal. A frequent use of this notation is for identifying a point on a graph, when in Cartesian coordinates the first element is the x-coordinate and the second is the y-coordinate of a point on the graph. Thus in set notation the graph of the function $y = f(x)$, for which x lies between a and b with $a < b$, which can also be written $a < x < b$, becomes

$$S = \{(x, f(x)) | a < x < b\}.$$

Similarly, *ordered triples* of numbers p, q and r, written (p, q, r), can be used to identify a point in three-dimensional Cartesian coordinates by taking p as the x-coordinate, q as the y-coordinate and r as the z-coordinate of a point in space. Here also ordering the numbers differently corresponds to identifying a different point in space and, for example, the ordered triples (p, q, r) and (r, q, p) will only be equal if $p=r$.

1.2 Integers, rationals and arithmetic laws

The reader will already be familiar with the fact that if the arithmetic operation of addition is performed on the natural (positive) numbers the result will be another natural number. Written symbolically, this statement becomes $a, b \in \mathbf{N} \Rightarrow (a+b) \in \mathbf{N}$, where the symbol $\Rightarrow$ may be read as 'the result to the left implies the result to the right'. However, the arithmetic operation of subtraction is less simple: If $a, b \in \mathbf{N}$ it is not necessarily true that $a-b \in \mathbf{N}$, because $a-b$ may be negative or zero.

Thus an attempt always to express the result of subtraction of natural numbers as natural numbers themselves must fail. This fact is usually expressed by saying that the system of natural numbers is not *closed* with respect to subtraction. The difficulty is, of course, resolved by supplementing the set $\mathbf{N}$ of natural numbers by the set $\mathbf{N}^* = \{\ldots, -3, -2, -1, 0\}$ of negative integers and zero. If, now, we combine the sets $\mathbf{N}$ and $\mathbf{N}^*$ to form the new set $\mathbf{Z} = \{\ldots, -3, -2, -1, 0, 1, 2, 3, \ldots\}$, the statements $a, b \in \mathbf{Z} \Rightarrow (a+b) \in \mathbf{Z}$ and $(a-b) \in \mathbf{Z}$ become unconditionally true.

The need to generalize numbers still further by extending the set $\mathbf{Z}$ is easily seen when division is involved, because the quotient a/b will only belong to $\mathbf{Z}$ if $a=kb$, where $k \in \mathbf{Z}$. Thus we see that the set $\mathbf{Z}$ is *not* closed with respect to division. To remedy this we further supplement the set $\mathbf{Z}$ by appending to it the rational numbers a/b, where $a, b \in \mathbf{Z}$. As the number 2 is represented by the rational numbers 2/1, 6/3, 10/5, . . ., as indeed are all other positive and negative integers, it follows that the set of rational numbers $\mathbf{Q}$ contains every element of $\mathbf{Z}$.

Numerous though the rational numbers obviously are, we now show that they may be arranged in a definite order and counted so that in fact, surprising as it may seem, they are no more numerous than the natural numbers. One way to show this is indicated in the following array which recognizes as different all rational representations in which cancelling of common factors has not been performed. Thus, for example, in this scheme 4/2, 6/3, 8/4, . . ., are counted as different rational numbers, despite the fact that they all represent 2. If desired these repetitions may be omitted from the resulting sequence of rational numbers, though the matter is not important. The counting or enumeration of the rationals proceeds in the order indicated by the arrows:

$$
\begin{array}{ccccccccccccccc}
 & & & & & & & 0 & & & & & & & \\
 & & & & & & & \searrow & & & & & & & \\
\cdots & \frac{-1}{3} & & \frac{-1}{2} & \leftarrow & \frac{-1}{1} & & \frac{1}{1} & \rightarrow & \frac{1}{2} & & \frac{1}{3} & \rightarrow & \frac{1}{4} & \cdots \\
 & \uparrow & & \downarrow & & \uparrow & & & & \downarrow & & \uparrow & & \downarrow & \\
\cdots & \frac{-2}{3} & & \frac{-2}{2} & & \frac{-2}{1} & \leftarrow & \frac{2}{1} & \leftarrow & \frac{2}{2} & & \frac{2}{3} & & \frac{2}{4} & \cdots \\
 & \uparrow & & \downarrow & & & & & & & & \uparrow & & \downarrow & \\
\cdots & \frac{-3}{3} & & \frac{-3}{2} & \rightarrow & \frac{-3}{1} & \rightarrow & \frac{3}{1} & \rightarrow & \frac{3}{2} & \rightarrow & \frac{3}{3} & & \frac{3}{4} & \cdots \\
\cdots & \vdots & & \vdots & & \vdots & & \vdots & & \vdots & & \vdots & & \vdots & \cdots
\end{array}
$$

If this form of counting is adopted then, after cancellations, the first few rationals to be specified are

$$0, 1, 1/2, 1, 2, -2, -1, -1/2, -1, -3/2, -3, \ldots .$$

If the repetitions are deleted the modified sequence becomes

$$0, 1, 1/2, 2, -2, -1, -1/2, -3/2, -3, \ldots .$$

Clearly all rationals are included somewhere in this scheme, so starting to count from the first term in the sequence, which is zero, we see that each rational may be put into coincidence with a natural number. This justifies our assertion that the rationals are countable.

At first sight it might seem that the rational numbers comprising the set **Q** represent all possible numbers, but in fact this is far from the truth since it is possible to show that other numbers exist that are infinitely more numerous than the rationals. These are the *irrational* numbers mentioned earlier. We will show the existence of only one such number.

Let us show that $\sqrt{2}$ is irrational or, to phrase this more precisely, that there is no rational number of which the square is 2. The argument is simple, starting from a given assumption and leading to a contradiction, thereby showing that the initial assumption must be false. Such an approach is called **proof by contradiction**, and it is a device that is frequently used in mathematics.

Suppose, if possible, that m/n, with m and n natural numbers having no common factor, is such that $(m/n)^2 = 2$. Then $m^2 = 2n^2$, which shows that m^2 and hence m must be even, so that we may set $m = 2r$ for some natural number r. Our assumption now becomes $(2r)^2 = 2n^2$, or $2r^2 = n^2$, which now shows that n^2 and hence n must be even, so we may write $n = 2s$ for some natural number s. Thus, contrary to our assumption, the numbers m and n have a common factor 2, contradicting the initial assumption and thereby showing that $\sqrt{2}$ must be irrational.

The set of numbers represented by **Q**, together with all the irrational numbers, can be shown to be *complete* in the sense that there are no other types of number that have been omitted. The set of all these numbers is called the *real number system* and is denoted by **R**. For future reference we summarize the basic properties of real numbers, taking full advantage of the mathematical shorthand that we have introduced.

It is often necessary to deal with integers that are required to be either even or odd, so some simple device must be introduced to identify such numbers. This is done by recalling that an integer is *even* if the result of dividing it by 2 is to produce another integer. Thus the number $2r$ is *even* for $r = 1, 2, 3, \ldots$. If 1 is added to an even number the result is an odd number, so $2r + 1$ is *odd* for $r = 1, 2, 3, \ldots$. Thus the numbers $2r$ and $(2r + 1)$ are respectively even and odd when r is an integer.

It is also often necessary to introduce into expressions a factor that switches between 1 and -1 as n takes successive integral values. Such a factor is $(-1)^n$, because this becomes 1 when n is even and -1 when n is odd.

Additive properties

A.1 $a, b \in \mathbf{R} \Rightarrow (a+b) \in \mathbf{R}$; $\mathbf{R}$ is *closed* with respect to addition.
A.2 $a,\ b \in \mathbf{R} \Rightarrow a+b = b+a$; addition is *commutative*.
A.3 $a,\ b,\ c \in \mathbf{R} \Rightarrow (a+b)+c = a+(b+c) = a+b+c$; addition is *associative*.
A.4 For every $a \in \mathbf{R}$ there exists a number $0 \in \mathbf{R}$ such that $0+a=a$; there is a *zero* element in $\mathbf{R}$.
A.5 If $a \in \mathbf{R}$ then there exists a number $-a \in \mathbf{R}$ such that $-a+a=0$; each number has a *negative*.

Multiplicative properties

M.1 $a, b \in \mathbf{R} \Rightarrow ab \in \mathbf{R}$: $\mathbf{R}$ is *closed* with respect to multiplication.
M.2 $a,\ b \in \mathbf{R} \Rightarrow ab = ba$: multiplication is *commutative*.
M.3 $a,\ b,\ c \in \mathbf{R} \Rightarrow (ab)c = a(bc)$: multiplication is *associative*.
M.4 There exists a number $1 \in \mathbf{R}$ such that $1.a = a$ for all $a \in \mathbf{R}$: there is a *unit* element in $\mathbf{R}$.
M.5 Let a be a non-zero number in $\mathbf{R}$. Then there exists a number $a^{-1} \in \mathbf{R}$ such that $a^{-1}a = 1$; each non-zero number has an *inverse*.
Usually we shall write $1/a$ in place of a^{-1}, so these two expressions are to be taken as being synonymous.

Distributive property

D.1 $a,\ b,\ c \in \mathbf{R} \Rightarrow a(b+c) = ab+ac$: multiplication is *distributive*.

The above results are self-evident for real numbers and are usually called the *real number axioms*. For future reference, we mention *matrices*, for which M.1 to M.5 are not generally true, and *vectors*, for which two forms of multiplication exist and for which M.5 has no meaning. It is an immediate consequence of the real number axioms that the familiar arithmetic operations of addition, subtraction, multiplication and division may be performed on real numbers any finite number of times. Henceforth we shall accept these axioms without further examination.

So far our list of properties of real numbers has been concerned only with equalities. The valuable property of real numbers that they can be arranged according to size, or *ordered*, has so far been overlooked. Ordering is achieved by utilizing the concept 'greater than' which, as we have already seen, when used in the form 'a greater than b', is denoted by $a > b$. Hence to the other real number axioms must be added:

Order properties

O.1 If $a \in \mathbf{R}$ then exactly one of the following is true; either $a > 0$ or $a = 0$ or $-a > 0$.
O.2 $a, b \in \mathbf{R}$, $a > 0$, $b > 0 \Rightarrow a+b > 0$, and $ab > 0$.

We now define $a > b$ and $a < b$, the latter being read 'a less than b' by $a > b \Rightarrow a-b > 0$ and $a < b \Rightarrow b-a > 0$. The following results are obvious consequences of the real number system and are called *inequalities*. In places they also involve the symbol $\geq$ which is to be read 'greater than or equal to'.

Elementary inequalities in R

I.1 $a > b$ and $c \geq d \Rightarrow a + c > b + d$.
I.2 $a > b \geq 0$ and $c \geq d > 0 \Rightarrow ac > bd$.
I.3 $k > 0$ and $a > b \Rightarrow ka > kb$; $k < 0$ and $a > b \Rightarrow ka < kb$.
I.4 $0 < a < b \Rightarrow a^2 < b^2$ and $a < b < 0 \Rightarrow a^2 > b^2$.
I.5 $a \lessgtr 0 \Rightarrow a^2 > 0$.
I.6 $a > b \Rightarrow -a < -b$.
I.7 $a < 0,\ b > 0 \Rightarrow ab < 0;\ \ a < 0,\ b < 0 \Rightarrow ab > 0$.
I.8 $a > 0 \Rightarrow a^{-1} > 0;\ \ a < 0 \Rightarrow a^{-1} < 0$.
I.9 $a > b > 0 \Rightarrow b^{-1} > a^{-1} > 0;\ \ a < b < 0 \Rightarrow b^{-1} < a^{-1} < 0$.

Inequalities provide a convenient and concise way of defining *intervals* on a line or *regions* in a plane, both of which are used to define the conditions under which functions are valid. To understand how this is achieved we need to use the *order property* of numbers and their correspondence with points on a line to identify a segment of the line whose left-hand end point is the number a, say, and whose right-hand end point is some other number b. Three cases arise according to whether (a) both end points are included in the interval, (b) both end points are excluded from the interval, or (c) one is included and one is excluded. These are called, respectively, (a) a *closed* interval, (b) an *open* interval, and (c) a *semi-open* interval. Thus an interval is closed at an end which contains the end point, otherwise it is open at that end. In terms of a left-hand end point a, and a right-hand end point b and the variable x representing an arbitrary point on the line, these inequalities are written:

(a) $a \leq x \leq b$ (closed interval),
(b) $a < x < b$ (open interval),
(c) $a < x \leq b$ (semi-open interval).

Thus, for example, $1 \leq x < 2$ defines the semi-open interval containing the point $x = 1$ and extending up to, but not including, the point $x = 2$. The different types of intervals are illustrated graphically in Fig. 1.1, using the convention that a thick line represents points included in the interval, with a solid dot representing an end point that is included and a circle representing an end point that is excluded.

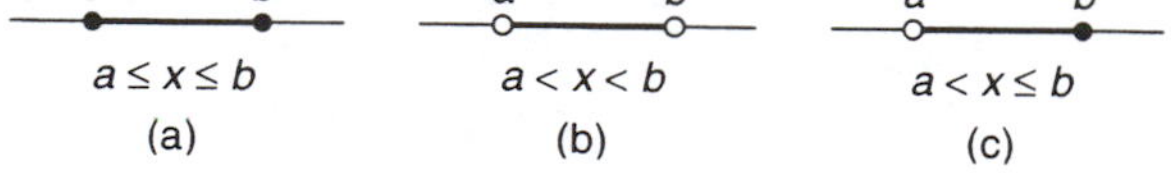

Fig. 1.1 Intervals on a line: (a) closed interval; (b) open interval; (c) semi-open interval.

Special cases occur when one or both of the end points of the interval are at infinity. The intervals $-\infty < x < a$ and $b < x < \infty$ are called *semi-infinite* intervals and $-\infty < x < \infty$ is called an *unbounded* interval or, more simply the *real line*. Since infinity is not a number but a limiting process, an end point at infinity is regarded as being open.

We illustrate the corresponding definition of a region in the (x, y)-plane by considering the three inequalities $x^2 + y^2 \leq a^2$, $y < x$, $x \geq 0$. The first defines the interior and boundary points of a circle of radius a centred on the origin, the second defines points below, but not on, the straight line $y = x$, and the third defines points in the right half of the (x, y)-plane including the points on the y-axis itself. These curves represent boundaries of the regions in question and the boundary points are only to be included in the region when possible equality is indicated by use of the signs $\geq$ or $\leq$. The three regions are indicated in Fig. 1.2(a), in which a full line indicates that points on it are to be included, a dotted line indicates that points on it are to be excluded, and shading indicates the side of the line on which the region in question must lie. Figure 1.2(b) indicates the region in which all the inequalities are satisfied.

The elementary inequalities may often be used to advantage to simplify complicated algebraic expressions by yielding helpful qualitative information as the following example indicates.

Example 1.1 Prove that if $a_1, a_2, \ldots, a_n$ and $b_1, b_2, \ldots, b_n$ are positive real numbers, then

$$\min_{1 \leq r \leq n} \left(\frac{a_r}{b_r}\right) \leq \frac{a_1 + a_2 + \ldots + a_n}{b_1 + b_2 + \ldots + b_n} \leq \max_{1 \leq r \leq n} \left(\frac{a_r}{b_r}\right).$$

Solution Here the left-hand side of the inequality is to be interpreted as meaning the minimum value of the expression (a_r/b_r), with r assuming any of the integral values from 1 to n, and the right-hand side is to be similarly

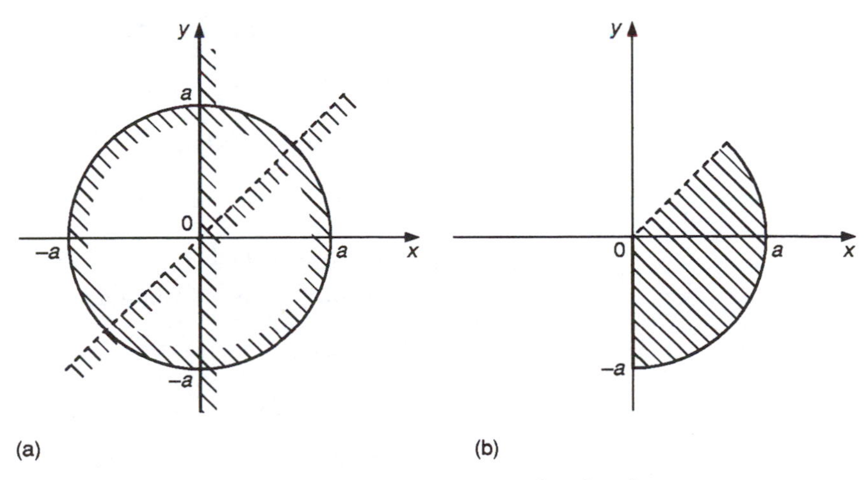

Fig. 1.2 Regions in a plane: (a) region boundaries $x^2 + y^2 = a^2$, $y = x$ and $x = 0$; (b) region $x^2 + y^2 \leq a^2$, $y < x$ and $x \geq 0$.

interpreted, reading maximum in place of minimum. The result follows by noticing that

$$\frac{a_1 + a_2 + \ldots + a_n}{b_1 + b_2 + \ldots + b_n} = \frac{1}{\sum_{r=1}^{n} b_r}\left[b_1\left(\frac{a_1}{b_1}\right) + b_2\left(\frac{a_2}{b_2}\right) + \ldots + b_n\left(\frac{a_n}{b_n}\right)\right],$$

where $\sum_{r=1}^{n} b_r = b_1 + b_2 + \ldots + b_n$. For if each of the expressions (a_1/b_1), $(a_2/b_2), \ldots, (a_n/b_n)$ is replaced by the smallest of these ratios, which could be the value taken by all the expressions if $a_1 = a_2 = \ldots = a_n > 0$ and $b_1 = b_2 = \ldots = b_n > 0$, then

$$\frac{a_1 + a_2 + \ldots + a_n}{b_1 + b_2 + \ldots + b_n} \geq \min_{1 \leq r \leq n}\left(\frac{a_r}{b_r}\right)\left[\frac{(b_1 + b_2 + \ldots + b_n)}{\sum_{r=1}^{n} b_r}\right]$$

$$= \min_{1 \leq r \leq n}\left(\frac{a_r}{b_r}\right),$$

which is the left half of the inequality. The right half follows by identical reasoning if maximum is written in place of minimum. ■

Example 1.2 Prove

(i) that if $a \geq 0$, $b \geq 0$ then

$$\tfrac{1}{2}(a + b) \geq \sqrt{ab} \quad \text{(Cauchy inequality)}$$

(ii) that if $ab > 0$, then

$$\frac{a}{b} + \frac{b}{a} \geq 2.$$

Solution (i) We shall consider the difference

$$\tfrac{1}{2}(a + b) - \sqrt{ab}$$

and determine its sign. We have

$$\tfrac{1}{2}(a + b) - \sqrt{ab} = \tfrac{1}{2}(a - 2\sqrt{ab} + b) = \tfrac{1}{2}(\sqrt{a} - \sqrt{b})^2.$$

For any non-negative values of a and b the right-hand side is non-negative. Equality is only possible when $a = b$. Thus the difference

$$\tfrac{1}{2}(a + b) - \sqrt{ab} \geq 0,$$

but this is equivalent to

$$\tfrac{1}{2}(a + b) \geq \sqrt{ab},$$

and we have proved the Cauchy inequality

$$\tfrac{1}{2}(a+b) \geq \sqrt{ab}.$$

(ii) We have

$$\frac{a}{b}+\frac{b}{a}-2=\frac{a^2+b^2-2ab}{ab}=\frac{(a-b)^2}{ab}.$$

As $ab > 0$ the right-hand side must be non-negative and it can only vanish when $a = b$. Thus the result has been proved, and equality is only possible when $a = b$. ■

Example 1.3 For what values of x is it true that

$$\frac{x+2}{x} > \frac{x-1}{x-2}.$$

Solution Care must be taken with inequalities of this type when clearing the quotients by multiplying by $x(x-2)$, because if $x(x-2) > 0$ then by I.3 ($k > 0$) the inequality sign remains unchanged, whereas if $x(x-2) < 0$ then by I.3 ($k < 0$) the inequality sign must be reversed. Suppose first that $x(x-2) > 0$. This is possible if either (i) $x > 0$ and $x > 2$, which follows if $x > 2$, or if (ii) $x < 0$ and $x < 2$, which follows if $x < 0$. The result of multiplication by the factor $x(x-2) > 0$ leads to the result

$$(x+2)(x-2) > x(x-1) \quad \text{or} \quad x^2-4 > x^2-x.$$

Cancelling x^2 on either side of the inequality reduces it to

$$-4 > -x, \quad \text{or by I.3}\,(k<0) \text{ to } x > 4.$$

However, the conditions $x > 0$, $x > 2$ and $x > 4$ imply that $x > 4$, and so must be rejected.

The other alternative to be considered is the possibility that $x(x-2) < 0$, which follows if (i) $x < 0$ and $x > 2$ which is impossible, or if (ii) $x > 0$ and $x < 2$ from which we have $0 < x < 2$. Multiplication by the factor $x(x-2) < 0$, followed by reversal of the inequality sign gives

$$x^2-4 < x^2-x, \quad \text{or} \quad -4 < -x,$$

so by I.3 ($k < 0$) we conclude that $x < 4$. The conditions $0 < x < 2$ and $x < 4$ are compatible provided $0 < x < 2$, so this last condition is seen to be the one that is necessary for the inequality to hold, together with $x > 4$.

A simple alternative geometrical approach, as opposed to the analytical one just used, is possible if the terms on either side of the inequality are first divided out to give

$$1+\frac{2}{x} > 1+\frac{1}{x-2}, \text{ which is equivalent to } \frac{2}{x}-\frac{1}{x-2} > 0.$$

Graphing these functions, it is easy to see the inequality is only true if $0 < x < 2$, or $x > 4$, though this task is left to the reader. ■

The last example we shall consider involves maximizing a simple expression subject to constraints imposed by linear inequalities.

Example 1.4 Find the maximum value of $x+y$, given that $0 \leq x-y \leq 2$, $2x+y \geq 1$, $0 \leq x \leq 4$, $0 \leq y \leq 5$.

Solution By drawing the boundaries indicated by these inequalities it is easily seen that the region R comprising points satisfying all of the inequalities is the shaded polygon shown in Fig. 1.3.

To maximize $x+y$, all that is now necessary is to locate the point inside R at which both x and y attain their greatest values. It is easily seen that this occurs on the boundary at P, which is the point (4, 4), so the maximum value of $x+y$ satisfying the inequalities is 8.

An analytical way of locating P which is useful when the polygon has a less convenient shape is the following. Set $x+y=c$, then $y=c-x$, showing that the value of c is the intercept on the y-axis of a straight line with gradient -1. Such a line is shown in Fig. 1.3 as a dotted line.

Thus, if we consider straight lines with gradient -1 through all points of region R, the maximum value of $x+y$ will be the largest value of the intercept on the y-axis of all such lines. In the case of Fig. 1.3 this occurs at point P, corresponding to $x+y=8$. This line through P is shown as a chain dotted line. This type of approach can also be used when seeking to maximize (or minimize) an expression such as $ax+by$.

Problems of the type just discussed, in which a linear combination of the variables must be maximized (or minimized) subject to a set of linear inequalities or constraints are called *linear programming problems*, or *LP*

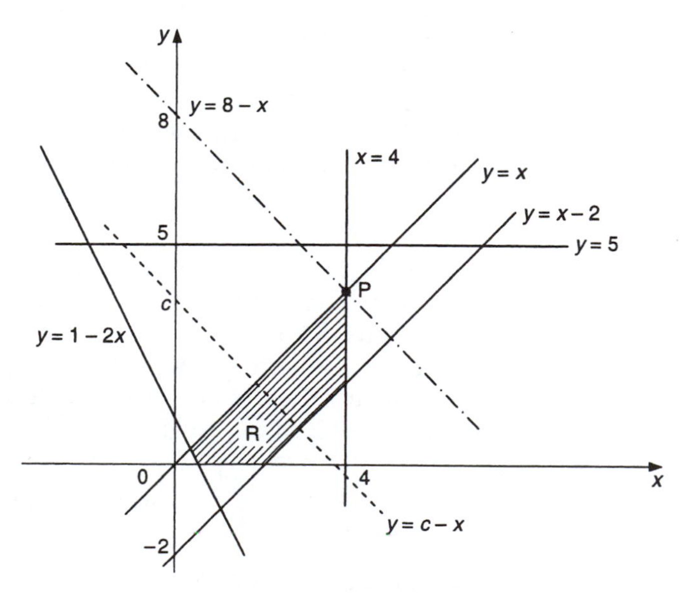

Fig. 1.3 The region R.

problems. Region R is then called the *feasible region* for solutions, and the solution corresponding to P is called the *optimum solution*.

Because of the number of dimensions involved, the graphical method used here fails when there are more than two variables, and other methods of solution must then be employed. These are discussed in more advanced accounts of engineering mathematics under the general heading of the *simplex method*.

We conclude this discussion of a simple linear programming problem by illustrating how this same problem might have arisen from an engineering production process. The mathematical formulation of the following problem is precisely the one given above.

Processes A and B are used simultaneously by a factory to produce a certain product. The outputs per day from processes A and B are $x(\geq 0)$ and $y(\geq 0)$ units, respectively. Production by process A cannot exceed 4 units, and production by process B cannot exceed 5 units. Constraints placed on the production due to the fact that certain parts of the two processes use the same equipment may be expressed in the form of the inequalities $0 \leq x - y \leq 2$ and $2x + y \geq 1$. Find the production by processes A and B which will jointly yield the maximum output of the product. ■

1.3 Absolute value of a real number

Definition 1.1 The *absolute value* $|a|$ of the real number a provides a measure of its size without regard to sign, and is defined as follows:

$$|a| = \begin{cases} a \text{ when } a \geq 0 \\ -a \text{ when } a < 0. \end{cases}$$

Thus if $a = 3$, then $|a| = 3$ and if $a = -5.6$ then $|a| = 5.6$.

There are three immediate consequences of this definition which we now enumerate as follows.

Theorem 1.1 If $a, b \in \mathbf{R}$ then

(a) $ab \leq |ab| = |a||b|$,
(b) $|a + b| \leq |a| + |b|$,
(c) $|a - b| \geq ||a| - |b||$.

Proof The proof is simply a matter of enumerating the possible combinations of positive and negative a and b, and then making a direct application of the definition of the absolute value. We shall only give the proof of (a) and (b).

There are three cases to be considered in (a): firstly, $a \geq 0, b \geq 0$; secondly, $a \geq 0, b < 0$; and thirdly, $a < 0, b \geq 0$. If $a \geq 0, b \geq 0$ then $ab \geq 0$ and so $|ab| = ab = |a||b|$. The second and third situations are essentially

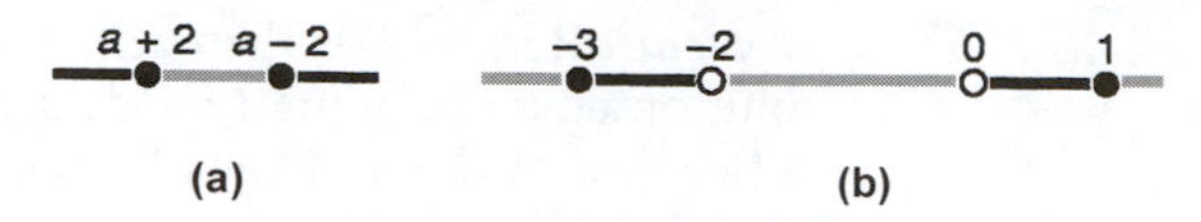

Fig. 1.4 Intervals on a line: (a) $|a-x|\geq 2$; (b) $|x+1|\leq 2$.

similar so we shall discuss only the second. As $a\geq 0$, $b<0$ we have $ab\leq 0$, whence $|ab|=-ab=a(-b)=|a||b|$, establishing (a). To prove (b), notice that $|a+b|^2=(a+b)^2=a^2+2ab+b^2=|a|^2+2ab+|b|^2$, and that $(|a|+|b|)^2=|a|^2+2|a||b|+|b|^2$. Thus since by (a) $ab\leq|a||b|$ we see that $|a+b|^2\leq(|a|+|b|)^2$. The result follows because both $|a+b|$ and $|a|+|b|$ are non-negative. For reason we give later, result (b) is usually called the *triangle inequality*. ■

The absolute value may also be used to define intervals since an expression of the form $|a-x|\geq 2$ implies two inequalities according to whether $a-x$ is positive or negative. If $a-x>0$ then $|a-x|=a-x$, and we have $a-x\geq 2$ or $x\leq a-2$. However, if $a-x<0$, then by the definition of the absolute value of $a-x$ we must have $|a-x|=-(a-x)$, showing that $-(a-x)\geq 2$, or, $x\geq a+2$. Taken together the results require that x may be equal to or greater than $2+a$ or equal to or less than $a-2$. x may not lie in the intervening interval of length 4 between $x=a-2$ and $x=a+2$. This is illustrated in Fig. 1.4(a), where a thick line is again used to indicate points in the interval satisfied by $|a-x|\geq 2$, and the solid dots are to be included at the ends of the appropriate interval.

By exactly similar reasoning, we see that if we consider the inequality $1<|x+1|\leq 2$, then if $x+1>0$, $|x+1|=x+1$ and the inequality becomes $1<x+1\leq 2$. Hence the intervals are $0<x\leq 1$. However, if $x+1<0$, then $|x+1|=-x-1$ and so the inequality becomes $1<-x-1\leq 2$, giving rise to the interval $-3\leq x<-2$. These intervals are shown in Fig. 1.4(b), with circles indicating points excluded from the end of the solid-line intervals and dots indicating points to be included.

In words, both $|x-a|$ and $|a-x|$ measure the (positive) distance of x from a. Thus $|a-x|<3$ means all points x whose distance from a are less than 3; this is just the interval $a-3<x<a+3$. Similarly, $|a-x|\geq 2$ means all points x whose distance from a is greater than or equal to 2.

On occasions the absolute value arises in an equation which is to be solved for the permissible values of some variable, say x. The approach is then usually either to consider the different cases according to whether the entry inside the absolute value is positive or negative or, if it is convenient, to remove the absolute value by squaring the expression. The following example illustrates these methods.

Example 1.5 For what values of x are the following equations satisfied?

(i) $|2x-3|=x+2$.
(ii) $|2x-3|=|x+1|$.
(iii) $|x-3|+|x-4|=x-7$.

Solution (i) Method 1. We consider the cases according to whether $2x-3$ is positive or negative. If $2x-3\geq 0$ then $|2x-3|=2x-3$, and the equation becomes

$$2x-3=x+2, \text{ so } x=5.$$

If $2x-3<0$ then $|2x-3|=3-2x$, and the equation becomes

$$3-2x=x+2, \text{ so } x=\tfrac{1}{3}.$$

Thus the equation is satisfied by $x=\frac{1}{3}$ and $x=5$.

Method 2. The absolute value may be removed by squaring to obtain the equation

$$(2x-3)^2=(x+2)^2, \text{ or } 3x^2-16x+5=0,$$

with the solutions $x=\frac{1}{3}$ and $x=5$.

(ii) Method 2 is the simplest in this case, because squaring gives

$$(2x-3)^2=(x+1)^2, \text{ or } 3x^2-14x+8=0,$$

with the solutions $x=\frac{2}{3}$ and $x=4$.

(iii) Squaring will not help here, so method 1 would seem to be necessary. However a little thought provides a simpler approach in this case. As the left-hand side of

$$|x-3|+|x-4|=x-7$$

is non-negative, we must have $x\geq 7$. Thus when $x\geq 7$ we have $x-3>0$ and $x-4>0$, so the equation reduces to

$$x-3+x-4=x-7 \text{ and } x\geq 7.$$

These two conditions are equivalent to the two contradictory conditions

$$x=0 \text{ and } x\geq 7,$$

so the equation has no solution. ■

1.4 Mathematical induction

Mathematical propositions often involve some fixed integer n, say, in a special role, and it is desirable to infer the form taken by the proposition for arbitrary integral n from the form taken by it for the specific value $n=n_1$. The logical method by which the proof of the general proposition, if true, may be established, is based on the properties of natural numbers and is called *mathematical induction*.

In brief, it depends for its success on the obvious fact that if A is some set of natural numbers and $1\in A$, then the statement that whenever integer $n\in A$, so also does its successor $n+1$, implies that $A=\mathbf{N}$, the set of natural numbers.

The formal statement of the process of mathematical induction is expressed by the following theorem where, for simplicity, the mathematical proposition corresponding to integer n is denoted by $S(n)$.

***Theorem 1.2* (mathematical induction)** If it can be shown that

(a) when $n = n_1$, the proposition $S(n_1)$ is true, and
(b) if for $n \geq n_1$, when $S(n)$ is true then so also is $S(n+1)$,

then the proposition $S(n)$ is true for all natural numbers $n \geq n_1$. ■

Example 1.6 Prove inductively that the sum of the first n natural numbers is given by $n(1+n)/2$. In other words, in this example the proposition denoted by $S(n)$ is that the following result is true:

$$1 + 2 + \ldots + n = n(1 + n)/2.$$

Proof (a) First the proposition must be shown to be true for some specific value $n = n_1$. Any integral value n_1 will suffice but if we set $n_1 = 1$ the proposition corresponding to $S(1)$ is immediately obvious. If, instead, we had chosen $n_1 = 3$, then it is easily verified that proposition $S(3)$ is true, namely that $1+2+3 = 3(1+3)/2$.
(b) We must now assume that proposition $S(n)$ is true and attempt to this implies that the proposition $S(n+1)$ is true. If $S(n)$ is true then

$$1 + 2 + \ldots + n = n(1 + n)/2$$

and, adding $(n+1)$ to both sides, we obtain

$$\begin{aligned} 1 + 2 + \ldots + n + (n + 1) &= n(1 + n)/2 + (n + 1) \\ &= (n + 1)(2 + n)/2. \end{aligned}$$

However, this is simply a statement of proposition $S(n+1)$ obtained by replacing n by $n+1$ in proposition $S(n)$. Hence $S(1)$ is true and $S(n) \Rightarrow S(n+1)$ so, by the conditions of Theorem 1.2, we have established that $S(n)$ is valid for all n. ■

The summing of terms, as in Example 1.6, arises repeatedly throughout mathematics, and so requires the introduction of a concise notation. This is accomplished by using the *summation symbol* Σ (*sigma*) and writing

$$\sum_{r=m}^{n} a_r \text{ in place of } a_m + a_{m+1} + \ldots + a_n, \text{ so that}$$

$$\sum_{r=m}^{n} a_r = a_m + a_{m+1} + \ldots + a_n.$$

In this notation the term a_r is called the *general term* in the summation, the subscript r is called the *summation index*, and the numbers m and n are called the *lower* and *upper summation limits*, respectively. Thus the summation starts with the term with summation index m and finishes with the term with summation index n.

In this concise notation, the proposition $S(n)$ in Example 1.6 would become

$$\sum_{r=1}^{n} r = n(1+n)/2.$$

For the next example we introduce the number called *factorial* n, written $n!$, which will occur frequently throughout the remainder of the book. Factorial n is *defined* as the product of all the natural numbers from 1 to n so that $n! = 1.2.3\ldots(n-2)(n-1)n$. Later, when working with series, it will be convenient to adopt the convention that 0! is *defined* as $0! = 1$. The first few factorials are listed below:

$$\begin{aligned} &0! = 1 \\ &1! = 1 \\ &2! = 1.2 = 2 \\ &3! = 1.2.3 = 6 \\ &4! = 1.2.3.4 = 24 \\ &5! = 1.2.3.4.5 = 120 \\ &6! = 1.2.3.4.5.6 = 720 \end{aligned}$$

On occasions, expressions like $(m+r)!/m!$ arise with r an integer. These are easily simplified by using the definition of $n!$, because

$$\begin{aligned} \frac{(m+r)!}{m!} &= \frac{1.2.3\ldots m(m+1)(m+2)\ldots(m+r)}{1.2.3\ldots m} \\ &= (m+1)(m+2)\ldots(m+r). \end{aligned}$$

For example, when applied to $8!/5!$ we see that $m = 5$ and $r = 3$, so $8!/5! = 6.7.8 = 336$.

Care must always be taken when working expressions like factorial $2n$ and factorial $2n+1$, because brackets are necessary as these mean $(2n)!$ and $(2n+1)!$, and not $2n!$ and $2n!+1$. So, for example, $(2n)! = 1.2.3\ldots n(n+1)(n+2)\ldots(2n-1)(2n)$, and $(2n+1)! = 1.2.3\ldots n(n+1)(n+2)\ldots(2n-1)(2n)(2n+1)$.

Example 1.7 Prove that $3^n < n!$ for all $n \geq 7$.

Proof Let the proposition $S(n)$ be that $3^n < n!$ for all $n \geq 7$. Then the result is true for $n = 7$, because $3^7 = 2187 < 7! = 5040$. Finally, to prove that $S(n) \Rightarrow S(n+1)$ we use the fact that

$$\begin{aligned} 3^{n+1} = 3.3^n &< 3.n!, \text{ because of proposition } S(n), \\ &< (n+1).n! \text{ provided } n \geq 7, \\ &= (n+1)!, \end{aligned}$$

and the result is proved. ■

We remark that this same form of proof will establish that it is always true that $m^n < n!$ for n greater than some suitably large number N, say.

It is a direct consequence of this result that for any $m > 0$, the quotient $m^n/n!$ will vanish as $n \to \infty$.

Later we shall use this form of proof in cases less trivial than Example 1.7.

Example 1.8 (A first-order difference equation)
Find by trial and error in terms of u_0 the nth term in the sequence of numbers $u_0, u_1, u_2, \ldots$, defined sequentially by the equation

$$u_n = 2u_{n-1} + 1, \tag{1.1}$$

and then verify the result by means of induction.

Proof Before proceeding to solve this problem we remark first that equations of this form which define a sequence of numbers u_n in terms of u_0 are called *first-order linear difference equations*.

We start by writing out the first few terms of the sequence, which are

$$\begin{aligned}
u_1 &= 2u_0 + 1 \\
u_2 &= 2u_1 + 1 = 4u_0 + 3 \\
u_3 &= 2u_2 + 1 = 8u_0 + 7 \\
u_4 &= 2u_3 + 1 = 16u_0 + 15 \\
&\ldots
\end{aligned}$$

Noticing that the coefficient of u_0 in the expression for u_n can be written 2^n, and that the constant term is always one less than the coefficient of u_0, we conclude that the solution should be

$$u_n = 2u_0 + (2^n - 1). \tag{1.2}$$

The initial term u_0 of the sequence is arbitrary and on account of this fact such a solution is called the *general solution* of the first-order difference equation (1.1). Once u_0 is specified by requiring that $u_0 = C$, say, then the solution is said to be a *particular solution*. Having deduced this solution intuitively, it is now necessary to use induction to verify that it is indeed a solution for all n.

The proof of Eqn (1.2) by induction again proceeds in two parts, with the proposition $S(n)$ being that Eqn (1.2) is the solution of Eqn (1.1).

(a) If $n_1 = 1$, then $u_1 = 2u_0 + (2 - 1) = 2u_0 + 1$, showing that the proposition $S(1)$ is true.
(b) Assuming the proposition $S(n)$ is true, then

$$\begin{aligned}
2u_n + 1 &= 2[2^n u_0 + (2^n - 1)] + 1 \\
&= 2^{n+1}u_0 + (2^{n+1} - 1) \\
&= u_{n+1},
\end{aligned}$$

showing that $S(n) \Rightarrow S(n+1)$. The result is thus true for all n. ■

Having introduced the notion of a difference equation, let us take the concept a little further so that it can be used in more general circumstances. A *homogeneous* linear difference equation of order 2 is a relationship of the form

$$u_n + au_{n-1} + bu_{n-2} = 0, \qquad (1.3)$$

where a and b are real constants and u_{n-2}, u_{n-1}, u_n are three consecutive members of a sequence of numbers. Given any two consecutive members of the sequence, say u_0 and u_1, then Eqn (1.3), provides an algorithm (rule) by which any other member of the sequence may be computed.

If we seek a solution u_n of the form

$$u_n = A\lambda^n, \qquad (1.4)$$

where A and λ are real constants, then substitution into Eqn (1.3) shows that

$$\lambda^2 + a\lambda + b = 0. \qquad (1.5)$$

This is called the *characteristic equation* associated with the difference equation (1.3), and shows that solutions of the form of Eqn (1.4) are only possible when λ is equal to one of two roots λ_1 and λ_2 of Eqn (1.5), which we assume to be real numbers. If $\lambda_1 \neq \lambda_2$, then $A\lambda_1^n$ and $B\lambda_2^n$ are both solutions of Eqn (1.3) and

$$u_n = A\lambda_1^n + B\lambda_2^n \qquad (1.6)$$

is also a solution, where A and B are arbitrary real constants (check this by substitution). This result is the general solution of Eqn (1.3). Given specific values for u_0 and u_1, A and B can be deduced by substituting into Eqn (1.6) and hence a particular solution can be found.

Example 1.9 Solve the difference equation

$$u_n - u_{n-1} - u_{n-2} = 0,$$

given that $u_0 = u_1 = 1$.

Solution The characteristic equation is

$$\lambda^2 - \lambda - 1 = 0,$$

with the two roots $\lambda_1 = (1+\sqrt{5})/2$ and $\lambda_2 = (1-\sqrt{5})/2$. Hence the general solution has the form

$$u_n = A\left(\frac{1+\sqrt{5}}{2}\right)^n + B\left(\frac{1-\sqrt{5}}{2}\right)^n.$$

To deduce the values of A and B particular to our problem we use the *initial conditions* $u_0 = 1$ and $u_1 = 1$ which when substituted into the above result give

$$1 = A + B \qquad \text{(case } n = 0,\ u_0 = 1)$$
$$1 = A\left(\frac{1 + \sqrt{5}}{2}\right) + B\left(\frac{1 - \sqrt{5}}{2}\right) \qquad \text{(case } n = 1,\ u_1 = 1).$$

Solving these equations for A and B we find

$$A = \frac{\sqrt{5} + 1}{2\sqrt{5}} \qquad B = \frac{\sqrt{5} - 1}{2\sqrt{5}}$$

whence the particular solution is

$$u_n = \left(\frac{\sqrt{5} + 1}{2\sqrt{5}}\right)\left(\frac{1 + \sqrt{5}}{2}\right)^n + \left(\frac{\sqrt{5} - 1}{2\sqrt{5}}\right)\left(\frac{1 - \sqrt{5}}{2}\right)^n.$$

The first few numbers $u_0, u_1, u_2, \ldots$, of the sequence generated by this algorithm are

$$1,\ 1,\ 2,\ 3,\ 5,\ 8,\ 13,\ 21,\ 34,\ 55,\ \ldots,$$

and comprise the well-known *Fibonacci sequence* of numbers. This sequence of numbers occurs naturally in the study of regular solids and in numerous other parts of mathematics. Naturally if only the first few members of the sequence are required then they are most easily found by use of the algorithm itself, which in the form

$$u_n = u_{n-1} + u_{n-2}$$

states that each member of the sequence is the sum of its two predecessors. The name of this sequence which has various applications in mathematics is due to the Italian Leonardo of Pisa, known as Fibonacci, who around 1200 AD arrived at the sequence by asking how many pairs of rabbits will exist n months after a single pair begins breeding. ■

It is not difficult to see that if the roots of the characteristic equation (1.5) are equal so that $\lambda_1 = \lambda_2 = \mu$, say, then $A\mu^n$ is a solution of Eqn (1.3). In terms of Eqn (1.3) this is equivalent to saying that $a^2 = 4b$ and $\mu = -a/2$. However, $A\mu^n$ cannot be the general solution since it only involves one arbitrary constant A, and it is necessary to have two such constants in the general solution to allow the specification of the initial conditions u_0 and u_1. The difficulty is easily resolved once we notice that $nB\mu^n$, with B an arbitrary real constant, is also a solution of Eqn (1.3). This is easily verified by direct substitution. For then we have for the general solution in the case of equal roots in the characteristic equation,

$$u_n = (A + nB)\mu^n. \tag{1.7}$$

To illustrate this situation, suppose that we are required to solve the difference equation

$$u_n = 6u_{n-1} - 9u_{n-2}$$

subject to the initial conditions $u_0 = 1$, $u_1 = 2$. Then the characteristic equation becomes

$$\lambda^2 - 6\lambda + 9 = 0,$$

with the double root $\lambda = 3$. From Eqn (1.7) the general solution must thus be

$$u_n = (A + nB).3^n.$$

Using the initial conditions $u_0 = 1$, $u_1 = 3$ then shows that

$$1 = A \text{ and } 2 = 3(A + B)$$

so that the solution to the problem in question is

$$u_n = (1 - \tfrac{1}{3}n)3^n.$$

When, as may happen in Eqn (1.3), $a^2 - 4b < 0$, the general solution can easily be shown to contain sines and cosines, but we shall not discuss this here as it necessitates the use of complex numbers.

One last important and very useful result that can be proved by means of mathematical induction is the *binomial theorem* for positive integral n, though the proof will not be given here.

Theorem 1.3 (Binomial theorem for positive integral n) If a, b are real numbers and n is a positive integer, then

$$(a + b)^n = a^n + na^{n-1}b + \frac{n(n-1)}{2!}a^{n-2}b^2 + \frac{n(n-1)(n-2)}{3!}a^{n-3}b^3 + \ldots + b^n$$

or, equivalently

$$(a + b)^n = a^n\left[1 + n\left(\frac{b}{a}\right) + \frac{n(n-1)}{2!}\left(\frac{b}{a}\right)^2 + \frac{n(n-1)(n-2)}{3!}\left(\frac{b}{a}\right)^3 + \ldots + \left(\frac{b}{a}\right)^n\right].$$

The expansion, which contains only $n+1$ terms when n is a positive integer, can be shown to be true for any real n, positive or negative. However, if n is not a positive integer then the expansion contains an infinite number of terms, and for the sum of the right-hand side to be finite the numbers a, b must be such that $|b/a| < 1$. When n is real and positive the result is also true for $|b/a| = 1$.

When $|b/a|$ is very much less than unity, indicated by writing $|b/a| \ll 1$, the right-hand side of this expansion is often approximated by retaining only the first two terms. Thus if $|b/a| \ll 1$ we have

$$(a+b)^n \approx a^n\left(1+n\frac{b}{a}\right),$$

where the sign $\approx$ is to be read 'is approximately equal to'.

The coefficients in the expansion for positive integral n are called *binomial coefficients*, and the coefficient of $(b/a)^r$ is often written $\binom{n}{r}$, where

$$\binom{n}{r} = \frac{n!}{(n-r)!r!} \quad \text{and} \quad \binom{n}{0} = \binom{n}{n} = 1.$$

In this notation the binomial expansion for positive integral n takes either the form

$$(a+b)^n = a^n + \binom{n}{1}a^{n-1}b + \binom{n}{2}a^{n-2}b^2 + \binom{n}{3}a^{n-3}b^3 + \ldots + b^n$$

or, equivalently,

$$(a+b)^n = a^n\left[1 + \binom{n}{1}\left(\frac{b}{a}\right) + \binom{n}{2}\left(\frac{b}{a}\right)^2 + \binom{n}{3}\left(\frac{b}{a}\right)^3 + \ldots + \left(\frac{b}{a}\right)^n\right].$$

The binomial expansions of $(a+b)^n$ for $n = 1, 2, \ldots, 4$ are

$$\begin{aligned}
(a+b)^1 &= a+b \\
(a+b)^2 &= a^2 + 2ab + b^2 \\
(a+b)^3 &= a^3 + 3a^2b + 3ab^2 + b^3 \\
(a+b)^4 &= a^4 + 4a^3b + 6a^2b^2 + 4ab^3 + b^4.
\end{aligned}$$

For reference, the binomial coefficients $\binom{n}{k}$ for $n = 1, 2, \ldots, 6$, $k \leq n$, are listed in the table below.

n \ k	0	1	2	3	4	5	6
1	1	1					
2	1	2	1				
3	1	3	3	1			
4	1	4	6	4	1		
5	1	5	10	10	5	1	
6	1	6	15	20	15	6	1

It is possible to generate the binomial coefficients in a very simple manner by using the following triangular array called *Pascal's triangle*.

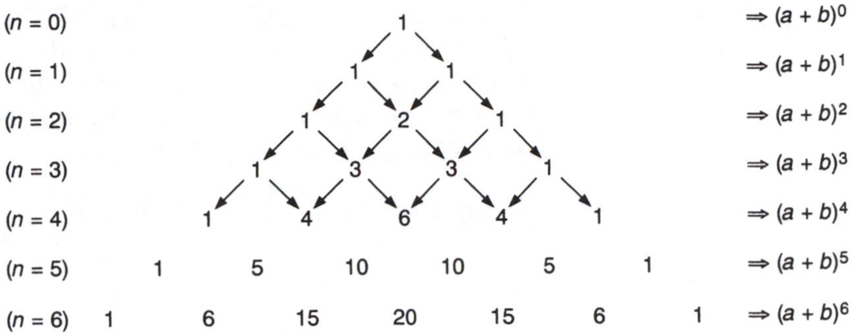

$(n = 0)$							1							$\Rightarrow (a+b)^0$
$(n = 1)$						1		1						$\Rightarrow (a+b)^1$
$(n = 2)$					1		2		1					$\Rightarrow (a+b)^2$
$(n = 3)$				1		3		3		1				$\Rightarrow (a+b)^3$
$(n = 4)$			1		4		6		4		1			$\Rightarrow (a+b)^4$
$(n = 5)$		1		5		10		10		5		1		$\Rightarrow (a+b)^5$
$(n = 6)$	1		6		15		20		15		6		1	$\Rightarrow (a+b)^6$

The entries in the nth row are the binomial coefficients $\binom{n}{k}$ ($k = 0, 1, 2, \ldots, n$). Each entry inside the triangle is obtained by summing the entries to its immediate left and right in the row above, as indicated by the arrows.

In general, if α is any real number, the binomial expansion of $(a+b)^\alpha$ for $|b/a| < 1$ is given by

$$(a+b)^\alpha = a^\alpha\left(1 + \alpha\left(\frac{b}{a}\right) + \frac{\alpha(\alpha-1)}{2!}\left(\frac{b}{a}\right)^2 + \frac{\alpha(\alpha-1)(\alpha-2)}{3!}\left(\frac{b}{a}\right)^3 + \cdots\right)$$

or, using the summation notation, by

$$(a+b)^\alpha = a^\alpha \sum_{r=0}^{\infty} \frac{\alpha(\alpha-1)(\alpha-2)\ldots(\alpha-r+1)}{r!}\left(\frac{b}{a}\right)^r.$$

Here the upper limit ∞ in the summation symbol indicates that, in general, the summation will contain an infinite number of terms, while when $r = 0$ the coefficient

$$\frac{\alpha(\alpha-1)(\alpha-2)\ldots(\alpha-r+1)}{r!}$$

is defined to be unity.

Some useful results obtained by applying the binomial expansion when n is not an integer are as follows:

$$(1 \pm x)^{1/4} = 1 \pm \frac{1}{4}x - \frac{1.3}{4.8}x^2 \pm \frac{1.3.7}{4.8.12}x^3 - \ldots, \qquad \text{for} \quad x^2 \le 1,$$

$$(1 \pm x)^{1/2} = 1 \pm \frac{1}{2}x - \frac{1.1}{2.4}x^2 \pm \frac{1.1.3}{2.4.6}x^3 - \ldots, \qquad \text{for} \quad x^2 \le 1,$$

$$(1 \pm x)^{-1/2} = 1 \mp \frac{1}{2}x + \frac{1.3}{2.4}x^2 \mp \frac{1.3.5}{2.4.6}x^3 + \ldots, \qquad \text{for} \quad x^2 < 1,$$

$$(1 \pm x)^{-1} = 1 \mp x + x^2 \mp x^3 + \ldots, \qquad \text{for} \quad x^2 < 1.$$

In these expansions the $\pm$ sign has been used to combine into a single expression the separate cases corresponding to the $+$ sign and the $-$ sign. When interpreting an equation using this symbol, wherever it occurs either all the upper signs or all the lower signs must be taken. Thus, for example,

$$(1+x)^{1/2} = 1 + \frac{1}{2}x - \frac{1.1}{2.4}x^2 + \frac{1.1.3}{2.4.6}x^3 - \ldots, \qquad \text{for} \quad x^2 \leq 1,$$

and

$$(1+x)^{-1/2} = 1 - \frac{1}{2}x + \frac{1.3}{2.4}x^2 - \frac{1.3.5}{2.4.6}x^3 + \ldots, \qquad \text{for} \quad x^2 < 1.$$

In these expansions the dots following the terms in x^3 indicate that the expansion does not terminate after a finite number of terms (it is an *infinite series*).

1.5 Review of trigonometric properties

This section lists the most frequently used trigonometric properties and identities. Various forms of proof of the identities are possible, but as the results are most easily established by using complex numbers we postpone until later the task of showing how they may be obtained.

In mathematics, when working with trigonometric functions the argument of the function is always in *radians* and never in degrees. As 2π radians are equivalent to 360 degrees, the conversion from θ degrees to x radians is accomplished by using the result

$$x = \frac{\pi\theta}{180} \tag{1.8}$$

from which it follows that the conversion from x radians to θ degrees is given by

$$\theta = \frac{180x}{\pi}. \tag{1.9}$$

The following results are often useful.

θ degrees	0	15	30	45	60	75	90	105	120	135	150	180
x radians	0	$\pi/12$	$\pi/6$	$\pi/4$	$\pi/3$	$5\pi/12$	$\pi/2$	$7\pi/12$	$2\pi/3$	$3\pi/4$	$5\pi/6$	π

Angles measured in degrees are denoted by the symbol $^\circ$, while those measured in radians are denoted by adding *rad*. Thus, for example, $30^\circ = \pi/6$ rad, and $45^\circ = \pi/4$ rad. Since radians will be used almost exclusively the notation rad will usually be omitted.

As shown in Fig. 1.5, the x- and y-axes divide the (x, y)-plane into four equal parts called *quadrants*. By convention, an angle θ to a line drawn through the origin O is measured *positively* in the *anticlockwise* sense

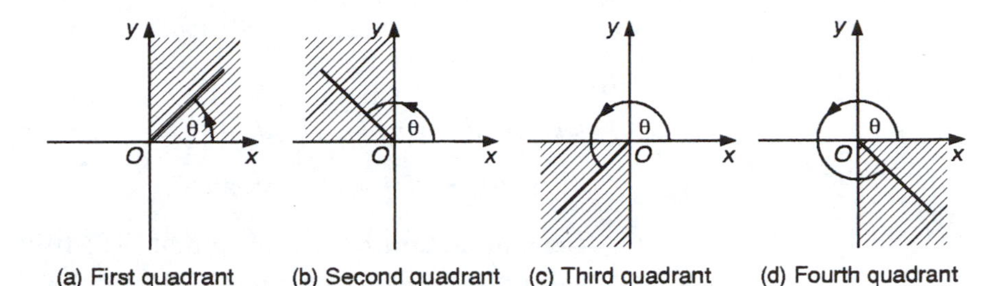

Fig. 1.5 The four quadrants of the (x, y)-plane: (a) first quadrant; (b) second quadrant; (c) third quadrant; (d) fourth quadrant.

from the x-axis, as indicated by the arrows in Fig. 1.5. The four quadrants correspond to:

first quadrant:	$0 \leq \theta \leq \pi/2$,
second quadrant:	$\pi/2 \leq \theta \leq \pi$,
third quadrant:	$\pi \leq \theta \leq 3\pi/2$,
fourth quadrant:	$3\pi/2 \leq \theta \leq 2\pi$.

The fundamental trigonometric functions are the *sine*, *cosine* and *tangent* functions. These are most conveniently defined in terms of the coordinates of a representative point P on a circle of unit radius, called a *unit circle*, centred on the origin. Consider the unit circle in Fig. 1.6 with the point $P(x, y)$ located on its circumference, and the line OP making an angle θ that is measured anticlockwise from the x-axis, as shown. Then, for $0 \leq \theta \leq 2\pi$, we will *define* the sine, cosine and tangent functions as follows:

(i) the *sine* of θ, written $\sin\theta$, is the y-coordinate of point P,
(ii) the *cosine* of θ, written $\cos\theta$, is the x-coordinate of point P,
(iii) the *tangent* of θ, written $\tan\theta$, is given by $\tan\theta = \dfrac{\sin\theta}{\cos\theta}$.

Applying Pythagoras' theorem to the triangle OPQ in Fig. 1.6, and using the fact that the radius of the unit circle is constant, shows the equation of the circle to be

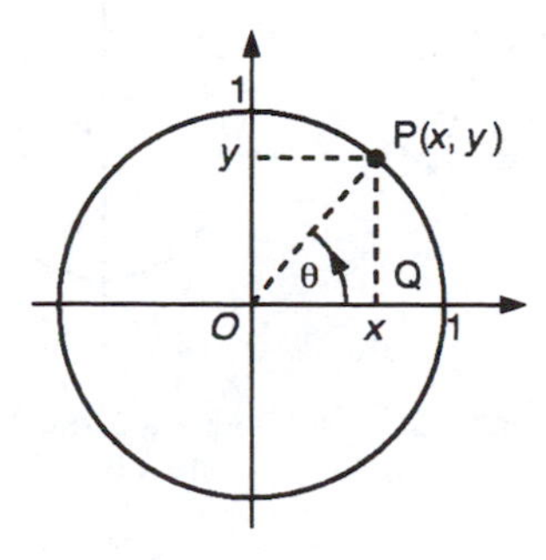

Fig. 1.6 Determination of the sine and cosine of an angle θ by means of the unit circle.

$$x^2 + y^2 = 1. \tag{1.10}$$

When combined with the definitions of $\sin\theta$ and $\cos\theta$ this result gives the *fundamental trigonometric identity*

$$\sin^2\theta + \cos^2\theta = 1, \quad \text{for all } \theta. \tag{1.11}$$

Here, by an *identity*, we mean that the left-hand side of the expression equals the right-hand side for *all* values of the variable involved (θ in this case), and not just for special values as happens with the roots of an equation.

The signs of the trigonometric functions in each quadrant can be determined by examination of Fig. 1.7. The results of Fig. 1.7 are summarized in the following diagram, in each quadrant of which are stated the trigonometric functions that are positive.

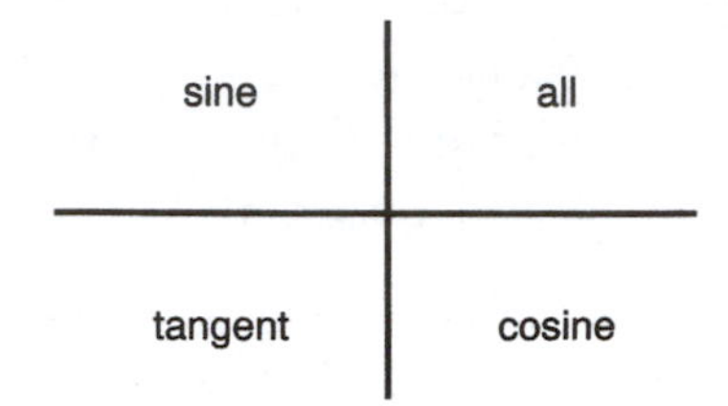

If instead of measuring θ positively from the x-axis (anticlockwise), it is measured negatively (clockwise) we arrive at the situation shown in Fig. 1.8. An application of the definitions of the trigonometric functions to the coordinates of the point $Q(x, -y)$ and to the triangle OQR yields the important and useful results:

$$\sin(-\theta) = -\sin\theta, \tag{1.12}$$

$$\cos(-\theta) = \cos\theta, \tag{1.13}$$

$$\tan(-\theta) = -\tan\theta, \tag{1.14}$$

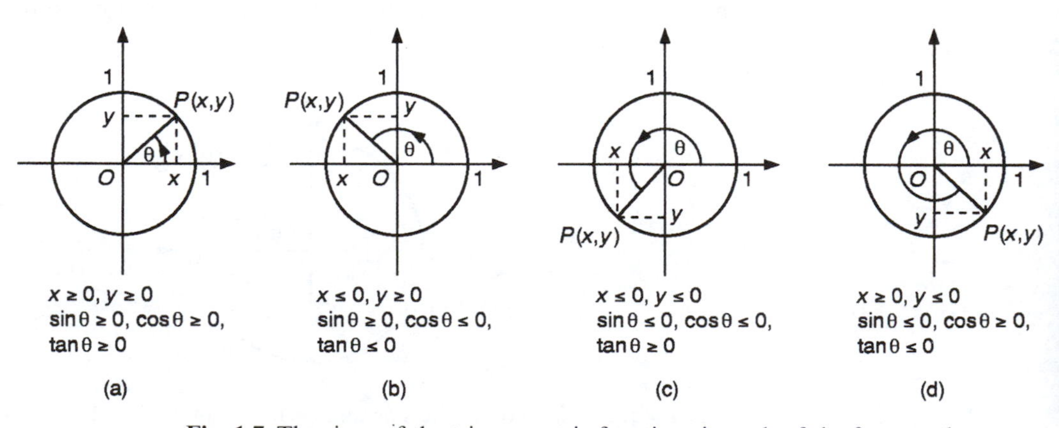

Fig. 1.7 The signs of the trigonometric functions in each of the four quadrants.

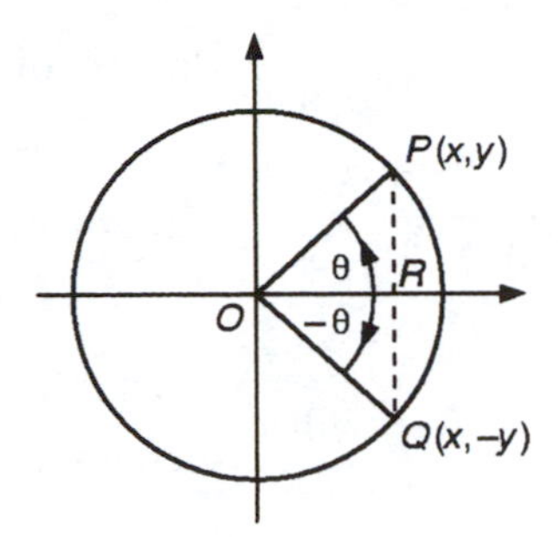

Fig. 1.8 Determination of trigonometric ratios for negative angles.

$$\sin\theta = \cos\left(\frac{\pi}{2} - \theta\right), \tag{1.15}$$

$$\cos\theta = \sin\left(\frac{\pi}{2} - \theta\right). \tag{1.16}$$

Later, in Section 2.3, we will see that functions like the sine and tangent functions which possess the property exhibited in (1.12) and (1.14) that changing the sign of the argument (θ in this case) leaves the magnitude of the function unchanged, but reverses its sign, are called *odd* functions. Conversely, functions like the cosine function which possess the property exhibited in (1.13) that changing the sign of the argument leaves both the magnitude and sign of the function unchanged are called *even* functions.

In general, functions are neither even nor odd, but when a function possessing one of these properties enters into an equation, the property can usually be used to advantage to simplify any subsequent manipulation that may be involved. For example, an expression like $2\sin\theta + \sin(-\theta)$ entering into an equation can be simplified, because $2\sin\theta + \sin(-\theta) = 2\sin\theta - \sin\theta = \sin\theta$, while an expression like $\cos\theta - \cos(-\theta)$ entering into an equation simplifies to zero. The important geometrical properties of functions which are either even or odd will be discussed in Section 2.3.

By scaling the triangle in Fig. 1.7(a), and considering the definitions of the three fundamental trigonometric functions, we arrive at the following alternative definitions for the sine, cosine and tangent functions in terms of the geometry of the right-angled triangle shown in Fig. 1.9:

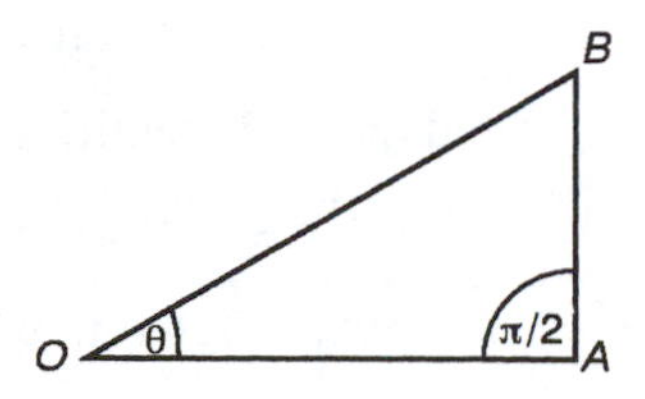

Fig. 1.9 A right-angled triangle.

$$\sin\theta = \frac{\text{length of side opposite the angle } \theta}{\text{length of the hypotenuse}} = \frac{AB}{OB},$$

$$\cos\theta = \frac{\text{length of side adjacent to the angle } \theta}{\text{length of the hypotenuse}} = \frac{OA}{OB},$$

$$\tan\theta = \frac{\text{length of side opposite the angle } \theta}{\text{length of side adjacent to the angle } \theta} = \frac{AB}{OA}.$$

Summary of properties of the trigonometric functions
Having defined the sine and cosine of an angle, we can use these ratios to define the other trigonometric functions:

$$\tan x = \frac{\sin x}{\cos x}, \quad \cot x = \frac{1}{\tan x} = \frac{\cos x}{\sin x},$$

$$\sec x = \frac{1}{\cos x}, \quad \csc x = \frac{1}{\sin x}.$$

A function $f(x)$ is *periodic* with *period* L if $f(x+L)=f(x)$ for every value of x, and L is the smallest value for which this is true. The functions sin x and cos x are periodic with period 2π, so that

$$\sin(x+2k\pi) = \sin x \text{ and } \cos(x+2k\pi) = \cos x, \quad \text{for all } x \text{ and } k = \pm 1, \pm 2, \ldots,$$

while the function tan x is periodic with period π, so that

$$\tan(x+k\pi) = \tan x, \text{ for } x \neq (2n+1)\pi/2 \text{ with } n = 0, 1, 2, 3, \ldots \text{ and } k = \pm 1, \pm 2, \ldots,$$

where the symbol $\neq$ is to be read 'not equal to'. The condition on x in the result concerning tan x simply excludes all odd multiples of $\pi/2$ at which tan x becomes infinite. Graphs of the fundamental trigonometric functions together with a statement of the values of x for which they are defined (their *domain of definition*) and the values assumed by the functions (their *range*) are shown in Fig. 1.10, from which their periodic properties can be seen.

Change of sign of argument

$$\sin(-\theta) = -\sin\theta$$

$$\cos(-\theta) = \cos\theta$$

$$\tan(-\theta) = -\tan\theta \tag{1.17}$$

Fundamental identities

$$\sin^2 x + \cos^2 x = 1 \tag{1.18}$$

$$\sec^2 x = 1 + \tan^2 x \tag{1.19}$$

$$\csc^2 x = 1 + \cot^2 x \tag{1.20}$$

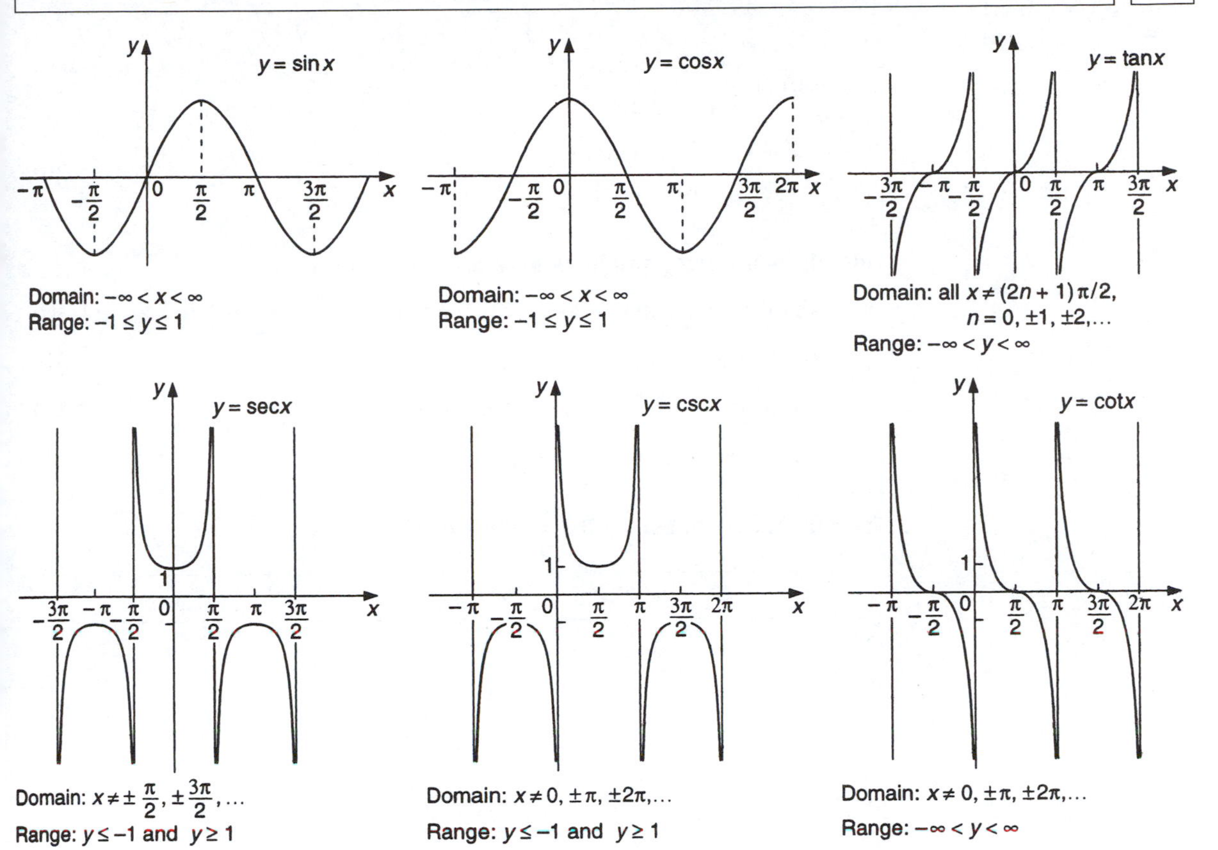

Fig. 1.10 Graphs of the trigonometric functions.

Sums and differences of angles

$$\sin(x+y) = \sin x \cos y + \cos x \sin y \tag{1.21}$$

$$\sin(x-y) = \sin x \cos y - \cos x \sin y \tag{1.22}$$

$$\cos(x+y) = \cos x \cos y - \sin x \sin y \tag{1.23}$$

$$\cos(x-y) = \cos x \cos y + \sin x \sin y \tag{1.24}$$

$$\tan(x+y) = \frac{\tan x + \tan y}{1 + \tan x \tan y} \tag{1.25}$$

$$\tan(x-y) = \frac{\tan x - \tan y}{1 + \tan x \tan y} \tag{1.26}$$

Identities involving double angles

$$\sin 2x = 2 \sin x \cos x \tag{1.27}$$

$$\begin{aligned} \cos 2x &= \cos^2 x - \sin^2 x \\ &= 1 - 2\sin^2 x \\ &= 2\cos^2 x - 1 \end{aligned} \tag{1.28}$$

$$\tan 2x = \frac{2\tan x}{1 - 2\tan^2 x}$$

$$= \frac{2\sin x \cos x}{\cos^2 x - \sin^2 x} \tag{1.29}$$

Identities involving products of sines and cosines

$$\sin(x+y) + \sin(x-y) = 2\sin x \cos y \tag{1.30}$$

$$\sin(x+y) - \sin(x-y) = 2\cos x \sin y \tag{1.31}$$

$$\cos(x+y) + \cos(x-y) = 2\cos x \cos y \tag{1.32}$$

$$\cos(x-y) - \cos(x+y) = 2\sin x \cos y \tag{1.33}$$

Special values of sin x, cos x and tan x

x	$\sin x$	$\cos x$	$\tan x$
0	0	1	0
$\pi/12$	$\frac{1}{2}\sqrt{2-\sqrt{3}}$	$\frac{1}{2}\sqrt{2+\sqrt{3}}$	$2-\sqrt{3}$
$\pi/6$	$\frac{1}{2}$	$\frac{1}{2}\sqrt{3}$	$1/\sqrt{3}$
$\pi/4$	$1/\sqrt{2}$	$1/\sqrt{2}$	1
$\pi/3$	$\frac{1}{2}\sqrt{3}$	$\frac{1}{2}$	$\sqrt{3}$
$5\pi/12$	$\frac{1}{2}\sqrt{2+\sqrt{3}}$	$\frac{1}{2}\sqrt{2-\sqrt{3}}$	$2+\sqrt{3}$
$\pi/2$	1	0	undefined ($\pm\infty$)
$7\pi/12$	$\frac{1}{2}\sqrt{2+\sqrt{3}}$	$-\frac{1}{2}\sqrt{2-\sqrt{3}}$	$-2-\sqrt{3}$
$2\pi/3$	$\frac{1}{2}\sqrt{3}$	$-\frac{1}{2}$	$-\sqrt{3}$
$3\pi/4$	$1/\sqrt{2}$	$-1/\sqrt{2}$	-1
$5\pi/6$	$\frac{1}{2}$	$-\frac{1}{2}\sqrt{3}$	$-1/\sqrt{3}$
$11\pi/12$	$\frac{1}{2}\sqrt{2-\sqrt{3}}$	$-\frac{1}{2}\sqrt{2+\sqrt{3}}$	$-2+\sqrt{3}$
π	0	-1	undefined ($\pm\infty$)

In the table the results for $\pi/6$ and $\pi/3$ follow by applying the alternative definitions of the trigonometric functions in terms of a right-angled triangle to the triangle OAB in the equilateral triangle OCB shown in Fig. 1.11(a), while those for $\pi/4$ follow in similar fashion by considering right-angled triangle OAB in Fig. 1.11(b). The results for $\pi/12$ were obtained by using the identity for $\sin(x-y)$ with $x = \pi/3$, $y = \pi/4$, and those for $5\pi/12$ follow from the identity for $\sin(x+y)$ with $x = \pi/3$, $y = \pi/12$. The results for the other angles may be obtained in similar fashion.

Finally, to close this section on trigonometric functions, we record the following simple results involving the sine and cosine functions, all of which follow directly from the elementary properties of the functions. These results will be useful in Chapter 16 when working with Fourier series.

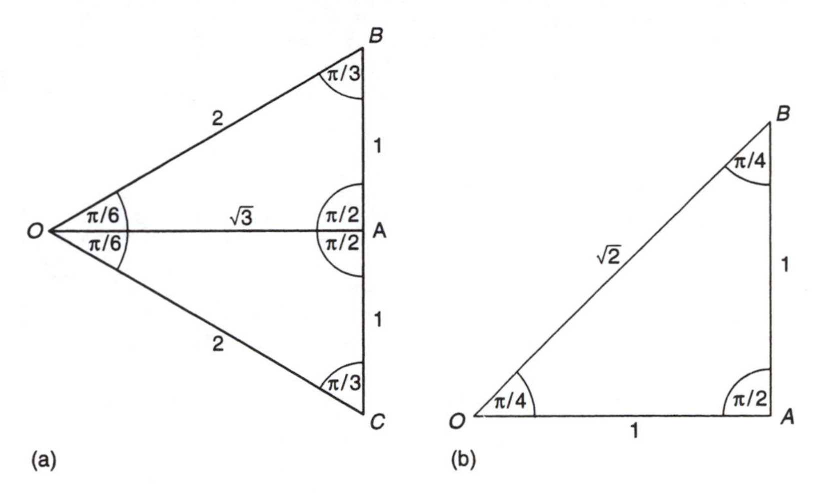

Fig. 1.11 Determination of trigonometric ratios for (a) $\pi/6$ and $\pi/6$; (b) $\pi/4$.

(i) $\sin(n\pi) = 0, \quad \text{for } n = 0, 1, 2, 3, \ldots$

(ii) $\cos(n\pi) = (-1)^n, \quad \text{for } n = 0, 1, 2, 3, \ldots$

(iii) $$\cos(n\pi/2) = \begin{cases} 1, & n = 0, 4, 8, 12, \ldots \\ 0, & n = 1, 3, 5, \ldots \\ -1, & n = 2, 4, 6, \ldots \end{cases}$$

(iv) $$\sin(n\pi/2) = \begin{cases} 1, & n = 1, 5, 9, \ldots \\ 0, & n = 2, 4, 6, \ldots \\ -1, & n = 3, 7, 11, \ldots \end{cases}$$

(v) $\sin[(2n+1)\pi/2)] = (-1)^n, \quad \text{for } n = 0, 1, 2, 3, \ldots$

(vi) $\cos[(2n+1)\pi/2)] = 0, \quad n = 0, 1, 2, 3, \ldots$

1.6 Cartesian geometry

This section offers a brief discussion of some elementary topics from Cartesian coordinate geometry that will prove useful elsewhere in this book. Hereafter a point P with x-coordinate x_0 and y-coordinate y_0 will either be written $P(x_0, y_0)$ to emphasize both the point P and its coordinates or, more simply, as (x_0, y_0) when only the coordinates are of importance.

1.6.1 Distance between two points

Let $P(x_0, y_0)$ and $Q(x_1, y_1)$ be any two points in the (x, y)-plane. Then the distance d between them, which is understood to be a non-negative quantity, is defined by means of Pythagoras' theorem as (see Fig. 1.12):

$$d^2 = |x_1 - x_0|^2 + |y_1 - y_0|^2.$$

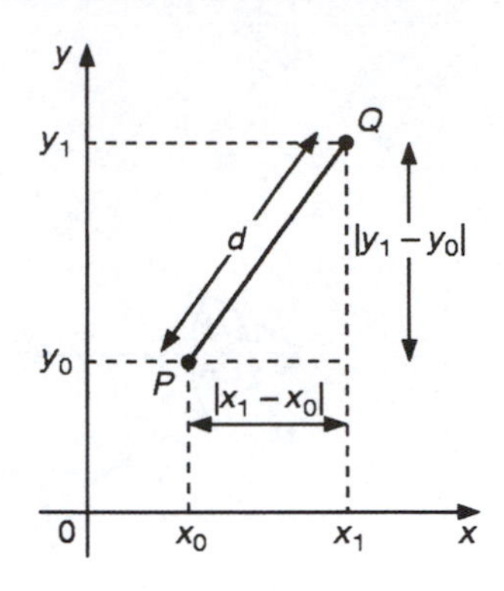

Fig. 1.12 The distance d between P and Q.

We have written $|x_1 - x_0|$ and $|y_1 - y_0|$, because these are the lengths of the projection of d on the x- and y-axes respectively, irrespective of whether $x_1 - x_0$ and $y_1 - y_0$ are positive or negative. However,

$$|x_1 - x_0|^2 = (x_1 - x_0)^2 \text{ and } |y_1 - y_0|^2 = (y_1 - y_0)^2,$$

so

$$d = \sqrt{(x_1 - x_0)^2 + (y_1 - y_0)^2} \geq 0.$$

Example 1.10 Find the distance between the points (2, 6) and (5, 3).

Solution Identifying (x_0, y_0) with (5, 3) and (x_1, y_1) with (2, 6) gives $x_0 = 5$, $y_0 = 3$, $x_1 = 2$ and $y_1 = 6$, and thus

$$d = \sqrt{(2-5)^2 + (6-3)^2} = 3\sqrt{2}. \quad \blacksquare$$

1.6.2 Straight line

When working with the usual orthogonal (mutually perpendicular) Cartesian x- and y-axes, the equation of a straight line in the (x, y)-plane has the form

$$y = mx + c. \tag{1.34}$$

Such a line L is shown in Fig. 1.13(a) in terms of which the constants m and c have a simple geometrical interpretation, as we now show.

Setting $x = 0$ in Eqn (1.34) shows that $y = c$ is the point of intersection or *intercept* of the line L with the y-axis. Furthermore, if α is the angle made by L with the x-axis, the angle being measured anticlockwise from the axis, and points P and Q on L correspond to $x = a$ and $x = b$, respectively, then

$$\tan\alpha = \frac{QR}{PR} = \frac{(mb + c) - (ma + c)}{b - a} = m$$

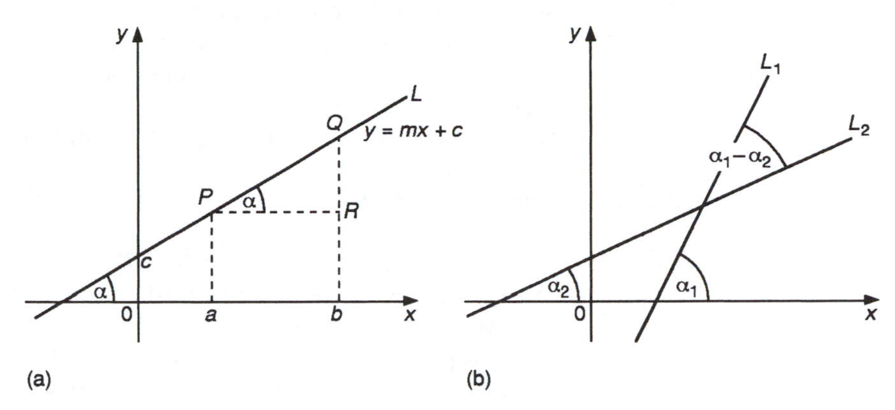

Fig. 1.13 (a) Straight line relative to Cartesian axes; (b) two intersecting straight lines L_1 and L_2.

or, equivalently,

$$\alpha = \arctan m. \tag{1.35}$$

Here the notation arctan is used to denote the inverse tangent in preference to the notation $\tan^{-1}$ that is also encountered in the literature (see section 2.2). The number m is called the *gradient* of line L, also its *slope*.

A useful consequence of result (1.35) is the determination of the condition that two straight lines L_1 and L_2 with equations

$$y = m_1 x + c_1 \quad \text{and} \quad y = m_2 x + c_2, \tag{1.36}$$

should be *orthogonal* (perpendicular). Two typical lines are shown in Fig. 1.13(b), and they will be orthogonal when $\alpha_1 - \alpha_2 = \pi/2$. Using the trigonometric identity (see Eqn (1.26))

$$\tan(\alpha_1 - \alpha_2) = \frac{\tan\alpha_1 - \tan\alpha_2}{1 + \tan\alpha_1 \tan\alpha_2},$$

and setting $m_1 = \tan\alpha_1$ and $m_2 = \tan\alpha_2$, gives

$$\tan(\alpha_1 - \alpha_2) = \frac{m_1 - m_2}{1 + m_1 m_2}.$$

Now if $\alpha_1 - \alpha_2 = \pi/2$, the expression on the right-hand side must be infinite, since $\tan \pi/2 = \pm\infty$, but as $m_1 \neq m_2$ this can only occur if

$$m_1 m_2 = -1. \tag{1.37}$$

Expressed in words, this condition for orthogonality states that the product of the two gradients m_1 and m_2 of the lines L_1 and L_2 must equal minus one. Naturally this condition is independent of the constants c_1 and c_2. A more general discussion of the straight line in three dimensions is given in section 4.8.

It follows directly from Eqn (1.34) that as there are two constants m and c involved in the determination of L, the line will be fixed by the specification of two independent conditions from which m and c may be determined. These may be either:

(a) the specification of two distinct points (x_1, y_1) and (x_2, y_2) on the line; or

(b) the specification of the gradient m and one point (x_1, y_1) on the line.

Example 1.11 Find the equation of the straight line through the points (1, 2) and (3, 7).

Solution Identifying the points (1, 2) and (3, 7) with the respective points P and Q in Fig. 1.13(a) we see that the gradient of the line $m = \tan \alpha = OR/PR = (7-2)/(3-1) = 2.5$, so that $y = 2.5x + c$. To determine c we use the fact that the points (1, 2) and (3, 7) lie on the line. Using the first of these in our equation gives $2 = 2.5 + c$, or $c = -0.5$ so that the required equation is $y = 2.5x - 0.5$. Had the second point (3, 7) been used instead we should, of course, have found the same value for c. ■

1.6.3 Circle

Consider a circle of radius r and centre $P(\alpha, \beta)$ as shown in Fig. 1.14(a). Then by Pythagoras' theorem

$$PQ^2 = PR^2 + QR^2$$

or, equivalently,

$$r^2 = (x_Q - \alpha)^2 + (y_Q - \beta)^2. \tag{1.38}$$

As Q is an arbitrary point on the circle we may drop the suffix Q to obtain the *standard form* of the equation for a circle of radius r with centre at (α, β),

$$(x - \alpha)^2 + (y - \beta)^2 = r^2. \tag{1.39}$$

When expanded this becomes

$$x^2 + y^2 - 2\alpha x - 2\beta y + \alpha^2 + \beta^2 - r^2 = 0. \tag{1.40}$$

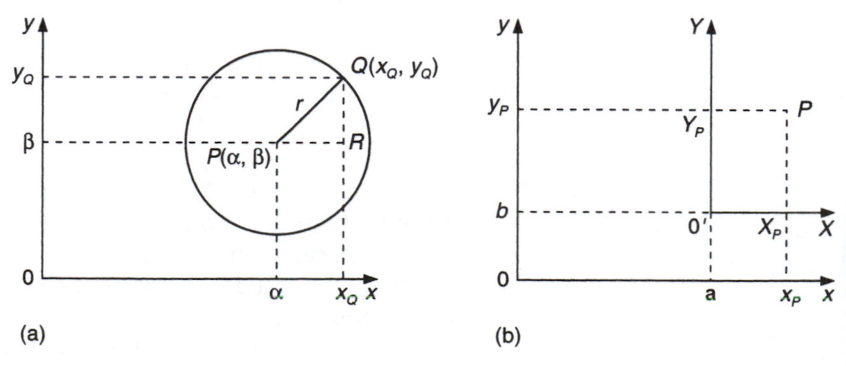

Fig. 1.14 (a) Circle centre (α, β) with radius r; (b) shift of origin.

If, now, this is compared with a general equation of the same form

$$x^2 + y^2 + ax + by + c = 0, \tag{1.41}$$

which may also describe a circle, we can deduce its radius and centre. By comparing corresponding coefficients in Eqns (1.40) and (1.41) we find

$$a = -2\alpha, \quad b = -2\beta, \quad c = \alpha^2 + \beta^2 - r^2. \tag{1.42}$$

Thus the circle represented by (1.41) has its centre at the point $(-\frac{1}{2}a, -\frac{1}{2}b)$ and a radius $r > 0$ that satisfies $r^2 = \frac{1}{4}a^2 + \frac{1}{4}b^2 - c$ if $a^2 + b^2 - 4c > 0$. Equation (1.41) does not represent a circle if $a^2 + b^2 - 4c < 0$.

Example 1.12 Find the equation of the straight line tangent to the circle

$$(x-2)^2 + (y-1)^2 = 4$$

at the point $(3, 1+\sqrt{3})$.

Solution The circle is shown in Fig. 1.15, with its centre P at (2, 1) and radius 2, and the point Q on its circumference at which the tangent line is required is located at $(3, 1+\sqrt{3})$. From section 1.6.2, the gradient m_R of the radius PQ is seen to be

$$m_R = \frac{y_Q - y_P}{x_Q - x_P} = \frac{(1+\sqrt{3}) - 1}{3-2} = \sqrt{3}.$$

As the gradient m_T of the tangent line T to the circle at Q must be such that $m_R m_T = -1$ (they are mutually perpendicular), it follows that $m_T = -1/\sqrt{3}$. Thus the tangent line T must have an equation of the form

$$y = -\frac{1}{\sqrt{3}}x + c.$$

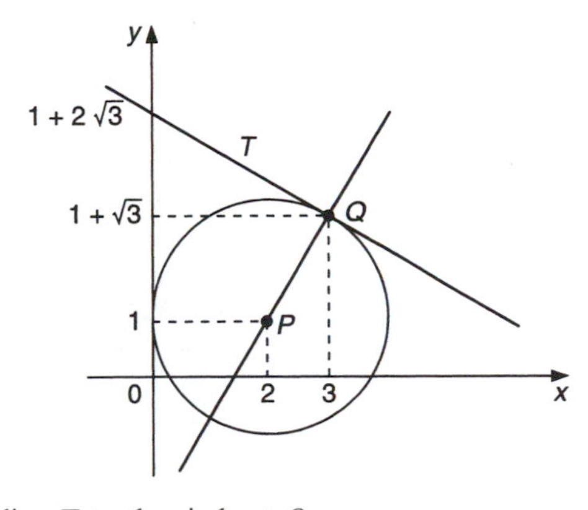

Fig. 1.15 Tangent line T to the circle at Q.

To determine c we use the fact that T passes through Q located at $x = 3$, $y = 1 + \sqrt{3}$. Consequently

$$1 + \sqrt{3} = -\frac{1}{\sqrt{3}} \cdot 3 + c, \quad \text{or } c = 1 + 2\sqrt{3}.$$

The required equation is thus

$$y = -\frac{1}{\sqrt{3}} x + 1 + 2\sqrt{3},$$

which shows the tangent line intersects the y-axis at $y = 1 + 2\sqrt{3}$. ■

1.6.4 Shift of origin

It is sometimes useful to *shift the origin* of a set of Cartesian axes from O to a point O' at (a, b) without rotating them, as shown in Fig. 1.14(b). If the new coordinate axes are $O'(X, Y)$ it becomes necessary to know the relationship that exists between the coordinate of a representative point P in terms of the original $O(x, y)$ axes, and the new $O'(X, Y)$ axes. Inspection of Fig. 1.14(b) shows that $X_P = x_P - a$ and $Y_P = y_p - b$, so that dropping the suffix P we arrive at the general result

$$X = x - a, \quad Y = y - b. \tag{1.43}$$

This is the required coordinate transformation. If, for example, the axes in Fig. 1.14(a) were shifted to the point $P(\alpha, \beta)$ in this manner, the equation of the circle (1.39) would become $X^2 + Y^2 = r^2$.

1.6.5 Rotation of axes

In certain circumstances it is necessary to *rotate* Cartesian axes about the origin through an angle θ as shown in Fig. 1.16. If the new coordinate axes are $O(X, Y)$ it is necessary to know the relationship that exists between the coordinates of the same representative point P in each

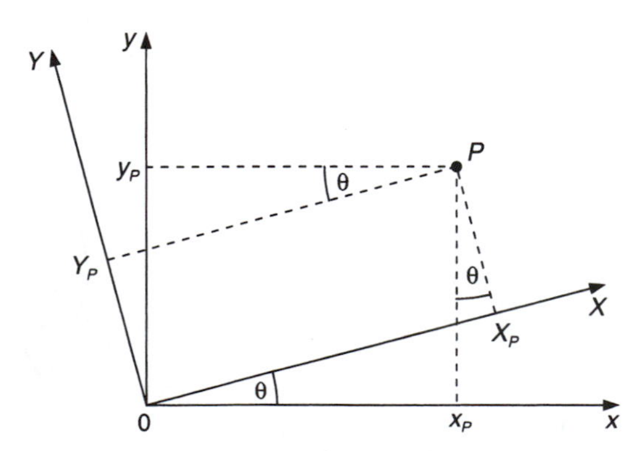

Fig. 1.16 Rotation of axes.

system of coordinates. Simple geometrical considerations in relation to the typical point P in Fig. 1.16, followed by dropping the suffix, show that the required result is

$$X = x\cos\theta + y\sin\theta, \quad Y = -x\sin\theta + y\cos\theta \tag{1.44a}$$

or, equivalently,

$$x = X\cos\theta - Y\sin\theta, \quad y = X\sin\theta + Y\cos\theta. \tag{1.44b}$$

The idea will be encountered again in section 10.9.

1.6.6 Parabola, ellipse and hyperbola

The simplest form of the equation for these three important curves, called their standard form, is as follows:

$$\text{Parabola} \quad y^2 = ax, \tag{1.45}$$

$$\text{Ellipse} \quad \frac{x^2}{a^2} + \frac{y^2}{b^2} = 1, \tag{1.46}$$

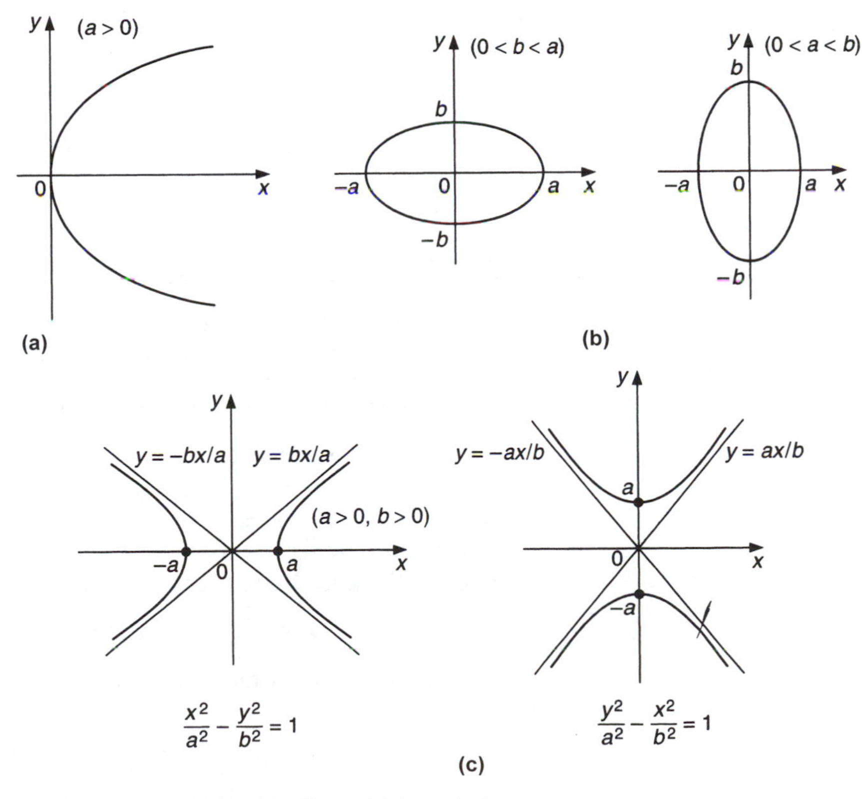

Fig. 1.17 (a) Parabola; (b) ellipse; (c) hyperbola.

Hyperbola $\dfrac{x^2}{a^2} - \dfrac{y^2}{b^2} = 1.$ (1.47)

The graphs of these functions are shown in Fig. 1.17.

In the ellipse (1.46) the longest dimension $2a$, which in the figure happens to lie along the x-axis, is called the *major axis* of the ellipse and the shortest dimension $2b$ which lies along the y-axis, is called the *minor axis*. Appropriately, the quantities a and b are called, respectively, the *semi-major* and the *semi-minor* axes of the ellipse, which is symmetrical about both the major and minor axes. In the case of the hyperbola (1.47) the two axes of symmetry $0x$ and $0y$ are called the *principal axes*, and the origin is called the *centre* of the hyperbola. The line of length $2a$ joining the points $(-a, 0)$ and $(a, 0)$ is called the *transverse axis* of the hyperbola. The two straight lines $y = \pm bx/a$ that become tangent to the hyperbola at infinity are called the *asymptotes* of the hyperbola (see section 3.7). The second diagram in Fig. 1.17(c) shows the form of the hyperbola when the number 1 on the right of the Eqn (1.47) is replaced by -1.

1.6.7 Sphere

A direct extension of the argument that led to the standard equation of the circle (1.39), shows that the standard equation of a sphere of radius r with its centre at the point (α, β, γ) is

$$(x-\alpha)^2 + (y-\beta)^2 + (z-\gamma)^2 = r^2. \tag{1.48}$$

This topic will be discussed again from the point of view of vectors in section 4.8.

1.6.8 General quadratic curves

The circle, parabola, ellipse and hyperbola all have equations which are special cases of the general quadratic equation

$$Ax^2 + Bxy + Cy^2 + Dx + Ey + F = 0, \tag{1.49}$$

in which the coefficients A, B, C, D, E and F are constants, with A, B and C not all zero. For example, the hyperbola (1.47) corresponds to $A = 1/a^2$, $C = -1/b^2$, $F = -1$ and $B = D = E = 0$.

If the x- and y-axes are rotated about the origin through an angle θ to become the X- and Y-axes respectively, it follows from (1.44b) that

$$x = X\cos\theta - Y\sin\theta, \quad y = X\sin\theta + Y\cos\theta. \tag{1.50}$$

The effect of such a rotation on the general quadratic equation (1.49) is to convert it to

$$A'X^2 + B'XY + C'Y^2 + D'X + E'Y + F' = 0, \tag{1.51}$$

where

$$
\begin{aligned}
A' &= A\cos^2\theta + B\cos\theta\sin\theta + C\sin^2\theta,\\
B' &= B(\cos^2\theta - \sin^2\theta) + 2(C - A)\sin\theta\cos\theta,\\
C' &= A\sin^2\theta - B\sin\theta\cos\theta + C\cos^2\theta,\\
D' &= D\cos\theta + E\sin\theta,\\
E' &= -D\sin\theta + E\cos\theta,\\
F' &= F.
\end{aligned} \tag{1.52}
$$

Now the form of (1.51) is simplified if the cross-product term XY vanishes, and this occurs when $B' = 0$. Using the trigonometric identities

$$\cos^2\theta - \sin^2\theta = \cos 2\theta, \quad 2\sin\theta\cos\theta = \sin 2\theta,$$

the condition $B' = 0$ is seen to be equivalent to requiring θ to satisfy either of the equations

$$\cot 2\theta = \frac{A - C}{B}, \quad \text{or} \quad \tan 2\theta = \frac{B}{A - C}. \tag{1.53}$$

Thus the cross-product term in the general quadratic equation (1.49) may always be removed by making a rotation of coordinates through an angle θ determined by either one of the equations in (1.53). This observation makes possible the identification of the type of graph represented by the general quadratic equation (1.49). To see this, suppose θ is chosen so that $B' = 0$, so that in the new coordinates (1.51) becomes

$$\tilde{A}X^2 + \tilde{C}Y^2 + \tilde{D}X + \tilde{E}Y + \tilde{F} = 0, \tag{1.54}$$

where $\tilde{A}$, $\tilde{C}$, $\tilde{D}$, $\tilde{E}$ and $\tilde{F}$ are the values assumed by A', C', D', E' and F' for this value of θ.

Inspection of (1.54) then shows it represents

(a) parabola if $\tilde{A} = 0$ or $\tilde{C} = 0$,
(b) an ellipse if $\tilde{A}$ and $\tilde{C}$ are different but both of the same sign, and a circle if $\tilde{A} = \tilde{C}$. It is possible for the ellipse and circle to degenerate to a single point, and they may not exist at all if no real X, Y satisfy (1.54),
(c) a hyperbola if $\tilde{A}$ and $\tilde{C}$ are non-zero and of opposite signs. It is possible for the hyperbola to degenerate into a pair of straight lines. (This happens, for example, in the case $X^2 - 4Y^2 = 0$.)

A simple calculation using (1.52) shows that for any rotation θ

$$B^2 - 4AC = B'^2 - 4A'C'. \tag{1.55}$$

Thus the number $B^2 - 4AC$, called the *discriminant* of the general quadratic equation (1.49), is invariant with respect to a rotation of axes. Choosing the angle θ which makes $B' = 0$ shows that

$$B^2 - 4AC = -4\tilde{A}\tilde{C}. \tag{1.56}$$

Combining this last result with observations (a) to (c) then provides the following simple discriminant test for the type of graph represented by (1.49). The general quadratic equation (1.49) represents

(a) a parabola if $B^2 - 4AC = 0$,
(b) an ellipse if $B^2 - 4AC < 0$,
(c) a hyperbola if $B^2 - 4AC > 0$,

with the possible occurrence of degeneracies of the type just described. Notice that as this test is applied to (1.49), it does not require the elimination of the cross-product term xy.

A special degeneracy arises in (1.54) if the left-hand side can be factorised into two linear factors, and also if $\tilde{A} = \tilde{C} = 0$, but not both $\tilde{A}$ and $\tilde{E}$ are zero. In the first case (1.30) degenerates into a pair of straight lines, while in the second case it reduces to a single straight line.

Example 1.13 Determine the nature of the graph of the equation

$$7x^2 - 6\sqrt{3}xy + 13y^2 - 4 = 0,$$

and sketch it.

Solution Here $A = 7$, $B = -6\sqrt{3}$ and $C = 13$, so the discriminant $B^2 - 4AC = -256$, showing that the graph is an ellipse. Now

$$\cot 2\theta = \frac{A - C}{B} = \frac{1}{\sqrt{3}},$$

so that $2\theta = \pi/3$, or $\theta = \pi/6$. Using $\theta = \pi/6$ in (1.44b) and substituting into the given equation reduces it to the equation of an ellipse in standard form

$$X^2 + \frac{Y^2}{\left(\frac{1}{2}\right)^2} = 1.$$

The semi-major axis is thus 1, the semi-minor axis is $\frac{1}{2}$; the centre is at the origin and the X- and Y-axes are obtained from the x- and y-axes by means of an anticlockwise rotation about the origin through an angle $\pi/6$. The graph is shown in Fig. 1.18. ■

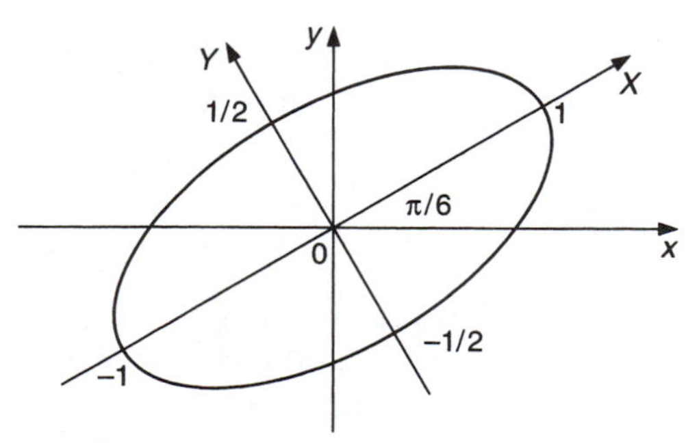

Fig. 1.18 Graph of ellipse $7x^2 - 6\sqrt{3}xy + 13y^2 = 4$.

1.6.9 Cone

Let Γ be a simple closed plane curve and V some fixed point not in the plane of Γ. Then a *cone* is the three-dimensional surface generated by a straight line passing through V which is allowed to move so that its point of intersection P with Γ traverses Γ. The point V is called the *vertex* of the cone, and each of the straight lines passing through V lying along the surface of the cone is called a *generator* of the cone. If there is a line passing through V which is not a generator of the cone, but about which the cone is symmetrical, this line is called the *axis* of the cone. A typical asymmetric cone is illustrated in Fig. 1.19(a) in which V is located at (x_0, y_0, z_0).

Each of the two separate parts of a cone which have the vertex V as their common point is called a *nappe* of the cone. Although in mathematics a cone comprises the two nappes, as in elementary geometry the name 'cone' is often given to a single nappe.

The equation of an *elliptic cone* with its vertex at the origin and its axis coincident with the z-axis is

$$\frac{x^2}{a^2} + \frac{y^2}{b^2} - \frac{z^2}{c^2} = 0. \tag{1.57}$$

This is so-called because the curve of cross-section in any plane normal to the axis (i.e., any plane $z = \text{const.}$) is an ellipse. An elliptic cone reduces to a *right circular cone* when $a = b$, for which the equation becomes

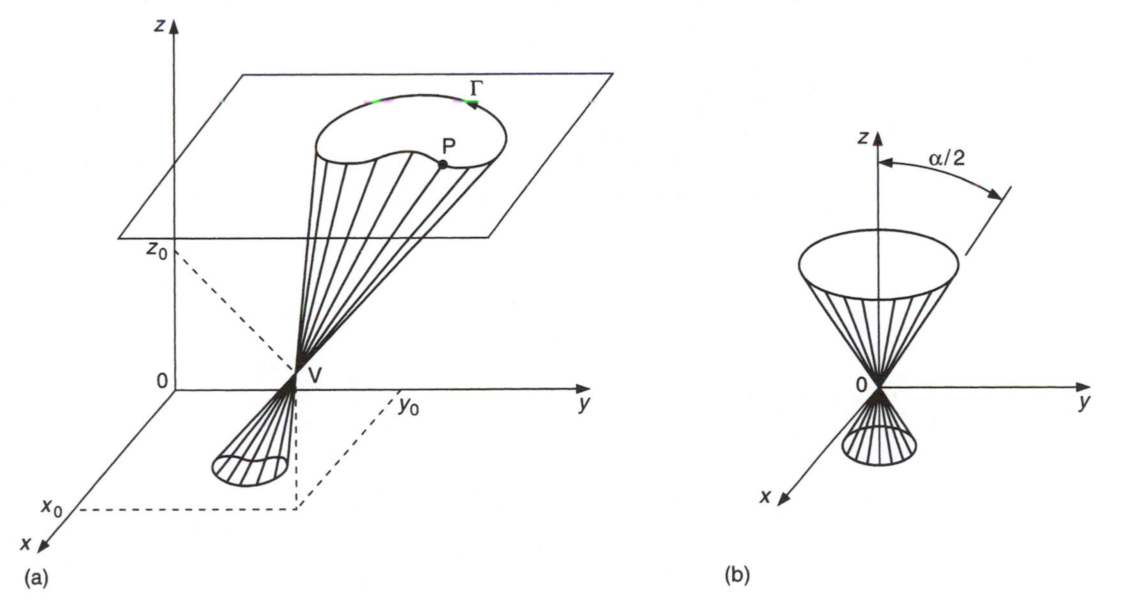

Fig. 1.19 (a) Asymmetric cone with vertex V at (x_0, y_0, z_0); (b) right circular cone.

$$\frac{x^2}{a^2}+\frac{y^2}{a^2}-\frac{z^2}{c^2}=0, \tag{1.58}$$

and in this case the curve of cross-section in any plane normal to the axis is a circle.

Because of rotational symmetry about the axis, the interior angle α at the vertex of this right circular cone is the angle between the two generators of the cone which lie in any plane containing the z-axis. Such a plane is $y=0$, when (1.58) reduces to

$$\frac{x^2}{a^2}-\frac{z^2}{c^2}=0, \quad \text{or} \quad x=\pm\left(\frac{a}{c}\right)z$$

showing $\alpha = 2 \arctan (a/c)$. The cone is shown in Fig. 1.19(b).

1.7 Polar coordinates

In many problems it is advantageous to use the polar coordinates (r, θ) to define points in a plane rather than the Cartesian coordinates (x, y). The *polar coordinates* (r, θ) of a point P in a plane are defined as follows.

We take an origin O in a plane and a fixed reference line through it also in the plane, as shown in Fig. 1.20. This line is sometimes called the *polar axis*. Then a point P in the plane is identified by specifying the length r of the line OP and the angle θ measured anticlockwise from the reference line to OP. The length r is called the *radial distance* and the angle θ the *polar angle*.

It is clear that the polar coordinates of a point are not unique, because although r is uniquely determined, the same point P will be identified if we replace θ by $\theta \pm 2n\pi$, with $n=0, 1, 2, \ldots$. Thus the polar angle θ is determined up to a multiple of 2π for every point in the plane apart from the origin at which it is not defined. Because of the necessity to use polar coordinates in problems involving differentiation, the angle θ will always be measured in radians and, for convenience, the angle θ will be chosen to lie in the interval $0 \le \theta < 2\pi$.

Now let the reference line be identified with the x-axis of a Cartesian coordinate system in the same plane with its origin also located at O.

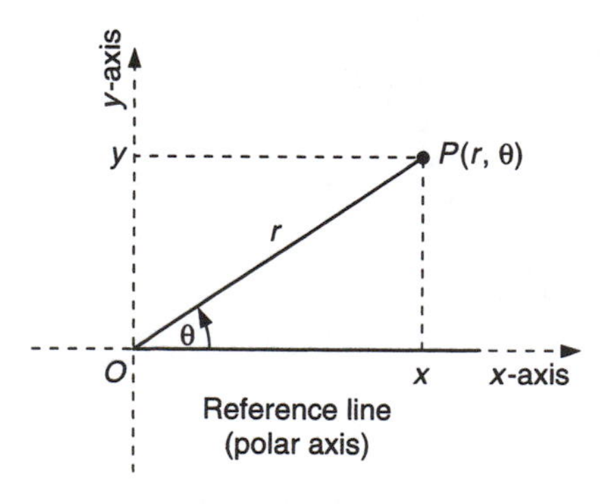

Fig. 1.20 Polar coordinates (r, θ).

Then the configuration is as shown in Fig. 1.20 and, if P has the Cartesian coordinates (x, y) and the polar coordinates (r, θ), we see that

$$x = r\cos\theta, \quad y = r\sin\theta. \tag{1.59}$$

These relationships are needed when converting from polar coordinates to Cartesian coordinates, and conversely, and it follows at once that

$$r = \sqrt{x^2 + y^2} \quad \text{and} \quad \theta = \arccos\left(\frac{x}{\sqrt{x^2 + y^2}}\right) = \arcsin\left(\frac{y}{\sqrt{x^2 + y^2}}\right). \tag{1.60}$$

Example 1.14

(a) Find the Cartesian coordinates of the point with polar coordinates $(4, 3\pi/4)$, and

(b) the polar coordinates of the point with Cartesian coordinates $(2, -2\sqrt{3})$.

Solution

(a) From Eqns (1.59) we have

$$x = 4\cos\frac{3\pi}{4}, \quad y = 4\sin\frac{3\pi}{4},$$

so $x = 2\sqrt{2}$, $y = 2\sqrt{2}$, as shown in Fig. 1.21(a).

(b) From Eqns (1.60) we have

$$r = \sqrt{2^2 + (2\sqrt{3})^2} = 4,$$

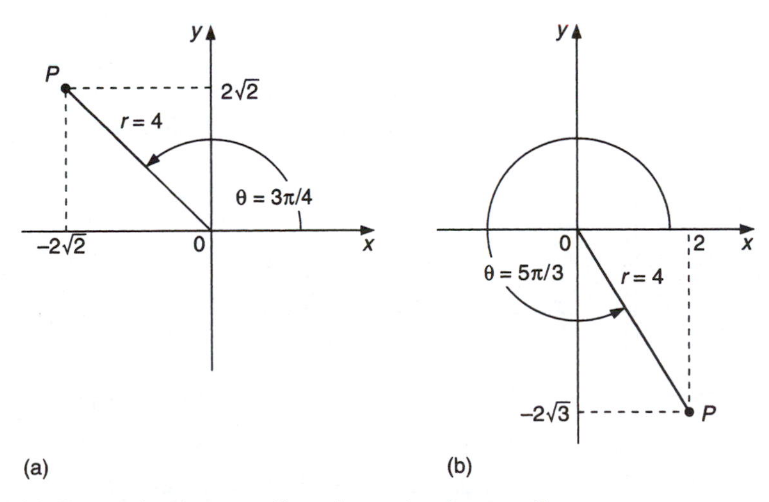

Fig. 1.21 Conversion between Cartesian and polar coordinates.

$$\theta = \arccos \tfrac{1}{2} = \arcsin\left(\frac{-\sqrt{3}}{2}\right).$$

Taken together, the last two results show that $\theta = 5\pi/3$. Thus the required polar coordinates are $(4, 5\pi/3)$, as shown in Fig. 1.21(b). We have used both expressions for θ given in Eqns (1.60) to ensure that we identify the correct quadrant in which the angle θ is located. Equivalently, we could have used the result $\theta = \arctan(y/x)$ and then identified the quadrant by using the signs of x and y. ■

Many interesting and useful geometrical figures are most easily specified in terms of polar coordinates. The following example illustrates a few such figures.

Example 1.15 Sketch the graphs of the following figures specified in terms of polar coordinates:

(a) $r = a(1 - \cos\theta)$ $(a > 0)$,
called a *cardioid* because of its heart shape;

(b) $r^2 = a \cos 2\theta$,
called a *lemniscate*;

(c) $r = \theta$, with θ allowed to increase without bound.

This is called an *Archimedean spiral*.

Solution The results are shown in Fig. 1.22. Points on the cardioid in Fig. 1.22(a) are generated by choosing a set of values of θ over the interval $0 \leq \theta \leq 2\pi$, computing the corresponding values of r, and plotting the points $P(r, \theta)$ using polar coordinates. As the values of the cosine function repeat themselves for $\theta > 2\pi$ (the period of the cosine function is 2π), taking θ outside this interval will simply generate points already on the cardioid.

The construction of the lemniscate in Fig. 1.22(b) requires a little more thought. We have $r = \sqrt{a \cos 2\theta}$, so the radius vector r, which must be real and non-negative, will only be defined when $\cos 2\theta \geq 0$. As the cosine function is an *even* function we have $\cos(-2\theta) = \cos 2\theta$, showing that the radius vector r must have the same values for θ and $-\theta$. Recalling

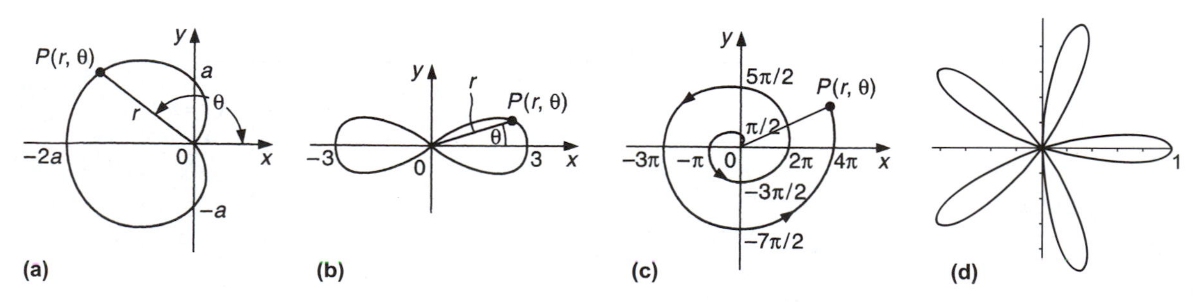

Fig. 1.22 (a) Cardioid; (b) lemniscate $r^2 = 9 \cos 2\theta$; (c) archimedean spiral; (d) five petal rose.

from Fig. 1.8 how negative angles are measured, this means that the curve above the reference line (the x-axis) in Fig. 1.20 will be repeated below the line as though the x-axis were a mirror, so to the right of the origin the curve must be symmetrical about the x-axis. As the cosine function has period 2π, it follows directly that for those values of θ for which $\cos 2\theta \geq 0$, the curve to the left of the origin must be a repeat of the curve to the right as though the y-axis were a mirror. Clearly, when $\cos 2\theta = 0$, the lemniscate must pass through the origin.

We have $\cos 2\theta \geq 0$ for $-\frac{1}{4}\pi \leq \theta \leq \frac{1}{4}\pi$, and plotting the points $P(r, \theta)$ in this interval gives the loop to the right of the y-axis in Fig. 1.22(b) which passes through the origin. The loop to the left follows after reflecting the curve to the right about the y-axis.

It is instructive to look more closely at the polar representation of the lemniscate $r = \sqrt{a \cos 2\theta}$. To cover the entire plane the polar angle θ must lie in an interval of length 2π, but in such an interval $\cos 2\theta$ will only be non-negative for $-\frac{1}{4}\pi \leq \theta \leq \frac{1}{4}\pi$ and $-\frac{3}{4}\pi \leq \theta \leq \frac{3}{4}\pi$, so a real curve will only be possible for θ in one of these two intervals. Values of θ in the first interval define the loop of the lemniscate to the right of the origin, while values of θ in the second interval define the loop of the lemniscate to the left of the origin. Thus there can be no real curve in the region of the (r, θ)-plane in the wedge shaped region $\frac{1}{4}\pi < \theta < \frac{3}{4}\pi$ (the one above the x-axis) and the wedge shaped region $-\frac{3}{4}\pi < \theta < -\frac{1}{4}\pi$ (the one below the x-axis). So, using these arguments without having first drawn the lemniscate, we have deduced that the lemniscate is symmetrical about both the x and y-axes, that it passes through the origin with the lines $x = \pm y$ as tangent lines, and it is obvious that it passes through the points with polar coordinates $(\sqrt{a}, 0)$ and $(\sqrt{a}, \pi)$, which are the points $(\sqrt{a}, 0)$ and $(-\sqrt{a}, 0)$ in the (x, y)-plane.

The construction of the Archimedean spiral in Fig. 1.22(c) is straightforward, and as θ increases with each revolution about the origin $r = 0$, so the spiral expands over the entire plane.

Figure 1.22(d) shows a polar plot of the curve $r = \cos 5\theta$ which has five lobes equally spaced around the origin. A plot like this, in which a lobe with one end located at the origin $r = 0$ is repeated symmetrically around the origin n times, is called an *n petal rose* because of its resemblance to a flower with n petals. Notice, however, that the general curve $r = a\cos n\theta$ has n lobes (petals) when n is odd, but $2n$ lobes (petals) when n is even. Sketch the curve $r = \cos 2\theta$ to convince yourself of this. ■

Finally, let us derive the equations of straight lines and circles in terms of polar coordinates. Inspection of Fig. 1.23(a) shows the line $y = a$ to have the equation

$$r \sin\theta = a, \tag{1.61}$$

while inspection of Fig. 1.23(b) shows the line $x = b$ to have the equation

$$r \cos\theta = b. \tag{1.62}$$

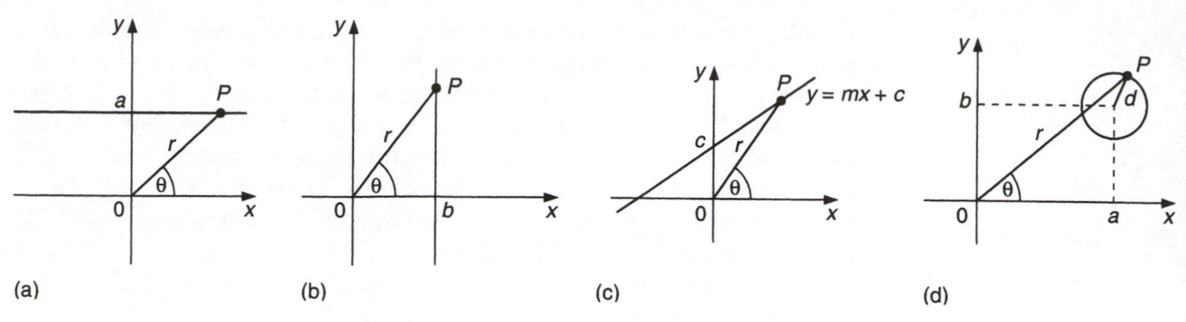

Fig. 1.23 Derivation of the equations in polar coordinates of (a) $y = a$; (b) $x = b$; (c) $y = mx + c$; (d) the circle centre (a, b) with radius d.

To obtain the polar coordinate form for the equation of the straight line $y = mx + c$ shown in Fig. 1.23(c) we substitute Eqns (1.59) and obtain

$$r \sin\theta = mr \cos\theta + c$$

or, equivalently,

$$r = \frac{c}{\sin\theta - m\cos\theta}. \tag{1.63}$$

To obtain the equation of the circle in Fig. 1.23(d) with radius d and its centre located at the point (a, b), which has the equation (see Eqn (1.39))

$$(x - a)^2 + (y - b)^2 = d^2,$$

we again substitute Eqns (1.59) and, after simplification, arrive at the equation

$$r^2 - 2r(a\cos\theta + b\cos\theta) + a^2 + b^2 - d^2 = 0. \tag{1.64}$$

1.8 Completing the square

Many situations arise in which it is necessary to express the quadratic expression

$$ax^2 + bx + c$$

in the alternative form

$$(px + q)^2 + r.$$

The method by which this is accomplished is called *completing the square*, and to see how it is carried out we first equate the two expressions to obtain

$$ax^2 + bx + c = (px + q)^2 + r. \tag{1.65}$$

Expanding the right-hand side gives

$$ax^2 + bx + c = p^2x^2 + 2pqx + q^2 + r, \tag{1.66}$$

which must be an *identity*; that is, it must be true for *all* values of x. This is only possible if the coefficients of corresponding powers of x on each side of this result are equal, so as $c = cx^0$ and $r = rx^0$,

$$a = p^2, \quad b = 2pq \quad \text{and} \quad c = q^2 + r,$$

showing that

$$p = \sqrt{a}, \quad q = \frac{b}{2\sqrt{a}} \quad \text{and} \quad r = c - \frac{b^2}{4a}. \tag{1.67}$$

We have now established that

$$ax^2 + bx + c = \left(\sqrt{a}x + \frac{b}{2\sqrt{a}}\right)^2 + c - \frac{b^2}{4a}$$

or, equivalently, that

$$ax^2 + bx + c = a\left(x + \frac{b}{2a}\right)^2 + c - \frac{b^2}{4a}. \tag{1.68}$$

Thus completing the square in the quadratic expression ax^2+bx+c has led us to the right-hand side of identity (1.68), which is the required alternative form for the quadratic.

It is, in fact, this process that leads to the familiar formula for finding the roots of the quadratic equation

$$ax^2 + bx + c = 0. \tag{1.69}$$

To see this, we rewrite Eqn (1.69) in the form

$$a\left(x + \frac{b}{2a}\right)^2 + \frac{4ac - b^2}{4a} = 0, \tag{1.70}$$

which is equivalent to

$$\left(x + \frac{b}{2a}\right)^2 + \frac{b^2 - 4ac}{4a^2},$$

so taking the square root and solving for x brings us to the familiar formula

$$x = \frac{-b \pm \sqrt{(b^2 - 4ac)}}{2a}. \tag{1.71}$$

Example 1.24 Complete the square to find an alternative expression for

$$4x^2 + 9x + 1.$$

Solution Here $a = 4$, $b = 9$ and $c = 1$, so substituting these values into the right-hand side of identity (1.68) gives

$$4x^2 + 9x + 1 = 4\left(x + \frac{9}{8}\right)^2 - \frac{65}{16}. \quad \blacksquare$$

Example 1.25 Find the Cartesian form of the equation involving the polar coordinates (r, θ) when

$$r^2 - 2r\cos\theta + 4r\sin\theta - 11 = 0.$$

Solution From Eqn (1.60) $r^2 = x^2 + y^2$, so substituting Eqns (1.59) gives

$$x^2 + y^2 - 2x + 4y - 11 = 0.$$

Completing the square this becomes

$$(x-1)^2 + (y+2)^2 = 16,$$

so the equation represents a circle of radius 4 with its centre at the point $(1, -2)$.

The same result could have been obtained without substituting for x and y and then completing the square, by comparison of the given equation and Eqn (1.64). This shows $-2 = -2a$ so $a = 1$, $-2b = 4$ so $b = -2$ and $a^2 + b^2 - d^2 = -11$, so $d^2 = 16$, whence $d = 4$. $\blacksquare$

1.9 Logarithmic functions

Let $a > 0$ be an arbitrary real constant. Then if $y > 0$ is any given real number, the equation

$$y = a^x \tag{1.72}$$

defines a number x called the *logarithm* of y to the *base* a, and to show this we write $x = \log_a(y)$. Conversely, y is called the *antilogarithm* of x, and this is shown by writing $y = \text{antilog}_a(x)$. For every base a, it follows that $\log_a(1) = 0$ and $\text{antilog}_a(0) = 1$.

Let the logarithm of y_1 to the base a be x_1 and the logarithm of y_2 to the base a be x_2. Then, using the properties of indices, it follows that the product

$$y_1 y_2 = a^{x_1} a^{x_2} = a^{(x_1 + x_2)}, \tag{1.73}$$

so the logarithm to the base a of the product $y_1 y_2$ is the *sum* of the two logarithms $x_1 + x_2$, each to the base a.

Similarly, the quotient

$$y_1 / y_2 = a^{x_1} / a^{x_2} = a^{(x_1 - x_2)}, \tag{1.74}$$

so the logarithm to the base a of the quotient y_1/y_2 is the *difference* of the two logarithms $x_1 - x_2$, each to the base a.

The logarithm is also related to the p/qth root of a number $y > 0$, because from Eqn (1.72)

$$y^{p/q} = (a^x)^{p/q} = a^{px/q}, \tag{1.75}$$

so the logarithm to the base a of $y^{p/q}$ is simply px/q.

Thus if for any given base a a table of logarithms is constructed, by using Eqns (1.72) to (1.75), the logarithms of y_1y_2, y_1/y_2 and $y^{p/q}$ can be found. Thereafter, if the table is used in reverse to find the antilogarithm y corresponding to a given logarithm x, it is possible to find the numbers y_1y_2, y_1/y_2 and $y^{p/q}$.

When logarithms were first introduced by John Napier, in a publication dated 1614, it was with the purpose of simplifying the task of multiplying and dividing arbitrary positive numbers, and the concept of a base as described above was not involved. Instead Napier constructed rods marked with numbers in geometric progression which could be used side by side to find the product y_1y_2 or the quotient y_1/y_2 of any two positive numbers y_1 and y_2. This computational device became known as *Napier's rods*, and it was the forerunner of the *slide rule.* It was not until later, as a result of a meeting between H. Briggs, the first Savilian Professor of geometry in Oxford, and Napier, that the idea of a base was introduced, when they agreed that the base should be ten, so $a = 10$. From that time, until the introduction of hand held calculators, slide rules and tables of logarithms to the base 10, called *common logarithms*, were used for arithmetic calculations. The symbol log y will always be used to denote $\log_{10} y$, so for simplicity we will write $x = \log y$ in place of $x = \log_{10} y$ and, correspondingly, we will write $y = \text{antilog } x$.

Although convenient for arithmetical calculations, and for plotting graphs when the dependent or independent variables vary over a large range of numerical values, logarithms to the base 10 are not convenient for mathematical work. It turns out that the appropriate base to use for mathematical analysis is Euler's constant $e = 2.7182818\ldots$, to be introduced later in Section 3.3. Logarithms to the base e, called *natural logarithms* and often *Naperian logarithms*, are always denoted by writing $x = \ln y$ without the necessity of writing $x = \ln_e y$. Although with natural logarithms $y = e^x$, for the time being we will write $y = \text{antiln } x$. Later, when Euler's constant e has been properly defined and the properties of the function e^x, called the *exponential function* and often written $\exp x$, have been developed, there will no longer be any use for the notation $y = \text{antiln } x$, because instead we will always write $y = e^x$. The relationship between logarithms to an arbitrary base a and $\ln x$ follows by recognizing that as $x = e^{\ln x}$, we can write $\log_a x = \log_a(e^{\ln x}) = \log_a(e) \ln x$. Setting $a = 10$ to find the relationship between $\ln x$ and $\log_{10} x$, the previous result becomes $\log_{10} x = \log_{10}(e) \ln x$, but $\log_{10}(e) = 0.434294$, so

$$\log_{10} x = 0.434294 \ln x \tag{1.76}$$

which is equivalent to,

$$\ln x = 2.302585 \log_{10} x. \tag{1.77}$$

Figure 1.24 shows superimposed graphs of $\log x$ and $\ln x$ for $0 < x \leq 10$, with each function tending to minus infinity as x approaches 0 through positive values, because the logarithmic functions are not defined for negative x.

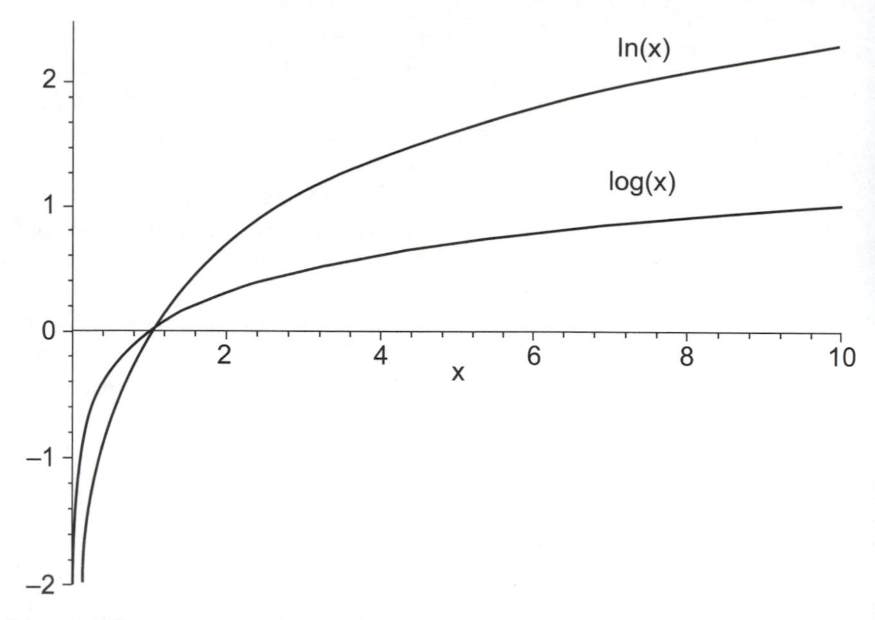

Fig. 1.24 Superimposed graphs of log x and ln x.

It will be seen later that in mathematical analysis the exponential function $y = e^x$ is a far more important function than ln x. Fig. 1.25 shows the superimposed graphs of the functions ln x and antiln x, and between them shown as a dot-dashed line the graph of $y = x$. To understand how the graph of antiln x is derived from the graph of ln x, consider a point on the graph of ln x corresponding to $x = 3$ when ln 3 is seen to be approximately 1. Reading this result in reverse means that antiln 3 is approximately equal to 1, which gives a point on the graph of antiln x. Repeating this process for all x, and superimposing the graphs of ln x and antiln x, gives the result of Fig. 1.25. A brief study of the way the graph of antiln x is derived from the graph of ln x shows that the curve for antiln x can be obtained from the curve for ln x by regarding the dot-dashed line $y = x$ as a mirror in which the graph of ln x is reflected.

When the concept of a function is considered in Chapter 2, it will become apparent from the considerations in Section 2.2 that antiln $x = e^y$ is the function *inverse* to the function ln x, and conversely.

Example 1.26 A measure of the magnitude of the noise generated by a machine rotating at x revolutions a second is given by the function

$$A(x) = \frac{1}{\sqrt{1 - 4x^2 + 16x^4}},$$

and it is required to graph the variation of $A(x)$ for x in the interval $0.01 \le x \le 100$.

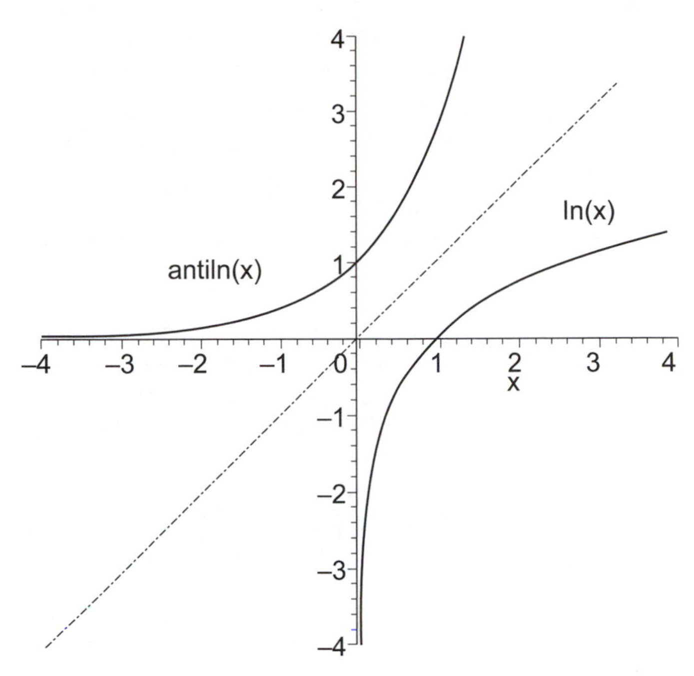

Fig. 1.25 Superimposed graphs of ln x and antiln x together with the graph of $y = x$.

Solution Inspection shows that most of the variation of $A(x)$ occurs around $x = 0$, with $A(x)$ approximately equal to 1 for $x < 0$ and 0 for $x > 0$. Thus a graph of $A(x)$ against x would be unhelpful, because the variation of $A(x)$ is confined to a narrow interval in x, and it is very small when compared with the variation of x. The difficulty is overcome by graphing $A(x)$ against $\log x$, because this will compress the variation in x to the interval $\log(0.01) \leq \log x \leq \log(100)$, which is equivalent to the interval $-2 \leq \log x \leq 2$. The resulting graph is shown in Fig. 1.26 from which it is seen that the maximum noise level occurs when $\log x \approx -0.42$, which corresponds to approximately 0.38 revolutions a second. The graph also

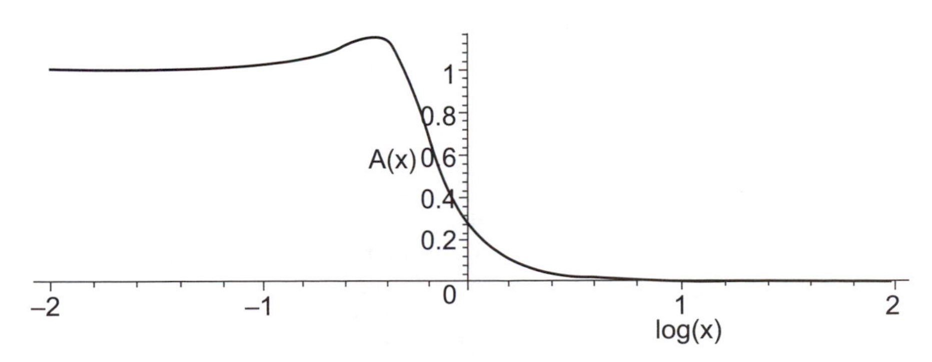

Fig. 1.26 The graph of $A(x)$ against $\log x$ for $0.01 \leq x \leq 100$.

shows how rapidly $A(x)$ approaches the value 1 for slower revolutions a second and the value 0 for faster revolutions a second. ■

To close this Section we derive a useful general result concerning logarithms by seeking a number y such that for logarithms to any base a

$$\log_a(x+y) = -\log_a(x-y).$$

As $\log_a(1/z) = \log_a(1) - \log_a(z)$, and $\log_a(1) = 0$, we see that

$$\log_a(1/z) = -\log_a(z), \tag{1.78}$$

which is itself a result that is often used to simplify equations. However, to proceed with our main task, we set $z = x - y$ and apply Eqn (1.78) to the term $\log_a(x-y)$ on the right of the original equation. Then, after taking the antilog to the base a of both sides of the equation, we arrive at the result

$$x + y = \frac{1}{x - y}.$$

This is equivalent to $x^2 - y^2 = 1$, and so $y = \sqrt{x^2 - 1}$. Finally, substituting for y in the original equation we arrive at the required identity

$$\log_a(x + \sqrt{x^2 - 1}) = -\log_a(x - \sqrt{x^2 - 1}). \tag{1.79}$$

Here, by an *identity*, we mean a result which is true for *all* values of x for which the logarithm is defined, and not merely for some specific value of x.

1.10 Greek symbols used in mathematics

The ratio of the circumference of a circle to its diameter, which is always represented by the Greek letter π, with the understanding that $\pi =$ 3.14159..., will already be familiar to the reader. However, in mathematics it is convenient to use many other Greek letters though, unlike π, they do not all have pre-assigned meanings. For reference purposes the following table lists first the lower- and then the upper-case letters in the Greek alphabet, together with their pronunciations.

α	A	alpha
β	B	beta
γ	Γ	gamma
δ	Δ	delta
ε	E	epsilon
ζ	Z	zeta
η	H	eta
θ	Φ	theta
ι	I	iota
κ	K	kappa
λ	Λ	lambda
μ	M	mu
ν	N	nu

ξ	Ξ	xi
o	O	omicron
π	Π	pi
ρ	P	rho
σ	Σ	sigma
τ	T	tau
υ	Y	upsilon
ϕ, φ	Φ	phi
χ	X	chi
ψ	Ψ	psi
ω	Ω	omega

Problems

Section 1.1

1.1 Enumerate the elements in the following sets in which **Z** signifies the set of positive and negative integers including zero:

(a) $S = \{n \mid n \in \mathbf{Z},\ 5 < n^2 < 47\}$;

(b) $S = \{n^3 \mid n \in \mathbf{Z},\ 15 < n^2 < 40\}$;

(c) $S = \{(m, n) \mid m,\ n \in \mathbf{Z},\ 15 < m^2 + n^2 < 17\}$;

(d) $S = \{(m, n, m+n) \mid m,\ n \in \mathbf{Z},\ 45 < m^2 + n^2,\ 3 < m+n < 9\}$;

(e) $S = \{x \mid x \in \mathbf{Z},\ x^2 + 0.1x - 1.1 = 0\}$.

1.2 Express the following sets in the notation of the previous question:
(a) the set of positive integers whose cubes lie between 7 and 126;
(b) the set of integers which are the squares of the integers lying between α and β $(0 < \alpha < \beta)$;
(c) the points in the plane that lie to the right of the y-axis and below the line $y = -1$.

1.3 Give an example of
(a) a finite set having numerical elements,
(b) a finite set having non-numerical elements, and in each case give an example of a proper subset (set within a set).

1.4 Give an example of
(a) a set of ordered triples involving numerical quantities,
(b) a set of ordered triples involving non-numerical quantities, and in each case give an example of a proper subset.

Section 1.2

1.5 Describe two different ways of defining N rational numbers between 1 and 2. Generalize one of these methods to interpolate N rationals between any two rationals a and b.

1.6 Prove that if α, β, are rational numbers, then so also are $\alpha+\beta$, $\alpha\beta$ and α/β.

1.7 Use the fact that $\sqrt{2}$ is irrational to prove that if α is a rational number, then $\alpha+\sqrt{2}$, $\alpha\sqrt{2}$ and $\sqrt{2}/\alpha$ are also irrational.

1.8. Given that $\sqrt{3}$ is irrational, prove that if α is a rational number then $\alpha+\sqrt{3}$, $\alpha\sqrt{3}$ and $\sqrt{3}/\alpha$ are also irrational.

1.9 Prove that if $a > b > 0$ and $k > 0$ then

$$\frac{b}{a} < \frac{b+k}{a+k} < 1 < \frac{a+k}{b+k} < \frac{a}{b}.$$

Section 1.3

1.10 Indicate by means of a diagram the intervals defined by the following expressions, using a dot to signify an end point belonging to an interval and a circle to indicate an end point excluded from the interval:

(a) $(x+2)(x+3) \leq (x-1)(x-2)$;

(b) $0 < | x-3 | \leq 1$;

(c) $| x | \leq 2$;

(d) $0 < | 2x+1 | < 1$;

(e) $| 3x+1 | \geq 2$;

(f) $\dfrac{x+2}{x-1} > \dfrac{x}{x+2}$.

1.11 Identify the regions in the (x, y)-plane determined by the following inequalities. Mark a boundary that belongs to the region by a full line; a boundary that does not by a dotted line; an end point that is included in an interval by a dot; an end point excluded from an interval by a circle:

(a) $x^2 + y^2 < 1$; $x < 0$; $y < -x$;

(b) $y \leq \sin x$; $x^2 + y^2 \geq p^2$; $y \leq \frac{1}{2}$;

(c) $\frac{1}{4}x^2 + y^2 > 1$; $| y | \leq \frac{1}{2}$;

(d) $y \geq x^2$; $| x-1 | \leq 1$; $y \leq 4$.

1.12 Prove that the inequality

$$\frac{x+3}{x} > \frac{x-2}{x-3}$$

is only satisfied for $0 < x < 3$ or $x > 9/2$.

1.13 Find the maximum of $x+2y$ subject to the constraint inequalities $0 \leq x \leq 4$, $y \geq 0$, $y \leq 3+\frac{1}{2}x$, $y \leq 6-x$.

Section 1.4

1.14 Give an inductive proof that

(a) $\sum_{r=0}^{n-1}(a+rd)=\frac{n}{2}[2a+(n-1)d]$, (Arithmetic Progression)

(b) $\sum_{r=1}^{n} r^2=\frac{n(n+1)(2n+1)}{6}$. (Sum of Squares)

1.15 Give an inductive proof of the results

(a) $\sum_{s=0}^{n-1} r^s=\frac{1-r^n}{1-r}$, (Geometric Progression)

(b) $\sum_{s=1}^{n} s^3=\left[\frac{n(n+1)}{2}\right]^2$. (Sum of Cubes)

1.16 Give an inductive proof that

$\sum_{r=1}^{n}(2r-1)=n^2$. (the sum of the first n odd numbers equals n^2)

1.17 Give an inductive proof that $(n!)^2>n^n$ for $n\geq 3$.

1.18 Give an inductive proof that

$$2^{n+5}.3^{4n}+5^{3n+1}$$

is divisible by 37 for any natural number n.

1.19 Give an inductive proof that

$$\frac{1}{1.2.3}+\frac{1}{2.3.4}+\ldots+\frac{1}{n(n+1)(n+2)}=\frac{1}{4}\left(1-\frac{2}{(n+1)(n+2)}\right).$$

1.20 Find the general solution to the difference equation

$$u_n-3u_{n-1}+2u_{n-2}=0.$$

Determining the particular solution corresponding to $u_1=3$, $u_2=7$.

1.21 Find the general solution to the difference equation

$$u_n+u_{n-1}-6u_{n-2}=0.$$

Determine the particular solution corresponding to $u_1=1$, $u_2=1$.

1.22 Find the solution to the difference equation

$$u_n-2u_{n-1}+u_{n-2}=0$$

given that $u_1=2$, $u_2=3$.

1.23 Use Pascal's triangle to find the expansion of $(a+b)^7$ and check the result against the one found from the binomial theorem.

1.24 Use the definition of $\binom{n}{r}$ to show that

$$\binom{n}{r} = \binom{n}{n-r}.$$

1.25 Find the coefficients of a^3b^5 and a^6b^2 in the expansion of $(a+b)^8$.

1.26 Use the general binomial expansion to verify the expansion of $(1+x)^{1/2}$ given at the end of section 1.4.

1.27 Use the general binomial expansion to verify the expansion of $(1-x)^{-1/2}$ given at the end of section 1.4.

1.28 Use the general binomial expansion to show that

$$(1-x)^{1/2} = 1 - \tfrac{1}{2}x^2 - \tfrac{1}{8}x^4 - \tfrac{1}{16}x^6 - \ldots.$$

1.29 Use the general binomial expansion to show that

$$\left(1 + \frac{1}{2}x\right)^{2/3} = 1 + \tfrac{1}{3}x - \tfrac{1}{36}x^2 + \tfrac{1}{162}x^3 - \ldots.$$

1.30 Use the general binomial expansion to show that

$$(1+2x^2)^{1/3} = 1 + \tfrac{2}{3}x^2 - \tfrac{4}{9}x^4 + \tfrac{40}{81}x^6 - \ldots.$$

Section 1.5

1.31 (a) Convert to radian measure the following angles expressed in degrees:

(i) 27°, (ii) 330°, (iii) 190°, (iv) 274°;

(b) Convert to degrees the following angles expressed in radians:

(i) 0.72 rad, (ii) 1.36 rad, (iii) 1.82 rad, (iv) 2.87 rad.

1.32 Deduce identities (1.19) and (1.20) by dividing identity (1.18) first by $\cos^2 x$ and then by $\sin^2 x$.

1.33 Deduce identities (1.27) and (1.28) from identities (1.21) and (1.23) by setting $y = x$.

1.34 Deduce identities (1.25) and (1.26) from identities (1.21) to (1.24) by using $\tan(x \pm y) = \sin(x \pm y)/\cos(x \pm y)$ and in each case $(\pm)$ dividing the numerator and denominator by a suitable factor.

1.35 By setting $y = 2x$ in (1.21) and then using (1.27) and (1.28), express $\sin 3x$ in terms of powers of $\sin x$ and $\cos x$.

1.36 By setting $y = 2x$ in (1.23) and then using (1.27) and (1.28) express $\cos 3x$ in terms of powers of $\sin x$ and $\cos x$.

Section 1.6

1.37 Find (a) the equation of the straight line that passes through the two points (x_1, y_1) and (x_2, y_2), (b) the equation of the straight line with slope 3/4 that passes through the point $(1, -2)$.

1.38 Find the condition on the coordinates of the points $P(x_1, y_1)$, $Q(x_2, y_2)$, $R(x_3, y_3)$ and $S(x_4, y_4)$ in order that the lines PQ and RS should be orthogonal.

1.39 Find the equation of the circle with centre at point (1, 2) that passes through the point (2, 4). What is the equation of the line that is tangent to its lowest point?

1.40 Express the equation of the straight line $y = 2x + 3$ when (a) the axes are shifted without rotation to the point $(-1, 2)$, (b) when the axes are rotated anticlockwise through an angle $\pi/4$.

1.41 Derive from first principles the equation of the sphere given in Eqn (1.24).

1.42 Show that if the coordinate axes are rotated anticlockwise through an angle $\pi/4$, the equation $xy = 1$ takes the form

$$\frac{X^2}{2} - \frac{Y^2}{2} = 1.$$

Sketch the graph and mark in the respective pairs of axes.

In the following three questions determine the nature of the graph of the given equation and sketch it.

1.43 $x^2 + 2\sqrt{3}xy - y^2 - 2 = 0$.

1.44 $5x^2 + 8xy + 5y^2 - 9 = 0$.

1.45 $6x^2 - 4\sqrt{3}xy + 2y^2 - 3x - 3\sqrt{3}y + 6 = 0$.

1.46 Show by making a suitable shift of origin of the form $x = X - a$, $y = Y - b$ to remove the terms in X and Y, that

$$x^2 - xy - 2y^2 + 3x + 2 = 0$$

is a degenerate hyperbola comprising a pair of straight lines, and find them.

1.47 Name the type of cone in each of the following cases, stating the location of the axis and the vertex.

(i) $$\frac{(x-1)^2}{4} + \frac{(y-2)^2}{4} - z^2 = 0,$$

(ii) $$\frac{(x+3)^2}{4} + \frac{(y-3)^2}{9} - z^2 = 0,$$

(iii) $$\frac{(x+2)^2}{9}+\frac{(y-4)^2}{16}-(z-3)^2=0,$$

(iv) $$\frac{(y-1)^2}{3}+\frac{(z-6)^2}{3}-x^2=0.$$

Supplementary computer problems

1.48 Use the difference equation

$$u_n = u_{n-1} + u_{n-2}$$

with $u_0 = u_1 = 1$ to find u_5, u_{10}, u_{15} and u_{25}.
Check your results against the solution

$$u_n = \left(\frac{\sqrt{5}+1}{2\sqrt{5}}\right)\left(\frac{1+\sqrt{5}}{2}\right)^n + \left(\frac{\sqrt{5}-1}{2\sqrt{5}}\right)\left(\frac{1-\sqrt{5}}{2}\right)^n$$

that was found in Example 1.9.

1.49 Use the difference equation

$$u_n = 6u_{n-1} - 9u_{n-2}$$

with $u_0 = 1$, $u_1 = 2$ to find u_5, u_{10}, u_{20} and u_{30}. Check your results against the solution

$$u_n = (1 - \tfrac{1}{3}n)3^n$$

that was found in section 1.4.

1.50 Use the difference equation

$$u_n = u_{n-1} + 3u_{n-2}$$

with $u_0 = 1$, $u_1 = 3$ to find u_6, u_{12}, u_{24} and u_{36}. Check your results against the solution

$$u_n = \tfrac{1}{3}[(-1)^{n+1} + 2^{n+2}].$$

1.51 Use the difference equation

$$u_n = 4u_{n-1} - 4u_{n-2}$$

with $u_0 = 3$, $u_1 = 2$ to find u_6, u_{15}, u_{25} and u_{40}. Check your results against the solution

$$u_n = (3 - 2n)2^n.$$

1.52 Verify the following summation formulae for $n = 10$, 15 and 20:

(i) $$\sum_{k=1}^{n} k = \tfrac{1}{2}n(n+1),$$

(ii) $$\sum_{k=1}^{n} k^2 = \tfrac{1}{6}n(n+1)(2n+1),$$

(iii) $\sum_{k=1}^{n} k^3 = \frac{1}{4}n^2(n+1)^2.$

1.53 The Cauchy–Schwarz inequality asserts that for any two sets of real numbers $a_1, a_2, \ldots, a_n$ and $b_1, b_2, \ldots, b_n$,

$$\left(\sum_{k=1}^{n} a_k b_k\right)^2 \leq \left(\sum_{k=1}^{n} a_k^2\right)\left(\sum_{k=1}^{n} b_k^2\right).$$

Choose any two sets of ten real numbers and take one set to represent a_1 to a_{10} and the other set to represent b_1 to b_{10}. Use these to verify the inequality for the case $n = 10$ by computing the values of the left- and right-hand sides of the inequality and then comparing them. Experiment and try to find the conditions under which the equality holds. (*Hint*: try simple relationships between the a_r and the b_r.)

1.54 The Minkowski inequality asserts that for any two sets of real numbers $a_1, a_2, \ldots, a_n$ and $b_1, b_2, \ldots, b_n$ and any number $p > 1$

$$\left[\sum_{k=1}^{n} (a_k + b_k)^p\right]^{1/p} \leq \left[\sum_{k=1}^{n} a_k^p\right]^{1/p} + \left[\sum_{k=1}^{n} b_k^p\right]^{1/p}.$$

Choose any two sets of real numbers to represent the a_1 to a_n and the b_1 to b_n and any number $p > 1$ and use them to verify the inequality by computing the values of the left- and right-hand sides of the inequality and then comparing them.

1.55 Let $a_1, a_2, \ldots, a_n$ and $b_1, b_2, \ldots, b_n$ be any two sets of real numbers such that either $a_1 \geq a_2 \geq \ldots \geq a_n$ and $b_1 \geq b_2 \geq \ldots \geq b_n$, or $a_1 \leq a_2 \leq \ldots \leq a_n$ and $b_1 \leq b_2 \leq \ldots \leq b_n$. Then the Chebyshev inequality asserts that

$$\left(\frac{a_1 + a_2 + \cdots + a_n}{n}\right)\left(\frac{b_1 + b_2 + \cdots + b_n}{n}\right) \leq \frac{1}{n}\sum_{k=1}^{n} a_k b_k.$$

Verify the inequality by taking any two suitable sets of 20 numbers for the a_r and the b_r, using them to compute the values of the left- and right-hand sides of the inequality and then comparing them.

1.56 Use a graphics package to generate the cardioid in Fig. 1.22(a).

1.57 Use a graphics package to generate the lemniscate in Fig. 1.22(b).

1.58 Use a graphics package to generate the Archimedean spiral in Fig. 1.22(c).

1.59 Use a graphics package to generate the three and eight petal roses

$$r = 2\cos 3\theta \quad \text{and} \quad r = 3\sin 4\theta.$$

2 Variables, functions and mappings

This chapter begins by introducing the related concepts of a variable and a function that together form the basis of analysis. The definition of a function of a single real variable as a relationship in which one value of the independent variable corresponds to only one value of the dependent variable, and conversely, then leads, by interchanging the roles of the dependent and independent variables, to the important concept of an inverse function. The graphical interpretation of the inverse of a function as the reflection of the graph of the function in the line $y = x$ simplifies its construction. Special care is taken with the discussion of the inverse trigonometric functions because their periodicity makes it necessary to introduce a principal branch for each trigonometric function.

A number of simple but important special functions such as the step function, even and odd functions, polynomials and periodic functions occur sufficiently frequently for it to be worth identifying and discussing them briefly before they are used in subsequent chapters. Two comparatively new but useful functions, called the *floor* and *ceiling* functions are also defined. These were introduced as recently as 1990 in connection with computing, and they are mentioned here because of their occurrence in readily available numerical computing packages and in their applications.

The class of curves in the plane described by means of the parametric representation $x = x(\alpha)$, $y = y(\alpha)$, with α as the parameter, is introduced because they are more general than the class of curves described by the usual explicit representation of the form $y = f(x)$. This is because a parametric representation can characterize a curve with multiple loops, for which the ordinary notion of a function becomes invalid due to there being more than one value of the dependent variable y corresponding to a given value of the independent variable x. The parametric representation, coupled with the concept of parametric differentiation that is introduced in Chapter 5, allows the derivative of the function at any point on its graph, with or without loops, to be defined and determined unambiguously.

The chapter ends with a brief discussion of the extension of the notion of a function of a single real variable to a function of several real variables which is essential for the development of mathematics and also for its applications. It is shown how, when a function of two real variables needs to be visualized, the use of level curves or of sectioning planes is often helpful, as these interpret a three-dimensional problem in terms of a set of two-dimensional curves.

2.1 Variables and functions

In the physical world the idea of one quantity depending on another is very familiar, a typical example being provided by the observed fact that the pressure of a fixed volume of gas depends on its temperature. This situation is reflected in mathematics by the notion of a *function*, which we shall now discuss in some detail.

The modern definition of a function in the context of real numbers is that it is a relationship, usually a formula, by which a correspondence is established between two sets A and B of real numbers in such a manner that to each number in set A there corresponds only one number in set B. The set A of numbers is called the *domain* of the function and the set B of numbers is the called *range* of the function.

If the function or rule by which the correspondence between numbers in sets A and B is established is denoted by f, and x denotes a typical number in the domain A of f, then the number in the range B to be associated with x by the function f is written $f(x)$ and is read 'f of x'. The numbers x and $f(x)$ are *variables*, with x being given the specific name *independent variable* and $f(x)$ the name *dependent variable*. The independent variable is also often called the *argument* of the function f.

It is often helpful to construct the graph of f which mathematically is the set of ordered number pairs $(x, f(x))$, where x belongs to the domain of f. Geometrically the graph of f is usually represented by a plane curve, drawn relative to an origin defined by the intersection of two perpendicular straight lines forming the axes. The process of construction is as follows. A distance proportional to x is measured along one axis and, using the same scale, a distance proportional to $f(x)$ along the other axis. Through each such point on an axis is then drawn a line parallel to the other axis and these two perpendicular lines intersect at a unique point in the plane of the axes. This point of intersection is the point $(x, f(x))$ and the graph of f is defined to be the locus or curve formed by joining up all such points corresponding to the domain of f, as illustrated by Fig. 2.1.

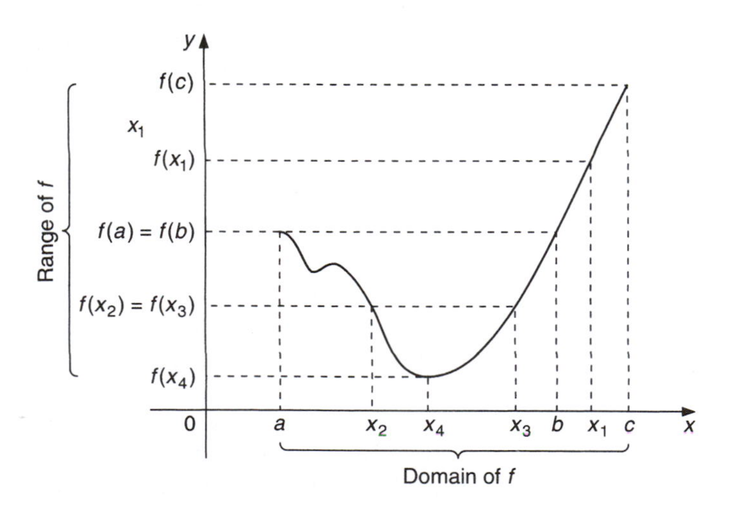

Fig. 2.1 Domain, range and graph of $f(x)$.

However, it is not necessary to use axes of this type, called rectangular Cartesian axes, and any other geometrical representation which gives a unique representation of the points $(x, f(x))$ would serve equally well. Thus the axes could be inclined at an angle $\alpha \neq \frac{1}{2}\pi$ and the scale of measurement along them need not be uniform. For example, it is often useful to plot the logarithm of x along the x-axis, rather than the x itself. This compresses the x scale so that large values of x may be conveniently displayed on the graph together with small values. Another possible representation involves the use of *polar coordinates* introduced in section 1.7. This will be taken up again later in connection with change of variables and double integrals.

Not every function can be represented in the form of an unbroken curve, and the function

$$f(x) = \begin{cases} 0 & \text{when } x \text{ is rational,} \\ 1 & \text{when } x \text{ is irrational,} \end{cases}$$

provides an extreme example of this situation. Here, although the graph would look like a line parallel to the x-axis on which all points have the value unity, in reality the infinity of points with rational x-coordinates would be missing since they lie on the x-axis itself. The domain is all the real numbers **R** and the range is just the two numbers zero and unity.

Because f transforms one set of real numbers into another set of real numbers a function is sometimes spoken of as a *transformation* between sets of real numbers. On account of the restriction to real numbers or, more explicitly, to real variables, the function $f(x)$ is called a *function of one real variable*. Another name that is often used for a function is a *mapping* of some set of real numbers into some other set of real numbers. This name is of course suggested by the geometrical illustration of the graph of a function, and we shall return more than once to the notion of a mapping. In this terminology, $f(x)$ is referred to as the *image* of x under the mapping f.

Since the domain and range of f occur as intervals on the x- and y-axes, it is convenient to use a simplified notation to identify the form of the interval that is involved. If the end point of an interval is included in the interval, that end of the interval is said to be *closed*. If the end point of an interval is excluded from the interval, that end of the interval is said to be *open*. We now adopt the almost standard notation summarized below in which a round bracket indicates an open end of an interval, and a square bracket indicates a closed end of an interval. Here, the symbol $\Leftrightarrow$ is to be read 'implies and is implied by', meaning that the expression to the left of the symbol implies the one to the right, and conversely:

$x \in (a, b) \Leftrightarrow a < x < b$; the *open* interval (a, b),

$x \in [a, b] \Leftrightarrow a \leq x \leq b$; the *closed* interval $[a, b]$,

$x \in (a, b] \Leftrightarrow a < x \leq b$; the *semi-open* interval $(a, b]$,

$x \in [a, b) \Leftrightarrow a \leq x < b$; the *semi-open* interval $[a, b)$,

$x \in (-\infty, a] \Leftrightarrow x \leq a$; the *semi-infinite* interval $(-\infty, a]$,

$x \in [a, \infty) \Leftrightarrow a \leq x;$ the *semi-infinite* interval $[a, \infty)$,

$x \in (-\infty, \infty) \Leftrightarrow \text{all } x \in \mathbf{R};$ the *infinite* interval $(-\infty, \infty)$.

An end point of an interval at $\pm\infty$ is shown as open, because ∞ is a limiting process and not a specific number.

As the definition of open and closed intervals is only a matter of considering the behaviour of the end points, we shall define the length of all the intervals (a, b), $[a, b)$, $(a, b]$, and $[a, b]$ to be the number $b - a$. This is consistent with regarding as zero the length of an 'interval' comprising only one point.

It may happen that each point x is associated with a unique image point $f(x)$ and, conversely, each image point $f(x)$ is associated with a unique point x. Such a mapping or function f is then said to be *one–one* in the domain in question. This means that to one x there corresponds only one $f(x)$, and conversely, as, for example, in Fig. 2.1 in the interval (b, c).

However, it is often the case that in some interval of the x-axis more than one point x may correspond to the same image point $f(x)$. This is again well illustrated by Fig. 2.1 if now we consider the interval $[a, b]$ and the points x_2 and x_3, both of which have the same image point since $f(x_2) = f(x_3)$. In such situations the mapping or function f is often said to be *many–one*.

A specific example might help here, and we choose for f the function $f(x) = x^2$ and the two different domains $[0, 3]$ and $[-1, 3]$. A glance at Fig. 2.2 shows that f maps the domain $[0, 3]$ onto the range $[0, 9]$ one–one, but that it maps the domain $[-1, 3]$ onto the same range $[0, 9]$ many–one. Expressed another way, the range $[0, 1]$ is mapped twice by points in the domain $[-1, 3]$; once by points in the subdomain $-1 \leq x < 0$ and once by points in the subdomain $0 < x \leq 1$. Again considering the domain $[-1, 3]$, the function $f(x) = x^2$ maps the subdomain $1 < x \leq 3$ onto the range $(1, 9]$ one–one.

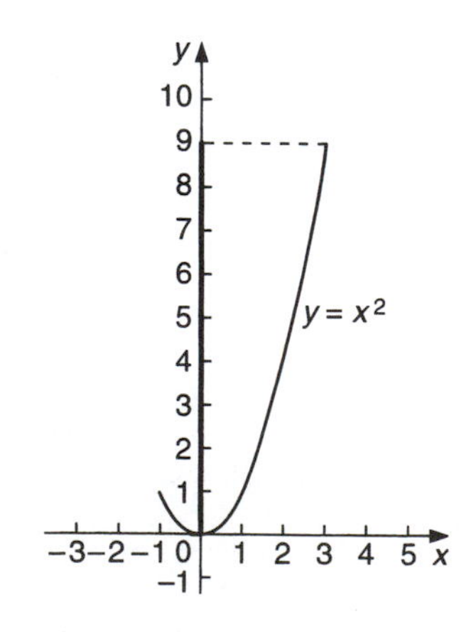

Fig. 2.2 A many–one mapping in the range [0, 1] and a one-one mapping in the range (1, 9].

In older books the term 'function' is used ambiguously in that it is sometimes applied to relationships which do not comply with our definition of a function. The most familiar example of this arises when considering the relationship $y^2 = x$, because for any given value of $x > 0$ there are two possible values of y, one of which is positive and the other negative because of the square root operation. Although it is permissible to describe this relationship as a mapping, it is incorrect to call it a function.

Nevertheless, the square root operation is fundamental to mathematics and we must find some way to make it and similar ones legitimate. The difficulty is easily resolved if we consider how the square root is used in applications. In point of fact, two different relationships are always considered which together are equivalent to $y^2 = x$. These are the functions $y_1 = +\sqrt{x}$ and $y_2 = -\sqrt{x}$, where the square root is always to be understood to denote the positive square root and the sign identifies the relationship being considered. Each of the mappings $y_1(x)$ and $y_2(x)$ of the domain $(0, \infty)$ is one–one, as Fig. 2.3 shows, so that each may be correctly termed a function, the particular one to be used in any application being determined by other considerations such as the result must either be positive or negative. These ideas will arise again later in connection with inverse functions.

In general, if the domain of function f is not specified it is understood to be the largest interval on the x-axis for which the function is defined. So if $f(x) = x^2 + 4$, then as this is defined for all x, the largest possible domain must be $(-\infty, \infty)$. Alternatively the function $f(x) = +\sqrt{(4 - x^2)}$ is only defined in terms of real numbers when $-2 \leq x \leq 2$, showing that the largest possible domain is $[-2, 2]$. Similarly, the function $f(x) = 1/(1 - x)$ is defined for all x with the sole exception of $x = 1$, so that the largest possible domain is the entire x-axis with the single point $x = 1$ deleted from it.

A function need not necessarily be defined for all real numbers on some interval and, as in probability theory, it is quite possible for the dependent and independent variables to assume only discrete values. Thus the rule which assigns to any positive integer n the number of positive integers whose squares are less than n, defines a perfectly good function. Denoting this function by f, we have for its first few values $f(1) = 0, f(2) = 1, f(3) = 1, f(4) = 1, f(5) = 2, f(6) = 2, f(7) = 2, f(8) = 2, f(9) = 2, f(10) = 3, \ldots$. Clearly, both its domain and its range are the set $\mathbf{N}$ of natural numbers and the mapping is obviously many–one.

Before examining some special functions, let us formulate our definition of a function in rather more general terms. This will be useful later since although in the above context the relationships discussed have always been between numbers, in future we shall establish relationships between quantities that are not simply real numbers. When we do so, it will be valuable if we can still utilize the notion of a function. This will occur, for example, when we establish correspondence between quantities called vectors and matrices which, although obeying algebraic laws, are not themselves real numbers.

The idea of a relationship between arbitrary quantities is one which we have already introduced in the previous chapter in connection with sets.

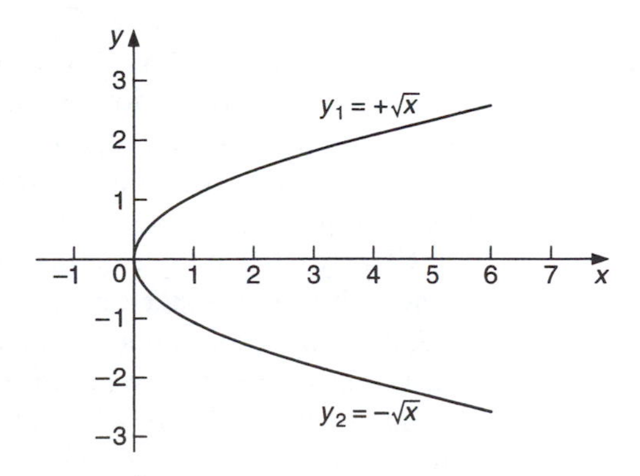

Fig. 2.3 The relationship $y^2 = x$.

As might be expected, set theory provides the natural language for the formulation and expression of general ideas associated with functions, and indeed we have already used the word 'set' quite naturally when thinking of a set of numbers. A more general definition follows.

Definition 2.1 A *function* f is a correspondence by which each element of set A which is called the domain of f, is associated with only one element of set B called the range of f.

To close this section we now provide a few examples illustrating some of the ideas just mentioned.

Example 2.1 The function $y = f(x)$ defined by the rule

$$f(x) = \frac{1}{(x-1)(x-2)}$$

is defined for all real x with the exception of the two points $x = 1$ and $x = 2$. The domain of f is thus the set of real numbers $\mathbf{R}$ with the two numbers 1 and 2 deleted. The range of f is $\mathbf{R}$ itself, with the understanding that $f(\infty) = 0$. ■

Example 2.2 A discrete-valued function may be defined by a table which is simply an arrangement of ordered number pairs in a sequence (see Table 2.1).

Table 2.1 A discrete-valued function

x	0	1	3	7
$f(x)$	2.1	4.2	1.0	6.3

■

Example 2.3 This example is a final illustration of our more general definition of a function. Take as the domain of the function f the set A of all people born in towns, and as the range B of the function f the set of all towns in the world. Then for the function f we propose the rule that assigns to every person his or her place of birth.

Clearly this example defines a many–one mapping of set A onto set B, since although a person can only be born in one place, many other people may have the same place of birth. This example also serves to distinguish clearly between the concept of a 'function' which is the rule of assignment, and the concept of the 'variables' associated with the function which here are people and places. ■

2.2 Inverse functions

In the previous section we remarked that a typical example of a correspondence between physical quantities was the observed fact that the pressure of a fixed volume of gas depends on its temperature. Expressed in this form, we are implying that the dependent variable is the pressure p and the independent variable is the temperature T, so that the law relating pressure to temperature has the general form

$$p = \phi(T), \tag{A}$$

where ϕ is some function that is determined by experiment.

However, we know from experience that it is often necessary to interchange these roles of dependent and independent variables and sometimes to regard the temperature T as the dependent variable and the pressure p as the independent variable, when the temperature–pressure law then has the form

$$T = \psi(p), \tag{B}$$

where, naturally, the function ψ is dependent on the form of the function ϕ. Indeed, formally, ϕ and ψ must obviously satisfy the identity $\phi[\psi(p)] \equiv p$ for all pressures p in the domain of ψ.

The relationships (A) and (B) are particular cases of the notion of a function and its inverse, and the idea is successful in this context because the correspondence between temperature and pressure is known to be one–one.

To examine the notion of an *inverse function* in more detail let us consider a general function

$$y = f(x) \tag{2.1}$$

that is one–one and defined on the domain $[a, b]$, together with its inverse

$$x = g(y) \tag{2.2}$$

which has for its domain the interval $[c, d]$ on the y-axis.

Graphically the process of inversion may be accomplished point by point as indicated in Fig. 2.4(a). This amounts to selecting a point y in $[c, d]$ and then finding the corresponding point x in $[a, b]$ by projecting a line horizontally from y until the graph of f is intercepted, after which a line is projected vertically downwards from this intercept to identify the required point on the x-axis.

The relationship between a one–one function and its inverse is represented in Fig. 2.4(b). In this diagram we have used the fact that when a function is represented as an ordered number pair, interchange of dependent and independent variables corresponds to interchange of numbers in the ordered number pair. The lower curve represents the function $y = f(x)$ and the upper curve represents the function $y = g(x)$, with the function g the inverse of f; both graphs being plotted using the same axes. The line $y = x$ is also shown on the graph to emphasize that geometrically the relationship between a one–one function and its inverse is obtained by reflecting the graph of either function in a mirror held along the line $y = x$. Henceforth such a process will simply be termed *reflection*

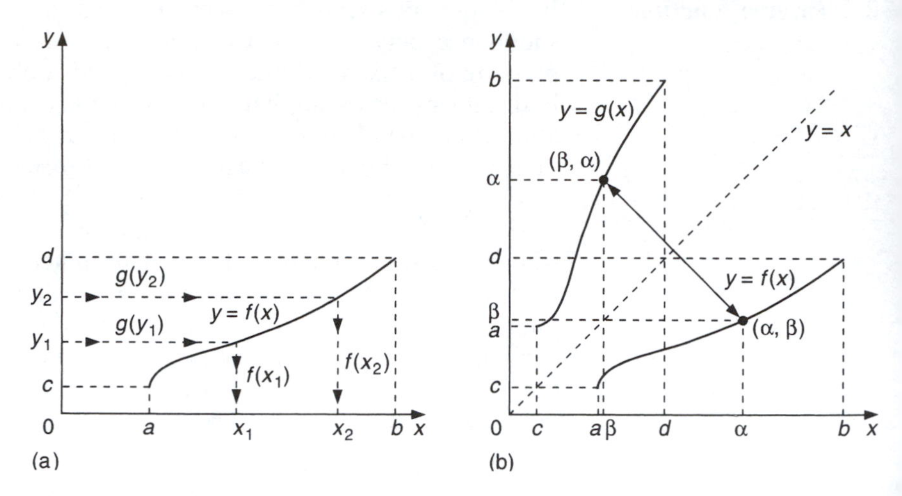

Fig. 2.4 Inversion (a) through the graph of $f(x)$; (b) by reflection in $y = x$.

in a line. Notice that when using this reflection property to construct the graph of an inverse function from the graph of the function itself, both functions are represented with y plotted vertically and x plotted horizontally. This follows because the range of f is the domain of g, and vice versa.

No difficulty can arise in connection with a one–one function and its inverse because of the one–one nature of the mapping. Expressed more precisely, we have used the obvious property illustrated by Fig. 2.4(a) that a one–one function f with domain $[a, b]$ is such that if $f(x_1) = f(x_2) \Rightarrow x_1 = x_2$ for all x_1 and x_2 in $[a, b]$.

In graphical terms, whenever the graph of f is unbroken, this result can only be true if the graph of f either increases or decreases steadily as x increases from a to b. When either of these properties is true of a function, it is said to be *strictly monotonic*. In particular, if a function f increases steadily as x increases from a to b, as in Fig. 2.4(a), then it is said to be *strictly monotonic increasing* and, conversely, if it decreases steadily then it is said to be *strictly monotonic decreasing*.

Notice that although the strict monotonic behaviour of a function will ensure that it is one–one, the converse will not be true unless the graph of the function is unbroken. To see this it is sufficient to consider $y = 1/x$ for all $x \neq 0$. Although this has a broken graph across the origin it is one–one, though not monotonic.

Slightly less stringent than the condition of strict monotonicity is the condition that a function f be just *monotonic*. This is the requirement that f be either non-decreasing or non-increasing, so that it is permissible for a function that is only monotonic to remain constant throughout some part of its domain of definition. The adjectives 'increasing' and 'decreasing' are again used to qualify the noun 'monotonic' in the obvious manner. Representative examples of monotonic and strictly monotonic functions, all with domain of definition $[a, b]$, are shown in Fig. 2.5.

The example of a strictly monotonic decreasing function shown in Fig. 2.5(b) has also been used to emphasize that a function need not be represented by an unbroken curve. The curve has a break at the single point $x = \alpha$ where it is defined to have the value $y = \beta$. However, as the value β lies between the functional values on adjacent sides of $x = \alpha$ the function is still strictly monotonic decreasing. Had we set $\beta = 0$, say, then the function would be neither strictly monotonic nor even monotonic on account of this one point!

It is sometimes useful to relate a function and its inverse by essentially the same symbol, and this is usually accomplished by adding the superscript minus one to the function. Thus the function inverse to f is often denoted by f^{-1}, which is not, of course, to be misinterpreted to mean $1/f$. Before examining some important special cases of inverse functions when many–one mappings are involved, let us formalize our previous arguments.

Definition 2.2 Let the set onto which the one–one function f with domain $[a, b]$ maps the set S of points be denoted by $f(S)$. Then we define the *inverse mapping* f^{-1} of $f(S)$ onto S by the requirement that $f^{-1}(y) = x$ if and only if $y = f(x)$ for all x in $[a, b]$.

It now only remains for us to consider how some important special mappings such as $y = x^2$, $y = \sin x$ and $y = \cos x$, together with other simple trigonometric functions which are all many–one mappings, may have unambiguous inverses defined.

Firstly, as we have already seen, the equation $y = x^2$ gives a many–one mapping of $[-a, a]$ onto $[0, a^2]$. Here the difficulty of defining an inverse is resolved by always taking the *positive* square root and defining two different inverse functions

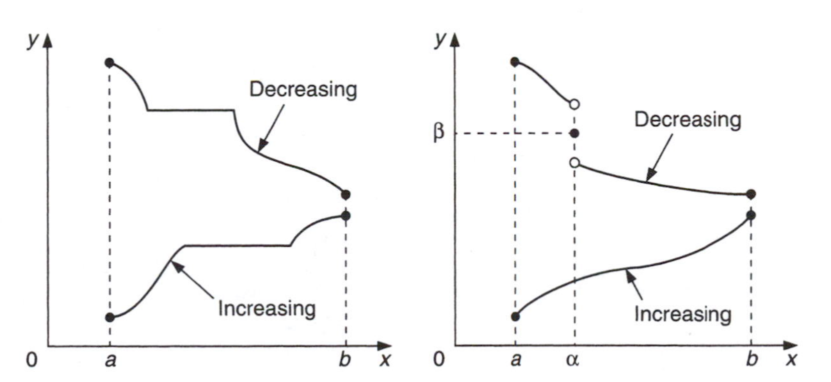

Fig. 2.5 Monotonic and strictly monotonic functions: (a) monotonic; (b) strictly monotonic.

$$x = +\sqrt{y} \quad \text{and} \quad x = -\sqrt{y},$$

which are then each a one–one mapping of $(0, a^2]$. The inversion must thus be regarded as having given rise to two different functions, the one to be selected depending on other factors as mentioned in connection with Fig. 2.3. If we recall that the domain of definition of a function forms an intrinsic part of the definition of that function, then $y = x^2$ may be regarded as two one–one mappings in accordance with the two inverses just introduced.

This is achieved by defining the many–one function $y = x^2$ on the domain $[-a, a]$ as the result of the two different one–one mappings

$$y = x^2 \text{ on } -a \leq x \leq 0 \quad \text{and} \quad y = x^2 \text{ on } 0 < x \leq a,$$

the difference here being only in the domains of definition $-a \leq x \leq 0$ and $0 < x \leq a$. The point 0 may be assigned to either domain since that single point maps one–one. The choice made here assigns it to the left-hand domain, so that $y = x^2$ then provides a one–one mapping between the domain $-a \leq x \leq 0$ and the range $[0, a^2]$. By means of this device we may, in general, reduce many–one mappings to a set of one–one mappings so that the inversion problem is always straightforward. Thus it is important to remember that both the domain and the range form part of the definition of a function.

It will suffice to discuss in detail only the inversion of the sine function, after which a summary of the results for the other elementary trigonometric functions will be presented in the form of a table. In general, the function $y = \sin x$ maps an argument x in the set **R** of real numbers onto $[-1, 1]$ many–one, but as shown in Fig. 2.6(a) the mapping is one–one onto $[-1, 1]$ in any of the restricted domains $[(2n-1)\frac{1}{2}\pi, (2n+1)\frac{1}{2}\pi]$, of

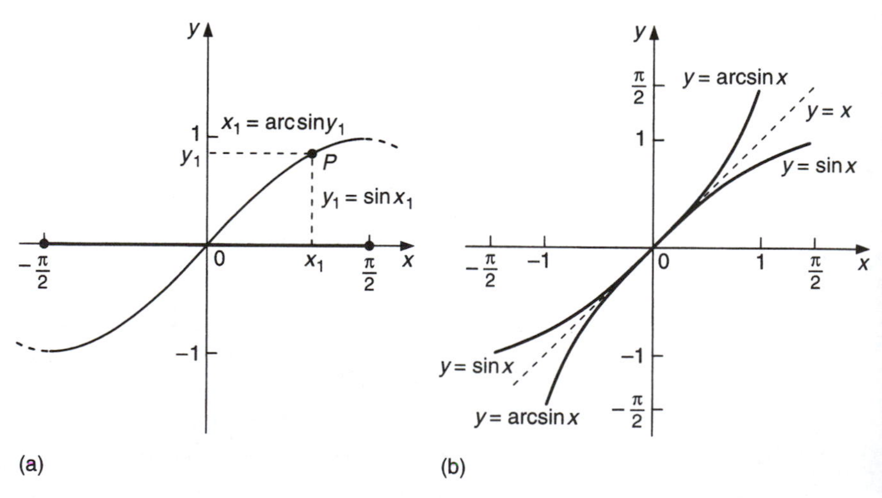

Fig. 2.6 Principal branch of sine function: (a) principal branch of sin x giving one-one mapping in $[-\frac{1}{2}\pi, \frac{1}{2}\pi]$; (b) inversion of sin x by reflection in $y = x$.

length π, for integral n, because in each of these the function is strictly monotonic.

Now in line with our approach to the inverse of the square root function, the ambiguity as regards the function inverse to sine may be completely resolved if we consider the many–one function $y = \sin x$ with $x \in \mathbf{R}$ as being replaced by an infinity of one–one functions $y = \sin x$, with domains $[(2n-1)\frac{1}{2}\pi, (2n+1)\frac{1}{2}\pi]$. For then in each domain corresponding to some integral value of n, because the mapping there is one–one, an appropriate inverse function may be defined without difficulty.

The intervals are all of length π and are often said to define different *branches* of the inverse sine function. In general, when no specific interval is named we shall write $x = \text{Arcsin } y$, whenever $y = \sin x$. The function Arcsine thus denotes an arbitrary branch of the inverse sine function. Because of the periodicity of the sine function, when considering the inverse function it is only necessary to study the behaviour of one branch of Arcsine. As is customary, we arbitrarily choose to work with the branch of the inverse sine function associated with the domain $[-\frac{1}{2}\pi, \frac{1}{2}\pi]$, calling this the *principal branch* and denoting the inverse function associated with this branch by arcsine. Hence for the inverse we shall always write $x = \arcsin y$ when $y = \sin x$ and $-\frac{1}{2}\pi \leq x \leq \frac{1}{2}\pi$. However, we shall need to work with inverse trigonometric functions in their own right, so we shall interchange the roles of x and y and henceforth consider the inverse trigonometric functions as functions of x. Thus we shall, for example, consider the function $y = \arcsin x$, with domain $-1 \leq x \leq 1$ and range $-\pi/2 \leq y \leq \pi/2$.

In Fig. 2.6(b) is shown in relation to the line $y = x$ the function $y = \sin x$ with domain of definition $[-\frac{1}{2}\pi, \frac{1}{2}\pi]$ and the associated function $y = \arcsin x$ with domain of definition $[-1, 1]$. The reflection property of inverse functions utilized in connection with Fig. 2.4(b) is again apparent here. It should perhaps again be emphasized that when an inverse function is obtained by reflection in the line $y = x$, then in both the curves representing the function and its inverse, the variable y is plotted as ordinate (i.e. vertically) and the variable x as abscissa (i.e. horizontally).

Table 2.2 summarizes information concerning the most important inverse trigonometric functions and should be studied in conjunction with Fig. 2.7. In general the notation for a function inverse to a named trigonometric function is obtained by adding the prefix *arc* when referring

Table 2.2 Trigonometric functions and their inverse functions

Function	*Inverse function*	*Branch*	*Domain for x*	*Range for y*
$y = \sin x$	$y = \arcsin x$	Principal	$[-1, 1]$	$[-\frac{1}{2}\pi, \frac{1}{2}\pi]$
$y = \cos x$	$y = \arccos x$	Principal	$[-1, 1]$	$[0, \pi]$
$y = \tan x$	$y = \arctan x$	Principal	$(-\infty, \infty)$	$(-\frac{1}{2}\pi, \frac{1}{2}\pi)$
$y = \csc x$	$y = \text{arccsc } x$	Principal	$\lvert x\rvert \geq 1$	$0 < \lvert y\rvert \leq \frac{1}{2}\pi$
$y = \sec x$	$y = \text{arcsec } x$	Principal	$\lvert x\rvert \geq 1$	$0 \leq y \leq \pi,\ y \neq \frac{1}{2}\pi$
$y = \cot x$	$y = \text{arccot } x$	Principal	$(-\infty, \infty)$	$(0, \pi)$

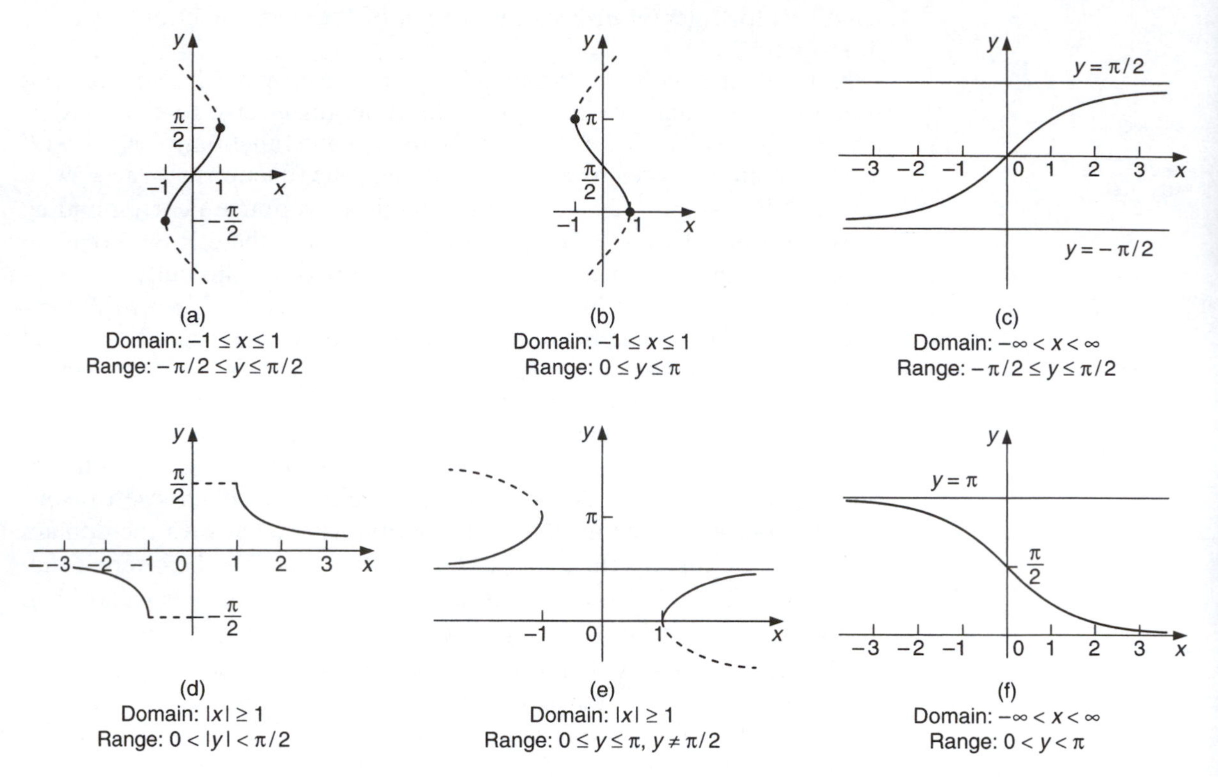

Fig. 2.7 Principal branches of the inverse trigonometric functions: (a) $y = \arcsin x$; (b) $y = \text{arcos}\ x$; (c) $y = \arctan x$; (d) $y = \text{arccsc}\ x$; (e) $y = \text{arcsec}\ x$; (f) $y = \text{arccot}\ x$.

to the principal branch and *Arc* otherwise. Other conventions are also used, the most common of which is to add the superscript minus one after the named function whose inverse is required and, for example, write $\sin^{-1}x$ in place of $\arcsin x$. Unfortunately notations are not uniform, so when using other books the reader would be well advised to check the notation in use.

There are two choices for the left-hand branch of $y = \text{arcsec}\ x$, but the one used here and shown in Fig. 2.7(e) is the most commonly accepted one. The following relationships between the inverse trigonometric functions are sometimes useful:

$$\text{arccsc}\ x = \arcsin(1/x), \quad \text{arcsec}\ x = \arccos(1/x),$$

$$\text{arccot}\ x = \tfrac{1}{2}\pi - \arctan x.$$

The inverse trigonometric functions may be related to angles in the first quadrant by means of the right-angled triangles in Fig. 2.8, obtained by using the definitions of $\sin x$, $\cos x$, $\tan x$, $\csc x$, $\sec x$ and $\cot x$. Thus $\sin(\arcsin x) = x$, as is confirmed by examination of the top left-hand triangle in Fig. 2.8, and the first of the three relationships above follows

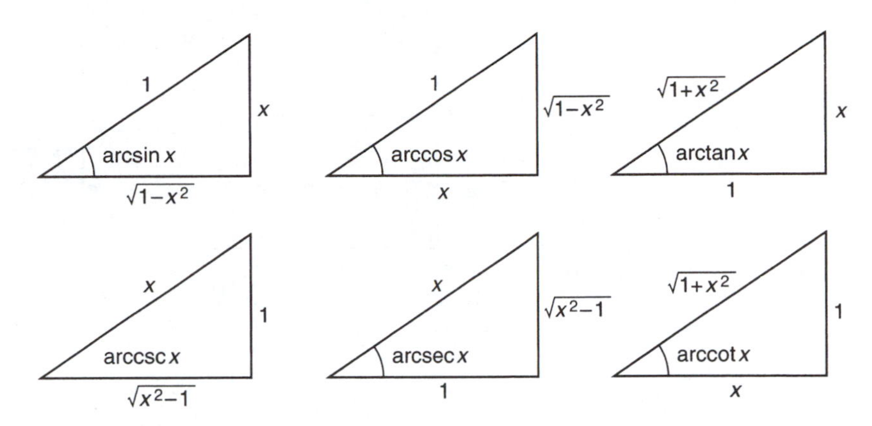

Fig. 2.8 Angles in the first quadrant and the inverse trigonometric functions.

from the bottom left-hand triangle in Fig. 2.8 because $\sin(\operatorname{arccsc} x) = 1/x$, so that taking the inverse sine gives $\operatorname{arcsec} x = \arcsin(1/x)$.

2.3 Some special functions

A number of special types of functions occur often enough to merit some comment. As the ideas involved in their definition are simple, a very brief description will suffice in all but a few cases. To clarify these descriptions, the functions are illustrated in Fig. 2.9.

2.3.1 Constant function

The *constant function* is a function $y = f(x)$ for which $f(x)$ is identically equal to some constant value for all x in the domain of definition $[a, b]$. Thus a constant function has the equation $y \equiv \text{constant}$, for $x \in [a, b]$ (see Fig. 2.9(a)).

2.3.2 Step and piecewise defined function

Consider some set of n sub-intervals or partitions $[a_0, a_1)$, $[a_1, a_2)$, $[a_2, a_3), \ldots, [a_{n-1}, a_n]$ of the interval $[a_0, a_n]$. Associate n constants $c_1, c_2, \ldots, c_n$ with these n sub-intervals. Then a *step function* defined on $[a_0, a_n]$ is the function $y = f(x)$ for which $f(x) \equiv c_r$, for all x in the rth sub-interval. The function will be properly defined provided a functional value is assigned to all points x in $[a_0, a_n]$ including end points of the intervals. Usually it is immaterial to which of two adjacent sub-intervals an end point is assigned and one possible assignment is indicated in Fig. 2.9(b), where a deleted end point is shown as a circle and an included end point as a dot.

A step function is a special case of a function defined in a *piecewise* manner, meaning that although the function is defined over some interval $a \le x \le b$, it is defined differently over different sub-intervals of $a \le x \le b$. A typical example of a function $f(x)$ defined in a piecewise manner

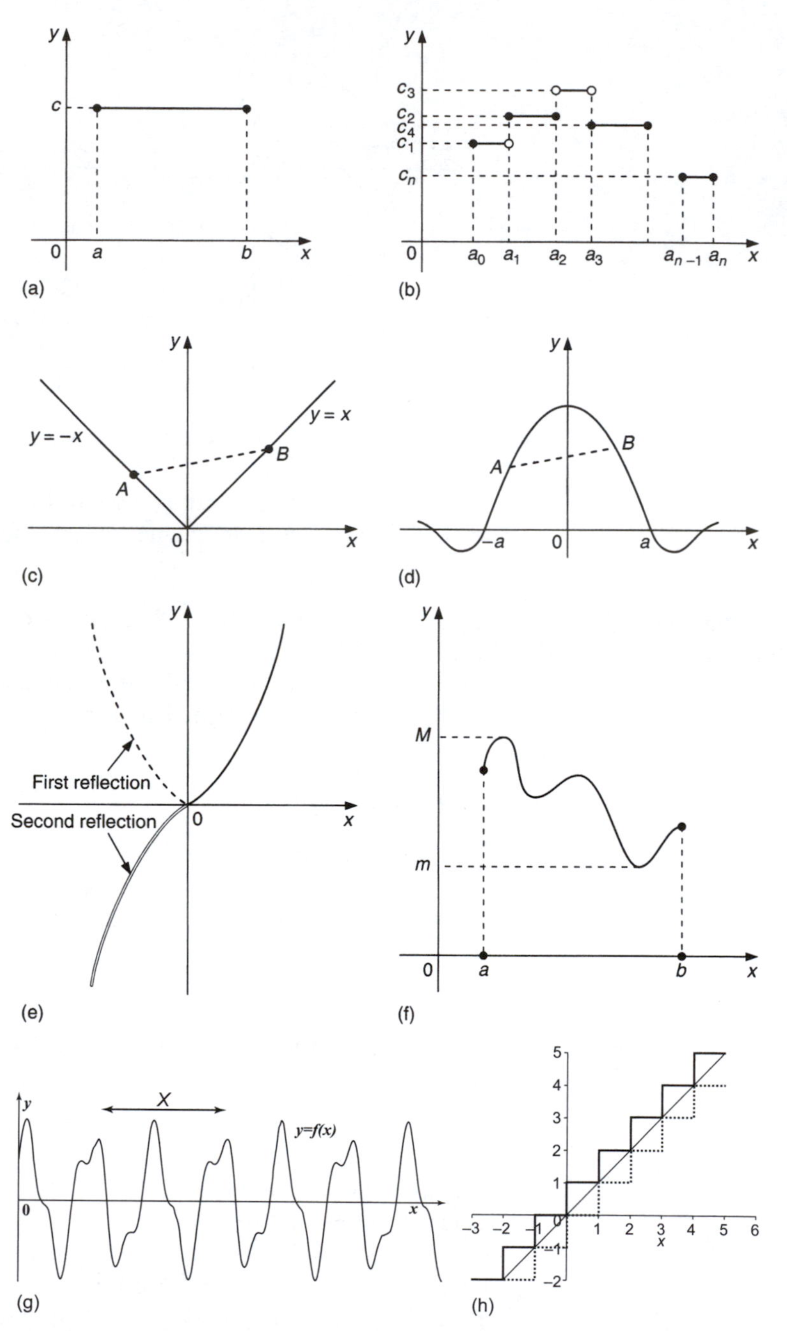

Fig. 2.9 Some special functions: (a) constant function; (b) step function; (c) $y = |x|$; (d) even function; (e) odd function; (f) bounded function on $[a, b]$; (g) periodic function; (h) floor and ceiling functions $\lfloor x \rfloor$ and $\lceil x \rceil$. Note $\lfloor x \rfloor$ and $\lceil x \rceil$ used in (h).

over the interval $0 \le x \le 2\pi$ is one for which $f(x) = \sin x$ for $0 \le x \le \pi$, with $f(x) = 2 - \sin x$ for $\pi < x \le 2\pi$. This is usually written more concisely as

$$f(x) = \begin{cases} \sin x, & 0 \le x \le \pi \\ 2 - \sin x, & \pi < x \le 2\pi \end{cases}.$$

The two curves in Fig. 2.5(a), and the top curve in Fig. 2.5(b), are typical graphs of functions defined in a piecewise manner. The two curves in Fig. 2.5(a) are both continuous for $a \le x \le b$, but the top curve in Fig. 2.5(b) experiences a "jump" across $x = \alpha$, where it is discontinuous.

2.3.3 The function $|x|$

From the definition of the absolute value of x it is easily seen that the graph of $y = |x|$ has the form shown in Fig. 2.9(c). It is composed of the line $y = x$ for $x \ge 0$ and the line $y = -x$ for $x < 0$.

2.3.4 Even function

An *even function* $y = f(x)$ is a function defined for positive and negative x for which $f(-x) = f(x)$ (see Fig. 2.9(d)). The geometrical implication of this definition is that the graph of an even function is symmetrical about the y-axis so that the graph for negative x is the reflection in the y-axis of the graph for positive x. Typical examples of even functions are $y = \cos x$, $y = 1/(1+x^2)$, the function $y = |x|$ just defined, and $y = \cos x/(1+x^2)$.

2.3.5 Odd function

An *odd function* $y = f(x)$ is a function defined for positive and negative x for which $f(-x) = -f(x)$. The geometrical implication of this definition is that the graph of an odd function is obtained from its graph for positive x by first reflecting the graph in the y-axis and then reflecting the result in the x-axis. In Fig. 2.9(e) the result of the first reflection is shown as a dotted curve and its reflection in the x-axis gives a second curve shown as a line in the third quadrant which, together with the original curve in the first quadrant, defines the odd function. By virtue of the definition we must have $f(0) = -f(0)$, showing that the graph of an odd function must either pass through the origin, or become $+\infty$ on one side of the origin and $-\infty$ on the other. Typical odd functions are $y = \sin x$, $y = x^3 - 3x$ and $y = 1/x$ for $x \ne 0$. Most functions are neither even nor odd. For example, $y = x^3 - 3x + 1$ is not even, since $y(-x) = (-x)^3 - 3(-x) + 1 = -x^3 + 3x + 1 \ne y(x)$, nor, by the same argument, is it odd, for $y(-x) \ne -y(x)$.

The following results concerning even and odd functions are often useful:

(i) the product of two even functions is an even function,
(ii) the product of two odd functions is an even function,

(iii) the product of an even function and an odd function is an odd function.

The proofs of these results follow directly from the definitions of even and odd functions. Result (i) follows from the fact that if $f(x)$ and $g(x)$ are both even functions, then $f(-x)g(-x)=f(x)g(x)$, showing that the product is an even function. Result (ii) follows from the fact that if $f(x)$ and $g(x)$ are both odd functions, then $f(-x)g(x)=[-f(x)][-g(x)]=f(x)g(x)$, showing that the product is an even function. Result (iii) follows from the fact that if $f(x)$ is an odd function and $g(x)$ is an even function, then $f(-x)g(-x)=[-f(x)]g(x)=-f(x)g(x)$, showing that the product is an odd function.

For two applications of these results see, for example, sections 7.6.4 and 16.3.2.

2.3.6 Bounded function

A function $y=f(x)$ is said to be *bounded* on an interval if it is never larger than some value M and never smaller than some value m for all values of x in the interval. An example is shown in Fig. 2.9(f). The numbers M and m are called, respectively, *upper* and *lower* bounds for the function $f(x)$ on the interval in question. It may of course happen that only one of these conditions is true, and if it never exceeds M then it is said to be bounded above, whereas if it is never less than m it is said to be bounded below. A bounded function is thus a function that is bounded both above and below. The bounds M and m need not be strict in the sense that the function ever actually attains them. Sometimes when the bounds are strict they are only attained at an end point of the domain of definition of the function.

Of all the possible upper bounds M that may be assigned to a function that is bounded above on some interval, there will be a smallest one M', say. Such a number M' is called the least upper bound or the *supremum* of the function on the interval, and the name is usually abbreviated to l.u.b. or to sup. Similarly, of all the possible lower bounds m that may be assigned to a function that is bounded below on some interval, there will be a largest one m', say. Such a number m' is called the greatest lower bound or the *infimum* of the function on the interval, and the name is usually abbreviated to g.l.b. or to inf.

Not all functions are bounded either above or below, as evidenced by the function $y=\tan x$ on $(-\frac{1}{2}\pi, \frac{1}{2}\pi)$, though it is bounded on any closed sub-interval not containing either end point. Typical examples of bounded functions on the interval $(-\infty, \infty)$ are $y=\sin x$ and $y=\cos x/(1+x^2)$. The function $y=1/(x-1)$ is bounded below by zero on the interval $(1, \infty)$ but is unbounded above, whereas the function $y=2-x^2$ is strictly bounded above by 2 but is unbounded below on the interval $(-\infty, \infty)$. The function $y=(x^2+1)/(3x^2+1)$ is bounded for all x, with a least upper bound 1 which is attained at $x=0$, and a greatest lower bound of $\frac{1}{3}$ which is not attained.

2.3.7 Convex and concave functions

A *convex* function is one which has the property that a chord joining any two points A and B on its graph always lies *above* the graph of the function contained between those two points. Similarly, a *concave* function is one which has the property that a chord joining any two points A and B on its graph always lies *below* the graph of the function contained between those two points. Thus the function $y = |x|$ shown in Fig. 2.9(c) is convex on the interval $(-\infty, \infty)$ whereas the function shown in Fig. 2.9(d) is only concave on the closed interval $[-a, a]$. In some books, instead of saying that a function is *convex*, they say it is *concave up* and, conversely, instead of saying a function is *concave*, they say it is *concave down*. Thus the function $y = |x|$ shown in Fig. 2.9(c) is concave up, while the function shown in Fig. 2.9(d) is concave down in $[-a, a]$.

2.3.8 Polynomial and rational functions

A *polynomial* of *degree n* is an algebraic expression of the form

$$y = a_n x^n + a_{n-1} x^{n-1} + \ldots + a_1 x + a_0,$$

where n is a positive integer. Polynomials are defined for all x.

Polynomials of low degree are named. Those of degree 1 are called *linear*, those of degree 2 *quadratic*, those of degree 3 *cubic*, those of degree 4 *quartic* and those of degree 5 *quintic*.

A number ζ will be said to be a *zero* of the polynomial

$$P_n(x) = a_n x^n + a_{n-1} x^{n-2} + \ldots + a_1 x + a_0$$

if $P_n(\zeta) = 0$. Equivalently, if we consider the equation $P_n(x) = 0$, then any number ζ such that $P_n(\zeta) = 0$ will be said to be a *root* of the equation. An important algebraic theorem we will often have cause to use, but which we shall not prove, is the *fundamental theorem of algebra*. This asserts that every polynomial $P_n(x)$ of degree n has n zeros $\zeta_1, \zeta_2, \ldots, \zeta_n$, and as a consequence can always be expressed as the product of n linear factors

$$P_n(x) = a_n(x - \zeta_1)(x - \zeta_2)\ldots(x - \zeta_n).$$

The zeroes $\zeta_1, \zeta_2, \ldots, \zeta_n$, which need not all be different, are not necessarily all real numbers. In section 4.2 we will prove that when the coefficients $a_0, a_1, \ldots, a_n$ of $P_n(x)$ are real, the zeroes will either be real or, if complex numbers, they will occur in complex conjugate pairs.

The simple linear polynomial equation

$$ax + b = 0 \quad (a \neq 0)$$

has the root $x = -b/a$. The quadratic polynomial equation

$$ax^2 + bx + c = 0$$

may always be solved for its two roots x_1 and x_2 by means of the well-known formulae

$$x_1 = \frac{-b + \sqrt{b^2 - 4ac}}{2a} \quad \text{and} \quad x_2 = \frac{-b - \sqrt{b^2 - 4ac}}{2a}.$$

The roots will be real when $b^2 - 4ac \geq 0$, and complex when $b^2 - 4ac < 0$ (see section 4.1).

A more complicated formula exists for the roots of a cubic, but for polynomials of degree greater than 3 it can be shown that no general formula can be found by which their roots may be determined. In general it is necessary to determine the roots of an arbitrary polynomial by means of numerical techniques.

An exception, in the form of a special case which occurs sufficiently frequently to merit comment, is the so-called *biquadratic equation*

$$ax^4 + bx^2 + c = 0.$$

The roots of this special quartic equation may be found by first setting $x^2 = z$, to reduce it to a quadratic equation in z, and using the formula for the roots of a quadratic to determine the roots

$$z_1 = \frac{-b + \sqrt{b^2 - 4ac}}{2a} \quad \text{and} \quad z_2 = \frac{-b - \sqrt{b^2 - 4ac}}{2a}.$$

Then, taking the square roots of z_1 and z_2 leads to the four roots of the biquadratic equation in the form

$$x_1 = \sqrt{z_1}, \ x_2 = -\sqrt{z_1}, \ x_3 = \sqrt{z_2} \text{ and } x_4 = -\sqrt{z_2}.$$

The roots will be real when $b^2 - 4ac > 0$ and complex when $b^2 - 4ac < 0$.

The number of times a root of an equation is repeated is called the *multiplicity* of the root. Thus if $P_7(x)$ is expressible in factorized form as

$$P_7(x) = (x - \zeta_1)^2(x - \zeta_2)(x - \zeta_3)^4,$$

it follows that ζ_1 is a root of multiplicity 2, ζ_2 is a single root (multiplicity 1) and ζ_3 is a root of multiplicity 4.

A *rational* function is a function which is capable of expression as the quotient of two polynomials and so has the form

$$y = \frac{b_m x^m + b_{m-1}x^{m-1} + \ldots + b_1 x + b_0}{a_n x^n + a_{n-1}x^{n-1} + \ldots + a_1 x + a_0},$$

and is defined for all values of x for which the denominator does not vanish.

An example of a polynomial of degree 2 is the quadratic function $y = x^2 - 3x + 4$; a typical rational function is

$$y = \frac{3x^2 - 2x - 1}{4x^3 + 11x^2 + 5x - 2},$$

which is defined for all values of x apart from $x = -2$, $x = -1$ and $x = \frac{1}{4}$, at which points the denominator vanishes but the numerator remains finite and non-zero. For this reason these values are called the *zeros* of the polynomial forming the denominator and they arise directly from its factorization

$$4x^3 + 11x^2 + 5x - 2 \equiv (4x - 1)(x + 2)(x + 1).$$

2.3.9 Algebraic function

An *algebraic* function arises when attempting to from the inverse of a rational function. The function $y = +\sqrt{x}$ for $x \geq 0$ provides a typical example here. More complicated examples are the functions:

$$y = x^{2/3} \quad y = x^2 + 2\sqrt{x} - 1 \quad y = x\sqrt{[x/(2-x)]}.$$

More precisely, we shall call the function $y = f(x)$ algebraic if it may be transformed into a polynomial involving the two variables x and y, the highest powers of x and y both being greater than unity. This criterion may easily be applied to any of the above examples. In the case of the last example, a simple calculation soon shows that it is equivalent to the polynomial $2y^2 - 2xy^2 - x^3 = 0$, which is of degree 2 in y and 3 in x.

2.3.10 Transcendental function

A function is said to be *transcendental* if it is not algebraic. A simple example is $y = x + \sin x$, which is defined for all x but is obviously not algebraic. A *transcendental number* is one which is the root of an equation that is not algebraic, such as a root of $x + \sin x = 0$.

2.3.11 The function $[x]$

On occasions when working with quantities that may only assume integral values it is useful to write $y = [x]$, meaning that we assign to every real number x the greatest integer y that is less than or equal to it. Thus, for example, we have $[-3] = -3$, $[-1.3] = -2$, $[0] = 0$, $[0.92] = 0$, $[\pi] = 3$ and $[17] = 17$.

2.3.12 The function sgn$\{f(x)\}$

The *signum function* $\operatorname{sgn}\{f(x)\}$ is equal to the value 1 when the sign of $f(x)$ is positive and to -1 when the sign of $f(x)$ is negative. It is undefined when $f(x) = 0$. In terms of the absolute value,

$$\operatorname{sgn}\{f(x)\} = \frac{f(x)}{|f(x)|}, \quad \text{for} \quad f(x) \neq 0.$$

2.3.13 A periodic function

A function $f(x)$ is said to be periodic in x if for some X it has the property that $f(x) = f(x + X)$ for all $x \in \mathbf{R}$ and the *period* of the function is then defined as the *smallest* value of X for which the periodicity condition is true. Fig. 2.9(g) shows a typical periodic function with period X. It is clear that the indicated value of X is the *period* of the function, because it is the *smallest* value for which periodicity occurs. Of course, the periodicity of $f(x)$ implies that $f(x)=(x+2X)=f(x+3X)=f(x+4X)=\ldots=f(x+nX)$ for $n=2, 3, \ldots$.

2.3.14 The floor and ceiling functions

The *floor* function, denoted by $\lfloor x \rfloor$, is defined as the largest integer smaller than or equal to x, and the *ceiling* function, denoted by $\lceil x \rceil$, is defined as the smallest integer larger than or equal to x. These two functions, which are used in computing and occur in most computer algebra languages, are illustrated in Fig. 2.9(h), with the *floor* function shown as the dotted 'staircase function' and the *ceiling* function as the solid "staircase function'. For purposes of comparison, the line $y=x$ shown as the solid straight line passing between the floor and ceiling functions, has been added for comparison. It should be mentioned that in some computer languages the *floor* function is called the *integer part* function and denoted by int(x), when the symbol [x] is then used to denote the *nearest integer* function because it lies between $\lfloor x \rfloor$ and $\lceil x \rceil$.

2.4 Curves and parameters

A parameter t may be associated with a curve in two quite different ways. In the first situation we shall discuss, the parameter α occurs as a constant in the equation describing the curve. Thus changing the value of t will change the curve that is described. This simple idea underlies the geometrical concept of an *envelope*, which is a curve that, when it exists, is tangent to each one of these curves for t in some given interval. The subject of envelopes will be taken up again later in connection with differentiation and with differential equations.

In the second situation, t will appear as a variable associated with two functions $s(t)$ and $t(t)$, which will describe separately the x- and y-coordinates of points on any unbroken curve. This use of a parameter is called the *parametrization of a curve* and is an alternative method of representing the equation of the curve.

2.4.1 Envelopes

This situation is best explained by means of an example. Consider the equation

$$(x-t)^2+y^2=\frac{t^2}{1+t^2},$$

which in this form is easily seen to describe a circle of radius $|t|/\sqrt{(1+t^2)}$ with its center on the x-axis at the point $x=t$. Obviously, changing t will both move the center of the circle and alter its radius, as shown in Fig. 2.10.

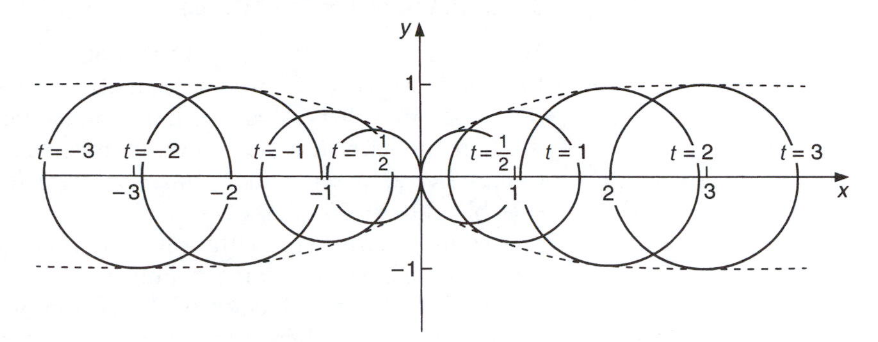

Fig. 2.10 Envelope shown as dotted line.

If t is allowed to vary in some interval, then the single equation will describe a set of circles, each one corresponding to a different value assumed by t in that interval. Collectively these circles are a *family* of circles with parameter t. If a curve exists that is tangent to members of a family of curves, but is not itself a member of the family, then it is an *envelope* of the family. An envelope can be curve of infinite length or on occasions it may reduce either to a curve of finite length or, in degenerate cases, to a single point.

In Fig. 2.10 the envelope is shown as a dotted curve and, as would be expected in this case, the envelope is symmetrical about both the x- and y-axes.

If the family of circles that led to this envelope is written in the form

$$(x - t)^2 + y^2 - \frac{t^2}{1 + t^2} = 0,$$

then it is seen to be a special case of an equation in three variables having the general form

$$f(x, y, t) = 0.$$

This is the standard form for an equation defining a family of curves with parameter t and it will be used later to determine the equation of the envelope when it exists.

However, it is easy to see that a family of curves does not always have an envelope associated with it, since the concentric circles $x^2 + y^2 = t^2$ form a perfectly good family with parameter t, but clearly there is no line that is tangent to each circle in the family.

Expression (2.3) is an *implicit* representation of a function in the sense that it is not directly obvious how and when it is possible to separate out y and re-express it in the more familiar *explicit* form

$$y = F(x, t). \tag{2.4}$$

2.4.2 Parametrization of a curve

We have seen that when a curve is represented by an explicit equation of the form $y = f(x)$, then for inversion reasons the mapping must be one–one. In other words, either f must be strictly monotonic in its domain of definition or, if not, it must be expressible piece by piece (piecewise) as a set of new functions which are strictly monotonic on suitably chosen domains.

A more general representation of a curve that overcomes the necessity for sub-division of the domain, and even allows curves with loops, may be achieved by the introduction of the notion of a *parametric representation of a curve*. The idea here is simple and is that instead of considering x and y to be directly related by some function f, we instead consider x and y separately to be one–one functions of the variable parameter t. Thus we arrive at the pair of equations.

$$x = s(t) \quad y = t(t), \tag{2.5}$$

with $a \le t \le b$, say, which together define a curve. For any value of t in $[a, b]$ we can use these equations to determine unique values of x and y, and hence to plot a single point on the curve represented parametrically by Eqns (2.5). The set of all points described by Eqn (2.5) then defines a curve.

As a simple example of a curve without loops we may consider the parametric equations

$$y = t^2 \quad x = t \quad \text{for} \quad -\infty < t < \infty.$$

In this case these parametric equations are simple enough for t to be eliminated, leading to the explicit equation $y = x^2$, a parabola that lies in the upper half plane and is symmetrical about the y-axis with its vertex passing through the origin. In more complicated cases the parameter cannot usually be eliminated and, indeed, this should not be expected since parametric representation is more general than explicit representation.

An important consequence of the parametric representation of a curve is that increasing the value of the parameter defines a sense of direction along the curve which is often very useful in more advanced applications of these ideas. An example of a curve containing a loop is provided by the parametric equations

$$x = t^3 - t \quad y = 4 - t^2 \quad \text{for} \quad -2 \le t \le 2,$$

which is shown in Fig. 2.11, together with the sense of direction defined by increasing t.

It is implicit in the concept of the parametric representation of a curve that a given curve may be parametrized in more than one way. Hence changing the variable in a parametrization will give different parametric representation of the same curve. Thus if in the example above we replace the parameter t by the parameter u using the relationship $t = u + 1$, then it is readily seen that

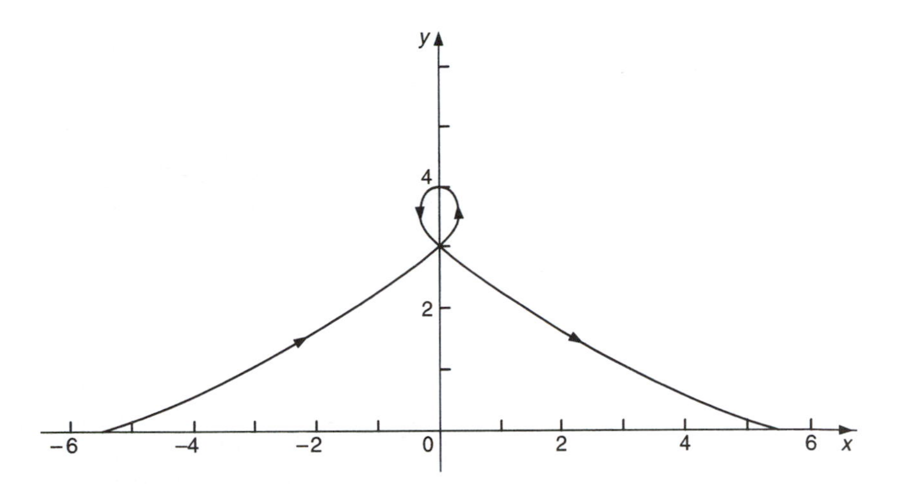

Fig. 2.11 Parametrization of a curve defining sense of direction.

$$x = u^3 + 3u^2 + 2u \quad y = 3 - 2u - u^2 \quad \text{for} \quad -3 \leq u \leq 1.$$

This is an alternative parametrization of the same curve shown in Fig. 2.11.

For a different example of a parametric representation, we now consider the lemniscate introduced in Example 1.15, a special case of which was shown in Fig. 1.22(b) using the polar representation $r^2 = 9\cos(2\theta)$. The geometrical definition of a lemniscate is the locus of points the product of whose distances from two fixed points called the *foci* is a constant. Let us derive the algebraic equation of a lemniscate, show how its representation in polar coordinates arises, and then how a parameterization simplifies the task of drawing the curve.

Figure 2.12(a) shows the two foci of a lemniscate located at the points A at $(-a, 0)$ and B at $(a, 0)$, and a typical point $P(x, y)$ on the lemniscate itself. Then from the Pythagoras theorem the length $PA = \sqrt{(x+a)^2 + y^2}$ and the length $PB = \sqrt{(x-a)^2 + y^2}$, so if the product $PA.PB = b^2$ (a constant), the algebraic equation of the lemniscate becomes $\sqrt{(x+a)^2 + y^2}\sqrt{(x-a)^2 + y^2} = b^2$, which is equivalent to

$$x^4 + 2x^2y^2 + y^4 = 2a^2(x^2 - y^2) + b^4 - a^4.$$

We will only consider the case $b = a$, which causes the algebraic equation of the lemniscase to reduce to

$$(x^2 + y^2)^2 = 2a^2(x^2 - y^2).$$

The polar coordinate representation now follows by setting $x = r\cos\theta$ and $y = r\sin\theta$, when the equation simplifies to

$$r^2 = 2a^2(\cos^2\theta - \sin^2\theta),$$

but $\cos^2\theta - \sin^2\theta = \cos 2\theta$, so the polar coordinate representation of the

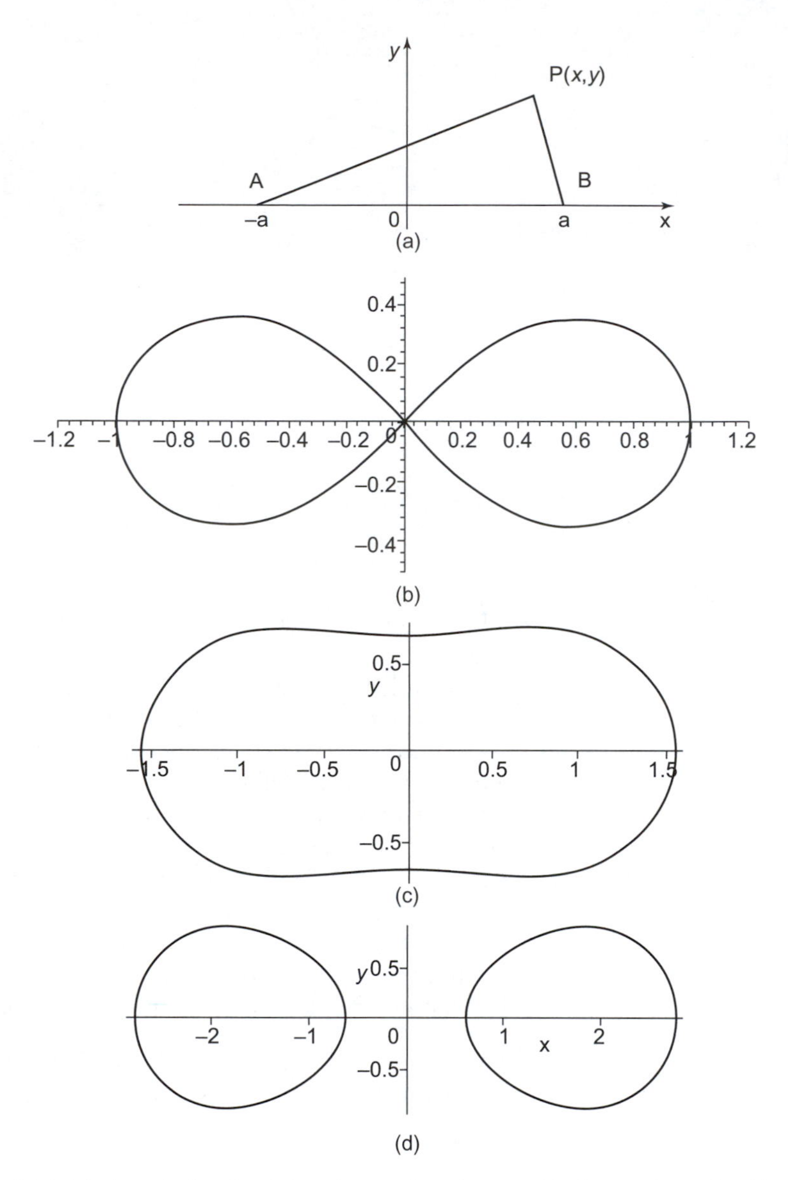

Fig. 2.12 (a) A typical point $P(x, y)$ on a lemniscate with foci at $(\pm a, 0)$; (b) A lemniscate with $a = 1$; (c) A Cassini oval with $a < b$; (d) Cassini ovals when $a > b$.

lemniscate becomes

$$r^2 = 2a^2 \cos 2\theta.$$

We have already seen how to plot this polar representation of the lemniscate, and it is clear that is far easier than attempting to plot it using the algebraic representation $(x^2 + y^2)^2 = 2a^2(x^2 - y^2)$ from which y cannot be found explicitly in terms of x (y is defined implicitly in terms of x). This representation is an even function of both x and y, so without actually plotting the curve it follows immediately that the lemniscate is symmetrical about both the x and y-axes, and inspection shows it must pass through the origin (0, 0) and the points $(\pm a\sqrt{2},\ 0)$, but this is all that can be deduced about the curve without actually plotting it.

However, the curve can be expressed in parametric form, one possible of which is given by

$$x = \frac{a\cos t}{1+\sin^2 t}, \quad y = \frac{a\sin t\cos t}{1+\sin^2 t}, \qquad 0 \le t < 2\pi.$$

The curve corresponding to $a = 1$ is shown in Fig. 2.12(b), where arrows have been added to show the sense of direction along the curve as t increases.

In fact the lemniscate considered here with $b = a$ is a special case of a family of curves called the *Cassini ovals*. When $a < b$ the curves form single loops enclosing the two foci, and the loop does not pass through the origin, though the oval shaped curves may be slightly pinched in along the y-axis as in Fig. 2.12(c), but when $a > b$ the curves separate into two distinct loops, each enclosing one of the foci as in Fig. 2.12(d). ■

2.5 Functions of several real variables

In physical situations, to say that a quantity depends only on one other quantity is usually a gross oversimplification. Indeed, this was so in the thermodynamic illustration used to introduce the notion of a function of one real variable, because we insisted on maintaining a *constant* volume of gas, thereby relating only the pressure and the temperature. In general the pressure p of a given gas will depend on both its temperature T and its volume v. Here we would say that there was a functional relationship between p, T and v which, in an implicit form, may be expressed by the equation

$$f(p,\ T,\ v) = 0. \tag{2.6}$$

The function f occurring here is a *function of three real variables* and obviously depends for its form on the particular gas involved.

Usually one of the three quantities, say p, is regarded as a dependent variable with the others, namely T and v, being regarded as independent variables. Solving Eqn (2.6) for p then gives rise to an explicit expression of the form

$$p = g(T,\ v), \tag{2.7}$$

with g then being called a *function of two real variables*. (This is a special case of the general idea of a function introduced earlier.)

Just as with a function of a single real variable, in addition to specifying the functional form it is also necessary to stipulate the domain of definition of the function. Thus Eqn (2.7), which in thermodynamic terms would be called the *equation of state* of the gas, would only be valid for some range of temperature and volume. In this case the reason for the restriction on the temperature and volume is a physical one, whereas in other situations it is likely to be a purely mathematical one.

Extending the ideas already introduced, we shall now let $\mathbf{R}^2$ denote the set of all ordered pairs (x, y) of real numbers and let S be some subset of $\mathbf{R}^2$.

Definition 2.3 We say f is a real-valued function of the real variables x and y defined in set S if, for every $(x, y) \in S$, there is defined a real number denoted by $f(x, y)$.

As is the case with a function of one variable, when the domain of definition of a real-valued function of two or more real variables is not specified it is to be understood to be the largest possible domain of definition that can be defined. Thus, for example, the largest subset S in $\mathbf{R}^2$ in which in the function $f(x, y) = \sqrt{(1 - x^2 - y^2)}$ is defined is given by

$$S = \{(x,\ y) \in \mathbf{R}^2 | x^2 + y^2 \leq 1\}.$$

This concept of a function immediately extends to include functions of more than two variables. Using $\mathbf{R}^n$ to denote the set of all ordered n-tuples $(x_1,\ x_2,\ \ldots,\ x_n)$ of real numbers of which S is some subset, this definition can be formulated.

Definition 2.4 We shall say that f is a real-valued function of the real variables x_1, x_2, $\ldots$, x_n defined in set S if, for every $(x_1, x_2, \ldots, x_n) \in S$, there is defined a real number denoted by $f(x_1,\ x_1,\ \ldots,\ x_n)$.

A typical example of a function of the three variables x, y, z is provided by $f(x, y, z) = \sqrt{(2 - x)} + \sqrt{(9 - y^2)} + \sqrt{(16 - z^4)}$. The largest subset S in $\mathbf{R}^3$ for which this function may be defined is obviously

$$S = \{(x,\ y,\ z) \in \mathbf{R}^3 | x \leq 2;\ \ -3 \leq y \leq 3;\ \ -2 \leq z \leq 2\}.$$

The geometrical idea underlying the graph of a function of a single variable also extends to real functions f of two real variables x, y. Denote the value of the function f at (x, y) by z, so that we may write $z = f(x, y)$. Then with each point of the (x, y)-plane at which f is defined we have associated a third number $z = f(x, y)$. Taking three mutually perpendicular straight lines with a common origin 0 as axes, we may then identify two of the axes with the independent variables x and y and the third with the dependent variable z. The ordered number triples $(x, y, z) \equiv (x, y, f(x, y))$ may then be plotted as points in a three-dimensional geometrical space. The set of points (x, y, z) corresponding to the domain of definition of the function $f(x, y, z)$ then defines a surface which, in practice, often turns out to be smooth. It is conventional to plot z vertically.

On account of the geometrical representation just described, even in $\mathbf{R}^n$ it is customary to speak of the ordered n-tuple of numbers $(x_1, x_2, \ldots, x_n)$ as defining a 'point' in the 'space' $\mathbf{R}^n$.

By way of illustration of a graph of a function of two variables we now consider

$$f(x, y) = \frac{x^2}{4} + \frac{y^2}{9} \quad \text{with} \quad \frac{x^2}{4} + \frac{y^2}{9} \leq 2,$$

where the inequality serves to define a domain of definition for the function. The surface described by this function has the equation $z = x^2/4 + y^2/9$ and the domain of definition is the interior and boundary of $x^2/4 + y^2/9 = 2$. If this latter expression is rewritten in the form $x^2/8 + y^2/18 = 1$ then it ca be seen that the domain of definition of f is in fact the interior of an ellipse in the (x, y)-plane having semi-minor axis $2\sqrt{2}$ and semi-major axis $3\sqrt{2}$, and being centered on the origin. As $f(x, y)$ is an essentially positive quantity it follows directly that $0 \leq z \leq 2$ in the domain of f.

To deduce the form of the surface, two further geometrical concepts are helpful. The first is the notion of the curve defined by taking a *cross-section* of the surface parallel to the z-axis. The second is the notion of a *contour line* or *level curve*, defined by taking a cross-section of the surface perpendicular to the z-axis.

To examine a cross-section of the surface by the plane $y = a$, say, we need only set $y \equiv a$ in $f(x, y)$ to obtain $z = x^2/4 + a^2/9$, showing that the curve so defined is a parabola in the plane $y = a$ with vertex at a height $z = a^2/9$ above the y-axis. A similar cross-section by the plane $x = b$ shows that the curve so defined is $z = b^2/4 + y^2/9$, which is also a parabola, but this time in the plane $x = b$ with its vertex at a height $z = b^2/4$ above the x-axis. (See Fig. 2.13(a).) If desired, sections by other planes parallel to the z-axis may also be used to assist visualization of the surface.

To assist in the visualization of the intersection of a plane with the surface $z = x^2/4 + y^2/9$, Fig. 2.13(b) shows a computer generated picture of the intersection of the plane $x = 1$ with this surface. The parabolic curves drawn on the curved surface in front of the plane $x = 1$ correspond to the curves of intersection of different planes $x = a$ for which $a > 1$.

As already stated, a curve defined by a section of the surface resulting from a cross-section taken perpendicular to the z-axis is called a contour line or level curve by direct analogy with cartography, where such lines are drawn on a map to show contours of constant altitude. Level curves are obtained by determining the curves in the (x, y)-plane for which $z =$ constant, and it is customary to draw them all on one graph in the (x, y)-plane, with the appropriate value of z shown against each curve. (See Fig. 2.13(c).)

Let us determine the level curve in our example corresponding to $z = \frac{1}{2}$ which is representative of z in the range $0 \leq z \leq 2$. We must thus find the curve with the equation $x^2/4 + y^2/9 = \frac{1}{2}$, which we choose to rewrite in the standard form $x^2/2 + y^2/(9/2) = 1$. This shows that it describes an ellipse centered on the origin with semi-minor axis $\sqrt{2}$ and semi-major axis

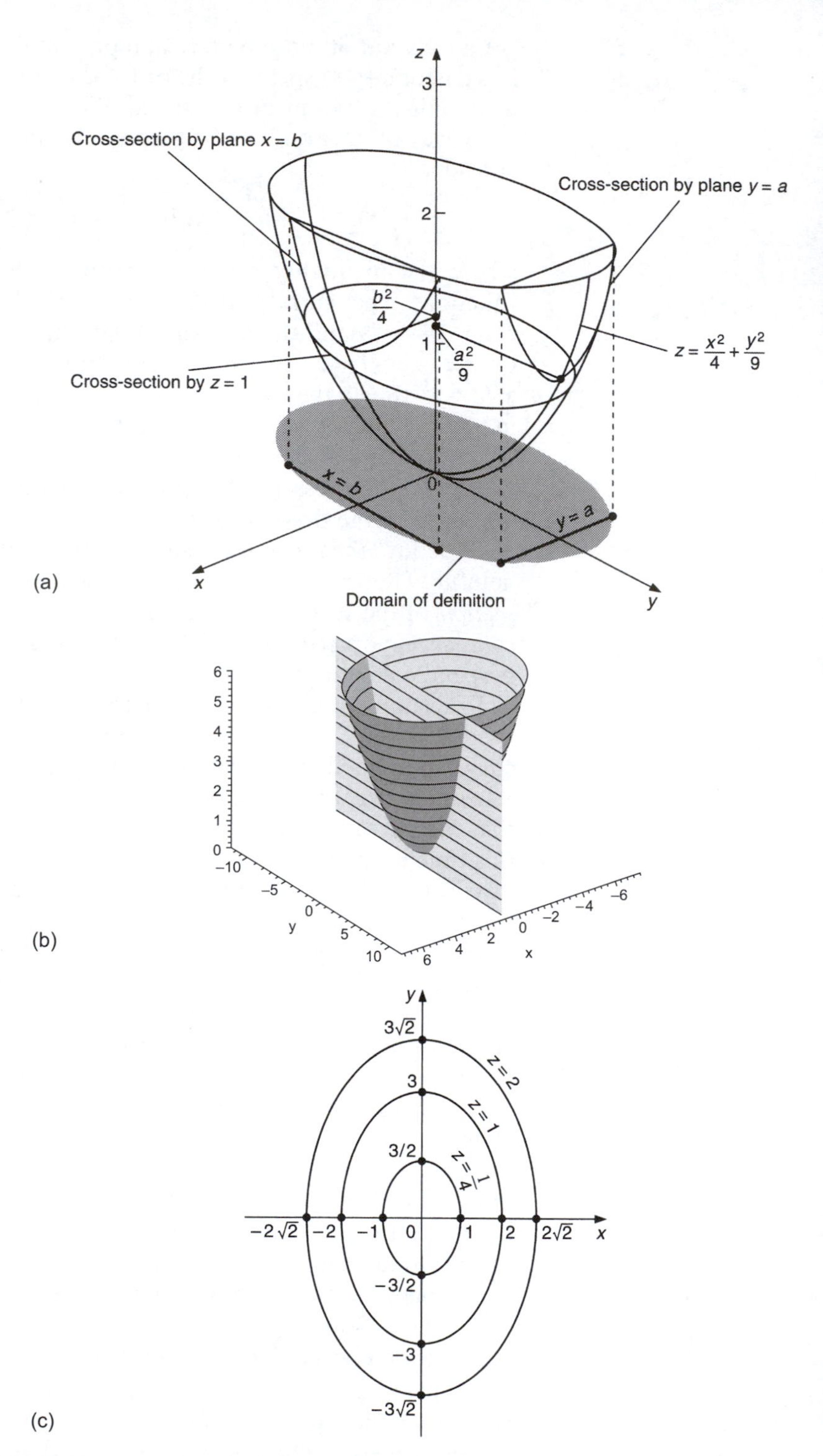

Fig. 2.13 Surface and level curves: (a) representation of surfaces; (b) computer plot of plane $x = 1$ intersecting the surface; (c) level curves.

$3/\sqrt{2}$. It is not difficult to see that all the level curves are ellipses, the one corresponding to $z = 2$ being the boundary of the domain of f and the one corresponding to $z = 0$ degenerating to the single point at the origin.

Not all graphs of functions of two or more variables are smooth, even though the surface of the graph may be unbroken. An example of this type is provided by the function

$$z = |x| + |y| - ||x| - |y||$$

with domain of definition $|x| \leq a$, $|y| \leq a$, whose graph is shown in perspective in Fig. 2.14(a). It is seen to comprise eight indentical plane

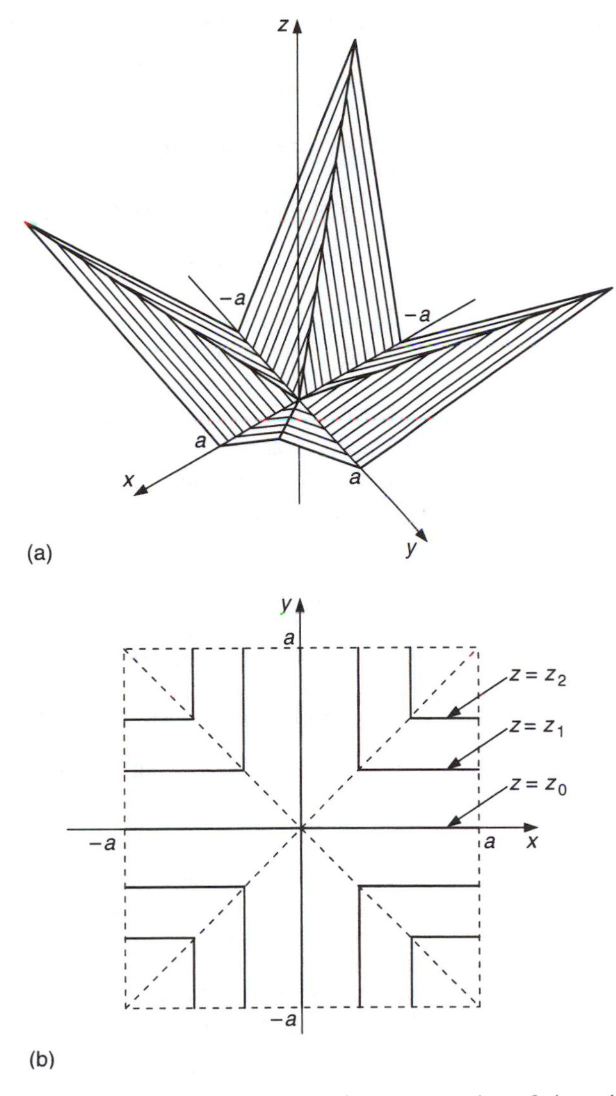

Fig. 2.14 Surface and level curves: (a) isometric representation of piecewise plane non-smooth surface; (b) level curves which are symmetric with respect to the x- and y-axes.

right-angled triangular shaped surfaces, fitted together so that four inverted ridges lie along the x- and y-axis, while four elevated ridges are symmetrically arranged about the (x, y)-plane, issuing out from the origin.

The only points on the graph which lie in the (x, y)-plane are those along the x- and y-axes themselves; all other points lie above the (x, y)-plane on the side of positive z. The level curves are shown in Fig. 2.14(b), where they are seen to be right-angled segments, symmetric with respect to the x- and y-axes. The level curves corresponding to the four inverted ridges lie along the axes. The four elevated ridges lie along the diagonals shown as dotted lines which are not, of course, level curves.

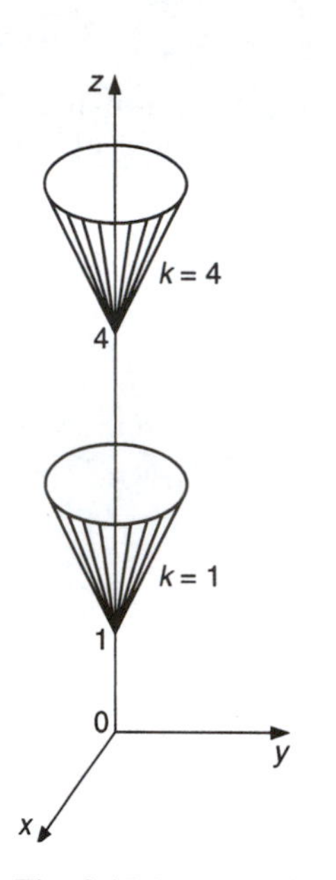

Fig. 2.15 Level surfaces for $w = z - \sqrt{x^2 + y^2}$.

If the functions involved have more than two independent variables it is necessary to generalize the notion of a level curve to that of a level surface–that is, a surface on which the dependent variable is constant. Thus the level surfaces of

$$w = z - \sqrt{x^2 + y^2}$$

are the surfaces $w = k$ (const.) given by

$$z - \sqrt{x^2 + y^2} = k.$$

It follows from section 1.6.9 that for any give k each of these is a right-circular cone, with its axis coincident with the z-axis and its vertex on the z-axis at $z = k$. Parts of these level surfaces are shown in Fig. 2.15.

Level surfaces play an important part in many physical problems. For example, a surface on which all points are at the same electrostatic potential is called an equipotential surface, and a surface on which all points are at the same temperature is called an isothermal surface.

Problems

Section 2.1

2.1 Sketch the graph of these functions:

(a) $f(x) = x^2 - 3x + 2 \quad (-1 \leq x \leq 3)$;

(b) $f(x) = x + \sin x \quad (-\pi/2 \leq x \leq \pi/2)$;

(c) $f(x) = x^3 \quad (-2 \leq x \leq 2)$;

(d) $f(x) = x^2 + 1/x \quad (0.2 \leq x \leq 2)$;

(e) $f(x) = x + 1/x^2 \quad (0.5 \leq x \leq 5)$.

2.2 Determine the domain and the range of each of functions (a) to (e) defined in Problem 2.1.

2.3 Determine the range of the function $f(x) = x^2 + 1$ corresponding each of the following domains and state when the mapping is one–one and when it is many–one:

(a) $[-1, 1]$; (b) $(2, 4)$; (c) $[-2, 4]$; (d) $[-3, 1]$.

2.4 Find the largest domain of definition for each of the following functions:

(a) $f(x) = x^3 + 3$; (b) $f(x) = x^2 + \surd(1 - x^2)$;

(c) $f(x) = x^2 + \surd(1 - x^3)$; (d) $f(x) = 1/(x^2 - 1)$;

(e) $f(x) = x + 1/x$; (f) $f(x) = x^2/(1 + x^2)$.

2.5 Let $f(n)$ denote the function that assigns to any positive integer n the number of positive integers whose square is less than or equal to $n+2$. By enumerating the first few values of $f(n)$ deduce the values of n for which $f(n) = 3$.

2.6 Sketch representative members of the two pencils of lines described by $y = \alpha(x - 1)+2$ and $y = \beta(x - 2)+3$, where α and β are parameters. Locate the singular points and suggest how α and β may be used as coordinates for points in the plane of the two pencils. When will the coordinates α and β fail to identify points? [*Note: A pencil* of straight lines is a family (set) of straight lines that all pass through a common point called a *singular point*.]

Section 2.2

2.7 Sketch the graphs of the following functions in their stated domains of definition and in each case use the process of reflection in the line $y = x$ to construct the graph of the inverse function:

(a) $f(x) = x^3$ with $x \in [-2, 2]$;

(b) $f(x) = x + \sin x$ with $x \in [0, \pi/2]$;

(c) $f(x) = x/(1 + x^2)$ with $x \in [-1, 2]$.

Section 2.3

2.8 Sketch these functions in their associated domains of definition:

(a) $f(x) = | 2x |$ for $x \in [-2, 2]$;

(b) $f(x) = x + | x |$ for $x \in [-2, 2]$;

(c) the step function assuming the values 1, 2, -3, 2, 4 on the x-intervals [0, 1), [1, 2], (2, 3.5), [3.5, 4], and (4, 5], respectively. Identify end points belonging to a line by a dot and end points deleted from a line by a circle.

(d) $$f(x) = \begin{cases} |x| \text{ for } x \in [0, 1) \\ |x - 1| \text{ for } x \in [1, 2) \\ |x - 2| \text{ for } x \in [2, 3]. \end{cases}$$

2.9 Where appropriate, classify the following functions as even or odd:

(a) $f(x) = x + |x|$;

(b) $f(x) = x + \sin 2x$;

(c) $f(x) = x^2 + \sin x$;

(d) $f(x) = 1/x$;

(e) $f(x) = x^2/(1+x^2)^2$;

(f) $f(x) = x^5 - x^3 + x$;

(g) $f(x) = 2\cos x + \sin x$.

It is obvious that any arbitrary function $f(x)$ which is defined in an interval $\mathscr{I}$ containing the original may be written in the form

$$f(x) = \tfrac{1}{2}(f(x) + f(-x)) + \tfrac{1}{2}(f(x) - f(-x)),$$

in any interval $\mathscr{I}$ lying within $\mathscr{I}$ that is symmetric about the origin. Such an interval $\mathscr{I}$ is said to be *interior* to $\mathscr{I}$. This shows that any such $f(x)$ is expressible as the sum of an even function $\frac{1}{2}(f(x) + f(-x))$, and an odd function $\frac{1}{2}(f(x) - f(-x))$ within $\mathscr{I}$. Apply this result to display the following functions as the sum of even and odd parts, in each case stating the largest interval $\mathscr{I}$ for which the result is true:

(h) $f(x) = 1 + x^3 + x\sin x$ for $-2\pi \le x \le 3\pi$;

(i) $f(x) = 1 + x + |x| \sin x$ for $-3\pi \le x \le 3\pi$;

(j) $f(x) = 1 - x + 2x^2 + 4x^3$ for $-4 \le x \le 3$.

2.10 Determine if upper and lower bounds exist for the following functions and, when appropriate, state their values and where they occur on the respective domains of definition:

(a) $f(x) = 1/x$ for $x \in [1, 4]$;

(b) $f(x) = 1/x$ for $x \in (0, 3]$;

(c) $f(x) = 1 + x^2$ for $x \in [-2, 1]$;

(d) $f(x) = \sin x$ for $x \in [0, 3\pi/2]$;

(e) $f(x) = \tan x$ for $x \in (-\pi/2, \pi/2)$.

2.11 The pairs of numbers enclosed by the curly brackets following each problem are upper and lower bounds for the associated function in its stated domain of definition. State whether or not each of these bounds is strict:

(a) $f(x) = x^3 + x + 1$ with $x \in [1, 2]$, $\{0, 11\}$;

(b) $f(x) = \sin x$ with $x \in [0, \pi/2]$, $\{0, 2\}$;

(c) $f(x) = 1/(1+x^2)$ with $x \in [0, 2], \{1/6, 2\}$;

(d) $f(x) = \sin(1/x)$ with $x \in [2/\pi, 30], \{0, 1\}$.

2.12 Determine by sketching whether the following functions are convex, concave or neither on their stated domains of definition:

(a) $f(x) = x^3$ for $x \in [1, 3]$;

(b) $f(x) = x^3$ for $x \in [-1, 1]$;

(c) $f(x) = (a^2 - x^2)^{1/2}$ for $x \in [-a/2, a]$;

(d) $f(x) = x + \sin x$ for $x \in [0, \pi/2]$;

(e) $f(x) = \sin x$ for $x \in [0, \pi]$.

2.13 (a) Superimpose on the graph of $2.5 \cos(\pi x)$ for $-1 < x < 1$ the graphs of the functions floor($2.5 \cos(\pi x)$) and ceiling($2.5 \cos(\pi x)$).

(b) Superimpose on the graph of $\sqrt{x}$ for $0 < x < 16$ the graphs of the functions floor($\sqrt{x}$) and ceiling($\sqrt{x}$).

2.14 (a) Superimpose on the graph of $2.5 \sin(\pi x)$ for $-1 < x < 1$ the graphs of the functions floor($2.5 \sin(\pi x)$) and ceiling($2.5 \sin(\pi x)$).

(b) Superimpose on the graph of 2^x for $-2 < x < 2$ the graphs of the functions floor(2^x) and ceiling(2^x).

Section 2.4

2.15 Draw the circles corresponding to $\alpha = \frac{1}{8}, \frac{1}{4}, \frac{1}{2}$, 1, and 2 in the equation $(x-1)^2 + (y-\alpha)^2 = \alpha^2/(1+\alpha^2)$ and sketch the envelope, indicating its asymptotes for large positive and negative α.

2.16 Draw the circles corresponding to $\alpha = \frac{1}{4}, \frac{1}{2}$, 1, 2, and 3 in the equation $(x-\alpha)^2 + y^2 = \frac{1}{2}\alpha^2$ and draw the envelope.

2.17 Deduce the envelope of the family of circles $(x-\alpha)^2 + y^2 = \alpha^2$, with parameter α.

2.18 Sketch representative ellipses belonging to the family $x^2/\alpha^2 + y^2/(4-\alpha)^2 = 1$, with parameter α and deduce the shape of the envelope.

2.19 Draw representative members of the family of straight lines $y = \alpha x + 2/\alpha$, with parameter α, and deduce the shape of the envelope.

2.20 Sketch the curve represented by the parametric equations $x = 2\cos\alpha$, $y = \sin\alpha$ for $-\pi/2 \le \alpha \le \pi/2$.

2.21 Sketch the curve represented by the parametric equations $x = \alpha^2 + 1$, $y = \alpha^3$ for $-2 \le \alpha \le 2$.

2.22 Sketch the curve represented by the parametric equations $x=\alpha^3+\alpha^2-2\alpha$, $y=5-\alpha^2$ for $-3\le\alpha\le 2$. Indicate by arrows on the curve the sense of direction corresponding to increasing α.

Section 2.5

2.23 The function $f(x, y)=x^2y$ has for its domain of definition the rectangle in the (x, y)-plane defined by $|x|\le 3$, $|y|\le 2$. Deduce the shape of the curves defined by cross-sections of the surface $z=f(x, y)$ taken by the three planes $x=-2$, $x=0$ and $x=2$ that are parallel to the (y, z)-axes and by the three planes $y=-2$, $y=0$ and $y=2$ that are parallel to the (x, z)-axes, using your results to sketch the surface. Sketch on one diagram the level curves corresponding to $z=-4$, $z=-2$, $z=0$, and $z=6$.

2.24 Sketch the surface $z=f(x, y)$ defined by the function $f(x, y)=1/(1+x^2+y^2)$ in the domain $|x|\le 4$, $|y|\le 4$. Draw the level curves corresponding to $z=1/9$, $z=1/3$, $z=2/3$, and $z=1$.

2.25 The surface $z=f(x, y)$ is defined by the function $f(x, y)=1/[(x-1)^2+(y-2)^2-1]$ with $2\le(x-1)^2+(y-2)^2\le 9$. Deduce the domain of definition of the function and then sketch the level curves corresponding to $z=\frac{1}{4}$, $z=\frac{1}{2}$, and $z=\frac{3}{4}$ on the same diagram. Use your result to sketch the surface. [*Hint:* Use the fact that the circle of radius ρ with center at (a, b) has the equation $(x-a)^2+(y-b)^2=\rho^2$.]

2.26 Sketch the level surfaces for the function

$$f(x, y, z)=\frac{x^2}{a^2}-\frac{y^2}{b^2}.$$

2.27 Sketch the level surfaces for the function

$$f(x, y, z)=\frac{x^2}{a^2}+\frac{y^2}{b^2}.$$

Supplementary computer problems

The problems in this section require the use of a graphics package.

2.28 Display simultaneously, using the same scales for x and y, the graphs of $y=x^2$ for $0\le x\le 3$, its inverse function $y=\sqrt{x}$ for $0\le x\le 9$ and the line $y=x$ for $0\le x\le 9$. Verify by inspection that the inverse function is the reflection of $y=x^2$ in the line $y=x$.

2.29 Display simultaneously, using the same scales for x and y, the graphs of $y=x^3$ for $0\le x\le 2$, its inverse function $y=x^{1/3}$ for $0\le x\le 8$ and the line $y=x$ for $0\le x\le 8$. Verify by inspection that the inverse function is the reflection of $y=x^3$ in the line $y=x$.

2.30 Display the function $y = x|x|$ for $-2 \le x \le 2$ and use the result to determine when it is a concave and when it is convex.

2.31 Display the function

$$y = 1 + \frac{2(x-1)}{|x-1|} - \frac{3(x-2)}{|x-2|} \quad \text{for} \quad -4 \le x \le 4,$$

and use the result to determine to which category of function described in section 2.3 it belongs.

2.32 Display the function

$$y = \begin{cases} 1 & \text{for} \quad -2 \le x \le 0 \\ \cos x & \text{for} \quad 0 \le x \le 2\pi \end{cases}$$

and use the result to describe the nature of the function on the intervals

(i) $-2 \le x \le 0$, (ii) $0 \le x \le \pi$, (iii) $\pi \le x \le 2\pi$.

2.33 Display the function $y = \sin x \cos 2x$ for $-\pi \le x \le \pi$ and verify by inspection that the product of an odd function (sine) and an even function (cosine) is an odd function.

2.34 Display the function $y = |x| \cos x$ for $-2\pi \le x \le 2\pi$ and verify by inspection that the product of two even functions is an even function.

2.35 Display the function $y = x^3 \sin x$ for $-\pi/2 \le x \le \pi/2$ and verify by inspection that the product of two odd functions is an even function.

2.36 Reproduce the eight circles shown in Fig. 2.10.

2.37 Display the family of circles

$$x^2 + (y - \alpha)^2 - \frac{8x}{1 + \alpha^2} = 0$$

corresponding to $\alpha = \pm\frac{1}{2}, \pm 1, \pm 2, \pm 3$, and use the result to sketch the envelope.

2.38 Reproduce the parametric curves shown in Fig. 2.11 and 2.12(b).

2.39 Display the parametrically defined curve $x = t - \sin t$, $y = 1 - \cos t$ for $0 \le t \le 2\pi$ (a cycloid).

2.40 Display the parametrically defined curve $x = 3\cos t - \cos 3t$, $y = 3\sin t - \sin 3t$ for $0 \le t \le \pi$.

2.41 Display the parametrically defined curve $x = 4\cos t + \cos 2t$, $y = 4\sin t - \sin 2t$ for $0 \le t \le \pi$.

2.42 Display the parametrically defined curve $x = \cos t + 3\cos 2t$, $y = 4\cos t - 3\sin 2t$ for $0 \le t \le \pi$.

2.43 The three-dimensional surface shown in Fig. 2.14(a) has the equation

$$z = |x| + |y| - ||x| - |y||.$$

A cross-section of this surface by the plane $y = c$ (const.) is obtained by setting $y = c$ in the equation for z. Display the form of the cross-section obtained by setting $c = 1, 2, 3$ and 4.

3 Sequences, limits and continuity

The study of a function of a real variable, and of the calculus in general, depends on the related concepts of a sequence and a limit, both of which are examined in some detail in the first two sections of the chapter. These ideas are then shown to lead in turn to notion of a continuous function and, in Chapter 5, to both the derivative of a function of a single real variable and to the partial derivatives of functions of several real variables.

The fundamental mathematical constant $\mathrm{e} = 2.718281\ldots$, can be defined in various different ways. The choice of definition used to introduce e in this chapter makes use of a special form of limit that can also be used to define the exponential function e^x that plays such an important role throughout mathematics. A different, but equivalent, definition of e that is useful for the numerical computation of the exponential function and for determining its differentiability properties is given later in Chapter 5. The exponential function and the related hyperbolic functions form topics for study in Chapter 6, in which the inverse of the exponential function, called the natural logarithmic function $\ln x$, is introduced.

The limit of a function of a single real variable is discussed in this chapter, followed by an explanation of the related notion of continuity. The extension of the discussion of limits to apply to functions of several real variables forms the next topic, and the idea is then used to extend the notion of continuity to such functions, preparatory to the discussion of partial differentiation given in Chapter 5.

Finally, a direct application of limits to the graphical interpretation of rational functions of a single real variable leads directly to the determination of any asymptote to the graph that may exist; that is, to the equation of a straight line that becomes tangent to the graph of the function at infinity. It is shown how asymptotes parallel to the x- and y-axes can be found and also how the equation of any oblique asymptote that exists can be determined.

3.1 Sequences

The notion of a 'sequence' is a constantly recurring one in everyday life, where it usually implies the ordering of some set of events with respect to time. The sets of events that are so ordered, or arranged, are very varied and may be either numerical or non-numerical in nature. Typical examples of commonplace sequences in these categories are these:

(a) the sequence of months in a year;
(b) the sequences of digits identifying a telephone subscriber;

(c) the sequence of machining operations required to make a certain component.

However, sequences are not necessarily decided by the chronological order of events, and they are often determined instead by some attribute possessed by the members of the set to be ordered. Thus, for example, two commonly occurring sequences to be found in any library are the entries in the alphabetic catalogues of authors and titles, neither of which are in the chronological order of acquisition of the books. Although these general ideas could be discussed at greater length, such an examination is inappropriate here, and it must suffice that these few examples show that sequences are commonplace in the world around us, and that they need not necessarily involve numbers.

These ideas find an immediate parallel in mathematics, where the natural order existing in **R**, combined with the arithmetic properties discussed in Chapter 1, enables us to deal very successfully and in great detail with questions relating to mathematical sequences. Our main preoccupation in this book will be with sequences of numbers and sequences of functions, so we must first make the mathematical notion of a sequence more precise. Before doing this, however, we must first issue a word of warning concerning the colloquial usage of the words *sequence* and *series*, and on their mathematical usage, which is quite different. Colloquially the words sequence and series are often used interchangeably, but in mathematics they have two quite different meanings which must never be confused. In brief, in mathematical terms a sequence is a set of quantities that is enumerated in a definite order, whereas a series involves the sum of a set of quantities. Thus 1, 3, 5, 7, 9,... is a sequence but $1+\frac{1}{2}+\frac{1}{4}+\frac{1}{8}+\frac{1}{16}+\ldots$ is a series.

If a sequence is composed of *elements* or *terms* u belonging to some set S, then it is conventional to indicate their order by adding a numerical suffix to each term. Consecutive terms in the sequence are usually numbered sequentially, starting from unity, so that the first few terms of a sequence involving u would be denoted by u_1, u_2, u_3,... . Rather than write out a number of terms in this manner this sequence is often represented by $\{u_n\}$, where u_n is the nth term, or general term, of the sequence. The sequence depends on the set chosen for S and the way suffixes are allocated to elements of S. A sequence will be said to be *infinite* or *finite* according to whether the number of terms it contains is infinite or finite and, unless explicitly stated, all sequences will be assumed to be infinite. The notation for a sequence is often modified to $\{u_n\}_{n=1}^{N}$ when only a finite number N of terms are involved, so that $\{u_n\}_{n=1}^{N} = u_1,\ u_2,\ \ldots, u_N$.

As an example of an infinite numerical sequence, let S be the set of real numbers and the rule by which suffixes are allocated be that to each integer suffix n we allocate the number $1/2^n$ which belongs to **R**. We thus arrive at the finite sequence $u_1 = 1/2$, $u_2 = 1/2^2$, $u_3 = 1/2^3$,..., which could either be written in the form

$$\frac{1}{2}, \frac{1}{2^2}, \frac{1}{2^3}, \ldots, \frac{1}{2^n}, \frac{1}{2^{n+1}}, \ldots,$$

or, more concisely, in the form

$$\{1/2^n\}.$$

Had the set S still been the set $\mathbf{R}$ of real numbers, but the rule of allocation of suffixes been changed, so that to each integer suffix n chosen from the first N natural numbers we allocated the number $1/(2n+1)$, then the finite sequence

$$\frac{1}{3}, \frac{1}{5}, \frac{1}{7}, \ldots, \frac{1}{(2N+1)}$$

would have resulted.

If we use the notion of a function $f(x)$ which is defined only for *integral* values of the argument x, the following concise definition can be formulated.

Definition 3.1 In mathematical terms a *sequence* is a function f defined only for integer values of its argument and having for its range an arbitrary set S.

Thus a sequence is a special kind of function whose domain is the set of positive integers. It follows that the first sequence displayed above could be regarded as resulting from the function $f(x) = 1/2^x$ with $u_n = f(n)$, where n is always a positive integer. By exactly similar reasoning, the second sequence can be derived from the function $f(x) = 1/(2x+1)$ by setting $u_n = f(n)$.

The connection between functions and sequences that is established in this definition makes it appropriate to describe numerical sequences in the same terms as would be used to describe the function giving rise to them. Thus if the terms of a sequence $\{u_n\}$ are such that $m < u_n < M$ for all values of n then the sequence is said to be *bounded*, while if $u_n + 1 > u_n$ for all n then the sequence is said to be *strictly monotonic increasing*. The terms *bounded above*, *bounded below*, *unbounded*, *strictly monotonic decreasing*, *monotonic*, and *oscillating*, etc., can also be used in the obvious manner as shown below where, by 'oscillating', we mean that alternate terms are of opposite sign.

Example 3.1 (a) $\{1/n\}_1^\infty$ is a bounded, strictly monotonic decreasing sequence. The upper bound 1 is strict because it is a member of the sequence, but the lower bound 0 is never actually attained.

(b) $\left\{\frac{1}{\sin(1/n)}\right\}_1^\infty$ is a strictly monotonic increasing sequence, strictly bounded below by $1/\sin 1$ but unbounded above.

(c) $\left\{\frac{(-1)^n}{n}\right\}_1^\infty$ is a bounded sequence with strict upper bound $\frac{1}{2}$, corresponding to $n=2$, and strict lower bound -1, corresponding to $n=1$, with all other terms lying between -1 and $\frac{1}{2}$, and the signs of the terms are alternately positive and negative.

(d) $\{u_n\}_1^\infty$, where $u_{2m-1}=m/(m+1)$ and $u_{2m}=u_{2m-1}$. The first six terms of this sequence are $\frac{1}{2}, \frac{1}{2}, \frac{2}{3}, \frac{2}{3}, \frac{3}{4}, \frac{3}{4}$ corresponding pairwise, respectively, to $m=1$, 2 and 3. The sequence is thus both bounded and monotonic increasing. It is not strictly monotonic increasing because pairs of terms are equal. The lower bound $\frac{1}{2}$ is strict, but the upper bound 1 is never actually attained.

(e) $\{(-1)^n\}$ is an oscillating but bounded sequence with strict upper bound 1 and strict lower bound -1. The only two numerical values in this sequence are 1 and -1, both of which occur infinitely many times. The expression $(-1)^n$ is often used as a multiplier of a more general term to switch its sign alternately between $+$ and $-$, according to whether n is even or odd.

(f) $\{(-2)^n\}$ is an oscillating but unbounded sequence. ■

Just as a graph proved to be useful when representing functions, so also may it be used to represent sequences. Exactly the same method of representation can be adopted, but this time, since the domain of the function defining the sequence is the set of natural numbers, the graph of a sequence will be a set of isolated points. A typical example is the graph of the first few terms of the sequence $\{u_n\}$ with $u_n=[n+(-1)^n]/n$, which are shown as dots in Fig. 3.1(a).

An obvious deficiency of this representation is that the horizontal axis must be made unreasonably long if a large number of terms are to be represented. This can be overcome by the following simple device which is sometimes of use since it compresses the representation of numbers 1 to infinity onto a line of finite length. The idea is illustrated in Fig. 3.1(b) where, on the horizontal axis, the integer n is associated with a point distant $1/n$ to the left of a fixed point P. The left end point of the line segment is then associated with the value 1, the mid-point with the value 2, and so on, with the point P itself corresponding to an infinite value of n.

An even simpler graphical representation than either of these is often used in which the values of successive terms in the sequence are plotted one-dimensionally as points on a straight line relative to some fixed origin. Because of the identification of the numerical value of a term of the sequence with a point on a line, the behaviour of a sequence is often spoken of in terms of the behaviour of the points in this representation (that is; there is a one–one mapping of $\{u_n\}$ onto the straight line). In terms of this representation, the same sequence that gave rise to Fig. 3.1 will appear as in Fig. 3.2. This could also have been obtained from Fig. 3.1(a) and (b) by projecting the points of the graphs horizontally across to meet the vertical axis.

In each of these three representations, the tendency for the points of the sequence $\{1+(-1)^n/n\}$ to cluster around the value unity as n increases is

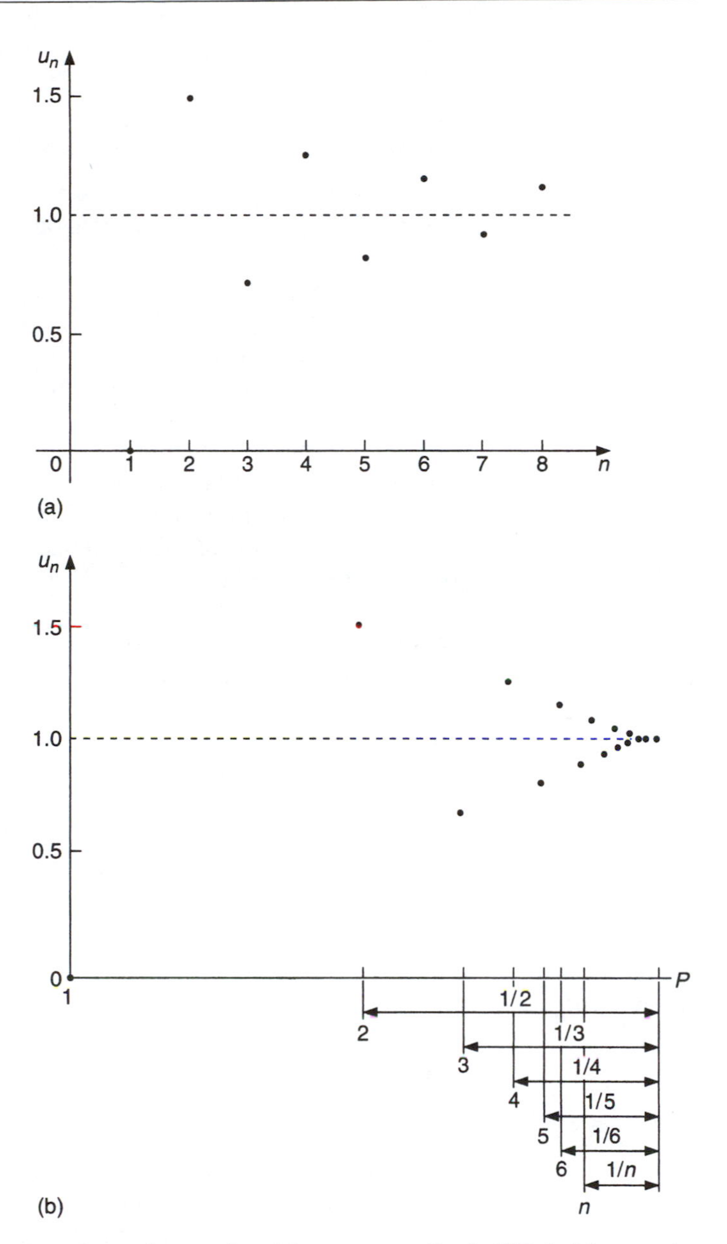

Fig. 3.1 Two alternative graphs of the sequence $\{1+(-1)^n/n\}$: (a) normal graph; (b) compressed horizontal axis.

obvious and clearly expresses an important property possessed by the sequence. We shall now explore this more fully.

In the sequence just discussed it is obvious that as n increases, so the points of the sequence cluster ever closer to the unit point in Fig. 3.2. If we adopt the convention of calling an *open* interval (a, b) containing some fixed point a *neighbourhood* of that point, then it is not difficult to see that any neighbourhood of the point unity will contain an infinite

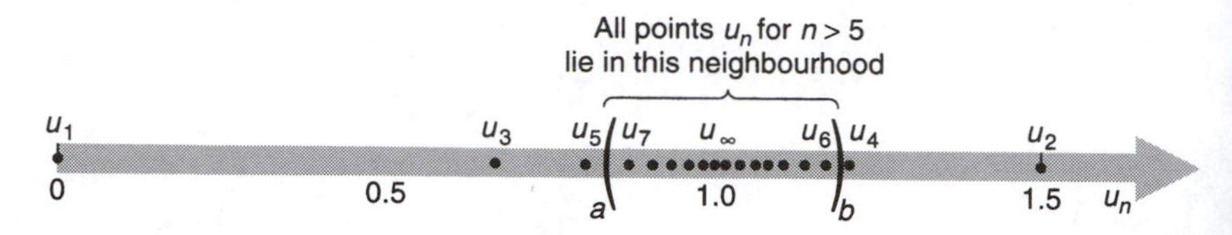

Fig. 3.2 The sequence $\{1+(-1)^n/n\}$ plotted on a line.

number of points of the sequence $\{u_n\}$. In fact in this case we can assert that no matter how small the length $b-a$ of the neighbourhood, there will always be an infinite number of points in (a, b), and there will always be a finite number of points outside (a, b). This is even true when $b-a$ shrinks virtually to zero!

The fact that any neighbourhood of the value unity has the property that an infinite number of points of the sequence are contained within it, whereas only a finite number of points lie without it, is recognized by saying that the *limit* of the sequence is unity. On account of this name the point corresponding to the value unity in Fig. 3.2 is called a *limit point* of the sequence. We shall examine the idea of a limit in the next section, and so for the moment will confine discussion to limit points. For this we shall require the notion of a sub-sequence. Henceforth, by a *sub-sequence* we shall mean a sequence $u_{n_1}, u_{n_2}, \ldots, u_{n_m}, \ldots,$ of terms belonging to the sequence $\{u_n\}$, where $n_1, n_2, \ldots, n_m, \ldots$ is some numerically ordered set of integers selected from the complete set of natural numbers. Thus u_2, u_9, u_{27}, $u_{31}, \ldots$ is a sub-sequence of u_1, u_2, u_3, ... and obviously the sub-sequence $\{u_2, u_9, u_{27}, u_{31}, \ldots\}$ is contained in the sequence $\{u_n\}$.

In terms of this we now give the following formal definition of a limit point of a sequence $\{u_n\}$.

Definition 3.2 A point u^* is said to be a *limit point* of the sequence $\{u_n\}$ if every neighbourhood of u^* contains an infinite number of elements of the sequence $\{u_n\}$.

Since we have not insisted that there be a *finite* number of points outside any neighbourhood of a limit point it follows that a sequence may have more than one limit point. We shall show by example that a limit point may or may not be a member of the sequence that defines it. This result, when applied to sequences with only one limit point, will later be seen to be very important, since it provides the justification for the approximation to irrational numbers in calculations by rational numbers. In sequences involving only *one* limit point the sequence will be said to *converge* to the value associated with the limit point. This value will be called the *limit* of the sequence.

Not all sequences have limit points and the following examples exhibit sequences having three, one, no and two limit points, respectively.

Example 3.2 (a) $\left\{\sin\left(\frac{n^2+1}{2n}\right)\pi\right\}$, has the three limit points -1, 0 and 1, of which 0 is a member of the sequence and the other two are not. The sequence does not converge.

(b) $\left\{\frac{1}{n}\sin\left(\frac{n\pi}{2}\right)\right\}$ has only one limit point at zero which is a member of the sequence. The sequence converges to zero.

(c) $\{n^2\}$ has no limit point and so the sequence does not converge.

(d) $\{1+(-1)^n+(-\frac{1}{2})^n\}$ has the two limit points 0 and 2, neither of which is a member of the sequence. The sequence does not converge. ■

One of the most important applications of the notion of a sequence is to the study of series. The difficulty here is to give a meaning to the sum of an infinite number of terms. What, for example, is the meaning of

$$\sum_{n=1}^{\infty}\frac{1}{n!}. \tag{A}$$

The solution is to be found in the behaviour of the sequence $\{S_m\}$ defined by

$$S_m=\sum_{n=1}^{m}\frac{1}{n!}.$$

The first few terms of the sequence $\{S_m\}$ are

$$S_1=1,\quad S_2=1+\frac{1}{2!},\quad S_3=1+\frac{1}{2!}+\frac{1}{3!},\quad S_4=1+\frac{1}{2!}+\frac{1}{3!}+\frac{1}{4!}$$

and obviously all such terms S_m will only involve the sum of a finitely many numbers. For obvious reasons S_m is called the mth *partial sum* of the series (A). The interpretation of the infinite sum (A) is to be found in the behaviour of the Nth term of the sequence $\{S_m\}$, namely the Nth partial sum S_N, as N tends to infinity. If $\{S_m\}$ has only one limit point at which S_m tends to some number S, then this will be called the *sum* of the series. If S is infinite, or $\{S_m\}$ has no limit, the series will be said to *diverge*. A moment's reflection will show the reader that this is the practical approach to the problem, since the term S_N is the sum of the first N terms of the infinite series (A), and it seems reasonable to assume that when the value of (A) is finite, it must be close to the value S_N, when N is suitably large. Thus the meaning of an infinite series is determined by the behaviour of the sequence $\{S_m\}$ of its partial sums.

These preliminary ideas on series must suffice for now, but we shall take them up again later and devise tests to determine whether series are convergent or divergent.

3.2 Limits of sequences

The term 'limit' was first introduced intuitively in the previous section in connection with a sequence $\{u_n\}$ which had only one limit point. As n increases so the points representing the terms u_n cluster ever closer to the limit point whose value L, say, is the limit of the sequence. This idea of a limit is correct in spirit, but it is not very satisfactory from the mathematical manipulative point of view since the phrase 'cluster ever closer to' is far too vague. The difficulty of making the expression 'limit' precise is connected with the exact meaning we give to this phrase.

Our difficulty can be resolved if we recall that any neighbourhood of a limit point will contain an infinite number of points of the sequence and, if there is only one limit point, will exclude only a finite number of points. Thinking in terms of numbers rather than points, a neighbourhood of a limit point is simply an open interval of the line on which the numbers u_n are plotted, and we already have a notation for representing such an interval. Suppose, for convenience, that the neighbourhood is symmetrical about the number L and of width 2ε, where ε, pronounced 'epsilon', is some arbitrarily small positive number. Then a variable u will be inside this neighbourhood if $L-\varepsilon<u<L+\varepsilon$. Recalling the definition of 'absolute value', this inequality can be rewritten concisely as $|u-L|<\varepsilon$. Different values of $\varepsilon>0$ determine different neighbourhoods of L, and if u is identified with the term u_n of the sequence, then L is the limit of the sequence if, no matter how small ε may become, only a finite number of terms u_n lie outside the neighbourhood and an infinite number lie within it.

We can now give a proper definition of a limit.

Definition 3.3 The sequence $\{u_n\}$ will be said to tend to the *limit* L if, and only if, for any arbitrarily small positive number ε, there exists an integer N such that

$$n>N \Rightarrow |u_n-L|<\varepsilon.$$

Let us test our definition on the sequence $\{u_n\}$ with $u_n=1+(-1)^n/n$. We already know that this sequence has only one limit point at the value unity, and consequently our definition should show that the limit is unity. Suppose, for the sake of argument, that we check to see that the definition is satisfied if $\varepsilon=1/100$. To do this we must find a number N such that when $n>N$ we have

$$\left|\left(1+\frac{(-1)^n}{n}\right)-1\right|<\frac{1}{100}.$$

This result is obviously equivalent to the requirement that $(1/n)<1/100$ which will be true for any value of n greater than 100. Hence if we take $N=100$, the conditions of the definition are satisfied. There are thus 100 terms outside the neighbourhood and an infinite number within it.

Had we demanded a much smaller value of ε, say $\varepsilon = 10^{-6}$, the identical argument would have shown that the definition is satisfied if $N = 10^6$. There would now be a very large number of terms outside the neighbourhood $0.999999 < u_n < 1.000001$, in fact 10^6 in all, but this is still a finite number, whereas the number of terms within the neighbourhood is still infinite. Clearly, however small the value of ε, the conditions of the definition will still apply, showing that it is in accord with our earlier intuitive ideas.

As a second example let us test our definition on the sequence $\{u_n\}$, with $u_n = 1 + (-\frac{1}{2})^n$, which by inspection has limit unity, because $\{u_n\}$ has only a single limit point, and it is equal to this value. Let us set $\varepsilon = 10^{-m}$, with m a positive integer, and show that for any choice of m there is an integer N such that when $n > N$ it is true that

$$|1 + (-\tfrac{1}{2})^n - 1| < \varepsilon = 1/10^m.$$

It may be seen at once that this condition is equivalent to $(\frac{1}{2})^n < 1/10^m$, or what is the same thing, $10^m < 2^n$. We see from this that for any choice of m it is always possible to find an integer N such that this last inequality is true when $n > N$. If, for example, $m = 2$ so that $\varepsilon = 1/100$, then $10^2 < 2^n$ provided $n \geq 7$, which corresponds to $N = 7$. If $m = 3$, so that $\varepsilon = 1/1000$, then $10^3 < 2^n$ provided $n \geq 10$, which corresponds to $N = 10$, whereas if $m = 4$, so that $\varepsilon = 1/10\,000$, then $10^4 < 2^n$ provided $n \geqslant 14$, which corresponds to $N = 14$. In each case the result is true for all n greater than the appropriate N, so that an infinite number of terms of the sequence will always lie within a distance ε from the limit 1, no matter how small ε may be; that is, no matter how large we make m. Thus our definition of a limit confirms that this sequence has the limit unity.

In general, when the sequence $\{u_n\}$ has a limit L, so that we say it converges to L, we shall write

$$\lim_{n \to \infty} u_n = L.$$

Whenever using this notation for a limit the reader must always keep in mind the underlying formal definition just given.

The definition and example just given above show that when a sequence has only one limit point, it must converge to the value associated with that limit point whenever the number of points outside any neighbourhood of the limit point is finite. Any sequence such as $\{u_n\}$ with $u_n = \sin\{\pi(n^2 + 1)/2n\}$ cannot have a limit, for it has three limit points at -1, 0 and 1 and any small neighbourhood taken about any one must, of necessity, exclude the infinitely many terms associated with the other two. Such a sequence does not converge.

Usually the limit of a sequence is of more importance than its individual terms, and in such circumstances the notation $\lim_{n \to \infty} u_n$ is advantageous in that it focuses attention on the general term u_n of the sequence. The result of the limiting operation is often readily deduced from the general term, as these examples indicate.

Example 3.3 Determine the limits in each of the following:

(a) $\lim_{n\to\infty}\left[\frac{(2n-1)(n+4)(n-2)}{n^3}\right]$;

(b) $\lim_{n\to\infty}\left[\frac{1}{n^2}+\frac{2}{n^2}+\cdots+\frac{n-1}{n^2}\right]$;

(c) $\lim_{n\to\infty}\left[\frac{5^{n+1}+7^{n+1}}{5^n-7^n}\right]$;

(d) $\lim_{n\to\infty}\left[\frac{1+2^2+3^2+\cdots+n^2}{n^2}\right]$.

Solution (a) The general term is $u_n=[(2n-1)(n+4)(n-2)]/n^3$, so that expanding the numerator and dividing by n^3 gives

$$u_n=2+\frac{3}{n}-\frac{18}{n^2}+\frac{8}{n^3}.$$

Obviously, as n increases, the last three terms comprising u_n approach zero, and in the limit we have

$$\lim_{n\to\infty}\left[\frac{(2n-1)(n+4)(n-2)}{n^3}\right]=2.$$

(b) The general term is $u_n=[1+2+\cdots+(n-1)]/n^2$, in which the numerator is the sum of an arithmetic progression. Now it is readily verified that $1+2+\cdots+(n-1)=n(n-1)/2$ so that

$$u_n=\left(\frac{n-1}{2n}\right)=\frac{1}{2}-\frac{1}{2n}.$$

Using the same argument as in (a) above, we see at once that as n increases so u_n approaches the value $\frac{1}{2}$, whence

$$\lim_{n\to\infty}\left[\frac{1}{n^2}+\frac{2}{n^2}+\cdots+\frac{n-1}{n^2}\right]=\frac{1}{2}.$$

(c) The general term here is $u_n=(5^{n+1}+7^{n+1})/(5^n-7^n)$ and by dividing numerator and denominator by 7^n it may be written:

$$u_n=\frac{5(5/7)^n+7}{(5/7)^n-1}.$$

Now $5/7<1$ so that $(5/7)^n$ will tend to zero as n increases. Thus u_n will approach the value -7. In this case we may write

$$\lim_{n\to\infty}\left[\frac{5^{n+1}+7^{n+1}}{5^n-7^n}\right]=-7.$$

(d) The general term is $u_n=[1^2+2^2+\cdots+n^2]/n^2$, in which the numerator is the sum of the squares of the first n natural numbers. Using the standard result Problem 1.14(b)

$$1^2 + 2^2 + \cdots + n^2 = \frac{n(n+1)(2n+1)}{6}$$

enables us to write

$$u_n = \frac{(n+1)(2n+1)}{6n} = \frac{1}{3}n + \frac{1}{2} + \frac{1}{6n}.$$

Hence as n increases without bound, so also will u_n. This sequence diverges and we write

$$\lim_{n\to\infty}\left[\frac{1^2 + 2^2 + \cdots + n^2}{n^2}\right] \to \infty.$$

Notice that we do not use the equality sign in connection with the symbol ∞, in accordance with the idea that infinity is not an actual number but essentially a limiting process. ■

An interesting sequence of historical importance is derived from what is called the *Archimedes recurrence formula*. To understand this, let a_n and b_n be the perimeters of the circumscribed and inscribed n sided regular polygons with respect to a circle of unit radius (a *unit circle*) as shown in Fig. 3.3.

Using purely geometrical arguments, Archimedes showed that the perimeters a_{2n} and b_{2n} of $2n$ sided circumscribed and inscribed polygons are related to the perimeters a_n and b_n of n sided polygons by the *recurrence formula*

$$a_{2n} = \frac{2a_nb_n}{a_n + b_n} \quad \text{and} \quad b_{2n} = \sqrt{a_{2n}b_n}.$$

Although at the time of Archimedes the mathematical constant π had not been introduced, we know that the circumference of a circle of radius 1 is 2π, so because the circumscribed and inscribed circles tend to the unit circle as $n \to \infty$, the limiting values of both a_n and b_n must each converge to 2π as $n \to \infty$. The sequence of the a_n and b_n converge remarkably rapidly, and starting from circumscribed and inscribed hexagons, as shown in Figs. 3.3, after using only five steps to arrive at a_{96} and b_{96}, Archimedes obtained a result which is equivalent to the following inequality for π, the upper limit of which is familiar to all school children as the approximation they should use for π in their calculations

$$\frac{223}{71} < \pi < \frac{22}{7}.$$

Routine calculations, starting with circumscribed and inscribed squares when $a_4 = 8$ and $b_4 = 8/\sqrt{2}$, leads to following results:

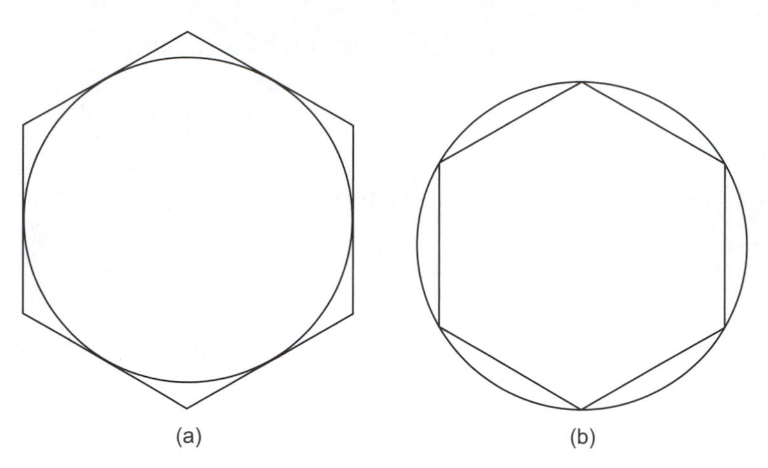

Fig. 3.3 (a) Circumscribed regular polygon on a circle of unit radius; (b) inscribed regular polygon in a circle of unit radius.

step	n	a_n	b_n
1	4	8	5.656954248
2	8	6.627416995	6.122934916
3	16	6.365195754	6.242890303
4	32	6.303449812	6.273096979
10	2048	6.283190224	6.273096979

So after only ten steps, the approximations to π given by $\frac{1}{2}a_{2048}$ and $\frac{1}{2}b_{2048}$ are 3.14159|5112 and 3.14159|1417, respectively, where a vertical rule has been inserted to show that the first five decimal digits are correct.

In this case we justified the limit 2π on the basis of a geometrical argument which suggests that the length of a curve can be approximated by the length of a polygonal approximation as the number of segments increases without bound. The justification for this will follow later when we use calculus to determine the length of an arc.

Before continuing our discussion of limits, let us introduce a useful notation. In the examples above it is apparent that the value of the limit of a sequence involving the ratio of two expressions as n increases, is entirely determined by the ratio of the most significant terms in the numerator and denominator. In the case of a polynomial involving x, the most significant term as x increases is obviously the highest-degree term in which it appears. Thus in (a), an inspection of the brackets in the numerator shows the most significant term to be $2n^3$, and as the denominator only involves n^3, it is at once obvious that for large n the ratio will approach $(2n^3/n^3) = 2$.

To streamline limiting arguments of this type, and yet to preserve something of the effect of the less significant terms, we now introduce the so-called 'big oh' notation appropriate to functions.

Definition 3.4 We say that function $f(x)$ is *of the order of* the function $g(x)$ as $x \to x_0$, written $f(x) = O(g(x))$, if

(a) $g(x) > 0$, and
(b) $|f(x)| < Mg(x)$,

where M is some constant.

The value of the constant M is usually unimportant as for most arguments it suffices that such as M should exist. We have these obvious results as $x \to \infty$:

$$2x^3 + 2x + 1 = O(x^3),$$

$$3x + \sin x = O(x),$$

$$\sin x = O(1),$$

where the symbol $O(1)$ is used here, as it will be elsewhere, to denote a constant.

In terms of this notation we may write the general term u_n in Example 3.3(a) in the simplified form

$$u_n = \frac{2n^3 + O(n^2)}{n^3} \qquad \text{whence} \quad u_n = 2 + \frac{O(n^2)}{n^3}. \tag{B}$$

By virtue of the definition of the symbol 'big oh', $O(n^2)$ implies an expression that is bounced above by Mn^2, so that $O(n^2)/n^3 \Rightarrow (Mn^2)/n^3$. However, $M/n \to 0$ as n increases without bound, so that

$$\lim_{n \to \infty} u_n = 2. \tag{C}$$

Normally the argument just outlined would be omitted, so that result (C) would be written down immediately after (B).

Implicit in the examples just examined are results which we now combine.

Theorem 3.1 If it can be shown that $u_1, u_2, u_3, \ldots$ and $v_1, v_2, v_3, \ldots$ are two sequences such that $\lim_{n \to \infty} u_n = L$ and $\lim_{n \to \infty} v_n = M$, then

(a) $u_1 + v_1, \quad u_2 + v_2, \quad u_3 + v_3, \quad \ldots$ is a sequence such that $\lim_{n \to \infty} (u_n + v_n) = L + M$;
(b) $u_1 v_1, u_2 v_2, u_3 v_3, \ldots$ is a sequence such that $\lim_{n \to \infty} u_n v_n = LM$;
(c) provided $M \neq 0$, $u_1/v_1, u_2/v_2, u_3/v_3, \ldots$ is a sequence such that $\lim_{n \to \infty} (u_n/v_n) = L/M$. ■

These assertions are intuitively obvious, though tedious to establish. By way of example we prove only result (a), which is the easiest, making full use of our definition of a limit and of the triangle inequality of Theorem 1.1(b).

Suppose ε is given. Then because $\{u_n\}$ converges to the limit L, there exists a number N_1 such that $n > N_1 \Rightarrow |u_n - L| < \frac{1}{2}\varepsilon$. By the same argument there exists another number N_2 such that $n > N_2 \Rightarrow |v_n - M| < \frac{1}{2}\varepsilon$. Now $|(u_n + v_n) - (L + M)| = |(u_n - L) - (v_n - M)| \leq |(u_n - L) - (v_n - M)|$, and so $n > \max(N_1, N_2) \Rightarrow |(u_n + v_n) - (L + M)| < \frac{1}{2}\varepsilon + \frac{1}{2}\varepsilon$. Thus, taking $N = \max(N_1, N_2)$ and given an arbitrarily small positive number ε, we have

$$n > N \Rightarrow |(u_n + v_n) - (L + M)| < \varepsilon$$

or

$$\lim_{n \to \infty} (u_n + v_n) = L + M.$$

In effect, this theorem justifies any argument in which it is asserted that, if a is close to A and b is close to B then $a+b$ is close to $A+B$, ab is close to AB, and, provided b and $B \neq 0$, a/b is close to A/B.

Theorem 3.2 Let $\{u_n\}$ and $\{v_n\}$ be two sequences which both converge to the same limit L, and suppose $\{w_n\}$ to be a third sequence. Then if for all n greater than some fixed value N, it is true that $u_n \leq w_n \leq v_n$, the sequence $\{w_n\}$ converges. Furthermore, the limit of the sequence $\{w_n\}$ is also L. ■

The proof of this theorem is not difficult and so is left to the reader as an exercise. In essence it involves two stages. The first is to establish that $\{u_n - w_n\}$ and $\{w_n - v_n\}$ are both *null sequences* in the sense that they converge to the limit zero. The second involves the use of Theorem 3.1 (a) to establish that these two null sequences imply $\lim_{n \to \infty} w_n = L$.

In applications, use of this theorem is often confined to proving that a given sequence $\{w_n\}$ converges, so that the sequences $\{u_n\}$ and $\{v_n\}$ then need to be devised to satisfy the conditions of the theorem.

Example 3.4 Given that

$$w_n = 1 + \frac{1}{2} + \frac{1}{2^2} + \cdots \frac{1}{2^{n-1}} + \frac{1}{3.2^n},$$

use Theorem 3.2 to prove that the sequence $\{w_n\}$ converges and to find the limit.

Solution Obviously, because $1/(3.2^n) < 1/2^n$,

$$1 + \frac{1}{2} + \frac{1}{2^2} + \cdots + \frac{1}{2^{n-1}} < w_n < 1 + \frac{1}{2} + \frac{1}{2^2} + \cdots + \frac{1}{2^{n-1}} + \frac{1}{2^n},$$

and so using the expression for the sum of a geometric progression given in Problem 1.15(a) we may write

$$2[1 - (\tfrac{1}{2})^n] < w_n < 2[1 - (\tfrac{1}{2})^{n+1}].$$

Thus for the sequence $\{u_n\}$ we take $u_n = 2[1 - (\frac{1}{2})^n]$ and for the sequence $\{v_n\}$ we take $v_n = 2[1 - (\frac{1}{2})^{n+1}]$. The conditions of the theorem are then satisfied, since $\lim_{n\to\infty} u_n = \lim_{n\to\infty} v_n = 2$. Hence the sequence $\{w_n\}$ converges and has for its limit the value 2. ■

At this stage in our discussion of sequences we prove the following fundamental theorem.

Theorem 3.3 (Fundamental theorem for sequences) Every increasing sequence which is bounded above tends to a limit and, conversely, every decreasing sequence which is bounded below tends to a limit.

Proof Let $\{a_n\}$ be an increasing sequence which is bounded above. Then there must be some number L that is the smallest of all possible upper bounds. The number L is called the *least upper bound* of the sequence. To prove the first part of the theorem we must show that $\{a_n\}$ converges to the limit L. From the definition of a least upper bound we know that given an arbitrarily small number $\varepsilon > 0$ there will be some integer N such that $L - a_n < \varepsilon$. This means that for all $n > N$ it follows that

$$a_N \leq a_n \text{ or, equivalently, that } L - a_n \leq L - a_N < \varepsilon.$$

However ε may be taken to be arbitrarily small, so this proves that

$$\lim_{n\to\infty} a_n = L.$$

The proof of the second part of the theorem is similar and so will be omitted. ■

It is this theorem that validates the usual arithmetic procedure for finding a square root. In the procedure an additional digit is added to the approximation at each stage, thereby giving rise to an increasing sequence that is bounded above. With a number such as $\sqrt{2}$ which we know to be irrational, this same theorem also justifies its successive approximation by the increasing sequence $\{u_n\}$ of rational numbers 1, 1.4, 1.41, 1.414, 1.4142,..., u_n,..., because this sequence is equivalent to 1, 7/5, 141/100, 707/500, 7071/5000,... . In this case an irrational number $\sqrt{2}$ is determined as the limit of a sequence of rationals. The implications are important, since although irrational numbers are of frequent occurrence, in the world in which we live we can only undertake practical calculations using rationals!

Not all sequences are defined explicitly by giving an expression for the general term u_n. Often a sequence is defined *recursively* by giving a formula relating the term u_n to its predecessor u_{n-1}, and then specifying the value of u_1. This is, of course, a difference equation, but in this context it is customary to call any rule of this kind a *recurrence relation*, or *algorithm*, and one of computational importance is

$$u_n = \frac{1}{2}\left\{u_{n-1} + \frac{a}{(u_{n-1})^{m-1}}\right\},$$

where m is an integer greater than unity and $a > 0$.

The particular significance of this recurrence relation stems from the fact that by using Theorem 3.2 it is not difficult to prove the rather surprising result that $\{u_n\}$ always converges to the limit $\sqrt[m]{a}$, that is to $a^{1/m}$, irrespective of the choice of u_1, provided only that it is positive. The value of the limit is obvious once convergence has been established, for denoting it by L and setting $x_{n-1} = x_n = L$, it follows directly from the recurrence relation that $L^m = a$, and so $L = a^{1/m}$.

Table 3.1 shows the effectiveness of this method as a computational procedure, or algorithm, for computing $\sqrt{2}$ to five decimal places, using three different starting values for u_1. To use the relation to compute $\sqrt{2}$ we must first set $m = 2$ and $a = 2$, to give

$$u_n = \frac{1}{2}\left[u_{n-1} + \frac{2}{u_{n-1}}\right].$$

Taking as representative the three starting values $u_1 = 1$, 1.4 and 5, we obtain Table 3.1, in which a dash signifies that no further change occurs in the last digit.

Obviously, convergence is most rapid when the value assumed for u_1 is a good approximation to the answer, and much effort may be spared by taking a sensible starting approximation. We mention here that a different form of recurrence relation has already been encountered in the Archimedes recurrence formula.

Table 3.1 Convergence of the algorithm for computing $\sqrt{2}$

n	u_n		
	$u_1 = 1$	$u_1 = 1.4$	$u_1 = 5$
1	1	1.4	5
2	1.5	1.41429	2.7
3	1.41667	1.41421	1.72037
4	1.41422	–	1.44146
5	1.41421	–	1.41447
6	–	–	1.41421

3.3 The number e

A number that is of fundamental importance throughout mathematics is *Euler's number*, denoted by the symbol e. This number is both *irrational* and *transcendental* and, to 15 decimal places, is given by

$$e = 2.718281828459045\ldots.$$

We will see later how the related *exponential function*

$$y = e^x \qquad \text{for} \qquad -\infty < x < \infty$$

enters into the definitions of the natural logarithmic function, the hyperbolic functions and the solution of differential equations. The exponential function is also denoted by $\exp(x)$, which is the notation used when the argument of the function is complicated as happens, for example, in

$$\exp[(1 + \sin x)/(1 + x^2)].$$

Graphs of $y = e^x$ and $y = e^{-x}$ are shown in Fig. 3.4, from which it should be noticed that e^x increases very rapidly as x increases and decreases very rapidly as x decreases, the converse being true for e^{-x}. The variation of e^x with x can be appreciated from the following numerical values: $e^{-15} = 3.059 \times 10^{-7}$, $e^{-10} = 4.540 \times 10^{-5}$, $e^{-5} = 6.738 \times 10^{-3}$, $e^5 = 1.484 \times 10^2$, $e^{10} = 2.203 \times 10^4$, $e^{15} = 3.269 \times 10^6$.

The Euler number can be defined in different ways, but because this chapter concerns limits of sequences we choose to give the one involving the limit

$$e = \lim_{n\to\infty}\left[\left(1 + \frac{1}{n}\right)^n\right]. \tag{3.1}$$

In section 5.3.7 this definition will be related to a geometrical property of the exponential function, and it will be shown that one way in which e^x may be defined is by

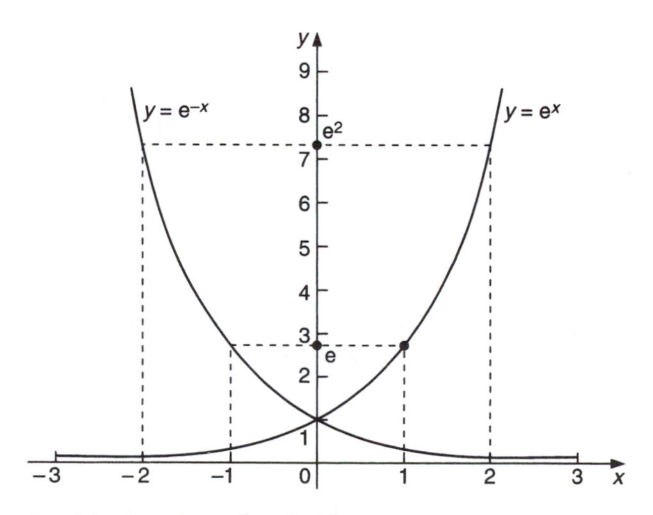

Fig. 3.4 Graph of the functions e^x and e^{-x}.

$$e^x = \lim_{n\to\infty}\left[\left(1+\frac{x}{n}\right)^n\right]. \tag{3.2}$$

To understand the nature of the sequence that is involved in Eqn (3.1), if we define the general term to be $u_n = [1+(1/n)]^n$, the first members of the sequence $\{u_n\}$ are $u_1 = 2$, $u_2 = 9/4$, $u_3 = 64/27, \ldots$. For the moment let us regard n as fixed and expand u_n by the binomial theorem to obtain

$$u_n = \left(1+\frac{1}{n}\right)^n = 1 + n\left(\frac{1}{n}\right) + \frac{n(n-1)}{2!}\left(\frac{1}{n}\right)^2 + \ldots$$
$$+ \frac{n(n-1)\cdots[n-(n-1)]}{n!}\left(\frac{1}{n}\right)^n. \tag{3.3}$$

Now rewrite this as

$$u_n = 1 + 1 + \frac{1}{2!}\left(1-\frac{1}{n}\right) + \frac{1}{3!}\left(1-\frac{1}{n}\right)\left(1-\frac{2}{n}\right) + \ldots$$
$$+ \frac{1}{n!}\left(1-\frac{1}{n}\right)\left(1-\frac{2}{n}\right)\cdots\left(1-\frac{n-1}{n}\right). \tag{3.4}$$

An examination of Eqn (3.4) suggests that as n increases, so the first few terms on the right appear to tend to

$$1 + 1 + \frac{1}{2!} + \frac{1}{3!} + \frac{1}{4!} + \cdots + \frac{1}{r!},$$

provided n is suitably large and r is much smaller than n. This would appear to indicate that in the limit as $n \to \infty$

$$e = \lim_{n\to\infty}\left(1+\frac{1}{n}\right)^n = 1 + 1 + \frac{1}{2!} + \frac{1}{3!} + \frac{1}{4!} + \cdots + \frac{1}{n!} + \cdots.$$

We will see later that this conjecture is correct, because we will show in more than one way that the Euler constant e can be defined as the infinite series

$$e = 1 + 1 + \frac{1}{2!} + \frac{1}{3!} + \frac{1}{4!} + \cdots + \frac{1}{r!} + \cdots = \sum_{n=0}^{\infty}\frac{1}{n!}, \tag{3.5}$$

where the standard convention (definition) $0! = 1$ has been used under the summation sign on the right. Next, setting

$$u_n(x) = \left(1+\frac{x}{n}\right)^n,$$

and expanding $u_n(x)$ as in Eqn (3.4), leads to the result

$$u_n(x) = 1 + x + \frac{x^2}{2!}\left(1 - \frac{1}{n}\right) + \frac{x^3}{3!}\left(1 - \frac{1}{n}\right)\left(1 - \frac{2}{n}\right) + \cdots$$
$$+ \frac{x^n}{n!}\left(1 - \frac{1}{n}\right)\left(1 - \frac{2}{n}\right)\cdots\left(1 - \frac{n-1}{n}\right).$$

Similar reasoning to that used above suggests that e^x is given by the infinite series

$$e^x = 1 + x + \frac{x^2}{2!} + \frac{x^3}{3!} + \frac{x^4}{4!} + \cdots + \frac{x^r}{r!} + \cdots = \sum_{n=0}^{\infty} \frac{x^n}{n!}, \qquad (3.6)$$

and later this also will be shown to be correct. So by observing the pattern of the coefficients as n becomes large, and using intuition, we have arrived at infinite series for the Euler constant e in Eqn (3.5), and an infinite series involving powers of x for the exponential function e^x in Eqn (3.6).

A discussion of the exponential function e^x will be given later, so now we only consider the question of the value of e defined in Eqn (3.4). Although we are unable to sum u_n, in order to show definition Eqn (3.1) makes sense we must at least show e has a finite value, and to do this we proceed as follows. The multipliers of $1/2!, 1/3!, \ldots, 1/n!$ in Eqn (3.4) are all less than unity, so an overestimate of the finite sum u_n is given by

$$u_n < 1 + 1 + 1/2! + 1/3! + \cdots + 1/n!.$$

The finite series on the right is still too difficult to sum explicitly, so some simplification must be made. The required result is achieved by noticing that if in $n! = 1.2.3\ldots(n-1)n$, each of the last $(n-1)$ numbers is replaced by 2, the resulting product $1.2.2\ldots2.2 = 2^{n-1}$ will be less than $n!$, so that $1/n! < 1/2^{n-1}$. Making use of this result and using the observation:

$$\frac{1}{3!} < \frac{1}{2^2}; \frac{1}{4!} < \frac{1}{2^3}; \ldots; \frac{1}{n!} < \frac{1}{2^{n-1}},$$

we further simplify the inequality for u_n to

$$u_n < 1 + 1 + \frac{1}{2} + \frac{1}{2^2} + \frac{1}{2^3} + \ldots + \frac{1}{2^{n-1}}. \qquad (3.6)$$

This can now be summed, since after the first term the remaining terms form a geometric progression. We arrive at the result

$$u_n < 1 + \frac{1 - (\frac{1}{2})^n}{1 - \frac{1}{2}},$$

and so

$$\lim_{n\to\infty} u_n < 3. \qquad (3.7)$$

All the terms in u_n and u_{n+1} are positive, and u_{n+1} has one more term than u_n. In addition, terms in u_{n+1} that are associated with factorials are larger than the corresponding terms in u_n so the sequence $\{u_n\}$ is increasing. Thus, as the sequence is bounded above by 3, it follows from the fundamental theorem for sequences that the limit e in Eqn (3.1) exists. The increasing nature of u_n, coupled with the fact that $u_1 = 2$, gives the following bounds for the Euler number

$$2 < \mathrm{e} < 3. \tag{3.8}$$

The definition of e in Eqn (3.1) is not suitable for the computation of its numerical value, nor is Eqn (3.2) suitable for the determination of e^x. An alternative and more convenient definition of e^x, justifying Eqn (3.6) and based on a geometrical property of the graph of $y = \mathrm{e}^x$, is given in section 5.3.7. This alternative definition also provides an efficient method for the computation of e by setting $x = 1$ in the expression determining e^x.

3.4 Limits of functions–continuity

The notion of the limit of a function $f(x)$ as x tends towards some value a is intuitively obvious in the case of functions whose graph is an unbroken curve. A typical function of this kind is illustrated in Fig. 3.5, from which it is easily seen that if x is considered to be a moving point on the x-axis, then $f(x)$ will approach the value $f(a)$ as x approaches a from either the left or the right. In this case $f(x)$ actually attains the value $f(a)$, and we shall speak of $f(a)$ as the 'limit of $f(x)$ as x tends to a' and write

$$\lim_{x\to a} f(x) = f(a).$$

Thus if $f(x) = x^3 - 2x^2 + x + 3$, then clearly in this case $\lim_{x\to 2} f(x) = 5 = f(2)$. A less obvious example involves finding $\lim_{x\to 1} f(x)$ when

$$f(x) = \frac{\sqrt{x} - 1}{x - 1},$$

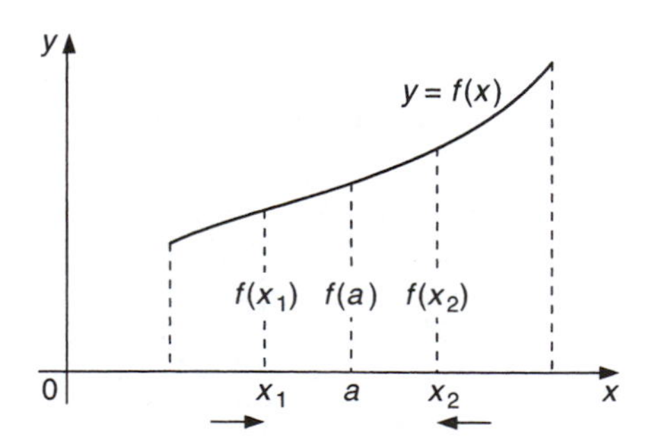

Fig. 3.5 Function $f(x)$ with unbroken graph.

since the formal substitution of $x = 1$ in $f(x)$ seems to yield 0/0, which is meaningless as it stands. This type of limit, which occurs throughout the calculus, is called an *indeterminate form*. The difficulty here is resolved by writing $x - 1 = (\sqrt{x} - 1)(\sqrt{x} + 1)$, and cancelling a factor $(\sqrt{x} - 1)$ in the numerator and denominator to give

$$f(x) = \frac{1}{\sqrt{x} + 1}, \qquad \text{for} \quad x \neq 1,$$

from which it is apparent that $\lim_{x \to 1} f(x) = \frac{1}{2}$, even though $f(x)$ is not defined $x = 1$.

In effect, the intuitive notion involved in the limit of a function is essentially the same as that for the limit of a sequence. Namely, we say that L is the limit of $f(x)$ as x tends to a if, for all x sufficiently close to a, $f(x)$ is close to L. In fact, the determination of the value of the limit L involves the behaviour of $f(x)$ *close* to $x = a$, but does not consider the actual value of $f(x)$ at $x = a$. Whether or not $f(a)$ is actually equal to L, or even defined, is immaterial. By only slightly modifying our definition of the limit of a sequence, we arrive at the following formal definition of the limit of a function, which is illustrated in Fig. 3.6, and will be used for our subsequent discussion of continuity.

Definition 3.5 The function $f(x)$ will be said to tend to the limit L as x tends to a if, and only if, for any arbitrarily small positive number ε, there exists a small positive number δ, pronounced 'delta', such that

$$0 < |x - a| < \delta \Rightarrow |f(x) - L| < \varepsilon.$$

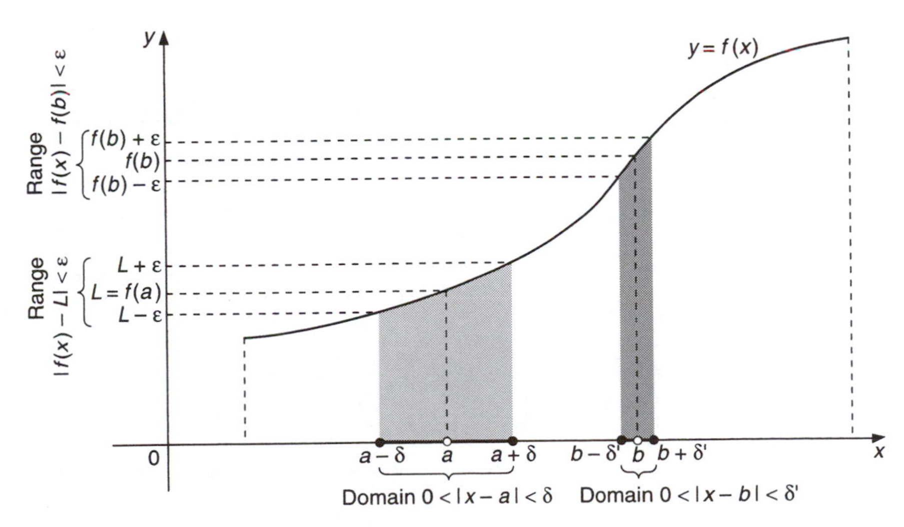

Fig. 3.6 Function $f(x)$ has a smooth graph and attains the limit L and $x = a$.

The significance of the condition $0 < |x - a| < \delta$, which in words says that the distance of x from a must be less than b, but that x must not equal a because then $|x - a| = 0$, may at first seem puzzling. However, it is present to ensure that the value $f(a)$ is specifically excluded from consideration as being irrelevant to the determination of the limit. Thus, if

$$f(x) = \begin{cases} 1 + x^2 & \text{for } x \neq 1, \\ 5 & \text{for } x = 1, \end{cases}$$

then $\lim_{x \to 1} f(x) = 2$, despite the fact that $f(1) = 5$.

If the graph of a function $f(x)$ is broken then more care must be exercised when discussing the notation of a limit. The reason can be seen after examination of Fig. 3.7 in which the graph has a break at $x = c$, at which point the functional value $f(c)$ has been allocated arbitrarily. This graph defines a perfectly satisfactory function, because for every x there is a unique $f(x)$, but as x approaches c from either the left or the right, so $f(x)$ approaches either the value L_- or L_+, values which are obviously limits in some sense. Furthermore $L_- \neq L_+$ and neither is equal to $f(c)$. To take account of this, we introduce the concepts of a limit from the left and a limit from the right.

To simplify the explanation we shall write $x \to a-$ in place of 'x tends to a from the left' and $x \to a+$ in place of 'x tends to a from the right'. In terms of this notation the function $f(x)$ in Fig. 3.7 has the property that $\lim_{x \to c-} f(x) = L_-$ and $\lim_{x \to c+} f(x) = L_+$ which is indicated in the diagram by means of arrows. Once again, in arriving at the limits from the left and right of a point, the functional value itself at that point is not involved. It may or may not equal one of the two limits so defined. These ideas may be expressed formally as a definition.

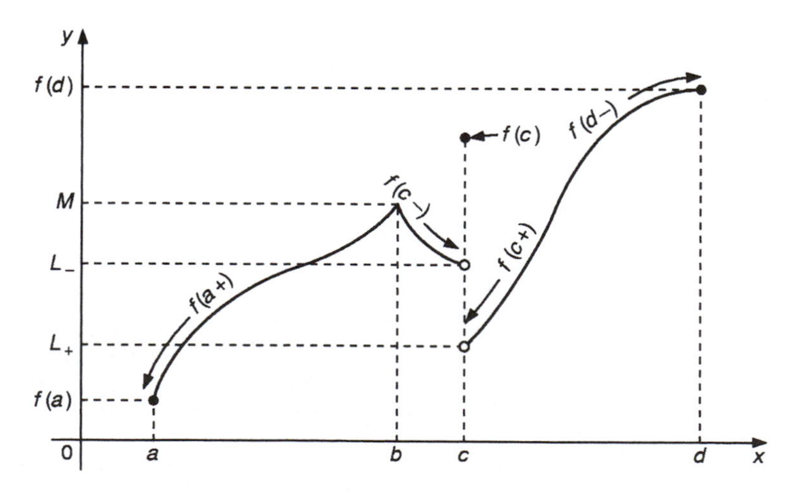

Fig. 3.7 Function $f(x)$ has broken graph.

Definition 3.6 The function $f(x)$ will be said to have the *left-hand limit*, or limit from the left, L_- as $x \to a-$ if, and only if, for any arbitrarily small positive number ε, there exists a small positive number δ such that

$$0 < a - x < \delta \Rightarrow |f(x) - L_-| < \varepsilon.$$

A corresponding definition exists for the *right-hand limit*, or limit from the right, as $x \to a+$ in which L_- is replaced by L_+ and $a - x$ by $x - a$.

Notice that the function $f(x)$ in Fig. 3.7 only has one-sided limits at $x = a$ and $x = d$ and, even though $f(x)$ has a cusp at $x = b$ and so is not smooth there, it nevertheless still has a limit in the ordinary sense at that point. This is because if $f(x)$ has identical left- and right-hand limits at a point $x = a$ so that $L_- = L_+ = L$, say, then $\lim_{x \to a} f(x)$ exists and is also equal to L.

We shall usually resolve limit problems of the type just discussed either intuitively or by appeal to a graph. An exception is the discussion of indeterminate forms which involve finding the limit of a quotient as x approaches some value at which both numerator and denominator vanish. This will be taken up again later as an application of calculus, though the reader should notice that we have already resolved one such simple problem involving a limit of the form 0/0 at the start of this section.

In the physical world, functional relationships are usually smoothly changing in the sense that a small change in the independent variable usually produces only a small change in the dependent variable. This smoothness-of-change property is given the mathematical name *continuity* and plays an important part throughout all mathematical analysis. A moment's thought will show that the following definition describes continuity in terms of the left- and right-hand limits.

Definition 3.7 The function $f(x)$ is said to be *continuous* at $x = x_0$ if

(a) $\lim_{x \to x_0-} f(x) = \lim_{x \to x_0+} f(x) = L$, and
(b) $f(x_0) = L$.

In this definition, (a) demands the equality of the left- and right-hand limits, and (b) ensures that there is no 'gap' in the graph of $f(x)$ at $x = x_0$. That is to say that the point $(x_0, f(x_0))$ lies on an unbroken curve and so coincides with the limits (a). An alternative, but equivalent, definition of continuity that is often used replaces (a) by the requirement that $\lim_{x \to x_0} f(x) = L$ but still retains (b). Either form of definition is equally good but we have chosen to emphasize the ideas of left- and right-hand limits since they find important applications in engineering and physics.

Continuity essentially describes a property of a function in a neighbourhood of a point of interest and not just at the point itself. Accordingly, a function will be said to be continuous in the interval (a, b) if it is continuous at all points x within (a, b). A point $x = a$ at which a function ceases to be continuous is a point of *discontinuity* of the function, which is said to be *discontinuous* at $x = a$.

Notice that the effect of condition (b) of our definition on a function such as

$$f(x) = \begin{cases} x^3 + 1 & \text{for } x \neq 1 \\ 6 & \text{for } x = 1 \end{cases}$$

is to show that $f(x)$ is continuous everywhere except at $x = 1$, at which point it is discontinuous.

The function $f(x) = (\sqrt{x} - 1)/(x - 1)$ considered previously had a discontinuity at $x = 1$ because $f(x)$ was not defined at the single point $x = 1$. However, it was shown that $\lim_{x \to 1} f(x) = \frac{1}{2}$, so if the value of $f(x)$ at $x = 1$ is *defined* as $f(1) = \frac{1}{2}$ the function will automatically become continuous for all x. Thus if a function is defined throughout some interval with the exception of a point $x = x_0$, but its limit exists as $x \to x_0$, the discontinuity at x_0 can be removed by *defining* the functional value at x_0 to be $f(x_0) = \lim_{x = x_0} f(x)$. Discontinuities like this are called *removable* discontinuities, and in the case just considered $f(x)$ can be defined as the *continuous* function

$$f(x) = \begin{cases} \dfrac{\sqrt{x} - 1}{x - 1}, & x \neq 1 \\ \frac{1}{2}, & x = 1 \end{cases}.$$

Examples of removable discontinuities occur throughout the calculus, and an important case will be encountered in Section 3.6 where the function $f(\theta) = \frac{\sin \theta}{\theta}$ is considered. This function is defined and continuous for all θ, except at $\theta = 0$. However it will be shown that $\lim_{\theta \to 0} (\frac{\sin \theta}{\theta}) = 1$, so $f(\theta)$ will become continuous for all θ if we adopt the definition

$$f(\theta) = \begin{cases} \dfrac{\sin \theta}{\theta}, & \theta \neq 0 \\ 1, & \theta = 0 \end{cases}.$$

In contrast, a discontinuity like the one at $x = c$ in Fig. 3.7 is not removable, because although in that case $f(c)$ is defined, the function $f(x)$ has only left and right hand limits as $x \to c$.

Let us paraphrase the notion of continuity. In effect, by requiring that a function $f(x)$ be continuous at $x = a$, we are insisting that if the variation of the function about the value $L = f(a)$ does not exceed $\pm\varepsilon$, where $\varepsilon > 0$ is arbitrary, then we can find an x-interval of width 2δ centred on $x = a$ within which this property is always true. This is illustrated by Fig. 3.6 which also indicates that in general the number δ depends on both ε and the value of x at which $f(x)$ is continuous. Thus for the same value of ε, the interval about $x = a$ is of width 2δ, whereas the interval about $x = b$ is of width $2\delta'$, with $\delta' \neq \delta$.

There are a number of immediate consequences of the definition of a limit of a function and of the definition of continuity which we now state as two important theorems.

***Theorem 3.4* (limits)** Suppose that $\lim_{x \to x_0} f(x) = L$ and $\lim_{x \to x_0} g(x) = M$. Then

(a) $\lim_{x \to x_0} \{f(x) + g(x)\} = L + M$;
(b) $\lim_{x \to x_0} f(x)g(x) = LM$;
(c) provided $M \neq 0$, $\lim_{x \to x_0} [f(x)/g(x)] = L/M$. ■

The proof of these results is similar in all respects to the proof of Theorem 3.1 so we shall not repeat the argument again.

***Theorem 3.5* (continuity)** If $f(x)$ and $g(x)$ are continuous at $x = x_0$, then so also are the functions

(a) $f(x) + g(x)$;
(b) $f(x)g(x)$;
(c) $f(x)/g(x)$, provided $g(x_0) \neq 0$.

If, furthermore, $f(x)$ is continuous at $x = x_0$ and $g(u)$ is continuous at $u = f(x_0)$, then the continuous function of a continuous function $g[f(x)]$ is continuous at $x = x_0$. ■

Once again the proof of this theorem is similar in all respects to the proof of Theorem 3.1 and so will be omitted.

Arguments involving continuity usually rely for their success on the knowledge that certain familiar functions are continuous. Once a small list of such functions has been established it can then be considerably enlarged by repeated applications of Theorem 3.5. Accordingly, we present in Table 3.2 a list of functions, in each case stating the intervals in which they are continuous. No proof will be given for most entries since the results would be obvious from their graphs, but for the sake of completeness we shall formally prove the first three entries.

Example 3.5 (a) Given that $C =$ constant, the function $f(x) = C$ is continuous everywhere.
The proof is trivial, since for any $x = x_0$, $f(x_0) \equiv C$ showing that the definition is always satisfied.

(b) The function $f(x) = x$ is continuous everywhere.
The proof is again trivial, but let us indicate how the alternative definition of continuity may be used. We must prove that for all x_0, $\lim_{x \to x_0} f(x)$ exists and is equal to $f(x_0)$. Now it is obvious from

Table 3.2 Short list of continuous functions

Function f(x)	*Interval over which f(x) is continuous*
C(constant)	$(-\infty, \infty)$
x	$(-\infty, \infty)$
$x^n (n>0)$	$(-\infty, \infty)$
$x^{-n} (n>0)$	$(-\infty, \infty)$ excluding point $x=0$
$\lvert x\rvert$	$(-\infty, \infty)$
$x^n + a_1 x^{n-1} + \ldots + a_n (n>0)$	$(-\infty, \infty)$
$\dfrac{x^n + a_1 x^{n-1} + \cdots + a_n}{x^m + b_1 x^{m-1} + \ldots + b_m}$	$(-\infty, \infty)$ excluding the zeroes of the denominator
$\sin x$	$(-\infty, \infty)$
$\cos x$	$(-\infty, \infty)$
$\tan x$	$(2n-1)\frac{\pi}{2} < x < (2n+1)\frac{\pi}{2}$, integral n
$\sec x$	$(2n-1)\frac{\pi}{2} < x < (2n+1)\frac{\pi}{2}$, integral n
$\operatorname{cosec} x$	$n\pi < x < (n+1)\pi$, integral n
$\cot x$	$n\pi < x < (n+1)\pi$, integral n

the definition of $f(x)$ that $f(x_0) = x_0$. Also, for any $x = x_0$ and given $\varepsilon > 0$, $|f(x) - f(x_0)| \equiv |x - x_0| < \varepsilon \Rightarrow |x - x_0| < \varepsilon$ so that in this case the quantity $\delta = \varepsilon$. The function is thus continuous at $x = x_0$ and, as x_0 was arbitrary, it finally follows that $f(x) = x$ is continuous everywhere.

(c) The function $f(x) = x^n$ with n a positive integer is continuous everywhere.

We give a proof by induction. Suppose the result is true for some n so that x^n is continuous at $x = x_0$ for all x_0. Now $x^{n+1} = x.x^n$, and we have just proved that x is continuous at x_0. Hence, using Theorem 3.4 (b), x^{n+1} is continuous. The result is true for $n = 1$ and so by the principle of induction it is true for all n. With a little more care this result can be shown to be true for *any* real positive n and not just for n a natural number. ■

The information contained in Table 3.2 is likely to be useful on many occasions and so should be memorized. Its application, together with Theorem 3.5, to questions of continuity is usually immediate. Thus, for example, the function $f(x) = 1/x + \sin x$ is continuous everywhere except at the point $x = 0$, and $f(x) = (x^m + a_1 x^{m-1} + \ldots + a_m)/\sin x$, with $m > 0$, is continuous everywhere except at the points $x = n\pi$ for which n is an integer, for there the denominator vanishes while the numerator does not.

Finally, in preparation for our use of limits in connection with the techniques of differentiation, we extend the O-notation to include functions of smaller order.

We shall write

$$f(x) = o(g(x)) \text{ as } x \to x_0$$

with the meaning that

$$\lim_{x \to x_0} \frac{f(x)}{g(x)} = 0.$$

The symbol o is read 'little oh' and in words the statement asserts that the function $f(x)$ is *of smaller order* than $g(x)$ as $x \to x_0$. For example, we may write $(1+x^2)^3 = 1+3x^2+o(x^3)$ as $x \to 0$, since $(1+x^2)^3 - 1 - 3x^2 = 3x^4 + x^6 = o(x^3)$ as $x \to 0$.

3.5 Functions of several variables–limits, continuity

The related concepts of a limit and the continuity of a function extend without difficulty to functions of more than one independent variable, provided only that the notion of the proximity of two points is suitably extended. The ideas involved here can best be appreciated if we confine attention to functions $f(x, y)$ of the two independent variables x and y.

Let us suppose that $f(x, y)$ has for its domain of definition some region D in the (x, y)-plane and that (x_0, y_0) is some point interior to D. Then, before considering $f(x, y)$, we must first make clear what is to be meant by $x \to x_0$, $y \to y_0$ in D.

An inspection of Fig. 3.8 shows that starting from the points P and Q in D, both the full curve and the dotted curve describe possible paths by which x and y may tend to the point (x_0, y_0) in D, and there are infinitely many other possible paths. In general, we shall write $x \to x_0$, $y \to y_0$, or, say that the point (x, y) tends to the point (x_0, y_0), if the positive number

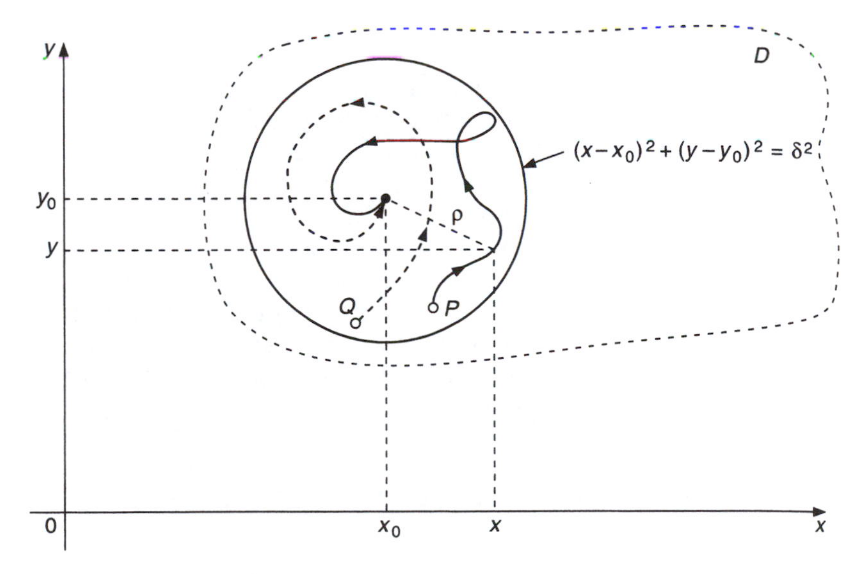

Fig. 3.8 Paths for which the point $(x, y) \to (x_0, y_0)$.

$\rho \to 0$, where $\rho = \sqrt{[(x-x_0)^2+(y-y_0)^2]}$ is the distance between the moving point (x, y) and the fixed point (x_0, y_0). Here ρ is pronounced 'rho'. This simple device then allows us to interpret a statement about the two variables x and y in terms of a statement about the single variable ρ. In words, this says that the *distance* ρ of (x, y) from the given point (x_0, y_0) must tend to zero. We may conveniently define a *neighbourhood* of the point (x_0, y_0) to be a circular region of arbitrary radius $\rho > 0$ centred on (x_0, y_0). Any rectangle or other simple closed geometrical curve containing (x_0, y_0) would, of course, serve equally well to define a neighbourhood of (x_0, y_0). When using such a neighbourhood it may or may not be necessary to exclude the boundary and the point (x_0, y_0) itself from the definition of the neighbourhood. If the point (x_0, y_0) is excluded from the neighbourhood we call the result a *punctured neighbourhood.*

Thus, for example, the square $x=0$, $y=0$, $x=1$, and $y=1$ defines a neighbourhood of the point $(\frac{1}{2}, \frac{1}{2})$. The function

$$f(x, y) = 1/\{xy(x-1)(y-1)(x-\tfrac{1}{2})(y-\tfrac{1}{2})\}$$

is defined *in* this neighbourhood, but not at $(\frac{1}{2}, \frac{1}{2})$, *on* the boundary or on the lines $x=\frac{1}{2}$, $y=\frac{1}{2}$.

Definition 3.8 is now proposed, with this interpretation of $x \to x_0$, $y \to y_0$ firmly in mind.

Definition 3.8 The function $f(x, y)$ will be said to tend to the *limit* L as $x \to x_0$ and $y \to y_0$, and we shall write

$$\lim_{\substack{x\to x_0 \\ y \to y_0}} f(x, y) = L,$$

if, and only if, the limit L is independent of the path followed by the point (x, y) as $x \to x_0$ and $y \to y_0$.

As before, we do not necessarily require that $f(x_0, y_0) = L$, or even that it is defined there, as the functional value actually at the limit point (x_0, y_0) is not involved in the limiting process. If it can be established that the result of the limiting operation depends on the path taken then, demonstrably, the function has no limit. The following examples make these ideas clear and, on account of their simplicity, are offered without proof.

Example 3.6 (a) If $f(x, y) = \dfrac{2x}{x^2+y^2+1}$, then $\displaystyle\lim_{\substack{x\to 1 \\ y\to 3}} \frac{2x}{x^2+y^2+1} = \frac{2}{11}$,

(b) If $f(x, y) = \dfrac{xy+1}{x^2+y^2}$, then $\displaystyle\lim_{\substack{x\to \infty \\ y\to 1}} \frac{xy+1}{x^2+y^2} = 0$,

(c) If $f(x, y) = \dfrac{\sin xy}{x^2 + y^2 + 1}$, then $\lim\limits_{\substack{x\to(1/2)\pi\\ y\to 1}} \dfrac{\sin xy}{x^2 + y^2 + 1} = \dfrac{4}{8 + \pi^2}$,

(d) If $f(x, y) = \dfrac{x(y-1)}{y(x-1)}$, then $\lim\limits_{\substack{x\to 1\\ y\to 1}} f(x, y)$ does not exist since $\lim\limits_{\substack{x\to 1\\ y\to 1}} f(x, y) = 1$ if taken along the line $y = x$, but $\lim\limits_{\substack{x\to 1\\ y\to 1}} f(x, y) = -1$ if taken along the line $y = 2 - x$. ■

As might be expected, the concept of continuity of a function $f(x, y)$ of two variables then follows as a direct extension of the definition of a limit.

Definition 3.9 The function $f(x, y)$ will be said to be continuous at the point (x_0, y_0) if:

(a) $\lim\limits_{\substack{x\to x_0\\ y\to y_0}} f(x, y) = L$ exists, and

(b) $f(x_0, y_0) = L$.

We shall say that $f(x, y)$ is continuous in a region if it is continuous at all points (x, y) belonging to that region. Notice that condition (a) demands that $f(x, y)$ has a unique limit as $x \to x_0$ and $y \to y_0$, and condition (b) then ensures that there is no 'hole' in the surface $z = f(x, y)$ at the point (x_0, y_0). The continuity of a function $f(x, y)$ is illustrated in Fig. 3.8, where a circular neighbourhood of the point (x_0, y_0) is shown in relation to the surface. In effect, continuity of $f(x, y)$ is simply requiring that a small change in location of the point (x, y) will cause only a small change in $z = f(x, y)$.

In Fig. 3.9 the point at (a, b) has been deliberately detached from the otherwise unbroken surface $z = f(x, y)$, so that the function $f(x, y)$ does

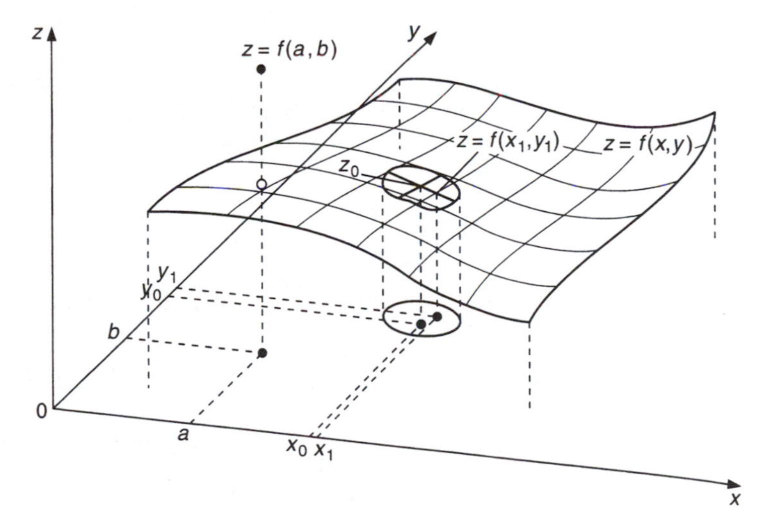

Fig. 3.9 Continuity of $f(x, y)$ at (x_0, y_0) and discontinuity at (a, b).

not satisfy the definition there and hence is not continuous at that single point. In general, a function of one or more variables which is not continuous at a point will be said to have a *discontinuity* at that point or, alternatively, to be *discontinuous* there. Thus the function of one variable shown in Fig. 3.7 has a discontinuity at $x=c$ and the function of two variables shown in Fig. 3.9 is discontinuous at $x=a$, $y=b$.

These ideas also extend to functions of several (more than two) real variables in an obvious manner once the 'distance' between two points has been defined satisfactorily. For functions $f(x, y, z)$ of the three independent variables x, y, z a suitable *distance function* between points (x_1, y_1, z_1) and (x_0, y_0, z_0) is the linear distance between them when plotted as points relative to three mutually perpendicular Cartesian axes. The distance ρ is then given by the Pythagoras rule as $\rho = \{(x_1-x_0)^2+(y_1-y_0)^2+(z_1-z_0)^2\}^{1/2}$.

The interpretation of distance in the so-called *finite-dimensional spaces* of n dimensions generated by functions of n independent variable is of considerable importance in mathematics. Essentially, of any function $\rho(P, Q)$ measuring the distance between point P and Q in such a space we require that for any points P, Q and R:

(a) $\rho(P, Q) \geq 0$,
(b) $\rho(P, Q) = 0$ if, and only if, $P = Q$,
(c) $\rho(P, Q) = \rho(Q, P)$,
(d) $\rho(P, R) \leq \rho(P, Q) + \rho(Q, R)$.

Here, condition (a) requires the distance to be non-negative, (b) requires two coincident points to be separated by a zero distance, (c) requires the distance from P to Q to be equal to that from Q to P, while (d) is the triangle inequality which asserts that the distance from P to R is less than or equal to the sum of the distance from P to Q and from Q to R. It is easy to check that the two distance functions already defined satisfy the above conditions, but this will be left as an exercise for the reader.

Again the determination of the regions in which any given function is continuous will usually be done either on an intuitive or on graphical basis. Thus, in Example 3.6 it is easily seen that:

(a) $f(x, y) = \dfrac{2x}{x^2+y^2+1}$ is continuous everywhere;

(b) $f(x, y) = \dfrac{xy+1}{x^2+y^2}$ is continuous everywhere except at $x=0$, $y=0$;

(c) $f(x, y) = \dfrac{\sin xy}{x^2+y^2+1}$ is continuous everywhere;

(d) $f(x, y) = \dfrac{x(y-1)}{y(x-1)}$ is continuous everywhere except at (0, 0) and (1, 1) and along $x=1$ and $y=0$.

3.6 A useful connecting theorem

By now it will have become apparent that there is a strong connection between theorems concerning limits of sequences and the corresponding theorems concerning limits of functions. In fact, with only trivial modification, most limit theorems that are true for sequences are also true for functions. Naturally this is no coincidence and the reason is explained by this connecting theorem.

Theorem 3.6 Let $f(x)$ be a function defined for all x in some interval $a \leq x \leq b$. Further, let $\{x_n\}$ be a numerical sequence defined in the same interval which converges to a limit α that is not a member of the sequence. Then if, and only if, $\lim_{n\to\infty} f(x_n) = L$ for each such sequence $\{x_n\}$, it follows that $\lim_{x\to a} f(x) = L$.

Proof The proof of this connecting theorem comprises two distinct parts. First it must be established that $\lim_{x\to\alpha} f(x) = L$, then sequences $\{x_n\}$ exist having the required property. Second, the converse result must be proved: that if the required sequences $\{x_n\}$ exist, then $\lim_{x\to\alpha} f(x) = L$. Together, these two results will ensure that the theorem works in both directions, so that corresponding function and sequence limit theorems satisfying the necessary conditions may be freely interchanged without further question.

The first part of the proof is a direct consequence of Definition 3.3 and 3.5. It follows from Definition 3.5 that when x is confined to some neighbourhood N_α of α, then $f(x)$ is confined to neighbourhood N_L of L. From Definition 3.3, since $\{x_n\}$ has the limit α, there must be some number n_0 such that for $n > n_0$ it follows that $f(x_n)$ will also be confined to the same neighbourhood N_L of L.

The second step is a little harder, since it involves an indirect proof by contradiction. It involves showing that if we assume that $\lim_{x\to\alpha} f(x) \neq L$, then a sequence $\{z_n\}$ can be found satisfying all the requirements of the theorem, for which $\lim_{n\to\infty} f(z_n) \neq L$. Hence the contradiction showing that the conclusion $\lim_{x\to\alpha} f(x) \neq L$ was false. We leave the details of the proof as an exercise for an interested reader. ■

To close this section, we shall use this theorem together with geometrical arguments to establish the three very useful limits:

$$\lim_{\theta\to 0}\left(\frac{\sin \alpha\theta}{\theta}\right) = \alpha; \tag{3.9}$$

$$\lim_{\theta\to 0}\left(\frac{1 - \cos \alpha\theta}{\theta}\right) = 0; \tag{3.10}$$

$$\lim_{\theta\to 0}\left(\frac{1 - \cos \alpha\theta}{\theta^2}\right) = \frac{\alpha^2}{2}, \tag{3.11}$$

where θ is pronounced 'theta', and α is pronounced 'alpha'.

These limits are all of the indeterminate variety mentioned earlier and, although this topic will receive special mention in a subsequent chapter, it is important for the development of our work that they be examined now. We shall establish that they are all related to the single limit

$$\lim_{\theta \to 0} \left(\frac{\sin \theta}{\theta} \right) = 1,$$

which we prove first.

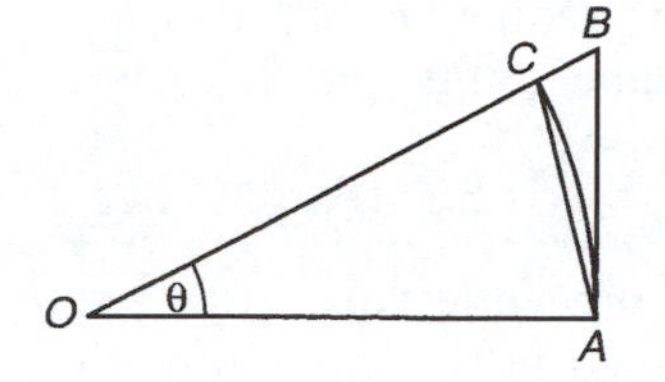

Fig. 3.10 Area inequalities.

Consider Fig. 3.10, which represents a circular arc of unit radius with its center at O, inscribed in the right-angled triangle OAB. It is obvious that

Area of triangle $OAC <$ Area of sector $OAC <$ Area of triangle OAB.

Expressed in terms of the angle θ, measured in radians, we have

$$\text{Area of triangle } OAC = \tfrac{1}{2} \text{ base} \times \text{height} = \tfrac{1}{2} \cdot 1 \cdot \sin \theta = \tfrac{1}{2} \sin \theta,$$

$$\text{Area of sector OAC} = \text{proportion of area of unit circle}$$

$$= \frac{\theta}{2\pi} \cdot \pi \cdot 1^2 = \tfrac{1}{2}\theta,$$

$$\text{Area of triangle } OAB = \tfrac{1}{2} \text{ base} \times \text{height} = \tfrac{1}{2} \cdot 1 \cdot \tan \theta = \tfrac{1}{2} \tan \theta,$$

so combining these results gives

$$\frac{1}{2} \sin \theta < \frac{1}{2} \theta < \frac{1}{2} \tan \theta.$$

Taking reciprocals, using inequality I.9 (section 1.2) which requires the reversal of the inequality signs, and multiplying by $\sin \theta$ which is positive, gives

$$\cos \theta < \frac{\sin \theta}{\theta} < 1. \tag{D}$$

This result must be true for all actual angles θ and, in particular, for the values of the sequence $\{\theta_n\}$ defined by $\theta_n = 1/n$. Thus (D) takes the form

$$\cos \theta_n < \frac{\sin \theta_n}{\theta_n} < 1 \tag{E}$$

and, since $\lim_{n \to \infty} \theta_n = 0$ where the limit is not a member of the sequence, we may combine Theorems 3.2 and 3.6 to deduce that

$$\lim_{\theta \to 0} \left(\frac{\sin \theta}{\theta} \right) = 1.$$

Stated simply, we have used the fact that as $\theta \to 0$, so that the required limit is squeezed between the values of $\cos \theta$ which tends to 1 and the number 1 itself, showing the limit must be 1.

To establish limit (3.9) it is only necessary to replace θ in this last result by $\alpha\theta$, giving rise to

$$\lim_{\alpha\theta\to 0}\left(\frac{\sin\alpha\theta}{\alpha\theta}\right)=1$$

or, equivalently, to

$$\lim_{\theta\to 0}\left(\frac{\sin\alpha\theta}{\theta}\right)=\alpha.$$

The limits (3.10) and (3.11) follow by using the identity $1-\cos\alpha\theta = 2\sin^2\frac{1}{2}\alpha\theta$ to form the expressions

$$\frac{1-\cos\alpha\theta}{\theta}=2\sin\frac{1}{2}\alpha\theta\left(\frac{\sin\frac{1}{2}\alpha\theta}{\theta}\right),$$

and

$$\frac{1-\cos\alpha\theta}{\theta^2}2\left(\frac{\sin\frac{1}{2}\alpha\theta}{\theta}\right)^2.$$

Applying result (3.9) to these we finally arrive at the required results

$$\lim_{\theta\to 0}\left(\frac{1-\cos\theta}{\theta}\right)\to 0.\alpha=0$$

and

$$\lim_{\theta\to 0}\left(\frac{1-\cos\alpha\theta}{\theta^2}\right)\to 2.\left(\frac{\alpha}{2}\right)^2=\frac{\alpha^2}{2}.$$

If (E) is divided by $\cos\theta_n$ which is positive, it becomes

$$1<\frac{\tan\theta_n}{\theta_n}<\sec\theta_n.$$

Taking the limit as $n\to 0$ this gives

$$\lim_{\theta\to 0}\left(\frac{\tan\theta}{\theta}\right)=1.$$

However, as

$$\lim_{\alpha\theta\to 0}\left(\frac{\tan\alpha\theta}{\alpha\theta}\right)=1,$$

we arrive at another useful limit,

$$\lim_{\theta\to 0}\left(\frac{\tan\alpha\theta}{\theta}\right)=\alpha. \tag{3.12}$$

The following general result is sometimes useful and, as we shall show by example, may be combined with Eqns (3.9) to (3.11) to give a number of interesting results.

Suppose $f(x)$ and $g(x)$ are two functions such that $\lim_{x\to a} f(x) = \alpha$ and $\lim_{x\to a} g(x) = \beta$, where α and β are both finite. Then clearly,

$$\lim_{x\to a} [f(x)]^{g(x)} = [\lim_{x\to a} f(x)]^{\lim_{x\to a}} = \alpha^{\beta}.$$

This result, which is true in general, is of course also true when one or more of the limits involved is of the form Eqns (3.9) to (3.11).

Example 3.7 Find

(a) $\lim_{\alpha\to 1}\left(\frac{x^3 + 2x^2 + x + 1}{x^2 + 2x + 3}\right)^{\{1-\cos[2(x-1)]\}/(x-1)^2}$,

(b) $\lim_{x\to 0}\left(\frac{1 - \cos 3x}{x^2}\right)^{(\sin 2x)/x}$.

Solution (a) Here $f(x) = (x^3+2x^2+x+1)/(x^2+2x+3)$, so that $\lim_{x\to 1} f(x) = 5/6$. Setting $u = x - 1$, so $x \to 1$ corresponds to $u \to 0$, we see that

$$\lim_{x\to 1} g(x) = \lim_{u\to 0}\left(\frac{1 - \cos 2u}{u^2}\right) = 2,$$

because of Eqn (3.11) with $\alpha = 2$. Hence $\lim_{x\to 1}[f(x)]^{g(x)} = (5/6)^2 = 25/36$.

(b) In this case $f(x) = (1 - \cos 3x)/x^2$ and $g(x) = (\sin 2x)/x$. A direct application of Eqns (3.9) and (3.11) then shows that $\lim_{x\to 0} f(x) = 9/2$ and $\lim_{x\to 0} g(x) = 2$ and thus $\lim_{x\to 0} [f(x)]^{g(x)} = (9/2)^2 = 81/4$. ■

3.7 Asymptotes

Some graphs have the property that one or more straight lines may be drawn that become tangent to them at infinity. Any straight line with this property is said to form an *asymptote* to the graph. There are, for example, three asymptotes to the graph shown in Fig. 3.11. They comprise the lines $x = c$ and $y = d$ that become tangent as $y \to \pm\infty$ and $x \to -\infty$, respectively, together with the oblique line QR with equation $y = ax + b$ that become tangent as $x \to +\infty$. The asymptote QR is called an *oblique asymptote* because it is parallel to neither the x-axis nor the y-axes.

To locate asymptote such as $x = c$ and $y = d$ that are parallel to the coordinate axes, all that is necessary in this case is to determine finite numbers $x = c$ and $y = d$ such that

$$\lim_{x\to c} f(x) = \pm\infty \qquad \text{and} \qquad \lim_{x\to -\infty} f(x) = d.$$

These asymptotes can usually be found by inspection of the function involved, so let us now consider the equation of the asymptote QR may be found.

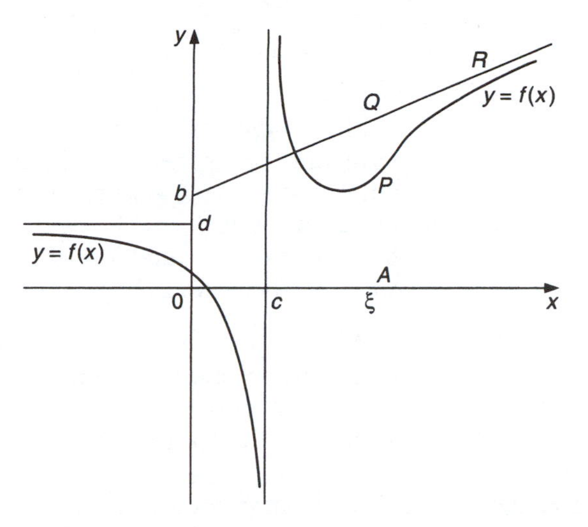

Fig. 3.11 Asymptotes.

When the functional form of $f(x)$ is simple, asymptotes which are not parallel to either axis can often be found by using elementary arguments. For example, in the simplest case $y = f(x)$ will have an asymptote which is not parallel to either axis whenever $f(x)$ is the quotient of two polynomials, and the degree of the numerator exceeds that of the denominator by unity. In this case the asymptote can be found by means of long division, followed by an examination of the form of the result for large x.

If a function has asymptotes, it is helpful to mark their positions before attempting to sketch its graph.

Example 3.8 Find the asymptotes to

$$y = \frac{2x^3 + 3x^2 + x + 1}{x^2 - 1}.$$

Solution The denominator has zeros at $x = \pm 1$, at which points the numerator is finite and non-vanishing, so the function will have the two vertical asymptotes $x = -1$ and $x = 1$. Since the degree of the numerator exceeds that of the denominator by unity there will also be an oblique asymptote. Carrying out the division we find

$$\begin{array}{r|l} & 2x + 3 \\ \hline x^2 - 1 & 2x^3 + 3x^2 + x + 1 \\ & 2x^3 \qquad\quad - 2x \\ \cline{2-2} & \qquad 3x^2 + 3x + 1 \\ & \qquad 3x^2 \qquad\;\; - 3 \\ \cline{2-2} & \qquad\qquad 3x + 4 \end{array}$$

which shows that

$$y = 2x + 3 + \left(\frac{3x + 4}{x^2 - 1}\right) = 2x + 3 + O\left(\frac{1}{x}\right),$$

so the oblique asymptote has the equation $y = 2x+3$, because the last term vanishes as $x \to +\infty$. Examination of the result of the long division not only shows the asymptotic behaviour, but also that for x close to ± 1 the graph is like $y = (3x+4)/(x^2 - 1)$. We say $y = 2x+3$ *dominates* the behaviour of the function for large x, while $y = (3x+4)/(x^2 - 1)$ *dominates* the behaviour for x close to ± 1. ■

When more complicated functions are involved a different form of argument becomes necessary.

The last examples illustrates how there may be a different oblique asymptote for positive and negative x.

Example 3.9 Find the asymptotes to

$$y = \frac{x^2 + 3x + 1}{|x|}.$$

Solution Inspection shows that $x = 0$ is one asymptote. To find the others, which will certainly exist because the degree of the numerator exceeds that of the denominator by unity, we first make use of the definition of $|x|$. It will be recalled that $|x|$ is, in fact, two statements: one for positive x and another for negative x. If $x > 0$, then $|x| = x$ and the equation becomes

$$y = \frac{x^2 + 3x + 1}{x} = x + 3 + \frac{1}{x},$$

showing that when $x > 0$ the oblique asymptote is $y = x+3$. If, however, $x < 0$, then $|x| = -x$ and the equation then becomes

$$y = \frac{x^2 + 3x + 1}{-x} = -x - 3 - \frac{1}{x},$$

showing that when $x < 0$ the oblique asymptote is $y = -x - 3$. We see from the long division that close to the origin the behaviour is dominated by $y = 1/|x|$. ■

Example 3.10 Find the asymptotes to

$$y = |x|\sin(3/x), \text{ with } y(0) = 0.$$

Solution It follows from (3.9) that for large x we may approximate $\sin(3/x)$ by $3/x$ and thus:

$$\text{as } x \to +\infty \quad \text{so } |x|\sin(|3|x) \to |x|\cdot\frac{3}{x} = x\cdot\frac{3}{x} = 3,$$

while

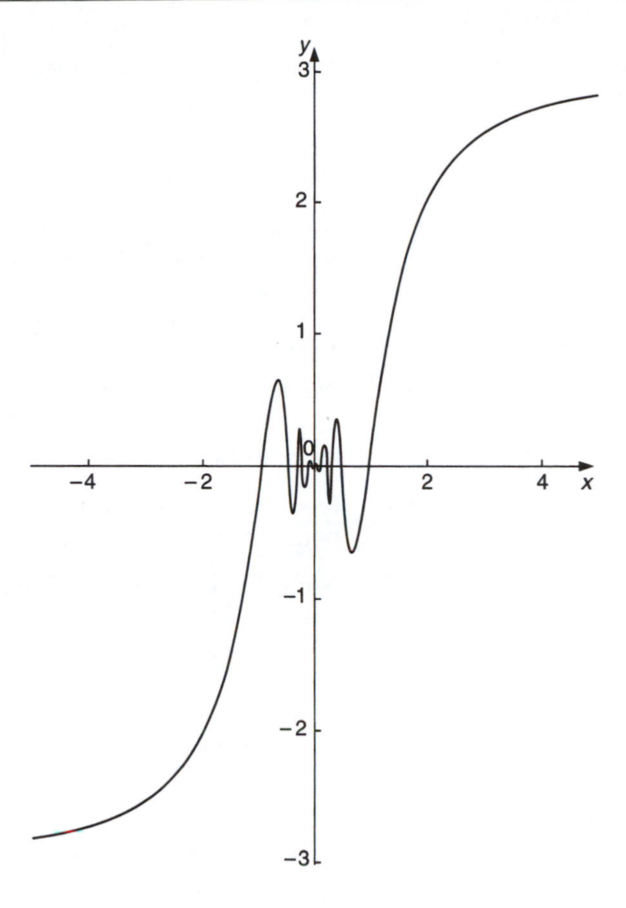

Fig. 3.12

$$\text{as } n \to -\infty \quad \text{so } |x|\sin(3/x) \to |x| \cdot \frac{3}{x} = -x \cdot \frac{3}{x} = -3.$$

So for $x > 0$ the asymptote is $y = 3$, whereas for $x < 0$, it is $y = -3$. The graph of this function is shown in Fig. 3.12, in which not all of the increasingly rapid oscillations that occur near the origin can be shown. ■

Problems

Section 3.1

3.1 Give an example of a numerical sequence and of a non-numerical sequence.

3.2 Use the terms bounded, unbounded, strictly monotonic increasing and strictly monotonic decreasing to classify the sequence $\{u_n\}$ which have the following general terms:

(a) $u_n = (-n)^{n+1}$; (b) $u_n = \left(n - \frac{1}{n}\right)^2$;

(c) $u_n = \sin(1/n)$; (d) $u_n = 2 + (-1)^n$;

(e) $u_n = \frac{n+1}{2n+3}$; (f) $u_n = \frac{2n+3}{n+1}$.

Section 3.2

3.3 Name the limit points of the sequence $\{u_n\}$ with the general term $u_n = \sin[(n^2 + n + 1)/2n]\pi$. Identify the sub-sequences that converge to these limit points.

3.4 Give examples of sequences having (a) no limit point, (b) one limit point, (c) two limit points.

3.5 Name the limit points of the sequence $\{u_n\}$ which has the general term

$$u_n = \begin{cases} 1 - \frac{1}{3^{2n}} & \text{for } n \text{ even} \\ \frac{1}{3^{2n+1}} & \text{for } n \text{ odd}. \end{cases}$$

State whether or not the limit points belong to the sequence.

3.6 Determine the following limits:

(a) $\lim_{n\to\infty} \frac{(3n+1)(2n-1)(n-1)}{n^3}$;

(b) $\lim_{n\to\infty} \frac{(2n^2+n-1)(n+2)}{(3n^2+7n+11)}$;

(c) $\lim_{n\to\infty} \frac{n+(-1)^n}{n-(-1)^n}$;

(d) $\lim_{n\to\infty} \frac{n+(-2)^n}{n-(-2)^n}$;

(e) $\lim_{n\to\infty} \left(\frac{1^2+2^2+3^2+\ldots+n^2}{2n^3}\right)$.

3.7 Starting with circumscribed and inscribed regular pentagons with respect to a circle of unit radius, and working to an accuracy of six decimal places, use the Archimedes recurrence formula to find a_{2560} and b_{2560}, and hence upper and lower estimates for π.

3.8 Determine the limits:

(a) $\lim_{n\to\infty}(\sqrt{(n+a)}-\sqrt{n})$, where $a>0$ is any real number;

(b) $\lim_{n\to\infty}\dfrac{n(2\sin n - 3\cos 2n)}{n^2+2n+1}$;

(c) $\lim_{n\to\infty}\left(\dfrac{3^{n+2}+5^{n+2}}{3^n-5^n}\right)$;

(d) $\lim_{n\to\infty}\sqrt[n]{(1+a^n)}(a\geq 0)$.

3.9 Use Theorem 3.2 to prove the convergence of the sequence $\{u_n\}$ in which

$$u_n=\frac{1}{n^2}\sin\left(1+\frac{1}{n}\right)\frac{\pi}{2}+\frac{2}{n^2}\sin\left(1+\frac{2}{n}\right)\frac{\pi}{2}+\frac{3}{n^2}\sin\left(1+\frac{3}{n}\right)\frac{\pi}{2}+\dots$$
$$+\frac{n-1}{n}\sin\left(1+\frac{n-1}{n}\right)\frac{\pi}{2}.$$

3.10 Using the algorithm $u_n=\frac{1}{2}\left\{u_{n-1}+\dfrac{11}{u_{n-1}^3}\right\}$, compute to four decimal places the first five terms of the sequence $\{u_n\}$ corresponding to the starting values (a) $u_1=1$, (b) $u_1=3$. Compare your result with the limiting value $11^{1/4}$.

3.11 Using the algorithm $u_n=\frac{1}{2}\left\{u_{n-1}+\dfrac{5}{u_{n-1}^2}\right\}$, compute to four decimal places the first five terms of the sequence $\{u_n\}$ corresponding to the starting values (a) $u_1=1$, (b) $u_1=2$. Compare your results with the limiting value $\sqrt[3]{5}$.

Section 3.3

The following two related problems show how the approximate behaviour of e^x in the interval $-2<x<2$ may be inferred directly from the sequence $\{v_n(x)\}$.

3.12 Define $v_n(x)$ by the expression

$$v_n(x)=\left(1+\frac{x}{n}\right)^n.$$

Use essentially the same arguments as those leading to Eqn (3.4) to prove that $\{v_n(x)\}$ is a strictly increasing sequence for any fixed positive x and then show that

$$v_n(x)\leq 1+x+\frac{x^2}{2}+\frac{x^3}{2^2}+\dots+\frac{x^n}{2^{n-1}}.$$

By summing this expression and taking the limit as $n\to\infty$ deduce that

$$1 \leq e^x < \frac{2+x}{2-x} \quad \text{for } 0 \leq x < 2.$$

Compare the result with Fig. 3.3.

3.13 Using the same definition of $v_n(x)$ as above, form the sub-sequences $\{v_{2m}(x)\}$ of even terms and $\{v_{2m+1}(x)\}$ of odd terms. Modify slightly the arguments used in the previous example to prove that both sub-sequences are strictly monotonic decreasing for negative x. Show that $\{v_{2m+1}(x) - v_{2m}(x)\}$ is a null sequence and hence deduce that both the even and odd sequence tends to the same limit. Modify $v_{2m}(x)$ to establish that

$$v_{2m}(x) \geq 1 - x + \frac{x^2}{2} - \frac{x^3}{2^2} + \ldots + \frac{x^{2m}}{2^{2m-1}}.$$

By summing this expression and taking the limit as $n \to \infty$ deduce that

$$\frac{2-x}{2+x} < e^{-x} \leq 1 \quad \text{for } 0 \leq x < 2.$$

Compare this result with Fig. 3.4.

Section 3.4

3.14 Determine the following limits of functions:

(a) $\lim_{x \to a} x^3 - x^2 + x + 1$; (b) $\lim_{x \to 3} \frac{x^2 + x + 1}{x^3 - 1}$;

(c) $\lim_{x \to 3} \frac{\sqrt{(x^2 - 6)}}{x^2 + 1}$; (d) $\lim_{x \to -2} \frac{x^3 + x^2 - x - 2}{(x+1)(x+2)}$;

(e) $\lim_{x \to 0} \frac{(x+h)^3 - x^3}{h}$;

(f) $\lim_{x \to \infty} \{\sqrt{(x^2 + 1000)} - \sqrt{(x^2 - 1000)}\}$;

(g) $\lim_{x \to \infty} x[\sqrt{(x^2 + 3)} - x]$.

3.15 Determine these limits when they exist:

(a) $\lim_{x \to 1} f(x)$ where $f(x) = \begin{cases} x^3 + x - 1 & \text{for } x \leq 1 \\ 1 + \sin(x-1) & \text{for } x > 1; \end{cases}$

(b) $\lim_{x \to 1} \frac{x-1}{x^2 - 1}$;

(c) $\lim_{x \to (1/2)\pi} f(x)$ where $f(x) = \begin{cases} x^2 + \sin\frac{1}{2}\pi x & \text{for } x \leq 3 \\ 4 + x^2 & \text{for } x > 3; \end{cases}$

(d) $\lim_{x \to 1} |x^2 - 1|$; (e) $\lim_{x \to (1/2)\pi} \dfrac{1 + \cos x}{1 - \sin x}$.

3.16 Determine the left- and right-hand limits of these functions at the stated points:

(a) $\lim_{x \to 2\pm} \dfrac{3^{x+1} + 5^{x+1}}{3^x + 5^x}$;

(b) $\lim_{x \to (1/2)\pi\pm} f(x)$ where $f(x) = \begin{cases} 1 + 2\sin x & \text{for } x \leq \frac{1}{2}\pi \\ \operatorname{cosec} x & \text{for } x > \frac{1}{2}\pi; \end{cases}$

(c) $\lim_{x \to 2\pm} |x^2 + x - 1|$;

(d) $\lim_{x \to 0\pm} f(x)$ where $f(x) = \begin{cases} -2 & \text{for } x \leq 0 \\ x + |x| & \text{for } x > 0; \end{cases}$

(e) $\lim_{x \to 3\pm} \dfrac{x}{3 - x}$.

3.17 Determine the domains of definition for which these functions are continuous:

(a) $f(x) = x + |x|$; (b) $f(x) = 1/(x^2 - 1)$;

(c) $f(x) = \dfrac{x^5 + x^2 - 1}{4 + \sin x - 2\cos x}$; (d) $f(x) = \dfrac{x^3 + 4x^2 + x - 6}{(x - 1)(x + 4)}$;

(e) $f(x) = \begin{cases} 2x + \sin x & \text{for } x \neq n\pi/2 \\ \dfrac{n^2 + 1}{2n^2 + 3} & \text{for } x = n\pi/2. \end{cases}$

3.18 Suppose it is known that a function $f(x)$ is continuous over the interval $x_0 \leq x \leq x_2$, and that $f(x_0) = y_0, f(x_1) = y_1$ and $f(x_2) = y_2$. Explain why it is reasonable to assume that when the functional values y_0, y_1 and y_2 are reasonably close together, $f(x)$ may in some sense be represented by the expression

$$f(x) \approx \frac{(x - x_1)(x - x_2)}{(x_0 - x_1)(x_0 - x_2)} y_0 + \frac{(x - x_0)(x - x_2)}{(x_1 - x_0)(x_1 - x_2)} y_1 + \frac{(x - x_0)(x - x_1)}{(x_2 - x_0)(x_2 - x_1)} y_2.$$

Any formula such as this, from which the behaviour of a function over an interval is inferred from its behaviour at specific points in that interval, is called an *interpolation formula*. This particular one is called the three-point Lagrangian iterpolation formula, and we shall see later that it gives *exact* results when applied to any linear or quadratic function $f(x)$. Considering $y = \sin x$ for $0 \leq x \leq 3\pi$, explain how this formula might give misleading results.

3.19 Apply the expressions given in Problem 3.18 to the function $y = \sin x$, taking as the points x_0, x_1 and x_2 the respective radian arguments 0.6, 0.9 and 1.2 and so find the appropriate three-point Lagrangian interpolation formula over the interval $0.6 \leq x \leq 1.2$. Use your result to deduce approximate values for sin 0.8 and sin 1.1 and compare these with the exact tabulated values.

3.20 The density of the material of a rod of length L is a function $f(x)$ of the distance x measured from one end. Describe, in physical terms, rods that are characterized by the following functions $f(x)$:

(a) $f(x) = \text{constant}$ for $0 \leq x \leq L$;

(b) $f(x) = \begin{cases} \rho_1 & \text{for } 0 \leq x < \frac{2}{3}L \\ \rho_2 & \text{for } \frac{2}{3}L < x \leq L; \end{cases}$

(c) $f(x) = \rho(1 + kx)\ 0 \leq x \leq L$.

Section 3.5

3.21 State the largest neighbourhood about the stated points P in which the following functions are defined. Also state if they are defined at P and on the boundary of the neighbourhood:

(a) $f(x, y) = 1/\{xy(2x - 1)(y + 2)(x + 1)(y - 2)\}$ taking point P as $(-1, 2)$

(b) $f(x, y) = \dfrac{1}{1 - x^2 - y^2}$ taking point P as $(0, 0)$;

(c) $f(x, y) = \dfrac{1}{1 + x^2 + y^2}$ taking point P as $(2, 3)$.

3.22 Find the points of lines of discontinuity of these functions:

(a) $f(x, y) = \begin{cases} 0 & \text{for } x^2 + y^2 = 1 \\ \dfrac{x \sin xy}{1 - x^2 - y^2} & \text{elsewhere}; \end{cases}$

(b) $f(x, y) = \begin{cases} 3 & \text{for } x = 1,\ y = 2 \\ \dfrac{xy}{1 + 2x^2 + y^2} & \text{elsewhere}; \end{cases}$

(c) $f(x, y) = \dfrac{x^3 + 2xy + 1}{y - 1}$;

(d) $f(x, y) = \dfrac{x^2 \sin y + y^2 \sin x + 2}{x^4 + 2x^2y^2 + y^4 + 1}$.

Section 3.6

3.23 Apply the results of section 3.6 to determine these limits:

(a) $\lim_{x \to 0} \dfrac{x}{\surd(1 - \cos x)}$; (b) $\lim_{x \to (1/4)\pi} \dfrac{1 - \surd 2 \cos x}{\surd[2 \sin(x - \frac{1}{4}\pi)]}$;

(c) $\lim_{h \to 0} \dfrac{\sin(x + h) - \sin x}{h}$; (d) $\lim_{x \to 0} \dfrac{2 \sin^3(x/4)}{x^3}$;

(e) $\lim_{x \to 0\pm} \dfrac{|\sin x|}{x}$.

3.24 Apply the results of section 3.6 to determine these limits:

(a) $\lim_{h \to 0} (x^2 + hx + 1)\left(\dfrac{\cos(x + h) - \cos x}{h}\right)$;

(b) $\lim_{x \to a} \dfrac{\sin x - \sin a}{x - a}$; (c) $\lim_{x \to 1} \left(\dfrac{\sin \pi x}{\sin \alpha \pi x}\right)$;

(d) $\lim_{x \to \infty} \left(x \sin \dfrac{1}{x}\right)$; (e) $\lim_{x \to 0} \left(\dfrac{x^2 - x + 4}{x^2 - x + 1}\right)^{(\sin 2x)/x}$;

(f) $\lim_{x \to 2} \left(\dfrac{x - 2}{x^2 - 4}\right)^{[\sin 3(x-2)]/(x-2)}$.

Section 3.7

3.25 Find the equations of the asymptotes, and use the results to sketch the graphs of the following functions:

(a) $f(x) = 1/(x^2 - 1)$;

(b) $f(x) = (x^3 + 3|x|^3 + x + 1)/(x + 1)^2$;

(c) $f(x) = (x^2 + 1)/(x^2 - 1)^{1/2}$.

3.26 Find the equations of the asymptotes and use the results to sketch the graphs of the following functions:

(a) $f(x) = 2x - 1 + 9x^2/(9x^2 + 2)^{1/2}$;

(b) $f(x) = |x| \sin(3/x)$ with $f(0) = 0$;

(c) $f(x) = 1 + \dfrac{x^3 + 2}{2(x^2 - 1)}$.

Supplementary computer problems

3.27 The general term of a sequence is

$$u_n = \cos\left(\frac{n^3 + 2}{n^2}\right)\pi.$$

Set $n = 2r + 1$ and compute u_{2r+1} for $r = 2, 10, 20, 30, 40, 50, 60, 70$ and 80. After consideration of the form of the general term and examination of the computed results, can it be concluded that this sequence converges to the limit -1?

3.28 The general term of a sequence is

$$u_n = (-1)^n \cos\left(\frac{n^2\pi}{n + 1}\right).$$

Compute u_n for $n = 10, 20, 30, 40, 50, 60, 70$ and 80. After consideration of the form of the general term and examination of the computed results, can it be concluded that the sequence converges to the limit -1?

3.29 The general term of a sequence is

$$u_n = \frac{1}{n}\sin\left[\frac{n^2\pi}{2(n + 1)}\right].$$

Compute u_n for $n = 10, 20, 30, 40, 50, 60, 70$ and 80. After consideration of the form of the general term and the computed results, can it be concluded that this sequence converges to the limit zero?

3.30 The general term of a sequence is

$$u_n = \frac{2n^2 + \cos(n\pi/2)}{3n^2 + 2}.$$

Compute u_n for $n = 10, 20, 30, 40, 50, 60, 70$ and 80. After consideration of the form of the general term and the computed results, can it be concluded that this sequence has a limit and, if so, what is its value?

3.31 The algorithm

$$u_n = \tfrac{1}{2}\left[u_{n-1} + \frac{a}{(u_{n-1})^{m-1}}\right],$$

with m an integer greater than unity and a a positive number, defines a sequence u_n that converges to $a^{1/m}$ for any initial value of $u_1 > 0$. Set $a = 3$, $m = 2$ and, starting with $u_1 = 1$, compute u_{10}, u_{20} and u_{30} and compare the results with the limit $\sqrt{3}$ to which the sequence converges. Repeat the calculations using the same values of a and m, but starting with (i) $u_1 = 0.2$, and (ii) $u_1 = 5$.

3.32 Using the algorithm in Problem 3.31 with $a = 5$, $m = 3$ and, starting with $u_1 = 1$, compute u_{10}, u_{20} and u_{30} and compare the results with the limit $5^{1/3}$ to which the sequence converges. Repeat the calculations using the same values of a and m, but starting with (i) $u_1 = 0.7$, and (ii) $u_1 = 3$.

3.33 It was shown in section 3.3 that

$$\mathrm{e} = \lim_{n\to\infty}\left[\left(1+\frac{1}{n}\right)^n\right] = 2.718281828\ldots.$$

Set $u_n = (1+\frac{1}{n})^n$, using $n = 10^r$, compute u_n for $r = 1, 2, 3, 4, 5, 6, 7, 8, 9$ and 10. Why does the result suddenly appear to converge to the value 1 instead of e?

3.34 It is shown later that e is also determined by the limit of the sequence

$$u_n = 1 + 1 + \frac{1}{2!} + \frac{1}{3!} + \ldots + \frac{1}{n!} = \sum_{r=0}^{n}\frac{1}{r!},$$

where, by definition, $0! = 1$. Compute u_5, u_{10}, u_{15}, u_{20}, u_{30} and u_{40}, and compare the results with the true limit given in Problem 3.33.

3.35 It is shown later that e^x is determined by the limit of the sequence

$$u_n = 1 + x + \frac{x^2}{2!} + \frac{x^3}{3!} + \ldots + \frac{x^4}{n!} = \sum_{r=0}^{n}\frac{x^n}{r!},$$

where, by definition, $0! = 1$. Set $x = 2.5$ and compute u_5, u_{10}, u_{15}, u_{20}, u_{30} and u_{40}, and compare the results with the exact value $\mathrm{e}^x = 12.18249396\ldots$.

3.36 Define the sequence

$$u_n = \frac{\sqrt{x_n} - 1}{x_n - 1},$$

where $x_n = 1 - (1/n)$ for $n = 2, 3, \ldots$. Compute u_{10}, u_{20}, u_{30}, u_{50}, u_{100}, u_{250} and u_{500}, and use the results to estimate the limit of the sequence. Relate the results to the indeterminate form

$$\lim_{x\to 1}\left(\frac{\sqrt{x} - 1}{x - 1}\right)$$

discussed in section 3.4.

3.37 It was shown in Example 3.7 that

$$\lim_{x\to 0}\left(\frac{1 - \cos 3x}{x^2}\right)^{((\sin 2x)/x)} = \frac{81}{4}.$$

Verify this by defining

$$u_n = \left(\frac{1 - \cos 3x_n}{x_n^2}\right), \quad v_n = \frac{\sin 2x_n}{x_n},$$

setting $x = 1/n$ for $n = 10$, 20, 30, 40, 50, 100 and 500, and computing $w_n = (u_n)^{u_n}$.

(*Note*: Remember that x_n is in radians.)

3.38 It was shown in Example 3.7 that

$$\lim_{x \to 1}\left(\frac{x^3 + 2x^2 + x + 1}{x^2 + 2x + 3}\right)^{[1 - \cos 2(x-1)]/(x-1)^2} = \frac{25}{36}.$$

Verify this by defining

$$u_n = \frac{x^3 + 2x^2 + x + 1}{x^2 + 2x + 3}, \quad v_n = \frac{1 - \cos 2(x - 1)}{(x - 1)^2},$$

setting $x_n = 1 + (1/n)$ for $n = 10$, 20, 30, 40, 50, 100 and 400, and computing

$$w_n = (u_n)^{v_n}.$$

(*Note*: Remember that x_n is in radians.)

4 Complex numbers and vectors

There are numerous operations in mathematics involving real variables whose outcome is a number that does not belong to the real number system. The simplest of these involves finding the square root of -1, denoted by $i = \sqrt{-1}$. The first part of this chapter is devoted to developing a general understanding of complex numbers. The study of complex numbers is the study of numbers that can be represented in the form $a+ib$, where a and b are real numbers. The number a is called the real part of the complex number, and the number b is called the *imaginary* part. Complex numbers provide the generalization of the real number system necessary to ensure that all numbers produced by mathematical operations are included in what is called the field of complex numbers. As would be expected, a real number is a special case of a complex number in which the imaginary part is zero.

It is shown how a complex number can be plotted as a point in a plane by regarding a as its x-coordinate and b as its y-coordinate, and when this is done the plane is called the complex plane, or sometimes the z-plane, because until the structure of a complex number needs to be known it is usually represented by the single letter z. Alternately, the line drawn from the origin to the point representing $a+ib$ may be taken to represent the complex number. This interpretation involves considering complex numbers as special two-dimensional vectors. The rules for the combination of complex numbers, and for performing basic algebraic operations upon them, are described in detail. These rules are necessary in order to work with complex numbers and, for example, for the study of roots of equations described in section 4.5.

The second part of the chapter describes vectors in three-dimensional space which are most general than the special two-dimensional vectors introduced in connection with complex numbers. A vector in three-dimensional space is defined as a quantity with both magnitude and direction. That is, it has a numerical measure of its magnitude (size) and also a direction in space along which it acts in a given sense. It is this direction and sense which distinguish a vector from a scalar quantity which only has a magnitude represented by a real number. These space vectors are most easily related to the rectangular Cartesian coordinate system $O\{x, y, z\}$, and when this is done it is shown how any vector can be represented as the sum of three components along x-, y- and z-axes associated with which are unit vectors $\mathbf{i}$, $\mathbf{j}$ and $\mathbf{k}$. The rules for the combination of vectors and scalars are shown to lead to an algebra of vectors that is widely used in applications of mathematics. The chapter ends with descriptions of a number of simple applications of vectors to mechanics.

4.1 Introductory ideas

A number of important properties of the real number system have already been considered, and we shall now examine to what extent quantities representable as displacements in space in a given direction may be incorporated into a number system. The name *vector quantity* is reserved for all quantities that are representable as a displacement in space or, more exactly as a directed line segment. Familiar vector quantities are force, magnetic field and velocity, which are all representable by a line whose length is proportional to their magnitude and whose direction is parallel to the direction of the original quantity. In addition, the line of action of a vector has a *sense* associated with it, which means that we must specify a direction along the line to indicate the way in which the vector acts.

Thus to represent a velocity of 3 m/s in an easterly direction we would first adopt a convenient length scale, say 1 cm to represent 1 m/s, and then, after marking the points of the compass on our paper, we would draw a line 3 cm long in an east–west direction. Finally, we would add an arrow to the line pointing eastwards to indicate the sense of the velocity. This line could be located anywhere on our paper since it does not represent a velocity that is associated with any particular point. Reversal of the arrow would correspond to a reversal of the direction of the velocity, so that the line would then represent a velocity of 3 m/s in a westerly direction.

Not all quantities are vectors, and another important group are called scalars. The word *scalar* describes any quantity that has magnitude but no direction. Typical scalar quantities which have units are temperature, mass and pressure. The real numbers are themselves scalars, and are used to describe the numerical magnitudes of both scalar and vector quantities, irrespective of whether units may be involved. The terms scalar and vector describe collectively two important groups of quantities in the real world. It should, however, be added that they do not jointly give a complete description of all possible physical quantities. Others exist that are neither scalar nor vector, though this need not be elaborated here.

In giving meaning to the square root operation when applied to negative numbers, we shall see that a special kind of two-dimensional vector arises. Its value in mathematics has proved to be so great that although such vectors are restricted to describing vector quantities in a plane, they have been given a special name, *complex numbers*. Because of this restriction, in addition to studying complex numbers, we shall need a more general theory of vectors so that we can describe the cited examples of vector quantities, and any others that may arise, in all possible situations and not just in a plane.

Despite this limitation of complex numbers, their vector properties are still important enough in special situations for them to be in this chapter. Their value elsewhere in mathematics, however, is even greater, and makes them a discipline in their own right. The main reason for this is to be found in their relationship to real numbers and in the consequences of their introduction into functional relationships in the roles of independent and dependent variables. In the meantime we shall develop the vector properties and algebra of complex numbers to the point of general usefulness in mathematics, postponing until the end of this chapter the alternative approach that is necessary for study of general three-dimensional vector quantities. As already mentioned, each is valuable as a separate discipline, though, as would be expected, each has a separate notation and, generally, a quite different field of application.

The following introduction to complex numbers is based only on a knowledge of elementary trigonometric identities, and not until after more study of the exponential and trigonometric functions will we unify our treatment of these two topics.

The origin of complex numbers was the desire of eighteenth-century mathematicians always to be able to compute the roots of polynomials, even when they are of the form

$$x^2 = -1. \qquad (4.1)$$

It was Leonhard Euler (1707–83) who first recognized that the real number system was deficient in respect of admitting solutions to all possible polynomials and, in connection with Eqn (4.1), he proposed that a new number i be introduced to extend the number system. In keeping with the mathematical beliefs of that period, he called i the *unit imaginary number* and related it to real numbers by requiring that

$$i^2 = -1. \qquad (4.2)$$

If we allow the use of this new symbol, then $i = \sqrt{-1}$ is the positive square root of minus one, whence Eqn (4.1) may be seen to have the two roots $x = i$ and $x = -i$. That $x = i$ is a root follows from the definition of i, while $x = -i$ is also a root since $(-i)^2 = (-1)^2.i^2 = 1 \cdot i^2 = -1$. With the introduction of i, equations such as

$$x^2 = -k, \quad (k > 0)$$

which are slightly more general than Eqn (4.1), can also be solved. The equation may be re-expressed in the form $x^2 = k.(-1)$, showing that its roots are $x = i\sqrt{k}$ and $x = -i\sqrt{k}$, where the positive square root is always taken. For example, if $x^2 = -9$, then the roots are $x = 3i$ and $x = -3i$.

The success of Euler's idea lies in the fact that only this one new number need be introduced to enable solutions to be found to all polynomials, irrespective of their degree. As a first step towards seeing this, consider the quadratic equation

$$ax^2 + bx + c = 0, \qquad (4.3)$$

and suppose that $b^2 - 4ac < 0$. Then, setting $4ac - b^2 = m^2$, and formally applying the usual formula for the roots of a quadratic, we obtain

$$x = \frac{-b \pm \sqrt{(-m^2)}}{2a} \qquad \text{or} \qquad x = \left(\frac{-b}{2a}\right) \pm i\left(\frac{m}{2a}\right).$$

Hence, denoting the two roots by x_1 and x_2, they take the form

$$x_1 = \left(\frac{-b}{2a}\right) + i\left(\frac{m}{2a}\right) \qquad \text{and} \qquad x_2 = \left(\frac{-b}{2a}\right) - i\left(\frac{m}{2a}\right). \tag{4.4}$$

The numbers x_1 and x_2 are not ordinary numbers since each comprises the sum of a real number and a multiple of the unit imaginary number i. On this basis it is reasonable to conjecture that each root of any arbitrary polynomial will be of the same form and, should the multiplier of i be zero, that root will reduce to a real number.

This conjecture is correct, but before we may verify it, we must see how to perform arithmetic on numbers of this special type. These are the complex numbers already mentioned and, henceforth, we shall always refer to them by this name. Unless the exact form of a complex number is needed, it is useful to denote it by a single symbol, usually z, so that an arbitrary complex number z is of the form

$$z = x + iy, \tag{4.5}$$

where x and y are real numbers. We call Eqn (4.5) the *real–imaginary form* of a complex number, and refer to x as the *real part* of z, and to y as the *imaginary part* of z. In symbolic form we write

$$x = \operatorname{Re} z, \quad y = \operatorname{Im} z. \tag{4.6}$$

Hence if $z = 4 - 7i$, then $\operatorname{Re} z = 4$ and $\operatorname{Im} z = -7$. We stress that $\operatorname{Re} z$ and $\operatorname{Im} z$ are real numbers. The *zero complex number* is denoted by 0 and represents the number $z = 0 + i.0$.

Already, and without proper justification, we have attributed some reasonable arithmetic properties to i. We have, for example, assumed results such as $\alpha i = i\alpha$ for all real α, and $\sqrt{(-\alpha)} = \sqrt{(-1)}.\ \sqrt{\alpha} = i\sqrt{\alpha}$. To proceed logically and rigorously it would be necessary to define addition, subtraction, multiplication and division for complex numbers and then to examine the applicability of the real number axioms of Chapter 1 in the case of complex numbers. This is necessary since whatever the arithmetic laws we now propose for complex numbers, they must obviously be in agreement with the real number axioms of Chapter 1, whenever the imaginary parts of complex numbers are zero. We shall not in fact justify the complex number axioms we now formulate, since this is a straightforward matter and provides good exercise for the student (see the problems at the end of the chapter). Instead, we simply summarize the results, pausing only to discuss in detail the most basic operations necessary for the manipulation of complex numbers.

4.2 Basic algebraic rules for complex numbers

First we shall agree to denote addition and subtraction of the complex numbers z_1 and z_2 in the usual manner by writing z_1+z_2 and z_1-z_2, respectively. Multiplication of the complex numbers z_1 and z_2 will be denoted by juxtaposition thus, z_1z_2. Before going on, and in order to work with equations, we must define the meaning of equality between two complex numbers, and then we can define the operations of addition, subtraction and multiplication. The following definitions are all phrased in terms of the arbitrary complex numbers $z_1=a+ib$ and $z_2=c+id$.

Definition 4.1 We shall say that the two complex numbers z_1 and z_2 are *equal*, and will write $z_1=z_2$ if, and only if, $a=c$ and $b=d$. That is if, and only if, their real parts and their imaginary parts are separately equal.

Example 4.1 Of the complex numbers z_1, z_2 and z_3 defined by $z_1=3-2i$, $z_2=1+3i$ and $z_3=3-2i$, it is obvious that $z_1=z_3$ but that $z_1\neq z_2$ and $z_3\neq z_2$. ■

Definiton 4.2 By the sum z_1+z_2 will be understood the single complex number which, written in real–imaginary form, has a real part that is the sum of the real parts of z_1 and z_2, and an imaginary part that is the sum of the imaginary parts of z_1 and z_2. Thus for the stated numbers z_1 and z_2 we have

$$z_1+z_2=(a+c)+i(b+d).$$

Example 4.2 If $z_1=2+i$ and $z_2=1-3i$, then $z_1+z_2=3-2i$. ■

Definiton 4.3 By the *difference* z_1-z_2 will be understood the single complex number which, written in real–imaginary form, has a real part that is the difference of the real parts of z_1 and z_2 and an imaginary part that is the difference between the imaginary parts of z_1 and z_2. Thus for the stated numbers z_1 and z_2 we have

$$z_1-z_2=(a-c)+i(b-d).$$

Example 4.3 If $z_1=5+6i$ and $z_2=4-2i$, then $z_1-z_2=1+8i$. ■

Using these definitions it is easily verified that axioms A.1 to A.5 of section 1.2 also apply to complex numbers. To proceed to an examination of the other axioms we must define the operation of multiplication.

Definition 4.4 The *product* z_1z_2, in which $z_1=a+ib$ and $z_2=c+id$, is a single complex number which may be written in real–imaginary form. The product is carried out algebraically as would be the ordinary product $(\alpha+\beta)(\gamma+\delta)$, and the final result is obtained by making the identifications $\alpha=a$, $\beta=ib$, $\gamma=c$, $\delta=id$ and using the result $i^2=-1$ to combine the four terms that result into a real part and an imaginary part. Thus we have

$$z_1z_2 = (a + ib)(c + id) = ac + iad + ibc + i^2bd$$
$$= (ac - bd) + i(ad + bc).$$

Example 4.4 If $z_1 = 2+3i$ and $z_2 = 1 - i$, then $z_1z_2 = 5+i$. As a more difficult example let us express $(1+i)^4+(1-i)^4$ in real–imaginary form.

Now $(1+i)^4 = (1+4i+6i^2+4i^3+i^4)$ and $(1-i)^4 = (1-4i+6i^2-4i^3+i^4)$, but as $i^2 = -1$, $i^3 = -i$, and $i^4 = 1$, these expressions become $(1+i)^4 = -4$ and $(1-i)^4 = -4$. Hence $(1+i)^4+(1-i)^4 = -8$. ■

The definitions of addition, subtraction and multiplication of complex numbers are used in the obvious manner for the solution of simple equations. Thus, if $2z-(2+i) = 4-3i$, then adding $(2+i)$ to both sides of the equation gives $2z+0 = (4-3i)+(2+i)$ or $2z = 6-2i$, whence $z = 3-i$.

In all cases, the reader should memorize the method employed in the definitions, rather than the quoted formulae.

With this definition of multiplication it is a simple matter to verify that axioms M.1 to M.4 and also axiom D.1 of section 1.2 apply to complex numbers. When one of the numbers z_1 or z_2 reduces to a real number, then the real and imaginary parts of the others are both scaled by the same factor. If the scale factor is -1 the sign of the complex number is reversed. To discuss axiom M.5 and division we need to proceed more carefully.

As it stands, an expression such as $(a+ib)/c$ is well defined as a complex number, for we may regard $(1/c)$ as a multiplier of $(a+ib)$ and, provided $c \neq 0$, Definition 4.4 will give the result. In this case a and b are both scaled by the factor $(1/c)$. However, it is not clear that the more general expression

$$z_3 = \frac{z_1}{z_2} = \frac{a + ib}{c + id} \tag{4.7}$$

is reducible to a complex number expressible in real–imaginary form. The key to this problem is to be found in M.5 itself when we recall that division is really defined as the operation inverse to multiplication. Hence, in effect, we must rewrite Eqn (4.7) in the equivalent form

$$z_3(c + id) = a + ib, \tag{4.8}$$

and then try to determine z_3. There is, however, an easier way in which to arrive at the result of this quotient. It is easily verified that any complex number $\alpha+i\beta$, when multiplied by the associated complex number $\alpha-i\beta$, gives the real number $\alpha^2+\beta^2$. Hence, if both sides of Eqn (4.8) are multiplied by $(c-id)$, the multiplier of z_3 will simply become the real number c^2+d^2. Carrying out this operation, Eqn (4.8) takes the form

$$z_3(c^2 + d^2) = (a + ib)(c - id)$$

whence, dividing by the real number (c^2+d^2) and expanding the product on the right, we find that

$$z_3 = \frac{(ac + bd) + i(bc - ad)}{c^2 + d^2}. \tag{4.9}$$

Equation (4.9) is now in the real–imaginary form of a complex number and is the result of the quotient (4.7). Many books take expression (4.9) as the formal definition of the quotient (4.7). The definition we shall propose shortly is equivalent to Eqn (4.9) in all respects, but its form is much easier to memorize. The simplification is achieved by the introduction of a new and useful operation called forming the complex conjugate of a complex number.

Definition 4.5 If $z = a + ib$ is an arbitrary complex number, then the complex number $\bar{z} = a - ib$ is the *complex conjugate* of z. The symbol $\bar{z}$ is read 'z bar'. Equivalently, we may state that the complex conjugate of a number is always obtained by changing the sign of the imaginary part of that number.

If $z = a + ib$, it follows directly from the definition of $\bar{z}$ that the product $z\bar{z}$ is always real, because

$$z\bar{z} = (a + ib)(a - ib) = a^2 + b^2. \tag{4.10}$$

With this definition and property in mind it is easy to show that the following definition of the quotient z_1/z_2 is equivalent to Eqn (4.9).

Definition 4.6 (division) The *quotient* z_1/z_2 of the two complex numbers z_1 and z_2 is the complex number $(z_1\bar{z}_2)/(z_2\bar{z}_2)$.

Using this definition it is a straightforward matter to verify axiom M.5 for complex numbers, provided only that $z_2 \neq 0$.

Example 4.5 We illustrate division by setting $z_1 = 2 + i$ and $z_2 = 3 - 2i$. Now $\bar{z}_2 = 3 + 2i$ and $z_1/z_2 = (z_1\bar{z}_2)(z_2\bar{z}_2)(2 + i)(3 + 2i)/(3 - 2i)(3 + 2i)$, whence $z_1/z_2 = (4 + 7i)/13$. By this same method, an equation of the form $2z(2 + i) = 1 + i$ is seen to have the solution $z = (1 + i)/(4 + 2i) = (3 + i)/10 = \frac{3}{10} + \frac{1}{10}i$. ■

On account of the fact that $\bar{z}$ is an ordinary complex number, its general properties are exactly the same as those of any other complex number. Hence the number axioms that apply to z, apply equally well to $\bar{z}$. The following specially useful results are easily proved, and are related to the arbitrary complex number $z = x + iy$, to its complex conjugate $\bar{z} = x - iy$ and to the real number $|z|$ associated with z and called the *modulus* of z and defined to be $|z| = (x^2 + y^2)^{1/2}$ (see Definition 4.6).

$$z = x + iy;$$

$$\bar{z} = x - iy;$$

$$z + \bar{z} = 2\text{Re}\, z = 2x;$$

$$z - \bar{z} = 2i\, \text{Im}\, z = 2iy;$$

$$z\bar{z} = x^2 + y^2 = |z|^2;$$

$$z = \overline{(\bar{z})}$$

$$\frac{1}{\bar{z}} = \overline{\left(\frac{1}{z}\right)};$$

$$\overline{(z^n)} = (\bar{z})^n;$$

$$\left|\frac{\bar{z}_1}{\bar{z}_2}\right| = \frac{|\bar{z}_1|}{|\bar{z}_2|};$$

$$\overline{(z_1 + z_2 + \ldots + z_n)} = \bar{z}_1 + \bar{z}_2 + \ldots + \bar{z}_n;$$

$$\overline{z_1 z_2 \ldots z_n} = \bar{z}_1 \bar{z}_2 \ldots \bar{z}_n.$$

We now utilize some of these simple properties of the complex conjugate operation to prove an important theorem concerning the roots of a polynomial and shall then deduce three very useful corollaries. In the process of doing so, we shall take as self-evident the fact that a polynomial $P(z)$ of degree n has n factors of the form $(z - \zeta)$, where ζ is pronounced 'zeta'. These are called *linear* factors because they are of degree 1, though some may be repeated. The numbers ζ may, or may not, be complex. The existence of n zeros of $P(z)$, corresponding to the n linear factors (some possibly repeated), is called the *fundamental theorem of algebra*.

Theorem 4.1 If the nth-degree polynomial

$$P(z) \equiv a_0 z^n + a_1 z^{n-1} + \ldots + a_n$$

has real coefficients $a_0, a_1, \ldots, a_n$, then if $z = \zeta$ is a complex zero of $P(z)$, so also is $z = \bar{\zeta}$ a zero of $P(z)$. Equivalently, if $(z - \zeta)$ is a factor of $P(z)$, then so also is $(z - \bar{\zeta})$.

Proof Suppose that $z = \zeta$ is a zero of $P(z)$. Then by definition

$$a_0 \zeta^n + a_1 \zeta^{n-1} + \ldots + a_n = 0.$$

Hence, taking the complex conjugate of this equation, we may write

$$\overline{(a_0 \zeta^n + a_1 \zeta^{n-1} + \ldots + a_n)} = 0.$$

However, the complex conjugate of a sum is the sum of the complex conjugates of the individual terms comprising the sum so that

$$\overline{(a_0 \zeta^n + a_1 \zeta^{n-1} + \ldots + a_n)} = \overline{a_0 \zeta^n} + \overline{a_1 \zeta^{n-1}} + \ldots + \bar{a}_n.$$

Now as the a_r, $r = 0, 1, \ldots, n$ are real, it follows that $\bar{a}_r = a_r$ and so

$$\overline{a_r\zeta^{n-r}} = a_r\overline{\zeta^{n-r}} = a_r(\overline{\zeta})^{n-r}, \quad \text{for } r = 0, 1, \ldots, n.$$

Hence,

$$a_0\overline{\zeta}^n + a_1\overline{\zeta^{n-r}} + \ldots + a_n = 0;$$

showing that $P(\overline{\zeta}) = 0$. Thus $z = \overline{\zeta}$ s also a zero of $P(z)$. ■

Paraphrased, Theorem 4.1 asserts that if a polynomial with real coefficients has complex zeros, then they must occur in complex conjugate pairs.

As any zero which is not complex must be real, it follows that we may formulate a corollary to Theorem 4.1.

Corollary 4.1(a) If a polynomial has real coefficients, then those of its zeros that are not real, occur in complex conjugate pairs. ■

If $z = \zeta$ and $z = \overline{\zeta}$ represent any pair of complex conjugate zeros in Theorem 4.1, then $(z - \zeta)$ and $(z - \overline{\zeta})$ must both be factors of $P(z)$. Hence their product $(z - \zeta)(z - \overline{\zeta})$ must also be a factor. Now

$$(z - \zeta)(z - \overline{\zeta}) = z^2 - (\zeta - \overline{\zeta})z + \zeta\overline{\zeta},$$

and as $\zeta + \overline{\zeta} = 2\text{Re}\,\zeta$ is a real number and $\zeta\overline{\zeta} = |\zeta|^2$ is also a real number, it follows that the pair of complex conjugate zeros correspond to a single quadratic factor with real coefficients. Hence Corollary 4.l(a) may be rephrased thus:

Corollary 4.1(b) Any polynomial with real coefficients may always be factorized into a set of factors which are linear or at most quadratic, each of which has real coefficients. Specifically, if the polynomial is of degree n and there are m pairs of complex conjugate zeros, then there will be $(n - 2m)$ linear factors with real coefficients and m quadratic factors with real coefficients. ■

Finally, as an obvious consequence of this last corollary:

Corollary 4.1(c) An odd-degree polynomial with real coefficients must have at least one real zero. ■

It must be emphasized that the zeros of a polynomial may be *repeated*. If the zero ζ of a polynomial $P(z)$ occurs m times it is said to have multiplicity m and, correspondingly, $(z-\zeta)^m$ is a factor of $P(z)$. Thus, for example, $P(z)=(z-2)(z-8)(z+4)^3$ has single zeros 2 and 8, and a zero -4 of multiplicity 3.

The significance of Theorem 4.1 and its corollaries is best illustrated by an example which shows how they may often be used to simplify a difficult problem to the point that the solution may be determined by familiar methods.

Example 4.6 A polynomial $P(z)$ of degree 5 is defined by the relationship

$$P(z) \equiv z^5 + 5z^4 + 10z^3 + 10z^2 + 9z + 5.$$

Given that $z=i$ is a zero, deduce the remaining four zeros and use the result to express $P(z)$ as the simplest possible product of factors having real coefficients.

Solution First, as the coefficients of $P(z)$ are all real, Theorem 4.1 is applicable. Hence if $z=i$ is a zero, then so also is $z=-i$. Thus $(z-i)$ and $(z+i)$ are factors, as is their product $(z-i)(z+i)=z^2+1$. Using ordinary long division to divide $P(z)$ by (z^2+1), we find that

$$P(z)/(z^2+1) = z^3 + 5z^2 + 9z + 5.$$

Hence to find the remaining factors we must now factorize this cubic polynomial. As the degree is odd, and the coefficients are real, Corollary 4.1(c) applies showing that it must have at least one real zero. At this point we have recourse to trial and error to find the real zero which for the purposes of this example has been made an integer.

Thus, setting

$$Q(z) = z^3 + 5z^2 + 9z + 5,$$

we must find a value $z=z_1$ such that $Q(z_1)=0$. By inspection we see that $Q(-1)=0$ showing that the real zero is $z=-1$. This corresponds to the linear factor with real coefficients $(z+1)$. Removing the factor $(z+1)$ from the cubic by long division, we then find that

$$\frac{P(z)}{(z^2+1)(z+1)} = \frac{Q(z)}{(z+1)} = z^2 + 4z + 5.$$

Finally we apply the standard formula for the roots of a quadratic to this expression to obtain the remaining two zeros. Completing the calculation, these are found to be $z=-2-i$ and $z=-2+i$. Thus the five zeros are $z=i$, $z=-i$, $z=-1$, $z=-2-i$ and $z=-2+i$. The required factorization involving only factors with real coefficients is thus

$$P(z) \equiv (z^2+1)(z+1)(z^2+4z+5). \quad \blacksquare$$

4.3 Complex numbers as vectors

So far we have discussed the basic arithmetic of complex numbers but have not mentioned their vector properties. To do this, and to give a geometrical representation of complex numbers, we plot them as points in a plane called the *complex plane* or, sometimes, the *z-plane*. Specifically, we shall use the real part of the complex number as its horizontal or x-coordinate and the imaginary part of the complex number as its vertical or y-coordinate. Thus to each complex number there corresponds just one point in the complex plane and, conversely, to each point in the complex plane there corresponds just one complex number. The relationship between points and complex numbers is one–one. In the complex plane, the x-axis is called the *real axis*, and the y-axis is called the *imaginary axis*. Other accounts of this subject often refer to this geometrical representation of complex numbers as the *Argand diagram*, in honour of its inventor.

In the complex plane, a complex number may either be considered as a point in the plane or, equivalently, as the directed straight-line element from the origin to the point in question. We shall remember this dual relationship between points and vectors but, for simplicity, will usually speak only of points in the complex plane.

This duality between points and vectors is indicated in Fig. 4.1, where the complex numbers $z=1$, $z=i$, $z=2+i$, $z=2-i$ and $z=-1-\frac{1}{4}i$ have been represented as points (Fig. 4.1(a)) and as vectors (Fig. 4.1(b)). In the case of the vector representation, arrows have been added to show that the vector is drawn from the origin to the point in question.

Notice that if a number, together with its complex conjugate, is plotted in the complex plane, as for example $2+i$ and $2-i$ in Fig. 4.1, then geometrically, in both the point and the vector representations, one is obtainable from the other by reflection in the x-axis as though it were a mirror.

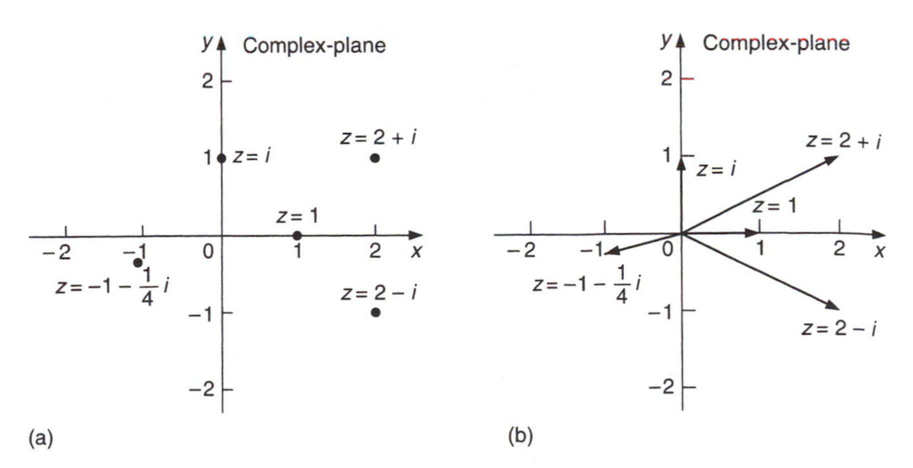

Fig. 4.1 Representation of complex numbers: (a) point representation; (b) vector representation.

Instead of adding and subtracting vectors analytically by use of Definitions 4.2 and 4.3, the same result may be achieved entirely geometrically, as we now indicate. Consider the sum of the vectors $z_1 = 2+i$ and $z_2 = 1+2i$. Analytically $z_1+z_2 = 3+3i$ and Fig. 4.2(a) show this result. The same result may be obtained geometrically by the following construction. If we wish to add vector z_2 to z_1, then for the purposes of addition we shall imagine vector z_2 to be freed from the origin, so that it is capable of translation anywhere in the complex plane, but we shall assume that wherever we relocate it in the complex plane it will always be kept parallel to its original position, and its length and sense will be preserved. The result of adding z_2 to z_1 is then achieved by translating z_2 in the manner described until its origin is located at the tip of vector z_1. The two arrows of vectors z_1 and z_2 then point in the same direction, and the vector z_1+z_2 is the line element directed from the origin 0 to the tip of the vector z_2 in its new position. In Fig. 4.2(a) this construction is represented by the lower triangle comprising the parallelogram. Such triangles are *vector triangles*.

A vector not attached to a specific origin or one which, for the purposes of combination with another vector, is freed from its origin to be relocated in some other part of the complex plane will be called a *free* vector. This is in contrast to a vector that is attached to a definite origin which we shall call a *bound* vector. In addition to z_2 to z_1 that we have just performed, z_1 was regarded as a bound vector and z_2 as a free vector.

Notice that by the same argument, z_1 may be freed and its origin translated to the tip of the bound vector z_2 to form the vector z_2+z_1, which is the line element directed from the origin to the tip of vector z_1 in

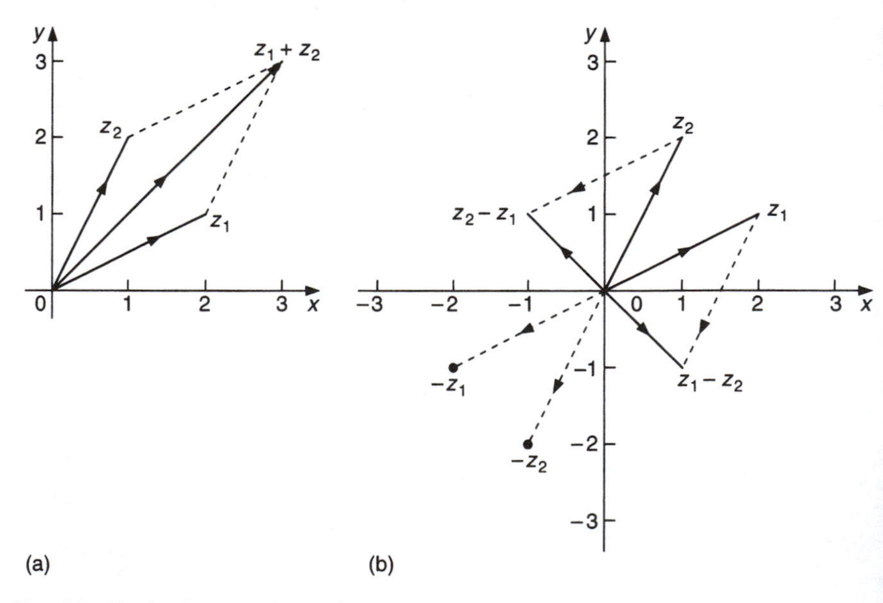

Fig. 4.2 Algebraic operations with complex numbers: (a) vector addition, z_1+z_2; (b) vector subtraction, z_1-z_2.

its new position. In Fig. 4.2(a) this construction is represented by the upper triangle comprising the parallelogram. The fact that both constructions gives rise to the same line representing on the one hand z_1+z_2, and on the other z_2+z_1, proves that vector addition is commutative, since $z_1+z_2=z_2+z_1$.

Before proceeding with the discussion of subtraction, we first observe that Definition 4.4 implies that multiplication of the bound vector z by -1 *reverses* its direction. That is to say, its origin remains fixed, but the line element representing the vector is rotated about the origin through the angle π. With this remark in mind, we see that subtraction of vector z_2 from z_1 (Fig. 4.2(b)), is just a special case of addition in which the vector to be added is $-z_2$. The vector $-z_2$ is obtained from z_2 by reversing the direction of z_2, as is indicated in Fig. 4.2(b) by the dotted line directed into the fourth quadrant. The vector z_1-z_2 is then the line element directed from the origin to the tip of the reversed vector z_2 in its new position. In Fig. 4.2(b) this construction is shown in the right-hand half of the plane. The same construction with the roles of z_1 and z_2 interchanged, is shown in the left-hand half of the plane, and, when compared with the first result, proves that $z_1-z_2=-(z_2-z_1)$. (Why?)

Thus far, complex numbers have been seen to obey the addition, multiplication and distributive axioms of real numbers, and the reader might be forgiven for wondering if there is any significant difference between them and the real numbers. The answer is *yes*. Whereas real numbers can be given a natural order according to their size, complex numbers cannot. A glance at Fig. 4.1(b) makes it clear that no natural order exists in the *field of complex numbers*, comprising all numbers in real–imaginary form, since even vectors of the same length may be differently directed, for instance the vectors drawn from the origin to points of a circle of radius 1 centered or the origin all have magnitude (length) 1, but each has a different direction. Whereas it makes sense to order the lengths of vectors, since these are scalar quantities and may be so ordered, the vectors themselves have no natural order. To further our argument we remind the reader of a result introduced earlier by now naming the length of a vector and introducing a notation whereby it may be manipulated in equations.

Definition 4.7 (modulus of a vector) The quantity

$$|z|=(x^2+y^2)^{1/2}=(z\bar{z})^{1/2}$$

is called the *modulus* of the vector $z=x+iy$. It is the length of the line element drawn from the origin to the point (x, y) in the complex plane (see Fig. 4.3(a)).

Notice that if $z=x+iy$ and $z_0=x_0+iy_0$, then $z-z_0=(x-x_0)+i(y-y_0)$, so that $|z-z_0|=[(x-x_0)^2+(y-y_0)^2]^{1/2}$. Thus from Pythagoras' theorem we see $|z-z_0|$ is the distance of z from z_0 in the complex

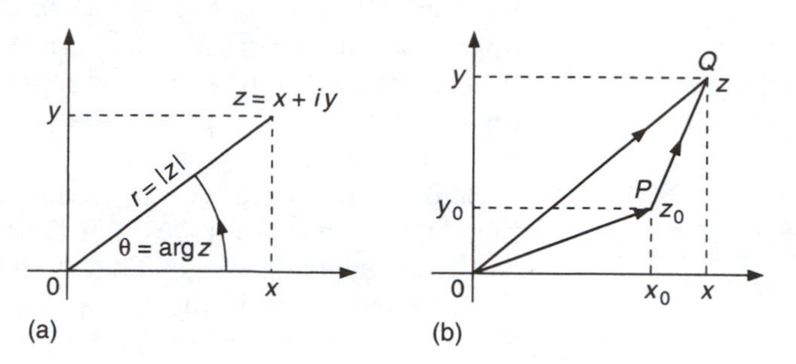

Fig. 4.3 (a) Modulus and argument representation; (b) distance of z from z_0.

plane. This is illustrated in Fig. 4.3(b) where P is the point z_0 and Q the point z.

Example 4.7 If $z = 3+4i$, then $|z| = (3^2+4^2)^{1/2} = 5 = |\bar{z}|$. ■

Notice that in the special case $\operatorname{Im} z = 0$, $|z|$ reduces to the absolute value of a real number since, as always, the positive square root is involved in the definition. The following useful results are easily verified:

$$z\bar{z} = |z|^2; \qquad |z_1 z_2| = |z_1| \cdot |z_2|; \qquad |z| = |\bar{z}|.$$

If either the upper or lower triangle comprising the parallelogram in Fig. 4.2(a) is considered, then clearly, when expressed in terms of the modulus, the Euclidean theorem 'the sum of the lengths of any two sides of a triangle exceeds or equals the length of the third side' becomes the following inequality relating moduli:

$$|z_1| + |z_2| \geq |z_1 + z_2|. \tag{4.11}$$

Equality will occur only when z_1 and z_2 are collinear and similarly directed. For obvious reasons Eqn (4.11) is called the *triangle inequality*, and it has already been encountered in simple form when we discussed the absolute value of the sum of two real numbers. An analytic proof of result (4.11) is set as a problem at the end of the chapter.

Another useful inequality relating the moduli of the complex numbers z_1 and z_2 is

$$|z_1 - z_2| \geq ||z_1| - |z_2||, \tag{4.12}$$

where again equality occurs only when z_1 and z_2 are collinear and similarly directed. The proof of this is also left to the reader as a problem.

Example 4.8 If $z_1 = 3+4i$ and $z_2 = 4+3i$, then $z_1 + z_2 = 7+7i$ and $z_1 - z_2 = -1+i$. Hence $|z_1| = (3^2+4^2)^{1/2} = 5$, $|z_2| = (4^2+3^2)^{1/2} = 5$ and $|z_1+z_2| = (7^2+7^2)^{1/2} = \surd 98$, $|z_1 - z_2| = (1^2+1^2)^{1/2} = \surd 2$, so that $|z_1| + |z_2| = 10$, $|z_1+z_2| = \surd 98$ and $||z_1| - |z_1|| = 0$. We have thus verified inequalities (4.11) and (4.12) in this special case, for they demand that for any z_1 and z_2

$$||z_1| - |z_2|| \leq |z_1 - z_2| \leq |z_1 + z_2| \leq |z_1| + |z_2|,$$

which in this case corresponds to the valid inequalities $0 < \sqrt{2} < \sqrt{98} < 10$. ■

4.4 Modulus–argument form of complex numbers

Referring again to Fig. 4.3, we see that the complex number z need not be specified in the standard form, for it may equally well be specified by giving both the value of $|z|$ and the angle θ which, by convention, is always measured positively in an anticlockwise direction from the x-axis to the line of the vector z. The angle θ is called the *argument* of z, and we shall write $\theta = \arg z$. The same point z will be identified if θ is increased or decreased by multiples of 2π, because angles θ and $\theta + 2k\pi$, where k is any integer, will give rise to the same line on Fig. 4.3. Later we shall see that this indeterminancy in θ plays an important role in the determination of the roots of complex numbers. When $\theta = \arg z$ is restricted to the interval $-\pi < \theta = \pi$, it will be termed the *principal value* of the argument.

If we define the real number r by the equation $r = |z|$, and still set $\theta = \arg z$, then the ordered number pair (r, θ) describes the *polar coordinates* of the point z in Fig. 4.3: that is, the radial distance of a point from the origin together with its bearing measured anticlockwise from a fixed line through the origin. The relationship between the Cartesian coordinates (x, y) and the polar coordinates (r, θ) of the same complex number z is immediate, since from Fig. 4.3 we have

$$x = r\cos\theta \quad y = r\sin\theta \tag{4.13}$$

or, equivalently,

$$r = (x^2 + y^2)^{1/2} \quad \cos\theta = \frac{x}{(x^2 + y^2)^{1/2}} \quad \sin\theta = \frac{y}{(x^2 + y^2)^{1/2}}. \tag{4.14}$$

Thus the complex number, or vector, $z = x + iy$ may also be written in what is called the *modulus–argument* form

$$z = r(\cos\theta + i\sin\theta). \tag{4.15}$$

Because $\arg z$ is indeterminate up to an angle $2k\pi$, we must phrase our definition of equality between two complex numbers carefully when it is to refer to complex numbers expressed in modulus–argument form.

Definition 4.8 The two numbers $z_1 = r(\cos\theta + i\sin\theta)$ and $z_2 = \rho(\cos\phi + i\sin\phi)$, expressed in modulus–argument form, will be said to be equal if, and only if, $r = \rho$ and $\theta = \phi + 2k\pi$, for integral values of k.

Equations (4.13) and (4.14) enable immediate interchange between the modulus–argument and the real–imaginary forms of z, as the following examples indicate.

Example 4.9
(a) Express $z = -4\sqrt{3} + 4i$ in modulus–argument form.
(b) Express $z = 2 + 5i$ in modulus–argument form.
(c) If $|z| = 3$ and $\arg z = -\pi/10$, express z in real–imaginary form.

Solution (a) From Eqn (4.14), $r=|z|=[(-4\sqrt{3})^2+4^2]^{1/2}=8$, while $\cos\theta=-(4\sqrt{3})/8=-(\sqrt{3})/2$ and $\sin\theta=4/8=\frac{1}{2}$, from which we deduce that the principal value of θ must lie in the second quadrant with $\theta=\arg z=5\pi/6$. Hence, in modulus–argument form

$$z=8\left(\cos\frac{5\pi}{6}+i\sin\frac{5\pi}{6}\right).$$

Notice that although we could have written $\theta=\arg z=\arctan(-1/\sqrt{3})$, it would not then have been clear in which quadrant θ must lie, and, consequently, we shall always specify $\sin\theta$ and $\cos\theta$ separately.

(b) Again from Eqn (4.14), $r=|z|=(2^2+5^2)^{1/2}=\sqrt{29}$, while this time $\cos\theta=2/\sqrt{29}$ and $\sin\theta=5/\sqrt{29}$, from which we deduce that the principal value of θ must lie in the first quadrant with $\theta=\arg z=1.1903$ rad. Hence, in modulus–argument form

$$z=\sqrt{29}(\cos 1.1903+i\sin 1.1903).$$

(c) The result is immediate, since Eqn (4.15) gives

$$z=3\left\{\cos\left(-\frac{\pi}{19}\right)+i\sin\left(-\frac{\pi}{10}\right)\right\}$$
$$=2.8533-0.9270i. \quad ■$$

We now examine the consequences of multiplication and division for complex numbers expressed in modulus–argument form. Let z_1 and z_2 be the two complex numbers:

$$z_1=r_1(\cos\theta_1+i\sin\theta_1) \quad \text{and} \quad z_2=r_2(\cos\theta_2+i\sin\theta_2). \tag{4.16}$$

Then by direct multiplication we find that

$$z_1z_2=r_1r_2[(\cos\theta_1\cos\theta_2-\sin\theta_1\sin\theta_2)$$
$$+i(\sin\theta_1\cos\theta_2+\cos\theta_1\sin\theta_2)],$$

and, using the trigonometric identities for $\cos(\theta_1+\theta_2)$ and $\sin(\theta_1+\theta_2)$, this may be written as

$$z_1z_2=r_1r_2[\cos(\theta_1+\theta_2)+i\sin(\theta_1+\theta_2)]. \tag{4.17}$$

We have thus proved that the result of the product z_1z_2 is a complex number with modulus $|z_1z_2|=r_1r_2$ and argument $\arg(z_1z_2)=\theta_1+\theta_2=\arg z_1+\arg z_2$. Thus the result of multiplying two complex numbers is to produce a complex number whose modulus is the product of the two separate moduli and whose argument is the sum of the two separate arguments (see Fig. 4.4(a)). A special case results if, remembering that

$\cos\frac{1}{2}\pi=0$, we write

$$i=\cos\tfrac{1}{2}\pi+i\sin\tfrac{1}{2}\pi. \tag{4.18}$$

It then follows from this that in the z-plane, multiplication by i corresponds geometrically to an anticlockwise rotation through $\frac{1}{2}\pi$ without any

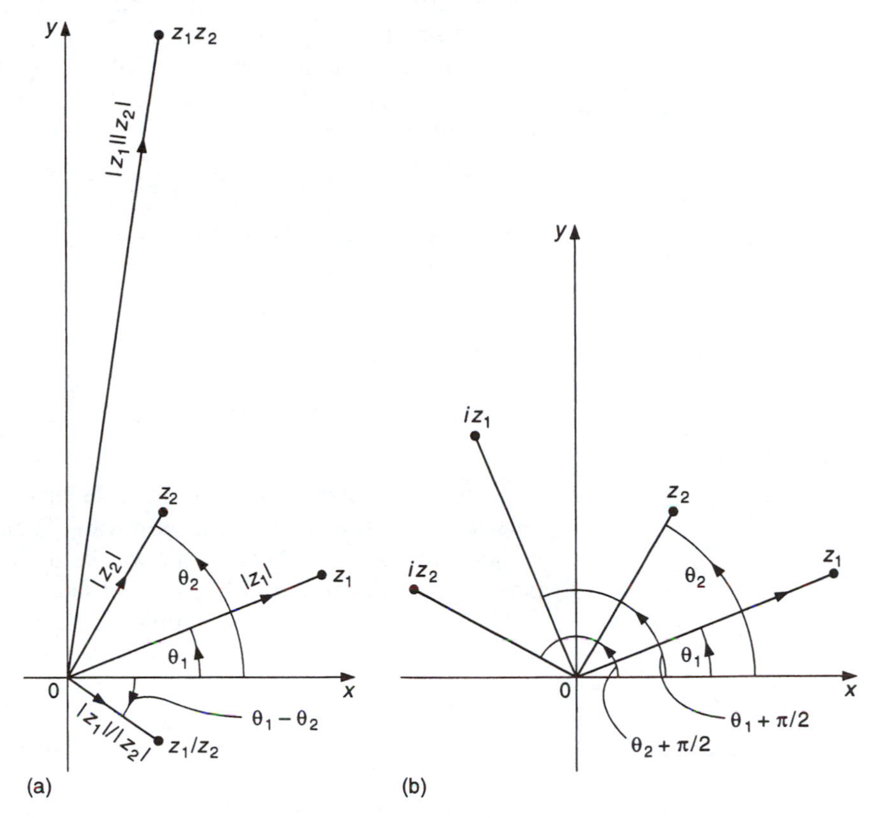

Fig. 4.4 (a) The product z_1z_2; (b) the effect of multiplication by i and the quotient z_1/z_2.

change of size. To illustrate this, the vectors iz_1 and iz_2 have been added to Fig. 4.4(b).

By repeated application of Eqn (4.17) it is easily proved that if $z_m = r_m(\cos\theta_m + i\,\sin\theta_m)$ for $m = 1, 2, \ldots, n$, then

$$z_1z_2\ldots z_n = r_1r_2\ldots r_n[\cos(\theta_1+\theta_2+\ldots+\theta_n) + i\sin(\theta_1+\theta_2+\ldots+\theta_n)]. \tag{4.19}$$

An argument essentially similar to that which gave rise to Eqn (4.17), but this time using the trigonometric identities for $\cos(\theta_1-\theta_2)$ and $\sin(\theta_1-\theta_2)$, establishes that whenever $z_2 \neq 0$, then with the same notation we have

$$\frac{z_1}{z_2} = \frac{r_1}{r_2}[\cos(\theta_1-\theta_2) + i\sin(\theta_1-\theta_2)]. \tag{4.20}$$

Obviously $|z_1/z_2| = r_1/r_2 = |z_1|/|z_2|$ and $\arg(z_1/z_2) = \theta_1 - \theta_2 = \arg z_1 - \arg z_2$. Expressed in words, this says that the result of dividing two complex numbers is to produce a complex number whose modulus is the quotient of the separate moduli and whose argument is the difference of the two separate arguments.

A most important special case of Eqn (4.19) occurs when all the z_1, $z_2, \ldots, z_n$ are equal to the same complex number $z = r(\cos\theta + i \sin\theta)$, say. The result then becomes

$$z^n = r^n(\cos n\theta + i \sin n\theta).$$

Substituting for z and cancelling a real factor r^n, we obtain the following important theorem.

Theorem 4.2 (de Moivre's theorem)

$$(\cos\theta + i\sin\theta)^n = \cos n\theta + i \sin n\theta. \quad \blacksquare$$

A more subtle argument would have yielded the fact that this remarkable result is true for all real values of n, and not just for the integral values utilized in our proof. This will be undertaken later when the complex exponential function has been discussed.

Theorem 4.2 provides a simple method by which certain forms of trigonometic identity may be established. One typical example is enough to illustrate this.

Example 4.10 Let us relate $\sin 4\theta$ and $\cos 4\theta$ to sums of powers of $\sin\theta$ and $\cos\theta$.

Solution Set $n = 4$ in Theorem 4.2 and expand the left-hand side by the binomial theorem, using the fact that $i^2 = -1$, $i^3 = -i$, $i^4 = 1$, etc., to obtain

$$\cos^4\theta + 4i\cos^3\theta\sin\theta - 6\cos^2\theta\sin^2\theta - 4i\cos\theta\sin^3\theta + \sin^4\theta$$
$$= \cos 4\theta + i\sin 4\theta.$$

Then, recalling that equality of complex numbers means equality of their real and imaginary parts considered separately, we have the two results:

$$\cos^4\theta - 6\cos^2\theta\sin^2\theta + \sin^4\theta = \cos 4\theta, \quad \text{(equality of real parts)}$$

and

$$4(\cos^3\theta\sin\theta - \cos\theta\sin^3\theta) = \sin 4\theta. \quad \text{(equality of imaginary parts)}$$

These are the desired results.

It is characteristic of complex numbers that any single complex equality implies two real equalities, and even if only one is sought the other will be generated automatically. The same method works for any positive integral value of n when it will connect $\sin n\theta$ and $\cos n\theta$ with sums of powers of $\sin\theta$ and $\cos\theta$.

We shall return to this idea in connection with the exponential function, and show that it is possible to use de Moivre's theorem to express $\sin^n\theta$ and $\cos^n\theta$ in terms of sums involving $\sin r\theta$.

Sometimes Theorem 4.2 can be used to reduce the labour of computation, as now shown.

Example 4.11 We shall evaluate z^{10} where $z = 1+i$. Rather than making repeated multiplications, or applying the binomial theorem, we write z in modulus–argument form as $z = \sqrt{2}(\cos \pi/4 + i \sin \pi/4)$, when we have $z^{10} = (\sqrt{2})^{10} (\cos \pi/4 + i \sin \pi/4)^{10}$. By de Moivre's theorem this becomes

$$z^{10} = 2^5\left(\cos \frac{5\pi}{2} + i \sin \frac{5\pi}{2}\right) = 32i. \quad \blacksquare$$

4.5 Roots of complex numbers

When performing algebra on real numbers the idea of the root of a number plays a fundamental part. The same is true when manipulating complex numbers, and we now discuss the general ideas involved in determining their roots.

Let p/q be any rational number, where p and q are integers with q supposed positive. We shall assume that p and q have no common factor.

Definition 4.9 We define $z^{p/q}$ by saying that:

$$w = z^{p/q} \Leftrightarrow w^q = z^p.$$

Let

$$w = \rho(\cos \phi + i \sin \phi) \quad \text{and} \quad z = r(\cos \theta + i \sin \theta). \tag{4.21}$$

Then from Definition 4.9 and de Moivre's theorem we have

$$\rho^q(\cos q\phi + i \sin q\phi) = r^p(\cos p\theta + i \sin p\theta). \tag{4.22}$$

Now from Definition 4.8 it follows that

$$\rho^q = r^p \quad \text{and} \quad q\phi = p\theta + 2k\pi, \quad \text{with } k \text{ an interger,} \tag{4.23}$$

and so

$$\rho = q^{p/q} \quad \text{and} \quad \phi = \frac{p\theta + 2k\pi}{q}. \tag{4.24}$$

The expression $w = z^{p/q}$ thus has the general form

$$w = z^{p/q} = r^{p/q}\left[\cos\left(\frac{p\theta + 2k\pi}{q}\right) + i \sin\left(\frac{p\theta + 2k\pi}{q}\right)\right]. \tag{4.25}$$

It is easily seen that only q different values $w_0, w_1, \ldots, w_{q-1}$ of w will result from Eqn (4.25) as the integer k increases or decreases through successive integral values. It is usual to give k the q successive values $k = 0, 1, 2, \ldots, q-1$. If k is allowed to increase beyond the value $q-1$, then the numbers $w_0, w_1, \ldots, w_{q-1}$ will simply be generated again because of the periodicity properties of the sine and cosine functions.

Example 4.12 We illustrate the use of Eqn (4.25) by determining the n numbers w satisfying the equation $w = (1)^{1/n}$. For obvious reasons these are called the *nth roots of unity*. Comparing this equation with the general expression $w = z^{p/q}$ that has just been discussed, we see that we must make the identifications $z = 1$, $p = 1$ and $q = n$. To proceed further we must write the number unity in its modulus–argument form

$$1 = 1.(\cos 0 + i \sin 0),$$

so that comparing this with z in Eqn (4.21) we see that the further identifications $r = 1$ and $\theta = 0$ must be made. Substitution of these quantities into Eqn (4.25) then gives the result

$$w_k = \cos \frac{2k\pi}{n} + i \sin \frac{2k\pi}{n} \quad \text{with} \quad k = 0, 1, 2, \ldots, n-1.$$

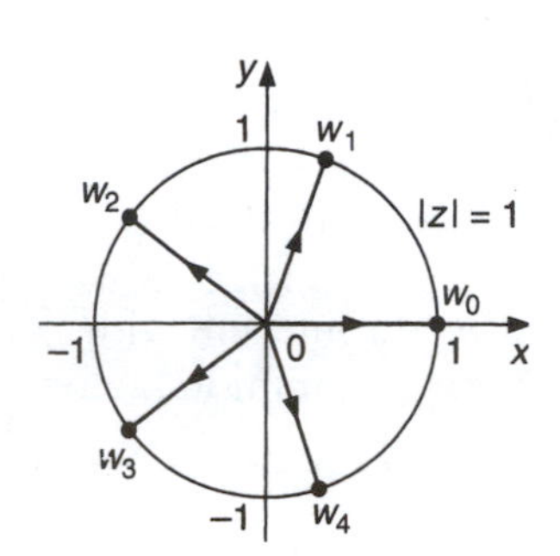

Fig. 4.5 Fifth roots of unity.

The result of this calculation with $n = 5$, for example, is to generate the fifth roots of unity. In Fig. 4.5 these roots are plotted as the numbers w_0, w_1, ..., w_4 in the complex plane. They are uniformly distributed around the unit circle centred on the origin. By making use of the vector properties of complex numbers we shall usually represent this circle by the convenient notation $|z| = 1$ and call it the *unit circle*. ■

Example 4.13 As a slightly more general example we now determine $z^{2/3}$, when $z = 1 + i$. In this case $p = 2$, $q = 3$, and, in modulus–argument form, $z = \sqrt{2}$ (cos $\pi/4$+i sin $\pi/4$), showing that $r = \sqrt{2}$ and $\theta = \pi/4$. Substitution into Eqn (4.25) gives

$$w = 2^{1/3}\left[\cos\left(\frac{1+4k}{6}\right)\pi + i \sin\left(\frac{1+4k}{6}\right)\pi\right] \quad \text{with } k = 0, 1, 2.$$

The three roots w_0, w_1 and w_2 are thus:

$$(k = 0){:}w_0 = 2^{1/3}\left(\cos\frac{\pi}{6} + i \sin\frac{\pi}{6}\right) = 2^{1/3}\left(\frac{\sqrt{3}}{2} + \frac{i}{2}\right),$$

$$(k = 1){:}w_1 = 2^{1/3}\left(\cos\frac{5\pi}{6} + i \sin\frac{5\pi}{6}\right) = 2^{1/3}\left(-\frac{\sqrt{3}}{2} + \frac{i}{2}\right),$$

$$(k = 2){:}w_2 = 2^{1/3}\left(\cos\frac{3\pi}{2} + i \sin\frac{3\pi}{2}\right) = -2^{1/3}i.$$

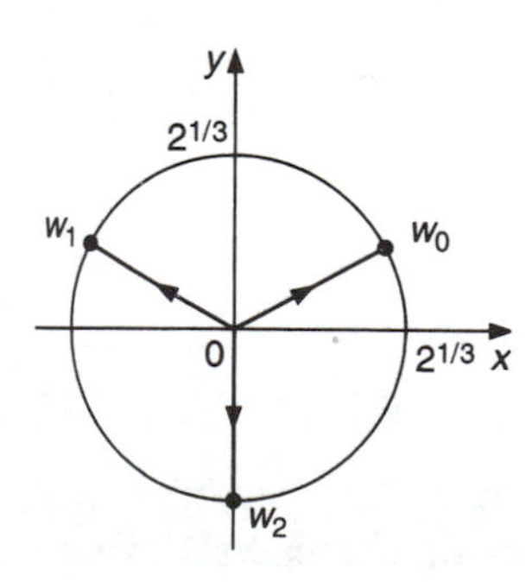

Fig. 4.6 Roots of $w = (1 + i)^{2/3}$.

These are plotted in the complex plane in Fig. 4.6, where they are seen to be uniformly distributed around the circle $|z| = 2^{1/3}$. Notice that $z^{2/3} = (z^{1/3})^2$, so the required roots could equally well have been obtained by finding the three roots of $z^{1/3}$ and then squaring them. ■

Example 4.14 As a final example let us find the roots of the equation

$$w = i^{-1/3}.$$

In terms of the notation of Eqns (4.21) and (4.25), and recalling that we have agreed always to take q as positive, we have $p = -1$, $q = 3$ and $z = i$. Now in modulus–argument form

$$i = 1.\left(\cos\frac{\pi}{2} + i\sin\frac{\pi}{2}\right),$$

so that $r = 1$ and $\theta = \pi/2$. Hence, substituting into Eqn (4.25), we find that

$$w = \cos\left[\frac{(-\pi/2) + 2k\pi}{3}\right] + i\sin\left[\frac{(-\pi/2) + 2k\pi}{3}\right] \quad \text{with } k = 0, 1, 2.$$

Hence the three roots w_0, w_1 and w_2 are:

$$(k = 0): w_0 = (\cos\pi/6 - i\sin\pi/6) = \tfrac{1}{2}(\sqrt{3} - i),$$

$$(k = 1): w_1 = (\cos\pi/2 - i\sin\pi/2) = i,$$

$$(k = 2): w_2 = (\cos 7\pi/6 - i\sin 7\pi/6) = -\tfrac{1}{2}(\sqrt{3} - i). \quad \blacksquare$$

This completes our preliminary encounter with complex numbers; our study will be resumed later in connection with the complex exponential function. The remainder of this chapter is devoted developing the foundations of our study of general vectors.

4.6 Introduction to space vectors

It is clear that any set of vector quantities that do not all lie in a plane cannot be represented vectorially in the form of complex numbers. For example, even the vectors describing the velocity of a vehicle as it is driven at constant speed past fixed points on a winding hill could not be so represented. Pairwise these velocity vectors define planes, and so could be represented by complex numbers in those planes, though different pairs of vectors would define different planes, thereby making any general representation impossible in terms of complex numbers. The trouble here is not hard to find. It is that complex numbers just happen to be capable of representation as planar vectors with their own appropriate descriptive language, though they were not developed with general vector representation in mind. In short, they are complex numbers first and vectors second; not the other way around.

To overcome this limitation and to be able to describe arbitrary vector quantities we must preserve the idea of a vector as a directed length, but rethink its description. This is best achieved using a diagram, so consider Fig. 4.7 which depicts mutually perpendicular Cartesian axes $O\{x, y, z\}$ with origin O. In more mathematical terms we describe these axes as being mutually *orthogonal*. This is a technical term that in a geometrical

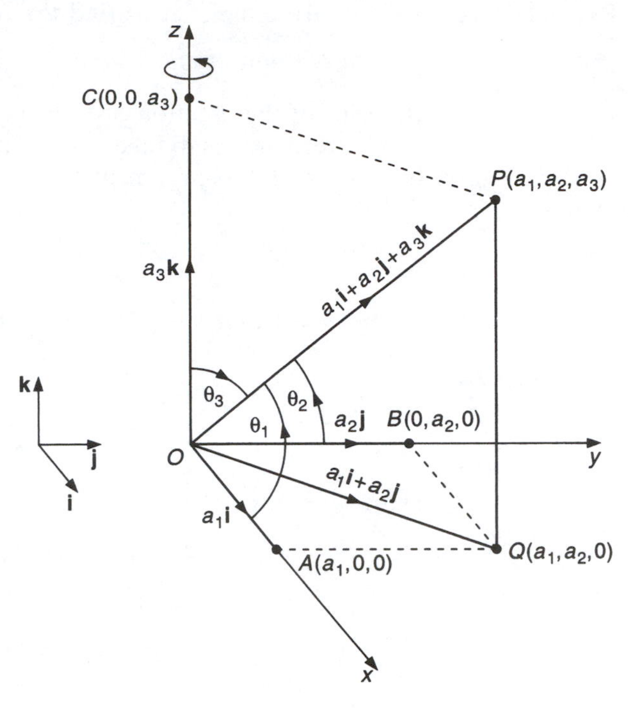

Fig. 4.7 Right-handed Cartesian axes.

context has the same meaning as perpendicular, though it is often used in a wider sense, when the word perpendicular would be inappropriate. Henceforth we shall almost always use the term 'orthogonal'.

The manner of identification of the x-, y- and z-coordinate axes is not arbitrary, but is made in such a manner that they form a right-handed system of axes. By this we mean that, having assigned axes for the variables x and y, together with the directions in which they increase positively, the direction of positive z is then chosen to be that in which a right-handed screw would advance were it aligned with the third axis and rotated in the sense x to y. This sense of rotation is indicated in Fig. 4.7. In the diagram the y- and z-axes are supposed to lie in the plane of the paper with the x-axis pointing out of the paper towards the viewer. Later we shall refer to this right-handed property in connection with axes which are not orthogonal, when right-handedness is still to be interpreted in exactly the same sense as above.

This right-handed property of the system of axes is shared by each pair of axes in turn, provided the senses of rotation are appropriately defined. Table 4.1 describes the convention that is always adopted. The table can easily be remembered in the concise form

$$\begin{matrix} x & y & z \\ y & z & x \\ z & x & y \end{matrix}$$

Table 4.1 Right-handed axes

Rotate		*Right handed screw advances in direction of positive*
From	*To*	
x	y	z
y	z	x
z	x	y

where the entry in any row is obtained from the entry in the row above by transferring the first letter of that entry to the last position. These entries are called *cyclic permutations* of the letters x, y and z, and further cyclic permutations will simply regenerate the table. These rules describe the right-handed symmetry of the $O\{x, y, z\}$ axes. If any two letters in an entry are interchanged, then by the same rule, the negative direction of the third axis is defined. Hence the set of letters yxz is to be interpreted 'rotate from y to x to make a right-handed screw aligned with the z-axis advance in the direction of negative z'.

If in the above argument a right-handed screw motion had been replaced by a left-handed screw motion, then a left-handed system of axes would have resulted. Although a left-handed system of axes is in all respects equivalent to a right-handed system for the purposes of vector representation, it is customary to work with a right-handed system.

Let P be the point with coordinates $x = a_1$, $y = a_2$ and $z = a_3$ illustrated in Fig. 4.7. We shall denote it by the more concise notation (a_1, a_2, a_3), where the first, second and third entries in this ordered number triple represent the x-, y- and z-coordinates, respectively. Then from the point of view of coordinate geometry it is the point P that is of interest, whereas from the point of view of vectors it is the directed line element from O to P that is of interest. To signify that it is the vector quantity that interests us here we shall write OP. Notice that by this convention the vector PO is the directed line from P to O and is opposite in sense to OP. In the future we will denote the length of the vector OP by $|OP|$, which is a scalar, and by definition this length will always be positive.

In Fig. 4.7 the lengths $OA = a_1$, $OB = a_2$ and $OC = a_3$ are called the *orthogonal projections* of OP onto the x-, y- and z-axes, and a simple application of Pythagoras' theorem gives the result

$$|OP|^2 = (OA)^2 + (OB)^2 + (OC)^2$$

or,

$$|OP|^2 = a_1^2 + a_2^2 + a_3^2.$$

Dividing by $|OP|^2$, this becomes

$$1 = \left(\frac{a_1}{|OP|}\right)^2 + \left(\frac{a_2}{|OP|}\right)^2 + \left(\frac{a_3}{|OP|}\right)^2,$$

which can then be rewritten in terms of the angles θ_1, θ_2, θ_3 as

$$1 = \cos^2\theta_1 + \cos^2\theta_2 + \cos^2\theta_3. \tag{4.26}$$

If the numbers l, m and n are defined by the relations

$$l = \cos\theta_1, \quad m = \cos\theta_2, \quad n = \cos\theta_3, \tag{4.27}$$

then Eqn (4.26) becomes

$$1 = l^2 + m^2 + n^2. \tag{4.28}$$

For obvious reasons l, m and n are called the *direction cosines* of OP with respect to the axes $O\{x, y, z\}$, and it is often convenient to write them in the form of an ordered number triple as $\{l, m, n\}$. The angles θ_1, θ_2 and θ_3 are indeterminate to within a multiple of 2π and, by convention, they will always be taken to lie in the $[0, \pi]$.

Consider the direction cosines l, m, n as defining a point P' in space with coordinates $x = l$, $y = m$ and $z = n$, then, by Pythagoras' theorem and Eqn (4.28), the vector OP' must have unit length. The direction and sense of OP' are the same as those of OP; only the lengths are different. Vectors of unit length in given directions prove to be extremely useful in vector analysis so they are appropriately called *unit vectors*.

Now by definition, the direction cosines l, m, n are proportional to the coordinates a_1, a_2, a_3 of the point P, and consequently the numbers a_1, a_2, and a_3 are often called the *direction ratios* of OP. To convert direction ratios to direction cosines it is necessary to *normalize* them by dividing by the square root of the sum of the squares of the direction ratios. This is, of course, equivalent to division by the quantity we have agreed to denote by $|OP|$.

Example 4.15 Find the direction ratios, the direction cosines and the angles θ_1, θ_2 and θ_3 of the vector OP, where P is the point $(1, -2, 4)$.

Solution The direction ratios are $1, -2, 4$ and $|OP|$, which is the square root of the sum of the squares of the direction ratios, is

$$|OP| = (1^2 + (-2)^2 + 4^2)^{1/2} = \sqrt{21}.$$

Hence the direction cosines of OP are $l = 1/\sqrt{21}$, $m = -2/\sqrt{21}$ and $n = 4/\sqrt{21}$, from which the angles θ_1, θ_2 and θ_3 are seen to be 1.351, 2.022 and 0.509 radians, respectively. As stated in Chapter 1, we shall always express angles in terms of radians, as here. ■

Example 4.16 Determine the angles of inclination θ_1, θ_2 and θ_3 of a vector to the x-, y- and z-axes, respectively, given that its direction cosines are:

(a) $\{\frac{1}{2}, -\sqrt{3}/2, 0\}$,
(b) $\{\frac{1}{2}, \frac{1}{4}, \sqrt{11}/4\}$.

Solution (a) Here $l = \cos\theta_1 = \frac{1}{2}$, $m = \cos\theta_2 = \sqrt{3}/2$, $n = \cos\theta_3 = 0$, so that $\theta_1 = \pi/3$, $\theta_2 = 5\pi/6$ and $\theta_3 = \pi/2$. Hence in this case the vector lies entirely in the (x, y)-plane.

(b) In this case $l = \cos\theta_1 = \frac{1}{2}$, $m = \cos\theta_2 = \frac{1}{4}$, $n = \cos\theta_3 = \sqrt{11}/4$, so that $\theta_1 = \pi/3$, $\theta_2 = 1.318$ and $\theta_3 = 0.593$. ■

Example 4.17 If a vector has direction cosines $\{\frac{1}{2}, m, \frac{1}{2}\}$ deduce the possible values of m. If, in addition, it is stated that the vector makes an obtuse angle θ_2 with the y-axis, determine the value of θ_2.

Solution We use Eqn (4.28), setting $l = \frac{1}{2}$ and $n = \frac{1}{2}$, to obtain

$$(\tfrac{1}{2})^2 + m^2 + (\tfrac{1}{2})^2 = 1.$$

Whence, $m^2 = \frac{1}{2}$ or $m = \pm 1/\sqrt{2}$. These values of m correspond to $\theta_2 = \pi/4$ for $m = 1/\sqrt{2}$, and to $\theta_2 = 3\pi/4$ for $m = -1/\sqrt{2}$. As the angle θ_2 is required to be obtuse we must select $\theta_2 = 3\pi/4$. ■

The idea of fixed origin is fundamental to coordinate geometry though it proves to be rather too restrictive in vector analysis. This is because it is only the magnitude, direction and sense of a vector that usually matter, and not the choice of origin and coordinate system in which the vector is represented. For example, when specifying a wind velocity it is normally sufficient to say 6 m/s due east, without identifying the particular points in space at which the air has this velocity.

In vector work this ambiguity as to the location of a vector in space is allowed by considering as equivalent any two vectors that may be represented by directed line elements of equal length which are parallel and have the same sense. In Fig. 4.8 we have depicted two vectors OP and $O'P'$ that are equivalent in the sense just defined. Another way to define

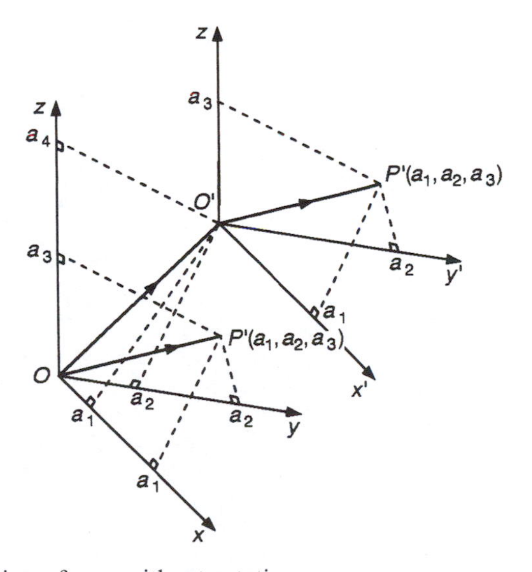

Fig. 4.8 Translation of axes without rotation.

this equivalence is to require that when the axes $O\{x, y, z\}$ are translated, without rotation, to the position $O'\{x, y', z\}$, the coordinates of P' with respect to the axes through O' are the same as those of P with respect to the axes through O. That is, if P is the point (a_1, a_2, a_3) in the system of axes $O'\{x', y', z'\}$. Do not get confused by this. If O' is the point $(\alpha_1, \alpha_2, \alpha_3)$ with respect to O $\{x, y, z\}$, then coordinates in the unprimed system are related to those in the primed system by the equations $x = \alpha_1 + x'$, $y = \alpha_2 + y'$, and $z = \alpha_3 + z'$.

This freedom to translate vectors now enables us to give direction cosines to any vector in space and not just to those having their base at O. Suppose, for example, that we require the length and direction cosines of the vector AB, where A is the point (a_1, a_2, a_3) and B is the point (b_1, b_2, b_3) when expressed relative to some set of axes $O\{x, y, z\}$. Then we see at once that the lengths of the projections of AB on the x-, y- and z-axes are $(b_1 - a_1)$, $(b_2 - a_2)$, and $(b_3 - a_3)$, respectively. Accordingly, by translating the vector AB until A in its new position A' coincides with O, we see that the tip B in its new position B' must be the point $((b_1 - a_1), (b_2 - a_2), (b_3 - a_3))$ (see Fig. 4.9). Hence $|AB|$, that is the length of AB, is

$$|AB| = [(b_1 - a_1)^2 + (b_2 - a_2)^2 + (b_3 - a_3)^2]^{1/2}. \tag{4.29}$$

The direction cosines of AB then follow as before and are

$$l = \frac{b_1 - a_1}{|AB|}, \quad m = \frac{b_2 - a_2}{|AB|}, \quad n = \frac{b_3 - a_3}{|AB|}. \tag{4.30}$$

Example 4.18 Find $|AB|$ and the direction cosines of the vector AB, if A has coordinates $(1, 2, 3)$ and B the coordinates $(4, 3, 6)$.

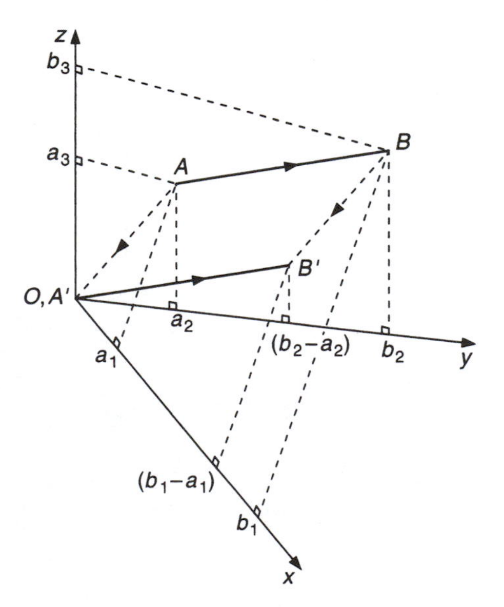

Fig. 4.9 Translation of a vector.

Solution From Eqn (4.29) we see that $|AB| = [(4-1)^2+(3-2)^2+(6-3)^2]^{1/2} = \sqrt{19}$, while from Eqn (4.30) it follows that $l = 3/\sqrt{19}$, $m = 1/\sqrt{19}$ and $n = 3/\sqrt{19}$. ■

It is now convenient to introduce a triad of unit vectors, denoted by **i**, **j** and **k**, that are parallel to and are directed in the positive senses of the x-, y- and z-axes, respectively. Here we remind the reader that these are called unit vectors because they are each of unit length on the x-, y- and z-length scales. Notice that the term 'right-handed' that was applied to the system of axes $O\{x, y, z\}$ also applies to the triad of vectors **i**, **j**, **k** when taken in this order. We shall use this idea again later.

An arbitrary vector in any one of the **i**, **j** or **k** directions may then be obtained by scaling the length of the appropriate unit vector by a multiplication factor μ. Thus a vector three times the size of the unit vector **i** will be written 3**i**, while a vector twice the size of the unit vector **k**, but oppositely directed, will be written $-2\mathbf{k}$.

Returning to Fig. 4.7, we see that in terms of **i**, **j**, **k**, the vectors OA, OB and OC may be written as

$$OA = a_1\mathbf{i}, \qquad OB = a_2\mathbf{j}, \qquad OC = a_3\mathbf{k}.$$

From our ideas of vector addition in a plane, the vector OQ lying in the (x, y)-plane is $OQ = OA + AQ$ or, because vectors may be translated, $OQ = OA + AB$. Now in terms of our unit vector notation this may be written $OQ = a_1\mathbf{i} + a_2\mathbf{j}$. Turning attention to the plane containing points O, Q and P, we see that by the same argument $OP = OQ + QP$. Again because vectors may be translated, $QP = OC$ so that, finally, on substituting for OQ and QP in the equation $OP = OQ + QP$, we obtain

$$OP = a_1\mathbf{i} + a_2\mathbf{j} + a_3\mathbf{k}. \tag{4.31}$$

For ease of notation, arbitrary vectors, like unit vectors, will usually be denoted by a single bold face symbol such as $\boldsymbol{\alpha}$, **a** or **r**. Thus a general point P in space with coordinate (x, y, z) will often be written

$$\mathbf{r} = x\mathbf{i} + y\mathbf{j} + z\mathbf{k}. \tag{4.32}$$

The almost universally accepted convention which we adopt here is to denote vector quantities by bold type and scalar quantities by italic type.

Because a vector such as **r** in Eqn (4.32) identifies a point P in space it is called a *position vector*. In the vector representation Eqn (4.31) the numbers a_1, a_2 and a_3 are called the *components* of OP.

Two vectors will only be said to be *equal* if, when written in the form of Eqn (4.31), their corresponding components are equal. The vector $\mathbf{a} = a_1\mathbf{i} + a_2\mathbf{j} + a_3\mathbf{k}$ will be said to be a scalar multiple λ of vector $\mathbf{b} = b_1\mathbf{i} + b_2\mathbf{j} + b_3\mathbf{k}$, and we will write $\mathbf{a} = \lambda\mathbf{b}$, if $a_1 = \lambda b_1$, $a_2 = \lambda b_2$ and $a_3 = \lambda b_3$. In the special case $\lambda = -1$ we have $\mathbf{a} = -\mathbf{b}$, showing that $|\mathbf{a}| = |\mathbf{b}|$, but that the senses of **a** and **b** are opposite. Thus in Fig. 4.7 we have $OP = -PO$.

The *zero* or *null* vector **0** is the vector whose three components are each identically zero. It is often denoted by 0 instead of **0**, since confusion is unlikely to arise on account of this simplification of the notation.

Following on from our first ideas of vectors, and in accordance with the derivation of Eqn (4.31), we now define the operations of addition and subtraction of vectors.

Definition 4.10 Let **a** and **b** be arbitrary vectors with components (a_1, a_2, a_3) and (b_1, b_2, b_3), respectively, so that they may be written $\mathbf{a} = a_1\mathbf{i} + a_2\mathbf{j} + a_3\mathbf{k}$ and $\mathbf{b} = b_1\mathbf{i} + b_2\mathbf{j} + b_3\mathbf{k}$. Then we define the *sum* $\mathbf{a} + \mathbf{b}$ of the two vectors **a** and **b** to be the vector $(a_1 + b_1)\mathbf{i} + (a_2 + b_2)\mathbf{j} + (a_3 + b_3)\mathbf{k}$. The *difference* $\mathbf{a} - \mathbf{b}$ of the two vectors **a** and **b** is defined to be the vector $(a_1 - b_1)\mathbf{i} + (a_2 - b_2)\mathbf{j} + (a_3 - b_3)\mathbf{k}$.

Because real numbers are commutative with respect to addition, it follows directly from this definition that the operation of vector addition is commutative. That is we have $\mathbf{a} + \mathbf{b} = \mathbf{b} + \mathbf{a}$ for all vectors **a** and **b**. When the subtraction operation is considered the properties of real numbers imply the result $\mathbf{a} - \mathbf{b} = -(\mathbf{b} - \mathbf{a})$ for all vectors **a** and **b**.

Example 4.19 If $\mathbf{a} = \mathbf{i} + \mathbf{j} + 2\mathbf{k}$ and $\mathbf{b} = 3\mathbf{i} - 3\mathbf{j} + \mathbf{k}$, then $\mathbf{a} + \mathbf{b} = (1 + 3)\mathbf{i} + (1 - 3)\mathbf{j} + (2 + 1)\mathbf{k}$, showing that $\mathbf{a} + \mathbf{b} = 4\mathbf{i} - 2\mathbf{j} + 3\mathbf{k}$. Reversal of the order of the vectors in the sum followed by the same argument proves the commutative property $\mathbf{a} + \mathbf{b} = \mathbf{b} + \mathbf{a}$ for these particular vectors. In the case of subtraction, we have $\mathbf{a} - \mathbf{b} = (1 - 3)\mathbf{i} + (1 - (-3))\mathbf{j} + (2 - 1)\mathbf{k}$, showing that $\mathbf{a} - \mathbf{b} = -2\mathbf{i} + 4\mathbf{j} + \mathbf{k}$. It is easily established that $\mathbf{a} - \mathbf{b} = -(\mathbf{b} - \mathbf{a})$. ■

Although these particular results could be illustrated diagrammatically, the vector triangles involved would look essentially the same as those used earlier in connection with addition and subtraction of complex numbers and would be arrived at by the same reasoning. Rather than illustrate this specific case, we present in Fig. 4.10 the results of addition and subtraction of arbitrary vectors **a** and **b**. Because a geometrical projection

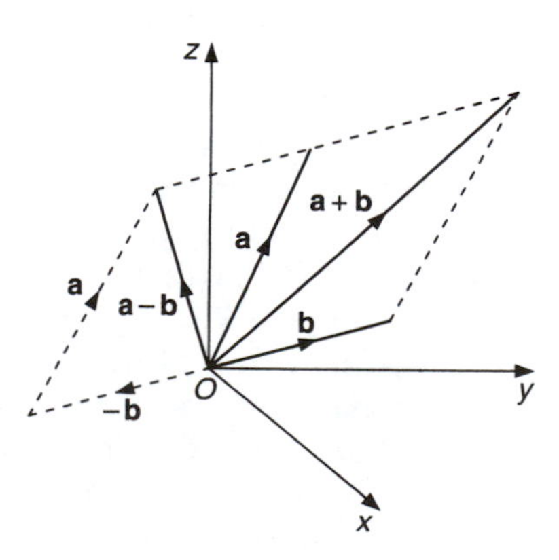

Fig. 4.10 Addition and subtraction of vectors.

method is necessary to illustrate three-dimensional problems on a sheet of paper, such diagrams are much less useful as a tool than was the case in a plane. Accordingly, we shall usually concentrate on an analytical approach to vectors, using diagrams only when they seem likely to be helpful.

Two terms worthy of note that are applied to vectors are the names 'parallel' and 'anti-parallel'. Two vectors will be said to be *parallel* when their lines of action are parallel and their senses are the same. Conversely, two vectors will be said to be *anti-parallel* when their lines of action are parallel but their senses are opposite. Thus if $\mathbf{a}$ is a vector and μ is a scalar, the vectors $\mathbf{a}$ and $\mu\mathbf{a}$ are parallel if $\mu > 0$ and are anti-parallel if $\mu < 0$. It follows than two vectors will be parallel if their corresponding direction cosines are equal, and they will be anti-parallel if their corresponding direction cosines are equal in magnitude but opposite in sign.

Example 4.20 The vectors $\mathbf{a} = \mathbf{i} + 2\mathbf{j} - 4\mathbf{k}$ and $\mathbf{b} = 3\mathbf{i} + 6\mathbf{j} - 12\mathbf{k}$ are such that we may write $3\mathbf{a} = \mathbf{b}$. Since the scalar $3 > 0$ it follows that $\mathbf{a}$ and $\mathbf{b}$ are parallel. However the vectors $\mathbf{c} = \mathbf{i} - 3\mathbf{j} + \mathbf{k}$ and $\mathbf{d} = -2\mathbf{i} + 6\mathbf{j} - 2\mathbf{k}$ are such that we may write $-2\mathbf{c} = \mathbf{d}$ and, as the scalar $-2 < 0$, it follows that $\mathbf{c}$ and $\mathbf{d}$ are anti-parallel. By the same argument, the two vectors $\mathbf{p} = 3\mathbf{i} - \mathbf{j} + 2\mathbf{k}$ and $\mathbf{q} = 6\mathbf{i} + 2\mathbf{j} + 4\mathbf{k}$ are neither parallel nor anti-parallel, since for no scalar μ is it true that $\mu\mathbf{p} = \mathbf{q}$. ■

The length of the vector $\mathbf{AB}$ which we have already denoted by $|\mathbf{AB}|$ is a useful quantity and, as with complex numbers, is called the modulus of the vector $\boldsymbol{AB}$. Its formal definition follows.

Definition 4.11 The *modulus* $|\mathbf{a}|$ of the vector $\mathbf{a} = a_1\mathbf{i} + a_2\mathbf{j} + a_3\mathbf{k}$ is the positive square root

$$|\mathbf{a}| = (a_1^2 + a_2^2 + a_3^2)^{1/2}.$$

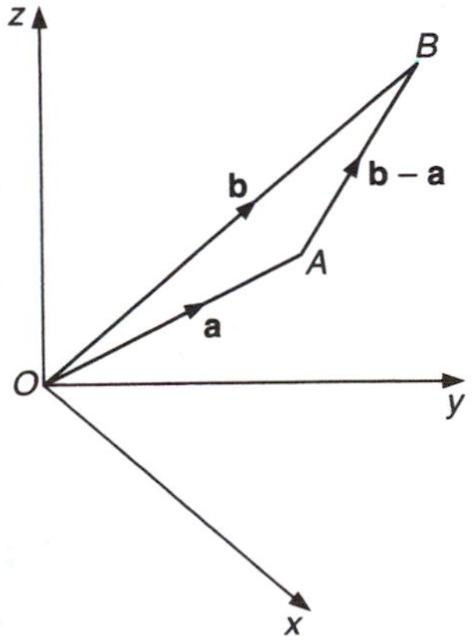

z
B
b
b − a
A
a
O
y
x

Fig. 4.11 Position vectors defining the vector AB.

It is an immediate consequence of this definition that any vector $\mathbf{r}$ with direction cosines $\{l, m, n\}$ may be written in the form

$$\mathbf{r} = |\mathbf{r}|(l\mathbf{i} + m\mathbf{j} + n\mathbf{k}). \tag{4.33}$$

The proof of this is obvious for, by definition, $l|\mathbf{r}|$ is the x-component of $\mathbf{r}$, $m|\mathbf{r}|$ is the y-component and $n|\mathbf{r}|$ is the z-component. The form of Eqn (4.33) shows that any vector may be expressed as the product of a scalar (its modulus) and a unit vector defining its direction and sense.

When it is necessary to define an arbitrary vector AB in space, this may easily be accomplished by using position vectors $\mathbf{a}$ and $\mathbf{b}$ to identify its end points A and B. This is illustrated in Fig. 4.11 from which, by the rules of vector addition, we may write

$$OA + AB = OB$$

or,

$$AB = OB - OA = \mathbf{b} - \mathbf{a}.$$

Examination of this simple but useful result suggests that an accurate name for the vector AB would be the 'position vector of B relative to A', since in this role it is A that plays the part of the origin. This more exact name is seldom used since the symbol AB is sufficiently clear as it stands.

Example 4.21 Let points A and B be identified by the position vectors $\mathbf{a} = -2\mathbf{i} - 3\mathbf{j} + \mathbf{k}$ and $\mathbf{b} = 3\mathbf{i} - \mathbf{j} + 4\mathbf{k}$, respectively. Find the vector AB together with its modulus and direction cosines.

Solution The diagram in Fig. 4.11 can be taken to represent this situation showing that vector $AB = \mathbf{b} - \mathbf{a}$. Substituting for the values of $\mathbf{a}$ and $\mathbf{b}$, we find $AB = (3\mathbf{i} - \mathbf{j} + 4\mathbf{k}) - (-2\mathbf{i} - 3\mathbf{j} + \mathbf{k})$, whence $AB = 5\mathbf{i} + 2\mathbf{j} + 3\mathbf{k}$. Then $|AB| = (5^2 + 2^2 + 3^2)^{1/2} = \sqrt{38}$, after which the usual argument establishes that $l = 5/\sqrt{38}$, $m = 2/\sqrt{38}$ and $n = 3/\sqrt{38}$. ■

By considering the plane containing the vectors $\mathbf{a}$, $\mathbf{b}$ and $\mathbf{b} - \mathbf{a}$ in Fig. 4.11, the arguments that established the triangle inequalities for complex numbers also establish them for arbitrary space vectors. Hence for arbitrary vectors $\mathbf{a}$ and $\mathbf{b}$ we have

$$||\mathbf{a}| - |\mathbf{b}|| \leq |\mathbf{a} - \mathbf{b}| \leq |\mathbf{a}| + |\mathbf{b}|.$$

Finally, to close this section, let us find the angle θ between two vectors $\mathbf{a}$ and $\mathbf{b}$ with the direction cosines $\{l_1, \mathrm{m}_1, n_1\}$ and $\{l_2, m_2, n_2\}$, respectively. When the lines of action of the vectors intersect the angle θ is well defined and, by convention, is always chosen to lie in the interval $[0, \pi]$. If the lines of action of two vectors do not intersect then they are merely translated until they do, when the angle θ is defined as above. It will suffice to consider the angle between two unit vectors directed along $\mathbf{a}$ and $\mathbf{b}$ since the length of the vectors will obviously not influence the angle between them. From Eqn (4.33), these unit vectors are seen to be $(l_1\mathbf{i} + m_1\mathbf{j} + n_1\mathbf{k})$ and $(l_2\mathbf{i} + m_2\mathbf{j} + n_2\mathbf{k})$. These are shown in Fig. 4.12. They have their tips P and Q at the respective points (l_1, m_1, n_1) and (l_2, m_2, n_2).

Now, by the cosine rule

$$|PQ|^2 = |OP|^2 + |OQ|^2 - 2|OP|.|OQ|\cos\theta, \tag{4.35}$$

but $|OP| = |OQ| = 1$, and by Eqn (4.29), $|PQ|^2 = (l_2 - l_1)^2 + (m_2 - m_1)^2 + (n_2 - n_1)^2$, while by Eqn (4.28), $l_1^2 + m_1^2 + n_1^2 = l_2^2 + m_2^2 + n_2^2 = 1$. Consequently, substituting into Eqn (4.35) and simplifying, we find the desired result

$$\cos\theta = l_1 l_2 + m_1 m_2 + n_1 n_2. \tag{4.36}$$

The angle of inclination θ follows directly from this equation. The restriction of the angle between the vectors to the interval $[0, \pi]$ means that the angle θ that is selected is as shown in Fig. 4.12.

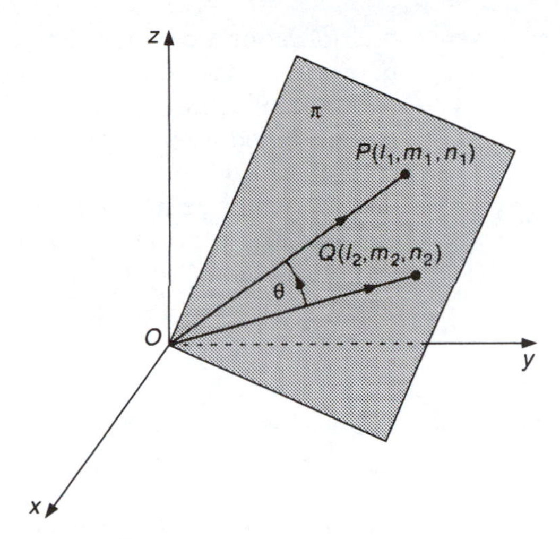

Fig. 4.12 Angle between two lines.

As a particular case, if

$$l_1 l_2 + m_1 m_2 + n_1 n_2 = 0, \tag{4.37}$$

then the two vectors **a** and **b** must be orthogonal.

Example 4.22 Find the angle of inclination θ between the vectors $\mathbf{a} = \mathbf{i} + 2\mathbf{j} + 3\mathbf{k}$ and $\mathbf{b} = 2\mathbf{i} - \mathbf{j} - \mathbf{k}$.

Solution Here $|\mathbf{a}| = \sqrt{14}$, $|\mathbf{b}| = \sqrt{6}$, so that the direction cosines $\{l_1, m_1, n_1\}$ of **a** are $l_1 = 1/\sqrt{14}$, $m_1 = 2/\sqrt{14}$, $n_1 = 3/\sqrt{14}$ while the direction cosines $\{l_2, m_2, n_2\}$ of **b** are $l_2 = 2/\sqrt{6}$, $m_2 = -1/\sqrt{6}$, $n_2 = -1/\sqrt{6}$. Hence by Eqn (4.36), the angle θ is the solution of the equation

$$\cos\theta = \left(\frac{1}{\sqrt{14}}\right)\left(\frac{2}{\sqrt{6}}\right) + \left(\frac{2}{\sqrt{14}}\right)\left(\frac{-1}{\sqrt{6}}\right) + \left(\frac{3}{\sqrt{14}}\right)\left(\frac{-1}{\sqrt{6}}\right),$$

or

$$\theta = \arccos\left(\frac{-3}{2\sqrt{21}}\right).$$

On account of the restriction of θ to the interval $[0, \pi]$ it finally follows that $\theta = 1.904$. ■

4.7 Scalar and vector products

If $\mathbf{a} = a_1\mathbf{i} + a_2\mathbf{j} + a_3\mathbf{k}$ is an arbitrary vector and λ is a scalar, then we have already defined the product $\lambda\mathbf{a}$ to be the vector $\lambda\mathbf{a} = \lambda a_1\mathbf{i} + \lambda a_2\mathbf{j} + \lambda a_3\mathbf{k}$. Hence the effect of multiplying a vector by a scalar is to magnify the vector without changing its direction. The result of this product is to generate a vector. We must now discuss the multiplication of two vectors.

Here three-dimensional vector algebra differs radically from the vector algebra of complex numbers. With complex numbers there is only one multiplication operation defined, and the product of two complex numbers is always a complex number. In the case of vectors we shall see that two multiplication operations are defined for a pair of vectors. One operation called a scalar product generates a scalar, whereas the other operation called a vector product generates a vector. The operation of division is not defined for vectors.

The scalar product of two vectors is a generalization of the notion of the orthogonal projection of a line element onto another line and is suggested by Eqn (4.36). Its definition follows.

Definition 4.12 The *scalar product* of the two vectors $\mathbf{a} = a_1\mathbf{i} + a_2\mathbf{j} + a_3\mathbf{k}$ and $\mathbf{b} = b_1\mathbf{i} + b_2\mathbf{j} + b_3\mathbf{k}$ is written $\mathbf{a}.\mathbf{b}$ and is defined to be the scalar quantity

$$\mathbf{a}.\mathbf{b} = a_1 b_1 + a_2 b_2 + a_3 b_3.$$

Because of the notation used, the scalar product is also called the *dot product*. Some books favour the notation $(\mathbf{a}, \mathbf{b})$ for the scalar product when it is then usually called the *inner product* of vectors **a** and **b**. To exhibit the relation of $\mathbf{a}.\mathbf{b}$ to Eqn (4.36) we first divide $\mathbf{a}.\mathbf{b}$ by the product of the moduli $|\mathbf{a}||\mathbf{b}|$ to get

$$\frac{\mathbf{a}.\mathbf{b}}{|\mathbf{a}||\mathbf{b}|} = \left(\frac{a_1}{|\mathbf{a}|}\right)\left(\frac{b_1}{|\mathbf{b}|}\right) + \left(\frac{a_2}{|\mathbf{a}|}\right)\left(\frac{b_2}{|\mathbf{b}|}\right) + \left(\frac{a_3}{|\mathbf{a}|}\right)\left(\frac{b_3}{|\mathbf{b}|}\right).$$

Then, from the definition of direction cosines, we recognize that this may be written

$$\frac{\mathbf{a}.\mathbf{b}}{|\mathbf{a}||\mathbf{b}|} = l_1 l_2 + m_1 m_2 + n_1 n_2, \tag{4.38}$$

where $\{l_1, m_1, n_1\}$ are the direction cosines of **a** and $\{l_2, m_2, n_2\}$ are the direction cosines of **b**. If θ is the angle of inclination between **a** and **b** then, by virtue of Eqn (4.36), expression (4.38) becomes

$$\mathbf{a}.\mathbf{b} = |\mathbf{a}||\mathbf{b}|\cos\theta. \tag{4.39}$$

This may be taken as an alternative definition of the scalar product $\mathbf{a}.\mathbf{b}$.

Alternative Definition 4.13 The scalar product of the two vectors **a** and **b** is written $\mathbf{a}.\mathbf{b}$ and is defined to be the scalar quantity

$$\mathbf{a}.\mathbf{b} = |\mathbf{a}|\ |\mathbf{b}|\cos\theta,$$

where θ is the angle between the vectors, with $0 \leq \theta \leq \pi$.

Notice that it is a direct consequence of the definition that the scalar product of two vectors is *commutative*. That is, we have $\mathbf{a}.\mathbf{b} = \mathbf{b}.\mathbf{a}$ for any two vectors $\mathbf{a}$ and $\mathbf{b}$.

Because of this property we shall sometimes, and without confusion, write $\mathbf{a}^2$ with the understanding that $\mathbf{a}^2 = \mathbf{a}.\mathbf{a}$. In practice Definition 4.12 is most used to find the scalar product since it relates the scalar product directly to the components of the vectors involved. The alternative form set out in Definition 4.13 is used to find the angle between the two vectors once the scalar product is known.

Example 4.23 Find the scalar product of the vectors $\mathbf{a} = -2\mathbf{i} - 3\mathbf{j} + \mathbf{k}$ and $\mathbf{b} = -\mathbf{i} + \mathbf{j} + 3\mathbf{k}$, and use the result to find the angle between $\mathbf{a}$ and $\mathbf{b}$.

Solution From Definition 4.12 we have $\mathbf{a}.\mathbf{b} = (-2)(-1) + (-3)(1) + (1)(3) = 2$. Now $|\mathbf{a}| = \surd 14$ and $|\mathbf{b}| = \surd 11$, so that substituting in Definition 4.13 we have $2 = \surd 14 . \surd 11 \cos\theta$ and hence $\cos\theta = 2/\surd 154$, or $\theta = \arccos(2/\surd 154) = 1.4089$ rad. ■

Consider the scalar products of the unit vectors $\mathbf{i}$, $\mathbf{j}$ and $\mathbf{k}$. Since these are mutually orthogonal the angle between any two is $\pi/2$. It follows from Definition 4.13 that the scalar product of any two different unit vectors from this triad is zero. As each of the vectors $\mathbf{i}$, $\mathbf{j}$ and $\mathbf{k}$ is parallel to itself, when forming the scalar product of one of these vectors with itself we must set $\theta = 0$. Thus as $|\mathbf{i}| = |\mathbf{j}| = |\mathbf{k}| = 1$, it follows from Definition 4.13 that $\mathbf{i}.\mathbf{i} = \mathbf{j}.\mathbf{j} = \mathbf{k}.\mathbf{k} = 1$. In summary, we have these important results, which should be memorized since they are fundamental to everything that follows:

$$\mathbf{i}.\mathbf{i} = \mathbf{j}.\mathbf{j} = \mathbf{k}.\mathbf{k} = 1,$$

$$\mathbf{i}.\mathbf{j} = \mathbf{j}.\mathbf{i} = 0,$$

$$\mathbf{i}.\mathbf{k} = \mathbf{k}.\mathbf{i} = 0,$$

$$\mathbf{j}.\mathbf{k} = \mathbf{k}.\mathbf{j} = 0.$$

These results are conveniently combined in Table 4.2 Each entry is to be interpreted as the scalar product of the vector at the left of the row of the entry, with the vector at the top of the column of the entry.

The scalar product of two vectors may be deduced, using Table 4.2, by simple algebraic manipulation without the use of Definition 4.12. To see

Table 4.2 Table of scalar products of **i**, **j** and **k**

First member	*Second member*		
	i	**j**	**k**
i	1	0	0
j	0	1	0
k	0	0	1

this, consider the vectors $\mathbf{a}=a_1\mathbf{i}+a_2\mathbf{j}+a_3\mathbf{k}$ and $\mathbf{b}=b_1\mathbf{i}+b_2\mathbf{j}+b_3\mathbf{k}$. First form their scalar product

$$\mathbf{a}.\mathbf{b}=(a_1\mathbf{i}+a_2\mathbf{j}+a_3\mathbf{k}).(b_1\mathbf{i}+b_2\mathbf{j}+b_3\mathbf{k})$$

and then expand the right-hand side as though ordinary algebraic quantities were involved to obtain

$$\begin{aligned}\mathbf{a}.\mathbf{b}&=(a_1\mathbf{i}).(b_1\mathbf{i})+(a_1\mathbf{i}).(b_2\mathbf{j})+(a_1\mathbf{i}).(b_3\mathbf{k})+(a_2\mathbf{j}).(b_1\mathbf{i})\\&\quad+(a_2\mathbf{j}).(b_2\mathbf{j})+(a_2\mathbf{j}).(b_3\mathbf{k})+(a_3\mathbf{k}).(b_1\mathbf{i})+(a_3\mathbf{k}).(b_2\mathbf{j})\\&\quad+(a_3\mathbf{k}).(b_3\mathbf{k}).\end{aligned}$$

Next, recognizing that the scalars a_i, b_i may be taken to the front of each scalar product involved, rewrite the result thus:

$$\begin{aligned}\mathbf{a}.\mathbf{b}&=a_1b_1\mathbf{i}.\mathbf{i}+a_1b_2\mathbf{i}.\mathbf{j}+a_1b_3\mathbf{i}.\mathbf{k}+a_2b_1\mathbf{j}.\mathbf{i}+a_2b_2\mathbf{j}.\mathbf{j}\\&\quad+a_2b_2\mathbf{j}.\mathbf{k}+a_3b_1\mathbf{k}.\mathbf{i}+a_3b_2\mathbf{k}.\mathbf{j}+a_3b_3\mathbf{k}.\mathbf{k}.\end{aligned}$$

Finally, using Table 4.2, this reduces to the desired result

$$\mathbf{a}.\mathbf{b}=a_1b_1+a_2b_2+a_3b_3.$$

In practice the intermediate working is always omitted and the result of a scalar product is written on sight by retaining only the products involving **i**.**i**, **j**.**j** and **k**.**k**.

Example 4.24 Determine the scalar products of these pairs of vectors:

(a) $\mathbf{a}=\mathbf{i}-3\mathbf{j}+\mathbf{k}$, $\mathbf{b}=-\mathbf{i}+\mathbf{j}-3\mathbf{k}$;
(b) $\mathbf{a}=2\mathbf{i}+\mathbf{j}-\mathbf{k}$, $\mathbf{b}=-\mathbf{i}+\mathbf{j}-\mathbf{k}$;
(c) $\mathbf{a}=2\mathbf{i}-\mathbf{j}+3\mathbf{k}$, $\mathbf{b}=-2\mathbf{i}+\mathbf{j}-3\mathbf{k}$;
(d) $\mathbf{a}=\mathbf{i}+2\mathbf{j}-\mathbf{k}$, $\mathbf{b}=\mathbf{i}+2\mathbf{j}-\mathbf{k}$.

Solutions To show the application of scalar products of unit vectors we shall retain the notation **i**.**i**, **j**.**j** and **k**.**k** in the first part of each calculation to indicate the origin of the terms involved. The terms involving products such as **i**.**j**, **i**.**k**,..., will be omitted as these scalar products are zero. The result will usually be written down on sight without any intermediate working.

(a) $\mathbf{a}.\mathbf{b}=(\mathbf{i}-3\mathbf{j}+\mathbf{k}).(-\mathbf{i}+\mathbf{j}-3\mathbf{k})$
$=(1)(-1)\,\mathbf{i}.\mathbf{i}+(-3)(1)\mathbf{j}.\mathbf{j}+(1)(-3)\mathbf{k}.\mathbf{k}$
$=-1-3-3=-7.$

(b) $\mathbf{a}.\mathbf{b}=(2\mathbf{i}+\mathbf{j}-\mathbf{k}).(-\mathbf{i}+\mathbf{j}-\mathbf{k})$
$=(2)(-1)\mathbf{i}.\mathbf{i}+(1)(1)\,\mathbf{j}.\mathbf{j}+(-1)(-1)\mathbf{k}.\mathbf{k}$
$=-2+1+1=0.$

Thus **a** and **b** are orthogonal.

(c) $\mathbf{a}.\mathbf{b}=(2\mathbf{i}-\mathbf{j}+3\mathbf{k}).(-2\mathbf{i}+\mathbf{j}-3\mathbf{k})$
$=(2)(-2)\mathbf{i}.\mathbf{i}+(-1)(1)\,\mathbf{j}.\mathbf{j}+(3)(-3)\mathbf{k}.\mathbf{k}$
$=-4-1-9=-14.$

(d) $\mathbf{a}.\mathbf{b} = (\mathbf{i}+2\mathbf{j}-\mathbf{k}).(\mathbf{i}+2\mathbf{j}-\mathbf{k})$
$= (1)(1)\ \mathbf{i}.\mathbf{i}+(2)(2)\mathbf{j}.\mathbf{j}+(-1)(-1)\mathbf{k}.\mathbf{k}$
$= 1+4+1 = 6.$ ■

Example 4.24(d) is a special case of the scalar product of a vector with itself and either from Definition 4.12 or 4.13 we see that for an arbitrary vector $\mathbf{a}$,

$$\mathbf{a}.\mathbf{a} = |\mathbf{a}|^2. \tag{4.40}$$

In words, 'the scalar product of a vector with itself is equal to the square of the modulus of that vector'. This simple result is often valuable when finding a unit vector parallel to a given arbitrary vector $\boldsymbol{\alpha}$. To see how this comes about, if we divide $\boldsymbol{\alpha}$ by its modulus $|\boldsymbol{\alpha}|$ to form the vector $\hat{\boldsymbol{\alpha}} = \boldsymbol{\alpha}/|\boldsymbol{\alpha}|$, then result (4.40) shows that $\hat{\boldsymbol{\alpha}}.\hat{\boldsymbol{\alpha}} = 1$ and so $\hat{\boldsymbol{\alpha}}$ is a unit vector.

Example 4.25 Find a unit vector $\hat{\boldsymbol{\alpha}}$ parallel to the vector $\boldsymbol{\alpha} = 3\mathbf{i}-\mathbf{j}-2\mathbf{k}$. Use the result to determine the projection of the vector $\mathbf{b} = 2\mathbf{i}+3\mathbf{j}+\mathbf{k}$ in the direction of $\boldsymbol{\alpha}$.

Solution Here $|\boldsymbol{\alpha}| = \sqrt{14}$ so that the desired unit vector $\hat{\boldsymbol{\alpha}} = \boldsymbol{\alpha}/\sqrt{14} = (3/\sqrt{14})\mathbf{i}-(1/\sqrt{14})\mathbf{j}-(2/\sqrt{14})\mathbf{k}$. Now the projection of vector $\mathbf{b}$ along $\boldsymbol{\alpha}$ is by definition the length l of vector $\mathbf{b}$ when projected normally onto the line determined by $\boldsymbol{\alpha}$. Thus it is $l = |\mathbf{b}|\cos\theta$, where θ is the angle between $\mathbf{b}$ and $\boldsymbol{\alpha}$. Since $|\hat{\boldsymbol{\alpha}}| = 1$ we may write this as $l = |\mathbf{b}||\hat{\boldsymbol{\alpha}}|\cos\theta$ or, by Definition 4.13, as $l = \mathbf{b}.\hat{\boldsymbol{\alpha}}$. Hence in this problem $l = (2\mathbf{i}+3\mathbf{j}+\mathbf{k}).\ \hat{\boldsymbol{\alpha}} = 1/\sqrt{14}$. ■

It follows from the definition of a scalar product of two vectors and from the properties of real numbers, that if $\mathbf{a}$, $\mathbf{b}$ and $\mathbf{c}$ are three arbitrary vectors, then

$$\mathbf{a}.(\mathbf{b}+\mathbf{c}) = \mathbf{a}.\mathbf{b}+\mathbf{a}.\mathbf{c}.$$

This is the distributive law for the scalar product of vectors.

Expressions of the form $\mathbf{a}.\mathbf{b}.\mathbf{c}$, $\mathbf{a}.\mathbf{b}.\mathbf{c}.\mathbf{d},\ldots$, are meaningless since the scalar product is only defined between a pair of vectors. Note also that division by vectors is not defined, since although we may write $\mathbf{a}.\mathbf{b} = n$, it makes no sense to write either $\mathbf{a} = n/.\mathbf{b}$ or $\mathbf{a}. = n/\mathbf{b}$.

The other form of product of two vectors is the vector product. We shall denote the vector product of vectors $\mathbf{a}$ and $\mathbf{b}$ by $\mathbf{a}\times\mathbf{b}$. Again because of the notation this is also called the *cross product* of two vectors. Other notations in use for the vector product are $[\mathbf{a}, \mathbf{b}]$ and $\mathbf{a}\wedge\mathbf{b}$. In preparation for the definition of $\mathbf{a}\times\mathbf{b}$ we now introduce a unit vector $\hat{\mathbf{n}}$ that is normal (i.e., orthogonal) to the plane defined by the vectors $\mathbf{a}$ and $\mathbf{b}$, and whose sense is such that $\mathbf{a}$, $\mathbf{b}$ and $\hat{\mathbf{n}}$, in this order, form a right-handed set of vectors. Here, although $\mathbf{a}$, $\mathbf{b}$ and $\hat{\mathbf{n}}$ are not necessarily mutually orthogonal, we use right-handedness exactly as was defined as the start of section 4.6.

Definition 4.14 The *vector product* of vectors $\mathbf{a}$ and $\mathbf{b}$ will be written $\mathbf{a} \times \mathbf{b}$ and is defined to be the vector quantity

$$\mathbf{a} \times \mathbf{b} = |\mathbf{a}|\,|\mathbf{b}|(\sin\theta)\hat{\mathbf{n}}$$

where θ is the angle between vectors $\mathbf{a}$ and $\mathbf{b}$ with $0 \le \theta \le \pi$, and $\hat{\mathbf{n}}$ is a unit vector normal to the plane of $\mathbf{a}$ and $\mathbf{b}$ such that $\mathbf{a}$, $\mathbf{b}$ and $\hat{\mathbf{n}}$, in this order, form a right-handed set of vectors.

This shows that the vector $\mathbf{a} \times \mathbf{b}$ is normal to both $\mathbf{a}$ and $\mathbf{b}$ and has magnitude $|\mathbf{a}||\mathbf{b}| \sin\theta$. The first interesting and unusual feature of this form of product is that it is not commutative. If $\mathbf{a}$, $\mathbf{b}$, $\hat{\mathbf{n}}$, in this order, form a right-handed set for the definition of $\mathbf{a} \times \mathbf{b}$, then for the definition of $\mathbf{b} \times \mathbf{a}$ it is necessary to take for the right-handed set the vectors $\mathbf{b}$, $\mathbf{a}$, $-\hat{\mathbf{n}}$, in the stated order. The immediate consequence is the important general result that if $\mathbf{a}$ and $\mathbf{b}$ are arbitrary vectors, then

$$\mathbf{a} \times \mathbf{b} = -(\mathbf{b} \times \mathbf{a}), \tag{4.41}$$

so the vector product is not commutative.

In contrast with the scalar product, it is easily seen that the vector product of parallel vectors is identically zero, whereas the vector product of orthogonal vectors is non-zero. A simple calculation gives Table 4.3 of vector products of the unit vectors $\mathbf{i}$, $\mathbf{j}$ and $\mathbf{k}$. The left-hand column identifies the first member of the vector product, and the top row identifies the second member of the vector product. The corresponding entry in the table gives the result of the vector product. The entries along the diagonal are all seen to be the zero or null vector.

If we take, for example, the first element in the left-hand column and the last element in the top row, we see that $\mathbf{i} \times \mathbf{k} = -\mathbf{j}$. In many respects it is easier to memorize these three results:

$$\mathbf{i} \times \mathbf{j} = \mathbf{k}, \quad \mathbf{j} \times \mathbf{k} = \mathbf{i}, \quad \mathbf{k} \times \mathbf{i} = \mathbf{j}, \tag{4.42}$$

and then the use property (4.41), than to remember Table 4.3 complete. The order of the vectors occurring in these key relations can be remembered by making the cyclic permutations

Table 4.3 Table of vector products of **i**, **j** and **k**

First member	*Second member*		
	i	**j**	**k**
i	**0**	**k**	$-\mathbf{j}$
j	$-\mathbf{k}$	**0**	**i**
k	**j**	$-\mathbf{i}$	**0**

$$\begin{array}{ccc} \mathbf{i} & \mathbf{j} & \mathbf{k} \\ \mathbf{j} & \mathbf{k} & \mathbf{i} \\ \mathbf{k} & \mathbf{i} & \mathbf{j} \end{array}$$

As with scalar products, this table of vector products may be used to calculate the vector product of any two vectors expressed in component form. Consider the vector product $\mathbf{a} \times \mathbf{b}$, where $\mathbf{a} = a_1\mathbf{i} + a_2\mathbf{j} + a_3\mathbf{k}$ and $\mathbf{b} = b_1\mathbf{i} + b_2\mathbf{j} + b_3\mathbf{k}$. Proceeding as though ordinary algebric quantities were involved, we write

$$\begin{aligned} \mathbf{a} \times \mathbf{b} &= (a_1\mathbf{i} + a_2\mathbf{j} + a_3\mathbf{k}) \times (b_1\mathbf{i} + b_2\mathbf{j} + b_3\mathbf{k}) \\ &= (a_1\mathbf{i}) \times (b_1\mathbf{i}) + (a_1\mathbf{i}) \times (b_2\mathbf{j}) + (a_1\mathbf{i}) \times (b_3\mathbf{k}) \\ &\quad + (a_2\mathbf{j}) \times (b_1\mathbf{i}) + (a_2\mathbf{j}) \times (b_2\mathbf{j}) + (a_2\mathbf{j}) \times (b_3\mathbf{k}) \\ &\quad + (a_3\mathbf{k}) \times (b_1\mathbf{i}) + (a_3\mathbf{k}) \times (b_2\mathbf{j}) + (a_3\mathbf{k}) \times (b_3\mathbf{k}), \end{aligned}$$

working on the assumption that vector multiplication is distributive over addition. Next we recognize that the scalars a_i, b_j may be taken out in front of each vector product that is involved, so that the expression becomes

$$\begin{aligned} \mathbf{a} \times \mathbf{b} = {} & a_1b_1\mathbf{i} \times \mathbf{i} + a_1b_2\mathbf{i} \times \mathbf{j} + a_1b_3\mathbf{i} \times \mathbf{k} + a_2b_1\mathbf{j} \times \mathbf{i} + a_2b_2\mathbf{j} \times \mathbf{j} \\ & + a_2b_3\mathbf{j} \times \mathbf{k} + a_3b_1\mathbf{k} \times \mathbf{i} + a_3b_2\mathbf{k} \times \mathbf{j} + a_3b_3\mathbf{k} \times \mathbf{k}. \end{aligned}$$

Finally, using Table 4.3 and collecting together the **i**, **j** and **k** terms, we obtain

$$\mathbf{a} \times \mathbf{b} = (a_2b_3 - a_3b_2)\mathbf{i} + (a_3b_1 - a_1b_3)\mathbf{j} + (a_1b_2 - a_2b_1)\mathbf{k}. \qquad (4.43)$$

This is often taken as definition of the vector product $\mathbf{a} \times \mathbf{b}$ in place of our Definition 4.14. Expression (4.43) may be considerably simplified if the concept of a determinant is used. Before showing this, we must digress slightly to define this term. A full account of determinants is given in section 9.3 which may be studied independently of the other material contained in Chapter 9.

Definition 4.15 Let a, b, c and d be any four numbers. Consider the two-row by two-column array of these numbers

$$\begin{array}{cc} a & b \\ c & d. \end{array} \qquad (A)$$

Define the expression

$$\begin{vmatrix} a & b \\ c & d \end{vmatrix} \qquad (B)$$

that is associated with this array by the identity

$$\begin{vmatrix} a & b \\ c & d \end{vmatrix} \equiv (ad - cb). \qquad (C)$$

We define the *second-order determinant* associated with the array (A) to be the number represented in symbols by (B) and having the value defined by (C). The process of expressing the left-hand side of (C) in the form of right-hand side is called *expanding* the determinant.

Example 4.26 Evaluate the second-order determinants

(a) $\begin{vmatrix} 1 & 7 \\ 3 & 9 \end{vmatrix}$;

(b) $\begin{vmatrix} 0 & -1 \\ 4 & 2 \end{vmatrix}$;

(c) $\begin{vmatrix} 2 & 6 \\ 1 & 3 \end{vmatrix}$.

Solution The values of the determinants follow directly from the definition:

(a) $\begin{vmatrix} 1 & 7 \\ 3 & 9 \end{vmatrix} \equiv (1)(9) - (3)(7) = 9 - 21 = -12;$

(b) $\begin{vmatrix} 0 & -1 \\ 4 & 2 \end{vmatrix} \equiv (0)(2) - (4)(-1) = 0 + 4 = 4;$

(c) $\begin{vmatrix} 2 & 6 \\ 1 & 3 \end{vmatrix} \equiv (2)(3) - (1)(6) = 6 - 6 = 0.$ ■

Definition 4.16 Let a_i, b_i and c_i, with $i = 1, 2, 3$, be any set of nine numbers. Consider the three-row by three-column array of these numbers

$$\begin{matrix} a_1 & a_2 & a_3 \\ b_1 & b_2 & b_3 \\ c_1 & c_2 & c_3. \end{matrix} \tag{D}$$

Define the expression

$$\begin{vmatrix} a_1 & a_2 & a_3 \\ b_1 & b_2 & b_3 \\ c_1 & c_2 & c_3 \end{vmatrix} \tag{E}$$

that is associated with this array to be the single number that is determined by the identity

$$\begin{vmatrix} a_1 & a_2 & a_3 \\ b_1 & b_2 & b_3 \\ c_1 & c_2 & c_3 \end{vmatrix} \equiv a_1 \begin{vmatrix} b_2 & b_3 \\ c_2 & c_3 \end{vmatrix} - a_2 \begin{vmatrix} b_1 & b_3 \\ c_1 & c_3 \end{vmatrix} + a_3 \begin{vmatrix} b_1 & b_2 \\ c_1 & c_2 \end{vmatrix}. \tag{F}$$

We define the *third-order determinant* associated with the array (D) to be the number represented in symbols by (E) and having the value defined by (F).

Example 4.27 Evaluate the third-order determinant

$$\Delta = \begin{vmatrix} 3 & -2 & -7 \\ 2 & 1 & 2 \\ 2 & 1 & 1 \end{vmatrix}.$$

Solution From definition,

$$\begin{vmatrix} 3 & -2 & -7 \\ 2 & 1 & 2 \\ 2 & 1 & 1 \end{vmatrix} = (3)\begin{vmatrix} 1 & 2 \\ 1 & 1 \end{vmatrix} - (-2)\begin{vmatrix} 2 & 2 \\ 2 & 1 \end{vmatrix} + (-7)\begin{vmatrix} 2 & 1 \\ 2 & 1 \end{vmatrix}.$$

Expanding the three second-order determinants and adding, we obtain the desired result

$$\Delta = 3(1-2) + 2(2-4) - 7(2-2) = -7. \quad \blacksquare$$

It is helpful to classify determinants in some simple way, which the next definition achieves.

Definition 4.17 We define the *order* of a determinant to be the number of terms that lie on a diagonal drawn from the top left-hand corner to the bottom right-hand corner. The values of these terms are immaterial.

Thus in Example 4.26 the determinants are second-order, whereas in Example 4.27 the determinant is third-order, and is evaluated in terms of three second-order determinants.

We are now able to give the promised alternative definition of a vector product, which is far more convenient to use.

Alternative Definition 4.18 We define the vector product $\mathbf{a} \times \mathbf{b}$ of the two vectors $\mathbf{a} = a_1\mathbf{i} + a_2\mathbf{j} + a_3\mathbf{k}$ and $\mathbf{b} = b_1\mathbf{i} + b_2\mathbf{j} + b_3\mathbf{k}$ to be the formal expansion of the determinant

$$\mathbf{a} \times \mathbf{b} \equiv \begin{vmatrix} \mathbf{i} & \mathbf{j} & \mathbf{k} \\ a_1 & a_2 & a_3 \\ b_1 & b_2 & b_3 \end{vmatrix}.$$

In this definition we have used the word ‘formal’ because, although the a_i and b_i are real numbers, the **i**, **j** and **k** are unit vectors. Aside from this, the expansion of the third-order determinant is performed exactly as in Example 4.27.

Example 4.28 Determine the vector product $\mathbf{a} \times \mathbf{b}$ where $\mathbf{a} = \mathbf{i} + \mathbf{j} - 2\mathbf{k}$ and $\mathbf{b} = -2\mathbf{i} + 3\mathbf{j} + \mathbf{k}$.

Solution To apply Definition 4.18 we first notice that the components a_1, a_2 and a_3 of **a** are 1, 1 and −2, while the components b_1, b_2 and b_3 of **b** are −2, 3 and 1. Hence

$$\mathbf{a}\times\mathbf{b}=\begin{vmatrix}\mathbf{i} & \mathbf{j} & \mathbf{k}\\ 1 & 1 & -2\\ -2 & 3 & 1\end{vmatrix}=\mathbf{i}\begin{vmatrix}1 & -2\\ 3 & 1\end{vmatrix}-\mathbf{j}\begin{vmatrix}1 & -2\\ -2 & 1\end{vmatrix}+\mathbf{k}\begin{vmatrix}1 & 1\\ -2 & 3\end{vmatrix}$$

and so

$$\mathbf{a}\times\mathbf{b}=7\mathbf{i}+3\mathbf{j}+5\mathbf{k}. \quad \blacksquare$$

This effectively demonstrates that for most practical purposes Definition 4.18 involves the least manipulation.

It is easily proved that the vector product is distributive, so that for any three vectors **a**, **b** and **c** we always have

$$\mathbf{a}\times(\mathbf{b}+\mathbf{c}) = \mathbf{a}\times\mathbf{b}+\mathbf{a}\times\mathbf{c}.$$

Indeed this is implied by the way in which Eqn (4.43) was derived.

With the introduction of the vector product, mixed products of the form $\mathbf{a}.(\mathbf{b}\times\mathbf{c})$ become possible. This type of product is known as a *triple scalar product*, and as it involves the scalar product of **a** with $(\mathbf{b}\times\mathbf{c})$ it is seen to be a scalar. If $\mathbf{a}=a_1\mathbf{i}+a_2\mathbf{j}+a_3\mathbf{k}$, $\mathbf{b}=b_1\mathbf{i}+b_2\mathbf{j}+b_3\mathbf{k}$ and $\mathbf{c}=c_1\mathbf{i}+c_2\mathbf{j}+c_3\mathbf{k}$, then, by combination of Definitions 4.12 and 4.18, we have

$$\mathbf{a}.(\mathbf{b}\times\mathbf{c})=(a_1\mathbf{i}+a_2\mathbf{j}+a_3\mathbf{k}).\begin{vmatrix}\mathbf{i} & \mathbf{j} & \mathbf{k}\\ b_1 & b_2 & b_3\\ c_1 & c_2 & c_3\end{vmatrix},$$

or

$$\mathbf{a}.(\mathbf{b}\times\mathbf{c})=a_1(b_2c_3-c_2b_3)-a_2(b_1c_3-c_1b_3)+a_3(b_1c_2-c_1b_2).$$

The terms on the right-hand side of this expression are the result of expanding (F) in Definition 4.16. When this result is recombined into a determinant we obtain the following general result that should always be used to determine a triple scalar product.

$$\mathbf{a}.(\mathbf{b}\times\mathbf{c})=\begin{vmatrix}a_1 & a_2 & a_3\\ b_1 & b_2 & b_3\\ c_1 & c_2 & c_3\end{vmatrix}. \tag{4.44}$$

By interchanging rows of the determinant it is readily shown that the dot. and the cross × in a triple scalar product may be interchanged so that

$$\mathbf{a}.(\mathbf{b}\times\mathbf{c})=(\mathbf{a}\times\mathbf{b}).\mathbf{c}. \tag{4.45}$$

Notice that the brackets in Eqns (4.44) and (4.45) are essential, because without them the expressions become meaningless. For example, $\mathbf{a}\times\mathbf{b}.\mathbf{c}$ could mean the vector product of **a** with the scalar **b.c**, which is not defined or the scalar product of $\mathbf{a}\times\mathbf{b}$ with **c**.

Example 4.29 Evaluate the triple scalar product $\mathbf{a}.(\mathbf{b} \times \mathbf{c})$, given that $\mathbf{a} = 2\mathbf{i}+\mathbf{k}$, $\mathbf{b} = \mathbf{i}+\mathbf{j}+2\mathbf{k}$ and $\mathbf{c} = -\mathbf{i}+\mathbf{j}$, and verify that $\mathbf{a}.(\mathbf{b} \times \mathbf{c}) = (\mathbf{a}.\mathbf{b}) \times \mathbf{c}$.

Solution The components of **a**, **b** and **c** are, respectively, (2, 0, 1), (1, 1, 2) and (–1, 1, 0). Hence

$$\mathbf{a}.(\mathbf{b} \times \mathbf{c}) = \begin{vmatrix} 2 & 0 & 1 \\ 1 & 1 & 2 \\ -1 & 1 & 0 \end{vmatrix} = 2.(-2) - 0.(2) + 1.(2) = -2.$$

Now

$$(\mathbf{a} \times \mathbf{b}).\mathbf{c} = \mathbf{c}.(\mathbf{a} \times \mathbf{b}) = \begin{vmatrix} -1 & 1 & 0 \\ 2 & 0 & 1 \\ 1 & 1 & 2 \end{vmatrix} = (-1).(-1) - 1.(3) + 0.2 = -2,$$

which after comparison with the previous result verifies that

$$\mathbf{a}.(\mathbf{b} \times \mathbf{c}) = (\mathbf{a} \times \mathbf{b}).\mathbf{c}. \quad \blacksquare$$

As our next generalization, we notice that vector products of more than two vectors are defined provided the order in which these products are to be carried out is specified by bracketing. As a special case we have the triple vector product $\mathbf{a} \times (\mathbf{b} \times \mathbf{c})$ of the three vectors **a**, **b** and **c** which differs from the triple vector product $(\mathbf{a} \times \mathbf{b}) \times \mathbf{c}$. The first expression signifies thevector product of **a** and $(\mathbf{b} \times \mathbf{c})$, while the second signifies the vector product of $(\mathbf{a} \times \mathbf{b})$ and **c**, and in general these are different vectors.

A straightforward application of Definition 4.18 establishes the following useful identity, from which some interesting results may be derived.

$$\mathbf{a} \times (\mathbf{b} \times \mathbf{c}) = (\mathbf{a}.\mathbf{c})\mathbf{b} - (\mathbf{a}.\mathbf{b})\mathbf{c}. \tag{4.46}$$

The details of the proof are left to the reader.

Example 4.30 Demonstrate the difference between the triple vector products $\mathbf{a} \times (\mathbf{b} \times \mathbf{c})$ and $(\mathbf{a} \times \mathbf{b}) \times \mathbf{c}$ by making the identifications $\mathbf{a} = \mathbf{i}$, $\mathbf{b} = \mathbf{i}+\mathbf{j}$, $\mathbf{c} = \mathbf{k}$.

Solution By direct substitution we find that $\mathbf{a} \times (\mathbf{b} \times \mathbf{c}) = \mathbf{i} \times [(\mathbf{i}+\mathbf{j})] \times \mathbf{k}]$ and so expanding this result by using Eqn (4.42) gives $\mathbf{a} \times (\mathbf{b} \times \mathbf{c}) = \mathbf{i} \times [-\mathbf{j}+\mathbf{i}] = -\mathbf{k}$. Similarly, in the second case, $(\mathbf{a} \times \mathbf{b}) \times \mathbf{c} = [\mathbf{i} \times (\mathbf{i}+\mathbf{j})] \times \mathbf{k} = \mathbf{k} \times \mathbf{k} = \mathbf{0}$. ■

4.8 Geometrical applications

This section illustrates something of the application of vectors to elementary geometry, and gives some simple but useful results. First we consider the representation of a straight line in vector form, and then show how the single vector equation may be reduced to the more familiar set of three Cartesian equations.

4.8.1 The straight line

Consider the problem of determining the equation of a straight line given that it passes through the point A with position vector $\mathbf{a}$ relative to O, and is parallel to vector $\mathbf{b}$. We shall denote the position vector of a general point P on the line by $\mathbf{r}$ as shown in Fig. 4.13.

By the rules of vector addition we have

$$OP = OA + AP$$

or

$$\mathbf{r} = \mathbf{a} + AP.$$

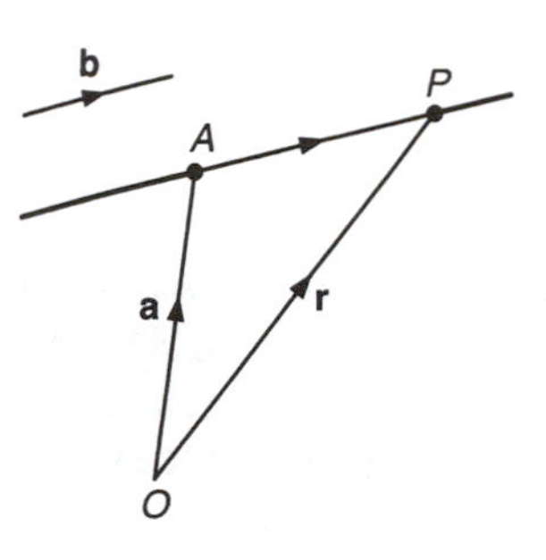

Fig. 4.13 Straight line through A parallel to $\mathbf{b}$.

However, as the straight line through A is parallel to the free vector $\mathbf{b}$, it follows that for any point P on the line there is a scalar λ such that we can write $AP = \lambda\mathbf{b}$. Applying this result to the equation above, we see that the vector equation for the straight line becomes

$$\mathbf{r} = \mathbf{a} + \lambda\mathbf{b}. \tag{4.47}$$

The scalar λ in this equation is simply a parameter, and different values of λ will determine different points on the line. To express this result in Cartesian form, set $\mathbf{r} = x\mathbf{i}+y\mathbf{j}+z\mathbf{k}$, $\mathbf{a} = a_1\mathbf{i}+a_2\mathbf{j}+a_3\mathbf{k}$ and $\mathbf{b} = b_1\mathbf{i}+b_2\mathbf{j}+b_3\mathbf{k}$, so that Eqn (4.47) becomes

$$x\mathbf{i}+y\mathbf{j}+z\mathbf{k} = a_1\mathbf{i} + a_2\mathbf{j} + a_3\mathbf{k} + \lambda(b_1\mathbf{i} + b_2\mathbf{j} + b_3\mathbf{k}).$$

This vector equation implies three scalar equations by virtue of the equality of its **i**-, **j**- and **k**-components. Hence we arrive at the three scalar equations

$$\begin{aligned} x &= a_1 + \lambda b_1 && (\mathbf{i}\text{-component}) \\ y &= a_2 + \lambda b_2 && (\mathbf{j}\text{-component}) \\ z &= a_3 + \lambda b_3 && (\mathbf{k}\text{-component}). \end{aligned}$$

If these are each solved for λ and equated, we obtain the more familiar result

$$\frac{x - a_1}{b_1} = \frac{y - a_2}{b_2} = \frac{z - a_3}{b_3} = \lambda. \tag{4.48}$$

Equations (4.48) are the *standard Cartesian form* for the equations of a straight line. Notice that the coefficients of x, y and z in Eqn (4.48) are all unity; that b_1, b_2 and b_3 are then the direction ratios of $\mathbf{b}$ and a_1, a_2 and a_3 define a point on the line. Equations (4.48) are sometimes expressed in the form of three simultaneous equations relating x and y, x and z and y and

z. This follows by cross-multiplying different pairs of expressions in Eqn (4.48).

Example 4.31 Find the vector equation of the line through the point with position vector $\mathbf{i}+3\mathbf{j}-\mathbf{k}$ which is parallel to the vector $2\mathbf{i}+3\mathbf{j}+4\mathbf{k}$. Determine the point on the line corresponding to $\lambda=2$ in the resulting equation. Also express the vector equation of the line in standard Cartesian form.

Solution From Eqn (4.47) we have

$$\mathbf{r}=(\mathbf{i}+3\mathbf{j}-\mathbf{k})+\lambda(2\mathbf{i}+3\mathbf{j}+4\mathbf{k}),$$

or

$$\mathbf{r}=(1+2\lambda)\mathbf{i}+3(1+\lambda)\mathbf{j}+(4\lambda-1)\mathbf{k}.$$

This is the vector equation of the line, and setting $\lambda=2$ determines the point $\mathbf{r}=5\mathbf{i}+9\mathbf{j}+7\mathbf{k}$. To express the equation of the line in Cartesian form we appeal to Eqns (4.48) and use the fact that $\mathbf{a}=\mathbf{i}+3\mathbf{j}-\mathbf{k}$ and $\mathbf{b}=2\mathbf{i}+3\mathbf{j}+4\mathbf{k}$. Hence $a_1=1$, $a_2=3$, $a_3=-1$ and $b_1=2$, $b_2=3$, $b_3=4$, so that the desired Cartesian equations are

$$\frac{x-1}{2}=\frac{y-3}{3}=\frac{z+1}{4}=\lambda.$$

As a check, we can also use these equations to determine the point corresponding to $\lambda=2$. We must solve the three equations

$$\frac{x-1}{2}=2, \qquad \frac{y-3}{3}=\frac{z+1}{4}=2,$$

which give $x=5$, $y=9$ and $z=7$. These are of course the coordinates of the tip of the position vector $\mathbf{r}=5\mathbf{i}+9\mathbf{j}+7\mathbf{k}$ which confirms our previous results. ■

The same approach may be used if the line is required to pass through the two points A and B with position vectors $\boldsymbol{\alpha}$ and $\boldsymbol{\beta}$, respectively. For then the line passes through $\boldsymbol{\alpha}$ and is parallel to the vector $\boldsymbol{\beta}-\boldsymbol{\alpha}$ which is just a segment of the line itself. Hence we identify $\mathbf{a}$ and $\boldsymbol{\alpha}$ and $\boldsymbol{b}$ with $\boldsymbol{\beta}-\boldsymbol{\alpha}$, after which the argument proceeds as before.

In the next example we illustrate how the non-standard Cartesian equations of a straight line may be reinterpreted in vector form.

Example 4.32 The equations

$$\frac{2x-1}{3}=\frac{y+2}{3}=\frac{-z+4}{2}$$

determine a straight line. Express them in vector form and find the direction ratios of the line.

Solution To express the equations in standard Cartesian form we must first make the coefficients of x, y and z each equal to unity. Hence we rewrite the equations:

$$\frac{x-\frac{1}{2}}{\frac{3}{2}}=\frac{y+2}{3}=\frac{z-4}{-2}.$$

The vector **a** then has components $a_1=\frac{1}{2}$, $a_2=-2$, $a_3=4$ and the vector **b** has the components $b_1=\frac{3}{2}=3$, $b_3=-2$. The latter three numbers are the desired direction ratios. The vector equation of the straight line itself follows by substituting $\mathbf{a}=\frac{1}{2}\mathbf{i}-2\mathbf{j}+4\mathbf{k}$ and $\mathbf{b}=\frac{3}{2}\mathbf{i}+3\mathbf{j}-2\mathbf{k}$ into Eqn (4.47) to obtain

$$\mathbf{r}=\tfrac{1}{2}(1+3\lambda)\mathbf{i}+(3\lambda-2)\mathbf{j}+2(2-\lambda)\mathbf{k}. \quad \blacksquare$$

On occasion it is necessary to determine the perpendicular distance p from a point C with position vector **c** to the line L with equation $\mathbf{r}=\mathbf{a}+\lambda\mathbf{b}$. This can be done by applying Pythagoras' theorem in Fig. 4.14.

We have the obvious result

$$p^2=(AC)^2-(AB)^2$$

but $AC=\mathbf{c}-\mathbf{a}$ so that $(AC)^2=|\mathrm{AC}|^2=(\mathbf{c}-\mathbf{a}).(\mathbf{c}-\mathbf{a})$, while length AB is the projection of AC onto the line L. Now the unit vector along L is $\mathbf{b}/|\mathbf{b}|$, so that $AB=(\mathbf{c}-\mathbf{a}).\mathbf{b}/|\mathbf{b}|$ and thus

$$(AB)^2=\left(\frac{(\mathbf{c}-\mathbf{a}).\mathbf{b}}{|\mathbf{b}|}\right)^2.$$

Combining these results gives

$$p^2=(\mathbf{c}-\mathbf{a}).(\mathbf{c}-\mathbf{a})-\left(\frac{(\mathbf{c}-\mathbf{a}).\mathbf{b}}{|\mathbf{b}|}\right)^2, \tag{4.49}$$

from which p may be deduced.

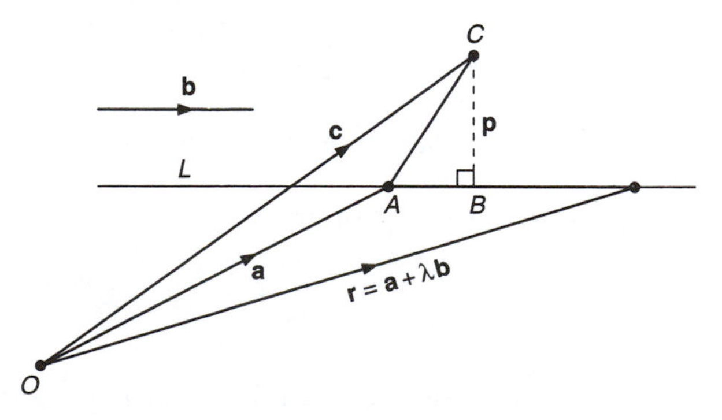

Fig. 4.14 Perpendicular distance of point from line.

Example 4.33 Find the distance of the point with position vector $\mathbf{i}+\mathbf{j}+\mathbf{k}$ from the line $\mathbf{r}=(\mathbf{i}+2\mathbf{j}+\mathbf{k})+\lambda(\mathbf{i}-2\mathbf{j}+\mathbf{k})$.

Solution In the notation leading to Eqn (4.49) we have $\mathbf{a}=\mathbf{i}+2\mathbf{j}+\mathbf{k}$, $\mathbf{b}=\mathbf{i}-2\mathbf{j}+\mathbf{k}$ and $\mathbf{c}=\mathbf{i}+\mathbf{j}+\mathbf{k}$. Hence $\mathbf{c}-\mathbf{a}=-\mathbf{j}$ and thus $(\mathbf{c}-\mathbf{a}).(\mathbf{c}-\mathbf{a})=(-\mathbf{j}).(-\mathbf{j})=1$. Also $(\mathbf{c}-\mathbf{a}).\mathbf{b}=-\mathbf{j}.(\mathbf{i}-2\mathbf{j}+\mathbf{k})=2$ so that $[(\mathbf{c}-\mathbf{a}).\mathbf{b}]^2=4$, while $|\mathbf{b}|^2=6$. Hence

$$\left(\frac{(\mathbf{c}-\mathbf{a}).\mathbf{b}}{|\mathbf{b}|}\right)^2=\frac{4}{6}=\frac{2}{3},$$

and so from Eqn (4.49), $p^2=1-\frac{2}{3}=\frac{1}{3}$ or $p=1/\sqrt{3}$ as p is essentially positive. ■

4.8.2 The plane

The equation of a plane is easily determined once it is recognized that a plane Π is specified when one point on it is known, together with any vector perpendicular to it. Such a vector, when normalized, is a *unit normal* to the plane Π and is unique except for its sign. The ambiguity as to the sign of the normal is, of course, because a plane has no preferred side. To derive its equation, consider Fig. 4.15.

Let $\mathbf{r}$ be the position vector relative to O of a point P on the plane Π, and $\mathbf{n}$ be a vector normal to the plane directed through the plane away from O so that the corresponding unit normal is $\hat{\mathbf{n}}=\mathbf{n}/|\mathbf{n}|$. Further, let the perpendicular distance ON from their origin O to the plane be p. Then for all points P we have $(OP)\cos\theta=p$, where $\angle NOP=\theta$. In terms of vectors this is

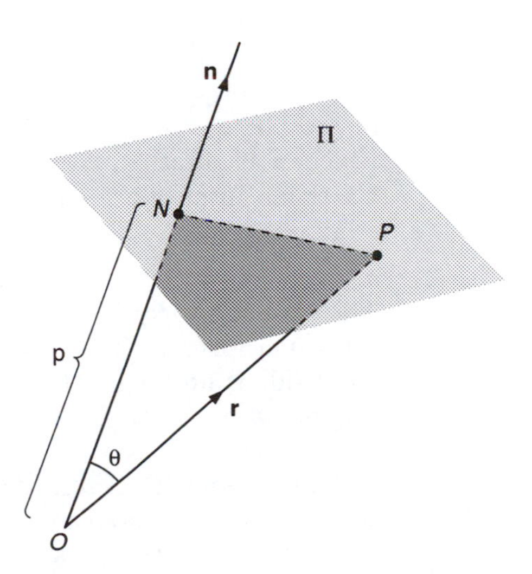

Fig. 4.15 Vector equation of a plane $\mathbf{n}.\mathbf{r}=|\mathbf{n}|p$.

$$\frac{\mathbf{r}.\mathbf{n}}{|\mathbf{n}|} = p, \tag{4.50}$$

which is just the vector equation of a plane. If the number p in Eqn (4.50) is positive then the plane lies on the side of the origin towards which $\mathbf{n}$ is directed, otherwise it lies on the opposite side.

To express result (4.50) in Cartesian form let $\mathbf{r} = x\mathbf{i} + y\mathbf{j} + z\mathbf{k}$ and the unit normal $\hat{\mathbf{n}} = \mathbf{n}/|\mathbf{n}| = l\mathbf{i} + m\mathbf{j} + n\mathbf{k}$, where of course $l^2 + m^2 + n^2 = 1$. Equation (4.50) becomes

$$lx + my + nz = p. \tag{4.51}$$

This is the *standard Cartesian form of the equation of a plane* in which $l^2 + m^2 + n^2 = 1$. Any equation of this form represents a plane having for its unit normal the vector $l\mathbf{i} + m\mathbf{j} + n\mathbf{k}$ and lying at a perpendicular distance p from the origin. If $p = 0$ the plane passes through the origin.

Instead of specifying a plane by giving $\mathbf{n}$ and the perpendicular distance p of the plane from the origin, it may also be specified by giving three points $\mathbf{a}$, $\mathbf{b}$, $\mathbf{c}$ on the plane that do not all lie on a straight line. As $\mathbf{a} - \mathbf{b}$ and $\mathbf{a} - \mathbf{c}$ lie in the plane, $\mathbf{n} = (\mathbf{a} - \mathbf{b}) \times (\mathbf{a} - \mathbf{c})$ is normal to the plane. We have $p = \mathbf{a}.\hat{\mathbf{n}}$, after which the argument proceeds as before. Of course, p is also given by $p = \mathbf{b}.\hat{\mathbf{n}} = \mathbf{c}.\hat{\mathbf{n}}$.

Example 4.34 Find the Cartesian equation of the plane containing the point (1, 2, 3) which is normal to the vector $\mathbf{i} + 2\mathbf{j} + 2\mathbf{k}$.

Solution First we use Eqn (4.50) to determine p. Since the point (1, 2, 3) lies in the plane, $\mathbf{r} = \mathbf{i} + 2\mathbf{j} + 3\mathbf{k}$ is the position vector of a point in the plane. The vector normal to the plane in this case is $\mathbf{n} = \mathbf{i} + 2\mathbf{j} + 2\mathbf{k}$, so that $|\mathbf{n}| = 3$ and the unit normal $\hat{\mathbf{n}} = \mathbf{n}/|\mathbf{n}| = (\mathbf{i} + 2\mathbf{j} + 2\mathbf{k})/3$. This shows that $l = \frac{1}{3}$, $m = \frac{2}{3}$, $n = \frac{2}{3}$. Hence, substituting into Eqn (4.50),

$$p = \frac{(\mathbf{i} + 2\mathbf{j} + 3\mathbf{k}).(\mathbf{i} + 2\mathbf{j} + 2\mathbf{k})}{3},$$

or $p = 11/3$. As $p > 0$, the plane must lie on the side of the origin towards which $\hat{\mathbf{n}}$ is directed. Substituting in Eqn (4.51), we find the desired Cartesian form of the equation of the plane:

$$\tfrac{1}{3}x + \tfrac{2}{3}y + \tfrac{2}{3}z = \tfrac{11}{3}.$$

This equation could equally well be written in the non-standard Cartesian form $x + 2y + 2z = 11$, though then the constant on the right-hand side is no longer the perpendicular distance of the plane from the origin. ■

Simple geometrical considerations similar to those set out above, when coupled with the scalar and vector product, enable various useful results to be derived very quickly. For example, as the angle θ between two planes is defined to be the angle between their unit normals $\hat{\mathbf{n}}_1$ and $\hat{\mathbf{n}}_2$

it follows that θ may be obtained from the scalar product $\hat{\mathbf{n}}_1.\hat{\mathbf{n}}_2 = \cos\theta$. Also the line of intersection of these two planes is perpendicular to both normals $\hat{\mathbf{n}}_1$ and $\hat{\mathbf{n}}_2$ and so is parallel to the vector $\mathbf{t}$ determined by the vector product $\mathbf{t} = \hat{\mathbf{n}}_1 \times \hat{\mathbf{n}}_2$. Rather than elaborate on these ideas here, a number of problems are given at the end of the chapter.

4.8.3 The sphere

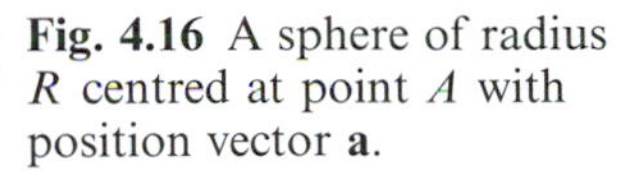

Fig. 4.16 A sphere of radius R centred at point A with position vector **a**.

Consider a sphere of radius R with its centre at the point A with the position vector **a** (Fig. 4.16). Then if **r** is the position vector of any point on the surface of the sphere, the modulus of the vector $\mathbf{r} - \mathbf{a}$ must equal R. In terms of vectors the equation of the sphere is

$$|\mathbf{r} - \mathbf{a}| = R$$

or, alternatively,

$$(\mathbf{r} - \mathbf{a}).(\mathbf{r} - \mathbf{a}) = R^2.$$

If, now, we expand this equation to get

$$\mathbf{r}.\mathbf{r} - 2\mathbf{r}.\mathbf{a} = R^2 - \mathbf{a}.\mathbf{a},$$

and then set $\mathbf{r} = x\mathbf{i} + y\mathbf{j} + z\mathbf{k}$, $\mathbf{a} = a_1\mathbf{i} + a_2\mathbf{j} + a_3\mathbf{k}$ and $R^2 - \mathbf{a}.\mathbf{a} = q$, we obtain the standard Cartesian form of the equation of a sphere

$$x^2 + y^2 + z^2 - 2a_1x - 2a_2y - 2a_3z = q. \tag{4.53}$$

Example 4.35 Find the Cartesian form of equation of the sphere of radius 2 having its centre at $\mathbf{a} = \mathbf{i} + \mathbf{j} + 2\mathbf{k}$.

Solution As $\mathbf{r} = x\mathbf{i} + y\mathbf{j} + z\mathbf{k}$ and $\mathbf{a} = \mathbf{i} + \mathbf{j} + 2\mathbf{k}$ we have $\mathbf{r} - \mathbf{a} = (x-1)\mathbf{i} + (y-1)\mathbf{j} + (z-2)\mathbf{k}$, while $R = 2$. Hence Eqn (4.52) becomes

$$(x-1)^2 + (y-1)^2 + (z-2)^2 = 4,$$

which is the desired Cartesian form of the equation. ■

4.9 Applications to mechanics

This section briefly introduces some of the many situations in mechanics that are best described vectorially. First is one of the simplest applications of vectors, that may already be familiar to the reader.

4.9.1 Polygon of forces – resultant

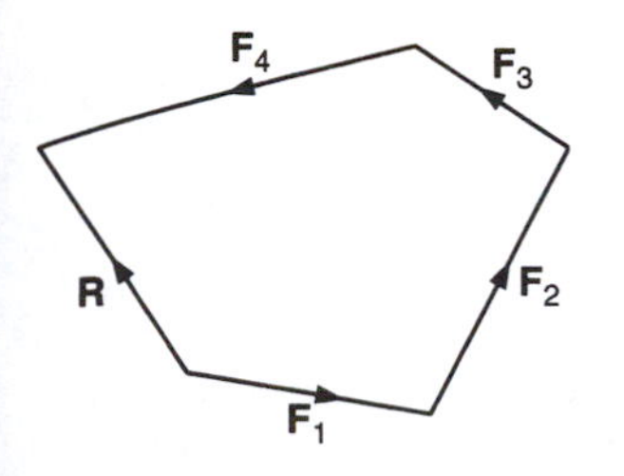

Fig. 4.17 Vector polygon.

It is known from experiment that when forces $\mathbf{F}_1, \mathbf{F}_2, \ldots, \mathbf{F}_n$ act on a rigid body through a single point O, their combined effect is equivalent to that of a single force **R**, their *resultant*, which acts through the same point O and is equal to their vector sum. Such a system of forces acting through a single point is a *concurrent* system of forces. Thus we have

$$\mathbf{R} = \mathbf{F}_1 + \mathbf{F}_2 + \ldots + \mathbf{F}_n. \tag{4.54}$$

These forces are often represented in the form of a *vector polygon* of forces as shown in Fig. 4.17, in which the senses of the forces $\mathbf{F}_i$ are all similarly directed and are opposite to the sense of $\mathbf{R}$.

Conversely, the vector polygon shows that the vector $-\mathbf{R}$ is the additional force that is required to act through O in order to maintain the system for forces in equilibrium.

Example 4.36 Forces $\mathbf{F}_1$, $\mathbf{F}_2$ and $\mathbf{F}_3$ have magnitudes $3\sqrt{3}$, $\sqrt{14}$ and $2\sqrt{6}$N and act concurrently through a point O along the lines of the vector $\mathbf{i}+\mathbf{j}+\mathbf{k}$, $3\mathbf{i}-\mathbf{j}+2\mathbf{k}$ and $-\mathbf{i}+2\mathbf{j}+\mathbf{k}$, respectively. Find force $\mathbf{Q}$ that must act through O for the system to remain in equilibrium.

Solution This is a direct application of the last remark about the vector polygon of forces, and the only problem is one of scaling. Let us agree that a vector of unit modulus represents a force of 1 N. From the conditions of the question we see that $\mathbf{F}_1$, $\mathbf{F}_2$ and $\mathbf{F}_3$ are respectively directed along the unit vectors

$$\hat{\mathbf{f}}_1 = \frac{1}{\sqrt{3}}(\mathbf{i}+\mathbf{j}+\mathbf{k}),$$

$$\hat{\mathbf{f}}_2 = \frac{1}{\sqrt{14}}(3\mathbf{i}-\mathbf{j}+2\mathbf{k}),$$

$$\hat{\mathbf{f}}_3 = \frac{1}{\sqrt{6}}(-\mathbf{i}+2\mathbf{j}+\mathbf{k}).$$

Using the scale factor we can use these to write

$$\mathbf{F}_1 = 3\sqrt{3}\hat{\mathbf{f}}_1 = 3\mathbf{i}+3\mathbf{j}+3\mathbf{k},$$

$$\mathbf{F}_2 = \sqrt{14}\hat{\mathbf{f}}_2 = 3\mathbf{i}-\mathbf{j}+2\mathbf{k},$$

$$\mathbf{F}_3 = 2\sqrt{6}\hat{\mathbf{f}}_3 = -2\mathbf{i}+4\mathbf{j}+2\mathbf{k}.$$

Hence the resultant $\mathbf{R}=\mathbf{F}_1+\mathbf{F}_2+\mathbf{F}_3=4\mathbf{i}+6\mathbf{j}+7\mathbf{k}$. The force necessary for equilibrium is $\mathbf{Q}=-\mathbf{R}$, showing that $\mathbf{Q}=-4\mathbf{i}-6\mathbf{j}-7\mathbf{k}$.

As $|\mathbf{Q}|=\sqrt{101}$, it follows immediately that the desired force is $\sqrt{101}$ N and acts in the direction of the unit vector $\hat{\mathbf{q}}$, where

$$\hat{\mathbf{q}} = \frac{-1}{\sqrt{101}}(4\mathbf{i}+6\mathbf{j}+7\mathbf{k}). \quad \blacksquare$$

In many problems of statics the centroid or the centre of mass of a system of particles is of importance. We now define this concept in terms of vectors.

Definition 4.19 The *centre of mass*, or *centroid*, of the system of masses $m_1, m_2, \ldots, m_n$ whose position vectors are $\mathbf{a}_1, \mathbf{a}_2, \ldots, \mathbf{a}_n$ is at the point G, where G has the position vector $\mathbf{g}$ determined by

$$\mathbf{g} = \frac{m_1\mathbf{a}_1 + m_2\mathbf{a}_2 + \ldots + m_n\mathbf{a}_n}{m_1 + m_2 + \ldots + m_n}.$$

Next we discuss simple problems about relative motions and relative velocity.

4.9.2 Relative velocity

Problems involving the motion of one point relative to another, which is itself moving, occur frequently in mechanics and easily lend themselves to vector treatment. They are best illustrated by example, but first we define relative velocity.

Definition 4.20 The relative velocity of a point P with velocity $\mathbf{u}$, relative to the point Q with velocity $\mathbf{v}$, is defined to be the velocity $\mathbf{u} - \mathbf{v}$.

Example 4.37 A man walks due east at 4 km/h and his dog runs north-east at 12 km/h. Find the velocity and speed of the man relative to his dog.

Solution Let a unit vector denote a velocity of magnitude 1 km/h and take $\mathbf{j}$ pointing due north and $\mathbf{i}$ pointing due east.

Unit vectors in the directions of motion of the man and dog are then $\mathbf{i}$ and $(\mathbf{i}+\mathbf{j})/\sqrt{2}$. The velocity $\mathbf{u}$ of the man is thus $\mathbf{u} = 4\mathbf{i}$, and the velocity $\mathbf{v}$ of the dog is $\mathbf{v} = 6\sqrt{2}(\mathbf{i}+\mathbf{j})$. Hence the velocity of the man relative to his dog is

$$\mathbf{u} - \mathbf{v} = 2(2 - 3\sqrt{2})\mathbf{i} - 6\sqrt{}\mathbf{j}.$$

His relative speed is $|\mathbf{u} - \mathbf{v}| = (160 - 48\sqrt{2})^{1/2}$ km/h.

4.9.3 Work done by a force

The scalar product can be used to give a convenient representation of the work W done by a force $\mathbf{F}$ that produces a displacement d of the particle on which it acts. The *work* done by a force of magnitude $|\mathbf{F}|$ when it displaces a particle through a distance $|\mathbf{d}|$ is defined as the product of the distance moved and the component of force in the direction of the displacement.

Hence we have

$$W = |\mathbf{F}||\mathbf{d}|\cos\theta,$$

where θ is the angle of inclination between **F** and **d**. So the final result is:

$$W = \mathbf{F}.\mathbf{d}. \tag{4.55}$$

Example 4.38 Calculate the work W done by a force **F** of 12 N whose line of action is parallel to $2\mathbf{i}+3\mathbf{j}-2\mathbf{k}$ when it moves its point of application through a displacement **d** of 4 m in a direction parallel to $-2\mathbf{i}+\mathbf{j}-3\mathbf{k}$.

Solution The unit vectors parallel of the force **F** and displacement **d** are $\hat{\mathbf{f}} = (2\mathbf{i} + 3\mathbf{j} - 2\mathbf{k})/\sqrt{17}$ and $\hat{\mathbf{d}} = (-2\mathbf{i} + \mathbf{j} - 3\mathbf{k})/\sqrt{14}$, respectively. Let $\hat{\mathbf{f}}$ denote a force of 1 N and $\hat{\mathbf{d}}$ a displacement of 1 m so that $\mathbf{F} = 12\hat{\mathbf{f}} = (24\mathbf{i} + 36\mathbf{j} - 24\mathbf{k})/\sqrt{17}$ and $\mathbf{d} = 4\hat{\mathbf{d}} = (-8\mathbf{i} + 4\mathbf{j} - 12\mathbf{k})/\sqrt{14}$. Then the work W that is done is

$$\begin{aligned} W &= \mathbf{F}.\mathbf{d} \ \text{Nm} \\ &= [(24)(-8) + (36)(4) + (-24)(-12)]/\sqrt{17}.\sqrt{14} \\ &= 240/\sqrt{238} \ \text{N m}. \quad \blacksquare \end{aligned}$$

We now turn to applications of the vector product. One of the easiest occurs in the determination of the angular velocity of a point rotating about a fixed axis.

4.9.4 Angular velocity

Consider a rigid body rotating with a constant spin Ω rad/s about a fixed axis L. Fig. 4.18 represents a point P in such a body, having the position vector **d** relative to a point O on the spin axis L. Point Q is the foot of the perpendicular from P to the line L.

The vector $\boldsymbol{\Omega}$ parallel to L with magnitude Ω and sense determined by a right-hand screw rule with respect to L and the direction of the spin Ω is called the *angular velocity* of the body. The *instantaneous linear velocity* **v** of point P with position vector **d** is obviously $\Omega.(QP)$ in a direction tangent to the dotted circle in Fig. 4.18. It is easily seen that we may rewrite this as

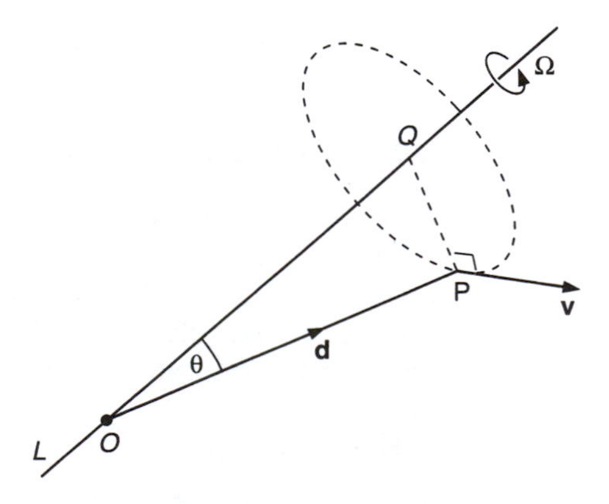

Fig. 4.18 Angular velocity.

$$|\mathbf{v}| = |\boldsymbol{\Omega}||\mathbf{d}|\sin\theta,$$

and in terms of the vector product the result follows from

$$\mathbf{v} = \boldsymbol{\Omega} \times \mathbf{d}. \tag{4.56}$$

The final two applications of the vector product involve the concept of the moment of a vector which is the first defined, and they require the use of a bound vector.

Definition 4.20 We define $\mathbf{M} = \mathbf{d} \times \mathbf{Q}$ to be the *moment of vector* $\mathbf{Q}$ about the point O, where $\mathbf{d}$ is the position vector relative to O of any point on the line of action of the bound vector $\mathbf{Q}$.

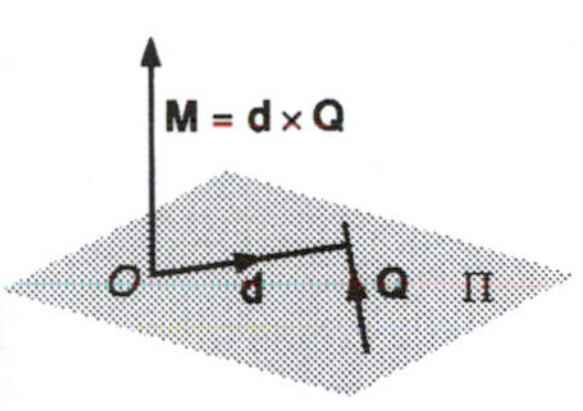

Fig. 4.19 Moment of a vector Q in plane Π about a point O in the plane, with $\mathbf{d}$ in plane Π the position vector of any point on the line of action of Q relative to O.

This definition is illustrated in Fig. 4.19 in which the plane Π contains the vectors $\mathbf{d}$ and $\mathbf{Q}$ and, by virtue of the definition of the moment, $\mathbf{M}$ is normal to Π.

The natural mechanical applications of this definition are to the moment of a force and to the moment of momentum about a fixed point. In both situations the line of action of the vector whose moment is to be found is important, as is its point of application in some circumstances.

If $\mathbf{Q}$ is identified with a force $\mathbf{F}$, then the expression

$$\mathbf{M} = \mathbf{d} \times \mathbf{F} \tag{4.57}$$

is the moment or *torque* of the force $\mathbf{F}$ about O. If the force is expressed in newtons and the displacement vector in metres, the units of torque are newton-metres. Similarly, if $\mathbf{Q}$ is identified with the *momentum* $m\mathbf{v}$ of a particle of mass m moving with velocity $\mathbf{v}$, then the vector

$$\mathbf{M} = \mathbf{d} \times (m\mathbf{v}) \tag{4.58}$$
$$= m\mathbf{d} \times \mathbf{v}$$

is the *moment of momentum* or the *angular momentum* of the particle about O.

Problems

Section 4.1

4.1 By setting $x^2 = w$, reduce the following quartic equations to quadratic equations, and hence obtain their roots:

(a) $x^4 + x^2 - 2 = 0$; (b) $x^4 + 5x^2 + 6 = 0$; (c) $x^4 - 5x^2 + 6 = 0$.

4.2 Find the real and imaginary parts of each of these complex numbers:

(a) $z = 9 - 6i$; (b) $z = 32$; (c) $z = 14 + 2i$; (d) $z = 17i$;

(e) $z = -3 + i$.

4.3 Write the following numbers in real–imaginary form, given that their real and imaginary parts are:

(a) $\text{Re}\, z = -11$, $\text{Im}\, z = 1$; (b) $\text{Re}\, z = 0$, $\text{Im}\, z = -3$;

(c) $\text{Re}\, z = 0$, $\text{Im}\, z = 0$; (d) $\text{Re}\, z = 4$, $\text{Im}\, z = 17$.

Section 4.2

4.4 Given that the following complex numbers are equal, deduce the values of a and b;

(a) $2 - 3i = 2 + ib$; (b) $a + 4i = 1 + ib$;

(c) $3 + 7i = a + ib$; (d) $5 + ia = b + 6i$.

4.5 Form the sums $z_1 + z_2$, given that:

(a) $z_1 = 3 - i$, $z_2 = 4 + 7i$;

(b) $z_1 = -2 - 4i$, $z_2 = 2 + 3i$;

(c) $z_1 = 5 + 6i$, $z_2 = -5 - 6i$;

(d) $z_1 = 4 - 3i$, $z_2 = 2 + 3i$.

4.6 Form the differences $z_1 - z_2$, given that:

(a) $z_1 = 2 + 6i$, $z_2 = 4 + 2i$;

(b) $z_1 = -2 + i$, $z_2 = -2 + 2i$;

(c) $z_1 = 4 + 7i$, $z_2 = 2 + 7i$;

(d) $z_1 = 3i$, $z_2 = 1 + 3i$.

4.7 Form the products $z_1 z_2$, given that:

(a) $z_1 = 1 + i$, $z_2 = 2 + 3i$;

(b) $z_1 = 3 - 5i$, $z_2 = 2 + 5i$;

(c) $z_1 = i$, $z_2 = 4 - 3i$;

(d) $z_1 = 2$, $z_2 = 9 - i$.

4.8 Evaluate $(1 + i)^5 - (1 - i)^5$.

4.9 Evaluate $\dfrac{1}{(1 + i)^4} + \dfrac{1}{(1 - i)^4}$.

4.10 Solve these equations for z:

(a) $2z(3 + i) = 2 + 3i$; (b) $3z(1 - 2i) = 1 + 4i$;

(c) $4z(1 - i) = 1 + i$; (d) $2z(4 + i) = 1 + 4i$.

4.11 Form the quotients z_1/z_2, given that:

(a) $z_1 = 3 + 2i$, $z_2 = 1 - i$; (b) $z_1 = 9 + 3i$, $z_2 = 3 + i$;

(c) $z_1 = 8 + 4i$, $z_2 = 2 - 4i$.

4.12 State which of the following polynomials have at least one real root and which, if they have complex roots, will have them occur in complex conjugate pairs. If no deductions can be made about the nature of the roots, then say so.

(a) $P(z) \equiv z^5 + 16z^4 + z^2 + 3z + 1;$

(b) $P(z) \equiv z^4 + 3z^3 + 2z^2 + 1;$

(c) $P(z) \equiv z^7 + 5z^5 - 2z^2 + z + i;$

(d) $P(z) \equiv z^3 - 6z^2 + 2z + 4.$

4.13 Given that $z = 2 + 3i$ is a root of the polynomial

$$P(z) \equiv z^4 - 4z^3 + 12z^2 + 4z - 13,$$

deduce the values of the other three roots. Factorize $P(z)$ into linear and quadratic factors with real coefficients.

4.14 Given that $z = i$ is a root of the polynomial

$$P(z) \equiv z^5 - 2z^4 + 10z^3 - 20z^2 + 9z - 18,$$

deduce the values of the other four roots. Factorize $P(z)$ into linear and quadratic factors with real coefficients.

Section 4.3

4.15 Plot the following vectors z_1 and z_2 in the complex plane and use geometrical methods to form their sum $z_1 + z_2$ and their difference $z_1 - z_2$:

(a) $z_1 = 2 + 3i, z_2 = -1 + 2i;$ (b) $z_1 = 3, z_2 = 4 - i;$
(c) $z_1 = 4i, z_2 = 3 - 4i;$ (d) $z_1 = -1 - 2i, z_2 = -1 + 2i.$

4.16 Find the modulus of each of these vectors:

(a) $4 - 3i;$ (b) $-2 + 3i;$ (c) $2 - 3i;$ (d) $3 + 4i;$ (e) $5i.$

4.17 Use the properties of the complex conjugate operation to prove that for any two complex numbers z_1 and z_2,

$$z_1\bar{z}_2 + \bar{z}_1 z_2 = 2 \text{ Re } z_1\bar{z}_2.$$

Then, using this result, together with the obvious inequality

$$|\text{Re}\, z_1\bar{z}_2| \leq |z_1\bar{z}_2|$$

and the identity

$$|z_1 + z_2|^2 = (z_1 + z_2)(\overline{z_1 + z_2}),$$

prove the triangle inequality,

$$|z_1 + z_2| \leq |z_1| + |z_2|.$$

4.18 Use the same form of argument as in Problem 4.17, together with the obvious inequality Re $z_1\bar{z}_2 \geq -|z_1\bar{z}_2|$, to prove the inequality

$$||z_1| - |z_2|| \leq |z_1 - z_2|.$$

Section 4.4

4.19 Express these numbers in modulus–argument form;

(a) $z = -3 + 4i$; (b) $z = -3 - 4i$;

(c) $z = -3 + 3i$; (d) $z = 2\sqrt{3} - 2i$.

4.20 Express the following numbers z in real–imaginary form, given that:

(a) $|z| = 4$, $\arg z = \dfrac{\pi}{3}$; (b) $|z| = 2$, $\arg z = \dfrac{-\pi}{6}$;

(c) $|z| = 6$, $\arg z = \dfrac{3\pi}{2}$; (d) $|z| = 3$, $\arg z = \dfrac{4\pi}{3}$.

4.21 Form the products $z_1 z_2$ and the quotients z_1/z_2 of the following numbers expressed in modulus–argument form:

(a) $z_1 = 3(\cos\frac{1}{6}\pi + i\sin\frac{1}{6}\pi)$; $z_2 = \frac{1}{2}(\cos\frac{1}{3}\pi + i\sin\frac{1}{3}\pi)$;

(b) $z_1 = 4(\cos\frac{1}{3}\pi - i\sin\frac{1}{3}\pi)$; $z_2 = 2(\cos\frac{1}{4}\pi + i\sin\frac{1}{4}\pi)$;

(c) $z_1 = \frac{1}{3}(\cos\frac{1}{4}\pi - i\sin\frac{1}{4}\pi)$; $z_2 = 6(\cos 3\pi/2 - i\sin 3\pi/2)$.

4.22 The second-order difference equation

$$au_n + bu_{n-1} + cu_{n-2} = 0$$

has for its general solution the expression

$$u_n = A\lambda_1^n + B\lambda_2^n$$

whenever the characteristic equation

$$a\lambda^2 + b\lambda + c = 0$$

has the distinct real roots λ_1 and λ_2. If $b^2 - 4ac < 0$, so that the characteristic equation has the complex conjugate roots λ and $\bar{\lambda}$, show that if u_n is to real, then the constants A and B must also be complex conjugates. Hence show that if $b^2 - 4ac < 0$ and $|\lambda| = r$, $\arg\lambda = \theta$, then the general solution is expressible in the form

$$u_n = r^n(C\cos n\theta + D\sin n\theta),$$

where C and D are real arbitrary constants.

Find the general solution of the following difference equation, and hence determine the particular solution appropriate to the stated initial conditions:

$$u_n - 3\sqrt{2}u_{n-1} + 9u_{n-2} = 0 \quad \text{with } u_0 = 1, \ u_1 = 3.$$

Section 4.5

4.23 Use de Moivre's theorem to express $\sin 7\theta$ and $\cos 7\theta$ in terms of powers of $\sin\theta$ and $\cos\theta$.

4.24 Evaluate z^{20} when $z = \sqrt{3} + i$.

4.25 Calculate the seventh roots of unity.

4.26 Find the roots of the equation $w = (-i)^{2/3}$.

4.27 Find the roots of the equation $w = (1 + i\sqrt{3})^{1/4}$.

Section 4.6

4.28 Determine the lengths $|OP|$ of the vectors OP given that O is the origin and the points P are:

(a) $(1, 1, 1)$; (b) $(-2, 1, 3)$; (c) $(-1, -1, -1)$; (d) $(3, -2, -4)$.

4.29 Find the lengths $|OP|$, the direction cosines and the angles θ_1, θ_2, θ_3 of the vectors OP, where the points P are:

(a) $(2, -1, -1)$; (b) $(4, 0, 2)$; (c) $(-1, 2, 1)$.

4.30 Find the direction ratios, the direction cosines and the angles θ_1, θ_2, θ_3 of the vectors OP, where the points P are:

(a) $(1, 1, 1)$; (b) $(-1, 1, 1)$; (c) $(2, 1, -1)$.

4.31 Determine the angles θ_1, θ_2, θ_3 for the vectors with the direction cosines:

(a) $\left\{\frac{\sqrt{3}}{2}, 0, \frac{1}{2}\right\}$; (b) $\left\{\frac{1}{\sqrt{3}}, \frac{1}{\sqrt{3}}, \frac{1}{\sqrt{3}}\right\}$; (c) $\left\{\frac{1}{3}, -\frac{1}{3}, \frac{\sqrt{7}}{3}\right\}$.

4.32 Determine the lengths $|AB|$ of the vectors AB, given that the end points A and B are:

(a) $A = (1, 1, 1)$, $B = (2, 0, 1)$;

(b) $A = (2, -1, 1)$, $B = (-2, 2, 2)$;

(c) $A = (-1, 3, 1)$, $B = (-2, -1, 0)$.

Use your results to determine the direction cosines for each of these vectors.

4.33 Write down the position vectors OP in terms of the unit vectors **i**, **j**, **k**, given that O is the origin and the points P are:

(a) $(1, 1, 1)$; (b) $(-2, 3, 7)$; (c) $(3, -1, 11)$; (d) $(0, 1, 0)$.

4.34 Determine the values of α, β and γ in order that:

$(1 - \alpha)\mathbf{i} + \beta(1 - \alpha^2)\mathbf{j} + (\gamma - 2)\mathbf{k} = \frac{1}{2}\mathbf{i} + 3\mathbf{j} + 2\mathbf{k}$.

4.35 Form the sum $\mathbf{a} + \mathbf{b}$ and difference $\mathbf{a} - \mathbf{b}$ of the vectors:

(a) $\mathbf{a} = 3\mathbf{i} - 2\mathbf{j} + \mathbf{k}$, $\mathbf{b} = -\mathbf{i} - 2\mathbf{j} + 3\mathbf{k}$;

(b) $\mathbf{a} = -\mathbf{i} + 2\mathbf{j} - \mathbf{k}$, $\mathbf{b} = 2\mathbf{i} - 4\mathbf{j} + 2\mathbf{k}$;

(c) $\mathbf{a} = 2\mathbf{j} - 3\mathbf{k}$, $\mathbf{b} = 2\mathbf{i} - \mathbf{j} + \mathbf{k}$.

4.36 State which of the following pairs of vectors **a** and **b** are parallel and which are anti-parallel:

(a) $\mathbf{a} = \mathbf{i} - 3\mathbf{j} + \mathbf{k}$, $\mathbf{b} = -4\mathbf{i} + 12\mathbf{j} - 4\mathbf{k}$;

(b) $\mathbf{a} = -2\mathbf{i} + 3\mathbf{j} - \mathbf{k}$, $\mathbf{b} = 2\mathbf{i} - 3\mathbf{j} + \mathbf{k}$;

(c) $\mathbf{a} = 4\mathbf{i} - \mathbf{j} - 3\mathbf{k}$, $\mathbf{b} = 8\mathbf{i} - 2\mathbf{j} - 6\mathbf{k}$;

(d) $\mathbf{a} = \mathbf{i} + 7\mathbf{j} + \mathbf{k}$, $\mathbf{b} = 3\mathbf{i} + 21\mathbf{j} + 3\mathbf{k}$.

Section 4.7

4.37 Express the following vectors **a** as the product of a scalar and a unit vector:

(a) $\mathbf{a} = 2\mathbf{i} - \mathbf{j} + 3\mathbf{k}$; (b) $\mathbf{a} = 3\mathbf{i} - 3\mathbf{j} + 3\mathbf{k}$;

(c) $\mathbf{a} = -\dfrac{\sqrt{71}}{9}\mathbf{i} + \tfrac{1}{3}\mathbf{j} - \tfrac{1}{9}\mathbf{k}$.

4.38 Find the vectors AB, and their direction cosines, given that A and B have position vectors **a** and **b**, respectively, where

(a) $\mathbf{a} = 3\mathbf{i} - 3\mathbf{j} + 5\mathbf{k}$, $\mathbf{b} = \mathbf{i} + 2\mathbf{j} - \mathbf{k}$;

(b) $\mathbf{a} = 2\mathbf{i} + 2\mathbf{j} + \mathbf{k}$, $\mathbf{b} = \mathbf{i} + 3\mathbf{j} - 2\mathbf{k}$.

4.39 Find the scalar products **a.b**, and hence find the angle between the vectors **a** and **b**, given that:

(a) $\mathbf{a} = 7\mathbf{i} - 3\mathbf{j} + \mathbf{k}$, $\mathbf{b} = -\mathbf{i} + 2\mathbf{j} + 2\mathbf{k}$;

(b) $\mathbf{a} = 2\mathbf{i} - 2\mathbf{j} + \mathbf{k}$, $\mathbf{b} = -3\mathbf{i} - 3\mathbf{j} + 4\mathbf{k}$;

(c) $\mathbf{a} = \mathbf{i} + 2\mathbf{j} + 3\mathbf{k}$, $\mathbf{b} = -2\mathbf{i} - 4\mathbf{j} - 6\mathbf{k}$.

4.40 Find unit vectors parallel to the vectors **a** where:

(a) $\mathbf{a} = 2\mathbf{i} - 2\mathbf{j} + \mathbf{k}$; (b) $\mathbf{a} = -3\mathbf{i} + \mathbf{j} + 2\mathbf{k}$; (c) $\mathbf{a} = 7\mathbf{i} - 2\mathbf{j} - 3\mathbf{k}$.

4.41 Prove the distributive law for the scalar product by using either definition of the scalar product.

4.42 Form the vector products $\mathbf{a} \times \mathbf{b}$ if:

(a) $\mathbf{a} = \mathbf{i} - 2\mathbf{j} - 4\mathbf{k}$, $\mathbf{b} = 2\mathbf{i} - 2\mathbf{j} + 3\mathbf{k}$;

(b) $\mathbf{a} = -\mathbf{i} + 4\mathbf{j} - \mathbf{k}$, $\mathbf{b} = 3\mathbf{i} + 2\mathbf{j} + 4\mathbf{k}$;

(c) $\mathbf{a} = -2\mathbf{i} + 4\mathbf{k}$, $\mathbf{b} = 3\mathbf{j} - 2\mathbf{k}$.

4.43 Evaluate the determinants:

(a) $\begin{vmatrix} 2 & 1 \\ 4 & 6 \end{vmatrix}$; (b) $\begin{vmatrix} 4 & 16 \\ -2 & 6 \end{vmatrix}$; (c) $\begin{vmatrix} 2 & 0 \\ 0 & 16 \end{vmatrix}$; (d) $\begin{vmatrix} 3 & 9 \\ 1 & 3 \end{vmatrix}$.

4.44 Find the values of λ, if any, for which these determinants vanish:

(a) $\begin{vmatrix} \lambda & 2 \\ 3 & 1 \end{vmatrix}$; (b) $\begin{vmatrix} \lambda & 2 \\ 3 & 2\lambda \end{vmatrix}$; (c) $\begin{vmatrix} 3 & \lambda \\ 0 & 2 \end{vmatrix}$; (d) $\begin{vmatrix} 3\lambda & 4 \\ 2 & -\lambda \end{vmatrix}$.

4.45 Evaluate the determinants:

(a) $\begin{vmatrix} 2 & 1 & 1 \\ 1 & 2 & 1 \\ 1 & 1 & 1 \end{vmatrix}$; (b) $\begin{vmatrix} 3 & 4 & 5 \\ 2 & 2 & 1 \\ 1 & 0 & 2 \end{vmatrix}$; (c) $\begin{vmatrix} 3 & 4 & 5 \\ 3 & 1 & 2 \\ 6 & 5 & 7 \end{vmatrix}$.

4.46 Evaluate the vector products $\mathbf{b} \times \mathbf{a}$, given that:

(a) $\mathbf{a} = 2\mathbf{i} - \mathbf{j} + 2\mathbf{k}$, $\mathbf{b} = -3\mathbf{i} + 2\mathbf{j} + \mathbf{k}$;

(b) $\mathbf{a} = -\mathbf{i} + \mathbf{j} + \mathbf{k}$, $\mathbf{b} = 4\mathbf{i} + 2\mathbf{j} + 3\mathbf{k}$;

(c) $\mathbf{a} = -\mathbf{i} - \mathbf{j} - \mathbf{k}$, $\mathbf{b} = 2\mathbf{i} + 2\mathbf{j} + 2\mathbf{k}$.

4.47 Determine unit vectors that are normal to both vectors $\mathbf{a}$ and $\mathbf{b}$ when:

(a) $\mathbf{a} = 3\mathbf{i} + 5\mathbf{j} - 2\mathbf{k}$, $\mathbf{b} = \mathbf{i} + \mathbf{j} + \mathbf{k}$;

(b) $\mathbf{a} = -4\mathbf{i} + 2\mathbf{k}$, $\mathbf{b} = \mathbf{j} - 3\mathbf{k}$.

State whether the results are unique and, if not, in what way they are indeterminate.

4.48 Use Definition 4.18 to evaluate the vector products $\mathbf{a} \times \mathbf{b}$, given that:

(a) $\mathbf{a} = -\mathbf{i} + 4\mathbf{j} - 2\mathbf{k}$, $\mathbf{b} = 2\mathbf{i} + 3\mathbf{j} + \mathbf{k}$;

(b) $\mathbf{a} = -2\mathbf{i} - 3\mathbf{j} + \mathbf{k}$, $\mathbf{b} = 6\mathbf{i} + 9\mathbf{j} - 3\mathbf{k}$;

(c) $\mathbf{a} = 3\mathbf{i} - \mathbf{k}$, $\mathbf{b} = 2\mathbf{j}$.

Evaluate these same vector products using Table 4.3 and compare the effort involved.

4.49 Verify the distributive property of the vector product

$$\mathbf{a} \times (\mathbf{b} + \mathbf{c}) = \mathbf{a} \times \mathbf{b} + \mathbf{a} \times \mathbf{c},$$

given that $\mathbf{a} = 2\mathbf{i} + \mathbf{j} - \mathbf{k}$, $\mathbf{b} = \mathbf{i} - 2\mathbf{j} + \mathbf{k}$ and $\mathbf{c} = 3\mathbf{i} - 2\mathbf{j} + 3\mathbf{k}$.

4.50 Evaluate the triple scalar products $\mathbf{a}.(\mathbf{b} \times \mathbf{c})$ and $(\mathbf{b} \times \mathbf{a}).\mathbf{c}$ given that:

(a) $\mathbf{a} = 2\mathbf{i} - \mathbf{j} - 3\mathbf{k}$, $\mathbf{b} = 3\mathbf{k}$, $\mathbf{c} = \mathbf{i} + 2\mathbf{j} + 2\mathbf{k}$;

(b) $\mathbf{a} = \mathbf{i} + 2\mathbf{j} + \mathbf{k}$, $\mathbf{b} = 2\mathbf{i} + \mathbf{j} + \mathbf{k}$, $\mathbf{c} = 4\mathbf{i} + 2\mathbf{j} + 2\mathbf{k}$.

4.51 Prove that if $\mathbf{a}$, $\mathbf{b}$ and $\mathbf{c}$ form three edges of a parallelepiped all meeting at a common point, then the volume of this solid figure is given by $|\mathbf{a}.(\mathbf{b} \times \mathbf{c})|$. Deduce that the vanishing of the triple scalar product implies that the vectors $\mathbf{a}$, $\mathbf{b}$ and $\mathbf{c}$ are co-planar (that is, all lie in a common plane).

Section 4.8

4.52 Find the vector equation of the line through the point with position vector $2\mathbf{i} - \mathbf{j} - 3\mathbf{k}$ which is parallel to the vector $\mathbf{i}+\mathbf{j}+\mathbf{k}$. Determine the points corresponding to $\lambda = -3, 0, 2$ in the resulting equation.

4.53 Find the vector equation of the line through the points A and B with position vectors $\mathbf{a} = 2\mathbf{i}+\mathbf{j}-\mathbf{k}$ and $\mathbf{b} = -\mathbf{i}+\mathbf{j}+2\mathbf{k}$. Determine the direction cosines of this line.

4.54 The equations

$$\frac{3x+3}{2} = \frac{-2y+1}{7} = \frac{2z+6}{3}$$

determine a straight line. Express them in vector form and find the direction cosines of the line.

4.55 If the points A and B have position vectors $\mathbf{a}$ and $\mathbf{b}$, and point C divides the line AB in the ratio $\lambda{:}\mu$, show that C has the position vector

$$\frac{\mu\mathbf{a} + \lambda\mathbf{b}}{\lambda + \mu} \qquad \text{provided } \lambda + \mu \neq 0.$$

4.56 Find the vector equation of the line that passes through the point A with position vector $\mathbf{a} = -2\mathbf{i} - \mathbf{j} + \mathbf{k}$ and is normal to both the vectors $\mathbf{b}$ and $\mathbf{c}$, where $\mathbf{b} = \mathbf{i}+2\mathbf{j}+3\mathbf{k}$ and $\mathbf{c} = -\mathbf{i}+\mathbf{j}-\mathbf{k}$.

4.57 Find the perpendicular distance of the point $2\mathbf{i}+\mathbf{j}+\mathbf{k}$ from the line $\mathbf{r} = (1+\lambda)\mathbf{i} + (2-3\lambda)\mathbf{j} + (2+\lambda)\mathbf{k}$.

4.58 Find the Cartesian equation of the plane containing the point $3\mathbf{i} - \mathbf{k}$ and also containing the two vectors $\boldsymbol{\alpha}$, $\boldsymbol{\beta}$ where $\boldsymbol{\alpha} = \mathbf{i}+2\mathbf{j}+\mathbf{k}$ and $\boldsymbol{\beta} = -\mathbf{i}+2\mathbf{j}+2\mathbf{k}$.

4.59 Find the angle between the two planes

$$\frac{x-1}{2} = \frac{y+2}{1} = \frac{z-3}{3}$$

and

$$\frac{2x-1}{2} = \frac{y-1}{3} = \frac{2z+1}{3}.$$

4.60 A line may be uniquely determined as the intersection of two planes

$$\mathbf{r}.\mathbf{n}_1 = p_1 \qquad \text{and} \qquad \mathbf{r}.\mathbf{n}_2 = p_2 \tag{G}$$

where $\mathbf{n}_1$ and $\mathbf{n}_2$ are not necessarily unit vectors. The direction of the line is normal to both $\mathbf{n}_1$ and $\mathbf{n}_2$ and so is parallel to $\mathbf{n}_1 \times \mathbf{n}_2$. Hence the line has the equation $\mathbf{r} = \mathbf{a} + \lambda(\mathbf{n}_1 \times \mathbf{n}_2)$, where λ is a parameter, and $\mathbf{a}$ is some point common to the two planes in (G). Apply these arguments to obtain the vector equation of the line determined by the planes

$$x + 2y - z = 3 \qquad \text{and} \qquad 2x + y + 2z = 1.$$

4.61 Find the Cartesian equation of the sphere of radius 3 about the centre $\mathbf{a} = 2\mathbf{i} + 3\mathbf{j} + \mathbf{k}$.

4.62 Construct the Cartesian equation of the sphere of radius 4 that lies on the side $z > 0$ of the plane $z = 0$ and is tangent to the point (3, 1, 0).

4.63 The inward drawn normal to a sphere of radius 2 at the point (1, 1, 2) on its surface is $\mathbf{n} = 2\mathbf{i} - \mathbf{j} + \mathbf{k}$. Deduce its equation in Cartesian form.

Section 4.9

4.64 Forces $\mathbf{F}_1$, $\mathbf{F}_2$, $\mathbf{F}_3$ and $\mathbf{F}_4$ have magnitudes $2\sqrt{6}$, $3\sqrt{5}$, 3 and 15N and act concurrently through a point O along the lines of the vectors $-\mathbf{i} + 2\mathbf{j} - \mathbf{k}$, $2\mathbf{i} + \mathbf{k}$, $2\mathbf{j}$ and $4\mathbf{i} + 3\mathbf{j}$, respectively. Find the resultant of these forces, and determine its magnitude in newtons.

4.65 Forces 1, 2 and 3 act at one corner of a cube along the diagonals of the faces meeting at that corner. Find the magnitude of their resultant and its inclination to the edges of the cube.

4.66 Find the centre of mass of the masses 1, 3, 4 and 2N situated at points with the respective position vectors $3\mathbf{i} - \mathbf{j} + \mathbf{k}$, $2\mathbf{i} + 2\mathbf{j} + 2\mathbf{k}$, $-\mathbf{i} + 7\mathbf{j} - \mathbf{k}$ and $4\mathbf{i} - 10\mathbf{k}$.

4.67 Prove that the centre of mass of a system of masses is independent of the choice of origin. (*Hint*: Choose a new origin O' with position vector $\mathbf{b}$ relative to the original origin O, and apply the definition of centre of mass.)

4.68 The velocity of a boat relative to the water is represented by $4\mathbf{i} + 3\mathbf{j}$, and that of the water relative to the earth by $2\mathbf{i} - \mathbf{j}$. What is the velocity of the boat relative to the earth if $\mathbf{i}$ and $\mathbf{j}$ represent velocities of 1 km/h to the east and north, respectively?

4.69 The point of application of the force $9\mathbf{i} + 6\mathbf{j} + 7\mathbf{k}$ moves a distance 5 m in the direction of the vector $3\mathbf{i} + \mathbf{j} + 4\mathbf{k}$. If the modulus of the force vector is equal to the magnitude of the force in newton, find the work done.

4.70 A body spins about a line through the origin parallel to the vector $2\mathbf{i}-\mathbf{j}+\mathbf{k}$ at 15 rad/s. Find the angular velocity vector $\boldsymbol{\Omega}$ for the body, and find the instantaneous linear velocity of a point in the body with position vector $\mathbf{i}+2\mathbf{j}+3\mathbf{k}$.

4.71 Find the torque of a force represented by $3\mathbf{i}+6\mathbf{j}+\mathbf{k}$ about point O given that it acts through the point with position vector $-\mathbf{i}+\mathbf{j}+2\mathbf{k}$ relative to O.

4.72 Mass 1, 3 and 2 units at the points specified by the position vectors $3\mathbf{i}-\mathbf{k}$, $2\mathbf{i}-3\mathbf{j}+\mathbf{k}$ and $\mathbf{i}+\mathbf{j}+\mathbf{k}$ relative to point O have velocities represented by $2\mathbf{j}+\mathbf{k}$, $3\mathbf{i}+\mathbf{j}+2\mathbf{k}$ and $\mathbf{i}-\mathbf{j}+\mathbf{k}$, respectively. Determine the vector sum of the moments of momentum of each of these masses about O.

Supplementary computer problems

4.73 Choose any two complex numbers z_1 and z_2, and use them to verify by substitution:

(a) $\left(\dfrac{z_1}{z_2}\right)^3 = \dfrac{z_1^3}{z_2^3}$;

(b) $(z_1^2 z_2)z_2 = (z_1 z_2^2)z_1$;

(c) $\left[\overline{\left(\dfrac{1}{z_1}\right)}\right]^5 = \dfrac{1}{(\bar{z}_1)^5}$.

4.74 Choose any four complex numbers z_1, z_2, z_3, and z_4, and use them to verify by substitution.

(a) $\overline{z_1 z_2 z_3 z_4} = \bar{z}_1 \bar{z}_2 (\overline{z_3 z_4}) = \bar{z}_1 \bar{z}_2 \bar{z}_3 \bar{z}_4$;

(b) $z_1^2(z_2 + z_3 z_4) = z_1^2 z_2 + z_1^2 z_3 z_4$.

4.75 Given that $\theta = 0.712$ rad and $n = 7$, verify de Moivre's theorem

$$(\cos\theta + i\sin\theta)^n = \cos n\theta + i\sin n\theta,$$

by substituting θ and n and evaluating each side of the identity.

4.76 Given that $\theta = 1.314$ rad and $n = -6$, verify de Moivre's theorem, stated in Problem 4.75, by substituting θ and n and evaluating both sides of the identity.

4.77 Given that $\theta = -2.567$ rad and $n = 8$, verify de Moivre's theorem, stated in Problem 4.75, by substituting θ and n and evaluating both sides of the identity.

4.78 Given

$$P(z) = \frac{2z^3 - 3iz + 1}{z^2 + i},$$

find $\text{Re}\{P(z)\}$ and $\text{Im}\{P(z)\}$ when $z = 1 + 2i$.

4.79 Given

$$P(z) = \frac{(3z + 4)^3 - 6iz + 2}{z^2 + z + 1},$$

find $\text{Re}\{P(z)\}$ and $\text{Im}\{P(z)\}$ when $z = 3 - 2i$.

4.80 Verify by subtitution that $z = -1 - 3i$, $z = -1 + 3i$, $z = -2i$ and $z = 3$ are the roots of the equation

$$z^4 - (1 - 2i)z^3 + (4 - 2i)z^2 - (30 - 8i)z - 60i = 0.$$

4.81 Verify by substitution that $z = -2 - 2i$, $z = 2 - 2i$, $z = 2$ and $z = i$ are the roots of the equation

$$z^4 + (3i - 2)z^3 - (4 + 6i)z^2 + (8 + 8i)z - 16i = 0.$$

4.82 Verify by substitution that $z = -2 - 2i$, $z = -2 + 2i$, $z = i$, $z = -i$ and $z = -3$ are the roots of the equation

$$z^5 + 7z^4 + 21z^3 + 31z^2 + 20z + 24 = 0.$$

4.83 Verify by substitution that $z = 4 - i$, $z = 4 + i$, $z = -1 - i$, $z = -1 + i$ and $z = -3$ are the roots of the equation

$$z^5 - 3z^4 - 15z^3 + 27z^2 + 88z + 102 = 0.$$

4.84 Choose any four complex numbers z_1, z_2, z_3 and z_4 and use them to verify the inequality

$$|z_1 + z_2 + z_3 + z_4| \leq |z_1| + |z_2| + |z_3| + |z_4|.$$

5 Differentiation of functions of one or more real variables

The calculus has two interrelated branches, called the differential and integral calculus. The fundamental concept in the differential calculus that forms the subject of study in the first part of this chapter is the definition and interpretation of the derivative df/dx of a function $f(x)$ of a single real variable x. The derivative is constructed and interpreted using the notions of a continuous function and a limit. The rules for performing differentiation on combinations of functions are then derived in order to enable the derivatives of more complicated functions to be determined in terms of a table of derivatives of elementary functions.

Many applications of differentiation help to resolve frequently occurring problems that arise elsewhere. Among the ones considered here are the identification of extrema of functions of a single real variable (maxima and minima), the resolution of limits called indeterminate forms first encountered in Chapter 3, and the result of repeated differentiation of the product of two functions (Leibniz's theorem).

The extension of the concept of a derivative to functions of several real variables forms the next topic to be considered, and leads to the study of partial differentiation. The basic rules for partial differentiation are shown to follow directly from those for ordinary differentiation by regarding all but one of the independent variables involved as constants when performing differentiation. The consequence is that for a suitably well-behaved function, a partial derivative may be defined with respect to each independent variable. To avoid ambiguity, the notation for partial derivatives differs from that for ordinary derivatives. Whereas if $f(x)$ is a function of a single real variable x, its first and second derivatives are written df/dx and d^2f/dx^2 respectively, if $f(x, y)$ is a function of the two real variables x and y, its first- and second-order partial derivatives, of which there are more than one, are written $\partial f/\partial x$, $\partial f/\partial y$ and $\partial^2 f/\partial x^2$, $\partial^2 f/\partial x\partial y$, $\partial^2 f/\partial y\partial x$ and $\partial^2 f/\partial y^2$, respectively.

The interrelationship between the mixed partial derivatives $\partial^2 f/\partial x\partial y$ and $\partial^2 f/\partial y\partial x$, the derivation and consequences of the chain rule and the effect of a change of independent variables, together with some techniques of partial differentiation, are considered at the end of the chapter.

5.1 The derivative

The important branch of mathematics known as the calculus is concerned with two basic operations called differentiation and integration. These operations are related and both rely for their definition on the use of limits.

The calculus was founded jointly, and independently, by Newton in England, and by his contemporary Leibniz in Germany to whom we owe the essentials of our present-day notation. In introducing the ideas underlying the derivative we shall make use of a simple dynamical problem in very much the same way that Newton did when first formulating his early ideas on differentiation. However, we have the advantage of understanding the nature of a limit more clearly than was the case in his day, so that after presenting our heuristic argument, we shall quickly formalize it in terms of the ideas set down in Chapter 3.

We shall consider how to define and determine the instantaneous speed of a point P moving in a non-uniform manner along a straight line. To be precise, we shall suppose that a fixed point O on the line has been selected, and that the distance s of point P from O at time t is determined by the equation

$$s = f(t),$$

where $f(t)$ is some suitable continuous function of t defined on some interval $\mathscr{I}$. Thus we know the position of P at a general time t, and are required to use this information to define and find the speed of P at any given instant of time. When the motion of P is uniform, so that its displacement is proportional to the elapsed time, the familiar definition of speed as distance per unit time can be used. However, if the motion is non-uniform we must consider the situation more carefully. We shall use intuition here and first consider the *difference quotient*

$$\frac{f(t_2) - f(t_1)}{t_2 - t_1} \tag{5.1}$$

in which t_1 and t_2 are two different times belonging to $\mathscr{I}$.

It seems reasonable to suppose that if t_2 were to be taken sufficiently close to t_1 then expression (5.1), which is the quotient of the finite distance travelled and the elapsed time, would in some sense provide a measure of the average speed of P in the small time interval $t_2 - t_1$. Even better would be the idea that we compute the difference quotient (5.1) not for one time t_2 close to t_1, but for a monotonic sequence $\{\tau_i\}$ of times having for its limit the time t_1 which is not a member of the sequence. This last condition is necessary because Eqn (5.1) is not defined if $t_2 = t_1$. Then if the sequence of difference quotients corresponding to Eqn (5.1) has a limit we propose to call the value of this limit the *instantaneous speed* $u(t_1)$ of P at time t_1.

Expressed in the symbolic form of Chapter 3 we may write

$$u(t_1) = \lim_{i \to \infty} \left[\frac{f(\tau_i) - f(t_1)}{\tau_i - t_1} \right]. \tag{5.2}$$

This definition is obviously consistent with the case of the uniform motion of P, for then every difference quotient involved in the determination of the limit (5.2) would give the same constant value u, say. We will call this value u the constant speed of P.

As the function $f(t)$ is continuous it is clearly desirable that we define it not in terms of the discrete variable τ_i but in terms of a continuous variable τ. Fortunately we can do this easily, for the conditions of the connection Theorem 3.6 are satisfied and allow us to rewrite Eqn (5.2) thus:

$$u(t) = \lim_{\tau - t}\left[\frac{f(\tau) - f(t)}{\tau - t}\right]. \tag{5.3}$$

The limit on the right is written $\mathrm{d}f/\mathrm{d}t$, so Eqn (5.3) becomes $u(t) = \mathrm{d}f/\mathrm{d}t$.

It should be appreciated that the limit $u(t)$ in Eqn (5.3) is a number and not a ratio of quantities as were the members of the sequence used to define the limit. The instantaneous speed $u(t)$ can be interpreted as the distance through which P would move in unit time if, during that time, it were to move at a constant speed equal to the value $u(t)$. Because Eqn (5.3) is consistent with the notion of a constant speed, it is customary to omit the adjective 'instantaneous' and to speak only of the speed of P.

The limit involved in Eqn (5.3) is of the indeterminate type, and it will be our object to devise techniques for evaluating such limits for a wide class of functions $f(t)$. In trivial cases these may be determined by simple algebraic considerations, as the following example shows.

Example 5.1 Suppose that the distance of a point P from a fixed origin at time t is determined by the equation $f(t) = kt^3$, where k is a constant with dimensions (Length) (Time) $^{-3}$. Find the functional form of the speed $u(t)$ at time t, and determine its value when $t = 4$.

Solution We are here required to evaluate the limit

$$u(t) = \lim_{\tau \to t}\left[\frac{k(\tau^3 - t^3}{\tau - t}\right]$$

which is the form assumed by Eqn (5.3) when $f(t) \equiv kt^3$.

Using the identity $\tau^3 - t^3 = (\tau - t)(\tau^2 + \tau t + t^2)$, we may write

$$\begin{aligned} u(t) &= \lim_{\tau \to t}\left[\frac{k(\tau - t)(\tau^2 + \tau t + t^2}{(\tau - t)}\right] \\ &= \lim_{\tau - t} k(\tau^2 + \tau t + t^2) \\ &= 3kt^2. \end{aligned}$$

Thus the functional form of the speed is $u(t) = 3kt^2$, so that at $t = 4$ the speed has the value $u(4) = 48k$. ■

It is often helpful to check the form of a result by means of *dimensional analysis.* This is achieved by representing the fundamental quantities of mass, length and time occurring in expressions and equations by the symbols M, L and T, and ignoring any purely numerical multipliers that may be involved. The equations then become identities between expressions of the form $L^p M^r T^s$, where p, r and s are real numbers. Quantities other than length, mass and time are represented as suitable combinations of these fundamental quantities. Thus speed and acceleration would be written LT^{-1} and LT^{-2}, respectively, with no account being taken of their magnitudes. We illustrate this approach with Example 5.1. As $f(t)$ is a distance L, it follows that in terms of dimensions $f(t) = kt^3$ becomes $L = kT^3$, showing that k has dimensions LT^{-3}, so that from the form of the solution we see that $u(t)$ must have the dimensions $kT^2 = (LT^{-3})T^2 = LT^{-1}$ which are the dimensions of speed, as required.

There is a valuable graphical interpretation of the limit (5.3) shown in Fig. 5.1 which is the graph of an arbitrary function $f(t)$ together with the chord PQ, where P is the point $(t, f(t))$ and Q the point $(\tau, f(\tau))$.

The difference quotient within the brackets of Eqn (5.3) before the limit is taken is the tangent of the angle QPR. In the limit as $\tau \to t$, the point Q approaches the point P and the chord PQ approaches the tangent PS to the curve $y = f(t)$ at P. The value $u(t)$ arrived at by considering the limit of the difference quotient (5.3) is thus the tangent of the angle SPR and so is equal to the *gradient* or slope of the curve $y = f(t)$ at P. The number $u(t_1)$ evaluated at any specific time $t = t_1$ is the *derivative* of $f(t)$ with respect to t at $t = t_1$. The limit $u(t)$ as a function of t is simply called the derivative of $f(t)$ with respect to t, and the operation of computing the derivative of a function is called *differentiation*. A function that possesses a derivative at each point of an interval is said to be *differentiable* in that interval. Hence in Example 5.1, the derivative of kt^3 with respect to t at $t = 4$ is $48\,k$, whereas the derivative of kt^3 with respect to t is the function $3kt^2$. The function kt^3 is obviously differentiable in any finite interval.

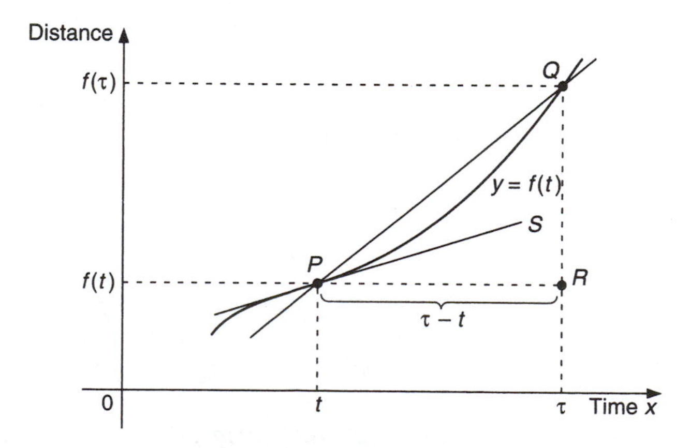

Fig. 5.1 Speed interpreted as a derivative.

This heuristic approach has served to introduce the limiting arguments underlying the concept of a derivative, and we must now carefully reformulate these arguments and express them in general terms. We shall use the following key definitions.

Definition 5.1 A function $f(x)$ of the real variable x will be said to be *differentiable* at x_0 if, and only if,

$$\lim_{x \to x_0} \frac{f(x) - f(x_0)}{x - x_0}$$

exists and is independent of the side from which x approaches x_0. More generally, $f(x)$ will be said to be *differentiable in an interval* $\mathscr{I}$ if it is differentiable at each point of $\mathscr{I}$. At any points of $\mathscr{I}$ for which the limit is not defined the function $f(x)$ will be said to be *non-differentiable*.

Definition 5.2 If $f(x)$ is a differentiable function of the real variable x at x_0, then the value of the expression

$$\lim_{x \to x_0} \frac{f(x) - f(x_0)}{x - x_0}$$

will be denoted by $f'(x_0)$ or $\left.\frac{\mathrm{d}f}{\mathrm{d}x}\right|_{x=x_0}$, and we shall say that it is the *derivative* of $f(x)$ at $x = x_0$. If, further, we define y by the equation $y = f(x)$, then we can also write the derivative of $f(x)$ at x_0 in the form $\left.\frac{\mathrm{d}y}{\mathrm{d}x}\right|_{x=x_0}$.

These definitions merely express in a more sophisticated way, what is often put as follows. Let $y = f(x)$. Then if δy is the increment in y occasioned by an increment δx in x, we have $y + \delta y = f(x + \delta x)$ and hence

$$\frac{\delta y}{\delta x} = \frac{f(x + \delta x) - f(x)}{\delta x}.$$

Thus at $x = x_0$,

$$\frac{\delta y}{\delta x} = \frac{f(x_0 + \delta x) - f(x_0)}{\delta x}$$

and so

$$\left.\frac{\mathrm{d}y}{\mathrm{d}x}\right|_{x=x_0} = \lim_{\delta x \to 0} \frac{f(x_0 + \delta x) - f(x_0)}{\delta x}.$$

To obtain the formulation of Definition 5.2 above, first write h in place of δx to obtain

$$\left.\frac{\mathrm{d}y}{\mathrm{d}x}\right|_{x=x_0} = \lim_{h \to 0} \frac{f(x_0 + h) - f(x_0)}{h},$$

and then write x in place of x_0+h, so that $h = x - x_0$.

What does the requirement, that $\lim_{x \to x_0} \{[f(x) - f(x_0)]/(x - x_0)\}$ should exist, actually mean? It is this. There is a number $f'(x_0)$ such that the left- and right-hand limits of the function $\varphi(x) = [f(x) - f(x_0)]/(x - x_0)$ as x approaches x_0 exist and are both equal to $f'(x_0)$. The function $\varphi(x)$ itself is defined near but not at $x = x_0$ but has the property that $\lim_{x \to x_0} \varphi(x) = f'(x_0)$. We shall use this idea, together with Theorem 3.4, when we discuss the general properties of derivatives of combinations of functions.

If in Definition 5.2 we write x_0+h in place of x, and replace x_0 by x in the subsequent result, we may formulate this definition.

Definition 5.3 If $y = f(x)$ is a differentiable function of the real variable x at all points of an interval $\mathcal{I}$, then the *derivative* of $f(x)$ in $\mathcal{I}$ is the function denoted either by $f'(x)$ or $\mathrm{d}y/\mathrm{d}x$ and defined by

$$f'(x) = \frac{\mathrm{d}y}{\mathrm{d}x} = \lim_{h \to 0} \frac{f(x + h) - f(x)}{h}.$$

The operation of computing the derivative of a function is *differentiation*.

An important consequence of Definition 5.3 is that

$$\frac{\mathrm{d}y}{\mathrm{d}x} = 1 \Big/ \left[\frac{\mathrm{d}x}{\mathrm{d}y}\right]. \tag{5.4}$$

This is easily proved by noticing that

$$\frac{f(x + h) - f(x)}{h} = 1 \Big/ \left[\frac{h}{f(x + h) - f(x)}\right],$$

because taking the limit as $h \to 0$, this reduces to result (5.4). The meaning of (5.4) is as follows. The derivative $\mathrm{d}y/\mathrm{d}x$ is the rate of change of y with respect to x, in which x is considered to be the independent variable and y the dependent variable. However, the derivative $\mathrm{d}x/\mathrm{d}y$ is the rate of change of x with respect of y, so here it is y that is now regarded as the independent variable with x in the role of the dependent variable. Thus Eqn (5.4) asserts that $\mathrm{d}y/\mathrm{d}x$ is the reciprocal of $\mathrm{d}x/\mathrm{d}y$.

Let us now apply exactly the same arguments to Fig. 5.2 as were used in connection with the speed at a point of the particle trajectory in Fig. 5.1. This time the graph represents any function $y=f(x)$ satisfying the conditions of Definition 5.3. Then if P is any point in the interval within which $f(x)$ is differentiable, and Q is an adjacent point, the chord PQ is, in some sense, an approximation to the tangent line to the curve PR at P. The limiting position of the chord PQ will lie along the tangent line to the curve at P and in terms of angles we have $\lim_{Q \to P} \theta = \alpha$. However,

$$\frac{f(x+h) - f(x)}{h} = \tan\theta$$

so that

$$\lim_{h \to 0} \frac{f(x+h) - f(x)}{h} = \lim_{h \to 0} \tan\theta$$

whence, finally,

$$f'(x) = \tan\alpha,$$

or, equivalently,

$$\frac{\mathrm{d}y}{\mathrm{d}x} = \tan\alpha.$$

This result shows that we may interpret the derivative of a differentiable function at a point as the *gradient* of the tangent line drawn to the curve at that point. It is implicit in the definition that the tangent line so defined should be independent of whether Q approaches P from the left or right.

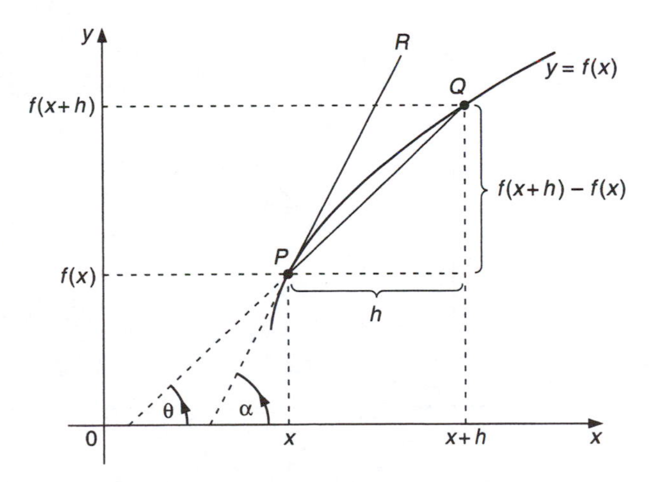

Fig. 5.2 Derivative interpreted as a gradient.

The geometrical interpretation of a derivative allows us to see quite clearly that in addition to the function needing to be continuous in the neighbourhood of a point at which it is required to be differentiable, it also needs a special kind of smoothness. Specifically, the left- and right-hand tangents to the curve at the point in question must be one and the same. Indeed, we could rephrase our definition of differentiability in terms of the equality of the left- and right-hand tangents at a point on the curve, just as we did when dealing with continuity.

Consider the function $f=f(x)$ shown in Fig. 5.3 and defined on the interval $[x_0, x_3]$, but only continuous in the semi-open intervals $[x_0, x_2)$ and $(x_2, x_3]$. Then, despite the fact that the function $f(x)$ is continuous in $[x_0, x_2)$ and $(x_2, x_3]$, it is only possible to assert that tangent lines in the sense implied by Definition 5.3 can be constructed for points in the open intervals (x_0, x_1), (x_1, x_2) and (x_2, x_3). No tangent line can be constructed at x_2 because of the discontinuity; two tangent lines l_1 and l_2 can be constructed at point P according as A and B approach P from the left and the right; while only right- and left- hand tangents l_3 and l_4 can be constructed at the end points x_0 and x_3 because the function $f(x)$ is not defined outside $[x_0, x_3]$.

We shall now show how Definition 5.3 may be used to determine the derivative of a function and also to prove its non-differentiability at a certain point. Our example is a continuous function whose behaviour is clear at all points other than the origin, at which the existence, or otherwise, of a tangent line to the curve cannot be deduced by inspection of its graph.

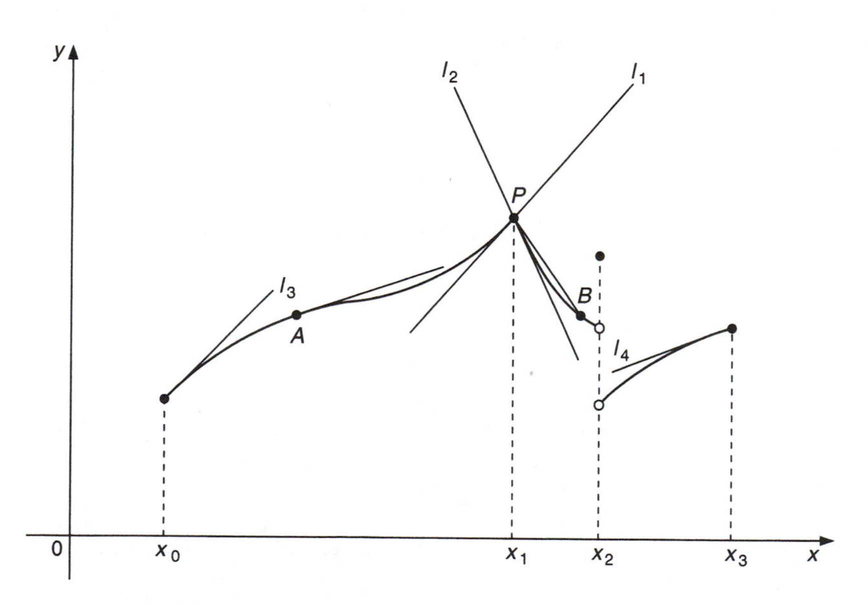

Fig. 5.3 Non-differentiable function at $x=x_1$ and $x=x_2$.

Example 5.2 Prove that the function f, defined by $f(x) = x\sin(1/x)$ for $x \neq 0$ and $f(0) = 0$, is continuous in $(-\infty, \infty)$ and sketch its graph. Find its derivative by use of Definition 5.3, and show that it is not differentiable at the origin.

Solution Only the behaviour of f in the vicinity of the origin is in doubt here. When $x \neq 0$ we may write $f(x) = [\sin(1/x)]/(1/x)$, showing that for large x, $f(x)$ behaves like $\lim_{h\to 0}(\sin h)/h = 1$. Conversely, as the origin is approached, so $x \to 0$ and because $\sin(1/x)$ is bounded by ± 1 it follows that $\lim_{x\to 0} f(x) = 0$. The limit of the function $f(x)$ at the origin is thus equal to the functional value itself, and so $f(x)$ is continuous at the origin. It is clearly continuous elsewhere since it is the product of two continuous functions. Hence it is everywhere continuous and Fig. 5.4 shows its graph, which is symmetric about the y-axis because $f(x)$ is an even function.

We shall approach the differentiability question in two stages: first for $x \neq 0$, and then for $x = 0$. Assuming $x \neq 0$ and making a direct application of Definition 5.3, we obtain

$$f'(x) = \lim_{h\to 0}\left[\frac{(x+h)\sin\left(\dfrac{1}{x+h}\right) - x\sin\dfrac{1}{x}}{h}\right],$$

which we re-express as

$$f'(x) = \lim_{h\to 0}\left[\frac{(x+h)\sin\left[\dfrac{1}{x}\left(1+\dfrac{h}{x}\right)^{-1}\right] - x\sin\dfrac{1}{x}}{h}\right].$$

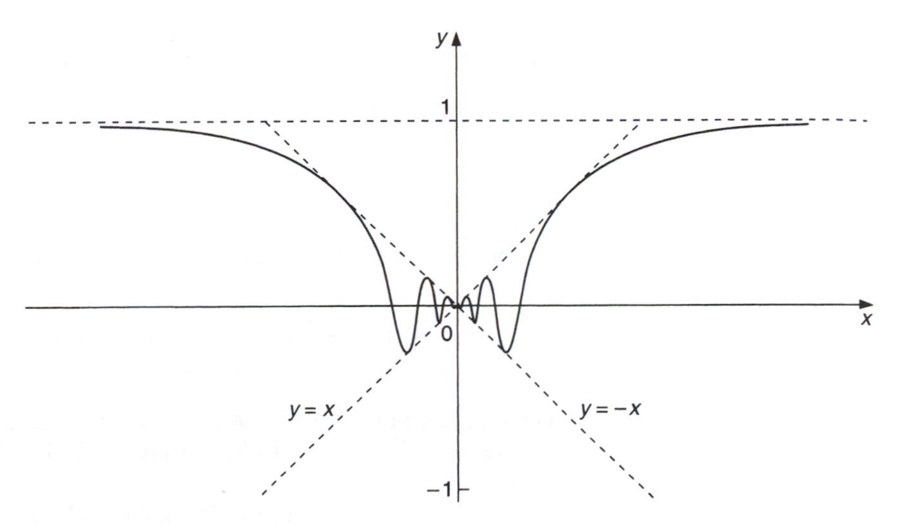

Fig. 5.4 The function $y = x\sin(1/x)$.

Now for h close to zero we may use the binomial theorem, together with our 'little oh' notation of section 3.4, to write $[1+(h/x)]^{-1}=1-(h/x)+o(h)$ as $h\to 0$, and hence

$$f'(x)=\lim_{h\to 0}\left[\frac{(x+h)\sin\left[\frac{1}{x}\left(1-\frac{h}{x}+o(h)\right)-x\sin\frac{1}{x}\right]}{h}\right].$$

Notice that the 'little oh' notation provides us with the essential information we need about the rest of the expansion of $[1+(h/x)]^{-1}$ after the two important terms $1-(h/x)$, without going into unnecessary detail. The imprecise notation $1-(h/x)+\ldots$ so often used is vague and uninformative. Next we write the argument of the sine function as

$$[(1/x)-(h/x^2)+[o(h)]/x],$$

and use the trigonometric expansion for the difference of two angles to obtain

$$f'(x)=\lim_{h\to 0}\left[\frac{(x+h)\left[\sin\frac{1}{x}\cos\left(\frac{h}{x^2}-\frac{o(h)}{x}\right)-\cos\frac{1}{x}\sin\left(\frac{h}{x^2}-\frac{o(h)}{x}\right)\right]-x\sin\frac{1}{x}}{h}\right].$$

Consider the behaviour of the terms comprising this quotient. If the first and last terms are taken together, then in the limit as $h\to 0$, they reduce to the single term $\sin(1/x)$. The remaining term in the centre is

$$-(x+h)\cos\frac{1}{x}\left[\frac{\sin\left(\frac{h}{x^2}-\frac{o(h)}{x}\right)}{h}\right];$$

and since $x\neq 0$ is fixed, it follows from limit (3.9) that this reduces to

$$-\frac{1}{x}\cos\frac{1}{x}$$

as $h\to 0$.

Combing these two results we find that the derivative $f'(x)$ is

$$f'(x)=\sin\frac{1}{x}-\frac{1}{x}\cos\frac{1}{x}\qquad\text{for}\qquad x\neq 0.$$

Thus we have used Definition 5.3 to compute the derivative, and as this is defined for all $x\neq 0$ it follows that $y=x\sin(1/x)$ is differentiable for all such x.

Finally we must examine the behaviour of the derivative at the origin using Definition 5.3. Setting $x=0$ we obtain

$$f'(0) = \lim_{h \to 0} \frac{h \sin(1/h) - 0}{h}$$
$$= \lim_{h \to 0} \sin(1/h).$$

As $\sin(1/h)$ oscillates boundedly with ever increasing frequency when $h \to 0$, it follows that $f'(0)$ is not defined. This establishes the non-differentiability of $f(x)$ at the origin as was required. ■

We close this section by deducing the derivatives of some important elementary functions, and stating them as theorems.

Theorem 5.1 The derivative of a constant function is zero.

Proof Let k be any constant, and consider the function $f(x)$ where $f(x) = k$ for all x. Then

$$\frac{f(x+h) - f(x)}{h} = \frac{k - k}{h} = 0 \text{ for all } x.$$

Hence

$$\lim_{h \to 0} \frac{f(x+h) - f(x)}{h} \equiv 0 \text{ for all } x. \quad ■$$

Theorem 5.2 If n is any positive integer, then the real function $y = x^n$ is differentiable everywhere and has the derivative $\mathrm{d}y/\mathrm{d}x = nx^{n-1}$. If m is any negative integer, then the function $y = x^m$ is differentiable everywhere except at the origin and has the derivative $\mathrm{d}y/\mathrm{d}x = mx^{m-1}$.

Proof We must first consider the limit of the difference quotient $[(x+h)^n - x^n]/h$. By the binomial theorem we have

$$\frac{(x+h)^n - x^n}{h}$$
$$= \frac{x^n + nx^{n-1}h + \dfrac{n(n-1)}{2!}x^{n-2}h^2 + \ldots + \dbinom{n}{r}x^{n-r}h^r + \ldots + h^n - x^n}{h}$$
$$= nx^{n-1} + \frac{n(n-1)}{2!}x^{n-2}h + \ldots + \binom{n}{r}x^{n-r}h^{r-1} + \ldots + h^{n-1}.$$

Now $\lim_{x \to 0} h = 0$ so $\lim_{h \to 0} h^r = 0$ for $1 \le r \le n-1$ and so

$$\lim_{h \to 0} \binom{n}{r} a^{n-r}h^{r-1} = 0.$$

Consequently,

$$\lim_{h\to 0}\frac{(x+h)^n - x^n}{h} = nx^{n-1}.$$

This is defined for all finite x including $x=0$ and so proves the first part of the theorem.

Next let $m = -n$. Then

$$\frac{(x+h)^m - x^m}{h} = \frac{(x+h)^{-n} - x^{-n}}{h} = \left(\frac{x^n - (x+h)^n}{h}\right)\frac{1}{x^n(x+h)^n}.$$

Now, from our result above,

$$\lim_{h\to 0}\frac{x^n - (x+h)^n}{h} = nx^{n-1}$$

while

$$\lim_{h\to 0}(x+h) = x \quad \text{and so} \quad \lim_{h\to 0}(x+h)^n = x^n.$$

If $x \neq 0$,

$$\lim_{h\to 0}\frac{1}{x^n(x+h)^n} = \frac{1}{\lim\limits_{h\to 0} x^n \cdot \lim\limits_{h\to 0}(x+h)^n} = \frac{1}{x^{2n}}.$$

Thus

$$\lim_{h\to 0}\frac{(x+h)^m - x^m}{h} = -nx^{n-1}\cdot\frac{1}{x^{2n}} = -nx^{-n-1} = mx^{m-1}.$$

Hence we have proved that

$$\left.\frac{dy}{dx}\right|_{x=x_0} = \left.\frac{d}{dx}(x^n)\right|_{x=x_0} = nx_0^{n-1} \text{ for all } x_0$$

if n is a positive integer, and for all non-zero x_0 if n is a negative integer. ■

Later we shall prove this result for all real n, and not as here for integral n. Henceforth we shall use the result freely, irrespective of the value of n.

Theorem 5.3 The functions $\sin\alpha x$ and $\cos\alpha x$ of the real variable x, where α is any real number, are differentiable everywhere and

$$\frac{d}{dx}(\sin\alpha x) = \alpha\cos\alpha x \qquad \frac{d}{dx}(\cos\alpha x) = -\alpha\sin\alpha x.$$

Proof These results follow by applying Definition 5.3 and then using limits (3.9) and (3.10). Thus we have

$$\frac{d}{dx}(\sin \alpha x) = \lim_{h \to 0} \frac{\sin[\alpha(x+h)] - \sin \alpha x}{h}$$

$$= \lim_{h \to 0} \left[\frac{\sin \alpha x \cos \alpha h + \cos \alpha x \sin \alpha h - \sin \alpha x}{h}\right]$$

$$= (\sin \alpha x) \lim_{h \to 0} \left(\frac{\cos \alpha h - 1}{h}\right) + (\cos \alpha x) \lim_{h \to 0} \left(\frac{\sin \alpha h}{h}\right)$$

$$= 0 + \alpha \cos \alpha x.$$

As this function is defined for all finite x, the first part of the required result has been established. The remainder of the proof follows exactly similar lines, and so will be omitted. ■

Example 5.3 Find the derivatives of the following functions $y = f(x)$ stating any point at which they are not differentiable.

(a) $f(x) = \begin{cases} 3 & \text{for } -\infty < x < 1 \\ 2 & \text{for } 1 \le x \le \infty; \end{cases}$

(b) $f(x) = x^5$ for all x;

(c) $f(x) = \begin{cases} x^{-3} & \text{for } x \ne 0 \\ 1 & \text{for } x = 0; \end{cases}$

(d) $f(x) = \sin 4x$;

(e) $f(x) = \cos 7x$;

Solution (a) By virtue of Theorem 5.1, the function $f(x)$ has a zero derivative for all x except at the point $x = 1$ where it is not defined.

(b) From Theorem 5.2 we have $dy/dx = 5x^4$ for all x.

(c) From Theorem 5.2 we have $dy/dx = -3x^{-4}$ for $x \ne 0$, and the derivative is not defined at $x = 0$.

(d) From Theorem 5.3, we have

$$\frac{d}{dx}(\sin 4x) = 4 \cos 4x \quad \text{for all } x.$$

(e) Again from Theorem 5.3, we have,

$$\frac{d}{dx}(\cos 7x) = -7 \sin 7x \quad \text{for all } x. \quad ■$$

By now it is obvious that Definition 5.3 is a working definition that can be used. However, some better method than its direct application is obviously needed to compute derivatives of complicated functions. This requirement will be systematically pursued in the next section.

5.2 Rules of differentiation

The complicated functions that occur in mathematical and physical studies are frequently the result of forming sums, products and quotients of simple algebraic and trigonometric functions. This suggests that our next task should comprise a general study of the operation of differentiation when applied to sums, products and quotients of arbitrary differentiable functions. We will present our results in the form of basic theorems which must become thoroughly familiar to the reader.

Theorem 5.4 (differentiation of a sum) If $f(x)$ and $g(x)$ are real-valued functions of x, differentiable at x_0, and k_1 and k_2 are constants, then the linear combination $k_1f(x)+k_2g(x)$ is also differentiable at x_0. Furthermore,

$$\frac{\mathrm{d}}{\mathrm{d}x}(k_1 f(x)+k_2 g(x))\bigg|_{x=x_0} = k_1 f'(x_0)+k_2 g'(x_0).$$

Proof Here we must apply Definition 5.3 to the linear combination $k_1f(x)+k_2g(x)$. We obtain

$$\begin{aligned}
&\frac{\mathrm{d}}{\mathrm{d}x}(k_1 f(x)+k_2 g(x))\bigg|_{x=x_0}\\
&\quad=\lim_{h\to 0}\frac{k_1 f(x_0+h)+k_2 g(x_0+h)-[k_1 f(x_0)+k_2 g(x_0)]}{h}\\
&\quad=k_1\lim_{h\to 0}\frac{f(x_0+h)-f(x_0)}{h}+k_2\lim_{h\to 0}\frac{g(x_0+h)-g(x_0)}{h}\\
&\quad=k_1 f'(x_0)+k_2 g'(x_0).
\end{aligned}$$

If f and g are both differentiable in some common interval $\mathscr{I}$, then the above argument when applied to each point of $\mathscr{I}$ yields the result

$$\frac{\mathrm{d}}{\mathrm{d}x}[k_1 f(x)+k_2 g(x)] = k_1 f'(x)+k_2 g'(x),$$

where x is now any point of $\mathscr{I}$. The constants k_1 and k_2 are often absorbed into the functions f and g, when the result could be expressed 'the derivative of a sum of functions is equal to the sum of their derivatives'. The task of showing that this result is true for a linear combination of an arbitrary number of differentiable functions is left as an exercise involving proof by induction. ■

Example 5.4 Let us use Theorem 5.4 to compute the derivative of $f(x)=\sin^2 x$.

Solution As it stands, at this point we cannot differentiate $f(x)$. However, by using the trigonometric identity $\sin^2 x=\frac{1}{2}(1-\cos 2x)$ we may transform $f(x)$ to the form

$$f(x) = \frac{1}{2}(1 - \cos 2x),$$

so that Theorem 5.4 becomes applicable. Then, using our earlier results concerning the differentiation of a constant and of $\cos \alpha x$ we find that

$$\begin{aligned}
\frac{d}{dx}(\sin^2 x) &= \frac{d}{dx}\{\tfrac{1}{2}(1 - \cos 2x)\} \\
&= \frac{d}{dx}(\tfrac{1}{2}) - \frac{d}{dx}(\tfrac{1}{2}\cos 2x) \\
&= 0 - \tfrac{1}{2}\frac{d}{dx}(\cos 2x) \\
&= -\tfrac{1}{2}\cdot(-2)\sin 2x \\
&= 2\sin x \cos x. \quad \blacksquare
\end{aligned}$$

Theorem 5.5 If $f(x)$ and $g(x)$ are differentiable real-valued functions at x_0, then so also is the product function $f(x)g(x)$. Furthermore,

$$\left.\frac{d}{dx}(f(x)g(x))\right|_{x \to x_0} = f'(x_0)g(x_0) + f(x_0)g'(x_0).$$

Proof Again we consider a difference quotient but this time, for economy of expression, use the form of limit given in Definition 5.2. We have the identity

$$\frac{f(x)g(x) - f(x_0)g(x_0)}{x - x_0} \equiv \left(\frac{f(x) - f(x_0)}{x - x_0}\right)g(x) + \left(\frac{g(x) - g(x_0)}{x - x_0}\right)f(x_0). \quad \text{(A)}$$

Now we wish to show that $\lim_{x \to x_0} f(x) = f(x_0)$. This would be true if $f(x)$ were continuous but we only know that it is differentiable and as yet do not know that this implies continuity. We shall prove that it does. As $f(x)$ is differentiable at $x = x_0$ we must have

$$\frac{f(x) - f(x_0)}{x - x_0} = f'(x_0) + o(h) \qquad \text{as } x \to x_0,$$

where $h = x - x_0$. Hence

$$f(x) - f(x_0) = (x - x_0)[f'(x_0) + o(h)] \qquad \text{as } x \to x_0.$$

This implies that if x is taken sufficiently close to x_0 then the difference $f(x) - f(x_0)$ can be made arbitrarily small. This is just our definition of continuity, and so we have proved that differentiability of $f(x)$ at x_0 implies its continuity at that point. Thus we are permitted to write

$$\lim_{x \to x_0} f(x) = f(x_0)$$

and, similarly,

$$\lim_{x \to x_0} g(x) = g(x_0).$$

Now

$$\lim_{x \to x_0} \left(\frac{f(x) - f(x_0)}{x - x_0} \right) = f'(x_0), \quad \lim_{x \to x_0} \left(\frac{g(x) - g(x_0)}{x - x_0} \right) = g'(x_0),$$

so, finally, taking the limit of (A) as $x \to x_0$, we obtain the result

$$\left. \frac{\mathrm{d}}{\mathrm{d}x}(f(x)g(x)) \right|_{x=x_0} = f'(x_0)g(x_0) + f(x_0)g'(x_0).$$

Again, if f and g are both differentiable in some common interval $\mathscr{I}$ then, as before, we obtain the more general result

$$\frac{\mathrm{d}}{\mathrm{d}x}(f(x)g(x)) = f'(x)g(x) + f(x)g'(x) \quad \text{for} \quad x \in \mathscr{I}. \quad \blacksquare$$

As an incidental detail of this proof we have shown that differentiability at a point implies continuity. This result is worth stating formally.

Theorem 5.6 If a real-valued function $f(x)$ is differentiable at the point x_0, then it is also continuous there. The converse result is not true.

Proof It only remains to prove that the converse result is not true: namely, that continuity does not imply differentiability. This has already been seen in connection with Fig. 5.3, but let us give a specific example. Our final assertion in Theorem 5.6 will be valid even if we can produce only one example of a function that is continuous at a point but is not differentiable there. Such an example used to prove the falsity of an assertion is a *counter-example*, and in this case we choose the function $f(x) = |x|$. This is known to be continuous at $x = 0$, but the derivative as defined in Definition 5.3 is not defined at the origin so the function is not differentiable at that point. $\blacksquare$

Example 5.5 Differentiate the function $f(x) = \sin^2 x$ and compute $f'(\frac{1}{4}\pi)$.

Solution We express the function as a product and use Theorem 5.5.

$$\frac{\mathrm{d}}{\mathrm{d}x}(\sin^2 x) = \frac{\mathrm{d}}{\mathrm{d}x}(\sin x . \sin x)$$

$$= \left[\frac{\mathrm{d}}{\mathrm{d}x}(\sin x)\right] \sin x + \sin x \left[\frac{\mathrm{d}}{\mathrm{d}x}(\sin x)\right]$$

$$= 2\sin x\left[\frac{\mathrm{d}}{\mathrm{d}x}(\sin x)\right]$$

$$= 2\sin x\cos x.$$

As would be expected, this verifies the result of Example 5.4. Finally, using this expression we compute

$$\left.\frac{\mathrm{d}}{\mathrm{d}x}(\sin^2 x)\right|_{x=\pi/4} = 2\sin\frac{\pi}{4}\cos\frac{\pi}{4} = 1. \quad \blacksquare$$

Our next theorem is important and concerns the rule for differentiating what is called a *composite function* or, more simply, the rule for the differentiation of *a function of a function*.

Theorem 5.7 (differentiation of composite functions) If $g(x)$ is a real-valued differentiable function at $x = x_0$ and $f(u)$ is a real-valued differentiable function at $u = g(x_0)$, then $f[g(x)]$ is differentiable at $x = x_0$. Furthermore,

$$\left.\frac{\mathrm{d}}{\mathrm{d}x}\{f[g(x)]\}\right|_{x=x_0} = f'[g(x_0)]\cdot g'(x_0).$$

Proof We have the obvious result

$$\frac{f[g(x)] - f[g(x_0)]}{x - x_0} = \frac{f[g(x)] - f[g(x_0)]}{g(x) - g(x_0)}\cdot\frac{g(x) - g(x_0)}{x - x_0}.$$

Since $g(x)$ is differentiable at x_0 it is continuous there, and so $g(x) \to g(x_0)$ as $x \to x_0$. So, writing $g(x) = u$, $g(x_0) = a$ we have

$$\frac{f[g(x)] - f[g(x_0)]}{x - x_0} = \frac{f(u) - f(a)}{u - a}\cdot\frac{g(x) - g(x_0)}{x - x_0}.$$

Taking the limit as $x \to x_0$, when $u \to a$, we find

$$\left.\frac{\mathrm{d}}{\mathrm{d}x}\{f[g(x)]\}\right|_{x=x_0} = \lim_{x\to x_0}\frac{f[g(x)] - f[g(x_0)]}{x - x_0}$$

$$= \lim_{u\to a}\left[\frac{f(u) - f(a)}{u - a}\right]\cdot\lim_{x\to x_0}\left[\frac{g(x) - g(x_0)}{x - x_0}\right]$$

$$= f'(a)\cdot g'(x_0)$$

$$= f'[g(x_0)]\cdot g'(x_0).$$

If this last result is true at each point of some interval $\mathscr{I}$, then we have the general result

$$\frac{d}{dx}\{f[g(x)]\} = f'[g(x)]\cdot g'(x).$$

When the substitution $u = g(x)$ is made, this can be written:

$$\frac{d}{dx}[f(u)] = \frac{df}{du}\cdot\frac{du}{dx}. \tag{5.6}$$

In this form the theorem is known as the *chain rule* for differentiation, and it is this result that is most often found in textbooks. By repeated application, the chain rule readily extends to enable the differentiation of more complicated composite functions such as the triple composite function $f\{g[h(x)]\}$, always provided the functions f, g and h have suitable differentiability properties. In this case, setting $v = h(x)$ and $u = g(v)$ result (5.6) takes the form

$$\frac{d}{dx}[f(u)] = \frac{df}{du}\cdot\frac{du}{dv}\cdot\frac{dv}{dx}. \tag{5.7}$$

Further extensions of the same kind are obviously possible.

Example 5.6 Differentiate the following functions, and find the values of their derivatives at $x = 1$:

(a) $\sin(x^2 + 3)$;
(b) $(x^3 + x + 1)^{1/3}$;
(c) $\sin\sqrt{(1 + x^2)}$.

Solution (a) Set $u = x^2 + 3$ so that

$$\frac{d}{dx}[\sin(x^2 + 3)] = \frac{d}{dx}(\sin u).$$

From the chain rule:

$$\frac{d}{dx}[\sin(x^2 + 3)] = \frac{d}{du}(\sin u)\cdot\frac{du}{dx}.$$

Now $(d/du)(\sin u) = \cos u$, $du/dx = 2x$ so that

$$\frac{d}{dx}[\sin(x^2 + 3)] = (\cos u)\cdot 2x$$

$$= 2x\cos(x^2 + 3).$$

Hence at $x = 1$,

$$\left.\frac{d}{dx}[\sin(x^2 + 3)]\right|_{x=1} = 2\cos 4.$$

(b) This time set $u = x^3 + x + 1$,

$$\frac{d}{dx}[(x^3 + x + 1)^{1/3}] = \frac{d}{dx}(u^{1/3}).$$

From the chain rule:

$$\frac{d}{dx}[(x^3 + x + 1)^{1/3}] = \frac{d}{du}(u^{1/3})\cdot\frac{du}{dx}.$$

Hence as $(d/du)(u^{1/3}) = \frac{1}{3}u^{-2/3}$, $du/dx = 3x^2 + 1$, we obtain

$$\frac{d}{dx}[(x^3 + x + 3)^{1/3}] = (\tfrac{1}{3}u^{-2/3})\cdot(3x^2 + 1)$$

$$= \tfrac{1}{3}(3x^2 + 1)\cdot(x^3 + x + 1)^{-2/3}.$$

Thus when $x = 1$,

$$\frac{d}{dx}[(x^3 + x + 1)^{1/3}]\bigg|_{x=1} = \frac{4}{3^{5/3}}.$$

(c) We must use the extension of the chain rule given in Eqn (5.7). Set $v = 1 + x^2$ so that $\sin\surd(1 + x^2) = \sin\surd v$, and $u = \surd v$ so that $\sin\surd(1 + x^2) = \sin u$. Then

$$\frac{d}{dx}[\sin\surd(1 + x^2)] = \frac{d}{dx}(\sin u)$$

$$= \left[\frac{d}{dx}(\sin u)\right]\frac{du}{dv}\cdot\frac{dv}{dx}$$

$$= \cos u\frac{du}{dv}\cdot\frac{dv}{dx}.$$

However,

$$\frac{dv}{dx} = 2x \quad \text{and} \quad \frac{du}{dv} = \tfrac{1}{2}v^{-1/2} = \frac{1}{2\surd(1 + x^2)}$$

so that, combining all the results,

$$\frac{d}{dx}[\sin\surd(1 + x^2)] = \frac{x\cos\surd(1 + x^2)}{\surd(1 + x^2)}.$$

Hence at $x = 1$,

$$\frac{d}{dx}[\sin\surd(1 + x^2)]\bigg|_{x=1} = \frac{\cos\surd 2}{\surd 2}.$$

***Theorem 5.8* (differentiation of a quotient)** If $f(x)$ and $g(x)$ are real-valued differentiable functions at x_0 and $g(x_0) \neq 0$, then the quotient $f(x)/g(x)$ is differentiable at x_0. Furthermore,

$$\frac{\mathrm{d}}{\mathrm{d}x}\left[\frac{f(x)}{g(x)}\right]\bigg|_{x=x_0} = \frac{g(x_0)f'(x_0) - g'(x_0)f(x_0)}{[g(x_0)]^2}.$$

Proof If we consider the quotient $f(x)/g(x)$ to be the product of the two functions $f(x)$ and $1/g(x)$, we have by Theorem 5.5

$$\frac{\mathrm{d}}{\mathrm{d}x}\left[\frac{f(x)}{g(x)}\right]\bigg|_{x=x_0} = \frac{1}{g(x)}.f'(x) + f(x)\frac{\mathrm{d}}{\mathrm{d}x}\left[\frac{f(x)}{g(x)}\right]\bigg|_{x=x_0}.$$

Now we must compute $(\mathrm{d}/\mathrm{d}x)(1/g)$. We set $g(x) = u$ so that, from the chain rule,

$$\begin{aligned}\frac{\mathrm{d}}{\mathrm{d}x}\left[\frac{1}{g(x)}\right]\bigg|_{x=x_0} &= \frac{\mathrm{d}}{\mathrm{d}x}\left[\frac{1}{u}\right]\bigg|_{x=x_0} = \frac{\mathrm{d}}{\mathrm{d}u}\left[\frac{1}{u}\right]\frac{\mathrm{d}u}{\mathrm{d}x}\bigg|_{x=x_0} \\ &= -\frac{1}{u^2}\frac{\mathrm{d}u}{\mathrm{d}x}\bigg|_{x=x_0} \\ &= -\frac{g'(x_0)}{[g(x_0)]^2}.\end{aligned}$$

Hence, combining our results, we obtain the desired result

$$\frac{\mathrm{d}}{\mathrm{d}x}\left[\frac{f(x)}{g(x)}\right]\bigg|_{x=x_0} = \frac{g(x_0)f'(x_0) - g'(x_0)f(x_0)}{[g(x_0)]^2}.$$

As in the other cases the general result follows when the conditions of the theorem are satisfied throughout some interval $\mathscr{I}$. It has the obvious form

$$\frac{\mathrm{d}}{\mathrm{d}x}\left[\frac{f(x)}{g(x)}\right] = \frac{g(x)f'(x) - g'(x)f(x)}{[g(x)]^2}. \quad \blacksquare$$

Example 5.7 Differentiate $(3x+1)/(x^2-2)$ and determine the values of x for which the derivative is not defined.

Solution Set $f(x) = 3x+1$ and $g(x) = x^2 - 2$. Then $f'(x) = 3$ and $g'(x) = 2x$ for all x, while $g(x) = 0$ for $x = \pm\sqrt{2}$. Hence, applying Theorem 5.8, we have

$$\begin{aligned}\frac{\mathrm{d}}{\mathrm{d}x}\left[\frac{3x+1}{x^2-2}\right] &= \frac{(x^2-2)\cdot 3 - (2x)(3x+1)}{(x^2-2)^2} \\ &= -\left[\frac{3x^2+2x+6}{(x^2-2)^2}\right],\end{aligned}$$

provided $x \neq \pm\sqrt{2}$.

As an alternative to the use of Theorem 5.8 we may use Theorem 5.5 in conjunction with Theorem 5.7. The argument goes as follows. First we set

$$\frac{3x+1}{x^2-2} = (3x+1)(x^2-2)^{-1}.$$

Then, differentiating the product on the right-hand side (Theorem 5.5) gives

$$\frac{d}{dx}\left[\frac{3x+1}{x^2-2}\right] = \left[\frac{d}{dx}(3x+1)\right](x^2-2)^{-1} + (3x+1)\left[\frac{d}{dx}(x^2-2)^{-1}\right]$$

$$= \frac{3}{x^2-2} + (3x+1)\frac{d}{dx}(u^{-1}),$$

where $u = x^2 - 2$. Finally, from Theorem 5.7 we know that

$$\frac{d}{dx}(u^{-1}) = \frac{du}{dx}\frac{d}{du}(u^{-1}) = \frac{-1}{u^2}\frac{du}{dx} = \frac{-2x}{(x^2-2)^2},$$

Table 5.1 Summary of the rules for differentiation

Let $f(x)$ and $g(x)$ be suitably differentiable functions and k_1 and k_2 be arbitrary, constants. The following are the rules for differentiation:

Differentiation of a sum

$$\frac{d}{dx}[k_1 f(x) + k_2 g(x)] = k_1\frac{df}{dx} + k_2\frac{dg}{dx}$$

Differentiation of a product

$$\frac{d}{dx}[f(x)g(x)] = f'(x)g(x) + f(x)g'(x)$$

Differentiation of a composite function (function of a function)

$$\frac{d}{dx}\{f(g(x))\} = f'[g(x)]g'(x)$$

or, if $u = g(x)$,

$$\frac{d}{dx}[f(u)] = \frac{df}{du}\cdot\frac{du}{dx} \qquad \text{(chain rule)}$$

Differentiation of a quotient

$$\frac{d}{dx}\left[\frac{f(x)}{g(x)}\right] = \frac{g(x)f'(x) - g'(x)f(x)}{[g(x)]^2} \qquad (g(x) \neq 0)$$

Table 5.2 Derivatives of trigonometric functions

$$\frac{d}{dx}(\sin x) = \cos x \qquad \frac{d}{dx}(\cos x) = -\sin x \qquad \frac{d}{dx}(\tan x) = \sec^2 x$$

$$\frac{d}{dx}(\csc x) = -\csc x \cot x \qquad \frac{d}{dx}(\sec x) = \sec x \tan x \qquad \frac{d}{dx}(\cot x) = -\csc^2 x$$

If $u = u(x)$, an application of the chain rule produces the following more general results:

$$\frac{d}{dx}(\sin u) = \cos u \frac{du}{dx} \qquad \frac{d}{dx}(\cos u) = -\sin u \frac{du}{dn} \qquad \frac{d}{dx}(\tan u) = \sec^2 u \frac{du}{dx}$$

$$\frac{d}{dx}(\csc u) = -\csc u \cot u \frac{du}{dx} \qquad \frac{d}{dx}(\sec u) = \sec u \tan u \frac{du}{dx} \qquad \frac{d}{dx}(\cot u) = -\csc^2 u \frac{du}{dx}$$

so that

$$\frac{d}{dx}\left[\frac{3x+1}{x^2-2}\right] = \frac{3}{x^2-2} - \frac{2x(3x+1)}{(x^2-2)^2} = -\left[\frac{3x^2+2x+6}{(x^2-2)^2}\right]. \quad \blacksquare$$

To complete this section, Table 5.1 summarizes the general rules for differentiation, while Table 5.2 gives the derivatives of the trigonometric functions. Unfamiliar results may be deduced by directly applying Theorem 5.8 to the definitions of the functions concerned. Here, and elsewhere, we use the modern abbreviation csc x for cosec x.

5.3 Some important consequences of differentiability

We preface this section by proving a result that belongs more properly to Chapter 3 since it depends for its validity only on the property of continuity. Our sole reason for discussing it here is to present it in the context in which it will first be used. It is usually known as the intermediate value theorem, and we shall now show that the idea underlying it is extremely simple.

Consider the situation in which a recording thermometer attached to some piece of equipment records its temperature at pre-assigned times. Suppose, for instance, that at times t_1 and t_2 the temperatures recorded were T_1 and T_2, respectively. Then although there is no record of the variation of the temperature $T(t)$ at times t between t_1 and t_2, it may be safely inferred that the temperature will pass at least once through each intermediate value between T_1 and T_2. It is quite possible for the temperature to assume values that do not lie between T_1 and T_2, but no assertion can be made about such an event. The situation is illustrated in Fig. 5.5, where T^* is a typical temperature intermediate between T_1 and T_2, and the dotted and solid lines represent two possible temperature variations with time.

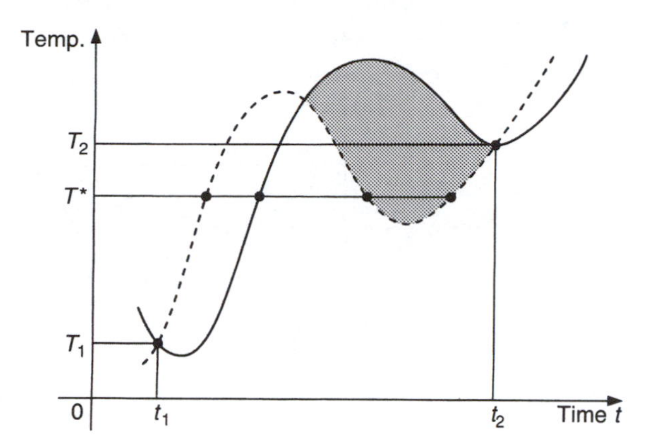

Fig. 5.5 Physical illustration of the intermediate value theorem.

This physical situation is an example of the operation of the intermediate value theorem in everyday life, and we are able to make our assertion because we know from experience that however rapidly a temperature may change, it can never undergo an abrupt jump. In mathematical terms, we are saying that temperature change must be a continuous process. Expressed like this the result seems obvious, but how may we prove it? Our simple proof relies on the postulate of section 3.2, which asserts that every bounded monotonic sequence tends to a limit, but first we state the formal result.

Theorem 5.9 (intermediate value theorem) Let the real-valued function $f(x)$ be continuous on the closed interval $[a, b]$ and such that $f(a) \neq f(b)$. Then if y^* is any number intermediate between $f(a)$ and $f(b)$, there exists at least one number x^* between a and b such that $y^* = f(x^*)$.

Proof Although a diagram is not essential for this proof, the representative situation shown in Fig. 5.6 will be of help.

First set $x_1 = \frac{1}{2}(a+b)$. Then if $f(x_1) = y^*$ the result is proved. If not, consider the intervals (a, x_1), (x_1, b). Then in one of these two intervals, y^* will lie between the functional values occurring at either end of the interval. Call this interval I_1 and let it be represented by the open interval (a_1, b_1). Thus in Fig. 5.6, I_1 is the right-hand interval and so in that case $a_1 = \frac{1}{2}(a+b)$, $b_1 = b$.

Next set $x_2 = \frac{1}{2}(a_1 + b_1)$. If $f(x_2) = y^*$ the result is proved. If not, consider the intervals (a_1, x_2), (x_2, b_1). Then in one of these two intervals, y^* will lie between the functional values occurring at either end of the interval. Call this interval I_2 and let it be represented by the open interval (a_2, b_2). In Fig. 5.6 the interval I_2 is the left-hand sub-interval of I_1, so that $a_2 = a_1$, $b_2 = \frac{1}{2}(a_1 + b_1)$.

We either prove the result directly for some x_n or we define an infinite sequence of open intervals $I_1, I_2, I_3, \ldots$, each contained within the

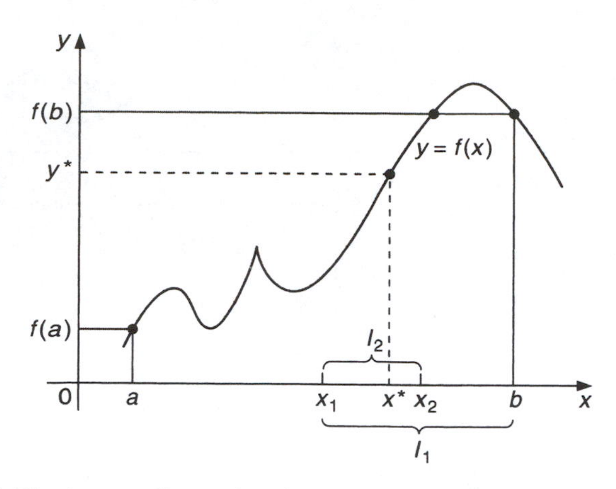

Fig. 5.6 The intermediate value theorem.

previous interval. Because each interval is contained by all its predecessors it then follows that the sequence of numbers $a_1, a_2, a_3, \ldots$ is monotonic increasing and bounded above, while the sequence of numbers $b_1, b_2, b_3, \ldots$ is monotonic decreasing and bounded below. Hence by the theorem of section 3.2, the sequences $\{a_i\}$ and $\{b_i\}$ both tend to a limit. That they both tend to the same limit follows from the fact that the length of the nth interval I_n is $(b-a)2^n$, which tends to zero as $n \to \infty$. Letting the common value of these two limits be denoted by x^*, we have $\lim_{n\to\infty} |f(a_n) - f(x^*)| = 0$, thereby showing the existence of the required number x^*.

The following result may be proved with the aid of the intermediate value theorem.

Corollary 5.9 Every function that is continuous in a closed interval attains both its greatest and least values at points of that interval. These values may occur at the end points of the interval.

Proof If a function f is continuous and strictly monotonic in $[a, b]$, then by Theorem 5.9 it attains each value between $f(a)$ and $f(b)$ precisely once for $x \in [a, b]$, and so its greatest and least values must be attained at the end of points $[a, b]$. Thus the result is true in this case. If f is continuous, but not monotonic in $[a, b]$, then $[a, b]$ may be divided into contiguous sub-intervals in each of which f is monotonic. Suppose it is strictly monotonic in each of these sub-intervals. Then its greatest and least values will occur at the ends of these sub-intervals at points where f is continuous, and each such point will belong to $[a, b]$. The largest of these values will be the greatest value of f on $[a, b]$, and the smallest will be the least value of f on $[a, b]$, and so the result is again true. Finally, if instead of being strictly monotonic in an interval (or intervals) f is merely monotonic, a slight modification of the above argument again shows the result to be true and the result is proved. The modification of the argument in this last case is left as an exercise.

5.3.1 Maxima and minima

One of the most familiar and useful applications of differentiation is to the problem of determining those points in some interval $[a, b]$ at which a function $f(x)$ assumes its maximum and minimum values. Collectively these values are known as the *extrema* of the function $f(x)$ on the interval $[a, b]$ and they are of various types, as this definition indicates.

Definition 5.4 (extrema) Let $f(x)$ be a continuous function defined on the interval $[a, b]$ so that it attains its greatest and least values at points of that interval. Then we say that the point x_0 belonging to $[a, b]$ is:

(a) an *absolute maximum* if $f(x_0) \geq f(x)$ for all points x in $[a, b]$;
(b) an *absolute minimum* if $f(x_0) \leq f(x)$ for all points x in $[a, b]$;
(c) a *local maximum* if $f(x_0+h) - f(x_0) \leq 0$ for $|h|$ sufficiently small;
(d) a *local minimum* if $f(x_0+h) - f(x_0) \geq 0$ for $|h|$ sufficiently small.

No assumption of differentiability has been made when formulating this definition, so that in Fig. 5.7 point P is an absolute maximum and both points R and T are local maxima. Point Q is an absolute minimum and point S a local minimum. Although the functional value at U lies intermediate between those at Q and S, it is not a local minimum in the sense of the definition, because it lies at the end of the domain of definition $[a, b]$ so that only the one-sided behaviour of the function is known there with respect to h.

If now, in addition to continuity, we also require of $f(x)$ that it be differentiable at the point x_0 occurring in Definition 5.4, we can easily devise a simple test to identify the points where extrema must occur. Consider point P in Fig. 5.7 as representative of a maximum at which the function is differentiable. The fact that P happens to be an absolute maximum is immaterial for the subsequent argument.

By supposition, if f is differentiable at P, the expression

$$f'(x_0) = \lim_{x \to x_0} \left[\frac{f(x) - f(x_0)}{x - x_0}\right]$$

must be independent of the manner of approach of x to x_0. Now for maxima of types (a) and (c) we have $f(x) - f(x_0) \leq 0$, and hence it follows that when $x < x_0$, $f'(x_0)$ is the limit of an essentially positive function; whereas when $x > x_0$, $f'(x_0)$ is the limit of an essentially negative function. Clearly this is only possible if $f'(x_0) = 0$. We have thus proved that if f is

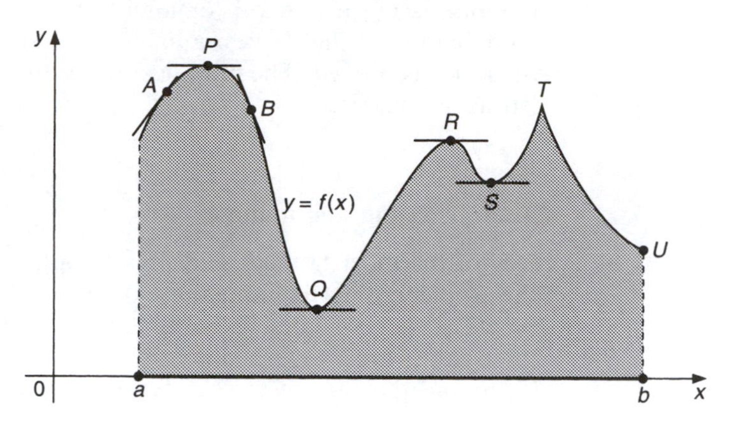

Fig. 5.7 Extrema of a function on $[a, b]$.

differentiable at x_0, then a necessary condition that f should have a maximum at x_0 is $f'(x_0) = 0$.

Similar reasoning establishes that the condition $f'(x_0) = 0$ is also a necessary condition for the differentiable function f to have a minimum at x_0. To show that the vanishing of the derivative f' at a point is not a *sufficient* condition for that point to be an extremum, we appeal to a counter example. The function $f = x^3$ has a continuous derivative $f' = 3x^2$ which vanishes at the origin. Nevertheless, f is negative for $x < 0$ and f is positive for $x > 0$, thereby showing that despite the vanishing of the derivative, neither a maximum nor a minimum of the function can occur at the origin. Later we shall identify behaviour of this nature as typical of a *point of inflection* with a horizontal tangent. Generally speaking, a point of inflection is a demarcation point on the graph of a differentiable function separating a region of convexity from a region of concavity. This simply says that at a point of inflection the tangent to the curve actually *crosses* the curve. Collectively the points at which the derivative vanishes, regardless of whether or not they are maxima, minima or points of inflection are called *critical points* or *stationary points* of the function, or sometimes *turning points*.

Combining the previous results, and recalling that the condition that f be differentiable at x_0 precludes behaviour of the type encountered at point T in Fig. 5.7, we are able to formulate the following general result.

Theorem 5.10 Let f be a real-valued differentiable function on some interval $[a, b]$. Then the *stationary points* of f are the numbers ξ (pronounced '*xi*') for which $f'(\xi) = 0$. ■

Once the stationary points of a function have been determined it is necessary to examine the functional behaviour in the vicinity of each one in order to determine the nature of the point involved. An absolute maximum is identified from among the local maxima by direct comparison of the functional values at the stationary points in question. A similar process identifies an absolute minimum.

Example 5.8 Without appealing to graphical ideas, find the location and nature of the extrema of the following two functions, and determine if they are differentiable at these points:

(a) $f(x) = \frac{1}{2}x^3 + 2x^2 + 3x + 1$;
(b) $f(x) = (2x - 5)x^{2/3}$.

Solution (a) The stationary points are determined by finding those values $x = \xi$ for which the derivative f' vanishes.

Now $f' = x^2 + 4x + 3$ and so the desired stationary points are given by the roots of the equation

$$\xi^2 + 4\xi + 3 = 0.$$

These roots are $\xi = -1$ and $\xi = -3$ and the functional values at the respective points are $f(-1) = -\frac{1}{3}$ and $f(-3) = 1$. As the derivative f' is the sum of continuous functions it is everywhere continuous, so that no cusp-like behaviour with associated extrema, as typified by point T in Fig. 5.7, can arise. So the two points $\xi = -1$ and $\xi = -3$ are the only ones at which stationary values can occur. An examination of the behaviour of the function near these points will determine if these stationary values correspond to maxima, minima or points of inflection.

A sketch graph would quickly show that in fact $\xi = -3$ corresponds to a local maximum and $\xi = -1$ to a local minimum, but we are specifically required to establish these results by analytical means. How, then, can we do this? The solution lies in a direct application of Definition 5.4, and we illustrate the argument by considering the stationary point $\xi = -1$. To find the behaviour of f close to $\xi = -1$ we shall set $x = -1 + h$, where h is small, and substitution in $f(x)$ to obtain

$$f(-1 + h) = \tfrac{1}{3}(-1 + h)^3 + 2(-1 + h)^2 + 3(-1 + h) + 1,$$

whence

$$f(-1 + h) = -\tfrac{1}{3} + h^2 + \frac{h^3}{3}.$$

Now $f(-1) = -\frac{1}{3}$ so that we may also write this result in the form

$$f(-1 + h) - f(-1) = h^2\left(1 + \frac{h}{3}\right).$$

Clearly for $|h|$ small, the right-hand side is essentially positive, and so we have succeeded in showing that close to $\xi = -1$,

$$f(\xi + h) - f(\xi) > 0,$$

and so by Definition 5.4 (d) the stationary point $\xi = -1$, at which $f(\xi) = -\frac{1}{3}$, is seen to be local minimum. An exactly similar argument will establish that the stationary point $\xi = -3$, at which $f(\xi) = 1$, is a local maximum. These are only local extrema because it is possible to find values of x for which $f > 1$ and $f < -\frac{1}{3}$.

(b) This case is more complicated. We have

$$\frac{df}{dx} = 2x^{2/3} + \frac{2(2x-5)}{3x^{1/3}}$$

showing that the stationary points of f are determined by the roots of the equation

$$0 = 2\xi^{2/3} + \frac{2(2\xi - 5)}{3\xi^{1/3}}, \quad \text{or} \quad \frac{10}{3}\frac{(\xi - 1)}{\xi^{1/3}} = 0.$$

This has the single root $\xi = 1$ at which $f(1) = -3$, showing that the function has only one stationary point. To determine the nature of this point let us set $x = 1 + h$, where $|h|$ is small, and substitute into $f(x)$ to find

$$f(1+h) = (2h-3)(1+h)^{2/3}.$$

Next we expand the factor $(1+h)^{2/3}$ by the binomial theorem as far as terms involving h^2, to obtain

$$f(1+h) = (2h-3)(1 + \tfrac{2}{3}h - \tfrac{1}{9}h^2 + O(h^3))$$

or

$$f(1+h) = -3 + \tfrac{4}{3}h^2 + O(h^3).$$

Using the fact that $f(1) = -3$ this becomes

$$f(1+h) - f(1) = \tfrac{4}{3}h^2 + O(h^3)$$

showing that close to $\xi = 1$, $f(\xi + h) - f(\xi) > 0$. Hence by Definition 5.4 (d), the stationary point $\xi = 1$ is seen to correspond to a local minimum. Again, it is only a local minimum because for large negative x we have $f < -3$.

We now observe that f' is defined for all x other than for $x = 0$, at which point $f(0) = 0$. The behaviour of the function in the vicinity of the origin needs examination since, as it is not differentiable there, Theorem 5.10 can provide no information about that point. Set $x = h$, where h is small, and substitute in f to get

$$f(h) = (2h-5)h^{2/3}.$$

Now $f(0) = 0$, so that we may rewrite this as

$$f(h) - f(0) = (2h - 5)h^{2/3},$$

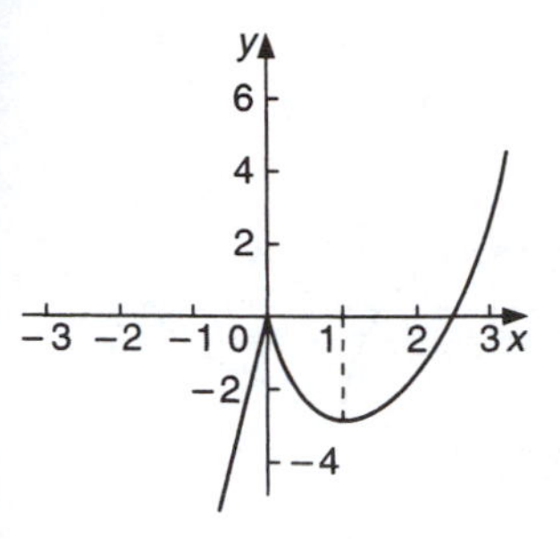

Fig. 5.8 $y = (2x - 5)x^{2/3}$.

thereby showing that as the right-hand side is essentially negative for suitably small h, close to $\xi = 0$ we have $f(\xi + h) - f(\xi) < 0$. From Definition 5.4(c) we now see that the origin is a local maximum, despite the fact that f is not differentiable at that point. It is only a local maximum because for large positive x we have $f > f(0)$. For reference purposes the function is shown in Fig. 5.8. ■

The method of classification of stationary points that we have just illustrated is always applicable, though it provides more information than is often required. This is so because not only does it discriminate between maxima and minima, but it also provides the approximate behaviour of the function close to the point in question. We shall return to this problem later to provide much simpler criteria by which the nature of stationary points may be identified.

5.3.2 Rolle's theorem

One form of this theorem may be stated as follows.

Theorem 5.11 (Rolle's theorem) Let f be a real-valued function that is continuous on the closed interval $[a, b]$ and differentiable at all points of the open interval (a, b). Then if $f(a) = f(b)$ there is at least one point $x = \xi$ interior to (a, b) at which $f'(\xi) = 0$.

Proof We know from Corollary 5.9 that a continuous function $f(x)$ defined on the closed interval $[a, b]$ must attain its maximum value M and its minimum value m at points of $[a, b]$. Then if $m = M$ on $[a, b]$, the function $f(x) =$ constant, and since the derivative of a constant is zero, the point $x = \xi$ at which $f'(\xi) = 0$ may be taken anywhere within the interval.

If $f(x)$ is not a constant function then $m \neq M$, and as $f(a) = f(b)$ it follows that at least one of the numbers m, M must differ from the value $f(a)$. We shall suppose that $M \neq f(a)$. Then clearly the value M must be attained at some point $x = \xi$ interior to (a, b). As f is assumed to be differentiable in (a, b) it follows that Theorem 5.10 must be applicable, showing that $f'(\xi) = 0$. A similar argument applies if $m \neq f(a)$. Geometrically this theorem simply asserts that the graph of any function satisfying the conditions of the theorem must have at least one point in the interval (a, b) at which the tangent to the curve is horizontal. ■

If f is not differentiable at even one interior point of (a, b) then Rolle's theorem cannot be applied. Our counter-example in this instance is the simple function $f(x) = |x|$ with $-1 \leq x \leq 1$. This function is everywhere

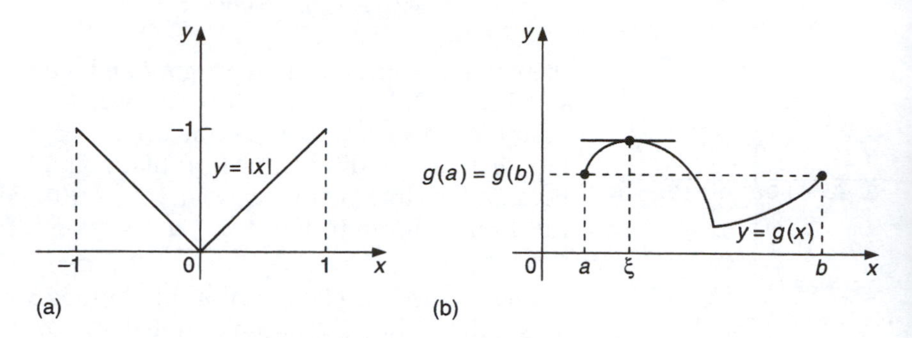

Fig. 5.9 Counter-example for Rolle's theorem: (a) Rolle's theorem does not apply – there is no point ξ for which $f'(\xi) = 0$; (b) $g'(\xi) = 0$, but Rolle's theorem does not apply, because although the function is continuous, its derivative is not.

continuous, and is differentiable at all points other than at the origin, but there is certainly no point $x = \xi$ on $[-1, 1]$ at which $f' = 0$. The graph of this function is shown in Fig. 5.9, with one of a function $g(x)$ not satisfying the conditions of the theorem but for which the result happens to be true.

5.3.3 Mean value theorems for derivatives

Our most important application of Rolle's theorem will be in the derivation of the mean value theorem for derivatives. It is difficult to indicate just how valuable and powerful this deceptively simple theorem really is as an analytical tool. However, something of its utility will, perhaps, be appreciated after studying the reminder of this chapter. We offer only a geometrical proof of the theorem.

Consider Fig. 5.10, which represents a graph of a differentiable function $f(x)$ on the open interval (a, b). Then as P and S are the points $(a, f(a))$ and $(b, f(b))$, the gradient m of the line PS is

$$m = \frac{f(b) - f(a)}{b - a}.$$

Now we may identify points Q and R, with respective x-coordinates and ξ and η (the latter pronounced 'eta') interior to (a, b), at which the tangent lines l_1 and l_2 to the graph are parallel to PS, and so must also have the same gradient m. Then because of the geometrical interpretation of the derivative f' as the gradient of the tangent line, at either P or Q we may equate m and f'. If we confine attention to point Q we have

$$\frac{f(b) - f(a)}{b - a} = f'(\xi),$$

where $a < \xi < b$. This is the form in which the mean value theorem for derivatives, also known as the *law of the mean*, is usually quoted. In

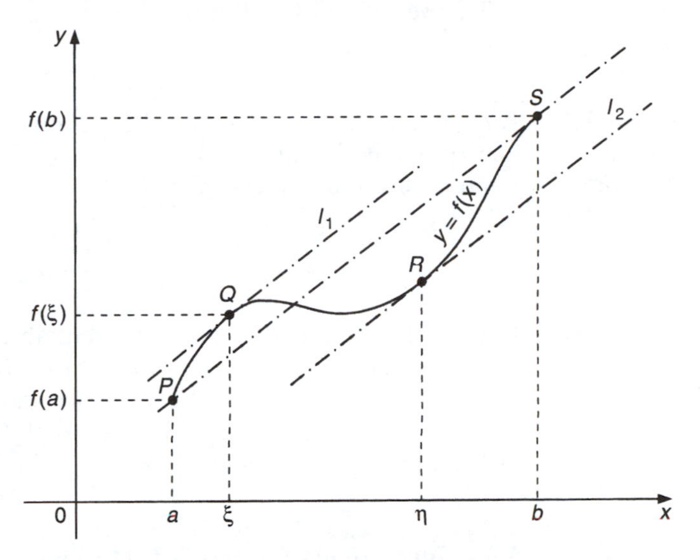

Fig. 5.10 Illustration of the mean value theorem.

geometrical terms the theorem asserts that there is always a point $(\xi, f(\xi))$ on the graph of the differentiable function, with $a < \xi < b$, at which the tangent to the curve is parallel to the secant line PS. The fact that the precise value of ξ is not usually known is, generally speaking, unimportant in the application of this theorem. This is because it is often used with some limiting argument in which $b \to a$, so that $\xi \to a$ also. A formal statement of the theorem is as follows.

Theorem 5.12 (mean value theorem for derivatives) If $f(x)$ is a real-valued function that is continuous in $[a, b]$ and differentiable in (a, b), then there exists a point ξ interior to (a, b) such that

$$\frac{f(b) - f(a)}{b - a} = f'(\xi). \quad \blacksquare$$

The existence of more than one point ξ in (a, b) at which this result is true is not precluded. This is so because it is only asserted that such a point exists, and not that there is necessarily only one such point. Such is the case, for example, in Fig. 5.10 since, as was remarked, $f'(\xi) = f'(\eta)$ with $\xi \neq \eta$, though both points ξ and η are interior to (a, b).

Corollary 5.12 If $g'(x) = h'(x)$ at all points $[a, b]$, then $g(x) = h(x) + \text{constant}$ in $[a, b]$.

Proof Set $f = g - h$ in Theorem 5.12 applied to the interval $[a, x]$. Then $g(x) - h(x) = g(a) - h(a) = \text{constant}$ and the result follows. $\blacksquare$

This result will be of importance when we discuss integration.

Theorem 5.13 **(Cauchy extended mean value theorem)** If $f(x)$ and $g(x)$ are real-valued functions that are continuous in $[a, b]$ and differentiable in (a, b) and $g'(x) \neq 0$ in (a, b), then there exists a point ξ interior to (a, b) such that

$$\frac{f(b) - f(a)}{g(b) - g(a)} = \frac{f'(\xi)}{g'(\xi)}.$$

Proof Define the continuous and differentiable function $F(x)$ as $F(x) = f(a)g(a) - f(b)g(a) + [g(a) - g(b)]f(x) - [f(a) - f(b)]g(x)$. Then $F(a) = F(b)$, so Rolle's theorem applies, and we have $F'(\xi) = 0$ for some $a < \xi < b$. Hence $[g(a) - g(b)]f'(\xi) - [f(a) - f(b)]g(\xi) = 0$, from which the result of the theorem then follows. ■

5.3.4 Indeterminate forms – L'Hospital's rule

Limits such as $\lim_{x \to 0}(\sin \alpha x)/x$ which apparently tend to the form 0/0 have already been encountered and given meaning in special cases. Closely related problems are those giving meaning to the limit of a quotient which apparently tends to ∞/∞, to a product that approaches $0.\infty$ or to a difference like $\infty - \infty$. These limit problems are called *indeterminate forms*. One of the most obvious applications of the extended mean value theorem is to resolve the value of the limit in either of these situations, and we now prove the simplest statement of a useful result generally known as L'Hospital's rule.

Theorem 5.14 **(first form of L'Hospital's rule)** If $f(x)$ and $g(x)$ are real-valued differentiable functions at $x = x_0$, and

(a) $f(x_0) = g(x_0) = 0$,

(b) $\lim_{x \to x_0} \frac{f'(x)}{g'(x)} = \lambda$, where λ is either a real number or infinity,

then

$$\lim_{x \to x_0} \frac{f(x)}{g(x)} = \lim_{x \to x_0} \frac{f'(x)}{g'(x)} = \lambda.$$

Proof Apply the extended mean value theorem to the functions $f(x)$ and $g(x)$ defined on the interval $[x, x_0]$ and use condition (a) to obtain

$$\frac{f(x)}{g(x)} = \frac{f'(\xi)}{g'(\xi)},$$

where $x < \xi < x_0$.

Now $x \to x_0$ implies that $\xi \to x_0$, so that by condition (b) we have the desired result

$$\lim_{x \to x_0} \frac{f(x)}{g(x)} = \lim_{\xi \to x_0} \frac{f'(\xi)}{g'(\xi)} = \lambda.$$

The fact that the variable ξ appears in the second limit in place of the x stated in the theorem is unimportant. Its function is simply that of a variable, and the symbol used to denote it is immaterial. ■

It may happen that the application of L'Hospital's rule gives rise to another indeterminate form. When this occurs the rule must be applied again, if necessary repeatedly, until the value of the limit has been determined. A modification of the rule taking this possibility into account is given in section 12.4.1.

It is essential to recognize that the rule may only be repeated until the indeterminate form is resolved; to continue to apply it beyond this stage will yield an incorrect result. For example,

$$\lim_{x \to 0} \left(\frac{\sin x}{x^2} \right) = \lim_{x \to 0} \left(\frac{\cos x}{2x} \right) = \infty,$$

but had the rule been applied again in error we would have obtained

$$\lim_{x \to 0} \left(\frac{\sin x}{x^2} \right) = \lim_{x \to 0} \left(\frac{-\sin x}{2} \right) = 0.$$

In general, when the symbol used to denote a variable is unimportant because it only appears in some intermediate calculation, the details of which do not concern us, we shall call it a *dummy variable*.

A useful extension of L'Hospital's rule is contained in the following corollary which allows examination of limits which tend to the form ∞/∞. We state it without proof.

Corollary 5.14 If $\varphi(x)$ and $\psi(x)$ are real-valued differentiable functions at $x = x_0$ and,

(a) $\lim_{x \to x_0} \varphi(x) \to \pm\infty, \quad \lim_{x \to x_0} \psi(x) \to \pm\infty,$

(b) $\lim_{x \to x_0} \dfrac{\varphi'(x)}{\psi'(x)} = \lambda$, where λ is either a real number or infinity, then

$$\lim_{x \to x_0} \frac{\varphi(x)}{\psi(x)} = \lim_{x \to x_0} \frac{\varphi'(x)}{\psi'(x)} = \lambda. \quad ■$$

Indeterminate forms that tend to $0.\infty$ or to $\infty - \infty$ are dealt with by rearranging the expression whose limit is required so that it becomes an indeterminate form to which either L'Hospital's rule or its corollary may be applied. This approach is illustrated in the following example.

Example 5.9 Determine the value of the following indeterminate forms.

(a) $\lim_{x\to 0}\dfrac{\sin \alpha x}{x}$;

(b) $\lim_{x\to 1}\dfrac{x^3+3x^2-2x-2}{2x^2-x-1}$;

(c) $\lim_{x\to 0}\dfrac{\sin 3x}{x^3}$;

(d) $\lim_{x\to(1/2)\pi-}\dfrac{\tan 3x}{\tan x}$;

(e) $\lim_{x\to 0-}\dfrac{\left(\dfrac{a}{x}\right)}{\cot bx}$;

(f) $\lim_{x\to 0}\dfrac{x\sin x+\cos x-1}{\cos x-1}$;

(g) $\lim_{x\to 0}\left(\dfrac{1}{\sin 2x}-\dfrac{1}{2x}\right)$;

(h) $\lim_{x\to 0+}(x\cot 2x)$.

Solution (a) This is of the form $\lim f/g\to 0/0$, with $f(x)=\sin\alpha x$ and $g(x)=x$. As $f'(x)=\alpha\cos\alpha x$ and $g'(x)=1$ it follows that

$$\lim_{x\to 0}\frac{\sin\alpha x}{x}=\lim_{x\to 0}\frac{\alpha\cos\alpha x}{1}=\alpha.$$

This confirms the limit that was obtained by a different method in Chapter 3.

(b) This is also of the form $\lim f/g\to 0/0$, but this time with $f(x)=x^3+3x^2-2x-2$ and $g(x)=2x^2-x-1$. It follows that $f'(x)=3x^2+6x-2$ and $g'(x)=4x-1$ so that

$$\lim_{x\to 1}\frac{x^3+3x^2-2x-2}{2x^2-x-1}=\lim_{x\to 1}\frac{3x^2+6x-2}{4x-1}=\frac{7}{3}.$$

(c) This is again of the form $\lim f/g\to 0/0$, with $f(x)=\sin 3x$ and $g(x)=x^3$. Hence $f'(x)=3\cos 3x$ and $g'(x)=3x^2$, so that

$$\lim_{x\to 0}\frac{\sin 3x}{x^3}=\lim_{x\to 0}\frac{\cos 3x}{x^2}\to+\infty.$$

(d) This is of the form $\lim f/g\to\infty/\infty$ with $f(x)=\tan 3x$ and $g(x)=\tan x$. Hence $f'(x)=3\sec^2 3x$ and $g'(x)=\sec^2 x$, and, by Corollary 5.14,

$$\lim_{x\to(1/2)\pi-}\frac{\tan 3x}{\tan x}=\lim_{x\to(1/2)\pi-}\frac{3\sec^2 3x}{\sec^2 x}=3\lim_{x\to(1/2)\pi-}\frac{\cos^2 x}{\cos^2 3x}.$$

This is again an indeterminate form, but now of the type 0/0. Applying Theorem 5.14 we have

$$3\lim_{x\to(1/2)\pi-}\frac{\cos^2 x}{\cos^2 3x}=3\lim_{x\to(1/2)\pi-}\frac{2\sin x\cos x}{6\sin 3x\cos 3x}=\lim_{x\to(1/2)\pi-}\left(\frac{\sin x}{\sin 3x}\right)$$

$$\times\lim_{x\to(1/2)\pi-}\left(\frac{\cos x}{\cos 3x}\right),$$

and hence

$$\lim_{x\to(1/2)\pi-}\frac{\tan 3x}{\tan x}=-\lim_{x\to(1/2)\pi-}\frac{\cos x}{\cos 3x}.$$

This last result is yet again an intermediate form of the type 0/0 so that a further application of Theorem 5.14 finally gives

$$\lim_{x\to(1/2)\pi-}\frac{\tan 3x}{\tan x}=-\lim_{x\to(1/2)\pi-}\frac{\sin x}{3\sin 3x}=\frac{1}{3}.$$

(e) This is of the form $\lim f/g\to\infty/\infty$, but it is easily seen that an application of Corollary 5.14 will not simplify the limit to be evaluated. Instead, we rewrite the limit in the form

$$\lim_{x\to 0-}\frac{\left(\frac{a}{x}\right)}{\cot bx}=\lim_{x\to 0-}a\frac{\tan bx}{x}$$

when it is seen that the alternative form is of the type $\lim f/g\to 0/0$ with $f(x)=a\tan bx$ and $g(x)=x$. Now $f'(x)=ab\sec^2 bx$ and $g'(x)=1$ so that by Theorem 5.14,

$$\lim_{x\to 0-}\frac{\left(\frac{a}{x}\right)}{\cot bx}=\lim_{x\to 0-}\frac{ab\sec^2 x}{1}=ab.$$

(f) This limit is of the form 0/0, so applying L'Hospital's Rule with

$$f(x)=x\sin x+\cos x-1 \text{ and } g(x)=\cos x-1$$

gives

$$\lim_{x\to 0}\frac{x\sin x+\cos x-1}{\cos x-1}=\lim_{x\to 0}\frac{x\cos x}{-\sin x},$$

which is again an indeterminate form. A further application of the rule, this time with $f(x)=x\cos x$ and $g(x)=-\sin x$, gives

$$\lim_{x\to 0}\frac{x\sin x+\cos x-1}{\cos x-1}=\lim_{x\to 0}\frac{x\cos x}{-\sin x}=\lim_{x\to 0}\frac{\cos x-x\sin x}{-\cos x}=-1.$$

Thus the required limit is -1, and it has been obtained by means of two applications of L'Hospital's rule.

(g) This limit is of the form $\infty-\infty$. Setting

$$I=\frac{1}{\sin 2x}-\frac{1}{2x}=\frac{2x-\sin 2x}{2x\sin 2x},$$

we see that the rearranged expression now becomes a limit of the form 0/0 as $x\to 0$. Thus applying L'Hospital's rule gives

$$\lim_{x\to o} I=\lim_{x\to 0}\frac{2-2\cos 2x}{2\sin 2x+4x\cos 2x},$$

which is again a limit of the form 0/0. A further application of the rule yields the result

$$\lim_{x\to o} I=\lim_{x\to 0}\frac{4\sin 2x}{8\cos 4x-8x\sin 2x}=0.$$

(h) This limit is of the form $0.\infty$. Setting

$$I=x\cot 2x=\frac{x}{\tan 2x},$$

we see that the rearranged expression now becomes a limit of the form 0/0 as $x\to 0$. An application of L'Hospital's rule gives

$$\lim_{x\to 0} I=\lim_{x\to 0}\left(\frac{1}{2\sec^2 2x}\right)=\frac{1}{2}.\quad\blacksquare$$

5.3.5 Identification of extrema

We return to the topic of extrema and, in particular, to the identification of functional behaviour at stationary values by means of the mean value theorem.

Suppose that a real-valued function $f(x)$ is differentiable in the interval (a, b) and has a maximum at an interior point x_0 of (a, b). Then if h is assumed to be positive and we consider the interval $[x_0-h, x_0]$ to the left of x_0, by the mean value theorem

$$\frac{f(x_0)-f(x_0-h)}{h}=f'(\xi),$$

where $x_0-h<\xi<x_0$.

Now by supposition $h>0$ and as x_0 is a maximum, the numerator of this expression will also be positive, showing that $f'(\xi)>0$. Hence by allowing h to tend to zero, it follows that $\xi\to x_0$, and we have shown that to the immediate left of the maximum we must have $f'>0$.

To the right of the maximum, and in the interval $[x_0, x_0+h]$, the same argument shows that

$$\frac{f(x_0 + h) - f(x_0)}{h} = f'(\eta),$$

where $x_0 < \eta < x_0+h$. This numerator is negative so that to the immediate right of the maximum we must have $f' < 0$.

Similar arguments, applied to a minimum and a point of inflection with a horizontal tangent, yield the following useful theorem, illustrated in Fig. 5.11.

Theorem 5.15 (identification of extrema using first derivative)

If $f(x)$ is a real-valued differentiable function in the neighbourhood of a point x_0 at which $f'(x_0) = 0$, then:

(a) the function has a maximum at x_0 if $f'(x) > 0$ to the left of x_0 and $f'(x) < 0$ to the right of x_0;
(b) the function has a minimum at x_0 if $f'(x) < 0$ to the left of x_0 and $f'(x) > 0$ to the right of x_0;
(c) the function has a point of inflection with zero gradient at x_0 if $f'(x)$ has the same sign to the left and right of x_0. ■

Sometimes these results are regarded as intuitively obvious deductions from the geometrical interpretation of a derivative in conjunction with the behaviour of the graph of the function. However, we have discussed them formally here as an illustration of an important consequence of the mean value theorem.

If, as in Fig 5.11(a), the graph of the function $y = f(x)$ lies above all chords drawn between any two points P and Q on the graph in some interval $a \leq x \leq b$ it will be recalled from section 2.3.7 that $f(x)$ is said to be *convex* in the interval. If, instead, of the graph lying above all chords PQ in an interval $a \leq x \leq b$ they lie below it, as in Fig. 5.11(b), the function $f(x)$ is said to be *concave* in the interval.

Example 5.10 We again consider the functions of Example 5.8.

(a) $f(x) = \frac{1}{3}x^3 + 2x^2 + 3x + 1$ with stationary points $x = \xi$ at $\xi = -1$ and $\xi = -3$. As $f'(x) = x^2+4x+3$ it follows that to the immediate left of $\xi = -1$ we have $f' < 0$, while to the immediate right $f' > 0$, showing that $\xi = -1$ corresponds to a minimum. This is most easily seen by computing $f'(x)$ at points to the immediate left and right of $x = -1$ at, say, $x = -1.05$ which is to the right, because $f'(-1.05) = -0.0975$ while $f'(-0.95) = 0.1025$. A similar argument shows that $\xi = -3$ corresponds to a maximum.

(b) $f(x) = (2x - 5)x^{2/3}$ with the one stationary point $x = \xi$ at $\xi = 1$. As $f'(x) = 2x^{2/3} + 2(2x - 5)/3x^{1/3}$, by computing $f'(x)$ to the immediate left and right of $x = 1$ as in (a), it follows that $f' < 0$ to the immediate left of

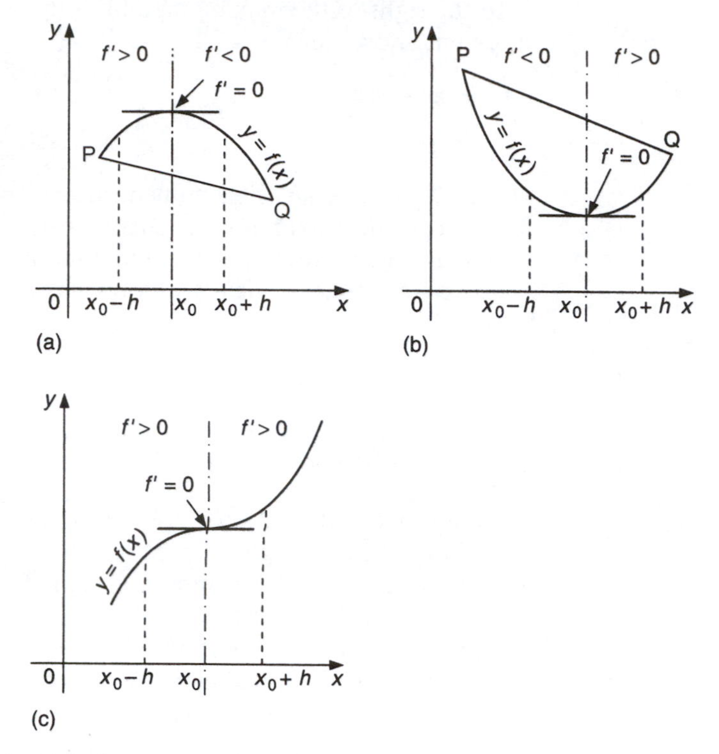

Fig. 5.11 Stationary values of $y = f(x)$: (a) local maximum and convex; (b) local minimum and concave; (c) point of inflection with zero gradient.

$\xi = 1$ and $f' > 0$ to the immediate right. Hence $\xi = 1$ corresponds to a minimum. As Theorem 5.15 stands, since f is not differentiable at the origin, the maximum that occurs there must be identified as in Example 5.8. However a trivial modification of the proof would show that results (a) and (b) of the theorem are still valid if f is not differentiable at x_0, so the sign of $f'(x)$ to the immediate left and right of $x = 0$ will again show it is a maximum (local).

5.3.6 Differentials

In using the notation $\mathrm{d}y/\mathrm{d}x$ to represent the derivative of the dependent variable y with respect to x, we have thus far been careful to emphasize that $\mathrm{d}y/\mathrm{d}x$ is simply a number defined by a limit. Although suggestive of increments, $\mathrm{d}y$ and $\mathrm{d}x$ taken separately have as yet no individual meaning. In many applications, particularly in differential equations which we encounter later, it is convenient to work with actual quantities $\mathrm{d}y$ and $\mathrm{d}x$ which we will call differentials.

However differentials must obviously be defined in a manner consistent with the notation $\mathrm{d}y/\mathrm{d}x$ when it is used to denote the derivative with respect to x of the function y defined by

$$y = f(x). \tag{5.8}$$

We achieve this by *defining* dy, the *first-order differential* of y, by

$$dy = f'(x).\Delta x, \tag{5.9}$$

where Δx is an increment in x of arbitrary size.

However, if, for the moment, we regard the independent variable x as a function of x we can write $x = g(x)$ with $g(x)=x$. Then by the above argument dx, the first-order differential of x, is defined by

$$dx = 1.\Delta x, \tag{5.10}$$

showing that we may meaningfully write Eqn (5.9) in the form

$$dy = f'(x)dx. \tag{5.11}$$

When needed, the actual increment in y consequent upon an increment Δx in x will be denoted by Δy. In general, the differential dy and the increment Δy are distinct quantities, and the interrelationship between them is indicated in Fig. 5.12.

In more advanced treatments the use of differentials is strictly avoided on account of logical difficulties encountered with their definition. However, they are so useful that we shall ignore these objections and use them freely whenever necessary.

It is an immediate consequence of this that if

$$y = k_1 f(x) + k_2 g(x)$$

then by Theorem 5.4, the *differential of a sum* is

$$dy = k_1 f'(x)dx + k_2 g'(x)dx$$

or, equivalently, in symbolic notation

$$d(k_1 f + k_2 g) = k_1\, df + k_2\, dg. \tag{5.12}$$

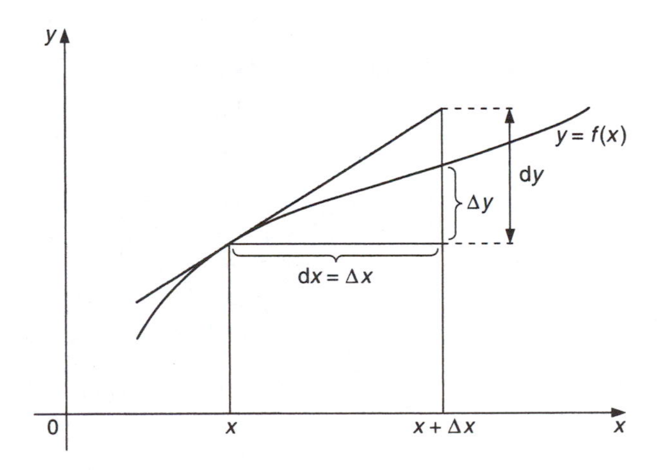

Fig. 5.12 Differentials dx and dy.

If we have

$$y = f(x)g(x)$$

then by Theorem 5.5, the *differential of a product* is

$$dy = g(x)f'(x)dx + f(x)g'(x)dx$$

or, equivalently, in symbolic notation

$$d(fg) = g df + f dg. \tag{5.13}$$

Finally, if

$$y = f(x)/g(x)$$

then by Theorem 5.8, the *differential of a quotient* is

$$dy = \frac{g(x)f'(x)dx - f(x)g'(x)dx}{g^2(x)}$$

or, equivalently, in symbolic notation

$$d\left(\frac{f}{g}\right) = \frac{g df - f dg}{g^2}. \tag{5.14}$$

Example 5.11 If $f(x) = \sin(x^2+4)$ and $g(x) = x^3$ find the differentials:

(a) $d(3f+g)$;
(b) $d(fg)$;
(c) $d\left(\frac{f}{g}\right)$.

Solution

(a)
$$\begin{aligned} d(3f+g) &= d[3\sin(x^2+4) + x^3] \\ &= 3\cos(x^2+4)d(x^2+4) + 3x^2dx \\ &= 6x\cos(x^2+4)dx + 3x^2dx. \end{aligned}$$

(b)
$$\begin{aligned} d(fg) &= d[x^3\sin(x^2+4)] \\ &= 3x^2\sin(x^2+4)dx + x^3\cos(x^2+4) + d(x^2+4) \\ &= 3x^2\sin(x^2+4)dx + 2x^4\cos(x^2+4) + dx. \end{aligned}$$

(c)
$$\begin{aligned} d\left(\frac{f}{g}\right) &= d\left[\frac{\sin(x^2+4)}{x^3}\right] \\ &= \frac{x^3\cos(x^2+4)d(x^2+4) - 3x^2\sin(x^2+4)dx}{x^6} \\ &= \frac{2x^2\cos(x^2+4)dx - 3\sin(x^2+4)dx}{x^4}. \end{aligned}$$ ■

For small values of $\mathrm{d}x$, the differential $\mathrm{d}y$ is a reasonable approximation to the actual increment Δy. This simple observation is often utilized to relate small changes in dependent and independent variables, as the next example shows.

Example 5.12 The pressure p of a polytropic gas is related to the density ρ by the expression

$$p = A\rho^{\gamma}$$

where A is a constant. Deduce the relationship connecting the differentials $\mathrm{d}p$ and $\mathrm{d}\rho$. Given that $\gamma = 3/2$ and $\rho = 4$, and taking $\mathrm{d}p$ as an approximation to the actual pressure change Δp, compute the approximate new pressure if ρ is increased by 0.1. Compare the approximate and exact results.

Solution In this case $p = f(\rho)$ with $f(\rho) = A\rho^{\gamma}$. Hence $f'(\rho) = \gamma A\rho^{\gamma-1}$ and thus the desired differential relation is

$$\mathrm{d}p = \gamma A\rho^{\gamma-1}\,\mathrm{d}\rho.$$

When $\gamma = 3/2$ and $\rho = 4$ it follows from the stated pressure–density law that the initial pressure p_0 is

$$p_0 = 4^{3/2}\,A = 8A.$$

Using the differential relation to compute the approximate pressure increase represented by the differential $\mathrm{d}p$, we find

$$\mathrm{d}p = (3/2).A.4^{1/2}.(0.1) = 0.3A.$$

Hence the approximate new pressure $p_0 + \mathrm{d}p = 8.3A$.

The exact new pressure $p_0 + \Delta p$ may be computed from the pressure–density law by setting $\rho = 4.1$ to obtain

$$p_0 + \Delta p = (4.1)^{3/2}\,A = 8.302A.$$

This shows that in this case the differential relation gives a good approximation to the pressure increase. ■

5.3.7 The exponential function

The *Euler number* e and the associated *exponential function* e^x can be defined in several different ways, one of which we saw in section 3.3. Our purpose here will be to establish a connection between the earlier definition, based on the limit of a sequence, and the derivative of the function

$$y = a^x\,(a > 1)$$

for a particular value of a.

Let us seek a number a with the property that the tangent line to its graph at $x = 0$ has gradient 1. Thus we define a to be such that

$$\left[\frac{d}{dx}[a^x]\right]_{x=0} = 1.$$

The behaviour of $y = a^x$ for different positive values of a is shown in Fig. 5.13, from which it can be seen that the gradient of the tangent line to the graph at $x = 0$ increases as a increases, starting from 0 when $a = 1$ and becoming arbitrarily large as $a \to \infty$.

By definition, for any choice of a, we have

$$\left[\frac{d}{dx}[a^x]\right]_{x=0} = \lim_{h\to 0}\left(\frac{a^h - 1}{h}\right).$$

The gradient of the tangent line to curve at P in Fig. 5.14 will be 1 if we set

$$\lim_{h\to 0}\left(\frac{a^h - 1}{h}\right) = 1.$$

Writing $h = 1/n$, so that $h \to 0$ corresponds to $n \to \infty$, causes the above limit to become

$$\lim_{n\to\infty}\left(\frac{a^{1/n} - 1}{1/n}\right) = 1.$$

Arguing intuitively, but with the notion of a limit of a sequence in mind, we recognize that this limit means that for n greater than some suitably large number N

$$\frac{a^{1/n} - 1}{1/n} \approx 1,$$

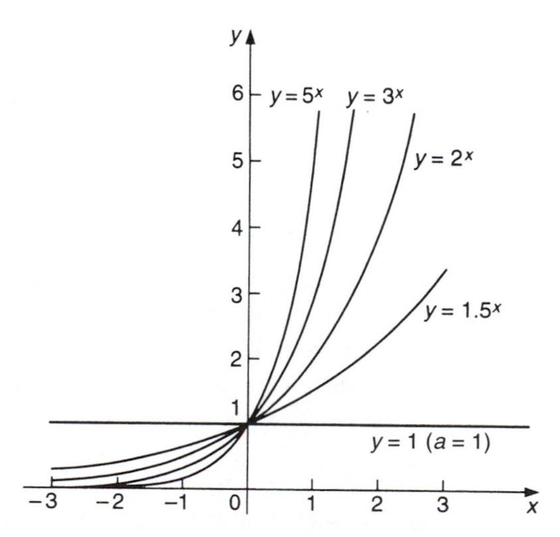

Fig. 5.13 Graphs of $y = a^x$, for different values of $a > 0$.

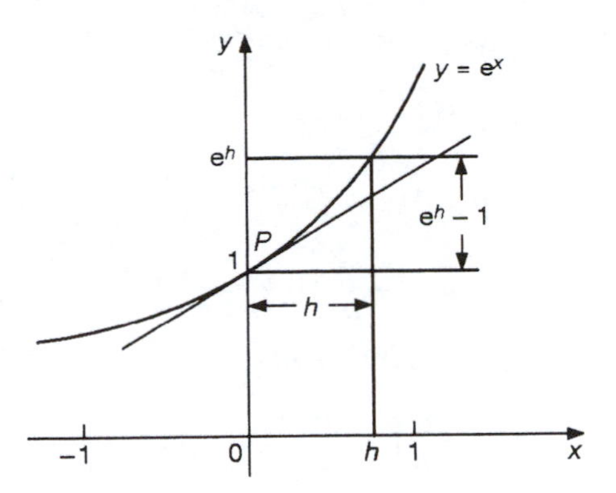

Fig. 5.14 Geometrical interpretation of $d[e^x]/dx$.

where the symbol $\approx$ is to be read 'is approximately equal to'. Thus

$$a^{1/n} \approx 1 + \frac{1}{n},$$

so after raising each side of this approximation to the power n we have

$$a \approx \left(1 + \frac{1}{n}\right)^n.$$

Proceeding to the limit, this becomes exact and shows that

$$a = \lim_{n\to\infty}\left(1 + \frac{1}{n}\right)^n,$$

which is precisely the definition of the Euler number given in Eqn (3.1), so that $a = e$. Thus we have established an important connection between the definition of e as the limit of the sequence in Eqn. (3.1), and as the number for which the function e^x is such that

$$\left[\frac{d}{dx}[e^x]\right]_{x=0} = 1.$$

If in the above argument we had set $h = x/n$, with $x = \text{const.}$, the same reasoning would have led to the more general result

$$e^x = \lim_{n\to\infty}\left(1 + \frac{x}{n}\right)^n.$$

It is useful to recall the following general properties of exponents, and hence of the exponential function e^x in particular. For any $a > 0$, $b > 0$ and for any real numbers x and y:

(1) $a^0 = 1$;

(2) $a^x a^y = a^{x+y}$;

(3) $\dfrac{a^x}{a^y} = a^{x-y}$;

(4) $(a^x)^y = a^{xy}$;

(5) $a^{-x} = \dfrac{1}{a^x}$;

(6) $(ab)^x = a^x b^x$;

(7) $\left(\dfrac{a}{b}\right)^x = \dfrac{a^x}{b^x}$.

For reference purposes, graphs of $y = \mathrm{e}^x$ and $y = \mathrm{e}^{-x}$ are shown in Fig. 5.15, from which it can be seen that each graph is the reflection of the other in the y-axis. This fact is, of course, a consequence of property 5 listed above.

Let us now determine the derivative of the exponential function

$$\frac{\mathrm{d}}{\mathrm{d}x}[\mathrm{e}^x]$$

when $x = x_0$, for any arbitrary value of x_0. By definition,

$$\left[\frac{\mathrm{d}}{\mathrm{d}x}[\mathrm{e}^x]\right]_{x=x_0} = \lim_{h\to 0}\left[\frac{\mathrm{e}^{x_0+h} - \mathrm{e}^{x_0}}{h}\right]$$

$$= \mathrm{e}^{x_0}\lim_{h\to 0}\left[\frac{\mathrm{e}^h - 1}{h}\right],$$

but this last limit is 1, so that

$$\left[\frac{\mathrm{d}}{\mathrm{d}x}[\mathrm{e}^x]\right]_{x=x_0} = \mathrm{e}^{x_0}.$$

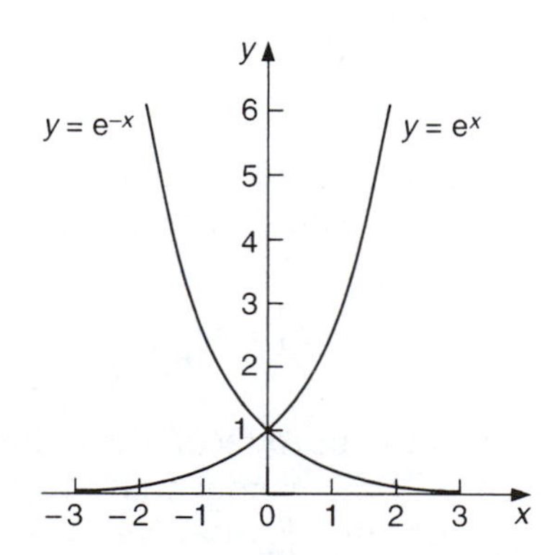

Fig. 5.15 The graphs of e^x and e^{-x}.

As x_0 was arbitrary the subscript zero may be omitted, causing this to become the fundamental result

$$\frac{d}{dx}[e^x] = e^x. \tag{5.15}$$

The derivative of $e^{f(x)}$ follows by applying the chain rule, for setting $u = f(x)$ we have

$$\frac{d}{dx}[e^{f(x)}] = \frac{d}{du}[e^u]\frac{du}{dx}$$
$$= e^{f(x)}f'(x),$$

where $f'(x)$ denotes the derivative of $f(x)$. Thus we have shown that

$$\frac{d}{dx}[e^{f(x)}] = f'(x)e^{f(x)}. \tag{5.16}$$

In particular, when $f(x) = kx$, it follows directly that

$$\frac{d}{dx}[e^{kx}] = ke^{kx}. \tag{5.17}$$

5.4 Higher derivatives–applications

We have seen how differentiation applied to a suitable function $f(x)$ yields as a result another function $f'(x)$, the derivative of $f(x)$ with respect to x. If the function $f'(x)$ is itself differentiable then a repetition of differentiation will result in a further function that we shall denote by $f''(x)$ and will call the *second derivative* of $f(x)$ with respect to x. We may usefully employ the dynamical problem that served to introduce the notion of a derivative to give meaning to the notion of a second derivative, for if $f'(x)$ represents a velocity, then $f''(x)$ represents an acceleration. If the function $f''(x)$ is itself differentiable then it is customary to denote the *third derivative* of $f(x)$ by $f'''(x)$, after which, if necessary, further derivatives are conventionally denoted by the use of bracketed superscript numerals. Hence the sixth derivative of a suitably differentiable function $f(x)$ would be written $f^{(6)}(x)$, and the nth derivative $f^{(n)}(x)$.

Alternative notations that are in use are

$$\frac{d^n y}{dx^n} \text{ or } D^n y.$$

These both represent the nth derivative with respect to x of $y = f(x)$ and for their determination require the successive application of differentiation n times. The number n is the *order* of the derivative and the symbol D symbolizes the operation of differentiation. Computationally the definition of the nth derivative of y with respect to x is equivalent to using either of these two equivalent algorithms

$$\frac{d}{dx}\left(\frac{d^{n-1}y}{dx^{n-1}}\right) = \frac{d^n x}{dx^n} \quad \text{or} \quad D[D^{n-1}y] \equiv D^n y. \tag{5.15}$$

These expressions are, of course, only meaningful when n is an integer, and we shall agree to the convention $D^0 y \equiv y$.

Geometrically, the function $d^n y/dx^n$ bears to the graph of $d^{n-1}y/dx^{n-1}$, the same relationship as does the function dy/dx to the graph of y. Namely, $d^n y/dx^n$ at $x = x_0$ is the gradient of the graph of $d^{n-1}y/dx^{n-1}$ as a function of x at the same point $x = x_0$. In particular, if $d^2y/dx^2 > 0$ this means dy/dx is *increasing*, whereas if $d^2y/dx^2 < 0$ it is *decreasing*.

Example 5.13 Determine dy/dx, d^2y/dx^2, d^3y/dx^3 and d^4y/dx^4, given that $y = f(x)$ with:

(a) $f(x) = \cos mx$;
(b) $f(x) = \tan x$;
(c) $f(x) = 1/(1 + x)$.

If possible make deductions about the nth derivative.

Solution

(a) $$\frac{dy}{dx} = f'(x) = \frac{d}{dx}(\cos mx) = -m \sin mx,$$

$$\frac{d^2y}{dx^2} = \frac{d}{dx}\left(\frac{dy}{dx}\right) = \frac{d}{dx}[-m \sin mx] = -m^2 \cos mx,$$

$$\frac{d^3y}{dx^3} = \frac{d}{dx}\left(\frac{d^2y}{dx^2}\right) = \frac{d}{dx}[-m^2 \cos mx] = m^3 \sin mx,$$

$$\frac{d^4y}{dx^4} = \frac{d}{dx}\left(\frac{d^3y}{dx^3}\right) = \frac{d}{dx}[m^3 \sin mx] = m^4 \cos mx.$$

Notice first that the power to which m is raised is equal to the order of the derivative involved, and that the sign of the derivative changes every two differentiations, starting with dy/dx and d^2y/dx^2 each with a negative sign. Secondly, after each differentiation, the function in the derivative alternates between a sine and the cosine, so if a general expression is to be found for the nth derivative, we must try to find a function with these characteristics. Consider the function

$$\cos[mx + (n\pi/2)] = \cos mx \cos(n\pi/2) - \sin mx \sin(n\pi/2)$$

as a possible candidate. When n is even, that is when $n = 2r$, this becomes

$$\cos[mx + (2r\pi/2)] = \cos[mx + r\pi] = \cos(mx)\cos(r\pi) - \sin(mx)\sin(r\pi).$$

However $\cos(r\pi) = (-1)^r$, which is a result that is often useful because $(-1)^r$ simply switches between 1 and -1, and $\sin(r\pi) = 0$, so we find that

$$\cos[mx + (2r\pi/2)] = (-1)^r \cos(mx),$$

showing that while the cosine function remains, its sign switches every time r increases by 1, that is after every two differentiations.

Similarly when n is odd, that is when $n = 2r + 1$, we have

$$\cos[mx + (2r+1)\pi/2] = \cos(mx)\cos[(2r+1)\pi/2] - \sin(mx)\sin[(2r+1)\pi/2].$$

But now $\cos[(2r+1)\pi/2] = 0$ and $\sin[(2r+1)\pi/2] = (-1)^r$, so we see that

$$\cos[mx + (2r+1)\pi/2] = -(-1)^r \sin[(2r+1)\pi/2] = (-1)^{r+1}\sin[(2r+1)\pi/2].$$

Now it is the sine function that remains with its sign switching every time r increases by 1, again after every two differentiations. Thus, after multiplying $\cos[mx + (n\pi/2)]$ by m^n, because a factor m occurs as a multiplier after each differentiation, we arrive at the following general expression for the nth derivative

$$(\mathrm{d}^n/\mathrm{d}x^n)(\cos mx) = m^n \cos[mx + (n\pi/2)],$$

for $n = 0, 1, 2, \ldots$, with $(\mathrm{d}^0/\mathrm{d}x^0)(\cos mx) = \cos mx$.

In respect of the function $y = \cos mx$, it is of importance to notice that the simple equation

$$\frac{\mathrm{d}^2y}{\mathrm{d}x^2} + m^2y = 0$$

connects the function and its second derivative. Because this equation involves derivatives it is a *differential* equation. Such equations are very important in both mathematics and the mathematical sciences; the later chapters of this book provide an introductory study of them.

(b) $$\frac{\mathrm{d}y}{\mathrm{d}x} = f'(x) = \frac{\mathrm{d}}{\mathrm{d}x}(\tan x) = \sec^2 x = 1 + \tan^2 x$$

$$\frac{\mathrm{d}^2y}{\mathrm{d}x^2} = \frac{\mathrm{d}}{\mathrm{d}x}\left(\frac{\mathrm{d}y}{\mathrm{d}x}\right) = \frac{\mathrm{d}}{\mathrm{d}x}(\sec^2 x) = 2\sec^2 x \tan x = 2\tan x(1 + \tan^3 x)$$

$$\frac{\mathrm{d}^3y}{\mathrm{d}x^3} = \frac{\mathrm{d}}{\mathrm{d}x}\left(\frac{\mathrm{d}^2y}{\mathrm{d}x^2}\right) = \frac{\mathrm{d}}{\mathrm{d}x}(2\sec^2 x \tan x)$$

$$= 2\sec^2 x(2\tan^2 x + \sec^2 x) = 2 + 8\tan^2 x + 6\tan^4 x$$

$$\frac{\mathrm{d}^4y}{\mathrm{d}x^4} = 16\tan x + 40\tan^3 x + 24\tan^5 x.$$

There is no simple rule by which $(\mathrm{d}^n/\mathrm{d}x^n)(\tan x)$ may be computed.

(c) $$\frac{\mathrm{d}y}{\mathrm{d}x} = f'(x) = \frac{\mathrm{d}}{\mathrm{d}x}\left(\frac{1}{1+x}\right) = \frac{-1}{(1+x)^2}$$

$$\frac{d^2y}{dx^2} = \frac{d}{dx}\left(\frac{dy}{dx}\right) = \frac{d}{dx}\left[\frac{-1}{(1+x)^2}\right] = \frac{2}{(1+x)^3}$$

$$\frac{d^3y}{dx^3} = \frac{d}{dx}\left(\frac{d^2y}{dx^2}\right) = \frac{d}{dx}\left[\frac{2}{(1+x)^3}\right] = \frac{-3!}{(1+x)^4}$$

$$\frac{d^4y}{dx^4} = \frac{4!}{(1+x)^5}.$$

From the pattern of the results we see that

$$\frac{d^n}{dx^n}\left(\frac{1}{1+x}\right) = \frac{(-1)^n n!}{(1+x)^{n+1}}. \quad \blacksquare$$

In general, functions are not capable of differentiation an indefinite number of times, because at some stage they may become non-differentiable.

5.4.1 Leibniz's theorem

This useful theorem is a consequence of Theorem 5.5 and facilitates the computation of high-order derivatives of the product $f(x)g(x)$ of the two functions $f(x)$ and $g(x)$, in terms of the derivatives of the individual functions $f(x)$ and $g(x)$ themselves.

The result is, perhaps, best expressed in terms of the symbolic differentiation operator D, and for our starting point we now re-express the result of Theorem 5.5 in terms of the operator D.

$$D(fg) = fDg + gDf.$$

Assuming functions $f(x)$ and $g(x)$ are suitably differentiable, a further application of the operator D together with Theorem 5.5 yields

$$\begin{aligned} D^2(fg) &= D(fDg + gDf) \\ &= Df.Dg + fD^2g + Dg.Df + gD^2f. \end{aligned}$$

However,

$$Df.Dg = \frac{df}{dx}\cdot\frac{dg}{dx} = \frac{dg}{dx}\frac{df}{dx} = Dg.Df,$$

so that

$$D^2(fg) = fD^2g + 2Df.Dg + gD^2f. \tag{5.16}$$

A repetition of the same argument shows that

$$D^3(fg) = fD^3g + 3Df.D^2g + 3D^2f.Dg + gD^3f. \tag{5.17}$$

The coefficients involved in $D^2(fg)$ and $D^3(fg)$ are seen to belong to the general pattern of binomial coefficients in the expansion of $(a+b)^n$, where n is the order of the derivative involved.

$$
\begin{array}{ll}
(n=0) & \binom{0}{0} \\
(n=1) & \binom{1}{0}\binom{1}{1} \\
(n=2) & \binom{2}{0}\binom{2}{1}\binom{2}{2} \\
(n=3) & \binom{3}{0}\binom{3}{1}\binom{3}{2}\binom{3}{3}
\end{array}
$$

or, equivalently, to the pattern

$$
\begin{array}{lllll}
(n=0) & 1 & & & \\
(n=1) & 1 & 1 & & \\
(n=2) & 1 & 2 & 1 & \\
(n=3) & 1 & 3 & 3 & 1 \\
 & \cdots & & &
\end{array}
$$

This indicates that in evaluating $D^n(fg)$, the coefficients arising belong to the $(n+1)$th row of either of these arrays, which are simply the rows of Pascal's triangle, introduced in section 1.4. We now state this conclusion in the form of a theorem.

***Theorem 5.16* (Leibniz's theorem)** If $f(x)$ and $g(x)$ are n times differentiable real-valued functions in the interval (a, b), then

$$D^n(fg) = \sum_{k=0}^{n} \binom{n}{k} D^{n-k}f.D^k g. \quad \blacksquare$$

The value and power of this theorem is best shown by an application.

Example 5.14 Use Leibniz's theorem to evaluate $(\mathrm{d}^3/\mathrm{d}x^3)(x^6 \sin x)$.

Solution Setting $n=3$ in the general result gives

$$D^3(fg) = gD^3f + 3D^2f.Dg + 3Df.D^2g + fD^3g.$$

Now we make the identifications $f(x)=x^6$ and $g(x)=\sin x$, whence it follows that $Df=6x^5$, $D^2f=30x^4$, $D^3f=120x^3$ and $Dg=\cos x$, $D^2g=-\sin x$, $D^3g=-\cos x$. Hence substitution into the above result gives

$$D^3(x^6 \sin x) = 120x^3 \sin x + 90x^4 \cos x - 18x^5 \sin x - x^6 \cos x.$$

5.4.2 Identification of extrema by second derivatives

An important application of the second derivative of a function $f(x)$ is to the identification of the nature of its extrema. Let us suppose that $f(x)$ is twice differentiable and that $f'(x_0) = 0$ and $f''(x_0) = L < 0$.

Then from Definition 5.2 and the notion of a second derivative, we must have that

$$f''(x_0) = \lim_{x \to x_0} \frac{f'(x) - f'(x_0)}{x - x_0} = L < 0.$$

By supposition $f'(x_0) = 0$, so that

$$f''(x_0) = \lim_{x \to x_0} \frac{f'(x)}{x - x_0} = L < 0.$$

This limit must be independent of the manner in which x approaches x_0 so that we must consider separately the cases that x lies to the left or to the right of x_0.

If x lies to the left of x_0 then $x - x_0 < 0$. Consequently, as the value L of the limit is negative, the expression defining $f''(x_0)$ implies that to the immediate left of x_0 it must be true that $f'(x) > 0$.

If x lies to the right of x_0 then $x - x_0 > 0$. Consequently, as the value L of the limit is negative, the expression defining $f''(x_0)$ implies that to the immediate right to x_0 it must be true that $f'(x) < 0$.

These results, in conjunction with Theorem 5.15(a), prove that at a stationary value x_0, for which $f''(x_0) < 0$, the function $f(x)$ attains a *maximum* value. An exactly similar argument proves that at a stationary value x_0, for which $f''(x_0) > 0$, the function $f(x)$ attains a *minimum* value.

To complete the argument, consider the situation in which $f''(x_0) = 0$. It might be conjectured that this corresponds to a point of inflection; and to establish the correctness of our intuition let us appeal to the geometrical interpretation of a derivative as a gradient.

Suppose that x_0 corresponds to a point of inflection with zero gradient. Then as x increases through the value x_0, either

(a) $f'(x)$ is initially positive and decreases to a minimum value $f'(x_0) = 0$, thereafter increasing again (cf. Fig. 5.11(c)); or
(b) $f'(x)$ is initially negative and increases to a maximum value $f'(x_0) = 0$, thereafter decreasing again.

In each case x_0 is a stationary value of the first derivatives $f'(x)$, so that, by an application of Theorem 5.10 to the function $f'(x)$, we find that $f''(x) = 0$ at a point of inflection.

We have proved the following theorem.

Theorem 5.17 (indentification of extrema using second derivatives) Let $f(x)$, be a real-valued twice differentiable function in (a, b) with a stationary point x_0 in (a, b) so that $f'(x_0) = 0$. Then, if

(a) $f''(x_0) < 0$ the function $f(x)$ has a maximum at x_0;
(b) $f''(x_0) > 0$ the function $f(x)$ has a minimum at x_0;
(c) $f''(x_0) = 0$ the function $f(x)$ has a point of inflection at x_0 with zero gradient provided that the sign of $f'(x)$ is the same to the immediate left and right of x_0. ■

The proof of this theorem shows clearly what was asserted earlier; namely, that a point of inflection on the graph of a function separates a region of convexity from a region of concavity, at which point the tangent to the curve actually crosses the curve. There is, of course, no necessity that this point should have associated with it a zero gradient.

Following this argument to its logical conclusion, we see that the proof of (c) above need only involve the sign of $f'(x)$ to the left and right of x_0 when $f'(x_0) = 0$, for then such arguments are needed to distinguish between an extremum and a point of inflection. If $f'(x_0) \neq 0$ such problems do not arise, and it is sufficient to look for those values ξ for which $f''(\xi) = 0$. We have thus proved the following general result.

Theorem 5.18 (location of points of inflection) If $f(x)$ is a real-valued twice differentiable function then its points of inflection, if any, occur at the values ξ for which $f''(\xi) = 0$, provided that $f'(\xi) \neq 0$. If, however, this is not so, and $f'(\xi) = 0$, then ξ corresponds to a point of inflection provided that the sign of $f'(x)$ is the same to the immediate left and right of ξ.

It is left as an exercise to prove that when $f'(x_0) = f''(x_0) = 0$, then provided $f'''(x_0)$ exists, our condition on $f'(x)$ may be replaced by the requirement $f'''(x_0) \neq 0$. The proof is essentially similar to that given for Theorem 5.17, though this time the starting point is the definition of $f''(x_0)$ expressed as a limit. We give this result as a corollary.

Corollary 5.18 If $f(x)$ is a real-valued thrice differentiable function and $f'(\xi) = f''(\xi) = 0$, then $f(x)$ has a point of inflection at $x = \xi$ if $f'''(\xi) \neq 0$.

Example 5.15 Locate and identify the stationary values of the following functions. Find any points of inflection they may have, together with the gradient of the tangent line at such points:

(a) $f(x) = x^3 - 12x + 1$ in $[-10, 10]$;
(b) $f(x) = \tan x$ in $[-\frac{1}{4}\pi, \frac{1}{4}\pi]$;

(c) $f(x) = (x - 1)^3$ in$(-\infty, \infty)$.

Solution (a) The stationary values are those values ξ for which $f'(\xi) = 0$. Hence as $f'(x) = 3x^2 - 12$, the stationary values are determined by the equation

$$3\xi^2 - 12 = 0.$$

This has roots $\xi = 2$, $\xi = -2$ which both lie in $[-10, 10]$ and are the desired stationary values. As $f''(x) = 6x$, it follows that $f''(2) = 12 > 0$ and $f''(-2) = -12 < 0$. Hence by Theorem 5.17, the point $\xi = 2$ is a minimum and the point $\xi = -2$ is a maximum. Since the function has no other stationary value there can be no point of inflection at which the tangent line has zero gradient. However, $f''(x) = 6x$ vanishes when $x = 0$, so that by Theorem 5.18 we see that $x = 0$ must correspond to a point of inflection. The gradient at $x = 0$ is $f'(0) = -12$ which is the gradient of the desired tangent line to the graph at the point of inflection.

(b) Here we have $f'(x) = \sec^2 x$ and clearly, since $\sec^2 x = 1 + \tan^2 x$, it follows that $f'(x) \neq 0$ in $[-\frac{1}{4}\pi, \frac{1}{4}\pi]$. The function $f(x) = \tan x$ thus has no stationary values in $[-\frac{1}{4}\pi, \frac{1}{4}\pi]$, though it assumes its greatest value at $\frac{1}{4}\pi$ and its least value at $-\frac{1}{4}\pi$. We have $f''(x) = 2 \sec^2 x \tan x$ which vanishes for $x = 0$. Hence by Theorem 5.18, the function $\tan x$ has a point of inflection at the origin at which the gradient of the tangent to the graph has the value $f'(0) = 1$.

(c) We see that $f'(x) = 3(x - 1)^2$ and so the condition $f'(\xi) = 0$ yields $\xi = 1$ as the single stationary value. However, $f''(x) = 6(x - 1)$ which shows that we also have $f''(1) = 0$. Appealing to the last part of Theorem 5.18 we see that, as $f'(x) = 3(x - 1)^2 > 0$ to both the left and right of $x = 1$, it follows that $f(x) = (x - 1)^3$ has a point of inflection at that point. The tangent line to the graph there has a zero gradient. Alternatively, as $f'''(x) \equiv 6 \neq 0$, the result also follows from Corollary 5.18. ■

5.4.3 Graph sketching

When sketching the graph of a function $y = f(x)$ the following features possessed by the graph can help if they can be identified:

(i) Locate and mark any points at which the function crosses the x- and y-axes.
(ii) Locate and draw any asymptotes that exist.
(iii) Locate, identify the nature of and mark points at which the function has a stationary point, remembering that the tangent to the graph at such points is parallel to the x-axis.
(iv) Locate and mark any points of inflection, remembering that a point of inflection occurs when $f''(x) = 0$, and that if such a point occurs at $x = c$, say, then the tangent line at the point $(c, f(c))$ has gradient $f'(c)$.
(v) Use the fact that the function is convex at a maximum and concave at a minimum.

A general test for *convexity* in an interval $a \le x \le b$ in which $f(x)$ is twice differentiable is easily seen to be that $f''(x) < 0$, for $a \le x \le b$, because $f'(x)$ decreases throughout the interval. Similarly, a general test for *concavity* in an interval $a \le x \le b$ is that $f''(x) > 0$, for $a \le x \le b$, because $f'(x)$ increases throughout the interval.

To illustrate matters, let us apply these arguments to the function $y = (2x - 5)x^{2/3}$ shown in Fig. 5.8.

(i) The function vanishes at the points $x = 0$ and $x = 5/2$ on the x-axis.
(ii) The function has no asymptotes.
(iii) $f'(x) = 0$ at $x = 1$, at which point $y = -3$, so the only stationary point is located at $(1, -3)$.
(iv) As $f''(x) = 10(2x+1)/(9x^{4/3})$, we have $f''(x) = 0$ when $x = -1/2$, corresponding to a point of inflection at $(-1/2, -6(1/4)^{1/3})$ with gradient $5(1/2)^{-1/3}$, though this cannot be seen in Fig. 5.8 because of its small size.
(v) From the form of $f''(x)$ in (iv), we see that $f''(x) > 0$ for $x > 0$ so that $y = f(x)$ must be concave for $x > 0$. As $f''(x)$ changes sign across $x = -1/2$, being positive for $-1/2 < x < 0$ and negative for $-\infty < x < -1/2$, the graph is concave in this first interval and convex in the second one.

5.5 Partial differentiation

The notion of continuity has already been extended so that it is meaningful in the context of functions of several independent variables. It is now appropriate to extend the notion of a derivative in a similar fashion. For simplicity of argument we shall work with the function $f(x, y)$ of two independent variables, and in order to visualize its behaviour geometrically, we will define a dependent variable by the equation

$$u = f(x, y). \tag{5.18}$$

The function may then be represented as a surface in three-dimensional space.

A typical surface generated by a function of the form of Eqn (5.18) is shown in Fig. 5.16 and, unlike functions of one independent variable, it is necessary to define more than one first-order derivative. The idea involved is simple: by holding one of the independent variables in f constant at some value of interest, the function f then becomes a function of the single remaining independent variable. We may then differentiate f as though it were a function only of that one variable. By holding first x and then y constant in this manner, two different derivatives may be defined which, because of their manner of computation, will be called *partial derivatives* to distinguish them from our earlier use of the term 'derivative'. We shall now express these ideas formally as a definition and set down the standard notation to be used.

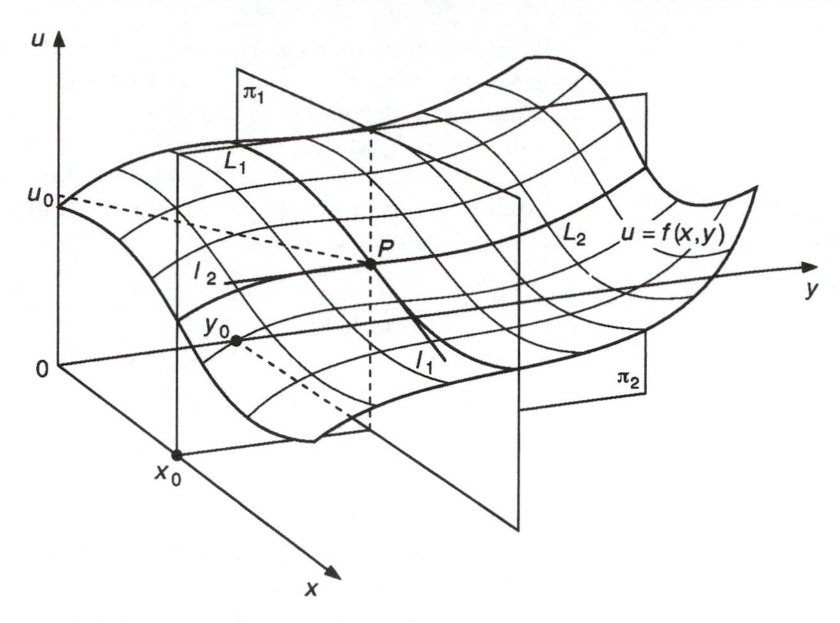

Fig. 5.16 Geometrical interpretation of higher derivatives.

Definition 5.5 (partial derivatives)

Let $f(x, y)$ be a function defined near (x_0, y_0). Suppose that

$$\lim_{x \to x_0} \frac{f(x, y_0) - f(x_0, y_0)}{x - x_0} \tag{A}$$

exists and is independent of the direction of approach of x to x_0, that is from either the left or right. Then f is differentiable *partially* with respect to x at (x_0, y_0). The value of the limit is denoted by $f_x(x_0, y_0)$ or by $\partial f/\partial x|_{(x_0, y_0)}$ and called the *first-order partial derivative* of f with respect to x at (x_0, y_0).

Similarly, suppose that

$$\lim_{y \to y_0} \frac{f(x_0, y) - f(x_0, y_0)}{y - y_0} \tag{B}$$

exists and is independent of the direction of approach of y to y_0. Then f is differentiable partially with respect to y at (x_0, y_0). The limit is denoted by $f_y(x_0, y_0)$ or by $\partial f/\partial y|_{(x_0, y_0)}$ and called the first-order partial derivative of f with respect to y at (x_0, y_0).

By analogy with ordinary derivatives, if $f(x, y)$ is differentiable partially with respect to x and y at all points of some region in the (x, y)-plane and these derivatives are continuous, then we say f is *differentiable* in that region. The operations of partial differentiation with respect to x and y are usually denoted by the differentiation operators $\partial/\partial x$ and $\partial/\partial y$, respectively.

Let us now interpret these definitions in terms of Fig. 5.16. The function $f(x, y_0)$ occurring in the numerator of limit (A) in Definition 5.5 is represented in that figure by the intersection of the surface $u = f(x, y)$ with the plane $y = y_0$ which has been labelled π_1. It is the curve L_1. The number $f_x(x_0, y_0)$ defined by limit (A) is the gradient of the tangent line l_1 to this curve at point P. By requiring the limit to be independent of the direction of approach of x to x_0, we have ensured that the tangent lines drawn to the curve at P, whether from the left or the right, will have the same gradient. In simpler terms this ensures that the curve L_1 is smooth and has no kink at P.

The number $f(x_0, y)$ occurring in the numerator of limit (B) in the definition is represented in Fig. 5.16 by the intersection of the surface $u = f(x, y)$ with the plane $x = x_0$ which has been labelled π_2. It is the curve L_2. The number $f_y(x_0, y_0)$ defined by limit (B) is the gradient of the tangent line l_2 to this curve at point P.

Thus by differentiating partially we mean that, during the process of differentiation, with respect to one independent variable, the other independent variable is to be regarded as a constant. In consequence, all the rules of differentiation developed for functions of a single variable are also rules of partial differentiation, provided only that the functions involved are suitably differentiate. On account of this, when, for example, the operator $\partial/\partial x$ acts on a function only of y, say $g(y)$, that function is to be regarded as a constant with respect to this operator and so $(\partial/\partial x)[g(y)] = 0$. Similarly $(\partial/\partial y)[h(x)] = 0$.

Example 5.16 In each of the following cases compute f_x and f_y as functions of x and y. Use the result to determine the numerical value of these derivatives at the stated points:

(a) $f(x, y) = x^3 + 2xy + 2y^2$; $(1, 2)$;
(b) $f(x, y) = x \sin xy + 3$; $(1, \frac{1}{2}\pi)$;
(c) $f(x, y) = x/(x^2 + y^2)$; $(1, 0)$.

Solution

$$\text{(a)}\quad f_x = \frac{\partial}{\partial x}[x^3 + 2xy + 2y^2]$$

$$= \frac{\partial}{\partial x}[x^3] + 2y\frac{\partial}{\partial x}[x] + 2y^2\frac{\partial}{\partial x}[1],$$

whence

$$\frac{\partial f}{\partial x} = 3x^3 + 2y.$$

At the point $(1, 2)$ we find that $\partial f/\partial x|_{(1,\,2)} = 7$. Similarly,

$$f_y = \frac{\partial}{\partial y}[x^3 + 2xy + 2y^2]$$
$$= x^3 \frac{\partial}{\partial y}[1] + 2x \frac{\partial}{\partial y}[y] + 2 \frac{\partial}{\partial y}[y^2],$$

whence

$$\frac{\partial f}{\partial y} = 2x + 4y.$$

At the point (1, 2) we find that $\partial f/\partial y|_{(1,\ 2)} = 10$.

(b) $$f_x = \frac{\partial}{\partial x}[x \sin xy + 3]$$
$$= x \frac{\partial}{\partial x}[\sin xy] + \sin xy \frac{\partial}{\partial x}[x] + \frac{\partial}{\partial x}[3],$$

whence

$$\frac{\partial f}{\partial x} = xy \cos xy + \sin xy.$$

At the point $(1, \frac{1}{2}\pi)$ we find that $\partial f/\partial x|_{(1,\ (1/2)\pi)} = 1$. Similarly,

$$f_y = \frac{\partial}{\partial y}[x \sin xy + 3]$$
$$= x \frac{\partial}{\partial y}[\sin xy] + \frac{\partial}{\partial y}[3],$$

whence

$$\frac{\partial f}{\partial y} = x^2 \cos xy$$

and

$$\left.\frac{\partial f}{\partial y}\right|_{(1,\ (1/2)\pi)} = 0.$$

(c) $$f_x = \frac{\partial}{\partial x}\left[\frac{x}{x^2 + y^2}\right]$$
$$= \frac{1}{x^2 + y^2} \frac{\partial}{\partial x}[x] + x \frac{\partial}{\partial x}[(x^2 + y^2)^{-1}]$$
$$= \frac{1}{x^2 + y^2} - \frac{x}{(x^2 + y^2)^2} \frac{\partial}{\partial x}[x^2 + y^2],$$

whence

$$\frac{\partial f}{\partial x}=\frac{1}{x^2+y^2}-\frac{2x^2}{(x^2+y^2)^2}=\frac{y^2-x^2}{(x^2+y^2)^2}.$$

At the point (1, 0) we find that $\partial f/\partial x|_{(1,\ 0)}=-1$. Similarly,

$$f_y=\frac{\partial}{\partial y}\left[\frac{x}{x^2+y^2}\right]$$

$$=x\frac{\partial}{\partial y}[(x^2+y^2)^{-1}]$$

$$=\frac{-x}{(x^2+y^2)^2}\frac{\partial}{\partial y}[x^2+y^2],$$

whence

$$\frac{\partial f}{\partial y}=\frac{-2xy}{(x^2+y^2)^2}$$

and so

$$\left.\frac{\partial f}{\partial y}\right|_{(1,\ 0)}=0. \quad \blacksquare$$

The notion of partial differentiation extends to functions of more than two independent variables in an obvious manner. Suppose that the function $f(x, y, z)$ is defined near the point (x_0, y_0, z_0). Then, provided the limits exist, we define the three first-order partial derivatives f_x, f_y and f_z by the expressions

$$\left.\frac{\partial f}{\partial x}\right|_{(x_0,\ y_0,\ z_0)}=\lim_{x\to x_0}\frac{f(x, y_0, z_0)-f(x_0, y_0, z_0)}{x-x_0},$$

$$\left.\frac{\partial f}{\partial y}\right|_{(x_0,\ y_0,\ z_0)}=\lim_{y\to y_0}\frac{f(x_0, y, z_0)-f(x_0, y_0, z_0)}{y-y_0},$$

$$\left.\frac{\partial f}{\partial z}\right|_{(x_0,\ y_0,\ z_0)}=\lim_{z\to z_0}\frac{f(x_0, y_0, z)-f(x_0, y_0, z_0)}{z-z_0}.$$

Clearly a function of n independent variables will have n different first-order derivatives, one with respect to each of the independent variables. The actual computation of these partial derivatives is carried out exactly as before.

Example 5.17 Find the first-order partial derivatives of

$$f(x, y, z)=x^3y^2+3\sin yz+2.$$

Solution This function has three independent variables so we must obtain three first-order partial derivatives, f_x, f_y and f_z. First we have

$$\frac{\partial f}{\partial x} = \frac{\partial}{\partial x}[x^3y^2 + 3\sin yz + 2]$$

$$= y^2\frac{\partial}{\partial x}[x^3] + 3\sin yz\frac{\partial}{\partial x}[1] + \frac{\partial}{\partial x}[2],$$

so

$$\frac{\partial f}{\partial x} = 3x^2y^2.$$

Next,

$$\frac{\partial f}{\partial y} = \frac{\partial}{\partial y}[x^3y^2 + 3\sin yz + 2]$$

$$= x^3\frac{\partial}{\partial y}[y^2] + 3\frac{\partial}{\partial y}[\sin yz] + \frac{\partial}{\partial y}[2],$$

so

$$\frac{\partial f}{\partial y} = 2x^3y + 3z\cos yz.$$

Finally,

$$\frac{\partial f}{\partial z} = \frac{\partial}{\partial z}[x^3y^2 + 3\sin yz + 2]$$

$$= x^3y^2\frac{\partial}{\partial z}[1] + 3\frac{\partial}{\partial z}[\sin yz] + \frac{\partial}{\partial z}[2],$$

so

$$\frac{\partial f}{\partial z} = 3y\cos yz. \quad \blacksquare$$

5.6 Total differentials

The idea of a differential, which was useful in ordinary differentiation, may also be developed to advantage in connection with partial differentiation. We first approach this problem from the geometrical standpoint, and then indicate how an analytical counterpart of these arguments can be produced.

Let us consider Eqn (5.18) and its geometrical representation in Fig. 5.16. The conditions for differentiability at P ensure that the surface has a tangent plane π at that point, and it is to this plane that we now confine our attention. An element of this tangent plane defined by the lines l_1 and l_2 through P is depicted in Fig. 5.17. Obviously points on π close to P must also be close to those points on the surface $u = f(x, y)$ that lie vertically below them. This suggests that for such points, the element

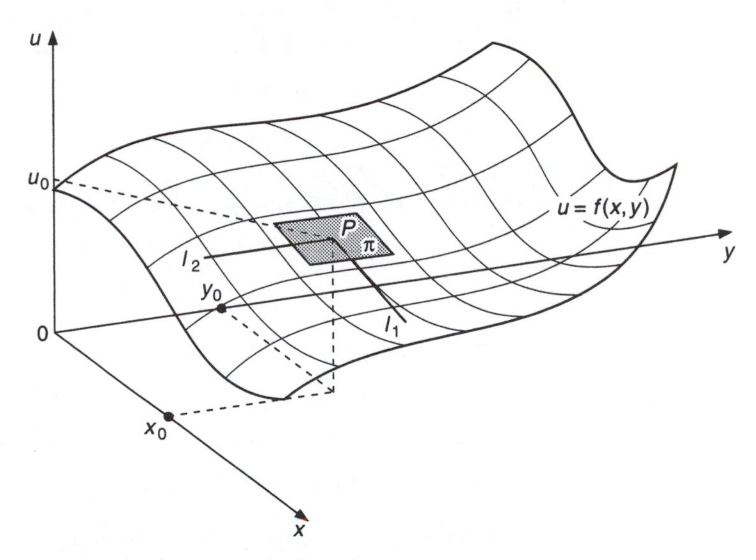

Fig. 5.17 Tangent plane p to surface $u = f(x, y)$ at point P.

of plane π neighbouring P represents a good approximation to the element of the curved surface defining the function u near to P. Thus variations of u close to P may, with propriety, be approximated by the variations of the corresponding points on π.

Since we are interested in variations of u about the point P at which $u_0 = f(x_0, y_0)$, we shall start by translating our coordinate axes without rotation to the point P. In this position the new x-, y- and u-coordinate axes will be denoted by x', y' and u', respectively, as shown in Fig. 5.18.

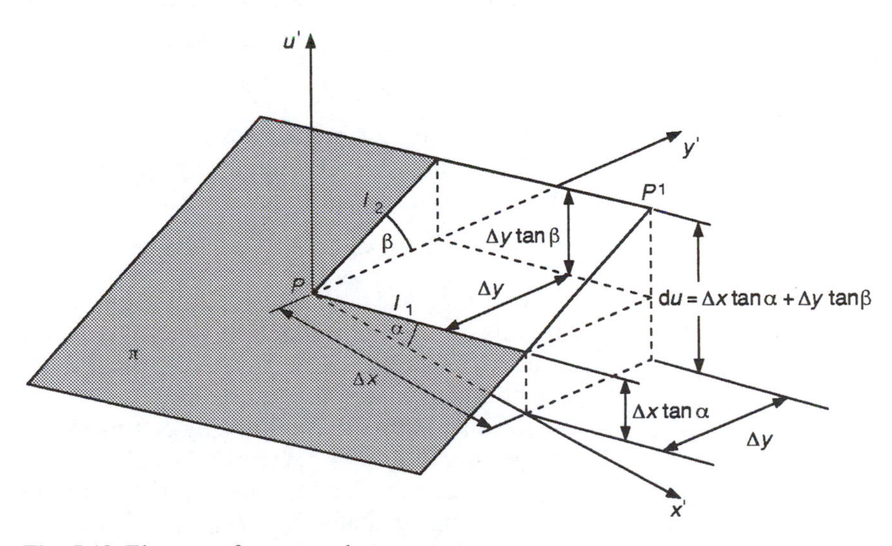

Fig. 5.18 Element of tangent plane.

If, relative to P, the x'- and y'-coordinates of a point P' are Δx and Δy, then it is obvious from Fig. 5.18 that the increment $\mathrm{d}u$ must be

$$\mathrm{d}u = \Delta x \tan\alpha + \Delta y \tan\beta,$$

where α and β are the angles between the lines l_1 and l_2 and the x'- and y'-axes, respectively.

However, by the definition of f_x and f_y, we have

$$f_x(x_0, y_0) = \tan\alpha, \quad f_y(x_0, y_0) = \tan\beta,$$

so that

$$\mathrm{d}u = f_x(x_0, y_0)\Delta x + f_y(x_0, y_0)\Delta y. \tag{5.19}$$

We now define differentials $\mathrm{d}x$ and $\mathrm{d}y$ in the independent variables x and y by setting $\mathrm{d}x = \Delta x$ and $\mathrm{d}y = \Delta y$. Expression (5.19) then becomes

$$\mathrm{d}u = f_x(x_0, y_0)\mathrm{d}x + f_y(x_0, y_0)\mathrm{d}y, \tag{5.20}$$

which is the relationship by which we define the *total differential* $\mathrm{d}u$ of the function $u = f(x, y)$. This is so called because it takes account of the total effect, on u, of the changes $\mathrm{d}x$ in x and $\mathrm{d}y$ in y. The additive effect of these changes is clearly apparent in Fig. 5.18 and results from using a tangent-plane approximation to the surface near P. As before, when $\mathrm{d}x$ and $\mathrm{d}y$ are suitably small, $\mathrm{d}u$ is a reasonable approximation to the true change Δu given by

$$\Delta u = f(x_0 + \mathrm{d}x, y_0 + \mathrm{d}y) - f(x_0, y_0). \tag{5.21}$$

An analytic rather than geometric justification of the tangent plane approximation used to define $\mathrm{d}u$ in Eqn (5.20) can be based on Theorem 5.12.

Equation (5.21), which is exact, is taken to be the starting point and by addition and subtraction of a term $f(x_0, y_0 + \Delta y)$, is written

$$\Delta u = [f(x_0 + \Delta x, y_0 + \Delta y) - f(x_0, y_0 + \Delta y)] \\ + [f(x_0, y_0 + \Delta y) - f(x_0, y_0)],$$

where in the first bracket only x changes, and in the second only y.

Then Theorem 5.12 may be applied to the first bracket with respect to x and to the second bracket with respect to y to yield

$$\Delta u = \Delta x f_x(x_0 + \xi\Delta x, y_0 + \Delta y) + \Delta y f_y(x_0, y_0 + \eta\Delta y), \tag{5.22}$$

where $0 < \xi < 1$ and $0 < \eta < 1$. Partial derivatives have been used here because, although in the first bracket it is only x that varies while in the second bracket it is only y that varies, both brackets are nevertheless functions of x and y.

Result (5.20) then follows by letting Δx and Δy become small. The continuity of $f_x(x_0 + \xi\Delta x, y_0 + \Delta y)$ allows it to be approximated by $f_x(x_0, y_0)$ with an error ε_1 and, similarly, the continuity of $f_y(x_0, y_0 + \eta\Delta y)$ allows it to be approximated by $f_y(x_0, y_0)$ with an error ε_2. Then, as $\Delta x, \Delta y \to 0$, so also do ε_1 and ε_2. It is left as an exercise for the interested reader to supply the details necessary to make this argument rigorous. If Eqn (5.20) is defined for all points (x_0, y_0) of some region in the (x, y)-plane, then the suffix zero may be discarded, and Eqn (5.20) can then be regarded as a functional relationship rather than a result that is true only near one point.

We have thus proved a special case of the following more general result whose proof differs in no significant detail.

Theorem 5.19 (total differential) Let $f(x_1, x_2, \ldots, x_n)$ be a real-valued function of n real variables, and let its first-order partial derivatives exist and be continuous in some region $\mathscr{R}$. Then the total differential $\mathrm{d}u$ of the function $u=f(x_1, x_2, \ldots, x_n)$ in the region $\mathscr{R}$ is given by

$$\mathrm{d}u = \frac{\partial f}{\partial x_1}\mathrm{d}x_1 + \frac{\partial f}{\partial x_2}\mathrm{d}x_2 + \ldots + \frac{\partial f}{\partial x_n}\mathrm{d}x_n. \quad \blacksquare$$

If we consider the surface generated by setting $u=$ constant, then on that surface $\mathrm{d}u \equiv 0$. Theorem 5.19 then takes the form

$$0 = \frac{\partial f}{\partial x_1}\mathrm{d}x_1 + \frac{\partial f}{\partial x_2}\mathrm{d}x_2 + \ldots \frac{\partial f}{\partial x_n}\mathrm{d}x_n, \tag{5.23}$$

showing that the differentials $\mathrm{d}x_1, \mathrm{d}x_2, \ldots, \mathrm{d}x_n$ are no longer independent since this *constraint* condition has been imposed on them. This is of course to be expected, since we have imposed the single condition $f(x_1, x_2, \ldots, x_n)=$ constant on the independent variables $x_1, x_2, \ldots x_n$ so that we are no longer free to change them arbitrarily. Indeed, if differentials $\mathrm{d}x_1, \mathrm{d}x_2, \ldots, \mathrm{d}x_{n-1}$ are chosen arbitrarily, then the remaining differential $\mathrm{d}x_n$ is uniquely determined by Eqn (5.23). If we call the number of independent variables the numbers of *degrees of freedom* associated with the equation $u=f(x_1, x_2, \ldots, x_n)$, then Eqn (5.23) implies the loss of a single degree of freedom.

Example 5.18 In thermodynamics, the pressure p of an ideal gas, its volume V, its absolute temperature T and the gas constant R are related by the *ideal gas law* $pV = RT$. Find the expression relating the total differential $\mathrm{d}p$ and the differentials $\mathrm{d}V$ and $\mathrm{d}T$.

Solution We have $p = RT/V$, and so $p=f(T, V)$ with $f(T, V)=RT/V$. Hence $\partial f/\partial T = R/V$ and $\partial f/\partial V = -RT/V^2$. Now interpreting Theorem 5.19 in this case we find

$$\mathrm{d}p = \left(\frac{\partial f}{\partial T}\right)\mathrm{d}T + \left(\frac{\partial f}{\partial V}\right)\mathrm{d}V, \tag{A}$$

and so

$$dp = \frac{R}{V}dT - \frac{RT}{V^2}dV.$$

Notice that the use of the symbol f in the total differential relation (A) to bring it into accord with the notation of Theorem 5.19 is not strictly necessary since $p \equiv f$. We could equally well have written equation (A) as

$$dp = \left(\frac{\partial p}{\partial T}\right)dT + \left(\frac{\partial p}{\partial V}\right)dV,$$

and used the immediately obvious result that

$$\frac{\partial p}{\partial T} = \frac{R}{V} \quad \text{and} \quad \frac{\partial p}{\partial V} = \frac{RT}{V^2}. \quad \blacksquare$$

Let us now consider the function $u = f(x, y)$ and, as a special case, set $u = 0$ so that the equation

$$f(x, y) = 0$$

defines y implicitly in terms of x. How, then, may we compute the derivative dy/dx without solving for y in terms of x? The solution to this problem is provided by Eqn (5.23), which in this case takes the form

$$0 = \frac{\partial f}{\partial x}dx + \frac{\partial f}{\partial y}dy.$$

We saw in connection with the definition of the differentials dy and dx in Eqn (5.11), that the function (dy/dx), namely the derivative of y with respect to x, is the ratio $dy{:}dx$ of the differentials. Hence dividing by the differential dx, assuming that $\partial f/\partial y \neq 0$, and rearranging gives the result

$$\frac{dy}{dx} = \frac{-(\partial f/\partial x)}{(\partial f/\partial y)}.$$

We state this as a corollary to Theorem 5.19.

Corollary 5.19(a) If the real variables x and y are related implicitly by the equation $f(x, y) = 0$, and the partial derivatives $\partial f/\partial x$ and $\partial f/\partial y$ exist and are continuous, then

$$\frac{dy}{dx} = -\left(\frac{\partial f}{\partial x}\right)\bigg/\left(\frac{\partial f}{\partial y}\right),$$

whenever $\partial f/\partial y \neq 0$. Insistence on this latter condition may be avoided by writing the result in the alternative form

$$\left(\frac{\partial f}{\partial y}\right)\frac{\mathrm{d}y}{\mathrm{d}x}+\frac{\partial f}{\partial x}=0. \quad \blacksquare$$

The situation is slightly different if three variables x, y, z are involved and z, say, is defined implicitly in terms of the independent variables x and y by the equation

$$f(x, y, z) = 0.$$

In these circumstances it is frequently necessary to compute $\partial z/\partial x$ and $\partial z/\partial y$ from this implicit relationship. To do so, notice that an obvious modification of Eqn (5.23) gives

$$\frac{\partial f}{\partial x}\mathrm{d}x+\frac{\partial f}{\partial y}\mathrm{d}y+\frac{\partial f}{\partial z}\mathrm{d}z=0,$$

but if z could be obtained explicitly, so that $z=z(x, y)$, it would also follow from Theorem 5.19 that

$$\mathrm{d}z=\frac{\partial z}{\partial x}\mathrm{d}x+\frac{\partial z}{\partial y}\mathrm{d}y.$$

Substitution of this result into the above expression gives

$$\left(\frac{\partial f}{\partial x}+\frac{\partial f}{\partial z}\cdot\frac{\partial z}{\partial x}\right)\mathrm{d}x+\left(\frac{\partial f}{\partial y}+\frac{\partial f}{\partial z}\cdot\frac{\partial z}{\partial y}\right)\mathrm{d}y=0,$$

and as x and y are independent variables, $\mathrm{d}x$ and $\mathrm{d}y$ are arbitrary so that this expression can only be true if

$$\frac{\partial f}{\partial x}+\frac{\partial f}{\partial z}\frac{\partial z}{\partial x}=0 \quad \text{and} \quad \left(\frac{\partial f}{\partial y}+\frac{\partial f}{\partial z}\frac{\partial z}{\partial y}\right)=0.$$

Hence, we find that provided $\partial f/\partial z \neq 0$,

$$\frac{\partial z}{\partial x}=-\left(\frac{\partial f}{\partial x}\right)\Big/\left(\frac{\partial f}{\partial z}\right) \quad \text{and} \quad \frac{\partial z}{\partial y}=-\left(\frac{\partial f}{\partial y}\right)\Big/\left(\frac{\partial f}{\partial z}\right).$$

We state this in the form of a further corollary.

Corollary 5.19(b) If the real variables x, y and z are related by the implicit equation $f(x, y, z)=0$ and the first-order derivatives of f exist and are continuous, then

$$\frac{\partial z}{\partial x}=-\left(\frac{\partial f}{\partial x}\right)\Big/\left(\frac{\partial f}{\partial z}\right) \quad \text{and} \quad \frac{\partial z}{\partial y}=-\left(\frac{\partial f}{\partial y}\right)\Big/\left(\frac{\partial f}{\partial z}\right),$$

when $\partial f/\partial z \neq 0$.

Example 5.19 (a) Find $\mathrm{d}y/\mathrm{d}x$ given that $x^2y+\sin xy=0$;
(b) Prove that $(\mathrm{d}/\mathrm{d}x)(x^r)=rx^{r-1}$ when r is rational;
(c) Find $\partial z/\partial y$ given that $f(x, y, z)=x^2+2xyz+z^3$.

Solution (a) We must apply Corollary 5.19(a). As, in this case,

$$f(x, y) = x^2y + \sin xy$$

it follows that

$$\frac{\partial f}{\partial x} = 2xy + y\cos xy$$

and

$$\frac{\partial f}{\partial y} = x^2 + x\cos xy.$$

Hence, by Corollary 5.19(a),

$$\frac{\mathrm{d}y}{\mathrm{d}x} = \frac{-(\partial f/\partial x)}{\partial f/\mathrm{d}y} = -\left(\frac{2xy + y\cos xy}{x^2 + x\cos xy}\right),$$

whenever $x^2+x\cos xy \neq 0$.

(b) This example may be omitted at a first reading because it merely justifies extending the result $\mathrm{d}[x^n]/\mathrm{d}x = nx^{n-1}$, already proved for integral n, to the case in which n may be an arbitrary rational number r. We have already shown in Theorem 5.2 that if $y=x^n$, then $\mathrm{d}y/\mathrm{d}x=nx^{n-1}$ for n a positive or negative integer. Now we must show this result is still true if the power involved is rational.

Let $y=x^r$ with $r=p/q$, where p and q are integers without any common factor. Then $y=x^{p/q}$ implies, and is implied by, $y^q=x^p$. Let $f(x, y)=y^q-x^p$ so that our equation corresponds to $f(x, y)=0$. Then there clearly exist pairs of real numbers (x, y) for which $y^q=x^p$, and by Theorem 5.2, $\partial f/\partial y=qy^{q-1}\neq 0$ when $y\neq 0$ (that is, when $x\neq 0$), and both $\partial f/\partial y$ and $\partial f/\partial x = -px^{p-1}$ are continuous functions. Hence the conditions of Corollary 5.19(a) are satisfied so that by the second form of its statement we may write

$$qy^{q-1}\frac{\mathrm{d}y}{\mathrm{d}x} - px^{p-1} = 0.$$

Thus

$$\frac{\mathrm{d}y}{\mathrm{d}x} = \frac{p}{q}\frac{x^{p-1}}{y^{q-1}} = \frac{p}{q}\frac{x^{p-1}}{(x^{p/q})^{q-1}} = \frac{p}{q}x^{p/q-1} = rx^{r-1},$$

when $x\neq 0$. In the event that $x=0$ we have

$$\left.\frac{\mathrm{d}}{\mathrm{d}x}(x^{p/q})\right|_{x=0} = \lim_{x\to 0}\frac{x^{p/q}-0}{x},$$

whenever this limit exists, which it does when $p/q > 1$, and is then equal to zero. This establishes our desired result for all x.

(c) Here,

$$f(x, y, z) = x^2 + 2xyz + z^3$$

and so

$$\frac{\partial f}{\partial x} = 2x + 2yz, \quad \frac{\partial f}{\partial y} = 2xz, \quad \frac{\partial f}{\partial z} = 2xy + 3z^2.$$

Thus by Corollary 5.19(b),

$$\frac{\partial z}{\partial x} = -\left(\frac{2x + 2yz}{2xy + 3z^2}\right) \quad \text{and} \quad \frac{\partial z}{\partial y} = \frac{-2xz}{2xy + 3z^2}.$$

5.7 Envelopes

A simple and useful application of the total differential is to the problem of the determination of envelopes already touched upon in section 2.4. Before proceeding with this application we now formally define an envelope.

Definition 5.6 Let a family of curves Γ in the (x, y)-plane with parameter α be defined by the implicit equation

$$f(x, y, \alpha) = 0.$$

Then the *envelope* of the family Γ, when it exists, is that curve $\mathscr{E}$ which is tangent to every member of the family.

Figure 5.19(a) shows some representative members of the family Γ corresponding to values α_1, α_2, α_3 and α_4 of the parameter α_1. Figure 5.19(b) shows the same situation on closely neighbouring curves C_1 and C_2 when the parametric value for C_2 is $\alpha_0 + \mathrm{d}\alpha$, which differs only by the differential $\mathrm{d}\alpha$ from the parametric value α_0 appropriate to C_1. We shall assume that the curves C_1 and C_2 intersect at the point P with coordinates (x_0, y_0).

Setting $u = f(x, y, \alpha)$, and regarding x, y and α as variables, if follows from Theorem 5.19 that

$$\mathrm{d}u = \frac{\partial f}{\partial x}\mathrm{d}x + \frac{\partial f}{\partial y}\mathrm{d}y + \frac{\partial f}{\partial \alpha}\mathrm{d}\alpha,$$

and as the family is defined by setting $u = 0$ (constant) it then follows, as in Eqn (5.23), that

$$0 = \frac{\partial f}{\partial x}\mathrm{d}x + \frac{\partial f}{\partial y}\mathrm{d}y \frac{\partial f}{\partial \alpha}\mathrm{d}\alpha.$$

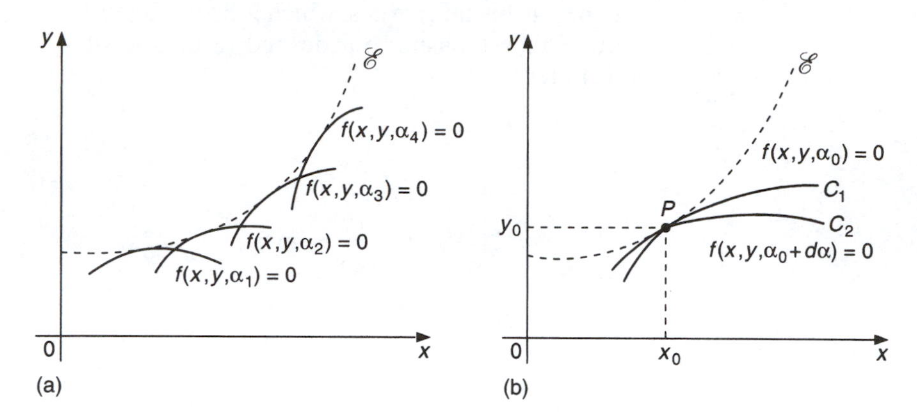

Fig. 5.19 Construction of envelope: (a) envelope of family of curves; (b) neighbouring members of the family.

This equation, which relates the differentials dx, dy and dα to the neighbouring curves C_1 and C_2, is, in particular, true at P. We signify this by writing

$$\left(\frac{\partial f}{\partial x}\right)_P \mathrm{d}x + \left(\frac{\partial f}{\partial y}\right)_P \mathrm{d}y + \left(\frac{\partial f}{\partial \alpha}\right)_P \mathrm{d}\alpha = 0, \tag{5.24}$$

where $(.)_P$ denotes that the associated quantity is to be evaluated at P. This equation is just the intersection condition for curves C_1 and C_2 at P.

As it is required of the envelope $\mathscr{E}$ that it be tangent to every member of the family Γ, it follows that as d$\alpha \to 0$ so curve C_2 must tend to C_1, and the gradient of the envelope $\mathscr{E}$ at P must tend to the gradient of the tangent to C_1 at P. To compute this we use the fact that $\alpha = \alpha_0$ is constant for curve C_1 so that the argument that gave rise to Eqn (5.24), when applied to $f(x,\ y,\ \alpha_0) = 0$, gives the tangency condition

$$0 = \left(\frac{\partial f}{\partial x}\right)_P \mathrm{d}x + \left(\frac{\partial f}{\partial y}\right)_P \mathrm{d}y. \tag{5.25}$$

Now both Eqns (5.24) and (5.25) must be simultaneously true for $\mathscr{E}$ and, consequently, we arrive at the condition

$$\left(\frac{\partial f}{\partial x}\right)_P \mathrm{d}\alpha = 0$$

which, since in general dα is a non-zero differential, can only be true if

$$\left(\frac{\partial f}{\partial \alpha}\right)_P = 0. \tag{5.26}$$

In addition to this result, the fact that P is a point on C_1 implies that $f(x_0,\ y_0,\ \alpha_0) = 0$ or, equivalently, that

$$[f(x,\ y,\ \alpha)]_P = 0. \tag{5.27}$$

Both conditions (5.26) and (5.27) must be satisfied if the envelope $\mathscr{E}$ is to pass through P and be tangent to C_1 at that point, so that, dropping the suffix P, we see that $\mathscr{E}$ is the locus of all points for which

$$f(x, y, \alpha) = 0 \quad \text{and} \quad \frac{\partial}{\partial x} f(x, y, \alpha) = 0. \tag{5.28}$$

Elimination of α between these two equations gives a relationship between x and y which is the desired equation of the envelope $\mathscr{E}$. We have thus proved the following result.

Theorem 5.20 (envelopes) When it exists, the equation of the envelope $\mathscr{E}$ of the family of curves

$$f(x, y, \alpha) = 0$$

with parameter α, is determined by the elimination of α between the equations

$$f(x, y, \alpha) = 0 \quad \text{and} \quad \frac{\partial}{\partial \alpha} f(x, y, \alpha) = 0. \quad \blacksquare$$

Example 5.20 Determine the envelope $\mathscr{E}$ of the family of curves

$$(x - \alpha)^2 + (y + \alpha)^2 = 1,$$

with parameter α.

Solution If we write the equation of this family of curves in the form

$$f(x, y, \alpha) = 0$$

then we must set

$$f(x, y, \alpha) = (x - \alpha)^2 + (y + \alpha)^2 - 1.$$

Hence the equation $\partial f/\partial \alpha = 0$ corresponds to

$$-(x - \alpha) + (y + \alpha) = 0$$

or, equivalently, to

$$\alpha = \tfrac{1}{2}(x - y).$$

To determine the envelope, the conditions of Theorem 5.20 require that $f(x, y, \alpha) = 0$ simultaneously with $\partial f/\partial \alpha = 0$. Hence substituting for the parameter α arrived at above from the condition $\partial f/\partial \alpha = 0$ into the family of the curves $f(x, y, \alpha) = 0$ gives

$$\tfrac{1}{4}(x + y)^2 + \tfrac{1}{4}(x + y)^2 = 1$$

or

$$x + y = \pm\sqrt{2}.$$

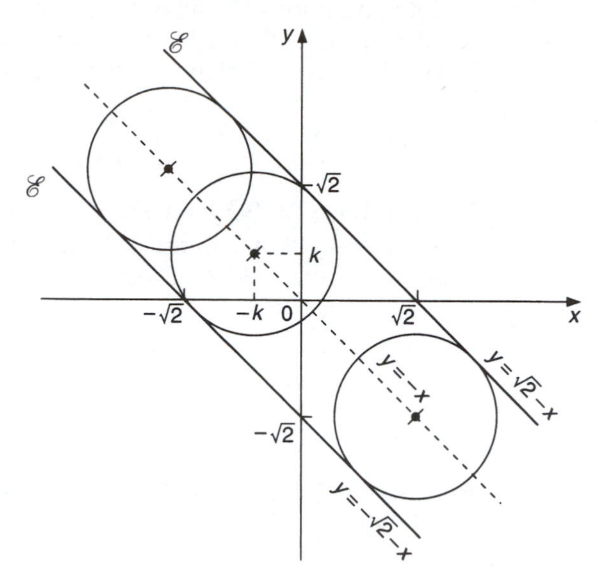

Fig. 5.20 Envelope of circles.

The desired envelope $\mathscr{E}$ thus comprises the two straight lines

$$y = \sqrt{2} - x \quad \text{and} \quad y = -\sqrt{2} - x.$$

This particular result could also have been deduced by geometrical arguments as follows. The original family of curves comprise circles of unit radius, each with its centre at $x = \alpha$, $y = -\alpha$. Consequently, the tangents to these circles which form their envelope $\mathscr{E}$ must be straight lines parallel to the line of centres $y = -x$ and separated from it by a unit distance (Fig. 5.20). ■

Although in this case it was possible to eliminate α from the equations arising from Theorem 5.20, this situation is not generally possible. It is usually necessary to express the envelope parametrically in terms of α.

5.8 The chain rule and its consequences

If, in Theorem 5.19, the variables $x_1, x_2, \ldots, x_n$ are specified in terms of a parameter t, say, then the result requires slight modification. Suppose that

$$x_1 = x_1(t),\ x_2 = x_2(t), \ldots,\ x_n = x_n(t),$$

which are all differentiate functions of t. Then the variable u becomes a function of the single real variable t, for we may write

$$u = F(t), \tag{5.29}$$

where $F(t) = f(x_1(t), x_2(t), \ldots, x_n(t))$.

Hence by an obvious adaptation of Eqn (5.11), defining differentials, we may write

$$\mathrm{d}u = F'(t)\mathrm{d}t, \tag{5.30}$$

where of course, $F'(t) = \mathrm{d}u/\mathrm{d}t$, the derivative of u with respect to t.

However, by a further application of Eqn (5.11) to each of the variables $x_1 = x_1(t),\ x_2 = x_2(t), \ldots,\ x_n = x_n(t)$, we have the result

$$\mathrm{d}x_1 = \left(\frac{\mathrm{d}x_1}{\mathrm{d}t}\right)\mathrm{d}t, \mathrm{d}x_2 = \left(\frac{\mathrm{d}x_2}{\mathrm{d}t}\right)\mathrm{d}t, \ldots, \mathrm{d}x_n = \left(\frac{\mathrm{d}x_n}{\mathrm{d}t}\right)\mathrm{d}t. \tag{5.31}$$

Substituting these expressions for the differentials $\mathrm{d}x_i$ in terms of the differential $\mathrm{d}t$ into the statement of Theorem 5.19 gives

$$\mathrm{d}u = \left(\frac{\partial f}{\partial x_1}\frac{\mathrm{d}x_1}{\mathrm{d}t} + \frac{\partial f}{\partial x_2}\frac{\mathrm{d}x_2}{\mathrm{d}t} + \ldots + \frac{\partial f}{\partial x_n}\frac{\mathrm{d}x_n}{\mathrm{d}t}\right)\mathrm{d}t. \tag{5.32}$$

Finally, a comparison of Eqns (5.30) and (5.32) shows that

$$F'(t) = \frac{\partial f}{\partial x_1}\frac{\mathrm{d}x_1}{\mathrm{d}t} + \frac{\partial f}{\partial x_2}\frac{\mathrm{d}x_2}{\mathrm{d}t} + \ldots + \frac{\partial f}{\partial x_n}\frac{\mathrm{d}x_n}{\mathrm{d}t}.$$

As $F'(t) = \mathrm{d}u/\mathrm{d}t$, this result facilitates the calculation of $\mathrm{d}u/\mathrm{d}t$ without the need for formal substitution into $u = f(x_1, x_2, \ldots, x_n)$ of the values

$$x_1 = x_1(t),\ x_2 = x_2(t), \ldots,\ x_n = x_n(t).$$

We have proved the following useful result.

Theorem 5.21 (chain rule for partial derivatives) Let $u = f(x_1,\ x_2, \ldots,\ x_n)$ be a real-valued function of n real variables and let its first-order partial derivatives exist and be continuous. Further, let each of the variables $x_1, x_2, \ldots, x_n$ be a differentiate function of the single real variable t, so that we may write

$$x_1 = x_1(t),\ x_2 = x_2(t), \ldots, x_n = x_n(t).$$

Then the total derivative of u with respect to t is given by

$$\frac{\mathrm{d}u}{\mathrm{d}t} = \frac{\partial f}{\partial x_1}\frac{\mathrm{d}x_1}{\mathrm{d}t} + \frac{\partial f}{\partial x_2}\frac{\mathrm{d}x_2}{\mathrm{d}t} + \ldots + \frac{\partial f}{\partial x_n}\frac{\mathrm{d}x_n}{\mathrm{d}t}. \quad \blacksquare$$

Two special cases of this theorem are of sufficient importance to merit recording as corollaries. The first arises when f is a function of only two variables between which an explicit relationship exists, and the parameter t is identified with one of these variables.

As only two variables are involved we shall avoid the use of numerical suffixes by agreeing to write $x_1 = x$ and $x_2 = y$ where, by supposition, $y = y(x)$ is some known explicit relation. The statement of Theorem 5.21 then becomes

$$\frac{\mathrm{d}u}{\mathrm{d}t} = \frac{\partial f}{\partial x}\frac{\mathrm{d}x}{\mathrm{d}t} + \frac{\partial f}{\partial y}\frac{\mathrm{d}y}{\mathrm{d}t}.$$

If, now, we identify t with x, then $t=x$ and $\mathrm{d}x/\mathrm{d}t=1$, $\mathrm{d}y/\mathrm{d}t=\mathrm{d}y/\mathrm{d}x$ so that the above result becomes

$$\frac{\mathrm{d}u}{\mathrm{d}x}=\frac{\partial f}{\partial x}+\frac{\partial f}{\partial y}\frac{\mathrm{d}y}{\mathrm{d}x}.$$

The expression on the right-hand side is the total derivative of u with respect to x. The first term on the right takes account of the change directly due to x, while the second term takes account of the fact that y is itself a function of x. This result enables $\mathrm{d}u/\mathrm{d}x$ to be obtained without needing to substitute $y=y(x)$ in the relation $u=f(x, y)$.

Corollary 5.21(a) If $u=f(x, y)$ is a real-valued function of the real variables x and y with continuous first-order derivatives, and y is related to x by the explicit equation $y=y(x)$, then

$$\frac{\mathrm{d}u}{\mathrm{d}x}=\frac{\partial f}{\partial x}+\frac{\partial f}{\partial y}\frac{\mathrm{d}y}{\mathrm{d}x}. \quad \blacksquare$$

More generally, suppose that $u=f(x, y)$, while x and y are related implicitly by the equation

$$g(x, y) = 0.$$

How must we modify our previous argument in order that we may compute the total derivative $\mathrm{d}u/\mathrm{d}x$? The result of Corollary 5.21(a), is still true but obviously $\mathrm{d}y/\mathrm{d}x$ now depends on the form of g. To find the form of $\mathrm{d}y/\mathrm{d}x$ we can use Corollary 5.19(a), writing $f=g$, to see that

$$\frac{\mathrm{d}y}{\mathrm{d}x}=-\left(\frac{\partial g}{\partial x}\right)\bigg/\left(\frac{\partial g}{\partial y}\right),$$

showing that

$$\frac{\mathrm{d}u}{\mathrm{d}x}=\frac{\partial f}{\partial x}-\left(\frac{\partial f}{\partial y}\right)\left(\frac{\partial g}{\partial x}\right)\bigg/\left(\frac{\partial g}{\partial y}\right),$$

provided $\partial g/\partial y \neq 0$. We state this as our next result.

Corollary 5.21(b) If $u=f(x, y)$ is a real-valued function of the real variables x and y with continuous first-order derivatives, and y is related implicitly to x by the equation $g(x, y)=0$, then

$$\frac{\mathrm{d}u}{\mathrm{d}x}=\frac{\partial f}{\partial x}-\left(\frac{\partial f}{\partial y}\right)\left(\frac{\partial g}{\partial x}\right)\bigg/\left(\frac{\partial g}{\partial y}\right),$$

provided $\partial g/\partial y \neq 0$.

Example 5.21 Determine the derivative $\mathrm{d}u/\mathrm{d}t$ given that

$$u = \sin(x^2 + y^2) \quad \text{with} \quad x = 3t,\ y = 1/(1 + t^2).$$

Solution We must apply Theorem 5.21, making the identifications $x_1 = x$, $x_2 = y$ and $f(x, y) = \sin(x^2 + y^2)$, with $x = 3t$ and $y = 1/(1 + t^2)$. Hence

$$\frac{\partial f}{\partial x} = 2x\cos(x^2 + y^2), \qquad \frac{\partial f}{\partial y} = 2y\cos(x^2 + y^2),$$

while

$$\frac{\mathrm{d}x}{\mathrm{d}t} = 3, \qquad \frac{\mathrm{d}y}{\mathrm{d}t} = \frac{-2t}{(1 + t^2)^2}.$$

Substituting in Theorem 5.21,

$$\frac{\mathrm{d}u}{\mathrm{d}t} = 2x\cos(x^2 + y^2).(3) + 2y\cos(x^2 + y^2).\left[\frac{-2t}{(1 + t^2)^2}\right]$$

or

$$\frac{\mathrm{d}u}{\mathrm{d}t} = 2\cos(x^2 + y^2)\left[3x - \frac{2yt}{(1 + t^2)^2}\right].$$

Using the known relationships between x, y and t, the derivative $\mathrm{d}u/\mathrm{d}t$ can thus be computed for any desired value of t. The details are left to the reader. ■

Example 5.22 Determine the total derivate $\mathrm{d}u/\mathrm{d}x$ in each case:

(a) $u = x\cos y + y\cos x$ when $y = 1 + x + x^3$;
(b) $u = x^2 + 2xy - y^2$ when $x^2 + y^2 + \cos xy = 0$.

Solution (a) This requires an application of Corollary 5.21(a). We set

$$f(x, y) = x\cos y + y\cos x \qquad \text{and} \quad y = 1 + x + x^3$$

so that

$$\frac{\partial f}{\partial x} = \cos y - y\sin x, \qquad \frac{\partial f}{\partial y} = -x\sin y + \cos x$$

and

$$\frac{\mathrm{d}y}{\mathrm{d}x} = 1 + 3x^2.$$

Hence, substituting into Corollary 5.21(a),

$$\frac{du}{dx} = \cos y - y \sin x + (\cos x - x \sin y)(1 + 3x^2).$$

(b) In this case we use Corollary 5.21(b), with

$$f(x, y) = x^2 + 2xy - y^2 \quad \text{and} \quad g(x, y) = x^2 + y^2 + \cos xy.$$

Hence

$$\frac{\partial f}{\partial x} = 2x + 2y, \quad \frac{\partial f}{\partial y} = 2x - 2y,$$

$$\frac{\partial g}{\partial x} = 2x - y \sin xy \quad \frac{\partial g}{\partial y} = 2y - x \sin xy.$$

Finally, applying Corollary 5.21(b),

$$\frac{du}{dx} = 2(x + y) - \frac{2(x - y)(2x - y \sin xy)}{(2y - x \sin xy)}. \quad \blacksquare$$

5.9 Change of variable

This section discusses a somewhat more complicated situation than that covered by Theorem 5.21, namely, the implications on partial differentiation of changing the independent variables in a function $u = f(x_1, x_2, \ldots, x_n)$ that is to be differentiated. This situation commonly occurs as a result of changing coordinate systems to suit physical problems, as the following example illustrates. Suppose that $p = p(x, y, z)$ is the pressure in a fluid flowing parallel to the z-axis. Then $\partial p/\partial z$ is the pressure gradient along the direction of flow and $\partial p/\partial x$, $\partial p/\partial y$ are the transverse pressure gradients in the plane $z = \text{constant}$.

Now, if the flow takes place in a rectangular duct with sides described by $x = \text{constant}$, $y = \text{constant}$, then the Cartesian coordinates $O(x, y, z)$ are obviously the natural ones to use. However, if the flow takes place in a cylindrical pipe, then the z-axis is still convenient as it can be aligned with the axis of the pipe, but the x- and y-axes are now less useful since the wall of the pipe becomes the curve $x^2 + y^2 = \text{constant}$. Clearly, a more sensible coordinate system would be the cylindrical polar coordinates r, θ, z' in which r and θ define a point in the plane $z' = \text{constant}$. Figure 5.21 illustrates this idea. The plane $z = z' = 0$ in both the $O\{x, y, z\}$ and $O\{r, \theta, z'\}$ systems of axes is denoted by Π. Relative to these two systems the point P has the coordinates $O\{x, y, z\}$ and $O\{r, \theta, z'\}$, respectively, where

$$x = r \cos \theta, \qquad y = r \sin\theta, \quad z = z'. \tag{5.33}$$

How can the pressure gradients described by the partial derivatives $\partial p/\partial r$, $\partial p/\partial \theta$ and $\partial p/\partial z'$ be determined from Eqn (5.33), and the known functions $\partial p/\partial x$, $\partial p/\partial y$ and $\partial p/\partial z$? The rest of this section is devoted to solving this type of problem. Notice that from the definition of partial differentiation, $\partial p/\partial z$ and $\partial p/\partial z'$ have essentially the same meaning,

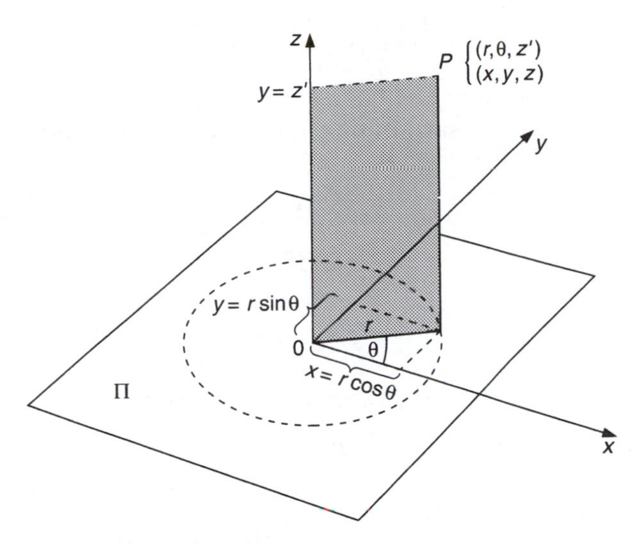

Fig. 5.21 Cylindrical polar coordinates.

whereas $\partial p/\partial r$ is the derivative of p computed along a radius with θ and z' held constant, whilst $\partial p/\partial \theta$ is the derivative of p tangential to a circle $r = \text{constant}$ drawn on the plane $z' = \text{constant}$.

Although the replacement of coordinate variables in this manner involves replacing a set of n independent variables by a new set also comprising n in number ($n = 3$ above), we shall first prove a more general result. Specifically, consider the implication of the situation in which

$$u = f(x_1, x_2, \ldots, x_n), \tag{5.34}$$

when the independent variables $x_1, x_2, \ldots, x_n$ are themselves differentiable functions of another set of variables which we denote by $\alpha_1, \alpha_2, \ldots, \alpha_m$. It is not necessary that m should equal n. Thus we have

$$\begin{aligned} x_1 &= x_1(\alpha_1, \alpha_2, \ldots, \alpha_m), \\ x_2 &= x_2(\alpha_1, \alpha_2, \ldots, \alpha_m), \\ &\ldots \\ x_n &= x_n(\alpha_1, \alpha_2, \ldots, \alpha_m). \end{aligned} \tag{5.35}$$

If the variables x_i in Eqn (5.34) were to be replaced by the equivalent functions (5.35) involving the variables α_i, then f would become some function $F(\alpha_1, \alpha_2, \ldots, \alpha_m)$ of $\alpha_1, \alpha_2, \ldots, \alpha_m$ so that by Theorem 5.19 we could write

$$du = \frac{\partial F}{\partial \alpha_1} d\alpha_1 + \frac{\partial F}{\partial \alpha_2} d\alpha_2 + \ldots + \frac{\partial F}{\partial \alpha_m} d\alpha_m. \tag{5.36}$$

Next, observe that by applying this same theorem to the equation for x_i in Eqn (5.35) we obtain

$$dx_i = \frac{\partial x_i}{\partial \alpha_1} d\alpha_1 + \frac{\partial x_i}{\partial \alpha_2} d\alpha_2 + \ldots + \frac{\partial x_i}{\partial \alpha_m} d\alpha_m, \tag{5.37}$$

for $i = 1, 2, \ldots, n$.

Substituting these expressions into the statements of Theorem 5.19 then gives

$$du = \frac{\partial f}{\partial x_1}\left[\frac{\partial x_1}{\partial \alpha_1} d\alpha_1 + \frac{\partial x_1}{\partial \alpha_2} d\alpha_2 + \ldots + \frac{\partial x_1}{\partial \alpha_m} d\alpha_m\right] + \ldots$$
$$+ \frac{\partial f}{\partial x_n}\left[\frac{\partial x_n}{\partial \alpha_1} d\alpha_1 + \frac{\partial x_n}{\partial \alpha_2} d\alpha_2 + \ldots + \frac{\partial x_n}{\partial \alpha_m} d\alpha_m\right]. \tag{5.38}$$

On rearrangement this becomes

$$du = \left[\frac{\partial f}{\partial x_1}\frac{\partial x_1}{\partial \alpha_1} + \frac{\partial f}{\partial x_2}\frac{\partial x_2}{\partial \alpha_1} + \ldots + \frac{\partial f}{\partial x_n}\frac{\partial x_n}{\partial \alpha_1}\right] d\alpha_1 + \ldots$$
$$+ \left[\frac{\partial f}{\partial x_1}\frac{\partial x_1}{\partial \alpha_m} + \frac{\partial f}{\partial x_2}\frac{\partial x_2}{\partial \alpha_m} + \ldots + \frac{\partial f}{\partial x_n}\frac{\partial x_n}{\partial \alpha_m}\right] d\alpha_m. \tag{5.39}$$

Since $f(x_1, x_2, \ldots, x_n) = F(\alpha_1, \alpha_2, \ldots, \alpha_m)$, it follows by a direct comparison of the ith terms of Eqns (5.36) and (5.39) that

$$\frac{\partial F}{\partial \alpha_i} = \frac{\partial f}{\partial x_1}\frac{\partial x_1}{\partial \alpha_i} + \frac{\partial f}{\partial x_2}\frac{\partial x_2}{\partial \alpha_i} + \ldots, + \frac{\partial f}{\partial x_n}\frac{\partial x_n}{\partial \alpha_i} \tag{5.40}$$

for $i = 1, 2, \ldots, m$.

We state this result in the form of a general theorem.

Theorem 5.22 (change of variable) Let $f(x_1, x_2, \ldots, x_n)$ be a real-valued function of real variables $x_1, x_2, \ldots, x_n$ whose first-order derivatives exist and are continuous. Further, let $x_1 = x_1(\alpha_1, \alpha_2, \ldots, \alpha_m), x_2 = x_2(\alpha_1, \alpha_2, \ldots, \alpha_m), \ldots, x_n = x_n(\alpha_1, \alpha_2, \ldots, \alpha_m)$ be differentiable functions of the real variables $\alpha_1, \alpha_2, \ldots, \alpha_m$. Then if $f(x_1, x_2, \ldots, x_n)$ becomes the function $F(\alpha_1, \alpha_2, \ldots, \alpha_m)$ as a result of this change of variables,

$$\frac{\partial F}{\partial \alpha_1} = \frac{\partial f}{\partial x_1}\frac{\partial x_1}{\partial \alpha_1} + \frac{\partial f}{\partial x_2}\frac{\partial x_2}{\partial \alpha_1} + \ldots + \frac{\partial f}{\partial x_n}\frac{\partial x_n}{\partial \alpha_1}$$
$$\frac{\partial F}{\partial \alpha_2} = \frac{\partial f}{\partial x_1}\frac{\partial x_1}{\partial \alpha_2} + \frac{\partial f}{\partial x_2}\frac{\partial x_2}{\partial \alpha_2} + \ldots + \frac{\partial f}{\partial x_n}\frac{\partial x_n}{\partial \alpha_2}$$
$$\ldots$$
$$\frac{\partial F}{\partial \alpha_m} = \frac{\partial f}{\partial x_1}\frac{\partial x_1}{\partial \alpha_m} + \frac{\partial f}{\partial x_2}\frac{\partial x_2}{\partial \alpha_m} + \ldots + \frac{\partial f}{\partial x_n}\frac{\partial x_n}{\partial \alpha_m}. \qquad \blacksquare$$

In the case of a function $f = f(x, y)$ of the two variables x and y in which x and y are related to the two new variables u and v by the known functions $x = x(u, v)$ and $y = y(u, v)$, the result of Theorem 5.22 reduces to

$$\frac{\partial F}{\partial u} = \frac{\partial f}{\partial x}\frac{\partial x}{\partial u} + \frac{\partial f}{\partial y}\frac{\partial y}{\partial u}$$

$$\frac{\partial F}{\partial v} = \frac{\partial f}{\partial x}\frac{\partial x}{\partial v} + \frac{\partial f}{\partial y}\frac{\partial y}{\partial v}.$$

Example 5.23 Let the differentiable function $f(z, y, z)$ become the function $F(r, \theta, z')$ under the change of variables from Cartesian to cylindrical polar coordinates $x = r\cos\theta$, $y = r\sin\theta$ and $z = z'$. Use Theorem 5.22 to find $\partial F/\partial r$, $\partial F/\partial\theta$ and $\partial F/\partial z'$ in terms of r, θ and z', given that

$$f(x, y, z) = x^2 + 3xy + y^2 + z^2.$$

Solution Although in this case the required results could be found directly by substituting into f to find F, and then differentiating, we use Theorem 5.22 by way of illustration. We apply Theorem 5.22 with $m = n = 3$ by making the identifications $x_1 = x$, $x_2 = y$, $x_3 = z$ and $\alpha_1 = r$, $\alpha_2 = \theta$, $\alpha_3 = z'$. As a result we find

$$\frac{\partial F}{\partial r} = \frac{\partial f}{\partial x}\frac{\partial x}{\partial r} + \frac{\partial f}{\partial y}\frac{\partial y}{\partial r} + \frac{\partial f}{\partial z}\frac{\partial z}{\partial r},$$

$$\frac{\partial F}{\partial \theta} = \frac{\partial f}{\partial x}\frac{\partial x}{\partial \theta} + \frac{\partial f}{\partial y}\frac{\partial y}{\partial \theta} + \frac{\partial f}{\partial z}\frac{\partial z}{\partial \theta},$$

$$\frac{\partial F}{\partial z'} = \frac{\partial f}{\partial x}\frac{\partial x}{\partial z'} + \frac{\partial f}{\partial y}\frac{\partial y}{\partial z'} + \frac{\partial f}{\partial z}\frac{\partial z}{\partial z'}.$$

However, from the change of variables it follows that

$$\frac{\partial x}{\partial r} = \cos\theta, \frac{\partial x}{\partial \theta} = -r\sin\theta, \frac{\partial x}{\partial z'} = 0, \frac{\partial y}{\partial r} = \sin\theta$$

$$\frac{\partial y}{\partial \theta} = r\cos\theta, \frac{\partial y}{\partial z'} = 0, \frac{\partial z}{\partial r} = \frac{\partial z}{\partial \theta} = 0, \frac{\partial z}{\partial z'} = 1.$$

Hence, substituting these expressions into the above transformation equations gives

$$\frac{\partial F}{\partial r} = \frac{\partial f}{\partial x}\cos\theta + \frac{\partial f}{\partial y}\sin\theta,$$

$$\frac{\partial F}{\partial \theta} = -\frac{\partial f}{\partial x}r\sin\theta + \frac{\partial f}{\partial y}r\cos\theta,$$

$$\frac{\partial F}{\partial z'} = \frac{\partial f}{\partial z}.$$

Next, using the fact that $f(x, y, z) = x^2 + 3xy + y^2 + z^2$, we see that

$$\frac{\partial f}{\partial x} = 2x + 3y = r(2\cos\theta + 3\sin\theta),$$

$$\frac{\partial f}{\partial y} = 3x + 2y = r(3\cos\theta + 2\sin\theta),$$

$$\frac{\partial f}{\partial z} = 2z = 2z'.$$

Finally, combining these results with those for $\partial F/\partial r$, $\partial F/\partial\theta$ and $\partial F/\partial z'$, substituting for x, y and z in terms of r, θ and z' and simplifying, we find

$$\frac{\partial F}{\partial r} = 2r(1 + 3\sin\theta\cos\theta), \frac{\partial F}{\partial\theta} = 3r^2(\cos^2\theta - \sin^2\theta), \frac{\partial F}{\partial z'} = 2z'. \quad \blacksquare$$

5.10 Some applications of $\frac{dy}{dx} = 1/\left(\frac{dx}{dy}\right)$

5.10.1 Differentiating the inverse trigonometric functions

The first application of the result $\dfrac{dx}{dy} = 1\Big/\left(\dfrac{dy}{dx}\right)$ established in Eqn (5.4) is to the differentiation of inverse trigonometric functions. In Table 2.2 of section 2.2, we agreed to write

$$y = \arcsin x$$

when $x = \sin y$ and $-\pi/2 \le y \le \pi/2$. Now,

$$\frac{d}{dy}(\sin y) = \cos y \ne 0 \qquad \text{for} \quad -\pi/2 < y < \pi/2;$$

that is, for $-1 < x < 1$ and so,

$$\frac{dy}{dx} = 1\Big/\left(\frac{dx}{dy}\right) = \frac{1}{\cos y} = \frac{1}{\sqrt{(1 - \sin^2 y)}} = \frac{1}{\sqrt{(1 - x^2)}}.$$

The *positive* square root has been taken here because the principal branch of the function $y = \arcsin x$ is a monotonic *increasing* function of x in its domain of definition $-1 \le x \le 1$. By this same argument, the *negative* square root is taken when differentiating the principal branch of the function $y = \arccos x$ which is a monotonic *decreasing* function of x in its domain of definition $-1 \le x \le 1$. Thus

$$\frac{d}{dx}(\arcsin x) = \frac{1}{\sqrt{(1 - x^2)}} \qquad \text{for} \quad -1 < x < 1.$$

Similar arguments establish Table 5.3. In the entries for the derivatives of arccsc and arcsec, the term $|x|$ has been introduced to take account of

Table 5.3 Derivatives of inverse trigonometric functions

$\dfrac{d}{dx}(\arcsin x/a) = \dfrac{1}{\sqrt{(a^2 - x^2)}}$ for $-a < x < a$	$\dfrac{d}{dx}(\arccos x/a) = \dfrac{-1}{\sqrt{(a^2 - x^2)}}$ for $-a < x < a$
$\dfrac{d}{dx}(\arctan x/a) = \dfrac{a}{a^2 + x^2}$ for all x	$\dfrac{d}{dx}(\text{arccsc}\, x/a) = \dfrac{-a}{\lvert x\rvert\sqrt{(x^2 - a^2)}}$ for $\lvert x\rvert > a$
$\dfrac{d}{dx}(\text{arcsec}\, x/a) = \dfrac{a}{\lvert x\rvert\sqrt{(x^2 - a^2)}}$ for $\lvert x\rvert > a$	$\dfrac{d}{dx}(\text{arccot}\, x/a) = \dfrac{-a}{a^2 + x^2}$ for all x

the two separate cases that need consideration when deriving these results; namely, when $x > a$ and when $x < -a$, as can be seen from Fig. 2.7(a). These same ideas will be encountered again in the next chapter in connection with Table 6.3, when they will be discussed in more detail.

5.10.2 Differentiating parametrically defined functions

In Chapter 2 we saw that curves may be described parametrically thus:

$$x = X(t), \quad y = Y(t),$$

where t is a parameter defined in some interval $\mathscr{J}$. The question that now arises is how may we find dy/dx in terms of the functions $X(t)$ and $Y(t)$.

Let us suppose that $X(t)$ and $Y(t)$ are differentiable functions of t with continuous derivatives and that $X'(t) \neq 0$. Then solving $x = X(t)$ in the form $t = f(x)$, we have $y = Y[f(x)]$. From Theorem 5.7 on the differentiation of composite functions (the chain rule) we have

$$\frac{dy}{dx} = \frac{d}{dx}\{Y[f(x)]\} = \frac{dY}{df}\frac{df}{dx}$$

or, equivalently,

$$\frac{dy}{dx} = \frac{dy}{dt}\cdot\frac{dt}{dx}. \tag{5.41}$$

However, $dt/dx = 1/(dx/dt)$ so that

$$\frac{dy}{dx} = \frac{dy}{dt}\Big/\frac{dx}{dt}. \tag{5.42}$$

Hence, like x and y, the derivative $\mathrm{d}y/\mathrm{d}x$ is now also known parametrically in terms of t.

This result is best remembered in symbolic form:

$$\frac{\mathrm{d}}{\mathrm{d}x} = \frac{1}{(\mathrm{d}x/\mathrm{d}t)}\frac{\mathrm{d}}{\mathrm{d}t}. \tag{5.43}$$

Higher-order derivatives with respect to x may be found either by a repetition of the argument leading to Eqn (5.42), or by successive application of Eqn (5.43).

Thus, using Eqn (5.43), we have

$$\frac{\mathrm{d}^2y}{\mathrm{d}x^2} = \frac{\mathrm{d}}{\mathrm{d}x}\left(\frac{\mathrm{d}y}{\mathrm{d}x}\right) = \frac{1}{(\mathrm{d}x/\mathrm{d}t)}\left[\frac{\mathrm{d}}{\mathrm{d}t}\left(\frac{\mathrm{d}y}{\mathrm{d}t}\Big/\frac{\mathrm{d}x}{\mathrm{d}t}\right)\right]. \tag{5.44}$$

Example 5.24

(a) Determine the relationship between $\dfrac{\mathrm{d}^2x}{\mathrm{d}y^2}$ and $\dfrac{\mathrm{d}^2y}{\mathrm{d}x^2}$.

(b) If $x = t + 2\sin t$, $y = \cos t$, determine $\mathrm{d}y/\mathrm{d}x$ and $\mathrm{d}^2y/\mathrm{d}x^2$, and hence deduce their values when $t = 0$.

Solution

(a)
$$\frac{\mathrm{d}^2x}{\mathrm{d}y^2} = \frac{\mathrm{d}}{\mathrm{d}y}\left(\frac{\mathrm{d}x}{\mathrm{d}y}\right) = \frac{\mathrm{d}}{\mathrm{d}y}\left(1\Big/\frac{\mathrm{d}y}{\mathrm{d}x}\right)$$
$$= \frac{\mathrm{d}x}{\mathrm{d}y}\frac{\mathrm{d}}{\mathrm{d}x}\left(1\Big/\frac{\mathrm{d}y}{\mathrm{d}x}\right) = \frac{1}{\left(\dfrac{\mathrm{d}y}{\mathrm{d}x}\right)}\,\frac{-1}{\left(\dfrac{\mathrm{d}y}{\mathrm{d}x}\right)^2}\,\frac{\mathrm{d}^2y}{\mathrm{d}x^2},$$

and so,

$$\frac{\mathrm{d}^2x}{\mathrm{d}y^2} = -\left(\frac{\mathrm{d}^2y}{\mathrm{d}x^2}\right)\Big/\left(\frac{\mathrm{d}y}{\mathrm{d}x}\right)^3.$$

This establishes the important fact that although

$$\frac{\mathrm{d}y}{\mathrm{d}x} = 1\Big/\left(\frac{\mathrm{d}x}{\mathrm{d}y}\right),$$

in general

$$\frac{\mathrm{d}^2x}{\mathrm{d}y^2} \neq 1\Big/\left(\frac{\mathrm{d}^2y}{\mathrm{d}x^2}\right).$$

(b) We have

$$\frac{\mathrm{d}x}{\mathrm{d}t} = 1 + 2\cos t, \quad \frac{\mathrm{d}y}{\mathrm{d}t} = -\sin t,$$

so that by Eqn (5.42)

$$\frac{dy}{dx} = \frac{dy}{dt} \bigg/ \frac{dx}{dt} = \frac{-\sin t}{1 + 2\cos t}.$$

When $t = 0$ we have $x = 0$, $y = 1$ and

$$\left.\frac{dy}{dx}\right|_{t=0} = \left.\frac{-\sin t}{1 + 2\cos t}\right|_{t=0} = 0.$$

Next, as

$$\frac{d^2y}{dx^2} = \frac{1}{(dx/dt)} \frac{d}{dt}\left(\frac{dy}{dx}\right)$$

we have

$$\frac{d^2y}{dx^2} = \frac{1}{1 + 2\cos t} \frac{d}{dt}\left[\frac{-\sin t}{1 + 2\cos t}\right].$$

Thus, performing the differentiation and simplifying,

$$\frac{d^2y}{dx^2} = -\left[\frac{2 + \cos t}{(1 + 2\cos t)^3}\right]$$

and so

$$\left.\frac{d^2y}{dx^2}\right|_{x=0} = -\left.\left[\frac{2 + \cos t}{(1 + 2\cos t)^3}\right]\right|_{t=0} = -\frac{1}{9}. \quad \blacksquare$$

5.11 Higher-order partial derivatives

If the function $f(x, y)$ is differentiable with continuous first-order derivatives f_x and f_y, then it can also happen that these partial derivatives which are functions of x and y are themselves differentiable. Thus we are led to consider the further partial derivatives

$$\frac{\partial}{\partial x}(f_x), \quad \frac{\partial}{\partial y}(f_x), \quad \frac{\partial}{\partial x}(f_y) \quad \text{and} \quad \frac{\partial}{\partial y}(f_y).$$

These functions, when they exist, are *second-order* partial derivatives of f and are respectively denoted by

$$\frac{\partial^2 f}{\partial x^2}, \quad \frac{\partial^2 f}{\partial y \partial x}, \quad \frac{\partial^2 f}{\partial x \partial y} \quad \text{and} \quad \frac{\partial^2 f}{\partial y^2}.$$

Notice the order in which differentiation is performed; $\partial^2 f/\partial y \partial x$ means first find $\partial f/\partial x$, and then differentiate it partially with respect to y.

We will adopt an alternative notation and write these same derivatives as

$$f_{xx}, \quad f_{yx}, \quad f_{xy} \quad \text{and} \quad f_{yy}.$$

In this notation the order of the suffixes and their meaning is the same as above; thus $f_{yx} = \dfrac{\partial}{\partial x}[\partial f/\partial x] = \partial^2 f/\partial y \partial x$. The centre pair of derivatives are *mixed second-order partial derivatives*. We have

$$\frac{\partial}{\partial y}(f_x) = \frac{\partial^2 f}{\partial y \partial x} = f_{yx} \quad \text{and} \quad \frac{\partial}{\partial x}(f_y) = \frac{\partial^2 f}{\partial x \partial y} = f_{xy}.$$

In terms of limiting operations we have

$$f_{xx}(x, y) = \lim_{h \to 0} \frac{f_x(x+h, y) - f_x(x, y)}{h},$$

$$f_{xy}(x, y) = \lim_{k \to 0} \frac{f_y(x, y+k) - f_y(x, y)}{k},$$

$$f_{yx}(x, y) = \lim_{h \to 0} \frac{f_y(x+h, y) - f_y(x, y)}{h},$$

$$f_{yy}(x, y) = \lim_{k \to 0} \frac{f_x(x, y+k) - f_x(x, y)}{k}.$$

It is important to notice that the double operations of partial differentiation that lead to the mixed derivatives f_{xy} and f_{yx} are performed in different orders. Consequently, we have no right to expect that the derivatives that result will be equal to one another. However, the following theorem can be proved, though we shall not give the details here.

Theorem 5.24 (equality of mixed derivatives) Let $f(x, y)$ be a real-valued function of the real variables x, y, and let f_x, f_y, f_{xy} exist and be continuous in the neighbourhood of the point (x_0, y_0). Then f_{yx} also exists at (x_0, y_0) and

$$\left.\frac{\partial^2 f}{\partial x \partial y}\right|_{(x_0, y_0)} = \left.\frac{\partial^2 f}{\partial y \partial x}\right|_{(x_0, y_0)}. \quad \blacksquare$$

Still higher-order derivatives can be defined by an obvious extension of the notation. Thus, for a suitably differentiable function f we may define the third-order partial derivatives

$$f_{xxx}, \quad f_{yyx}, \quad f_{xyx}, \quad f_{yyy}, \quad \text{etc.}$$

If the higher-order derivatives involved are continuous then, by an obvious extension of Theorem 5.24, the order of performing differentiations may be disregarded. In the case of the mixed third-order partial derivative f_{xyx} this would imply that

$$f_{xyx} = \frac{\partial}{\partial x}\left[\frac{\partial}{\partial y}(f_x)\right] = \frac{\partial}{\partial y}\left[\frac{\partial}{\partial x}(f_x)\right] = f_{xxy}.$$

Hence, under these conditions, it is proper to extend the ∂ notation by writing

$$\frac{\partial^3 f}{\partial x^3}, \quad \frac{\partial^3 f}{\partial x^2 \partial y}, \quad \frac{\partial^3 f}{\partial x \partial y^2}, \quad \frac{\partial^3 f}{\partial y^3}, \quad \text{etc.}$$

We draw attention here to the fact that other suffix conventions for higher order partial derivatives are also in use, and that in the most common one the order of the suffixes is reversed relative to the one used here. In this book, because it seems most natural, we use the convention that $f_{xy} = \partial/\partial x(\partial f/\partial x)$, whereas in the other convention this same partial derivative is denoted by f_{yx}. Similarly instead of, as here, writing $f_{xyy} = \partial/\partial x(\partial^2 f/\partial y^2)$, the other convention denotes this third order partial derivative on the right by f_{yyx}, with corresponding reversals of the order of suffixes in other higher order partial derivatives. The difference between the two conventions is only of importance on the rare occasions that a higher order partial derivative is not continuous. This is because when the higher order partial derivatives are continuous the equality of mixed derivatives means that the result is the same in both conventions so, for example, it then follows that $f_{xy} = f_{yx}, f_{xyy} = f_{yyx} = f_{yxy}$, and so on.

Example 5.25 If $f(x, y) = x^4 + 2x^2y^2 + xy^4$, find the second- and third-order partial derivatives of f.

Solution First-order derivatives:

$$f_x = 4x^3 + 4xy^2 + y^4, \quad f_y = 4x^2y + 4xy^3.$$

Second-order derivatives

$$f_{xx} = 12x^2 + 4y^2, \quad f_{yy} = 4x^2 + 12xy^2,$$

$$f_{yx} = \frac{\partial}{\partial y}(f_x) = 8xy + 4y^3.$$

This mixed derivative is continuous, and so $f_{xy} = f_{yx}$. As a check in this case we compute f_{xy} directly:

$$f_{xy} = \frac{\partial}{\partial x}(f_y) = 8xy + 4y^3.$$

Third-order derivatives:

$$f_{xxx} = 24x, \quad f_{yyy} = 24xy, \quad f_{yxy} = \frac{\partial}{\partial y}(f_{xy}) = 8x + 12y^2,$$

$$f_{yxx} = \frac{\partial}{\partial y}(f_{xx}) = 8y.$$

The continuity of the third-order derivatives we have computed ensures the existence and equality of the other corresponding third-order derivatives that may be defined. Thus, for example, as $f_{xxy} = 8y$ is continuous, there is no need to compute f_{xyx}, since it exists and is equal to f_{xxy}.

The final examples show how the approach of section 5.9 may be extended to enable the calculation of higher-order partial derivatives when a change of variable is involved.

Example 5.26 Under the transformation

$$x = r\cos\theta, \quad y = r\sin\theta$$

the twice differentiable and continuous function $f(x, y)$ becomes the function $F(r, \theta)$. Find $\partial^2 F/\partial r\partial\theta$ in terms of partial derivatives with respect to x and y.

Solution Proceeding as in Example 5.23, though here the change of variables is from (x, y) to (r, θ) so that z and z' are absent, we find

$$\frac{\partial F}{\partial r} = \cos\theta\frac{\partial f}{\partial x} + \sin\theta\frac{\partial f}{\partial y},$$

and

$$\frac{\partial F}{\partial \theta} = -r\sin\theta\frac{\partial \phi}{\partial \chi} + r\cos\theta\frac{\partial f}{\partial y}.$$

In terms of the *operations* of partial differentiation with respect to x, y and r, θ these expressions show that

$$\frac{\partial}{\partial r} \equiv \cos\theta\frac{\partial}{\partial x} + \sin\theta\frac{\partial}{\partial y}, \tag{A}$$

and

$$\frac{\partial}{\partial \theta} = -r\sin\theta\frac{\partial}{\partial x} + r\cos\theta\frac{\partial}{\partial y}. \tag{B}$$

Now, by definition, we have from (B) that

$$\frac{\partial^2 F}{\partial r\partial\theta} = \frac{\partial}{\partial r}\left(\frac{\partial F}{\partial \theta}\right)$$
$$= \frac{\partial}{\partial r}\left(-r\sin\theta\frac{\partial f}{\partial x} + r\cos\theta\frac{\partial f}{\partial y}\right),$$

or

$$\frac{\partial^2 F}{\partial r \partial \theta} = -\frac{\partial}{\partial r}\left(r \sin\theta \frac{\partial f}{\partial x}\right) + \frac{\partial}{\partial r}\left(r \cos\theta \frac{\partial f}{\partial y}\right).$$

Since r, θ are independent variables, but x and y depend on r and θ, by performing the indicated differentiations we arrive at

$$\frac{\partial^2 F}{\partial r \partial \theta} = -\sin\theta \frac{\partial f}{\partial x} - r\sin\theta \frac{\partial}{\partial r}\left(\frac{\partial f}{\partial x}\right) + \cos\theta \frac{\partial f}{\partial y} + r\cos\theta \frac{\partial}{\partial r}\left(\frac{\partial f}{\partial y}\right).$$

Using (A) to interpret $\partial/\partial r$ in terms of $\partial/\partial x$ and $\partial/\partial y$ we find

$$\begin{aligned}\frac{\partial^2 F}{\partial r \partial \theta} &= -\sin\theta \frac{\partial r}{\partial x} - r\sin\theta\left(\cos\theta \frac{\partial}{\partial x} + \sin\theta \frac{\partial}{\partial y}\right)\left(\frac{\partial f}{\partial x}\right) \\ &\quad + \cos\theta \frac{\partial f}{\partial y} + r\cos\theta\left(\cos\theta \frac{\partial}{\partial x} + \sin\theta \frac{\partial}{\partial y}\right)\left(\frac{\partial f}{\partial y}\right) \\ &= -\sin\theta \frac{\partial f}{\partial x} - r\sin\theta\cos\theta \frac{\partial^2 f}{\partial x^2} - r\sin^2\theta \frac{\partial^2 f}{\partial y \partial x} \\ &\quad + \cos\theta \frac{\partial f}{\partial y} + r\cos^2\theta \frac{\partial^2 f}{\partial x \partial y} + r\sin\theta\cos\theta \frac{\partial^2 f}{\partial y^2}.\end{aligned}$$

However, since f is twice differentiable and continuous, it follows from Theorem 5.24 that $\partial^2 f/\partial x \partial y = \partial^2 f/\partial y \partial x$, so that the result simplifies to

$$\frac{\partial^2 F}{\partial r \partial \theta} = -\sin\theta \frac{\partial f}{\partial x} + \cos\theta \frac{\partial f}{\partial y} + r(\cos^2\theta - \sin^2\theta)\frac{\partial^2 f}{\partial x \partial y} - r\sin\theta\cos\theta\left(\frac{\partial^2 f}{\partial x^2} - \frac{\partial^2 f}{\partial y^2}\right). \quad \blacksquare$$

In conclusion, we remark that other higher-order partial derivatives may be found in similar fashion. Thus we have

$$\frac{\partial^2 F}{\partial r^2} = \frac{\partial}{\partial r}\left(\frac{\partial F}{\partial r}\right), \quad \frac{\partial^2 F}{\partial \theta^2} = \frac{\partial}{\partial \theta}\left(\frac{\partial F}{\partial \theta}\right),$$

$$\frac{\partial^3 F}{\partial r^2 \partial \theta} = \frac{\partial}{\partial r}\left(\frac{\partial^2 F}{\partial r \partial \theta}\right), \ldots$$

The details of such calculations are left as an exercise for the reader, but the results for the second-order partial derivatives are as follows:

$$\frac{\partial^2 F}{\partial r^2} = \cos^2\theta \frac{\partial^2 f}{\partial x^2} + 2\sin\theta\cos\theta \frac{\partial^2 f}{\partial x \partial y} + \sin^2\theta \frac{\partial^2 f}{\partial y^2},$$

$$\frac{\partial^2 F}{\partial \theta^2} = -r\left(\cos\theta\frac{\partial f}{\partial x} + \sin\theta\frac{\partial f}{\partial y}\right) + r^2\sin^2\theta\frac{\partial^2 f}{\partial x^2}$$
$$-2r^2\sin\theta\cos\theta\frac{\partial^2 f}{\partial x\partial y} + r^2\cos^2\theta\frac{\partial^2 f}{\partial y^2}.$$

We remark in passing that it follows by combining the above results that

$$\frac{\partial^2 F}{\partial r^2} + \frac{1}{r}\frac{\partial F}{\partial r} + \frac{1}{r^2}\frac{\partial^2 F}{\partial \theta^2} = \frac{\partial^2 f}{\partial x^2} + \frac{\partial^2 f}{\partial y^2}.$$

The equation

$$\frac{\partial^2 f}{\partial x^2} + \frac{\partial^2 f}{\partial y^2} = 0 \tag{5.45}$$

satisfied by a function $f(x, y)$, and the corresponding result

$$\frac{\partial^2 F}{\partial r^2} + \frac{1}{r}\frac{\partial F}{\partial r} + \frac{1}{r^2}\frac{\partial^2 F}{\partial \theta^2} = 0 \tag{5.46}$$

satisfied by $F(r, \theta)$, where $f(x, y)$ is related to $F(r, \theta)$ by $x = r\cos\theta$ and $y = r\sin\theta$, are called *partial differential equations*.

Equation (5.45) is named and called the *Laplace equation* expressed in the Cartesian coordinates (x, y). The corresponding result in Eqn (5.46) is called Laplace's equation in the polar coordinates (r, θ).

Laplace's equation plays an important role in engineering and science because many quantities of physical interest are expressible in terms of the solution of this equation. For example, the steady-state temperature in a plate and the electrostatic potential in a cavity both satisfy Laplace's equation. The choice of coordinates best suited to describe the geometry of the physical problem (the shape of the plate or cavity) determines which form of the equation it is best to use.

Example 5.27 Under the transformation

$$x = s^3 + 2s, \quad y = st$$

the twice differentiable function $f(x, y)$ becomes $F(s, t)$. Find $\partial F/\partial s$ and $\partial F/\partial t$ in terms of $\partial f/\partial x$ and $\partial f/\partial y$, and show that

$$\frac{\partial^2 F}{\partial s\partial t} = \frac{\partial f}{\partial y} + (3s^3 + 2s)\frac{\partial^2 f}{\partial x\partial y} + st\frac{\partial^2 f}{\partial y^2}.$$

Solution Interpreting Theorem 5.22 in terms of the present problem, we have

$$\frac{\partial F}{\partial s} = \frac{\partial f}{\partial x}\frac{\partial x}{\partial s} + \frac{\partial f}{\partial y}\frac{\partial y}{\partial s}$$

$$\frac{\partial F}{\partial t} = \frac{\partial f}{\partial x}\frac{\partial x}{\partial t} + \frac{\partial f}{\partial y}\frac{\partial y}{\partial t},$$

but

$$x = s^3 + 2s, \quad y = st,$$

so

$$\frac{\partial x}{\partial s} = 3s^2 + 2, \quad \frac{\partial x}{\partial t} = 0 \quad \frac{\partial y}{\partial s} = t \quad \text{and} \quad \frac{\partial y}{\partial t} = s.$$

Consequently, substituting these results into the above equations gives

$$\frac{\partial F}{\partial s} = (3s^2 + 2)\frac{\partial f}{\partial x} + t\frac{\partial f}{\partial y} \quad \text{and} \quad \frac{\partial F}{\partial t} = s\frac{\partial f}{\partial y}.$$

We see from this that in terms of the operators $\partial/\partial x$ and $\partial/\partial y$, the operators $\partial/\partial s$ and $\partial/\partial t$ are

$$\frac{\partial}{\partial s} = (3s^2 + 2)\frac{\partial}{\partial x} + t\frac{\partial}{\partial y} \quad \text{and} \quad \frac{\partial}{\partial t} = s\frac{\partial}{\partial y}.$$

Thus

$$\begin{aligned}
\frac{\partial^2 F}{\partial s \partial t} &= \frac{\partial}{\partial s}\left[\frac{\partial F}{\partial t}\right] = \frac{\partial}{\partial s}\left[s\frac{\partial f}{\partial y}\right] \\
&= \frac{\partial f}{\partial y} + s\frac{\partial}{\partial s}\left[\frac{\partial f}{\partial y}\right] \\
&= \frac{\partial f}{\partial y} + s\left[(3s^2 + 2)\frac{\partial}{\partial x} + t\frac{\partial}{\partial y}\right]\frac{\partial f}{\partial y} \\
&= \frac{\partial f}{\partial y} + (3s^3 + 2s)\frac{\partial^2 f}{\partial x \partial y} + st\frac{\partial^2 f}{\partial y^2}.
\end{aligned}$$

The same result would have been obtained had we used the fact that

$$\frac{\partial^2 F}{\partial s \partial t} = \frac{\partial^2 F}{\partial t \partial s}$$

and started from

$$\frac{\partial^2 F}{\partial s \partial t} = \frac{\partial}{\partial t}\left[\frac{\partial F}{\partial s}\right]. \quad \blacksquare$$

Problems

Section 5.1

5.1 Give examples of four physical quantities that are essentially defined in terms of a derivative.

5.2 Use Definitions 5.1 and 5.2 to prove that the following functions are differentiable in the stated intervals and to compute their derivatives. Evaluate these derivatives for the stated values:

(a) $f(x) = 3x^2$ in $[0, 3]$, find $f'(2)$;

(b) $f(x) = 2x^3 + x + 1$ in $[-1, 4]$, find $f'(3)$;

(c) $f(x) = |x|$ in $(0, \infty)$, find $f'(1)$;

(d) $f(x) = |x|$ in $(-\infty, 0)$, find $f'(-3)$;

(e) $f(x) = 1/x$ in $[1, 5]$, find $f'(4)$;

(f) $f(x) = x^{1/4}$ in $(0, \infty)$, find $f'(2)$.

5.3 Prove that $f(x) = |x|$ is not differentiable at the origin.

5.4 Consider the graph of $f(x) = x^3 + x + 1$. Let x_1 and x_2 be two points on the x-axis with the property that the gradient dy/dx of the curve $y = f(x)$ at $x = x_2$ is four times the gradient at $x = x_1$. Derive the algebraic equation connecting x_1 and x_2, and deduce that $|x_2| \geq 1$.

5.5 Deduce the gradients of the functions $f(x)$ to the immediate left and right of $x = 1$ given that:

(a) $(x) = \begin{cases} x^3 + x + 1 & \text{for } x \geq 1 \\ 5 - x - x^2 & \text{for } x < 1; \end{cases}$

(b) $f(x) = \begin{cases} x^3 - x + 3 & \text{for } x \geq 1 \\ 2x + 1 & \text{for } x < 1. \end{cases}$

5.6 At which points in the stated intervals, if any, are the following functions $f(x)$ non-differentiable:

(a) $f(x) = x + \sin 2x \quad \text{for } 0 \leq x \leq \pi$;

(b) $f(x) = \left\{\begin{matrix} x + 1/x & \text{for } x \neq 0 \\ 0 & \text{for } x = 0 \end{matrix}\right\}$ in the interval $[-1, 1]$;

(c) $f(x) = \left\{\begin{matrix} 1 & \text{for } x \text{ rational} \\ 0 & \text{for } x \text{ irrational} \end{matrix}\right\}$ in the interval $[0, 1]$.

5.7 The function $f(x)$ is defined on the interval $0 \leq x \leq 1$ by the expression

$$f(x) = \begin{cases} \sin 2x & \text{for } 0 < x \leq \frac{1}{6}\pi \\ ax + b & \text{for } \frac{1}{6}\pi < x \leq 1. \end{cases}$$

Deduce the values of a and b in order that the function should be continuous and have a continuous derivative at $x = \frac{1}{6}\pi$. Interpret these conditions geometrically.

Section 5.2

5.8 By assuming Theorem 5.2 is also valid for rational n where necessary, find the derivatives of the following functions, stating at which points in their domains of definition, if any, they are non-differentiable:

(a) $f(x) = \begin{Bmatrix} x^{1/3} + \cos 3x, & \text{for } x \neq 0 \\ 0, & \text{for } x = 0 \end{Bmatrix}$ in the interval $-\frac{1}{2}\pi \leq x \leq \pi$;

(b) $f(x) = x \sin 2x + x^{5/3}$ for $-1 \leq x \leq 3$;

(c) $f(x) = |\cos x|$ for $0 \leq x \leq \pi$.

5.9 Differentiate the following functions:

(a) $y = x^{1/3} \sin x$;

(b) $y = (x^2 + 3x + 1)(1 + \cos 2x)$;

(c) $y = \sin 6x \cos 2x$;

(d) $y = (x^3 + 2x - 1) \cos 3x$.

5.10 Differentiate the following functions by making a repeated application of Theorem 5.5:

(a) $y = (1 + x^2) \sin 7x \cos 4x$;

(b) $y = (1 + 2x^2 + x^4)^3$;

(c) $y = \cos^3 2x$;

(d) $y = (1 + x^3)^2 \sin^2 3x$.

5.11 Differentiate these composite functions:

(a) $y = (x^2 + 2x + 1)^{3/2}$;

(b) $y = (a + bx^3)^{1/3}$;

(c) $y = (2 + 3 \sin 2x)^5$;

(d) $y = \sin(1 + 2x^3)$;

(e) $y = \sin [\sin (1 + x^2)]$;

(f) $y = \cos [(1 + x^4)^{1/2}]$.

5.12 Differentiate these quotients:

(a) $y = (x^2 + 3x + 7)/(x^4 + 1)$;

(b) $y = \dfrac{\sin(1 + x^2)}{x^4 + 2x^2 + 6}$;

(c) $y = \dfrac{1}{3 \cos^3 x} - \dfrac{1}{\cos x}$;

(d) $y = \dfrac{\tan(1 - x^2 + x^4)}{\sin(x^2 + 1)}$;

(e) $y = \dfrac{1 + \sqrt{x}}{1 - \sqrt{x}}$.

5.13 Differentiate these functions:

(a) $y = \dfrac{1}{(1 - 3\cos x)^2}$;

(b) $y = \dfrac{x}{a^2\sqrt{(b^2 + x^2)}}$;

(c) $y = \dfrac{\tan(1 + x^2 + x^4)}{\sin(1 + x^2)}$;

(d) $y = \csc^2(1 + 3x)$;

(e) $y = \dfrac{\sin x + 2\cos x}{\sin x - 2\cos x}$;

(f) $y = \cot\left(\dfrac{3x - 1}{x^2 + 4}\right)$.

5.14 If the functions $f_1(x)$, $f_2(x)$, $g_1(x)$, $g_1(x)$ and $g_2(x)$ are differentiable, show by direct expansion that the following result is true:

$$\frac{\mathrm{d}}{\mathrm{d}x}\begin{vmatrix} f_1(x) & f_2(x) \\ g_1(x) & g_2(x) \end{vmatrix} = \begin{vmatrix} f_1'(x) & f_2'(x) \\ g_1(x)) & g_2(x) \end{vmatrix} + \begin{vmatrix} f_1(x) & f_2(x) \\ g_1'(x) & g_2'(x) \end{vmatrix}.$$

Apply this result to differentiate the determinants:

(a) $\begin{vmatrix} x^2 & x\sin x \\ \cos x & 1 \end{vmatrix}$; (b) $\begin{vmatrix} (1 + x^2\cos x) & (2 - \sin^2 x) \\ (1 - x^2\cos x) & (2 + \sin^2 x) \end{vmatrix}$.

Section 5.3

5.15 Why is it not possible to conclude from the intermediate value theorem that if $f(x) = 1/(1 - |x|)$ for $|x| \neq 1$ and $f(1) = 0.5$ then
(a) there is no point $x = \xi$ in the interval $[0, 6]$ for which $f(\xi) = 0$;
(b) yet there is a point $x = \eta$ in the interval $[-11, -2]$ for which $f(\eta) = -0.5$? Identify the point on the x-axis giving rise to this functional value.

5.16 The function $f(x) = \frac{1}{3}x^3 - x + 2$ which is defined in the interval $(-\infty, \infty)$ has extrema at the points $x = 1$, $x = -1$. Identify their nature by considering the behaviour of the function close to these points. Are they relative or absolute extrema?

5.17 Find the critical points of the function $f(x)=x^3-x^2-4x+4$. Identify the nature of the extrema associated with them by considering the functional behaviour close to each of these points.

5.18 Find the critical point of the function $f(x)=(x-1)x^{2/3}$ and identify its nature. Do the points $x=-1$, $x=0$ correspond to extrema of the function and, if so, of what type are they?

5.19 Locate and identify the critical points of the function $f(x)=x^2(3-x)^2$.

5.20 Identify the critical points and extrema of the function

$$f(x)=\begin{cases} x^2-3x+2 & \text{for } 0\le x\le 2.5 \\ x^2-7x+12 & \text{for } 2.5<x\le 5. \end{cases}$$

5.21 Apply Rolle's theorem to the following functions where it is applicable, and hence determine at how many points in the stated intervals $[a,\ b]$ the functions satisfy the result of that theorem:

(a) $f(x)=x^2-1$ in $[-2,\ 2]$;

(b) $f(x)=1+\sin x$ in $[-2\pi,\ 3\pi]$;

(c) $f(x)=1/(1+|x|)$ in $[-1,\ 1]$;

(d) $f(x)=\begin{cases} x^2+3x+2 & \text{for } -1\le x\le 0 \\ x^2-3x+2 & \text{for } 0<x<1. \end{cases}$

5.22 Show that the conditions of the mean value theorem apply to $f(x)=x+\sin x$ for the interval $[0,\ \frac{1}{2}\pi]$. Find the value of ξ in the statement of the theorem.

The following four problems illustrate how the mean value theorem may be used to estimate the behaviour of functions in closed intervals.

5.23 Let $f(x)$ be a differentiable function having a monotonic increasing derivative in the intervals $[a,\ b]$. Then by writing the mean value theorem in the form $f(b)=f(a)+(b-a)f'(\xi)$, with $a<\xi<b$, prove that $f(a)+(x-a)f'(a)<f(x)<f(a)+(x-a)f'(b)$, for $a<x<b$. We shall agree to say that these inequalities define upper and lower estimates of $f(x)$ in $[a,\ b]$. Show also that if $f'(x)$ is monotonic decreasing, then the inequalities must be reversed in the above expression.

5.24 Apply the result of Problem 5.23 to the function $f(x)=\sin x$ in the interval $[0,\ \frac{1}{2}\pi]$ in order to prove that $0<(\sin x)/x<1$ for $0<x<\frac{1}{2}\pi$.

5.25 Apply the result of Problem 5.23 to the function $f(x)=(1+x^2)^{3/2}$ in the interval $[1,\ 2]$, thereby obtaining upper and lower estimates for it in that interval.

5.26 If $f(x) = 1 + x + (1/5)\sin^2 x$, show that $f'(x)$ is monotonic increasing in the interval $[-\frac{1}{4}\pi, \frac{1}{4}\pi]$. Hence apply the result of Problem 5.23 to $f(x)$ to obtain upper and lower estimates for $f(x)$ in that interval. Evaluate the inequalities for $x = 0$ and $x = \frac{1}{4}\pi$, and compare the estimates with the exact result.

5.27 Let the functions $f(x)$ and $g(x)$ be continuous in $[a, b]$ and differentiable in (a, b), with $g'(x)$ non-zero in (a, b). Show that under these conditions Rolle's theorem may be applied to the function $F(x)$ defined by $F(x) = f(a)g(a) - f(b)g(a) + [g(a) - g(b)]f(x) - [f(a) - f(b)]g(x)$, for $a \le x \le b$. Hence establish the Cauchy extended mean value theorem.

5.28 By repeatedly applying L'Hospital's rule where necessary, evaluate the following indeterminate forms of the type 0/0:

(a) $\lim_{x\to 0} \dfrac{\tan \alpha x}{x}$; (b) $\lim_{x\to 0} \dfrac{x\cos x - \sin x}{x^3}$;

(c) $\lim_{x\to 0} \dfrac{\tan x - \sin x}{x - \sin x}$; (d) $\lim_{x\to 1} \dfrac{x^3 - 2x^2 - x + 2}{x^3 - 7x + 6}$;

(e) $\lim_{x\to 0} \dfrac{x^2 - \sin^2 x}{x^2 \sin^2 x}$.

5.29 Evaluate the following indeterminate forms which are of the type ∞/∞:

(a) $\lim_{x\to 0} (\pi/x)/\cot(\pi x/2)$; (b) $\lim_{x\to(1/2)\pi} \tan x/\tan 5x$;

(c) $\lim_{x\to\infty} \dfrac{3x^2 + x - 1}{x^2 + 2}$; (d) $\lim_{x\to 0} \dfrac{\cot x}{x - \cot x}$.

5.30 Explain the fallacy in this argument. The limit

$$\lim_{x\to\infty} \frac{x^2 + x\sin x + \sin x}{x^2}$$

does not exist, because applying Corollary 5.14 to L'Hospital's rule gives

$$\lim_{x\to\infty} \frac{x^2 + x\sin x + \sin x}{x^2} = \lim_{x\to\infty} \frac{2x + \sin x + x\cos x + \cos x}{2x}$$

$$= \lim_{x\to\infty} \left[1 + \tfrac{1}{2}\cos x + \frac{\sin x + \cos x}{2x}\right] = 1 + \tfrac{1}{2}\lim_{x\to\infty} \cos x,$$

which is indeterminate. What is the true value of this limit?

5.31 Indeterminate limits of the form $\infty - \infty$, $0.\infty$ can be reduced to the types 0/0 or ∞/∞ by means of the following simple devices. If the limit is of the type $0.\infty$ set $\lim_{x\to a} f(x) = 0$ and $\lim_{x\to a} g(x) \to \infty$, then

$$\lim_{x\to a}[f(x)g(x)] = \lim_{x\to a}[f(x)/(1/g(x)] \text{ (type } 0/0)$$
$$= \lim_{x\to a}[g(x)/(1/f(x))] \text{ (type } \infty/\infty).$$

If the limit is of the type $\infty - \infty$ set $\lim_{x\to a} f(x) = 0$, $\lim_{x\to a} g(x) = 0$, then

$$\lim_{x\to a}\left[\frac{1}{f(x)} - \frac{1}{g(x)}\right] = \lim_{x\to a}\left[\frac{g(x) - f(x)}{f(x)g(x)}\right] \text{(type } 0/0)$$
$$= \lim_{x\to a}\left[\frac{1/f(x)g(x)}{1/(g(x) - f(x))}\right] \text{(type } \infty/\infty).$$

Apply these results to evaluate the following limits:

(a) $\lim_{x\to 0}\left(\frac{1}{\sin x} - \frac{1}{x}\right)$; (b) $\lim_{x\to 3}\left(\frac{1}{x-3} - \frac{5}{x^2 - x - 6}\right)$;

(c) $\lim_{x\to 0}(1 - \cos x)\cot x$; (d) $\lim_{x\to\infty} x \sin\frac{3}{x}$;

(e) $\lim_{x\to 1}(1 - x)\tan\frac{\pi x}{2}$; (f) $\lim_{x\to(1/2)\pi}\left(\frac{x}{\cot x} - \frac{\pi}{2\cos x}\right)$.

5.32 If $y = f(x)$, where f is a differentiable function, find the differential dy given that:

(a) $f(x) = x^6 + 3x^2 + x + 6$;

(b) $f(x) = x\sin(x^2 + 1)$;

(c) $f(x) = \frac{1 + x}{2 + \sin^2 x}$;

(d) $f(x) = (1 + x^2)^{1/2}$.

5.33 Metals A and B have coefficients of linear expansion α and β, respectively. That is to say, when the temperature changes by an amount t from the ambient value T_0, the linear dimensions of metal A change by a factor $(1 + \alpha t)$, while those of metal B change by a factor $(1 + \beta t)$. Suppose that a block of metal A contains a cylindrical cavity of height H_0 and radius R_0 at temperature T_0 which is empty apart from a cylinder of metal B which has height h_0 and radius r_0 at that same temperature. Obtain an approximate expression for the small volume change dV of the cavity between the cylinders consequent upon a small change of temperature dt.

Section 5.4

5.34 Compute the first and second derivatives of the functions $f(x)$ listed below:

(a) $f(x) = \tan x$;

(b) $f(x) = x^2 \sin x$;

(c) $f(x) = (1 + x)(3 \sin x + \cos 2x)$;

(d) $f(x) = (x^2 + 1)^{1/2}$;

(e) $f(x) = \sin(1 + x^2)$;

(f) $f(x) = \dfrac{1}{x}$.

5.35 Show that the function $f(x)$ defined below is continuous and has a continuous first derivative at $x = 1$, but that it has a discontinuous second derivative at that point:

$$f(x) = \begin{cases} x^4 + x^2 - x + 1 & \text{for } x \leq 1 \\ 2x^3 - x^2 + x & \text{for } x > 1. \end{cases}$$

5.36 Use Leibinz's theorem to evaluate the third derivatives of the following functions:

(a) $f(x) = \dfrac{x^7}{1 + x}$; (b) $f(x) = (x^7 - 1)\sin x$;

(c) $f(x) = \sin^2 x$; (d) $f(x) = x^3 \cos 2x$.

5.37 Apply Theorems 5.17 and 5.18 to locate and identify the extrema and points of infection of the following functions, using your results to determine the gradients at the points of inflection:

(a) $f(x) = 2x^3 + 3x^2 - 12x + 5$;

(b) $f(x) = \dfrac{x^2}{x^2 + 3}$;

(c) $f(x) = x^2(x - 12)^2$.

5.38 Determine the values of a and b in order that $f(x) = x^3 + ax^2 + bx + 1$ should have a point of inflection at $x = 2$ at which the gradient of the tangent to the graph is -3.

Section 5.5

5.39 Compute the derivatives f_x and f_y given that:

(a) $f(x, y) = x^2/y$;

(b) $f(x, y) = 3x^2y + (x + y)^2x + 1$;

(c) $f(x, y) = \sin(x^2 + y^2)$;

(d) $f(x, y) = x\cos(1 + x^2y^2)$.

5.40 Given that

$$f(x, y) = x^3 + 3x^2y + 4xy^2 + 2y^3$$

prove that $xf_x + yf_y = 3f$.

5.41 Compute the derivatives f_x, f_y, f_z given that:

(a) $f(x, y, z) = x^2yz + \dfrac{1}{xyz^2}$;

(b) $f(x, y, z) = x\cos yz + y\cos xz + z\cos xy$;

(c) $f(x, y, z) = \cos(x^2 + xy + yz)$.

5.42 Show that if

$$f(x, y, z) = \frac{x}{(x^2 + y^2 + z^2)^{3/2}}$$

then $xf_x + yf_y + zf_z = -2f$.

Section 5.6

5.43 Find the total differential du given that $u = f(x, y, z)$, where:

(a) $f(x, y, z) = \dfrac{1}{x^2yz} + xyz$;

(b) $f(x, y, z) = x\sin(y^2 + z^2)$;

(c) $f(x, y, z) = (1 - x^2 - y^2 - z^2)^{3/2}$.

5.44 Compute dy/dx from the following implicit relationships:

(a) $x^2 + y^2 = 4$;

(b) $x\sin xy = 1$;

(c) $x^2y + 2xy^2 + y^3 = 2$.

5.45 Compute dz/dx and dz/dy given that:

(a) $x^2 + y^2 + z^2 = 1$;

(b) $xyz + \sin xz^2 = 2$;

(c) $x^2 - 2y^2 + 3z^2 - yz + y = 0$;

(d) $x\cos y + y\cos z + z\cos x = 1$.

Section 5.7

5.46 Find the envelope of the family of curves with parameter α

$$(x-\alpha)^2 + y^2 = \alpha^2/2.$$

5.47 Find the envelope of the family of curves with parameter α

$$y = \alpha x + \frac{3}{2\alpha}.$$

5.48 When a particle is projected into the air from the origin with velocity V at an angle θ to the horizontal then, neglecting air resistance, its height y when distant x from the point of projection is given by

$$y = x \tan\theta - \frac{gx^2}{2V^2\cos^2\theta}$$

where g is acceleration due to gravity. By regarding θ as a parameter, show that the envelope of the family of trajectories for $0 \leq \theta \leq \pi$ is a parabola, and find its equation. This is usually called the parabola of safety because no projectile can penetrate beyond it.

5.49 Find the envelope of the family of curves with parameter α specified by

$$(x-\alpha)^2 + (y+\alpha)^2 - \alpha^2 = 0.$$

5.50 Show that the envelope of the family of curves with parameter α defined by

$$x\cos\alpha + y\sin\alpha = 2$$

is a circle. Find its center and radius. Interpret this family geometrically.

Section 5.8

5.51 Find $\mathrm{d}u/\mathrm{d}t$ given that:

(a) $u = xy + \sin(x^2 + y^2)$ with $x = 2t,\ y = (1+t^2)^{1/2}$;

(b) $u = (1 + x^2 + y^2)^{3/2}$ with $x = t(1+t),\ y = t^3$;

(c) $u = \dfrac{z}{(x^2 + y^2)^{1/2}}$ with $x = 3\cos t,\ y = 3\sin t,\ z = t^2$.

5.52 If $u = x^2 - xy + y^3$, compute $\mathrm{d}u/\mathrm{d}t$ at points on the curve specified parametrically by the equations $x = 2t+1,\ y = t^2 + t - 2$.

5.53 Prove that if $u = f(2x^2 + y^2)$, where f is a differentiable function, then

$$y\frac{\partial u}{\partial x} - 2x\frac{\partial u}{\partial y} = 0.$$

(*Hint:* Set $t = 2x^2 + y^2$).

5.54 If $u = f(x, y)$, compute $\mathrm{d}u/\mathrm{d}x$ given that:

(a) $f(x, y) = (1 + xy + x^2)$, where $y = \tan\left(\dfrac{x}{2}\right)$;

(b) $f(x, y) = (1 + x^2 - y^2)^{3/2}$, where $y = \cos 3x$;

(c) $f(x, y) = x\cos y + y\cos x - 1$, where $y = 1 + \sin^2 x$.

5.55 If $u = f(x, y)$ and $g(x, y) = 0$ are differentiable functions, compute $\mathrm{d}u/\mathrm{d}x$ given that:

(a) $f(x, y) = x^3 + 3xy + y^3$ and $g(x, y) = x\cos y + y\cos x - 2$;

(b) $f(x, y) = x^2y^2 + \sin xy$ and $g(x, y) = x^2 - 2y^2 - 3$.

Section 5.9

5.56 Use theorem 5.22 with $n = 2$, $m = 3$ to determine $\partial f/\partial u$, $\partial f/\partial v$ and $\partial f/\partial w$, given that:

$$f = x^2 + \tfrac{1}{2}y^2,$$

where $x = u^2 + v + w$ and $y = uvw$.

5.57 If u and v are functions of x and y which satisfy $u^2 - v^2 + 2x + 3y = 0$ and $uv + x - y = 0$, find $\partial u/\partial x$, $\partial u/\partial y$, $\partial v/\partial x$ and $\partial v/\partial y$ in terms of u and v.

5.58 Prove that if $z = f(u, v)$, where $u = x + 3t$, $v = y - 2t$, then

$$\frac{\partial z}{\partial t} = 3\frac{\partial z}{\partial u} - 2\frac{\partial z}{\partial v}.$$

5.59 Show that if $u = 1/r^n$, where $r^2 = x^2 + y^2 + z^2$, then

$$\frac{\partial^2 u}{\partial x^2} + \frac{\partial^2 u}{\partial y^2} + \frac{\partial^2 u}{\partial z^2} = \frac{n(n-1)}{r^n + 2}.$$

Section 5.10

5.60 Compute $\mathrm{d}x/\mathrm{d}y$ for each of the following relationships:

(a) $y = 1 + x^2 + x\sin x$;

(b) $y = (1 - x + x^2)^{1/2}$;

(c) $y = x + \tan x$.

5.61 Differentiate these functions:

(a) $f(x) = x^2 \operatorname{arcsec}(x/a)$;

(b) $f(x) = (x^2 + x + 1)/\arcsin(x^2 - 2)$;

(c) $f(x) = (1 + x + \arccos 2x)^{3/2}$.

5.62 Compute dy/dx and d^2y/dx^2 for each of the following parametrically defined curves:

(a) $x = t - 1,\ y = t^3$;

(b) $x = \cos^3 t, y = 2\sin^2 t$;

(c) $x = \arccos \dfrac{1}{\sqrt{(1 + t^2)}},\ y = \arcsin \dfrac{1}{\sqrt{(1 + t^2)}}$;

(d) $x = 2(\cos t + t \sin t),\ y = 2(\sin t - t \cos t)$.

5.63 Compute dy/dx and d^2y/dx^2 at $t = \frac{1}{2}\pi$ if $x = t - \sin t$ and $y = 2(1 - \cos t)$.

Section 5.11

5.64 Compute $\partial^2 z/\partial x^2$, $\partial^2 z/\partial x \partial y$, $\partial^2 z/\partial y \partial x$ and $\partial^2 z/\partial y^2$ for each of the following functions and hence show that $\partial^2 z/\partial x \partial y = \partial^2 z/\partial y \partial x$:

(a) $z = (x^2 + y^2)^{1/2}$;

(b) $z = x \cos y + y \cos x$;

(c) $z = \arctan(y/x)$.

5.65 Compute $f_{xx}(1, 1)$, $f_{xx}(1, 1)$ and $f_{xx}(1, 1)$ given that

$$f(x, y) = (1 + x)^4 (1 + y)^3.$$

Is $\partial^2 f/\partial x \partial y = \partial^2 f/\partial y \partial x$? Give reasons for your answer.

5.66 Given that

$$f(x, y) = \begin{cases} \dfrac{xy}{x^2 + y^2} & \text{for } x \neq 0,\ y \neq 0 \\ 1 & \text{for } x = 0,\ y = 0 \end{cases}$$

compute $\partial^2 f/\partial x \partial y$ starting, with reasons, when it is equal to $\partial^2 f/\partial y \partial x$. Is there any point at which this result is not true and, if so, what property of the function invalidates the result? (*Hint:* Consider limits taken along the line $y = mx$.)

5.67 Let

$$f(x,y) = \begin{cases} \dfrac{x^2y}{3x^4 + 2y^2} & \text{for } x \neq 0,\ y \neq 0 \\ 0 & \text{for } x = 0,\ y = 0. \end{cases}$$

By considering $\lim_{\substack{x\to 0\\ y\to 0}} f(x,y)$ (a) along the lines $y = mx$, and (b) along the parabolas $y = kx^2$, with m, k arbitrary constants, show $f(x, y)$ is non-differentiable at (0, 0), and hence that neither f_{xy} nor f_{yx} is defined at the origin.

5.68 Let $f(x, y) = |x| + |y| + ||x| - |y||$ for all x, y. Show that $f_{xy} = f_{yx} = 0$ everywhere except on the lines $x = 0$, $y = 0$, $y = x$ and $y = -x$.

In Problems 5.69 to 5.72 show that the following expressions w satisfy the equation

$$\frac{\partial^2 w}{\partial x^2} + \frac{\partial^2 w}{\partial y^2} + \frac{\partial^2 w}{\partial z^2} = 0.$$

5.69 $w = 3x^2 - 5y^2 + 2z^2$.

5.70 $w = z \arctan(x/y)$.

5.71 $w = (x^2 + y^2 + z^2)^{-1/2}$.

5.72 $w = 5z + \arctan\left(\dfrac{2x}{x^2 + y^2 - 1}\right)$.

5.73 Under the transformation $X = x - ct$, $Y = x + ct$ with $c =$ constant, the twice differentiable function $f(x, t)$ becomes $F(X, Y)$. Show that under this transformation the equation

$$\frac{\partial^2 f}{\partial t^2} = c^2 \frac{\partial^2 f}{\partial x^2} \text{ becomes } \frac{\partial^2 F}{\partial X \partial Y} = 0.$$

5.74 Under the transformation $x = s + t$, $y = t - 2s$ the twice differentiable function $f(x, y)$ becomes $F(s, t)$. Find the form of $\partial^2 F/\partial s \partial t$.

5.75 Under the transformation

$$x = u + v,\ y = uv$$

the twice differentiable and continuous function $f(x, y)$ becomes $F(u, v)$. By expressing $\partial F/\partial u$ and $\partial F/\partial v$ in terms of $\partial f/\partial x$ and $\partial f/\partial y$ show that

$$\frac{\partial^2 F}{\partial u \partial v} = \frac{\partial f}{\partial y} + \frac{\partial^2 f}{\partial x^2} + \frac{\partial^2 f}{\partial x \partial y} + \frac{\partial^2 f}{\partial y^2}.$$

(*Hint:* Find $\partial u/\partial x$ and $\partial u/\partial y$ by partial differentiation of the given transformation.)

5.76 Under the transformation

$$x = u^2 + v^2, \; y = uv$$

the twice differentiable and continuous function $f(x, y)$ becomes $F(u, v)$. By expressing $\partial F/\partial v$ in terms of $\partial f/\partial u$ and $\partial f/\partial y$ show that

$$\frac{\partial^2 F}{\partial u^2} = 2\frac{\partial f}{\partial x} + 4y\frac{\partial^2 f}{\partial x \partial y} + 4u^2\frac{\partial^2 f}{\partial x^2} + \frac{\partial^2 f}{\partial y^2}.$$

5.77 Under the transformation

$$x = u^2 + v^2, \; y = 2uv$$

the twice differentiable and continuous function $f(x, y)$ becomes $F(u, v)$. By expressing $\partial F/\partial v$ in terms of $\partial f/\partial x$ and $\partial f/\partial y$ show that

$$\frac{\partial^2 F}{\partial u^2} = 2\frac{\partial f}{\partial x} + 4y\frac{\partial^2 f}{\partial x \partial y} + 4x\frac{\partial^2 f}{\partial y^2}.$$

Supplementary computer problems

5.78 Given

$$f(x) = x^3 + 4x^2 - x + 1,$$

compute

$$F(h) = \frac{f(2 + h) - f(2)}{h}$$

for $h = -0.1, -0.05, -0.01, -0.005, 0.005, 0.01, 0.05, 0.1$, and confirm that as $h \to 0$ so $F(h)$ tends to $\mathrm{d}f/\mathrm{d}x$ at $x = 2$.

5.79 Given

$$f(x) = \sin(2x + 1), \; g(x) = (x^2 + x + 1)_1^{1/2},$$

compute

$$F(h) = \frac{f(h + 3)g(h + 3) - f(3)g(3)}{h}$$

for $h = -0.1, -0.05, -0.01, -0.005, 0.005, 0.01, 0.05, 0.1$, and confirm that as $h \to 0$ so $F(h)$ tends to $\mathrm{d}[f(x)g(x)]/\mathrm{d}x$ at $x = 3$.

5.80 Given

$$f(x) = (x^3 + 2x + 3)^{1/3}, \; g(x) = \cos(1 + x^2),$$

compute

$$F(h) = \left[\frac{f(h + 1)}{g(h + 1)} - \frac{f(1)}{g(1)}\right] \Big/ h$$

for $h = -0.1, -0.05, -0.01, -0.005, 0.005, 0.01, 0.05, 0.1$, and confirm that as $h \to 0$ so $F(h)$ tends to $d[f(x)g(x)]/dx$ at $x = 1$.

5.81 Compute

$$f(x) = \frac{x^3 + 3x^2 - 2x - 2}{2x^2 - x - 1}$$

for $x = 0.95, 0.96, 0.97, 0.98, 0.99, 0.995, 1.005, 1.01, 1.02, 1.03, 1.04, 1.05$, and confirm the result of Example 5.9(h) that $\lim_{x \to 1} f(x) = \frac{7}{3}$.

5.82 Find the envelope of the family of curves

$$\alpha x^2 + y^2 = 1 + \alpha^2$$

in which α is a parameter. Display the envelope together with the members of the family corresponding to $\alpha = 0.2, 0.5, 1.0, 1.25, 1.5$.

6 Exponential, logarithmic and hyperbolic functions and an introduction to complex functions

The exponential function e^x was introduced in Chapter 3 as an example of a function defined by a limit. Its occurrence throughout mathematics, together with its inverse, the natural logarithmic function $\ln x$, is sufficiently frequent to justify studying e^x and $\ln x$ separately. In the first part of this chapter the differentiability property of e^x is established by means of a definition of the exponential function which is different than but equivalent to the one used in Chapter 3 and makes use of that definition together with a geometrical argument and the idea that a function can be defined in terms of a series involving powers of x. The differentiability properties of the natural logarithmic function are then deduced from those of the exponential function by using the fact that $dy/dx = 1/(dx/dy)$.

The hyperbolic functions $\sinh x$, $\cosh x$ and $\tanh x$ are all defined in terms of the exponential function, so it is natural that they should be introduced and their differentiability properties established immediately after an account of the exponential function. The chapter then considers the exponential function with a complex argument, and shows how by making use of the series representations for the sine and cosine functions it is possible to establish de Moivre's theorem for any real x. The theorem is then used to show how various trigonometric identities can be deduced in a simple manner. The representation of functions as series involving powers of x, called power series, is taken up again and developed formally in Chapter 12. Finally a brief introduction to complex functions is given to justify the replacement of x by ix in the series for e^x.

6.1 The exponential function

The Euler number e and the exponential function e^x were first introduced in section 3.3 as the limits

$$e = \lim_{n\to\infty}\left(1 + \frac{1}{n}\right)^n \tag{6.1}$$

and

$$\mathrm{e}^x = \lim_{n \to \infty} \left(1 + \frac{x}{n}\right)^n \tag{6.2}$$

while in section 5.3.7 the graph of the exponential function was shown to have a unit gradient at the origin. Before seeking a more convenient and practical definition for e^x than the one in Eqn (6.2), let us first make use of the above definition to deduce the derivative of e^x.

Let x_0 be an arbitrary fixed number. Then, by definition, the derivative of e^x at $x = x_0$ is

$$\left[\frac{\mathrm{d}}{\mathrm{d}x}[\mathrm{e}^x]\right]_{x=x_0} = \lim_{h \to 0}\left[\frac{\mathrm{e}^{x_0+h} - \mathrm{e}^{x_0}}{h}\right]$$

$$= \mathrm{e}^{x_0} \lim_{h \to 0}\left[\frac{\mathrm{e}^h - 1}{h}\right], \tag{6.3}$$

but we saw in section 5.3.7 that this last limit is simply unity, so we have shown that

$$\left[\frac{\mathrm{d}}{\mathrm{d}x}[\mathrm{e}^x]\right]_{x=x_0} = \mathrm{e}^{x_0}. \tag{6.4}$$

As x_0 was arbitrary the subscript zero may be omitted, causing this to become the fundamental general result

$$\frac{\mathrm{d}}{\mathrm{d}x}[\mathrm{e}^x] = \mathrm{e}^x. \tag{6.5}$$

Setting $y = \mathrm{e}^x$, this last result is seen to be the equation

$$\frac{\mathrm{d}y}{\mathrm{d}x} = y. \tag{6.6}$$

This is an example of what is called a *differential equation*, because it relates the function $y(x)$ to its derivative dy/dx. Differential equations and their solutions will be studied later in chapters 13, 14 and 15.

However, let us anticipate one approach to the solution of differential equations and seek a series solution to the function $y = \mathrm{e}^x$ of the form

$$y = a_0 + a_1 x + a_2 x^2 + a_3 x^3 + \ldots = \sum_{r=0}^{\infty} a_r x^r, \tag{6.7}$$

where $y(0) = 1$. Assuming that this infinite series may be differentiated term by term, we have

$$\frac{\mathrm{d}y}{\mathrm{d}x} = a_1 + 2a_2 x + 3a_3 x^2 + \ldots = \sum_{r=0}^{\infty}(r+1)a_{r+1}x^r, \tag{6.8}$$

so that substituting for y and $\mathrm{d}y/\mathrm{d}x$ in Eqn (6.6) yields

$$\sum_{r=0}^{\infty}(r+1)a_{r+1}x^r = \sum_{r=0}^{\infty} a_r x^r. \tag{6.9}$$

For this result to be unconditionally true for all x, as it must be to satisfy our definition of e^x, it follows that it must be an *identity* in x. This can only be possible if the coefficients of the corresponding powers of x on each side of Eqn (6.9) are identical. Hence, equating the coefficients of the general term x^r, we find that

$$(r+1)a_{r+1} = a_r, \tag{6.10}$$

for $r = 0, 1, 2, \ldots$.

As we require that $y(0) = 1$, it follows by setting $x = 0$ in Eqn (6.7) that $a_0 = 1$. Using this result, together with Eqn (6.10), defines the coefficients a_r recursively. Setting $r = 0$, we have

$$a_1 = a_0, \qquad \text{so} \quad a_1 = 1;$$

setting $r = 1$, we have

$$2a_2 = a_1, \qquad \text{so } a_2 = 1/2!;$$

setting $r = 2$, we have

$$3a_3 = a_2, \qquad \text{so } a_3 = 1/3!;$$

and, in general,

$$a_r = 1/r! \tag{6.11}$$

for $r = 0, 1, 2, \ldots$, where, by definition, $0! = 1$.

Substituting these coefficients into the series in Eqn (6.7) then gives

$$e^x = 1 + x + \frac{x^2}{2!} + \frac{x^3}{3!} + \ldots + \frac{x^r}{r!} + \ldots = \sum_{n=0}^{\infty} \frac{x^n}{n!}, \tag{6.12}$$

which is the result (3.6) that was obtained intuitively.

Changing the sign of x in this series then shows that

$$e^{-x} = 1 - x + \frac{x^2}{2!} - \frac{x^3}{3!} + \ldots + (-1)^r \frac{x^r}{r!} + \ldots = \sum_{n=0}^{\infty}(-1)^n \frac{x^n}{n!}. \tag{6.13}$$

To arrive at a series representation for the Euler number e itself we only need to set $x = 1$ in the series for e^x to arrive at the result

$$e = 1 + 1 + \frac{1}{2!} + \frac{1}{3!} + \ldots + \frac{1}{r!} + \ldots = \sum_{n=0}^{\infty} \frac{1}{n!} \tag{6.14}$$

which, to 15 decimal places, has the numerical value

$$e = 2.718281828459045\ldots.$$

We have established the following theorem.

***Theorem 6.1* (exponential theorem)** Defining the number e by

$$\mathrm{e} = \sum_{n=0}^{\infty} \frac{1}{n!},$$

then for all x it follows that

$$\mathrm{e}^x = 1 + x + \frac{x^2}{2!} + \frac{x^3}{3!} + \ldots + \frac{x^r}{r!} + \ldots = \sum_{n=0}^{\infty} \frac{x^n}{n!}$$

and

$$\mathrm{e}^{-x} = 1 - x + \frac{x^2}{2!} - \frac{x^3}{3!} + \ldots + (-1)^r \frac{x^e}{r!} + \ldots = \sum_{n=0}^{\infty} (-1)^n \frac{x^n}{n!}. \quad \blacksquare$$

The graph of $y = \mathrm{e}^x$ is shown in Fig. 6.1, on which is drawn the tangent to the curve at the point $P(1, \mathrm{e})$ is drawn making an angle α with the x-axis where, because of Eqn (6.6) it follows that $\tan\alpha = \mathrm{e}$.

It is worth formally recording the differentiability properties of the exponential function. However, we first remark that if $f(x) = \mathrm{e}^{g(x)}$, where $g(x)$ is a differentiable function of x, then, setting $g(x) = u$ so that $f(x) = \mathrm{e}^u$ and using the chain rule, we obtain

$$\frac{\mathrm{d}f}{\mathrm{d}x} = \frac{\mathrm{d}f}{\mathrm{d}u} \cdot \frac{\mathrm{d}u}{\mathrm{d}x} = \mathrm{e}^u g'(x) = g'(x)\mathrm{e}^{g(x)}. \tag{6.15}$$

Theorem 6.2 If $f(x) = \mathrm{e}^{g(x)}$, where $g(x)$ is a differentiable function of x, then

$$\frac{\mathrm{d}}{\mathrm{d}x}\{\mathrm{e}^{g(x)}\} = g'(x)\mathrm{e}^{g(x)}.$$

In particular, if $g(x) = \alpha x$, where α is a constant, then,

$$\frac{\mathrm{d}}{\mathrm{d}x}(\mathrm{e}^{\alpha x}) = \alpha \mathrm{e}^{\alpha x}. \quad \blacksquare$$

Let us now establish an important property of e^x. Consider the quotient e^x/x^p, where p is any positive integer. Then from Eqn (6.12) it follows that

$$\frac{\mathrm{e}^x}{x^p} = \frac{1 + x + \dfrac{x^2}{2!} + \ldots + \dfrac{x^p}{p!} + \dfrac{x^{p+1}}{(p+1)!} + \ldots}{x^p} > \frac{x}{(p+1)!}.$$

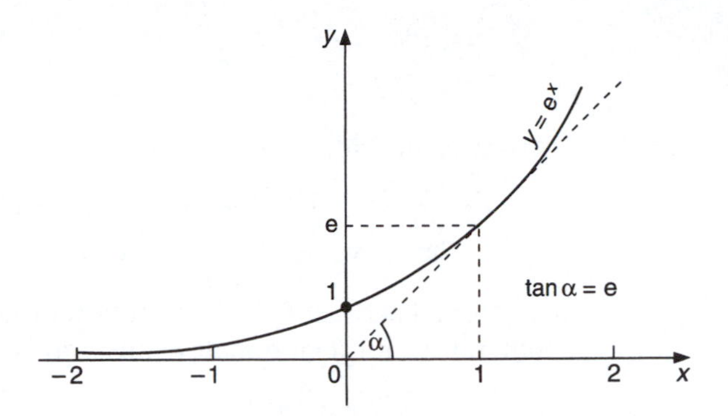

Fig. 6.1 The exponential function.

Hence we have shown that for any integer $p > 0$

$$\lim_{x\to\infty} \frac{e^x}{x^p} > \lim_{x\to\infty} \frac{x}{(p+1)!} \to \infty,$$

and, equivalently,

$$\lim_{x\to\infty} \frac{x^p}{e^x} = 0. \tag{6.16}$$

We have proved the following result.

Theorem 6.3 The function e^x increases more rapidly than any positive power of x as $x \to \infty$, so that

$$\lim_{x\to\infty} \frac{x^p}{e^x} = 0 \text{ for all } p > 0. \quad \blacksquare$$

We have already noted that $\lim_{x\to\infty} e^x \to \infty$, and as $e^x = 1/e^{-x}$ it follows that $\lim_{x\to-\infty} e^x = 0$ or, equivalently, $\lim_{x\to\infty} e^{-x} = 0$. From Theorem 6.1 it follows that the function e^x is everywhere positive and since, by virtue of its definition, its derivative is everywhere a strictly monotonic increasing function of x it must be a *convex* function.

These last properties are frequently of help when studying limiting problems involving the exponential function, as illustrated in the following examples.

Example 6.1 Deduce the values of the following limits:

(a) $\lim_{x\to\infty} \dfrac{3e^x + x^3 + 1}{2e^x + x^7}$;

(b) $\lim_{x\to\infty} \dfrac{2e^x + x^2 + 2}{3e^{3x} + 7}$;

(c) $\lim_{x\to 0} \dfrac{e^{ax} - e^{bx}}{2x}$.

Solution (a) After division of the numerators and denominator by e^x we have

$$\frac{3e^x + x^3 + 1}{2e^x + x^7} = \frac{3 + (x^3/e^x) + (1/e^x)}{2 + (x^7/e^x)},$$

and from Theorem 6.3 it then follows that all but the initial terms in the numerator and denominator must vanish as $x \to \infty$, so that

$$\lim_{x\to\infty} \frac{3e^x + x^3 + 1}{2e^x + x^7} = \frac{3}{2}.$$

(b) In this case, after division of the numerator and denominator by e^{3x} (the most significant term in the denominator), we have

$$\frac{2e^{2x} + x^2 + 2}{3e^{3x} + 7} = \frac{2e^{-x} + (x^2/e^{3x}) + (2/e^{3x})}{3 + (7/e^{3x})}.$$

However, this time as $x \to \infty$ all the numerator tends to zero while the denominator approaches the value 3. Hence we have

$$\lim_{x\to\infty} \frac{2e^{2x} + x^2 + 2}{3e^{3x} + 7} = 0.$$

(c) This limit involves an indeterminate form of the type 0/0, so we appeal to Theorem 5.14. Writing $f(x) = e^{ax} - e^{bx}$ and $g(x) = 2x$ we see that $f(0) = g(0) = 0$, and

$$\lim_{x\to 0} \frac{f'(x)}{g'(x)} = \lim_{x\to 0} \frac{ae^{ax} - be^{bx}}{2} = \frac{a-b}{2}.$$

Hence, by the conditions of Theorem 5.14,

$$\lim_{x\to 0} \frac{e^{ax} - e^{bx}}{2x} = \lim_{x\to 0} \frac{ae^{ax} - be^{bx}}{2} = \frac{a-b}{2}.$$

This result could have been obtained by use of the series definition of e^x, because

$$\begin{aligned}\lim_{x\to 0}\left[\frac{e^{ax} - e^{bx}}{2x}\right] &= \lim_{x\to 0}\left\{\frac{[1 + ax + O(a^2x^2)] - [1 + bx + O(b^2x^2)]}{2x}\right\}\\ &= \frac{a-b}{2} + \lim_{x\to 0}\left[\frac{O(a^2x) - O(b^2x)}{2}\right]\\ &= \frac{a-b}{2}.\end{aligned}$$

6.2 Differentiation of functions involving the exponential function

The exponential function occurs frequently in mathematics, and all of its differentiability properties follow from Theorem 6.2 combined with the fundamental differentiation theorems of Chapter 5. These results are straightforward and are best illustrated by examples. The first example illustrates the ordinary differentiation of simple combinations of functions.

Example 6.2 Differentiate the following functions $f(x)$:

(a) $f(x) = 2x^2 + 3e^{2x}$;

(b) $f(x) = x^2 e^{3x}$;

(c) $f(x) = 2\exp(x^3 + 2x + 1)$;

(d) $f(x) = e^{2x}/(1 + e^x)$;

(e) $f(x) = \sin(1 + e^{2x})$.

Solution (a)

$$f'(x) = \frac{d}{dx}(2x^2 + 3e^{2x}) = 4x + 3\frac{d}{dx}(e^{2x})$$

and so

$$f'(x) = 4 + 6e^{2x}.$$

(b)

$$f'(x) = e^{3x}\frac{d}{dx}(x)^2 + x^2\frac{d}{dx}(e^{3x})$$

so that

$$f'(x) = 2xe^{3x} + 3x^2e^{3x}.$$

(c) This is a more complicated example of a composite function or, more simply, of a function of a function. Set $u = x^3 + 2x + 1$ so that

$$f(x) = 2e^u.$$

Then, by the chain rule,

$$f'(x) = \frac{df}{du}\cdot\frac{du}{dx}$$

but

$$\frac{df}{du} = \frac{d}{du}(2e^u) = 2e^u = 2\exp(x^3 + 2x + 1) \text{ and } \frac{du}{dx} = 3x^2 + 2$$

so that, finally,

$$f'(x) = (6x^2 + 4)\exp(x^3 + 2x + 1).$$

(d) Writing $f(x)$ in the form

$$f(x) = e^{2x}(1 + e^x)^{-1}$$

we have

$$f'(x) = (1 + e^x)^{-1}\frac{d}{dx}(e^{2x}) + e^{2x}\frac{d}{dx}[(1 + e^x)^{-1}]$$

or

$$f'(x) = \frac{2e^{2x}}{(1 + e^x)} + e^{2x}\frac{d}{dx}[(1 + e^x)^{-1}].$$

To evaluate the last term set $1+e^x = u$, so that we then need to evaluate

$$\frac{d}{dx}\left(\frac{1}{u}\right)$$

which, by the chain rule, is

$$\frac{d}{dx}\left(\frac{1}{u}\right) = \frac{d}{du}\left(\frac{1}{u}\right)\cdot\frac{du}{dx}.$$

However, $du/dx = e^x$ and $(d/du)(1/u) = -(1/u^2) = -1/(1+e^x)^2$, showing that

$$\frac{d}{dx}[(1 + e^x)^{-1}] = \frac{-e^x}{(1 + e^x)^2}.$$

Hence, combining our results, we find

$$f'(x) = \frac{2e^{2x}}{(1 + e^x)} - \frac{e^{3x}}{(1 + e^x)^2}.$$

(e) This is another composite function. Set $u = 1+e^{2x}$, so that $f(x) = \sin u$. Proceeding as before we then see that

$$f'(x) = \frac{df}{du}\cdot\frac{du}{dx} = 2e^{2x}\cos(1 + e^{2x}). \quad ■$$

Higher-order derivatives are defined, as usual, by repeating the differentiation process the requisite number of times.

Example 6.3 Find $f''(x)$, given that:

(a) $f(x) = x^2e^{-2x}$;
(b) $f(x) = (x - 1)e^x$.

Solution (a) Proceeding as before we find that

$$f'(x) = 2xe^{-2x} - 2x^2e^{-2x},$$

and

$$f''(x) = 2e^{-2x} - 4xe^{-2x} - 4xe^{-2x} + 4x^2e^{-2x}.$$

Collecting terms we obtain

$$f''(x) = 2(1 - 4x + 2x^2)e^{-2x}.$$

(b) $f'(x) = e^x + (x-1)e^x = xe^x$

so that

$$f''(x) = e^x + xe^x. \quad ■$$

Partial differentiation of functions involving the exponential function is also straightforward, as the following example indicates.

Example 6.4 Determine f_x, f_y and f_{xy}, given that

$$f(x, y) = (x^2 + y^2)\exp(x^2 - y^2).$$

Solution

$$\begin{aligned}\frac{\partial f}{\partial x} &= 2x\exp(x^2 - y^2) + (x^2 + y^2)\frac{\partial}{\partial x}[\exp(x^2 - y^2)]\\ &= 2x\exp(x^2 - y^2) + 2x(x^2 + y^2)\exp(x^2 - y^2).\end{aligned}$$

Notice that $\partial f/\partial x$ comprises the sum of everywhere continuous functions and so it itself is everywhere continuous.

$$\begin{aligned}\frac{\partial f}{\partial y} &= 2y\exp(x^2 - y^2) + (x^2 + y^2)\frac{\partial}{\partial y}[\exp(x^2 - y^2)]\\ &= 2y\exp(x^2 - y^2) - 2y(x^2 + y^2)\exp(x^2 - y^2).\end{aligned}$$

The partial derivative $\partial f/\partial y$ is also seen to be everywhere continuous. Theorem 5.24 now tells us that $\partial^2 f/\partial x\partial y = \partial^2 f/\partial y\partial x$, so that we may differentiate either $\partial f/\partial x$ or $\partial f/\partial y$ to arrive at f_{xy}. We choose to differentiate f_x partially with respect to y:

$$\begin{aligned}\frac{\partial^2 f}{\partial y\,\partial x} &= -4xy\exp(x^2 - y^2) + 4xy\exp(x^2 - y^2)\\ &\quad -4xy(x^2 + y^2)\exp(x^2 - y^2),\end{aligned}$$

whence

$$\frac{\partial^2 f}{\partial x\,\partial y} = \frac{\partial^2 f}{\partial y\,\partial x} = -4xy(x^2 + y^2)\exp(x^2 - y^2).$$

6.3 The logarithmic function

Having introduced the exponential function there is now a need for an inverse function. Recalling section 1.9 we call it the *natural logarithmic function* and denote it by ln.

> ***Definition 6.1*** We define the natural logarithmic function $\ln x$ by the requirement that $y = \ln x \Leftrightarrow x = e^y$.

We may use this definition to compute the derivative of $\ln x$. As $dy/dx = 1/(dx/dy)$ and $x = e^y$, it follows that $dx/dy = e^y$, whence

$$\frac{dy}{dx} = \frac{1}{e^y} = \frac{1}{x}.$$

Now e^y is essentially positive, so that

$$\frac{d}{dx}(\ln x) = \frac{1}{x} \quad \text{for} \quad x > 0. \tag{6.17}$$

It is obvious that $\ln 1 = 0$ (because $e^0 = 1$) and, as x increases strictly monotonically with y, it also follows that $\ln x \to +\infty$ as $x \to +\infty$, and $\ln x \to -\infty$ as $x \to 0$ (because $e^{-\infty} = 0$).

Let us now prove that

$$\lim_{x\to\infty} \frac{\ln x}{x^\alpha} = 0 \qquad \text{for all } \alpha > 0.$$

As $x = e^y$ we have

$$\frac{\ln x}{x^\alpha} = \frac{y}{e^{\alpha y}}$$

and so

$$\lim_{x\to\infty} \frac{\ln x}{x^\alpha} = \lim_{y\to\infty} \frac{y}{e^{\alpha y}} = \frac{1}{\alpha} \lim_{y\to\infty} \frac{\alpha y}{e^{\alpha y}}.$$

Setting $u = ay$ we arrive at

$$\lim_{x\to\infty} \frac{\ln x}{x^\alpha} = \frac{1}{\alpha} \lim_{u\to\infty} \frac{u}{e^u} = 0,$$

by virtue of Theorem 6.3. This limiting result shows that $\ln x$ increases slower than any positive power of x, however small, as $x \to \infty$.

If in this last result we set $x = 1/u$ it becomes

$$\lim_{u\to\infty} u^\alpha \ln(1/u) = 0,$$

but $\ln(1/u) = \ln 1 - \ln u = -\ln u$, and so

$$\lim_{u\to\infty}(u^{\alpha}\ln u) = 0 \text{ for all } \alpha > 0.$$

Since the natural logarithmic function is defined as the function inverse to the exponential function we may use the arguments of section 2.2 to deduce the graph of $y = \ln x$. All that is necessary (cf. Fig. 2.4) is to reflect the graph of $y = e^x$ in the line $y = x$. Figure 6.2 shows the graph of $y = \ln x$ together with the graphs of $y = e^x$ and $y = x$.

Collecting the previous results we arrive at the following theorem.

Theorem 6.4 If $y = \ln x$, then

(a) $\dfrac{dy}{dx} = \dfrac{1}{x}$ for $x > 0$;

(b) $\lim_{x\to\infty} \dfrac{\ln x}{x^{\alpha}} = 0$ for all $\alpha > 0$;

(c) $\lim_{x\to 0}(x^{\alpha}\ln x) = 0$ for all $\alpha > 0$.

Logarithms to other bases can be used if convenient. They are defined as follows.

Definition 6.2 We define the logarithmic function to the base c, denoted by $\log_c x$ where $c \neq 1$ is a positive number, by the requirement that

$$y = \log_c x \Leftrightarrow x = c^y.$$

For reference purposes we record the following familiar properties of the logarithmic function. Let ln and $\log_c$ represent logarithms to the base e and c respectively, and a, b, r be real numbers. Then:

(a) $\ln ab = \ln a + \ln b$;

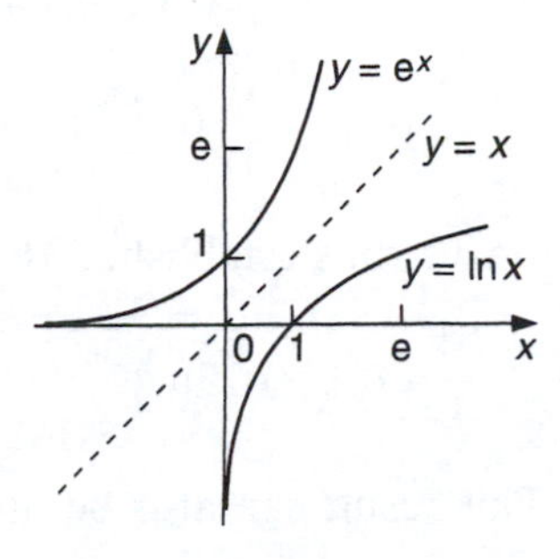

Fig. 6.2 The function $y = \ln x$ obtained by reflection of $y = e^x$ in the line $y = x$.

(b) $\ln a^r = r \ln a$;

(c) $\log_c a = \dfrac{\ln a}{\ln c}$;

(d) $\log_c \mathrm{e} = \dfrac{1}{\ln c}$;

(e) $a = \mathrm{e}^{\ln a}$;

(f) $\ln 1 = 0$;

(g) $\ln \mathrm{e} = 1$;

(h) $\log_c 1 = 0$;

(i) $\log_c c = 1$.

Results (c) and (d) quoted above are immediately useful if it is necessary to differentiate $\log_a x$. For we have

$$\log_a x = \frac{\ln x}{\ln a}$$

so that

$$\frac{\mathrm{d}}{\mathrm{d}x}(\log_a x) = \frac{1}{\ln a} \cdot \frac{\mathrm{d}}{\mathrm{d}x}(\ln x),$$

whence

$$\frac{\mathrm{d}}{\mathrm{d}x}(\log_a x) = \frac{1}{x \ln a} = \frac{\log_a \mathrm{e}}{x}.$$

Let us now find the derivative of the function a^x, where a is any positive number. Notice first that, by virtue of Definition 6.1,

$$a = \mathrm{e}^{\ln a}$$

so that

$$a^x = (\mathrm{e}^{\ln a})^x = \mathrm{e}^{x \ln a}.$$

Now $\ln a$ is simply a constant, so we have

$$\frac{\mathrm{d}}{\mathrm{d}x}(a^x) = \frac{\mathrm{d}}{\mathrm{d}x}(\mathrm{e}^{x \ln a}) = (\ln a)\mathrm{e}^{x \ln a} = a^x \ln a.$$

We have thus established the useful result

$$\frac{\mathrm{d}}{\mathrm{d}x}(a^x) = a^x \ln a.$$

This result can also be obtained in another manner. We set

$$y = a^x,$$

so that taking the natural logarithm gives

$$\ln y = x \ln a.$$

Differentiating this result with respect to x we obtain

$$\frac{\mathrm{d}}{\mathrm{d}x}(\ln y) = \frac{\mathrm{d}}{\mathrm{d}x}(x \ln a)$$

or

$$\frac{1}{y} \cdot \frac{\mathrm{d}y}{\mathrm{d}x} = \ln a$$

so that

$$\frac{\mathrm{d}y}{\mathrm{d}x} = y \ln a,$$

from which we obtain

$$\frac{\mathrm{d}}{\mathrm{d}x}(a^x) = a^x \ln a.$$

For our final general result we consider the differentiation of the function $y = \ln g(x)$, where $g(x)$ is a differentiable function. Setting $u = g(x)$ so that $y = \ln u$ and using the chain rule gives

$$\frac{\mathrm{d}y}{\mathrm{d}x} = \frac{\mathrm{d}y}{\mathrm{d}u} \cdot \frac{\mathrm{d}u}{\mathrm{d}x} = \frac{1}{u} g'(x)$$

so that, finally, we arrive at the result

$$\frac{\mathrm{d}}{\mathrm{d}x}[\ln g(x)] = \frac{g'(x)}{g(x)}.$$

We have thus proved the two basic results that

$$\frac{\mathrm{d}}{\mathrm{d}x}(a^x) = a^x \ln a, \text{ and}$$

$$\frac{\mathrm{d}}{\mathrm{d}x}[\ln g(x)] = \frac{g'(x)}{g(x)}.$$

Let us now examine some representative cases of limits involving logarithms.

Example 6.5 Evaluate the following limits:

(a) $\displaystyle\lim_{x \to \infty} \frac{\ln x^3}{x}$;

(b) $\lim_{x\to\infty} \frac{\ln a^x}{3x+1}$ with $a > 0$.

Solution (a) We have

$$\frac{\ln x^3}{x} = \frac{3\ln x}{x}$$

so that by Theorem 6.4(b) it follows at once that

$$\lim_{x\to\infty} \frac{\ln x^3}{x} = 0.$$

(b) We have

$$\frac{\ln a^x}{3x+1} = \frac{x\ln a}{3x+1}$$

and so

$$\lim_{x\to\infty} \frac{\ln a^x}{3x+1} = \lim_{x\to\infty} \frac{x\ln a}{3x+1} = \frac{1}{3}\ln a. \quad \blacksquare$$

Example 6.6 (a) Determine the derivative dy/dx for each of the following functions $y = f(x)$ where:

(i) $f(x) = 3^x x^2$;
(ii) $f(x) = (\sin x)^x$.

(b) If $u = x\ \ln[1+(x/y)] + y\ \ln[1+(y/x)]$, show that

$$x\frac{\partial u}{\partial x} + y\frac{\partial u}{\partial y} = u.$$

Solution (a)

(i) We have

$$\frac{d}{dx}(3^x x^2) = 3^x \frac{d}{dx}(x^2) = x^2 \frac{d}{dx}(3^x)$$

which, by virtue of Eqn (6.21), becomes

$$\frac{d}{dx}(3^x x^2) = 2x.3^x + x^2 3^x \ln 3$$

giving

$$\frac{d}{dx}(3^x x^2) = (2x + x^2 \ln 3)3^x.$$

(ii) We set $y = (\sin x)^x$ and take logarithms to get

$$\ln y = x \ln \sin x.$$

Now, differentiating, we find that

$$\frac{1}{y} \cdot \frac{dy}{dx} = \ln \sin x + x \frac{d}{dx}(\ln \sin x)$$

or

$$\frac{dy}{dx} = (\sin x)^x (\ln \sin x + x \cot x).$$

(b) We start by computing $\partial u / \partial x$. It is readily seen that

$$\begin{aligned}\frac{\partial u}{\partial x} &= \ln\left(1 + \frac{x}{y}\right) + x \frac{\partial}{\partial x} \ln\left(1 + \frac{x}{y}\right) + y \frac{\partial}{\partial x} \ln\left(1 + \frac{y}{x}\right) \\ &= \ln\left(1 + \frac{x}{y}\right) + x . \frac{1}{1 + x/y} . \frac{1}{y} + y . \frac{1}{1 + y/x}\left(\frac{-y}{x^2}\right),\end{aligned}$$

and so

$$\frac{\partial u}{\partial x} = \ln\left(1 + \frac{x}{y}\right) + \frac{x}{x + y} - \frac{y^2}{x(x + y)}.$$

The symmetry of x and y in u then allows us to interchange x and y in the above partial derivative in order to derive $\partial u / \partial y$ without further calculation. We obtain

$$\frac{\partial u}{\partial y} = \ln\left(1 + \frac{y}{x}\right) + \frac{y}{x + y} - \frac{x^2}{y(x + y)}.$$

Hereafter, direct substitution verifies that

$$x \frac{\partial u}{\partial x} + y \frac{\partial u}{\partial y} = u. \quad \blacksquare$$

Differentiation of complicated functions involving products and quotients may often be simplified by means of the natural logarithmic function. The process is called *logarithmic differentiation*, though it should more properly be called differentiation by means of the logarithmic function. It is best illustrated by example. Suppose we need to find dy/dx, where

$$y = \frac{f(x)}{g(x)\, h(x)}.$$

Taking natural logarithms and using their properties gives

$$\ln y = \ln f(x) - \ln g(x) - \ln h(x).$$

Differentiation with respect to x then gives

$$\frac{1}{y}\frac{dy}{dx}=\frac{f'(x)}{f(x)}-\frac{g'(x)}{g(x)}-\frac{h'(x)}{h(x)},$$

which when multiplied by y gives $\mathrm{d}y/\mathrm{d}x$. The simplification arises from the fact that in general the expression for $(1/y)\,(\mathrm{d}y/\mathrm{d}x)$ is easier to calculate than $\mathrm{d}y/\mathrm{d}x$.

Example 6.7 Find $\mathrm{d}y/\mathrm{d}x$ by means of logarithmic differentiation if

$$y=\frac{(2x-7)^5}{(3x+1)^{1/2}(2x-1)^{3/2}}.$$

Solution Taking the natural logarithm gives

$$\ln y = 5\ln(2x-7)-\tfrac{1}{2}\ln(3x+1)-\tfrac{3}{2}\ln(2x-1).$$

Differentiating with respect to x we find

$$\frac{1}{y}\frac{\mathrm{d}y}{\mathrm{d}x}=\frac{10}{2x-7}-\frac{3}{2(3x+1)}-\frac{3}{2x-1},$$

which after multiplication by y and simplification becomes

$$\frac{\mathrm{d}y}{\mathrm{d}x}=\frac{(72x^2+142x+1)(2x-7)^4}{2(3x+1)^{3/2}(2x-1)^{5/2}}. \quad \blacksquare$$

6.4 Hyperbolic functions

It is useful to define new functions called the *hyperbolic sine*, written $\sinh x$ and pronounced either as 'cinch x' or 'shine x', and the *hyperbolic cosine*, written $\cosh x$ and pronounced 'kosh x', which are related to the exponential function. This is achieved as follows.

Definition 6.3 (hyperbolic functions) For all real x we define $\sinh x$ and $\cosh x$ by the requirement that

$$\sinh x=\frac{\mathrm{e}^x-\mathrm{e}^{-x}}{2},\qquad \cosh x=\frac{\mathrm{e}^x+\mathrm{e}^{-x}}{2}.$$

It is an immediate consequence of combining the series for e^x and e^{-x} that

$$\sinh x = x+\frac{x^3}{3!}+\frac{x^5}{5!}+\frac{x^7}{7!}+\ldots+\frac{x^{2n+1}}{(2n+1)!}+\ldots=\sum_{n=0}^{\infty}\frac{x^{2n+1}}{(2n+1)!}, \tag{6.23}$$

and

$$\cosh x = 1+\frac{x^2}{2!}+\frac{x^4}{4!}+\frac{x^6}{6!}+\ldots+\frac{x^{2n}}{(2n)!}+\ldots=\sum_{n=0}^{\infty}\frac{x^{2n}}{(2n)!}. \tag{6.24}$$

Furthermore, it also follows from Definition 6.3 that $\sinh x$ is an odd function and $\cosh x$ is an even function, because $\sinh(-x) = -\sinh x$ and $\cosh(-x) = \cosh x$.

We now define the *hyperbolic tangent*, *cotangent*, *cosecant* and *secant*, denoted by $\tanh x$, $\coth x$, $\operatorname{csch} x$ and $\operatorname{sech} x$, as follows. The function $\tanh x$ is pronounced either as 'tansh x' or as 'than x', while the function $\coth x$ is pronounced as spelled, with the 'th' sounded in both cases as in 'thought'. The function $\operatorname{csch} x$ is pronounced as 'cosech x', while the function $\operatorname{sech} x$ is also pronounced as spelled, where in each case the 'ch' is sounded as in 'church'.

Definition 6.4

$$\tanh x = \frac{\sinh x}{\cosh x}; \quad \coth x = \frac{\cosh x}{\sinh x};$$

$$\operatorname{csch} x = \frac{1}{\sinh x}; \quad \operatorname{sech} x = \frac{1}{\cosh x}.$$

We illustrate how useful identities may be established directly from Definition 6.3. Let us prove that

$$\sinh a \cosh b + \cosh a \sinh b = \sinh(a+b).$$

Substituting for $\sinh a$ and $\cosh b$ from Definition 6.3, we obtain

$$\frac{e^a - e^{-a}}{2} \cdot \frac{e^b + e^{-b}}{2} + \frac{e^a + e^{-a}}{2} \cdot \frac{e^b - e^{-b}}{2} = \frac{e^{(a+b)} - e^{-(a+b)}}{2},$$

which proves our result since $[e^{(a+b)} - e^{-(a+b)}]/2 = \sinh(a+b)$. Similar manipulation establishes the validity of all the identities listed in Table 6.1.

Appeal to Definitions 6.3 and 6.4, together with the differentiability properties of the exponential function, establishes Table 6.2, the table of derivatives.

The behaviour of the hyperbolic sine and cosine functions is indicated graphically in Fig. 6.3(a) to which, for comparison, the graphs of $y = \frac{1}{2}e^x$ and $y = \frac{1}{2}e^{-x}$ have been added. Figures 6.3(b) – (e) show the behaviour of the other hyperbolic functions that are defined in terms of $\sinh x$ and $\cosh x$.

Functions inverse to the hyperbolic sine and cosine are introduced through the following definitions.

Table 6.1 Identities for hyperbolic functions

$\sinh(x \pm y) = \sinh x \cosh y \pm \cosh x \sinh y$	$\sinh 2x = 2 \sinh x \cosh x$
$\cosh(x \pm y) = \cosh x \cosh y \pm \sinh x \sinh y$	$\cosh 2x = \cosh^2 x + \sinh^2 x$
$\tanh(x \pm y) = \dfrac{\tanh x + \tanh y}{1 \pm \tanh x \tanh y}$	$\tanh 2x = \dfrac{2 \tanh x}{1 + \tanh^2 x}$
$\cosh^2 x - \sinh^2 x = 1$	$\cosh^2 x = \dfrac{(\cosh 2x) + 1}{2}$
$\tanh^2 x + \mathrm{sech}^2 x = 1$	$\sinh^2 x = \dfrac{(\cosh 2x) - 1}{2}$
$1 + \mathrm{csch}^2 x = \coth^2 x$	

Table 6.2 Derivatives of hyperbolic functions

$\dfrac{d}{dx}(\sinh x) = \cosh x$	$\dfrac{d}{dx}(\coth x) = -\mathrm{csch}^2 x$
$\dfrac{d}{dx}(\cosh x) = \sinh x$	$\dfrac{d}{dx}(\mathrm{csch}\, x) = -\mathrm{csch}\, x \coth x$
$\dfrac{d}{dx}(\tan hx) = \mathrm{sech}^2 x$	$\dfrac{d}{dx}(\mathrm{sech}\, x) = -\mathrm{sech}\, x \tanh x$

If $u = u(x)$, an application of the chain rule produces the following more general results:

$\dfrac{d}{dx}(\sinh u) = \cosh u \dfrac{du}{dx}$	$\dfrac{d}{dx}(\coth u) = -\mathrm{csch}^2 u \dfrac{du}{dx}$
$\dfrac{d}{dx}(\cosh u) = \sinh u \dfrac{du}{dx}$	$\dfrac{d}{dx}(\mathrm{csch}\, u) = -\mathrm{csch}\, u \coth u \dfrac{du}{dx}$
$\dfrac{d}{dx}(\tanh u) = \mathrm{sech}^2 u \dfrac{du}{dx}$	$\dfrac{d}{dx}(\mathrm{sech}\, u) = -\mathrm{sech}\, u \tanh u \dfrac{du}{dx}$

Definition 6.5

The *inverse hyperbolic sine*, $\operatorname{arcsinh} x$, and the *inverse hyperbolic cosine*, $\operatorname{arccosh} x$, are defined by the relationships:

(a) $y = \operatorname{arcsinh} x \Leftrightarrow x = \sinh y$;
(b) $y = \operatorname{arccosh} x \Leftrightarrow x = \cosh y$.

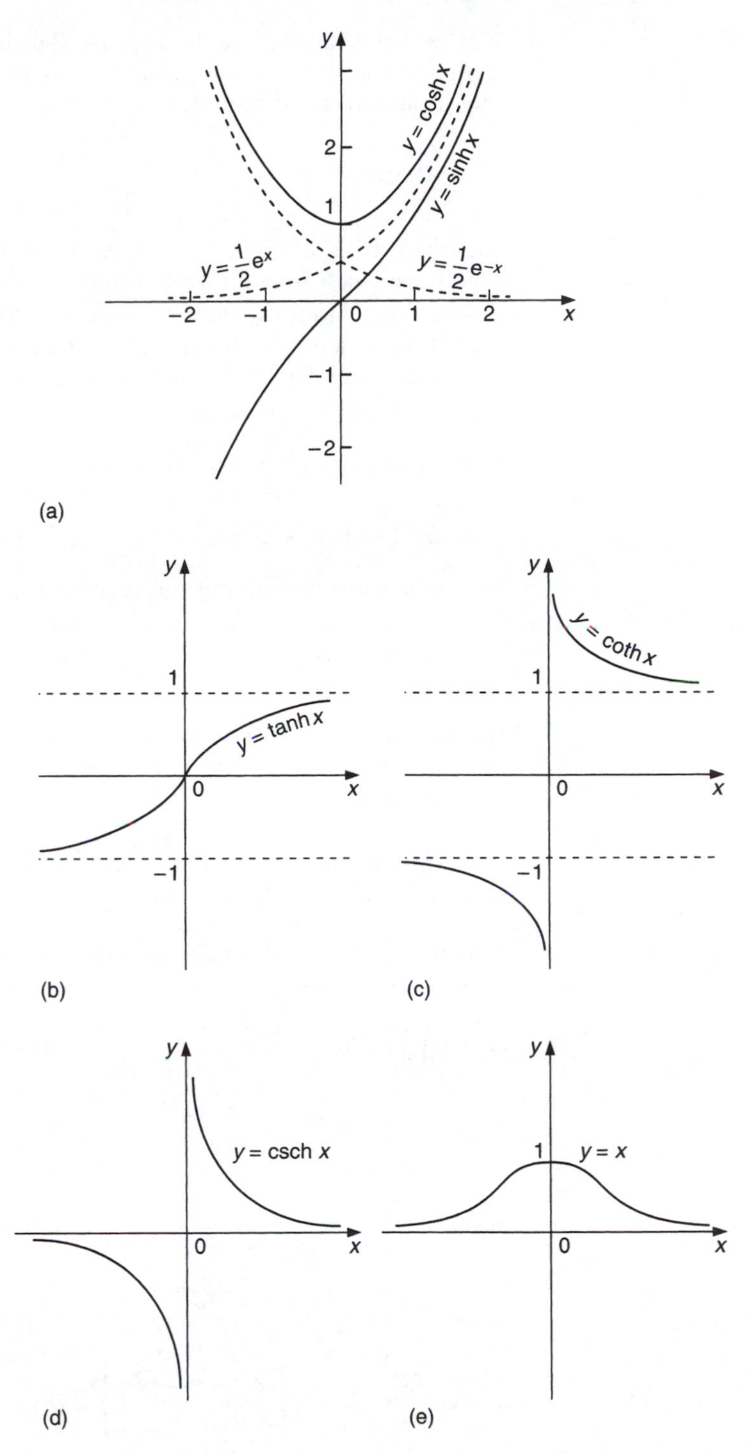

Fig. 6.3 Hyperbolic functions: (a) $y = \sinh x$ and $y = \cosh x$; (b) $y = \tanh x$; (c) $y = \coth x$; (d) $y = \operatorname{csch} x$; (e) $y = \operatorname{sech} x$.

It is often convenient to express the inverse hyperbolic functions in terms of the natural logarithmic function. To show how this may be accomplished, we derive the expression

$$\operatorname{arcsinh}\left(\frac{x}{a}\right) = \ln\left[\frac{x}{a} + \frac{\sqrt{a^2 + x^2}}{|a|}\right],$$

and then, in Table 6.3, list the results for the other inverse hyperbolic functions which may all be obtained in similar fashion.

The strange looking factor $x/|x|$ occurring on the right-hand side of the entry for arccsch (x/a) is simply to produce a factor of magnitude 1 whose sign is the sign of x, and which is undefined for $x = 0$; it is usually written either as sign (x), or as sgn (x).

If $y = \operatorname{arcsinh}\left(\frac{x}{a}\right)$, then $x = a \sinh y = (a/2)(\mathrm{e}^y - \mathrm{e}^{-y})$, and so

$$a\,\mathrm{e}^{2y} - 2x\,\mathrm{e}^y - a = 0.$$

Setting $u = \mathrm{e}^y$ and solving the resulting quadratic equation for u gives

$$u = \mathrm{e}^y = \left[\frac{x}{a} \pm \frac{\sqrt{x^2 + a^2}}{|a|}\right].$$

In this result the denominator in the second term is written as $|a|$, and not as a, because, by definition, the square root operation is taken to yield a positive quantity.

Adopting the positive sign, since e^y is non-negative, and then taking the natural logarithm of the result gives

Table 6.3 Inverse hyperbolic functions expressed in terms of the logarithmic function

$\operatorname{arcsinh}\left(\frac{x}{a}\right) = \ln\left[\frac{x}{a} + \frac{\sqrt{x^2 + a^2}}{\|a\|}\right],$	$\|x/a\| < \infty$
$\operatorname{arccosh}\left(\frac{x}{a}\right) = \ln\left[\frac{\|x\|}{\|a\|} + \frac{\sqrt{x^2 - a^2}}{\|a\|}\right],$	$\|x/a\| > 1$
$\operatorname{arctanh}\left(\frac{x}{a}\right) = \frac{1}{2}\ln\left[\frac{a + x}{a - x}\right],$	$\|x/a\| < 1$
$\operatorname{arcsech}\left(\frac{x}{a}\right) = \ln\left[\frac{x}{a} + \frac{\sqrt{x^2 + a^2}}{\|x\|}\right],$	$\|x/a\| < \infty, \quad x/a \neq 0$
$\operatorname{arccsch}\left(\frac{x}{a}\right) = \ln\left\{\frac{x}{\|x\|}\left[\frac{\|a\|}{x} + \frac{\sqrt{a^2 - x^2}}{x}\right]\right\},$	$0 < x/a < 1$
$\operatorname{arccoth}\left(\frac{x}{a}\right) = \frac{1}{2}\ln\left[\frac{x + a}{x - a}\right],$	$\|x/a\| > 1$

$$\operatorname{arcsinh}\left(\frac{x}{a}\right) = \ln\left[\frac{x}{a} + \frac{\sqrt{x^2 + a^2}}{|a|}\right], \qquad \text{for } -\infty < x/a < \infty.$$

The derivatives of the inverse hyperbolic functions are readily obtained by direct use of their definitions, and we illustrate the process by deriving $(\mathrm{d}/\mathrm{d}x)\,(\operatorname{arcsinh} x)$.

If $y = \operatorname{arcsinh} x$, then $x = \sinh y$ and so, differentiating with respect to x, we obtain

$$1 = \cosh y \frac{\mathrm{d}y}{\mathrm{d}x},$$

and so

$$\frac{\mathrm{d}y}{\mathrm{d}x} = \frac{1}{\cosh y} = \frac{1}{\sqrt{(1 + \sinh^2 y)}},$$

by virtue of the third identity in Table 6.1 and the fact that $\cosh y$ is essentially positive. Hence, using the fact that $x = \sinh y$, we find that

$$\frac{\mathrm{d}}{\mathrm{d}x}(\operatorname{arcsinh} x) = \frac{1}{\sqrt{(1 + x^2)}} \text{ for all } x.$$

In the case of $y = \operatorname{arccosh} x$ we must proceed with more care. If $y = \operatorname{arccosh} x$, so that $x = \cosh y$, then, as before, differentiating with respect to x gives

$$1 = \sinh y \frac{\mathrm{d}y}{\mathrm{d}x},$$

or

$$\frac{\mathrm{d}y}{\mathrm{d}x} = \frac{1}{\sinh y}.$$

Now from the graph of Fig. 6.3(a) we see that $\sinh y$ is positive if its argument $\operatorname{arccosh} x > 0$ and negative if $\operatorname{arccosh} x < 0$. Thus two different inverse functions must be defined.

If $\operatorname{arccosh} x > 0$, then

$$\frac{\mathrm{d}y}{\mathrm{d}x} = \frac{1}{\sinh y} = \frac{1}{\sqrt{(\cosh^2 y - 1)}} = \frac{1}{\sqrt{(x^2 - 1)}} \qquad \text{for } x > 1.$$

Conversely, if $\operatorname{arccosh} x < 0$, then

$$\frac{\mathrm{d}y}{\mathrm{d}x} = \frac{1}{\sinh y} = \frac{-1}{\sqrt{(\cosh^2 y - 1)}} = \frac{-1}{\sqrt{(x^2 - 1)}} \qquad \text{for } x > 1.$$

Other inverse hyperbolic functions are defined similarly, and it is left to the reader to verify the remaining entries in Table 6.4. (In many books the

Table 6.4 Derivatives of inverse hyperbolic functions with $u = u(x)$

$$\frac{d}{dx}(\operatorname{arcsinh} u) = \frac{1}{\sqrt{(1+u^2)}}\frac{du}{dx} \quad \text{for all } u$$

$$\frac{d}{dx}(\operatorname{arccosh} u) = \frac{1}{\sqrt{(u^2-1)}}\frac{du}{dx} \quad \text{for } \operatorname{arccosh} u > 0 \text{ and } u > 1$$

$$\frac{d}{dx}(\operatorname{arccosh} u) = \frac{-1}{\sqrt{(u^2-1)}}\frac{du}{dx} \quad \text{for } \operatorname{arccosh} u < 0 \text{ and } u > 1$$

$$\frac{d}{dx}(\operatorname{arctanh} u) = \frac{1}{1-u^2}\frac{du}{dx} \quad \text{for } |u| < 1$$

$$\frac{d}{dx}(\operatorname{arccoth} u) = \frac{1}{1-u^2}\frac{du}{dx} \quad \text{for } |u| > 1$$

$$\frac{d}{dx}(\operatorname{arccsch} u) = \frac{-1}{|u|\sqrt{(1+u^2)}}\frac{du}{dx} \quad \text{for all } u \neq 0$$

$$\frac{d}{dx}(\operatorname{arcsech} u) = \frac{-1}{u\sqrt{(1-u^2)}}\frac{du}{dx} \quad \text{for } \operatorname{arcsech} u > 0 \text{ and } 0 < u < 1$$

$$\frac{d}{dx}(\operatorname{arcsech} u) = \frac{-1}{u\sqrt{(1-u^2)}}\frac{du}{dx} \quad \text{for } \operatorname{arcsech} u < 0 \text{ and } 0 < u < 1$$

inverse function is denoted by a superscript -1, so that $\sinh^{-1} x$ is written in place of $\operatorname{arcsinh} x$, etc.)

The inverse hyperbolic functions satisfy the following identities:

$$\operatorname{arcsech} x = \operatorname{arccosh}\left(\frac{1}{x}\right), \quad \operatorname{arccsch} x = \operatorname{arcsinh}\left(\frac{1}{x}\right),$$

$$\operatorname{arccoth} x = \operatorname{arctanh}\left(\frac{1}{x}\right). \tag{6.25}$$

The following examples are representative of the limiting and differentiability problems encountered with hyperbolic functions.

Example 6.8

(a) Evaluate $\lim_{x\to\infty} \dfrac{5\sinh 3x + xe^x}{4e^{3x}}$;

(b) Find $f'(x)$ if $f(x) = \sinh(x^2+3x+1)^{1/2}$;
(c) Find $f'(x)$ given that $f(x) < 0$ is given by $f(x) = \text{arccosh}(\sin^2 x)$;
(d) Determine f_x and f_y given that $f(x, y) = xy\cosh(x^2+y^2)$.

Solution (a) From Definition 6.3

$$\sinh 3x = \tfrac{1}{2}(e^{3x} - e^{-3x})$$

so, substituting this into the required limit, after dividing both numerator and denominator by e^{3x} and applying the usual arguments, it follows at once that

$$\lim_{x\to\infty} \frac{5\sinh 3x + xe^x}{4e^{3x}} = \lim_{x\to\infty} \frac{\frac{5}{2}(e^{3x} - e^{-3x}) + xe^x}{4e^{3x}} = \frac{5}{8}.$$

(b) $f'(x) = [\cosh(x^2 + 3x + 1)^{1/2}].\dfrac{1}{2}.\dfrac{(2x+3)}{(x^2 + 3x + 1)^{1/2}}$

so that

$$f'(x) = \frac{(2x+3)}{2(x^2 + 3x + 1)^{1/2}}\cosh(x^2 + 3x + 1)^{1/2}.$$

(c) Set $y = \text{arccosh}(\sin^2 x)$, so that

$$\sin^2 x = \cosh y.$$

Differentiation with respect to x then gives

$$2\sin x\cos x = \sinh y\frac{dy}{dx},$$

or

$$\frac{dy}{dx} = \frac{2\sin x\cos x}{\sinh y}.$$

As we are told that $y = f(x) < 0$, it then follows that

$$\frac{dy}{dx} = \frac{-2\sin x\cos x}{\surd((\cosh^2 y) - 1)} = \frac{-2\sin x\cos x}{\surd((\sin^4 x) - 1)}$$

provided $\sin x \neq 1$.

(d) $$\frac{\partial f}{\partial x} = y \cosh(x^2 + y^2) + xy\partial/\partial x \cosh(x^2 + y^2)$$

$$= y \cosh(x^2 + y^2) + 2x^2 y \sinh(x^2 + y^2).$$

Similarly,

$$\frac{\partial f}{\partial x} = x \cosh(x^2 + y^2) + 2xy^2 \sinh(x^2 + y^2).$$

6.5 Exponential function with a complex argument

If we formally replace x by ix in the series expansion of e^x in Theorem 6.1, we obtain

$$e^{ix} = 1 + ix - \frac{x^2}{2!} - i\frac{x^3}{3!} + \frac{x^4}{4!} + i\frac{x^5}{5!} - \frac{x^6}{6!} + \ldots + i^n \frac{x^n}{n!} + \ldots.$$

Clearly e^{ix} is a complex number for any fixed real number x and, writing it in the form $e^{ix} = C(x) + iS(x)$, it follows by equating real and imaginary parts that

$$C(x) = 1 - \frac{x^2}{2!} + \frac{x^4}{4!} - \frac{x^6}{6!} + \ldots + (-1)^n \frac{x^{2n}}{(2n)!} + \ldots$$

and

$$S(x) = x - \frac{x^3}{3!} + \frac{x^5}{5!} + \frac{x^7}{7!} + \ldots + (-1)^n \frac{x^{2n+1}}{(2n+1)!} + \ldots.$$

Thus, in fact, if x is regarded as a variable, $S(x)$ and $C(x)$ are functions of x, and e^{ix} is, in some sense yet to be properly defined, a function of a complex variable.

Assuming that the series for $C(x)$ may be differentiated term by term, it is easily verified that

$$C'(x) = -x + \frac{x^3}{3!} - \frac{x^5}{5!} + \frac{x^7}{7!} + \ldots + (-1)^{n+1} \frac{x^{2n+1}}{(2n+1)!} + \ldots.$$

Next, differentiating $C'(x)$ again with respect to x yields

$$C''(x) = -1 + \frac{x^2}{2!} - \frac{x^4}{4!} + \frac{x^6}{6!} + \ldots + (-1)^{n+1} \frac{x^{2n}}{(2n)!} + \ldots,$$

showing that in fact

$$C''(x) = -C(x).$$

Now, setting $x = 0$ in the series for $C(x)$ and $C'(x)$, we find that

$$C(0) = 1 \quad \text{and} \quad C'(0) = 0.$$

Hence the function $C(x)$ is seen to be the solution of the special differential equation

$$\frac{d^2y}{dx^2} + y = 0$$

with $y(0) = 1$ and $y'(0) = 0$.

This same differential equation with the conditions on y was encountered in Example 5.13(a), where it was derived as the equation satisfied by $y = \cos x$ and its derivatives. Thus the function $C(x)$ is, in reality, the function $\cos x$. An analogous argument establishes that $S(x) \equiv \sin x$. On account of this identification of $C(x)$ and $S(x)$ we may write

$$e^{ix} = \cos x + i \sin x. \tag{6.26}$$

As a direct consequence of replacing x by $-x$ in Eqn (6.26) and using the fact that $\cos x$ is even, but $\sin x$ is odd, we find that

$$e^{-ix} = \cos x - i \sin x. \tag{6.27}$$

Combination of Eqns (6.26) and (6.27) leads to the following definitions of the sine and cosine functions.

Definition 6.6

$$\sin x = \frac{e^{ix} - e^{-ix}}{2i},$$

$$\cos x = \frac{e^{ix} + e^{-ix}}{2}.$$

Comparison of Eqns (4.15) and (6.26) shows that e^{ix} represents a complex number of unit modulus lying on the unit circle drawn about the origin. The *argument* of e^{ix} is x. Special values of e^{ix} that occur frequently are:

$$\begin{array}{ll} (x = 0) & e^{i0} = 1 \\ (x = \pi/2) & e^{i\pi/2} = i \\ (x = \pi) & e^{i\pi} = -1 \\ (x = 3\pi/2) & e^{i3\pi/2} = -i \\ (x = 2\pi) & e^{i2\pi} = 1. \end{array}$$

Slightly more general than Eqn (6.26) is the complex number $e^{(x+iy)}$ for, by the property of indices together with Eqn (6.26), we have

$$e^{(x+iy)} = e^x . e^{iy} = e^x(\cos y + i \sin y), \tag{6.28}$$

showing that

$$|e^{(x+iy)}| = e^x \quad \text{and} \quad \arg e^{(x+iy)} = y. \tag{6.29}$$

Thus the modulus–argument form of a general non-zero complex number z may be written

$$z = re^{i\theta},$$

where

$$r = |z| \text{ and } \theta = \arg z. \tag{6.30}$$

This is, of course, an alternative form of Eqn (4.15).

As it is true for any exponent α that $(a^x)^\alpha = a^{\alpha x}$, it follows that $(e^{ix})^\alpha = e^{i\alpha x}$, so that from Eqn (6.26) we arrive at the result

$$(\cos x + i \sin x)^\alpha = \cos \alpha x + i \sin \alpha x. \tag{6.31}$$

This is simply *de Moivre's theorem* (Theorem 4.2) for *any* exponent α and not just for the integral values used in the first proof of this important theorem.

To close, let us apply these results to give an alternative derivative of the results of Example 4.10, and also the express $\sin^n \theta$ and $\cos^n \theta$ in terms of sums involving $\sin r\theta$ and $\cos r\theta$, as promised in that example. As in Chapter 4, the argument is best presented by example.

Example 6.9

(a) Express $\sin n\theta$ and $\cos n\theta$ in terms of $\cos\theta$ and $\sin\theta$. Deduce the form taken by the result when $n = 4$.
(b) Express $\cos^7\theta$ in terms of $\cos r\theta$.
(c) Express $\sin^5\theta$ in terms of $\sin r\theta$.
(d) Prove that $\sin(A+B) = \sin A \cos B + \cos A \sin B$ and $\cos(A+B) = \cos A \cos B - \sin A \sin B$.

Solution

(a)

$$\cos n\theta = \text{Re}(e^{in\theta}) = \text{Re}[(e^{i\theta})^n] = \text{Re}[(\cos\theta + i \sin\theta)^n];$$

$$\sin n\theta = \text{Im}(e^{in\theta}) = \text{Im}[(e^{i\theta})^n] = \text{Im}[(\cos\theta + i \sin\theta)^n].$$

When $n = 4$ we have

$$(\cos\theta + i \sin\theta)^4 = \cos^4\theta + 4i \cos^3\theta \sin\theta - 6 \cos^2\theta \sin^2\theta - 4i \cos\theta \sin^3\theta + \sin^4\theta.$$

Hence

$$\cos 4\theta = \text{Re}\,[(\cos\theta + i \sin\theta)^4] = \cos^4\theta - 6\cos^2\theta \sin^2\theta + \sin^4\theta$$

and

$$\sin 4\theta = \text{Im}[(\cos\theta + i \sin\theta)^4] = 4(\cos^3\theta \sin\theta - \cos\theta \sin^3\theta).$$

(b) From Definition 6.6 we may write

$$\cos^7\theta = \left(\frac{e^{i\theta} + e^{-i\theta}}{2}\right)^7.$$

Expanding the right-hand side by the binomial theorem, simplifying and grouping terms, we obtain

$$\cos^7 \theta = \frac{1}{2^6}\left(\frac{e^{7i\theta} + e^{-7i\theta}}{2} + 7\frac{e^{5i\theta} + e^{-5i\theta}}{2} + 21\frac{e^{3i\theta} + e^{-3i\theta}}{2} + 35\frac{e^{i\theta} + e^{-i\theta}}{2}\right).$$

Again using Definition 6.6, we see that this immediately simplifies to

$$\cos^7 \theta = \frac{1}{64}(\cos 7\theta + 7\cos 5\theta + 21\cos 3\theta + 35\cos \theta).$$

(c) From Definition 6.6 we may write

$$\sin^5 \theta = \left(\frac{e^{i\theta} - e^{-i\theta}}{2i}\right)^5.$$

Expanding the right-hand side, simplifying and grouping terms gives

$$\sin^5 \theta = \frac{1}{2^4}\left(\frac{e^{5i\theta} - e^{-5i\theta}}{2i} - 5\frac{e^{3i\theta} - e^{-3i\theta}}{2i} + 10\frac{e^{i\theta} - e^{-i\theta}}{2i}\right).$$

Again appealing to Definition 6.6, we see that this immediately reduces to

$$\sin^5 \theta = \frac{1}{16}(\sin 5\theta - 5\sin 3\theta + 10\sin \theta).$$

(d) As $e^{iA}.e^{iB} = e^{i(A+B)}$ we have

$$(\cos A + i\sin A)(\cos B + i\sin B) = \cos(A+B) + i\sin(A+B).$$

The required results follow by equating the respective real and imaginary parts on either side of this equation. Thus, equating real parts, for example, gives

$$\cos A \cos B - \sin A \sin B = \cos(A+B). \quad \blacksquare$$

A variant of the method used here and in example (b) above is to be found outlined in Problems 6.25 and 6.26.

6.6 Functions of a complex variable, limits, continuity and differentiability

This section justifies the replacement of x by ix in the series for e^x, leading to a function of a complex variable, and it derives the conditions that must be satisfied if a complex function is to be differentiable. The replacement of the real variable x in the function $y=f(x)$ by the complex variable $z=x+iy$ leads naturally to the study of functions of a complex variable $w=f(z)$. Like functions of a real variable, $w=f(z)$ is defined in some region of the complex plane (the z-plane). To apply the calculus to complex functions it is necessary to extend the notions of continuity and differentiability, and this is shown to lead in turn to the requirement that in order for $w=f(z)$ to have a uniquely defined derivative, and so to be an analytic function, it is necessary that it satisfies certain conditions. These conditions are that the real part u and the imaginary part v of w satisfy the Cauchy–Riemann equations

$$\frac{\partial u}{\partial x}=\frac{\partial v}{\partial y} \quad \text{and} \quad \frac{\partial u}{\partial y}=-\frac{\partial v}{\partial x}.$$

In Chapter 2 we used the term 'a real-valued function of a real variable' to mean any rule that associates with each real number from the domain of definition of the function a unique real number from the range of that function. Symbolically, if D denotes the set of points in the domain of a function f, and R denotes the set of points in the range of f, this relationship or mapping is given by

$$R=f(D).$$

These ideas still hold good when the domain D and the range R include complex numbers. Thus if z is any point in D, and w is the unique number assigned to z by the function f, we write

$$w=f(z). \tag{6.32}$$

The number $z=x+iy$ is allowed to assume any value in D and so, if desired, could be called a complex independent variable, when w could then properly be called a complex dependent variable. Usually we shall simply refer to z and w as complex variables. It must be appreciated that, like z, the variable w has a real part and an imaginary part, both of which are in general dependent on x and y through the variable $z=x+iy$. We summarize these ideas formally as follows.

Definition 6.7 (function of a complex variable)

We shall say that f is a *function of the complex variable* $z=x+iy$, and write

$$w=f(z),$$

if f associates a unique complex number $w=u+iv$ with each complex number z belonging to some region D of the complex plane.

Specific examples of functions of a complex variable are:

(a) $w = iz + 1$; (b) $w = z\bar{z}$; (c) $w = z^2 + 2z + 1$;

(d) $w = 1/(z-2)$; (e) $w = \sin z$.

With the exception of (d), which is not defined for $z = 2$, these functions are defined for all z.

The difference between a function of a complex variable and a real-valued function of a real variable is made clear by expressing these examples in real and imaginary form. Thus writing $z = x + iy$ and $w = u + iv$ we find:

(a) $w = i(x+iy) + 1 = (1-y) + ix$, showing that $u = 1 - y,\ v = x$.
(b) $w = (x+iy)(x-iy) = x^2 + y^2$, showing that $u = x^2 + y^2,\ v = 0$. This is an example of a function that always maps a complex variable into a real variable.
(c) $w = (x+iy)^2 + 2(x+iy) + 1 = (x^2 + 2x - y^2 + 1) + i(2y + 2xy)$, showing that $u = x^2 + 2x - y^2 + 1,\ v = 2y(1+x)$.
(d) $w = 1/(x + iy - 2) = [(x-2) - iy]/(x^2 + y^2 - 4x + 4)$, showing that $u = (x-2)/(x^2 + y^2 - 4x + 4),\ v = -y/(x^2 + y^2 - 4x + 4)$, provided only that $x \neq 2$ and $y \neq 0$.
(e) $w = \sin z = \sin(x+iy) = \sin x \cos iy + \cos x \sin iy$, and so using the results of Problem 6.21, that $\cos iy = \cosh y, \sin iy = i \sinh y$, we arrive at $w = \sin x \cosh y + i \cos x \sinh y$. Thus in this case $u = \sin x \cosh y,\ v = \cos x \sinh y$.

Any function of x, y and complex constants that gives rise to a unique complex number when x and y are specified defines a function of the complex variable z by virtue of the relationship $z = x + iy$. For suppose that

$$(x + y + 1) + i(x - 2y) = f(z),$$

then to determine $f(z)$ when $z = 1 + 2i$ we simply write $x + iy = 1 + 2i$, showing that $x = 1,\ y = 2$, after it follows from the form of $f(z)$ that $f(1+2i) = 4 - 3i$.

The concept of a limit can also be extended to complex functions $f(z) = u(x, y) + iv(x, y)$. We shall say that $w = f(z)$ has the *limit* w_0 as $z \to z_0$ if each of the two real functions $u(x, y)$ and $v(x, y)$ of the two real variables x, y has a limit in the sense of Definition 3.8 as $(x, y) \to (x_0, y_0)$, where $z = x + iy$ and $z_0 = x_0 + iy_0$. That is to say, the complex limit w_0 of $w = f(z)$ must be *independent* of the path in the complex plane by which $z \to z_0$. By analogy with real functions of two real variables, we shall say that $w = f(z)$ is *continuous* at $z = z_0$ if

$$\lim_{z \to z_0} f(z) = w_0 \quad \text{and} \quad f(z_0) = w_0. \tag{6.33}$$

Using these ideas we now define the derivative dw/dz, or $f'(z)$, of $w = f(z)$ as the limit of the usual difference quotient $[f(z+h) - f(z)]/h$ as $h \to 0$, whenever this limit exists (remember that h is complex). Complex functions with this differentiability property throughout some region D are called *analytic* functions in D. Let us now examine the condition that this definition imposes on $u(x, y)$ and $v(x, y)$ on account of the required independence of the limit of the difference quotient with respect to the manner in which $h \to 0$. We start by writing out the difference quotient in terms of u, v, setting $z = x + iy$ and $h = \lambda + i\mu$, where λ, μ are real numbers:

$$\frac{f(z+h) - f(z)}{h} = \frac{u(x+\lambda, y+\mu) + iv(x+\lambda, y+\mu) - u(x, y) - iv(x, y)}{\lambda + i\mu}. \tag{6.34}$$

First, we choose to let $h \to 0$ through purely real values, so that $\mu = 0$, and we find that

$$\begin{aligned}\frac{dw}{dz} &= \lim_{\lambda \to 0}\left[\frac{u(x+\lambda, y) - u(x, y)}{\lambda} + i\frac{v(x+\lambda, y) - v(x, y)}{\lambda}\right] \\ &= \frac{\partial u}{\partial x} + i\frac{\partial v}{\partial x}.\end{aligned} \tag{6.35}$$

Next, we deduce the form of dw/dz by choosing to let $h \to 0$ through purely imaginary values, so that $\lambda = 0$, and the same form of argument then gives

$$\frac{dw}{dz} = \frac{1}{i}\left[\frac{\partial u}{\partial y} + i\frac{\partial v}{\partial y}\right]. \tag{6.36}$$

If the limit of difference quotient (6.34) exists, then it is unique, so that Eqns (6.35) and (6.36) must be alternative forms of the same result. Thus, equating real and imaginary parts, we obtain the two equations that must be satisfied simultaneously by the real and imaginary parts of $f(z)$:

$$\frac{\partial u}{\partial x} = \frac{\partial v}{\partial y} \quad \text{and} \quad \frac{\partial u}{\partial y} = -\frac{\partial v}{\partial x}. \tag{6.37}$$

These are called the *Cauchy–Riemann equations.*

Our argument has shown that a differentiable or analytic function $w = u + iv$ must satisfy the Cauchy–Riemann equations (6.37), and a similar argument establishes the converse result; namely, that if the Cauchy–Riemann equations are satisfied by a complex function, then it must have a unique derivative.

It is a direct consequence of our definitions of limit, continuity and differentiability that all the limit and continuity theorems (Theorems 3.4 and 3.5) and differentiation theorems (Theorems 5.4 to 5.8) for real functions apply also to analytic functions. Points at which $w = f(z)$ is not analytic are called *singularities* of $f(z)$. Thus the function $f(z) = 1/(z+1)$ is easily seen to be analytic everywhere except at the point $z = -1$, which is a singularity.

Provided $f(z)$ is an analytic function, when z is purely real the forms assumed by $f(z)$ and $f(x)$ will be identical. Thus we may deduce the form of f when expressed as $w = f(z) = u(x, y) + iv(x, y)$ from the following simple rule.

Rule 6.1 (Expression of f in terms of z)

An analytic function $w = u(x, y) + iv(x, y)$ may be expressed in terms of z by formally setting $y = 0$ in the right-hand side and then replacing x by z.

The application of the rule is well illustrated by applying it to examples (a), (c), (d) and (e) above. The function $w = z\bar{z} = u + iv = x^2 + y^2$ in (b) is not analytic because it does not satisfy the Cauchy–Riemann equations in (6.37). To see this, notice that $u = x^2 + y^2$, so $\partial u/\partial x = 2x$ and $\partial u/\partial y = 2y$, but $v \equiv 0$ so $\partial v/\partial x = \partial v/\partial y = 0$ showing that the Cauchy–Riemann equations are not satisfied. Were Rule 6.1 to be applied, then as $w = z\bar{z} = x^2 + y^2$, we would conclude that $w = f(z) = z^2$, which is incorrect. Results (6.35) and (6.36) may be used to deduce the form of $f'(z)$ by using the simple observation that when z is purely real, so that $z = x$, the forms assumed by $f'(z)$ and $f'(x)$ are identical. This gives the following straightforward rule for determining the derivative $f'(z)$ of the analytic function $f(z)$ which is sometimes helpful.

Rule 6.2 (Determination of the derivative of a complex function)

If $f(z) = u + iv$ satisfies the Cauchy–Riemann equations, then the derivative $f'(z)$ expressed in terms of z may be deduced from the result

$$f'(z) = \frac{\partial u}{\partial x} + i\frac{\partial v}{\partial x}$$

by formally setting $y = 0$, and then replacing x by z.

Example 6.10 Determine which of the following functions satisfy the Cauchy–Riemann equations and thus possess uniquely defined derivatives. Give the form of this derivative when it is defined.

(a) $w = z^2$;
(b) $w = \cos z$;
(c) $w = |z|$.

Solution

(a) If $w = z^2$, then $w = (x + iy)^2 = x^2 - y^2 + i2xy$ and so $u = x^2 - y^2$, $v = 2xy$. So $u_x = 2x$, $u_y = -2y$, $v_x = 2y$, and $v_v = 2x$. It is readily seen that these expressions satisfy the Cauchy–Riemann equations, and so we may conclude that $w = z^2$ possesses a unique derivative. It follows from Eqn (6.35) that

$$f'(z) = 2x + i2y = 2z.$$

This result was so simple that appeal to Rule 6.2 was not necessary.
(b) If $w = \cos z$, then $w = \cos(x + iy) = \cos x \cos iy - \sin x \sin iy$, or $w = \cos x \cosh y - i \sin x \sinh y$, and so $u = \cos x \cosh y$, $v = -\sin x \sinh y$. Hence, $u_x = -\sin x \cosh y$, $u_y = \cos x \cosh y$, $v_x = -\cos x \sinh y$ and $v_y = -\sin x \cosh y$. Here also it is immediately apparent that the expressions satisfy the Cauchy–Riemann equations, showing that $w = \cos z$ possesses a unique derivative.

Let us work with Rule 6.2 to determine $f'(z)$ in terms of z. We must therefore start with the equation

$$f'(z) = \frac{\partial u}{\partial x} + i\frac{\partial v}{\partial x}.$$

In this case we find

$$f'(z) = -\sin x \cosh y - i \cos x \sinh y.$$

Then, setting $y = 0$ and replacing x by z gives

$$f'(z) = -\sin z.$$

It is instructive to compare this rapid method with the direct approach we now indicate.

$$\begin{aligned} f'(z) &= -\sin x \cosh y - i \cos x \sinh y \\ &= -\sin x \cos iy - \cos x \sin iy \\ &= -\sin(x + iy) = -\sin z. \end{aligned}$$

(c) If $w = |z|$, then $w = (x^2 + y^2)^{1/2}$, showing that $u = (x^2 + y^2)^{1/2}$, $v = 0$. Then as $u_x = x/(x^2 + y^2)^{1/2}$, $u_y = y/(x^2 + y^2)^{1/2}$, $v_x = v_y = 0$, it is clear that $w = |z|$ can only satisfy the Cauchy–Riemann equations at $z = 0$ in the complex plane. We conclude that $w = |z|$ is not an analytic function. ■

Example 6.11 Determine the constants a and b in order that

$$w = x^2 + ay^2 - 2xy + i(bx^2 - y^2 + 2xy)$$

should satisfy the Cauchy–Riemann equations. Deduce the derivative of w.

Solution Here we have $u = x^2 + ay^2 - 2xy$, $v = bx^2 - y^2 + 2xy$ so that $u_x = 2x - 2y$, $u_y = 2ay - 2x$, $v_x = 2bx + 2y$, and $v_y = -2y + 2x$. It is certainly true that $u_x = v_y$, so that the first of the Cauchy–Riemann equations is automatically satisfied. For the second equation to be satisfied we must require that $u_y = -v_x$, or $2ay - 2x = -(2bx + 2y)$. This is only possible if $a = -1$, $b = 1$.

Now as $f'(z) = u_x + iv_x$, we have

$$f'(z) = 2x - 2y + i(2x + 2y).$$

Again, working with Rule 6.2 gives

$$f'(z) = 2(1 + i)z. \quad ■$$

Example 6.12 Show that the function $w = e^z$ with $z = x + iy$ is analytic, and hence justify the reasoning used in Section 6.5 to arrive at result (6.26).

Solution Setting $z = x + iy$ in $w = e^z$ gives $w = e^x e^{iy} = e^x(\cos y + i \sin y)$, so writing $w = u + iv$ we see that $u = e^x \cos y$ and $v = e^x \sin y$. Thus $\partial u/\partial x = e^x \cos y$ and $\partial u/\partial y = -e^x \sin y$, while $\partial v/\partial x = -e^x \sin y$ and $\partial v/\partial y = e^x \cos y$. These derivatives satisfy the Cauchy–Riemann equations so the function $w = e^z$ is analytic, and hence can be differentiated. In fact it is not difficult to show that an immediate consequence of the analyticity of a complex function is that it may be differentiated an arbitrary number of times with each derivative being analytic, though the justification of this statement will not be given here. The result due to de Moivre in Eqn (6.26) follows by setting $x = 0$ and replacing y by x. It is the fact that e^z is analytic, and so can be repeatedly differentiated, that justifies the conclusions reached in Section 6.5 concerning the functions $C(x)$ and $S(x)$. ■

Supposing that u_{xy}, v_{xy} exist and are continuous, it follows directly by partial differentiation of the Cauchy–Riemann equations $u_x = v_y$, $u_y = -v_x$ that

$$\frac{\partial^2 u}{\partial x^2} + \frac{\partial^2 u}{\partial y^2} = 0 \quad \text{and} \quad \frac{\partial^2 v}{\partial x^2} + \frac{\partial^2 v}{\partial y^2} = 0. \tag{6.38}$$

These equations are identical in form and are examples of an important partial differential equation called *Laplace's equation*, any solution of which is called a *harmonic function*. The harmonic functions u and v associated with an analytic function $f(z) = u + iv$ are called *conjugate harmonic functions*. For example,

$$\cos z = \cos x \cosh y - i \sin x \sin y$$

is an analytic function with $u = \cos x \cosh y$, $v = -\sin x \sinh y$. Now both u and v are such that u_{xy}, v_{xy} are continuous, so it follows immediately that u and v satisfy Eqns (6.38). Hence $u = \cos x \cosh y$, $v = -\sin x \sinh y$ are conjugate harmonic functions. The term conjugate is, of course, used here in a different sense than when discussing complex conjugates.

Notice that u and v will only be conjugate harmonic functions when they are, respectively, the real and imaginary parts of the same analytic function $f(z) = u + iv$; that is, when $u = \operatorname{Re} f(z)$ and $v = \operatorname{Im} f(z)$, satisfy the Cauchy–Riemann equations.

Thus as the functions $f_1(z)=u_1+iv_1=z^2=(x^2-y^2)+i2xy$ and $f_2(z)=u_2+iv_2=z^3=(x^3-3xy^2)+i(3x^2y-y^3)$ are analytic, $u_1=x^2-y^2$ and $v_1=2xy$ are conjugate harmonic functions, as are $u_2=x^3-3xy^2$ and $v_2=3x^2y-y^3$. However, the functions u_1 and v_2 are not conjugate harmonic functions, and neither are the functions u_2 and v_1.

Problems

Section 6.1

6.1 Solve the differential equation $\mathrm{d}y/\mathrm{d}x=y$, with $y(0)=c$, as in Section 6.1, by substituting

$$y=\sum_{r=0}^{\infty}a_rx^r.$$

Hence deduce that, provided $c\neq 0$, the differential equation has the non-trivial solution $y=ce^x$.

6.2 Evaluate the following limits:

(a) $\lim\limits_{x\to\infty}\dfrac{4e^{2x}+xe^x+3}{5xe^{3x}+e^x+1}$; (b) $\lim\limits_{x\to\infty}\dfrac{(x^2+1)e^{3x}+e^x+1}{(2x^2-3x+1)e^{3x}}$;

(c) $\lim\limits_{x\to-\infty}\dfrac{(2-x^2)e^x+3}{1+(2+x)e^{2x}}$; (d) $\lim\limits_{x\to\infty}\dfrac{3(2e^{-3x}+x^2+1)}{4e^{2x}+2x+1}$.

6.3 Make use of the series expansion of e^x to evaluate the following limits, and verify your result by using Theorem 5.14:

(a) $\lim\limits_{x\to 0}\dfrac{e^{2x}-1}{3x}$; (b) $\lim\limits_{x\to 0}\dfrac{1-e^{-x}}{\sin 4x}$; (c) $\lim\limits_{x\to 0}\dfrac{e^x-1-x}{3x^2}$.

6.4 Differentiate the following functions:

(a) $f(x)=2e^x\cos x$; (b) $f(x)=e^{3x}\arcsin x$;

(c) $f(x)=e^x/x^2$; (d) $f(x)=e^{x\sin x}$.

6.5 Differentiate the following functions:

(a) $f(x)=\arcsin e^{2x}$; (b) $f(x)=\surd(xe^x+x)$;

(c) $f(x)=\sin(xe^x+2)$; (d) $f(x)=(e^x-1)/(e^x+1)$.

Section 6.2

6.6 Differentiate the following functions:

(a) $f(x)=3\exp[-(x^2+x+1)]$; (b) $f(x)=\exp(\sin^2 x)$;

(c) $f(x)=\cos[\exp(x\sin x+2)]$.

6.7 Consider the function $f(x)$ defined as follows:

$$f(x)=\begin{cases}e^{-1/x^2} & \text{for } x\neq 0,\\ 0 & \text{for } x=0.\end{cases}$$

Clearly the differentiability properties of this function at the origin must be deduced directly from the definition of a derivative. To deduce these properties, show first that for $x \neq 0$, it follows that

$$f'(x) = \frac{2}{x^3} e^{-1/x^2}.$$

Then, by using Definition 5.2 together with Theorem 6.3, prove that $f'(0) = 0$, and hence deduce that

$$\lim_{x \to 0} f'(x) = f'(0) = 0.$$

Finally, deduce that in general,

$$f^{(n)}(x) = e^{-1/x^2} \times (\text{polynomial in } 1/x),$$

and hence by using an inductive argument prove that $f^{(n)}(0) = 0$ for *all* n. This is an example of a function which is capable of differentiation an arbitrary number of times for all x, and yet which has every derivative equal to zero at one point of its domain of definition.

6.8 Show that $u = xy + xe^{y/x}$ satisfies the equation

$$x\frac{\partial u}{\partial x} + y\frac{\partial u}{\partial y} = xy + u.$$

6.9 Find $\partial f/\partial x$ and $\partial f/\partial y$, given that

$$f(x, y) = e^{\sin(y/x)}.$$

6.10 Find $\partial f/\partial u$ and $\partial f/\partial v$ if

$$f(x, y) = 2 \arctan \frac{x}{y}$$

with $x = u \sin v$ and $y = u \cos v$.

Section 6.3

6.11 Evaluate the following limits:

(a) $\displaystyle \lim_{x \to \infty} \frac{(x - 1) \ln x^2}{x^2}$;

(b) $\displaystyle \lim_{x \to \infty} \frac{\ln(3 + 2^x)}{4x + 1}$;

(c) $\displaystyle \lim_{x \to 0} \frac{\ln(3 \sin x) - \ln[(1 + x) \sin x]}{2e^x - 1}$;

(d) $\displaystyle \lim_{x \to \infty} [\ln(3x + 1) - \ln(2x + 5)]$;

(e) $\lim_{x\to\infty} \dfrac{\ln(1+2e^x)}{x}$.

6.12 Differentiate the following functions:

(a) $f(x) = \ln(x^3 + 7x^2 + 2)$;

(b) $f(x) = \ln \sin 2x$;

(c) $f(x) = \ln \cos\left(\dfrac{x-1}{x}\right)$.

6.13 If $y = [f(x)]^{g(x)}$ then, taking the natural logarithm,

$$\ln y = g(x) \ln f(x).$$

Hence, differentiating with respect to x, it follows that

$$\frac{dy}{dx} = \left[g'(x)\ln f(x) + \frac{g(x)}{f(x)} f'(x)\right][f(x)]^{g(x)}.$$

Use this result to differentiate the following functions:

(a) $y = x^x$;

(b) $y = (\sin 2x)^x$;

(c) $y = x^{\sin x}$;

(d) $y = 10^{\ln \sin x}$.

6.14 If $u = x \ln(1 + x/y) + y \ln(1 + y/x)$,

$$\text{show that } x^2\frac{\partial^2 u}{\partial x^2} = y^2\frac{\partial^2 u}{\partial y^2}.$$

6.15 By taking logarithms, deduce $\partial u/\partial x$, $\partial u/\partial y$ and $\partial u/\partial z$ if $u = (xy)^z$.

Section 6.4

6.16 Prove by means of the definitions that

$$(\cosh x + \sinh x)^n = \cosh nx + \sinh nx.$$

6.17 Prove by means of the definitions that:

(a) $2 \sinh x \cosh y = \sinh(x + y) + \sinh(x - y)$;

(b) $2 \cosh x \cosh y = \cosh(x + y) + \cosh(x - y)$;

(c) $2 \sinh x \sinh y = \cosh(x + y) - \cosh(x - y)$,

6.18 Evaluate the following limits, using the series (6.27) and (6.28) where necessary:

(a) $\lim_{x\to\infty} \dfrac{x^3 \cosh 2x + e^x}{(2x^3 + x + 1)e^{2x} + x^3 e^{-2x}}$;

(b) $\lim_{x\to-\infty} \frac{x^3\cosh 2x + e^x}{(2x^3 + x + 1)e^{2x} + x^3e^{-2x}}$; (c) $\lim_{x\to 0} \frac{\sinh \alpha x}{x}$;

(d) $\lim_{x\to 0} \frac{1 - \cosh 2x}{3x^2}$.

6.19 Differentiate the following functions:

(a) $f(x) = \sinh 2x \cosh^2 x$;

(b) $f(x) = \exp(1 + \cosh 3x)$;

(c) $f(x) = \ln(\tanh x)$;

(d) $f(x) = \operatorname{arcsech}(x^2 + \frac{1}{2})$ if $f(x) > 0$;

(e) $f(x) = \cosh(\sin 2x)$.

6.20 Evaluate $\partial u/\partial x$ and $\partial u/\partial y$ given that:

(a) $u(x, y) = \sin x \cosh xy$;

(b) $u(x, y) = \sinh(x^2 + x\sin y + 3y^2)$;

(c) $u(x, y) = x^{\cosh(x^2+2y^2)}$.

Section 6.5

6.21 Establish by means of the definitions that:

(a) $\sin(iz) = i\sinh z$; (b) $\cos(iz) = \cosh z$;

(c) $\sinh(iz) = i\sin z$; (d) $\cosh(iz) = \cos z$.

6.22 Given that a, b are positive real numbers, deduce four trigonometric identities by equating real and imaginary parts in each of the following results

$e^{ia}\cdot e^{ib} = e^{i(a+b)}$ and $e^{ia}\cdot e^{-ib} = e^{i(a-b)}$.

6.23 Express the following complex numbers in the form $re^{i\theta}$:

(a) $1 + i$; (b) $1 - i$; (c) $-8(i\sqrt{3} - 1)$;

(d) $(-1 + i)^6$; (e) $(5 + 14i)/(4 + i)$.

6.24 Show by means of de Moivre's theorem that:

(a) $32\cos^6\theta = 10 + 15\cos 2\theta + 6\cos 4\theta + \cos 6\theta$;

(b) $\sin 7\theta = 7\sin\theta - 56\sin^3\theta + 112\sin^5\theta - 64\sin^7\theta$.

6.25 Verify that if $z = e^{i\theta}$, then

$$\cos\theta = \frac{1}{2}\left(z + \frac{1}{z}\right) \quad \text{and} \quad \sin\theta = -\frac{i}{2}\left(z - \frac{1}{z}\right)$$

and, more generally,

$$\cos r\theta = \frac{1}{2}\left(z^r + \frac{1}{z^r}\right) \quad \text{and} \quad \sin r\theta = -\frac{i}{2}\left(z^r - \frac{1}{z^r}\right).$$

By replacing $\cos\theta$ and $\sin\theta$ by their equivalent expressions involving z, make use of these results to express $\cos^2\theta \sin^3\theta$ in terms of $\sin n\theta$.

6.26 Use the method of Problem 6.25 to express $\sin^8\theta$ in terms of $\cos n\theta$.

6.27 Consider the function $\cosh z$, where $z = x + iy$. Then, using Definition 6.3, deduce that $\cosh z = 0$ when $z = (2n+1)\pi i/2$, with $n = 0$, ± 1, ± 2, ... Use the results of Problem 6.21 to deduce the zeros of $\cos z$.

6.28 Consider the function $\sin z$, where $z = x + iy$. Then, using Definition 6.6, deduce that $\sin z = 0$ when $z = n\pi$, with $n = 0$, ± 1, ± 2, ... Use the results of Problem 6.21 to deduce the zeros of $\sinh z$.

Supplementary computer problems

6.29 Use the approximation

$$e^x = \sum_{r=0}^{10} \frac{x^r}{r!},$$

in which $0! = 1$, to compute $\exp(\sin x)$ for $x = -2, -1, 1.5$ and 4, and compare your results with the actual values 0.402807, 0.431076, 2.711481 and 0.469164, respectively.

6.30 Display simulateneously, using the same scales for x and y, the function $y = e^x$ for $-2 = x = 2$, the-inverse function $y = \ln x$ for $0.2 = x = 7$ and the line $y = x$ for $-2 = x = 7$, and verify by inspection that the inverse function is the reflection of $y = e^x$ in the line $y = x$.

6.31 Display the function

$$f(x) = \frac{x^p}{e^x}$$

for $p = 1, 2, 3, 5$ and $0 \leq x \leq 15$, and verify by inspection that

$$\lim_{x\to\infty} \frac{x^p}{e^x} \to 0 \quad \text{for } p > 0.$$

6.32 Display simultaneously the functions

$$y = \sinh x \qquad \text{and} \quad y = \tfrac{1}{2}e^x$$

for $-4 \leq x \leq 4$, and confirm by inspection that $\sinh x$ is approximated by $\frac{1}{2}e^x$ when x is large and positive.

6.33 Display simultaneously the functions

$$y = \cosh x, \quad y = \tfrac{1}{2}e^{x} \qquad \text{and} \quad y = \tfrac{1}{2}e^{-x}$$

for $-4 \le x \le 4$, and confirm by inspection that $\cosh x$ is approximated by $\frac{1}{2}e^{x}$ when x is large and positive, and by $\frac{1}{2}e^{-x}$ when x is large and negative.

6.34 Display the function

$$y = \tanh x$$

for $-4 \le x \le 4$, and confirm by inspection that

$$\lim_{x\to\infty} \tanh x = 1 \qquad \text{and} \qquad \lim_{x\to-\infty} \tanh x = -1.$$

6.35 Choose any two numbers x and y and verify by substitution that

(i) $\sinh(x+y) = \sinh x \cosh y + \cosh x \sinh y$;

(ii) $\cosh(x-y) = \cosh x \cosh y - \sinh x \sinh y$;

(iii) $\tanh(x+y) = \dfrac{\tanh x + \tanh y}{1 + \tanh x \tanh y}$.

6.36 Using the definition in Table 6.3 display the function

$$y = \text{arcsinh}\, x \qquad \text{for} \quad -3 \le x \le 3.$$

6.37 Using the definition in Table 6.3 display the function

$$y = \text{arccosh}\, x \qquad \text{for} \quad 1 \le x \le 4.$$

6.38 Using the definition in Table 6.3 display the function

$$y = \text{arctanh}\, x \qquad \text{for} \quad -0.9 \le x \le 0.9.$$

6.39 Use the approximation

$$\sin x = x - \frac{x^3}{3!} + \frac{x^5}{5!} - \ldots + (-1)^n \frac{x^{2n+1}}{(2n+1)!} = \sum_{r=0}^{n} (-1)^r \frac{x^{2r+1}}{(2r+1)!},$$

with $n = 4$, to compute $\sin x$ for $x = 0.2$, 0.45, 0.75, 2.2, and compare the calculations with the exact results 0.198669, 0.434966, 0.681639 and 0.808496, respectively. Repeat the calculations with $n = 8$, and note any increase in accuracy.

6.40 Use the approximation

$$\cos x = 1 - \frac{x^2}{2!} + \frac{x^4}{4!} - \ldots + (-1)^n \frac{x^{2n}}{(2n)!} = \sum_{r_1=0}^{n} (-1)^r \frac{x^{2r}}{(2r)!},$$

where $0! = 1$, with $n = 4$, to compute $\cos x$ for $x = -0.3$, 0.5, 0.8, 2.5, and compare the calculations with the exact results 0.955336, 0.877583, 0.696707 and -0.801144, respectively. Repeat the calculations with $n = 8$, and note any increase in accuracy.

6.41 Use the approximation

$$\sinh x = x + \frac{x^3}{3!} + \frac{x^5}{5!} + \ldots + \frac{x^{2n+1}}{(2n+1)!} = \sum_{r=0}^{n} \frac{x^{2r+1}}{(2r+1)!},$$

with $n = 4$, to compute $\sinh x$ for $x = -1.2, 0.2, 0.75, 1.25, 3.4$, and compare the calculations with the exact results -1.509461, 0.201336, 0.822317, 1.601919, 14.965363, respectively. Repeat the calculations with $n = 8$, and note any increase in accuracy.

6.42 Use the approximation

$$\cosh x = 1 + \frac{x^2}{2!} + \frac{x^4}{4!} + \ldots + \frac{x^{2n}}{(2n)!} = \sum_{r=0}^{n} \frac{x^{2n}}{(2^n)!},$$

with $n = 4$, to compute $\cosh x$ for $x = 0.25, 0.55, 0.8, 1.9, 4.1$, and compare the calculations with the exact results 1.031413, 1.155101, 1.337435, 3.417732, 30.178430, respectively. Repeat the calculations with $n = 8$, and note any increase in accuracy.

Section 6.6

6.43 For what values of z are the following complex functions defined:

(a) $w = z^2 + iz + 1$; (b) $w = (z-1)/(z-2)$;

(c) $(z+1)(z-i)(z^2+4)$; (d) $w = \sinh z$.

6.44 If $f(z) = u + iv$, find the expressions for the functions u, v in terms of x, y given that:

(a) $f(z) = z^2 + z\bar{z} + 1$; (b) $f(z) = \dfrac{z+i}{z+2}$;

(c) $f(z) = \cosh z$; (d) $f(z) = \cos z$.

6.45 Given the following forms of $f(z)$, deduce their value if $z = 1 + 2i$:

(a) $f(z) = x^2 + 3xy + iy^2$;

(b) $f(z) = \dfrac{x^2 + 2iy + 1}{x^2 + y^2}$;

(c) $f(z) = \sin\dfrac{\pi x}{2}(x^2 - iy^2) + i\cos\dfrac{3\pi y}{2}(x^2 + iy^2)$.

6.46 Give reasons to justify the assertion that

$f(z) = z\sin(z^2 + 3z + 2) + 1/(z + 2 - i)$

is continuous everywhere except at $z = -2 + i$.

6.47 Determine which of the following functions $f(z)$ satisfy the Cauchy–Riemann equations:

(a) $f(z) = z^3 - iz^2 + 3$;

(b) $f(z) = \cosh(z + 3i)$;

(c) $f(z) = \sin z + z\bar{z}$;

(d) $f(z) = (x^3 - 3xy^2) + i(3x^2y - y^3)$;

(e) $f(z) = z(z + \bar{z})/2$;

(f) $f(z) = \sinh 3x \cos y + i \cosh 3x \sin y$.

6.48 Find the points, if any, at which the following functions are not analytic:

(a) $f(z) = 3z + \sinh z$;

(b) $f(z) = z/(z + 2)$;

(c) $f(z) = \cos(1/z)$;

(d) $f(z) = \dfrac{\sin z}{z^2 + 1}$.

7 Fundamentals of integration

The purpose of this chapter is to introduce the basic ideas underlying the definite integral, and then to prove a theorem that shows how definite integrals can be evaluated by using the idea of differentiation in reverse. The *definite integral* of a function $f(x)$ of the real variable x, evaluated between the limits $x = a$ and $x = b$, and written

$$\int_a^b f(x)\,dx,$$

is a real number. It is shown how this number, defined as the limit of a sum, may be interpreted as the area between the graph of $y = f(x)$, the x-axis and the lines $x = a$ and $x = b$, with areas above the x-axis being regarded as positive and those below it as negative. The first fundamental theorem of the integral calculus that is then proved provides the connection between integration and differentiation through the result

$$\frac{d}{dx}\int_a^x f(t)\,dt = f(x),$$

where the upper limit of integration is the variable x, and t is a dummy variable. The concept of an antiderivative is introduced, and it is shown in the fundamental theorem of differential calculus how by using an antiderivative it is possible to evaluate definite integrals.

Among the simple but useful geometrical applications of definite integrals that are discussed are the determination of arc length and the area and volume of surfaces of revolution. Applications to mechanics are also introduced, and include the location of centres of mass of bodies, their moments of inertia about specified axes and the line integral.

7.1 Definite integrals and areas

This chapter is concerned with the theory of the operation known as *integration*, which occupies a central position in the calculus. The connection between differentiation and integration is basic to the whole of the calculus and is contained in a result we shall prove later known as the fundamental theorem of calculus. Once again, limiting operations will play an essential part in the development of our argument. In fact we will show not only how they enable a satisfactory general theory of integration to be established, but also how they provide a tool, albeit a clumsy one, for the actual numerical integration of functions. However, aside from a number of simple but important examples, the practical details of the evaluation of integrals of specific classes of function will be deferred until Chapter 8.

We begin by seeking to determine the shaded area I of Fig. 7.1 which is interior to the region bounded above and below by the curve $y = f(x)$ and the x-axis, respectively, and to the left and right by the lines $x = a$, $x = b$. This approach will lead naturally to what is called the *definite integral* of $f(x)$ over the interval $a \leq x \leq b$, and it illustrates a valuable geometrical interpretation of the process of integration. Although we use the definite integral to give precise meaning to the notion of the area contained within a closed curve, this appeal to geometry is not actually necessary when defining a definite integral. Indeed, we shall also show how a purely analytical definition of a definite integral, quite independent of any geometrical arguments, may be formulated.

Let $f(x)$ be a non-negative continuous function defined in the closed interval $[a, b]$ and consider, for a moment, the conceptual problem that arises when trying to determine the area I defined by it in Fig. 7.1. The only simple plane geometrical figure for which the concept of area is defined in an elementary and unambiguous manner is the rectangle, so that we shall seek to define the area I in terms of the limit of a sum of rectangular areas.

We shall start our discussion from the postulates that (a) the area of a rectangle is given by the product length × breadth, (b) the area of two non-overlapping rectangles is the sum of their separate areas, and (c) if a rectangle is divided into two parts by a curve, then the sum of the separate non-rectangular areas comprising these two parts is equal to the area of the rectangle.

On the basis of postulate (c), we at once see that the area I in Fig. 7.1 exceeds the rectangular area $ABEF$, but is less than the rectangular area $ACDF$. Letting m, M denote, respectively, the minimum and maximum values attained by $f(x)$ in $[a, b]$, this result becomes

$$m(b-a) \leq I \leq M(b-a). \tag{7.1}$$

This inequality must obviously be refined if it is ever to lead to the actual value of I. In principle, our approach will be simple, for we shall

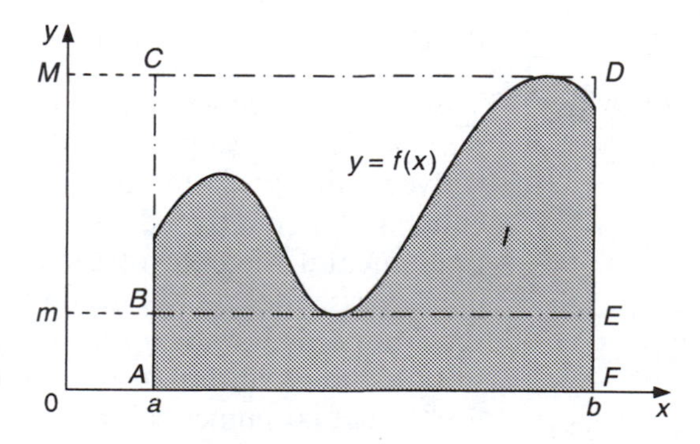

Fig. 7.1 Area I defined by $y = f(x)$.

begin by dividing $[a, b]$ into n adjacent sub-intervals in each of which an inequality of type (7.1) will apply, after which we shall use postulate (b) to find better upper and lower bounds for I.

Specifically, we start by choosing any sequence of $n+1$ numbers x_0, $x_1,\ldots,x_n$ subject only to the requirements that $x_0 = a$, $x_n = b$, and

$$x_0 < x_1 < \ldots < x_{n-1} < x_n.$$

The sequence $\{x_r\}_{r=0}^{n}$ so defined is called a *partition* P_n of the interval $[a, b]$, and for any given value of n it is obviously not unique. Next, on each sub-interval $[x_{i-1}, x_i]$, let the function $f(x)$ attain a minimum value m_i and a maximum value M_i, and denote the length of the ith sub-interval by Δ_i, so that

$$\Delta_i = x_i - x_{i-1}.$$

We now define numbers $\underline{S}_{P_n}$ and $\overline{S}_{P_n}$ called respectively, the *lower* and *upper sums* taken over the partition P_n, by the expressions

$$\underline{S}_{P_n} = m_1\Delta_1 + m_2\Delta_2 + \ldots + m_n\Delta_n = \sum_{r=1}^{n} m_r\Delta_r \tag{7.2}$$

and

$$\overline{S}_{P_n} = M_1\Delta_1 + M_2\Delta_2 + \ldots + M_n\Delta_n = \sum_{r=1}^{n} M_r\Delta_r. \tag{7.3}$$

Clearly, as Fig. 7.2 illustrates, $\underline{S}_{P_n}$ and $\overline{S}_{P_n}$ are, respectively, under- and overestimates of the area I.

The fact that $\underline{S}_{P_n} \leq \overline{S}_{P_n}$ is apparent on geometrical grounds, but it also follows without appeal to geometry by considering the difference

$$\overline{S}_{P_n} - \underline{S}_{P_n} = (M_1 - m_1)\Delta_1 + (M_2 - m_2)\Delta_2 + \ldots + (M_n - m_n)\Delta_n. \tag{7.4}$$

In this equation we have, by definition, $\Delta_r > 0$ and $M_r = m_r$ for $r = 1, 2, \ldots, n$, so the expression on the right is positive, showing that

$$\overline{S}_{P_n} - \underline{S}_{P_n} \geq 0, \quad \text{or} \quad \underline{S}_{P_n} \leq \overline{S}_{P_n},$$

and thus, by postulate (c),

$$\underline{S}_{P_n} \leq I \leq \overline{S}_{P_n}. \tag{7.5}$$

As the number n of partitions increases, and the lengths Δ_i of all the intervals in the partition tend to zero, so the inequality in (7.5) becomes sharper and sharper, and in the limit as $n \to \infty$ the value I of the definite integral will be determined either by $\underline{S}_{P_\infty}$ or by $\overline{S}_{P_\infty}$.

The upper and lower sums used in Eqns (7.2) and (7.3) are not really necessary when defining I, though we will see later that they do serve a practical purpose if this simple approach to the determination of a definite integral is implemented numerically. Instead of considering the greatest and least values M_i and m_i of $f(x)$ in each interval Δ_i, we can equally

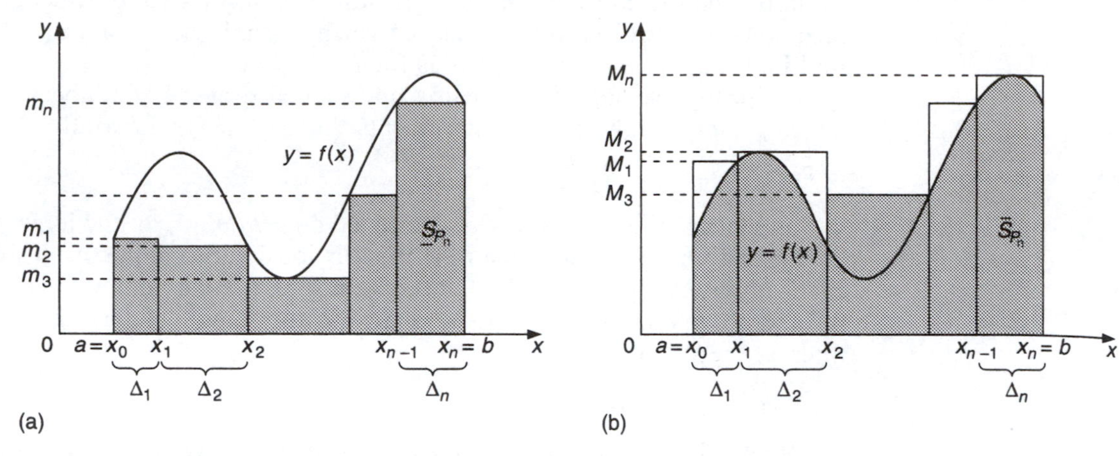

Fig. 7.2 (a) Shaded area represents lower sum $\overline{S}_{P_n}$ (b) Shaded area represents upper sum $\underline{S}_{P_n}$.

well arrive at the value of I by using the value of $f(x)$ at an arbitrary point ξ_i in Δ_i, and consider instead the limiting value of S_{P_n} as $n \to \infty$, where

$$S_{P_n} = f(\xi_1)\Delta_1 + f(\xi_2)\Delta_2 + \ldots + f(\xi_n)\Delta_n. \tag{7.6}$$

This is possible because as each interval Δ_i, decreases, so both m_i and M_i, tend to $f(\xi_i)$.

A partition P_N of $a \leq x \leq b$ is said to be a *refinement* of a partition P_n if $N > n$ and the largest interval in P_N is smaller than the largest interval in P_n. With this definition in mind we arrive at the following definition for the definite integral of a continuous non-negative function $f(x)$.

Definition 7.1 (definite integral of a continuous non-negative function)

Let $f(x)$ be a continuous non-negative function on the closed interval $a \leq x \leq b$, and let $P_1, P_2, \ldots, P_m, \ldots$ be a sequence of successive refinements of some partition P of $[a, b]$. Then, if ξ_i is any point in the ith sub-interval of length Δ_i generated by the partition P_m, the *definite integral* of $f(x)$ integrated over the interval $[a, b]$, and written symbolically

$$\int_a^b f(x)\,\mathrm{d}x,$$

is defined to be

$$\int_a^b f(x)\,\mathrm{d}x = \lim_{m\to\infty} \sum_{i=1}^{n} f(\xi_i)\Delta_i.$$

In the context of a definite integral, the function $f(x)$ is called the *integrand*, the numbers a, b are called the *lower and upper limits of integration*, respectively, and the sign $\int$ itself is called the *integral sign*.

In summary then, a definite integral of a non-negative continuous function $f(x)$ integrated over the interval $[a, b]$ is a positive number defined by means of a limiting process. It may be interpreted geometrically as the shaded area I below the curve $y = f(x)$ as shown in Fig. 7.1.

To use Definition 7.1 as a numerical method for approximating a definite integral it is convenient to make every interval Δ_i the same length, so if an interval $a \leq x \leq b$ is to be partitioned into n sub-intervals we have $\Delta_i = (b - a)/n$, for $i = 1, 2, \ldots, n$. The following Example shows the result of applying this approach to the determination of a definite integral involving trigonometric functions.

Example 7.1 Find the values of the upper and lower sums for the integral of $f(x) = 3 - 3\sin 2x + 2\cos x$ over the interval $0 \leq x \leq \pi$ using a partition with 8 intervals of equal length.

Solution The function $f(x)$ is shown in Fig. 7.3(a, b), where the respective lower and upper sums are to be interpreted as the sum of the rectangular areas in each case. The result of determining $\underline{S}_{P_8}$ and $\overline{S}_{P_8}$ numerically lead to the inequality

$$\underline{S}_{P_8} < \int_0^\pi (3 - 3\sin 2x + 2\cos x)\,\mathrm{d}x < \overline{S}_{P_8},$$

with $\underline{S}_{P_8} = 7.35679$ and $\overline{S}_{P_8} = 11.49277$. These estimates are wide apart, and to obtain a significant improvement using this simple approach necessitates increasing the number of intervals in the partition to over 100. There is no point in doing this because later the value of the integral will be determined analytically and shown to be $I = 9.42478$. Later, when the numerical solution of definite integrals is discussed, quite different methods will be developed which while requiring far less computational effort, nevertheless yield high accuracy. ■

To show that this is a working definition, in the sense that it can be used to yield a useful answer, let us now apply it to a simple function.

Example 7.2 Evaluate the definite integral

$$\int_a^b x^2\,\mathrm{d}x, \quad \text{where} \quad a < b.$$

Solution As x^2 is everywhere continuous and is non-negative on the stated interval, Definition 7.1 applies. Thus we start by considering a convenient partition P_n in which $[a, b]$ is divided into n equal sub-intervals, each of length $\Delta = (b - a)/n$. Then, if for convenience we identify ξ_i with the right-hand end point of the ith sub-interval, we have

$$\xi_1 = a + \Delta, \quad \xi_2 = a + 2\Delta, \quad \xi_3 = a + 3\Delta, \ldots, \quad \xi_n = a + n\Delta.$$

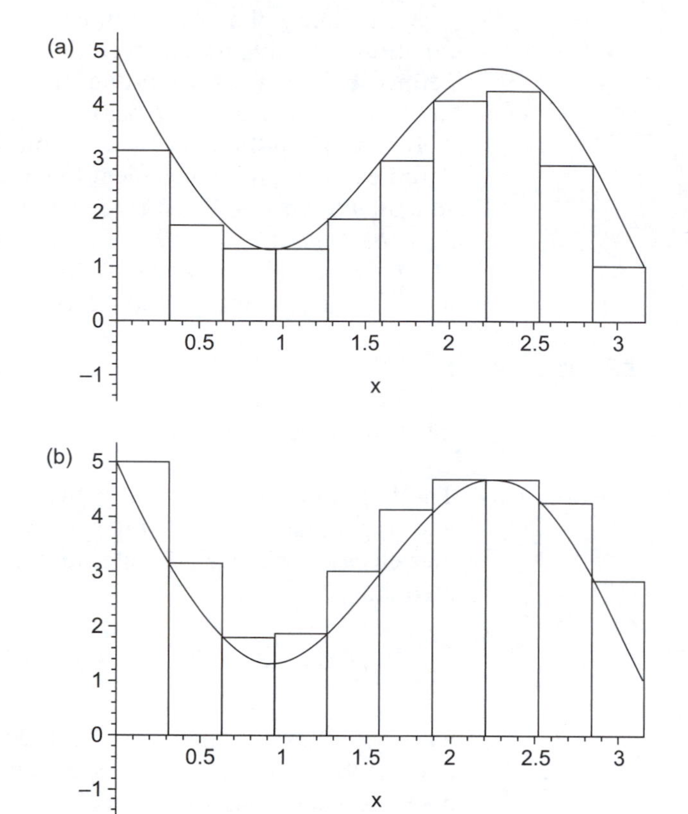

Fig. 7.3 (a) The lower sum approximation $\underline{S}_{P_8}$ represented by eight rectangles; (b) the upper sum approximation $\overline{S}_{P_8}$ represented by eight rectangles.

Hence, from Definition 7.1,

$$I = \lim_{n\to\infty}\sum_{i=1}^{n}(a + i\Delta)^2\Delta.$$

Expanding and grouping the terms of the summation then gives

$$I = \lim_{n\to\infty}[na^2\Delta + 2a\Delta^2(1 + 2 + 3 + \ldots + n) + \Delta^3(1^2 + 2^2 + 3^2 + \ldots + n^2)].$$

Using the fact that $\Delta = (b - a)/n$, together with the expressions for the sum of the first n natural numbers and for the sum of their squares (see Problem 1.14), namely

$$1 + 2 + 3 + \ldots + n = \frac{n}{2}(n + 1)$$

and

$$1^2 + 2^2 + 3^2 + \ldots + n^2 = \frac{n(n+1)(2n+1)}{6},$$

it follows that

$$I = \lim_{n \to \infty} \left\{ a^2(a-b) + a(b-a)^2 \left[\frac{n(n+1)}{n^2}\right] + (b-a)^3 \left[\frac{(n+1)(2n+1)}{6n^2}\right] \right\}.$$

Thus, taking the limit as $n \to \infty$, we find

$$I = \tfrac{1}{3}(b^3 - a^3),$$

and so

$$\int_a^b x^2\,\mathrm{d}x = \tfrac{1}{3}(b^3 - a^3).$$

If we let $a = 1$, $b = 2$, then

$$\int_1^2 x^2\,\mathrm{d}x = \tfrac{1}{3}(2^3 - 1^3) = \frac{7}{3}.$$

The same result would, of course, have been obtained had ξ_i been chosen in any other way. ■

When the behaviour of $f(x)$ is monotonic over the interval $a \le x \le b$, inequality (7.5), coupled with Definition 7.1, can often be used to derive interesting and useful series approximations to the definite integral, as the following example illustrates.

Example 7.3 Show that

$$\sum_{r=1}^{n} \left(\frac{1}{n+r-1}\right) \ge \int_1^2 \frac{\mathrm{d}x}{x} \ge \sum_{r=1}^{n} \left(\frac{1}{n+r}\right).$$

Solution In this case $f(x) = 1/x$, which is continuous, positive and monotonic decreasing on the interval [1, 2], so that Definition 7.1 applies. We again choose a partition P_n which divides the interval [1, 2] into n equal sub-intervals of length $\Delta = 1/n$. Now take point x_r in the partition P_n to be $x_r = 1 + r/n$, for $r = 0, 1, \ldots, n$, so that

$$f(x_r) = \frac{n}{n+r}.$$

Then as $f(x)$ is a monotonic decreasing function of x, it follows that on the interval $[x_{r-1}, x_r]$, $f(x)$ attains its maximum value M_r at x_{r-1} and its minimum value m_r, at x_r, where

$$M_r = \frac{n}{n+r-1} \quad \text{and} \quad m_r = \frac{n}{n+r}.$$

Hence

$$\underline{S}_{P_n} = \sum_{r=1}^{n} \left(\frac{n}{n+r} \right) \frac{1}{n} \quad \text{and} \quad \overline{S}_{P_n} = \sum_{r=1}^{n} \left(\frac{n}{n+r-1} \right) \frac{1}{n},$$

so, from Definition 7.1, we deduce that

$$\sum_{r=1}^{n} \left(\frac{1}{n+r-1} \right) \geq \int_1^2 \frac{\mathrm{d}x}{x} \geq \sum_{r=1}^{n} \left(\frac{1}{n+r} \right).$$

A few numbers might help here, so we show in the table below the behaviour of the upper and lower sums $\overline{S}_{P_n}$ and $\underline{S}_{P_n}$ as a function of n.

n	$\overline{S}_{P_n}$	$\underline{S}_{P_n}$
5	0.7456	0.6456
10	0.7188	0.6688
15	0.7101	0.6768
∞	0.6931	0.6931

We shall discover later that the exact result, which is shown in this table against the entry ∞, is in fact ln 2. ■

Before closing this section, let us give brief consideration to the effect of removing the condition of continuity imposed on the function $f(x)$ and substituting instead the condition that $f(x)$ is bounded. The argument proceeds as before until the stage at which $\underline{S}_{P_m}$ and $\overline{S}_{P_m}$ are defined. Then, without the continuity of $f(x)$ to ensure that $|M_r - m_r| \to 0$ as $|x_r - x_{r-1}| \to 0$, it is no longer possible to infer that when $\lim \underline{S}_{P_m}$ and $\lim \overline{S}_{P_m}$ exist, they are necessarily equal.

To see how this difficulty is overcome, consider a function $f(x)$ defined over the interval $0 \leq x \leq 3$ shown in Fig. 7.4 with a finite jump discontinuity at $x = 1$. Then, all that needs to be done, is to write

$$\int_0^3 f(x)\,\mathrm{d}x = \int_0^{1-} f(x)\,\mathrm{d}x + \int_{1+}^3 f(x)\,\mathrm{d}x,$$

apply Definition 7.1 to each integral on the right, and sum the results. Here the first integral on the right denotes the integral of $f(x)$ with values from $f(0)$ to $\lim_{x\uparrow 1} f(x)$, with this last result representing the limit of $f(x)$ as x *increases* to 1. Similarly, the second integral on the right denotes the integral of $f(x)$ with values from $\lim_{x\downarrow 1} f(x)$ to $f(3)$, with the first result representing the limit of $f(x)$ as x *decreases* to 1.

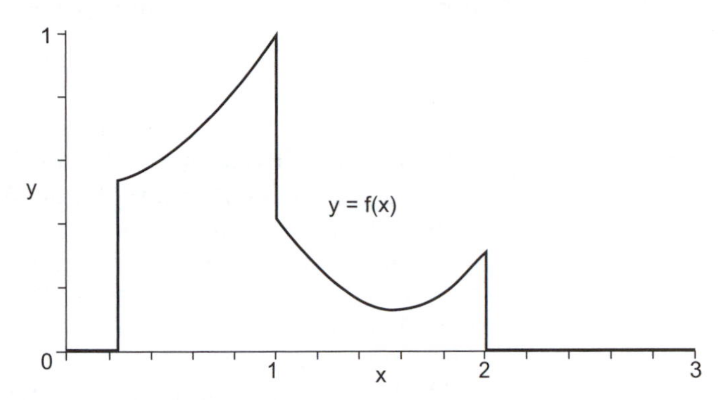

Fig. 7.4 A function $f(x)$ with a finite jump discontinuity at $x = 1$.

Definition 7.2 (Riemann integral of a non-negative function) Let $f(x)$ be a non-negative piecewise continuous function on the closed interval $a \le x \le b$, and let $P_1, P_2, \ldots, P_m, \ldots$ be a sequence of successive refinements of some partition P of $[a, b]$.

Then the *Riemann integral* of $f(x)$ integrated over the interval $[a, b]$, and written symbolically

$$\int_a^b f(x)\,\mathrm{d}x,$$

is defined to be the sum of the integrals, in the sense of Definition 7.1, over each interval in which $f(x)$ is continuous.

This form of definition was first given by B. Riemann (1826–66), whose idea it also was to place the value of the integral between upper and lower bounds that can be brought arbitrarily close together. The definite integral is known formally as the *Riemann integral*, though usually only the term 'definite integral' will be used. Because of this $\underline{S}_{P_m}$ and $\overline{S}_{P_m}$ are called, respectively, the lower and upper Riemann sums.

To show that not all bounded discontinuous functions are Riemann integrable, it is only necessary to consider the integral over the interval $0 \le x \le 1$ of the function

$$f(x) = \begin{cases} 1 & \text{for } x \quad \text{rational} \\ 0 & \text{for } x \quad \text{irrational.} \end{cases}$$

Clearly $f(x)$ is non-negative and bounded on [0, 1], but by a suitable choice of the numbers ξ_i in the approximating sum of Definition 7.2, the limit of the sums may be made to assume any value between zero and unity. This situation arises because the limits of the upper and lower sums are not properly defined. In more advanced accounts these difficulties are overcome by defining a more general form of integral known as the *Lebesgue integral*.

7.2 Integration of arbitrary continuous functions

As most functions assume both positive and negative values in their domain of definition, our notion of a definite integral as formulated so far is restrictive, for it requires that the integrand be non-negative. A brief examination of the introductory arguments used in the previous section shows that this restriction stems from our idea of an area as being an essentially positive quantity, although this was not stated explicitly at any stage in our argument.

Nothing in the limiting arguments that we used requires either the upper and lower sums themselves, or any of the terms comprising them, to be non-negative. Since a term in either of these sums will be negative when m_r or M_r is negative, that is, when $f(x)$ is negative, it follows that the interpretation of a definite integral as an area may be extended to continuous functions $f(x)$ which assume negative values provided that areas below the x-axis are regarded as negative. This is illustrated in Fig. 7.5, in which the positive and negative area contributions to the definite integral of $f(x)$ integrated over the interval $[a, b]$ are marked accordingly.

Thus using this convention when interpreting a definite integral as an area, we may remove the condition that the integrand $f(x)$ be non-negative throughout all of section 7.1. Because it simply amounts to the deletion of the word 'non-negative', we shall not trouble to reformulate our earlier definitions to take account of this result. We remark that had we introduced the definite integral via the upper and lower Riemann sums, without any appeal to graphs and areas, this artificial restriction would never have arisen.

The definition of the definite integral of an arbitrary continuous function $f(x)$ integrated over the interval $[a, b]$ immediately implies a number of important general results which we now state in the form of a theorem. No proofs will be offered since the results are virtually self-evident.

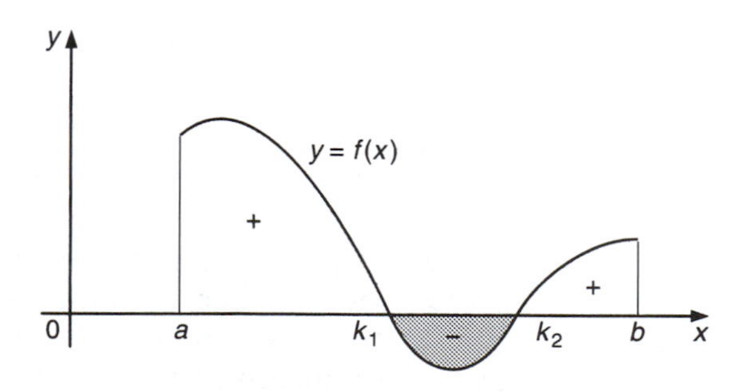

Fig. 7.5 Positive and negative areas defined by $y = f(x)$.

Theorem 7.1 (properties of definite integrals) Let $f(x)$, $g(x)$ be continuous functions defined on the closed interval $a \leq x \leq b$, and let c be a constant and k be such that $a \leq k \leq b$. Then

(a) $\int_a^b f(x)\,dx = \int_a^k f(x)\,dx + \int_k^b f(x)\,dx$ (Additivity with respect to interval of integration),

(b) $\int_a^b cf(x)\,dx = c\int_a^b f(x)\,dx$ (Homogeneity),

(c) $\int_a^b (f(x) + g(x))\,dx = \int_a^b f(x)\,dx + \int_a^b g(x)\,dx$ (Linearity). ■

By virtue of these results, the definite integral of the function $f(x)$ appropriate to Fig. 7.5 could, if desired, be written in terms of the sum of three integrals involving non-negative integrands. To achieve this, notice that $f(x)$ is negative for $k_1 \leq x \leq k_2$, so that for all x in this interval, $-f(x)$ is positive. Then, first expressing our integral as the sum of three separate integrals over adjacent intervals

$$\int_a^b f(x)\,dx = \int_a^{k_1} f(x)\,dx + \int_{k_1}^{k_2} f(x)\,dx + \int_{k_2}^b f(x)\,dx, \tag{7.6}$$

we can replace $-f(x)$ by $|f(x)|$ in the second of these integrals and change its sign to obtain

$$\int_a^b f(x)\,dx = \int_a^{k_1} f(x)\,dx - \int_{k_1}^{k_2} |f(x)|dx + \int_{k_2}^b f(x)\,dx. \tag{7.7}$$

Each of these integrals is now the definite integral of a non-negative function as required.

We must now take account of the fact that so far it has been implicit in our definition of a definite integral that x increases positively from a to b, where $b > a$. This *sense*, or *direction*, of integration is indicated in the definite integral by writing a at the bottom of the integral sign $\int$ to signify the lower limit of integration and by writing b at the top to signify the upper limit of integration. If, despite the fact that $b > a$, their positions as upper and lower limits of integration are reversed, this implies that integration is to be carried out in the direction in which x increases negatively. Because we are now allowing areas to have both magnitude and sign, to be consistent we must compensate for a reversal of the limits of integration by changing the sign of the integral. Hence we arrive at our next definition.

Definition 7.3 (reversal of limits of integration) If $a < b$, we define the definite integral

$$\int_b^a f(x)\,dx$$

of a continuous function $f(x)$ by the equation

$$\int_b^a f(x)\,dx = -\int_a^b f(x)\,dx.$$

Example 7.4 Evaluate the definite integral

$$\int_3^1 2x^2\,dx.$$

Solution From Definition 7.2 we have

$$\int_3^1 2x^2\,dx = -\int_1^3 2x^2\,dx.$$

Hence an application of Theorem 7.3(b), together with the result of Example 7.2, shows that

$$\int_3^1 2x^2\,dx = -2\int_1^3 x^2\,dx = -2(\tfrac{1}{3})(3^3 - 1^3) = -\frac{52}{3}. \quad ■$$

Since a definite integral is simply a number, the choice of symbol used to denote the argument of the function f forming the integrand is arbitrary, and often it is convenient, and sometimes essential, to replace x by some other variable, say t. Thus

$$\int_a^b f(x)\,dx \quad \text{and} \quad \int_a^b f(t)\,dt$$

are identical in meaning, so that

$$\int_a^b f(x)\,dx = \int_a^b f(t)\,dt. \tag{7.8}$$

On account of this fact, the variable in the integrand of a definite integral is called a *dummy* variable, and it is sometimes said to be 'integrated out' when the integral is evaluated. This fact is usually recognized in more advanced accounts of the theory of integration by simply writing

$$\int_a^b f$$

in place of either of the expressions in Eqn (7.8). The full significance of the symbol dx, which is suggestive of a differential, comes when changes of variable of the form $x = g(u)$ are made in Eqn (7.8), and it is for this reason that we choose to retain it. This matter will be taken up in detail in Chapter 8, where it is shown that because of the chain rule for differentiation, dx can indeed be interpreted as a differential.

Now that the definite integral has been extended to arbitrary continuous integrands, we are in a position to determine quite general areas. Consider, for example, the situation illustrated in Fig. 7.6(a) in which it is desired to determine the area I of the shaded region. Then obviously, referring to Fig. 7.6(b), (c) we have

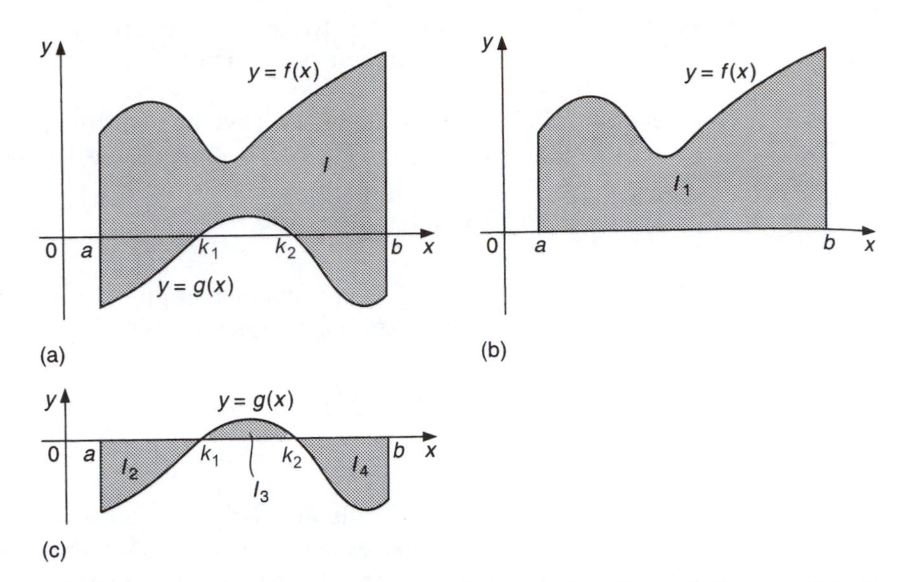

Fig. 7.6 (a) Area I bounded by curves $y = f(x)$ and $g(x)$; (b) area below $y = f(x)$; (c) positive and negative areas defined by $y = g(x)$.

$$I = I_1 + I_2 - I_3 + I_4,$$

where I_1 to I_4 represent the *positive* areas identified by these symbols. However, we know that

$$I_1 = \int_a^b f(x)\,\mathrm{d}x,$$

and from the form of argument leading to Eqn (7.15) we also know that

$$-I_2 = \int_a^{k_1} g(x)\,\mathrm{d}x, \quad I_3 = \int_{k_1}^{k_2} g(x)\,\mathrm{d}x, \quad -I_4 = \int_{k_2}^{b} g(x)\,\mathrm{d}x,$$

where k_1 and k_2 are the first and second points of intersection of $y = g(x)$ with the x-axis as x increases from a to b. However, by Theorem 7.1(a) we have

$$-I_2 + I_3 - I_4 = \int_a^b g(x)\,\mathrm{d}x,$$

so that combining these results we obtain

$$I = \int_a^b f(x)\,\mathrm{d}x - \int_a^b g(x)\,\mathrm{d}x.$$

From Theorem 7.1(b) it then finally follows that

$$I = \int_a^b (f(x) - g(x))\,\mathrm{d}x. \tag{7.9}$$

Example 7.5 Find the area I between the two curves $y = e^{2x}$ and $y = -x^2$ which is bounded to the left by the line $x = 1$ and to the right by the line $x = 3$.

Solution We start by making the obvious identifications $f(x) = e^{2x}$, $g(x) = -x^2$, $a = 1$ and $b = 3$. Then from Eqn (7.9) it follows that

$$I = \int_1^3 (e^{2x} + x^2)\,dx$$

whence, using the results of Example 7.1 and Problem 7.3, which shows how to integrate the exponential function, we find

$$I = \tfrac{1}{2}(e^6 - e^2) + \frac{26}{3}. \quad \blacksquare$$

The fact that a definite integral is additive with respect to its interval of integration enables a function to be integrated even when it has discontinuities. It is simply the sum of the integrals over each interval where the function is continuous.

Example 7.6 Evaluate the definite integral

$$I = \int_0^2 f(x)\,dx$$

where $f(x)$ in the piecewise function

$$f(x) = \begin{cases} x^2 & \text{for } 0 \le x \le 1 \\ e^{5x} & \text{for } 1 \le x \le 2. \end{cases}$$

Solution We have

$$I = \int_0^{1-} x^2\,dx + \int_{1+}^2 e^{5x}\,dx,$$

where $1-$ denotes the limit as x *increases* to 1, while $1+$ denotes the limit as x *decreases* to 1. Evaluating the integrals using the results of Example 7.1 and Problem 7.3, and taking the appropriate limits, gives

$$I = \tfrac{1}{3} + \tfrac{1}{5}(e^{10} - e^5). \quad \blacksquare$$

Sometimes a more difficult situation than this arises in which either the integrand tends to infinity at some point in the interval of integration or, perhaps, the interval of integration itself is infinite in length. Such definite integrals are called *improper integrals*, and the way in which to attribute a value to any such integral is as follows.

Let us illustrate something of the difficulty that can arise if ideas are not made precise. Consider the integral

$$\int_{-1}^{1} \frac{dx}{x^2}.$$

Then since $y = 1/x^2$ is essentially positive, the area under the curve must also be positive. Now if we apply the result of Problem 7.4 without questioning its applicability, we have

$$\int_{-1}^{1} \frac{dx}{x^2} = -\frac{1}{1} + \frac{1}{-1} = -2$$

which, since it is negative, contradicts our previous conclusion. What has gone wrong? The trouble is that $1/x^2$ tends to infinity as $x \to 0$, so that the arguments of Problem 7.4 are not applicable, for it was presupposed there that the interval of integration excluded the origin. When dealing with improper integrals of this type in which the integrand has an infinity within the interval of integration, we shall assign a value to the integral according to the following definition.

***Definition 7.4* (improper integral due to infinity of integrand)** Let the function $f(x)$ be continuous throughout the intervals $a \leq x < c$ and $c < x \leq b$, and suppose that $f(x)$ has a singularity at $x = c$ in the sense that $f(x)$ tends to infinity as $x \to c$. Then the integral of $f(x)$ over the interval of integration $[a, b]$ is said to be *improper*, and it is defined to have the value

$$I = \lim_{\varepsilon \to 0} \int_{a}^{c-\varepsilon} f(x)\,dx + \lim_{\delta \to 0} \int_{c+\delta}^{b} f(x)\,dx,$$

whenever both limits involved exist. Under these circumstances the improper integral will be said to *converge* to the value I. When either or both of the limits does not exist, the integral will be said to be *divergent*. If the point c coincides with an end point of the interval $[a, b]$, then I is defined to be equal to the limit of the single integral for which the interval of integration lies within $[a, b]$.

On the basis of this definition we are now able to determine the value to be attributed to the improper integral used as an illustration above. Let us do this in the form of an example.

Example 7.7 Evaluate the improper integrals:

$$\text{(a)}\quad I_1 = \int_{-1}^{1} \frac{dx}{x^2}; \quad \text{(b)}\quad I_2 = \int_{-1}^{0} \left(\frac{x^2+1}{x^2}\right) dx.$$

Solution The integrand $1/x^2$ tends to infinity as $x \to 0$, so that for case (a), when appealing to Definition 7.4, we need to make the identifications $a = -1$, $b = 1$, $c = 0$ and $f(x) = 1/x^2$. Thus,

$$I_1 = \lim_{\varepsilon \to 0} \int_{-1}^{-\varepsilon} \frac{dx}{x^2} + \lim_{\delta \to 0} \int_{\delta}^{1} \frac{dx}{x^2} \qquad (\delta,\ \varepsilon > 0).$$

Using the result of Problem 7.4, we find that

$$I_1 = \lim_{\varepsilon \to 0} \left(\frac{1}{\varepsilon} - 1 \right) + \lim_{\delta \to 0} \left(-1 + \frac{1}{\delta} \right) \to \infty.$$

Thus the improper integral (a) is divergent.

In case (b) the integrand is $(x^2+1)/x^2$, which again tends to infinity as $x \to 0$. However, in this case we must make the identifications $a = -1$, $b = 0$, $c = 0$ and $f(x) = 1 + 1/x^2$, so that this time the singularity in the integrand occurs at the right-hand end-point of the interval of integration $[-1, 0]$ (that is, at the upper limit of integration).

It then follows from Definition 7.4 that

$$I_2 = \lim_{\varepsilon \to 0} \int_{-1}^{-\varepsilon} \left(1 + \frac{1}{x^2} \right) dx, \qquad (\varepsilon > 0)$$

which, from the results of Problems 7.2 (b) and 7.4, becomes

$$I_2 = \lim_{\varepsilon \to 0} \left\{ (-\varepsilon + 1) + \left(\frac{1}{\varepsilon} - 1 \right) \right\} \to \infty.$$

Hence the improper integral (b) is also divergent. ■

The one remaining different form of improper integral requiring consideration occurs when the interval of integration is infinite. In these circumstances we shall assign a value to the integral according to the following definition.

Definition 7.5 (improper integral due to infinite interval of integration)

Let the function $f(x)$ be continuous on the interval $[a, \infty)$. Then the integral of $f(x)$ over the interval of integration $[a, \infty)$ is said to be *improper*, and it is defined to have the value

$$I_1 = \lim_{k \to \infty} \int_a^k f(x)\, dx,$$

whenever this limit exists. Under these circumstances the improper integral will be said to converge to the value I_1. When the limit does not exist or is infinite, the integral will be said to be *divergent*. Similarly, if the interval of integration is $(-\infty, b]$, then when the limit exists, the improper integral of $f(x)$ over the interval of integration $(-\infty, b]$ is defined to have the value

$$I_2 = \lim_{k \to \infty} \int_{-k}^{b} f(x)\, dx.$$

Symbolically, these improper integrals will be denoted, respectively, by

$$I_1 = \int_a^{\infty} f(x)\,\mathrm{d}x \quad \text{and} \quad I_2 = \int_{-\infty}^{b} f(x)\,\mathrm{d}x.$$

Sometimes this form of integral is called an *infinite integral*.

Example 7.8 Evaluate the improper integral

$$I = \int_3^{\infty} \frac{\mathrm{d}x}{x^2}.$$

Solution It follows at once from Definition 7.5 that

$$I = \lim_{k\to\infty} \int_3^k \frac{\mathrm{d}x}{x^2},$$

so that, by virtue of the result of Problem 7.4,

$$I = \lim_{k\to\infty}\left[\frac{-1}{k} + \frac{1}{3}\right] = \frac{1}{3}.$$

Hence this improper integral converges to the value 1/3. ■

7.3 Integral inequalities

A number of useful inequalities may be deduced concerning definite integrals, the simplest of which has already been stated in Eqn (7.1). Let us now derive our first result of this type, of which Eqn (7.1) represents a special case.

Suppose that the definite integrals of $f(x)$ and $g(x)$, taken over the interval $[a, b]$, both exist. In brief, let us agree to say that $f(x)$ and $g(x)$ are *integrable* over the interval $[a, b]$. Now suppose that $f(x) \le g(x)$ for $a \le x \le b$. Then if P_m is a partition of $[a, b]$, we have

$$\int_a^b g(x)\,\mathrm{d}x - \int_a^b f(x)\,\mathrm{d}x = \int_a^b (g(x) - f(x))\,\mathrm{d}x$$

$$= \lim_{m\to\infty} \sum_{i=1}^{n} (g(\xi_i) - f(\xi_i))\Delta_i,$$

where ξ_i is some point in the ith sub-interval of length Δ_i generated by the partition P_m. Now since by hypothesis $f(x) \le g(x)$, it follows that $f(\xi_i) \le g(\xi_i)$ so that the right-hand side must be non-negative. Thus we have proved the following theorem.

Theorem 7.2 (inequality between two definite integrals) Let $f(x) \le g(x)$ be two integrable functions over the interval $[a, b]$. Then,

$$\int_a^b f(x)\,\mathrm{d}x \le \int_a^b g(x)\,\mathrm{d}x. \quad ■$$

Equation (7.1) follows as a trivial consequence of this result, for the theorem implies that if $\phi(x) \leq f(x) \leq \psi(x)$ are three integrable functions over the interval $[a, b]$, then

$$\int_a^b \phi(x)\,\mathrm{d}x \leq \int_a^b f(x)\,\mathrm{d}x \leq \int_a^b \psi(x)\,\mathrm{d}x. \tag{7.10}$$

Hence, if m, M are, respectively, the minimum and maximum values of $f(x)$ on $[a, b]$, our required result follows by setting $\phi(x) = m$, $\psi(x) = M$, whence we obtain

$$m(b-a) \leq \int_a^b f(x)\,\mathrm{d}x \leq M(b-a). \tag{7.11}$$

This last simple result implies a more important result which we now derive by appeal to the intermediate value theorem of Chapter 5. Writing inequality (7.11) in the form

$$m \leq \frac{1}{b-a}\int_a^b f(x)\,\mathrm{d}x \leq M$$

shows that the number

$$\frac{1}{b-a}\int_a^b f(x)\,\mathrm{d}x$$

is intermediate between m and M, which are extreme values of the function $f(x)$ itself. Hence, provided $f(x)$ is continuous, it then follows from the intermediate value theorem that some number ξ exists, strictly between a and b, such that

$$f(\xi) = \frac{1}{b-a}\int_a^b f(x)\,\mathrm{d}x. \tag{7.12}$$

This result is called the *first mean value theorem for integrals*, and it constitutes our next theorem.

Theorem 7.3 (first mean value theorem for integrals) Let $f(x)$ be continuous on the interval $[a, b]$. Then there exists a number ξ, strictly between a and b, for which

$$\int_a^b f(x)\,\mathrm{d}x = (b-a)f(\xi). \quad \blacksquare$$

We remark in passing that integral inequality (7.10) is often useful in determining the convergence or divergence of improper integrals when the integrand involved is too complicated to be evaluated analytically. Consider, for example, the improper integral

$$I = \int_1^\infty \frac{\mathrm{d}x}{\mathrm{e}^{-x} + x^2} = \lim_{k\to\infty}\int_1^k \frac{\mathrm{d}x}{\mathrm{e}^{-x} + x^2}.$$

Then on the interval $1 \le x \le \infty$ we have

$$x^2 < e^{-x} + x^2 < 1 + x^2,$$

and so

$$\frac{1}{x^2} > \frac{1}{e^{-x} + x^2} > \frac{1}{1 + x^2}.$$

Setting $\phi(x) = (1+x^2)^{-1}$, $f(x) = (e^{-x}+x^2)^{-1}$ and $\psi(x) = 1/x^2$ in (7.10) gives the inequalities

$$\lim_{k\to\infty} \int_1^k \frac{dx}{1 + x^2} < I < \lim_{k\to\infty} \int_1^k \frac{dx}{x^2}.$$

Later the left-hand side of this inequality will be shown to be equal to $\pi/4$, while from Problem 7.4 in the problem section the right-hand side is equal to 1, so we have the result

$$\frac{\pi}{4} < \int_1^\infty \frac{dx}{e^{-x} + x^2} < 1.$$

Although this does not give the precise value of I it shows that it is bounded (between $\pi/4$ and 1), so this improper integral is convergent. Similar arguments can also be used to prove convergence or divergence of improper integrals of the type identified in Definition 7.4. Estimates of this type are useful in more advanced work when improper integrals have to be integrated using special numerical technique, since it is fruitless trying to compute the value of an integral that can be shown to be divergent.

7.4 The definite integral as a function of its upper limit—the indefinite integral

If the lower limit of a definite integral is held constant, but the upper limit is replaced by the variable x, then the numerical value of the integral will clearly depend on x. Another way of describing this situation is if we say that a definite integral with a variable upper limit x defines a *function* of x. In Fig. 7.7 this idea is illustrated in terms of areas, with the region marked $F(x)$ denoting the area below the curve $y = f(t)$ which is bounded on the left by the line $t = a$, and on the right by the line $t = x$.

In terms of the definite integral we have

$$F(x) = \int_a^x f(t)\,dt. \tag{7.13}$$

Now let us suppose that $f(t)$ is continuous in some interval $[a, b]$, with $a \le x \le b$. Notice here that for the first time it is necessary to use the dummy variable t, because x and t are fulfilling two different roles in Eqn (7.13). To be precise, x represents the upper limit of integration, while the dummy variable t represents the general variable in the interval of integration $a \le t \le x$.

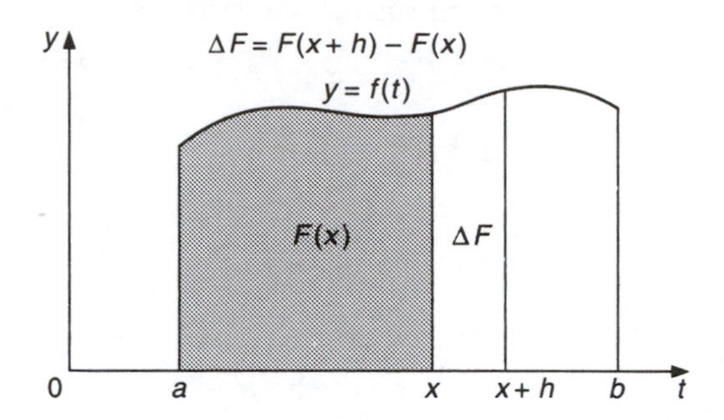

Fig. 7.7 Area below $y=f(t)$ as a function of the upper limit of integration x.

Consider the difference

$$F(x+h)-F(x)=\int_a^{x+h} f(t)\,\mathrm{d}t-\int_a^x f(t)\,\mathrm{d}t$$
$$=\int_x^{x+h} f(t)\,\mathrm{d}t. \tag{7.14}$$

Then the first mean value theorem for integrals allows us to rewrite Eqn (7.14) in the form

$$F(x+h)-F(x)=hf(\xi), \tag{7.15}$$

where $x<\xi<x+h$.

Now, forming the difference quotient $\{F(x+h)-F(x)\}/h$, we find

$$\frac{F(x+h)-F(x)}{h}=f(\xi),$$

so that taking the limit as $h\to 0$, which squeezes ξ arbitrarily close to x, gives

$$F'(x)=\lim_{h\to 0}\left\{\frac{F(x+h)-F(x)}{h}\right\}=f(x). \tag{7.16}$$

This important result shows that the integrand of integral (7.13) at the upper limit of integration $t=x$ is equal to the derivative of $F(x)$ with respect to x.

Suppose now that $G(x)$ is any function for which $G'(x)=f(x)$. Then,

$$G'(x)-F'(x)=\frac{\mathrm{d}}{\mathrm{d}x}\{G(x)-F(x)\}=0,$$

and so, from Corollary 5.12,

$$G(x)=F(x)+\text{constant}. \tag{7.17}$$

A function $F(x)$ such that $F'(x)=f(x)$ is called an *antiderivative* of $f(x)$ and Eqn (7.17) shows that antiderivatives of $f(x)$ can only differ by a constant.

Combining Eqns (7.13) and (7.17) shows that the most general function $G(x)$ whose derivative is equal to $f(x)$ must be of the form

$$G(x) = \int_a^x f(t)\,\mathrm{d}t + C, \qquad (7.18)$$

where C is an arbitrary constant.

We call Eqn (7.18) the *indefinite integral* of $f(x)$ with respect to x. Thus an antiderivative of $F(z)$ only differs from its indefinite integral by a constant. The name 'antiderivative' offers an accurate description of the process by which it is to be found. Namely, an antiderivative arises from the process of reversing the operation of differentiation, and the most frequent method of finding antiderivatives utilizes this idea by employing tables of derivatives in reverse – that is to say, by matching an integrand with an entry in a table of derivatives and thereby finding the functional form of $G(x)$ apart from the additive arbitrary constant.

Usually the *indefinite integral* of $f(x)$ with respect to x is written symbolically in the form

$$\int f(x)\,\mathrm{d}x = F(x) + C. \qquad (7.19)$$

Hence the constant C is called the *constant of integration*, and the operation of finding an indefinite integral is called *integration*. In this notation, the fact that an antiderivative is a *function* related to the operation of integration, and not just a number as is an ordinary definite integral, is indicated by again employing the integral sign, but this time without limits. On occasions an antiderivative is signified by the notation

$$\int^x f(x)\,\mathrm{d}x,$$

rather than the notation used in Eqn (7.19).

Table 7.1 lists a few of the antiderivatives which are of most frequent occurrence in mathematics. Other useful elementary antiderivatives that should be memorized together with an account of systematic methods for finding antiderivatives, are given in Chapter 8. These are easily checked, because from Eqn (7.16), $F'(x) = f(x)$. For reference purposes a more extensive list of antiderivatives is to be found at the end of the book.

Let us now return to Eqn (7.17) and notice that it follows from this that

$$G(b) - G(a) = F(b) - F(a) = F(b) = \int_a^b f(x)\,\mathrm{d}x. \qquad (7.20)$$

Hence we have proved that

$$\int_a^b f(x)\,\mathrm{d}x = G(b) - G(a), \qquad (7.21)$$

where $G'(x) = f(x)$. This provides an efficient method for the evaluation of definite integrals, for expressed in words it asserts that the definite integral of $f(x)$ taken over an interval $[a, b]$ is the difference between the value of any antiderivative of $f(x)$ at $x = b$ and $x = a$.

Table 7.1 Antiderivatives $F(x)$ of some of the more commonly occurring functions $f(x)$ (i.e., $\int f(x)\,dx = F(x) + C$)

	$f(x)$	$F(x)$
1	a(const)	ax
2	$x^n (n \neq -1)$	$\dfrac{x^{n+1}}{n+1}$
3	$e^{\lambda x}$	$\dfrac{1}{\lambda} e^{\lambda x}$
4	$\sin ax$	$-(1/a)\cos ax$
5	$\cos ax$	$(1/a)\sin ax$
6	$\sinh ax$	$(1/a)\cosh ax$
7	$\cosh ax$	$(1/a)\sinh ax$

Notice that the arbitrary constant C associated with $G(x)$ cancels out in Eqn (7.21) and so may be omitted when evaluating definite integrals.

It is now time to express results (7.16) and (7.21) in the form of two basic theorems known, respectively, as the *first* and *second fundamental theorems of calculus*.

Theorem 7.4 (first fundamental theorem of calculus) If $f(x)$ is continuous for $a \le x \le b$, and

$$F(x) = \int_a^x f(t)\,dt,$$

then $F'(x) = f(x)$ for all points x in $[a, b]$.

Alternatively expressed, this result may also be written

$$\frac{d}{dx}\int_a^x f(t)\,dt = f(x). \quad \blacksquare$$

Thus, by way of example, if

$$F(x) = \int_1^x \frac{1 + \sin t + 3\cos^2 t}{1 + 3t^2}\,dt,$$

then from Theorem 7.4 it follows that

$$\frac{dF(x)}{dx} = \frac{d}{dx}\int_1^x \frac{1 + \sin t + 3\cos^2 t}{1 + 3t^2}\,dt$$

$$= \frac{1 + \sin x + 3\cos^2 x}{1 + 3x^2}.$$

Theorem 7.5 (second fundamental theorem of calculus) If $f(x)$ is continuous for $a \le x \le b$ and $G(x)$ is any antiderivative of $f(x)$, then

$$\int_a^x f(t)\,dt = G(x) - G(a). \quad \blacksquare$$

The statement of Theorem 7.5 is often written in the form

$$\int_a^b f(x)\,\mathrm{d}x = G(x)|_{x=a}^{x=b},$$

with the understanding that

$$G(x)|_{x=a}^{x=b} = G(b) - G(a). \quad \blacksquare$$

It follows from Theorem 7.5 that the definite integral calculated so laboriously in Example 7.1 may be evaluated directly by appeal to entry no. 2 in Table 7.1. To see this set $n = 2$, so that $f(x) = x^2$, then $F(x) = x^3/3$, and by Theorem 7.5 we immediately deduce that

$$\int_a^b x^2\,\mathrm{d}x = \tfrac{1}{3}(b^3 - a^3).$$

The systematic employment of the fundamental theorems of calculus will be taken up in detail in Chapter 8, since our concern here is primarily with the theory rather than the practice of integration.

Finally, to emphasize that the indefinite integral is a function, we now give an example of such an integral which defines an important mathematical function. Since we have the relationship

$$\frac{\mathrm{d}}{\mathrm{d}x}[\ln x] = \frac{1}{x}, \quad \text{for } x > 0,$$

it follows from Theorem 7.5 that, provided $a > 0$,

$$\int_a^x \frac{\mathrm{d}t}{t} = \ln x - \ln a.$$

Hence, setting $a = 1$ gives the result

$$\ln x = \int_1^x \frac{\mathrm{d}t}{t}, \tag{7.22}$$

which is illustrated as the shaded area in Fig. 7.8. This result also shows that the entry corresponding to $n = \infty$ in the table in Example 7.2 is ln 2.

7.5 Differentiation of an integral containing a parameter

It can sometimes happen that an integrand, in addition to being a function of x, also depends on a parameter α, and so is of the form

$$I(\alpha) = \int_a^b f(x, \alpha)\,\mathrm{d}x, \tag{7.23}$$

where $f(x, \alpha)$ is a differentiable function of α.

To derive our next theorem let us differentiate $I(\alpha)$ with respect to α. First, notice that by the mean value theorem for derivatives

$$f(x, \alpha + h) = f(x, \alpha) + h\left(\frac{\partial f}{\partial \alpha}\right)_{(x,\xi)}, \quad \text{with } \alpha < \xi < \alpha + h, \tag{7.24}$$

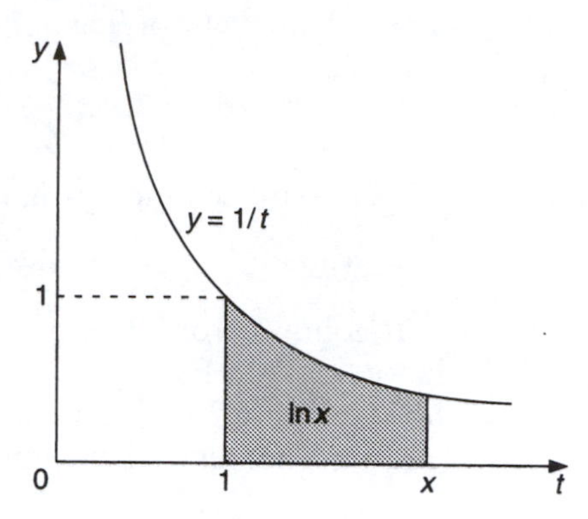

Fig. 7.8 Natural logarithm represented as an area.

where $(\partial f/\partial x)_{(x,\alpha)}$ means $\partial f/\partial \alpha$ evaluated at $\alpha = \xi$ while holding x constant. Thus

$$I(\alpha + h) = \int_a^b \left[f(x, \alpha) + h\left(\frac{\partial f}{\partial \alpha}\right)_{(x,\xi)} \right] \mathrm{d}x, \quad \text{with} \quad \alpha < \xi < \alpha + h,$$

so

$$I(\alpha + h) - I(\alpha) = h \int_a^b \left(\frac{\partial f}{\partial \alpha}\right)_{(x,\ \xi)} \mathrm{d}x, \quad \text{with} \quad \alpha < \xi < \alpha + h,$$

and thus

$$\frac{I(\alpha + h) - I(\alpha)}{h} = \int_a^b \left(\frac{\partial f}{\partial \alpha}\right)_{(x,\xi)} \mathrm{d}x, \quad \text{with} \quad \alpha < \xi < \alpha + h. \tag{7.25}$$

Taking the limit as $h \to 0$, the left-hand side of this last result becomes $I'(\alpha)$, while the integrand on the right-hand side reduces to $\partial f/\partial \alpha$. We have thus proved the following theorem concerning the differentiation of an integral with respect to a parameter contained in its integrand.

Theorem 7.6 (differentiation of an integral containing a parameter) If $f(x, \alpha)$ is both integrable with respect to x over the interval $[a, b]$ and differentiable with respect to α, then

$$\frac{\mathrm{d}}{\mathrm{d}\alpha} \int_a^b f(x, \alpha)\,\mathrm{d}x = \int_a^b \frac{\partial f}{\partial \alpha}\,\mathrm{d}x. \quad \blacksquare$$

This theorem has many uses as, for example, with the *Laplace transform* which will be introduced when we come to discuss differential equations. There, an improper integral of the form

$$\int_0^\infty \mathrm{e}^{-st} f(t)\,\mathrm{d}t$$

defines what is called the *Laplace transform* of the function $f(t)$ and, given the Laplace transform of a known function, the Laplace transform of a related one can be deduced by differentiation with respect to s. To illustrate matters, let us consider the Laplace transform of sin kt which will be deduced later and is given by

$$\int_0^\infty e^{-st}\sin kt\,dt = \frac{k}{k^2 + s^2}.$$

Applying Theorem 7.6, we have

$$\frac{d}{ds}\int_0^\infty e^{-st}\sin kt\,dt = \frac{d}{ds}\left[\frac{k}{k^2 + s^2}\right].$$

and so

$$\int_0^\infty - te^{-st}\sin kt\,dt = -\frac{2ks}{(k^2 + s^2)^2},$$

or

$$\int_0^\infty e^{-st}t\sin kt\,dt = \frac{2ks}{(k^2 + s^2)^2},$$

showing that the last result on the right is the Laplace transform of $t\sin kt$.

7.6 Other geometrical applications of definite integrals

This section offers a brief discussion of the application of the definite integral to the determination of arc length for plane curves, the area of a surface of revolution, and the volume of a solid of revolution. Each result will be derived by appeal to the basic definition of a definite integral, since it will first be necessary to define the precise meaning of the concepts that are involved.

7.6.1 Arc length of a plane curve

Consider the plane curve Γ with equation $y = f(x)$ joining points M and N, as illustrated in Fig. 7.9(a). Our task here will be first to define the meaning of the length s of the arc MN, and then to deduce a method by which it may be found once the equation of Γ has been given. Let $Q_0, Q_1, \ldots, Q_n$ represent any set of points on Γ, the first of which coincides with the left-hand end point M, and the last of which coincides with the right-hand end point N. Then if Δs_i denotes the length of the chord joining Q_{i-1} to Q_i, the length S_n of the polygonal line joining M to N is

$$S_n = \sum_{i=1}^{n} \Delta s_i. \tag{7.26}$$

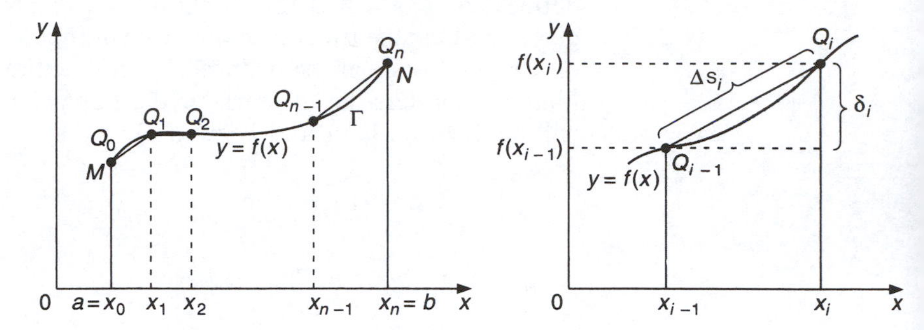

Fig. 7.9 (a) Arc length of curve; (b) element of arc length.

Now the projection of the set of points $Q_0, Q_1, \ldots, Q_n$ onto the x-axis defines a set of points $a = x_0 < x_1 < \cdots < x_n = b$ which form a partition P_n of the interval $[a, b]$. Setting $\|\Delta\|_{P_n} = \max(x_i - x_{i-1})$, we shall define the length s of the arc Γ from M to N to be

$$s = \lim_{n\to\infty} S_n = \lim_{\|\Delta\|_{P_n}\to 0} \sum_{i=1}^{n} \Delta s_i. \tag{7.27}$$

Now setting $\Delta_i = x_i - x_{i-1}$ and $\delta_i = f(x_i) - f(x_{i-1})$, it follows directly by an application of Pythagoras' theorem (Fig. 7.9(b)) that

$$\Delta s_i = \surd(\Delta_i^2 + \delta_i^2) = \left[\sqrt{\left(1 + \left(\frac{\delta_i}{\Delta_i}\right)^2\right)}\Delta_i\right].$$

However, by virtue of the mean value theorem for derivatives we may write, provided that $f(x)$ is differentiable on $[a, b]$,

$$\frac{\delta_i}{\Delta_i} = \frac{f(x_i) - f(x_{i-1})}{x_i - x_{i-1}} = f'(\xi_i),$$

where $x_{i-1} < \xi_i < x_i$, and so

$$\Delta s_i = [\surd(1 + \{f'(\xi_i)\}^2)]\Delta_i. \tag{7.28}$$

Thus the desired arc length s will be determined by evaluating

$$s = \lim_{n\to\infty} \sum_{i=1}^{n} [\surd(1 + \{f'(\xi_i)\}^2)]\Delta_i. \tag{7.29}$$

We see from Definition 7.1 that this is simply the definite integral of the function $\surd(1 + [f'(x)]^2)$ integrated from $x = a$ to $x = b$, and hence

$$s = \int_a^b \surd(1 + [f'(x)]^2)\,\mathrm{d}x = \int_a^b \sqrt{\left(1 + \left(\frac{\mathrm{d}y}{\mathrm{d}x}\right)^2\right)}\mathrm{d}x. \tag{7.30}$$

Theorem 7.7 *(arc length of plane curve)* Let $y = f(x)$ be a differentiable function on the interval $[a, b]$. Then the length s of the plane curve Γ defined by the graph of this function in the (x, y)-plane between the points $(a, f(a))$, $(b, f(b))$ is given by

$$s = \int_a^b \sqrt{(1 + [f'(x)]^2)}\, dx = \int_a^b \sqrt{\left(1 + \left(\frac{dy}{dx}\right)^2\right)} dx. \quad \blacksquare$$

Example 7.9 Determine the length of arc of the curve $y = \cosh x$ between the points $(1, \cosh 1)$ and $(3, \cosh 3)$.

Solution We have $a = 1$, $b = 3$, $y = \cosh x$, and so $dy/dx = \sinh x$, whence

$$s = \int_1^3 \sqrt{(1 + \sinh^2 x)}\, dx = \int_1^3 \cosh x\, dx,$$

where we have used the hyperbolic identity $\cosh^2 x = 1 + \sinh^2 x$. Now since $(d/dx)(\sinh x) = \cosh x$, it follows that $\sinh x$ is an antiderivative of $\cosh x$, so that by Theorem 7.5 we have

$$s = \int_1^3 \cosh x\, dx = \sinh x|_1^3 = \sinh 3 - \sinh 1. \quad \blacksquare$$

Remember that when evaluating a definite integral using Theorem 7.5 the arbitrary additive constant associated within indefinite integral is always omitted because it cancels out.

Theorem 7.7 will fail for curves Γ of the type shown in Fig. 7.10, because any representation of the function in the form $y = f(x)$ will not be single-valued on the interval $[\alpha, \beta]$, and so it will not be differentiable there.

The difficulty here is easily overcome by using the fact that each point on the curve Γ can be uniquely defined and a unique derivative assigned if the curve Γ is capable of parametric representation in the form

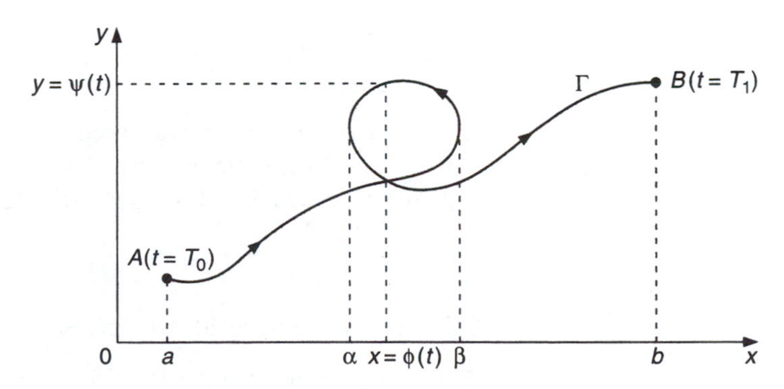

Fig. 7.10 Length of parametrically defined curve Γ.

$$x = \phi(t), \quad y = \psi(t) \quad \text{for } T_0 \le t \le T_1, \tag{7.31}$$

with $\phi(t)$, $\psi(t)$ differentiable on $[T_0, T_1]$.

Using the result for parametric differentiation

$$f'(x) = \frac{dy}{dx} = \frac{\psi'(t)}{\phi'(t)}$$

in Eqn (7.29), and then employing the differential relationship $\Delta_i = \phi'(t)\Delta t$ to define Δ_i in terms of Δt, we find that

$$s = \lim_{n\to\infty}\sum_{i=1}^{n}\sqrt{\left(1 + \left[\frac{\psi'(\xi_i)}{\phi'(\xi_i)}\right]^2\right)}\phi'(\xi_i)\Delta t,$$

where $t_{i-1} < \xi_i < t_i$.

Thereafter, the argument that gave rise to Eqn (7.30), now gives rise to

$$s = \int_{T_0}^{T_1}\sqrt{\left(1 + \left[\frac{\psi'(t)}{\phi'(t)}\right]^2\right)}\phi'(t)\,dt$$

$$= \int_{t_0}^{t_1}\sqrt{([\phi'(t)]^2 + [\psi'(t)]^2)}\,dt. \tag{7.32}$$

Theorem 7.8 (arc length of a parametrically defined curve) Let $\phi(t)$, $\psi(t)$ be differentiable functions in $T_0 \le t \le T_1$. Then the length s of the plane curve defined parametrically by $x = \phi(t)$, $y = \psi(t)$ between the points $(\phi(T_0), \psi(T_0))$, $(\phi(T_1), \psi(T_1))$ is given by

$$s = \int_{T_0}^{T_1}\sqrt{([\phi'(t)]^2 + [\psi'(t)]^2)}\,dt. \quad \blacksquare$$

By way of example, let us apply Theorem 7.8 to determine the length of the loop in Fig. 7.10 described parametrically by

$$x = t^3 - t \quad \text{and} \quad y = 4 - t^2 \quad \text{for} \quad -2 \le t \le 2.$$

Setting $\phi(t) = t^3 - t$ and $\psi(t) = 4 - t^2$ and substituting into Theorem 7.8, we have

$$s = \int_{-2}^{2}\sqrt{(3t^2 - 1)^2 + (2t)^2}\,dt = \int_{-2}^{2}\sqrt{9t^4 - 2t^2 + 1}\,dt.$$

This integral cannot be evaluated analytically in terms of simple functions, and so it must be determined by using a numerical method such as Simpson's rule which will be derived later, an application of which shows that $s = 17.1462$.

7.6.2 Area of surface of revolution

Let $f(x)$ be a non-negative function. Then the name *surface of revolution* is given to any surface which is generated by rotating a plane curve

$y = f(x)$ about either the x-axis or the y-axis. Since the determination of the area in either case is exactly similar, we shall discuss in detail only the case of the revolution of the curve $y = f(x)$ about the x-axis, as shown in Fig. 7.11.

A problem arises here as to how to define the area of a non-cylindrical curved surface. We propose to approach the problem by sectioning the surface into annular strips of width Δ_i as shown in Fig. 7.11, and then to approximate the area ΔS of each such annular strip by representing it by the conical area which is obtained by rotating the chord PQ of length Δs_i about the x-axis. Then if this element of area of cone between the planes $x = x_{i-1}$ and $x = x_i$ is ΔS_i this will be given by 2π times the product of the average radius of the conical strip and its slant length Δs_i, so that

$$\Delta S_i = 2\pi\left(\frac{y_{i-1} + y_i}{2}\right)\Delta s_i.$$

Similar elements of area may be defined for each of the other annular strips defined by some partition P_n of the interval $[a, b]$ by the set of points $a = x_0 < x_1 < \ldots < x_n = b$. Thus we shall define the area S of the surface of revolution generated by rotating $y = f(x)$ about the x-axis, and contained between the planes $x = a$ and $x = b$, to be

$$S = \lim_{n\to\infty}\sum_{i=1}^{n}\Delta S_i = \lim_{n\to\infty}\pi\sum_{i=1}^{n}(y_{i-1} + y_i)\Delta s_i.$$

Hence, if $f(x)$ is differentiable in $a \le x \le b$, by using result (7.28) we find

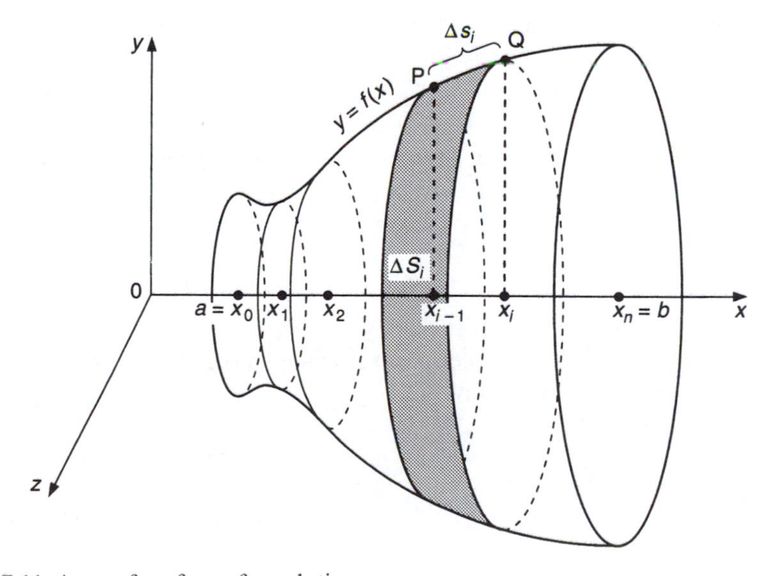

Fig. 7.11 Area of surface of revolution.

$$S = \lim_{n\to\infty} \pi \sum_{i=1}^{n} (y_{i-1} + y_i)\sqrt{(1 + [f'(\xi_i)]^2)}\Delta_i,$$

where $x_{i-1} < \xi_i < x_i$.

Once again our previous form of argument shows that this is just the definite integral of the function $2\pi f(x)\sqrt{(1 + [f'(x)]^2)}$ integrated from $x = a$ to $x = b$, and so

$$S = 2\pi \int_a^b f(x)\sqrt{(1 + [f'(x)]^2)}\mathrm{d}x. \tag{7.33}$$

***Theorem 7.9* (area of surface of revolution)** Let $f(x)$ be a non-negative and differentiable function on $a \leq x \leq b$. Then the area S of the surface of revolution generated by rotating the graph of the function $y = f(x)$ about the x-axis, and contained between the planes $x = a$ and $x = b$, is given by

$$S = 2\pi \int_a^b y\sqrt{[1 + (\mathrm{d}y/\mathrm{d}x)^2]}\,\mathrm{d}x = 2\pi \int_a^b f(x)\sqrt{(1 + [f'(x)]^2)}\,\mathrm{d}x.$$

Similarly, the area of the surface of revolution generated by rotating the non-negative differentiable function $x = g(y)$ about the y-axis, and contained between the planes $y = c$ and $y = d$, is given by

$$S = 2\pi \int_c^d x\sqrt{[1 + (\mathrm{d}x/\mathrm{d}y)^2]}\mathrm{d}y = 2\pi \int_c^d g(y)\sqrt{(1 + [g'(y)]^2)}\mathrm{d}y.$$

Example 7.10 Find the area contained between the planes $x = -1$ and $x = 2$ of the surface of revolution about the x-axis of the curve $y = \cosh x$.

Solution We have $a = -1$, $b = 2$ and $f(x) = \cosh x$, and so $f'(x) = \sinh x$, whence

$$S = 2\pi \int_{-1}^{2} \cosh x\sqrt{(1 + \sinh^2 x)}\,\mathrm{d}x = 2\pi \int_{-1}^{2} \cosh^2 x\,\mathrm{d}x,$$

where use has been made of the hyperbolic identity $\cosh^2 x = 1 + \sinh^2 x$. To evaluate this result we now use another hyperbolic identity $\cosh^2 x = \frac{1}{2}(1 + \cosh 2x)$ to obtain

$$S = \pi \int_{-1}^{2} (1 + \cosh 2x)\,\mathrm{d}x.$$

Then, as it is easily verified that $\frac{1}{2}\sin 2x$ is an antiderivative of $\cosh 2x$, we have

$$S = \pi \int_{-1}^{2} (1 + \cosh 2x)\,\mathrm{d}x = \pi(x + \tfrac{1}{2}\sinh 2x)|_{-1}^{2}$$

$$= \tfrac{1}{2}\pi(6 + \sinh 4 + \sinh 2) = 57.9888. \quad \blacksquare$$

7.6.3 Volume of revolution

Finally, let us determine the *volume of revolution* V of the volume shown in Fig. 7.11, in which it is supposed that $f(x)$ is a non-negative function. This time, to define the volume of such a figure, we consider cylindrical elements of volume of thickness Δ_i, and place upper and lower bounds on that element of volume by the obvious inequality:

$$\pi \times (\text{least radius of annulus})^2 \times \Delta_i \leq \text{element of volume}$$
$$\leq \pi \times (\text{greatest radius of annulus})^2 \times \Delta_i.$$

Then, if $x_{i-1} < \xi_i < x_i$, a volume element ΔV_i satisfying this inequality and bounded to the left by the plane $x = x_{i-1}$ and to the right by the plane $x = x_i$, is

$$\Delta V_i = \pi [f(\xi_i)]^2 \Delta_i.$$

The volume of revolution generated by rotating $y = f(x)$ about the x-axis, and contained between the planes $x = a$ and $x = b$, will then be defined to be

$$V = \lim_{n \to \infty} \pi \sum_{i=1}^{n} [f(\xi_i)]^2 \Delta_i.$$

A repetition of the previous form of argument then yields

$$V = \pi \int_a^b [f(x)]^2 \mathrm{d}x. \tag{7.34}$$

Notice that we have imposed no differentiability requirements on $f(x)$, so that result (7.34) is applicable even if $f(x)$ is only piecewise continuous.

***Theorem 7.10* (volume of solid revolution)** Let $f(x)$ be a piecewise continuous non-negative function on $a \leq x \leq b$. Then the volume of the solid of revolution generated by rotating the curve $y = f(x)$ about the x-axis, and contained between the planes $x = a$ and $x = b$, is given by

$$V = \pi \int_a^b [f(x)]^2 \mathrm{d}x.$$

Similarly, if $g(y)$ is a piecewise continuous non-negative function on $c \leq y \leq d$, the volume of the solid of revolution generated by rotating the curve $x = g(y)$ about the y-axis, and contained between $y = c$ and $y = d$, is given by

$$V = \pi \int_c^d [g(y)]^2 \mathrm{d}y. \quad \blacksquare$$

Example 7.11 Determine the volume of revolution generated by rotating the parabola $y = 1 + x^2$ about the x-axis, and contained between the planes $x = 1$ and $x = 2$.

Solution Here we have $a = 1$, $b = 2$ and $f(x) = 1 + x^2$, so that

$$V = \pi\int_1^2 (1 + x^2)^2 \mathrm{d}x = \pi\int_1^2 (1 + 2x^2 + x^4)\mathrm{d}x$$

$$= \pi\left(x + \frac{2x^3}{3} + \frac{x^5}{5}\right)\Bigg|_1^2 = \frac{178\pi}{15}. \quad \blacksquare$$

7.6.4 Integration of even and odd functions

Two simple but very useful results that often help when integrating even or odd functions over an interval $-a \leq x \leq a$ that is symmetrical about the origin are as follows:

(a) If $f(x)$ is an even function that is defined and integrable for $-a \leq x \leq a$, then

$$\int_{-a}^{a} f(x)\mathrm{d}x = 2\int_0^a f(x)\mathrm{d}x. \tag{7.35}$$

(b) If $f(x)$ is an odd function that is defined and integrable for $-a \leq x \leq a$, then

$$\int_{-a}^{a} f(x)\mathrm{d}x = 0. \tag{7.36}$$

The proofs of these results are simple. If $f(x)$ is an even function it is symmetrical about the y-axis. Thus the positive and negative areas between the graph and the x-axis for positive x are exactly matched by those for negative x, so that

$$\int_{-a}^{0} f(x)\mathrm{d}x = \int_0^a f(x)\mathrm{d}x,$$

from which it follows that

$$\int_{-a}^{a} f(x)\mathrm{d}x = \int_{-a}^{0} f(x)\mathrm{d}x + \int_0^a f(x)\mathrm{d}x = 2\int_0^a f(x)\mathrm{d}x.$$

If, however, $f(x)$ is an odd function, then every positive area between the graph and the x-axis for positive x is matched by a negative area for negative x, and conversely, so that

$$\int_{-a}^{0} f(x)\mathrm{d}x = -\int_0^a f(x)\mathrm{d}x,$$

from which it follows that

$$\int_{-a}^{a} f(x)\mathrm{d}x = \int_{-a}^{0} f(x)\mathrm{d}x + \int_0^a f(x)\mathrm{d}x = 0.$$

For example, it follows from Eqn (7.35) that $\int_{-\pi/2}^{\pi/2} \cos x \, dx = 2\int_0^{\pi/2} \cos x \, dx = 2$, while from Eqn (7.36) it follows that $\int_{-\pi}^{\pi} \sin x \, dx = 0$.

7.6.5 Areas in polar coordinates

Let an arc Γ be defined in terms of polar coordinates by the equations

$$r = f(\theta) \quad \text{for } \alpha \leq \theta \leq \beta.$$

Then the area OAB bounded by Γ in Fig. 7.12 may be approximated by dividing the angle AOB into n sub-angles δ_i, *for* $i = 1, 2, \ldots, n$, and summing the areas ΔA_i of all the triangular area elements such as ROS, with angle δ_i at O and radius $OP = r_i = f(\theta_i)$.

As δ_i is small we have

$$\Delta A_i = \text{area } ROS = \tfrac{1}{2}(OP)\cdot(RS) \approx \tfrac{1}{2}(OP)\cdot(OP\delta_i) = \tfrac{1}{2}[f(\theta_i)]^2\delta_i.$$

Proceeding to the limit as $n \to \infty$ and all $\delta_i \to 0$, we see that the area differential

$$dA = \tfrac{1}{2}[f(\theta)]^2 d\theta,$$

so the area bounded by the lines OA, OB and the arc Γ is seen to be given by integrating dA from $\theta = \alpha$ to $\theta = \beta$ to obtain

$$A = \int_{\alpha}^{\beta} \tfrac{1}{2}[f(\theta)]^2 \, d\theta. \tag{7.37}$$

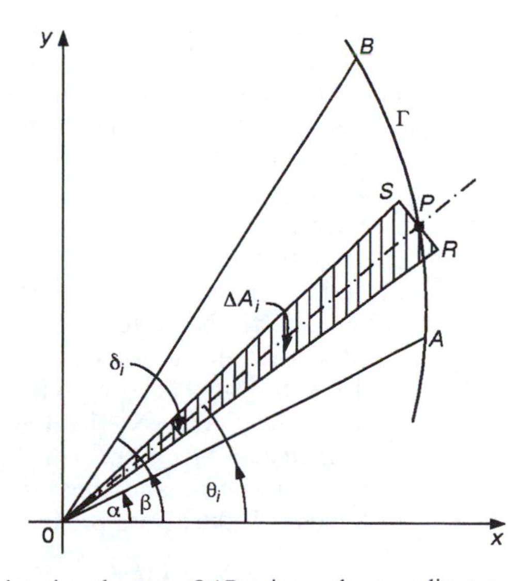

Fig. 7.12 Approximating the area OAB using polar coordinates.

Theorem 7.11 (Area in plane polar coordinates) Let an arc Γ be described in terms of plane polar coordinates by the equation $r = f(\theta)$ for $\alpha \leq \theta \leq \beta$. Then the area A between the arc Γ and the radial lines $\theta = \alpha$ and $\theta = \beta$ is given by

$$A = \tfrac{1}{2}\int_{\alpha}^{\beta} [f(\theta)]^2 \, d\theta.$$

If, for example, Eqn (7.37) is used to find the area inside the cardioid $r = a(1 - \cos\theta)$, $0 \leq \theta \leq 2\pi$ shown in Fig. 1.22(a), we have

$$\begin{aligned} A &= \tfrac{1}{2}a^2 \int_0^{2\pi} (1 - \cos\theta)^2 d\theta = \tfrac{1}{2}a^2 \int_0^{2\pi} (1 - 2\cos\theta + \cos^2\theta)\, d\theta \\ &= \tfrac{1}{2}a^2 \int_0^{2\pi} [1 - 2\cos\theta + \tfrac{1}{2}(1 + \cos 2\theta)]\, d\theta \\ &= \tfrac{1}{2}a^2 [\theta - 2\sin\theta + \tfrac{1}{2}\theta + \tfrac{1}{4}\sin 2\theta]\Big|_{\theta=0}^{2\pi} \\ &= (3/2)\pi a^2. \end{aligned}$$

7.7 Centre of mass and moment of inertia

In mechanics the concepts of the centre of mass and moment of inertia of a body are both of considerable importance, and in general they rely for their determination on the techniques of integration. If the body comprises a system of discrete masses, each rigidly positioned relative to the others, then only a summation is involved in the computation. For continuous bodies, however, it is necessary to extend the idea of an integral to more than one dimension in order to perform the necessary calculation. Despite this fact there are many useful cases in which some symmetry of the body can be utilized to reduce the calculation to the evaluation of an ordinary definite integral, and it is this situation that we shall consider here.

To illustrate ideas, let us examine how the *centre of mass* G (also called the *centre of gravity*) of an arbitrary plane lamina of mass M and uniform density may be determined. A *lamina* is a flat sheet of material so thin that only its two-dimensional shape need be considered. In terms of mechanics, the moments m_1 and m_2 measure the turning effect of the distributed mass of the lamina about the respective Y and X axes. If a horizontal lamina is supported at a single point located at its centre of mass G then $\bar{x} = \bar{y} = 0$, causing the turning moments m_1 and m_2 to vanish. This means that a horizontal lamina supported at its centre of mass G will be perfectly balanced. A typical lamina S is shown in Fig. 7.13(a) relative to axes $O\{X, Y\}$ that lie in its plane. Suppose that the element located at $P(x, y)$ has mass dM. Then, as the moment of a mass about the Y-axis is defined as the product of the mass and its perpendicular distance from the Y-axis, the moment of the element at P about the Y-axis is $x dM$, and the total moment m_1 of the lamina S about the Y-axis will thus be approximated by

$$m_1 \approx \sum x dM,$$

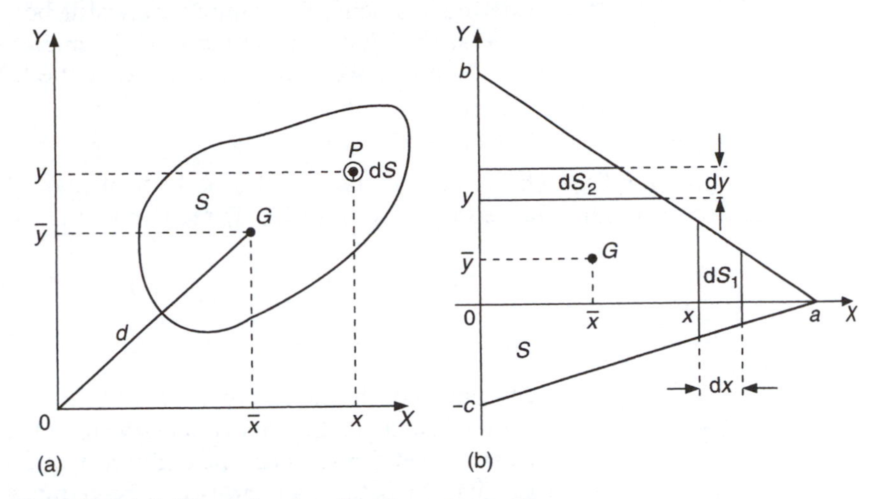

Fig. 7.13 (a) Arbitrary lamina; (b) triangular lamina.

where the summation of the quantities $x\mathrm{d}M$ is understood to be extended over all elements $\mathrm{d}M$ of the lamina. It is usual to modify the notation for integrals and to write the limiting form of this summation as the subdivision of S into elements is refined and $\mathrm{d}M \to 0$ as

$$m_1 = \int_S x\,\mathrm{d}M. \tag{7.38}$$

The natural extension of integration in this way to a region S in the (X, Y)-plane gives rise to what are called *double integrals*, though these will be considered in Chapter 9.

We now set

$$m_1 = M\bar{x}, \tag{7.39}$$

which serves to define the perpendicular distance $\bar{x}$ of the centre of mass G of the lamina from the Y-axis. Similarly, the perpendicular distance $\bar{y}$ of the centre of mass from the X-axis may be determined from the corresponding result for the moment about the X-axis

$$m_2 = M\bar{y}, \quad \text{where } m_2 = \int_S y\,\mathrm{d}M. \tag{7.40}$$

In this case the integral notation signifies that the contributions to the moment of the lamina about the X-axis of each different element $y\mathrm{d}M$ are to be integrated for all possible y belonging to the lamina. Results (7.39) and (7.40) then determine the point $(\bar{x}, \bar{y})$ at which is located the *centre of mass* G of the lamina.

This same form of argument extends immediately to three-dimensional bodies once we define in them a system of axes $O\{X, Y, Z\}$. Then, in addition to the moments m_1 and m_2 already discussed, there will be a third moment m_3 of the body about the Z-axis. The integral notation used to define moments for laminas may still be used provided it is reinterpreted as the limit of a summation of elements extending over the volume S of the body. We have thus arrived at the following general definition.

Definition 7.5 (centre of mass) The centre of mass G of a body S of mass M and uniform density is located at the point $(\bar{x}, \bar{y}, \bar{z})$ relative to the axes $O\{X, Y, Z\}$, where

$$\bar{x} = \frac{1}{M}\int_s x\,\mathrm{d}M, \quad \bar{y} = \frac{1}{M}\int_s y\,\mathrm{d}M, \quad \bar{z} = \frac{1}{M}\int_s z\,\mathrm{d}M.$$

As phrased, this definition is only of theoretical interest, because as yet we are not in a position to evaluate the integrals that are involved. Despite this fact, useful results can be deduced from this definition for the many cases in which these integrals can be reduced to ordinary definite integrals. We illustrate this by the following examples.

Example 7.12 Find the centre of mass of the uniform triangular lamina of mass M shown in Fig. 7.13(b).

Solution The upper side of the triangle between the points $(0, b)$ and $(a, 0)$ has the equation $y = b(1 - x/a)$, and the lower side from the point $(-c, 0)$ to $(a, 0)$ has the equation $y = c(x/a - 1)$. So the width $h(x)$ of strip $\mathrm{d}S_1$ at a distance x from the Y-axis is

$$h(x) = b(1 - x/a) - c(x/a - 1) = (b + c)(1 - x/a).$$

As the mass M of the lamina is uniformly distributed, and the area of the lamina is $\frac{1}{2}a(b + c)$, the mass per unit area is $2M/[a(b+c)]$. Thus the mass $\mathrm{d}M$ of element $\mathrm{d}S_1$ is

$$\mathrm{d}M = \frac{2M}{a(b + c)}h(x)\,\mathrm{d}x = \frac{2M}{a}(1 - x/a)\,\mathrm{d}x.$$

From the first result in Definition 7.6 we find that

$$\bar{x} = \frac{1}{M}\int_0^a \frac{2M}{a}x(1 - x/a)\,\mathrm{d}x = a/3.$$

To find $\bar{y}$ requires a little more work. The width of strip $\mathrm{d}S_2$ in the part of the triangle above the X-axis is $a(1 - y/b)\mathrm{d}y$, so its area is $a(1 - y/b)\mathrm{d}y$, and its mass $\mathrm{d}M$ is

$$\mathrm{d}M = \frac{2M}{(b + c)}(1 - y/b)\,\mathrm{d}y \quad \text{for } 0 \le y \le b.$$

Similar reasoning shows the mass $\mathrm{d}M$ of strip $\mathrm{d}S_2$ in the part of the triangle below the X-axis is

$$\mathrm{d}M = \frac{2M}{(b + c)}(1 + y/c)\,\mathrm{d}y \quad \text{for} \quad -c \leq y \leq 0.$$

Combining these results and using the second result in Definition 7.6 gives

$$\begin{aligned}\bar{y} &= \frac{1}{M}\left\{\frac{2M}{(b + c)}\int_0^b y(1 - y/b)\,\mathrm{d}y + \frac{2M}{(b + c)}\int_{-c}^0 y(1 + y/c)\,\mathrm{d}y\right\} \\ &= \tfrac{1}{3}(b - c).\end{aligned}$$

We have shown that the coordinates of the centre of mass G of the uniform lamina is located at $\bar{x} = a/3$ and $\bar{y} = \frac{1}{3}(b - c)$. This result is general, because the vertices of the triangle were located at arbitrary points on the X and Y-axes. Consequently, the centre of mass G of a uniform triangular lamina will be in the same position relative to *each* side of the triangle that is made to coincide with the Y-axis.

Geometrical reasoning, which we will omit, shows the point G is located on any median of the triangle, one third of the way along the median measured from its base. Remember, a median of a triangle is the line drawn from a vertex to the center of the opposite side of the triangle which forms its base, and that all three medians intersect at the same point. Thus G is located at the point where the medians intersect. This result is illustrated in Fig. 7.14 for a triangle with its vertices at the points

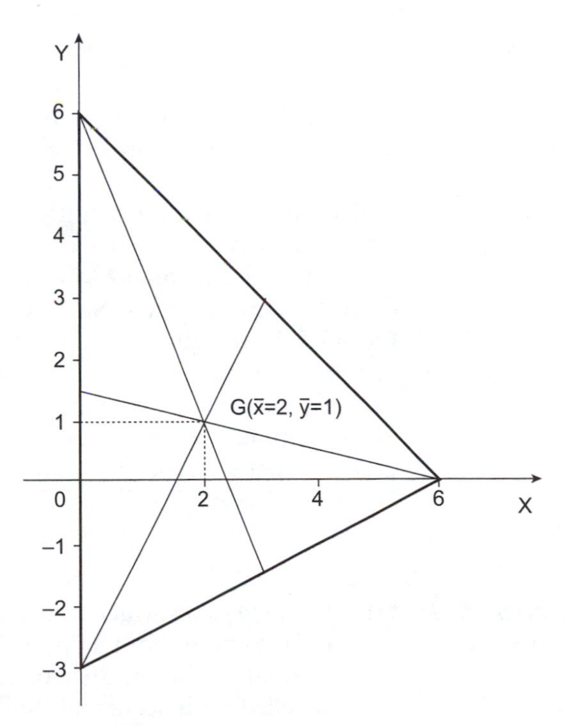

Fig. 7.14 The medians of the uniform triangular lamina all intersect at G.

(6, 0), (0, 6) and (0, −3). In terms of Fig. 7.13(b) this corresponds to $a = 6$, $b = 6$ and $-c = -3$, so $c = 3$. Thus $\bar{x} = a/3 = 2$ and $\bar{y} = \frac{1}{3}(b - c) = 1$. ■

Example 7.13 Find the centre of mass of a solid uniform right circular cone of mass M and height a when the radius of the base is b.

Solution As symmetry shows that the centre of mass must lie on the axis of the cone, we shall take the X-axis to coincide with the axis of the cone and the origin O to lie at the centre of the base, so the Y- and Z-axes lie in the base, and the vertex is at $X = a$. The volume $\mathrm{d}V$ of a cylindrical element of thickness $\mathrm{d}x$ at a distance x from the $\{X, Y\}$ plane is easily seen to be

$$\mathrm{d}V = \frac{\pi b^2 (a - x)^2}{a^2}\,\mathrm{d}x.$$

The volume of the cone is $\frac{1}{3}\pi ab^2$, so the mass $\mathrm{d}M$ of the cylindrical element follows after multiplication of its volume $\mathrm{d}V$ by the mass per unit volume $3M/\pi ab^2$ of the cone, giving

$$\mathrm{d}M = \frac{3M(a - x)^2}{a^3}\,\mathrm{d}x.$$

The moment of this about a diameter of the base is $x\,\mathrm{d}M$, so from the first result in Definition 7.5

$$\bar{x} = \frac{1}{M}\int_0^a \frac{3M(a - x)^2}{3}\,x\,\mathrm{d}x = a/4.$$

Thus the centre of mass of a uniform solid cone is on its axis one quarter of its height above the base. ■

If an arc of a curve is regarded as though it were a wire of uniform density its centre of mass may be calculated, although, as mass is not involved, it is customary to refer to this as the *centroid of the arc*, rather than its centre of mass. Similarly, if a closed plane curve is regarded as though it were a plane lamina of uniform density, its centre of mass may also be calculated, though here again this is referred to as the *centroid of the area* defined by the curve, rather than its centre of mass. With these ideas in mind we now mention two interesting and useful relationships that exist between the area and volume of revolution of either a plane arc or a closed curve and its centroid.

Theorem 7.12 (Pappus' theorem)

(a) The area of a surface of revolution generated by rotating the arc of a plane curve about a line in its plane that does not intersect the arc is equal to the product of the length of the arc and the distance travelled by its centroid during the rotation.

(b) The volume of revolution generated by rotating a closed plane curve about a line in its plane that does not intersect the curve is equal to the product of the area of the plane curve and the distance travelled by its centroid during the rotation.

Proof As the proofs of (a) and (b) are similar we shall only give that of (b), leaving the proof of (a) as an exercise for the reader. Suppose we wish to find the volume of the surface generated when area S in Fig. 7.13(a) is rotated about the X-axis. The element of volume generated by the rotation about the X-axis through an angle θ $(0 \leq \theta \leq 2\pi)$ of the plane element of area $\mathrm{d}S$ is $\mathrm{d}V = \theta y \mathrm{d}S$. The total volume V of revolution so formed is thus

$$V = \theta \int_s y \,\mathrm{d}S = \theta S \left\{ \frac{1}{S} \int_s y \,\mathrm{d}S \right\} = \theta \bar{y} S,$$

which is the required result. If S is rotated a full revolution then $\theta = 2\pi$, so

$$V = 2\pi \bar{y} S.$$

Example 7.14 (a) Find the surface area of a right circular cone of height h and semi-angle α at its vertex.

(b) Find the volume of revolution of the torus that is formed when a circle of radius a is rotated one complete revolution about a line in its plane at a perpendicular distance b from its centre, when $b > a$.

Solution (a) The side of the cone (a generator) is of length $h \sec \alpha$, and the centroid of this line, which from symmetry must lie at its mid-point, is at a perpendicular distance $\frac{1}{2} h \tan \alpha$ from the axis of the cone. From Theorem 7.11(a) it then follows that the area of the surface of the cone $S = (h \sec \alpha) \times 2\pi(\frac{1}{2} h \tan \alpha) = \pi h^2 \sec \alpha \tan \alpha$.

(b) By symmetry the centroid of a circular disc must lie at its centre. Thus in a full revolution the centroid will travel a distance $2\pi b$, so that by Theorem 7.11(b) the volume of the torus $V = \pi a^2 \times 2\pi b = 2\pi^2 a^2 b$. ■

The fundamental idea of a moment of inertia is best illustrated in its simplest form in terms of a discrete system of particles with masses m_1, $m_2, \ldots, m_N$ that are positioned rigidly in space with respect to one another. If a fixed line L is specified, and the masses $m_1, m_2, \ldots, m_N$ are at the respective perpendicular distances $x_1, x_2, \ldots, x_N$ from L, then the *moment of inertia* I_L of the mass system about the axis L is defined to be

$$I_L = \sum_{r=1}^{N} m_r x_r^2. \tag{7.41}$$

The suffix L is employed here to emphasize that the moment of inertia depends on the choice of axis L.

The dimensions of a moment of inertia are (mass) × (length)^2, so that when I_L is known it is possible to define a length k_L by setting

$$I_L = Mk_L^2,$$

where M is the total mass of the system. When this is done the length k_L is called the *radius of gyration* of the system relative to the axis L.

As with our discussion of centre of mass, so the definition of moment of inertia may also be extended to take account of a continuous body S. If such a body is subdivided into parts, with a representative part having a mass $\mathrm{d}M$ at a perpendicular distance x from the axis L, result (7.41) shows that the moment of inertia I_L of the body about axis L will be approximated by

$$I_L \approx \sum x^2\,\mathrm{d}M. \tag{7.42}$$

This summation is understood to be extended over all elements $\mathrm{d}M$ of the body. When the subdivision is refined so that $\mathrm{d}M \to 0$, we arrive at the limiting form of (7.42),

$$I_L = \int_S x^2 \mathrm{d}M. \tag{7.43}$$

We have thus motivated the following general definition.

***Definition 7.6* (moment of inertia and radius of gyration)** The moment of inertia I_L of a solid body S about any axis L is given by

$$I_L = \int_S x^2 \mathrm{d}M,$$

where x is the perpendicular distance of the element of mass $\mathrm{d}M$ from the axis L. The radius of gyration of the body relative to the axis L is $k_L = (I_L/M)^{1/2}$.

Example 7.15 Find the moment of inertia I_x, I_y about the X- and Y-axes, respectively, of the uniform triangular lamina of mass M shown in Fig. 7.13(b), and determine the radius of gyration r_Y.

Solution From Example 7.12 we know the mass of strip $\mathrm{d}S_2$ is

$$\mathrm{d}M = \begin{cases} \dfrac{2M}{(b+c)}(1 - y/b)\,\mathrm{d}y, & 0 \le y \le b \\ \dfrac{2M}{(b+c)}(1 + y/c)\,\mathrm{d}y, & -c \le y \le 0 \end{cases},$$

and the moment of inertia I_X about the X-axis is

$$I_\mathrm{X} = \int_{-c}^{b} y^2 \mathrm{d}M.$$

So substituting for $\mathrm{d}M$ this becomes

$$I_\mathrm{X} = \frac{2M}{(b+c)}\left\{\int_0^b y^2(1 - y/b)\,\mathrm{d}y + \int_{-c}^0 y^2(1 + y/c)\,\mathrm{d}y\right\},$$

and after integrating we have

$$I_X = \frac{M(b^3 + c^3)}{6(b + c)}.$$

The mass of strip dS_1 is

$$dM = \frac{2M}{a}(1 - x/a)dx \quad \text{for } 0 \leq x \leq a,$$

and the moment of inertia I_Y about the Y-axis is

$$I_Y = \int_0^a x^2 dM.$$

So substituting for dM and integrating we find that

$$I_Y = \frac{2M}{a}\int_0^a x^2(1 - x/a)dx = \frac{Ma^2}{6}.$$

The radius of gyration $k_Y = \sqrt{I_Y/M} = a/\sqrt{6}$. ■

It is appropriate to ask what is the moment of inertia of a lamina about an axis to its plane, and the following theorem provides the answer.

Theorem 7.13 (moment of inertia about an axis normal to a lamina) The moment of inertia I_Z of a lamina about an axis Z normal to its plane, point O in the plane, is equal to the sum of the moments of inertia I_X and I_Y about any two orthogonal axes OX and OY in its plane, so that

$$I_Z = I_X + I_Y.$$

Proof Let the lamina in question be represented by Fig. 7.13(a). We then have

$$I_Z = \int_S (OP)^2 dM = \int_S (x^2 + y^2)^2 dM = I_X + I_Y,$$

which was to be shown. ■

Example 7.16 Find the moment of inertia I_Z for the lamina shown in Fig. 7.13(b).

Solution From Theorem 7.12 and the results of Example 7.15 we have at once $I_Z = I_X + I_Y = \frac{1}{6}M(b^3 + c^3 + a^2b + a^2c)/(b + c)$. ■

One final result that is useful when calculating moments of inertia is the relationship that exists between the moment of inertia I_Z of a body about an arbitrary axis Z and the moment of inertia I_G of the body about an axis parallel to Z that passes through the centre of mass G.

***Theorem 7.13* (parallel axes theorem)** If I_Z is the moment of inertia of a body of mass M about an arbitrary axis Z, and I_G is the moment of inertia of the body about an axis parallel to Z that passes through the centre of mass G of the body, then

$$I_Z = I_G + Md^2,$$

where d is the perpendicular distance between the two axes.

Proof Suppose Fig. 7.13(a) represents the situation in which the relationship between the moments of inertia about the axis Z normal to the plane of the lamina passing through the origin and a parallel axis through the centre of mass G are to be established. Let $\mathrm{d}M$ represent the mass of the cylindrical element of the body, taken parallel to the Z-axis, that has cross-section $\mathrm{d}S$. Then the moment of inertia of this element about the axis through G parallel to the Z-axis is

$$\begin{aligned}(GP)^2\mathrm{d}M &= \{(x-\bar{x})^2 + (y-\bar{y})^2\}\mathrm{d}M \\ &= (x^2+y^2)\mathrm{d}M - 2(\bar{x}x + \bar{y}y)\mathrm{d}M + (\bar{x}^2+\bar{y}^2)\mathrm{d}M, \\ &= (x^2+y^2)\mathrm{d}M - 2\{\bar{x}[(x-\bar{x})+\bar{x}] + \bar{y}[(y-\bar{y})+\bar{y}]\}\mathrm{d}M \\ &\quad + (\bar{x}^2+\bar{y}^2)\mathrm{d}M.\end{aligned}$$

Thus, by definition

$$\begin{aligned}I_G &= \int_S (x^2+y^2)\,\mathrm{d}M - 2\bar{x}\int_S (x-\bar{x})\,\mathrm{d}M - 2\bar{y}\int_S (y-\bar{y})\,\mathrm{d}M \\ &\quad - (\bar{x}^2 - \bar{y}^2)\int_S \mathrm{d}M \\ &= I_Z - d^2M,\end{aligned}$$

because as G is the centre of mass, the second and third integrals must vanish. ■

Example 7.17 Find the moment of inertia of a uniform isosceles triangular lamina of mass M about a line normal to its plane that passes through its centre of mass.

Solution Setting $c = -b$ in Example 7.16 and interpreting the result in terms of Fig. 7.13(b), we find for an isoceles triangle with base $2b$ and height a that the moment of inertia about the Z-axis through O is $I_Z = M(a^2+b^2)/6$. The result of Example 7.12 shows that the centre of mass G must be at a distance $a/3$ along the median drawn from the mid-point of the base to the vertex. Thus, as Theorem 7.13 can be put in the form $I_G = I_Z - Md^2$, where in this case $d = a/3$, we find $I_G = M(a^2+b^2)/6 - Ma^2/9 = M(a^2+3b^2)/18$. ■

7.8 Line integrals

A special form of definite integral involving a function $F(x, y, z)$ that reduces to an ordinary definite integral is what is called the *line integral*. Suppose that a curve Γ is defined parametrically in terms of t by

$$x = f(t), \quad y = g(t), \quad z = h(t) \quad \text{for } a \le t \le b, \tag{7.44}$$

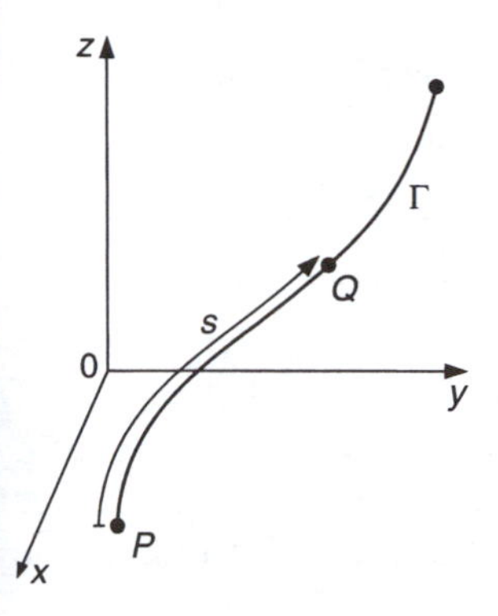

Fig. 7.15 Curves Γ in space, with s the arc length along Γ.

and that s is the arc length measured along Γ from its intial point P located at $(f(a), g(a), h(a))$ to the point Q at $(f(t), g(t), h(t))$ as in Fig. 7.15. Then, provided $F(x, y, z)$ is defined at each point of Γ, its integral along Γ from P to Q is

$$\lim_{n\to\infty} \sum_{i=1}^{n} f(\xi_i, \eta_i, \zeta_i)\Delta s_i$$

where P_n is a partition of the curve Γ, and (ξ_i, η_i, ζ_i) is a point in the ith element Δs_i of the arc length along Γ. Denoting the *line integral* of $F(x, y, z)$ along Γ by $\int_\Gamma F(x, y, z)\,ds$, by proceeding to the limit and substituting the parametrizations for x, y and z in terms of t, we obtain the following integral involving $f(t)$, $g(t)$ and $h(t)$

$$\int_\Gamma F(x, y, z)\,ds = \int_{t=a}^{t=b} F(f(t), g(t), h(t))\,ds. \tag{7.45}$$

However, extending the argument concerning the arc length of a curve in section 7.6.1 to three dimensions shows that

$$\left(\frac{ds}{dt}\right)^2 = [f'(t)]^2 + [g'(t)]^2 + [b'(t)]^2,$$

and so

$$ds = \{[f'(t)]^2 + [g'(t)]^2 + [b'(t)^2]\}^{1/2}\,dt, \tag{7.46}$$

Finally, combining Eqns (7.45) and (7.46), we arrive at the following *rule for the determination of a line integral* in terms of an ordinary definite integral in terms of t:

$$\int_\Gamma F(x, y, z)\,ds = \int_{t=a}^{t=b} F(f(t), g(t), h(t))\{[f'(t)]^2 + [g'(t)]^2 + [h'(t)]^2\}^{1/2}\,dt. \tag{7.47}$$

Example 7.18 Find the line integral

$$\int_\Gamma F(x, y, z)ds,$$

given that $F(x, y, z) = x^2 + y^2 + z^2$, and the path Γ of the line integral is defined as the helix $x = \cos t$, $y = \sin t$ and $z = t$, for $0 \le t \le 2\pi$.

Solution Substituting for x, y and z in $F(x, y, z)$ gives

$$F(x, y, z) = \cos^2 t + \sin^2 t^2 + t^2 = 1 + t^2$$

and in the notation of Eqn (7.47) we have $a = 0$, $b = 2\pi$, $f(t) = \cos t$, $g(t) = \sin t$ and $h(t) = t$ so

$$\{[f'(t)]^2 + [g'(t)]^2 + [h'(t)]^2\}^{1/2} = [\sin^2 t + \cos^2 t + 1]^{1/2} = \sqrt{2}.$$

Thus, substituting into Eqn (7.47) yields

$$\int_\Gamma F(x, y, z)\,\mathrm{d}s = \int_0^{2\pi} (1 + t^2)\sqrt{2}\,\mathrm{d}t = \sqrt{2}(2\pi + 8\pi^3/3). \quad \blacksquare$$

Problems **Section 7.1**

7.1 Let $f(x) = \lambda x$ on some closed interval $a \le x \le b$ lying in the positive part of the x-axis, where $\lambda > 0$ is a constant. Then, if P_n is a partition of $[a, b]$ into n sub-intervals of equal length, determine the form of the lower and upper sums $\underline{S}_{P_n}$, $\overline{S}_{P_n}$ for $f(x)$ taken over this partition, and prove directly by taking the limit that

$$\lim_{n\to\infty} \underline{S}_{P_n} = \lim_{n\to\infty} \overline{S}_{P_n}.$$

Hence deduce that

$$\int_a^b \lambda x\,\mathrm{d}x = \frac{\lambda}{2}(b^2 - a^2).$$

7.2 Let λ, $\mu > 0$ be constants, and set $f(x) = \mu + \lambda x$ on some closed interval $a \le x \le b$ lying in the positive part of the x-axis. Show, using the method of Problem 7.1, that

$$\int_a^b (\mu + \lambda x)\,\mathrm{d}x = \mu(b - a) + \frac{\lambda}{2}(b^2 - a^2). \qquad \text{(A)}$$

Show also by this method that

$$\int_a^b \mu\,\mathrm{d}x = \mu(b - a), \qquad \text{(B)}$$

and deduce from (A), (B) and the result of Problem 7.1 that

$$\int_a^b (\mu + \lambda x)\,\mathrm{d}x = \int_a^b \mu\,\mathrm{d}x + \int_a^b \lambda x\,\mathrm{d}x.$$

This provides a direct proof of the linearity of the operation of integration in the special case that $f(x) = \mu + \lambda x$.

7.3 Let $f(x) = \mathrm{e}^{\lambda x}$, and take P_n to be a partition of the closed interval $[a, b]$ into n sub-intervals of equal length. By taking the numbers ξ_i of Definition 7.1 to be at the left-hand end points of the sub-intervals, compute the approximating sum S_{P_n} corresponding to $f(x) = \mathrm{e}^{\lambda x}$, and by finding its limit prove that

$$\int_a^b \mathrm{e}^{\lambda x}\mathrm{d}x = \frac{1}{\lambda}(\mathrm{e}^{\lambda b} - \mathrm{e}^{\lambda a}).$$

7.4 Let $[a, b]$ be any closed interval not containing the origin, and denote by P_m the partition of this interval into m equal sub-intervals each of length $(b-a)/m$. Denote by x_r the point $x_r = a + (r/m)(b-a)$ lying at the right-hand end point of the rth interval. Then, setting $\xi_r = \sqrt{(x_{r-1}x_r)}$, show, by considering $x_{r-1} - \xi_r$ and $x_r - \xi_r$, that $x_{r-1} < \xi_r < x_{r+1}$. By writing $f(x) = 1/x^2$ in Definition 7.2 and taking P_m and the points ξ_r in that definition to be as defined above, prove that

$$\int_a^b \frac{\mathrm{d}x}{x^2} = \left(\frac{1}{a} - \frac{1}{b}\right).$$

$$\left[\textit{Hint}: \text{Use the fact that } \sum_{r=1}^{n} \frac{1}{x_{r-1}x_r} = \sum_{r=1}^{n}\left(\frac{1}{x_r - x_{r-1}}\right)\left(\frac{1}{x_{r-1}} - \frac{1}{x_r}\right).\right]$$

7.5 Determine the lower bounds m_r and the upper bounds M_r of the function $f(x) = 1/(1+x^2)$ in each of the n adjacent sub-intervals of length $1/n$ comprising a partition P_n of the closed interval $[0, 1]$. Use these results to deduce the form taken by the upper and lower sums $\overline{S}_{P_n}$, $\underline{S}_{P_n}$, and show that

$$\lim_{n\to\infty}(\overline{S}_{P_n} - \underline{S}_{P_n}) = 0.$$

Deduce from this that

$$\int_0^1 \frac{\mathrm{d}x}{1+x^2} = \lim_{n\to\infty} n\left\{\frac{1}{n^2+1^2} + \frac{1}{n^2+2^2} + \frac{1}{n^2+3^2} + \ldots + \frac{1}{n^2+n^2}\right\}$$

or, equivalently,

$$= \lim_{n\to\infty} n\left\{\frac{1}{n^2} + \frac{1}{n^2+1^2} + \frac{1}{n^2+2^2} + \ldots + \frac{1}{n^2+(n-1)^2}\right\}.$$

We shall see later that this integral has the value $\frac{1}{4}\pi$, and so each of these different expressions has this same interesting limit.

Section 7.2

7.6 Find the area I between the curves $y = x^2 + 2$ and $y = -x + 1$, which is bounded to the left by the line $x = -1$ and to the right by the line $x = 2$.

7.7 Discuss, without attempting to evaluate any integrals that are involved, the problem of determining the area between the curves $y = 1 + \sin x$ and $y = 1 + \cos x$ which is bounded to the left by the line $y = 0$ and to the right by the line $y = 2\pi$.

7.8 Find the area I between the two curves $y = 1/x^2$ and $y = e^{0.5x} - 3$, which is bounded to the left by the line $x = 1$ and to the right by the line $x = 2$.

7.9 Evaluate the integral

$$I = \int_0^3 f(x)\mathrm{d}x,$$

given that

$$f(x) = \begin{cases} x & \text{for } 0 \leq x \leq 1; \\ 2 + 2x & \text{for } 1 \leq x \leq 2; \\ x - 1 & \text{for } 2 \leq x \leq 3. \end{cases}$$

7.10 On the assumption that the definite integral

$$\int_a^b \frac{\mathrm{d}x}{\sqrt{(1 - x^2)}} = \arcsin b - \arcsin a,$$

prove that the improper integral

$$I = \int_0^1 \frac{\mathrm{d}x}{\sqrt{(1 - x^2)}}$$

is convergent, and determine its value.

7.11 Sketch the area bounded below by the positive x-axis, and above by the line $y = x$ on the interval $0 \leq x \leq 1$, and by the curve $y = 1/x^2$ on the interval $1 \leq x < \infty$. Determine this area I by the use of an improper integral combined with elementary geometrical arguments.

Section 7.3

7.12 Use Theorem 7.4 to place bounds on the value of the definite integral

$$I = \int_{\pi/4}^{\pi/3} e^{-x^2} \cos^3 x \,\mathrm{d}x.$$

7.13 Evaluate the definite integral

$$\int_{-1}^3 x^2 \,\mathrm{d}x,$$

and use the result to determine the number ξ in Theorem 7.5 when it is applied to this definite integral. Is the number ξ unique? Repeat the argument, but this time applying it to the definite integral

$$\int_{-2}^2 x^2 \,\mathrm{d}x.$$

Is there a unique number ξ in this case?

7.14 Prove the following result, which is a restricted form of the second mean value theorem for integrals. Let $f(x) \geq 0$ be continuous and monotonic decreasing on $[a, b]$, and let $g(x) \geq 0$ be continuous on $[a, b]$. Then

$$\int_a^b f(x)g(x)\,\mathrm{d}x = f(a)\int_a^\xi g(x)\,\mathrm{d}x,$$

where $a < \xi < b$. State the corresponding form of the theorem when $f(x) \geq 0$ is continuous and monotonic increasing on $[a, b]$. [*Hint*: Consider the integrand $f(a)\{f(x)g(x)/f(a)\}$ and use Theorem 7.4.]

7.15 The requirement of continuity for $f(x)$ in Theorem 7.5 is essential, for without it the result of the theorem may, or may not, be true. Illustrate this by considering step functions $f(x)$ defined on the interval [1, 4], and show that it is possible to define ones for which,

(a) no number ξ exists which satisfies Theorem 7.5;
(b) an infinity of numbers ξ exist which satisfy Theorem 7.5.

Section 7.4

7.16 Use Theorem 7.7 to evaluate the following definite integrals:

(a) $\int_a^b (x^{5/2} + 3\mathrm{e}^x)\,\mathrm{d}x$, (b) $\int_0^\pi \sin x\,\mathrm{d}x$, (c) $\int_0^{2\pi} \sin x\,\mathrm{d}x$,

(d) $\int_0^{2\pi} |\sin x|\,\mathrm{d}x$.

7.17 Use Theorem 7.7 to determine the area contained between the x-axis and the curve $y = 1 + x^3 + 2\sin x$, which is bounded to the left by the line $x = 0$ and to the right by the line $x = \pi$.

Section 7.6

7.18 Express in terms of a definite integral the arc length of the curve $y = 1 + x^2 + \sin 2x$, that lies between the points on the curve corresponding to $x = 1$ and $x = 4$.

7.19 Prove that the circumference of a circle of radius a is $2\pi a$ by using the parametric equations of a circle $x = a\cos t$, $y = a\sin t$ with $0 \leq t \leq 2\pi$.

7.20 Find the area contained between the planes $x = -2$ and $x = 3$ of the surface of revolution about the x-axis generated by the curve $y = 2 + \cosh x$. [*Hint*: An antiderivative of $\cosh x$ is $\sinh x + C$.]

7.21 If the curve $y = f(x)$ has an inverse $x = \phi(y)$, state the form taken by Theorem 7.12 when the curve $y = f(x)$ between the points $(a, f(a))$ and $(b, f(b))$ is rotated about the y-axis.

7.22 Determine the volume of revolution contained between the parabola $y = 2 + x + x^2$ and the cubic $y = 5 + 2x + x^3$, which lies between the planes $x = 1$ and $x = 2$.

7.23 If the curve $y = f(x)$ has an inverse $x = \phi(y)$, state the form taken by Theorem 7.13 when the curve $y = f(x)$ between the points $(a, f(a))$ and $(b, f(b))$ is rotated about the y-axis.

Section 7.7

7.24 Show by taking axes that coincide with two adjacent edges of a uniform rectangular lamina that its centre of mass coincides with its mid-point.

7.25 Show that the location of the centre of mass of a uniform wire of mass M that is bent into a semi-circular arc of radius a lies at a distance $2a/\pi$ from the centre of the circle along the radius down to the mid-point of the arc.

7.26 Show that the centre of mass of the part of the solid paraboloid of revolution generated by rotating the curve $y^2 = 4ax$ about the x-axis that lies between the planes $x = 0$ and $x = a$ lies on the x-axis at $x = 2a/3$ from the origin.

7.27 Prove Theorem 7.11(a).

7.28 Show that the moment of inertia about a diameter of a uniform disc of mass M and radius a is $\frac{1}{4}Ma^2$. What is its moment of inertia about an axis normal to its plane that passes through its centre?

7.29 Show that the moment of inertia about its axis of a solid right circular cone of mass M, height h and diameter of base a is $3Ma^2/10$.

7.30 Show that the moment of inertia of a solid hemisphere of mass M and radius a about an axis at right angles to its plane face passing through the boundary of the face is $7Ma^2/5$.

Supplementary computer problems

Using upper and lower sums when estimating the value of a definite integral of $f(x)$ over the interval $a \le x \le b$ has been shown to place bounds on the value of the integral. However, the bounds are likely to be far apart unless the number of sub-intervals used is suitably large, so the actual value of the integral is uncertain. A different approach to approximating the value of such an integral using n sub-intervals Δ_i over the interval of integration $a \le x \le b$ starts by evaluating $f(x)$ at the *mid-point* u_i of each Δ_i. Next a rectangle is constructed over each Δ_i, with height $f(u_i)$, so the graph of $f(x)$ passes through the mid-point of the top of each rectangle. Setting $\xi_i = u_i$ in Definition 7.1, gives the result

$$\int_a^b f(x)\,dx = \lim_{n\to\infty}\sum_{i=1}^{n} f(u_i)\Delta_i.$$

To simplify the subsequent computation, we now take all n sub-intervals to be of equal length, so that $\Delta_i = (b-a)/n$ for $i = 1, 2, \ldots, n$. Then, for any given choice of n, the definite integral is approximated by

$$\int_a^b f(x)\,dx \approx \left(\frac{b-a}{n}\right)\sum_{i=1}^{n} f(u_i). \tag{A}$$

The purpose of the computer problems that follow is to apply approximation (A) to some given functions $f(x)$ to estimate the values of the definite integrals. As estimate (A) lies between the upper and lower sums obtained using the same sub-intervals, in general result (A) can be expected to provide a reasonable approximation to the value of a definite integral.

In Problems 7.31 through 7.34 plot the function $f(x)$ over the appropriate interval for a small number n, say $n \le 8$, superimpose the n rectangles of width $\Delta = (b-a)/n$ on the graph of $f(x)$ to illustrate the relationship of the approximation to the required definite integral. Use approximation (A) to estimate the definite integral for the chosen value of n, while ensuring that if $f(x)$ is discontinuous the discontinuity is located at the *end* of a rectangle. Where possible, evaluate the definite integral analytically, and compare the result with your numerical approximation.

7.31 $f(x) = \dfrac{x\sin x}{1+\sqrt{2x}}, \quad 0 \le x \le \pi.$

7.32 $f(x) = 3 + x\sin 2x, \quad 0 \le x \le \pi.$

7.33 $f(x) = \begin{cases} 2 + 2\sin x, & 0 \le x \le \pi/2 \\ 2 + \sin x, & \pi/2 < x \le \pi. \end{cases}$

7.34 $f(x) = \begin{cases} \frac{1}{4}(3 + 5x^2), & 0 \le x \le 1 \\ \frac{1}{4}(3 + x^2), & 1 < x \le 3. \end{cases}$

In Problems 7.35 through 7.38 plot the graph of the function $f(x)$ to visualize the curve or surface that is involved, and use approximation (A) with $n = 6$ to estimate the required geometrical property.

7.35 Approximate the length of the parabola $f(x) = x^2$ from $x = 0$ to $x = 2$.

7.36 Approximate the area of the surface of revolution between the planes $x = 1$ and $x = 3$ generated by rotating the parabola $y = x^2$ about the x-axis.

7.37 Approximate the volume of the surface of revolution generated by rotating the curve $y = x + \cos x$ about the x-axis that lies between the planes $x = 0$ and $x = \pi/2$.

7.38 Approximate the area of the surface of revolution between the planes $y = \pi/2$ and $y = 3\pi/2$ generated by rotating the curve $x = 2y + \sin 2y$ about the y-axis.

Approximation (A) gives surprisingly good results when used with the above problems, because the functions involved vary comparatively little over the intervals of integration, and they do not change rapidly. If a function oscillates rapidly over its interval of integration, or if it experiences a very large change of magnitude, the approximation will give very poor results unless the value of n is large enough to take account of the variations. The last two problems have been chosen to illustrate these effects.

7.39 Plot the function $f(x) = 2 + \sin(4x^2)$ over the interval $x = 0$ to $x = \pi$ to show its rapid oscillatory behaviour. Use approximation (A) with $n = 6$, 10 and 20 to illustrate the slow convergence of the approximations to the integral over the interval, to the true value 6.6043 obtained by using an accurate numerical integration scheme.

7.40 Confirm that the function $f(x) = \exp(2x^2)$ experiences a large change of magnitude over the interval $x = 0$ to $x = 2$, with most of the change being confined to the last part of the interval. Use approximation (A) with $n = 6$, 10 and 20 to illustrate the slow convergence of the approximations to the integral over the interval, to the true value 403.0200 obtained by using an accurate numerical integration scheme.

8 Systematic integration

To make effective use of integration it is necessary to develop a systematic approach to the integration of standard forms of integral. This is accomplished in this chapter, which discusses integration by substitution, integration by parts, the use of reduction formulae and the use of partial fractions to simplify the task of integrating rational functions. It is shown how these same techniques may be used to find antiderivatives (indefinite integrals) which are just functions, and definite integrals which are numbers. The chapter ends with a discussion of differentiation under the integral sign and the integration of trigonometric functions involving multiple angles.

8.1 Integration of elementary functions

The main objective of this chapter is to explore some of the systematic methods for determining an *antiderivative*, that is, a function $F(x)$ whose derivative is equal to some given function $f(x)$. As described in the previous chapter, we shall denote the antiderivative of the function f by $\int f(x)\,dx$, with the understanding that

$$\int f(x)\,\mathrm{d}x = F(x) + C \tag{8.1}$$

with C an arbitrary constant.

Alternatively, as any *indefinite integral* of f must also be an antiderivative of f, we may identify $F(x)$ in Eqn (8.1) with $\int_a^x f(t)\,\mathrm{d}y$, where a is arbitrary, and incorporate the constant C into the constant resulting from the arbitrary lower limit to obtain the equivalent expression

$$\int f(x)\,\mathrm{d}x = \int_a^x f(t)\,\mathrm{d}t. \tag{8.2}$$

Remember that the symbol $\int f(x)\,\mathrm{d}x$ is derived from differentiation and denotes the most general *function* whose derivative is f. The allied symbol $\int_a^b f(x)\,\mathrm{d}x$, denoting a definite integral of f, derives from integration and is simply a real *number*. Considering the definition of an antiderivative, we shall say that two antiderivatives are equal if they differ only by a constant.

It should be recalled that the connection between the concepts of an antiderivative and a definite integral is provided by the fundamental theorem of calculus, which when related to Eqn (8.1) asserts that

$$\int_a^b f(x)\,\mathrm{d}x = F(b) - F(a). \tag{8.3}$$

Unfortunately, the theorems for the differentiation of wide classes of functions seldom have any counterpart for determining antiderivatives. Ultimately, success in finding an antiderivative depends on whether or not the function f can be so simplified that one may be recognized by using tables of derivatives in reverse: that is, matching the desired derivative f with one in the table, and reading backwards to deduce an antiderivative. Thus, to find the antiderivative of $3 \sec x \tan x$, we first glean from Table 5.2 that

$$\frac{\mathrm{d}}{\mathrm{d}x}(\sec x) = \sec x \tan x$$

or, equivalently,

$$\frac{\mathrm{d}}{\mathrm{d}x}(3 \sec x) = 3 \sec x \tan x$$

showing that the antiderivative is

$$\int 3 \sec x \tan x \,\mathrm{d}x = 3 \sec x + C.$$

In colloquial terms, the process of finding the most general antiderivative of the function $f(x)$ is called the 'integration of $f(x)$'.

Table 8.1 gives a preliminary working list of important integrals which has been compiled from the tables of derivatives in Chapters 5 and 6; a more extensive table is given at the end of the book.

The two separate results shown against number 3 are usually contracted to

$$\int \frac{\mathrm{d}x}{x} = \ln|x| + C,$$

with the tacit understanding that $x \neq 0$, and that the arbitrary constant C differs according to whether x is positive or negative. With obvious modifications, this convention will be extended to include all integrals involving the logarithmic function.

The following statement is equivalent to Eqn (8.1) and it arises as a direct consequence of the definition of an antiderivative. We formulate it as a general theorem.

Theorem 8.1

$$\frac{\mathrm{d}}{\mathrm{d}x}\int f(x)\,\mathrm{d}x = f(x). \quad \blacksquare$$

Table 8.1 Basic table for integrals

	Integral	Condition
1.	$\int x^n \,\mathrm{d}x = \dfrac{x^{n+1}}{n+1} + C$	$n \neq -1$
2.	$\int a^x \,\mathrm{d}x = \dfrac{a^x}{\ln a} + C$	$a \neq 1,\ a > 0$
3.	$\int \dfrac{\mathrm{d}x}{x} = \begin{cases} \ln x + C \\ \ln(-x) + C \end{cases}$	for $x > 0$ for $x < 0$
4.	$\int \mathrm{e}^{ax} \,\mathrm{d}x = \dfrac{1}{a}\mathrm{e}^{ax} + C$	$a \neq 0$
5.	$\int \cos ax \,\mathrm{d}x = \dfrac{1}{a}\sin ax + C$	$a \neq 0$
6.	$\int \sin ax \,\mathrm{d}x = -\dfrac{1}{a}\cos ax + C$	$a \neq 0$
7.	$\int \dfrac{\mathrm{d}x}{\sqrt{(a^2 - x^2)}} = \arcsin\dfrac{x}{a} + C$	for $\lvert x\rvert < \lvert a\rvert$
8.	$\int \dfrac{\mathrm{d}x}{a^2 + x^2} = \dfrac{1}{a}\arctan\dfrac{x}{a} + C$	$a \neq 0$
9.	$\int \dfrac{\mathrm{d}x}{\sqrt{a^2 + x^2}} = \begin{cases} \operatorname{arcsinh}(x/a), \\ \ln(x + \sqrt{a^2 + x^2}) \end{cases}$	$a \neq 0$ $a \neq 0$
10.	$\int \dfrac{\mathrm{d}x}{\sqrt{x^2 - a^2}} = \begin{cases} \operatorname{arccosh}(x/a), \\ \ln\lvert x + \sqrt{x^2 - a^2}\rvert, \end{cases}$	$\lvert x\rvert > a > 0$ $\lvert x\rvert > a > 0$
11.	$\int \dfrac{\mathrm{d}x}{a^2 - x^2} = \begin{cases} \dfrac{1}{a}\operatorname{arctanh}(x/a), \\ \dfrac{1}{2a}\ln\left(\dfrac{x+a}{x-a}\right), \end{cases}$	$\lvert x\rvert < \lvert a\rvert$ $\lvert x\rvert > \lvert a\rvert$
12.	$\int \dfrac{\mathrm{d}x}{x^2 - a^2} = \begin{cases} -\dfrac{1}{a}\operatorname{arccoth}(x/a), \\ \dfrac{1}{2a}\ln\left(\dfrac{x-a}{x+a}\right), \end{cases}$	$\lvert x\rvert < \lvert a\rvert$ $\lvert x\rvert > \lvert a\rvert$

In words, this general result merely asserts the obvious fact that the derivative of the antiderivative of a function $f(x)$ is the function $f(x)$ itself. Its most frequent application is probably to the verification of antiderivatives. For example, let us use the theorem to verify the antiderivative

$$\int \frac{g'\,\mathrm{d}x}{\sqrt{(a^2 - g^2)}} = \arcsin\left(\frac{g}{a}\right) + C, \tag{A}$$

where $g = g(x)$ is some differentiable function of x and $\lvert g\rvert < a$.

$$\frac{d}{dx}\int \frac{g'\,dx}{\sqrt{(a^2 - g^2)}} = \frac{g'}{\sqrt{(a^2 - g^2)}}. \tag{B}$$

Now, differentiating the right-hand side of (A), we find

$$\frac{d}{dx}\left[\arcsin\left(\frac{g}{a}\right) + C\right] = \frac{1}{\sqrt{(1-(g/a)^2}}\cdot\frac{g'}{a}$$

$$= \frac{g'}{\sqrt{(a^2 - g^2)}},$$

which is identical with (B). Thus, (A) is verified.

A final general result of great value is the fact that the derivative of a linear combination of functions is equal to the same linear combination of their derivatives (Theorem 5.4). Expressed in terms of antiderivatives, this implies the following general theorem.

Theorem 8.2 (linearity of integrals) If $f(x)$ and $g(x)$ are two integrable functions and k_1 and k_2 are constants, then

$$\int (k_1 f + k_2 g)\,dx = k_1\int f\,dx + k_2\int g\,dx. \quad \blacksquare$$

It is, of course, this theorem that permits us to simplify many expressions to the point that antiderivatives may be deduced from tables of standard integrals (antiderivatives) such as Table 8.1. Hence we have

$$\int (5x^2 - 2\cos x)\,dx = 5\int x^2\,dx - 2\int \cos x\,dx$$

$$= \frac{5x^3}{3} - 2\sin x + C.$$

The separate arbitrary constants associated with each of the antiderivatives on the right-hand side have been combined into the single arbitrary constant C.

The remaining sections of this chapter are concerned with outlining the details of the main techniques available for finding antiderivatives.

8.2 Integration by substitution

Possibly the most frequently used technique of integration is that in which the variable under the integral sign is changed in a manner which simplifies the task of finding the antiderivative. This process is known as *integration by substitution* or *integration by change of variable*. It is in this technique that the full significance of the symbol dx in Eqn (8.1) is first realized. Indeed, by making a straightforward application of the chain rule for differentiation (Theorem 5.7) we shall shortly establish a simple mechanical rule for effecting a variable change by using differentials.

Integration by substitution depends for its success on transforming an integrand in one variable into one of simpler form in another variable. The choice of the substitution is dictated by experience, as there are no definite rules which can be formulated. As a general rule, a substitution is chosen which leads to the simplification of some difficult feature of the integrand, like a square or a square root. Sometimes more than one substitution is necessary in order to determine an antiderivative, and often integration by substitution needs to be combined with a different integration technique, such as partial fractions or integration by parts, each of which will be discussed later.

To illustrate the method, suppose

$$F(x)=\int f(x)\,\mathrm{d}x, \tag{8.4}$$

and that we wish to change the variable x to a new variable u, where $x=x(u)$ is a given differentiable function of u. From the chain rule (Eqn (5.6)) we have

$$\frac{\mathrm{d}F}{\mathrm{d}u}=\frac{\mathrm{d}F}{\mathrm{d}x}\frac{\mathrm{d}x}{\mathrm{d}u}, \tag{8.5}$$

but differentiation of Eqn (8.4) gives

$$\frac{\mathrm{d}F}{\mathrm{d}x}=f(x), \quad \text{or} \quad \frac{\mathrm{d}F}{\mathrm{d}x}=f(x(u)), \tag{8.6}$$

so combining Eqns (8.5) and (8.6) we find that

$$\frac{\mathrm{d}F}{\mathrm{d}u}=f(x(u))\frac{\mathrm{d}x}{\mathrm{d}u}. \tag{8.7}$$

Consequently, in terms of antiderivatives, this is seen to be equivalent to

$$F(x)=\int f(x)\,\mathrm{d}x=\int f(x(u))\left(\frac{\mathrm{d}x}{\mathrm{d}u}\right)\mathrm{d}u. \tag{8.8}$$

Thus the change of variable from x to u may be accomplished by means of the following simple rule in which:

(i) x is replaced by the corresponding function of u,

(ii) $\mathrm{d}x$ is replaced by $\left(\frac{\mathrm{d}x}{\mathrm{d}u}\right)\mathrm{d}u$, and

(iii) after integrating Eqn (8.8) the variable u is changed back to x.

Although Eqn (8.8) looks more complicated than Eqn (8.4), a suitable choice of the change of variable from x to $x=x(u)$ can make Eqn (8.8) easier to integrate.

Example 8.1 Evaluate the antiderivatives

$$\text{(a)} \quad I = \int \cos(4x+5)\,\mathrm{d}x, \text{ and} \quad \text{(b)} \quad I = \int \frac{3}{(2x-1)^2}\,\mathrm{d}x.$$

Solution (a) Let us set $u = 4x + 5$ to simplify the argument of the cosine function. Differentiating $u = 4x + 5$ with respect to u to find $\mathrm{d}x/\mathrm{d}u$ gives

$$1 = 4\frac{\mathrm{d}x}{\mathrm{d}u}, \qquad \text{so} \qquad \frac{\mathrm{d}x}{\mathrm{d}u} = \frac{1}{4}.$$

Then transforming to the variable u, either by substituting into Eqn (8.8), or by using the rule, gives

$$I = \int \cos(4x+5)\,\mathrm{d}x = \int \cos u \left(\frac{\mathrm{d}x}{\mathrm{d}u}\right)\mathrm{d}u = \frac{1}{4}\int \cos u\,\mathrm{d}u$$
$$= \frac{1}{4}\sin u + C.$$

Transforming back to the original variable x this becomes

$$I = \int \cos(4x+5)\,\mathrm{d}x = \frac{1}{4}\sin(4x+5) + C.$$

(b) Set $u = 2x - 1$ to simplify the integrand. Differentiating $u = 2x - 1$ with respect to u to find $\mathrm{d}x/\mathrm{d}u$ gives

$$1 = 2\frac{\mathrm{d}x}{\mathrm{d}u}, \qquad \text{so} \qquad \frac{\mathrm{d}x}{\mathrm{d}u} = \frac{1}{2}.$$

Again substituting into Eqn (8.8), or using the rule, gives

$$I = \int \frac{3}{(2x-1)^2}\,\mathrm{d}x = \int \frac{3}{u^2}\left(\frac{\mathrm{d}x}{\mathrm{d}u}\right)\mathrm{d}u = \frac{3}{2}\int \frac{1}{u^2}\,\mathrm{d}u$$
$$= -\frac{3}{2}\frac{1}{u} + C.$$

Transforming back to the original variable x, this becomes

$$I = \int \frac{3}{(2x-1)^2}\,\mathrm{d}x = -\frac{3}{2}\frac{1}{(2x-1)} + C. \quad \blacksquare$$

The examples considered above are both special cases of Eqn (8.8) in which the argument of the function f depends on x through the *linear combination* $ax+b$, so that $f = f(ax+b)$.

To evaluate the integral $\int f(ax+b)\,\mathrm{d}x$, we proceed as follows. Setting

$$u = ax + b$$

differentiation with respect to u yields

$$1 = a\frac{dx}{du} \quad \text{or} \quad \frac{dx}{du} = \frac{1}{a},$$

and so Eqn (8.8) reduces to the useful special case

$$\int f(ax+b)\,dx = \frac{1}{a}\int f(u)\,du, \quad \text{with } u = ax+b. \tag{8.9}$$

Example 8.2 Evaluate the antiderivatives

(a) $I = \int \sin^6 x \cos x \, dx$, and

(b) $I = \int (3x^3 + 4x + 1)^8 \left(\frac{9}{2}x^2 + 2\right) dx.$

Solution (a) Setting $u = \sin x$, differentiation with respect to u gives

$$1 = \cos x \frac{dx}{du},$$

or in terms of differentials,

$$du = \cos x \, dx.$$

Thus substitution into Eqn (8.8), or use of the rule, gives

$$I = \int \sin^6 x \cos x \, dx = \int u^6 \, du = \frac{1}{7}u^7 + C = \frac{1}{7}\sin^7 x + C.$$

(b) Set $u = 3x^3 + 4x + 1$ to simplify the integrand. Then differentiation with respect to u gives

$$1 = (9x^2 + 4)\frac{dx}{du},$$

or, in terms of differentials,

$$du = (9x^2 + 4)\,dx.$$

Substitution into Eqn (8.8), or by applying the rule, gives

$$I = \int (3x^3 + 4x + 1)^8 \left(\frac{9}{2}x^2 + 2\right) dx = \frac{1}{2}\int u^8 du = \frac{1}{18}u^9 + C$$

$$= \frac{1}{18}(3x^3 + 4x + 1)^9 + C.$$

Example 8.3 Evaluate the antiderivatives

(a) $I = \int e^{\sin 2x}(1 - 2\sin^2 x)\,dx$, (b) $I = \int \tan x \, dx$,

(c) $I = \int \frac{4x+1}{4x^2+2x-1}\,dx$, and (d) $I = \int \frac{x+1}{x^2+1}\,dx$.

Solution (a) Set $u = \sin 2x$ to simplify the exponential function in the integrand, and make use of the identity $\cos 2x = 1 - 2\sin^2 x$ to relate the factor $(1 - 2\sin^2 x)$ to the cosine of the double angle in x.

Differentiating $u = \sin 2x$ with respect to u gives

$$1 = 2\cos 2x \frac{dx}{du},$$

so in terms of differentials

$$dx = \frac{1}{2\cos 2x}\,du.$$

Substituting into Eqn (8.8), or applying the rule, gives

$$I = \int e^{\sin 2x}(1 - 2\sin^2 x)\,dx = \int e^{\sin 2x}\cos 2x\,dx$$

$$= \frac{1}{2}\int e^u\,du = \frac{1}{2}e^u + C = \frac{1}{2}e^{\sin 2x} + C \text{ (with } u = \sin 2x).$$

(b) Write

$$I = \int \frac{\sin x}{\cos x}\,dx$$

and set $u = \cos x$ to simplify the integrand. Differentiating the substitution with respect to u gives

$$1 = -\sin x \frac{dx}{du},$$

or, in terms of differentials,

$$dx = -\frac{du}{\sin x}.$$

Substituting into Eqn (8.8), or applying the rule, gives

$$I = \int \tan x\,dx = \int \frac{\sin x}{u}\left(\frac{dx}{du}\right)du = -\int \frac{du}{u}$$

$$= -\ln|u| + C = -\ln|\cos x| + C = \ln|\sec x| + C.$$

The form of the last result follows from the fact that $\ln|\sec x| = \ln|1/\cos x| = \ln 1 - \ln|\cos x| = -\ln|\cos x|$.

(c) Set $u = 4x^2 + 2x - 1$ to simplify the integrand. Differentiating the substitution with respect to u gives

$$1 = (8x + 2)\frac{dx}{du},$$

or in terms of differentials

$$dx = \left(\frac{1}{8x + 2}\right)du.$$

Substituting into Eqn (8.8), or applying the rule, gives

$$I = \int \frac{4x + 1}{4x^2 + 2x - 1}\,dx = \int \frac{4x + 1}{u}\left(\frac{dx}{du}\right)du = \frac{1}{2}\int \frac{1}{u}\,du$$

$$= \frac{1}{2}\ln|u| + C = \frac{1}{2}\ln|4x^2 + 2x - 1| + C.$$

(d) Set $u = x^2 + 1$ in an attempt to simplify the integrand. Differentiating the substitution with respect to u gives

$$1 = 2x\frac{dx}{du},$$

or in terms of differentials

$$x\,dx = \frac{1}{2}\,du.$$

Now

$$I = \int \frac{x + 1}{x^2 + 1}\,dx = \int \frac{x}{x^2 + 1}\,dx + \int \frac{1}{x^2 + 1}\,dx,$$

so using Eqn (8.8), or the rule, to simplify the first antiderivative, while recognizing that the second antiderivative is arctan x, gives

$$I = \frac{1}{2}\int \frac{du}{u} + \int \frac{1}{x^2 + 1}\,dx = \frac{1}{2}\ln|u| + \arctan x + C$$

$$= \frac{1}{2}\ln(1 + x^2) + \arctan x + C. \quad \blacksquare$$

Rather more general than the form considered in Eqn (8.4) is the antiderivative

$$I = \int k(x)f[g(x)]\,dx, \tag{8.10}$$

in which it is required to change from the variable x to the variable u by means of the relationship $g(x) = h(u)$, where $h(u)$ is some given differentiable function of u, and $g(x)$ is a differentiable function of x. Then a direct extension of the arguments given above leads to the following rule for the change of variable.

Rule 8.1 (Integration by substitution)

We suppose that in the antiderivative

$$I = \int k(x).f[g(x)]\mathrm{d}x$$

it is required to change from the variable x to the variable u by means of the relationship $g(x) = h(u)$, where g and h are differentiable functions, with $g'(x) \neq 0$. The result may be deduced from I above by:

(a) replacing $g(x)$ in $f[g(x)]$ by $h(u)$;
(b) solving $g(x) = h(u)$ for x and then replacing x in $k(x)$ by this result;
(c) replacing $\mathrm{d}x$ by $\mathrm{d}u$, where $\mathrm{d}u$ is obtained from the differential relationship $g'(x)\,\mathrm{d}x = h'(u)\,\mathrm{d}u$;
(d) replacing x in $g'(x)$ by the result obtained in (b).

We now illustrate the application of Rule 8.1 in a series of examples. Unfortunately, although the rule tells us how to change the variable, it offers us no information on the type of variable change that should be made if a simplification is to be achieved. That is to say, it does not tell us the functional forms to choose for g and h. Only experience can help here.

Example 8.4 Evaluate the antiderivative

$$I = \int x^3\sqrt{(1 + x^2)}\,\mathrm{d}x.$$

Solution This antiderivative is of the general type given in Rule 8.1. First we make the obvious identification $k(x) = x^3$ and then, to remove the square root function which is difficult to manipulate, we shall try setting

$$1 + x^2 = u^2.$$

That is to say, in the hope that it will lead to a simpler expression, we make the further identifications

$$g(x) = 1 + x^2 \quad \text{and} \quad h(u) = u^2.$$

The function f in Rule 8.1 then becomes the square root function, with $\sqrt{(1+x^2)} = u$. Rather than trying to solve for x, for the moment we shall use the result $x^3 = x.x^2 = x(u^2 - 1)$, which gives $x^3\sqrt{(1+x^2)} = xu(u^2 - 1)$.

Now $g'(x) = 2x$ and $h'(u) = 2u$, so that the differential relation $g'(x)\,\mathrm{d}x = h'(u)\,\mathrm{d}u$ gives rise to $x\mathrm{d}x = u\mathrm{d}u$. Hence, in differential form,

$$x^3\sqrt{(1 + x^2)}\,\mathrm{d}x = u(u^2 - 1)x\,\mathrm{d}x = u^2(u^2 - 1)\,\mathrm{d}u,$$

and so by the rule

$$I = \int x^3\surd(1 + x^2)\,\mathrm{d}x = \int u^2(u^2 - 1)\,\mathrm{d}u.$$

The antiderivative on the right-hand side is now straightforward and may be integrated on sight to give

$$I = \frac{u^5}{5} - \frac{u^3}{3} + C$$

or

$$I = \frac{(1 + x^2)^{5/2}}{5} - \frac{(1 + x^2)^{3/2}}{3} + C.$$

Example 8.5 Evaluate the antiderivatives

$$\text{(a)}\quad I = \int \surd(1 + x^2)\,\mathrm{d}x, \quad \text{and} \quad \text{(b)}\quad I = \int \sqrt{1 + 4x^2}\,\mathrm{d}x.$$

Solution (a) In this antiderivative $k(x) \equiv 1$, but it is not immediately clear how best to change the variable. It is left to the reader to see why neither of the possible substitutions $u^2 = 1 + x^2$ or $u = 1 + x^2$ brings about any effective simplification. Instead, let us seek to remove the square root by making the substitution $x = \sinh u$. Then $1 + x^2 = 1 + \sinh^2 u = \cosh^2 u$, so that $\surd(1 + x^2) = \cosh u$. Next, as $g(x) = x$ and $h(u) = \sinh u$, $g'(x) = 1$, $h'(u) = \cosh u$ and so $\mathrm{d}x = \cosh u\ u\mathrm{d}u$. Applying the rule then gives

$$\surd(1 + x^2)\,\mathrm{d}x = \cosh u.\cosh u\,\mathrm{d}u = \cosh^2 u\,\mathrm{d}u,$$

whence

$$I = \int \cosh^2 u\,\mathrm{d}u.$$

Now use the identity $\cosh^2 u = \frac{1}{2}(\cosh 2u + 1)$ to give

$$I = \tfrac{1}{2}\int (\cosh 2u + 1)\,\mathrm{d}u$$

$$= \tfrac{1}{4}\sinh 2u + \frac{u}{2} + C.$$

To return to the variable x it is necessary to use the results $u = \operatorname{arcsinh} x$ and $\cosh u = \surd(1 + x^2)$, together with the identity $\sinh 2u = 2\sinh u \cosh u$, to obtain

$$I = \tfrac{1}{2}[x\surd(1 + x^2) + \operatorname{arcsinh} x] + C.$$

In Section 6.4 we saw that

$$\arcsin x = \ln\,[x + \sqrt{1 + x^2}],$$

so

$$I = \tfrac{1}{2}\{x\sqrt{1+x^2} + \ln[x + \ln\sqrt{1+x^2}]\} + C.$$

(b) This integral can be related to the one in (a) by making the substitution $x = \frac{1}{2}u$, because $dx = \frac{1}{2}du$ and $1+4x^2 = 1+u^2$, so that

$$\int \sqrt{1+4x^2}\,dx = \tfrac{1}{2}\int \sqrt{1+u^2}\,du,$$

and from result (a)

$$\tfrac{1}{2}\int \sqrt{1+u^2}\,du = \tfrac{1}{4}\{u\sqrt{1+u^2} + \ln[u + \sqrt{1+u^2}]\} + C,$$

where because the constant C in (a) is arbitrary, the same symbol will be used for the arbitrary constant in this last result. (Remember that arithmetic is not performed on arbitrary constants, because if C is arbitrary, so also is $\frac{1}{2}C$.)

Returning to the original variable by writing $u = 2x$, we find that

$$\int \sqrt{1+4x^2}dx = \tfrac{1}{2}x\sqrt{1+4x^2} + \tfrac{1}{4}\ln\,[2x + \sqrt{1+4x^2}] + C.$$

Example 8.6 Evaluate the antiderivative

$$I = \int 2x\surd(1+x^2)\,dx.$$

Solution Setting $u = 1+x^2$, it follows that $du = 2x dx$, so that

$$2x\surd(1+x^2)\,dx = \surd u\,du,$$

whence

$$I = \int \surd u\,du = \tfrac{2}{3}u^{3/2} + C$$
$$= \tfrac{2}{3}(1+x^2)^{3/2} + C. \quad \blacksquare$$

A useful special case of Eqn (8.10) arises when $k(x) = g'(x)$, and we set $u = g(x)$. For then, after applying Rule 8.1, we find that

$$\int f[g(x)]g'(x)\,dx = \int f(u)\,du. \tag{8.11}$$

Three cases in which Eqn (8.11) leads to specially simple results occur in the following standard forms which are often useful: (i) $f(u) = u^n$, (ii) $f(u) = \dfrac{1}{u}$, and (iii) $f(u) = e^u$. In case (i) Eqn (8.11) becomes

$$\int [g(x)]^n g'(x)\,dx = \int u^n\,du = \frac{u^{n+1}}{n+1} + C,$$

and thus

$$\int [g(x)]^n g'(x)\,\mathrm{d}x = \frac{1}{(n+1)}[g(x)]^{n+1} + C. \tag{8.12}$$

In case (ii) Eqn (8.11) becomes

$$\int \frac{g'(x)}{g(x)}\,\mathrm{d}x = \int \frac{\mathrm{d}u}{u} = \ln|u| + C,$$

and thus

$$\int \frac{g'(x)}{g(x)}\,\mathrm{d}x = \ln|g(x)| + C. \tag{8.13}$$

In case (iii) Eqn (8.11) becomes

$$\int \mathrm{e}^{g(x)} g'(x)\,\mathrm{d}x = \int \mathrm{e}^u\,\mathrm{d}u = \mathrm{e}^u + C,$$

and thus

$$\int \mathrm{e}^{g(x)} g'(x)\,\mathrm{d}x = \mathrm{e}^{g(x)} + C. \tag{8.14}$$

It will be recognized that the antiderivative in Example 8.2(b) is closely related to the form given in Eqn (8.12) with $g(x) = 3x^3 + 4x + 1$ and $n = 8$. The slight modification obtained by writing it as

$$I = \frac{1}{2}\int (3x^3 + 4x + 1)^8 (9x^2 + 4)\,\mathrm{d}x$$

brings it into direct correspondence with Eqn (8.12), so the antiderivative is seen to be

$$I = \frac{1}{2}\cdot\frac{1}{9}(3x^3 + 4x + 1)^9 + C.$$

Correspondingly, the antiderivatives in Example 8.3(b) to (d) are seen to be of the form given in Eqn (8.13). Thus Example 8.3(c) may be brought into the form of Eqn (8.13) with $g(x) = 4x^2 + 2x - 1$ if it is rewritten as

$$\int \frac{4x+1}{4x^2+2x-1}\,\mathrm{d}x = \frac{1}{2}\int \frac{8x+2}{4x^2+2x-1}\,\mathrm{d}x,$$

from which it follows at once that

$$\int \frac{4x+1}{4x^2+2x-1}\,\mathrm{d}x = \frac{1}{2}\ln|4x^2 + 2x - 1| + C.$$

Apart from a factor, Example 8.3(a) is seen to be of the form given in Eqn (8.14). Using the identity $\cos 2x = 1 - 2\sin^2 x$ to transform the antiderivative we have, as before,

$$I = \int e^{\sin 2x} \cos 2x \, dx = \frac{1}{2} \int e^{\sin 2x}(2 \cos 2x) \, dx.$$

It now follows directly from Eqn (8.14) that

$$\int e^{\sin 2x}(1 - 2 \sin^2 x) \, dx = \frac{1}{2} \int e^{\sin 2x}(2 \cos 2x) \, dx = e^{\sin 2x} + C.$$

It is an immediate consequence of Eqn (8.3) that Rule 8.1 applies to definite integrals provided that the limits are also transformed by the same transformation law. The restatement of Rule 8.1 in terms of definite integrals is as follows.

Rule 8.2 (Integrating definite integrals by substitution)

We suppose that in the definite integral

$$I = \int_a^b k(x).f[g(x)]dx$$

it is required to change from the variable x to the variable u by means of the relationship $g(x) = h(u)$, where g and h are differentiable functions, with $g'(x) \neq 0$. The result may be deduced from I above by:

(a) transforming the differential expression $k(x).\, f[g(x)]dx$ as indicated in Rule 8.1;
(b) solving $g(x) = h(u)$ for u and using the result to replace the limits for x in the definite integral by the corresponding limits for u.

Example 8.7 Evaluate the definite integral

$$I = \int_0^1 x^2 \sqrt{(1 - x^2)} \, dx.$$

Solution Let us make the substitution $x = \sin u$, so that $dx = \cos u \, du$, whence

$$\begin{aligned} x^2 \sqrt{(1 - x^2)} \, dx &= \sin^2 u \cos u \cos u \, du \\ &= \sin^2 u \cos^2 u \, du. \end{aligned}$$

Then, as $u = \arcsin x$, using the principal branch of the sine function, we find from Rule 8.2 that

$$\begin{aligned} I = \int_0^1 x^2 \sqrt{(1 - x^2)} \, dx &= \int_{\arcsin 0}^{\arcsin 1} \sin^2 u \cos^2 u \, du \\ &= \int_0^{\pi/2} \sin^2 u \cos^2 u \, du \end{aligned}$$

To evaluate this last definite integral we use a technique from Chapter 6 which is often helpful. From Definition 6.6 we may write

$$\sin^2 u \cos^2 u = \left(\frac{e^{iu} - e^{-iu}}{2i}\right)^2 \left(\frac{e^{iu} + e^{-iu}}{2}\right)^2$$
$$= \left(\frac{e^{2iu} - 2 + e^{-2iu}}{-4}\right)\left(\frac{e^{2iu} + 2 + e^{-2iu}}{4}\right)$$
$$= \frac{e^{4iu} + e^{-4iu} - 2}{-16},$$

and thus

$$\sin^2 u \cos^2 u = \tfrac{1}{8}(1 - \cos 4u).$$

Using this result in the definite integral, which may then be evaluated on sight, we finally obtain

$$I = \tfrac{1}{8}\int_0^{\pi/2} (1 - \cos 4u)\, du$$
$$= \tfrac{1}{8}\left[u - (\tfrac{1}{4}\sin 4u)\right]\Big|_0^{\pi/2} = \tfrac{1}{16}\pi,$$

and so

$$\int_0^1 x^2\sqrt{(1 - x^2)}\, dx = \tfrac{1}{16}\pi. \quad \blacksquare$$

Example 8.8 Evaluate the definite integral

$$I = \int_0^1 (2x + 5)\cosh(x^2 + 5x + 1)\, dx.$$

Solution Inspection shows that this example is of the form of Eqn (8.11) with the function $f \equiv \cosh$ and $g(x) = x^2 + 5x + 1$.

As $g(0) = 1$, $g(1) = 7$, by setting $u = g(x)$ we at once obtain

$$I = \int_1^7 \cosh u\, du = (\sinh 7 - \sinh 1). \quad \blacksquare$$

Whenever a substitution is chosen, care must always be exercised to ensure that it complies with any mathematical constraints imposed by the integrand. Thus, for example, if $(1 - x^2)^{1/2}$ enters into an integrand it is possible to remove the square root, and so simplify the expression, by using either of the substitutions $x = \sin u$ or $x = \cos u$. This is permissible because for $(1 - x^2)^{1/2}$ to remain real x must be restricted so that $|x| \le 1$, and with these substitutions $|x| \le 1$ for all u. However, these same substitutions would be inappropriate for an integrand in which the expression $(x^2 - 1)^{1/2}$ occurs, because if this is to remain real we must require that $|x| \ge 1$. An appropriate substitution in this case would be $x = \cosh u$, because $\cosh u \ge 1$ for all u.

Table 8.2 Simple substitutions to simplify integrands

Algebraic form occurring in the integrand	*Suggested substitution*
ax	$u=ax$
$ax+b$	$u=ax+b$
a^2+x^2	$x=a\tan u$
$(ax+b)^{1/2}\quad (x\geq -b/a)$	$u^2=ax+b$
$(a^2+x^2)^{1/2}$	$x=a\sinh u$ or, possibly, $x=a\tan u$
$(a^2-x^2)^{1/2}$	$x=a\cos u$ or $x=a\sin u$
$(x^2-a^2)^{1/2}$	$x=a\sec u$ or $x=a\cosh u$

Table 8.2 lists are some simple substitutions which often prove useful. Two special substitutions not given in this list are discussed in sections 8.6.1 and 8.6.2.

8.3 Integration by parts

This most valuable technique is based on Theorem 5.5, concerning the derivative of the product of two functions. That theorem asserts that f, g are two differentiable functions of x, then

$$\frac{\mathrm{d}}{\mathrm{d}x}[f(x)g(x)] = f(x)g'(x)+f'(x)g(x).$$

Taking the antiderivative of this result gives

$$f(x)g(x)=\int f(x)g'(x)\,\mathrm{d}x+\int g(x)f'(x)\,\mathrm{d}x$$

which, on rearrangement, becomes

$$\int f(x)g'(x)\,\mathrm{d}x=f(x)g(x)-\int g(x)f'(x)\,\mathrm{d}x.$$

This is one form of the required result. Using the differential notation $\mathrm{d}f=f'(x)\mathrm{d}x$, $\mathrm{d}g=g'(x)\mathrm{d}x$ enables this to be contracted to the equivalent and easily remembered alternative form

$$\int f\,\mathrm{d}g=fg-\int g\,\mathrm{d}f.$$

These results are now formulated as our next theorem.

Theorem 8.3 (integration by parts) If f, g are differentiable functions of x, then

$$\int f(x)g'(x)\,\mathrm{d}x=f(x)g(x)-\int g(x)f'(x)\,\mathrm{d}x$$

or, expressed in differential notation,

$$\int f\,\mathrm{d}g=fg-\int g\,\mathrm{d}f. \quad \blacksquare$$

This useful theorem is the nearest possible approach to a general theorem for finding the antiderivative of the product of two functions. It depends on the fact that often the antiderivative $\int g\,df$ is easier to determine than the antiderivative $\int f\,dg$. Naturally the technique of integration by substitution can also be employed when evaluating $\int g\,df$.

When definite integrals are involved it is not difficult to see that the result is still valid provided the limits are also applied to the product fg. The general result is as follows.

Theorem 8.4 (integration by parts: definite integral) If f, g are differentiable functions of x in $[a, b]$, then

$$\int_a^b f(x)g'(x)\,dx = f(x)g(x)\big|_a^b - \int_a^b g(x)f'(x)\,dx$$

$$= f(b)g(b) - f(a)g(a) - \int_a^b g(x)f'(x)\,dx. \quad \blacksquare$$

As before, we illustrate both of these theorems by means of examples. These have been carefully chosen to demonstrate a variety of situations in which integration by parts is useful.

Example 8.9 Evaluate the antiderivative

$$I = \int x^k \ln x\,dx \qquad \text{for} \quad x > 0, \; k \neq -1.$$

Solution The problem here, as with all applications of the technique of integration by parts, is to decide upon the functions f and g. A little experimentation will soon convince the reader that I will only simplify if we set $f(x) = \ln x$ and $g(x) = x^{k+1}/(k+1)$, for then $g'(x) = x^k$ and $f'(x) = 1/x$. Accordingly, using the differential notation of Theorem 8.3, we write I in the form

$$I = \int \ln x\,d\left[\frac{x^{k+1}}{k+1}\right].$$

Applying Theorem 8.3 gives

$$I = \frac{x^{k+1}\ln x}{k+1} - \int \frac{x^{k+1}}{k+1}\cdot\frac{1}{x}\,dx$$

$$= \frac{x^{k+1}\ln x}{k+1} - \frac{x^{k+1}}{(k+1)^2} + C.$$

Had the differential notation not been used, the integral I would have been written

$$I = \int \ln x\left\{\frac{d}{dx}\left[\frac{x^{k+1}}{k+1}\right]\right\}dx,$$

after which the argument would have proceeded as before with $f(x) = \ln x$ and $g(x) = x^{k+1}/(k+1)$. ■

Example 8.10 Evaluate the definite integral

$$\int_0^{1/2} \arcsin x \, dx.$$

Solution This time we make the identifications $f(x) = \arcsin x$ and $g(x) = x$ and write

$$\int_0^{1/2} \arcsin x \, d[x] = x \arcsin x \Big|_0^{1/2} - \int_0^{1/2} \frac{x \, dx}{\sqrt{(1-x^2)}}. \tag{C}$$

We have

$$x\arcsin x \Big|_0^{1/2} = \tfrac{1}{2}\arcsin\tfrac{1}{2} - 0 = \pi/12 - 0 = \pi/12,$$

but the definite integral on the right-hand side is still not recognizable. To simplify it let us now set $u = 1 - x^2$ so that $x\,dx = -\frac{1}{2}du$; using Rule 8.2, we obtain

$$\int_0^{1/2} \frac{x\,dx}{\sqrt{(1-x^2)}} = -\tfrac{1}{2}\int_1^{3/4} \frac{du}{\sqrt{u}} = -\tfrac{1}{2}.2u^{1/2}\Big|_1^{3/4} = 1 - \frac{\sqrt{3}}{2}.$$

Combining this result with (C) gives

$$\int_0^{1/2} \arcsin x \, dx = \frac{\pi}{12} + \frac{\sqrt{3}}{2} - 1. \quad ■$$

Example 8.11 Evaluate the antiderivative

$$I = \int e^{ax} \sin bx \, dx.$$

Solution This time we choose to make the identification $f(x) = \sin bx$, $g(x) = (1/a)e^{ax}$ and to write I in the differential form

$$I = \int \sin bx \, d\left(\frac{1}{a} e^{ax}\right).$$

Integrating by parts we find

$$\int e^{ax} \sin bx \, dx = \frac{1}{a} e^{ax} \sin bx - \frac{b}{a}\int e^{ax} \cos bx \, dx.$$

As the integral on the right appears no simpler than the one on the left, let us use this same device on the second term above to obtain

$$\int e^{ax}\sin bx\,dx = \frac{1}{a}e^{ax}\sin bx - \frac{b}{a}\int \cos bx\, d\left(\frac{1}{a}e^{ax}\right)$$

$$= \frac{1}{a}e^{ax}\sin bx - \frac{b}{a^2}e^{ax}\cos bx - \frac{b^2}{a^2}\int e^{ax}\sin bx\,dx + C.$$

Combining terms gives

$$\left(1+\frac{b^2}{a^2}\right)\int e^{ax}\sin bx\,dx = \frac{e^{ax}(a\sin bx - b\cos bx)}{a^2} + C,$$

and so

$$\int e^{ax}\sin bx\,dx = \frac{e^{ax}(a\sin bx - b\cos bx)}{a^2+b^2} + C^*$$

where C^* is related to C by $C^* = a^2C/(a^2+b^2)$, there is no necessity to distinguish between C and C^*, because as C was an arbitrary constant of integration, as is C^*.

8.4 Reduction formulae

It not infrequently happens that an antiderivative I involving a parameter m may be reduced by means of the technique of integration by parts to an expression of similar form in which the parameter has a value differing by an integer k from its original value. If we denote such an antiderivative by I_m, then a typical situation is the one in which we arrive at an expression of the form

$$I_m = A(m) + kI_{m-1}, \tag{8.15}$$

where $A(m)$ is some known function, and k is a constant.

Expressions of this form provide an algorithm for the computation of any antiderivative of the given type once one of them is known, because the I_m are then defined recursively by this relation in terms of I_1, say. It is customary to refer to expressions of the general form of Eqn (8.15) as *reduction formulae*. The same idea is equally applicable, without essential modification, to definite integrals.

Example 8.12 Determine the reduction formula for

$$I_m = \int \cos^m \theta\, d\theta.$$

Use the result to determine I_7.

Solution We rewrite I_m as follows and use integration by parts.

$$I_m = \int \cos^{m-1}\theta \, d(\sin\theta)$$

$$= \cos^{m-1}\theta \sin\theta - \int \sin\theta.(m-1)\cos^{m-2}\theta(-\sin\theta)\, d\theta$$

$$= \cos^{m-1}\theta \sin\theta + (m-1)\int \cos^{m-2}\theta \sin^2\theta \, d\theta$$

$$= \cos^{m-1}\theta \sin\theta + (m-1)\int \cos^{m-2}\theta(1-\cos^2\theta)\, d\theta$$

$$= \cos^{m-1}\theta \sin\theta + (m-1)\int \cos^{m-2}\theta \, d\theta - (m-1)\int \cos^m\theta \, d\theta.$$

Recalling the definition of I_m we discover that this may be re-expressed in terms of I_m and I_{m-2} as

$$I_m = \cos^{m-1}\theta.\sin\theta + (m-1)I_{m-2} - (m-1)I_m,$$

whence we arrive at the required reduction formula

$$I_m = \frac{\cos^{m-1}\theta \sin\theta}{m} + \left(\frac{m-1}{m}\right) I_{m-2}.$$

Setting $m = 7$ gives

$$I_7 = \frac{\cos^6\theta \sin\theta}{7} + \frac{6}{7} I_5$$

$$= \frac{\cos^6\theta \sin\theta}{7} + \frac{6}{7}\left(\frac{\cos^4\theta \sin\theta}{5} + \frac{4}{5} I_3\right)$$

$$= \frac{\cos^6\theta \sin\theta}{7} + \frac{6}{35}\cos^4\theta \sin\theta + \frac{24}{35}\left(\frac{\cos^2\theta \sin\theta}{3} + \frac{2}{3} I_1\right).$$

As $I_1 = \int \cos\theta \, d\theta = \sin\theta + C$ this gives the result

$$\int \cos^7\theta \, d\theta = \frac{1}{7}\cos^6\theta \sin\theta + \frac{6}{35}\cos^4\theta \sin\theta$$

$$+ \frac{8}{35}\cos^2\theta \sin\theta + \frac{16}{35}\sin\theta + C. \quad \blacksquare$$

Example 8.13 Evaluate the definite integral

$$J_m = \int_0^{\pi/2} \cos^m\theta \, d\theta,$$

and deduce its relationship to $\int_0^{\pi/2} \sin^m\theta \, d\theta$.

Solution We can make use of the reduction formula determined in the previous example. It follows from

$$I_m = \frac{\cos^{m-1}\theta \sin\theta}{m} + \left(\frac{m-1}{m}\right) I_{m-2},$$

by imposing the limits of integration on the first term on the right, that the definite integral J_m obeys the reduction formula

$$J_m = \frac{\cos^{m-1}\theta \sin\theta}{m}\bigg|_0^{(1/2)\pi} + \left(\frac{m-1}{m}\right) J_{m-2} = \left(\frac{m-1}{m}\right) J_{m-2}.$$

We must now consider separately even and odd values of m. Firstly, if m is *even*, so that we may write $m = 2n$, then

$$J_{2n} = \frac{2n-1}{2n} \cdot \frac{2n-3}{2n-2} \cdots \frac{1}{2} J_0.$$

Secondly, if m is *odd*, so that we may write $m = 2n+1$, then

$$J_{2n+1} = \frac{2n}{2n+1} \cdot \frac{2n-2}{2n-1} \cdots \frac{2}{3} J_1.$$

So, using the fact that $J_0 = \int_0^{\pi/2} 1\, d\theta = \frac{1}{2}\pi$ and $J_1 = \int_0^{\pi/2} \cos\theta\, d\theta = 1$, we obtain:

$$J_{2n} = \frac{1.3.5\ldots(2n-1)}{2.4.6\ldots 2n} \cdot \tfrac{1}{2}\pi;$$

$$J_{2n+1} = \frac{2.4.6\ldots 2n}{3.5.7\ldots(2n+1)}.$$

Finally, let us prove that

$$J_m = \int_0^{\pi/2} \cos^m x\, dx = \int_0^{\pi/2} \sin^m x\, dx.$$

To achieve this make the variable change $x = \frac{1}{2}\pi - u$ in J_m to obtain

$$\int_0^{\pi/2} \cos^m x\, dx = -\int_{\pi/2}^{0} \cos^m (\tfrac{1}{2}\pi - u)\, du = \int_0^{\pi/2} \cos^m (\tfrac{1}{2}\pi - u)\, du$$

$$= \int_0^{\pi/2} \sin^m u\, du. \quad ■$$

This last example is of some interest historically, as it provided the first infinite product representation for π. One form of the argument used to derive this result proceeds as follows.

It is readily seen from the expressions for J_{2n} and J_{2n+1} that

$$\tfrac{1}{2}\pi = \left[\frac{2.4\ldots 2n}{3.5\ldots(2n-1)}\right]^2 \frac{1}{2n+1}\frac{J_{2n}}{J_{2n+1}}. \qquad (8.16)$$

Now in the interval $(0, \tfrac{1}{2}\pi)$ the following inequalities hold:

$$\sin^{2n-1}x > \sin^{2n}x > \sin^{2n+1}x > 0,$$

so that as

$$J_m = \int_0^{\pi/2} \sin^m x\,\mathrm{d}x,$$

it follows at once that

$$J_{2n-1} \geq J_{2n} \geq J_{2n+1}.$$

This is equivalent to

$$\frac{J_{2n-1}}{J_{2n+1}} \geq \frac{J_{2n}}{J_{2n+1}} \geq 1, \qquad (8.17)$$

but as

$$\frac{J_{2n-1}}{J_{2n+1}} = \frac{2n+1}{2n},$$

we must have

$$\lim_{n\to\infty} \frac{J_{2n-1}}{J_{2n+1}} = 1. \qquad (8.18)$$

By virtue of Eqns (8.17) and (8.18) it also follows that

$$\lim_{n\to\infty} \frac{J_{2n}}{J_{2n+1}} = 1.$$

So, taking the limit of Eqn (8.16) as $n \to \infty$, we arrive at the expression

$$\tfrac{1}{2}\pi = \lim_{n\to\infty}\left(\frac{2}{1}\cdot\frac{2}{3}\cdot\frac{4}{3}\cdot\frac{4}{5}\cdot\frac{6}{5}\cdots\frac{2n-2}{2n-1}\cdot\frac{2n}{2n-1}\cdot\frac{2n}{2n+1}\right). \qquad (8.19)$$

This famous result, called an *infinite product*, was first obtained by the sixteenth-century mathematician John Wallis. If S_n denotes that nth partial product

$$S_n = \frac{2}{1}\cdot\frac{2}{3}\cdot\frac{4}{3}\cdot\frac{4}{5}\cdots\frac{2n-2}{2n-1}\cdot\frac{2n}{2n-1}\cdot\frac{2n}{2n+1}, \qquad (8.20)$$

then the limit in Eqn (8.19) is to be interpreted to mean that $|\tfrac{1}{2}\pi - S_n| \to 0$ as $n \to \infty$.

Reduction formulae may involve more than one parameter, as the final example illustrates.

Example 8.14 Show that

$$I_{m,n} = \int \sin^m x \cos^n x \, dx$$

satisfies the reducion formula

$$(m+n)I_{m,n} = -\sin^{n-1} x \cos^{n+1} x + (m-1)I_{m-2,n}.$$

Solution Write $I_{m,n}$ in the form shown below and integrate by parts.

$$I_{m,n} = \int \sin^{m-1} x \cos^n x \, d(-\cos x)$$

$$= -\sin^{m-1} x \cos^{n+1} x - \int (-\cos x)[(m-1)\sin^{m-2} x \cos^{n+1} x$$

$$-n \sin^m x \cos^{n-1} x] dx$$

$$= -\sin^{m-1} x \cos^{n+1} x + (m-1)I_{m-2,n+2} - nI_{m,n}.$$

Next reduce $I_{m-2,n+2}$ to a simpler form by writing

$$I_{m-2,n+2} = \int \sin^{m-2} x \cos^{n+2} x \, dx$$

$$= \int \sin^{m-2} x \cos^n x (1 - \sin^2 x) \, dx$$

which shows that

$$I_{m-2,n+2} = I_{m-2,n} - I_{m,n}.$$

Using this to eliminate $I_{m-2,n+2}$ from the previous result gives

$$I_{m,n} = -\sin^{m-1} x \cos^{n+1} x + (m-1)I_{m-2,n}$$

$$-(m-1)I_{m,n} - nI_{m,n},$$

or

$$(m+n)I_{m,n} = -\sin^{m-1} x \cos^{n+1} x + (m-1)I_{m-2,n}.$$

8.5 Integration of rational functions – partial fractions

It will be recalled from Chapter 2 that a rational function is a quotient $N(x)/D(x)$, in which $N(x)$ and $D(x)$ are polynomials. Antiderivatives of rational functions are often required, and in this section we indicate ways of expressing rational functions as the sum of simpler expressions, the antiderivatives of which are either known or may be found by standard methods. Our approach to the general problem of finding the antiderivative

$$I = \int \frac{N(x)}{D(x)} \, \mathrm{d}x \tag{8.21}$$

will be first to consider some important special cases. However, before proceeding with our discussion we first recall that the *degree* of a polynomial is the highest power to appear in it. Thus a polynomial of degree 2 is a quadratic, while a polynomial in x of degree 0 is a constant.

(a) Suppose that $N(x)$ is a polynomial of degree 0, and $D(x)$ is a polynomial of degree 1, and consider the case in which

$$\frac{N(x)}{D(x)} = \frac{1}{cx + d}.$$

Then, making the substitution $u = cx + d$, we find

$$\int \frac{\mathrm{d}x}{cx + d} = \frac{1}{c} \int \frac{\mathrm{d}u}{u} = \frac{1}{c} \ln |u| + C$$

and so

$$\int \frac{\mathrm{d}x}{cx + d} = \frac{1}{c} \ln |cx + d| + C.$$

A similar argument establishes that

$$\int \frac{\mathrm{d}x}{(cx + d)^n} = \frac{-1}{c(n - 1)} \cdot \frac{1}{(cx + d)^{n-1}} + C.$$

(b) Suppose $N(x)$ is of degree 0, and $D(x)$ is of degree 2, and consider the case in which

$$\frac{N(x)}{D(x)} = \frac{1}{ax^2 + bx + c}.$$

Then completing the square in the denominator $D(x)$ gives

$$ax^2 + bx + c = a\left[\left(x + \frac{b}{2a}\right)^2 + \left(\frac{c}{a} - \frac{b^2}{4a^2}\right)\right] = a\left[\left(x + \frac{b}{2a}\right)^2 + \alpha\right],$$

where $\alpha = (c/a) - (b^2/4a^2)$ may be positive, negative or zero. Making the variable change $u = x + (b/2a)$ then shows that

$$I = \int \frac{\mathrm{d}x}{ax^2 + bx + c} = \frac{1}{a} \int \frac{\mathrm{d}u}{u^2 + \alpha}.$$

This is a standard integral which may be identified from Table 8.1 once the sign of α has been determined. It will involve either the function arctan or the function arctanh.

(c) Suppose $N(x)$ is of degree 1, and $D(x)$ is of degree 2, so that

$$\frac{N(x)}{D(x)} = \frac{px + q}{ax^2 + bx + c}.$$

Then we can write

$$I = \int \frac{px + q}{ax^2 + bx + c}\,\mathrm{d}x = \int \frac{(p/2a)(2ax + b) + [q - (pb/2a)]}{ax^2 + bx + c}\,\mathrm{d}x,$$

from which we find

$$I = \frac{p}{2a}\int \frac{2ax + b}{ax^2 + bx + c}\,\mathrm{d}x + \left(\frac{2aq - pb}{2a}\right)\int \frac{\mathrm{d}x}{ax^2 + bx + c}.$$

The second antiderivative is the one discussed in (b) above, and by setting $u = ax^2 + bx + c$, the first antiderivative reduces to

$$\int \frac{2ax + b}{ax^2 - bx + c}\,\mathrm{d}x = \int \frac{\mathrm{d}u}{u} = \ln|u| + C = \ln|ax^2 + bx + c| + C.$$

Combining this result with that of case (b) then leads to the desired antiderivative I.

(d) Suppose $N(x)$ is of degree 1, and $D(x)$ is a quadratic raised to the power $n > 1$, so that

$$\frac{N(x)}{D(x)} = \frac{px + q}{(ax^2 + bx + c)^n}.$$

Then, using the identity

$$px + q = \left(\frac{p}{2a}\right)(2ax + b) + \left(q - \frac{pb}{2a}\right)$$

enables us to write

$$\int \frac{px + q}{(ax^2 + bx + c)^n}\,\mathrm{d}x = \left(\frac{p}{2a}\right)\int \frac{2ax + b}{(ax^2 + bx + c)^n}\,\mathrm{d}x + \left(q - \frac{pb}{2a}\right)\int \frac{\mathrm{d}x}{(ax^2 + bx + c)^n}.$$

Setting $u = ax^2 + bx + c$ in the first antiderivative on the right-hand side then leads to

$$\int \frac{2ax + b}{(ax^2 + bx + c)^n}\,\mathrm{d}x = \int \frac{\mathrm{d}u}{u^n} = \left(\frac{-1}{n - 1}\right)\frac{1}{u^{n-1}} + C$$
$$= \left(\frac{-1}{n - 1}\right)\frac{1}{(ax^2 + bx + c)^{n-1}} + C.$$

The second antiderivative on the right-hand side must be evaluated by means of a reduction formula.

In the case $n = 1$ we have the obvious result

$$\int \frac{2ax + b}{ax^2 + bx + c}\,\mathrm{d}x = \ln|ax^2 + bx + c| + C.$$

Having considered a number of special cases, we must now examine how we should proceed when $D(x)$ is any polynomial with real coefficients, and the degree of the polynomial $N(x)$ is less than that of $D(x)$. The coefficient a_0 of the highest power of x in $D(x)$ will be assumed to be unity, since if this is not the case it can always be made so by division of $N(x)$ and $D(x)$ by a_0. Now we know from Corollary 4.1(b) that $D(x)$ may be factorized into real factors of the form

$$D(x) = (x - a)^k (x - b)^l \ldots (x^2 + px + q)^m, \tag{8.22}$$

where $x = a, b, \ldots,$ are real roots with multiplicities $k, l, \ldots,$ and $(x^2 + px + q)^m$ represents an m-fold repeated pair of complex conjugate roots.

Then from elementary algebraic considerations it may be shown that when the degree of $N(x)$ is less than that of $D(x)$ we may always set

$$\begin{aligned}\frac{N(x)}{D(x)} &= \frac{A_1}{(x-a)} + \frac{A_2}{(x-a)^2} + \ldots + \frac{A_k}{(x-a)^k} + \frac{B_1}{(x-b)} \\ &+ \frac{B_2}{(x-b)^2} + \ldots + \frac{B_l}{(x-b)^l} + \ldots \frac{P_1 x + Q_1}{(x^2 + px + q)} \\ &+ \frac{P_2 x + Q_2}{(x^2 + px + q)^2} + \ldots + \frac{P_m x + Q_m}{(x^2 + px + q)^m}.\end{aligned} \tag{8.23}$$

That is to say, every rational fraction may be expressed as a sum of simple functions of the types whose antiderivatives were obtained in cases (a) to (d).

The expression on the right-hand side of Eqn (8.23) is called a *partial fraction* expansion of the rational fraction $N(x)/D(x)$, and the coefficients $A_1, A_2, \ldots P_m, Q_m$ are called *undetermined coefficients*. The undetermined coefficients may be found by cross-multiplication of this expression, followed by equating the coefficients of equal powers of x. Antiderivatives of rational functions $N(x)/D(x)$ may thus be found by a combination of the method of partial fractions and the results of cases (a) to (d).

If the degree of $N(x)$ equals or exceeds that of $D(x)$ by n, where $n = 0, 1, \ldots,$ then the situation may be reduced to the one just described by simply adding to the partial fraction expansion (8.23) the extra terms

$$R_0 + R_1 x + R_2 x^2 + \ldots + R_n x^n.$$

This result is often achieved with less effort by first dividing $N(x)$ by $D(x)$ by long division. The circumstances usually dictate which approach is the easier.

The above results may be combined to form the following set of rules to determine the form of a partial fraction expansion.

Steps for determining the form of a partial fraction expansion

Let it be required to determine the form of the partial fraction expansion of the rational function $N(x)/D(x)$, where $N(x)$ and $D(x)$ are polynomials in x with real coefficients.

1. Factorize $D(x)$ into real linear factors like $(x-a)^r$, if $x=a$ is a real root of $D(x)$ with multiplicity r, and real quadratic factors like $(x^2+px+q)^s$ if the factor x^2+px+q has multiplicity s, and x^2+px+q has complex conjugate factors.
2. For each factor $(x-a)^r$ include the terms

$$\frac{A_1}{(x-a)}+\frac{A_2}{(x-a)^2}+\ldots+\frac{A_r}{(x-a)^r}$$

   in the partial fraction expansion, with $A_1, A_2, \ldots, A_r$ undetermined coefficients.
3. For each factor $(x^2+px+q)^s$ include the terms

$$\frac{P_1x+Q_1}{(x^2+px+q)}+\frac{P_2x+Q_2}{(x^2+px+q)^2}+\ldots+\frac{P_sx+Q_s}{(x^2+px+q)^s}$$

   in the partial fraction expansion, with the P_i and Q_i undetermined coefficients.
4. If the degree of the numerator $N(x)$ equals or exceeds the degree of the denominator $D(x)$ by $r(r=0, 1, 2, \ldots)$, include the terms

$$R_0+R_1x+\ldots+R_rx^r$$

   in the partial fraction expansion, with $R_0, R_1, \ldots, R_r$ undetermined coefficients.
5. The form of the partial fraction expansion appropriate to $N(x)/D(x)$ is then the sum of all the terms produced by steps 2 to 4.
6. Find the undetermined coefficients by equating $N(x)/D(x)$ to the sum of all the partial fractions obtained in step 5, cross-multiplying the result to remove all rational functions and identifying the undetermined coefficients by equating the coefficients of corresponding powers of x on either side of the identity.

Example 8.15 Evaluate

$$I = \int \left(\frac{x^3 + 5x^2 + 9x + 5}{x^2 + 3x + 1} \right) dx.$$

Solution Here, as the degree of $N(x)$ only exceeds that of $D(x)$ by one, we shall start by dividing the integrand to get

$$\frac{x^3 + 5x^2 + 9x + 5}{x^2 + 3x + 1} = x + 2 + \frac{2x + 3}{x^2 + 3x + 1},$$

so that

$$I = \int (x + 2)\, dx + \int \frac{2x + 3}{x^2 + 3x + 1}\, dx.$$

The first antiderivative is trivial, while the second is of the form discussed in case (c), so that

$$I = \frac{x^2}{2} + 2x + \ln |x^2 + 3x + 1| + C.$$

Example 8.16 Evaluate

$$I = \int \frac{x\, dx}{(x + 2)^2 (x - 1)}.$$

Solution In this case we must adopt the partial fraction expansion

$$\frac{x}{(x + 2)^2 (x - 1)} = \frac{A}{x + 2} + \frac{B}{(x + 2)^2} + \frac{C}{x - 1}.$$

Cross-multiplication gives

$$x = A(x + 2)(x - 1) + B(x - 1) + C(x + 2)^2$$

or

$$x = A(x^2 + x - 2) + B(x - 1) + C(x^2 + 4x + 4).$$

Equating coefficients of equal powers of x gives:

coefficient of x^2: $0 = A + C,$
coefficient of x : $1 = A + B + 4C,$
coefficient of x^0: $0 = -2A - B + 4C,$

showing that $A = -1/9$, $B = 2/3$ and $C = 1/9$. We may thus write

$$\int \frac{x\, dx}{(x + 2)^2 (x - 1)} = -\frac{1}{9} \int \frac{dx}{x + 2} + \frac{2}{3} \int \frac{dx}{(x + 2)^2} + \frac{1}{9} \int \frac{dx}{x - 1}.$$

These antiderivatives were all discussed in case (a), so that using those results we obtain

$$I = -\frac{1}{9}\ln|x+2| - \frac{2}{3}\frac{1}{(x+2)} + \frac{1}{9}\ln|x-1| + C.$$

Example 8.17 Find the antiderivative

$$I = \int \frac{x^4 - x^3 + 5x^2 + x + 3}{(x+1)(x^2 - x + 1)^2}\,\mathrm{d}x.$$

Solution Here $N(x) = x^4 - x^3 + 5x^2 + x + 3$ and $D(x) = (x+1)(x^2 - x + 1)^2$, so that the degree of $N(x)$ is 4, and the degree of $D(x)$ is 5. Following on from our earlier reasoning, we must set

$$\frac{x^4 - x^3 + 5x^2 + x + 3}{(x+1)(x^2 - x + 1)^2} = \frac{A}{x+1} + \frac{Bx + C}{x^2 - x + 1} + \frac{Dx + E}{(x^2 - x + 1)^2}.$$

Cross-multiplication gives the identity

$$\begin{aligned} x^4 - x^3 + 5x^2 + x + 3 = A(x^2 - x + 1)^2 \\ + (Bx + C)(x+1)(x^2 - x + 1) \\ + (Dx + E)(x+1). \end{aligned}$$

Instead of expanding the right-hand side and then equating coefficients of equal powers of x as in the previous example, we shall use the fact that $(x+1)$ is a factor of $D(x)$ to simplify this expression. Setting $x = -1$ we find that $9 = 9A$, or $A = 1$ and so

$$\begin{aligned} x^4 - x^3 + 5x^2 + x + 3 = (x^2 - x + 1)^2 + (Bx + C)(x^3 + 1) \\ + (Dx + E)(x + 1), \end{aligned}$$

whence

$$x^3 + 2x^2 + 3x + 2 = (Bx + C)(x^3 + 1) + (Dx + E)(x + 1).$$

Having eliminated A we now proceed as before and equate coefficients of equal powers of x to find B, C, D and E:

coefficient of x^4: $0 = B$,
coefficient of x^3: $1 = C$,
coefficient of x^2: $2 = D$,
coefficient of x : $3 = B + E + D$,
coefficient of x^0: $2 = C + E$.

Thus, $B = 0$, $C = 1$, $D = 2$, $E = 1$ and so

$$\begin{aligned} I &= \int \frac{\mathrm{d}x}{x+1} + \int \frac{\mathrm{d}x}{x^2 - x + 1} + \int \frac{2x + 1}{(x^2 - x + 1)^2}\,\mathrm{d}x \\ &= I_1 + I_2 + I_3. \end{aligned}$$

Now

$$I_1 = \int \frac{dx}{x+1} = \ln|x+1| + C_1$$

and

$$I_2 = \int \frac{dx}{(x-\frac{1}{2})^2 + (\sqrt{3}/2)^2} = \frac{2}{\sqrt{3}} \arctan\left(\frac{2x-1}{\sqrt{3}}\right) + C_2.$$

To evaluate I_3 write

$$\begin{aligned} I_3 &= \int \frac{2x-1}{(x^2-x+1)^2}\, dx + \int \frac{2\, dx}{(x^2-x+1)^2} \\ &= \frac{-1}{(x^2-x+1)} + \int \frac{2\, dx}{[(x-\frac{1}{2})^2 + (\sqrt{3}/2)^2]^2}. \end{aligned}$$

Next, setting $x - \frac{1}{2} = (\sqrt{3}/2)\tan\theta$, so that $dx = (\sqrt{3}/2)\sec^2\theta\, d\theta$, gives

$$\int \frac{2\, dx}{[(x-\frac{1}{2})^2 + (\sqrt{3}/2)^2]^2} = \int \frac{\sqrt{3}\sec^2\theta\, d\theta}{(\frac{3}{4}\sec^2\theta)^2} = \frac{16\sqrt{3}}{9} \int \cos^2\theta\, d\theta.$$

Using the identity $\cos^2\theta = \frac{1}{2}(1+\cos 2\theta)$ this may be evaluated to give

$$\begin{aligned} \int \frac{2\, dx}{[(x-\frac{1}{2})^2 + (\sqrt{3}/2)^2]^2} &= \frac{8\sqrt{3}}{9}[\theta + \tfrac{1}{2}\sin 2\theta] + C_3 \\ &= \frac{8\sqrt{3}}{9}\left\{\arctan\left(\frac{2x-1}{\sqrt{3}}\right) + \frac{\sqrt{3}}{4} \cdot \frac{2x-1}{(x^2-x+1)}\right\} + C_3. \end{aligned}$$

Hence we have shown that

$$I_3 = \frac{-1}{(x^2-x+1)} + \frac{8\sqrt{3}}{9}\left\{\arctan\left(\frac{2x-1}{\sqrt{3}}\right) + \frac{\sqrt{3}}{4} \cdot \frac{2x-1}{(x^2-x+1)}\right\} + C_3.$$

Adding I_1, I_2 and I_3 to find I finally gives

$$I = \ln|x+1| + \frac{14\sqrt{3}}{9} \arctan\frac{2x-1}{\sqrt{3}} + \frac{4x-5}{3(x^2-x+1)} + C.$$

A factor $(x^2-x+1)^3$ in the denominator would have led to $\int\cos^4\theta\, d\theta$ and so, in general, we would obtain antiderivatives of the form $\int\cos^{2n}\theta\, d\theta$. ■

8.6 Other special techniques of integration

A great variety of different methods exist for evaluating particular types of antiderivative, and in this final section we illustrate only a few specially useful ones with the help of some examples. Extensive tables of integrals are readily available and, where possible, they should be used to minimize tedious manipulation.

8.6.1 Substitution $t = \tan(x/2)$

If we write $t = \tan(x/2)$, using the identity

$$\tan 2A = \frac{2\tan A}{1 - \tan^2 A}$$

and setting $A = \frac{1}{2}x$ it follows that

$$\tan x = \frac{2t}{1 - t^2}.$$

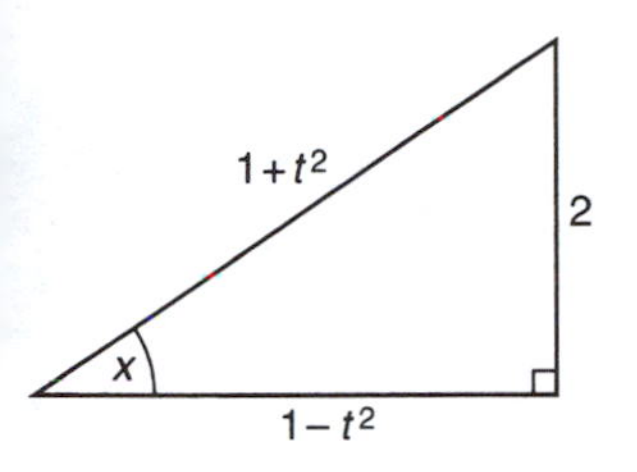

Fig. 8.1 Relationship between x and t in terms of a right-angled triangle.

Consideration of Fig. 8.1 then shows that

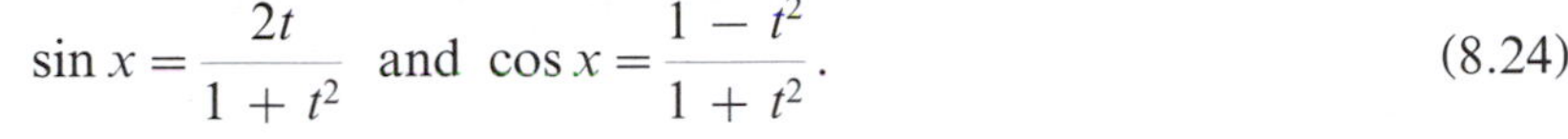

$$\sin x = \frac{2t}{1 + t^2} \quad \text{and} \quad \cos x = \frac{1 - t^2}{1 + t^2}. \tag{8.24}$$

Differentiation of the first result followed by use of the second one establishes the differential relationship

$$\mathrm{d}x = \frac{2\,\mathrm{d}t}{1 + t^2}. \tag{8.25}$$

Consequently, in principle, any rational function $R(\sin x, \cos x)$ that involves only the sine and cosine functions may be transformed by means of (8.24) into a rational function involving t. On account of this result and (8.25), it then follows that

$$I = \int R(\sin x, \cos x)\,\mathrm{d}x = \int R\left[\frac{2t}{1 + t^2}, \frac{1 - t^2}{1 + t^2}\right]\frac{2\mathrm{d}t}{1 + t^2}.$$

Thus I has been transformed into an antiderivative of a rational function involving t.

Example 8.18 Evaluate

$$I = \int \frac{\cos x\,\mathrm{d}x}{1 + \sin x}.$$

Solution Transforming to the variable t as indicated above gives

$$I = \int \frac{2(1 - t^2)}{(1 + t^2)(1 + t)^2}\,\mathrm{d}t.$$

It is readily established that

$$\frac{2(1-t^2)}{(1+t^2)(1+t)^2} = \frac{2}{1+t} - \frac{2t}{1+t^2},$$

showing that

$$I = \int \frac{2\,\mathrm{d}t}{1+t} - \int \frac{2t}{1+t^2}\,\mathrm{d}t$$
$$= \ln(1+t)^2 - \ln(1+t^2) + C.$$

Thus

$$I = \ln\left[\frac{(1+t)^2}{1+t^2}\right] + C = \ln\left[1 + \frac{2t}{1+t^2}\right] + C$$

whence from (8.24),

$$I = \ln(1 + \sin x) + C. \quad \blacksquare$$

8.6.2 Integration of $R[x, \sqrt{(ax^2+bx+c)}]$

We define $R[x, \sqrt{(ax^2+bx+c)}]$ to be a rational function involving x and $\sqrt{(ax^2+bx+c)}$. Special cases of this general type in which $b=0$ have been encountered in Examples 8.5 and 8.7 where it was shown that the substitutions $x = \sin u$ and $x = \sinh u$ can be used to reduce the integrand to one involving only trigonometric or hyperbolic functions. If it is of trigonometric type then the technique of section 8.6.1 may be used to reduce the integrand further to a rational function. If the integrand is of hyperbolic type then the following substitution

$$t = \tanh(x/2),$$

together with

$$\sinh x = \frac{2t}{1-t^2} \quad \text{and} \quad \cosh x = \frac{1+t^2}{1-t^2} \tag{8.26}$$

and the differential relation

$$\mathrm{d}x = \frac{2\,\mathrm{d}t}{1-t^2} \tag{8.27}$$

will again reduce the integrand to a rational function.

The substitutions for $\sinh x$ and $\cosh x$ follow from the result

$$\tanh x = \frac{2\tanh(\frac{1}{2}x)}{1+\tanh^2(\frac{1}{2}x)},$$

and use of the other hyperbolic identities to find expressions for $\sinh x$ and $\cosh x$ in terms of t. The result for $\mathrm{d}x$ follows by differentiation of $\sinh x$ followed by use of the substitution for $\cosh x$.

If $b \neq 0$, then completing the square under the square root sign gives

$$\sqrt{(ax^2+bx+c)} = \sqrt{a\left[\left(x+\frac{b}{2a}\right)^2 + \left(\frac{c}{a}-\frac{b^2}{4a^2}\right)\right]}.$$

The substitution $u = x + (b/2a)$ will then reduce the problem to one of the two special cases just discussed, according to the sign of a and $[(c/a) - (b^2/4a^2)]$.

Example 8.19 Evaluate

$$I = \int \frac{\mathrm{d}x}{\sqrt{(2-3x-4x^2)}}.$$

Solution First we complete the square under the square root sign to obtain

$$I = \int \frac{\mathrm{d}x}{\sqrt{\{4[41/64-(x+3/8)^2]\}}}.$$

Then, setting $u = x + \frac{3}{8}$, this becomes

$$I = \tfrac{1}{2}\int \frac{\mathrm{d}u}{\sqrt{[(41/64)^2-u^2]}} = \tfrac{1}{2}\arcsin\frac{8u}{\sqrt{41}} + C$$

and thus

$$I = \tfrac{1}{2}\arcsin\left(\frac{8x+3}{\sqrt{41}}\right) + C$$

or

$$I = \tfrac{1}{2}\ln[u + \sqrt{1+u^2}] + C,$$

with

$$u = \tfrac{1}{\sqrt{41}}(8x+3). \quad \blacksquare$$

8.6.3 Integration by means of differentiation under integral sign

This approach utilizes the idea of differentiation under the integral sign with respect to a parameter. It relies on finding a known antiderivative involving a parameter α say, with the property that the derivative of its integrand with respect to this parameter α is capable of being simply related to the integrand of the desired antiderivative. Specifically, the method uses the result of Theorem 7.8 that is also true for indefinite integrals, for which it takes the following if

$$F(x,\alpha) = \int f(x,\alpha)\,\mathrm{d}x,$$

then

$$\frac{\partial F(x,\alpha)}{\partial \alpha} = \int \frac{\partial f(x,a)}{\partial \alpha}\,\mathrm{d}x.$$

Example 8.20 Evaluate by means of differentiation under integral sign the antiderivative

$$I = \int \frac{\mathrm{d}x}{(x^2 + a^2)^{3/2}}.$$

Solution We first note that the integrand $1/(x^2+a^2)^{3/2}$ is simply related to the derivative

$$\frac{\partial}{\partial a}\left[\frac{1}{(x^2 + a^2)^{1/2}}\right].$$

Accordingly, let us consider the familiar antiderivative

$$\int \frac{\mathrm{d}x}{(x^2 + a^2)^{1/2}} = \operatorname{arcsinh}\frac{x}{a} + C.$$

Then

$$\frac{\partial}{\partial a}\int \frac{\mathrm{d}x}{(x^2 + a^2)^{1/2}} = \frac{\partial}{\partial a}\left[\operatorname{arcsinh}\frac{x}{a} + C\right],$$

and so

$$-\frac{1}{2}\int \frac{2a\,\mathrm{d}x}{(x^2 + a^2)^{3/2}} = -\left(\frac{x}{a^2}\right)\frac{1}{((x/a)^2 + 1)^{1/2}},$$

or

$$\int \frac{\mathrm{d}x}{(x^2 + a^2)^{3/2}} = \frac{x}{a^2(x^2 + a^2)^{1/2}} + C'.$$

The arbitrary constant C' has been added since we are deducing an antiderivative and not just an indefinite integral. ■

8.6.4 Integration of trigonometric functions involving multiple angles

Antiderivatives of products of trigonometric functions involving multiple angles are of considerable importance and the most frequently occurring ones are:

$$I_1 = \int \sin mx \cos nx\,\mathrm{d}x, \tag{8.28}$$

$$I_2 = \int \sin mx \sin nx\,\mathrm{d}x, \tag{8.29}$$

$$I_3 = \int \cos mx \cos nx\,\mathrm{d}x. \tag{8.30}$$

These are easily evaluated by appeal to the trigonometric identities:

$$\sin mx \cos nx = \tfrac{1}{2}[\sin(m+n)x + \sin(m-n)x], \tag{8.31}$$

$$\sin mx \sin nx = \tfrac{1}{2}[\cos(m-n)x - \cos(m+n)x], \tag{8.32}$$

$$\cos mx \cos nx = \tfrac{1}{2}[\cos(m+n)x + \cos(m-n)x]. \tag{8.33}$$

Substitution of these identities into the above antiderivatives produces:

$$I_1 = \begin{cases} -\dfrac{1}{2}\left[\dfrac{\cos(m-n)x}{(m-n)} + \dfrac{\cos(m+n)x}{(m+n)}\right] + C & \text{for } m^2 \neq n^2 \\ -\dfrac{1}{4m}\cos 2mx + C & \text{for } m = n, \end{cases} \tag{8.34}$$

$$I_2 = \begin{cases} \dfrac{1}{2}\left[\dfrac{\sin(m-n)x}{(m-n)} - \dfrac{\sin(m+n)x}{(m+n)}\right] + C & \text{for } m^2 \neq n^2 \\ \dfrac{1}{2m}(mx - \sin mx \cos mx) + C & \text{for } m = n, \end{cases} \tag{8.35}$$

$$I_3 = \begin{cases} \dfrac{1}{2}\left[\dfrac{\sin(m-n)x}{(m-n)} + \dfrac{\sin(m+n)x}{(m+n)}\right] + C & \text{for } m^2 \neq n^2 \\ \dfrac{1}{2m}(mx + \sin mx \cos mx) + C & \text{for } m = n. \end{cases} \tag{8.36}$$

Example 8.21 Evaluate the following two antiderivatives:

$$I_1 = \int \sin 3x \cos 5x \, dx, \quad I_2 = \int \sin^2 3x \, dx.$$

Solution The antiderivatives follow immediately by substitution in (8.34) and (8.35):

$$I_1 = \frac{\cos 2x}{4} - \frac{\cos 8x}{16} + C, \quad I_2 = \frac{x}{2} - \frac{\sin 3x \cos 3x}{6} + C.$$

8.7 Integration by means of tables

The task of integration can be greatly simplified by the use of extensive tables of integrals which are readily available. These tables list antiderivatives of many combinations of the functions which arise throughout engineering and science. Entries in such tables are usually arranged in order of increasing complexity of the functions forming the integrands. Starting with algebraic functions, they progress through transcendental functions such as the exponential, logarithmic, trigonometric and hyperbolic functions, to combinations of such functions and then to the so-called higher transcendental functions (e.g., Bessel functions). Unless otherwise stated, all parameters in the integrals listed in such tables may be specified arbitrarily.

A reader who has mastered the simple methods of integration outlined in this chapter should proceed to use these same techniques when attempting to reduce a desired integral to a combination of known ones listed in such tables. To assist in this task a short table of integrals has been provided at the end of the book.

Two remarks should be made at this point. The first is that not every function, even though it may appear to be simple, has an antiderivative expressible in terms of known functions. This is because an antiderivative may be considered as another way of defining a function in terms of a given function (the integrand), and it does not necessarily follow that this new function is related to known elementary functions. A typical example of this type is

$$\int x^\alpha \sin^2 x \, \mathrm{d}x,$$

where α is an arbitrary real number. This has no antiderivative which is expressible in terms of known functions.

The second remark is that it is well for the reader to remember that even when an antiderivative is known, it may have more than one functional form. This can have the effect of making the same result appear quite different, depending on which form is used. A simple example of this kind is

$$\int \frac{\mathrm{d}x}{a^2 - x^2} = \begin{cases} \dfrac{1}{a} \operatorname{arctanh} \dfrac{x}{a} + C \\ \dfrac{1}{2a} \ln \left| \dfrac{x + a}{x - a} \right| + C. \end{cases}$$

The first form follows from the arguments used when hyperbolic functions and their inverses were first discussed, and it is valid for $|x| < |a|$. The second form follows by use of partial fractions and is valid for $|x| \neq |a|$, so it is the more general result of the two.

In point of fact, if the first result is supplemented by the result

$$\int \frac{\mathrm{d}x}{a^2 - x^2} = -\frac{1}{a} \operatorname{arccoth} \frac{x}{a} + C,$$

which is valid for $|x| > |a|$, then, taken together, these two results are equivalent to the second one in logarithmic form.

Finally, as an example of a condition imposed on a parameter in an antiderivative, we need only consider

$$\int x^n \, \mathrm{d}x = \frac{x^{n+1}}{n+1} + C \qquad \text{for } n \neq -1.$$

This condition on n is necessary because although the integrand is well defined for $n = -1$, the antiderivative is not then equal to the expression on the right-hand side. For the single case $n = -1$ the antiderivative ceases to be algebraic and becomes logarithmic, when the right-hand side needs to be replaced by $\ln x + C$.

Example 8.22 Use Reference Table 4 at the end of the book to evaluate

$$I = \int \frac{7x^2 + 4x}{(1 + 3x)^2}\,\mathrm{d}x.$$

Solution We have

$$I = 7\int \frac{x^2}{(1 + 3x)^2}\,\mathrm{d}x + 4\int \frac{x}{(1 + 3x)^2}\,\mathrm{d}x.$$

The first integral is of the form given in entry 23 with $a = 1$, $b = 3$, and the second integral is of the form given in entry 22, with the same values for a and b. Thus, making these identifications for a and b, we find

$$I = \frac{7}{27}\left[1 + 3x - \frac{1}{1 + 3x} - 2\ln|1 + 3x|\right] + \frac{4}{9}\left[\frac{1}{1 + 3x} + \ln|1 + 3x|\right] + C.$$

Here, as is usual, the arbitrary constants from each integration have been combined into a single arbitrary constant C. ■

Problems

Section 8.1

8.1 Find the following antiderivatives:

(a) $\int \frac{3\,\mathrm{d}x}{4x^2 - 16}$; (b) $\int \sin 3x\,\mathrm{d}x$; (c) $\int \frac{\mathrm{d}x}{9 - x^2}$;

(d) $\int \frac{\mathrm{d}x}{4 + x^2}$; (e) $\int \frac{1}{3}\cos 4x\,\mathrm{d}x$; (f) $\int 3^x\,\mathrm{d}x$.

8.2 Verify by means of differentiation that

$$\int \frac{\mathrm{d}x}{\sqrt{(x^2 - a^2)}} = \ln|x + \sqrt{(x^2 - a^2)}| + C.$$

Compare this form of result with that shown against entry 10 of Table 8.1.

8.3 Verify by means of differentiation that

$$\int \frac{\mathrm{d}x}{a^2 - b^2x^2} = \frac{1}{2ab}\ln\left|\frac{a + bx}{a - bx}\right| + C.$$

Compare this more general result with those shown against entries 11 and 12 of Table 8.1.

8.4 Verify by means of differentiation that

$$\int \frac{dx}{\sqrt{(a^2 + x^2)}} = \ln[x + \sqrt{(a^2 + x^2)}] + C.$$

Compare this form of result with that shown against entry 9 of Table 8.1.

8.5 Evaluate the following antiderivatives:

(a) $\int (x^2 + 3\sin x + 1)\,dx$; (b) $\int (4^x + 2\cos 2x)\,dx$,

(c) $\int (4\sinh x + \sin x)\,dx$; (d) $\int (e^{ax} + 3)\,dx$.

8.6 Use the following identities to evaluate the four antiderivatives listed below:

$$\sinh mx \cosh mx = \tfrac{1}{2}[\sinh(m+n)x + \sinh(m-n)x]$$

$$\sinh mx \sinh nx = \tfrac{1}{2}[\cosh(m+n)x - \cosh(m-n)x]$$

$$\cosh mx \cosh nx = \tfrac{1}{2}[\cosh(m+n)x + \cosh(m-n)x]$$

(a) $\int \sinh 4x \cosh 2x\,dx$; (b) $\int \sinh x \sinh 3x\,dx$;

(c) $\int \cosh 4x \cosh 2x\,dx$; (d) $\int \cosh^2 2x\,dx$.

Section 8.2

Use the indicated substitutions to evaluate the following antiderivatives.

8.7 $\int \frac{dx}{x\sqrt{(x^2 - 4)}}$, $x = 1/u$.

8.8 $\int \sqrt{(1 - x^2)}\,dx$, $x = \sin u$.

8.9 $\int \frac{\tanh x\,dx}{2\sqrt{(\cosh x - 1)}}$, $\cosh x = 1 + u^2$.

8.10 $\int \cos x \sqrt{\sin x}\,dx$, $\sin x = u$.

8.11 $\int x(3x^2 + 1)^5 dx$, $3x^2 + 1 = u$.

Evaluate the following antiderivatives by means of a suitable trigonometric substitution.

8.12 $\int \frac{\sqrt{(x^2 + 1)}}{x}\,dx$.

8.13 $\int \frac{\sqrt{(x^2-1)}}{x}\,\mathrm{d}x.$

Evaluate the following definite integrals.

8.14 $\int_0^1 (3x^2+1)\sinh(x^3+x+3)\,\mathrm{d}x.$

8.15 $\int_0^1 x^5\sqrt{(1+x^2)}\,\mathrm{d}x.$

8.16 $\int_2^6 \sqrt{(x-2)}\,\mathrm{d}x.$

8.17 $\int_1^2 \left(\frac{4x+6}{x^2+3x+1}\right)\mathrm{d}x.$

Section 8.3

Evaluate the following antiderivatives using the technique of integration by parts.

8.18 $\int \mathrm{e}^{ax}\sin x\,\mathrm{d}x.$

8.19 $\int x\,\mathrm{e}^{ax}\,\mathrm{d}x.$

8.20 $\int \frac{x\,\mathrm{d}x}{\sin^2 x}.$

8.21 $\int \sin x \sinh x\,\mathrm{d}x.$

8.22 $\int x7^x\,\mathrm{d}x.$

8.23 $\int \ln^2 x\,\mathrm{d}x.$

8.24 $\int x \arcsin x\,\mathrm{d}x.$

Section 8.4

8.25 Given that $I_n = \int(1-x^3)^n\,\mathrm{d}x$, where n is an integer, show that

$$(3n+1)I_n = x(1-x^3)^n + 3nI_{n-1}.$$

Hence prove that

$$\int_0^1 (1-x^3)^5 = 3^6/(2^4.7.13).$$

8.26 The integral I_m is defined by

$$I_m = \int_0^{\infty} \frac{x^{2m-1}}{(x^2+1)^{m+3}} \, dx \text{ for integral } m \geq 0.$$

Show that

$$I_{m-1} = \frac{m+2}{m-1} I_m,$$

and by using the substitution $x = \tan\theta$ prove that

$$\int_0^{\pi/2} \sin^7\theta \cos^5\theta \, d\theta = \frac{1}{120}.$$

8.27 If

$$T_n = \int \tan^n\theta \, d\theta,$$

where $n \neq 1$ is a positive integer, show that

$$T_n = \frac{\tan^{n-1}\theta}{n-1} - T_{n-2}.$$

Use this result to evaluate

$$\int_0^{\pi/4} \tan^6\theta \, d\theta.$$

8.28 The function $I_{m,n}$ is defined by

$$I_{m,n} = \int x^m (a+bx)^n \, dx,$$

in which m, n are positive integers. Prove that

$$b(m+n+1)I_{m,n} + maI_{m-1,n} = x^m(a+bx)^{n+1}.$$

Section 8.5

Evaluate the following antiderivatives by means of partial fractions.

8.29 $$\int \frac{x^2 - 5x + 9}{x^2 - 5x + 6} \, dx.$$

8.30 $$\int \frac{dx}{x^3 - 2x^2 + x}.$$

8.31 $\displaystyle\int \frac{x^2 - 8x + 7}{(x^2 - 3x - 10)^2}\,\mathrm{d}x.$

8.32 $\displaystyle\int \left(\frac{6x^4 - 5x^3 + 4x^2}{2x^2 - x + 1}\right)\mathrm{d}x.$

Section 8.6

Evaluate the following antiderivatives by means of the substitution $t = \tan(x/2)$.

8.33 $\displaystyle\int \frac{\mathrm{d}x}{8 - 4\sin x + 7\cos x}.$

8.34 $\displaystyle\int \frac{\sin x}{(1 - \cos x)^3}\,\mathrm{d}x.$

Evaluate the following antiderivatives by means of one or more suitable substitutions.

8.35 $\displaystyle\int \frac{\mathrm{d}x}{\sqrt{(2 + 3x - 2x^2)}}.$

8.36 $\displaystyle\int \frac{3x - 6}{\sqrt{(x^2 - 4x + 5)}}\,\mathrm{d}x.$

8.37 $\displaystyle\int \frac{\mathrm{d}x}{x\sqrt{(1 - x^2)}}.$

Evaluate the following trigonometric antiderivatives.

8.38 $\displaystyle\int \sin ax \sin(ax + \varepsilon)\,\mathrm{d}x,$ a, ε non-zero constants.

8.39 $\displaystyle\int \cos x \cos^2 3x\,\mathrm{d}x.$

8.40 $\displaystyle\int \sin x \sin 2x \sin 3x\,\mathrm{d}x.$

Use the results of this chapter, together with Definitions 7.4 and 7.5 of Chapter 7, to classify the following improper integrals as convergent or divergent. Determine the value of all improper integrals that are convergent, stating any conditions that must be imposed to ensure this.

8.41 $\displaystyle\int_0^1 \frac{dx}{x^\lambda}$.

8.42 $\displaystyle\int_1^\infty \frac{dx}{1+x^2}$.

8.43 $\displaystyle\int_0^\infty \frac{dx}{(1+x)\sqrt{x}}$.

8.44 $\displaystyle\int_0^\infty \cos x \, dx$.

9 Double integrals in Cartesian and plane polar coordinates

9.1 Double integrals in Cartesian coordinates

Double integrals arise in connection with many practical problems, one of the simplest of which is the determination of the volume between a plane area S in the (x, y)-plane and a surface $z = f(x, y)$ lying directly above it. Others involve finding the mass and moment of inertia of a variable density plane lamina, finding the amount of heat flowing through a plane area S in a unit time (the heat flux through S), and the analogous problem of finding the amount of fluid flowing through a plane area S in a unit time when the speed of flow normal to S at each point of S is a function of position in S (the flux of fluid through S).

When introducing the ordinary definite integral in Chapter 7, it was natural to start by using the convenient interpretation of $\int_a^b f(x)\mathrm{d}x$ as the area between the curve $y = f(x)$, the x-axis and the two ordinates $x = a$ and $x = b$. A corresponding geometrical analogy can be used when introducing the double integral written

$$\iint_S f(x, y)\mathrm{d}A, \quad \text{or} \quad \text{sometimes} \int_S f(x, y)\mathrm{d}A, \tag{9.1}$$

where the integrand $f(x, y)$ is a continuous function of the two variables x and y, S is a plane area in the (x, y)-plane over which the integration is to be extended, and $\mathrm{d}A$ is the differential element of area in S.

Let us begin by interpreting a double integral as a volume in order to give one possible physical interpretation of Eqn (9.1). We will consider a volume to be an essentially non-negative quantity, so when determining the volume between the part of the surface $z = f(x, y)$ vertically above an area S in the (x, y)-plane it is necessary to confine attention to surfaces for which $f(x, y) \geq 0$, because if $f(x, y) < 0$ volumes will be negative. Consider Fig. 9.1(a) showing a surface $z = f(x, y)$ lying vertically above a plane area S in the (x, y)-plane bounded by a closed curve Γ in the form of a single loop. Then the volume ΔV_{ij} of the small prism-like volume above the rectangle $x_i \leq x \leq x_{i+1}$ and $y_j \leq y \leq y_{j+1}$ containing an arbitrary point P in the (x, y)-plane with coordinates (ξ_i, η_j) is approximately

$$\Delta V_{ij} \approx f(\xi_i, \eta_j)\Delta A_{ij}, \tag{9.2}$$

where ΔA_{ij} is the area of the rectangle containing P.

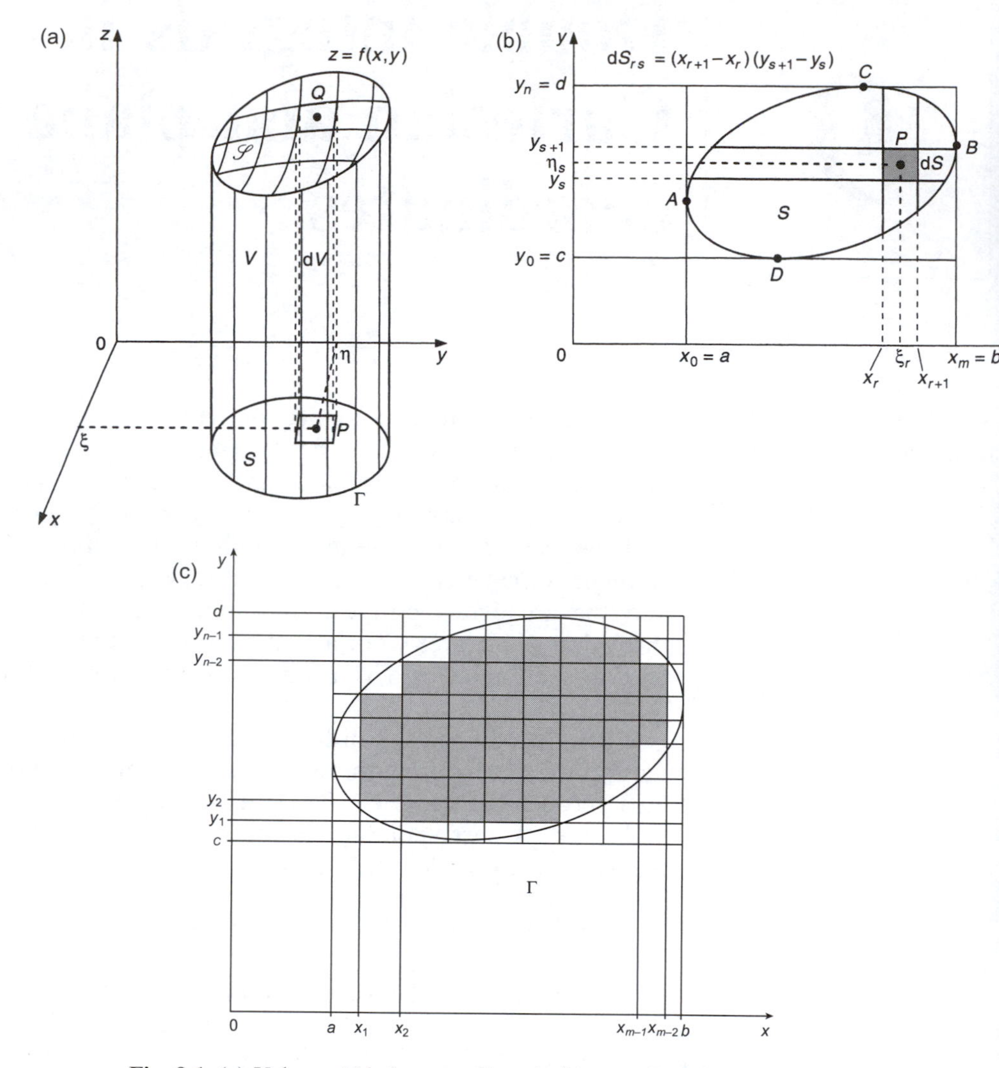

Fig. 9.1 (a) Volume V below $z = f(x, y)$; (b) area S and area element ΔA_{ij}; (c) sub-division of S into rectangles lying entirely inside S.

To proceed further, we enclose area S inside the rectangle $a \le x \le b$ and $c \le y \le d$, as shown in Fig. 9.1(b), and subdivide the interval $a \le x \le b$ into m sub-intervals and the interval $c \le y \le d$ into n sub-intervals by means of the two arbitrary sets of numbers (partitions)

$$a = x_0 < x_1 < \cdots < x_{m-1} < x_m = b$$

and

$$c = y_0 < y_1 < \cdots < y_{n-1} < y_n = d.$$

Drawing lines parallel to the x and y-axes through all points (x_r, y_s) for

$0 \le r \le m$ and $0 \le s \le n$ will define mn sub-rectangles inside the rectangle $a \le x \le b$, $c \le y \le d$. With (ξ_i, η_j) the arbitrary point in the rectangle $x_i \le x \le x_{i+1}$, $y_j \le y \le y_{j+1}$ as in Fig. 9.1(b), by setting $\Delta A_{ij} = (x_{i+1} - x_i)(y_{j+1} - y_j)$, the approximation (9.2) becomes

$$\Delta V_{ij} \approx (x_{i+1} - x_i)(y_{j+1} - y_j) f(\xi_i, \eta_j). \tag{9.3}$$

The volume V lying vertically above area S and below the surface $z = f(x, y)$ is approximated by

$$V \approx \sum_{i=1}^{m} \sum_{j=1}^{m} (x_{i+1} - x_i)(y_{j+1} - y_j) f(\xi_i, \eta_j), \tag{9.4}$$

where the summation only extends over those sub-rectangles lying entirely within S, as shown in Fig. 9.1(c).

The double Riemann integral now follows from approximation (9.4) by proceeding to the limit as $m, n \to \infty$, while allowing the lengths of the subdivisions in the x and y directions to tend to zero. So far, as a double integral has been interpreted as a volume, it has been necessary to confine attention to non-negative integrands $f(x, y)$, but this is an unnecessary restriction when considering double integrals in general. So from now on the integrand $f(x, y)$ in a double integral may be either positive or negative.

We now give the formal definition of a double Riemann integral, usually called more concisely a double integral.

Definition 9.1 (Double Riemann integral in Cartesian coordinates)

The double Riemann integral of a continuous function $f(x, y)$ over an area S in the (x, y)-plane is defined as

$$\iint_S f(x, y)\,\mathrm{d}A = \lim_{m \to \infty,\, n \to \infty} \sum_{i=1}^{m} \sum_{j=1}^{n} (x_{i+1} - x_i)(y_{j+1} - y_j) f(\xi_i, \eta_j),$$

where the lengths of all the x and y- partitions tend to zero as $m, n \to \infty$ and $f(\xi_i, \eta_j)$ is set equal to zero for every sub-rectangle not lying entirely inside S.

Definition 9.1 suffices to define a double integral in analytical terms, but it is not a practical way of evaluating double integrals. However, before discussing how to evaluate double integrals we first draw attention to properties shared by both ordinary integrals and double integrals. These follow directly from Definition 9.1, so they will be stated in the form of a theorem without proof.

Theorem 9.1 (Properties of double integrals) If the functions $f(x, y)$ and $g(x, y)$ are continuous in a finite region S of the (x, y)-plane bounded by a closed curve Γ in the form of a single loop, then

(a) $$\iint_S \{\alpha f(x, y) + \beta g(x, y)\}\mathrm{d}A = \alpha\iint_S f(x, y)\mathrm{d}A + \beta\iint_S g(x, y)\mathrm{d}A,$$

where α and β are arbitrary constants;

(b) if region S is divided into N parts $S_1, S_2, \ldots, S_N$, such that $S = S_1 + S_2 + \ldots + S_N$ with each part bounded by a closed curve in the form of a single loop, then

$$\iint_S f(x, y)\mathrm{d}A = \iint_{S_1} f(x, y)\mathrm{d}A + \iint_{S_2} f(x, y)\mathrm{d}A + \cdots + \iint_{S_N} f(x, y)\mathrm{d}A.$$

These results are both important and useful when evaluating double integrals, as can be seen from the following observations. Setting $\beta = 0$ in result (a) gives

$$\iint_S \alpha f(x, y)\mathrm{d}A = \alpha\iint_S f(x, y)\mathrm{d}A, \tag{9.5}$$

showing that the double integral of a constant multiple of an integrand $f(x, y)$ is equal to the value of the double integral of $f(x, y)$ multiplied by the constant.
Next, setting $\alpha = 1$ and $\beta = -1$ in result (a) shows that

$$\iint_S \{f(x, y) - g(x, y)\}\mathrm{d}A = \iint_S f(x, y)\mathrm{d}A - \iint_S g(x, y)\mathrm{d}A. \tag{9.6}$$

So, for example, if $f(x, y) \geq g(x, y) \geq 0$ everywhere over some area S, result (9.6) implies that the volume between the surfaces $f(x, y)$ and $g(x, y)$ lying vertically above S is the difference between the two double integrals on the right of Eqn (9.6).

The result of Theorem 9.1(b) is useful when a function $f(x, y)$ experiences a finite jump discontinuity across a curve γ that divides S into two parts S_1 and S_2. Because then

$$\iint_S f(x, y)\mathrm{d}A = \iint_{S_1} f(x, y)\mathrm{d}A + \iint_{S_2} f(x, y)\mathrm{d}A.$$

The simplest type of double integral is one in which the area S in the (x, y)-plane is the rectangle $a \leq x \leq b$, $c \leq y \leq d$. To see why this is, suppose y is held constant at $y = \eta$ causing the integrand to reduce to $f(x, \eta)$, which is now only a function of x. Now let δ be a small increment in y and consider the double integral of $f(x, y)$ over the area $a \leq x \leq b$ and $\eta \leq y \leq \eta + \delta$, while for the moment interpreting the double integral as a volume. Then the contribution V_η to the slice of volume V between the planes $y = \eta$ and $y = \eta + \delta$ shown in Fig. 9.3 can be approximated by the product of the thickness δ of the slice and the area in the plane $y = \eta$

beneath the curve $z=f(x, \eta)$, lying above the x-axis, and between the lines $x=a$ and $x=b$.

We may write

$$\Delta V_\eta \approx \delta \int_a^b f(x, \eta)\mathrm{d}x, \tag{9.7}$$

so if the contributions from all such slices in the interval $c \leq y \leq d$ are summed, and we proceed to the limit as the slices become arbitrarily thin, we arrive at the result

$$V = \iint_S f(x, y)\mathrm{d}A = \int_c^d \left[\int_a^b f(x, y)\mathrm{d}x\right]\mathrm{d}y. \tag{9.8}$$

This now tells us how to evaluate the double integral, because first we evaluate the integral $\int_a^b f(x, y)\mathrm{d}x$ while regarding y as a constant, and then we evaluate the resulting integral, which is now only a function of y, over the interval $c \leq y \leq d$.

Clearly, the result would have been the same if we had set $x=\xi$, approximated the volume V_ξ between the planes $x=\xi$ and $x=\xi+\varepsilon$ by the product of the thickness of the slice ε and the area in the plane $x=\xi$ beneath the curve $z=f(\xi, y)$, above y-axis and between the lines $y=c$ and $y=d$.

This gives the equivalent result

$$V = \iint_S f(x, y)\mathrm{d}A = \int_a^b \left[\int_c^d f(x, y)\mathrm{d}y\right]\mathrm{d}x, \tag{9.9}$$

and in this form of the result we first evaluate the integral $\int_c^d f(x, y)\mathrm{d}y$ while regarding x as a constant, and then integrate the result with respect to x over the interval $a \leq x \leq b$.

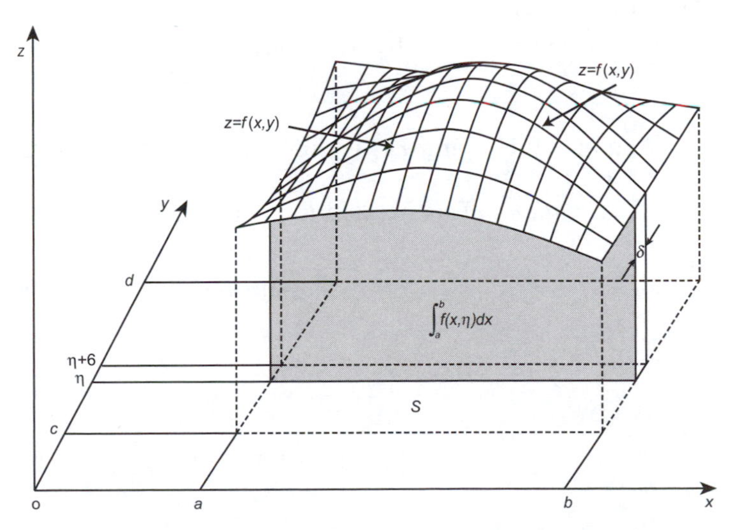

Fig. 9.2 The volume V_η of the slice between the planes $y=\eta$ and $y=\eta+\delta$.

There is no need to restrict these results by requiring that $f(x, y)$ to be non-negative, because this restriction was only introduced to enable us to interpret the double integral as a (positive) volume, so from now on $f(x, y)$ can be any continuous function defined over the area S in the (x, y)-plane.

By convention, the double integral (9.8) is written

$$V = \int_c^d \int_a^b f(x, y) \mathrm{d}x \mathrm{d}y, \tag{9.10}$$

and the double integral (9.9) is written

$$V = \int_a^b \int_c^d f(x, y) \mathrm{d}y \mathrm{d}x. \tag{9.11}$$

When interpreting integrals written in this more compact form it is only necessary to remember that the integral of $f(x, y)$ is evaluated first with respect to the variable occurring in the differential closest to the integrand $f(x, y)$, between the limits closest to the integrand, after which the result is integrated with respect to the other variable between the limits furthest from the integrand.

The double integrals (9.10) and (9.11) are called *iterated integrals*, or sometimes *repeated integrals*, and we have seen that when integration is over a rectangle S, the order in which the integration is performed does not change the result provided the two integrations are evaluated between appropriate limits. We are now in a position to formulate the following theorem.

Theorem 9.2 (Evaluation of iterated integrals in Cartesian coordinates)

If $f(x, y)$ is a continuous function, and integration is to be performed over the rectangle S defined as $a \leq x \leq b$, $c \leq y \leq d$, then

$$\iint_S f(x, y) \mathrm{d}A = \int_c^d \int_a^b f(x, y) \mathrm{d}x \mathrm{d}y = \int_a^b \int_c^d f(x, y) \mathrm{d}y \mathrm{d}x.$$

Example 9.1 Find $I = \iint_S \sin\{\pi(x + y)\} \mathrm{d}A$, if S is the rectangle $0 \leq x \leq 1$, $0 \leq y \leq \frac{1}{2}$.

Solution This iterated integral can be written

$$I = \int_0^1 \int_0^{1/2} \sin\{\pi(x + y)\} \mathrm{d}y \mathrm{d}x = \int_0^1 \left[\int_0^{1/2} \sin\{\pi(x + y)\} \mathrm{d}y \right] \mathrm{d}x.$$

However

$$\int_0^{1/2} \sin\{\pi(x + y)\} \mathrm{d}y = \frac{1}{\pi} \{\cos \pi x - \cos(\pi(x + \tfrac{1}{2}))\},$$

so

$$I = \int_0^1 \frac{1}{\pi}\{\cos \pi x - \cos(\pi(x + \tfrac{1}{2}))\}\mathrm{d}x = \frac{1}{2\pi^2}.$$

In this case, because of the symmetry of the integrand with respect to x and y, reversing the order of integration this double integral would not have simplified the calculation. ■

Example 9.2 Find $I = \iint_S (x + y)\mathrm{e}^{(x+y)}\mathrm{d}A$, if S is the rectangle $0 \le x \le 1$, $1 \le y \le 3$.

Solution This iterated integral can be written

$$I = \int_1^3 \int_0^1 (x + y)\mathrm{e}^{(x+y)}\mathrm{d}x\mathrm{d}y = \int_1^3 \left[\int_0^1 (x + y)\mathrm{e}^{(x+y)}\mathrm{d}x\right]\mathrm{d}y.$$

However,

$$\int_0^1 (x + y)\mathrm{e}^{(x+y)}\mathrm{d}x = \mathrm{e}^y(1 - y) + y\mathrm{e}^{(1+y)},$$

so

$$I = \int_1^3 \{\mathrm{e}^y(1 - y) + y\mathrm{e}^{(1+y)}\}\mathrm{d}y = 2\mathrm{e}^4 - \mathrm{e}^3 - \mathrm{e}.$$

Notice that, by way of illustration, this iterated integral was evaluated with the integrations in the reverse order to those used in Example 9.1. On occasions the evaluation of one of the two equivalent forms of double integral in Theorem 9.2 is easier than the other. ■

The next double integral is one where only one form of integral in Theorem 9.2 can be used, because the other form involves an integrand for which there is no known antiderivative.

Example 9.3 Find $\iint_S x\cos(xy)\mathrm{d}A$ if S is the area $0 \le x \le \pi/2$, $0 \le y \le 1$.

Solution Writing the double integral as $\int_0^{\pi/2}\mathrm{d}x\int_0^1 x\cos(xy)\mathrm{d}y$, and using the fact that when x is regarded as a constant $\frac{\mathrm{d}}{\mathrm{d}y}(\sin(xy)) = x\cos(xy)$, we find that

$$\int_0^{\pi/2}\mathrm{d}x\int_0^1 x\cos(xy)\mathrm{d}y = \int_0^{\pi/2}[\sin(xy)]_{y=0}^1\mathrm{d}x = \int_0^{\pi/2}\sin x\,\mathrm{d}x = 1.$$

Had the integration been performed in the reverse order it would have given

$$\int_0^1 \mathrm{d}y\int_0^{\pi/2} x\cos(xy)\mathrm{d}x = \int_0^1 \left[\frac{2\cos(\pi y/2) + \pi y\sin(\pi y/2) - 2}{2y^2}\right]\mathrm{d}y,$$

and this integrand has no known antiderivative. ■

We are now in a position to lift the restriction that the region S over which integration is to be performed is a rectangle. Fist though, it is important to make quite clear the nature of the region over which integrations are to be performed. Consider Fig. 9.3(a), which shows a region S bounded by a closed curve Γ in the form of a single loop. Curves such as these are called *simple curves*, to distinguish them from curves which intersect themselves called a *non-simple curves*. A typical non-simple curve is one shaped like a figure eight, inside each loop of which lies a different region.

A region S bounded by a closed simple curve Γ is said to be *simply connected* if every closed simple curve drawn inside Γ only contains points of S. This distinguishes it from what is called a *multiply connected region*, like the region inside an annulus between two concentric circles. Such a region is *not* simply connected because any closed simple curve in S drawn around the inner circular boundary of the annulus will contain both points of S and points inside the inner circle that do not belong to S.

Returning now to Fig. 9.3(a), and considering the double integral of a continuous function $f(x, y)$, we see that curve Γ is double valued, because for each x in the interval $a < x < b$ there are two values of y. To overcome this ambiguity, let Γ be described in terms of the arcs ADB and ACB as follows:

$$\Gamma = \begin{cases} y = g_1(x) & \text{for } a \le x \le b \text{ defining arc } ADB \\ y = g_2(x) & \text{for } a \le x \le b \text{ defining arc } ACB. \end{cases} \tag{9.12}$$

We now consider the strip dS in Fig. 9.3(a) taken parallel to the y-axis. Remembering how a double integral was defined over a rectangle, the generalization to region S in Fig. 9.3(a) is seen to be as follows. First integrate $f(x, y)$ over strip dS with respect to y, holding x constant at $x = \xi$, and then free the variable x and integrate the result that is now only a function of x with respect to x over the interval $a \le x \le b$. The difference now is that when integrating over dS the limits for y are $y = g_1(\xi)$ and $y = g_2(\xi)$. Thus the double integral of a continuous function $f(x, y)$ over S becomes

$$\iint_S f(x, y)\mathrm{d}A = \int_a^b \left[\int_{g_1(x)}^{g_2(x)} f(x, y)\mathrm{d}y \right] \mathrm{d}x. \tag{9.13}$$

This is not the only way this double integral can be evaluated, as can be seen from Fig. 9.3(b) where now the strip dS is parallel to the x-axis, and Γ is described in terms of the arcs DAC and DBC as follows:

$$\Gamma = \begin{cases} x = h_1(y) & \text{for } c \le y \le d \text{ defining arc } DAC \\ x = h_2(y) & \text{for } c \le y \le d \text{ defining arc } DBC. \end{cases} \tag{9.14}$$

This time integration over strip dS is first performed with respect to x, holding $y = \eta$ constant, and the result which is a function of y is integrated with respect to y over the interval $c \le y \le d$. This leads directly to the second form of the same double integral

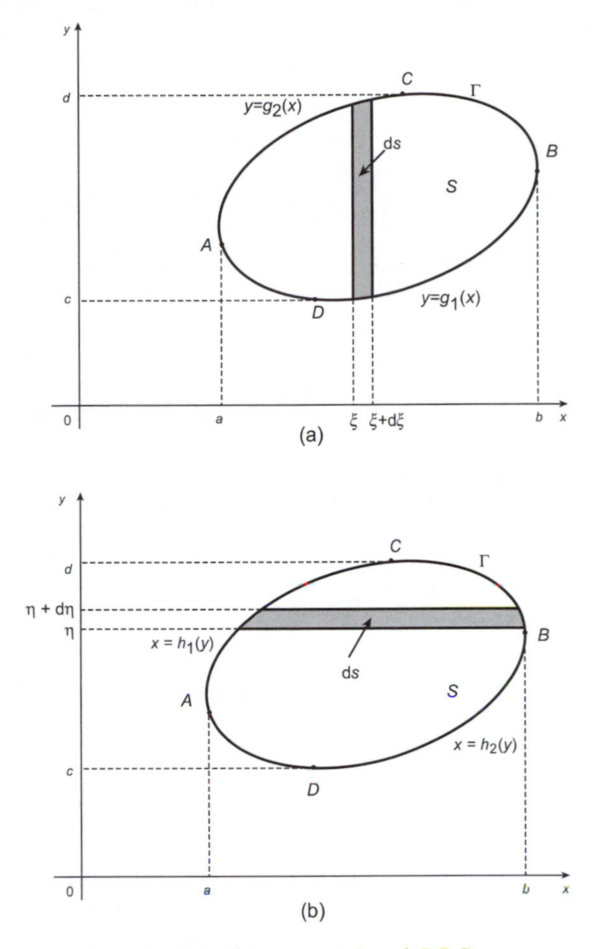

Fig. 9.3 (a) Arcs ADB and ACB; (b) arcs DAC and DBC.

$$\iint_S f(x, y)\mathrm{d}A = \int_c^d \left[\int_{h_1(y)}^{h_2(y)} f(x, y)\mathrm{d}x \right] \mathrm{d}y. \tag{9.15}$$

Alternative more concise conventions used for the iterated integral (9.13) are

$$\int_a^b \int_{g_1(x)}^{g_2(x)} f(x, y)\mathrm{d}y\mathrm{d}x \quad \text{and} \quad \int_a^b \mathrm{d}x \int_{g_1(x)}^{g_2(x)} f(x, y)\mathrm{d}y, \tag{9.16}$$

with corresponding versions for (9.15), but in what follows we will only use the second form as it makes quite clear the order in which the integrations are to be performed while holding the second variable constant. We use the second form of notation in (9.16) and combine results (9.13) and (9.15) to form the following theorem.

> ***Theorem 9.3*** (iterated integrals with variable limits) If the function $f(x, y)$ is continuous in a finite region S bounded by a simple curve Γ described as in (9.12) or (9.14), the double integral and the two iterated integrals are all equal, so
>
> $$\iint_S f(x, y)\mathrm{d}A = \int_a^b \mathrm{d}x \int_{g_1(x)}^{g_2(x)} f(x, y)\mathrm{d}y = \int_c^d \mathrm{d}y \int_{h_1(y)}^{h_2(y)} f(x, y)\mathrm{d}x. \quad \blacksquare$$

Example 9.4 Evaluate the double integral $\iint_S (x^2 + y^2)\mathrm{d}A$ when S is the area between the x-axis, the parabola $y = x^2$ and the line $x = 2$.

Solution The function $f(x, y)$ is continuous in the region S bounded by a simple curve shown in Fig. 9.4.

From the first result in Theorem 9.3 we find that

$$\iint_S (x^2 + y^2)\mathrm{d}A = \int_0^2 \mathrm{d}x \int_0^{x^2} (x^2 + y^2)\mathrm{d}y = \int_0^2 \left(x^4 + \frac{x^6}{3}\right)\mathrm{d}x = \frac{1312}{105}.$$

Alternatively, using the second result in Theorem 9.3, we have

$$\iint_S (x^2 + y^2)\mathrm{d}A = \int_0^4 \mathrm{d}y \int_{\sqrt{y}}^2 (x^2 + y^2)\mathrm{d}x$$

$$= \int_0^1 \left(\frac{8}{3} + 2y^2 - \frac{1}{3}y^{3/2} - y^{5/2}\right)\mathrm{d}y = \frac{1312}{105}.$$

This is an example where the first form of iterated integral happens to be slightly easier to evaluate than the second. So, in general, it is best to examine both forms of the integral to find the simplest before proceeding with the calculation. $\blacksquare$

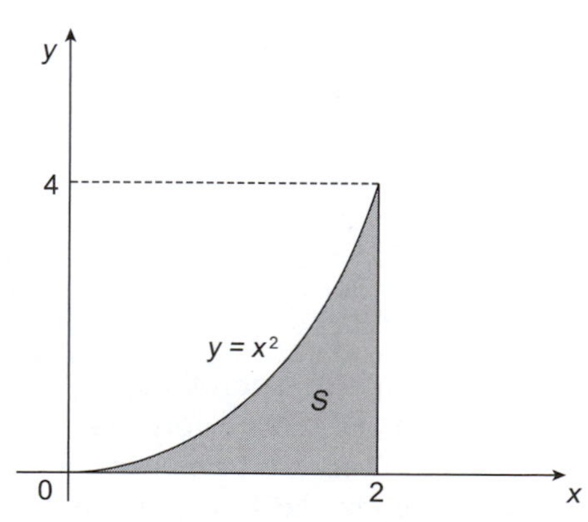

Fig. 9.4 The area S between the parabola $y = x^2$, the x-axis and the line $x = 2$.

Example 9.5 Find the volume V_+ of the part of the ellipsoid

$$\frac{x^2}{a^2}+\frac{y^2}{b^2}+\frac{z^2}{c^2}=1$$

that lies above the (x, y)-plane, and use the result to find the volume V of the complete ellipsoid.

Solution The surface of the ellipsoid above the (x, y)-plane is shown in Fig. 9.5. We have

$$V_+ = \iint_S f(x, y)\mathrm{d}S$$

where S is the closed elliptical region given by

$$\frac{x^2}{a^2}+\frac{y^2}{b^2}\leq 1, \quad \text{and} \quad f(x, y)=c\sqrt{1-\frac{x^2}{a^2}-\frac{y^2}{b^2}}.$$

Thus

$$V_+ = c\int_{-a}^{a}\mathrm{d}x\int_{-b\sqrt{1-(x/a)^2}}^{b\sqrt{1-(x/a)^2}}\sqrt{1-\frac{x^2}{a^2}-\frac{y^2}{b^2}}\mathrm{d}y.$$

To perform the integration with respect to y, while holding x constant, we simplify the integral by making the substitutions

$$y=b\sqrt{1-(x/a)^2}\sin t \quad \text{and} \quad \mathrm{d}y=b\sqrt{1-(x/a)^2}\cos t\,\mathrm{d}t,$$

with $-\frac{1}{2}\pi \leq t \leq \frac{1}{2}\pi$.

As a result we obtain

$$V_+ = bc\int_{-a}^{a}\mathrm{d}x\int_{-\pi/2}^{\pi/2}\sqrt{\left(1-\frac{x^2}{a^2}\right)-\left(a-\frac{x^2}{a^2}\right)\sin^2 t}\sqrt{1-\frac{x^2}{a^2}}\cos t\,\mathrm{d}t,$$

which after simplification becomes

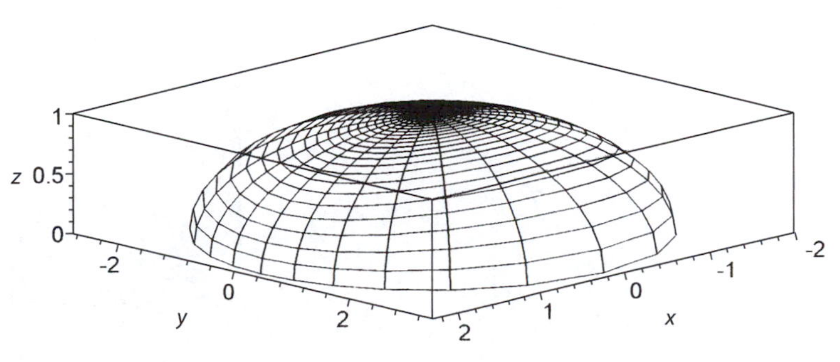

Fig. 9.5 The half-ellipsoid above the (x, y)-plane with $a=2$, $b=3$, $c=1$.

$$V_+ = bc\int_{-a}^{a}\left(1-\frac{x^2}{a^2}\right)\mathrm{d}x\int_{-\pi/2}^{\pi/2}\cos^2 t = \left(\frac{4abc}{3}\right)\left(\frac{\pi}{2}\right) = \frac{2}{3}\pi abc.$$

From symmetry, the volume V of the complete ellipsoid is $2V_+$, so

$$V = \frac{4}{3}\pi abc.$$

When $a = b = c$ the ellipsoid becomes a sphere, and the above expression reduces to the well known result that the volume of a sphere of radius a is

$$V = \frac{4}{3}\pi a^3. \quad ■$$

Example 9.6 Find the double integral

$$\iint_S xy^2 \mathrm{d}A,$$

where S is the area in the first quadrant inside the circle $x^2 + y^2 = 1$.

Solution The region S is shown in Fig. 9.6. We choose to integrate first with respect to y, regarding x as a constant, so that y is to be integrated over the interval $0 \le y \le \sqrt{1-x^2}$, after which x is to be integrated over the interval $0 \le x \le 1$. This gives

$$\iint_S xy^2 \mathrm{d}A = \int_0^1 \mathrm{d}x \int_0^{\sqrt{1-x^2}} xy^2 \mathrm{d}y = \int_0^1 \left[x\frac{y^3}{3}\right]_{y=0}^{y=\sqrt{1-x^2}} \mathrm{d}x$$

$$\int_0^1 x(1-x^2)^{3/2}\mathrm{d}x = \frac{1}{15},$$

where the substitution $u^2 = 1 - x^2$ was used when evaluating the last

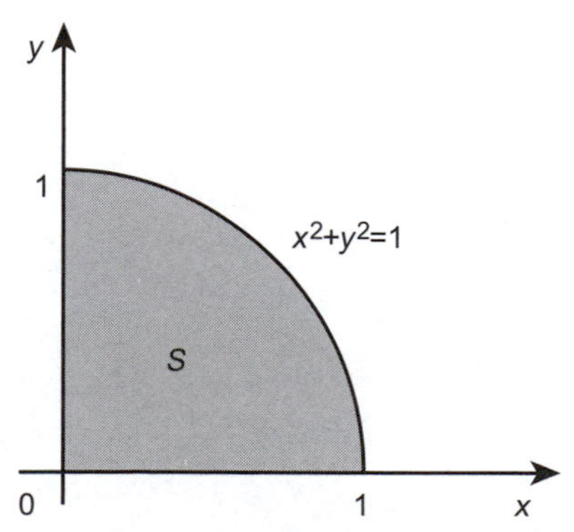

Fig. 9.6 The region S in the first quadrant inside the circle $x^2 + y^2 = 1$.

integral. When we come to consider double integrals in terms of polar coordinates we will see this same result can be obtained very simply. ■

Example 9.7 Use a double integral to find the moment of inertia I_Y about the Y-axis of the triangular lamina considered in Example 7.12 and illustrated in Fig. 7.13(b).

Solution In Example 7.12 the top side of the triangular lamina was found to have the equation $y = b(1 - x/a)$, and the lower side the equation $y = c(x/a - 1)$. The mass per unit area was found to be $2M/[a(b+c)]$, so the mass $\mathrm{d}M$ of an element of area $\mathrm{d}x\mathrm{d}y$ will be $\mathrm{d}M = 2M\,\mathrm{d}x\mathrm{d}y/[a(b+c)]$. Thus

$$I_Y = \iint_S x^2 \mathrm{d}M = \frac{2M}{a(b+c)} \int_0^a \mathrm{d}x \int_{c(x/a-1)}^{b(1-x/a)} x^2 \mathrm{d}y,$$

so

$$I_Y = \frac{2M}{a(b+c)} \int_0^a x^2 \left[\int_{c(x/a-1)}^{b(1-x/a)} 1 \cdot \mathrm{d}y \right] \mathrm{d}x$$

$$\frac{2M}{a} \int_0^a \left(x^2 - \frac{x^3}{a} \right) \mathrm{d}x = \frac{2M}{a} \cdot a^3 \left(\frac{1}{3} - \frac{1}{4} \right) = \frac{Ma^2}{6}.$$

This is the result that was obtained in Example 7.12 using a method that avoided double integrals. ■

To close this brief introduction to double integrals in Cartesian coordinates, it is necessary that something should be said about double integrals involving regions bounded by non-simple curves Γ. A typical example is shown in Fig. 9.7.

It will be sufficient to consider the annular region S shown in Fig. 9.8 which is non-simple because not every closed simple curve drawn in S will only contain points that are in S. This can be seen from the dashed closed

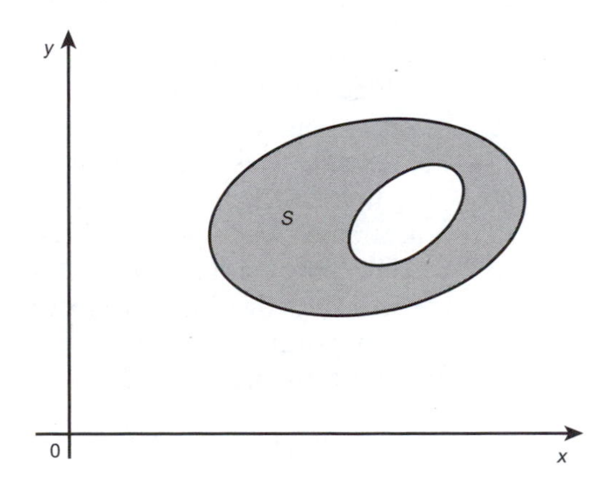

Fig. 9.7

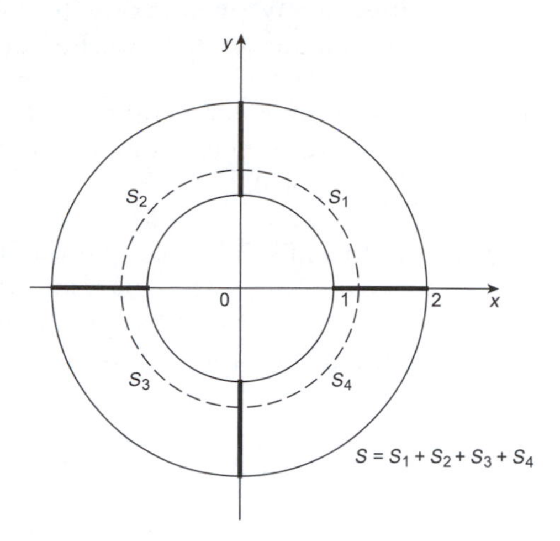

Fig. 9.8 The sub-division of an annular region S into four simply connected regions.

simple curve in Fig. 9.8 which, although drawn entirely in S, contains points both in S and points inside the inner circle that do not belong to S.

The difficulty when working with multiply connected regions is overcome by sub-dividing them into regions that are simply connected, integrating over each in turn, and adding the results. Such a sub-division involving four simply connected regions S_1, S_2, S_3 and S_4 belonging to the annular region contained between the circles $x^2+y^2=1$ and $x^2+y^2=4$ is shown in Fig. 9.8.

Example 9.8 Find the double integral

$$I = \iint_S xy^2 \mathrm{d}A,$$

where S is the area in the upper half of the (x, y)-plane contained between the circles $x^2+y^2=1$ and $x^2+y^2=4$.

Solution The area S is the part of the annulus containing regions S_1 and S_2 shown in Fig. 9.8. Consideration of the symmetry of the annulus about the y-axis, coupled with the fact that the integrand xy^2 is an odd function of x while y^2 is non-negative, shows that the integral over S_1 will be the negative of the integral over S_2, so the value of the definite integral over the area $S=S_1+S_2$ must be zero. Let us prove this by direct integration.

The simplest approach is to use the reasoning in Example 9.6, and to regard I_1 as the difference between the integral of xy^2 over the area in the first quadrant inside $x^2+y^2=4$, and the integral of xy^2 over the area in the first quadrant inside $x^2+y^2=1$, so that

$$I_1 = \iint_{S_1} xy^2 \mathrm{d}A = \int_0^2 \mathrm{d}x \int_0^{\sqrt{4-x^2}} xy^2 \mathrm{d}y - \int_0^1 \mathrm{d}x \int_0^{\sqrt{1-x^2}} xy^2 \mathrm{d}y$$

$$= \int_0^2 \left[x\frac{y^3}{3} \right]_{y=0}^{y=\sqrt{4-x^2}} \mathrm{d}x - \int_0^1 \left[x\frac{y^3}{3} \right]_{y=0}^{\sqrt{1-x^2}} \mathrm{d}x = \tfrac{32}{15} - \tfrac{1}{15} = \tfrac{31}{15}.$$

The integral I_2 over S_2 follows in similar fashion as

$$I_2 = \iint_{S_2} xy^2 \mathrm{d}A = \int_{-2}^{0} \mathrm{d}x \int_0^{\sqrt{4-x^2}} xy^2 \mathrm{d}y - \int_{-1}^{0} \mathrm{d}x \int_0^{\sqrt{1-x^2}} xy^2 \mathrm{d}y$$

$$= \int_{-2}^{0} \left[x\frac{y^3}{3} \right]_{y=0}^{y=\sqrt{4-x^2}} \mathrm{d}x - \int_{-1}^{0} \left[x\frac{y^3}{3} \right]_{y=0}^{\sqrt{1-x^2}} \mathrm{d}x = -\tfrac{32}{15} + \tfrac{1}{15} = -\tfrac{31}{15}.$$

As $S = S_1 + S_2$, adding I_1 and I_2 gives the expected result

$$I = \iint_S xy^2 \mathrm{d}A = I_1 + I_2 = 0.$$

This double integral could, of course, have been evaluated differently as the sum of the integral of xy^2 over the part of the annulus in the upper-half plane between $x = -1$ and $x = 1$, and the integrals of xy^2 over the parts of the annulus in the upper-half plane between $x = -2$ and -1, and $x = 1$ and $x = 2$. This approach gives

$$I = \int_{-1}^{1} \mathrm{d}x \int_{\sqrt{1-x^2}}^{\sqrt{4-x}} xy^2 \mathrm{d}y + \int_{-2}^{-1} \mathrm{d}x \int_0^{\sqrt{4-x^2}} xy^2 \mathrm{d}y + \int_1^2 \mathrm{d}x \int_0^{\sqrt{4-x^2}} xy^2 \mathrm{d}y.$$

The details of these integrations are left as an exercise, but the value of the first integral turns out to be zero, and although last two integrals are non-zero their values cancel, so once again we find that $I = 0$. The next section will show that double integrals over annular regions are more easily evaluated using polar coordinates, because the upper circular boundary of the annulus is $r = 2$ for $0 \le \theta \le \pi$, while the lower circular boundary of the annulus is $r = 1$ for $0 \le \theta \le \pi$, so the integrals involved will not have variable limits. ■

9.2 Double integrals using polar coordinates

For some regions S, as in Example 9.8, the Cartesian coordinate system is inconvenient, and it is better to use another coordinate system. The only one to be considered here is the *plane polar coordinate system* (r, θ), where r and θ are related to Cartesian coordinates (x, y) by

$$x = r\cos\theta, \quad y = r\sin\theta. \tag{9.17}$$

Here $r \geq 0$, and because of the periodicity of x and y with respect to θ, the angle θ is restricted to the interval $0 \leq \theta \leq 2\pi$. Fig. 9.9 shows a typical region S for which integration in terms of polar coordinates is appropriate. The two straight line boundaries of region S are the radial lines $\theta = \alpha$ and $\theta = \beta$, while the lower boundary has the equation $r = g(\theta)$ and the upper boundary the equation $r = h(\theta)$, where for later use we set $R_1 = \min\, g(\theta)$ and $R_2 = \max\, h(\theta)$.

To find a double integral of a continuous function $f(r, \theta)$ over S it is necessary to consider the form of an element of area dA in polar coordinates, and to do this we need to examine Fig. 9.10 showing an enlargement of a typical element of area $ABCD$. The curves AB and CD are circular arcs with the respective radii r_1 and $r_1 + \delta_r$, where δ_r is a small increment in r, and the straight line boundaries BC and AD are the respective radial lines $\theta = \alpha$ and $\theta = \alpha + \delta_\theta$, where δ_θ is a small increment in θ.

The area of the annulus $r_1 \leq r \leq r_1 + \delta_r$, with r_1 chosen arbitrarily inside this interval, is $\pi((r_1 + \delta_r)^2 - r_1^2) = \pi(2r_1\delta_r + \delta_r^2)$, so the element of area dA between the radial lines $\theta = \alpha$ and $\theta = \alpha + \delta_\theta$ is

$$dA = \pi(2r_1\delta_r + \delta_r^2)(\delta_\theta/2\pi) = r_1\delta_r\delta_\theta + \delta_r^2\delta_\theta.$$

The magnitude of last term on the right is third order in the small increments δ_r and δ_θ, and so is less than the magnitude of the first term which is only of second order in these increments. Consequently, when defining a double Riemann integral in terms of polar coordinates which involves

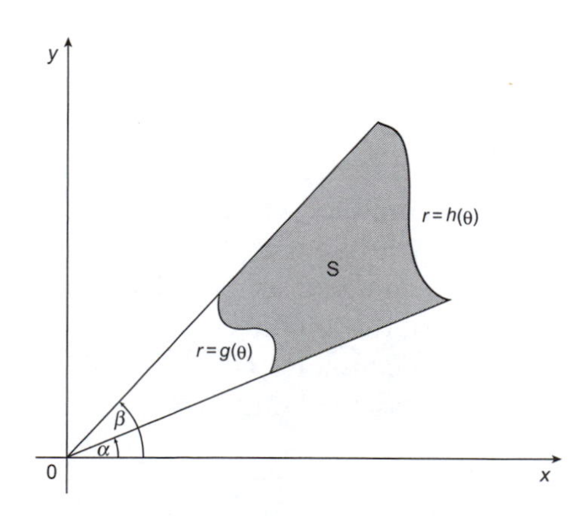

Fig. 9.9 A typical region S in defined using polar coordinates.

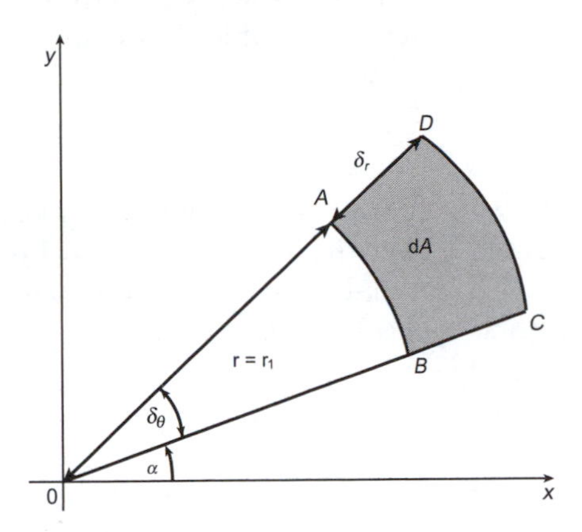

Fig. 9.10 An enlargement of an element of area dA in polar coordinates.

allowing δ_r and δ_θ tend to zero, the last term may be neglected. Thus the area dA will be approximated by

$$\Delta A = r\delta_r\delta_\theta, \tag{9.18}$$

where because r_1 was taken arbitrarily the suffix 1 has been dropped.

For convenience, we now divide the interval $R_1 \leq r \leq R_2$ into m sub-intervals each of length $\delta_r = (R_1 - R_2)/m$, and the angle $\alpha \leq \theta \leq \beta$ into n sub-intervals each of length $\delta_\theta = (\alpha - \beta)/n$. Next we denote by ΔA_{ij} the element of area in the ith sub-interval in the r coordinate and the jth sub-interval in the θ coordinate, and take an arbitrary point (ξ_r, η_j) inside this element of area. Then, in terms of polar coordinates, the approximate contribution ΔV_{ij} made by the double integral over ΔA_{ij} to the double Riemann integral over S is

$$\Delta V_{ij} \approx f(\xi_i, \eta_j)\Delta A_{ij} = f(\xi_i, \eta_j)\xi_i\delta_r\delta_\theta, \tag{9.19}$$

where we set $f(\xi_i, \eta_j) = 0$ for every element ΔA_{ij} that does not lie entirely within S. The double Riemann integral of $f(r, \theta)$ over S is now defined as follows.

Definition 9.2 (Double Riemann integral in polar coordinates)

The double Riemann integral of a continuous function $f(r, \theta)$ over an area S in the (r, θ)-plane is defined as

$$\iint_S f(r,\theta)\mathrm{d}A = \lim_{m\to\infty, n\to\infty} \sum_i^m \sum_j^n \xi_i\delta_r\delta_\theta f(\xi_i, \eta_j),$$

where the lengths δ_r and δ_θ tend to zero as $m, n \to \infty$ and $f(\xi_i, \eta_j)$ is set equal to zero for every area element ΔA_{ij} not lying completely inside S.

This result leads directly to the following theorem for the evaluation of iterated double Riemann integrals with variable limits over the region S shown in Fig. 9.9.

Theorem 9.4 (Evaluation of iterated double integrals with variable limits in polar coordinates)

If $f(r, \theta)$ is a continuous function, and integration is to be performed over the area S shown in Fig. 9.9 comprising the area between the two radial lines $\theta = \alpha$ and $\theta = \beta$, and between the lower boundary $r = g(\theta)$ and the upper boundary $r = h(\theta)$, then

$$\iint_S f(r, \theta)\mathrm{d}A = \int_\alpha^\beta \left[\int_{g(\theta)}^{h(\theta)} rf(r, \theta)\mathrm{d}r\right]\mathrm{d}\theta = \int_\alpha^\beta \mathrm{d}\theta \int_{g(\theta)}^{h(\theta)} rf(r, \theta)\mathrm{d}r.$$

The connection between double integrals in Cartesian coordinates and polar coordinates is provided by the following obvious result in which results (9.17) have been used to replace x and y and $\mathrm{d}A$ has been replaced by $r\mathrm{d}r\mathrm{d}\theta$.

$$\iint_S f(x, y)\mathrm{d}A = \iint_D f(r\cos\theta, r\sin\theta)r\mathrm{d}r\mathrm{d}\theta, \tag{9.20}$$

where S is the area over which integration is to be performed expressed in terms of Cartesian coordinates, and D is the corresponding area in terms of polar coordinates. Take note to include the extra r in the integral on the right.

Example 9.9 Find the integral

$$\iint_S xy^2\mathrm{d}A,$$

where S is the area in the first quadrant inside the circle $x^2 + y^2 = 1$.

Solution This is the double integral considered in Example 9.5, where because Cartesian coordinates were used to describe area S it was necessary to use an iterated integral with variable limits. However, in polar coordinates the specification of area S is much simpler, as it becomes $0 \le r \le 1$ with $0 \le \theta \le \pi/2$, so this time the limits of the iterated integral are constants. Setting $x = r\cos\theta$ and $y = r\sin\theta$ and substituting into the double integral using Theorem 9.3, with $\alpha = 0$, $\beta = \pi/2$, $g(\theta) = 0$ and $h(\theta) = 1$, gives

$$\iint_S xy^2\mathrm{d}A = \int_0^{\pi/2}\mathrm{d}\theta\int_0^1 (r^3\cos\theta\sin^2\theta)r\mathrm{d}r$$

$$= \left(\int_0^1 r^4\mathrm{d}r\right)\left(\int_0^{\pi/2}\cos\theta\sin^2\theta\, d\theta\right).$$

We have $\int_0^1 r^4 \mathrm{d}r = \frac{1}{5}$, and as $\frac{\mathrm{d}}{\mathrm{d}\theta}(\sin^3\theta) = 3\cos\theta\sin^2\theta$, it follows that

$$\int_0^{\pi/2} \cos\theta \sin^2\theta \,\mathrm{d}\theta = \frac{1}{3},$$

and so

$$\iint_S xy^2 \mathrm{d}A = \int_0^{\pi/2} \mathrm{d}\theta \int_0^1 (r^3\cos\theta\sin^2\theta) r\mathrm{d}r = \frac{1}{5}\cdot\frac{1}{3} = \frac{1}{15}.$$

This was the result obtained in Example 9.5 after a more complicated calculation. ■

Example 9.10 Find the volume V contained between the planes $x+y+3z=6$ and $z=0$ above the sector S in the (r, θ)-plane defined by $0 \le r \le 2$, $0 \le \theta \le \pi/4$.

Solution As z is non-negative, the volume V is given by

$$V = \iint_S z\,\mathrm{d}A = \iint_S (2 - \tfrac{1}{3}x - \tfrac{1}{3}y)\mathrm{d}A.$$

Using Eqn (9.20) to convert this to polar coordinates we find that

$$V = \int_0^{\pi/4} \mathrm{d}\theta \int_0^2 (2 - \tfrac{1}{3}r\cos\theta - \tfrac{1}{3}r\sin\theta) r\mathrm{d}r$$

$$= \int_0^{\pi/4} (4 - \tfrac{8}{9}\cos\theta - \tfrac{8}{9}\sin\theta)\mathrm{d}\theta = \pi - \tfrac{8}{9}. \quad ■$$

Example 9.11 Find $\iint_S f(r, \theta)\mathrm{d}A$ given that $f(r, \theta) = r\theta$, and S is the area between the radial lines $\theta = 0$ and $\theta = \pi/2$, and between the lower radial boundary $r = 1$ and the upper radial boundary $r = 2+\theta$.

Solution This integral involves variable limits, with $\alpha = 0$, $\beta = \pi/2$, $g(\theta) = 1$ and $h(\theta) = 2+\theta$. So from Theorem 9.3

$$\begin{aligned}\iint_S f(r,\theta)\mathrm{d}A &= \int_0^{\pi/2} \mathrm{d}\theta \int_1^{2+\theta} (r\theta) r\mathrm{d}r \\ &= \int_0^{\pi/2} \left[\int_1^{2+\theta} (r^2\theta)\mathrm{d}r\right] \mathrm{d}\theta \\ &= \int_0^{\pi/2} \tfrac{1}{3}(7\theta + 12\theta^2 + 6\theta^3 + \theta^4)\mathrm{d}\theta \\ &= \frac{7}{24}\pi^2 + \frac{1}{6}\pi^3 + \frac{1}{32}\pi^4 + \frac{1}{480}\pi^5 \approx 11.728. \quad ■\end{aligned}$$

Example 9.12 Find the volume of the cylinder $x^2+y^2=4$ bounded above and below by the spheroidal surface $x^2+y^2+4z^2=25$.

Solution The equation of the spheroidal surface can be written $z = \pm\frac{1}{2}\sqrt{25 - x^2 - y^2}$, so above the plane $z = 0$ (that is when $z > 0$) this becomes $z = \frac{1}{2}\sqrt{25 - x^2 - y^2}$.
Consequently the volume V_+ of the cylinder above the plane $z = 0$ is given by

$$V_+ = \iint_S z\,\mathrm{d}x\mathrm{d}y = \iint_S \tfrac{1}{2}\sqrt{25 - x^2 - y^2}\,\mathrm{d}x\mathrm{d}y,$$

where S is the area of intersection of the cylinder with the plane $z = 0$, and so is the area inside the circle $x^2 + y^2 = 4$.

Converting to polar coordinates, using the facts that then $x^2 + y^2 = r^2$, the equation of the cylinder becomes $r = 2$ and the symmetry of the surface about the z-axis means that the integral with respect to r is independent of θ for $0 \le \theta \le 2\pi$, this becomes

$$V_+ = \int_0^{2\pi} \mathrm{d}\theta \int_0^2 \tfrac{1}{2} r\sqrt{25 - r^2}\,\mathrm{d}r = \int_0^{2\pi} \left(\tfrac{125}{6} - \tfrac{7\sqrt{21}}{2}\right)\mathrm{d}\theta = \tfrac{1}{3}(125 - 21\sqrt{21})\pi.$$

The cylinder is symmetrical above and below the plane $z = 0$, so the volume of the complete cylinder with spheroidal caps is $V = 2V_+$, showing that

$$V = \tfrac{2}{3}(125 - 21\sqrt{21})\pi. \quad \blacksquare$$

Example 9.13 Find the moment of inertia about the z-axis of a thin semicircular disk in the upper half of the (x, y)-plane with boundary $x^2 + y^2 = 4$ and constant density ρ.

Solution As the density is constant, the moment of inertia I_0 about the z-axis is

$$I_0 = \iint_S \rho(x^2 + y^2)\mathrm{d}x\mathrm{d}y,$$

where S is the area inside $x^2 + y^2 = 4$ and $y > 0$. Converting to polar coordinates when $x^2 + y^2 = 4$ becomes $r = 2$ and $\mathrm{d}x\mathrm{d}y$ becomes $r\mathrm{d}r\mathrm{d}\theta$ we find that

$$I_0 = \int_0^{\pi} \mathrm{d}\theta \int_0^2 (\rho r^2) r\mathrm{d}r = \int_0^{\pi} \left(\frac{\rho r^4}{4}\right)\Bigg|_{r=0}^{2} \mathrm{d}\theta = 4\pi\rho.$$

Notice that I_0 would have been far more difficult to compute had Cartesian coordinates been used instead of polar coordinates. $\blacksquare$

Example 9.14 Find the mass m of a thin plate inside the cardioid $r = 1 + \cos\theta$ and outside the circle $r = 1$, if its density is ρr, with ρ a constant.

Solution The shape of the plate is shown in Fig. 9.11. Examination of Fig. 9.11 shows that the mass m of the plate will be given by

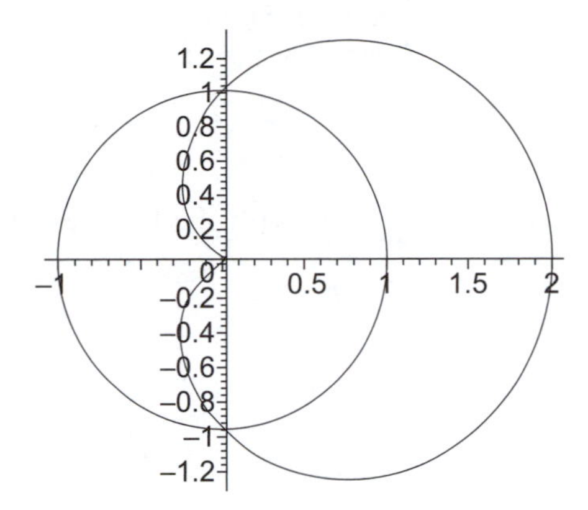

Fig. 9.11 The area of the thin plate inside the cardioid $r = 1 + \cos\theta$ and outside the circle $r = 1$.

$$m = \int_{-\pi/2}^{\pi/2} d\theta \int_{1}^{1+\cos\theta} (\rho r) r dr = \int_{-\pi/2}^{\pi/2} \rho(\cos\theta + \cos^2\theta + \tfrac{1}{3}\cos^3\theta) d\theta$$
$$= \rho(\tfrac{22}{9} + \tfrac{1}{2}\pi). \quad \blacksquare$$

The last example shows how a double integral can sometimes be used to evaluate an ordinary definite integral for which the integrand has no known antiderivative. The definite integral concerned is

$$\int_{-\infty}^{\infty} e^{-x^2} dx = \sqrt{\pi},$$

and the integrand e^{-x^2} has no known antiderivative. The definite integral occurs frequently in the theory of probability, throughout statistics, in the study of the heat flow, and elsewhere.

Example 9.15 Use a double integral to show that $\int_{-\infty}^{\infty} e^{-x^2} dx = \sqrt{\pi}$.

Solution The approach will be to take the limit of an integral inequality for double integrals that is similar to the one in Eqn (7.10). We start by considering the integral of $e^{-(x^2+y^2)}$ over the circle S_R defined as $x^2 + y^2 = R^2$, which in polar coordinates becomes the integral of e^{-r^2} over the circle $0 \leq r \leq R$ with $0 \leq \theta \leq 2\pi$. From Eqn (9.20) we have

$$\iint_{S_R} e^{-(x^2+y^2)} dx dy = \int_0^{2\pi} d\theta \int_0^R r e^{-r^2} dr$$
$$= \int_0^{2\pi} \tfrac{1}{2}(1 - e^{-R^2}) d\theta = \pi(1 - e^{-R^2}).$$

Now let D_R be the square $-R \leq x \leq R$, $-R \leq y \leq R$, then the circular area S_R is less than area of square D_R, which in turn is less than the circular area S_{2R}, so as e^{-r^2} is positive we can write the obvious double integral inequality

$$\iint_{S_R} e^{-(x^2+y^2)}dxdy \le \iint_{D_R} e^{-(x^2+y^2)}dxdy \le \iint_{S_{2R}} e^{-(x^2+y^2)}dxdy.$$

Together with the first result this yields the inequality

$$\pi(1-e^{-R^2}) \le \iint_{S_R} e^{-x^2}e^{-y^2}dxdy = \pi(1-e^{-4R^2}).$$

However,

$$\iint_{S_R} e^{-x^2}e^{-y^2}dxdy = \left(\int_{-R}^{R} e^{-x^2}dx\right)\left(\int_{-R}^{R} e^{-y^2}dy\right) = \left(\int_{-R}^{R} e^{-x^2}dx\right)^2,$$

where we have used the fact that in the second definite integral y is a dummy variable, and so can be replaced by x.

Combining results shows that

$$\pi(1-e^{-R^2/4}) \le \left(\int_{-R}^{R} e^{-x^2}dx\right)^2 \le \pi(1-e^{-R^2}).$$

The required result now follows by taking the limit of this last inequality as $R\to\infty$, when it reduces the equality

$$\pi = \left(\int_{-\infty}^{\infty} e^{-x^2}dx\right)^2,$$

from which we find that $\sqrt{\pi} = \int_{-\infty}^{\infty} e^{-x^2}dx$. ■

Problems **Section 9.1**

9.1 Find $\iint_S (x+y^2)dA$ where S is the area $0 \le x \le 2$, $1 \le y \le 3$.

9.2 Find $\iint_S xydA$ where S is the area $0 \le x \le 1$, $0 \le y \le 3$.

9.3 Find $\iint_S x^2\sin 2y dA$ where S is the area $0 \le x \le 2$, $1 \le y \le 3$.

9.4 Find $\iint_S \frac{1}{2x+3y}dA$ where S is the area $1 \le x \le 2$, $2 \le y \le 4$.

9.5 Find $\iint_S \sin(2x-3y)dA$ where S is the area $0 \le x \le \pi/4$, $0 \le y \le \pi/3$.

9.6 Use Cartesian coordinates to find $\iint_S y(1+2x^2)dA$ where S is the area in the first quadrant inside the circle $x^2+y^2=1$.

9.7 Sketch the area of integration involved in the iterated integral $\int_0^3 dy\int_{y^2}^9 dx$. Evaluate this iterated integral, write down an equivalent iterated integral, and evaluate it to show that it yields the same result.

9.8 Evaluate the iterated integral $\int_0^1 dy \int_0^{\sqrt{1-y^2}} (2x+y)dx$.

9.9 Evaluate the iterated integral $\int_0^2 dx \int_0^{\sqrt{9-y^2}} \frac{1}{\sqrt{9-y^2}} dy$.

9.10 Use an iterated integral to find the area above the x-axis, that lies below $y = 9 - x^2$ for $0 \le x \le 3$, and above the line $y = 2 - x$.

9.11 Evaluate the iterated integral $\int_0^3 dx \int_x^{2x^2} xy^2 dy$.

9.12 Use an iterated integral to evaluate $\iint_S x^2 y \, dA$, where S is the area below the parabola $y = \sqrt{x}$, above the line $y = \frac{1}{2}x$, and between the lines $x = 1$ and $x = 4$.

9.13 Use an iterated integral to find $\iint_S x^2 y dA$, where S is the area in the first quadrant that lies inside the circle $x^2 + y^2 = 4$.

9.14 Evaluate $\int_0^2 dx \int_0^{\sqrt{x}} xy^2 dy$. Sketch the area S over which the integration is performed. Use the sketch to reverse the order of integration, and integrate the result to confirm that value of the new integral is the same as before.

9.15 $\int_0^1 dx \int_0^{\sqrt{1-x^2}} \sqrt{1-x^2-y^2} dy$ (Hint: Evaluate the integral with respect to y over the interval $0 \le y \le a$, and then take the limit as $a \to \sqrt{1-x^2}$ from below.)

9.16 $\int_{-3}^3 dy \int_{y^2-4}^5 (x+2y)dx$.

9.17 $\int_1^2 dx \int_{1/x}^x x^2/y^2 dy$.

9.18 $\int_0^2 dx \int_{\sqrt{2x-x^2}}^1 (2x+3y^2)dy$.

Section 9.2

9.19 Use polar coordinates to find $\iint_S x^2 y dA$ where S is the area in the first quadrant inside the circle $x^2 + y^2 = 4$.

9.20 Find $\iint_S r(1+\theta)dA$, where S is the area $1 \le r \le 3$, $0 \le \theta \le \pi/2$.

9.21 Find $\iint_S r \sin(r\theta)dA$, where S is the area $1 \le r \le 4$, $0 \le \theta \le \pi$.

9.22 Find $\iint_S r^2 \theta \cos(r\theta)dA$, where S is the area $0 \le r \le 2$, $0 \le \theta \le \pi$.

9.23 Use polar coordinates to find the volume between the parallel planes $2x + y + 2z = 4$ and $2x + y + 2z = 2$ above the area $0 \le r \le 1$, $0 \le \theta \le \pi/4$.

9.24 Sketch the loop $r = 4\cos(4\theta)$ for $-\pi/8 \le \theta \le \pi/8$. Use an iterated integral in polar coordinates to find the area inside the loop.

9.25 Determine the area of integration for the iterated integral $\int_{-2}^2 dx \int_0^{\sqrt{4-x^2}} (x^2+y^2)^{3/2} dy$, and then use polar coordinates to evaluate it.

9.26 By converting to polar coordinates, find

$$\int_0^2 \mathrm{d}x \int_0^{\sqrt{4-x^2}} \frac{1}{(1+2x^2+2y^2)^{3/2}}\,\mathrm{d}y.$$

Compare the effort involved when the iterated integral is evaluated directly using Cartesian coordinates.

9.27 Use polar coordinates to evaluate the integral of $\exp[-(x^2+y^2)]$ over the part of the circle $x^2+y^2=1$ that lies in the upper half of the (x, y)-plane.

9.28 Use polar coordinates to find the volume of the cylinder $x^2+y^2=1$ bounded above by the spherical surface $x^2+y^2+z^2=4$ and below by the plane $z=0$. Hence find the volume of the hemisphere $x^2+y^2+z^2=4$ above the plane $z=0$ when the part inside the cylinder $x^2+y^2=1$ is removed.

9.29 Find the mass m of a thin plate inside the part of the cardioid $r=1+\cos\theta$ in the upper half of the (x, y)-plane and outside the circle $r=2$ if its mass per unit area is ρ/r, with ρ a constant.

9.30 Find the moment of inertia about the z-axis of a thin quarter circular disk in the first quadrant of the (x, y)-plane with boundary $x^2+y^2=1$ if its mass per unit area is $\rho(1+r)$.

9.31 Find the moment of inertia I_0 of a thin plate in the shape of the cardioid $r=2(1+\cos\theta)$ about a line through the origin $r=0$ drawn perpendicular to the plate if the mass per unit area of the plate is constant and equal to ρ.

9.32 Sketch the three petal rose $r=1+\cos 3\theta$. A thin plate with this shape has a constant mass per unit area ρ. Find the mass of the plate and its moment of inertia I_0 about a line through the origin $r=0$ drawn perpendicular to the plate.

9.33 $\int_{-\pi/2}^{\pi/2} \mathrm{d}\theta \int_0^{3\cos\theta} r^2 \sin^2\theta\,\mathrm{d}r.$

10 Matrices and linear transformations

The study of matrices that is developed in this chapter forms part of the branch of mathematics called linear algebra. A matrix is an array of quantities, usually numbers, but sometimes functions, in m rows and n columns. Systems of linear simultaneous equations can be expressed in matrix form and then manipulated as a single entity, and it is partly for this reason that matrices are so valuable in applications to engineering and science.

In many respects the algebraic rules for the combination of matrices parallel the usual algebraic rules for the combination of numbers. However, there are significant differences, one of which is that the order in which matrices are multiplied is important and cannot be altered without altering or even invalidating the resulting product. A further difference is that division of matrices is not defined, though it is shown how to find a multiplicative inverse called an inverse matrix for special square matrices. There are many other differences that are discussed as and when they occur.

Arising out of the study of inverse matrices comes the next topic to be included, which is the study of determinants. A determinant is a uniquely defined number derived in a special way from an array of numbers involving n rows and n columns. Rules for the manipulation of determinants simplify the task of their evaluation and are valuable in establishing special properties of determinants that are useful in their application.

The solution of systems of linear equations by Gaussian elimination is a practical method that is described, whereas their solution by means of inverse matrices is mainly of theoretical importance, though the two methods are equivalent. It is shown that Gaussian elimination is essential when a system is homogeneous, and it is necessary to establish first whether or not it has a non-trivial solution, and second the form of the solution when it exists.

The chapter closes with an introductory study of the eigenvalues and eigenvectors of square matrices which represents an important branch of linear algebra. Eigenvalue problems arise in a great variety of ways which are often related to oscillation problems. In such cases the eigenvalues characterize the frequencies of oscillation of a system, and the eigenvectors then describe its corresponding modes of oscillation. The last section gives a brief introduction to some of the applications of matrices, including the use of linear transformations in computer graphics and the numerical solution of the Laplace equation in the context of the temperature distribution in a metal.

10.1 Matrix algebra

In this section we introduce the fundamental ideas connected with matrices and their algebra, and proceed quickly through the basic definitions and theorems which will be illustrated by examples. The solution of systems of linear equations, an introduction to linear transformations and a discussion of some of their applications will be presented in subsequent sections.

Definition 10.1 (matrix and its order)

A *matrix* is a rectangular array of *elements* or *entries* a_{ij} involving m rows and n columns. The first suffix i in element a_{ij} is called the *row index* of the element and the second suffix j is called the *column index* of the element. These indices specify the row number and column number in which the element is located, with row 1 occurring at the top of the array and column 1 at the extreme left. A matrix with m rows and n columns is said to be of *order* m by n, and this is written $m \times n$. Alternative names that are also used instead of *order* are *dimension*, *shape* and *size* of a matrix.

Thus we denote an $m \times n$ matrix $\mathbf{A}$ with elements a_{ij} by writing

$$\mathbf{A} = \underbrace{\begin{bmatrix} a_{11} & a_{12} & \cdots & a_{1n} \\ a_{21} & a_{22} & \cdots & a_{2n} \\ \cdot & \cdot & \cdots & \cdot \\ a_{m1} & a_{m2} & \cdots & a_{mn} \end{bmatrix}}_{n \text{ columns}} \left.\vphantom{\begin{bmatrix} a \\ a \\ a \\ a \end{bmatrix}}\right\} \quad m \text{ rows.} \tag{10.1}$$

The convention used here is that a matrix is shown as a single bold-face letter, while its elements are shown by using the same letter, but in lowercase italic form with suffixes.

Elements in a matrix may be real or complex numbers, or even functions, as illustrated below:

$$\begin{bmatrix} 1 & 4 & 3 \\ 0 & 1 & 2 \\ 2 & 2 & 1 \end{bmatrix} \quad \text{(3 × 3 matrix with real numbers as elements)}$$

$$\begin{bmatrix} 1+3i & 2 & -3i \\ 2i & 2-4i & 3 \end{bmatrix} \quad \text{(2 × 3 matrix with complex numbers as elements)}$$

$$\begin{bmatrix} \cos\theta & -\sin\theta \\ \sin\theta & \cos\theta \end{bmatrix} \quad \text{(2 × 2 matrix with functions as elements)}$$

$$\begin{bmatrix} x \\ y \\ z \end{bmatrix}, \quad [2 \quad 1 \quad 4 \quad 0]$$ (a 3×1 matrix with variables x, y and z as elements and a 1×4 matrix with real numbers as elements).

Example 10.1 Construct the 2×3 matrix **A** whose elements are given by

$$a_{ij} = i + 2j.$$

Solution As **A** is a 2×3 matrix it has two rows and three columns, so that $i = 1, 2$ and $j = 1, 2, 3$. Thus

$$a_{11} = 1 + 2 \times 1 = 3, \quad a_{12} = 1 + 2 \times 2 = 5, \quad a_{13} = 1 + 2 \times 3 = 7,$$

$$a_{21} = 2 + 2 \times 1 = 4, \quad a_{22} = 2 + 2 \times 2 = 6, \quad a_{23} = 2 + 2 \times 3 = 8,$$

so entering these entries in

$$\mathbf{A} = \begin{bmatrix} a_{11} & a_{12} & a_{13} \\ a_{21} & a_{22} & a_{23} \end{bmatrix}$$

gives

$$\mathbf{A} = \begin{bmatrix} 3 & 5 & 7 \\ 4 & 6 & 8 \end{bmatrix}. \quad \blacksquare$$

Example 10.2 Find the elements a_{12}, a_{31} and a_{23} in the matrix

$$\mathbf{A} = \begin{bmatrix} 1 & 4 & 2 & 3 \\ 2 & 6 & -1 & 0 \\ 2 & 1 & -4 & 7 \end{bmatrix}$$

Solution It follows at once that $a_{12} = 4$, $a_{31} = 2$ and $a_{23} = -1$. ■

Special names are given to certain types of matrices, and we now describe and give examples of some of the more frequently used terms.

(a) A *row matrix* or *row vector* is any matrix of order $1 \times n$. The following is an example of a row vector of order 1×4:

$$[3 \quad 0 \quad 7 \quad 2]. \tag{10.2}$$

(b) A *column matrix* or *column vector* is any matrix of order $n \times 1$. The following is an example of a column vector of order 3×1:

$$\begin{bmatrix} 11 \\ 2 \\ 5 \end{bmatrix}. \tag{10.3}$$

(c) A *square matrix* is any matrix of order $n \times n$. The following is an example of a square matrix of order 3×3:

$$\begin{bmatrix} 1 & 2 & 4 \\ 3 & 0 & 2 \\ 5 & 1 & 3 \end{bmatrix}. \tag{10.4}$$

Three particular cases of square matrices that are worthy of note are the diagonal matrix, the symmetric matrix and the skew-symmetric matrix. Of these, the *diagonal matrix* has non-zero elements only on what is called the *principal diagonal*, which runs from the top left of the matrix to the bottom right. The principal diagonal is also often referred to as the *leading diagonal.* The following is an example of a diagonal matrix of order 4×4:

$$\begin{bmatrix} 3 & 0 & 0 & 0 \\ 0 & 0 & 0 & 0 \\ 0 & 0 & 2 & 0 \\ 0 & 0 & 0 & 5 \end{bmatrix}. \tag{10.5}$$

This example illustrates the fact that although all the *off-diagonal* elements of a diagonal matrix are zero, some (but not all) of the elements on the leading diagonal may be zero.

The diagonal matrix in which every element of the leading diagonal is unity is called either the *unit matrix* or the *identity matrix*, and it is usually denoted by **I**. The unit matrix of order 3×3 thus has the form

$$\begin{bmatrix} 1 & 0 & 0 \\ 0 & 1 & 0 \\ 0 & 0 & 1 \end{bmatrix}. \tag{10.6}$$

A *symmetric matrix* is one is which the elements obey the rule $a_{ij} = a_{ji}$, so that the pattern of numbers has a reflection symmetry about the leading diagonal. A typical symmetric matrix of order 3×3 is:

$$\begin{bmatrix} 5 & 1 & 3 \\ 1 & 2 & -2 \\ 3 & -2 & 7 \end{bmatrix}. \tag{10.7}$$

A *skew-symmetric matrix* is one in which the elements obey the rule $a_{ij} = -a_{ji}$, so that the leading diagonal must contain zeros, while the pattern of numbers has a reflection symmetry about the principal diagonal but with a reversal of sign. A typical skew-symmetric matrix of order 3×3 is:

$$\begin{bmatrix} 0 & 1 & 5 \\ -1 & 0 & -3 \\ -5 & 3 & 0 \end{bmatrix}. \tag{10.8}$$

(d) A *null matrix* is the name given to a matrix of any order which contains only zero elements. It is usually denoted by the symbol **0**. The null matrix of order 2×3 has the form

$$\mathbf{0} = \begin{bmatrix} 0 & 0 & 0 \\ 0 & 0 & 0 \end{bmatrix}. \quad (10.9)$$

(e) An *upper-triangular matrix* is a square matrix in which every element below the leading diagonal is zero, and a *lower-triangular matrix* is a square matrix in which every element above the leading diagonal is zero. Typical examples of such matrices of order 3×3 are:

$$\begin{bmatrix} 1 & 3 & 7 \\ 0 & 2 & -3 \\ 0 & 0 & 4 \end{bmatrix} \quad \text{(upper triangular)}$$

$$\begin{bmatrix} 2 & 0 & 0 \\ 4 & -1 & 0 \\ 7 & 1 & 3 \end{bmatrix} \quad \text{(lower triangular).} \quad (10.10)$$

(f) A *banded matrix* is one in which all the non-zero elements occur on the leading diagonal and on one or more lines parallel to the leading diagonal, but above and below it. The most commonly occurring banded matrix is the *tridiagonal matrix* in which all the non-zero elements occur on the leading diagonal, on the diagonal immediately above it called the *superdiagonal*, and on the diagonal immediately below it called the *sub-diagonal*. A typical tridiagonal matrix of order 5×5 has the form

$$\begin{bmatrix} a_{11} & a_{12} & 0 & 0 & 0 \\ a_{21} & a_{22} & a_{23} & 0 & 0 \\ 0 & a_{32} & a_{33} & a_{34} & 0 \\ 0 & 0 & a_{43} & a_{44} & a_{45} \\ 0 & 0 & 0 & a_{54} & a_{55} \end{bmatrix}. \quad (10.11)$$

***Definition 10.2* (equality of matrices)** Two matrices **A** and **B** with general elements a_{ij} and b_{ij}, respectively, are *equal* only when they are both of the same order and $a_{ij} = b_{ij}$ for all possible pairs of indices (i, j).

Example 10.3 It is possible for the matrices

$$\begin{bmatrix} 5 & a^3 \\ a^2 & 1 \end{bmatrix} \quad \text{and} \quad \begin{bmatrix} 5 & -27 \\ 9 & 1 \end{bmatrix}$$

to be equal and, if so, for what value of a does equality occur?

Solution The matrices are both of the same order, and hence they will be equal when their corresponding elements are equal. As corresponding elements on the principal diagonal are indeed equal, we need only confine attention to the off-diagonal elements. Thus the matrices will be equal if there is a common solution to the two equations $a^2=9$ and $a^3=-27$. Obviously, equality will occur if $a=-3$. ■

Definition 10.3 (addition of matrices) Two matrices **A** and **B** with general elements a_{ij} and b_{ij}, respectively, will be said to be *conformable for addition* only if they are both of the same order. Their sum $\mathbf{C}=\mathbf{A}+\mathbf{B}$ is the matrix **C** with elements $c_{ij}=a_{ij}+b_{ij}$.

As addition of real numbers is commutative we have $a_{ij}+b_{ij}=b_{ij}+a_{ij}$. This shows that addition of conformable matrices must also be commutative, whence

$$\mathbf{A}+\mathbf{B}=\mathbf{B}+\mathbf{A}.$$

Now addition of real numbers is also associative, so that $(a_{ij}+b_{ij})+c_{ij}=a_{ij}+(b_{ij}+c_{ij})$. Hence if a_{ij}, b_{ij} and c_{ij} are general elements of matrices **A**, **B** and **C** which are conformable for addition, then this also implies that addition of matrices is associative, whence

$$(\mathbf{A}+\mathbf{B})+\mathbf{C}=\mathbf{A}+(\mathbf{B}+\mathbf{C}).$$

These two results comprise our first theorem.

Theorem 10.1 (matrix addition is both commutative and associative) If **A**, **B** and **C** are matrices which are conformable for addition, then

(a) $\mathbf{A}+\mathbf{B}=\mathbf{B}+\mathbf{A}$ (matrix addition is commutative);

(b) $(\mathbf{A}+\mathbf{B})+\mathbf{C}=\mathbf{A}+(\mathbf{B}+\mathbf{C})$ (matrix addition is associative). ■

Example 10.4 Determine the constants a, b, c and d is order that the following matrix equation should be valid:

$$\begin{bmatrix} 0 & a & 3 \\ b & 2 & 2 \end{bmatrix}+\begin{bmatrix} c & 1 & 2 \\ 1 & 1 & d \end{bmatrix}=\begin{bmatrix} 4 & 3 & 5 \\ 7 & 3 & 5 \end{bmatrix}.$$

Solution Adding the two matrices on the left-hand side we arrive at the matrix equation

$$\begin{bmatrix} c & (a+1) & 5 \\ (b+1) & 3 & (d+2) \end{bmatrix}=\begin{bmatrix} 4 & 3 & 5 \\ 7 & 3 & 5 \end{bmatrix}.$$

Equating corresponding elements shows that $a=2$, $b=6$, $c=4$ and $d=3$. ■

***Definition 10.4* (multiplication by scalar)** If k is a scalar and the matrix $\mathbf{A}$ has elements a_{ij}, then the matrix $\mathbf{B} = k\mathbf{A}$ is of the same order as $\mathbf{A}$ and has element ka_{ij}.

Example 10.5 Determine $2\mathbf{A}+5\mathbf{B}$, given that:

$$\mathbf{A} = \begin{bmatrix} 1 & 2 \\ 3 & 4 \end{bmatrix} \quad \text{and} \quad \mathbf{B} = \begin{bmatrix} -1 & 3 \\ 4 & 2 \end{bmatrix}.$$

Solution

$$2\mathbf{A} + 5\mathbf{B} = 2\begin{bmatrix} 1 & 2 \\ 3 & 4 \end{bmatrix} + 5\begin{bmatrix} -1 & 3 \\ 4 & 2 \end{bmatrix},$$

or

$$2\mathbf{A} + 5\mathbf{B} = \begin{bmatrix} 2 & 4 \\ 6 & 8 \end{bmatrix} + \begin{bmatrix} -5 & 15 \\ 20 & 10 \end{bmatrix},$$

whence

$$2\mathbf{A} + 5\mathbf{B} = \begin{bmatrix} -3 & 19 \\ 26 & 18 \end{bmatrix}. \quad \blacksquare$$

***Definition 10. 5* (difference of two matrices)** If the matrices $\mathbf{A}$ and $\mathbf{B}$ are both of the same order, then the *difference* $\mathbf{A} - \mathbf{B}$ is defined by the relation

$$\mathbf{A} - \mathbf{B} = \mathbf{A} + (-1)\mathbf{B}.$$

Example 10.6 Determine $\mathbf{A} - \mathbf{B}$, given that:

$$\mathbf{A} = \begin{bmatrix} 1 & 3 \\ 4 & -2 \\ 1 & 6 \end{bmatrix} \quad \text{and} \quad \mathbf{B} = \begin{bmatrix} 4 & 2 \\ 3 & 1 \\ 0 & -2 \end{bmatrix}.$$

Solution

$$\mathbf{A} - \mathbf{B} = \begin{bmatrix} 1 & 3 \\ 4 & -2 \\ 1 & 6 \end{bmatrix} + (-1)\begin{bmatrix} 4 & 2 \\ 3 & 1 \\ 0 & -2 \end{bmatrix},$$

and so

$$\mathbf{A} - \mathbf{B} = \begin{bmatrix} 1 & 3 \\ 4 & -2 \\ 1 & 6 \end{bmatrix} + \begin{bmatrix} -4 & -2 \\ -3 & -1 \\ 0 & 2 \end{bmatrix} = \begin{bmatrix} -3 & 1 \\ 1 & -3 \\ 1 & 8 \end{bmatrix}.$$

***Definition 10.6* (matrix multiplication)** The two matrices $\mathbf{A}$ and $\mathbf{B}$ with general elements a_{ij} and b_{ij} are said to be *conformable for matrix multiplication* provided that the number of columns in $\mathbf{A}$ equals the number of rows in $\mathbf{B}$. If $\mathbf{A}$ is of order $m \times n$

and **B** is of order $n \times r$, then the matrix product **AB** is the matrix **C** of order $m \times r$ with elements c_{ij}, where

$$c_{ij} = a_{i1}b_{1j} + a_{i2}b_{2j} + \ldots + a_{in}b_{nj}.$$

The number c_{ij} is called the *inner product* of the ith row of **A** with the jth column of **B**.

The definition of matrix multiplication can be simplified by recognizing that the product of the n-element row vector

$$\mathbf{r} = [r_1 \quad r_2 \quad r_3 \ldots r_n]$$

and the n-element column vector

$$\mathbf{c} = \begin{bmatrix} c_1 \\ c_2 \\ \vdots \\ c_n \end{bmatrix}$$

is

$$\mathbf{rc} = [r_1 \quad r_2 \quad \ldots r_n] \begin{bmatrix} c_1 \\ c_2 \\ \vdots \\ c_n \end{bmatrix} = r_1c_1 + r_2c_2 + \ldots r_nc_n. \tag{10.12}$$

For if $\mathbf{a}_i$ is the ith row of matrix **A** and $\mathbf{b}_j$ is the jth column of matrix **B** in Definition 9.6, then the element c_{ij} in the matrix product **AB** is

$$c_{ij} = \boldsymbol{a}_i\boldsymbol{b}_j.$$

Thus the matrix product **AB** may be written

$$\mathbf{AB} = \begin{bmatrix} \mathbf{a}_1\mathbf{b}_1 & \mathbf{a}_1\mathbf{b}_2 & \cdots & \mathbf{a}_1\mathbf{b}_r \\ \mathbf{a}_2\mathbf{b}_1 & \mathbf{a}_2\mathbf{b}_2 & \cdots & \mathbf{a}_2\mathbf{b}_r \\ \cdots & \cdots & \cdots & \cdots \\ \mathbf{a}_m\mathbf{b}_1 & \mathbf{a}_m\mathbf{b}_2 & \cdots & \mathbf{a}_m\mathbf{b}_r \end{bmatrix}. \tag{10.13}$$

This emphasizes the fact that the element c_{ij} in the matrix product **AB** is obtained by multiplying the ith row of **A** and the jth column of **B** in the order $\mathbf{a}_i\mathbf{b}_j$.

Example 10.7 Determine **A**+**BC**, given that:

$$\mathbf{A} = \begin{bmatrix} 1 & 4 \\ 2 & 3 \end{bmatrix}, \qquad \mathbf{B} = \begin{bmatrix} 1 & 4 & 2 \\ 2 & 1 & 1 \end{bmatrix} \qquad \text{and} \qquad \mathbf{C} = \begin{bmatrix} 3 & 4 \\ 1 & 0 \\ 0 & 2 \end{bmatrix}.$$

Solution Matrix **B** is of order 2×3 and matrix **C** is of order 3×2, showing that **B** and **C** are conformable for multiplication. We have

$$\mathbf{BC} = \begin{bmatrix} 1 & 4 & 2 \\ 2 & 1 & 1 \end{bmatrix} \begin{bmatrix} 3 & 4 \\ 1 & 0 \\ 0 & 2 \end{bmatrix} = \begin{bmatrix} 1\cdot 3 + 4\cdot 1 + 2\cdot 0 & 1\cdot 4 + 4\cdot 0 + 2\cdot 2 \\ 2\cdot 3 + 1\cdot 1 + 1\cdot 0 & 2\cdot 4 + 1\cdot 0 + 1\cdot 2 \end{bmatrix}$$

$$= \begin{bmatrix} 7 & 8 \\ 7 & 10 \end{bmatrix},$$

and so

$$\mathbf{A} + \mathbf{BC} = \begin{bmatrix} 1 & 4 \\ 2 & 3 \end{bmatrix} + \begin{bmatrix} 7 & 8 \\ 7 & 10 \end{bmatrix} = \begin{bmatrix} 8 & 12 \\ 9 & 13 \end{bmatrix}. \quad \blacksquare$$

It is an immediate consequence of the definition of matrix multiplication that, although the matrix product **AB** may be defined, it is not generally true that $\mathbf{AB} = \mathbf{BA}$ and, indeed, the product **BA** may not even be defined. This fact is emphasized by the following remark.

If **A** and **B** are any two matrices, the matrix product **AB** is only defined if the number of columns in **A** equals the number of rows in **B**. If **A** is an $m \times n$ matrix and **B** is an $n \times r$ matrix, then **AB** is defined. We may show this symbolically as follows:

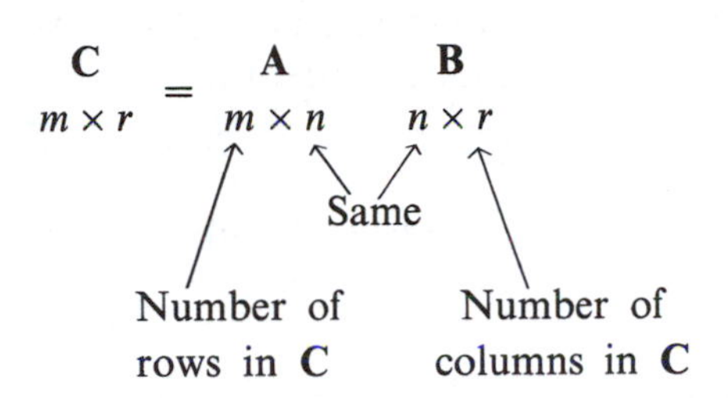

For example, if

$$\mathbf{A} = \begin{bmatrix} 1 & 2 \\ 3 & 1 \end{bmatrix}, \quad \mathbf{B} = \begin{bmatrix} 1 & -1 \\ 1 & 2 \end{bmatrix},$$

then $\mathbf{AB} = \begin{bmatrix} 3 & 3 \\ 4 & -1 \end{bmatrix}$ but $\mathbf{BA} = \begin{bmatrix} -2 & 1 \\ 7 & 4 \end{bmatrix}$, so $\mathbf{AB} \neq \mathbf{BA}$. However, if $\mathbf{C} = [1, 3]$ the product **AC** is not defined, but $\mathbf{CA} = [10, 5]$.

On account of the fact that matrix multiplication is not normally commutative, it is important to use a terminology that distinguishes between matrix multipliers that appear on the left or the right in a matrix product. This is achieved by adopting the convention that when matrix **B** is multiplied by matrix **A** from the *left* to form the product **AB**, we shall say that **B** is *pre-multiplied* by **A**. Conversely, when the matrix **B** is multiplied by **A** from the *right* to form the product **BA**, we shall say that **B** is *post-multiplied* by **A**.

The most important results concerning matrix multiplication are contained in the following theorem, which asserts that matrix multiplication is distributive with respect to addition, and that it is also associative.

Theorem 10.2 (matrix multiplication is distributive and associative)

If matrices **A**, **B** and **C** are conformable for multiplication, then:

(a) matrix multiplication is distributive with respect to addition, so that

$$\mathbf{A}(\mathbf{B}+\mathbf{C}) = \mathbf{AB}+\mathbf{AC};$$

(b) matrix multiplication is associative, so that

$$\mathbf{A}(\mathbf{BC}) = (\mathbf{AB})\mathbf{C};$$

(c) if **A** is a square matrix of order $n \times n$ and **I** is the unit matrix of order $n \times n$, then

$$\mathbf{AI} = \mathbf{IA} = \mathbf{A}.$$

Proof To establish result (a), let **B** and **C** be of order $m \times n$, and denote their general elements by b_{ij} and c_{ij}, respectively, so that the general element of $\mathbf{B}+\mathbf{C}$ is $b_{ij}+c_{ij}$. Then if **A** is of order of $r \times m$ with general element a_{ij}, and d_{ij} is the general element of $\mathbf{D} = \mathbf{A}(\mathbf{B}+\mathbf{C})$ which is of order $r \times n$, we have from Definition 10.6 that

$$d_{ij} = a_{i1}(b_{1j}+c_{1j}) + a_{i2}(b_{2j}+c_{2j}) + \ldots + a_{im}(b_{mj}+c_{mj}).$$

Performing the indicated multiplication and regrouping, we have

$$d_{ij} = (a_{i1}b_{1j} + a_{i2}b_{2j} + \ldots + a_{im}b_{mj}) + (a_{i1}c_{1j} + a_{i2}c_{2j} + \ldots + a_{im}c_{mj}).$$

However, from Definition 10.6 this is seen to be equivalent to

$$\mathbf{D} = \mathbf{AB}+\mathbf{AC},$$

which was to be proved.

Result (b) may be established in similar fashion, and to achieve this we assume **A**, **B** and **C** to be respectively of order $p \times q$, $q \times m$ and $m \times n$ with general elements a_{ij}, b_{ij} and c_{ij}.

From Definition 10.6 we know that the general element occurring in row i, column j of the product **BC** has the form

$$b_{i1}c_{1j} + b_{i2}c_{2j} + \ldots + b_{im}c_{mj},$$

so that the general element d_{ij} occurring in row i column j of the product $\mathbf{D} = \mathbf{A}(\mathbf{BC})$, which is of order $p \times n$, must have the form

$$\begin{aligned} d_{ij} &= a_{i1}(b_{11}c_{1j} + b_{12}c_{2j} + \ldots + b_{1m}c_{mj}) \\ &\quad + a_{i2}(b_{21}c_{1j} + b_{22}c_{2j} + \ldots + b_{2m}c_{mj}) \\ &\quad + \ldots + a_{iq}(b_{q1}c_{1j} + b_{q2}c_{2j} + \ldots + b_{qm}c_{mj}). \end{aligned}$$

Regrouping of the terms then gives

$$\begin{aligned} d_{ij} &= (a_{i1}b_{11} + a_{i2}b_{21} + \ldots + a_{iq}b_{q1})c_{1j} \\ &\quad + (a_{i1}b_{12} + a_{i2}b_{22} + \ldots + a_{iq}b_{q2})c_{2j} \\ &\quad + \ldots + (a_{i1}b_{1m} + a_{i2}b_{2m} + \ldots + a_{iq}b_{qm})c_{mj}. \end{aligned}$$

Appealing once more to Definition 10.6 we find that this is equivalent to

$$\mathbf{D} = (\mathbf{AB})\mathbf{C},$$

which was to be proved.

Result (c) may be established very easily by introducing the symbol

$$\delta_{ij} = \begin{cases} 1, & i = j \\ 0, & i \neq j \end{cases}.$$

Called the *Kroneker delta*. Then, if the general element of the unit matrix is c_{ij}, it follows that $c_{ij} = \delta_{ij}$, $i, j = 1, 2, \ldots, n$. Using this definition of c_{ij} in each of the matrix products **AI** and **IA**, it follows immediately that $\mathbf{AI} = \mathbf{IA} = \mathbf{A}$. ■

Because of Theorem 10.2(b), no confusion arises if we write

$$\mathbf{AA} = \mathbf{A}^2, \ \mathbf{AAA} = \mathbf{A}^3, \ldots,$$

whence it follows that for p, q positive integers

$$\mathbf{A}^p\mathbf{A}^q = \mathbf{A}^{p+q}, \qquad (10.14)$$

and

$$(\mathbf{A}^p)^q = \mathbf{A}^{pq}. \qquad (10.15)$$

Example 10.8 If

$$\mathbf{A} = [1 \quad 2], \quad \mathbf{B} = \begin{bmatrix} 1 & 3 \\ -1 & 2 \end{bmatrix}, \quad \mathbf{C} = \begin{bmatrix} 2 & 1 \\ 3 & 1 \end{bmatrix},$$

verify that
(a) $\mathbf{A}(\mathbf{B} + \mathbf{C}) = \mathbf{AB} + \mathbf{AC}$,
(b) $\mathbf{A}(\mathbf{BC}) = (\mathbf{AB})\mathbf{C}$.

Solution (a) We have

$$\mathbf{B} + \mathbf{C} = \begin{bmatrix} 3 & 4 \\ 2 & 3 \end{bmatrix},$$

so that

$$\mathbf{A}(\mathbf{B} + \mathbf{C}) = [1 \quad 2]\begin{bmatrix} 3 & 4 \\ 2 & 3 \end{bmatrix} = [7 \quad 10];$$

whereas

$$\mathbf{AB} = [-1 \quad 7] \qquad \text{and} \qquad \mathbf{AC} = [8 \quad 3],$$

so that

$$\mathbf{AB} + \mathbf{AC} = [7 \quad 10].$$

(b) We have

$$\mathbf{BC} = \begin{bmatrix} 1 & 3 \\ -1 & 2 \end{bmatrix} \begin{bmatrix} 2 & 1 \\ 3 & 1 \end{bmatrix} = \begin{bmatrix} 11 & 4 \\ 4 & 1 \end{bmatrix}$$

so that

$$\mathbf{A}(\mathbf{BC}) = [1 \quad 2] \begin{bmatrix} 11 & 4 \\ 4 & 1 \end{bmatrix} = [19 \quad 6];$$

whereas

$$\mathbf{AB} = [1 \quad 2] \begin{bmatrix} 1 & 3 \\ -1 & 2 \end{bmatrix} = [-1 \quad 7],$$

whence

$$(\mathbf{AB})\mathbf{C} = [-1 \quad 7] \begin{bmatrix} 2 & 1 \\ 3 & 1 \end{bmatrix} = [19 \quad 6]. \quad \blacksquare$$

An important matrix operation involves the interchange of rows and columns of a matrix, thereby changing a matrix of order $m \times n$ into one of order $n \times m$. Thus a row vector is changed into a column vector and a matrix of order 3×2 is changed into a matrix of order 2×3. This operation is called the operation of *transposition* and is denoted by the addition of a prime to the matrix in question.

***Definition 10.7* (transposition operation)** If $\mathbf{A}$ is a matrix of order $m \times n$, then its *transpose* $\mathbf{A}'$, sometimes written $\mathbf{A}^T$, is the matrix of order $n \times m$ which is derived from $\mathbf{A}$ by the interchange of rows and columns. Symbolically, if a_{ij} is the element in the ith row and jth column of $\mathbf{A}$, then a_{ji} is the element in the corresponding position in $\mathbf{A}'$. Thus row 1 of $\mathbf{A}$ becomes column 1 of $\mathbf{A}'$, row 2 of $\mathbf{A}$ becomes column 2 of $\mathbf{A}'$, ..., and row m of $\mathbf{A}$ becomes column m of $\mathbf{A}'$.

Example 10.9 Find $\mathbf{A}'$ and $(\mathbf{A}')'$, given that:

$$\mathbf{A} = \begin{bmatrix} 1 & 4 & 7 & 3 \\ 2 & -1 & 4 & -1 \end{bmatrix}.$$

Solution Writing the first row in place of the first column and the second row in place of the second column, as is required by Definition 10.7, we find that

$$\mathbf{A}' = \begin{bmatrix} 1 & 2 \\ 4 & -1 \\ 7 & 4 \\ 3 & -1 \end{bmatrix}.$$

The same argument shows that

$$(\mathbf{A}')' = \begin{bmatrix} 1 & 4 & 7 & 3 \\ 2 & -1 & 4 & -1 \end{bmatrix}. \quad \blacksquare$$

It is obvious from the definition of the transpose operation that $(\mathbf{A}')' = \mathbf{A}$, as was indeed illustrated in the last example. It is also obvious from Definitions 10.3 and 10.5 that if $\mathbf{A}$ and $\mathbf{B}$ are conformable for addition, then

$$(\mathbf{A} \pm \mathbf{B})' = \mathbf{A}' \pm \mathbf{B}'. \tag{10.16}$$

Now if $\mathbf{A}$ is of order $m \times n$ and $\mathbf{B}$ is of order $n \times r$, and the general matrix elements are a_{ij} and b_{ij}, respectively, the element c_{ij} in the ith row and jth column of the matrix product $\mathbf{C} = \mathbf{AB}$ is

$$c_{ij} = a_{i1}b_{1j} + a_{i2}b_{2j} + \ldots + a_{in}b_{nj}.$$

By definition, this is the element that will appear in the jth row and ith column of $(\mathbf{AB})'$.

Applying the transpose operation separately to $\mathbf{A}$ and $\mathbf{B}$ we find that $\mathbf{A}'$ is of order $n \times m$ and $\mathbf{B}'$ is of order $r \times n$, so that only the matrix product $\mathbf{B}'\mathbf{A}'$ is conformable.

Now the elements of the jth row of $\mathbf{B}'$ are the elements of the jth column of $\mathbf{B}$, and the elements of the ith column of $\mathbf{A}'$ are the elements of the ith row of $\mathbf{A}$, so that the element d_{ji} in the jth row and the ith column of the product $\mathbf{D} = \mathbf{B}'\mathbf{A}'$ must be

$$d_{ji} = b_{1j}a_{i1} + b_{2j}a_{i2} + \ldots + b_{nj}a_{in},$$

or equivalently

$$d_{ji} = a_{i1}b_{1j} + a_{i2}b_{2j} + \ldots + a_{in}a_{nj}.$$

However, equating elements in the jth row and ith column of $(\mathbf{AB})'$ and $\mathbf{B}'\mathbf{A}'$ we find that $c_{ij} = d_{ji}$, and so

$$(\mathbf{AB})' = \mathbf{B}'\mathbf{A}'. \tag{10.17}$$

We summarize these results into a final theorem.

Theorem 10.3 (properties of the transposition operation)

If $\mathbf{A}$ and $\mathbf{B}$ are conformable for addition or multiplication, as required, then:

(a) $(\mathbf{A}')' = \mathbf{A}$ (transposition is *reflexive*);
(b) $(\mathbf{A}+\mathbf{B})' = \mathbf{A}'+\mathbf{B}'$;
(c) $(\mathbf{A}-\mathbf{B})' = \mathbf{A}'-\mathbf{B}'$;
(d) $(\mathbf{AB})' = \mathbf{B}'\mathbf{A}'$. $\blacksquare$

Example 10.10 Verify that $(\mathbf{AB})' = \mathbf{B}'\mathbf{A}'$, given that:

$$\mathbf{A} = \begin{bmatrix} 1 & 3 \\ 2 & 4 \end{bmatrix} \quad \text{and} \quad \mathbf{B} = \begin{bmatrix} 2 & -1 \\ 3 & 1 \end{bmatrix}.$$

Solution We have

$$\mathbf{AB} = \begin{bmatrix} 1 & 3 \\ 2 & 4 \end{bmatrix} \begin{bmatrix} 2 & -1 \\ 3 & 1 \end{bmatrix} = \begin{bmatrix} 11 & 2 \\ 16 & 2 \end{bmatrix},$$

so that

$$(\mathbf{AB})' = \begin{bmatrix} 11 & 16 \\ 2 & 2 \end{bmatrix}.$$

However,

$$\mathbf{B}'\mathbf{A}' = \begin{bmatrix} 2 & 3 \\ -1 & 1 \end{bmatrix} \begin{bmatrix} 1 & 2 \\ 3 & 4 \end{bmatrix} = \begin{bmatrix} 11 & 16 \\ 2 & 2 \end{bmatrix},$$

which is equal to $(\mathbf{AB})'$. ■

10.2 Determinants

The notion of a determinant, when first introduced in Chapter 4, was that of a single *number* associated with a square array of numbers. In its subsequent application in that chapter it was used in a subsidiary role to simplify the manipulation of the vector product, and in that capacity it gave rise to a *vector*. The notion of a determinant is, in fact, implicit in Theorem 5.22 when, in connection with the change of variable in partial differentiation, it can be introduced with functions as elements. It is then called a *Jacobian*, and in this role it is often called a *functional determinant* and gives rise to a *function* that is closely related to the one–one nature of the change of variables involved.

These are but two of the situations in which determinants occur in different branches of mathematics, and it is the object of this section to examine some of the most important algebraic properties of determinants. The *order* of a determinant is the number of elements in its leading diagonal. Our results will only be proved for determinants of order 3 but they are, in fact, all true for determinants of any order.

We begin by rewriting Definition 4.16 using the matrix element notation as follows.

Definition 10.8 (third-order determinant) Let $\mathbf{A}$ be the square matrix of order 3×3,

$$\mathbf{A} = \begin{bmatrix} a_{11} & a_{12} & a_{13} \\ a_{21} & a_{22} & a_{23} \\ a_{31} & a_{32} & a_{33} \end{bmatrix}.$$

Then the expression

$$|\mathbf{A}| = \begin{vmatrix} a_{11} & a_{12} & a_{13} \\ a_{21} & a_{22} & a_{23} \\ a_{31} & a_{32} & a_{33} \end{vmatrix}$$

is called the *third-order determinant* associated with the square matrix **A**, and it is defined to be the number

$$|\mathbf{A}| = a_{11}\begin{vmatrix} a_{22} & a_{23} \\ a_{32} & a_{33} \end{vmatrix} - a_{12}\begin{vmatrix} a_{21} & a_{23} \\ a_{31} & a_{33} \end{vmatrix} + a_{13}\begin{vmatrix} a_{21} & a_{22} \\ a_{31} & a_{32} \end{vmatrix};$$

where we recall from Chapter 4 that, for any a, b, c and d, the second-order determinant is defined by

$$\begin{vmatrix} a & b \\ c & d \end{vmatrix} \equiv ad - bc.$$

The notation det **A** is also frequently used in place of $|\mathbf{A}|$ to signify the determinant of **A**.

This definition has a number of consequences of considerable value in simplifying the manipulation of determinants. Let us confine attention to the third-order determinant which is typical of all orders of determinant, and expand the last expression for $|\mathbf{A}|$ in Definition 10.8. We have

$$\begin{vmatrix} a_{11} & a_{12} & a_{13} \\ a_{21} & a_{22} & a_{23} \\ a_{31} & a_{32} & a_{33} \end{vmatrix} = a_{11}a_{22}a_{33} - a_{11}a_{23}a_{32} + a_{12}a_{23}a_{31} - a_{12}a_{21}a_{33} - a_{13}a_{21}a_{32} - a_{13}a_{22}a_{31}, \tag{10.18}$$

showing that one, and only one, element of each row and each column of the determinant appears in each of the products on the right-hand side defining $|\mathbf{A}|$. Hence, if any row or column of a determinant is multiplied by a factor λ, then the value of the determinant is multiplied by λ, because a factor λ will then appear in each product on the right-hand side of Eqn (10.18). Conversely, if any row or column of a determinant is divided by a factor λ, then the value of the determinant will be divided by λ. It is also obvious from Eqn (10.18) that $|\mathbf{A}| = 0$ if all the elements of a row or column of $|\mathbf{A}|$ are zero, or if all the corresponding elements of two rows or columns of $|\mathbf{A}|$ are equal, because then all of the products in Eqn (10.18) cancel.

Suppose, for example, that $\lambda = 3$ and

$$|\mathbf{A}| = \begin{vmatrix} 1 & 2 & 3 \\ 2 & 1 & 1 \\ 4 & 1 & 2 \end{vmatrix}.$$

Then it is easily shown that $|\mathbf{A}| = -5$, so that $3|\mathbf{A}| = -15$. Now this result could have been obtained equally well by using the above argument and multiplying any row or any column of $|\mathbf{A}|$ by 3. Thus, if the first row of $|\mathbf{A}|$ is multiplied by 3 we have

$$3|\mathbf{A}| = \begin{vmatrix} 3 & 6 & 9 \\ 2 & 1 & 1 \\ 4 & 1 & 2 \end{vmatrix} = -15$$

or, alternatively, if the third column is multiplied by 3 we have

$$3|\mathbf{A}| = \begin{vmatrix} 1 & 2 & 9 \\ 2 & 1 & 3 \\ 4 & 1 & 6 \end{vmatrix} = -15.$$

It is readily verified from Eqn (10.18) that interchanging any two rows or columns of $|\mathbf{A}|$ changes the sign of the determinant. Thus if

$$\mathbf{A} = \begin{vmatrix} 1 & 4 & 3 \\ 2 & 1 & 4 \\ 9 & 4 & -6 \end{vmatrix},$$

then

$$|\mathbf{A}| = \begin{vmatrix} 1 & 4 & 3 \\ 2 & 1 & 4 \\ 9 & 4 & -6 \end{vmatrix} = 167,$$

but interchanging the first two rows gives

$$\begin{vmatrix} 2 & 1 & 4 \\ 1 & 4 & 3 \\ 9 & 4 & -6 \end{vmatrix} = -167.$$

Similarly, interchanging the first and last rows in $|\mathbf{A}|$ gives

$$\begin{vmatrix} 9 & 4 & -6 \\ 2 & 1 & 4 \\ 1 & 4 & 3 \end{vmatrix} = -167.$$

A particularly simple case arises when $|\mathbf{A}|$ is the determinant associated with a diagonal matrix $\mathbf{A}$, for then all off-diagonal elements are automatically zero. This implies that Eqn (10.18) reduces to $|\mathbf{A}| = a_{11}a_{22}a_{33}$, which is just the product of the elements of the leading diagonal. Thus if

$$|\mathbf{A}| = \begin{vmatrix} 3 & 0 & 0 \\ 0 & -2 & 0 \\ 0 & 0 & 4 \end{vmatrix},$$

then $|\mathbf{A}| = (3)(-2)(4) = -24$.

Another useful result is that the value of a determinant is unchanged when elements of a row (or column) have added to them some multiple of the corresponding elements of some other row (or column). We prove this result by direct expansion in the following general case. Consider the determinant $|\mathbf{D}|$ obtained from $|\mathbf{A}|$ by adding to the elements of column 3 of $|\mathbf{A}|$, λ times the corresponding elements in column 2 of $|\mathbf{A}|$ to obtain:

$$|\mathbf{D}| = \begin{vmatrix} a_{11} & a_{12} & a_{13} + \lambda a_{12} \\ a_{21} & a_{22} & a_{23} + \lambda a_{22} \\ a_{31} & a_{32} & a_{33} + \lambda a_{32} \end{vmatrix}.$$

Then at once Definition 9.8 asserts that

$$|\mathbf{D}| = a_{11}\begin{vmatrix} a_{22} & a_{23} \\ a_{32} & a_{33} \end{vmatrix} + a_{11}\begin{vmatrix} a_{22} & \lambda a_{22} \\ a_{32} & \lambda a_{32} \end{vmatrix} - a_{12}\begin{vmatrix} a_{21} & a_{23} \\ a_{31} & a_{33} \end{vmatrix}$$

$$- a_{12}\begin{vmatrix} a_{21} & \lambda a_{22} \\ a_{31} & \lambda a_{32} \end{vmatrix} + a_{13}\begin{vmatrix} a_{21} & a_{22} \\ a_{31} & a_{32} \end{vmatrix} + \lambda a_{12}\begin{vmatrix} a_{21} & a_{22} \\ a_{31} & a_{32} \end{vmatrix}.$$

Now the second term on the right-hand side is zero, while the fourth and last terms cancel, leaving only three remaining terms. These are seen to comprise the definition of $|\mathbf{A}|$, so that we have proved that $|\mathbf{D}| = |\mathbf{A}|$ or, in symbols, that

$$\begin{vmatrix} a_{11} & a_{12} & a_{13} + \lambda a_{12} \\ a_{21} & a_{22} & a_{23} + \lambda a_{22} \\ a_{31} & a_{32} & a_{33} + \lambda a_{32} \end{vmatrix} = \begin{vmatrix} a_{11} & a_{12} & a_{13} \\ a_{21} & a_{22} & a_{23} \\ a_{31} & a_{32} & a_{33} \end{vmatrix}.$$

A similar result would have been obtained had different columns been used or, indeed, had rows been used instead of columns.

An obvious implication of this result is that if a row (or column) of a determinant is expressible as the sum of multiples of other rows (or columns) of the determinant, then the value of the determinant must be zero. This is so because by subtraction of this sum of multiples of other rows (or columns) from the row (or column) in question, it is possible to produce a row (or column) containing only zero elements.

Let us illustrate how a determinant may be simplified by means of this result. Consider the determinant

$$|\mathbf{A}| = \begin{vmatrix} 7 & 18 & 8 \\ 1 & 5 & 7 \\ 3 & 9 & 4 \end{vmatrix}.$$

Subtracting twice the third row from the first row we find

$$|\mathbf{A}| = \begin{vmatrix} 1 & 0 & 0 \\ 1 & 5 & 7 \\ 3 & 9 & 4 \end{vmatrix},$$

whence

$$|\mathbf{A}| = 1 \times \begin{vmatrix} 5 & 7 \\ 9 & 4 \end{vmatrix} = -43.$$

We now summarize these findings in the form of a theorem.

Theorem 10.4 (properties of determinants)

(a) A determinant in which all the elements of a row or column are zero, itself has the value zero.
(b) A determinant in which all corresponding elements in two rows (or columns) are equal has the value zero.
(c) If the elements of a row (or column) of a determinant are multiplied by a factor λ, then the value of the determinant is multiplied by λ.
(d) The value of a determinant associated with a diagonal matrix is equal to the product of the elements on the leading diagonal.
(e) The value of a determinant is unaltered by adding to the elements of any row (or column) a constant multiple of the corresponding elements of any other row (or column).
(f) If a row (or column) of a determinant is expressible as the sum of multiples of other rows (or columns) or the determinant, then its value is zero.
(g) If any two rows or columns of a determinant are interchanged, the sign of the determinant is changed.
(h) The value of a determinant in upper- or lower-triangular form is equal to the product of the elements on the leading diagonal. ■

Higher-order determinants can be defined with exactly similar properties to those enumerated in the theorem above. Thus the determinant $|\mathbf{A}|$ of order n associated with the square matrix $\mathbf{A}$ of order $n \times n$ has $n!$ terms in its expansion, each of which contains one, and only one, element from each row and column of $\mathbf{A}$.

Definition 10.9 (fourth-order determinant)

If $\mathbf{A}$ is the square matrix of order (4×4)

$$\mathbf{A} = \begin{bmatrix} a_{11} & a_{12} & a_{13} & a_{14} \\ a_{21} & a_{22} & a_{23} & a_{24} \\ a_{31} & a_{32} & a_{33} & a_{34} \\ a_{41} & a_{42} & a_{43} & a_{44} \end{bmatrix},$$

then the expression

$$|\mathbf{A}| = \begin{bmatrix} a_{11} & a_{12} & a_{13} & a_{14} \\ a_{21} & a_{22} & a_{23} & a_{24} \\ a_{31} & a_{32} & a_{33} & a_{34} \\ a_{41} & a_{42} & a_{43} & a_{44} \end{bmatrix},$$

is called the forth-order determinant associated with the square matrix $\mathbf{A}$, and it is defined to be the number

$$|\mathbf{A}| = a_{11} \begin{vmatrix} a_{22} & a_{23} & a_{24} \\ a_{32} & a_{33} & a_{34} \\ a_{42} & a_{43} & a_{44} \end{vmatrix} - a_{12} \begin{vmatrix} a_{21} & a_{23} & a_{24} \\ a_{31} & a_{33} & a_{34} \\ a_{41} & a_{43} & a_{44} \end{vmatrix}$$

$$+a_{13}\begin{vmatrix} a_{21} & a_{22} & a_{24} \\ a_{31} & a_{32} & a_{34} \\ a_{41} & a_{42} & a_{44} \end{vmatrix} - a_{14}\begin{vmatrix} a_{21} & a_{22} & a_{23} \\ a_{31} & a_{32} & a_{33} \\ a_{41} & a_{42} & a_{43} \end{vmatrix}.$$

An inductive argument applied to Definitions 10.8 and 10.9 shows one way in which higher-order determinants may be defined, but clearly our notation needs some simplification to avoid unwieldy expressions of the type given above. This is achieved by the introduction of the *minor* and the *cofactor* of an element of a square matrix.

Definition 10.10 (minors and cofactors) Let $\mathbf{A}$ be a square matrix of order $n \times n$ with general element a_{ij}, and let $|\mathbf{A}|$ be the determinant of order n associated with $\mathbf{A}$. Denote by M_{ij} the determinant of order $n-1$ associated with the matrix of order $(n-1)\times(n-1)$ derived from $\mathbf{A}$ by the deletion of row i and column j. Then M_{ij} is called the *minor* of the element a_{ij} of $\mathbf{A}$, and $A_{ij} = (-1)^{i+j} M_{ij}$ is called the *cofactor* of the element a_{ij} of $\mathbf{A}$.

Notice that the factor $(-1)^{i+j}$ entering into the definition of a cofactor of $\mathbf{A}$ is simply a device for multiplying the associated minor by a factor $+1$ when $i+j$ is an *even* number, and by a factor -1 when $i+j$ is an *odd* number. Thus minors and cofactors are equal when $i+j$ is even, and they have equal magnitude but opposite sign when $i+j$ is odd.

Example 10.11 Find the minors and cofactors of the matrix

$$\mathbf{A} = \begin{bmatrix} 1 & 0 & 3 \\ 2 & 1 & 4 \\ 1 & 2 & 1 \end{bmatrix}.$$

Solution The minor M_{11} is derived from $\mathbf{A}$ by deleting row 1 and column 1 and equating M_{11} to the determinant formed by the remaining elements. That is,

$$M_{11} = \begin{vmatrix} 1 & 4 \\ 2 & 1 \end{vmatrix} = -7.$$

Similarly, minor M_{12} is derived from $\mathbf{A}$ by deleting row 1 and column 2 and equating M_{12} to the determinant formed by the remaining elements. That is,

$$M_{12} = \begin{vmatrix} 2 & 4 \\ 1 & 1 \end{vmatrix} = -2.$$

Identical reasoning then shows that $M_{13}=3$, $M_{21}=-6$, $M_{22}=-2$, $M_{23}=2$, $M_{31}=-3$, $M_{32}=-2$ and $M_{33}=1$. As the cofactors A_{ij} are given by $(-1)^{i+j}M_{ij}$, it follows that $A_{11}=-7$, $A_{12}=2$, $A_{13}=3$, $A_{21}=6$, $A_{22}=-2$, $A_{23}=-2$, $A_{31}=-3$, $A_{32}=2$ and $A_{33}=1$. ■

If $\mathbf{A}$ is a square matrix with general element a_{ij} and corresponding cofactor A_{ij}, it is easily seen from the above definitions that:

(a) if $\mathbf{A}$ is of order 2×2, then $|\mathbf{A}| = a_{11}A_{11} + a_{12}A_{12}$;
(b) if $\mathbf{A}$ is of order 3×3, then $|\mathbf{A}| = a_{11}A_{11} + a_{12}A_{12} + a_{13}A_{13}$;
(c) if $\mathbf{A}$ is of order 4×4, then $|\mathbf{A}| = a_{11}A_{11} + a_{12}A_{12} + a_{13}A_{13} + a_{14}A_{14}$.

This suggests that if $\mathbf{A}$ is of order $n \times n$, then for $|\mathbf{A}|$ we could adopt the definition

$$|\mathbf{A}| = a_{11}A_{11} + a_{12}A_{12} + \ldots + a_{1n}A_{1n}. \tag{10.19}$$

This is a true statement and could be accepted as a definition, but it is not the most general one which may be adopted. To see this we return to Eqn (10.18) and rearrange the terms on the right-hand side to give

$$|\mathbf{A}| = a_{31}(a_{12}a_{23} - a_{13}a_{22}) - a_{32}(a_{11}a_{23} - a_{13}a_{21}) + a_{33}(a_{11}a_{22} - a_{12}a_{21}).$$

Hence, working backwards, we have

$$|\mathbf{A}| = a_{31}\begin{vmatrix} a_{12} & a_{13} \\ a_{22} & a_{23} \end{vmatrix} - a_{32}\begin{vmatrix} a_{11} & a_{13} \\ a_{21} & a_{23} \end{vmatrix} + a_{33}\begin{vmatrix} a_{11} & a_{12} \\ a_{21} & a_{22} \end{vmatrix},$$

thereby showing that it is also true that

$$|\mathbf{A}| = a_{31}A_{31} + a_{32}A_{32} + a_{33}A_{33}. \tag{10.20}$$

We now have two equivalent but different-looking expressions for $|\mathbf{A}|$, either of which could be taken as the definition of $|\mathbf{A}|$. The expression in (b) above involves the elements and cofactors of the first row of $\mathbf{A}$ and the expression in Eqn (10.20) involves the elements and cofactors of the third row of $\mathbf{A}$. A repetition of this argument involving other rearrangements of the terms of Eqn (10.18) shows that $|\mathbf{A}|$ may be evaluated as the sum of the products of the elements and their cofactors of *any* row or column of $\mathbf{A}$. This very valuable and the general result is known as the *Laplace expansion theorem*, and it is true for determinants of any order, though we have only proved it for a third-order determinant. Let us state the result formally as it would apply to a determinant of order n. The proof, which follows inductively from the definition of a determinant of order 2, will be omitted.

Theorem 10.5 (Laplace expansion theorem) The determinant $|\mathbf{A}|$ associated with any $n \times n$ square matrix $\mathbf{A}$ is obtained by summing the products of the elements and their cofactors in any row or column of $\mathbf{A}$. If $\mathbf{A}$ has the general element a_{ij} and the corresponding cofactor is A_{ij}, then this result is equivalent to:

$$|\mathbf{A}| = \sum_{j=1}^{n} a_{ij}A_{ij}$$

for $i = 1, 2, \ldots, n$ (which is the expansion by elements of the ith row); or

$$|\mathbf{A}| = \sum_{i=1}^{n} a_{ij}A_{ij}$$

for $j = 1, 2, \ldots, n$ (the expansion by elements of the jth column). ■

Example 10.12 Evaluate the determinant

$$|\mathbf{A}| = \begin{vmatrix} 1 & 4 & 2 \\ 3 & -2 & 1 \\ 1 & 5 & 2 \end{vmatrix}$$

by expanding it (a) in terms of the elements of row 2, and (b) in terms of the elements of column 3.

Solution (a) Using the first expression in Theorem 9.5, we have

$$|\mathbf{A}| = -3\begin{vmatrix} 4 & 2 \\ 5 & 2 \end{vmatrix} -2\begin{vmatrix} 1 & 2 \\ 1 & 2 \end{vmatrix} -1\begin{vmatrix} 1 & 4 \\ 1 & 5 \end{vmatrix} = 5.$$

(b) Using the second expression in Theorem 9.5 gives

$$|\mathbf{A}| = 2\begin{vmatrix} 3 & -2 \\ 1 & 5 \end{vmatrix} -1\begin{vmatrix} 1 & 4 \\ 1 & 5 \end{vmatrix} +2\begin{vmatrix} 1 & 4 \\ 3 & -2 \end{vmatrix} = 5.$$ ■

An important extension of Theorem 10.5 asserts that the sum of the products of the elements of any row (or column) of a square matrix **A** with the cofactors corresponding to the elements of a *different* row (or column) is zero. This is easily proved as follows.

Let **A** be a matrix of order $n \times n$, and let **B** be obtained from **A** by replacing row q of **A** by row p. Then **B** has the elements of rows p and q equal, so that by Theorem 10.4(b) it follows that $|\mathbf{B}| = 0$. Expanding $|\mathbf{B}|$ in terms of elements of row q by Theorem 10.5 we then find

$$|\mathbf{B}| = a_{p1}A_{q1} + a_{p2}A_{q2} + \ldots + a_{pn}A_{qn} = 0,$$

which was to be proved. A similar argument establishes the corresponding result for columns, and so we have proved our assertion.

Theorem 10.6 The sum of the products of the elements of any row (or column) of a square matrix **A** with the cofactors corresponding to the elements of a different row (or column) is zero. Symbolically, if a_{ij} is the general element of **A** and $\mathbf{A}_{ij}$ is its cofactor, then using the expansion by elements of a row and cofactors of a different row,

$$\sum_{i=1}^{n} a_{pi}A_{qi} = 0$$

if $p \neq q$; and, using the expansion by elements of a column and cofactors of a different column,

$$\sum_{i=1}^{n} a_{ip}A_{iq} = 0$$

if $p \neq q$. ■

***Corollary 10.5* (product of two determinants)** If $|\mathbf{A}|$, $|\mathbf{B}|$ are any two determinants of equal order, then $|\mathbf{A}|\,|\mathbf{B}| = |\mathbf{AB}|$. ■

We shall omit the proof of this result, which is based on a generalization of Theorem 10.5 and the rule for the multiplication of matrices.

Example 10.13 Verify that the sum of the products of the elements of column 1 and the corresponding cofactors of column 2 of the following matrix is zero:

$$\mathbf{A} = \begin{bmatrix} 1 & 3 & 2 \\ 4 & 1 & 2 \\ 3 & 1 & 3 \end{bmatrix}.$$

Solution The elements of column 1 are $a_{11} = 1$, $a_{21} = 4$, $a_{31} = 3$. The cofactors corresponding to the elements of the second column are $A_{12} = -6$, $A_{22} = -3$, $A_{32} = 6$. Hence

$$a_{11}A_{12} + a_{21}A_{22} + a_{31}A_{32} = (1)(-6) + (4)(-3) + (3)(6) = 0. \quad ■$$

The effort involved in evaluating determinants using the Laplace expansion theorem increases rapidly as the order increases. The practical method for the evaluation of a determinant of order much greater than 3 is to use Theorem 10.4(e) to reduce the determinant to upper-triangular form, and then to use Theorem 10.4(h) to find its value by forming the product of the elements in the leading diagonal of the equivalent upper-triangular determinant.

Rule 10.1 (The practical evaluation of high-order determinants)

Let the determinant to be evaluated be of order n.

1. Subtract suitable multiples of row 1 from rows 2 to n to make zero the first element in each of the new rows 2 to n.

2. Working with the determinant produced in Step 1, subtract suitable multiples of row 2 from rows 3 to n to make zero the first element in each of the new rows 3 to n.
3. Repeat this process until all elements below the leading diagonal are zero.
4. The value of the original determinant will then be given by the product of all the elements in the leading diagonal of the equivalent upper-triangular determinant.

During this process interchange of rows may become necessary, with the consequent change of sign of the determinant, if at the mth stage the mth element in row m becomes zero.

Example 10.14 Evaluate the following determinant by reducing it to upper-triangular form

$$|\mathbf{A}| = \begin{vmatrix} 1 & 2 & 1 & 1 \\ 1 & 4 & 2 & 1 \\ 2 & 0 & 1 & 2 \\ 2 & 6 & 0 & 1 \end{vmatrix}.$$

Solution Introduce the notation $R^1_{m+1} = R_{m+1} + kR_m$, with the meaning that the new $(m+1)$th row R^1_{m+1} is obtained by adding to the original $(m+1)$th row R_{m+1}, k times the original mth row R_m. An application of Rule 10.1 produces the following result:

$$|\mathbf{A}| = \begin{vmatrix} 1 & 2 & 1 & 1 \\ 1 & 4 & 2 & 1 \\ 2 & 0 & 1 & 2 \\ 2 & 6 & 0 & 1 \end{vmatrix}$$

$$= \begin{matrix} R^1_1 = R_1 \\ R^1_2 = R_2 - R_1 \\ R^1_3 = R_3 - 2R_1 \\ R^1_4 = R_4 - 2R_1 \end{matrix} \begin{vmatrix} 1 & 2 & 1 & 1 \\ 0 & 2 & 1 & 0 \\ 0 & -4 & -1 & 0 \\ 0 & 2 & -2 & -1 \end{vmatrix}$$

$$= \begin{matrix} R^1_1 = R_1 \\ R^1_2 = R_2 \\ R^1_3 = R_3 + 2R_1 \\ R^1_4 = R_4 - R_1 \end{matrix} \begin{vmatrix} 1 & 2 & 1 & 1 \\ 0 & 2 & 1 & 0 \\ 0 & 0 & 1 & 0 \\ 0 & 0 & -3 & -1 \end{vmatrix}$$

$$= \begin{matrix} R^1_1 = R_1 \\ R^1_2 = R_2 \\ R^1_3 = R_3 \\ R^1_4 = R_4 + 3R_3 \end{matrix} \begin{vmatrix} 1 & 2 & 1 & 1 \\ 0 & 2 & 1 & 0 \\ 0 & 0 & 1 & 0 \\ 0 & 0 & 0 & -1 \end{vmatrix}$$

$$= 1 \times 2 \times 1 \times (-1) = -2. \quad \blacksquare$$

10.3 Linear dependence and linear independence

We are now in a position to discuss the important idea of linear independence. This concept has already been used implicitly in Chapter 4 when the three mutually orthogonal unit vectors **i**, **j** and **k** were introduced comprising what in linear algebra is called a *basis* for the vector space. By this we mean that all other vectors are expressible in terms of the vectors comprising the basis through the operations of scaling and vector addition, but that no member of the basis itself is expressible in terms of the other members of the basis. Thus *no* choice of the scalars λ, μ can ever make the vectors **i** and $\lambda\mathbf{j}+\mu\mathbf{k}$ equal. It is in this sense that the unit vectors **i**, **j**, **k** comprising the basis for ordinary vector analysis are *linearly independent*, and obviously any other set of unit vectors **a**, **b**, **c** which are not co-planar, and no two of which are parallel, would serve equally well as a basis for this space.

The same idea carries across to matrices when the term 'vector' is interpreted to mean either a matrix row vector or a matrix column vector. Thus the three column vectors

$$\mathbf{C}_1 = \begin{bmatrix} 1 \\ 3 \\ -2 \end{bmatrix}, \quad \mathbf{C}_2 = \begin{bmatrix} 2 \\ 1 \\ 4 \end{bmatrix} \text{ and } \mathbf{C}_3 = \begin{bmatrix} 5 \\ 5 \\ 6 \end{bmatrix}$$

are *not* linearly independent because $\mathbf{C}_3 = \mathbf{C}_1 + 2\mathbf{C}_2$, whereas the three row vectors

$$\mathbf{R}_1 = [1 \quad 0 \quad 0], \ \mathbf{R}_2 = [0 \quad 1 \quad 0] \text{ and } \mathbf{R}_3 = [0 \quad 0 \quad 1]$$

are obviously linearly independent, because no choice of the scalars λ, μ can ever make the vectors $\mathbf{R}_1$ and $\lambda\mathbf{R}_2+\mu\mathbf{R}_3$ equal. It is these ideas that underlie the formulation of the following definition.

Definition 10.11 (linear dependence and linear independence)

The set of n matrix row or column vectors $\mathbf{V}_1, \mathbf{V}_2, \ldots, \mathbf{V}_n$ which are conformable for addition will be said to be *linearly dependent* if there exist n scalars $\alpha_1, \alpha_2, \ldots, \alpha_n$, not all zero, such that

$$\alpha_1\mathbf{V}_1 + \alpha_2\mathbf{V}_2 + \ldots + \alpha_n\mathbf{V}_n = 0.$$

When no such set of scalars exists, so that this relationship is only true when $\alpha_1 = \alpha_2 = \ldots = \alpha_n = 0$, then the n matrix vectors $\mathbf{V}_1, \mathbf{V}_2, \ldots$, will be said to be *linearly independent*.

In the event that the n matrix vectors in Definition 10.11 represent the rows or columns of a rectangular matrix **A**, the linear dependence or independence of the vectors $\mathbf{V}_1, \mathbf{V}_2, \ldots, \mathbf{V}_n$ becomes a statement about the linear dependence or independence of the rows or columns of **A**. In particular, if **A** is a square matrix, and linear dependence exists between its rows (or columns), then by definition it is possible to express at least one row (or column) or **A** as the sum of multiples of the other rows (or columns). Thus from Theorem 10.4(f), we see that linear dependence among the rows or columns of a square matrix $\mathbf{A}$ implies the condition $|\mathbf{A}| = 0$. Similarly, if $|\mathbf{A}| \neq 0$ then the rows and columns of $\mathbf{A}$ cannot be linearly dependent.

Theorem 10.7 (test for linear independence) The rows and columns of a square matrix $\mathbf{A}$ are linearly independent if, and only if, $|\mathbf{A}| \neq 0$. Conversely, linear dependence is implied between rows or columns of a square matrix $\mathbf{A}$ if $|\mathbf{A}| = 0$.

Example 10.15 Test the following matrices for linear independence between rows or columns:

$$\mathbf{A} = \begin{bmatrix} 1 & 4 & 3 \\ -2 & 18 & 7 \\ 4 & -6 & 1 \end{bmatrix} \quad \text{and} \quad \mathbf{B} = \begin{bmatrix} 1 & 1 & 0 \\ 3 & 2 & 1 \\ 1 & 1 & 3 \end{bmatrix}.$$

Solution We shall apply Theorem 10.7 by examining $|\mathbf{A}|$ and $|\mathbf{B}|$. A simple calculation shows that $|\mathbf{A}| = 0$, so that linear dependence exists between either the rows or the columns of $\mathbf{A}$. In fact, denoting the columns of $\mathbf{A}$ by $\mathbf{C}_1$, $\mathbf{C}_2$ and $\mathbf{C}_3$, we have $\mathbf{C}_2 = 2(\mathbf{C}_3 - \mathbf{C}_1)$. As $|\mathbf{B}| = -3$ the rows and columns of $\mathbf{B}$ are linearly independent. ■

Let us now give consideration to any linear independence that may exist between the rows or columns of a rectangular matrix $\mathbf{A}$ of order $m \times n$. If r rows (or columns) of $\mathbf{A}$ are linearly independent, where $r \leq \min(m, n)$, then Theorem 10.7 implies that there is at least one determinant of order r that may be formed by taking these r rows (or columns) which is *non-zero*, but that all determinants of order greater than r must of necessity vanish. This number r is called the *rank* of the matrix $\mathbf{A}$, and it represents the greatest number of linearly independent rows or columns existing in $\mathbf{A}$. If, for example, $\mathbf{A}$ is a square matrix of order $n \times n$ and $|\mathbf{A}| \neq 0$, this implies that the rank of $\mathbf{A}$ must be n.

Definition 10.12 (rank of a matrix) The *rank* r of a matrix $\mathbf{A}$, written $r = \text{rank}\ \mathbf{A}$, is the greatest number of linearly independent rows or columns that exist in the matrix $\mathbf{A}$. Numerically, r is equal to the order of the largest non-vanishing determinant $|\mathbf{B}|$ associated with any square matrix $\mathbf{B}$ which can be constructed from $\mathbf{A}$ by combination of r rows and r columns.

The task of determining the rank of a matrix $\mathbf{A}$ may be greatly simplified by using the fact that the addition of a multiple of one row of $\mathbf{A}$ to another will not alter the linear dependence of the rows of $\mathbf{A}$, and so will not alter rank $\mathbf{A}$. This property may be used to simplify the structure of $\mathbf{A}$ until its rank becomes obvious. The simplification is accomplished by reducing $\mathbf{A}$ by successive row operations of this type to what is called a *row-equivalent echelon matrix*, in which all elements below a line drawn through the leading diagonal of the original elements $a_{11}, a_{22}, a_{33}, \ldots$ are zero.

The process is illustrated in the following example, in which, as before, the notation $R_4^1 = 3R_4 - 2R_2$ means that the new fourth row R_4^1 is obtained by subtracting twice the original second row R_2 from three times the original fourth row R_4, and here the symbol $\sim$ means 'is row equivalent to'.

Example 10.16 Find the rank of the following matrix

$$\mathbf{A} = \begin{bmatrix} 1 & 0 & 0 & 1 & 0 \\ -1 & 1 & 1 & -1 & 1 \\ -3 & 0 & 1 & -1 & 0 \\ -2 & 0 & 1 & 0 & 0 \end{bmatrix}.$$

Solution

$$\mathbf{A} = \begin{bmatrix} 1 & 0 & 0 & 1 & 0 \\ -1 & 1 & 1 & -1 & 1 \\ -3 & 0 & 1 & -1 & 0 \\ -2 & 0 & 1 & 0 & 0 \end{bmatrix} \sim \mathbf{R}_2 + \mathbf{R}_1 \begin{bmatrix} 1 & 0 & 0 & 1 & 0 \\ 0 & 1 & 1 & 0 & 1 \\ -3 & 0 & 1 & -1 & 0 \\ -2 & 0 & 1 & 0 & 0 \end{bmatrix}$$

$$\sim \mathbf{R}_3 + 3\mathbf{R}_1 \begin{bmatrix} 1 & 0 & 0 & 1 & 0 \\ 0 & 1 & 1 & 0 & 1 \\ 0 & 0 & 1 & 2 & 0 \\ -2 & 0 & 1 & 0 & 0 \end{bmatrix} \sim \mathbf{R}_4 + 2\mathbf{R}_1 \begin{bmatrix} 1 & 0 & 0 & 1 & 0 \\ 0 & 1 & 1 & 0 & 1 \\ 0 & 0 & 1 & 2 & 0 \\ 0 & 0 & 1 & 2 & 0 \end{bmatrix}$$

$$\sim \mathbf{R}_4 - \mathbf{R}_3 \begin{bmatrix} 1 & 0 & 0 & 1 & 0 \\ 0 & 1 & 1 & 0 & 1 \\ 0 & 0 & 1 & 2 & 0 \\ 0 & 0 & 0 & 0 & 0 \end{bmatrix}.$$

Now the largest non-zero determinant which is possible in this last matrix is of order 3 because the last row contains only zeros. There is certainly one non-zero determinant of order 3 to be found among the first three rows of this last matrix, namely the determinant

$$\begin{bmatrix} 1 & 0 & 0 \\ 0 & 1 & 1 \\ 0 & 0 & 1 \end{bmatrix} = 1,$$

so rank $\mathbf{A} = 3$.

10.4 Inverse and adjoint matrices

The operation of division is not defined for matrices, but a multiplicative inverse matrix denoted by $\mathbf{A}^{-1}$ can be defined for any square matrix $\mathbf{A}$ for which $|\mathbf{A}| \neq 0$. This multiplicative inverse $\mathbf{A}^{-1}$ is unique and has the property that

$$\mathbf{A}^{-1}\mathbf{A} = \mathbf{A}\mathbf{A}^{-1} = \mathbf{I}$$

where $\mathbf{I}$ is the unit matrix, and it is defined in terms of what is called the matrix *adjoint* to $\mathbf{A}$. The uniqueness follows from the fact that if $\mathbf{B}$ and $\mathbf{C}$ are each inverse to $\mathbf{A}$, then $\mathbf{B}(\mathbf{AC})\mathbf{C} = (\mathbf{BA})\mathbf{C}$, so that $\mathbf{BI} = \mathbf{IC}$, or $\mathbf{B} = \mathbf{C}$.

Definition 10.13 (adjoint matrix) Let $\mathbf{A}$ be a square matrix, then the transpose of the matrix of cofactors of $\mathbf{A}$ is called the matrix *adjoint* to $\mathbf{A}$, and it is denoted by adj $\mathbf{A}$. A square matrix and its adjoint are both of the same order.

Example 10.17 Find the matrix adjoint to:

$$\mathbf{A} = \begin{bmatrix} 1 & 2 & 1 \\ 3 & 1 & 0 \\ 2 & 1 & 2 \end{bmatrix}.$$

Solution The cofactors A_{ij} of $\mathbf{A}$ are: $A_{11} = 2$, $A_{12} = -6$, $A_{13} = 1$, $A_{21} = -3$, $A_{22} = 0$, $A_{23} = 3$, $A_{31} = -1$, $A_{32} = 3$ and $A_{33} = -5$. Hence the matrix of cofactors $\mathbf{C}$ has the form

$$\mathbf{C} = \begin{bmatrix} 2 & -6 & 1 \\ -3 & 0 & 3 \\ -1 & 3 & -5 \end{bmatrix},$$

so that its transpose, which by definition is adj $\mathbf{A}$, is

$$\mathbf{C}' = \text{adj } \mathbf{A} = \begin{bmatrix} 2 & -3 & -1 \\ -6 & 0 & 3 \\ 1 & 3 & -5 \end{bmatrix}. \quad \blacksquare$$

Now from Theorem 10.5 and 10.6, we see that the effect of forming either the product (adj $\mathbf{A}$)$\mathbf{A}$ or the product $\mathbf{A}$(adj $\mathbf{A}$) is to produce a diagonal matrix in which each element of the leading diagonal is $|\mathbf{A}|$. That is, we have shown that

$$(\text{adj } \mathbf{A})\mathbf{A} = \mathbf{A}(\text{adj } \mathbf{A}) = \begin{bmatrix} |\mathbf{A}| & 0 & 0 & \cdots & 0 \\ 0 & |\mathbf{A}| & 0 & \cdots & 0 \\ \cdot & \cdot & \cdot & \cdots & 0 \\ 0 & 0 & 0 & \cdots & |\mathbf{A}| \end{bmatrix},$$

whence

$$(\text{adj } \mathbf{A})\mathbf{A} = \mathbf{A}(\text{adj } \mathbf{A}) = |\mathbf{A}|\mathbf{I}. \tag{10.21}$$

Thus, provided $|\mathbf{A}| \neq 0$, by writing

$$\mathbf{A}^{-1} = \frac{\operatorname{adj} \mathbf{A}}{|\mathbf{A}|}, \tag{10.22}$$

we arrive at the result

$$\mathbf{A}^{-1}\mathbf{A} = \mathbf{A}\mathbf{A}^{-1} = \mathbf{I}. \tag{10.23}$$

The matrix $\mathbf{A}^{-1}$ is called the matrix *inverse* to $\mathbf{A}$, and it is only defined for square matrices $\mathbf{A}$ for which $|\mathbf{A}| \neq 0$. A square matrix whose associated determinant is non-vanishing is called a *non-singular* matrix. A square matrix whose associated determinant vanishes is said to be *singular*. Although the inverse matrix is only defined for non-singular square matrices, the adjoint matrix is defined for any square matrix, irrespective of whether or not it is non-singular.

Definition 10.14 (inverse matrix) If $\mathbf{A}$ is a square matrix for which $|\mathbf{A}| \neq 0$, the matrix *inverse* to $\mathbf{A}$, which is denoted by $\mathbf{A}^{-1}$, is defined by the relationship

$$\mathbf{A}^{-1} = \frac{\operatorname{adj} \mathbf{A}}{|\mathbf{A}|}.$$

Example 10.18 Find the matrix inverse to the matrix $\mathbf{A}$ of Example 10.17 above.

Solution It is easily found from the cofactors already computed that $|\mathbf{A}| = -9$. This follows, for example, by expanding $|\mathbf{A}|$ in terms of elements of the first row to obtain $|\mathbf{A}| = (1)(2) + (2)(-6) + (1)(1) = -9$. Hence from Definition 10.14, we have

$$\mathbf{A}^{-1} = \frac{\operatorname{adj}\mathbf{A}}{|\mathbf{A}|} = (-1/9)\begin{bmatrix} 2 & -3 & -1 \\ -6 & 0 & 3 \\ 1 & 3 & -5 \end{bmatrix} = \begin{bmatrix} -2/9 & 1/3 & 1/9 \\ 2/3 & 0 & -1/3 \\ -1/9 & -1/3 & 5/9 \end{bmatrix}. \quad \blacksquare$$

The steps in the determination of an inverse matrix are perhaps best remembered in the form of a rule.

Rule 10.2 (Determination of inverse matrix)

To determine the matrix $\mathbf{A}^{-1}$ which is inverse to the square matrix $\mathbf{A}$ by means of Definition 10.14 proceed as follows:

(a) Construct the matrix of cofactors of $\mathbf{A}$.
(b) Transpose the matrix of cofactors of $\mathbf{A}$ to obtain adj $\mathbf{A}$.
(c) Calculate $|\mathbf{A}|$ and, if it is not zero, divide adj $\mathbf{A}$ by $|\mathbf{A}|$ to obtain $\mathbf{A}^{-1}$.
(d) If $|\mathbf{A}| = 0$, then $\mathbf{A}^{-1}$ is not defined.

It is a trivial consequence of Definition 10.14 and the fact that for any square matrix $\mathbf{A}$, $|\mathbf{A}| = |\mathbf{A}'|$ that

$$(\mathbf{A}^{-1})' = (\mathbf{A}')^{-1}. \qquad (10.24)$$

Also, if $\mathbf{A}$ and $\mathbf{B}$ are non-singular matrices of the same order, then

$$(\mathbf{B}^{-1}\mathbf{A}^{-1})\mathbf{AB} = \mathbf{I} = \mathbf{AB}(\mathbf{B}^{-1}\mathbf{A}^{-1}),$$

showing that

$$(\mathbf{AB})^{-1} = \mathbf{B}^{-1}\mathbf{A}^{-1}. \qquad (10.25)$$

By making use of Corollary 10.5, which asserts that for any two square matrices $\mathbf{A}$, $\mathbf{B}$ of the same order $|\mathbf{AB}| = |\mathbf{A}||\mathbf{B}|$, we are able to prove another useful result concerning the inverse matrix. If $|\mathbf{A}| \neq 0$, then $\mathbf{AA}^{-1} = \mathbf{I}$ showing that $|\mathbf{AA}^{-1}| = 1$, or $|\mathbf{A}||\mathbf{A}^{-1}| = 1$. It follows from this that:

$$|\mathbf{A}| = 1/|\mathbf{A}^{-1}|. \qquad (10.26)$$

One final result follows directly from the obvious fact that $(\mathbf{A}^{-1})^{-1}\mathbf{A}^{-1} = \mathbf{I}$, which is always true provided $|\mathbf{A}^{-1}| \neq 0$. If we post-multiply this result by $\mathbf{A}$ we find

$$(\mathbf{A}^{-1})^{-1}\mathbf{A}^{-1}\mathbf{A} = \mathbf{IA}$$

giving

$$(\mathbf{A}^{-1})^{-1}\mathbf{I} = \mathbf{A},$$

whence

$$(\mathbf{A}^{-1})^{-1} = \mathbf{A}. \qquad (10.27)$$

***Theorem 10.8* (properties of inverse matrix)** If $\mathbf{A}$ and $\mathbf{B}$ are non-singular square matrices of the same order, then:

(a) $\mathbf{AA}^{-1} = \mathbf{A}^{-1}\mathbf{A} = \mathbf{I}$;
(b) $(\mathbf{AB})^{-1} = \mathbf{B}^{-1}\mathbf{A}^{-1}$;
(c) $(\mathbf{A}^{-1})' = (\mathbf{A}')^{-1}$;
(d) $(\mathbf{A}^{-1})^{-1} = \mathbf{A}$;
(e) $|\mathbf{A}| = 1/|\mathbf{A}^{-1}|$. ■

Example 10.19 Verify that $(\mathbf{A}^{-1})' = (\mathbf{A}')^{-1}$, given that

$$\mathbf{A} = \begin{bmatrix} 1 & 3 \\ 2 & 4 \end{bmatrix}.$$

Solution We have

$$\mathbf{A}^{-1} = \begin{bmatrix} -3 & 3/2 \\ 1 & -1/2 \end{bmatrix},$$

so that

$$(\mathbf{A}^{-1})' = \begin{bmatrix} -2 & 1 \\ 3/2 & -1/2 \end{bmatrix}.$$

However,

$$\mathbf{A}' = \begin{bmatrix} 1 & 2 \\ 3 & 4 \end{bmatrix},$$

so that

$$(\mathbf{A}')^{-1} = \begin{bmatrix} -2 & 1 \\ 3/2 & -1/2 \end{bmatrix},$$

confirming that $(\mathbf{A}^{-1})' = (\mathbf{A}')^{-1}$. ■

It follows from the definition of matrix multiplication that if $\mathbf{A}$ is an $n \times n$ matrix, pre-multiplication of $\mathbf{A}$ by an $n \times n$ matrix $\mathbf{R}$ to form the matrix product $\mathbf{RA}$ is equivalent to performing operations on the rows of $\mathbf{A}$.

Suppose, if possible, that m matrices $\mathbf{R}_1, \mathbf{R}_2, \ldots, \mathbf{R}_m$ can be found such that

$$\mathbf{R}_m\mathbf{R}_{m-1}\ldots\mathbf{R}_2\mathbf{R}_1\mathbf{A} = \mathbf{I}.$$

Then post multiplication by $\mathbf{A}^{-1}$ (provided it exists) gives the result

$$\mathbf{R}_m\mathbf{R}_{m-1}\ldots\mathbf{R}_2\mathbf{R}_1 = \mathbf{A}^{-1},$$

which we may rewrite as

$$\mathbf{A}^{-1} = \mathbf{R}_m\mathbf{R}_{m-1}\ldots\mathbf{R}_2\mathbf{R}_1\mathbf{I}.$$

Thus the sequence of elementary row operations performed on a matrix $\mathbf{A}$ which possesses an inverse ($\mathbf{A}$ is non-singular), will also transform the unit matrix $\mathbf{I}$ into $\mathbf{A}^{-1}$.

This result provides an alternative method for the computation of an inverse matrix which does not involve finding $|\mathbf{A}|$ and the cofactors of $\mathbf{A}$. The method is formulated as the following rule.

Rule 10.3 (Determination of an inverse matrix by elementary row transformations)

Let

$$\mathbf{A} = \begin{bmatrix} a_{11} & a_{12} & \cdots & a_{1n} \\ a_{21} & a_{22} & \cdots & a_{2n} \\ \cdot & \cdot & \cdots & \cdot \\ a_{n1} & a_{n2} & \cdots & a_{nn} \end{bmatrix} \quad \text{and} \quad \mathbf{I} = \begin{bmatrix} 1 & 0 & \cdots & 0 \\ 0 & 1 & \cdots & 0 \\ 0 & 0 & \cdots & 0 \\ 0 & 0 & \cdots & 1 \end{bmatrix}$$

both be $n \times n$ matrices. Then if a sequence of elementary row operations can be found which when performed on **A** reduces it to **I**, the same sequence of operations when performed on **I** in the same order will transform it to $\mathbf{A}^{-1}$.

Example 10.20 Use row transformations to find $\mathbf{A}^{-1}$ given that

$$\mathbf{A} = \begin{bmatrix} 1 & 0 & 1 \\ 1 & 2 & 0 \\ 3 & 0 & 4 \end{bmatrix}.$$

Solution We start by writing **A** and **I** side by side as follows:

$$\begin{bmatrix} 1 & 0 & 1 \\ 1 & 2 & 0 \\ 3 & 0 & 4 \end{bmatrix} \begin{bmatrix} 1 & 0 & 0 \\ 0 & 1 & 0 \\ 0 & 0 & 1 \end{bmatrix}.$$

Next we perform elementary row operations on **A** and **I** simultaneously, so that **A** is reduced to **I**, when **I** is transformed into $\mathbf{A}^{-1}$:

$$\begin{matrix} R_1^1 = R_1 \\ R_2^1 = R_2 - R_1 \\ R_3^1 = R_3 - 3R_1 \end{matrix} \begin{bmatrix} 1 & 0 & 1 \\ 0 & 2 & -1 \\ 0 & 0 & 1 \end{bmatrix} \begin{bmatrix} 1 & 0 & 0 \\ -1 & 1 & 0 \\ -3 & 0 & 1 \end{bmatrix}$$

$$\begin{matrix} R_1^1 = R_1 - R_3 \\ R_2^1 = R_2 - R_3 \\ R_3^1 = R_3 \end{matrix} \begin{bmatrix} 1 & 0 & 0 \\ 0 & 2 & 0 \\ 0 & 0 & 1 \end{bmatrix} \begin{bmatrix} 4 & 0 & -1 \\ -4 & 1 & 1 \\ -3 & 0 & 1 \end{bmatrix}$$

$$\begin{matrix} R_1^1 = R_1 \\ R_2^1 = \frac{1}{2}R_2 \\ R_3^1 = R_3 \end{matrix} \begin{bmatrix} 1 & 0 & 0 \\ 0 & 1 & 0 \\ 0 & 0 & 1 \end{bmatrix} \begin{bmatrix} 4 & 0 & -1 \\ -2 & 1/2 & 1/2 \\ -3 & 0 & 1 \end{bmatrix}.$$

The reduction of **A** to the matrix **I** is now complete, so the transformed unit matrix on the right is $\mathbf{A}^{-1}$, and thus

$$\mathbf{A}^{-1} = \begin{bmatrix} 4 & 0 & -1 \\ -2 & 1/2 & 1/2 \\ -3 & 0 & 1 \end{bmatrix}.$$

A simple calculation verifies that $\mathbf{A}\mathbf{A}^{-1} = \mathbf{I}$. ■

10.5 Matrix functions of a single variable

All the matrix results that have been obtained so far are equally valid whether applied to matrices whose elements are numerical constants, or to matrices whose elements are functions of a single variable t. When the latter is the case it is convenient to copy the notation for a function used hitherto, and to represent the matrix by writing $\mathbf{A}(t)$. In many respects it is convenient to regard all matrices in this manner, since matrices with constant number elements correspond to the subset of all possible matrices $\mathbf{A}(t)$ in which all elements are constant functions.

When the elements of $\mathbf{A}(t)$ are all differentiable with respect to t in some interval, it is reasonable to define a derivative of $\mathbf{A}(t)$ with respect to t, and for this purpose we shall work with the following definition.

Definition 10.15 (derivative of a matrix) Let $\mathbf{A}(t)$ be a matrix of order $m \times n$ whose elements $a_{ij}(t)$ are all differentiable functions of t in some common interval $t_0 < t < t_1$. Then the *derivative* of $\mathbf{A}(t)$ with respect to t in $t_0 < t < t_1$, written $\mathrm{d}\mathbf{A}/\mathrm{d}t$, is defined to be the matrix of order $m \times n$ with elements $\mathrm{d}a_{ij}/\mathrm{d}t$. The matrix $\mathbf{A}(t)$ will be said to be *differentiable* for $t_0 < t < t_1$. Symbolically this result becomes:

$$\frac{\mathrm{d}}{\mathrm{d}t}\begin{bmatrix} a_{11}(t) & a_{12}(t) & \cdots & a_{1n}(t) \\ a_{21}(t) & a_{22}(t) & \cdots & a_{2n}(t) \\ \cdot & \cdot & \cdot & \cdot \\ a_{m1}(t) & a_{m2}(t) & \cdots & a_{mn}(t) \end{bmatrix} = \begin{bmatrix} \frac{\mathrm{d}a_{11}}{\mathrm{d}t} & \frac{\mathrm{d}a_{12}}{\mathrm{d}t} & \cdots & \frac{\mathrm{d}a_{1n}}{\mathrm{d}t} \\ \frac{\mathrm{d}a_{21}}{\mathrm{d}t} & \frac{\mathrm{d}a_{22}}{\mathrm{d}t} & \cdots & \frac{\mathrm{d}a_{2n}}{\mathrm{d}t} \\ \frac{\mathrm{d}a_{m1}}{\mathrm{d}t} & \frac{\mathrm{d}a_{m2}}{\mathrm{d}t} & \cdots & \frac{\mathrm{d}a_{mn}}{\mathrm{d}t} \end{bmatrix},$$

for $t_0 < t < t_1$.

Example 10.21 Find $\mathrm{d}\mathbf{A}/\mathrm{d}t$ given that:

$$\mathbf{A}(t) = \begin{bmatrix} \cosh t & \sin t & \cosh 2t \\ \sinh t & \cos t & \sinh 2t \end{bmatrix}.$$

Solution From Definition 10.15 we have at once:

$$\frac{\mathrm{d}\mathbf{A}}{\mathrm{d}t} = \begin{bmatrix} \sinh t & \cos t & 2\sinh 2t \\ \cosh t & -\sin t & 2\cosh 2t \end{bmatrix},$$

for all t. ■

If $a_{ij}(t)$ and $b_{ij}(t)$ are differentiable functions in some common interval $t_0 < t < t_1$, then we know from the work of Chapter 5 that

$$\frac{\mathrm{d}}{\mathrm{d}t}(a_{ij} \pm b_{ij}) = \frac{\mathrm{d}a_{ij}}{\mathrm{d}t} \pm \frac{\mathrm{d}b_{ij}}{\mathrm{d}t},$$

and also

$$\frac{\mathrm{d}}{\mathrm{d}t}(a_{i1}b_{1j}+a_{i2}b_{2j}+\ldots+a_{in}b_{nj})$$
$$=\left(\frac{\mathrm{d}a_{i1}}{\mathrm{d}t}b_{1j}+\frac{\mathrm{d}a_{i2}}{\mathrm{d}t}b_{2j}+\ldots+\frac{\mathrm{d}a_{in}}{\mathrm{d}t}b_{nj}\right)$$
$$+\left(a_{i1}\frac{\mathrm{d}b_{1j}}{\mathrm{d}t}+a_{i2}\frac{\mathrm{d}b_{2j}}{\mathrm{d}t}+\ldots+a_{in}\frac{\mathrm{d}b_{nj}}{\mathrm{d}t}\right).$$

Consequently, it then follows directly from Definitions 10.3 to 10.6 that for suitably conformable matrices **A** and **B**:

$$\frac{\mathrm{d}}{\mathrm{d}t}(\mathbf{A}\pm\mathbf{B})=\frac{\mathrm{d}\mathbf{A}}{\mathrm{d}t}\pm\frac{\mathrm{d}\mathbf{B}}{\mathrm{d}t}, \tag{10.28}$$

$$\frac{\mathrm{d}}{\mathrm{d}t}(\lambda\mathbf{A})=\lambda\frac{\mathrm{d}\mathbf{A}}{\mathrm{d}t},\ \text{for any constant scalar } \lambda \tag{10.29}$$

and

$$\frac{\mathrm{d}}{\mathrm{d}t}(\mathbf{AB})=\frac{\mathrm{d}\mathbf{A}}{\mathrm{d}t}\mathbf{B}+\mathbf{A}\frac{\mathrm{d}\mathbf{B}}{\mathrm{d}t}. \tag{10.30}$$

Notice that in general $\mathrm{d}\mathbf{A}^2/\mathrm{d}t \neq 2\mathbf{A}(\mathrm{d}\mathbf{A}/\mathrm{d}t)$, for setting $\mathbf{B}=\mathbf{A}$ in Eqn (10.30) yields

$$\frac{\mathrm{d}\mathbf{A}^2}{\mathrm{d}t}=\frac{\mathrm{d}\mathbf{A}}{\mathrm{d}t}\mathbf{A}+\mathbf{A}\frac{\mathrm{d}\mathbf{A}}{\mathrm{d}t}, \tag{10.31}$$

where $(\mathrm{d}\mathbf{A}/\mathrm{d}t)\mathbf{A}$ and $\mathbf{A}(\mathrm{d}\mathbf{A}/\mathrm{d}t)$ are not necessarily equal (section 10.1). It also follows that if **K** is a constant matrix in the sense that its elements are constant functions of t, then

$$\frac{\mathrm{d}\mathbf{K}}{\mathrm{d}t}=0. \tag{10.32}$$

Using the results of Theorems 10.3(d) and 10.8(b), together with Eqn (10.30), we can derive two useful results. The first result applies to any two matrices **A**, **B** which are conformable for multiplication and is

$$\frac{\mathrm{d}}{\mathrm{d}t}(\mathbf{AB})'=\frac{\mathrm{d}}{\mathrm{d}t}(\mathbf{B}'\mathbf{A}')=\frac{\mathrm{d}\mathbf{B}'}{\mathrm{d}t}\mathbf{A}'+\mathbf{B}'\frac{\mathrm{d}\mathbf{A}'}{\mathrm{d}t}; \tag{10.33}$$

the second result applies to any two non-singular square matrices which are conformable for multiplication and is

$$\frac{\mathrm{d}}{\mathrm{d}t}(\mathbf{AB})^{-1}=\frac{\mathrm{d}}{\mathrm{d}t}(\mathbf{B}^{-1}\mathbf{A}^{-1})=\frac{\mathrm{d}\mathbf{B}^{-1}}{\mathrm{d}t}\mathbf{A}^{-1}+\mathbf{B}^{-1}\frac{\mathrm{d}\mathbf{A}^{-1}}{\mathrm{d}t}. \tag{10.34}$$

We now summarize these results in the form of a general theorem.

Theorem 10. 9 (properties of matrix differentiation) Let $\mathbf{A}(t)$ and $\mathbf{B}(t)$ be suitably conformable matrices which are differentiable in some common interval $t_0 < t < t_1$, and let $\mathbf{K}$ be a constant matrix and λ a scalar. Then throughout the interval $t_0 < t < t_1$:

(a) $\dfrac{\mathrm{d}}{\mathrm{d}t}(\mathbf{A}+\mathbf{B}) = \dfrac{\mathrm{d}\mathbf{A}}{\mathrm{d}t} + \dfrac{\mathrm{d}\mathbf{B}}{\mathrm{d}t}$;

(b) $\dfrac{\mathrm{d}}{\mathrm{d}t}(\mathbf{A}-\mathbf{B}) = \dfrac{\mathrm{d}\mathbf{A}}{\mathrm{d}t} - \dfrac{\mathrm{d}\mathbf{B}}{\mathrm{d}t}$;

(c) $\dfrac{\mathrm{d}}{\mathrm{d}t}(\lambda\mathbf{A}) = \lambda\dfrac{\mathrm{d}\mathbf{A}}{\mathrm{d}t}$ (λ a constant scalar);

(d) $\dfrac{\mathrm{d}}{\mathrm{d}t}(\mathbf{A}\mathbf{B}) = \dfrac{\mathrm{d}\mathbf{A}}{\mathrm{d}t}\mathbf{B} + \mathbf{A}\dfrac{\mathrm{d}\mathbf{B}}{\mathrm{d}t}$;

(e) $\dfrac{\mathrm{d}\mathbf{K}}{\mathrm{d}t} = 0$ ($\mathbf{K}$ a constant matrix);

(f) $\dfrac{\mathrm{d}}{\mathrm{d}t}(\mathbf{A}\mathbf{B})' = \dfrac{\mathrm{d}\mathbf{B}'}{\mathrm{d}t}\mathbf{A}' + \mathbf{B}'\dfrac{\mathrm{d}\mathbf{A}'}{\mathrm{d}t}$;

(g) $\dfrac{\mathrm{d}}{\mathrm{d}t}(\mathbf{A}\mathbf{B})^{-1} = \dfrac{\mathrm{d}\mathbf{B}^{-1}}{\mathrm{d}t}\mathbf{A}^{-1} + \mathbf{B}^{-1}\dfrac{\mathrm{d}\mathbf{A}^{-1}}{\mathrm{d}t}$,

where $\mathbf{A}$ and $\mathbf{B}$ are non-singular matrices. ■

Example 10.22 Verify Eqn (10.33) for the matrices

$$\mathbf{A}(t) = \begin{bmatrix} t & 1 \\ -1 & t^2 \end{bmatrix} \quad \text{and} \quad \mathbf{B}(t) = \begin{bmatrix} 2 & t^2 \\ t^3 & 1 \end{bmatrix}.$$

Solution We have

$$\mathbf{AB} = \begin{bmatrix} 2t+t^3 & 1+t^3 \\ t^5-2 & 0 \end{bmatrix},$$

so that

$$(\mathbf{AB})' = \begin{bmatrix} 2t+t^3 & t^5-2 \\ 1+t^3 & 0 \end{bmatrix},$$

and thus

$$\frac{\mathrm{d}}{\mathrm{d}t}(\mathbf{AB})' = \begin{bmatrix} 2+3t^2 & 5t^4 \\ 3t^2 & 0 \end{bmatrix}.$$

Now,

$$\mathbf{A}'(t) = \begin{bmatrix} t & -1 \\ 1 & t^2 \end{bmatrix} \quad \text{and} \quad \mathbf{B}'(t) = \begin{bmatrix} 2 & t^2 \\ t^3 & 1 \end{bmatrix}$$

so that

$$\frac{\mathrm{d}\mathbf{A}'}{\mathrm{d}t} = \begin{bmatrix} 1 & 0 \\ 0 & 2t \end{bmatrix} \quad \text{and} \quad \frac{\mathrm{d}\mathbf{B}'}{\mathrm{d}t} = \begin{bmatrix} 0 & 3t^2 \\ 2t & 0 \end{bmatrix}.$$

Using these results we have

$$\frac{d\mathbf{B}'}{dt}\mathbf{A}' + \mathbf{B}'\frac{d\mathbf{A}'}{dt} = \begin{bmatrix} 0 & 3t^2 \\ 2t & 0 \end{bmatrix}\begin{bmatrix} t & -1 \\ 1 & t^2 \end{bmatrix} + \begin{bmatrix} 2 & t^3 \\ t^2 & 1 \end{bmatrix}\begin{bmatrix} 1 & 0 \\ 0 & 2t \end{bmatrix}$$

$$= \begin{bmatrix} 3t^2 & 3t^4 \\ 2t^2 & -2t \end{bmatrix} + \begin{bmatrix} 2 & 2t^4 \\ t^2 & 2t \end{bmatrix}$$

$$= \begin{bmatrix} 2+3t^2 & 5t^4 \\ 3t^2 & 0 \end{bmatrix} = \frac{d}{dt}(\mathbf{AB})'. \quad \blacksquare$$

10.6 Solution of systems of linear equations

A system of m linear inhomogeneous equations in the n variables x_1, x_2, ..., x_n has the general form

$$\begin{aligned} a_{11}x_1 + a_{12}x_2 + \ldots + a_{1n}x_n &= k_1 \\ a_{21}x_1 + a_{12}x_2 + \ldots + a_{2n}x_n &= k_2 \\ \ldots \\ a_{m1}x_1 + a_{m2}x_2 + \ldots + a_{mn}x_n &= k_m, \end{aligned} \tag{10.35}$$

where the term *inhomogeneous* refers to the fact that not all of the numbers k_1, k_2, ..., k_m are zero. Defining the matrices

$$\mathbf{A} = \begin{bmatrix} a_{11} & a_{12} & \cdots & a_{1n} \\ a_{21} & a_{22} & \cdots & a_{2n} \\ \cdots & \cdots & \cdots & \cdots \\ a_{m1} & a_{m2} & \cdots & a_{mn} \end{bmatrix}, \quad \mathbf{X} = \begin{bmatrix} x_1 \\ x_2 \\ \cdots \\ x_n \end{bmatrix} \quad \text{and} \quad \mathbf{K} = \begin{bmatrix} k_1 \\ k_2 \\ \cdots \\ k_m \end{bmatrix},$$

this system can be written

$$\mathbf{AX} = \mathbf{K}. \tag{10.36}$$

Here $\mathbf{A}$ is called the *coefficient matrix*, $\mathbf{X}$ the *solution vector* and $\mathbf{K}$ the *inhomogeneous vector*.

In the event that $m = n$ and $|\mathbf{A}| \neq 0$ it follows that $\mathbf{A}^{-1}$ exists, so that premultiplication of Eqn (10.36) by $\mathbf{A}^{-1}$ gives for the solution vector,

$$\mathbf{X} = \mathbf{A}^{-1}\mathbf{K}. \tag{10.37}$$

This method of solution is of more theoretical than practical interest because the task of computing $\mathbf{A}^{-1}$ becomes prohibitive when n is much greater than 3. However, one useful method of solution for small systems of such equations ($n = 3$) known as *Cramer's rule* may be deduced from Eqn (10.37).

Consideration of Eqn (10.37) and Definition 10.14 shows that x_i, the ith element in the solution vector $\mathbf{X}$, is given by

$$x_i = \frac{1}{|\mathbf{A}|}(k_1 A_{1i} + k_2 A_{2i} + \ldots + k_n A_{ni}) \tag{10.38}$$

for $i = 1, 2, \ldots, n$, where A_{ij} is the cofactor of **A** corresponding to element a_{ij}. Using Laplace's expansion theorem we then see that the numerator of Eqn (10.38) is simply the expansion of $|\mathbf{A}_i|$, where $\mathbf{A}_i$ denotes the matrix derived from **A** by replacing the ith column of **A** by the column vector **K**. Thus we have derived the simple result

$$x_i = \frac{|\mathbf{A}_i|}{|\mathbf{A}|}$$

for $i = 1, 2, \ldots, n$, which expresses the elements of the solution vector **X** of Eqn (10.35) in terms of determinants.

Rule 10.4 (Cramer's rule)

To solve n linear inhomogeneous equations in n variables proceed as follows:

(a) Compute $|\mathbf{A}|$, the determinant of the coefficient matrix, and, if $|\mathbf{A}| \neq 0$, proceed to the next step.
(b) Compute the modified coefficient determinants $|\mathbf{A}_i|$, $i = 1, 2, \ldots, n$, where $\mathbf{A}_i$ is derived from **A** by replacing the ith column of **A** by the inhomogeneous vector **K**;
(c) Then the solutions $x_1, x_2, \ldots, x_n$ are given by

$$x_i = \frac{|\mathbf{A}_i|}{|\mathbf{A}|}$$

for $i = 1, 2, \ldots, n$.
(d) If $|\mathbf{A}| = 0$ the method fails.

Example 10.23 Use Cramer's rule to solve the equations:

$$\begin{aligned} x_1 + 3x_2 + x_3 &= 8 \\ 2x_1 + x_2 + 3x_3 &= 7 \\ x_1 + x_2 - x_3 &= 2. \end{aligned}$$

Solution The coefficient matrix **A** and the modified coefficient matrices $\mathbf{A}_1$, $\mathbf{A}_2$ and $\mathbf{A}_3$ are obviously:

$$\mathbf{A} = \begin{bmatrix} 1 & 3 & 1 \\ 2 & 1 & 3 \\ 1 & 1 & -1 \end{bmatrix}, \quad \mathbf{A}_1 = \begin{bmatrix} 8 & 3 & 1 \\ 7 & 1 & 3 \\ 2 & 1 & -1 \end{bmatrix},$$

$$\mathbf{A}_2 = \begin{bmatrix} 1 & 8 & 1 \\ 2 & 7 & 3 \\ 1 & 2 & -1 \end{bmatrix} \quad \text{and} \quad \mathbf{A}_3 = \begin{bmatrix} 1 & 3 & 8 \\ 2 & 1 & 7 \\ 1 & 1 & 2 \end{bmatrix}.$$

Hence $|\mathbf{A}| = 12$, $|\mathbf{A}_1| = 12$, $|\mathbf{A}_2| = 24$, and $|\mathbf{A}_3| = 12$, so that

$$x_1 = \frac{|\mathbf{A}_1|}{|\mathbf{A}|} = 1, \quad x_2 = \frac{|\mathbf{A}_2|}{|\mathbf{A}|} = 2, \quad x_3 = \frac{|\mathbf{A}_3|}{|\mathbf{A}|} = 1. \quad \blacksquare$$

In the special case in which $m = n$, but $|\mathbf{A}| = 0$, the inverse matrix does not exist and so any method using $\mathbf{A}^{-1}$ must fail. In these circumstances we must consider more carefully what is meant by a solution. In general, when a solution vector $\mathbf{X}$ exists whose elements simultaneously satisfy all the equations in the system, the equations will be said to be *consistent*. If no solution vector exists having this property then the equations will be said to be *inconsistent*. Consider the following equations:

$$\begin{aligned} x_1 + x_2 + 2x_3 &= 9 \\ 4x_1 - 2x_2 + x_3 &= 4 \\ 5x_1 - x_2 + 3x_3 &= 1. \end{aligned}$$

These equations are obviously inconsistent, because the left-hand side of the third equation is just the sum of the left-hand sides of the first two equations, whereas the right-hand sides are not so related (that is, $1 \neq 9 + 4$). In effect, what we are saying is that there is a linear dependence between the rows of the left-hand side of the equations which is not shared by the inhomogeneous terms. The row linear dependence in the coefficient matrix $\mathbf{A}$ is determined by the rank of $\mathbf{A}$, for if rank $\mathbf{A} = r$, then there are only r linearly independent rows in $\mathbf{A}$. We now offer a brief discussion of one way in which the general problem of consistency may be approached.

Obviously, when working conventionally with the individual equations comprising (9.35) we know that: (a) equations may be scaled, (b) equations may be interchanged, and (c) multiples of one equation may be added to another. This implies that if we consider the coefficient matrix $\mathbf{A}$ of the system and supplement it on the right by the elements of the inhomogeneous vector $\mathbf{K}$ to form what is called the *augmented matrix*, then these same operations are valid for the rows of the augmented matrix. Clearly, rank will not be affected by these operations. If the ranks of $\mathbf{A}$ and of the augmented matrix denoted by $(\mathbf{A}, \mathbf{K})$ are the same, then the equations must be consistent; otherwise they must be inconsistent.

***Definition 10.16* (augmented matrix and elementary row operations)** Suppose that $\mathbf{AX} = \mathbf{K}$, where

$$\mathbf{A} = \begin{bmatrix} a_{11} & a_{12} & \cdots & a_{1n} \\ a_{21} & a_{22} & \cdots & a_{2n} \\ \cdots & \cdots & \cdots & \cdots \\ a_{n1} & a_{n2} & \cdots & a_{nn} \end{bmatrix}, \quad \mathbf{X} = \begin{bmatrix} x_1 \\ x_2 \\ \cdots \\ x_n \end{bmatrix} \quad \text{and} \quad \mathbf{K} = \begin{bmatrix} k_1 \\ k_2 \\ \cdots \\ k_n \end{bmatrix}.$$

Then the *augmented matrix*, written (**A**, **K**), is defined to be the matrix

$$(\mathbf{A}, \mathbf{K}) = \begin{bmatrix} a_{11} & a_{12} & \cdots & a_{1n} & k_1 \\ a_{21} & a_{22} & \cdots & a_{2n} & k_2 \\ \cdots & \cdots & \cdots & \cdots & \cdots \\ a_{n1} & a_{n2} & \cdots & a_{nn} & k_n \end{bmatrix}.$$

An *elementary row operation* performed on an augmented matrix is any one of the following:

(a) scaling of all elements in a row by a factor λ;
(b) interchange of any two rows;
(c) addition of a multiple of one row to another row.

An augmented matrix will be said to have been reduced to echelon form by elementary row operations when the first non-zero element in any row is unity, and it lies to the right of the unity in the row above.

Example 10.24 Perform elementary row operations on the augmented matrix corresponding to the inconsistent equations above to reduce them to echelon form. Find the ranks of **A** and (**A**, **K**).

Solution The augmented matrix is given by

$$(\mathbf{A}, \mathbf{K}) = \begin{bmatrix} 1 & 1 & 2 & 9 \\ 4 & -2 & 1 & 4 \\ 5 & -1 & 3 & 1 \end{bmatrix}.$$

Leaving rows 1 and 2 unchanged, subtract from the elements of row 3 the sum of the corresponding elements in rows 1 and 2, or, using the notation introduced previously, form the matrix with rows $R_1' = R_1$, $R_2' = R_2$ and $R_3' = R_3 - R_1 - R_2$ to obtain

$$\begin{bmatrix} 1 & 1 & 2 & 9 \\ 4 & -2 & 1 & 4 \\ 0 & 0 & 0 & -12 \end{bmatrix}.$$

Form the matrix with rows $R_1' = R_1$, $R_2' = R_2 - 4R_1$, $R_3' = R_3$ to obtain

$$\begin{bmatrix} 1 & 1 & 2 & 9 \\ 0 & -6 & -7 & -32 \\ 0 & 0 & 0 & -12 \end{bmatrix}.$$

Form the matrix with rows $R_1' = R_1$, $R_2' = (-1/6)R_2$, $R_3' = (-1/12)R_3$ to obtain

$$\begin{bmatrix} 1 & 1 & 2 & 9 \\ 0 & 1 & 7/6 & 16/3 \\ 0 & 0 & 0 & 1 \end{bmatrix}.$$

This is now in echelon form and the rank of the matrix comprising the first three columns and corresponding to matrix **A** is 2, which must be the rank of the coefficient matrix **A**. The rank of the augmented matrix (**A**, **K**) must be the rank of the echelon equivalent of the augmented matrix, which is clearly 3. Thus rank $\mathbf{A} \neq$ rank $(\mathbf{A}, \mathbf{K})$, showing that the system of equations is inconsistent.

The general conclusion that may be reached from the echelon form of an augmented matrix (**A**, **K**) is that equations are consistent only when the ranks of **A** and (**A**, **K**) are the same. If the equations are consistent, and **A** is of order $n \times n$ and the rank $r < n$, we shall have fewer equations than variables. In these circumstances we may solve for any r of the variables x_i in terms of the $n - r$ remaining ones which can then be assigned arbitrary values. The solution will not be unique, since it will then depend on the $n - r$ values x_i that may be assigned arbitrarily.

Theorem 10.10 (solution of inhomogeneous systems) The system of equations

$$\mathbf{AX} = \mathbf{K},$$

where **A** is of order $n \times n$ and **X**, **K** are of order $n \times 1$, has a unique solution if $|\mathbf{A}| \neq 0$. If $|\mathbf{A}| = 0$, then the equations are only consistent when the ranks of **A** and (**A**, **K**) are equal. In this case, if the rank $r < n$, it is possible to solve for r variables in terms of the $n - r$ remaining variables which may then be assigned arbitrary values. ■

Example 10.25 Solve the following equations by reducing the augmented matrix to echelon form:

$$x_1 + 3x_2 - x_3 = 6$$

$$8x_1 + 9x_2 + 4x_3 = 21$$

$$2x_1 + x_2 + 2x_3 = 3.$$

Solution The augmented matrix is

$$(\mathbf{A}, \ \mathbf{K}) = \begin{bmatrix} 1 & 3 & -1 & 6 \\ 8 & 9 & 4 & 21 \\ 2 & 1 & 2 & 3 \end{bmatrix}.$$

Form the matrix with rows $R_1' = R_1$, $R_2' = R_2 - 3R_3 - 2R_1$, $R_3' = R_3$ to obtain.

$$\begin{bmatrix} 1 & 3 & -1 & 6 \\ 0 & 0 & 0 & 0 \\ 2 & 1 & 2 & 3 \end{bmatrix}.$$

Form the matrix with rows $R_1' = R_1$, $R_2' = R_3$, $R_3' = R_2$ to obtain

$$\begin{bmatrix} 1 & 3 & -1 & 6 \\ 2 & 1 & 2 & 3 \\ 0 & 0 & 0 & 0 \end{bmatrix}.$$

Form the matrix with rows $R_1' = R_1$, $R_2' = (-1/5)(R_2 - 2R_1)$ to obtain

$$\begin{bmatrix} 1 & 3 & -1 & 6 \\ 0 & 1 & -4/5 & 9/5 \\ 0 & 0 & 0 & 0 \end{bmatrix}.$$

This is now in echelon form and clearly the ranks of $\mathbf{A}$ and $(\mathbf{A}, \mathbf{K})$ are both 2, showing that the equations are consistent. However, only two equations exist between the three variables x_1, x_2 and x_3 because re-interpreting the echelon form of the augmented matrix as a set of scalar equations in x_1, x_2 and x_3 gives

$$x_1 + 3x_2 - x_3 = 6 \quad \text{and} \quad x_2 - \frac{4}{5}x_3 = \frac{9}{5}.$$

Hence, assigning x_3 arbitrarily, we find that

$$x_1 = \frac{3}{5} - x_3 \quad \text{and} \quad x_2 = \frac{9}{5} + \frac{4}{5}x_3.$$

This is the most general solution possible for this system. ■

When the inhomogeneous vector $\mathbf{K} = \mathbf{0}$, the resulting system of equations $\mathbf{AX} = \mathbf{0}$ is said to be *homogeneous.* Consider the case of a homogeneous system of n equations involving the n variables x_1, x_2, $\ldots$, x_n. Then it is obvious that a *trivial solution* $x_1 = x_2 = \ldots = x_n = 0$ corresponding to $\mathbf{X} = \mathbf{0}$ always exists, but a *non-trivial solution*, in the sense that not all $x_1, x_2, \ldots, x_n$ are zero, can only occur if $|\mathbf{A}| = 0$. To see this, notice that if $|\mathbf{A}| \neq 0$ then $\mathbf{A}^{-1}$ exists, so that premultiplication of $\mathbf{AX} = \mathbf{0}$ by $\mathbf{A}^{-1}$ gives at once the trivial solution $\mathbf{X} = \mathbf{0}$ as being the only possible solution. Conversely, if $|\mathbf{A}| = 0$, then certainly at least one row of $\mathbf{A}$ is linearly dependent upon the other rows, showing that not all of the variables $x_1, x_2, \ldots, x_n$ can be zero.

When a non-trivial solution exists to a homogeneous system of n equations involving n variables it cannot be unique, for if $\mathbf{X}$ is a solution vector, then so also is $\lambda\mathbf{X}$, where λ is a scalar. As in our previous discussion, if the rank of $\mathbf{A}$ which is of order $n \times n$ is r, then we may solve for r of the variables $x_1, x_2, \ldots, x_n$ in terms of the $n - r$ remaining ones which can then be assigned arbitrary values.

Theorem 10.11 (solution of homogeneous systems)

The homogeneous system of equations

$$\mathbf{AX} = 0,$$

where $\mathbf{A}$ is of order $n \times n$ and $\mathbf{X}$, $\mathbf{0}$ are of order $n \times 1$, always has the trivial solution $\mathbf{X} = \mathbf{0}$. It has a non-trivial solution only when $|\mathbf{A}| = 0$. If $\mathbf{A}$ is of rank $r < n$, it is possible to solve for r variables in terms of the $n - r$ remaining variables which may then be assigned arbitrary values. If $\mathbf{X}$ is a non-trivial solution, so also is $\lambda\mathbf{X}$, where λ is an arbitrary scalar. ■

Example 10.26 Solve the equations

$$\begin{aligned} x_1 - \ x_2 + \ x_3 &= 0 \\ 2x_1 + \ x_2 - \ x_3 &= 0 \\ x_1 + 5x_2 - 5x_3 &= 0. \end{aligned}$$

Solution There is the trivial solution $x_1 = x_2 = x_3 = 0$ and, since the determinant associated with the coefficient matrix vanishes, there are also non-trivial solutions. The augmented matrix is now

$$(\mathbf{A}, \mathbf{0}) = \begin{bmatrix} 1 & -1 & 1 & 0 \\ 2 & 1 & -1 & 0 \\ 1 & 5 & -5 & 0 \end{bmatrix},$$

which is easily reduced by elementary row transformations to the echelon form

$$\begin{bmatrix} 1 & -1 & 1 & 0 \\ 0 & 1 & -1 & 0 \\ 0 & 0 & 0 & 0 \end{bmatrix}.$$

This shows that there are only two equations between the three variables x_1, x_2 and x_3, for the echelon form of the augmented matrix is seen to be equivalent to the two scalar equations

$$x_1 - x_2 + x_3 = 0 \qquad \text{and} \qquad x_2 - x_3 = 0.$$

Hence, assigning x_3 arbitrarily, we have for our solution $x_1 = 0$ and $x_2 = x_3 = k$ (say). ■

A practical numerical method of solution called *Gaussian elimination* is usually used when dealing with inhomogeneous systems of n equations involving n variables. This is essentially the same method as the one described above for the reduction of an augmented matrix to echelon form. The only difference is that it is not necessary to make the first non-zero element appearing in any row in the position corresponding to the leading diagonal equal to unity.

During the process of Gaussian elimination it may happen that an element on the leading diagonal that is to be used to eliminate all elements in the column below it becomes zero. Should this occur the difficulty is overcome by interchanging the row with one below it in which the corresponding element is non-zero. In fact, in computer packages, the solution of n inhomogeneous linear equations in n unknowns is based on this approach.

Thus if, for example, the following augmented matrix arose during Gaussian elimination

$$\begin{bmatrix} 1 & 3 & 6 & 1 & 2 \\ 0 & 0 & 3 & 1 & 4 \\ 0 & 4 & 1 & 0 & 1 \\ 0 & 1 & 2 & 1 & 3 \end{bmatrix},$$

the occurrence of the zero in the second element of the leading diagonal would terminate ordinary Gaussian elimination. In this case the problem is overcome by interchanging rows 2 and 3 to obtain the equivalent augmented matrix,

$$\begin{bmatrix} 1 & 3 & 6 & 1 & 2 \\ 0 & 4 & 1 & 0 & 1 \\ 0 & 0 & 3 & 1 & 4 \\ 0 & 1 & 2 & 1 & 3 \end{bmatrix},$$

after which the elimination can proceed as before.

The element on the leading diagonal that is used to eliminate all elements in the column below it is called the *pivot element.* In general, at each stage of Gaussian elimination, whenever necessary, the row with the pivotal element is interchanged with a row *below* it to bring to the leading diagonal the element with the largest absolute value. This strategy prevents the introduction of unnecessary errors by using a pivot with small absolute value to eliminate elements below it with much greater absolute values.

Example 10.27 Solve the following equations by Gaussian elimination:

$$\begin{aligned} x_1 - x_2 - x_3 &= 0 \\ 3x_1 + x_2 + 2x_3 &= 6 \\ 2x_1 + 2x_2 + x_3 &= 1. \end{aligned}$$

Solution The augmented matrix

$$(\mathbf{A}, \mathbf{K}) = \begin{bmatrix} 1 & -1 & -1 & 0 \\ 3 & 1 & 2 & 6 \\ 2 & 2 & 1 & 2 \end{bmatrix}.$$

Form the matrix with the rows $R_1' = R_1$, $R_2' = R_2 - 3R_1$, $R_3' = R_3 - 2R_1$ to obtain

$$\begin{bmatrix} 1 & -1 & -1 & 0 \\ 0 & 4 & 5 & 6 \\ 0 & 4 & 3 & 2 \end{bmatrix}.$$

Form the matrix with the rows $R_1' = R_1$, $R_2' = R_2$, $R_3' = R_3 - R_2$ to obtain

$$\begin{bmatrix} 1 & -1 & -1 & 0 \\ 0 & 4 & 5 & 6 \\ 0 & 0 & -2 & -4 \end{bmatrix}.$$

The solution is now found by the process of *back-substitution,* using the scalar equations corresponding to this modified augmented matrix. That is, the equations

$$x_1 - x_2 - x_3 = 0$$
$$4x_2 + 5x_3 = 6$$
$$-2x_3 = -4.$$

The last equation gives $x_3 = 2$ and, using this result in the second then gives $x_2 = -1$. Combination of these results in the first equation then gives $x_1 = 1$. ■

It is not proposed to offer more than a few general remarks about the solutions of m equations involving n variables. If the equations are consistent, but there are more equations than variables so that $m > n$, it is clear that there must be linear dependence between the equations. In the case that the rank of the coefficient matrix is equal to n there will obviously be a unique solution for, despite appearances, there will be only n linearly independent equations involving n variables. If, however, the rank is less than n we are in the situation of solving for r variables $x_1, x_2, \ldots$ in terms of the remaining $n - r$ variables whose values may be assigned arbitrarily. In the remaining case where there are fewer equations than variables we have $m < n$. When this system is consistent it follows that at least $n - m$ variables must be assigned arbitrary values.

10.7 Eigenvalues and eigenvectors

Let us examine the consequence of requiring that in the system

$$\mathbf{AX} = \mathbf{K}, \tag{10.40}$$

where $\mathbf{A}$ is of order $n \times n$ and $\mathbf{X}$, $\mathbf{K}$ are of order $n \times 1$, the vector $\mathbf{K}$ is proportional to the vector $\mathbf{X}$ itself. That is, we are requiring that $\mathbf{K} = \lambda\mathbf{X}$, where λ is some scalar multiplier as yet unknown. This requires us to solve the system

$$\mathbf{AX} = \lambda\mathbf{X}, \tag{10.41}$$

which is equivalent to the homogeneous system

$$(\mathbf{A} - \lambda\mathbf{I})\mathbf{X} = 0, \tag{10.42}$$

where $\mathbf{I}$ is the unit matrix.

Now we know from Theorem 10.11 that Eqn (10.42) can only have a non-trivial solution when the determinant associated with the coefficient matrix vanishes, so that we must have

$$|\mathbf{A} - \lambda\mathbf{I}| = 0. \tag{10.43}$$

When expanded, this determinant gives rise to an algebraic equation of degree n in λ of the form

$$\lambda^n + \alpha_1\lambda^{n-1} + \alpha_2\lambda^{n-2} + \ldots + \alpha_n = 0. \tag{10.44}$$

The determinant (10.43) is called the *characteristic determinant* associated with $\mathbf{A}$ and Eqn (10.44) is called the *characteristic equation.* It has n roots $\lambda_1, \lambda_2, \ldots, \lambda_n$, each of which is called either an *eigenvalue*, a *characteristic root*, or, sometimes, a *latent root* of $\mathbf{A}$.

Example 10.28 Find the characteristic equation and the eigenvalues corresponding to

$$\mathbf{A} = \begin{bmatrix} 1 & 2 \\ 3 & 0 \end{bmatrix}.$$

Solution We have

$$\mathbf{A} - \lambda\mathbf{I} = \begin{bmatrix} 1 & 2 \\ 3 & 0 \end{bmatrix} - \lambda \begin{bmatrix} 1 & 0 \\ 0 & 1 \end{bmatrix} = \begin{bmatrix} 1-\lambda & 2 \\ 3 & -\lambda \end{bmatrix},$$

so that

$$|\mathbf{A} - \lambda\mathbf{I}| = \begin{bmatrix} 1-\lambda & 2 \\ 3 & -\lambda \end{bmatrix} = \lambda^2 - \lambda - 6.$$

Thus the characteristic equation is

$$\lambda^2 - \lambda - 6 = 0,$$

and its roots, the eigenvalues of $\mathbf{A}$, are $\lambda = 3$ and $\lambda = -2$. ■

For any 2×2 matrix

$$\mathbf{A} = \begin{bmatrix} a_{11} & a_{12} \\ a_{21} & a_{22} \end{bmatrix},$$

the characteristic equation of $\mathbf{A}$ may always be written

$$|\mathbf{A} - \lambda\mathbf{I}| = \lambda^2 - (\mathrm{tr}\,\mathbf{A})\lambda + |\mathbf{A}|,$$

where $\mathrm{tr}\,\mathbf{A} = a_{11} + a_{22}$ is called the *trace* of $\mathbf{A}$.

No consideration will be given here to the interpretation that is to be placed on the appearance of repeated roots of the characteristic equation, and henceforth we shall always assume that all the eigenvalues (roots) are distinct.

Returning to Eqn (10.42) and setting $\lambda = \lambda_i$, where λ_i is any one of the eigenvalues, we can then find a corresponding solution vector $\mathbf{X}_i$ which, because of Theorem 10.11, will only be determined to within an arbitrary scalar multiplier. This vector $\mathbf{X}_i$ is called either an *eigenvector*, a *characteristic vector*, or sometimes a *latent vector* of $\mathbf{A}$ corresponding to λ_i. The eigenvectors of a square matrix $\mathbf{A}$ are of fundamental importance in both the theory of matrices and in their application, and some indication of this will be given later in, Section 15.5.

Example 10.29 Find the eigenvectors of the matrix $\mathbf{A}$ in Example 10.28.

Solution Use the fact that the eigenvalues have been determined as being $\lambda = 3$ and $\lambda = -2$ and make the identifications $\lambda_1 = 3$ and $\lambda_2 = -2$. Now let the eigenvectors $\mathbf{X}_1$ and $\mathbf{X}_2$, corresponding to λ_1 and λ_2, be denoted by

$$\mathbf{X}_1 = \begin{bmatrix} x_1^{(1)} \\ x_2^{(1)} \end{bmatrix} \quad \text{and} \quad \mathbf{X}_2 = \begin{bmatrix} x_1^{(2)} \\ x_2^{(2)} \end{bmatrix}.$$

Then for the case $\lambda = \lambda_1$, Eqn (9.42) becomes

$$\begin{bmatrix} (1-3) & 2 \\ 3 & (0-3) \end{bmatrix} \begin{bmatrix} x_1^{(1)} \\ x_2^{(1)} \end{bmatrix} = \mathbf{0},$$

whence

$$-2x_1^{(1)} + 2x_2^{(1)} = 0 \qquad \text{and} \quad 3x_1^{(1)} - 3x_2^{(1)} = 0.$$

These are automatically consistent by virtue of their manner of definition, so that we find that $x_1^{(1)} = x_2^{(2)}$. So, arbitrarily assigning to $x_1^{(1)}$ the value $x_1^{(1)} = 1$, we find that the eigenvector $\mathbf{X}_1$ corresponding to $\lambda_1 = 3$ is

$$\mathbf{X}_1 = \begin{bmatrix} 1 \\ 1 \end{bmatrix}.$$

A similar argument for $\lambda = \lambda_2$ gives

$$\begin{bmatrix} (1+2) & 2 \\ 3 & (0+2) \end{bmatrix} \begin{bmatrix} x_1^{(2)} \\ x_2^{(2)} \end{bmatrix} = \mathbf{0},$$

whence

$$3x_1^{(2)} + 2x_2^{(2)} = 0.$$

Again, arbitrarily assigning to $x_1^{(2)}$ the value $x_1^{(2)} = 1$, we find that $x_2^{(2)} = -3/2$. Thus the eigenvector $\mathbf{X}_2$ corresponding to $\lambda_2 = -2$ is

$$\mathbf{X}_2 = \begin{bmatrix} 1 \\ -\frac{3}{2} \end{bmatrix}.$$

Obviously $\mu\mathbf{X}_1$ and $\mu\mathbf{X}_2$ are also eigenvectors for any arbitrary scalar μ. This fact is often used to *normalize* an eigenvector with components $x_1, x_2, \ldots, x_n$ by setting $\mu = 1/\surd(x_1^2 + x_2^2 + \ldots + x_n^2)$. Thus for $\mathbf{X}_1$ above $\mu = 1/\surd 2$, and for $\mathbf{X}_2$, $\mu = \frac{1}{2}\surd 13$. ■

Example 10.30 Find the eigenvalues, eigenvectors and normalized eigenvectors of

$$\mathbf{A} = \begin{bmatrix} 0 & 2 & 4 \\ 1 & 1 & -2 \\ -2 & 0 & 5 \end{bmatrix}.$$

Solution The characteristic determinant from which the eigenvalues are to be determined is

$$|\mathbf{A} - \lambda\mathbf{I}| = \mathbf{0}.$$

In terms of the given matrix $\mathbf{A}$ this becomes

$$\begin{vmatrix} -\lambda & 2 & 4 \\ 1 & (1-\lambda) & -2 \\ -2 & 0 & (5-\lambda) \end{vmatrix} = 0,$$

which, after expanding, reduces to the characteristic equation

$$\lambda^3 - 6\lambda^2 + 11\lambda - 6 = 0.$$

This has the three roots

$$\lambda_1 = 1, \quad \lambda_2 = 2 \quad \text{and} \quad \lambda_3 = 3$$

which are the required eigenvalues of **A**.

To determine the eigenvector $\mathbf{X}_i$ corresponding to the eigenvalue λ_i we must now solve the matrix equation

$$\mathbf{AX}_i = \lambda_i \mathbf{X}_i$$

or, equivalently, the homogeneous system

$$(\mathbf{A} - \lambda_i \mathbf{I})\mathbf{X}_i = 0.$$

First consider the case where $\lambda = \lambda_1 = 1$. Setting

$$\mathbf{X}_1 = \begin{bmatrix} x_1^{(1)} \\ x_2^{(1)} \\ x_3^{(1)} \end{bmatrix},$$

with $\lambda = \lambda_1 = 1$, it follows that the elements $x_1^{(1)}$, $x_2^{(1)}$ and $x_3^{(1)}$ of the eigenvector $\mathbf{X}_1$ corresponding to $\lambda = \lambda_1$ must satisfy the homogeneous system of equations

$$\begin{aligned} -x_1^{(1)} + 2x_2^{(1)} + 4x_3^{(1)} &= 0 \\ x_1^{(1)} - 2x_3^{(1)} &= 0 \\ -2x_1^{(1)} + 4x_3^{(1)} &= 0. \end{aligned}$$

Thus $x_3^{(1)} = \frac{1}{2}x_1^{(1)}$ and $x_2^{(1)} = -\frac{1}{2}x_1^{(1)}$, with $x_1^{(1)}$ an arbitrary number. Setting $x_1 = 1$ shows that the eigenvector $\mathbf{X}_1$ corresponding to $\lambda = \lambda_1$ is

$$\mathbf{X}_1 = \begin{bmatrix} 1 \\ -\frac{1}{2} \\ \frac{1}{2} \end{bmatrix}.$$

We now consider the case where $\lambda = \lambda_2 = 2$. Setting

$$\mathbf{X}_2 = \begin{bmatrix} x_1^{(2)} \\ x_2^{(2)} \\ x_3^{(2)} \end{bmatrix},$$

with $\lambda = \lambda_2 = 2$, it follows in similar fashion that the elements $x_1^{(2)}$, $x_2^{(2)}$ and $x_3^{(2)}$ of the eigenvector $\mathbf{X}_2$ corresponding to $\lambda = \lambda_2$ must satisfy the homogeneous system of equations

$$\begin{aligned} -2x_1^{(2)} + 2x_2^{(2)} + 4x_3^{(2)} &= 0 \\ x_1^{(2)} - x_2^{(2)} - 2x_3^{(2)} &= 0 \\ -2x_1^{(2)} + 3x_3^{(2)} &= 0. \end{aligned}$$

Thus $x_3^{(2)} = \frac{2}{3}x_1^{(2)}$ and $x_2^{(2)} = -\frac{1}{3}x_1^{(2)}$, with $x_1^{(2)}$ an arbitrary number. Setting $x_1^{(2)} = 1$ shows that the eigenvector $\mathbf{X}_2$ corresponding to $\lambda = \lambda_2$ is

$$\mathbf{X}_2 = \begin{bmatrix} 1 \\ -\frac{1}{3} \\ \frac{2}{3} \end{bmatrix}.$$

In analogous fashion the eigenvector $\mathbf{X}_3$ corresponding to $\lambda = \lambda_3$ can be shown to be

$$\mathbf{X}_3 = \begin{bmatrix} 1 \\ -\frac{1}{2} \\ 1 \end{bmatrix}.$$

A normalized vector $\hat{\mathbf{X}}$ corresponding to

$$\mathbf{X} = \begin{bmatrix} x_1 \\ x_2 \\ x_3 \end{bmatrix}$$

is, by definition,

$$\hat{\mathbf{X}} = \frac{1}{\sqrt{x_1^2 + x_2^2 + x_3^3}} \begin{bmatrix} x_1 \\ x_2 \\ x_3 \end{bmatrix}.$$

Consequently the normalized eigenvectors are

$$\hat{\mathbf{X}}_1 = \frac{1}{\sqrt{6}} \begin{bmatrix} 2 \\ -1 \\ 1 \end{bmatrix}, \quad \hat{\mathbf{X}}_2 = \frac{1}{\sqrt{14}} \begin{bmatrix} 3 \\ -1 \\ 2 \end{bmatrix}, \quad \hat{\mathbf{X}}_3 = \frac{1}{3} \begin{bmatrix} 2 \\ -1 \\ 2 \end{bmatrix}. \quad \blacksquare$$

10.8 Matrix interpretation of change of variables in partial differentiation

The change of variable formulae for partial derivatives given in Theorem 5.22 are best written as the matrix product

$$\begin{bmatrix} \dfrac{\partial F}{\partial \alpha_1} \\ \dfrac{\partial F}{\partial \alpha_2} \\ \cdots \\ \dfrac{\partial F}{\partial \alpha_m} \end{bmatrix} = \begin{bmatrix} \dfrac{\partial x_1}{\partial \alpha_1} & \dfrac{\partial x_2}{\partial \alpha_1} & \cdots & \dfrac{\partial x_n}{\partial \alpha_1} \\ \dfrac{\partial x_1}{\partial \alpha_2} & \dfrac{\partial x_2}{\partial \alpha_2} & \cdots & \dfrac{\partial x_n}{\partial \alpha_2} \\ \cdots & \cdots & \cdots & \cdots \\ \dfrac{\partial x_1}{\partial \alpha_m} & \dfrac{\partial x_2}{\partial \alpha_m} & \cdots & \dfrac{\partial x_n}{\partial \alpha_m} \end{bmatrix} \begin{bmatrix} \dfrac{\partial f_1}{\partial \alpha_1} \\ \dfrac{\partial f_1}{\partial \alpha_2} \\ \cdots \\ \dfrac{\partial f_1}{\partial \alpha_n} \end{bmatrix}.$$

In this matrix product the $m \times n$ matrix on the right-hand side is called the *Jacobian* matrix of the transformation. The most important case arises when $m = n$, for then the Jacobian matrix becomes an $n \times n$ matrix which will have an inverse provided it is non-singular.

As an example of this, we consider the transformation of first-order partial derivatives from Cartesian to cylindrical polar coordinates discussed in Example 5.23. The results may be written

$$\begin{bmatrix} \dfrac{\partial F}{\partial r} \\ \dfrac{\partial F}{\partial \theta} \\ \dfrac{\partial F}{\partial z'} \end{bmatrix} = \begin{bmatrix} \cos\theta & \sin\theta & 0 \\ -r\sin\theta & r\cos\theta & 0 \\ 0 & 0 & 1 \end{bmatrix} \begin{bmatrix} \dfrac{\partial f}{\partial x} \\ \dfrac{\partial f}{\partial y} \\ \dfrac{\partial f}{\partial z} \end{bmatrix}, \tag{10.45}$$

where

$$\mathbf{J} = \begin{bmatrix} \cos\theta & \sin\theta & 0 \\ -r\sin\theta & r\cos\theta & 0 \\ 0 & 0 & 1 \end{bmatrix} \tag{10.46}$$

is the Jacobian matrix of this transformation.

A simple calculation shows $|\mathbf{J}| = r$, so that $\mathbf{J}$ will be non-singular for finite r provided only that $r \neq 0$. So, for r finite but non-zero, that is away from the origin of the polar coordinates, the inverse matrix $\mathbf{J}^{-1}$ exists and is easily shown to be

$$\mathbf{J}^{-1} = \begin{bmatrix} \cos\theta & -\dfrac{1}{r}\sin\theta & 0 \\ \sin\theta & \dfrac{1}{r}\cos\theta & 0 \\ 0 & 0 & 1 \end{bmatrix}.$$

Pre-multiplication of the original result (9.45) by $\mathbf{J}^{-1}$ then shows the inverse transformation from (r, θ, z') to (x, y, z) is given by

$$\begin{bmatrix} \dfrac{\partial f}{\partial x} \\ \dfrac{\partial f}{\partial y} \\ \dfrac{\partial f}{\partial z} \end{bmatrix} = \begin{bmatrix} \cos\theta & -\dfrac{1}{r}\sin\theta & 0 \\ \sin\theta & \dfrac{1}{r}\cos\theta & 0 \\ 0 & 0 & 1 \end{bmatrix} \begin{bmatrix} \dfrac{\partial F}{\partial r} \\ \dfrac{\partial F}{\partial \theta} \\ \dfrac{\partial F}{\partial z'} \end{bmatrix}. \tag{10.47}$$

Thus the transformation of first-order partial derivatives from Cartesian to polar coordinates, and back from polar coordinates to Cartesian coordinates, is uniquely determined provided only that r is finite and non-zero. Similar situations apply to other standard coordinate transformations, though we shall not develop these matters further.

10.9 Linear transformations

An important way in which matrices arise is in connection with what are called *linear transformations.* The idea involved is best understood if described in terms of coordinate transformations, and for this purpose we now examine a special change of coordinates in a plane.

Let us consider the effect of a rotation on points in a plane, and in doing so adopt the standard convention that a rotation in the *positive* sense is one in the *anticlockwise* direction. Suppose the point P in Fig. 10.1, with Cartesian coordinates (x, y), is rotated anticlockwise about the origin O through an angle θ to a point Q with coordinates (x', y') in such a way that its distance r from the origin remains constant. Then if the line OP makes an angle α with the positive x-axis, the line OQ makes an angle $\alpha+\theta$ with the axis, and so

$$x = r\cos\alpha, \quad y = r\sin\alpha, \tag{10.48}$$

while

$$x' = r\cos(\alpha+\theta) = r\cos\alpha\cos\theta - r\sin\alpha\sin\theta$$

$$y' = r\sin(\alpha+\theta) = r\sin\alpha\cos\theta + r\cos\alpha\sin\theta.$$

Combining these results shows that after the rotation the coordinates of Q become

$$\begin{aligned} x' &= x\cos\theta - y\sin\theta \\ y' &= x\sin\theta + y\cos\theta. \end{aligned} \tag{10.49}$$

In matrix form these results can be written

$$\tilde{\mathbf{X}} = \mathbf{AX}, \tag{10.50a}$$

where

$$\tilde{\mathbf{X}} = \begin{bmatrix} x' \\ y' \end{bmatrix}, \quad \mathbf{X} = \begin{bmatrix} x \\ y \end{bmatrix} \quad \text{and} \quad \mathbf{A} = \begin{bmatrix} \cos\theta & -\sin\theta \\ \sin\theta & \cos\theta \end{bmatrix}. \tag{10.50b}$$

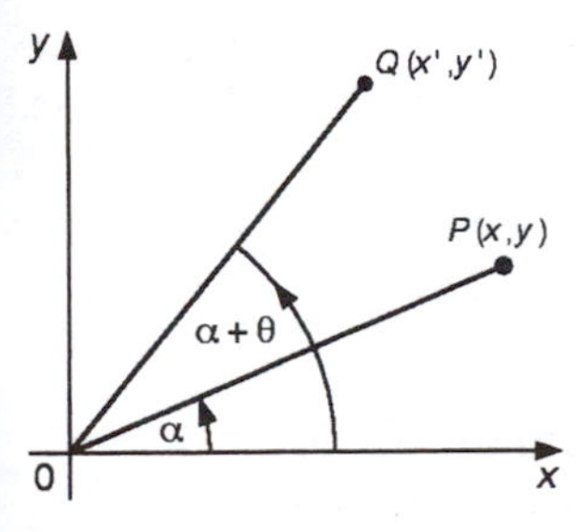

Fig. 10.1 Anticlockwise rotation from P to Q.

Thus the effect of an anticlockwise rotation on points in the (x, y)-plane is described by transformation (10.50), and as this is a linear result it is an example of a *linear transformation.* It is important to recognize that the *rotation matrix* $\mathbf{A}$ describes how *all* points in the (x, y)-plane are transformed, and not just the point P. It is usual to say that $\tilde{\mathbf{X}}$ is the *image* of $\mathbf{X}$ under the transformation, and still to use this term when $\tilde{\mathbf{X}}$ is a vector with an arbitrary number of elements.

It is instructive to examine the effect of two successive rotations on a point, the first through an angle θ and the second through an angle ϕ. Using Eqns (10.50) we see that the point (x, y) has as its image the point (x', y'), where

$$\begin{bmatrix} x' \\ y' \end{bmatrix} = \begin{bmatrix} \cos\phi & -\sin\phi \\ \sin\phi & \cos\phi \end{bmatrix} \begin{bmatrix} \cos\theta & -\sin\theta \\ \sin\theta & \cos\theta \end{bmatrix} \begin{bmatrix} x \\ y \end{bmatrix}$$

so

$$\begin{bmatrix} x' \\ y' \end{bmatrix} = \begin{bmatrix} \cos\phi\cos\theta - \sin\phi\sin\theta & -(\cos\phi\sin\theta + \sin\phi\cos\theta) \\ \cos\phi\sin\theta + \sin\phi\cos\theta & \cos\phi\cos\theta - \sin\phi\sin\theta \end{bmatrix} \begin{bmatrix} x \\ y \end{bmatrix}$$

$$= \begin{bmatrix} \cos(\theta+\phi) & -\sin(\theta+\phi) \\ \sin(\theta+\phi) & \cos(\theta+\phi) \end{bmatrix} \begin{bmatrix} x \\ y \end{bmatrix}.$$

This shows that the matrix transformation representing two successive rotations is described by the rotation matrix

$$\begin{bmatrix} \cos(\theta+\phi) & -\sin(\theta+\phi) \\ \sin(\theta+\phi) & \cos(\theta+\phi) \end{bmatrix}.$$

The last result is to be expected, because the effect of two successive rotations, first by θ and then by ϕ, is equivalent to replacing θ by $\theta + \phi$ in Eqns (10.50). If the first rotation is anticlockwise through an angle θ and the second is clockwise through the same angle, that is through the angle $-\theta$, this last result reduces to the unit matrix showing, correctly, that points have been returned to their original positions.

This last result involving two successive arbitrary rotations is an example of a *composite* linear transformation; that is, a linear transformation produced as a result of the combination of other linear transformations.

We now adopt the following simplified definition of a linear transformation which is sufficiently general for our purposes.

Definition 10.17 (linear transformation) A general *linear transformation* or point *transformation* of the vector $\mathbf{X}$ of order $m \times 1$ into an image vector $\tilde{\mathbf{X}}$ of order $n \times 1$ is defined as a transformation of the form

$$\tilde{\mathbf{X}} = \mathbf{AX} + \mathbf{K},$$

where matrix $\mathbf{A}$ is of order $n \times m$, and $\mathbf{K}$ is a vector of order $n \times 1$.

The special case just considered involved a mapping of points of the plane brought about solely by an anticlockwise rotation of the plane through an angle θ about the origin. In that case the transformation corresponded to $\mathbf{K} = \mathbf{0}$, and

$$\mathbf{A} = \begin{bmatrix} \cos\theta & -\sin\theta \\ \sin\theta & \cos\theta \end{bmatrix}.$$

This is a pure rotation that leaves all distances from the origin unchanged. Aside from the rotation transformation characterized by Eqns (9.50), there are three other simple but important transformations we will need to use that are listed below.

1. The *identity transformation*. This is the transformation $\tilde{\mathbf{X}} = \mathbf{X}$, and it corresponds to the case $\mathbf{K} = \mathbf{0}$ and $\mathbf{A} = \mathbf{I}$. Under this transformation $\mathbf{X}$ and its image $\tilde{\mathbf{X}}$ are coincident.
2. *The translation transformation*. This is the transformation $\tilde{\mathbf{X}} = \mathbf{X} + \mathbf{K}$, and it corresponds to an arbitrary non-zero vector $\mathbf{K}$ and $\mathbf{A} = \mathbf{I}$. The effect of the transformation is to translate $\mathbf{X}$ to its image $\tilde{\mathbf{X}}$, without rotation or change of scale.
3. *The dilation transformation*. This is a transformation $\tilde{\mathbf{X}} = \mathbf{AX}$, in which $\mathbf{A}$ is a non-singular diagonal matrix. Its effect when mapping $\mathbf{X}$ into $\tilde{\mathbf{X}}$ is to change the scale of the different elements of $\mathbf{X}$ without translation or rotation. In the special case that all the diagonal elements are equal, say to λ, its effect is one of a magnification of $\mathbf{X}$ when $\lambda > 1$, and to a contraction when $0 < \lambda < 1$.

Example 10.31 If

$$\mathbf{X} = \begin{bmatrix} x \\ y \end{bmatrix}, \quad \tilde{\mathbf{X}} = \begin{bmatrix} x' \\ y' \end{bmatrix} \quad \text{and} \quad \mathbf{A} = \begin{bmatrix} 5 & 0 \\ 0 & 2 \end{bmatrix},$$

deduce the image of the curve $y = \sinh x$ under the transformation

$$\tilde{\mathbf{X}} = \mathbf{AX}.$$

Solution We have

$$\begin{bmatrix} x' \\ y' \end{bmatrix} = \begin{bmatrix} 5 & 0 \\ 0 & 2 \end{bmatrix} \begin{bmatrix} x \\ y \end{bmatrix},$$

so that

$$x' = 5x \quad \text{and} \quad y' = 2y.$$

Thus the image curve of $y = \sinh x$ is given parametrically by

$$x' = 5x \quad \text{and} \quad y' = 2\sinh x$$

or, equivalently, by

$$y' = 2\sin(x'/5).$$

10.10 Applications of matrices and linear transformations

It is the object of this final section to indicate a few of the diverse applications of the work of this chapter. Of necessity, we will be able to do no more than outline this large and fruitful field of study, and for our first example we shall determine the currents flowing in an electrical network.

10.10.1 Electrical networks

Matrices play an important part in many branches of electrical engineering. One of these branches deals with the analysis of electrical networks, and the simplest of these involves only resistors and batteries. We now illustrate this by determining the currents flowing in such a network. In order to accomplish this we will need to make use of *Kirchhoff's laws*:

(i) the algebraic sum of the potential drops around a closed circuit is zero, and
(ii) the current entering any junction (node) of the circuit must equal the algebraic sum of the currents leaving that junction.

The latter law is automatically satisfied if cyclic currents are used, and these are indicated in Fig. 10.2, in which the circuit to be analysed is shown. The number adjacent to a battery is its electromotive force (e.m.f.), in volts, and the number adjacent to a resistor is its resistance in ohms. The cyclic currents I_1, I_2, I_3 and I_4 are shown flowing in the four branches according to the convention that an anticlockwise flow is positive. Thus, considering the closed circuit with cyclic current I_1, the current flowing through the resistor of 6 ohms is $I_1 - I_3$, while the current flowing through the resistor of 10 ohms is I_1. It should be remembered that the currents may be either positive or negative, and that a negative current means that its indicated direction of flow must be reversed.

Applying the first law to each of the four closed circuits associated with the cyclic currents, and using Ohm's law $E = IR$ relating the current I, the voltage E and the resistance R in a resistor, we arrive at the equations

$$\begin{aligned}
12 &= 10I_1 + 4(I_1 - I_2) + 6(I_1 - I_3) + 2(I_1 - I_4) \\
8 &= 4(I_2 - I_1) + 3(I_2 - I_3) \\
9 &= 6(I_3 - I_1) + 3(I_3 - I_2) + 5(I_3 - I_4) \\
10 &= 2(I_4 - I_1) + 5(I_4 - I_3) + 7I_4.
\end{aligned}$$

After grouping terms these equations become

$$\begin{aligned}
22I_1 - 4I_2 - 6I_3 - 2I_4 &= 12 \\
-4I_1 + 7I_2 - 3I_3 &= 8 \\
-6I_1 - 3I_2 + 14I_3 - 5I_4 &= 9 \\
-2I_1 - 5I_3 + 14I_4 &= 10,
\end{aligned} \tag{10.52}$$

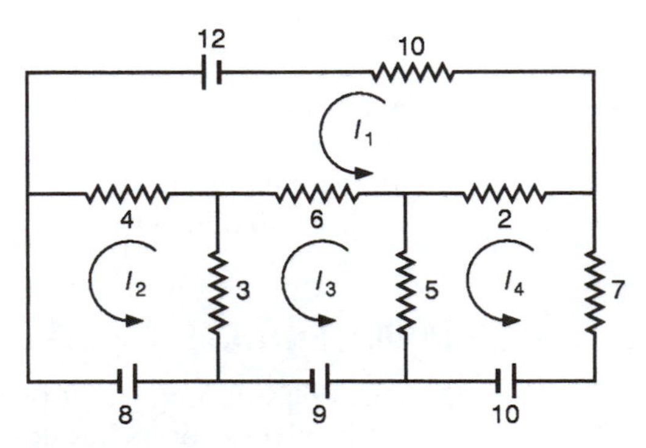

Fig. 10.2 Simple electrical network.

which may be written in the matrix form

$$\mathbf{Ri} = \mathbf{E}, \tag{10.53}$$

where

$$\mathbf{R} = \begin{bmatrix} 22 & -4 & -6 & -2 \\ -4 & 7 & -3 & 0 \\ -6 & -3 & 14 & -5 \\ -2 & 0 & -5 & 14 \end{bmatrix}, \quad \mathbf{i} = \begin{bmatrix} I_1 \\ I_2 \\ I_3 \\ I_4 \end{bmatrix} \quad \text{and} \quad \mathbf{E} = \begin{bmatrix} 12 \\ 8 \\ 9 \\ 10 \end{bmatrix}.$$

Equation (10.53) may be considered as a generalization of Ohm's law, in which **R** is the resistance matrix, **i** is the current vector and **E** is the applied e.m.f. vector.

In matrix form the solution to Eqn. (10.53) may be written

$$\mathbf{i} = \mathbf{R}^{-1}\mathbf{E}, \tag{10.54}$$

provided $|\mathbf{R}| \neq 0$. Alternatively, Eqns (10.52) may be solved by Gaussian elimination, which is usually the most convenient method.

10.10.2 Forces in a truss

A structure formed from N rigid rods, which are all co-planar and pin-jointed together at M joints to form a single rigid body within the plane, is called a *truss*. If it is possible to remove a rod, and for the remaining structure still to be rigid in the plane, the rod is said to be *redundant*.

If the truss is in equilibrium, the algebraic sum of all the forces acting on the truss in both the horizontal and vertical directions must be zero, as must be the algebraic sum of all the moments. These results comprise three linear conditions which must always be satisfied.

Now, equating to zero the algebraic sum of the horizontal and vertical components of the forces acting at each joint will give rise to $2M$ linear equations. However, these will imply the three overall equilibrium conditions just mentioned, so only $2M - 3$ of the equations at the joints will be linearly independent. Since each of the N rods will have an unknown force acting within it there will be N unknowns in all.

We conclude that there will be as many linearly independent equations as there are unknowns if

$$N = 2M - 3.$$

The truss is then said to be *statically determinate*, and a unique force acts in each rod.

If $N > 2M - 3$ there will be fewer linearly independent equations than there are unknowns, and an infinity of solutions will be possible. Such a system is said to be *statically indeterminate*.

Finally, if $N < 2M - 3$, there will be more linearly independent equations than there are unknowns, corresponding to the presence of redundant rods in the truss.

To illustrate matters, consider the loaded truss shown in Fig. 10.3 in which the rods are all of equal length and of negligible weight relative to the vertical loads W at B and 2W at D. It is assumed that the truss rests on smooth supports at A and E. The truss has seven rods and five joints, so $N=7$, $M=5$ and we see that $N=2M-3$, so the system is statically determinate.

Taking moments about A and E shows that the reaction at A is $5W/4$ and the reaction at E is $7W/4$. Equating the algebraic sum of the vertical and horizontal components of force acting at each joint gives:

at A

$$F_1 \sin\frac{\pi}{3} - \frac{5W}{4} = 0, \quad F_1 \cos\frac{\pi}{3} + F_2 = 0,$$

at B

$$F_1 \sin\frac{\pi}{3} + F_3 \sin\frac{\pi}{3} - W = 0, \quad F_1 \cos\frac{\pi}{3} - F_3 \cos\frac{\pi}{3} - F_4 = 0,$$

at C

$$F_3 \sin\frac{\pi}{3} + F_5 \sin\frac{\pi}{3} = 0, \quad F_2 + F_3 \cos\frac{\pi}{3} - F_5 \cos\frac{\pi}{3} - F_6 = 0,$$

at D

$$F_5 \sin\frac{\pi}{3} + F_7 \sin\frac{\pi}{3} - 2W = 0, \quad F_4 + F_5 \cos\frac{\pi}{3} - F_7 \cos\frac{\pi}{3} = 0$$

at E

$$F_7 \sin\frac{\pi}{3} - \frac{7W}{4} = 0, \quad F_6 + F_7 \cos\frac{\pi}{3} = 0. \tag{10.55}$$

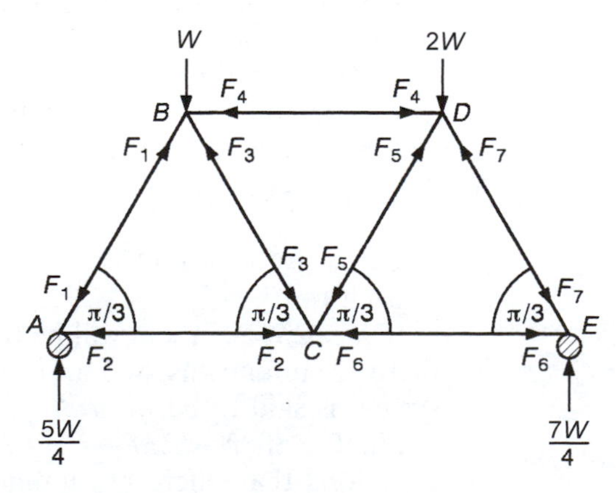

Fig. 10.3 Simple loaded truss.

This is a system of ten linear equations for the seven unknown forces, and it can only have a unique solution if three of these equations are linearly dependent on the remaining seven equations. In fact the last three equations depend linearly on the first seven. The forces are most easily found by solving these equations sequentially. A positive force will indicate that the arrow in the diagram is correctly oriented, and a negative force will indicate that its sense must be reversed.

A simple calculation shows

$$F_1 = 1.4434W,\ F_2 = -0.7217W,\ F_3 = -0.2887W,$$

$$F_4 = 0.8660W,\ F_5 = 0.2887W,\ F_6 = -1.0104W,$$

$$F_7 = 2.0208W.$$

To formulate this as a matrix problem, all that is necessary is to write Eqns (10.55) in the form

$$\mathbf{AF} = \mathbf{K}, \tag{10.56}$$

where

$$\mathbf{A} = \begin{bmatrix} \sin\frac{\pi}{3} & 0 & 0 & 0 & 0 & 0 & 0 \\ \cos\frac{\pi}{3} & 1 & 0 & 0 & 0 & 0 & 0 \\ \sin\frac{\pi}{3} & 0 & \sin\frac{\pi}{3} & 0 & 0 & 0 & 0 \\ \cos\frac{\pi}{3} & 0 & -\cos\frac{\pi}{3} & -1 & 0 & 0 & 0 \\ 0 & 0 & \sin\frac{\pi}{3} & 0 & \sin\frac{\pi}{3} & 0 & 0 \\ 0 & 1 & \cos\frac{\pi}{3} & 0 & -\cos\frac{\pi}{3} & -1 & 0 \\ 0 & 0 & 0 & 0 & \sin\frac{\pi}{3} & 0 & \sin\frac{\pi}{3} \\ 0 & 0 & 0 & 1 & \cos\frac{\pi}{3} & 0 & -\cos\frac{\pi}{3} \\ 0 & 0 & 0 & 0 & 0 & 0 & \sin\frac{\pi}{3} \\ 0 & 0 & 0 & 0 & 0 & 1 & \cos\frac{\pi}{3} \end{bmatrix} \quad \mathbf{F} = \begin{bmatrix} F_1 \\ F_2 \\ F_3 \\ F_4 \\ F_5 \\ F_6 \\ F_7 \end{bmatrix} \quad \mathbf{K} = \begin{bmatrix} 5W/4 \\ 0 \\ W \\ 0 \\ 0 \\ 0 \\ 2W \\ 0 \\ 7W/4 \\ 0 \end{bmatrix}.$$

It follows from section 10.6 that there will be a unique solution provided rank $\mathbf{A} = \text{rank}\ (\mathbf{A}, \mathbf{K}) = 7$. This is easily verified by reducing $(\mathbf{A}, \mathbf{K})$ to echelon form by means of elementary row operations. The forces may then be found by back-substitution from the echelon form of Eqn. (10.56).

10.10.3 Differentials as linear transformations

We now consider a generalization of the total differential as described in Theorem 5.19 and subsequently used in Theorem 5.22. Let us suppose that

$$\left.\begin{aligned} u_1 &= f_1(x_1, x_2, \ldots, x_n) \\ u_2 &= f_2(x_1, x_2, \ldots, x_n) \\ &\ldots \\ u_n &= f_n(x_1, x_2, \ldots, x_n) \end{aligned}\right\}. \tag{10.57}$$

then it follows from Theorem 5.19 and the properties of matrices that

$$\begin{bmatrix} \mathrm{d}u_1 \\ \mathrm{d}u_2 \\ \vdots \\ \mathrm{d}u_n \end{bmatrix} = \begin{bmatrix} \frac{\partial f_1}{\partial x_1} & \frac{\partial f_1}{\partial x_2} & \cdots & \frac{\partial f_1}{\partial x_n} \\ \frac{\partial f_2}{\partial x_1} & \frac{\partial f_2}{\partial x_2} & \cdots & \frac{\partial f_2}{\partial x_n} \\ \vdots & \vdots & \vdots\vdots\vdots & \vdots \\ \frac{\partial f_n}{\partial x_1} & \frac{\partial f_n}{\partial x_2} & \cdots & \frac{\partial f_n}{\partial x_n} \end{bmatrix} \begin{bmatrix} \mathrm{d}x_1 \\ \mathrm{d}x_2 \\ \vdots \\ \mathrm{d}x_n \end{bmatrix}. \tag{10.58}$$

This can be written

$$\mathrm{d}\mathbf{u} = \mathbf{A}\mathrm{d}\mathbf{x} \tag{10.59}$$

by identifying $\mathrm{d}\mathbf{u}$, $\mathrm{d}\mathbf{x}$ with the $n \times 1$ column vectors in Eqn (10.58) in the obvious manner, and $\mathbf{A}$ with the $n \times n$ matrix of partial derivatives. Viewed in this light, Eqn (10.59) may be seen to be a *local* linear transformation mapping $\mathrm{d}\mathbf{x}$ into $\mathrm{d}\mathbf{u}$. The adjective 'local' is used here because the transformation will only be a *linear transformation* when $\mathbf{A}$ is a constant matrix, and as the elements of A are functions of $x_1, x_2, \ldots, x_n$. They can only be approximated by constants in the neighbourhood of any fixed point P with coordinates $\{x_1^P, x_2^P, \ldots, x_n^P\}$. For different points P, the transformation $\mathbf{A}$ will be different, showing that Eqn (10.59) represents a more general type of transformation than a general linear point transformation.

Transformation (10.59) will be one – one provided that $\mathbf{A}^{-1}$ exists, for then a unique inverse mapping

$$\mathrm{d}\mathbf{x} = \mathbf{A}^{-1}\mathrm{d}\mathbf{x} \tag{10.60}$$

will exist. The condition for this is, of course, that $|\mathbf{A}| \neq 0$ at the point P. This will be recognized as the non-vanishing Jacobian condition already encountered in Chapter 5.

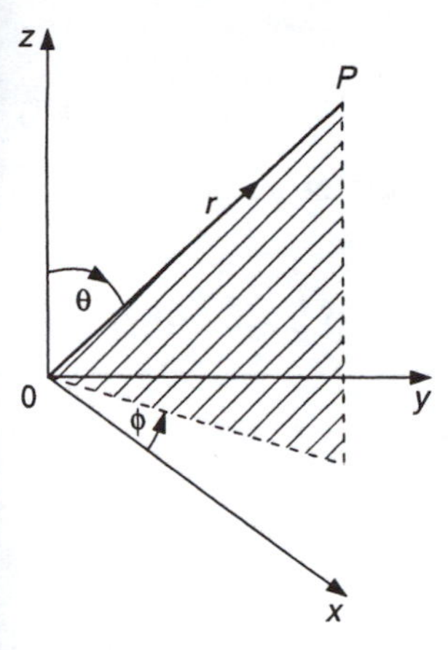

Fig. 10.4 Spherical polar coordinates.

By way of example, consider the relationship between the spherical polar coordinates (r, ϕ, θ) and the Cartesian coordinates (x, y, z) illustrated in Fig. 10.4 and described by

$$x = r \sin\theta \cos\phi$$

$$y = r \sin\theta \sin\phi$$

$$z = r \cos\theta.$$

Making the identifications $u_1 = x$, $u_2 = y$, $u_3 = z$ and $x_1 = r$, $x_2 = \theta$, $x_3 = \phi$, a simple calculation shows that Eqn (9.58) will take the form

$$\begin{bmatrix} \mathrm{d}x \\ \mathrm{d}y \\ \mathrm{d}z \end{bmatrix} = \begin{bmatrix} \sin\theta\cos\phi & r\cos\theta\cos\phi & -r\sin\theta\sin\phi \\ \sin\theta\sin\phi & r\cos\theta\sin\phi & r\sin\theta\cos\phi \\ \cos\theta & -r\sin\theta & 0 \end{bmatrix} \begin{bmatrix} \mathrm{d}r \\ \mathrm{d}\theta \\ \mathrm{d}\phi \end{bmatrix}. \tag{10.61}$$

Denoting the square matrix in Eqn (10.61) by $\mathbf{A}$, it is easily established that the Jacobian determinant $|\mathbf{A}| = r^2 \sin\theta$. Calculating the inverse matrix $\mathbf{A}^{-1}$ and using it to deduce the inverse mapping we have, provided $r^2 \sin\theta \neq 0$ that

$$\begin{bmatrix} \mathrm{d}r \\ \mathrm{d}\theta \\ \mathrm{d}\phi \end{bmatrix} = \frac{1}{r^2 \sin\theta} \begin{bmatrix} r^2\sin^2\theta\cos\phi & r^2\sin^2\theta\sin\phi & r^2\sin^2\theta\cos\theta \\ r\sin\theta\cos\theta\cos\phi & r\sin\theta\cos\theta\sin\phi & -r\sin^2\theta \\ -r\sin\phi & r\cos\phi & 0 \end{bmatrix} \begin{bmatrix} \mathrm{d}x \\ \mathrm{d}y \\ \mathrm{d}z \end{bmatrix}. \tag{10.62}$$

10.10.4 Application of rank to dimensional analysis – Buckingham pi theorem

In many branches of engineering and science, a valuable method of approach to difficult problems is via the method of dimensional analysis touched on briefly at the start of Chapter 5. In essence, this method seeks first to characterize a physical situation by forming dimensionless groups from the variables involved, and then to determine the functional relationships which relate these dimensionless groups. Our contribution will be to the first part of this process, for we shall determine how many dimensionless groups exist.

Let us suppose that a physical situation is described by n variables $u_1, u_2, \ldots, u_n$, each of which corresponds to a physical quantity. Suppose also that each of these quantities is capable of expression dimensionally in terms of length [L], mass [M] and time [T], and that u_i has dimensions

$$[L]^{a_i}[M]^{b_i}[T]^{c_i}.$$

Then the product of powers

$$u_1^{k_1} u_2^{k_2} \ldots u_n^{k_n}, \tag{10.63}$$

where $k_1, k_2, \ldots, k_n$ are real numbers, must have dimensions

$$[L]^{a_1k_1+\ldots+a_nk_n}[M]^{b_1k_1+\ldots+b_nk_n}[T]^{c_1k_1+\ldots+c_nk_n}.$$

Such products of powers will be dimensionless, in the sense that they are pure numbers having dimensions

$$[L]^0[M]^0[T]^0,$$

only if

$$a_1k_1 + a_2k_2 + \ldots + a_nk_n = 0$$
$$b_1k_1 + b_2k_2 + \ldots + b_nk_n = 0$$
$$c_1k_1 + c_2k_2 + \ldots + c_nk_n = 0$$

or, equivalently, if

$$\begin{bmatrix} a_1 & a_2 & \ldots & a_n \\ b_1 & b_2 & \ldots & b_n \\ c_1 & c_2 & \ldots & c_n \end{bmatrix} \begin{bmatrix} k_1 \\ k_2 \\ \vdots \\ k_n \end{bmatrix} = \mathbf{0}. \tag{10.64}$$

Now if the rank of the coefficient matrix of order $3 \times n$ in Eqn (10.64) is r, then we know from the work of section 9.6 that it is possible to express $n-r$ of the variables $k_1, k_2, \ldots, k_n$ in terms of the remaining r variables.That is to say, it will be possible to form $n-r$ dimensionless quantities $\pi_1, \pi_2, \ldots, \pi_{n-r}$ from the n variables $u_1, u_2, \ldots, u_n$. The dimensionless variables π_i, are called *π-variables* or *pi-variables.* Hence we have established the following result.

Theorem 10.12 (Buckingham pi theorem) Let a physical situation be capable of description in terms of n physical quantities $u_1, u_2, \ldots, u_n$, where u_i has dimensions $[L]^{a_i}[M]^{b_i}[T]^{c_i}$. Then, if r is the rank of the matrix

$$\begin{bmatrix} a_1 & a_2 & \ldots & a_n \\ b_1 & b_2 & \ldots & b_n \\ c_1 & c_2 & \ldots & c_n \end{bmatrix},$$

the physical situation is capable of description in terms of $n-r$ dimensionless variables $\pi_1, \pi_2, \ldots, \pi_{n-r}$ formed from the variables u_1, $u_2, \ldots, u_n$. ■

This is best illustrated by example. In the slow viscous flow of a fluid between parallel planes, some functional relationships of the form

$$V = f(k, d, \eta)$$

exist between the average flow velocity V, the pressure gradient k along the flow, the distance d between the planes and the viscosity η. The dimensions of these quantities which will form the matrix in the Buckingham pi theorem are shown in the table below.

	V	k	d	η
L	1	−2	1	−1
M	0	1	0	1
T	−1	−2	0	−1

The rank of the 3×4 matrix whose elements comprise the entries in this table is 3, as may be seen, for example, by using elementary row operations to reduce it to its echelon equivalent form

$$\begin{bmatrix} 1 & -2 & 1 & -1 \\ 0 & 1 & 0 & 1 \\ 0 & 0 & 1 & 0 \end{bmatrix},$$

in which the determinant formed from the first three columns is non-zero.

Thus, from the conditions of the theorem, the number of π-variables is $4 - 3 = 1$. A dimensionless grouping in this case is $kd^2/\eta V$, and any product of powers of the form shown in (9.63) must be a power of this one dimensionless group. Hence this physical problem is capable of description in terms of the one dimensionless grouping $\pi = kd^2/\eta V$. As the velocity profile across the flow only depends on the distance x from one of the walls, our result implies that all such flows will be characterized by one curve describing the variation of π with the dimensionless distance x/d.

10.10.5 Computer graphics

A major application of linear transformations occurs in the computer graphics routines used in all computer-aided design (CAD) packages. In their simplest two-dimensional form these routines generate diagrams by combining straight-line segments with arcs of circles and ellipses, by scaling and moving basic geometrical shapes about the (x, y)-plane, and by using any symmetry that may exist.

We will only consider the simplest cases involving two-dimensional geometrical shapes in the (x, y)-plane. The basic operations involved are as follows:

(i) *Scaling* a shape to adjust its size, and by using different scale factors in the x- and y-directions to adjust its shape.
(ii) *Translating* a basic shape, either to move it to a different position in the plane or to duplicate it if it occurs more than once in the finished diagram.
(iii) *Rotating* a shape.

(iv) *Reflecting* a basic diagram about a given straight line. This operation can be used for the rapid construction of a shape that is symmetrical about the given line.
(v) *Shearing* operations which may be regarded as an extension of scaling, in that the linear transformation in the x-direction is influenced by the linear transformation in the y-direction, and conversely.

These operations have, in effect, already been introduced in section 10.9, but we shall now interpret them in terms of diagrams and the matrix transformations that are involved.

Scaling

The scaling of a shape need not be the same in both the x- and y-directions. Thus if k_x and k_y are the scale factors in the respective x- and y-directions, the simple linear transformation involved is given by

$$x' = k_x x \quad \text{and} \quad y' = k_y y. \tag{10.65}$$

In matrix form these equations become

$$\tilde{\mathbf{X}} = \mathbf{AX}, \tag{10.66}$$

where

$$\tilde{\mathbf{X}} = \begin{bmatrix} x' \\ y' \end{bmatrix}, \quad \mathbf{X} = \begin{bmatrix} x \\ y \end{bmatrix} \quad \text{and} \quad \mathbf{A} = \begin{bmatrix} k_x & 0 \\ 0 & k_y \end{bmatrix}. \tag{10.67}$$

Setting $k_x = 3$ and $k_y = 2$, the unit square in Fig. 10.5(a) is seen to be transformed into the rectangle in Fig. 10.5(b). Similarly, the unit circle $x^2 + y^2 = 1$ in Fig. 10.5(c) becomes the ellipse

$$\frac{(x')^2}{9} + \frac{(y')^2}{4} = 1$$

in Fig. 10.5(d).

Translation

The translation of a shape is accomplished by means of the linear transformation

$$x' = a + x, \quad y' = b + y, \tag{10.68}$$

corresponding to a shift of origin. In matrix form this linear transformation becomes

$$\tilde{\mathbf{X}} = \mathbf{K} + \mathbf{AX}, \tag{10.69}$$

where

$$\tilde{\mathbf{X}} = \begin{bmatrix} x' \\ y' \end{bmatrix}, \quad \mathbf{K} = \begin{bmatrix} a \\ b \end{bmatrix} \quad \text{and} \quad \mathbf{A} = \begin{bmatrix} 1 & 0 \\ 0 & 1 \end{bmatrix}. \tag{10.70}$$

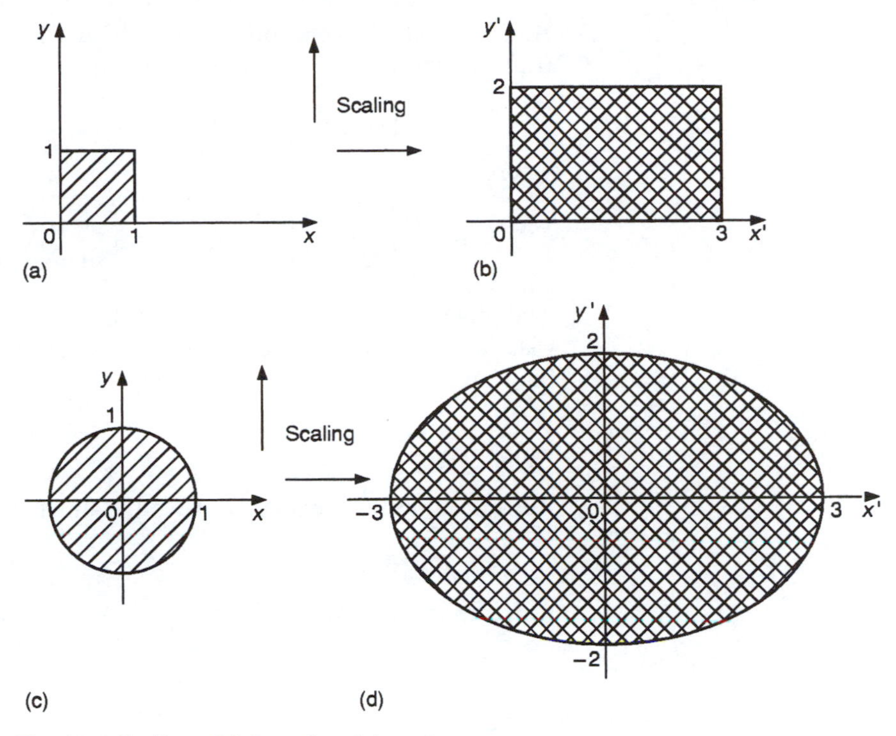

Fig. 10.5 Scaling with $k_x = 3$ and $k_y = 2$.

The effect of this transformation on the unit square is shown in Fig. 10.6 where, for convenience, both the original square and the result of its translation are shown in the same diagram.

Rotation

In section 10. 9 we saw that when a pure rotation θ about the origin in the anticlockwise direction is regarded as a rotation in the positive sense, the

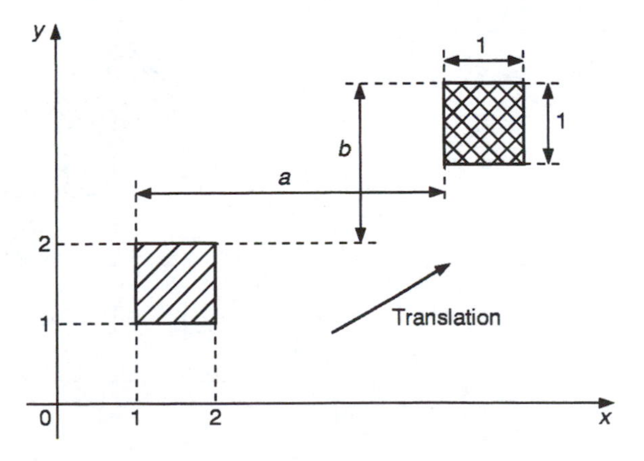

Fig. 10.6 Translation.

effect of such a rotation about the origin is governed by the matrix transformation

$$\tilde{\mathbf{X}} = \mathbf{AX}, \tag{10.71}$$

where

$$\tilde{\mathbf{X}} = \begin{bmatrix} x' \\ y' \end{bmatrix}, \quad \mathbf{X} = \begin{bmatrix} x \\ y \end{bmatrix} \quad \text{and} \quad \mathbf{A} = \begin{bmatrix} \cos\theta & -\sin\theta \\ \sin\theta & \cos\theta \end{bmatrix}. \tag{10.72}$$

The effect this rotation has on a rectangle is shown in Fig. 10.7 for the case $\theta = \pi/4$, where again both rectangles are shown on the same diagram.

Reflection

Reflection about the x-axis obeys the linear transformation

$$x = x, \quad y = -y, \tag{10.73}$$

which in matrix form becomes

$$\tilde{\mathbf{X}} = \mathbf{AX}, \tag{10.74}$$

where

$$\tilde{\mathbf{X}} = \begin{bmatrix} x' \\ y' \end{bmatrix}, \quad \mathbf{X} = \begin{bmatrix} x \\ y \end{bmatrix} \quad \text{and} \quad \mathbf{A} = \begin{bmatrix} 1 & 0 \\ 0 & -1 \end{bmatrix}. \tag{10.75}$$

Reflection about the y-axis obeys the same linear transformation as Eqn (10.74), but this time

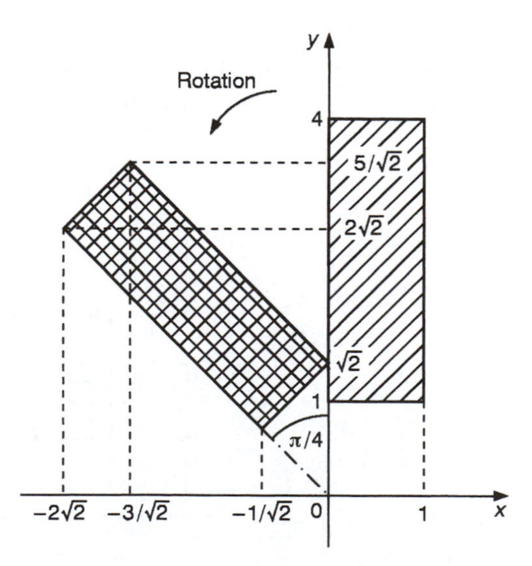

Fig. 10.7 Anticlockwise rotation through angle $\theta = \pi/4$.

$$\mathbf{A} = \begin{bmatrix} -1 & 0 \\ 0 & 1 \end{bmatrix}. \tag{10.76}$$

The effect of these reflections on a semi-circle are shown in Fig. 10.8; Fig. 10.8(a) shows the effect of a reflection in the x-axis, while Fig. 10.8(b) shows the effect of a reflection in the y-axis.

The effect of a reflection in a line through the origin inclined to the x-axis at an angle θ is obtained by means *of a composite transformation*, involving rotation and reflection. This is illustrated in Fig. 10.9. The first step involves a clockwise rotation of the semi-circle in Fig. 10.9(a) to bring the semi-circle into alignment with the x-axis, in Fig. 10.9(b) A reflection in the x-axis is then performed in Fig. 10.9(c), after which an anticlockwise rotation is made to restore the semi-circle to its original position, together with its reflection, to form a complete circle as in Fig. 10.9(d).

The transformation for the first step is obtained from the rotational transformation by replacing θ by $-\theta$, because θ is measured positively in the anticlockwise direction and this rotation is in the clockwise sense. Thus the transformation involved is

$$\mathbf{A}_1 = \begin{bmatrix} \cos(-\theta) & -\sin(-\theta) \\ \sin(-\theta) & \cos(-\theta) \end{bmatrix} = \begin{bmatrix} \cos\theta & \sin\theta \\ -\sin\theta & \cos\theta \end{bmatrix},$$

and the result is shown in Fig. 10.9(b).

Next, to obtain a reflection in the x-axis, we use the reflection transformation in Eqn (10.74), so the transformation to be performed involved the reflection matrix

$$\mathbf{A}_2 = \begin{bmatrix} 1 & 0 \\ 0 & -1 \end{bmatrix}.$$

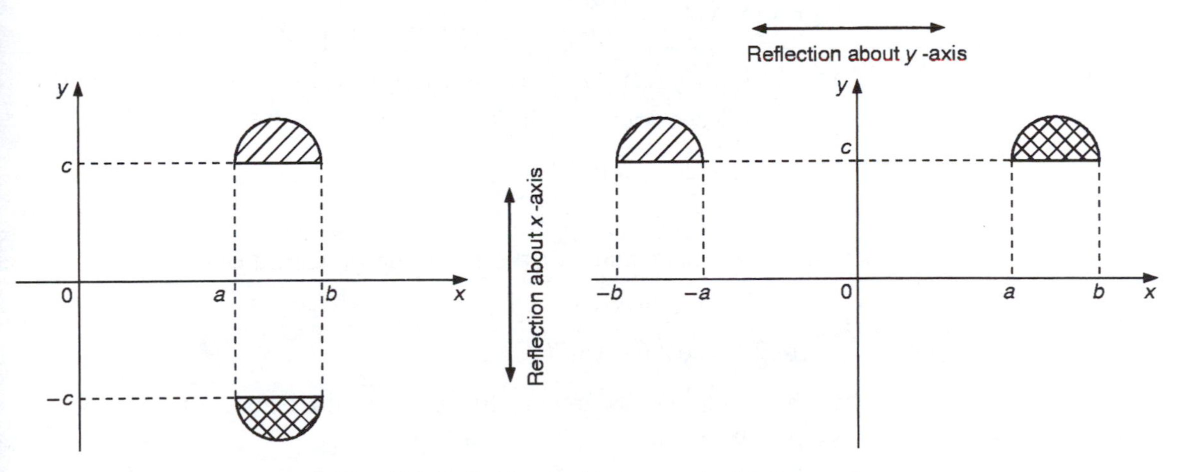

Fig. 10.8 Reflection in (a) the x-axis; (b) the y-axis.

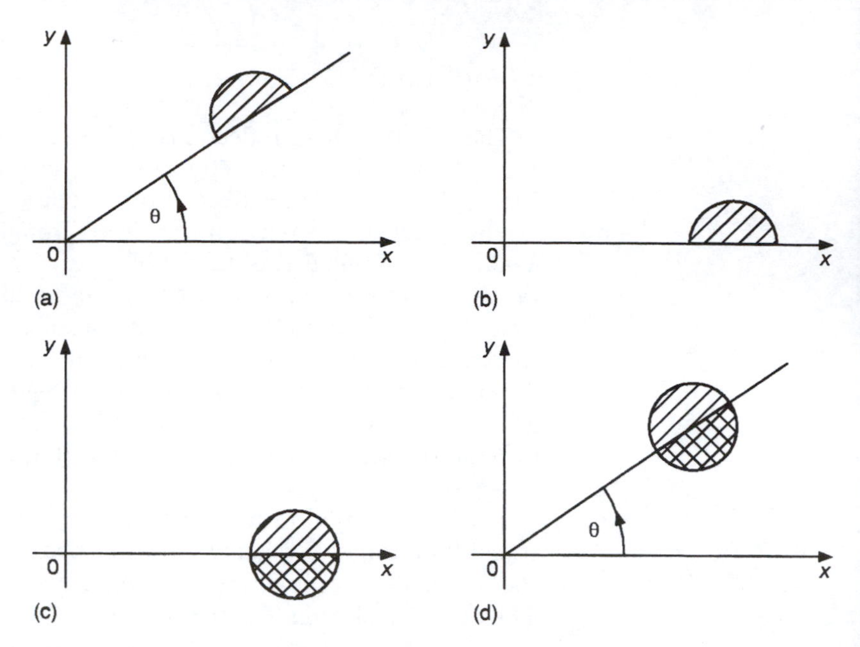

Fig. 10.9 Reflection in a line inclined at angle θ to the x-axis.

Finally, to restore the semi-circle to its original position, we must make a rotation through an angle θ in the anticlockwise direction, so we use the rotational transformation described by the matrix

$$\mathbf{A}_3 = \begin{bmatrix} \cos\theta & -\sin\theta \\ \sin\theta & \cos\theta \end{bmatrix}.$$

Thus, combining these transformations in the order in which they have been performed, we find that the composite transformation required is given by

$$\begin{aligned} \mathbf{T} = \mathbf{A}_3\mathbf{A}_2\mathbf{A}_1 &= \begin{bmatrix} \cos\theta & -\sin\theta \\ \sin\theta & \cos\theta \end{bmatrix} \begin{bmatrix} 1 & 0 \\ 0 & -1 \end{bmatrix} \begin{bmatrix} \cos\theta & \sin\theta \\ -\sin\theta & \cos\theta \end{bmatrix} \\ &= \begin{bmatrix} \cos^2\theta - \sin^2\theta & 2\cos\theta\sin\theta \\ 2\cos\theta\sin\theta & \sin^2\theta - \cos^2\theta \end{bmatrix} \\ &= \begin{bmatrix} \cos 2\theta & \sin 2\theta \\ \sin 2\theta & -\cos 2\theta \end{bmatrix}, \end{aligned}$$

where use has been made of the trignometric identities

$$\sin 2\theta = 2\sin\theta\cos\theta$$

$$\cos 2\theta = \cos^2\theta - \sin^2\theta.$$

Thus the required reflection in a line inclined at an angle θ to the x-axis is described by

$$\tilde{\mathbf{X}} = \mathbf{TX},$$

where

$$\tilde{\mathbf{X}} = \begin{bmatrix} x' \\ y' \end{bmatrix}, \quad \mathbf{X} = \begin{bmatrix} x \\ y \end{bmatrix} \quad \text{and} \quad \mathbf{T} = \begin{bmatrix} \cos 2\theta & \sin 2\theta \\ \sin 2\theta & -\cos 2\theta \end{bmatrix}$$

In the special case in which reflection is in the line $y = x$, so that $\theta = \pi/4$, the transformation matrix $\mathbf{T}$ becomes

$$\mathbf{T} = \begin{bmatrix} 0 & 1 \\ 1 & 0 \end{bmatrix}.$$

Shear

Shear occurs when the scaling of the coordinate in the x-direction is also influenced by the coordinate in the y-direction, and conversely. Thus the most general shearing transformation is of the form

$$x' = ax + by, \quad y' = cx + dy, \tag{10.77}$$

so the matrix transformation involved becomes

$$\tilde{\mathbf{X}} = \mathbf{AX}, \tag{10.78}$$

where

$$\tilde{\mathbf{X}} = \begin{bmatrix} x' \\ y' \end{bmatrix}, \quad \mathbf{X} = \begin{bmatrix} x \\ y \end{bmatrix} \quad \text{and} \quad \mathbf{A} = \begin{bmatrix} a & b \\ c & d \end{bmatrix}. \tag{10.79}$$

The effect this has on the rectangle in Fig. 10.10(a) is shown in Fig. 10.10(b), for the case in which $a = 1$, $b = 0$, $c = 1$ and $d = 1$.

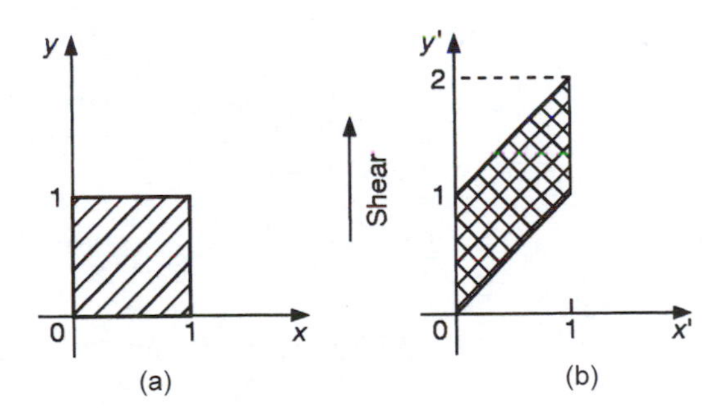

Fig. 10.10 Shear.

10.10.6 Simultaneous equations and the numerical solution of Laplace's equation

The partial differential equation

$$\frac{\partial^2 u}{\partial x^2} + \frac{\partial^2 u}{\partial y^2} = 0 \tag{10.80}$$

satisfied by a twice differentiable function $u(x, y)$ in a region D of the (x, y)-plane is called the *Laplace equation*. This equation is of special importance in engineering and physics because its solution governs the mathematical description of many different physical phenomena. Typical amongst these is the temperature at any point in a long rectangular metal rod whose sides are maintained at different temperatures, the electric potential inside a long rectangular cavity whose walls are maintained at different potentials, and what is called the *velocity potential* in an incompressible inviscid fluid (one without viscosity), from which the velocity at any point in the fluid can be found.

In fact the Laplace equation was already encountered in Section 6.6 in a different context. There it was seen to be the equation satisfied by the real and imaginary parts of a differentiable complex function, though as its significance there belongs to the theory of complex analysis this aspect of the Laplace equation will not be considered.

From the brief descriptions of the first two physical applications, it is apparent that to determine, for example, the temperature distribution $u(x, y)$ in a cross-section of long rectangular bar with uniform cross-section and the z-axis directed along the length of the bar, it is necessary that $u(x, y)$ should satisfy the specified temperature distributions on each plane face of the bar. Conditions like this are called *boundary conditions*, and to find the temperature distribution in a cross-section of the bar, it is necessary to solve the Laplace equation subject to these boundary conditions. A typical problem involves solving the Laplace equation

$$\frac{\partial^2 u}{\partial x^2} + \frac{\partial^2 u}{\partial y^2} = 0, \tag{10.81}$$

subject to the boundary conditions

$$u(x, y) = \begin{cases} u(x, 0) = \sin x, \; 0 \leq x \leq \pi \\ u(0, y) = 0, \; 0 \leq y \leq \pi \\ u(\pi, y) = 0, \; 0 \leq y \leq \pi \\ u(x, \pi) = 0, \; 0 \leq x \leq \pi. \end{cases} \tag{10.82}$$

If the bar is long, and the temperature distribution in a cross-section well away from the ends of the bar is considered, it will not be influenced by conditions at the ends of the bar, so the variable z can be ignored and we can work with the two-dimensional problem represented by Eqn (10.81) and its boundary conditions (10.82). For obvious reasons this problem is called a two-dimensional *boundary value problem* for the Laplace equation, and it is illustrated in Fig. 10.11.

Using a more advanced approach that cannot be described here, it can be shown that this boundary value problem has the analytical solution

$$u(x, y) = \frac{\sin x \sinh(\pi - y)}{\sinh \pi}. \tag{10.83}$$

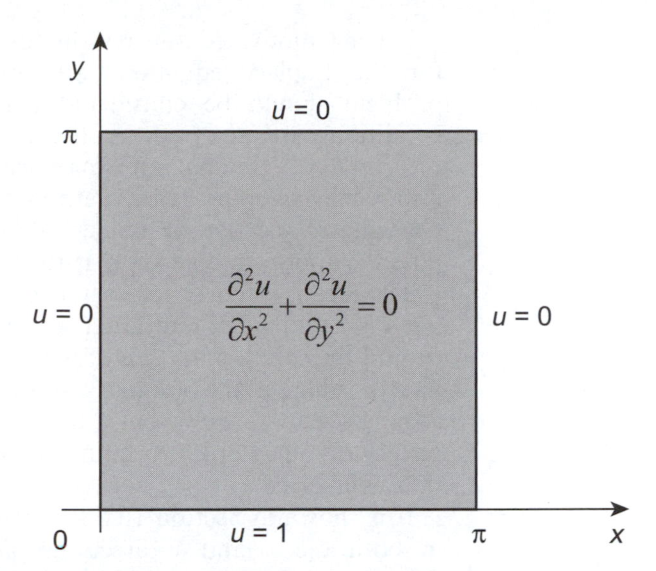

Fig. 10.11 The boundary value problem for a rectangular cross-section of a long metal bar.

It is easy to check that Eqn (10.83) is the exact solution, because it satisfies both the Laplace equation (10.81) and the boundary conditions (10.82). Fig. 10.12 shows a computer plot of this analytical solution from which it can be seen that, as would be expected, the solution is symmetrical about the line $x = \pi/2$, and it decays to zero on three sides of the rectangle in Fig. 10.11.

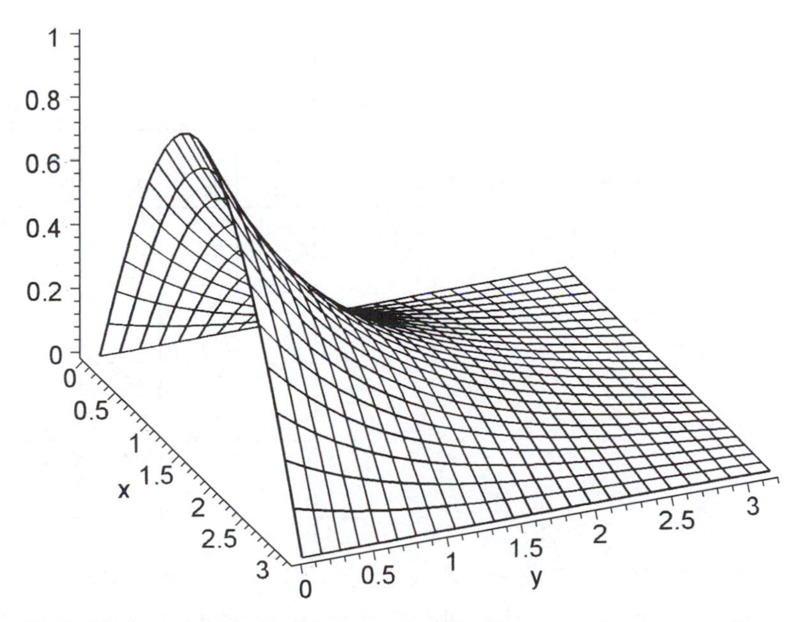

Fig. 10.12 The exact theoretical temperature distribution in the cross-section.

Having motivated the reason for solving boundary value problems for the Laplace equation, we now come to the reason why this problem should be considered here. When more general boundary conditions are involved, or the region D is not a simple rectangle, the task of finding an analytical solution is either difficult or impossible, so some other method of solution becomes necessary. This necessitates the use of numerical methods, and we now outline one possible approach and apply it to the above boundary value problem.

The idea is to cover the region D, in this case the rectangle $0 \le x \le \pi$, $0 \le y \le \pi$ with a 6×6 rectangular grid, as shown in Fig. 10.13, and to try to find the solution at each of the 16 internal grid points in terms of the known values on the boundary. A much finer grid could be used to obtain a more accurate result, but a 6×6 grid has been chosen for illustrative purposes and in order to limit the amount of algebra and computation that is involved.

It is shown in Section 12.6 that if the separation of the grid lines is h in both the x and y-directions, and $u(x, y)$ is twice differentiate, then the value u_0 of $u(x, y)$ satisfying Laplace's equation at point P_0 in Fig. 10.14 is related to the values u_1, u_2, u_3 and u_4 at the adjacent points P_1, P_2, P_3 and P_4 by the equation

$$u_1 + u_2 + u_3 + u_4 - 4u_0 = 0. \tag{10.84}$$

Result (10.84) is called a *finite difference* approximation for the Laplace equation at P_0 using equal spacing between points in the x and

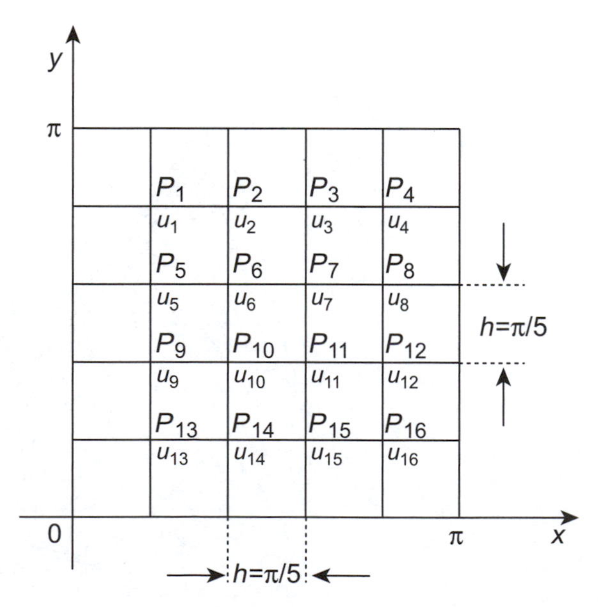

Fig. 10.13 The grid to be used with boundary value problem (10.81) and (10.82).

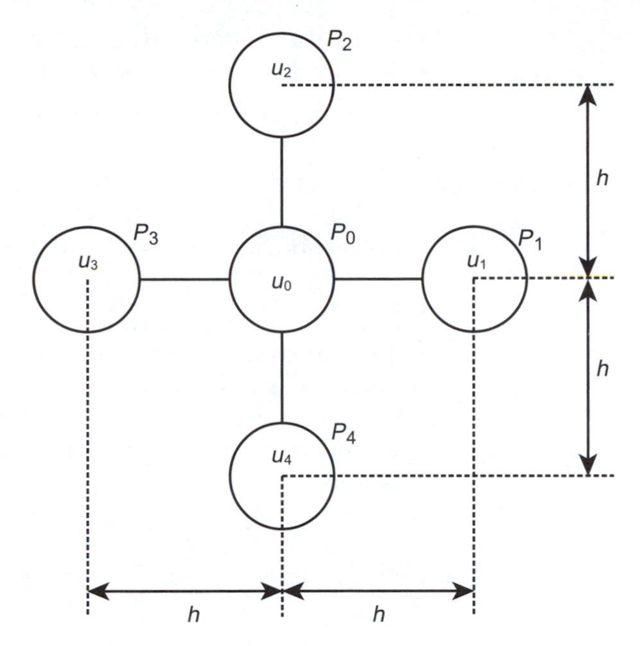

Fig. 10.14 The finite difference approximation for the Laplace equation at P_0.

y-directions. Later, it will be seen in Section 12.6 that the error involved when the Laplace equation is replaced by this finite difference approximation is of the order of h^3. If the separation between points in the x and y-directions is different, result (10.84) must be modified in the way described in Section 12.6.

The approach we now adopt is to apply the finite difference approximation (10.84) to each of the 16 grid points in Fig. 10.13 at which the $u(x, y)$ is unknown, with the value of $u(x, y)$ at point P_i being denoted by u_i, for $i = 1, 2, \ldots, 16$. This will lead to 16 linear algebraic equations, some of which will involve the specified values of $u(x, y)$ (temperatures) at points on the rectangular boundary. Solving these simultaneous equations will determine the unknown temeperature at each grid point. Using the numbering of grid points shown in Fig. 10.13 leads to the following system of finite difference equations (linear simultaneous equations):

$$u_5 + u_2 - 4u_1 = 0,\ \ u_1 + u_3 + u_6 - 4u_2 = 0,\ \ u_2 + u_7 + u_4 - 4u_3 = 0,$$

$$u_3 + u_8 - 4u_4 = 0,\ \ u_1 + u_9 + u_6 - 4u_5 = 0,\ \ u_2 + u_5 + u_{10} + u_7 - 4Pu_6 = 0$$

$$u_3 + u_6 + u_8 + u_{11} - 4u_7 = 0,\ \ u_4 + u_7 + u_{12} - 4u_8 = 0,\ \ u_5 + u_{10} + u_{13} - 4u_9 = 0$$

$$u_6 + u_9 + u_{14} + u_{11} - 4u_{10} = 0,\ \ u_7 + u_{12} + u_{15} + u_{10} - 4u_{11} = 0,$$

$$u_8 + u_{11} + u_{16} - 4u_{12} = 0, \quad u_9 + u_{14} + \sin(\pi/5) - 4u_{13} = 0,$$

$$u_{10} + u_{15} + \sin(2\pi/5) + u_{13} - 4u_{14} = 0, \quad u_{11} + u_{16} + \sin(3\pi/5) + u_{14} - 4u_{15} = 0,$$

$$u_{12} + \sin(4\pi/5) + u_{15} - 4u_{16} = 0.$$

This system can be solved by computer using Gaussian elimination to find the unknown temperatures P_1 through P_{16}. The result obtained in this way is shown in the array below, where the number in bold type is the exact result calculated from Eqn (10.83), and the number below it in italic type is the corresponding value calculated from the set of finite difference equations.

y	$x=0$	$x=\pi/5$	$x=2\pi/5$	$x=3\pi/5$	$x=4\pi/5$	$x=\pi$
	0	**0**	**0**	**0**	**0**	**0**
π	*0*	*0*	*0*	*0*	*0*	*0*
	0	**0.0363**	**0.0588**	**0.0588**	**0.0363**	**0**
$4\pi/5$	*0*	*0.0341*	*0.0552*	*0.0552*	*0.0341*	*0*
	0	**0.0866**	**0.1401**	**0.1401**	**0.0866**	**0**
$3\pi/5$	*0*	*0.0822*	*0.1330*	*0.1330*	*0.0822*	*0*
	0	**0.1699**	**0.2749**	**0.2749**	**0.1699**	**0**
$2\pi/5$	*0*	*0.1637*	*0.2649*	*0.2649*	*0.1637*	*0*
	0	**0.3181**	**0.5147**	**0.5147**	**0.3181**	**0**
$\pi/5$	*0*	*0.3121*	*0.5050*	*0.5050*	*0.3121*	*0*
	0	**0.5878**	**0.9511**	**0.9511**	**0.5878**	**0**
0	*0*	*0.5878*	*0.9511*	*0.9511*	*0.5878*	*0*

Fig. 10.15 gives a visual indication of the accuracy with which the exact solution is approximated by the finite difference approximation, by superimposing on a plot of the analytical solution the calculated values u_i, represented by small boxes.

To obtain a better approximation it is necessary to increase the number of grid points. When the number of grid points becomes very large, as can happen when the shape of region D is complicated, the resulting system of finite difference equations needs to be represented in matrix form, after which an iterative scheme can be used to find the value u_i, of the solution at each point P_i. ■

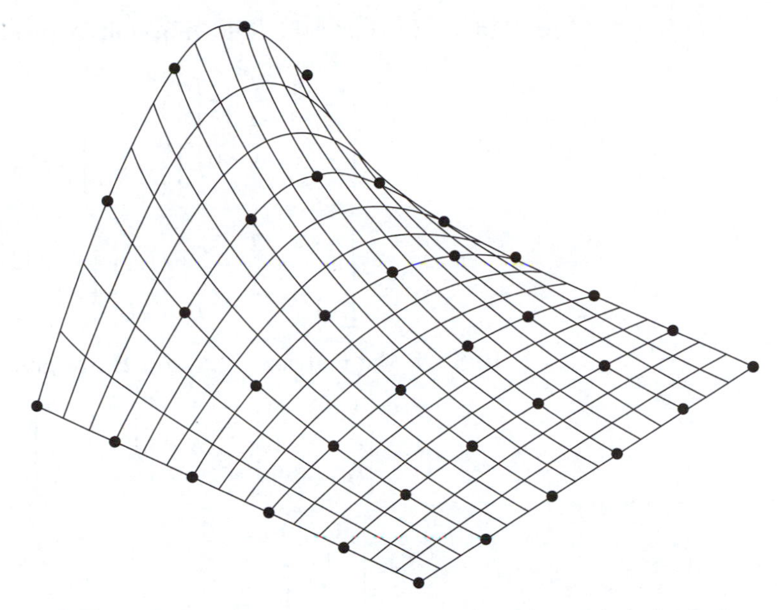

Fig. 10.15 The calculated values u_i, superimposed on the analytical solution.

Problems

Section 10.1

10.1 Suggest two physical situations in which the outcomes may be displayed in the form of a matrix.

10.2 Find the sum $\mathbf{A}+\mathbf{B}$ and difference $\mathbf{A}-\mathbf{B}$ of the matrices

$$\mathbf{A} = \begin{bmatrix} 1 & 2 & 3 & 4 \\ 2 & 1 & 2 & 2 \\ 1 & 2 & 0 & 0 \end{bmatrix}, \quad \mathbf{B} = \begin{bmatrix} 2 & 3 & 1 & 2 \\ 0 & 2 & 2 & 0 \\ 1 & -2 & 1 & 1 \end{bmatrix}.$$

10.3 Evaluate the following inner products:

(a) $\begin{bmatrix} 2 & 1 & 1 & 3 \end{bmatrix} \begin{bmatrix} 1 \\ 2 \\ 2 \\ 1 \end{bmatrix}$; (b) $\begin{bmatrix} 1 & -2 & 7 & 4 \end{bmatrix} \begin{bmatrix} 2 \\ 3 \\ 0 \\ 1 \end{bmatrix}$;

(c) $\begin{bmatrix} 2 & -1 & 3 & 1 \end{bmatrix} \begin{bmatrix} 2 \\ -1 \\ 3 \\ 1 \end{bmatrix}$.

10.4 Evaluate the following matrix products:

(a) $$\begin{bmatrix} 0 & 3 & 1 & 2 \\ 1 & 2 & 2 & 2 \\ 1 & 1 & 1 & 0 \end{bmatrix} \begin{bmatrix} 1 & 2 \\ 1 & -1 \\ 0 & 0 \\ 1 & 1 \end{bmatrix};$$

(b) $$\begin{bmatrix} -1 & 2 & 1 & 3 \\ 1 & -1 & 1 & -1 \\ 1 & 0 & 0 & 1 \end{bmatrix} \begin{bmatrix} -1 & 1 \\ 2 & 2 \\ 1 & 1 \\ 3 & 2 \end{bmatrix}.$$

10.5 If the matrices **I**, **A** and **B** are given by

$$\mathbf{I} = \begin{bmatrix} 1 & 0 & 0 \\ 0 & 1 & 0 \\ 0 & 0 & 1 \end{bmatrix}, \quad \mathbf{A} = \begin{bmatrix} 2 & 1 & 3 \\ 1 & 2 & 1 \\ 5 & 1 & 4 \end{bmatrix}, \quad \text{and}$$

$$\mathbf{B} = \begin{bmatrix} 1 & 0 & 2 & 1 \\ -1 & 3 & 1 & 0 \\ 2 & 1 & 2 & 1 \end{bmatrix},$$

show that
(a) $\mathbf{IA} = \mathbf{AI} = \mathbf{A}$;
(b) $\mathbf{IB} = \mathbf{B}$ but that $\mathbf{BI}$ is not defined.

10.6 Display each of the following sets of simultaneous equations in matrix form:

(a) $$\begin{aligned} x + 4y + z &= 9 \\ x - 3y + 2z &= -4 \\ x + y - z &= 1; \end{aligned}$$

(b) $$\begin{aligned} w + 2x - y &= 4 \\ x - 3y + 2z &= -1 \\ 2w + 5x - 3z &= 0 \\ 4w - y + 4z &= 2; \end{aligned}$$

(c) $$\begin{aligned} 3w + x - 2y + 4z &= 1 \\ w - 3x + y - 3z &= -4 \\ w + 7x + 2y + 5z &= 2; \end{aligned}$$

(d) $$\begin{aligned} 2x + y - z &= \lambda x \\ 3x + 2y + 4z &= \lambda y \\ x - 3y + 2z &= \lambda z. \end{aligned}$$

Section 10.2

10.7 State which of the following pairs of matrices can be made equal by assigning suitable values to the constants a, b and c. Where appropriate, determine what these values must be.

(a) $$\begin{bmatrix} 1 & 2 & 1 & 0 \\ 3 & a & b & 2 \\ 1 & 2 & c & 1 \end{bmatrix} \quad \text{and} \quad \begin{bmatrix} 1 & 2 & 1 & 0 \\ 3 & 1 & 2 & 2 \\ 1 & 2 & 4 & 1 \end{bmatrix};$$

(b) $\begin{bmatrix} 1 & 5 & a & 2 \\ 2 & a^2 & 3 & b \\ 4 & 3 & 2 & c \end{bmatrix}$ and $\begin{bmatrix} 1 & 5 & 1 & 2 \\ 2 & 4 & 3 & 4 \\ 4 & 3 & 2 & 1 \end{bmatrix}$;

(c) $\begin{bmatrix} 1 & a+b & 3 \\ a+c & 2 & 4 \\ 1 & 2 & b+c \end{bmatrix}$ and $\begin{bmatrix} 1 & 4 & 3 \\ 0 & 2 & 4 \\ 1 & 2 & 2 \end{bmatrix}$.

10.8 Determine $3\mathbf{A}+2\mathbf{B}$ and $2\mathbf{A}-6\mathbf{B}$ given that

$$\mathbf{A} = \begin{bmatrix} 1 & 3 & 7 \\ 2 & -1 & 6 \end{bmatrix} \quad \text{and} \quad \mathbf{B} = \begin{bmatrix} 2 & -1 & 4 \\ 3 & -3 & 2 \end{bmatrix}.$$

10.9 If

$$\mathbf{A} = \begin{bmatrix} 2 & 1 & 0 \\ 3 & 2 & 0 \\ 1 & 0 & 1 \end{bmatrix}, \quad \mathbf{B} = \begin{bmatrix} 1 & 1 & 1 & 0 \\ 2 & 1 & 1 & 0 \\ 2 & 3 & 1 & 2 \end{bmatrix},$$

$$\mathbf{C} = \begin{bmatrix} 2 & 3 & 4 \\ 1 & 5 & 6 \end{bmatrix}, \quad \mathbf{D} = \begin{bmatrix} 1 \\ 2 \\ 3 \end{bmatrix},$$

find the matrix products **AB** and **CD**.

10.10 This example shows that the matrix product $\mathbf{AB} = \mathbf{0}$ does not necessarily imply either that $\mathbf{A} = \mathbf{0}$ or that $\mathbf{B} = \mathbf{0}$. If

$$\mathbf{A} = \begin{bmatrix} 1 & -1 & 1 \\ -3 & 2 & -1 \\ -2 & 1 & 0 \end{bmatrix} \quad \text{and} \quad \mathbf{B} = \begin{bmatrix} 1 & 2 & 3 \\ 2 & 4 & 6 \\ 1 & 2 & 3 \end{bmatrix},$$

find **AB** and **BA** and show that $\mathbf{AB} \neq \mathbf{BA}$.

10.11 Show that the matrix equation

$$\mathbf{AX} = \mathbf{K},$$

where

$$\mathbf{A} = \begin{bmatrix} 1 & 3 & 1 \\ 1 & 1 & 2 \\ 2 & 2 & 0 \end{bmatrix}, \quad \mathbf{X} = \begin{bmatrix} x_1 \\ x_2 \\ x_3 \end{bmatrix} \quad \text{and} \quad \mathbf{K} = \begin{bmatrix} 1 \\ 2 \\ 3 \end{bmatrix},$$

may be solved for x_1, x_2 and x_3 by pre-multiplication by **B**, where

$$\mathbf{B} = \begin{bmatrix} -\frac{1}{2} & \frac{1}{4} & \frac{5}{8} \\ \frac{1}{2} & -\frac{1}{4} & -\frac{1}{8} \\ 0 & \frac{1}{2} & -\frac{1}{4} \end{bmatrix}.$$

10.12 Use matrix multiplication to verify the results of Theorem 10.2 when **A**, **B** and **C** are of the form

$$\mathbf{A} = \begin{bmatrix} 1 & 3 & 2 \\ 0 & 1 & 4 \\ 2 & 3 & 1 \end{bmatrix}, \quad \mathbf{B} = \begin{bmatrix} -1 & 2 & 1 \\ 3 & -2 & -1 \\ 1 & 4 & 2 \end{bmatrix}, \quad \text{and}$$

$$\mathbf{C} = \begin{bmatrix} -2 & 2 & 1 \\ 0 & 2 & 4 \\ 1 & 3 & 1 \end{bmatrix}.$$

10.13 If $\mathbf{A}$ is a square matrix, then the associative property of matrices allows us to write $\mathbf{A}^n$ without ambiguity because, for example, $\mathbf{A}^3 = \mathbf{A}(\mathbf{A}\mathbf{A}) = (\mathbf{A}\mathbf{A})\mathbf{A}$. If

$$\mathbf{A} = \begin{bmatrix} \cosh x & \sinh x \\ \sinh x & \cosh x \end{bmatrix},$$

use the hyperbolic identities to express $\mathbf{A}^2$ and $\mathbf{A}^3$ in their simplest form and use induction to deduce the form of $\mathbf{A}^n$.

10.14 Transpose the following matrices:

(a) $[1 \quad 4 \quad 17 \quad 3]$; (b) $\begin{bmatrix} 2 & 5 & 7 & 9 \\ 4 & 3 & 0 & 1 \end{bmatrix}$; (c) $\begin{bmatrix} 1 & 4 & 19 \\ 4 & 0 & 2 \\ 19 & 2 & 4 \end{bmatrix}$;

(d) $\begin{bmatrix} 0 & 3 & -2 \\ -3 & 0 & 1 \\ 2 & -1 & 0 \end{bmatrix}$; (e) $\begin{bmatrix} 4 \\ 3 \\ 1 \\ 0 \end{bmatrix}$.

10.15 Use Definition 10.7 and Theorem 10.3 to prove that:
(a) the sum of a square matrix and its transpose is a *symmetric* matrix;
(b) the difference of a square matrix and its transpose is *skew-symmetric* matrix.
Illustrate each of these results by an example.

10.16 Verify that $(\mathbf{AB})' = \mathbf{B}'\mathbf{A}'$, given that

$$\mathbf{A} = \begin{bmatrix} 1 & 4 & 7 \\ 9 & -3 & 1 \end{bmatrix} \quad \text{and} \quad \mathbf{B} = \begin{bmatrix} -4 & 2 \\ 3 & 1 \\ -5 & 6 \end{bmatrix}.$$

Section 10.3

10.17 Evaluate the determinants

(a) $\begin{vmatrix} 1 & 2 \\ 4 & 7 \end{vmatrix}$; (b) $\begin{vmatrix} 1 & 0 & 3 \\ 2 & 0 & 5 \\ 1 & 3 & 7 \end{vmatrix}$; (c) $\begin{vmatrix} 1 & 2 & 5 \\ 3 & 1 & 5 \\ -5 & 0 & -5 \end{vmatrix}$.

10.18 Without expanding the determinant, prove that

$$\begin{vmatrix} 1+a_1 & a_1 & a_1 \\ a_2 & 1+a_2 & a_2 \\ a_3 & a_3 & 1+a_3 \end{vmatrix} = (1 + a_1 + a_2 + a_3).$$

10.19 Use Theorem 9.4 to simplify the following determinants before expansion:

(a) $|\mathbf{A}|=\begin{vmatrix}42 & 61 & 50\\ 3 & 0 & 2\\ 4 & 6 & 5\end{vmatrix}$; (b) $|\mathbf{A}|=\begin{vmatrix}0 & 9 & 3\\ 2 & 16 & 4\\ 1 & 2 & 1\end{vmatrix}$;

(c) $|\mathbf{A}|=\begin{vmatrix}2 & 1 & 5\\ 5 & 17 & 56\\ 4 & 1 & 7\end{vmatrix}$.

10.20 Without expanding, prove that

$$\begin{vmatrix}x^2+a_1^2 & a_1a_2 & a_1a_3\\ a_2a_1 & x^2+a_2^2 & a_2a_3\\ a_3a_1 & a_3a_2 & x^2+a_3^2\end{vmatrix}=x^4(x^2+a_1^2+a_2^2+a_3^2).$$

10.21 Show without expansion that

$$\begin{vmatrix}a^2 & b^2 & c^2\\ a & b & c\\ 1 & 1 & 1\end{vmatrix}=(a-b)(a-c)(b-c).$$

This determinant is called an *alternate* determinant. Illustrate the result by means of a numerical example and verify it by direct expansion.

10.22 Prove that

$$|\mathbf{A}|=\begin{vmatrix}\sin(x+\tfrac{1}{4}\pi) & \sin x & \cos x\\ \sin(x+\tfrac{1}{4}\pi) & \cos x & \sin x\\ 1 & a & 1-a\end{vmatrix}$$

is independent of a, and express it as a function of x.

10.23 Using any two 3×3 matrices $\mathbf{A}$ and $\mathbf{B}$ of your choice, find $|\mathbf{A}|$, $|\mathbf{B}|$ and $|\mathbf{AB}|$, and hence verify Corollary 10.5, which asserts that for any two determinants $|\mathbf{A}|$ and $|\mathbf{B}|$ of equal order,

$|\mathbf{A}||\mathbf{B}|=|\mathbf{AB}|$.

Section 10.4

10.24 Which of the following sets of vectors are linearly independent? Where linear dependence exists determine its form:

(a) $\mathbf{C}_1=\begin{bmatrix}3\\0\\0\end{bmatrix}$, $\mathbf{C}_2=\begin{bmatrix}0\\-7\\0\end{bmatrix}$, $\mathbf{C}_3=\begin{bmatrix}0\\0\\15\end{bmatrix}$;

(b) $\mathbf{R}=[1\quad 9\quad -2\quad 14]$, $\mathbf{R}_2=[-2\quad -18\quad 4\quad -28]$;

(c) $\mathbf{C}_1 = \begin{bmatrix} 2 \\ 1 \\ 0 \end{bmatrix}$, $\mathbf{C}_2 = \begin{bmatrix} 1 \\ 1 \\ 2 \end{bmatrix}$, $\mathbf{C}_3 = \begin{bmatrix} 1 \\ 2 \\ 1 \end{bmatrix}$, $\mathbf{C}_4 = \begin{bmatrix} 5 \\ 6 \\ 4 \end{bmatrix}$.

10.25 Test the following matrices for linear independence between their rows or columns:

(a) $\begin{bmatrix} 1 & 2 & -1 & 0 \\ 2 & 3 & 1 & 1 \\ -1 & 1 & 0 & 2 \\ 0 & 1 & 2 & 3 \end{bmatrix}$; (b) $\begin{bmatrix} 0 & 2 & 3 & 1 \\ -2 & 0 & -1 & 2 \\ -3 & 1 & 0 & -2 \\ -1 & -2 & 2 & 0 \end{bmatrix}$;

(c) $\begin{bmatrix} 1 & 2 & 1 & 5 \\ 2 & 1 & 2 & 0 \\ 1 & 0 & 2 & 1 \\ 5 & 3 & 7 & 7 \end{bmatrix}$.

Section 10.5

10.26 Show that adj $\mathbf{A} = \mathbf{A}$ when

$$\mathbf{A} = \begin{bmatrix} -4 & -3 & -3 \\ 1 & 0 & 1 \\ 4 & 4 & 3 \end{bmatrix}.$$

10.27 Find the matrix adjoint to each of the following matrices:

(a) $\begin{bmatrix} 1 & 2 & 3 \\ 2 & 3 & 2 \\ 3 & 3 & 4 \end{bmatrix}$; (b) $\begin{bmatrix} 1 & 2 & 3 \\ 1 & 3 & 4 \\ 1 & 4 & 3 \end{bmatrix}$; (c) $\begin{bmatrix} a & b \\ c & d \end{bmatrix}$.

10.28 Set

$$\begin{bmatrix} 1 & 2 \\ 3 & 4 \end{bmatrix} \begin{bmatrix} a & b \\ c & d \end{bmatrix} = \begin{bmatrix} 1 & 0 \\ 0 & 1 \end{bmatrix}$$

and equate corresponding elements to determine the inverse of

$$\begin{bmatrix} 1 & 2 \\ 3 & 4 \end{bmatrix}.$$

10.29 Find the inverse of

$$\mathbf{A} = \begin{bmatrix} 3 & -2 & -1 \\ -4 & 1 & -1 \\ 2 & 0 & 1 \end{bmatrix}.$$

Verify that:
(a) $\mathbf{A}^{-1}\mathbf{A} = \mathbf{A}\mathbf{A}^{-1} = \mathbf{I}$;
(b) $(\mathbf{A}^{-1})^{-1} = \mathbf{A}$.

10.30 Given that **A** and **B** are

$$\mathbf{A}=\begin{bmatrix}1 & 2 & 1\\ 1 & 4 & 2\\ 0 & 3 & 2\end{bmatrix} \quad \text{and} \quad \mathbf{B}=\begin{bmatrix}1 & -1 & 2\\ 0 & 2 & 4\\ 1 & 0 & 3\end{bmatrix},$$

verify that $(\mathbf{AB})^{-1}=\mathbf{B}^{-1}\mathbf{A}^{-1}$.

Section 10.6

10.31 Find $\mathrm{d}\mathbf{A}/\mathrm{d}t$ and determine the largest interval about the origin in which it is defined, given that

$$\mathbf{A}(t)=\begin{bmatrix}2t^3 & \tan t & \cos t\\ 3 & 4-t^2 & 1+t\end{bmatrix}.$$

10.32 Given that

$$\mathbf{A}(t)=\begin{bmatrix}\cosh t & \sinh t\\ \sinh t & \cosh t\end{bmatrix} \quad \text{and} \quad \mathbf{B}(t)=\begin{bmatrix}t & 1\\ 2t & t^2\end{bmatrix},$$

verify results (d), (f) and (g) of Theorem 9.9.

10.33 Show that for the matrix

$$\mathbf{A}(t)=\begin{bmatrix}\cos t & -\sin t\\ \sin t & \cos t\end{bmatrix}$$

it is true that $(\mathrm{d}/\mathrm{d}t)\mathbf{A}^2=2\mathbf{A}(\mathrm{d}\mathbf{A}/\mathrm{d}t)$, but that this is not true for the matrix

$$\mathbf{A}(t)=\begin{bmatrix}1 & t\\ 2 & t^2\end{bmatrix}.$$

10.34 Show that if $\mathbf{A}(t)$ is a non-singular matrix, then

$$\frac{\mathrm{d}\mathbf{A}^{-1}}{\mathrm{d}t}=-\mathbf{A}^{-1}\frac{\mathrm{d}\mathbf{A}}{\mathrm{d}t}\mathbf{A}^{-1}.$$

Verify this result when

$$\mathbf{A}=\begin{bmatrix}\cos t & -\sin t\\ \sin t & \cos t\end{bmatrix}.$$

Section 10.7

10.35 Solve the following equations using Cramer's rule:

$$x_1+x_2+x_3=7$$

$$2x_1-x_2+2x_3=8$$

$$3x_1+2x_2-x_3=11.$$

10.36 Solve the equations of the previous example using the inverse matrix method, and compare the task with the previous method.

10.37 Solve the following equations using Cramer's rule:

$$\begin{aligned} x_1 - x_2 + x_3 - x_4 &= 1 \\ 2x_1 - x_2 + 3x_3 + x_4 &= 2 \\ x_1 + x_2 + 2x_3 + 2x_4 &= 3 \\ x_1 + x_2 + x_3 + x_4 &= 3. \end{aligned}$$

10.38 Write down the augmented matrix corresponding to the equations:

$$\begin{aligned} 2x_1 - x_2 + 3x_3 &= 1 \\ 3x_1 + 2x_2 - x_3 &= 4 \\ x_1 - 4x_2 + 7x_3 &= 3. \end{aligned}$$

Show, by reducing this matrix to its echelon equivalent, that these equations are inconsistent.

10.39 Write down the augmented matrix corresponding to the equations:

$$\begin{aligned} 3x_1 + 2x_2 - x_3 &= 4 \\ 2x_1 - 5x_2 + 2x_3 &= 1 \\ 5x_1 + 16x_2 - 7x_3 &= 10. \end{aligned}$$

Show, by reducing this matrix to its echelon equivalent, that these equations are consistent and solve them.

10.40 Solve the following equations using Gaussian elimination:

$$\begin{aligned} 1.202x_1 - 4.371x_2 + 0.651x_3 &= 19.447 \\ -3.141x_1 + 2.243x_2 - 1.626x_3 &= -13.702 \\ 0.268x_1 - 0.876x_2 + 1.341x_3 &= 6.849. \end{aligned}$$

10.41 Discuss briefly, but do not solve, the following sets of equations:

(a) $\begin{aligned} x_1 + x_2 &= 1 \\ 2x_1 - x_2 &= 5; \end{aligned}$ (b) $\begin{aligned} x_1 + x_2 &= 1 \\ 2x_1 - x_2 &= 5 \\ x_1 - x_2 &= 0; \end{aligned}$

(c) $\begin{aligned} x_1 + x_2 &= 1 \\ 2x_1 - x_2 &= 5 \\ -x_1 - 2x_2 &= 0; \end{aligned}$ (d) $\begin{aligned} x_1 + x_2 - x_3 &= 0 \\ 2x_1 - x_2 - 5x_3 &= 0. \end{aligned}$

Section 10.8

10.42 Write down the characteristic equations for the following matrices:

$$\text{(a)}\quad \mathbf{A} = \begin{bmatrix} 1 & 4 \\ 3 & 7 \end{bmatrix}; \quad \text{(b)}\quad \mathbf{A} = \begin{bmatrix} 1 & 0 & 2 \\ 2 & 1 & 1 \\ 0 & 2 & 1 \end{bmatrix}.$$

10.43 Find the eigenvalues and engenvectors of

$$\mathbf{A} = \begin{bmatrix} 1 & -1 \\ -2 & 0 \end{bmatrix}.$$

10.44 Prove that the eigenvalues of a diagonal matrix of any order are given by the elements on the leading diagonal. What form do the eigenvectors take?

10.45 Find the eigenvalues, eigenvectors and normalized eigenvectors of

$$\mathbf{A} = \begin{bmatrix} 3 & 0 & 0 \\ 5 & 4 & 0 \\ 3 & 6 & 1 \end{bmatrix}.$$

Supplementary computer problems

10.46 Define any two 5×5 matrices **A** and **B** and use them to verify that

$$\mathbf{I}(\mathbf{A} + 2\mathbf{B}) = (2\mathbf{B} + \mathbf{A})\mathbf{I} = \mathbf{A} + 2\mathbf{B}.$$

10.47 Define any 5×5 matrix **A**, and use it to verify that

$$\mathbf{A}(\mathbf{A}^2) = (\mathbf{A}^2)\mathbf{A}.$$

10.48 Define three 5×5 matrices **A**, **B** and **C**, and use them to verify that

(a) $\mathbf{A}(\mathbf{B} + \mathbf{C}) = \mathbf{AB} + \mathbf{AC}$;

(b) $\mathbf{A}(\mathbf{BC}) = (\mathbf{AB})\mathbf{C}$.

10.49 Define any two 5×5 matrices **A** and **B** and use them to verify that

$$(\mathbf{AB})' = \mathbf{B}'\mathbf{A}'.$$

10.50 Given

$$\mathbf{A} = \begin{bmatrix} 1 & 3 & 4 & -2 \\ 6 & 1 & 1 & 2 \\ -1 & 0 & 2 & 1 \\ 2 & 1 & 3 & 4 \end{bmatrix}, \quad \text{show that } \det \mathbf{A} = \det \mathbf{A}' = -84.$$

10.51 Given

$$\mathbf{A} = \begin{bmatrix} 3 & -2 & 1 & 4 \\ 1 & 6 & -1 & -2 \\ 0 & 2 & 1 & 4 \\ 2 & 1 & 4 & 1 \end{bmatrix}, \quad \text{show that } \det \mathbf{A} = \det \mathbf{A}' = -394.$$

10.52 Given

$$\mathbf{A} = \begin{bmatrix} 0 & 3 & 1 & -3 \\ 1 & 4 & -2 & 3 \\ 2 & 0 & -5 & 1 \\ 3 & 0 & 1 & 2 \end{bmatrix},$$

show $\det \mathbf{A} \neq 0$ and use Rule 10.2 to find $\mathbf{A}^{-1}$. Verify your result by showing that $\mathbf{A}\mathbf{A}^{-1} = \mathbf{I}$.

10.53 Given

$$\mathbf{A} = \begin{bmatrix} 1 & 2 & -1 & 3 \\ 2 & 1 & 2 & 1 \\ -1 & 2 & 2 & 1 \\ 4 & -1 & -1 & 0 \end{bmatrix},$$

show $\det \mathbf{A} \neq 0$ and use Rule 10.2 to find $\mathbf{A}^{-1}$. Verify your result by showing that $\mathbf{A}\mathbf{A}^{-1} = \mathbf{I}$.

10.54 Define any two non-singular 4×4 matrices **A** and **B** and use them to verify that

$$(\mathbf{AB})^{-1} = \mathbf{B}^{-1}\mathbf{A}^{-1}.$$

10.55 Solve by any method

$$\begin{aligned} 1.3x_1 + 2.2x_2 + x_3 + 3x_4 &= 1.2 \\ 4.2x_1 - 1.5x_2 + 2x_3 - x_4 &= 6.5 \\ x_1 - 7.2x_2 + 3x_3 + 2x_4 &= 1.4 \\ 3.1x_1 + 2x_2 - 2.7x_3 + 1.6x_4 &= 3.1. \end{aligned}$$

10.56 Solve by any method

$$\begin{aligned} 3.2x_1 + 1.4x_2 - 2.3x_3 + 1.2x_4 &= -3.4 \\ 1.4x_1 - 2.6x_2 + 4.7x_3 - 1.1x_4 &= 1.7 \\ 0.9x_1 + 1.8x_2 - 3.6x_3 + 2.2x_4 &= 4.6 \\ 2.1x_1 - 1.9x_2 - 5.4x_3 + 2.9x_4 &= 7.5. \end{aligned}$$

10.57 Solve by any method

$$\begin{aligned} -3.98x_1 + 1.12x_2 + 0.14x_3 + 1.12x_4 &= 3.47 \\ 1.11x_1 - 4.13x_2 + 1.04x_3 + 0.14x_4 &= -3.42 \\ 0.23x_1 + 1.14x_2 - 3.87x_3 + 1.41x_4 &= -2.37 \\ 1.02x_1 - 0.17x_2 + 0.98x_3 - 3.58x_4 &= -1.98. \end{aligned}$$

10.58 Solve by any method

$$\begin{aligned} -4.98x_1 + 1.32x_2 + 0.54x_3 + 1.43\ x_4 &= 2.69 \\ 1.41x_1 - 3.98x_2 + 1.37x_3 + 0.156x_4 &= -4.39 \end{aligned}$$

$$0.43x_1 + 0.79x_2 - 4.27x_3 + 1.53\ x_4 = -3.14$$

$$2.02x_1 - 1.26x_2 + 1.26x_3 - 4.41\ x_4 = -2.43.$$

10.59 Verify by multiplication that if

$$\mathbf{A} = \begin{bmatrix} -1 & 0 & 1 & 2 \\ -1 & 1 & 3 & 3 \\ 1 & 0 & 1 & 1 \\ -2 & 1 & 6 & 9 \end{bmatrix},$$

then $\mathbf{A}$ can be factorized as $\mathbf{A} = \mathbf{LU}$, with a lower-triangular matrix $\mathbf{L}$ and an upper-triangular matrix $\mathbf{U}$ given by

$$\mathbf{L} = \begin{bmatrix} 1 & 0 & 0 & 0 \\ 1 & 1 & 0 & 0 \\ -1 & 0 & 1 & 0 \\ 2 & 1 & 1 & 1 \end{bmatrix}, \quad \mathbf{U} = \begin{bmatrix} -1 & 0 & 1 & 2 \\ 0 & 1 & 2 & 1 \\ 0 & 0 & 2 & 3 \\ 0 & 0 & 0 & 1 \end{bmatrix}.$$

10.60 Verify by multiplication that if

$$\mathbf{A} = \begin{bmatrix} 1 & 0 & 0 & 1 \\ 1 & 2 & 1 & 0 \\ 0 & 1 & 0 & 1 \\ 0 & 0 & 1 & 2 \end{bmatrix},$$

then $\mathbf{A}$ can be factorized as $\mathbf{A} = \mathbf{LU}$, with a lower-triangular matrix $\mathbf{L}$ and upper-triangular matrix $\mathbf{U}$ given by

$$\mathbf{L} = \begin{bmatrix} 1 & 0 & 0 & 0 \\ 1 & 1 & 0 & 0 \\ 0 & \frac{1}{2} & 1 & 0 \\ 0 & 0 & -2 & 1 \end{bmatrix}, \quad \mathbf{U} = \begin{bmatrix} 1 & 0 & 0 & 1 \\ 0 & 2 & 1 & -1 \\ 0 & 0 & -\frac{1}{2} & \frac{3}{2} \\ 0 & 0 & 0 & 5 \end{bmatrix}.$$

10.61 Write the equation of a circle of radius 2 centred at the origin in the parametric form

$x = 2\cos t, \quad y = 2\sin t$ and plot the graph for $0 \le t \le 2\pi$.

Use the transformation of Eqn (10.66) with

$$\mathbf{A} = \begin{bmatrix} 3 & 0 \\ 0 & 2 \end{bmatrix}$$

to find $\tilde{\mathbf{X}}$, and hence x' and y'. Use the result to reproduce Fig. 10.5(c) illustrating the scaling effect.

10.62 Parametrize the sine function by writing

$x = t, \quad y = \sin t$ for $0 \le t \le 2\pi$

and plot its graph. Use the transformation of Eqn (10.69) with

$$\mathbf{A} = \begin{bmatrix} 1 & 0 \\ 0 & 1 \end{bmatrix}, \quad \mathbf{K} = \begin{bmatrix} 1 \\ 2 \end{bmatrix}$$

to find $\tilde{\mathbf{X}}$, and hence x' and y'. Use to result to illustrate the translation of a graph.

10.63 Parametrize the cosine function writing

$$x = t, \quad y = \cos t \qquad \text{for} \quad 0 \le t \le 2\pi,$$

and use the transformation of Eqn (10.74) with

$$\mathbf{A} = \begin{bmatrix} 1 & 0 \\ 0 & -1 \end{bmatrix}$$

to illustrate the reflection of the graph of a function in the x-axis.

10.64 Write the equation of a parabola in the parametric form

$$x = t, \quad y = t^2 \qquad \text{for} \quad -3 \le t \le 3,$$

and use the transformation of Eqn (10.71) with

$$\mathbf{A} = \begin{bmatrix} \cos \pi/4 & -\sin \pi/4 \\ \sin \pi/4 & \cos \pi/4 \end{bmatrix}$$

to illustrate the rotation of the graph of a function through an angle $\pi/4$.

10.65 Write the equation of a circle of radius 3 centred on the origin in the parametric form

$$x = 3\cos t, \quad y = 3\sin t \qquad \text{for} \quad 0 \le t \le 2\pi,$$

and use the transformation of Eqn (10.78) with

$$\mathbf{A} = \begin{bmatrix} 1 & 0 \\ 0 & 1 \end{bmatrix}$$

to illustrate the effect of shear on a circle.

10.66 Parametrize the sine function by writing

$$x = t, \quad y = \sin t \qquad \text{for} \quad 0 \le t \le 2\pi,$$

and use the transformation

$$\tilde{\mathbf{X}} = \mathbf{ABX}$$

with

$$\mathbf{A} = \begin{bmatrix} \cos \pi/4 & -\sin \pi/4 \\ \sin \pi/4 & \cos \pi/4 \end{bmatrix}, \quad \mathbf{B} = \begin{bmatrix} 1 & 0 \\ 1 & 1 \end{bmatrix}$$

to illustrate the effect on the graph of a function by first performing a shear and then a rotation. Repeat the calculations, but this time using the transformation

$$\tilde{\mathbf{X}} = \mathbf{BAX}$$

to illustrate the effect of performing the shear and rotational operations in the reverse order.

10.67 Use 2×2 matrices of your own to illustrate the effect of a shear.

10.68 Use 2×2 matrices of your own to illustrate the effect of first performing a shear, and then a rotation.

Problems 10.69 and 10.70 involve longer calculations and require the use of an important result called the *Cayley-Hamilton* theorem. This theorem states that every $n \times n$ matrix $\mathbf{A}$ satisfies its own characteristic equation so if, for example, $\mathbf{A}$ is a 4×4 matrix and its characteristic equation is

$$a_1\lambda^4 + a_2\lambda^3 + a_3\lambda^2 + a_4\lambda + a_5 = 0, \qquad (*)$$

then the Cayley-Hamilton theorem asserts that

$$a_1\mathbf{A}^4 + a_2\mathbf{A}^3 + a_3\mathbf{A}^2 + a_4\mathbf{A} + a_5\mathbf{I} = 0.$$

10.69 Given that

$$\mathbf{A} = \begin{bmatrix} 1 & 1 & -1 & 0 \\ 1 & -1 & 1 & 0 \\ 1 & 3 & 3 & 0 \\ -1 & 2 & -1 & 1 \end{bmatrix},$$

find the characteristic equation (*) and verify that $\mathbf{A}$ satisfies the Cayley-Hamilton theorem. Confirm that det $\mathbf{A} \neq 0$, so that the inverse matrix $\mathbf{A}^{-1}$ exists. Pre-multiply the Cayley-Hamilton equation by $\mathbf{A}^{-1}$, and rearrange the resulting matrix equation to find $\mathbf{A}^{-1}$ using only the operations of matrix multiplication and the multiplication of a matrix by a scalar. Confirm the result obtained in this way by showing that $\mathbf{A}\,\mathbf{A}^{-1} = \mathbf{I}$.

10.70 Repeat Problem 10.69, but this time using

$$\mathbf{A} = \begin{bmatrix} -1 & 2 & 0 & 1 \\ -1 & 3 & -1 & 1 \\ -1 & -2 & 1 & -1 \\ 0 & -2 & 1 & 1 \end{bmatrix}.$$

In Problems 10.71 and 10.72, neither of which has a simple analytical solution, use a computer to solve the stated boundary value problem in the unit square $0 \leq x \leq 1, 0 \leq y \leq 1$. Use the grid pattern shown in Fig. 10.13, but this time with the interval between grid points equal to 0.2 in both the x and y-directions. Display your results in a form that allows the decay of the solution as a function of y to be seen for any fixed value of x.

10.71 $u(x, 0) = \begin{cases} u(x, 0) = 10x^2(1-x), \ 0 \le x \le 1 \\ u(0, y) = 0, \ 0 \le y \le 1 \\ u(1, y) = 0, \ 0 \le y \le 1 \\ u(x, 1) = 0, \ 0 \le x \le 1. \end{cases}$

10.72 $u(x, y) = \begin{cases} u(x, 0) = 10x(1-x)^3, \ 0 \le x \le 1 \\ u(0, y) = 0, \ 0 \le y \le 1 \\ u(1, y) = 0, \ 0 \le y \le 1 \\ u(x, 1) = 0, \ 0 \le x \le 1. \end{cases}$

10.73 Solve Laplace's equation in the L-shaped region shown in Fig. 10.16 using a square mesh in which $h = 1$ and the boundary conditions are as given in the figure, and number the mesh points as shown.

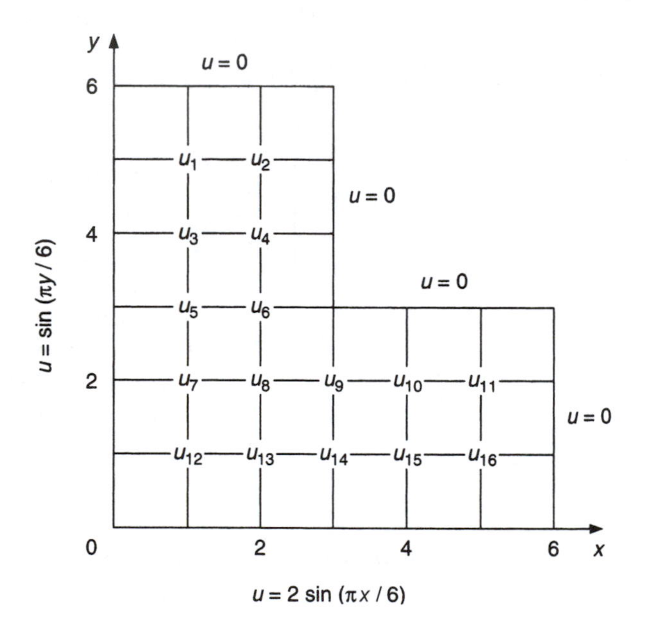

Fig. 10.16

10.74 Solve Laplace's equation in the square with the internal boundary shown in Fig. 10.17. Use a square mesh in which $h = 1$, and the boundary conditions are as given in the figure, and number the mesh points as shown.

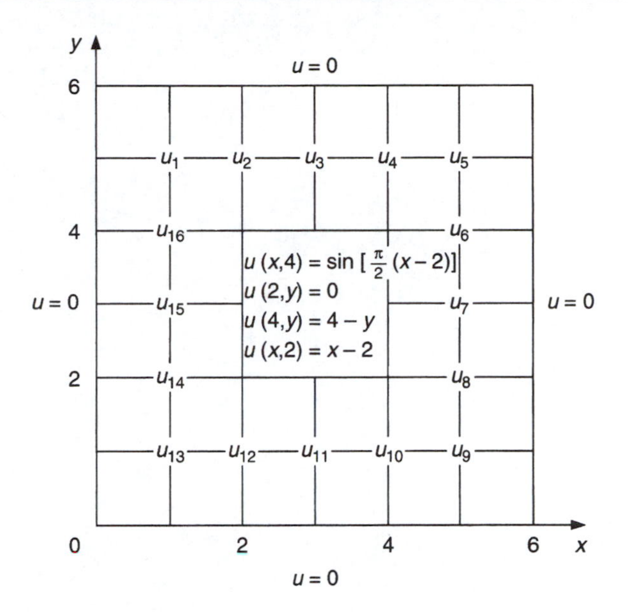
y
u = 0
6
u_1 u_2 u_3 u_4 u_5
4
u_{16} u_6
$u(x,4) = \sin[\frac{\pi}{2}(x-2)]$
$u(2,y) = 0$
u = 0
u_{15} u_7
u = 0
$u(4,y) = 4 - y$
$u(x,2) = x - 2$
2
u_{14} u_8
u_{13} u_{12} u_{11} u_{10} u_9
0
2
4
6
x
u = 0

Fig. 10.17

11 Scalars, vectors and fields

This chapter starts by showing how the theory of curves in space, which has many applications, can be developed to great advantage by using vectors. This is accomplished by regarding the general point on a curve as a function of a parameter t, and writing the position vector of the general point as

$$\mathbf{r}(t) = f(t)\mathbf{i} + g(t)\mathbf{j} + h(t)\mathbf{k},$$

where f, g and h are functions of t and $\mathbf{i}$, $\mathbf{j}$ and $\mathbf{k}$ are the usual fixed unit vectors. It is then shown how differentiation of the position vector leads to the definition of the unit tangent vector $\mathbf{T}$ to the curve at a point corresponding to a given value of t.

In section 11.2 integrals of vector functions are introduced by analogy with ordinary integrals, and then in section 11.4 the ideas of scalar and vector fields are developed, together with the notion of a directional derivative and the gradient operator that play important roles in continuum mechanics, electromagnetic theory and elsewhere. The chapter ends with the definition of the divergence and curl vector operators followed by an introduction to conservative fields and potential functions.

11.1 Curves in space

If the coordinates (x, y, z) of a point P in space are described by

$$x = f(t), \quad y = g(t), \quad z = h(t), \tag{11.1}$$

where f, g, h are continuous functions of t, then as t increases so the point P moves in space tracing out some curve. It follows that Eqns (11.1) represent a *parametric* description of a curve Γ in space and, furthermore, that they define a direction along the curve Γ corresponding to the direction in which P moves as t increases. For example, the parametric equations

$$x = 2\cos 2\pi t, \quad y = 2\sin 2\pi t, \quad z = 2t,$$

for $0 \leq t \leq 1$, describe one turn of a helix, as may be seen by noticing that the projection of the point P on the (x, y)-plane traces one revolution of the circle $x^2+y^2=4$ as t increases from $t=0$ to $t=1$, while the z-coordinate of P steadily increases from $z=0$ to $z=2$ (see Fig. 11.1(b)).

If we now denote by $\mathbf{r}$ the position vector OP of a point P on Γ relative to the origin O of our coordinate system, and introduce the triad of orthogonal unit vectors $\mathbf{i}$, $\mathbf{j}$, $\mathbf{k}$ used in Chapter 4, it follows

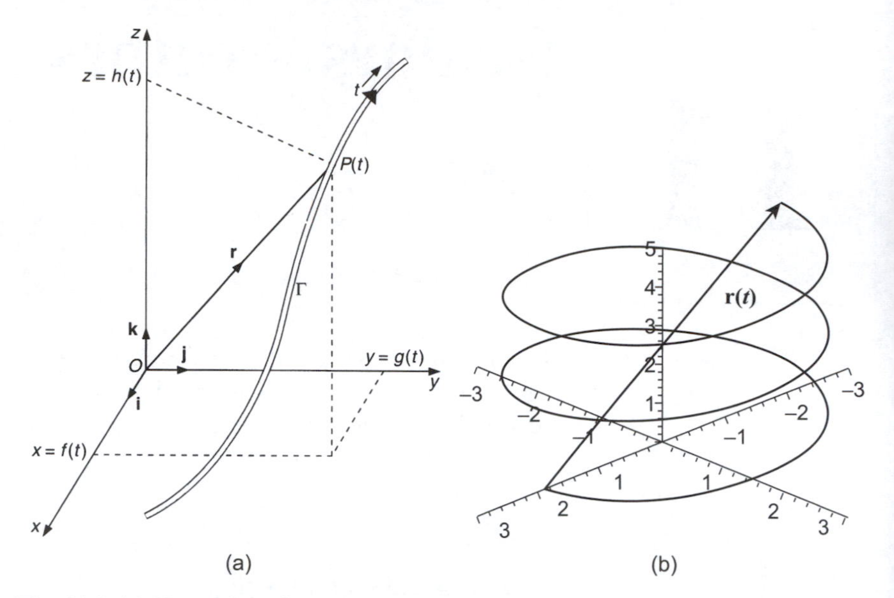

Fig. 11.1 (a) Vector function of one variable interpreted as a curve in space; (b) the helix $\mathbf{r}(t) = 2\,(\cos 2\pi t\boldsymbol{i} + \sin 2\pi t\boldsymbol{j} + t\boldsymbol{k})$.

(Fig. 11.1(a)) that

$$\mathbf{r} = f(t)\mathbf{i} + g(t)\mathbf{j} + h(t)\mathbf{k}. \tag{11.2}$$

Expressions of this form are called *vector functions* of one real varable, in which the dependence on the parameter t is often displayed concisely by writing $\mathbf{r} = \mathbf{r}(t)$. The name 'vector function' arises because $\mathbf{r}$ is certainly a vector and, as it depends on the real independent variable t, it must also be a function in the sense that to each t there corresponds a vector $\mathbf{r}(t)$. Knowledge of the vector function $\mathbf{r}(t)$ implies knowledge of the three scalar functions f, g and h, and conversely.

The geometrical anaogy used here to interpret a general vector function $\mathbf{r}(t)$ is particularly valuable in dynamics where the point $P(t)$ with position vector $\mathbf{r}(t)$ usually represents a moving particle, and the curve Γ its trajectory in space. Under these conditions it is frequently most convenient if the parameter t is identified with time, though in some circumstances identification with the distance s to P measured along Γ from some fixed point on Γ is preferable. Useful though these geometrical and dynamical analogies are, we shall in the main use them only to help further our understanding of general vector functions.

The name 'vector function' suggests, correctly, that it is possible to give satisfactory meaning to the terms limit, continuity and derivative when applied to $\mathbf{r}(t)$. As in the ordinary calculus, the key concept is that of a limit. Intuitively the idea of a limit is clear: when we say $\mathbf{u}(t)$ tends to a limit $\mathbf{v}$ as $t \to t_0$, we mean that when t is close to t_0, the vector function $\mathbf{u}(t)$ is in some sense close to the vector $\mathbf{v}$. In what sense, though, can the two vectors $\mathbf{u}(t)$ and $\mathbf{v}$ be said to be close to one another? Ultimately, all that is necessary is to interpret this as meaning that $|\mathbf{u}(t)-\mathbf{v}|$ is small.

So, we shall say that $\mathbf{u}(t)$ tends to the limit $\mathbf{v}$ as $t \to t_0$ if, by taking t sufficiently close to t_0, it is possible to make $|\mathbf{u}(t)-\mathbf{v}|$ arbitrarily small. As with our previous notion of continuity we shall then say that $\mathbf{u}(t)$ is continuous at t_0 if $\lim_{t\to t_0}\mathbf{u}(t)=\mathbf{v}$ and, in addition, $\mathbf{u}(t_0)=\mathbf{v}$. We incorporate these ideas into a formal definition as follows.

Definition 11.1 (vector function – limits and continuity)

Let $\mathbf{u}(t)=u_1(t)\mathbf{i}+u_2(t)\mathbf{j}+u_3(t)\mathbf{k}$ and $\mathbf{v}=v_1\mathbf{i}+v_2\mathbf{j}+v_3\mathbf{k}$. Then if, for any $\varepsilon>0$, it is possible to make

$$|\mathbf{u}(t)-\mathbf{v}|<\varepsilon \quad \text{when} \quad |t-t_0|<\delta,$$

by taking t sufficiently close to t_0, we shall write

$$\lim_{t\to t_0}\mathbf{u}(t)=v,$$

and say, v is the *limit* of $\mathbf{u}(t)$ as $t\to t_0$. If in addition $\mathbf{u}(t_0)=\mathbf{v}$, then $\mathbf{u}(t)$ will be said to be continuous at $t=t_0$. A vector function that is continuous at all points in the interval $a\le t\le b$ will be said to be continuous throughout that interval.

As usual, a vector function that is not continuous at $t=t_0$ will be said to be *discontinuous*. It is obvious from this definition that $\mathbf{u}(t)$ can only tend to the limit $\mathbf{v}$ as $t\to t_0$ if the limit of each component of $\mathbf{u}(t)$ is equal to the corresponding component of the vector $\mathbf{v}$. Thus the limit of a vector function of one variable is directly related to the limits of the three scalar functions of one variable $u_1(t)$, $u_2(t)$ and $u_3(t)$. This is proved by setting $\mathbf{v}=v_1\mathbf{i}+v_2\mathbf{j}+v_3\mathbf{k}$ and writing

$$|\mathbf{u}(t)-\mathbf{v}|=[(u_1(t)-v_1)^2+(u_2(t)-v_2)^2+(u_3(t)-v_3)^2]^{1/2},$$

showing that $|\mathbf{u}(t)-\mathbf{v}|<\varepsilon$ as $t\to t_0$ is only possible if

$$\lim_{t\to t_0}(u_i(t)-v_i)=0 \quad \text{for} \quad i=1,\,2,\,3,$$

or

$$\lim_{t\to t_0}u_1(t)=v_1,\quad \lim_{t\to t_0}u_2(t)=v_2,\quad \lim_{t\to t_0}u_3(t)=v_3.$$

A systematic application of these arguments enables the following theorem to be proved.

Theorem 11.1 (continuous vector functions)

If the vector functions $\mathbf{u}(t)$, $\mathbf{v}(t)$ are defined and continuous throughout the interval $a\le t\le b$, then the vector functions $\mathbf{u}(t)+\mathbf{v}(t)$, $\mathbf{u}(t)\times\mathbf{v}(t)$ and the scalar function $\mathbf{u}(t).\mathbf{v}(t)$ are also defined and continuous throughout that same interval.

Example 11.1 At what point are the vector functions $\mathbf{u}(t)$, $\mathbf{v}(t)$ discontinuous if

$$\mathbf{u}(t) = (\sin t)\mathbf{i} + (\sec t)\mathbf{j} + \frac{1}{t-1}\mathbf{k},$$

$$\mathbf{v}(t) = t\mathbf{i} + (1+t^2)\mathbf{j} + \mathrm{e}^t\mathbf{k}.$$

Verify by direct calculation that $\mathbf{u}(t)+\mathbf{v}(t)$, $\mathbf{u}(t).\mathbf{v}(t)$ and $\mathbf{u}(t)\times\mathbf{v}(t)$ are continuous functions in any interval not containing a point of discontinuity of $\mathbf{u}(t)$ or $\mathbf{v}(t)$.

Solution The **i**-component of $\mathbf{u}(t)$ is defined and continuous for all t, whereas the **j**-component is discontinuous for $t = (2n+1)\frac{1}{2}\pi$ with $n = 0, \pm 1, \pm 2, \ldots$ and the **k**-component is discontinuous for the single value $t = 1$. All three components of $\mathbf{v}(t)$ are continuous for all t. We have by vector addition

$$\mathbf{u}(t) + \mathbf{v}(t) = (t + \sin t)\mathbf{i} + (1 + t^2 + \sec t)\mathbf{j} + \left(\mathrm{e}^t + \frac{1}{t-1}\right)\mathbf{k},$$

showing that the components of $\mathbf{u}(t)+\mathbf{v}(t)$ give rise to the same points of discontinuity as the function $\mathbf{u}(t)$. We may thus conclude that the vector sum is continuous throughout any interval not containing one of these points. For example, $\mathbf{u}(t)+\mathbf{v}(t)$ is continuous in both the open interval $(\frac{1}{2}\pi, 3\pi/3)$ and the closed interval $[5, 7]$ but it is discontinuous in $(0, \pi)$.

The scalar product $\mathbf{u}(t).\mathbf{v}(t)$ is given by

$$\mathbf{u}(t).\mathbf{v}(t) = t\sin t + (1+t^2)\sec t + \frac{\mathrm{e}^t}{(t-1)},$$

which is, of course, a scalar. Again we see by inspection that the scalar product is continuous in any interval not containing a point of discontinuity of $\mathbf{u}(t)$.

The vector product $\mathbf{u}(t)\times\mathbf{v}(t)$ is

$$\mathbf{u}(t)\times\mathbf{v}(t) = \begin{vmatrix} \mathbf{i} & \mathbf{j} & \mathbf{k} \\ \sin t & \sec t & 1/(t-1) \\ t & 1+t^2 & \mathrm{e}^t \end{vmatrix},$$

giving

$$\mathbf{u}(t)\times\mathbf{v}(t) = \left(\mathrm{e}^t\sec t - \frac{1+t^2}{t-1}\right)\mathbf{i} + \left(\frac{t}{t-1} - \mathrm{e}^t\sin t\right)\mathbf{j}$$
$$+[(1+t^2)\sin t - t\sec t]\mathbf{k}.$$

Here also inspection of the components shows that the vector product is continuous in any interval not containing a point of discontinuity of $\mathbf{u}(t)$. ■

The following definition (interpreted later) shows that, as might be expected, the idea of a derivative can also be applied to vector functions of one variable.

Definition 11.2 (derivative of vector function) Let $\mathbf{u}(t)$ be a continuous vector function throughout some interval $a \leq t \leq b$ at each point of which the limit

$$\lim_{\Delta t \to 0} \frac{\mathbf{u}(t + \Delta t) - \mathbf{u}(t)}{\Delta t}$$

is defined. Then $\mathbf{u}(t)$ is said to be *differentiable* throughout that interval with derivative

$$\frac{\mathrm{d}\mathbf{u}}{\mathrm{d}t} = \lim_{\Delta t \to 0} \frac{\mathbf{u}(t - \Delta t) - \mathbf{u}(t)}{\Delta t}.$$

The geometrical interpretation of the derivative of a vector function of a real variable is apparent in Fig. 11.2. In that figure, as t increases, so the curve Γ is described by a point $P(t)$ with position vector $\mathbf{u}(t)$ relative to O. The point denoted by $P'(t+\Delta t)$ is the position assumed by $\mathbf{u}$ at time $t+\Delta t$, so that $OP = \mathbf{u}(t)$, $OP' = \mathbf{u}(t+\Delta t)$, and $PP' = \Delta\mathbf{u}$ is the increment in $\mathbf{u}(t)$ consequent upon the increment Δt in t.

It can be seen that as $\Delta t \to 0$, so the vector $\Delta\mathbf{u}$ tends to the line of the tangent to the curve Γ at $P(t)$ with $\Delta\mathbf{u}$ being directed from P to P'. To interpret $\mathrm{d}\mathbf{u}/\mathrm{d}t$ in terms of components when $\mathbf{u}(t) = u_1(t)\mathbf{i} + u_2(t)\mathbf{j} + u_3(t)\mathbf{k}$, we need only observe that

$$\frac{\mathrm{d}\mathbf{u}}{\mathrm{d}t} = \lim_{\Delta t \to 0} \frac{\mathbf{u}(t + \Delta t) - \mathbf{u}(t)}{\Delta t}$$

$$= \lim_{\Delta t \to 0} \left[\frac{u_1(t + \Delta t) - u_1(t)}{\Delta t}\right]\mathbf{i} + \lim_{\Delta t \to 0} \left[\frac{u_2(t + \Delta t) - u_2(t)}{\Delta t}\right]\mathbf{j}$$

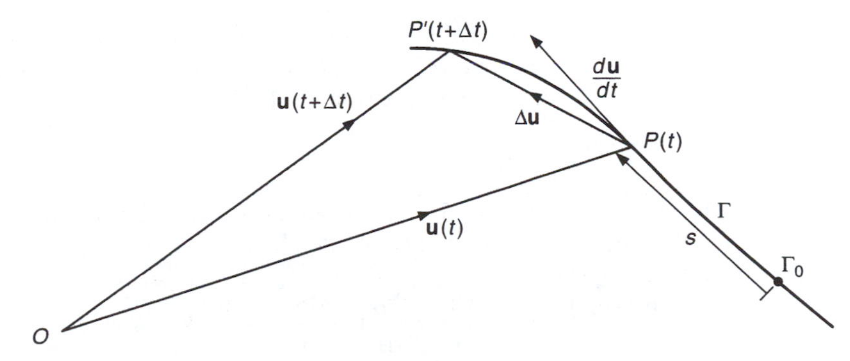

Fig. 11.2 Geometrical interpretation of $\mathrm{d}\mathbf{u}/\mathrm{d}t$.

$$+\lim_{\Delta t\to 0}\left[\frac{u_3(t+\Delta t)-u_3(t)}{\Delta t}\right]\mathbf{k},$$

from which it follows that

$$\frac{d\mathbf{u}}{dt}=\frac{du_1}{dt}\mathbf{i}+\frac{du_2}{dt}\mathbf{j}+\frac{du_3}{dt}\mathbf{k}. \tag{11.3}$$

The unit vector $\mathbf{T}$ that is tangent to Γ at $P(t)$ and points in the direction in which $P(t)$ will move with increasing t is obviously

$$\mathbf{T}=\frac{d\mathbf{u}}{dt}\Big/\left|\frac{d\mathbf{u}}{dt}\right|. \tag{11.4}$$

If s is the distance to P measured positively in the sense P to P' along Γ from some fixed point on that curve (Fig. 11.2), then we know from our work with differentials that $du_1=u_1'dt$, $du_2=u_2'dt$, $du_3=u_3'dt$. Now as the differentials du_1, du_2, du_3 are mutually orthogonal and represent the increments in the coordinates $(u_1(t), u_2(t), u_3(t))$ of P to an adjacent point distant ds away along Γ with coordinates $(u_1(t+dt), u_2(t+dt), u_3(t+dt))$, we may apply Pythagoras' theorem to obtain

$$(ds)^2=(u_1'\,dt)^2+(u_2'\,dt)^2+(u_3'\,dt)^2,$$

whence

$$\frac{ds}{dt}=\sqrt{\left[\left(\frac{du_1}{dt}\right)^2+\left(\frac{du_2}{dt}\right)^2+\left(\frac{du_3}{dt}\right)^2\right]}. \tag{11.5}$$

Comparison of Eqns (11.3) and (11.5) then gives the result

$$\left|\frac{d\mathbf{u}}{dt}\right|=\frac{ds}{dt}, \tag{11.6}$$

from which we see that if t is regarded as time, then the vector function $\mathbf{v}=d\mathbf{u}/dt$ is the *velocity vector* of $P(t)$ as it moves with *speed* ds/dt along Γ in the direction of $\mathbf{T}$. These results merit recording as a theorem.

Theorem 11.2 (derivative of a vector function of a single variable)

Let $\mathbf{u}(t)=u_1(t)\mathbf{i}+u_2(t)\mathbf{j}+u_3(t)\mathbf{k}$ be a differentiable vector function of the real variable t. Then

$$\frac{d\mathbf{u}}{dt}=\frac{du_1}{dt}\mathbf{i}+\frac{du_2}{dt}\mathbf{j}+\frac{du_3}{dt}\mathbf{k}. \quad\blacksquare$$

As a consequence of this theorem we may write

$$\frac{\mathrm{d}u}{\mathrm{d}t} = \frac{\mathrm{d}s}{\mathrm{d}t}\frac{\mathrm{d}\mathbf{u}}{\mathrm{d}t}\bigg/\left|\frac{\mathrm{d}\mathbf{u}}{\mathrm{d}t}\right| = \frac{\mathrm{d}s}{\mathrm{d}t}\mathbf{T}, \tag{11.7}$$

which is a result of considerable use in dynamics when t is identified with time.

Higher-order derivatives such as $\mathrm{d}^2\mathbf{u}/\mathrm{d}t^2$ and $\mathrm{d}^3\mathbf{u}/\mathrm{d}t^3$ may also be defined in the obvious fashion as $\mathrm{d}^2\mathbf{u}/\mathrm{d}t^2 = (\mathrm{d}/\mathrm{d}t)(\mathrm{d}\mathbf{u}/\mathrm{d}t)$, $\mathrm{d}^3\mathbf{u}/\mathrm{d}t^3 = (\mathrm{d}/\mathrm{d}t)(\mathrm{d}^2\mathbf{u}/\mathrm{d}t^2)$, provided only that the components of $\mathbf{u}(t)$ have suitable differentiability properties. Thus, for example, if the second derivatives of the components of $\mathbf{u}(t)$ exist we have

$$\frac{\mathrm{d}^2\mathbf{u}}{\mathrm{d}t^2} = \frac{\mathrm{d}^2u_1}{\mathrm{d}t^2}\mathbf{i} + \frac{\mathrm{d}^2u_2}{\mathrm{d}t^2}\mathbf{j} + \frac{\mathrm{d}^2u_3}{\mathrm{d}t^2}\mathbf{k}. \tag{11.8}$$

We have seen that if t is identified with time, and $\mathbf{u}(t)$ is the position vector of a point P, then $\mathrm{d}\mathbf{u}/\mathrm{d}t$ is the velocity vector of P. It follows from this same argument that $\mathrm{d}^2u/\mathrm{d}t^2$ is the *acceleration vector* of P.

Example 11.2 The position vector r of a particle at time t is given by

$$\mathbf{r} = a(\cos\omega t)\mathbf{i} + a(\sin\omega t)\mathbf{j} + \alpha t^2\mathbf{k},$$

where $\mathbf{i}$, $\mathbf{j}$, $\mathbf{k}$ have their usual meanings and a, ω and α are constants. Find the acceleration vector at time t, and deduce the times at which it will be perpendicular to the position vector. Hence deduce the unit tangent to the particle trajectory at these times.

Solution By making the identifications $\mathbf{u} = \mathbf{r}$, $u_1(t) = a\cos\omega t$, $u_2(t) = a\sin\omega t$ and $u_3(t) = \alpha t^2$ and then applying Theorem 11.2, we find that the velocity vector is

$$\frac{\mathrm{d}\mathbf{r}}{\mathrm{d}t} = -a\omega(\sin\omega t)\mathbf{i} + a\omega(\cos\omega t)\mathbf{j} + 2\alpha t\mathbf{k}.$$

A further differentiation yields the required acceleration vector

$$\frac{\mathrm{d}^2\mathbf{r}}{\mathrm{d}t^2} = -a\omega^2(\cos\omega t)\mathbf{i} - a\omega^2(\sin\omega t)\mathbf{j} + 2\alpha\mathbf{k}.$$

Expressed vectorially, the condition that $\mathbf{r}$ and $\mathrm{d}^2\mathbf{r}/\mathrm{d}t^2$ should be perpendicular is simply that $\mathbf{r}.(\mathrm{d}^2\mathbf{r}/\mathrm{d}t^2) = 0$. Hence to find the time at which this condition is satisfied, we must solve the equation

$$[a(\cos\omega t)\mathbf{i} + a(\sin\omega t)\mathbf{j} + \alpha t^2\mathbf{k}].$$
$$[-a\omega^2(\cos\omega t)\mathbf{i} - a\omega^2(\sin\omega t)\mathbf{j} + 2\alpha\mathbf{k}] = 0.$$

Forming the required scalar product gives

$$-a^2\omega^2\cos^2\omega t - a^2\omega^2\sin^2\omega t + 2\alpha^2 t^2 = 0$$

which immediately simplifies to

$$a^2\omega^2 = 2\alpha^2 t^2,$$

showing that the desired times are

$$t = \pm \frac{a\omega}{\alpha\sqrt{2}}.$$

To deduce the unit tangent **T** at these times we use the fact that

$$\mathbf{T} = \frac{\mathrm{d}\mathbf{r}}{\mathrm{d}t} \Big/ \left|\frac{\mathrm{d}\mathbf{r}}{\mathrm{d}t}\right|,$$

where here

$$\left|\frac{\mathrm{d}\mathbf{r}}{\mathrm{d}t}\right| = \sqrt{[(a\omega \sin \omega t)^2 + (a\omega \cos \omega t)^2 + (2\alpha t)^2]}$$
$$= \sqrt{(a^2\omega^2 + 4\alpha^2 t^2)}.$$

Denoting by $\mathbf{T}_{\pm}$, the unit tangent to the trajectory at $t = \pm a\omega/\alpha\sqrt{2}$, we find by substitution of these values of t in the above expression that

$$\mathbf{T}_{+} = \frac{1}{\sqrt{3}}\left[-\left(\sin \frac{a\omega^2}{\alpha\sqrt{2}}\right)\mathbf{i} + \left(\cos \frac{a\omega^2}{\alpha\sqrt{2}}\right)\mathbf{j} + \sqrt{2}\mathbf{k}\right]$$

and

$$\mathbf{T}_{-} = \frac{1}{\sqrt{3}}\left[\left(\sin \frac{a\omega^2}{\alpha\sqrt{2}}\right)\mathbf{i} + \left(\cos \frac{a\omega^2}{\alpha\sqrt{2}}\right)\mathbf{j} - \sqrt{2}\mathbf{k}\right). \quad \blacksquare$$

With the obvious differentiability requirements, if $\mathbf{u}(t)$ and $\mathbf{v}(t)$ are differentiable vector functions with respect to t, then so also are $\mathbf{u}+\mathbf{v}$, $\mathbf{u}.\mathbf{v}$, $\mathbf{u} \times \mathbf{v}$ and $\phi\mathbf{u}$, where $\phi = \phi(t)$ is a scalar function of t. As the following theorem is easily proved by resolution of the vector functions involved into component form, it is stated without proof.

Theorem 11.3 (differentiation of sums and products of vector functions)

If $\mathbf{u}(t)$ and $\mathbf{v}(t)$ are differentiable vector functions throughout some interval $a \le t \le b$ and $\phi(t)$ is a differentiable scalar function throughout that same interval, then

(a) $\dfrac{\mathrm{d}}{\mathrm{d}t}(\mathbf{u}+\mathbf{v}) = \dfrac{\mathrm{d}\mathbf{u}}{\mathrm{d}t} + \dfrac{\mathrm{d}\mathbf{v}}{\mathrm{d}t}$;

(b) $\dfrac{\mathrm{d}}{\mathrm{d}t}(\phi\mathbf{u}) = \phi\dfrac{\mathrm{d}\mathbf{u}}{\mathrm{d}t} + \mathbf{u}\dfrac{\mathrm{d}\phi}{\mathrm{d}t}$;

(c) $\dfrac{\mathrm{d}}{\mathrm{d}t}(\mathbf{u}\cdot\mathbf{v}) = \mathbf{u}\cdot\dfrac{\mathrm{d}\mathbf{v}}{\mathrm{d}t} + \dfrac{\mathrm{d}\mathbf{u}}{\mathrm{d}t}\cdot\mathbf{v}$;

(d) $\dfrac{\mathrm{d}}{\mathrm{d}t}(\mathbf{u}\times\mathbf{v}) = \mathbf{u}\times\dfrac{\mathrm{d}\mathbf{v}}{\mathrm{d}t} + \dfrac{\mathrm{d}\mathbf{u}}{\mathrm{d}t}\times\mathbf{v}$;

and, if **c** is a constant vector,

(e) $\frac{d}{dt}\mathbf{c} = \mathbf{0};$

where the order of the vector products on the right-hand side of (d) must be strictly observed. ■

11.2 Antiderivatives and integrals of vector functions

The notion of an antiderivative, already encountered in Chapter 8, extends naturally to a vector function of a real variable.

Definition 11.3 (antiderivative – vector function) The vector function $\mathbf{F}(t)$ of the real variable t will be said to be the *antiderivative* of the vector function $\mathbf{f}(t)$ if

$$\frac{d}{dt}\mathbf{F}(t) = \mathbf{f}(t).$$

Naturally, an antiderivative $\mathbf{F}(t)$ is indeterminate so far as an additive arbitrary constant vector $\mathbf{C}$ is concerned, because by Theorem 11.3(e), $d\mathbf{C}/dt = \mathbf{0}$. Continuing the convention adopted in Chapter 8, the operation of antidifferentiation with respect to a vector function of the single real variable t will be denoted by $\int$, so that

$$\int \mathbf{f}(t)\,dt = \mathbf{F}(t) + \mathbf{C}, \tag{11.9}$$

where $\mathbf{C}$ is an arbitrary constant vector.

It is obvious that Eqn (11.9), when taken in conjunction with Theorem 11.2, implies the following result.

Theorem 11.4 (antiderivative of a vector function) If

$$\int \mathbf{f}(t)\,dt = \mathbf{F}(t) + \mathbf{C},$$

where $\mathbf{f}(t) = f_1(t)\mathbf{i} + f_2(t)\mathbf{j} + f_3(t)\mathbf{k}$, $\mathbf{F}(t) = F_1(t)\mathbf{i} + F_2(t)\mathbf{j} + F_3(t)\mathbf{k}$ and $\mathbf{C} = C_1\mathbf{i} + C_3\mathbf{j} + C_3\mathbf{k}$ is an arbitrary constant vector, then

$$\int f_i(t)\,dt = F_i(t) + C_i, \quad i = 1, 2, 3$$

with

$$\frac{dF_i}{dt} = f_i(t). \quad ■$$

Expressed in words, the antiderivative of $\mathbf{f}(t)$ has components equal to the antiderivatives of the components of $\mathbf{f}(t)$.

Example 11.3 Find the antiderivative of $\mathbf{f}(t)$ given that

$$\mathbf{f}(t) = (\cos t)\mathbf{i} + (1 + t^2)\mathbf{j} + e^{-t}\mathbf{k}.$$

Solution It follows immediately from Theorem 11.4 that,

$$\int \mathbf{f}(t)\,dt = \mathbf{i}\int \cos t\,dt + \mathbf{j}\int (1 + t^2)\,dt + \mathbf{k}\int e^{-t}dt$$

$$= (\sin t)\mathbf{i} + \left(t + \frac{t^3}{3}\right)\mathbf{j} - e^{-t}\mathbf{k} + \mathbf{C}. \quad \blacksquare$$

The obvious modification to Theorem 11.4 to enable us to work with definite integrals of vector functions of a single real variable comprises the next theorem. Because it is strictly analogous to the scalar case, it is offered without proof.

Theorem 11.5 (definite integral of a vector function) If $\mathbf{F}(t)$ is an antiderivative of $\mathbf{f}(t)$, then

$$\int_a^b \mathbf{f}(t)\,dt = \mathbf{F}(b) - \mathbf{F}(a).$$

Example 11.4 Evaluate the definite integral

$$\int_0^{\pi/4} [t^2\mathbf{i} + (\sec^2 t)\mathbf{j} + \mathbf{k}]dt.$$

Solution From Theorem 11.5 we have the result

$$\int_0^{\pi/4} [t^2\mathbf{i} + (\sec^2 t)\mathbf{j} + \mathbf{k}]dt = \left(\frac{t^3}{3}\mathbf{i} + (\tan t)\mathbf{j} + t\mathbf{k}\right)\Bigg|_0^{\pi/4}$$

$$= \frac{\pi^3}{192}\mathbf{i} + \mathbf{j} + \tfrac{1}{4}\pi\mathbf{k}. \quad \blacksquare$$

A slightly more interesting application of a definite integral is provided by the following example concerning the motion of a particle in space.

Example 11.5 A point moving in space has acceleration

$$(\sin 2t)\mathbf{i} - (\cos 2t)\mathbf{k}.$$

Find the equation of its path if it passes through the point with position vector $\mathbf{r}_0 = \mathbf{j} + 2\mathbf{k}$ with velocity $2\mathbf{j}$ at time $t = 0$.

Solution If $\mathbf{r}$ is the general position vector of the point at time t, then the velocity is $\mathbf{v}(t) = d\mathbf{r}/dt$ and the acceleration is $\mathbf{a}(t) = d^2\mathbf{r}/dt^2$. Hence

$$\frac{d^2\mathbf{r}}{dt^2} = (\sin 2t)\mathbf{i} - (\cos 2t)\mathbf{k},$$

so that integrating the acceleration equation from 0 to t and replacing t in the integrand by the dummy variable τ gives

$$\int_0^t \left(\frac{d^2\mathbf{r}}{d\tau^2}\right) d\tau = \int_0^t [(\sin 2\tau)\mathbf{i} - (\cos 2\tau)\mathbf{k}]d\tau.$$

Hence

$$\left(\frac{d\mathbf{r}}{d\tau}\right)\bigg|_0^t = -\tfrac{1}{2}[(\cos 2\tau)\mathbf{i} + (\sin 2\tau)\mathbf{k}]|_0^t,$$

and so

$$\mathbf{v}(t) = \mathbf{v}_0 + \tfrac{1}{2}(1 - \cos 2t)\mathbf{i} - \tfrac{1}{2}(\sin 2t)\mathbf{k}.$$

Now setting $t = 0$ and using the initial conditions of the problem we find that $\mathbf{v}_0 = 2\mathbf{j}$, so the velocity equation becomes

$$\mathbf{v}(t) = \tfrac{1}{2}(1 - \cos 2t)\mathbf{i} + 2\mathbf{j} - \tfrac{1}{2}(\sin 2t)\mathbf{k}.$$

To find the equation of the path a further integration is required, so setting $\mathbf{v}(t) = d\mathbf{r}/dt$, integrating the velocity equation from 0 to t gives

$$\int_0^t \left(\frac{d\mathbf{r}}{d\tau}\right) d\tau = \int_0^t [\tfrac{1}{2}(1 - \cos 2\tau)\mathbf{i} + 2\mathbf{j} - \tfrac{1}{2}(\sin 2\tau)\mathbf{k}]d\tau.$$

Hence

$$\mathbf{r}(\tau)|_0^t = [\tfrac{1}{2}(\tau - \tfrac{1}{2}\sin 2\tau)\mathbf{i} + 2\tau\mathbf{j} + \tfrac{1}{4}(\cos 2\tau)\mathbf{k}]|_0^t,$$

and so

$$\mathbf{r}(t) = \mathbf{r}_0 + \tfrac{1}{2}(t - \tfrac{1}{2}\sin 2t)\mathbf{i} + 2t\mathbf{j} + \tfrac{1}{4}(\cos 2t - 1)\mathbf{k},$$

where $\mathbf{r}_0$ is an arbitrary vector. Again appealing to the initial conditions of the problems, we find that $\mathbf{r}_0 = \mathbf{j} + 2\mathbf{k}$, so that, finally, the particle path must be

$$\mathbf{r}(t) = \tfrac{1}{2}(t - \tfrac{1}{2}\sin 2t)\mathbf{i} + (1 + 2t)\mathbf{j} + \tfrac{1}{4}(7 + \cos 2t)\mathbf{k}. \quad \blacksquare$$

The form of definite integral of a vector function so far considered is itself a vector. We now discuss one final generalization of the notion of a definite integral involving a vector function that generates a scalar.

Let a curve Γ defined parametrically in terms of the arc length s have the general position vector $\mathbf{r} = \mathbf{r}(s)$ and unit tangent vector $\mathbf{T}(s)$, and let $\mathbf{F}(s)$ be a vector function of s. Then at any point of Γ the scalar $\phi(s) = \mathbf{F}(s). \mathbf{T}(s)$ represents the component of $\mathbf{F}(s)$ tangential to Γ. If the scalar function $\phi(s)$ is then integrated from $s = a$ to $s = b$, this is obviously equivalent to integrating the tangential component of $\mathbf{F}(s)$ along Γ from the point $\mathbf{r} = \mathbf{r}(a)$ to the point $\mathbf{r} = \mathbf{r}(b)$. An integral of this form is therefore called either a *line integral* or a *curvilinear integral* of the vector function $\mathbf{F}(s)$ taken along the curve Γ, which is sometimes referred to as the *path of integration*.

Definition 11.4 (line integral of vector function)

The line integral of the vector function $\mathbf{F}(s)$ taken along the curve Γ between the points A and B with position vectors $\mathbf{r} = \mathbf{r}(a)$ and $\mathbf{r} = \mathbf{r}(b)$, respectively, is the quantity

$$J = \int_a^b \phi(s)\,\mathrm{d}s = \int_a^b \mathbf{F}.\mathbf{T}\,\mathrm{d}s.$$

where $\phi(s) = \mathbf{F}(s).\mathbf{T}(s)$, s denotes arc length along Γ and $\mathbf{T}(s)$ is the unit tangent vector to Γ.

In terms of the general position vector $\mathbf{r}$ of a point on the curve and the fact that s is the arc length along Γ, we obviously have the relationship $\mathrm{d}\mathbf{r} = \mathbf{T}\mathrm{d}s$, so that the line integral may also be written

$$J = \int_A^B \mathbf{F}.\mathrm{d}\mathbf{r}$$

or, more simply still if Γ denotes part of a curve, as

$$J = \int_\Gamma \mathbf{F}.\mathrm{d}\mathbf{r}.$$

In component form, setting the differential $\mathrm{d}\mathbf{r} = \mathrm{d}x\mathbf{i} + \mathrm{d}y\mathbf{j} + \mathrm{d}z\mathbf{k}$ and $\mathbf{F} = F_1\mathbf{i} + F_2\mathbf{j} + F_3\mathbf{k}$, we have at once

$$\int_\Gamma \mathbf{F}.\mathrm{d}\mathbf{r} = \int_\Gamma F_1\,\mathrm{d}x + F_2\,\mathrm{d}y + F_3\,\mathrm{d}z. \tag{11.10}$$

If desired, the integral (11.10) may be defined vectorially in terms of the limit of a sum in a manner strictly analogous to the definition of an ordinary definite integral. To achieve this, let the interval $a \le d \le b$ be divided into n sub-intervals $s_{i-1} \le s \le s_i$, with $i = 1, 2, \ldots, n$, where $s_0 = a$ and $s_n = b$. Then setting $\mathrm{d}\mathbf{r}_i = \mathbf{r}(s_i) - \mathbf{r}(s_{i-1})$ as in Fig. 11.3, the line integral (11.10) may be approximated by the sum

$$J_n = \sum_{i=1}^{n} \mathbf{F}(s_i).\mathrm{d}\mathbf{r}_i.$$

If the number of sub-divisions n is now allowed to tend to infinity in such a manner that the lengths of all the sub-divisions tend to zero then, as with an ordinary definite integral, we arrive at the result

$$\int_A^B \mathbf{F}.\mathrm{d}\mathbf{r} = \lim_{n\to\infty} \sum_{i=1}^{n} \mathbf{F}(s_i).\mathrm{d}\mathbf{r}_i.$$

When used in this context, the differential $\mathrm{d}\mathbf{r}_i$ is usually called a *line element* of the curve Γ joining A to B.

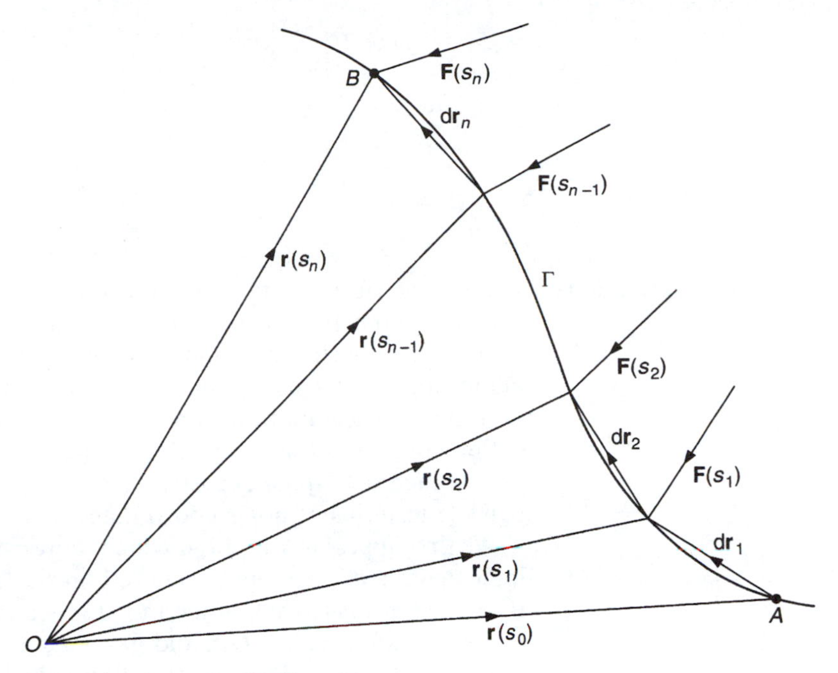

Fig. 11.3 Line integral of **F** along Γ.

Example 11.6 Evaluate the line integral

$$\int_{\Gamma} \mathbf{F}.\mathbf{dr},$$

given that $\mathbf{F} = yz\mathbf{i} + xz\mathbf{j} + 2xy\mathbf{k}$, and Γ is that part of the circular helix $x = a\cos t$, $y = a\sin t$, $z = kt$ that corresponds to the interval $0 \le t \le 2\pi$.

Solution First we use Eqn (11.10) to write the line integral as

$$\int_{\Gamma} \mathbf{F}.\mathbf{dr} = \int_{\Gamma} yz\,\mathrm{d}x + 2xz\,\mathrm{d}y + xy\,\mathrm{d}z.$$

Now along the path Γ we have the relationships

$$x = a\cos t, \quad y = a\sin t, \quad z = kt,$$

which imply the differential relationships

$$\mathrm{d}x = -a\sin t\,\mathrm{d}t, \quad \mathrm{d}y = a\cos t\,\mathrm{d}t, \quad \mathrm{d}z = k\mathrm{d}t.$$

Hence

$$\int_{\Gamma} \mathbf{F}.\mathbf{dr} = \int_0^{2\pi} (-a^2kt\sin^2 t + 2a^2kt\cos^2 t + a^2k\sin t\cos t)\,\mathrm{d}t$$

$$= -a^2k\left[\frac{t^2}{4} - \frac{t\sin 2t}{4} - \frac{\cos 2t}{8}\right]_0^{2\pi} + 2a^2k\left[\frac{t^2}{4} + \frac{t\sin 2t}{4}\right.$$

$$+\frac{\cos 2t}{8}\Big]_0^{2\pi} - \frac{a^3k}{4}\Big[\cos 2t\Big]_0^{2\pi}$$

$$= a^2\pi^2 k. \quad ■$$

11.3 Some applications

Kinematics, an important branch of mechanics, is essentially concerned with the geometrical aspect of the motion of particles along curves. Of particular importance is that class of motions that occur entirely in one plane, and so are called *planar motions*. In many of these situations – for example, particle motion in an orbit – the position of a particle is best defined in terms of the polar coordinates (r, θ) in the plane of the motion. Let us then determine expressions for the velocity and acceleration of a particle in terms of polar coordinates.

We first appeal to Fig. 11.4, which represents a particle P moving in the indicated direction along the curve Γ in the (x, y)-plane. The unit vectors $\mathbf{R}$, $\mathbf{\Theta}$ are normal to each other and are such that $\mathbf{R}$ is directed from O to P along the radius vector OP, and $\mathbf{\Theta}$ points in the direction of increasing θ. Then clearly $\mathbf{R}$ and $\mathbf{\Theta}$ are vector functions of the single variable θ, with

$$\mathbf{R} = (\cos\theta)\mathbf{i} + (\sin\theta)\mathbf{j} \quad \text{and} \quad \mathbf{\Theta} = \mathbf{k}\times\mathbf{R} = -(\sin\theta)\mathbf{i} + (\cos\theta)\mathbf{j}. \tag{11.11}$$

where use has been made of the results in Eqn (4.42). It follows from these relationships that

$$\frac{d\mathbf{R}}{d\theta} = \mathbf{\Theta} \quad \text{and} \quad \frac{d\mathbf{\Theta}}{d\theta} = -\mathbf{R}. \tag{11.12}$$

In terms of the unit vectors $\mathbf{R}$, $\mathbf{\Theta}$ the point P has the position vector

$$\mathbf{r} = r\mathbf{R}, \tag{11.13}$$

so that the velocity $d\mathbf{r}/dt$ must be

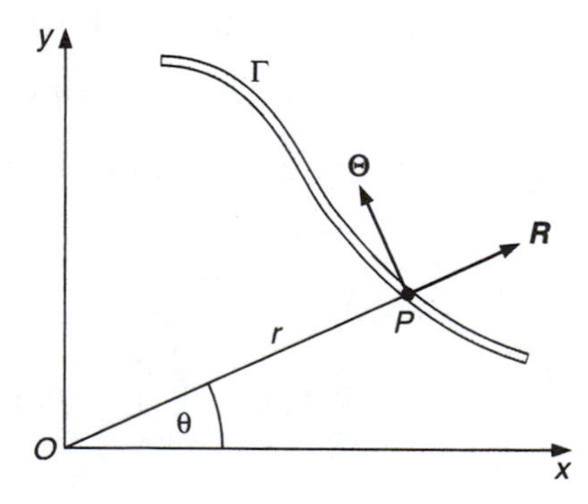

Fig. 11.4 Planar motion of particle in terms of polar coordinates.

$$\frac{d\mathbf{r}}{dt} = \frac{dr}{dt}\mathbf{R} + r\frac{d\mathbf{R}}{dt}$$

$$= \frac{dr}{dt}\mathbf{R} + r\frac{d\mathbf{R}}{d\theta}\frac{d\theta}{dt},$$

showing that the velocity vector of P is

$$\dot{\mathbf{r}} = \dot{r}\mathbf{R} + r\dot{\theta}\mathbf{\Theta}, \quad (11.14)$$

where differentiation with respect to time has been denoted by a dot.

Here the quantity $\dot{r}$ is called the *radial component of velocity*, and $r\dot{\theta}$ is called the *transverse component of velocity*. A further differentiation with respect to time yields for the acceleration vector $\ddot{\mathbf{r}} = d^2\mathbf{r}/dt$ the expression

$$\ddot{\mathbf{r}} = \ddot{r}\mathbf{R} + \dot{r}\dot{\mathbf{R}} + \dot{r}\dot{\theta}\mathbf{\Theta} + r\ddot{\theta}\mathbf{\Theta} + r\dot{\theta}\dot{\mathbf{\Theta}}$$

or

$$\ddot{\mathbf{r}} = \ddot{r}\mathbf{R} + \dot{r}\dot{\theta}\frac{d\mathbf{R}}{d\theta} + (\dot{r}\dot{\theta} + r\ddot{\theta})\mathbf{\Theta} + r\dot{\theta}^2\frac{d\mathbf{\Theta}}{d\theta}.$$

Hence by Eqn (11.21) this is seen to be equivalent to

$$\ddot{\mathbf{r}} = (\ddot{r} - r\dot{\theta}^2)\mathbf{R} + (2\dot{r}\dot{\theta} + r\ddot{\theta})\mathbf{\Theta}. \quad (11.15)$$

The quantity $\ddot{r} - r\dot{\theta}^2$ is called the *radial component of acceleration*, and $2\dot{r}\dot{\theta} + r\ddot{\theta}$ is called the *tranverse component of acceleration*.

Example 11.7 A particle is constrained to move with constant speed v along the cardioid $r = a(1+\cos\theta)$. Prove that

$$v = 2a\dot{\theta}\cos(\tfrac{1}{2}\theta),$$

and show that the radial component of the acceleration is constant.

Solution From Eqn (11.14) and the expression $r = a(1+\cos\theta)$, it follows that the velocity vector r is given by

$$\dot{\mathbf{r}} = -a(\sin\theta)\dot{\theta}\mathbf{R} + a(1+\cos\theta)\dot{\theta}\mathbf{\Theta}.$$

Now as $v^2 = \dot{\mathbf{r}}^2 = \dot{\mathbf{r}}.\dot{\mathbf{r}}$, we have

$$v^2 = a^2\dot{\theta}^2\sin^2\theta + a^2\dot{\theta}^2(1+\cos\theta)^2 = 2a^2\dot{\theta}^2(1+\cos\theta).$$

Using the identity $1+\cos\theta = 2\cos^2(\theta/2)$ in this expression and taking the square root yields the required result

$$v = 2a\dot{\theta}\cos(\theta/2).$$

To complete the problem we now make appeal to the fact that the radial acceleration component is $\ddot{r} - r\dot{\theta}^2$, while by supposition $v =$ constant. From our previous working we know that

$$v^2 = 2a^2\dot{\theta}^2(1+\cos\theta),$$

so that differentiating with respect to t, using the fact that $v = \text{constant}$ and cancelling $\dot{\theta}$ gives

$$\ddot{\theta} = \frac{\dot{\theta}^2 \sin\theta}{2(1 + \cos\theta)}$$

or, as

$$\dot{\theta}^2 = \frac{v^2}{2a^2(1 + \cos\theta)},$$

$$\ddot{\theta} = \frac{v^2 \sin\theta}{4a^2(1 + \cos\theta)^2}.$$

Differentiation of $r = a(1 + \cos\theta)$ shows that $\ddot{r} = -a[(\cos\theta)\dot{\theta}^2 + (\sin\theta)\ddot{\theta}]$, so substituting for r, $\dot{\theta}^2$ and $\ddot{\theta}$ in the radial component of acceleration, we find, as required, that

$$\ddot{r} - r\dot{\theta}^2 = \frac{-3v^2}{4a} = \text{constant.} \quad \blacksquare$$

A vector treatment of particle dynamics follows quite naturally from the ideas presented so far. Thus a particle of variable mass m moving with velocity $\mathbf{v}$ has, by definition, the *linear momentum* $\mathbf{M}$, where

$$\mathbf{M} = m\mathbf{v}.$$

Now by Newton's second law of motion we know that, with a suitable choice of units, we may equate the force $\mathbf{F}$ to the rate of change of momentum, so it follows that we may write

$$\mathbf{F} = \frac{\mathrm{d}\mathbf{M}}{\mathrm{d}t}.$$

However,

$$\frac{\mathrm{d}\mathbf{M}}{\mathrm{d}t} = \frac{\mathrm{d}m}{\mathrm{d}t}\mathbf{v} + m\frac{\mathrm{d}\mathbf{v}}{\mathrm{d}t}$$

and hence

$$\mathbf{F} = \frac{\mathrm{d}m}{\mathrm{d}t}\mathbf{v} + m\frac{\mathrm{d}\mathbf{v}}{\mathrm{d}t}. \tag{11.16}$$

In the case of a particle of constant mass m, we have $\mathrm{d}m/\mathrm{d}t = 0$, reducing this to the familiar equation of motion

$$\mathbf{F} = m\mathbf{a}, \tag{11.17}$$

where $\mathbf{a} = \mathrm{d}\mathbf{v}/\mathrm{d}t$ is the acceleration.

Similarly, the angular momentum of a particle of fixed mass m about the origin is defined by the relation $\mathbf{\Omega} = \mathbf{r} \times m\mathbf{v}$, where $\mathbf{r}$ is the position vector of the particle relative to the origin, and $\mathbf{v} = \mathrm{d}\mathbf{r}/\mathrm{d}t$ is its velocity.

Then the rate of change of angular momentum about the origin is

$$\frac{\mathrm{d}\boldsymbol{\Omega}}{\mathrm{d}t} = m\mathbf{v} \times \mathbf{v} + m\mathbf{r} \times \frac{\mathrm{d}\mathbf{v}}{\mathrm{d}t}$$

$$= \mathbf{r} \times \mathbf{F},$$

by virtue of Eqn (11.17) and the fact that $\boldsymbol{v} \times \boldsymbol{v} = 0$. This is the vector form of the *principle of angular momentum*, which asserts that the rate of change of angular momentum about the origin is equal to the momentum about the origin of the force acting on the particle.

The line integral

$$J = \int_{\Gamma} \mathbf{F}.\mathrm{d}\mathbf{r}$$

also occurs naturally in many contexts, perhaps the simplest of which is in connection with the work done by a force. If $\mathbf{F}$ is identified with a force, and $\mathrm{d}\mathbf{r}$ is a displacement along some specific curve Γ joining points A and B, then J represents the work done by the varying force $\mathbf{F}$ as it moves its point of application along the curve Γ from A to B (cf. Fig. 11.4).

In the special case that $\mathbf{F}$ is a constant force and Γ is a straight-line segment with end points at $s = a$ and $s = b$, this simplifies to an already familiar result. Suppose that $\mathbf{F} = F\boldsymbol{\alpha}$ and $\mathrm{dr} = \mathrm{d}s\,\boldsymbol{\beta}$, where $\boldsymbol{\alpha}$, $\boldsymbol{\beta}$ are constant unit vectors inclined at an angle θ. Then

$$J = \int_A^B \mathbf{F}.\mathrm{d}\mathbf{r} = F(\boldsymbol{\alpha}.\boldsymbol{\beta}) \int_a^b \mathrm{d}s$$

$$= F(b - a)\cos\theta.$$

Thus, as would be expected in these circumstances, the work done by $\mathbf{F}$ is the product of the component $F\cos\theta$ of the force $\mathbf{F}$ along the line of motion and the total displacement $(b - a)$.

The line integral also occurs in fields other than particle dynamics; in fluid mechanics, for example, if $\mathbf{F}$ is identified with the fluid velocity $\mathbf{v}$ and Γ is some closed curve drawn in the fluid, then the scalar quantity γ defined by the line integral

$$\gamma = \int_{\Gamma} \mathbf{v}.\mathrm{d}\mathbf{r}$$

is called the *circulation* around the curve Γ. In books on fluid mechanics it is shown that γ provides a measure of the degree of rotational motion present in a fluid. For a special class of fluid flow known as *potential flow* the circulation is everywhere zero, irrespective of the choice of Γ. These flows are said to be *irrotational* and are of fundamental importance. Line integrals around closed curves are generally denoted by the symbol $\oint$ with the conventional that the path of integration is taken anticlockwise, so that for the circulation γ we would write

$$\gamma = \oint_\Gamma \mathbf{v}.\mathbf{dr}.$$

A reversal of the direction of integration around Γ would change the sign of γ.

An exactly similar application of the line integral occurs in electromagnetic theory, where the electromotive force (e.m.f.) between the ends A and B of a wire coinciding with a curve Γ is related to the electric field vector $\mathbf{E}$ by the line integral

$$\text{e.m.f.} = \int_A^B \mathbf{E}.\mathbf{dr}.$$

Example 11.8 Find the work done by a force $\mathbf{F} = yz\mathbf{i} + x\mathbf{j} + z\mathbf{k}$ in moving its point of application along the curve Γ defined by $x = t$, $y = t^2$, $z = t^3$ from the point with parameter $t = 1$ to the point with parameter $t = 2$.

Solution

$$\text{Work done} = \int_\Gamma \mathbf{F}.\mathbf{dr} = \int_\Gamma (yz\mathbf{i} + x\mathbf{j} + z\mathbf{k}).(\mathrm{d}x\mathbf{i} + \mathrm{d}y\mathbf{j} + \mathrm{d}z\mathbf{k})$$
$$= \int_\Gamma yz\,\mathrm{d}x + x\,\mathrm{d}y + z\,\mathrm{d}z.$$

Now as $x = t$, $y = t^2$, $z = t^3$, it follows that

$$\mathrm{d}x = \mathrm{d}t, \quad \mathrm{d}y = 2t\,\mathrm{d}t, \quad \mathrm{d}z = 3t^2\,\mathrm{d}t$$

and so, substituting in the above expression, we find

$$\text{Work done} = \int_1^2 (4t^5 + 2t^2)\,\mathrm{d}t = 140/3 \text{ units.} \quad \blacksquare$$

Example 11.9 If the fluid velocity $\mathbf{v} = x^2y\mathbf{i}$, determine the circulation γ of $\mathbf{v}$ around the contour Γ comprising the boundary of the rectangle $x = \pm a$, $y = \pm b$.

Solution By definition, the circulation γ is

$$\gamma = \oint_\Gamma \mathbf{v}\cdot\mathbf{dr} = \oint_\Gamma x^2y\mathbf{i}.(\mathrm{d}x\mathbf{i} + \mathrm{d}y\mathbf{j} + \mathrm{d}z\mathbf{k}) = \int_\Gamma x^2y\,\mathrm{d}x,$$

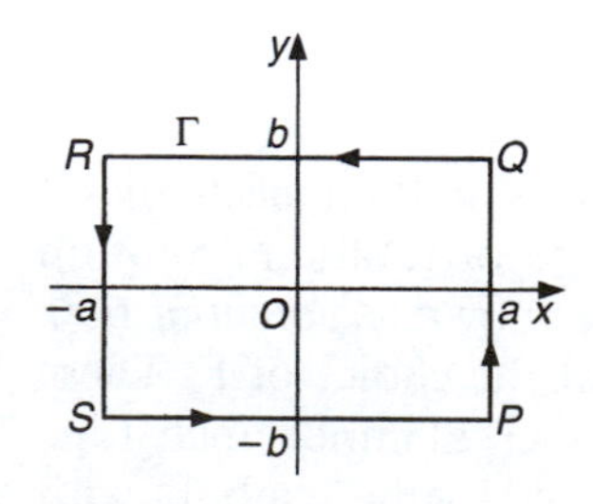

Fig. 11.5 Circulation around a rectangle.

where the direction of integration is anticlockwise around Γ. Now the line integral around Γ may be represented as the sum of four integrals as follows,

$$\gamma = \int_P^Q x^2y\,\mathrm{d}x + \int_Q^R x^2y\,\mathrm{d}x + \int_R^S x^2y\,\mathrm{d}x + \int_S^P x^2y\,\mathrm{d}x,$$

where the limits refer to the corners of the rectangle in Fig. 11.5.

The first and third integrals vanish since x is constant along PQ and RS, with the consequence that $\mathrm{d}x = 0$. Along QR, $y = b$ and along SP $y = -b$, so that

$$\gamma = \int_{a}^{-a} bx^2\,\mathrm{d}x + \int_{-a}^{a} -bx^2\,\mathrm{d}x = \frac{-4a^3b}{3}. \quad \blacksquare$$

11.4 Fields, gradient and directional derivative

The scalar function $\phi = \sqrt{(1-x^2)} + \sqrt{(1-y^2)} + \sqrt{(1-z^2)}$ is defined within and on the cube-shaped domain $|x| \leq 1$, $|y| \leq 1$, $|z| \leq 1$ and assigns a specific number ϕ to every point within that region. In the language of vector analysis, ϕ is said to define a *scalar field* throughout the cube. In general, any scalar function ϕ of position will define a scalar field within its domain of definition. A typical physical example of a scalar field is provided by the temperature at each point of a body.

Similarly, if $\mathbf{F}$ is a vector function of position, we say that $\mathbf{F}$ defines a *vector field* throughout its domain of definition in the sense that it assigns a specific vector to each point. Thus the vector function $\mathbf{F} = (\sin x)\mathbf{i} + xy\mathbf{j} + ye^z\mathbf{k}$ defines a vector field throughout all space.

As heat flows in the direction of decreasing temperature, it follows that associated with the scalar temperature field within a body there must also be a vector field which assigns to each point a vector describing the direction and maximum rate of flow of heat. Other physical examples of vector fields are provided by the velocity field $\mathbf{v}$ throughout a fluid, and the magnetic field $\mathbf{H}$ throughout a region.

To examine more closely the nature of a scalar field, and to see one way in which a special type of vector field arises, we must now define what is called the *gradient* of a scalar function. This is a vector differentiation operation that associates a vector field with every continuously differentiable scalar function.

***Definition 11.5* (gradient of scalar function)** If the scalar function $\phi(x, y, z)$ is a continuously differentiable function with respect to the independent variables x, y and z, then the *gradient* of ϕ, written grad ϕ, is defined to be the vector

$$\operatorname{grad}\phi = \frac{\partial\phi}{\partial x}\mathbf{i} + \frac{\partial\phi}{\partial y}\mathbf{j} + \frac{\partial\phi}{\partial z}\mathbf{k}.$$

For the moment let it be understood that $\mathbf{r} = x\mathbf{i} + y\mathbf{j} + z\mathbf{k}$ is a specific point, and consider a displacement from it, $\mathrm{d}\mathbf{r} = \mathrm{d}x\mathbf{i} + \mathrm{d}y\mathbf{j} + \mathrm{d}z\mathbf{k}$. Then it follows from the definition of grad ϕ that

$$\mathrm{d}\mathbf{r}.\operatorname{grad}\phi = \frac{\partial\phi}{\partial x}\mathrm{d}x + \frac{\partial\phi}{\partial y}\mathrm{d}y + \frac{\partial\phi}{\partial z}\mathrm{d}z,$$

in which it is supposed that grad ϕ is evaluated at $\mathbf{r}=x\mathbf{i}+y\mathbf{j}+z\mathbf{k}$. Theorem 5.19 then asserts that the right-hand side of this expression is simply the total differential $\mathrm{d}\phi$ of the scalar function ϕ, so that we have the result

$$\mathrm{d}\phi = \mathrm{d}\mathbf{r}.\mathrm{grad}\,\phi. \tag{11.18}$$

If we set $\mathrm{d}s = |\mathrm{d}\mathbf{r}|$, then $\mathrm{d}\mathbf{r}/\mathrm{d}s$ is the unit vector in the direction of $\mathrm{d}\mathbf{r}$. Writing $\mathbf{a} = \mathrm{d}\mathbf{r}/\mathrm{d}s$, Eqn (11.18) is thus seen to be equivalent to

$$\frac{\mathrm{d}\phi}{\mathrm{d}s} = \mathbf{a}.\mathrm{grad}\,\phi. \tag{11.19}$$

Because $\mathbf{a}$.grad ϕ is the projection of grad ϕ along the unit vector $\mathbf{a}$, expression (11.19) is called the *directional derivative* of ϕ in the direction of $\mathbf{a}$.

In other words, $\mathbf{a}$.grad ϕ is the rate of change of ϕ with respect to distance measured in the direction of $\mathbf{a}$. We have already utilized the notion of a directional derivative in connection with the derivation of the Cauchy–Riemann equations, though at that time neither the term nor vector notation was employed.

As the largest value of the projection $\mathbf{a}$.grad ϕ at a point occurs when $\mathbf{a}$ is taken in the same direction as grad ϕ, it follows that grad ϕ points in the direction in which the maximum change of the directional derivative of ϕ occurs.

In more advanced treatments of the gradient operator it is this last property that is used to define grad ϕ, since it is essentially independent of the coordinate system that is utilized. From this more general point of view our Definition 11.5 then becomes the interpretation of grad ϕ in terms of rectangular Cartesian coordinates.

The *vector differential operator* ∇, pronounced either 'del' or 'nabla', is defined in terms of rectangular Cartesian coordinates as

$$\nabla \equiv \mathbf{i}\frac{\partial}{\partial x} + \mathbf{j}\frac{\partial}{\partial y} + \mathbf{k}\frac{\partial}{\partial z}. \tag{11.20}$$

Note that ∇ is a vector differential operator, not a vector. It only generates a vector when it acts on a suitably differentiable scalar function. We have the obvious result that

$$\mathrm{grad}\,\phi \equiv \frac{\partial\phi}{\partial x}\mathbf{i} + \frac{\partial\phi}{\partial y}\mathbf{j} + \frac{\partial\phi}{\partial z}\mathbf{k} \equiv \left(\mathbf{i}\frac{\partial}{\partial x} + \mathbf{j}\frac{\partial}{\partial y} + \mathbf{k}\frac{\partial}{\partial z}\right)\phi \equiv \nabla\phi. \tag{11.21}$$

Example 11.10 Determine grad ϕ if $\phi = z^2\cos(xy - \frac{1}{4}\pi)$, and hence deduce its value at the point $(1, \frac{1}{2}\pi, 1)$.

Solution We have

$$\frac{\partial \phi}{\partial x} = -yz^2\sin(xy - \tfrac{1}{4}\pi), \quad \frac{\partial \phi}{\partial y} = -xz^2\sin(xy - \tfrac{1}{4}\pi),$$

$$\frac{\partial \phi}{\partial z} = 2z\cos(xy - \tfrac{1}{4}\pi).$$

Hence,

$$\text{grad}\,\phi = \frac{\partial \phi}{\partial x}\mathbf{i} + \frac{\partial \phi}{\partial y}\mathbf{j} + \frac{\partial \phi}{\partial z}\mathbf{k}$$

$$= -yz^2[\sin(xy - \tfrac{1}{4}\pi)]\mathbf{i} - xz^2[\sin(xy - \tfrac{1}{4}\pi)]\mathbf{j} + 2z[\cos(xy - \tfrac{1}{4}\pi)]\mathbf{k}.$$

At the point $(1, \frac{1}{2}\pi, 1)$ we thus have

$$(\text{grad}\,\phi)_{(1,(1/2)\pi,1)} = \frac{1}{\sqrt{2}}[-(\tfrac{1}{2}\pi)\mathbf{i} - \mathbf{j} + 2\mathbf{k}]. \quad \blacksquare$$

Example 11.11 If $\mathbf{r} = x\mathbf{i} + y\mathbf{j} + z\mathbf{k}$, and $r = |\mathbf{r}|$, deduce the form taken by grad r^n.

Solution As $r = (x^2 + y^2 + z^2)^{1/2}$, it follows from Eqn (11.21) and the chain rule that

$$\text{grad}\,r^n = \left(\mathbf{i}\frac{\partial}{\partial x} + \mathbf{j}\frac{\partial}{\partial y} + \mathbf{k}\frac{\partial}{\partial z}\right)r^n$$

$$= \left(\mathbf{i}\frac{\partial r}{\partial x}\cdot\frac{\partial}{\partial r} + \mathbf{j}\frac{\partial r}{\partial y}\cdot\frac{\partial}{\partial r} + \mathbf{k}\frac{\partial r}{\partial z}\cdot\frac{\partial}{\partial r}\right)r^n$$

$$= nr^{n-1}\left(\frac{\partial r}{\partial x}\mathbf{i} + \frac{\partial r}{\partial y}\mathbf{j} + \frac{\partial r}{\partial z}\mathbf{k}\right).$$

However,

$$\frac{\partial r}{\partial x} = \frac{x}{r}, \quad \frac{\partial r}{\partial y} = \frac{y}{r}, \quad \frac{\partial r}{\partial z} = \frac{z}{r}$$

and so

$$\text{grad}\,r^n = nr^{n-2}(x\mathbf{i} + y\mathbf{j} + z\mathbf{k}) = nr^{n-2}\mathbf{r}. \quad \blacksquare$$

The following theorem is an immediate consequence of the definition of the gradient operator and of the operation of partial differentiation.

Theorem 11.6 (properties of the gradient operator) If ϕ and ψ are two continuously differentiable scalar functions in some domain D, and a, b are scalar constants, then

(a) $\text{grad } a = \mathbf{0}$;
(b) $\text{grad}(a\phi + b\psi) = a \text{ grad } \phi + b \text{ grad } \psi$;
(c) $\text{grad}(\phi\psi) = \phi \text{ grad } \psi + \psi \text{ grad } \phi$. ■

The surfaces $\phi(x, y, z) = \text{constant}$ associated with a scalar function ϕ are called *level surfaces* of ϕ. If we form the total differential of ϕ at a point on a special level surface $\phi = \text{constant}$ then $\mathrm{d}\phi = 0$ and, as in Eqn (5.23), we obtain the result

$$\frac{\partial \phi}{\partial x}\mathrm{d}x + \frac{\partial \phi}{\partial y}\mathrm{d}y + \frac{\partial \phi}{\partial z}\mathrm{d}z = 0.$$

This is equivalent to

$$\mathrm{d}\mathbf{r}.\text{grad } \phi = 0, \tag{11.22}$$

where now $\mathrm{d}\mathbf{r}$ is constrained to lie in the level surface.

This vector condition shows that grad ϕ must be normal to $\mathrm{d}\mathbf{r}$, and as $\mathrm{d}\mathbf{r}$ is constrained to be an arbitrary tangential vector to the level surface at the point in question, it follows that the vector grad ϕ must be normal to the level surface. The unit normal $\mathbf{n}$ to the surface is thus $\mathbf{n} = \text{grad } \phi / |\text{grad } \phi|$. Notice that this normal is unique apart from its sign, since it may be directed towards either side of the surface. This simple argument has proved the following general result.

Theorem 11.7 (normal to a level surface) If is a continuous differentiable scalar function, the unit normal $\mathbf{n}$ to any point of the level surface $\phi = \text{constant}$ is determined by

$$\mathbf{n} = \frac{\text{grad } \phi}{|\text{grad } \phi|}.$$

Example 11.12 If $\phi = x^2 + 3xy^2 + yz^3 - 12$, find the unit normal $\mathbf{n}$ to the level curve $\phi = 3$ at the point (1, 2, 1). Deduce the equation of the tangent plane to the level surface at this point.

Solution The level surface $\phi = 3$ is defined by the equation $\psi = 0$, where

$$\psi = \phi - 3 = x^2 + 3xy^2 + yz^3 - 15 = 0.$$

Hence

$$\text{grad } \psi = (2x + 3y^2)\mathbf{i} + (6xy + z^3)\mathbf{j} + 3yz^2\mathbf{k}$$

which, at (1, 2, 1), becomes

$$(\text{grad } \psi)_{(1,2,1)} = 14\mathbf{i} + 13\mathbf{j} + 6\mathbf{k}.$$

As $\psi = 0$ is the desired level surface (it corresponds to $\phi = 3$), it follows from Theorem 11.7 that the unit normal to this surface at the point (1, 2, 1) must be,

$$\mathbf{n} = \frac{14\mathbf{i} + 13\mathbf{j} + 6\mathbf{k}}{\sqrt{(14^2 + 13^2 + 6^2)}} = \frac{14\mathbf{i} + 13\mathbf{j} + 6\mathbf{k}}{\sqrt{401}}.$$

Now the equation of a plane is $\mathbf{n}.\mathbf{r} = p$, where $\mathbf{r} = x\mathbf{i}+y\mathbf{j}+z\mathbf{k}$ is a general point on the plane, $\mathbf{n}$ is the unit normal to the plane, and p is its perpendicular distance from the origin. The point $\mathbf{r}_0 = \mathbf{i}+2\mathbf{j}+\mathbf{k}$ is a point on the plane so that $\mathbf{n}.\mathbf{r} = \mathbf{n}.\mathbf{r}_0(=p)$, which after substituting for $\mathbf{n}$, $\mathbf{r}$ and $\mathbf{r}_0$ becomes

$$\left(\frac{14\mathbf{i} + 13\mathbf{j} + 6\mathbf{k}}{\sqrt{401}}\right)\cdot(x\mathbf{i} + y\mathbf{j} + z\mathbf{k}) = \left(\frac{14\mathbf{i} + 23\mathbf{j} + 6\mathbf{k}}{\sqrt{401}}\right)\cdot(\mathbf{i} + 2\mathbf{j} + \mathbf{k}),$$

showing that the required equation is

$$14x + 13y + 6z = 46.$$

We have seen how the gradient operator associates a vector field grad ϕ with every continuously differentiable scalar field ϕ. Any vector field $\mathbf{F} = \text{grad}\ \phi$ which is expressible as the gradient of a scalar field ϕ is called a *conservative* vector field, and ϕ is then referred to as the *scalar potential* associated with the vector field.

This has an important implication when line integrals involving conservative vector fields are considered. Let us suppose that $\mathbf{F} = \text{grad}\ \phi$, and that

$$J = \int_A^B \mathbf{F}.\mathrm{d}\mathbf{r}. \tag{11.23}$$

Then

$$J = \int_A^B \text{grad}\ \phi.\mathrm{d}\mathbf{r}, \tag{11.24}$$

and by virtue of Eqn (11.25) this can be written

$$J = \int_A^B \mathrm{d}\phi = \phi(\mathbf{B}) - \phi(\mathbf{A}). \tag{11.25}$$

Hence when $\mathbf{F}$ belongs to a conservative vector field, results (11.23) and (11.25) show that the line integral J of $\mathbf{F}$ depends only on the end points of the path of integration, and not on the path itself.

This fundamental result has far reaching consequences and forms the basis of many important developments, of which potential fluid flows, steady-state heat flows and gravitational potential theory are typical examples.

Suppose, for example, that $\mathbf{F}$ is identified with a conservative force field. Then result (11.25) represents the change in the potential energy of a particle as it moves from point A to point B. That J depends only on the difference $\phi(B)$ and $\phi(A)$ and not on the path joining A to B explains why, when using potential energy arguments in mechanics, no consideration need be given to the path that is followed by particles.

11.5 Divergence and curl of a vector

In this section we introduce two special operations involving differentiation of vectors called the *divergence* and *curl* of a vector. The divergence is an operation that when performed on a vector produces a *scalar* quantity, while curl is an operation that when performed on a vector produces another *vector*. The motivation for these operations will appear during their derivation. For the sake of simplicity, we first introduce these operations in terms of two-dimensional vectors, and then generalize the results to three dimensions.

We consider the divergence operation first, and set

$$\mathbf{F}(x, y) = F_1(x, y)\mathbf{i} + F_2(x, y)\mathbf{j}, \tag{11.26}$$

so that $\mathbf{F}$ is a two-dimensional vector function that depends on position in the (x, y)-plane, with $F_1(x, y)$ and $F_2(x, y)$ differentiable functions with respect to their arguments. Let us now consider the small rectangle $ABCD$ shown in Fig. 11.6 with sides AB and CD each of length h parallel to the x-axis, and sides AD and BC each of length k parallel to the y-axis, where h, k are assumed to be small.

A measure of the amount of vector $\mathbf{F}$ directed out of the rectangle through side AD is provided by the product of the average value of the component of $\mathbf{F}$ in the direction of the outward drawn normal to AD and the length k of side AD.

Because k is small, the average value of $\mathbf{F}$ along AD may be approximated by its value $\mathbf{F}(x, y)$ at A. The outward drawn unit normal to AD is $-\mathbf{i}$, so the product of the average value of the component of $\mathbf{F}$ in the direction of the outward drawn normal to AD and the length k of AD is approximately

$$[\mathbf{F}(x, y)].(-\mathbf{i}k) = -F_1(x, y)k.$$

A similar argument applied to side BC shows the amount of vector $\mathbf{F}$ directed out of the rectangle through side BC is, approximately,

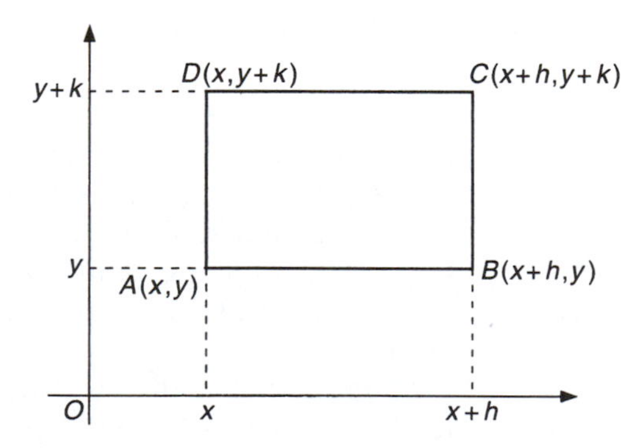

Fig. 11.6 Geometrical interpretation of div **F**.

$$[\mathbf{F}(x+h, y)].(\mathbf{i}k) = F_1(x+h, y)k.$$

Thus if Q_h is the amount of $\mathbf{F}$ directed out of the rectangle in the *horizontal* direction through the parallel sides AD and BC, it follows that Q_h is given approximately by

$$Q_h = [F_1(x+h, y) - F_1(x, y)]k. \tag{11.27}$$

The mean value theorem asserts that

$$F_1(x+h, y) = F_1(x, y) + h\left(\frac{\partial F_1}{\partial x}\right)_{(\xi,y)},$$

where $(\partial F_1/\partial x)_{(\xi,y)}$ is the value of $\partial F_1/\partial x$ at the point (ξ, y), for some ξ in the interval $x < \xi < x+h$. Thus Eqn (11.27) may be rewritten as

$$Q_h = hk\left(\frac{\partial F_1}{\partial x}\right)_{(\xi,y)}. \tag{11.28}$$

The same form of argument applied to the sides AB and CD shows that Q_v, the amount of $\mathbf{F}$ directed out of the rectangle *vertically* through these sides, is given approximately by

$$Q_v = hk\left(\frac{\partial F_2}{\partial y}\right)_{(x,\eta)}, \tag{11.29}$$

where $(\partial F_2/\partial y)_{(x,y)}$ is the value of $\partial F_2/\partial y$ at the point (x, η) for some η in the interval $y < \eta < y+k$.

Thus the total amount of $\mathbf{F}$ leaving the rectangle per unit area is approximated by $Q_h + Q_v$ divided by the area hk of rectangle $ABCD$. Letting $h, k \to 0$, so that $\xi \to x$ and $\eta \to y$, in the limit the scalar quantity $(Q_h + Q_v)/hk$, called the *divergence* of $\mathbf{F}$, and written div $\mathbf{F}$ is given by

$$\text{div } \mathbf{F} = \frac{\partial F_1}{\partial x} + \frac{\partial F_2}{\partial y}. \tag{11.30}$$

When this argument is generalized to a vector $\mathbf{F}$ of the form

$$\mathbf{F}(x, y, z) = F_1(x, y, z)\mathbf{i} + F_2(x, y, z)\mathbf{j} + F_3(x, y, z)\mathbf{k},$$

we arrive at the following definition.

Definition 11.6 (divergence of a vector) Let the vector

$$\mathbf{F}(x, y, z) = F_1(x, y, z)\mathbf{i} + F_2(x, y, z)\mathbf{j} + F_3(x, y, z)\mathbf{k}$$

be such that F_1, F_2 and F_3 are differentiable functions of their arguments. Then the *divergence* of $\mathbf{F}$ is defined as the scalar quantity

$$\text{div } \mathbf{F} = \frac{\partial F_1}{\partial x} + \frac{\partial F_2}{\partial y} + \frac{\partial F_3}{\partial z}.$$

A useful physical interpretation of the divergence of **F** follows by identifying **F** with the velocity vector in a fluid. In such a case div **F** at a point $P(x, y, z)$ measures the flow of fluid into or out of a unit volume per unit time in a unit volume surrounding the point P. The quantity div **F** will be positive when fluid flows into the volume and negative when it flows out of it.

Example 11.13 Find the divergence of the vector

$$\mathbf{F}(x, y, z) = (x^2 + 3yz)\mathbf{i} + 3xyz\mathbf{j} + 2yz^3\mathbf{k}.$$

Solution Making the identifications $F_1 = x^2 + 3yz$, $F_2 = 3xyz$ and $F_3 = 2yz^3$, and using Definition 11.6, gives

$$\begin{aligned}\text{div } \mathbf{F} &= \frac{\partial}{\partial x}[x^2 + 3yz] + \frac{\partial}{\partial y}[3xyz] + \frac{\partial}{\partial z}[2yz^3] \\ &= 2x + 3xz + 6yz^2. \quad \blacksquare\end{aligned}$$

Next we motivate the definition of the curl of a vector **F** and, as with the divergence, start by considering a two-dimensional vector of the form

$$\mathbf{F}(x, y) = F_1(x, y)\mathbf{i} + F_2(x, y)\mathbf{j}. \tag{11.31}$$

This time, however, by traversing the sides of the rectangle in Fig. 11.6 in an anticlockwise direction, we shall determine the sum of the products of the average changes in the component of **F** along each side of the rectangle and the lengths of the sides. Finally, by dividing this quantity by the area hk of the rectangle and letting $h, k \to 0$, we will arrive at a quantity that measures the rotation of **F** around this limiting rectangle. The result is a scalar quantity with magnitude determined by this limiting operation and a direction of rotation about the unit vector **k** normal to the (x, y)-plane. This quantity, called the *curl* of **F**, and written curl **F**, is a vector, because it has both magnitude and direction.

Along the line AB vector **F** may be approximated by $\mathbf{F}(x, y)$, its value at A, but the unit vector along AB is **i**, so the product R_{AB} of the average of the component of **F** and the length h of AB is approximately

$$\mathbf{R}_{AB} = [\mathbf{F}(x, y)].(\mathbf{i}h) = F_1(x, y)h.$$

Similarly, along CD, **F** may be approximated by $\mathbf{F}(x, y+k)$, its value at D, though now the unit vector along CD is $-\mathbf{i}$ so the product $R_{C,D}$ of the average of the component of **F** and the length h of CD is approximately

$$\mathbf{R}_{CD} = [\mathbf{F}(x, y + k)].(-\mathbf{i}h) = -F_1(x, y + k)h,$$

and so

$$R_{AB} + R_{CD} = -[F_1(x, y + k) - F_1(x, y)]h. \tag{11.32}$$

An application of the mean value theorem, as in the derivation of div **F**, reduces Eqn (11.32) to

$$R_{AB} + R_{CD} = -hk\left(\frac{\partial F_1}{\partial y}\right)_{(x,\eta)}, \tag{11.33}$$

where $(\partial F_1/\partial y)_{(x,\eta)}$ is the value of $\partial F_1/\partial y$ at the point (x, η), for some η in the interval $y < \eta < y+k$.

A similar argument applied to the sides BC and DA in the directions indicated by the order of the letters yields

$$R_{BC} + R_{DA} = hk\left(\frac{\partial F_2}{\partial x}\right)_{(\xi,y)}, \tag{11.34}$$

where $(\partial F_2/\partial x)_{(\xi,y)}$ is the value of $\partial F_2/\partial x$ at the point (ξ, y), for some ξ in the interval $x < \xi < x+h$.

Forming the sum $R_{AB} + R_{CD} + R_{BC} + R_{DA}$, dividing by the area of the rectangle hk and letting $h, k \to 0$, so that $\xi \to x$ and $\eta \to y$, shows that

$$\lim_{\substack{h\to 0\\ k\to 0}}\left(\frac{R_{AB} + R_{CD} + R_{BC} + R_{DA}}{h.k}\right) = \frac{\partial F_2}{\partial x} - \frac{\partial F_1}{\partial y}.$$

As this quantity was obtained by moving around the sides of the rectangle in an anticlockwise direction, that is from the direction of the unit vector **i** to the direction of the unit vector **j**, and $\mathbf{i} \times \mathbf{j} = \mathbf{k}$, it follows that it has associated with it the unit vector **k**. We define this new vector quantity to be the *curl* of **F**, written curl **F**, where

$$\operatorname{curl}\ \mathbf{F} = \left(\frac{\partial F_2}{\partial x} - \frac{\partial F_1}{\partial y}\right)\mathbf{k}. \tag{11.35}$$

When these arguments are generalized to the vector

$$\mathbf{F}(x, y, z) = F_1(x, y, z)\mathbf{i} + F_2(x, y, z)\mathbf{j} + F_3(x, y, z)\mathbf{k},$$

we arrive at the following definition.

Definition 11.7 (curl of a vector) Let the vector

$$\mathbf{F}(x, y, z) = F_1(x, y, z)\mathbf{i} + F_2(x, y, z)\mathbf{j} + F_3(x, y, z)\mathbf{k},$$

be such that F_1, F_2 and F_3 are differentiable functions of their arguments. Then the *curl* of **F** is defined as the vector

$$\operatorname{curl}\ \mathbf{F} = \left(\frac{\partial F_3}{\partial y} - \frac{\partial F_2}{\partial z}\right)\mathbf{i} + \left(\frac{\partial F_1}{\partial z} - \frac{\partial F_3}{\partial x}\right)\mathbf{j} + \left(\frac{\partial F_2}{\partial x} - \frac{\partial F_1}{\partial y}\right)\mathbf{k}.$$

A useful physical interpretation of curl **F** at a point $P(x, y, z)$ is provided by identifying **F** with the velocity vector in a fluid, for then curl **F** describes how a small particles suspended in the fluid at point **P** will be rotated by the fluid as it moves with the fluid.

Example 11.14 Find the curl of the vector

$$\mathbf{F}(x, y, z) = (x^2 - y^2)\mathbf{i} + xy^3z\mathbf{j} + (x^2 + y^2 - z^2)\mathbf{k}.$$

Solution Identifying **F** with the vector in Definition 11.7 shows that

$$F_1 = x^2 - y^2, \quad F_2 = xy^3z \quad \text{and} \quad F_3 = x^2 + y^2 - z^2,$$

so

$$\begin{aligned}\text{curl } \mathbf{F} &= \left[\frac{\partial}{\partial y}(x^2 + y^2 - z^2) - \frac{\partial}{\partial z}(xy^3z)\right]\mathbf{i} + \left[\frac{\partial}{\partial z}(x^2 - y^2)\right. \\ &\quad \left. - \frac{\partial}{\partial x}(x^2 + y^2 - z^2)\right]\mathbf{j} + \left[\frac{\partial}{\partial x}(xy^3z) - \frac{\partial}{\partial y}(x^2 - y^2)\right]\mathbf{k} \\ &= (2y - xy^3)\mathbf{i} - 2x\mathbf{j} + (y^3z + 2y)\mathbf{k}. \quad \blacksquare\end{aligned}$$

The notations for the operations of grad ϕ, div F and curl F can be unified by defining the vector operator ∇, called 'del', as

$$\nabla = i\frac{\partial}{\partial x} + j\frac{\partial}{\partial y} + k\frac{\partial}{\partial z}. \tag{11.36}$$

It is important to recognize that ∇ is *not* a vector, but a *vector operator* that can act on either a scalar or a vector. It follows that

$$\text{grad } \phi = \nabla\phi, \tag{11.37}$$

$$\text{div } \mathbf{F} = \nabla.\mathbf{F}, \tag{11.38}$$

$$\text{curl } \mathbf{F} = \nabla \times \mathbf{F}. \tag{11.39}$$

Examination of Definition 11.7 shows that curl **F** can be written in the following convenient form

$$\text{curl } \mathbf{F} = \begin{vmatrix} \mathbf{i} & \mathbf{j} & \mathbf{k} \\ \dfrac{\partial}{\partial x} & \dfrac{\partial}{\partial y} & \dfrac{\partial}{\partial z} \\ F_1 & F_2 & F_3 \end{vmatrix}. \tag{11.40}$$

Example 11.15 Given that

$$\phi = x^2 + 3xy + y^2z^2 \quad \text{and} \quad \mathbf{F} = x^2\mathbf{i} + xyz\mathbf{j} + yz^2\mathbf{k},$$

find $\nabla\phi$, $\nabla\cdot\mathbf{F}$ and $\nabla \times \mathbf{F}$.

Solution

$$\begin{aligned}\text{grad } \phi = \nabla\phi &= \left(\mathbf{i}\frac{\partial}{\partial x} + \mathbf{j}\frac{\partial}{\partial y} + \mathbf{k}\frac{\partial}{\partial z}\right)(x^2 + 3xy + yz^2) \\ &= (2x + 3y)\mathbf{i} + (3x + 2yz)\mathbf{j} + 2yz\mathbf{k},\end{aligned}$$

$$\text{div}F = \left(\mathbf{i}\frac{\partial}{\partial x} + \mathbf{j}\frac{\partial}{\partial y} + \mathbf{k}\frac{\partial}{\partial z}\right) \cdot (x^2\mathbf{i} + xyz\mathbf{j} + yz^2\mathbf{k})$$

$$= 2x + xz + 2yz,$$

$$\text{curl } \mathbf{F} = \begin{vmatrix} \mathbf{i} & \mathbf{j} & \mathbf{k} \\ \dfrac{\partial}{\partial x} & \dfrac{\partial}{\partial y} & \dfrac{\partial}{\partial z} \\ x^2 & xyz & yz^2 \end{vmatrix}$$

$$= (z^2 - xy)\mathbf{i} + yz\mathbf{k}. \quad \blacksquare$$

11.6 Conservative fields and potential functions

We conclude this chapter by mentioning an important connection that exists between line integrals and the gradient of a scalar function. First we give two definitions.

Definition 11.8 (conservative field) A vector field $\mathbf{F}$ is called a *conservative field* if the line integral $\int_A^B \mathbf{F}.\mathbf{dr}$ is independent of the path joining points A and B.

Definition 11.9 (potential function) If a vector field $\mathbf{F}$ is such that $\mathbf{F} = \nabla\phi$, then ϕ is called a *potential function* for $\mathbf{F}$.

Conservative fields are fundamental to the study of the mechanics of bodies moving in gravitational fields in which dissipative effects such as friction can be ignored, because a gravitational field is a conservative field and due to this energy is *conserved*. The result is that the sum of the potential and kinetic energies remains constant (hence the name *conservative field*). In a gravitational field the change in potential energy as a body moves from point A at height h_1 to point B at height h_2 is simply the difference between the potential energies at heights h_1 and h_2, and it is independent of the way in which the body moves from point A to point B.

We now state the following important theorem without proof.

Theorem 11.8 (fundamental theorem for conservative fields) A vector field $\mathbf{F}$ is a conservative field if a differentiable scalar function ϕ exists such that $\mathbf{F} = \nabla\phi$. For such a vector field $\mathbf{F}$ it follows that

$$\int_A^B \text{F.d}r = \int_A^B \nabla\phi.\text{d}r = \phi(B) - \phi(A). \quad \blacksquare$$

The final theorem, which we again offer without proof, provides a test by which to determine whether or not a vector field **F** is conservative.

Theorem 11.9 (test for a conservative field) The vector field

$$\mathbf{F} = P(x, y, z)\mathbf{i} + Q(x, y, z)\mathbf{j} + R(x, y, z)\mathbf{k}$$

is a conservative field if

$$\frac{\partial R}{\partial y} = \frac{\partial Q}{\partial z}, \quad \frac{\partial P}{\partial z} = \frac{\partial R}{\partial x}, \quad \frac{\partial Q}{\partial x} = \frac{\partial P}{\partial y}. \quad \blacksquare$$

Example 11.16 Show that $\mathbf{F} = y\mathbf{i} + (x + z^2)\mathbf{j} + 2yz\mathbf{k}$ is a conservative field, and find a potential function for **F**.

Solution Making the identification $P = y$, $Q = x + z^2$ and $R = 2yz$, a routine calculation shows that the conditions of Theorem 11.9 are satisfied, so that **F** is a conservative field. Because **F** is conservative it follows that a function ϕ exists such that

$$\nabla\phi = y\mathbf{i} + (x + z^2)\mathbf{j} + 2yz\mathbf{k}.$$

Substituting $\nabla\phi = (\partial\phi/\partial x)\mathbf{i} + (\partial\phi/\partial y)\mathbf{j} + (\partial\phi/\partial z)\mathbf{k}$ in the above result and equating the components of **i**, **j** and **k** then shows that

$$\frac{\partial\phi}{\partial x} = y, \quad \frac{\partial\phi}{\partial y} = x + z^2 \quad \text{and} \quad \frac{\partial\phi}{\partial z} = 2yz.$$

Integrating the first equation with respect to x, and because partial differentiation is involved regarding y and z as constants, gives

$$\phi = xy + f(y, z),$$

where, for the moment, $f(y, z)$ is an arbitrary function of y and z that will be determined later. Using this result to compute $\dfrac{\partial\phi}{\partial y}$ and equating the result to the expression for $\dfrac{\partial\phi}{\partial y}$ given above shows that

$$\frac{\partial\phi}{\partial y} = x + \frac{\partial f}{\partial y} = x + z^2,$$

showing that $\dfrac{\partial f}{\partial y} = z^2$, from which we see by integration with respect to y that

$$f(y, z) = yz^2 + g(z).$$

Using this result in our earlier expression for ϕ gives

$$\phi = xy + yz^2 + g(z).$$

Using this to compute $\partial\phi/\partial z$ and equating it to $\partial\phi/\partial z = 2yz$ then shows that

$$\frac{\partial\phi}{\partial z} = 2yz + \frac{dg}{dz} = 2yz,$$

and so

$$\frac{dg}{dz} = 0 \quad \text{showing that } g \equiv C(\text{const.}).$$

Thus the required potential is

$$\phi = xy + yz^2 + C,$$

showing that there are infinitely many potential functions for **F**, though they only differ one from the other by a constant. ■

In fluid mechanics a potential function plays an important part in the study of inviscid, irrotational, incompressible flow. This is because it is possible to introduce a potential function, called the *velocity potential*, with the property that the gradient of the velocity potential at any point gives the fluid velocity at that point.

Problems

Section 11.1

11.1 Sketch and give a brief description of the curves described by the following vector functions of a single real variable t:

(a) $\mathbf{r} = a(\cos 2\pi t)\mathbf{i} + b(\sin 2\pi t)\mathbf{j} + t\mathbf{k}$;

(b) $\mathbf{r} = a(\cos 2\pi t)\mathbf{i} + b(\sin 2\pi t)\mathbf{j} + t^2\mathbf{k}$;

(c) $\mathbf{r} = t\mathbf{i} + t^2\mathbf{j} + t^3\mathbf{k}$.

11.2 Form the vector functions $\mathbf{u}(t)+\mathbf{v}(t)$, $\mathbf{u}(t) \times \mathbf{v}(t)$, and the scalar function $\mathbf{u}(t).\mathbf{v}(t)$ given that:

$$\mathbf{u}(t) = t^2\mathbf{i} + (\sinh t)\mathbf{j} + \left(\frac{1 - t^2}{1 + t^2}\right)\mathbf{k}$$

and

$$\mathbf{v}(t) = 2t\mathbf{i} + (\cosh t)\mathbf{j} + (\sin t)\mathbf{k}.$$

11.3 Determine $d\mathbf{u}/dt$ and $d^2\mathbf{u}/dt^2$ for the vectors $\mathbf{u}$ defined in (a) and (c) of Problem 11.1, and find $d\mathbf{u}/dt$ for the vector

$$\mathbf{u} = \frac{\mathbf{r}}{r^2} + (\mathbf{a}.\mathbf{r})\mathbf{b} + \mathbf{a} \times \frac{d^2\mathbf{r}}{dt^2},$$

where $\mathbf{r} = \mathbf{r}(t)$, $r = |\mathbf{r}|$ and $\mathbf{a}$, $\mathbf{b}$ are constant vectors.

11.4 The position vector of a particle at time t is

$$\mathbf{r} = [\cos(t-1)]\mathbf{i} + [\sinh(t-1)]\mathbf{j} + \alpha t^3\mathbf{k}.$$

Find the condition imposed on α by requiring that at time $t = 1$ the acceleration vector is normal to the position vector.

11.5 Find the unit tangent $\mathbf{T}$ to the curve

$$\mathbf{r} = t\mathbf{i} + t^2\mathbf{j} + t^3\mathbf{k}$$

at the points corresponding to $t = 0$ and $t = 1$.

11.6 Find the unit tangent vector to the helix

$$\mathbf{r} = \cos t\mathbf{i} + \sin t\mathbf{j} + t\mathbf{k}$$

as a function of t.

Section 11.2

11.7 Find the antiderivative of the following two functions $\mathbf{f}(t)$:

(a) $\mathbf{f}(t) = (\cosh 2t)\mathbf{i} + \dfrac{1}{t}\mathbf{j} + t^3\mathbf{k}$;

(b) $\mathbf{f}(t) = t^2(\sin t)\mathbf{i} + e^t\mathbf{j} + (\ln t)\mathbf{k}$.

11.8 Verify the following antiderivatives using Definition 11.3:

(a) $\displaystyle\int\left(\mathbf{r}.\frac{d\mathbf{r}}{dt}\right)dt = \tfrac{1}{2}(\mathbf{r}.\mathbf{r}) + C = \tfrac{1}{2}\mathbf{r}^2 + C$;

(b) $\displaystyle\int\left(\frac{d\mathbf{r}}{dt}.\frac{d^2\mathbf{r}}{dt^2}\right)dt = \frac{1}{2}\frac{d\mathbf{r}}{dt}.\frac{d\mathbf{r}}{dt} + C = \frac{1}{2}\left(\frac{d\mathbf{r}}{dt}\right)^2 + C$;

(c) $\displaystyle\int\mathbf{r}\times\frac{d^2\mathbf{r}}{dt^2} = \mathbf{r}\times\frac{d\mathbf{r}}{dt} + \mathbf{C}$,

where C, $\mathbf{C}$ are arbitrary constants.

11.9 Use the result of Problem 11.7 to express $d\mathbf{r}/dt$ in terms of $\mathbf{r}$, given that $\mathbf{r}$ satisfies the vector differential equation

$$\frac{d^2\mathbf{r}}{dt^2} + \Omega^2\mathbf{r} = \mathit{0}.$$

11.10 Evaluate the definite integral

$$\int_1^2 [t^2 e^t\mathbf{i} + t(\log t)\mathbf{j} + t^2\mathbf{k}]dt.$$

11.11 The displacement of a particle P is given in terms of the time t by

$$\mathbf{r} = (\cos 2t)\mathbf{i} + (\sin 2t)\mathbf{j} + t^2\mathbf{k}.$$

If v and f are the magnitudes of the velocity and acceleration respectively, show that

$$5v^2 = f^2(1 + t^2).$$

11.12 A point moving in space has acceleration $(\cos t)\mathbf{i}+(\sin t)\mathbf{j}$. Find the equation of its path if it passes through the point $(-1, 0, 0)$ with velocity $-\mathbf{j}+\mathbf{k}$ at time $t = 0$.

11.13 Evaluate the line integral of $\mathbf{F} = xy\mathbf{i}+yz\mathbf{j}+z\mathbf{k}$ along the contour defined by $\mathbf{r} = t\mathbf{i}+t^2\mathbf{j}+t^3\mathbf{k}$ from $t = 0$ to $t = 1$.

11.14 Evaluate the line integral of $\mathbf{F} = 4xy\mathrm{i} - 2x^2\mathbf{j}+3z\mathbf{k}$ from the origin to the point (2, 1, 0) along the straight lines:

(a) from the origin to the point (2, 0, 0) and then from the point (2, 0, 0) to the point (2, 1, 0);

(b) from the origin to the point (0, 1, 0) and then from the point (0, 1, 0) to the point (2, 1, 0);

(c) from the origin to the point (2, 1, 0) along the straight line joining these two points.

(*Hint*: the contours (a), (b), (c) all lie in the plane $z = 0$.)

11.15 Evaluate the line integral $\mathbf{F} = 4xy\mathbf{i}+2x^2\mathbf{j}+3z\mathbf{k}$ from the origin to the point (2, 1, 0) along the contours of Problem 11.14.

Section 11.3

11.16 A particle moves in a curve given by

$$r = a(1 - \cos\theta) \quad \text{with} \quad \frac{d\theta}{dt} = 3.$$

Find the components of velocity and acceleration. Show that the velocity is zero when $\theta = 0$. Find the acceleration when $\theta = 0$.

11.17 A particle moves on that portion of the curve $r = ae^{\theta}\cos\theta$ (a = constant) for which $0 < \theta < \frac{1}{4}\pi$, so that its radial velocity u remains constant. Find its transverse velocity and its radial and transverse components of acceleration as functions of r and θ.

11.18 If the fluid velocity $\mathbf{v} = y\mathbf{i}+2x\mathbf{j}$, determine the circulation γ by integrating anticlockwise around the rectangular contour $x = \pm a$, $y = \pm b$. Show that the sign of γ is reversed if the direction of integration is taken clockwise around the same contour.

11.19 Consider three rectangular regions (a) $0 \le x \le 1$, $-1 \le y \le 1$, (b) $0 \le x \le 1$, $1 \le y \le 2$ and (c) $0 \le x \le 1$, $-1 \le y \le 2$, and denote their boundary curves by Γ_1, Γ_2 and Γ_3. If $\mathbf{F} = 2y\mathbf{i}+x\mathbf{j}$, evaluate the three line integrals

$$J_1 = \int_{\Gamma_1} \mathbf{F}.\mathbf{dr}, \quad J_2 = \int_{\Gamma_2} \mathbf{F}.\mathbf{dr}, \quad J_3 = \int_{\Gamma_3} \mathbf{F}.\mathbf{dr},$$

and hence show that $J_1 + J_2 = J_3$.

11.20 Given that $\mathbf{F} = (\cos y)\mathbf{i} + (\sin x)\mathbf{j}$, evaluate the line integral of $\mathbf{F}$ taken anticlockwise around the triangle with vertices at the points $(0, 0)$, $(\frac{1}{2}\pi, 0)$, $(\frac{1}{2}\pi, \frac{1}{2}\pi)$.

11.21 A vector field $\mathbf{F}$ is said to be *irrotational* if its line integral around any closed curve Γ is zero. By integrating around two conveniently chosen contours, deduce which of the following vector fields are irrotational:

(a) $\mathbf{F} = y(\sinh z)\mathbf{i} + x(\sinh z)\mathbf{j} + xy(\cosh z)\mathbf{k}$;

(b) $\mathbf{F} = x\mathbf{i} + y\mathbf{j} + z\mathbf{k}$;

(c) $\mathbf{F} = xyz^2\mathbf{i} + x^2z^2\mathbf{j} + x^2yz\mathbf{k}$.

Section 11.4

11.22 Find the gradient of the following function ϕ:

(a) $\phi = \cosh xyz$;

(b) $\phi = x^2 + y^2 + z^2$;

(c) $\phi = xy \tanh(x - z)$.

11.23 Find the directional derivative off the following functions ϕ in the direction of the vector $(\mathbf{i} + 2\mathbf{j} - 2\mathbf{k})$:

(a) $\phi = 3x^2 + xy^2 + yz$;

(b) $\phi = x^2yz + \cos y$;

(c) $\phi = 1/xyz$.

11.24 If $\mathbf{a}$ is a constant vector and $\mathbf{r} = x\mathbf{i} + y\mathbf{j} + z\mathbf{k}$, $r = |\mathbf{r}|$, prove that

(a) $\operatorname{grad}(\mathbf{a}.\mathbf{r}) = \mathbf{a}$;

(b) $\operatorname{grad} r = \mathbf{r}/r$;

(c) $\operatorname{grad}\left(\frac{1}{r}\right) = -\frac{\mathbf{r}}{r^3}$.

11.25 By using the Cartesian representation of grad ϕ as expressed in Definition 11.5, prove that

(a) $\operatorname{grad}(a\phi + b\psi) = a \operatorname{grad} \phi + b \operatorname{grad} \psi$;

(b) $\operatorname{grad}(\phi\psi) = \phi \operatorname{grad} \psi + \psi \operatorname{grad} \phi$,

where a, b are scalar constants and ϕ, ψ are suitably differentiable functions.

11.26 A vector field $\mathbf{F}$ will be *irrotational* if it is expressible in the form $\mathbf{F} = \operatorname{grad} \phi$, with ϕ a scalar potential. Find the most general scalar potential ϕ that will give rise to the irrotational vector field

$$\mathbf{F} = (x + \tfrac{1}{2}y^2z^2)\mathbf{i} + xyz^2\mathbf{j} + xy^2z\mathbf{k}.$$

11.27 Find the unit normal $\mathbf{n}$ to the surface $x^2+2y^2-z^2-8=0$ at the point (1, 2, 1). Deduce the equation of the tangent plan to the surface at this point.

11.28 Find the unit normal $\mathbf{n}$ to the surface $x^2-4y^2+2z^2=6$ at the point (2, 2, 3). Deduce the equation of the plane which has $\mathbf{n}$ as its normal and which passes through the origin.

11.29 If (x_0, y_0, z_0) is a point on the conic surface

$$\frac{x^2}{a}+\frac{y^2}{b}+\frac{z^2}{c}=1,$$

show that the tangent plane to the surface at that point is

$$\frac{xx_0}{a}+\frac{yy_0}{b}+\frac{zz_0}{c}=1.$$

11.30 The vector field $\mathbf{F}$ is generated by the scalar potential $\phi = x^2y$. Verify directly by integration that the line integral of $\mathbf{F}$ along each of the three paths of Problem 11.14 is equal to 4.

Section 11.5

11.31 Find div $\mathbf{F}$ if $\mathbf{F} = e^x(\sin y)\mathbf{i}+xy^2z\mathbf{j}+x^2z^2\mathbf{k}$, and hence determine its value at the point $(1, \pi/2, 2)$.

11.32 Find div $\mathbf{F}$ if $\mathbf{F} = [\sinh(x+y+z)]\mathbf{i}+(x+2y+z)\mathbf{j}+yz\mathbf{k}$, and hence determine its value at the point (0, 0, 0).

11.33 Find div $\mathbf{F}$ if $\mathbf{F} = \text{grad}\,\phi$, and $\phi = x^2+3yz+xz^2$.

11.34 Use the definitions of div and grad to prove that, provided ϕ is suitably differentiable, then

$$\text{div grad}\ \phi = \frac{\partial^2\phi}{\partial x^2}+\frac{\partial^2\phi}{\partial y^2}+\frac{\partial^2\phi}{\partial z^2}.$$

11.35 Use the definition of div to prove the following general result in which k_1 and k_2 are constants and $\mathbf{F}_1$ and $\mathbf{F}_2$ are suitably differentiable vector fields

$$\text{div}\,(k_1\mathbf{F}_1 + k_2\mathbf{F}_2) = k_1\text{div}\mathbf{F}_1 + k_2\text{div}\mathbf{F}_2.$$

11.36 Use the definition of div to prove that

$$\text{div}\,(\phi\mathbf{F}) = \phi\ \text{div}\ \mathbf{F}+(\text{grad}\ \phi).\mathbf{F},$$

where ϕ is a suitably differentiable scalar function, and $\mathbf{F}$ is a suitably differentiable vector field.

11.37 Find curl $\mathbf{F}$, given that $\mathbf{F} = yz\mathbf{i}+2xz\mathbf{j}+xy^2\mathbf{k}$.

11.38 Find curl **F**, given that $\mathbf{F} = y(\sin z)\mathbf{i} + xyz(\cos z)\mathbf{j} + x(\cos y)\mathbf{k}$.

11.39 Use the definition of curl **F** in the form (11.40) to prove that $\operatorname{div}\operatorname{curl}\mathbf{F} = 0$.

11.40 Use the definition of curl to prove that

$$\operatorname{curl}(k_1\mathbf{F}_1 + k_2\mathbf{F}_2) = k_1\mathbf{F}_1 + k_2\operatorname{curl}\ \mathbf{F}_2,$$

where k_1 and k_2 are constants and $\mathbf{F}_1$ and $\mathbf{F}_2$ are suitably differentiable vector fields.

11.41 Use the definition of curl to prove that if ϕ is a suitably differentiable scalar function

$$\operatorname{curl}(\operatorname{grad}\ \phi) = \mathbf{0}.$$

Section 11.6

11.42 Show that

$$\mathbf{F} = e^y(\cos x)\mathbf{i} + (e^y \sin x + z^2)\mathbf{j} + 2yz\mathbf{k}$$

is a conservative field, and find a potential function for **F**.

11.43 Show that

$$\mathbf{F} = (z + \cos z)\mathbf{i} + (\cos z)\mathbf{i} + [x - (x + y)\sin z]\mathbf{k}$$

is a conservative field, and find a potential function for **F**.

11.44 Show that

$$\mathbf{F} = (2x + y^2 z)\mathbf{i} + 2xyz\mathbf{j} + x^2 y\mathbf{k}$$

is a conservative field, and find a potential function for **F**.

11.45 Show that

$$\mathbf{F} = 2xyz^3\mathbf{i} + x^2 z^3\mathbf{j} + 3x^2 yz^2\mathbf{k}$$

is a conservative field, and find a potential function for **F**.

11.46 Find a and b such that

$$\mathbf{F} = (y^2 z + z^2)\mathbf{i} + axyz\mathbf{j} + (xy^2 + bxz)\mathbf{k}$$

is conservative, and using these values of a and b find a potential function for **F**.

11.47 Find a and b such that

$$\mathbf{F} = 4xy\mathbf{i} + (ax^2 + byz^2)\mathbf{j} + by^2 z\mathbf{k}$$

is conservative, and using these values of a and b find a potential function for **F**.

12 Series, Taylor's theorem and its uses

The purpose of this chapter is first to develop and make precise the ideas underlying the convergence of series. It then proceeds to examine the way in which functions may be represented in terms of series, after which it shows some of the most important ways in which series are used. Functions arising in mathematics are often defined in terms of Maclaurin or Taylor series. The conditions under which these power series converge to a finite number for any given value of their argument, and the error made when such a series is truncated, are developed from the tests for convergence that are established in the first part of the chapter.

The applications of series are numerous and varied, and included in this chapter are the use of series to evaluate indeterminate forms (L'Hospital's rule), the location and identification of extrema of functions of one and several real variables, the analysis of constrained extrema (Lagrange multipliers), and the least-squares fitting of data. Series are used in many parts of numerical analysis, as illustrated in Chapter 17, and also in the solution of variable-coefficient differential equations and partial differential equations that are described in more advanced accounts of mathematics.

12.1 Series

The term *series* denotes the sum of the members of a sequence of numbers $\{a_n\}$, in which a_n represents the *general term*. The number of terms added may be finite or infinite, according to whether the sequence used is finite or infinite in the sense of Chapter 3. The sum to N terms of the infinite sequence $\{a_n\}$ is written

$$a_1 + a_2 + \ldots + a_N = \sum_{n=1}^{N} a_n,$$

and is called a *finite series* because the number of terms involved in the summation is finite. The so-called *infinite series* derived from the infinite sequence $\{a_n\}$ by the addition of all its term is written

$$a_1 + a_2 + \ldots a_r + \ldots = \sum_{n=1}^{\infty} a_n.$$

The following are specific examples of numerical series of essentially different types:

(a) $\sum_{n=1}^{N} n^2 = 1^2 + 2^2 + \ldots + N^2,$

in which the general term $a_n = n^2$;

(b) $\sum_{n=0}^{\infty} \frac{1}{n!} = 1 + 1 + \frac{1}{2!} + \frac{1}{3!} + \ldots + \frac{1}{r!} + \ldots,$

in which the general term $a_n = 1/n!$, and we define $0! = 1$;

(c) $\sum_{n=1}^{\infty} \frac{1}{n} = 1 + \frac{1}{2} + \frac{1}{3} + \ldots + \frac{1}{r} + \ldots,$

in which the general term $a_n = 1/n$;

(d) $\sum_{n=1}^{\infty} \frac{2n^2 + 1}{4n + 2} = \frac{1}{2} + \frac{9}{10} + \frac{19}{14} + \ldots + \frac{2r^2 + 1}{4r + 2} + \ldots,$

in which the general term $a_n = (2n^2 + 1)/(4n + 2)$;

(e) $\sum_{n=1}^{\infty} (-1)^{n+1} = 1 - 1 + 1 - 1 + \ldots + (-1)^{r+1} + \ldots,$

in which the general term $a_n = (-1)^{n+1}$.

Only (a) is a finite series; the remainder are infinite.

There is obviously no difficulty in assigning a sum to a finite series, but how are we to do this in the case of an infinite series? A practical approach would be an attempt to approximate the infinite series by means of a finite series comprising only its first N terms. To justify this, it would be necessary to show in some way that the sum of the remainder R_N of the series after N terms tends to zero as N increases and, even better if possible, to obtain an upper bound for R_N. This was, of course, the approach adopted in Chapter 6 when discussing the exponential series which comprises example (b) above. In the event of an upper bound for R_N being available, this could be used to deduce the number of terms that need to be taken in order to determine the sum to within a specified accuracy.

The spirit of this practical approach to the summation of series is exactly what is adopted in a rigrous discussion of series. The first question to be determined is whether or not a given series has a unique sum; the estimation of the remainder term follows afterwards, and usually proves to be more difficult.

To assist us in our formal discussion of series we use the already familiar notion of the nth partial sum S_n of the series $\sum_{n=1}^{\infty} a_n$, which is defined to be the finite sum

$$S_n = \sum_{r=1}^{n} a_r = a_1 + a_2 + \ldots + a_n.$$

Then, in terms of S_n, we have the following definition of convergence, which is in complete agreement with the approach we have just outlined.

Definition 12.1 (convergence of series) The series $\sum_{m=1}^{\infty} a_m$ will be said to be *convergent* to the finite sum S if its nth partial sum S_n is such that

$$\lim_{n\to\infty} S_n = S.$$

If the limit of S_n is not defined, or is infinite, the series will be said to be *divergent*.

The *remainder* after n terms, R_n, is given by

$$R_n = a_{n+1} + a_{n+2} + \ldots + a_{n+r} + \ldots = \sum_{r=n+1}^{\infty} a_n,$$

so that if $\{S_n\}$ converges to the limit S, then $R_n = S - S_n$, and Definition 12.1 is obviously equivalent to requiring that

$$\lim_{n\to\infty}(S - S_n) = \lim_{n\to\infty} R_n = 0.$$

Example 12.1 Find the nth partial sum of the series

$$1 + \frac{1}{3} + \frac{1}{9} + \frac{1}{27} + \ldots + \frac{1}{3^n} + \ldots,$$

and hence show that it converges to the sum 3/2. Find the remainder after n terms and deduce how many terms need be summed in order to yield a result in which the error does not exceed 0.01.

Solution This series is a geometric progression with initial term unity and common ratio 1/3. Its sum to n terms, which is the desired nth partial sum S_n, may be determined by the standard formula (see Problem 12.2) which gives

$$S_n = \frac{1 - (1/3)^n}{1 - 1/3} = \frac{3}{2}[1 - (1/3)^n].$$

We have

$$\lim_{n\to\infty} S_n = \lim_{n\to\infty} \frac{3}{2}\left[1 - \left(\frac{1}{3}\right)^n\right] = 3/2,$$

showing that the series is convergent to the sum 3/2.

As S_n is the sum to n terms, the remainder after n terms, R_n, must be given by $R_n = 3/2 - S_n$, and so

$$R_n = \frac{1}{2}\left(\frac{1}{3}\right)^{n-1}.$$

If the remainder must not exceed 0.01, $R_n \leq 0.01$, from which it is easily seen that the number n of terms needed is $n \geq 5$. The determination of R_n was simple in this instance because we were fortunate enough to have available an explicit formula for S_n. In general such a formula is seldom available. ■

The definition of convergence has immediate consequences as regards the addition and subtraction of series. Suppose Σa_n and Σb_n are convergent series with sums α, β. (It is customary to omit summation limits when they are not important.) Let their respective partial sums be $S_n = a_1+a_2+\ldots+a_n$, $S'_n = b_1+b_2+\ldots+b_n$ and consider the series $\Sigma(a_n+b_n)$ which has the partial sum $S''_n = S_n + S'_n$. Then

$$\begin{aligned}\lim_{n\to\infty} S''_n &= \lim_{n\to\infty}(S_n + S'_n)\\ &= \lim_{n\to\infty} S_n + \lim_{n\to\infty} S'_n = \alpha + \beta,\end{aligned}$$

showing that

$$\sum_{n=1}^{\infty}(a_n + b_n) = \alpha + \beta.$$

A corresponding result for the difference of two series may be proved in similar fashion. We have established the following general result.

Theorem 12.1 (sum and difference of convergent series) If the series $\sum_{n=1}^{\infty} a_n$ and $\sum_{n=1}^{\infty} b_n$ are convergent to the respective sums α and β, then

$$\sum_{n=1}^{\infty}(a_n + b_n) = \alpha + \beta; \quad \sum_{n=1}^{\infty}(a_n - b_n) = \alpha - \beta. \quad ■$$

Example 12.2 Suppose that $a_n = (1/2)^n$ and $b_n = (1/3)^n$, so that the series involved are again geometric progressions with $\sum_{n=1}^{\infty}(1/2)^n = 1$ and $\sum_{n=1}^{\infty}(1/3)^n = 1/2$. Then it follows from Theorem 12.1 that

$$\sum_{n=1}^{\infty}[(1/2)^n + (1/3)^n] = 3/2 \quad \text{and} \quad \sum_{n=1}^{\infty}[(1/2)^n - (1/3)^n] = 1/2. \quad ■$$

Let us now derive a number of standard tests by which the convergence or divergence of a series may be established. We begin with a test for divergence.

Suppose first that a series Σa_n with nth partial sum S_n converge to the sum S. Then from our discussion of the convergence of a sequence to a limit given in Definition 3.3, we know that for any arbitrarily small $\varepsilon \geq 0$ there must exist some integer N such that

$$|S_n - S| < \varepsilon \qquad \text{for} \quad n > N.$$

This immediately implies the additional result

$$|S_{n+1} - S| < \varepsilon.$$

Hence adding these two results and using the triangle inequality $|a|+|b| \geq |a+b|$, with $|a| = |S_n - S|$ and $|b| = |S_{n+1} - S|$, we find that

$$\varepsilon + \varepsilon > |S_{n+1} - S| + |S_n - S| = |S_{n+1} - S| + |S - S_n| \geq |S_{n+1} - S_n|.$$

However, as $S_{n+1} - S_n = a_{1+1}$, we have proved that

$$|a_{n+1}| < 2\varepsilon \qquad \text{for} \quad n > N.$$

As ε was arbitrary, this shows that for a series to be convergent, it is necessary that

$$\lim_{n\to\infty} |a_n| = 0$$

or, equivalently,

$$\lim_{n\to\infty} a_n = 0.$$

If this is not the case then the series $\sum_{n=1}^{\infty} a_n$ must diverge. This condition thus provides us with a positive, test for divergence.

Theorem 12.2(a) (test for divergence) The series $\sum_{n=1}^{\infty} a_n$ diverges if

$$\lim_{n\to\infty} a_n \neq 0. \quad \blacksquare$$

This theorem shows, for example, that the series (d) is divergent, because $a_n = (2n^2 + 1)/(4n + 2)$, and hence it increases without bound as n increases. It is important to take note of the fact that this theorem gives no information in the event that $\lim_{n\to\infty} a_n = 0$. Although we have shown that this is a *necessary* condition for convergence, it is not a *sufficient* condition because divergent series exist for which the condition is true.

Theorem 12.2(a) gives no information about either series (a) or (c) as in each case $\lim_{n\to\infty} a_n = 0$. In fact, by using another argument, we have already proved that the series representation for e in (b) is convergent, whereas we shall prove shortly that (c), called the *harmonic series* is divergent. Series (e) is divergent by our definition, because a_n oscillates finitely between 1 and -1, and also S_n does not tend to any limit.

The terms of series are not always of the same sign, and so it is useful to associate with the series Σa_n the companion series $\Sigma|a_n|$. If this latter series is convergent, then the series Σa_n is said to be *absolutely convergent*. It can happen that although Σa_n is convergent, $\Sigma|a_n|$ is divergent. When this occurs the series Σa_n is said to be *conditionally convergent*. Now when terms of differing signs are involved, the sum of the absolute values of the terms of a series clearly exceeds the sum of the terms of the series, and so it seems reasonable to expect that absolute convergence implies convergence. Let us prove this fact.

Theorem 12.2(b) (absolute convergence implies convergence) If the series $\sum_{n=1}^{\infty} |a_n|$ is convergent, then so also is the series $\sum_{n=1}^{\infty} a_n$.

Proof The proof of this result is simple. Let $S_n = |a_1| + |a_2| + \ldots + |a_n|$ and $S_n' = a_1 + a_2 + \ldots + a_n$ be the nth partial sums, respectively, of the series. Then, as $a_r + |a_r|$ is either zero or $2|a_r|$, it follows that

$$0 \leq S_n + S_n' \leq 2S_n'.$$

Now by supposition $\lim_{n\to\infty} S_n' = S'$ exists, so that taking limits we arrive at

$$0 \leq \lim_{n\to\infty}(S_n + S_n') \leq 2S'.$$

This implies that the series with nth term $a_n + |a_n|$ must be convergent and hence, using Theorem 12.1, that $\sum_{n=1}^{\infty} a_n$ must be convergent. ■

Example 12.3 Consider the series

$$\sum_{n=0}^{\infty} \frac{(-1)^n}{n!} = 1 - 1 + \frac{1}{2!} - \frac{1}{3!} + \ldots.$$

As $a_n = (-1)^n/n!$, we have $|a_n| = 1/n!$, which is the general term of the exponential series defining e. Thus Theorem 12.2, and the convergence of the exponential series, together imply the convergence of $\sum_{n=0}^{\infty} (-1)^n/n!$ In fact this is the series representation of $1/\mathrm{e}$. ■

Suppose Σb_n is a convergent series of positive terms, and that Σa_n is a series with the property that if N is some positive integer, then $|a_n| \leq b_n$ for $n > N$. Then clearly the convergence of Σb_n implies the convergence of $\Sigma |a_n|$ and, by Theorem 12.2, also the convergence of Σa_n. By a similar argument, if for $n > N$, $0 \leq b_n \leq a_n$ and Σb_n is known to be divergent, then clearly Σa_n must also be divergent. We incorporate these results into a useful comparison test.

Theorem 12.3 (comparison test) (a) Convergence test. Let Σb_n be a convergent series of positive terms, and let Σa_n be a series with the property that there exists a positive integer N such that

$$|a_n| \le b_n \quad \text{for} \quad n > N.$$

Then Σa_n is an absolutely convergent series.

(b) Divergence test. Let Σb_n be a divergent series of positive terms, and let Σa_n be a series of positive terms with the property that there exists a positive interger N such that

$$0 \le b_n \le a_n \quad \text{for} \quad n > N.$$

Then Σa_n is a divergent series. ■

Example 12.4 (a) Consider the series $\sum_{n=1}^{\infty} [2+(-1)^n]/2^n$. We have

$$a_n = \frac{2+(-1)^n}{2^n} \le \frac{3}{2^n},$$

and as $\sum_{n=1}^{\infty} 3/2^n = 3\sum_{n=1}^{\infty} 1/2^n = 3$, the conditions of Theorem 12.3(a) are satisfied if we set $b_n = 3/2^n$. It thus follows that the series Σa_n is convergent.

(b) Consider the series $\sum_{n=1}^{\infty} (n+1)/n^2$. Here we have

$$a_n = \frac{n+1}{n^2} = \frac{1}{n}\left(\frac{n+1}{n}\right) > \frac{1}{n},$$

and as the harmonic series $\Sigma 1/n$ is divergent (this will be proved later), the conditions of Theorem 12.3(b) are satisfied when we set $b_n = 1/n$. Hence Σa_n is divergent. ■

A powerful test for the convergence or divergence of a series Σa_n of positive terms follows by a comparison of the rectangles in Fig. 12.1.

Let $f(x)$ be a non-increasing function defined for $1 \le x < \infty$ which decreases to zero as x tends to infinity, and let $f(n) = a_n$, where n is an integer. Then we have the obvious inequality

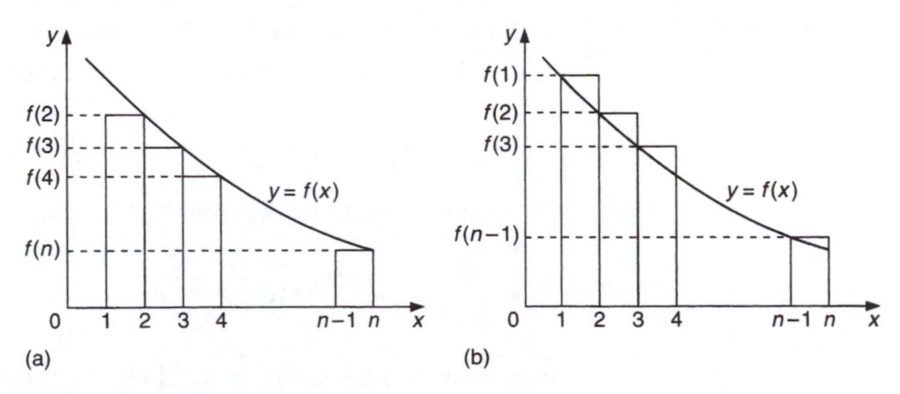

Fig. 12.1 Comparison between series and integral.

$$\sum_{r=2}^{n} f(r) \leq \int_{1}^{n} f(x)\,\mathrm{d}x \leq \sum_{r=1}^{n-1} f(r) \leq \sum_{r=1}^{n} f(r).$$

However, if we replace the upper summation limit $n-1$ on the right-hand side by n the inequality is strengthened because $f(n) > 0$, so we have

$$\sum_{r=2}^{n} f(r) \leq \int_{1}^{n} f(x)\,\mathrm{d}x \leq \sum_{r=1}^{n} f(r),$$

which is equivalent to

$$\sum_{r=2}^{n} a_r \leq \int_{1}^{n} f(x)\,\mathrm{d}x \leq \sum_{r=1}^{n} a_r.$$

As the right-hand side of this inequality only exceeds the left-hand side by the single positive term a_1, it must follow that in the limit, the infinite series Σa_r and the integral

$$\lim_{n\to\infty} \int_{1}^{n} f(x)\,\mathrm{d}x$$

converge or diverge together. This conclusion may be incorporated into a test as follows.

Theorem 12.4 (integral test) Let $f(x)$ be a positive non-increasing function defined on $1 \leq x < \infty$ with $\lim_{x\to\infty} f(x) = 0$. Then, if $a_n = f(n)$, the series $\sum_{n=1}^{\infty} a_n$ converges if

$$\lim_{n\to\infty} \int_{1}^{n} f(x)\,\mathrm{d}x$$

is finite and diverges if it is infinite.

Corollary 12.4 (R_N deduced from integral test) Let $f(x)$ be a positive non-increasing function defined on $1 \leq x < \infty$ with $\lim_{n\to\infty} f(x) = 0$ and let $\sum_{n=1}^{\infty} a_n$ be convergent, where $a_n = f(n)$. Then the remainder R_N after N terms satisfies the inequality

$$R_N \leq \int_{N}^{\infty} f(x)\,\mathrm{d}x.$$

Proof The result follows at once from the obvious inequality

$$\sum_{r=N+1}^{N'} a_r \leq \int_{N}^{N'} f(x)\,\mathrm{d}x \leq \sum_{r=N}^{N'} a_r$$

by taking the limit as $N' \to \infty$. This is possible because, by hypothesis, Σa_n is convergent so that the improper integral involved exists. ■

Example 12.5 (a) Consider the series $\sum_{n=1}^{\infty} 1/n^k$ where $k > 0$. Then the function $f(x) = 1/x^k$ satisfies the conditions of Theorem 12.4. Hence this series converges or diverges depending on whether

$$\lim_{n\to\infty} \int_1^n \frac{dx}{x^k}$$

is finite or infinite. If $k \neq 1$ we have

$$\lim_{n\to\infty} \int_1^n \frac{dx}{x^k} = \left(\frac{1}{1-k}\right) \lim_{n\to\infty} \left[\frac{1}{n^{k-1}} - 1\right].$$

Hence for $0 < k < 1$ this limit is infinite, showing that the series is divergent for k in this range, whereas for $k > 1$ this limit has the finite value $1/(k-1)$, showing that the series is convergent for $k > 1$.

When $k = 1$ we obtain the harmonic series encountered as Example (c) at the start of this chapter, and this series must be treated separately. As it follows that

$$\lim_{n\to\infty} \int_1^n \frac{dx}{x} = \lim_{n\to\infty} \log n \to \infty,$$

we have proved that the harmonic series is divergent.

(b) Consider the series $\sum_{n=1}^{\infty} n/(1+n^2)$. Here we set $f(x) = x/(1+x^2)$, so we must examine

$$L = \lim_{n\to\infty} \int_1^n \frac{x\,dx}{1+x^2}.$$

Setting $x^2 = u$ we find

$$L = \lim_{n\to\infty} \tfrac{1}{2}[\log(1+n^2) - \log 2] \to \infty.$$

Hence the series is divergent. ■

Two other useful tests known as the ratio test and the nth root test may be derived from Theorem 12.3, essentially using a geometric progression for purposes of comparision. The idea involved in these tests is that a series is tested against itself, and that its convergence or divergence is then deduced from the rate at which successive terms decrease or increase.

Suppose that Σa_n is a series for which the ratio a_{n+1}/a_n is always defined and that $\lim_{n\to\infty} |a_{n+1}/a_n| = L$, where $L < 1$. Let r be some fixed number such that $L < r < 1$. Then the existence of the limit L implies that there exists an integer N such that

$$|a_n + 1| < r|a_n| \text{ for } n > N.$$

Hence it follows that

$$|a_{N+2}| < r|a_{N+1}|,\ |a_{N+3}| < r|a_{N+2}| < r^2|a_{N+1}|,\ \ldots,$$

and in general

$$|a_{N+m+1}| < r^m |a_{N+1}|.$$

Thus if R_N is the remainder after N terms we have

$$R_N = \sum_{n=N+1}^{\infty} a_n \leq \sum_{n=N+1}^{\infty} |a_n| < |a_{N+1}|(1 + r + r^2 + \ldots). \tag{A}$$

The expression in brackets is a convergent geometric progression because, by hypothesis, $r < 1$. As the remainder term R_N is finite, and is less than the sum of the absolute values of the terms comprising the tail of the series, it is easily seen that the series Σa_n must be absolutely convergent. If $L > 1$ the terms grow in size, and the series Σa_n is divergent. Nothing may be deduced if $L = 1$ for then the series may either be convergent or divergent as illustrated by Example 12.5(a). In that case $a_{n+1}/a_n = n^k/(n+1)^k$, giving $\lim |a_{n+1}/a_n| = 1$; and the series was seen to be divergent for $0 < k \leq 1$ and convergent for $k > 1$.

Expressed formally, as follows, these results are called the *ratio test.*

Theorem 12.5 (ratio test) If the series $\sum_{n=1}^{\infty} a_n$ is such that $a_n \neq 0$ and $\lim_{n\to\infty} \left|\frac{a_{n+1}}{a_n}\right| = L$, then

(a) the series Σa_n converges absolutely if $L < 1$,
(b) the series Σa_n diverges if $L > 1$,
(c) the test fails if $L = 1$. ■

Example 12.6 (a) Consider the series

$$\sum_{n=1}^{\infty} \frac{(-1)^n n!}{n^n}.$$

Then $a_n \neq 0$ and

$$\frac{a_{n+1}}{a_n} = (-1)^{2n+1} \frac{(n+1)!n^n}{(n+1)^{n+1}n!} = (-1)^{2n+1}\left(1 + \frac{1}{n}\right)^{-n}.$$

Hence

$$\lim_{n\to\infty} \left|\frac{a_{n+1}}{a_n}\right| = \lim_{n\to\infty} 1 \bigg/ \left(1 + \frac{1}{n}\right)^n = 1/\mathrm{e},$$

where the final result follows by virtue of the work of section 3.3. As $\mathrm{e} > 1$, the ratio test proves the absolute convergence of this series.

(b) Consider the series $\sum_{n=1}^{\infty} 1/n!$. Here $a_n = 1/n! \neq 0$ and

$$\frac{a_{n+1}}{a_n} = \frac{n!}{(n+1)!} = \frac{1}{n+1} = \left|\frac{a_{n+1}}{a_n}\right|.$$

Hence

$$\lim_{n\to\infty}\left|\frac{a_{n+1}}{a_n}\right| = \lim_{n\to\infty}\frac{1}{n+1} = 0,$$

and as $0 < 1$ the ratio test proves the series to be convergent.

(c) Consider the series $\sum_{n=1}^{\infty} 3^n/n$.

Then $a_n \neq 0$ and

$$\frac{a_{n+1}}{a_n} = \left(\frac{3^{n+1}}{n+1}\right)\left(\frac{n}{3^n}\right) = 3\left(\frac{n}{n+1}\right) = \left|\frac{a_{n+1}}{a_n}\right|.$$

Now

$$\lim_{n\to\infty}\left|\frac{a_{n+1}}{a_n}\right| = \lim_{n\to\infty}\frac{3n}{n+1} = 3,$$

and as $3 > 1$ the ratio test proves the series to be divergent.

(d) Consider the series $\sum_{n=1}^{\infty} 1/(2n+1)^2$.

Then $a_n \neq 0$ and

$$\frac{a_{n+1}}{a_n} = \left(\frac{2n+1}{2n+3}\right)^2 = \left|\frac{a_{n+1}}{a_n}\right|.$$

Now

$$\lim_{n\to\infty}\left|\frac{a_{n+1}}{a_n}\right| = \lim_{n\to\infty}\left(\frac{2n+1}{2n+3}\right)^2 = 1,$$

so that the ratio test fails in this case. In fact the series is convergent, as may readily be proved by use either of the comparison test, with $b_n = 1/n^2$, or the integral test. ■

In the so-called nth root test, appeal is also made to the geometric progression to prove convergence. Suppose the series Σa_n is such that

$$\lim_{n\to\infty} \sqrt[n]{|a_n|} = L,$$

and that $L < 1$. Then if r is some definite number such that $L < r < 1$, the existence of the limit implies that there exists an integer N such that

$$\sqrt[n]{|a_n|} < r \quad \text{for} \quad n > N.$$

Hence $|a_n| < r^n$ for $n > N$. Thus, as with the ratio test, the remainder after N terms may be overestimated by the sum of the absolute values of the remaining terms, and the result still further overestimated in terms of $|a_{N+1}|$ and a geometric progression with common ratio r. As $r < 1$ this remainder is finite, thereby establishing that Σa_n is absolutely convergent. If $L > 1$, then successive terms grow and the series is divergent. As with the ratio test, the nth root test fails when $L = 1$, for then Σa_n may be either convergent or divergent. Stated formally we have:

Theorem 12.6 (*nth root test*) If the series $\sum_{n=1}^{\infty} a_n$ is such that

$$\lim_{n\to\infty} \sqrt[n]{|a_n|} = L,$$

then

(a) the series Σa_n is absolutely convergent if $L < 1$,
(b) the series Σa_n is divergent if $L > 1$,
(c) the test fails if $L = 1$. ■

Example 12.7 (a) Consider the series

$$\sum_{n=1}^{\infty} \left(\frac{nk}{3n+1}\right)^n,$$

where k is a constant. Then

$$a_n = \left(\frac{nk}{3n+1}\right)^n \quad \text{and} \quad \lim_{n\to\infty} \sqrt[n]{|a_n|} = \lim_{n\to\infty} \frac{nk}{3n+1} = \frac{k}{3}.$$

Thus the nth root test shows that the series will be convergent if $k < 3$ and divergent if $k > 3$. It fails if $k = 3$, though Theorem 12.2 then shows the series to be divergent.

(b) Consider the series $\sum_{n=1}^{\infty} n/2^n$. Then $a_n = n/2^n = |a_n|$ and $\sqrt[n]{|a_n|} = \frac{1}{2}\sqrt[n]{n}$. Taking logarithms we find

$$\ln [\sqrt[n]{|a_n|}] = \ln \tfrac{1}{2} + \frac{1}{n} \ln n.$$

Now by Theorem 6.4 (b) we know that $\lim_{n\to\infty} (\ln n)/n = 0$, so that

$$\lim_{n\to\infty} \ln [\sqrt[n]{|a_n|}] = \ln \tfrac{1}{2},$$

whence

$$\lim_{n\to\infty} \sqrt[n]{n} = \tfrac{1}{2}.$$

As $\frac{1}{2} < 1$ the test thus proves convergence. In this instance it would have been simpler to use the ratio test to prove convergence. ■

For our final result we prove that all series in which the signs of terms alternate while the absolute values of successive terms decrease monotonically to zero, are convergent. Such series are called *alternating series* and are of the general form

$$\sum_{n=1}^{\infty}(-1)^{n+1}a_n = a_1 - a_2 + a_3 - a_4 + \ldots,$$

where $a_n > 0$ for all n.

To prove our assertion of convergence we assume $a_1 > a_2 > a_3 > \ldots$, and $\lim a_n = 0$, and first consider the partial sum S_{2r} corresponding to an even number of terms $2r$. We write S_{2r} in the form

$$S_{2r} = (a_1 - a_2) + (a_3 - a_4) + \ldots + (a_{2r-1} - a_{2r}).$$

Then, because $a_1 > a_2 > a_3 > \ldots$, it follows that $S_{2r} > 0$. By a slight rearrangement of the brackets we also have

$$S_{2r} = a_1 - (a_2 - a_3) - (a_4 - a_5) - \ldots - (a_{2r-2} - a_{2r-1}) - a_{2r},$$

showing that as all the brackets and quantities are positive, $S_{2r} < a_1$. Hence, as $\{S_{2r}\}$ is a bounded monotonic decreasing sequence, we know from the fundamental theorem for sequences in Chapter 3 that it must tend to a limit S, where

$$0 < S < a_1.$$

Next consider the partial sum S_{2r+1} corresponding to an odd number of terms $2r+1$. We may write $S_{2r+1} = S_{2r} + a_{2r+1}$. Then, taking the limit of S_{2r+1} we have

$$\lim_{r\to\infty} S_{2r+1} = \lim_{r\to\infty} S_{2r} + \lim_{r\to\infty} a_{2r+1} = S,$$

because by supposition $\lim a_{2r+1} = 0$. Thus both the partial sums S_{2r} and the partial sums S_{2r+1} tend to the same limit S, and so we have proved that for n both even and odd

$$\lim_{n\to\infty} S_n = S,$$

thereby showing that the series converges.

***Theorem 12.7* (alternating series test)** The series $\sum_{n=1}^{\infty}(-1)^{n+1}a_n$ converges if $a_n > 0$ and $a_{n+1} \le a_n$ for all n and, in addition,

$$\lim_{n\to\infty} a_n = 0.$$

Example 12.8 (a) Consider the alternating series

$$\sum_{n=0}^{\infty}\frac{(-1)^n}{2^n} = 1 - \frac{1}{2} + \frac{1}{2^2} - \frac{1}{2^3} + \ldots,$$

in which the absolute value of the general term $a_n = (1/2)^n$. Then, as it is true that $a_{n+1} < a_n$ and $\lim a_n = 0$, the test shows that the series is convergent.

(b) Consider the alternating series

$$\sum_{n=1}^{\infty}(-1)^{n+1}2^{1/(n+1)} = 2^{1/2} - 2^{1/3} + 2^{1/4} + \ldots,$$

in which the absolute value of the general term $a_n = 2^{1/(1+n)}$. Now it is true that $a_{n+1} < a_n$, but $\lim a_n = 1$, so that the last condition of the theorem is violated, rendering it inapplicable. Theorem 12.2 shows the series to the divergent. ■

The form of argument that was used to show $0 < S_{2r} < a_1$ also shows that

$$0 < \sum_{r=2m+1}^{\infty} (-1)^{n+1} a_r < a_{2m+1}$$

and, by a slight modification, that

$$-a_{2m} \leq \sum_{r=2m}^{\infty}(-1)^{n+1} a_r < 0.$$

As $R_{2m} = \sum_{r=2m+1}^{\infty} (-1)^r a_r$ is the remainder after an even number $2m$ of terms, and $R_{2m-1} = \sum_{r=2m}^{\infty} (-1)^r a_r$ is the remainder after an odd number $2m-1$ of terms, it follows that if N is either even or odd, then

$$0 < |R_N| < a_{N+1}.$$

Expressed in words, this asserts that when an alternating series is terminated after the Nth term, the absolute value of the error involved is less than the magnitude a_{N+1} of the next term.

Corollary 12.7 (R_N for alternating series) If the alternating series $\sum_{n=1}^{\infty} (-1)^{n+1} a_n$ converges, and R_N is the remainder after N terms then

$$0 < |R_N| < a_{N+1}. \quad ■$$

Using the convergent alternating series in Example 12.8(a) for purposes of illustration we see that $a_n = 1/2^n$, and so the remainder R_N must be such that

$$0 < |R_N| < 1/2^{N+1}.$$

For example, termination of the summation of this series after five terms would result in an error whose absolute magnitude is less than 1/64.

A calculation involving the summation of a finite number of terms is often facilitated by grouping and interchanging their order. Although these operations are legitimate when the number of terms involved is finite, we must question their validity when dealing with an infinite number of terms. Later we shall show that the grouping of terms is permissible for any convergent series, but that rearrangement of terms is only permissible in a series when it is absolutely convergent, for only then does this operation leave the sum unaltered.

An example will help here to indicate the dangers of manipulating a series without first questioning the legitimacy of the operations to be performed upon it. Consider the alternating series

$$1 - \tfrac{1}{2} + \tfrac{1}{3} - \tfrac{1}{4} + \tfrac{1}{5} - \tfrac{1}{6} + \dots,$$

which is seen to be convergent by virtue of our last theorem, and denote its sum by S. Then we have

$$S = 1 - \tfrac{1}{2} + \tfrac{1}{3} - \tfrac{1}{4} + \tfrac{1}{5} - \tfrac{1}{6} + \tfrac{1}{7} - \tfrac{1}{8} + \tfrac{1}{9} - \tfrac{1}{10} + \tfrac{1}{11} - \tfrac{1}{12} + \dots$$

or, on rearranging the terms,

$$\begin{aligned} S &= 1 - \tfrac{1}{2} - \tfrac{1}{4} + \tfrac{1}{3} - \tfrac{1}{6} - \tfrac{1}{8} + \tfrac{1}{5} - \tfrac{1}{10} - \tfrac{1}{12} + \dots \\ &= (1 - \tfrac{1}{2}) - \tfrac{1}{4} + (\tfrac{1}{3} - \tfrac{1}{6}) - \tfrac{1}{8} + (\tfrac{1}{5} - \tfrac{1}{10}) - \tfrac{1}{12} + \dots \\ &= \tfrac{1}{2}(1 - \tfrac{1}{2} + \tfrac{1}{3} - \tfrac{1}{4} + \tfrac{1}{5} - \tfrac{1}{6} + \dots) \\ &= \tfrac{1}{2}S. \end{aligned}$$

This can only be true if $S = 0$, but clearly this is impossible because Corollary 12.7 above shows that the error in the summation after only one term is less than $\frac{1}{2}$, and therefore S is certainly positive with $\frac{1}{2} < S < 1$.

What has gone wrong? The answer is that in a sense we are 'robbing Peter to pay Paul'. This occurs because both the series $\Sigma 1/(2n+1)$ and the series $\Sigma 1/2n$ from which are derived the positive and negative terms in our series are divergent, and we have so rearranged the terms that they are weighted in favour of the negative ones. Other rearrangements could in fact be made to yield any sum that was desired. In other words, we are working with a series that is only conditionally convergent, and not absolutely convergent. It would seem from this that perhaps if a series Σa_n is absolutely convergent, then its terms should be capable of rearrangement and grouping without altering the sum. Let us prove the truth of this conjecture, but first we prove the simpler result that the grouping or bracketing of the terms of a convergent series leaves its sum unaltered.

Suppose that Σa_n is a convergent series with sum S. Take as representative of the possible groupings of its terms the series derived from Σa_n by the insertion of brackets as indicated below:

$$(a_1 + a_2) + (a_3 + a_4 + a_5) + a_6 + (a_7 + a_8) + \dots.$$

Now denote the bracketed terms by $b_1, b_2, \dots$, where $b_1 = a_1 + a_2$, $b_2 = a_3 + a_4 + a_5, \dots$, so that we have associated a new series Σb_n with the original series Σa_n. If the nth partial sums of Σa_n and Σb_n are S_n and S'_n, respectively, then the partial sums $S'_1, S'_2, S'_3, S'_4, \dots$ of Σb_n are, in reality, the partial sums $S_2, S_5, S_6, S_8, \dots$ of Σa_n. As Σa_n is convergent to

S by hypothesis, any sub-sequence of its partial sums $\{S_n\}$ must also converge to S. In particular this applies to the sequence $S_2, S_5, S_6, S_8, \ldots$, derived by the inclusion of parentheses. Hence Σb_n is also convergent to the sum S, which proves our result.

We now examine the effect of rearranging the terms of a series. Let Σa_n be absolutely convergent so that $\Sigma|a_n|$ must be convergent, and let Σb_n be a rearrangement of Σa_n. Then, as the terms of $\Sigma|b_n|$ are in one–one correspondence with those of $\Sigma|a_n|$, it is clear that $\Sigma|b_n| = \Sigma|a_n|$, from which we deduce that Σb_n is also absolutely convergent.

Next we must show that Σa_n and Σb_n have the same sum. If S_n is the nth partial sum of Σa_n which has the sum S, then by taking n sufficiently large we may make $|S_n - S|$ as small as we wish; say less than an arbitrarily small positive number ε. Now let S'_m be the mth partial sum of Σb_n. Then, as S_n contains the first n terms of Σa_n, with their suffixes in sequential order, by taking m large enough we can obviously make S'_m contain all the terms of S_n together with $m - n$ additional terms $a_p, q_q, \ldots, q_r$, where $n < p < q < \ldots < r$.

Hence we may write

$$S'_m = S_n + a_p + a_q + \ldots + a_r,$$

whence

$$S'_m - S = S_n - S + a_p + a_q + \ldots + a_r.$$

Taking absolute values gives

$$|S'_m - S| \leq |S_n - S| + |a_p| + |a_q| + \ldots + |a_r|.$$

Now, n was chosen such that $|S_n - S| < \varepsilon$, so that

$$|S'_m - S| \leq \varepsilon + |a_p| + |a_q| + \ldots + |a_r|.$$

However, the remaining terms on the right-hand side of this inequality all occur after a_n in the series Σa_n, and as $|S_n - S| < \varepsilon$, it must follow that their total contribution cannot exceed ε, and thus

$$|S'_m - S| < 2\varepsilon.$$

This shows that the mth parial sum of Σb_n converges to the sum S, so that rearrangement of the terms of an absolutely convergent series is permissible and does not affect its sum.

Theorem 12.8 (grouping and rearrangement of series)

If the series $\sum_{n=1}^{\infty} a_n$ is convergent, then brackets may be inserted into the series without affecting its sum. If, in addition, the series $\sum_{n=1}^{\infty} a_n$ is absolutely convergent, then its terms may be rearranged without altering its sum. ■

Example 12.9 (a) Consider the series

$$\sum_{m=1}^{\infty} \frac{1}{m(m+1)},$$

which is easily seen to be convergent by use of the comparison test with $b_m = 1/m^2$. The first part of Theorem 12.8 asserts that we may group terms by inserting brackets as we wish. So, using the identity

$$\frac{1}{m(m+1)} = \frac{1}{m} - \frac{1}{m+1},$$

we find for the nth partial sum S_n the expression

$$S_n = \sum_{m=1}^{n} \left(\frac{1}{m} - \frac{1}{m+1} \right).$$

Now successive terms in this summation cancel, or *telescope* as the process is sometimes called, leaving only the first and the last. This is best seen by writing out the expression for S_n in full as follows:

$$\begin{aligned} S_n &= \left(\frac{1}{1} - \frac{1}{2} \right) + \left(\frac{1}{2} - \frac{1}{3} \right) + \ldots + \left(\frac{1}{n-1} - \frac{1}{n} \right) + \left(\frac{1}{n} - \frac{1}{n+1} \right) \\ &= 1 - \frac{1}{n+1}. \end{aligned}$$

Hence, if the sum of the series is S, we have

$$S = \lim_{n \to \infty} S_n = \lim_{n \to \infty} \left[1 - \frac{1}{n+1} \right] = 1.$$

(b) Consider the series

$$-\frac{1}{2} + \frac{1}{3} + \frac{1}{2^2} + \frac{1}{3^2} - \frac{1}{2^3} + \frac{1}{3^3} + \frac{1}{2^4} + \frac{1}{3^4} - \ldots,$$

which can be shown to be absolutely convergent by an extension of the nth root test. The second part of Theorem 12.8 is applicable, so that we may arrange terms and, denoting the sum by S, we obtain

$$\begin{aligned} S &= \sum_{n=1}^{\infty} \frac{(-1)^n}{2^n} + \sum_{n=1}^{\infty} \frac{1}{3^n} \\ &= -\frac{1}{2} \left(\frac{1}{1 + \frac{1}{2}} \right) + \frac{1}{3} \left(\frac{1}{1 - \frac{1}{3}} \right) = -\tfrac{1}{3} + \tfrac{1}{2} = \tfrac{1}{6}. \end{aligned}$$ ■

The use of brackets in a divergent series can sometimes produce a convergent series and, conversely, when attempting to alter the form of a convergent series a divergent series may sometimes be produced inadvertently.

For instance, taking Example 12.9(b), we could have written

$$\sum_{n=1}^{\infty} \frac{1}{n(n+1)} = \sum_{1}^{\infty}\left(\frac{n+1}{n} - \frac{n+2}{n+1}\right)$$

$$= \sum_{1}^{\infty} \frac{n+1}{n} - \sum_{1}^{\infty} \frac{n+2}{n+1}$$

$$= 2 + \sum_{2}^{\infty} \frac{n+1}{n} - \sum_{2}^{\infty} \frac{n+1}{n} = 2,$$

which we know to be an incorrect result. The error is, of course, contained in the first line in which we attempt to equate an absolutely convergent series with the difference between two divergent series.

12.2 Power series

Up to now we have been concerned entirely with series that did not contain the variable x. A more general type of series called a *power series* in $(x - x_0)$ has the general form

$$\sum_{n=0}^{\infty} a_n(x - x_0)^n = a_0 + a_1(x - x_0) + a_2(x - x_0)^2 + \ldots, \tag{12.1}$$

in which the coefficients a_0, a_1, ..., a_n, ... are constants. When x is assigned some fixed value c, say, the power series Eqn (12.1) reduces to an ordinary series of the kind discussed in section 12.1, and so may be tested for convergence by any appropriate test mentioned there.

For simplicity we now apply the *ratio test* to series Eqn (12.1), allowing x to remain a free variable, in order to try to deduce the interval of x in which the series is absolutely convergent. If $\alpha_n(x)$ is the absolute value of the ratio of the $(n+1)$th term to the nth term as a function of x, we have

$$\alpha_n(x) = \left|\frac{a_{n+1}(x - x_0)^{n+1}}{a_n(xx - x_0)^n}\right| = \left|\frac{a_{n+1}}{a_n}\right| |x - x_0|.$$

Now for any specific value of x, the ratio test asserts that the series will be convergent if $\lim_{n\to\infty} \alpha_n(x) < 1$, whence we must require

$$\lim_{n\to\infty} \left|\frac{a_{n+1}}{a_n}\right| |x - x_0| < 1.$$

Thus, taking reciprocals, we see that the largest value r, say, of $|x - x_0|$ for which this is true is given by

$$r = \lim_{n\to\infty} \left|\frac{a_n}{a_{n+1}}\right|, \tag{12.2}$$

provided that this limit exists.

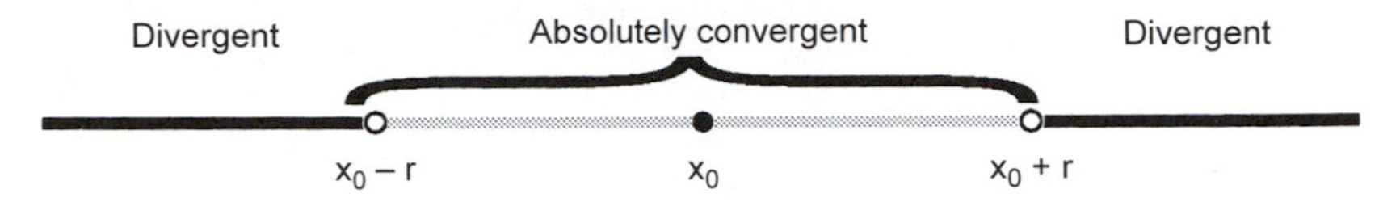

Fig. 12.2 Interval of convergence.

The inequality

$$|x - x_0| < r \tag{12.3}$$

thus defines the x-interval $(x_0 - r, x_0 + r)$ within which the power series Eqn (12.1) is absolutely convergent. For x outside this interval the ratio test shows that the power series must be divergent (see Fig. 12.2). The interval itself is called the *interval of convergence* of the power series, and the number r is called the *radius of convergence* of the power series. The interval of convergence has been deliberately displayed in the form of an open interval because the ratio test can offer no information about the behaviour of the series at the end points. In fact the power series may either be convergent or divergent at these points.

The radius of convergence of a power series can also be deduced from the nth root test, where it is easily seen that

$$r = \lim_{n\to\infty} \frac{1}{\sqrt[n]{|a_n|}}, \tag{12.4}$$

provided that this limit exists.

Definition 12.2 (radius of convergence of power)

The radius of convergence r of the power series $\sum_{n=0}^{\infty} a_n(x - x_0)^n$ is defined either as:

$$r = \lim_{n\to\infty} \left| \frac{a_n}{a_{n+1}} \right|,$$

or

$$r = \lim_{n\to\infty} \frac{1}{\sqrt[n]{|a_n|}},$$

provided that these limits exist.

Example 12.10 (a) Let us show that the series for the exponential function is absolutely convergent for all real x. We have

$$e^x = 1 + x + \frac{x^2}{2!} + \frac{x^3}{3!} + \ldots + \frac{x^n}{n!} + \ldots,$$

in which the general term $a_n = 1/n!$.

Now

$$\left|\frac{a_n}{a_{n+1}}\right| = \left|\frac{(n+1)!}{n!}\right| = (n+1),$$

so that

$$r = \lim_{n\to\infty}(n+1) \to \infty.$$

We have thus proved that the power series for e^x is absolutely convergent for all real x. This is an example of a power series with an infinite radius of convergence.

(b) Consider the series

$$x - \frac{x^2}{2} + \frac{x^3}{3} - \frac{x^4}{4} + \ldots,$$

which reduces to the illustrative example following Corollary 12.7, when $x = 1$. We shall see later that this is the power series expansion of $\ln(1+x)$.

Then, again applying limit (12.2), we have $a_n = (-1)^{n+1}/n$, and so

$$\left|\frac{a_n}{a_{n+1}}\right| = \left(\frac{n+1}{n}\right).$$

Thus we have

$$r = \lim_{n\to\infty}\left(\frac{n+1}{n}\right) = 1.$$

Hence, the series is absolutely convergent for $|x| < 1$. As we already know the series is convergent for $x = 1$, and divergent for $x = -1$ for then it becomes the harmonic series with the signs of all terms reversed, we have proved that the power series for $\ln(1+x)$ is absolutely convergent for $-1 < x \leq 1$. This was an example of a power series with radius of convergence unity.

(c) Consider the series

$$1 + x + (2x)^2 + (3x)^3 + \ldots + (nx)^n + \ldots,$$

then $a_n = n^n$ so that

$$\frac{1}{\sqrt[n]{|a_n|}} = \frac{1}{n}.$$

Hence, from Eqn (12.4),

$$r = \lim_{n\to\infty}\frac{1}{n} = 0.$$

This series has zero radius of convergence and so is absolutely convergent *only* when $x = 0$. That is to say this power series has a finite sum, and so is convergent, only at the one point $x = 0$ on the real line. ■

As a power series is yet another example of the representation of a function of the variable x, it is reasonable to enquire how we may differentiate and integrate functions that are so defined. The answers to these questions are provided by the following theorem which we state without proof, because the details involve the notion of *uniform convergence* which will not be discussed.

Theorem 12.9 (differentiation and integration of power series)

Let the function $f(x)$ be defined by the power series

$$f(x) = \sum_{n=0}^{\infty} a_n x^n,$$

with radius of convergence $r > 0$. Then, within the common interval of convergence $-r < x < r$,

(a) $f(x)$ is a continuous function;

(b) $f'(x) = \sum_{n=1}^{\infty} n a_n x^{n-1}$;

(c) $\int_0^x f(t)\,dt = \sum_{n=0}^{\infty} \frac{a_n}{n+1} x^{n+1}$.

Example 12.11 Find the radius and interval of convergence of

$$f(x) = \sum_{n=1}^{\infty} \frac{x^n}{n(n+1)}.$$

Deduce $f'(x)$ and find its interval of convergence.

Solution The nth coefficient a_n of the power series for $f(x)$ is $a_n = 1/n(n+1)$, and so the radius of convergence r is given by

$$r = \lim_{n\to\infty} \left| \frac{a_n}{a_{n+1}} \right| = \lim_{n\to\infty} \left| \frac{n+2}{n} \right| = 1.$$

To specify the complete interval of convergence it remains to examine the behaviour of the power series at the end points of the interval $-1 < x < 1$.

The series may be seen to be convergent at $x = 1$ by using the comparison test with $b_n = 1/n^2$. When $x = -1$ the series becomes an alternating series and is seen to be convergent by Theorem 12.7. Thus the complete interval of convergence for $f(x)$ is $-1 \leq x \leq 1$.

Under the conditions of Theorem 12.9(b) we may differentiate the power series for $f(x)$ term by term within $-1 < x < 1$, so that

$$f'(x) = \sum_{n=1}^{\infty} \frac{x^{n-1}}{n+1}.$$

To specify the complete interval of convergence for this new series which, by Theorem 12.9(b), is certainly convergent in $-1 < x < 1$, we must again examine the end points of the interval $-1 < x < 1$. The series for $f'(x)$ becomes an alternating series when $x = -1$, and is convergent by Theorem 12.7. At $x = 1$ it becomes the harmonic series, and so is divergent. The complete interval of convergence for $f'(x)$ is thus $-1 \leq x < 1$. The effect of termwise differentiation has been to produce divergence of the differentiated series at the right-hand end point of an interval of convergence at which $f(x)$ is convergent. ∎

Example 12.12 Find the power series representation of $\arctan x$ by considering the integral

$$\arctan x = \int_0^x \frac{dt}{1+t^2}.$$

Deduce a series expansion for $\frac{1}{4}\pi$.

Solution An application of the binomial theorem to the function $(1+a)^{-1}$ gives the result

$$\frac{1}{1+a} = 1 - a + a^2 - a^3 + a^4 - \ldots,$$

for $-1 < a < 1$. Setting $a = t^2$ we arrive at the power series representation of $(1+t^2)^{-1}$,

$$\frac{1}{1+t^2} = 1 - t^2 + t^4 - t^6 + t^8 - \ldots. \tag{B}$$

The conditions of Theorem 12.9(c) apply, and we may integrate this power series term by term to obtain

$$\arctan x = \int_0^x \frac{dt}{1+t^2} = \int_0^x (1 - t^2 + t^4 - t^6 + t^8 - \ldots)\,dt,$$

or

$$\arctan x = x - \frac{x^3}{3} + \frac{x^5}{5} - \frac{x^7}{7} + \ldots. \tag{12.5}$$

This is the desired power series for arctan x and by the conditions of Theorem 12.9(b) it is certainly convergent within the interval $-1 < x < 1$, which is the interval of convergence of the original power series Eqn (B).

At each of the points $x = \pm 1$ of this interval, the power series Eqn (12.5) becomes an alternating series which is seen to be convergent by

Theorem 12.7. Hence the interval of convergence of the integrated series Eqn (12.5) is $-1 \leq x \leq 1$. Using the fact that $\arctan 1 = \frac{1}{4}\pi$, we find a representation for $\pi/4$ in the form

$$\tfrac{1}{4}\pi = 1 - \tfrac{1}{3} + \tfrac{1}{5} - \tfrac{1}{7} + \ldots,$$

which is a very *slowly convergent* numerical series, so it is not a useful way of calculating π.

12.3 Taylor's theorem

So far we have discussed the convergence properties of a function $f(x)$ which is defined by a given power series. Let us now reverse this idea and enquire how, when given a specific function $f(x)$, its power series representation may be obtained. Otherwise expressed, we are asking how the coefficients a_n in the power series

$$f(x) = \sum_{n=0}^{\infty} a_n x^n \tag{12.6}$$

may be determined when $f(x)$ is some given function.

First, by setting $x = 0$, we discover that $f(0) = a_0$. Then, on the assumption that the power series Eqn (12.6) has a radius of convergence $r > 0$, differentiate it term by term to obtain

$$f'(x) = \sum_{n=1}^{\infty} n a_n x^{n-1}, \tag{12.7}$$

for $-r < x < r$.

Again setting $x = 0$ shows that $f'(0) = a_1$. Differentiating Eqn (12.7) again with respect to x yields

$$f''(x) = \sum_{n=2}^{\infty} n(n-1) a_n x^{n-2}, \tag{12.8}$$

for $-r < x < r$, from which we conclude $f''(0) = 2!a_2$.

Proceeding systematically in this manner gives the general result

$$f^{(m)}(x) = \sum_{n=m}^{\infty} m(m-1)\ldots(m-n+1) a_m x^{n-m}, \tag{12.9}$$

for $-r < x < r$, so that $f^{(n)}(0) = n!a_n$, where we have set $f^{(n)}(x) = \mathrm{d}^n f/\mathrm{d}x^n$.

Thus the coefficients in power series Eqn (12.6) are determined by the formula

$$a_n = \frac{f^{(n)}(0)}{n!} \tag{12.10}$$

for $n \geq 1$ and $a_0 = f(0)$.

Substituting these coefficients into Eqn (12.6), we finally arrive at the power series

Maclaurin series

$$f(x) = f(0) + xf'(0) + \frac{x^2}{2!}f''(0) + \ldots + \frac{x^n}{n!}f^{(n)}(0) + \ldots = \sum_{n=0}^{\infty} \frac{x^n f^{(n)}(0)}{n!}. \tag{12.11}$$

The expression on the right-hand side of this equation is known as the *Maclaurin series* for $f(x)$, and it presupposes that $f(x)$ is differentiable an infinite number of times. To justify the use of the equality sign in Eqn (12.11) it is, of course, necessary to test the series for convergence to verify that its radius of convergence $r > 0$, and to show that $|f(x) - S_n(x)| \to 0$ as $n \to \infty$, where $S_n(x)$ is the sum of the first n terms of the Maclaurin series. We shall return to this matter later.

To transform Eqn (12.11) into a power series in $(x - x_0)$ we set $x = x_0 + h$ and let $f(x_0 + h) = \phi(h)$. Then $\phi'(h) = f'(x_0 + h)$, $\phi''(h) = f''(x_0 + h)$, $\ldots$, $\phi^{(n)}(h) = f^{(n)}(x_0 + h)$, $\ldots$. It thus follows that $\phi^{(n)}(0) = f^{(n)}(x_0)$ for $n \geq 1$ and $\phi(0) = f(x_0)$. The Maclaurin series for $\phi(h)$ is

$$\phi(h) = \phi(0) + h\phi'(0) + \frac{h^2}{2!}\phi''(0) + \ldots + \frac{h^n}{n!}\phi^{(l)}(0) + \ldots,$$

or, reverting to the function f,

Taylor series

$$f(x) = f(x_0) + (x - x_0)f'(x_0) + \frac{(x - x_0)^2}{2!}f''(x_0) + \ldots$$
$$+ \frac{(x - x_0)^n}{n!}f^{(n)}(x_0) + \ldots . \tag{12.12}$$

Expressed in this form, the expression on the right-hand side is called the *Taylor series* for $f(x)$ about the point $x = x_0$.

This result could equally well have been obtained had we started with the assumption that

$$f(x) = \sum_{n=0}^{\infty} a_n(x - x_0)^n$$

in place of Eqn (12.6) and item proceeded on above, only this time setting $x = x_0$ instead of $x = 0$ at each stage of the argument.

Example 12.13 Find the Maclaurin series for $\ln(1 + x)$ and $\ln(1 - x)$. Deduce the expansion for $\ln[(1 + x)/(1 - x)]$.

Solution Setting $f(x) = \ln(1+x)$, we find

$$f'(x) = \frac{1}{1+x}, \quad f''(x) = \frac{-1}{(1+x)^2}, \ldots, \quad f^{(n)}(x) = \frac{(-1)^{n-1}(n-1)!}{(1+x)^n}, \ldots,$$

and so

$$f^{(n)}(0) = (-1)^{n-1}(n-1)!$$

for $n \geq 1$, while $f(0) = 0$. Combining this expression for $f^{(n)}(0)$ with Eqn (12.11) gives for the Maclaurin series for $\ln(1+x)$,

$$\ln(1+x) = x - \frac{x^2}{2} + \frac{x^3}{3} - \frac{x^4}{4} + \ldots.$$

This has already been examined for convergence in Example 12.10(b) and found to be absolutely convergent in the interval $-1 < x \leq 1$.

In the case of the function $\ln(1-x)$, the same argument shows that

$$f^{(n)}(0) = -(n-1)!$$

for $n \geq 1$ and $f(0) = 0$, so that the Maclaurin series for $\ln(1-x)$ has the form

$$\ln(1-x) = -x - \frac{x^2}{2} - \frac{x^3}{3} - \frac{x^4}{4} - \ldots.$$

This can readily be seen to have $-1 \leq x < 1$ for its interval of convergence. Using the fact that $\ln\{(1+x)/(1-x)\} = \ln(1+x) - \ln(1-x)$ gives the desired result

$$\ln\left(\frac{1+x}{1-x}\right) = 2\left\{x + \frac{x^3}{3} + \frac{x^5}{5} + \frac{x^7}{7} + \ldots\right\},$$

for $-1 < x < 1$. ■

Strictly speaking, we are not yet entitled to use the equality sign between the function and its Maclaurin series, as we have not yet established the convergence of the nth partial sum of the series to the function it represents. We will do this later.

Example 12.14 Use Taylor's series to express the polynomial

$$P(x) = x^4 + 3x^3 + x^2 + 2x + 1$$

in terms of powers of $(x-1)$.

Solution To utilize the Taylor series in Eqn (2.12) we must set $x_0 = 1$ and $f(x) = P(x)$. Then a simple calculation shows that

$$P(1) = 8, P'(1) = 17, P''(1) = 32, PP'''(1) = 42, P^{(4)}(1) = 24$$

and $P^{(n)}(1) \equiv 0$ for $n \geq 5$.

Hence we arrive at the finite power series

$$P(x) = 8 + (x-1) \times 17 + \frac{(x-1)^2}{2!} \times 32 + \frac{(x-1)^3}{3!} \times 42 + \frac{(x-1)^4}{4!} \times 24,$$

or

$$P(x) = 8 + 17(x-1) + 16(x-1)^2 + 7(x-1)^3 + (x-1)^4.$$

A routine but tedious expansion and combination of these terms verifies that this is equivalent to the original polynomial $P(x)$. The use of the equality sign is fully justified here since we are dealing with a finite power series. ■

It can happen that the derivatives of a function $f(x)$ are not defined at $x = 0$ so that its formal Maclaurin series expansion cannot be obtained. In this case, provided the function is infinitely differentiable at the point $x = x_0$, then $f(x)$ may be expanded in a Taylor series about that point. Such a case is illustrated by the following simple example.

Example 12.15 Derive the nth derivative $f^{(n)}(x)$ of the function $f(x) = x \ln x$, and show that $f^{(n)}(0)$ is not defined. Deduce the Taylor series expansion of $f(x)$ about the point $x = 1$.

Solution Direct differentiation shows that $f^{(1)}(x) = 1 + \ln x$, $f^{(2)}(x) = 1/x$, $f^{(3)}(x) = -1/x^2$, $f^{(4)}(x) = 2!/x^3$, $f^{(5)}(x) = -3!/x^4$, ..., and in general

$$f^{(n)}(x) = \frac{(-1)^n (n-2)!}{x^{n-1}}$$

for $n \geq 2$. Hence it is clear that $f^{(n)}(0)$ is not defined for any n. However, the numbers $f^{(n)}(1)$ are defined for all n and $f^{(n)}(1) = (-1)^n (n-2)!$ for $n \geq 2$, while $f(1) = 0$, $f^{(1)}(1) = 1$. The Taylor series for $x \ln x$ can now be obtained from Eqn (12.12) by making the identification $x_0 = 1$ and then using the derivatives $f^{(n)}(1)$ which have just been computed. We find

$$x \ln x = (x-1) + \frac{(x-1)^2}{1.2} - \frac{(x-1)^3}{2.3} + \frac{(x-1)^4}{3.4} - \frac{(x-1)^5}{4.5} + \ldots,$$

which is the desired result. Again, we have used the equality sign without first showing that the nth partial sum of the Taylor series converges to $x \ln x$ as $n \to \infty$.

Regarding this as a power series in the variable $t = (x-1)$, we find that the coefficient a_n of the power t^n is $a_n = (-1)^n / n(n-1)$, whence the radius of convergence

$$r = \lim_{n\to\infty}\left|\frac{a_n}{a_{n+1}}\right| = \lim_{n\to\infty}\left|\frac{n(n+1)}{(n-1)n}\right| = 1.$$

The power series is thus absolutely convergent in the interval $-1 < t < 1$ or, equivalently, in $0 < x < 2$. The series is convergent when $x = 2$, because then it becomes an alternating series. It is also convergent when $x = 0$ by comparison with the series with the general term $b_n = 1/n^2$. In fact we can do better than this when $x = 0$, for then we can actually sum the series. Aside from the first term, which becomes -1, the sum of the remaining terms must be $+1$ by virtue of Example 12.9(a), showing that if the equality sign may be believed, then

$$\lim_{x\to 0}(x \ln x) = 0.$$

This is encouraging, because it is in agreement with the result which can be obtained from Theorem 6.4(b) by replacing x by $1/x$. This would strongly suggest that our series is in fact equal to $x \ln x$ in the complete interval of convergence $0 \le x \le 2$. ■

We have attempted to emphasize that although we have indicated how a Maclaurin or Taylor series may be associated with a function $f(x)$ that is infinitely differentiable, the general question of just exactly when the series is equal to the function with which it is associated still remains open. To indicate that an infinitely differentiable function need not be represented by its Maclaurin series at more than a single point, despite the fact the series is convergent for *all* x, we examine the function

$$f(x) = \begin{cases} e^{-1/x2} & \text{for} \quad x \neq 0 \\ 0 & \text{for} \quad x = 0. \end{cases}$$

This function is easily seen to be infinitely differentiable, and to be such that $f^{(n)}(0) = 0$ for all n. The Maclaurin series for $f(x)$ is thus

$$f(x) = 0 + 0 + 0 + \ldots,$$

which is clearly convergent for all x, yet it is only equal to the function $f(x)$ at the single point $x = 0$. Such behaviour is quite exceptional, yet the fact that it is associated with a seemingly simple function justifies the caution with which we must approach the question of equality between a function and its power series expansion.

On occasions, the computation of the nth derivative $f^{(n)}(x)$ is simplified by employing Leibniz's theorem, as we now illustrate (although not difficult, this example may be omitted at a first reading).

Example 12.16 If $f(x) = \cos(k \arccos x)$, and $f^{(n)}(x)$ denotes the nth derivative of $f(x)$, show that

$$(1 - x^2)f^{(n+2)}(x) - (2n+1)xf^{(n+1)}(x) - (n^2 - k^2)f^{(n)}(x) = 0,$$

for $n = 0, 1, \ldots$, where $f^{(0)}(x) \equiv f(x)$. Deduce the Maclaurin series for $f(x)$.

Solution As $f(x) = \cos(k \arccos x)$, it follows by differentiation that

$$f'(x) = \frac{k \sin(k \arccos x)}{\sqrt{(1 - x^2)}}$$

and

$$f''(x) = \frac{-k^2 \cos(k \arccos x)}{1 - x^2} + \frac{x k \sin(k \arccos x)}{(1 - x^2)^{3/2}}.$$

A little manipulation shows that $f(x)$ satisfies the differential equation

$$(1 - x^2)f''(x) - xf'(x) + k^2 f(x) = 0,$$

or

$$(1 - x^2)f^{(2)}(x) - xf^{(1)}(x) + k^2 f^{(0)}(x) = 0.$$

Now differentiating this equation n times, and using the symbolic differentiation operator D, gives

$$D^n[(1 - x^2)f^{(2)}(x) - xf^{(1)}(x) + k^2 f^{(0)}(x)] = 0,$$

or

$$D^n[(1 - x^2)f^{(2)}(x)] - D^n[xf^{(1)}(x)] + D^n[k^2 f^{(0)}(x)] = 0.$$

Employing Leibniz's theorem (Theorem 5.16), this becomes

$$(1 - x^2)f^{(n+2)}(x) + n(-2x)f^{(n+1)}(x) + \frac{n(n-1)}{2!}(-2)f^{(n)}(x)$$
$$- xf^{(n+1)}(x) - nf^{(n)}(x) + k^2 f^{(n)}(x) = 0,$$

showing that

$$(1 - x^2)f^{(n+2)}(x) - (2n + 1)xf^{(n+1)}(x) - (n^2 - k^2)f^{(n)}(x) = 0.$$

This is a differential equation, but setting $x = 0$ it reduces to a recurrence relation for $f^{(n)}(0)$:

$$f^{(n+2)}(0) = (n^2 - k^2)f^{(n)}(0) \qquad \text{for} \quad n = 0, 1, 2, \ldots$$

As $f^{(0)}(0) = f(0) = \cos(k \arccos 0) = \cos(\frac{1}{2}k\pi)$ and $f^{(1)}(0) = f'(0) = k \sin(k \arccos 0) = k \sin(\frac{1}{2}k\pi)$, we have

$$f^{(2)}(0) = -k^2 f^{(0)}(0) = -k^2 \cos\frac{k\pi}{2},$$

$$f^{(4)}(0) = (2^2 - k^2)f^{(2)}(0) = -k^2(2^2 - k^2)\cos\frac{k\pi}{2},$$

$$f^{(6)}(0) = (4^2 - k^2)f^{(4)}(0) = -k^2(2^2 - k^2)(4^2 - k^2)\cos\frac{k\pi}{2},$$

and

$$f^{(3)}(0) = (1^2 - k^2)f^{(1)}(0) = k(1^2 - k^2)\sin\frac{k\pi}{2},$$

$$f^{(5)}(0) = (3^2 - k^2)f^{(3)}(0) = k(1^2 - k^2)(3^2 - k^2)\sin\frac{k\pi}{2},$$

$$f^{(7)}(0) = (5^2 - k^2)f^{(5)}(0) = k(1^2 - k^2)(3^2 - k^2)(5^2 - k^2)\sin\frac{k\pi}{2},$$

and so on. The general expressions are

$$f^{(2m-1)}(0) = k(1^2 - k^2)(3^2 - k^2)\ldots[(2m-3)^2 - k^2]\sin\frac{k\pi}{2},$$

$$f^{(2m)}(0) = -k^2(2^2 - k^2)(4^2 - k^2)\ldots[(2m-2)^2 - k^2]\cos\frac{k\pi}{2},$$

from which we conclude that the Maclaurin series for $\cos(k \arccos x)$ has the form

$$\cos(k\arccos x) = \cos\frac{k\pi}{2} + xk\sin\frac{k\pi}{2} - \frac{x^2}{2!}k^2\cos\frac{k\pi}{2}$$
$$+\frac{x^3}{3!}k(1^2 - k^2)\sin\frac{k\pi}{2} - \frac{x^4}{4!}k^2(2^2 - k^2)\cos\frac{k\pi}{2} + \ldots.$$

■

To make further progress it now becomes necessary for us to settle the question of when a Maclaurin or Taylor series is really equal to the function with which it is associated. Let the function $f(x)$ be infinitely differentiable and have the Taylor series representation Eqn (12.12), and let $P_{n-1}(x)$ be the sum of the first n terms of the series terminating at the power $(x - x_0)^{n-1}$, so that

$$P_{n-1}(x) = f(x_0) + (x - x_0)f(x_0) + \frac{(x - x_0)^2}{2!}f''(x_0) + \ldots$$
$$+ \frac{(x - x_0)^{n-1}}{(n-1)!}f^{(n-1)}(x_0).$$

Then a necessary and sufficient condition that the Taylor series should converge to $f(x)$ is obviously that

$$\lim_{n\to\infty} |f(x) - P_{n-1}(x)| = 0.$$

As would be expected, this suggests that to establish convergence we must examine the behaviour of the remainder of the series after n terms. To achieve this we now prove *Talyor's theorem*, one form of which is stated below.

Theorem 12.10 (Taylor's theorem with a remainder) Let $f(x)$ be a function which is differentiable n times in the interval $a \leq x \leq b$. Then there exists a number ξ, strictly between a and x, such that

$$f(x) = f(a) + (x-a)f'(a) + \frac{(x-a)^2}{2!}f''(a) + \ldots + \frac{(x-a)^{n-1}}{(n-1)!}f^{(n-1)}(a) + \frac{(x-a)^n}{n!}f^{(n)}(\xi).$$

Proof The proof of Taylor's theorem we now offer will be based on Rolle's theorem. Let k be defined such that

$$f(b) = f(a) + (b-a)f'(a) + \ldots + \frac{(b-a)^{n-1}}{(n-1)!}f^{(n-1)}(a) + \frac{(b-a)^n}{n!}k,$$

and define the function $F(x)$ by the expression

$$F(x) = f(b) - f(x) - (b-x)f'(x) - \ldots - \frac{(b-x)^{n-1}}{(n-1)!}f^{(n-1)}(x) - \frac{(b-x)^n}{n!}k.$$

The $F(b) = F(a) = 0$, and a simple calculation shows that

$$F'(x) = \frac{(b-x)^{n-1}}{(n-1)!}\{k - f^{(n)}(x)\}.$$

Since, by hypothesis, $f^{(n-1)}(x)$ is differentiable in $a \leq x \leq b$, the function $F(x)$ satisfies the conditions of Rolle's theorem, which asserts that there must be a number ξ, strictly betwen a and b, for which $F'(\xi) = 0$. As $a < \xi < b$, the factor $(b-\xi)^{n-1} \neq 0$, so that we must have $k = f^{(n)}(\xi)$. Theorem 12.10 then follows by replacing b by x in $f(b)$ and $a < \xi < b$. ■

Corollary 12.10 (uniqueness of power series) If a function $f(x)$ has two power series expansions about the point $x = x_0$, each with the same radius of convergence $r > 0$, then the series are identical.

Proof The result is almost immediate, because if the two power series with the same radius of convergence are

$$f(x) = \sum_{n=0}^{\infty} a_n(x-x_0)^n \quad \text{and} \quad f(x) = \sum_{n=0}^{\infty} b_n(x-x_0)^n,$$

the form of argument used to establish Eqn (12.10) shows that $n!a_n = f^{(n)}(x_0) = n!b_n$ for all n. So as $a_n = b_n$, for all n, the power series expansion of $f(x)$ about $x = x_0$ is unique. ■

If we identify b with x and a with x_0, Taylor's theorem with a remainder takes the form

$$f(x) = (x_0) + (x - x_0)f'(x_0) + \ldots + \frac{(x - x_0)^{n-1}}{(n - 1)!}f^{(n-1)}(x_0)$$

$$+ \frac{(x - x_0)^n}{n!}f^{(n)}(\xi), \qquad (12.13)$$

where $x_0 < \xi < x$. For obvious reasons the last term of this expression is called the *remainder term* and is usually denoted by $R_n(x)$. The form stated here in which

$$R_n(x) = \frac{(x - x_0)^n}{n!}f^{(n)}(\xi), \qquad (12.14)$$

with $x_0 < \xi < x$, is known as the *Lagrange* form of the remainder term.

When $x_0 = 0$ Eqn (12.13) reduces to Maclaurin's theorem with a Lagrange remainder,

$$f(x) = f(0) + xf'(0) + \frac{x^2}{2!}f''(0) + \ldots + \frac{x^{n-1}}{(n - 1)!}f^{(n-1)}(0)$$

$$+ \frac{x^n}{n!}f^{(n)}(\xi), \qquad (12.15)$$

where $0 < \xi < x$.

Example 12.17 Find the Lagrange remainders $R_n(x)$ after n terms in the Maclaurin series expansion of e^x, $\sin x$ and $\cos x$. By showing that in each case $R_n(x) \to 0$ as $n \to \infty$, prove that these functions are equal to their Maclaurin series expansions.

Solution If $f(x) = e^x$, it is easily shown that Eqn (12.15) takes the form

$$e^x = 1 + x + \frac{x^2}{2!} + \frac{x^3}{3!} + \ldots + \frac{x^{n-1}}{(n - 1)!} + R_n(x),$$

where $R_n(x) = (x^n/n!)e^{\xi}$ and $0 < \xi < x$. Now $e^{\xi} < e^{|x|}$, and if R is an integer greater than $2x$, then $x/n < \frac{1}{2}$ for $n \geq R$, and so

$$\frac{x^n}{n!} = \frac{x}{1}\cdot\frac{x}{2}\cdots\frac{x}{R-1}\cdots\frac{x}{R}\cdots\frac{x}{n} < \frac{x^{R-1}}{(R - 1)!}\left(\tfrac{1}{2}\right)^{n-R+1}.$$

Hence for any fixed x, $e^{|x|}$ is a finite constant, and we already know that $x^n/n! \to 0$ as $n \to \infty$. It follows from this that $R_n(x) \to 0$ as $n \to \infty$. This provides an alternative verification of the results of section 6.1.

If $f(x) = \sin x$, then the Maclaurin series with a Lagrange remainder Eqn (12.15) becomes

$$\sin x = x - \frac{x^3}{3!} + \frac{x^5}{5!} - \ldots + \frac{x^n}{n!}\sin\left(\xi + \frac{n\pi}{2}\right),$$

where $0 < \xi < x$. The Lagrange remainder Eqn (12.14) is the last term, so

$$R_n(x) = \frac{x^n}{n!}\sin\left(\xi + \frac{n\pi}{2}\right).$$

Since $|\sin[\xi + (n\pi/2)]| \leq 1$, it follows that

$$|R_n(x)| \leq \left|\frac{x^n}{n!}\right|,$$

showing that $R_n(x) \to 0$ as $n \to \infty$. This establishes the convergence of $\sin x$ to its Maclaurin series, and the argument for the cosine function is exactly similar. ■

Example 12.18 Establish that $\ln(1+x)$ converges to its Maclaurin series in the interval $-1 < x \leq 1$.

Solution The Maclaurin series with a remainder is (see Example 12.13)

$$\ln(1+x) = x - \frac{x^2}{2} + \frac{x^3}{3} - \frac{x^4}{4} + \ldots + \frac{(-1)^{n-2}x^{n-1}}{n-1} + R_n(x),$$

where the Lagrange remainder is

$$R_n(x) = \frac{(-1)^{n-1}x^n}{n(1+\xi)^n},$$

with $0 < \xi < x$. For the interval $0 \leq x \leq 1$, we must have $0 < \xi < 1$ so that $1 + \xi > 1$, and hence $(1+\xi)^n > 1$. Thus $|R_n(x)| < x^n/n < 1/n \to 0$ as $n \to \infty$, thereby proving convergence of the Maclaurin series to $\ln(1+x)$ for $0 \leq x \leq 1$.

We must proceed differently to prove convergence for the interval $-1 < x < 0$. Set $y = -x$ and consider the interval $0 < y < 1$, in which we may write

$$\ln(1+x) = \ln(1-y) = -\int_0^y \frac{dt}{1-t}.$$

Using the binomial theorem

$$\frac{1}{1-t} = 1 + t + t^2 + \ldots + t^{n-1} + \frac{(-t)^n}{1-t},$$

we have, after integration,

$$\ln(1-y) = -y - \frac{y^2}{2} - \frac{y^3}{3} - \ldots - \frac{y^n}{n} + \int_0^y \frac{(-t)^n dt}{1-t}.$$

Thus our remainder term is now expressed in the form of the integral

$$R_n(y) = (-1)^n \int_0^y \frac{t^n}{1-t}\,\mathrm{d}t.$$

Now,

$$|R_n(y)| = \int_0^y \frac{t^n}{1-t}\,\mathrm{d}t < \left(\frac{1}{1-y}\right)\int_0^y t^n \mathrm{d}t = \frac{y^{n+1}}{(1-y)(n+1)}$$

$$< \frac{1}{(1-y)(n+1)},$$

showing that $|R_n(y)| \to 0$ as $n \to \infty$. This establishes convergence in the interval $-1 < x < 0$. Taken together with the first result, we have succeeded in showing that the Maclaurin series of $\ln(1+x)$ converges to the function itself in the interval $-1 < x \leq 1$. This provides the justification for our final result in Example 12.13. ■

When performing numerical calculations with Taylor series, the remainder term provides information on the number of terms that must be retained in order to attain any specified accuracy. Suppose, for example, we wished to calculate sin 31° correct to five decimal places by means of Eqn (12.13). Then first we would need to set $f(x) = \sin x$ to obtain

$$\sin x = \sin x_0 + (x - x_0)\cos x_0 - \frac{(x-x_0)^2}{2!}\sin x_0 + \ldots$$

$$+ \frac{(x-x_0)^{n-1}}{(n-1)!}\sin\left(x_0 + \frac{n\pi}{2}\right) + R_n(x),$$

where the remainder

$$R_n(x) = \frac{(x-x_0)^n}{n!}\sin\left(\xi + \frac{n\pi}{2}\right),$$

with $x_0 < \xi < x$.

As the arguments of trigonometric functions must be specified in radian measure, it is necessary to set x equal to the radian equivalent of 31° and then to choose a convenient value for x_0. Now 31° is equivalent to $\pi/6 + \pi/180$ radians, so that a convenient value for x_0 would be $x_0 = \pi/6$. This is, of course, the radian equivalent of 30°. The remainder term $R_n(x)$ now becomes

$$R_n(x) = \left(\frac{\pi}{180}\right)^n \frac{1}{n!}\sin\left(\xi + \frac{n\pi}{2}\right),$$

whence

$$|R_n(x)| \leq \left(\frac{\pi}{180}\right)^n \cdot \frac{1}{n!}.$$

For our desired accuracy we must have $|R_n(x)| < 5 \times 10^{-6}$. Hence n must be such that

$$\left(\frac{\pi}{180}\right)^n \cdot \frac{1}{n!} < 5 \times 10^{-6}.$$

A short calculation soon shows this condition is satisfied for $n \geq 3$, so that the expansion need only contain powers as far as $(x - x_0)^2$.

The polynomial

$$P_{n-1}(x) = f(x_0) + (x - x_0)f'(x_0) + \ldots + \frac{(x - x_0)^{n-1}}{(n-1)!} f^{(n-1)}(x_0) \tag{12.16}$$

associated with Taylor's theorem as expressed in Eqn (12.13) is called a *Taylor polynomial* of degree $n - 1$ about the point $x = x_0$. It is obviously an *approximating polynomial* for the function $f(x)$ in the sense that $|f(x) - P_{n-1}(x)| \to 0$ as $n \to \infty$ for all x within the interval of convergence. Hence $P_{n-1}(x)$ is strictly analogous to the nth partial sum used in the previous section. By way of example, the Taylor polynomial $P_3(x)$ for the exponential function e^x about the point $x = 0$ is

$$P_3(x) = 1 + x + \frac{x^2}{2!} + \frac{x^3}{3!},$$

while its general Taylor polynomial $P_n(x)$ about the point $x = 0$ is

$$P_n(x) = 1 + x + \frac{x^2}{2!} + \ldots + \frac{x^n}{n!}.$$

Sometimes Taylor and Maclaurin series can be found without appeal to Theorem 12.10 as the following Example shows, where the known series for e^x and $\cos x$ are used to find the Maclaurin series expansions of more complicated functions.

Example 12.19 Find the Maclaurin series for (a) $e^{(1-x^2)}$, (b) $\sin^4 x$ and (c) by representing e^{-x^2} by its Taylor polynomial $P_2(x)$ with the remainder term, evaluate the definite integral

$$I = \int_0^{0.2} e^{-x^2} dx,$$

and estimate the error involved.

Solution (a) The Maclaurin series for e^{-x^2} follows from the Maclaurin series for e^x by replacing x by $-x^2$, so

$$e^{-x^2} = \sum_{n=0}^{\infty}(-1)^n x^{2n}/n!.$$

As $e^{(1-x^2)} = e.e^{-x^2}$ it follows at once that the required Maclaurin series is

$$e^{(1-x^2)} = e\sum_{n=0}^{\infty}(-1)^n x^{2n}/n!.$$

(b) To find this Maclaurin series expansion without direct appeal to Theorem 12.10 we will use the trigonometric identity $\sin^4 x = \frac{1}{8}(3 - 4\cos 2x + \cos 4x)$. This identity can be derived by writing $\sin x = (e^{ix} - e^{-ix})/(2i)$, raising the expression on the right to the fourth power, grouping terms, and then re-writing the result in terms of $\cos 2x$ and $\cos 4x$. As the Maclaurin series for $\cos x$ is

$$\cos x = \sum_{n=0}^{\infty}(-1)^n \frac{x^{2n}}{(2n)!},$$

by replacing x by $2x$ and then by $4x$, we can write

$$\sin^4 x = \frac{1}{8}\left(3 - 4\sum_{n=0}^{\infty}(-1)^n \frac{(2x)^{2n}}{(2n)!} + \sum_{n=0}^{\infty}(-1)^n \frac{(4x)^{2n}}{(2n)!}\right).$$

After simplification this becomes

$$\sin^4 x = \frac{1}{8}\left(3 + \frac{(-1)^n 2^{2n}(2^{2n} - 4)}{(2n)!}x^{2n}\right) = x^4 - \tfrac{2}{3}x^6 + \tfrac{1}{5}x^8 - \tfrac{34}{945}x^{10} + \ldots.$$

(c) Although the Taylor polynomial $P_2(x)$ associated with $f(x) = e^{-x^2}$ can be deduced directly from the series for e^x as in (a), the remainder term will be required so we apply Taylor's theorem with a remainder term to $f(x)$, when we find that $f(0) = 1, f'(0) = 0, f''(0) = -2$ and $f'''(0) = 4x(3 - 2x^2)e^{-x^2}$. Thus $P_2(x) = 1 - x^2$, and so from Theorem 12.10 we obtain

$$e^{-x^2} = P_2(x) + \frac{x^3}{3!}f'''(\xi),$$

where $0 < \xi < 0.2$.

Now we have

$$I \approx \int_0^{0.2} P_2(x)\,dx = \int_0^{0.2}(1 - x^2)\,dx = 0.19733,$$

which is our approximate value for the integral. To assess the error E we use the fact that

$$\begin{aligned}E &= \int_0^{0.2} e^{-x^2}\,dx - \int_0^{0.2} P_2(x)\,dx \\ &= \int_0^{0.2} (e^{-x^2} - P_2(x))\,dx \\ &= \int_0^{0.2} \frac{x^3}{3!} f'''(\xi)\,dx.\end{aligned}$$

In this expression, $\xi = \xi(x)$, because $0 < \xi < x$ and x is itself integrated over the interval $0 \leq x \leq 0.2$. Although the functional form of $\xi(x)$ is unknown, we may obtain an overestimate of E by replacing $f'''(\xi)$ by its greatest value in the interval $0 < x < 0.2$. Using the fact that $f'''(x) = 4x(3 - 2x^2)e^{-x}$ and $\max|f'''(\xi)| = \max|f'''(x)|$ we estimate this latter quantity by assigning to each of the three factors in $f'''(x)$ its maximum value. We find that

$$\max|f'''(x)| \leq 0.8 \times 3 \times 1,$$

whence

$$E \leq \frac{2.4}{3!}\int_0^{0.2} x^3 dx = 0.00016.$$

The true value rounded to five decimal places is 0.19737. ■

Theorem 12.10 is called the *generalized mean value theorem*, since when $n = 1$ it reduces to the already familiar mean value theorem derived in Chapter 5 (Theorem 5.12). Let us now derive the analogue of Taylor's theorem with a remainder for a function of two variables.

Suppose that $f(x, y)$ has continuous partial derivatives up to those of nth order, and consider the function

$$F(t) = f(a + ht,\ b + kt), \tag{12.17}$$

in which a, b, h and k are constants. Then $F(t) = f(x, y)$, where $x = a + ht$, $y = b + kt$, and in the neighbourhood of (a, b) we have

$$\begin{aligned}\frac{dF}{dt} &= \frac{df}{dt} = \frac{\partial f}{\partial x}\frac{dx}{dt} + \frac{\partial f}{\partial y}\frac{dy}{dt} \\ &= h\frac{\partial f}{\partial x} + k\frac{\partial f}{\partial y}.\end{aligned}$$

Write this result in the form

$$\frac{df}{dt} = \left(h\frac{\partial}{\partial x} + k\frac{\partial}{\partial y}\right)f,$$

where the expression in brackets is a partial differential operator with respect to x and y and not a function. It only generates a function when it acts on a suitably differentiable function f. In consequence, differentiating r times, we have

$$\left(\frac{\mathrm{d}}{\mathrm{d}t}\right)^r f = \left(h\frac{\partial}{\partial x} + k\frac{\partial}{\partial y}\right)^r f \qquad \text{for} \quad r = 1, 2, \ldots, \tag{12.18}$$

with the understanding that:

$$\left(h\frac{\partial}{\partial x} + k\frac{\partial}{\partial y}\right)f = h\frac{\partial f}{\partial x} + k\frac{\partial f}{\partial y},$$

$$\left(h\frac{\partial}{\partial x} + k\frac{\partial}{\partial y}\right)^2 f = h^2\frac{\partial^2 f}{\partial x^2} + 2hk\frac{\partial^2 f}{\partial x \partial y} + k^2\frac{\partial^2 f}{\partial y^2},$$

$$\left(h\frac{\partial}{\partial x} + k\frac{\partial}{\partial y}\right)^2 f = h^3\frac{\partial^3 f}{\partial x^3} + 3h^2k\frac{\partial^3 f}{\partial x^2 \partial y} + 3hk^2\frac{\partial^3 f}{\partial x \partial y^2} + k^3\frac{\partial^3 f}{\partial y^3}.$$

Now $F(0) = f(a, b)$, $F(1) = f(a+h, b+k)$ and $F(t)$ is differentiable n times for $0 \leq t \leq 1$. Consequently, by applying Theorem 12.10 to the function $F(t)$ we obtain

$$F(1) = F(0) + F'(0) + \frac{1}{2!}F''(0) + \ldots + \frac{1}{(n-1)!}F^{(n-1)}(0) + \frac{1}{n!}F^{(n)}(\xi), \tag{12.19}$$

where $0 < \xi < 1$.

However, we also have

$$F^{(r)}(0) = \left(h\frac{\partial}{\partial x} + k\frac{\partial}{\partial y}\right)^r f\bigg|_{\substack{x=a\\y=b}}$$

and

$$F^{(n)}(\xi) = \left(h\frac{\partial}{\partial x} + k\frac{\partial}{\partial y}\right)^n f\bigg|_{\substack{x=a+\xi h,\\y=b+\xi k}} \tag{12.20}$$

whence by substitution of Eqn (12.20) into Eqn (12.19) we obtain:

$$f(a+h, b+k) = f(a, b) + hf_x(a, b) + kf_y(a, b) + \frac{1}{2!}\left(h\frac{\partial}{\partial x} + k\frac{\partial}{\partial y}\right)^2 f\bigg|_{\substack{x=a\\y=b}} + \ldots + \frac{1}{(n-1)!}\left(h\frac{\partial}{\partial x} + k\frac{\partial}{\partial y}\right)^{n-1} f\bigg|_{\substack{x=a\\y=b}} + \frac{1}{n!}\left(h\frac{\partial}{\partial x} + k\frac{\partial}{\partial y}\right)^n f\bigg|_{\substack{x=a+\xi h,\\y=b+\xi k}} \tag{12.21}$$

where $0 < \xi < 1$.

This result is Taylor's theorem for a function $f(x, y)$ of two variables, and it is terminated with a Lagrange remainder term involving nth partial derivatives. The result is also often known as the *generalized mean value theorem for a function of two variables*. In particular, by taking $n = 1$ we obtain the result

$$f(a+h, b+k) = f(a,b) + hf_x(a+\xi h,\ b+\xi k)$$
$$+ kf_y(a+\xi h,\ b+\xi k), \tag{12.22}$$

where $0 < \xi < 1$. This is the two-variable analogue of Theorem 5.12 to which it obviously reduces when $f = f(x)$, for then $f_y \equiv 0$. Result Eqn (12.21) is of such importance that it merits stating in the form of a theorem.

Theorem 12.11 (generalized mean value theorem in two variables)

Let $f(x, y)$ have continuous partial derivatives up to those of order n in some neighbourhood of the point (a, b). Then if (x, y) is any point within this neighbourhood,

$$f(x,y) = f(a,\ b) + \left((x-a)\frac{\partial}{\partial x} + (y-b)\frac{\partial}{\partial y}\right)f\bigg|_{(a,b)}$$
$$+ \frac{1}{2!}\left((x-a)\frac{\partial}{\partial x} + (y-b)\frac{\partial}{\partial y}\right)^2 f\bigg|_{(a,b)} + \dots$$
$$+ \frac{1}{(n-1)!}\left((x-a)\frac{\partial}{\partial x} + (y-b)\frac{\partial}{\partial y}\right)^{n-1} f\bigg|_{(a,b)} + R_n(x,y),$$

where the Lagrange remainder

$$R_n(x,y) = \frac{1}{n!}\left((x-a)\frac{\partial}{\partial x} + (y-b)\frac{\partial}{\partial y}\right)^n f\bigg|_{(\eta,\xi)},$$

in which $\eta = a + \xi(x-a)$, $\zeta = b + \xi(y-b)$, and $0 < \xi < 1$. ■

So, for example, with $n = 2$ we have

$$f(a,\ b) + (x-a)\left(\frac{\partial f}{\partial x}\right)_{(a,b)} + (y-b)\left(\frac{\partial f}{\partial y}\right)_{(a,b)} + R_2(x,y),$$

while with $n = 3$ we have

$$f(x,y) = f(a,b) + (x-a)\left(\frac{\partial f}{\partial x}\right)_{(a,b)} + (y-b)\left(\frac{\partial f}{\partial y}\right)_{(a,b)}$$
$$+ \frac{1}{2!}\left[(x-a)^2\left(\frac{\partial^2 f}{\partial x^2}\right)_{(a,b)} + 2(x-a)(y-b)\left(\frac{\partial^2 f}{\partial x\,\partial y}\right)_{(a,b)}\right.$$
$$\left. + (y-b)^2\left(\frac{\partial^2 f}{\partial y^2}\right)_{(a,b)}\right] + R_3(x,y),$$

where the suffix (a, b) signifies that the associated expression is to be evaluated with $x = a$ and $y = b$.

Example 12.20 Use the generalized mean value theorem in two variables to expand the function

$$f(x,y) = e^{x+2xy}$$

about the point (0, 0). Terminate the expansion with the Lagrange remainder term $R_3(x,y)$ and display its form.

Solution As the expansion is required about the point (0, 0) we must set $a = 0$, $b = 0$ in Theorem 12.11 and take $n = 3$. Routine calculation shows that:

$$f(0,0) = 1, \quad f_x(0,0) = 1, \quad f_y(0,0) = 0, \quad f_{xx}(0,0) = 1,$$

$$f_{xy}(0,0) = 2, \quad f_{yy}(0,0) = 0,$$

while

$$f_{xxx}(x,y) = (1+2y)^3 e^{x+2xy},$$

$$f_{xxy}(x,y) = 2(1+2y)[2 + x(1+2y)]e^{x+2xy},$$

$$f_{yyx}(x,y) = 4x[2 + x(1+2y)]e^{x+2xy},$$

$$f_{yyy}(x,y) = 8x^3 e^{x+2xy}.$$

From Theorem 12.11 we find

$$e^{x+2xy} = 1 + x + \tfrac{1}{2}x^2 + 2xy + R_3(x,y),$$

where

$$R_3(x,y) = \frac{1}{3!}\left(x^3 f_{xxx}(x,y) + 3x^2 y f_{xxy}(x,y) + 3xy^2 f_{yyx}(x,y) + y^3 f_{yyy}(x,y)\right)_{(\eta,\zeta)}$$

with $\eta = \xi x$, $\zeta = \xi y$, and $0 < \xi < 1$.

12.4 Applications of Taylor's theorem

The applications of Taylor's theorem with a remainder are so numerous that we can do no more here than describe some that occur most frequently. It is intended that these illustrations will indicate the power of this theorem and the fact that its use is not confined exclusively to the estimation of errors in the series expansions of functions.

12.4.1 Indeterminate forms

The form of L'Hospital's rule given in Theorem 5.14 is capable of immediate extension as follows.

Theorem 12.12 (extended L'Hospital's rule) Let $f(x)$ and $g(x)$ be n times differentiable functions which are such that $f(a) = g(a) = 0$ and $f^{(r)}(a) = g^{(r)}(a) = 0$ for $r = 1, 2, \ldots, n-1$, but $\lim_{x \to a} f^{(n)}(x)$ and $\lim_{x \to a} g^{(n)}(x)$ are not both zero. Then

$$\lim_{x\to a}\frac{f(x)}{g(x)}=\frac{\lim_{x\to a} f^{(n)}(x)}{\lim_{x\to a} g^{(n)}(x)}.$$

Proof Using Taylor's theorem with a remainder to expand numerator and denominator separately gives

$$\frac{f(a+h)}{g(a+h)}=\frac{f(a)+hf'(a)+\ldots+\dfrac{h^n}{n!}f^{(n)}(\xi_1)}{g(a)+hg'(a)+\ldots+\dfrac{h^n}{n!}g^{(n)}(\xi_2)}=\frac{f^{(n)}(\xi_1)}{g^{(n)}(\xi_2)},$$

where $a<\xi_1<a+h, a<\xi_2<a+h$. If now $h\to 0$, then $\xi_1, \xi_2\to a$ obtain the result of the theorem

$$\lim_{x\to a}\frac{f(x)}{g(x)}=\lim_{h\to 0}\frac{f(a+h)}{g(a+h)}=\frac{\lim_{x\to a} f^{(n)}(x)}{\lim_{x\to a} g^{(n)}(x)}.$$

This is, of course, the result which would be obtained by the repeated application of L'Hospital's rule referred to in section 5.3.4. ■

Example 12.21 Find the value of the expression

$$\lim_{x\to 0}\frac{x\sin x}{(a^x-1)(b^x-1)}.$$

Solution This is an indeterminate form. Setting $f(x)=x\sin x$, $g(x)=(a^x-1)(b^x-1)$, we first compute $f'(x)$ and $g'(x)$. We find $f'(x)=\sin x+x\cos x$ and $g'(x)=a^x(\ln a)(b^x-1)+b^x(\ln b)(a^x-1)$, and clearly $\lim_{x\to 0} f'(x)=\lim_{x\to 0} g'(x)=0$. The earlier form of L'Hospital's rule thus fails, so we appeal to Theorem 12.12 and compute $f''(x)$ and $g''(x)$. We find $f''(x)=2\cos x-x\sin x$ and $g''(x)=2a^xb^x\ln a\ln b+a^x(\ln a)^2(b^x-1)+b^x(\ln b)^2(a^x-1)$, from which we see that $\lim_{x\to 0} f''(x)=2$, $\lim_{x\to 0} g''(x)=2\ln a\ln b$. By the conditions of Theorem 12.12 we have

$$\lim_{x\to 0}\frac{x\sin x}{(a^x-1)(b^x-1)}=\frac{1}{\ln a\ln b}.$$ ■

12.4.2 Local behaviour of functions of one variable

In Chapter 5 we repeatedly turned to the problem of the local behaviour of a function of one variable in order to identify local maxima, local minima and points of inflection. Here again Taylor's theorem with a remainder helps to identify such points when not only the first derivative, but also successive higher-order derivatives vanish at a point.

Suppose that $f(x)$ is n times differentiable near $x=a$ and that $f^{(1)}(a)=f^{(2)}(a)=\ldots=f^{(n-1)}(a)=0$, but that $f^{(n)}(a)\neq 0$. Then by Taylor's theorem

$$f(a+h) = f(a) + hf^{(1)}(a) + \ldots + \frac{h^{n-1}}{(n-1)!}f^{(n-1)}(a) + \frac{h^n}{n!}f^{(n)}(\xi),$$

where $a < \xi < a+h$, but because of the vanishing of the first $(n-1)$ derivatives at $x=a$ this simplifies to

$$f(a+h) - f(a) = \frac{h^n}{n!}f^{(n)}(\xi).$$

The behaviour of the left-hand side of this expression was used in Chapter 5 to identify the nature of the extrema involved so that we see its sign is now determined solely by the sign of $h^n f^{(n)}(\xi)$ or, for suitably small h, by the sign of $h^n f^{(n)}(a)$. It is left to the reader to verify that the following theorem is an immediate consequence of this simple result when taken in conjunction with Definition 5.4.

Theorem 12.13 (identification of local extrema – one independent variable) A necessary and sufficient condition that a suitably differentiable function $f(x)$ have a local maximum (minimum) at $x=a$ is that the first derivative $f^{(n)}(x)$ with a non-zero value at $x=a$ shall be of even order and $f^{(n)}(a) < 0$ $(f^{(n)}(a) > 0)$. If the first derivative other than $f^{(1)}(a)$ with a non-zero value at $x=a$ is of odd order, then $f(x)$ has a point of inflection with an associated zero gradient at $x=a$. ■

12.5 Applications of the generalized mean value theorem

The applications of the extension of Taylor's theorem to functions of two or more variables are perhaps even more extensive than those of Taylor's theorem itself. This section illustrates a few of the simplest and most used, connected mainly with functions of two variables. The final application, connected with the least-squares fitting of a polynomial, involves functions with an arbitrary number of independent variables.

12.5.1 Stationary points of functions of two variables

Consider the function $z=f(x,y)$ of the two real independent variables x, y which is defined in some region D of the (x,y)-plane bounded by the curve γ. The notion of its graph is already familiar to us, and it comprises a surface S with points $(x,y,f(x,y))$, the projection of the boundary Γ of which onto the (x,y)-plane is the curve γ. A typical situation is shown in Fig. 12.3, where the point P is obviously a maximum and the point Q is obviously a minimum.

Intuitively, and by analogy with the single-variable case, it would seem that all that is necessary to locate extrema such as P, Q is to find those points (x_0, y_0) at which $f_x(x_0, y_0) = f_y(x_0, y_0) = 0$. This is, in effect, saying that the tangent plane at either a maximum or a minimum must be parallel to the (x,y)-plane. Unfortunately, this is not a sufficiently strin-

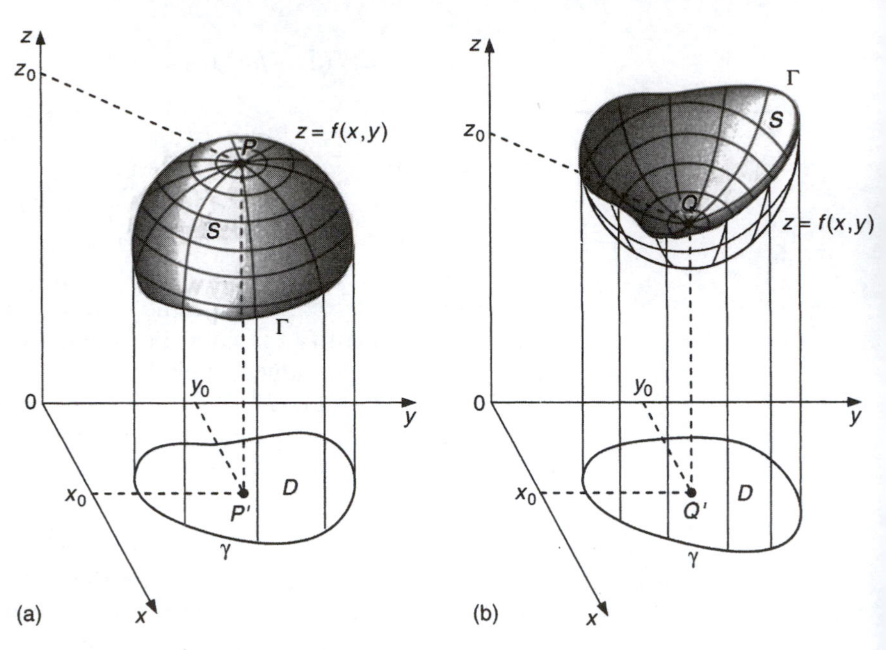

Fig. 12.3 (a) Surface having maximum at P; (b) surface having minimum at Q.

gent condition, for the point R in Fig. 12.4 is neither a maximum nor a minimum, yet the tangent plane at that point is certainly parallel to the (x, y)-plane. Because of the shape of the surface it is called a *saddle point*. It is characterized by the fact that if the surface is sectioned through R by different planes parallel to the z-axis, then for some the curve of section has a minimum at R and for others a maximum.

Each of these points P, Q, R is called a *stationary point* of the function $z = f(x, y)$ because f_x and f_y vanish at these points.

Definition 12.3 (stationary points of $f(x, y)$) Let $f(x, y)$ be a differentiate function in some region of the (x, y)-plane. Then any point (x_0, y_0) in D for which $f_x(x_0, y_0) = 0$ and $f_y(x_0, y_0) = 0$ is called a *stationary point* of the function $f(x, y)$ in D.

If for all (x, y) near (x_0, y_0) it is true that $f(x, y) < f(x_0, y_0)$, then $f(x, y)$ will be said to have a *local maximum* at (x_0, y_0). If for all (x, y) near to (x_0, y_0) it is true that $f(x, y) > f(x_0, y_0)$, then $f(x, y)$ will be said to have a *local minimum* at (x_0, y_0). In the event that $f(x, y)$ assumes values both greater and less than $f(x_0, y_0)$ for (x, y) near to a stationary point (x_0, y_0), then $f(x, y)$ will be said to have a *saddle point* at (x_0, y_0).

We now use the generalized mean value theorem to prove the following result.

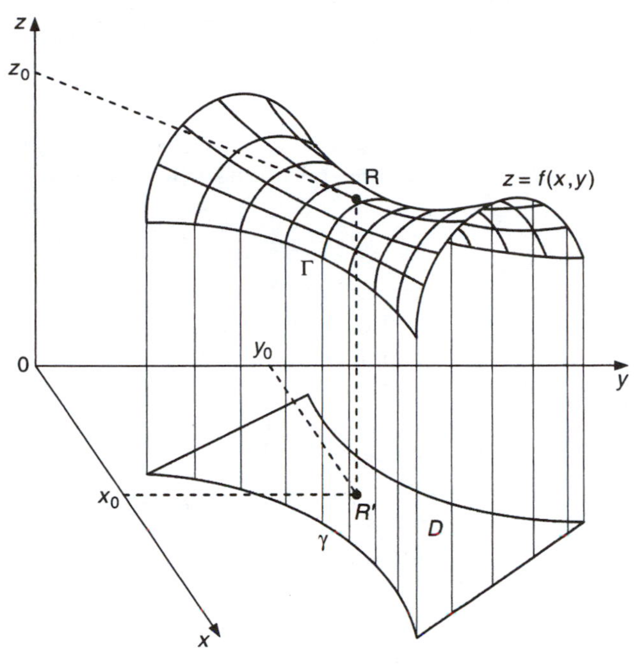

Fig. 12.4 Saddle point.

Theorem 12.14 (identification of extrema of $f(x,y)$)

Let $f(x,y)$ be a function with continuous first- and second-order partial derivatives. Then a sufficient condition for (x_0, y_0) to be a local maximum (minimum) for $f(x,y)$ is that:

(a) $f_x(x_0, y_0) = f_y(x_0, y_0) = 0$;
(b) $f_{xx}(x_0, y_0)f_{yy}(x_0, y_0) > f_{xy}^2(x_0, y_0)$;
(c) $f_{xx}(x_0, y_0) < 0$ for max. and $f_{xx}(x_0, y_0) > 0$ for min.

A sufficient condition for $f(x,y)$ to have a saddle point at (x_0, y_0) is that in addition to condition (a) above being satisfied condition (b) is replaced by

(b′) $f_{xx}(x_0, y_0)f_{yy}(x_0, y_0) < f_{xy}^2(x_0, y_0)$

and condition (c) is discarded.

Proof Notice first that (b) implies either that $f_{xx}(x_0, y_0) > 0$ and $f_{yy}(x_0, y_0) > 0$ or that $f_{xx}(x_0, y_0) < 0$ and $f_{yy}(x_0, y_0) < 0$. Consider the case $f_{xx}(x_0, y_0) > 0$. Then by the generalized mean value theorem with $n = 2$,

$$f(x_0 + h, y_0 + k) - f(x_0, y_0) = \tfrac{1}{2}[h^2 f_{xx}(\eta, \zeta) + 2hk f_{xy}(\eta, \zeta) + k^2 f_{yy}(\eta, \zeta)],$$

where $\eta = x_0 + \xi h$, $\zeta = y_0 + \xi k$, with $0 < \xi < 1$. Now as f_{xx}, f_{xy} and f_{yy} are assumed continuous, it follows from (b) that for sufficiently small h and

k, $f_{xx}(\xi, \eta)f_{yy}(\xi, \eta) - f_{xy}^2(\xi, \eta) > 0$. Thus we have

$$f(x_0 + h, y_0 + k) - f(x_0, y_0) = \tfrac{1}{2}(Ah^2 + 2Bhk + Ck^2),$$

where $A = f_{xx}(\eta, \zeta)$, $B = f_{xy}(\eta, \zeta)$, $C = f_{yy}(\eta, \zeta)$. Completing the square on the right-hand side of this equation allows us to write it as

$$f(x_0 + h, y_0 + k) - f(x_0, y_0) = \tfrac{1}{2}A\left[\left(h + \frac{B}{A}k\right)^2 + \left(\frac{AC - B^2}{A^2}\right)k^2\right].$$

Clearly, $(h + (B/A)k)^2 \geq 0$ and $[(AC - B^2)/A^2]k^2 > 0$ if $k \neq 0$ since, by hypothesis, $AC - B^2 > 0$. In the event $k = 0$, then $Ah^2 + 2Bhk + Ck^2 = Ah^2 > 0$ provided $h \neq 0$.

Thus, if at most one of h and k is zero, since we are assuming $A > 0$ we have shown that

$$f(x_0 + h, y_0 + k) - f(x_0, y_0) > 0$$

for small h, k or, equivalently,

$$f(x, y) > f(x_0, y_0),$$

for all (x, y) near (x_0, y_0). This is the condition that $f(x, y)$ should have a local minimum at (x_0, y_0).

The verification of the condition for a local maximum at (x_0, y_0) follows from the above argument by setting $g(x, y) = -f(x, y)$ and then supposing that $f_{xx}(x_0, y_0) < 0$. This establishes that $g(x, y)$ has a local minimum at (x_0, y_0) so that $f(x, y)$ must have a local maximum at that point.

The verification of the condition for a saddle point follows directly from consideration of the result

$$f(x_0 + h, y_0 + k) - f(x_0, y_0) = \tfrac{1}{2}A\left[\left(h + \frac{B}{A}k\right)^2 + \left(\frac{AC - B^2}{A^2}\right)k^2\right]$$

which was derived above. For now, by hypothesis, $AC - B^2 < 0$, so that the terms within the large parentheses are of opposite signs. This implies that $f(x_0 + h, y_0 + k) - f(x_0, y_0)$ can be made either positive or negative near (x_0, y_0) by a suitable choice of h, k. This is the condition for a saddle point and completes the proof of the theorem. ■

Example 12.22 Find the stationary points of the function

$$f(x, y) = 2x^3 - 9x^2y + 12xy^2 - 60y,$$

and identify their nature.

Solution We have

$$f_x = 6x^2 - 18xy + 12y^2 \quad \text{and} \quad f_y = -9x^2 + 24xy - 60.$$

The conditions $f_x = f_y = 0$ are equivalent to

$$(f_x = 0) \quad (x - y)(x - 2y) = 0$$
$$(f_y = 0) \quad 3x^2 - 8xy + 20 = 0.$$

From the first condition we may either have $x = y$ or $x = 2y$. Substituting $x = y$ in the second condition gives rise to the equation $y^2 = 4$, so that the stationary points corresponding to $x = y$ are (2, 2) and $(-2, -2)$. Substituting $x = 2y$ in the second condition gives rise to the condition $y^2 = 5$, so that the stationary points corresponding to $x = 2y$ are $(2\sqrt{5}, \sqrt{5})$ and $(-2\sqrt{5}, -\sqrt{5})$.

There are thus four stationary points associated with the function in question, and we must apply the tests given in Theorem 12.14 to identify their nature. We have

$$f_{xx} = 12x - 18y, \quad f_{xy} = -18x + 24y, \quad f_{yy} = 2x,$$

and it is easily verified that $f_{xx} f_{yy} - f_{xy}^2 < 0$ at both of the points (2, 2) and (–2, –2), showing that they must be saddle points. A similar calculation shows that $f_{xx} f_{yy} - f_{xy}^2 > 0$ at each of the other stationary points, though $f_{xx} > 0$ at $(2\sqrt{5}, \sqrt{5})$, showing that it must be a minimum, whereas $f_{xx} < 0$ at $(-2\sqrt{5}, -\sqrt{5})$, showing that it must be a maximum. ■

Example 12.23 Find the stationary points of the function

$$f(x, y) = x^3 + y^3 - 2(x^2 + y^2) + 3xy,$$

and identify their nature.

Solution We have

$$f_x = 3x^2 - 4x + 3y \quad \text{and} \quad f_y = 3y^2 - 4y + 3x.$$

The conditions $f_x = f_y = 0$ are equivalent to

$$(f_x = 0) \quad 3x^2 - 4x + 3y = 0$$

$$(f_y = 0) \quad 3y^2 - 4y + 3x = 0.$$

Subtracting these equations gives

$$3(x^2 - y^2) - 4(x - y) - 3(x - y) = 0,$$

or

$$(x - y)(3x + 3y - 7) = 0.$$

Hence, either

$$x = y \quad \text{or} \quad 3x + 3y - 7 = 0.$$

Setting $x = y$ in $f_x = 0$ reduces it to

$$x(3x - 1) = 0,$$

so that either $x = 0$ or $x = \frac{1}{3}$, and as $x = y$ it follows that the stationary points corresponding to this case are (0, 0) and $(\frac{1}{3}, \frac{1}{3})$.

We must now consider the other possibility that $3x + 3y - 7 = 0$. Using this last result in $f_x = 0$ to eliminate y reduces it to

$$3x^2 - 7x + 7 = 0.$$

As this quadratic equation has complex roots the condition $3x+3y-7 = 0$ cannot lead to any further stationary points. Thus the only stationary points possessed by the function are located at (0, 0) and $(\frac{1}{3}, \frac{1}{3})$.

There is no necessity to examine the effect of the conditions $x = y$ and $3x+3y-7 = 0$ on $f_y = 0$, for these will simply lead to the same conclusions, because these conditions were derived by solving the simultaneous equations $f_x = 0$ and $f_y = 0$.

Now to identify the nature of the stationary points, it is necessary to make use of the last part of Theorem 12.14. As

$$f_{xx} = 6x - 4, \quad f_{xy} = 3 \quad \text{and } f_{yy} = 6y - 4$$

it follows that

$$f_{xx} f_{yy} - f_{xy}^2 = (6x - 4)(6y - 4) - 9.$$

Now at (0, 0) we see that $f_{xx} f_{yy} - f_{xy}^2 = 7 > 0$, so as $f_{xx} = -4 < 0$ we conclude that the point (0, 0) is a local maximum.

At $(\frac{1}{3}, \frac{1}{3})$ we see that $f_{xx} f_{yy} - f_{xy}^2 = -5 < 0$, so we conclude that the point $(\frac{1}{3}, \frac{1}{3})$ is a saddle point. ■

12.5.2 Constrained extrema

A rather more difficult problem involving the location of the extrema of a function $z = f(x, y)$ of two variables occurs when the points (x, y) are constrained to lie on some curve $g(x, y) = 0$. This is illustrated in Fig. 12.5, in which $f(x, y)$ is defined at points in the region D of the (x, y)-plane contained within the curve γ defined by $g(x, y) = 0$. The boundary points on the surface $z = f(x, y)$ corresponding to the boundary curve γ of D form the closed space curve Γ. Our task is to locate the maximum and minimum values P and Q assumed by $z = f(x, y)$ on the curve Γ. These correspond to the points P' and Q' on γ in Fig. 12.5.

In principle this is a problem of locating the extrema for a function of one variable, because solving $g(x, y) = 0$ explicitly for y in the form $y = h(x)$ shows that we must find and identify the stationary points of $z = F(x)$, where $F(x) = f(x, h(x))$. However, this is usually an impossible task because $g(x, y) = 0$ cannot, as a general rule, be solved explicitly for y. Instead we proceed as follows.

We have

$$z = f(x, y) \tag{12.23}$$

and

$$g(x, y) = 0, \tag{12.24}$$

so that forming the total derivatives of these with respect to x gives

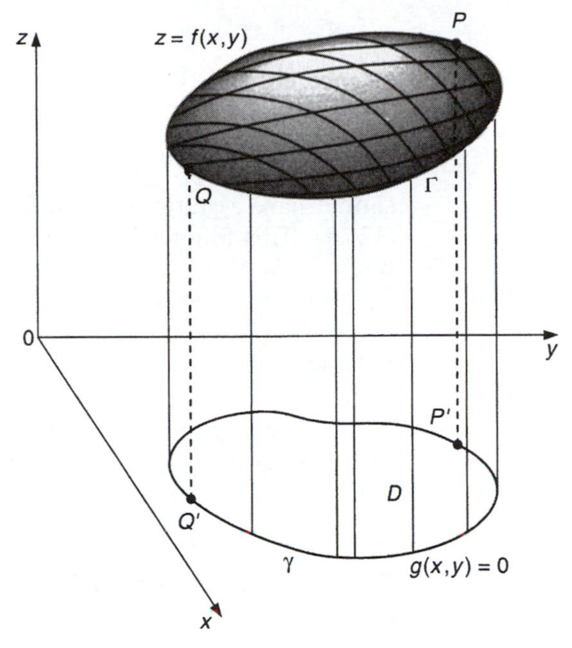

Fig. 12.5 Constrained extrema.

$$\frac{\mathrm{d}z}{\mathrm{d}x} = \frac{\partial f}{\partial x} + \frac{\partial f}{\partial y}\frac{\mathrm{d}y}{\mathrm{d}x} \tag{12.25}$$

and

$$0 = \frac{\partial g}{\partial x} + \frac{\partial g}{\partial y}\frac{\mathrm{d}y}{\mathrm{d}x}. \tag{12.26}$$

Thus on γ we have, provided $\partial g/\partial y \neq 0$,

$$\frac{\mathrm{d}y}{\mathrm{d}x} = -\frac{\partial g}{\partial x}\bigg/\frac{\partial g}{\partial y},$$

whence on γ Eqn (12.25) becomes

$$\frac{\mathrm{d}z}{\mathrm{d}x} = \frac{\partial f}{\partial x} - \frac{\partial f}{\partial y}\frac{\partial g}{\partial x}\bigg/\frac{\partial g}{\partial y}.$$

As already remarked, on γ the function $z = f(x, y)$ is effectively only a function of x, so that its stationary points will be determined by the condition $\mathrm{d}z/\mathrm{d}x = 0$.Thus the solution to our problem lies in solving the equation

$$\frac{\partial f}{\partial x}\frac{\partial g}{\partial y} - \frac{\partial f}{\partial y}\frac{\partial g}{\partial x} = 0 \tag{12.27}$$

subject to the constraint condition

$$g(x, y) = 0. \quad (12.28)$$

Algebraically, this is equivalent to determining the value of the parameter λ in order that the function of two variables

$$w = f(x, y) + \lambda g(x, y) \quad (12.29)$$

should have a stationary point subject to the constraint condition Eqn (12.28). This follows because for w to have a stationary point we need both $w_x = f_x + \lambda g_x = 0$ and $w_y = f_y + \lambda g_y = 0$, and these homogeneous equations have a solution only when condition (12.27) is satisfied. The parameter λ occurring in Eqn (12.29) is called a *Lagrange multiplier*. Solving these equations locates the stationary points but does not identify their nature. This must be undertaken by an examination of the conditions in the neighbourhood of the stationary points and possibly, as in the following examples, by other considerations implicit in the problem.

We now summarize this argument in the form of a rule.

Rule 12.1 (Lagrange multiplier method for functions of variables with one constraint)

It is required to find the stationary points of $f(x, y)$ subject to the constraint condition $g(x, y) = 0$.

1. Form the function

$$w = f(x, y) + \lambda g(x, y)$$

where λ is a parameter (the Lagrange multiplier).
2. Find the values of λ for which

$$\frac{\partial w}{\partial x} = \frac{\partial f}{\partial x} + \lambda \frac{\partial g}{\partial x} = 0,$$

$$\frac{\partial w}{\partial y} = \frac{\partial f}{\partial y} + \lambda \frac{\partial g}{\partial y} = 0.$$

3. Use these values of λ in the constraint condition $g(x, y) = 0$ to find the required stationary points.
4. Identify the nature of the stationary points.

Example 12.24 Determine the dimensions of the rectangle of maximum area whose sides are parallel to the coordinate axes and whose corners are constrained to lie on the ellipse $x^2 + 2y^2 = 1$.

Solution From the symmetry of the ellipse it follows that if (x, y) is a point on $x^2 + 2y^2 = 1$, then the rectangle having a corner at this point must have a side of length $2x$ parallel to the x-axis and a side of length $2y$ parallel to the y-axis. The area z of the rectangle is thus $z = 4xy$ and is, by definition, positive. The constraint condition corresponding to $g(x, y) = 0$ is

$x^2 + 2y^2 - 1 = 0$. So, making the identifications $f(x, y) = 4xy$ and $g(x, y) = x^2 + 2y^2 - 1$, we next form the function

$$w = 4xy + \lambda(x^2 + 2y^2 - 1)$$

corresponding to Eqn (12.29).

We have $w_x = 4y + 2\lambda x$ and $w_y = 4x + 4\lambda y$, and as the stationary points of w occur when $w_x = w_y = 0$, this is equivalent to requiring

$$\lambda x + 2y = 0 \qquad \text{and} \qquad x + \lambda y = 0.$$

For these homogeneous equations to have a solution, the determinant of their coefficients must vanish, giving rise to the condition

$$\begin{vmatrix} \lambda & 2 \\ 1 & \lambda \end{vmatrix} = 0 \qquad \text{or} \qquad \lambda^2 - 2 = 0.$$

Hence $\lambda = \pm\sqrt{2}$. When $\lambda = \sqrt{2}$ we have $x + \sqrt{2}y = 0$, and as this is subject to the constraint condition $x^2 + 2y^2 = 1$, it follows that the two possible solutions are $(-1/\sqrt{2}, 1/2)$, $(1\sqrt{2}, -1/2)$. When $\lambda = -\sqrt{2}$ we have $x - \sqrt{2}y = 0$ and the same reasoning leads to the two other solutions $(1/\sqrt{2}, 1/2)$, $(-1/2, -1/2)$. The extrema of $z = 4xy$ on the curve $x^2 + 2y^2 = 1$ thus occur at the four stated points. As both the area of the rectangle and the lengths of its sides must be positive, the only solution we may accept as being physically real is $(1\sqrt{2}, 1/2)$, for this implies sides of length $\sqrt{2}$ and 1. ■

The method may be extended to take account of more than two independent variables, and more than one constraint condition. First we consider the case of finding the extrema of the function of three variables

$$\Phi = f(x, y, z) \tag{12.30}$$

subject to the constraint

$$g(x, y, z) = 0. \tag{12.31}$$

Forming the total differential of (12.30) we have

$$\mathrm{d}\Phi = \frac{\partial f}{\partial x}\mathrm{d}x + \frac{\partial f}{\partial y}\mathrm{d}y + \frac{\partial f}{\partial z}\mathrm{d}z,$$

but at a stationary point $\mathrm{d}\Phi = 0$ *so* f must satisfy the condition

$$\frac{\partial f}{\partial x}\mathrm{d}x + \frac{\partial f}{\partial y}\mathrm{d}y + \frac{\partial f}{\partial z}\mathrm{d}z = 0, \tag{12.32}$$

which constrains $\mathrm{d}x$, $\mathrm{d}y$ and $\mathrm{d}z$. However, the constraint condition (12.31) imposes a further constraint on $\mathrm{d}x$, $\mathrm{d}y$ and $\mathrm{d}z$, for forming the total differential of (12.31) we find

$$\frac{\partial g}{\partial x}\mathrm{d}x + \frac{\partial g}{\partial y}\mathrm{d}y + \frac{\partial g}{\partial z}\mathrm{d}z = 0. \tag{12.33}$$

For (12.32) and (12.33) to be compatible the coefficients of dx, dy and dz must be proportional, so introducing a constant of proportionality α we see that

$$\frac{\left(\frac{\partial f}{\partial x}\right)}{\left(\frac{\partial g}{\partial x}\right)} = \frac{\left(\frac{\partial f}{\partial y}\right)}{\left(\frac{\partial g}{\partial y}\right)} = \frac{\left(\frac{\partial f}{\partial z}\right)}{\left(\frac{\partial g}{\partial z}\right)} = \alpha. \tag{12.34}$$

Results (12.34) may be rewritten in a more familiar form in which, for convenience, we set $\alpha = -\lambda$ to give

$$\frac{\partial f}{\partial x} + \lambda\frac{\partial g}{\partial x} = 0, \quad \frac{\partial f}{\partial y} + \lambda\frac{\partial g}{\partial y} = 0, \quad \frac{\partial f}{\partial z} + \lambda\frac{\partial g}{\partial z} = 0. \tag{12.35}$$

These are seen to be the conditions that the function

$$w = f(x, y, z) + \lambda g(x, y, z) \tag{12.36}$$

has a stationary point subject to the constraint $g(x, y, z) = 0$. Thus we have arrived at an immediate extension of Rule 12.1 for functions of two variables. However, just as before, once the stationary points have been located their nature must be determined by some other form of reasoning.

Rule 12.2 (Lagrange multiplier method for functions of three variables with one constraint)

It is required to find the stationary points of $f(x, y, z)$ subject to the constraint condition $g(x, y, z) = 0$.

1. Form the function

$$w = f(x, y, z) + \lambda g(x, y, z),$$

where λ is a parameter (the Lagrange multiplier).

2. Find the values of λ for which

$$\frac{\partial w}{\partial x} = \frac{\partial f}{\partial x} + \lambda\frac{\partial g}{\partial x} = 0,$$

$$\frac{\partial w}{\partial y} = \frac{\partial f}{\partial y} + \lambda\frac{\partial g}{\partial y} = 0,$$

$$\frac{\partial w}{\partial z} = \frac{\partial f}{\partial z} + \lambda\frac{\partial g}{\partial z} = 0.$$

3. Use these values of λ in the constraint condition $g(x, y, z) = 0$ to find the required stationary points.
4. Identify the nature of the stationary points.

Example 12.25 Locate and identify the stationary point of $f(x, y, z) = x^2 + 4y^2 + 3z^2$ subject to the constraint $x + y + 2z = 4$.

Solution We consider the function

$$w = f(x, y, z) + \lambda g(x, y, z),$$

where $g(x, y, z) = x + y + 2z - 4$ and find the λ for which $\partial w/\partial x = \partial w/\partial y = \partial w/\partial z = 0$ subject to $g(x, y, z) = 0$. The vanishing of the three partial derivatives leads to the three equations

$$2x + \lambda = 0, \quad 8y + \lambda = 0 \quad \text{and} \quad 6z + 2\lambda = 0.$$

These are equivalent to

$$\lambda = -2x = -8y = -3z,$$

showing that

$$x = 4y \quad \text{and} \quad z = \frac{8}{3}y.$$

Using these results in the constraint condition $x + y + 2z = 4$ then shows that the extremum occurs when $x = 48/31$, $y = 12/31$ and $z = 32/31$. The nature of this extremum is most easily identified by means of geometrical arguments. Notice first that the function f is the sum of squares, and so is non-negative. Then, since the constraint condition defines a plane, we are required to find the extremum of this non-negative function f for points on this plane. This in turn implies that the extremum of f must be a minimum, because f will increase without bound as we move arbitrarily far from the stationary point. Thus f has a minimum at $\left(\frac{48}{31}, \frac{12}{31}, \frac{32}{31}\right)$. ■

We conclude this section by mentioning the case of two constraint conditions. The full details of the proof will be omitted as they are based on the approach used to establish Rule 12.2.

Rule 12.3 (Lagrange multiplier method for functions of three variables with two constraints)

It is required to find the stationary points of $f(x, y, z)$ subject to the two constraints $g(x, y, z) = 0$ and $h(x, y, z) = 0$.

1. Form the function

$$w = f(x, y, z) + \lambda g(x, y, z) + \mu h(x, y, z)$$

where λ, μ are parameters (the two Lagrange multipliers).

2. Find the values of λ, μ for which

$$\frac{\partial w}{\partial x} = \frac{\partial f}{\partial x} + \lambda \frac{\partial g}{\partial x} + \mu \frac{\partial h}{\partial x} = 0,$$

$$\frac{\partial w}{\partial y} = \frac{\partial f}{\partial y} + \lambda \frac{\partial g}{\partial y} + \mu \frac{\partial h}{\partial y} = 0,$$

$$\frac{\partial w}{\partial z} = \frac{\partial f}{\partial z} + \lambda \frac{\partial g}{\partial z} + \mu \frac{\partial h}{\partial z} = 0.$$

3. Use these values of λ, μ in the constraint conditions $g(x, y, z) = 0$ and $h(x, y, z) = 0$ to find the required stationary points.
4. Identify the nature of the stationary points.

To see why the rule is true, notice that, arguing as for Rule 12.2, we arrive at the three equations

$$\frac{\partial f}{\partial x}\mathrm{d}x + \frac{\partial f}{\partial y}\mathrm{d}y + \frac{\partial f}{\partial z}\mathrm{d}z = 0,$$

$$\frac{\partial g}{\partial x}\mathrm{d}x + \frac{\partial g}{\partial y}\mathrm{d}y + \frac{\partial g}{\partial z}\mathrm{d}z = 0,$$

$$\frac{\partial h}{\partial x}\mathrm{d}x + \frac{\partial h}{\partial y}\mathrm{d}y + \frac{\partial h}{\partial z}\mathrm{d}z = 0.$$

The proportionality of coefficients then shows, for example, that:

$$\frac{\partial f}{\partial x} = \alpha \frac{\partial g}{\partial x}, \quad \frac{\partial f}{\partial x} = \beta \frac{\partial h}{\partial x},$$

for some constants α and β, with similar results holding for the y and z partial derivatives. Adding these results and setting $\alpha = 2\lambda$, $\beta = 2\mu$, we find

$$\frac{\partial f}{\partial x} + \lambda \frac{\partial g}{\partial x} + \mu \frac{\partial h}{\partial x} = 0,$$

which is the first of the results in step 2. The other two follow in similar fashion.

Example 12.26 Locate and identify the stationary points of the function $x^2 + y^2 + z^2$ subject to the constraints $2x^2 + 2y^2 - z^2 = 0$ and $x + y + z - 3 = 0$.

Solution Here $f(x, y, z) = x^2 + y^2 + z^2$, and we shall set $g(x, y, z) = 2x^2 + 2y^2 - z^2$ and $h(x, y, z) = x + y + z - 3$. Then by Rule 2.3 we must find the stationary points of the function

$$w = x^2 + y^2 + z^2 + \lambda(2x^2 + 2y^2 - z^2) + \mu(x + y + z - 3).$$

The equations $\partial w/\partial x = \partial w/\partial y = \partial w/\partial z = 0$ become, respectively,

$$2x + 4\lambda x + \mu = 0,$$

$$2y + 4\lambda y + \mu = 0,$$

$$2z - 2\lambda z + \mu = 0.$$

Subtracting the first two equations gives

$$2(x - y) + 4\lambda(x - y) = 0,$$

showing that either $\lambda = -\frac{1}{2}$ or $x = y$, while subtracting the first and last equations gives

$$2(x - z) + 2\lambda(2x + z) = 0.$$

Now from this last equation $\lambda = -\frac{1}{2}$ corresponds to $z = 0$, which, taken with the constraint condition $2x^2 + 2y^2 - z^2 = 0$, shows that $x = 0$ and $y = 0$. However, the point (0, 0, 0) does not satisfy the other constraint condition $x + y + z - 3 = 0$, so we must reject the condition $\lambda = -\frac{1}{2}$. If the remaining condition $x = y$ is used in the constraint condition $2x^2 + 2y^2 - z^2 = 0$ it leads to $z = \pm 2x$, while if it used in the constraint condition $x + y + z - 3 = 0$ it leads to $2x + z = 3$. Only the case $z = 2x$ can lead to a solution of $2x + z = 3$, which then gives $4x = 3$ or $x = 3/4$. As $x = y$ we also have $y = 3/4$, while as $x + y + z - 3 = 0$ we have $z = 3/2$. Thus the stationary point occurs at $(\frac{3}{4}, \frac{3}{4}, \frac{3}{2})$.

To identify the nature of this single stationary point we again use geometrical arguments. The constraint $2x^2 + 2y^2 - z^2 = 0$ is the equation of a cone, and the constraint $x + y + z - 3 = 0$ is the equation of a plane, so these intersect in a conic section (ellipse, hyperbola or parabola). The distance d from the origin of a point P in space with coordinates (x, y, z) is $d = (x^2 + y^2 + z^2)^{1/2}$, so our function $f(x, y, z) = d^2$. Since $d(x, y, z) \geq 0$, both d and d^2 attain their extrema at the same points, so the problem requires us to locate the points on a conic section whose distance from the origin is either a maximum and a minimum. As there is only one such point, namely $(\frac{3}{4}, \frac{3}{4}, \frac{3}{2})$, the curve of intersection must be a parabola, so the point will be a minimum. If the curve of intersection had been a different conic section there would have been two stationary points, one a maximum and one a minimum. ■

Example 12.27 Find the extrema of $V = x + y + z^2 + 10$ on the curve of intersection of the sphere $x^2 + y^2 + z^2 = 1$ and the plane $x + y + z = 1$, and hence identify the location and absolute maximum and minimum values of V on this curve.

Solution This problem involves two constraints, so we set

$$w = x + y + z^2 + 10 + \lambda(x^2 + y^2 + z^2 - 1) + \mu(x + y + z - 1).$$

The equations $\partial w/\partial x = 0$, $\partial w/\partial y = 0$, $\partial w/\partial z = 0$ become

$$1 + 2\lambda x + \mu = 0$$

$$1 + 2\lambda y + \mu = 0$$

$$2z + 2\lambda z + \mu = 0.$$

Subtracting the second equation from the first one gives

$$\lambda(x - y) = 0,$$

showing that either $\lambda = 0$ or $x = y$.

Considering the condition $\lambda = 0$, we see from the first equation that $\mu = -1$, when it follows from the third equation that $z = 1/2$. The first constraint condition now becomes $x^2 + y^2 = 3/4$ while the second becomes $x + y = 1/2$. Solving these equations gives $x = \frac{1}{4}(1 - \sqrt{5})$ or $x = \frac{1}{4}(1 + \sqrt{5})$. The corresponding values of y are $y = \frac{1}{4}(1 + \sqrt{5})$ and $y = \frac{1}{4}(1 - \sqrt{5})$, so one extremum occurs at the point P_1 located at $(\frac{1}{4}(1 - \sqrt{5}), \frac{1}{4}(1 + \sqrt{5}), \frac{1}{2})$ and another at the point P_2 located at $(\frac{1}{4}(1 + \sqrt{5}), \frac{1}{4}(1 - \sqrt{5}), \frac{1}{2})$.

Let us now consider the case $\lambda \neq 0$, $x = y$, when the first constraint condition becomes $2x^2 + z^2 = 1$, and the second becomes $2x + z = 1$. Solving these equations gives $x = y = 0$ or $x = y = 2/3$, when the corresponding values of z become $z = 1$ and $z = -1/3$. Thus two additional extrema occur at the points P_3 located at $(0, 0, 1)$ and P_4 located at $(\frac{2}{3}, \frac{2}{3}, -\frac{1}{3})$.

Substitution into the expression for V gives $V(P_1) = V(P_2) = 43/4 = 10.75$, while $V(P_3) = 11$ and $V(P_4) = 103/9 = 11.444$. So on the curve of intersection the greatest value of V occurs at P_4 while two identical minima occur at P_1 and P_2. ■

12.5.3 Stationary points of functions of several variables

By direct analogy with the concept of stationary points of a differentiable function of two variables, the stationary points of a differentiable function $w = f(x_1, x_2, \ldots, x_n)$ of the n independent variables $x_1, x_2, \ldots x_n$ are defined to be those points at which

$$\frac{\partial f}{\partial x_1} = \frac{\partial f}{\partial x_2} = \ldots = \frac{\partial f}{\partial x_n} = 0. \tag{12.37}$$

The concepts of maximum and minimum also extend in an obvious manner, for if P is a stationary point of $w = f(x_1, x_2, \ldots, x_n)$ and Q is a neighbouring point, we shall say that P is a local minimum of f if $f(Q) - f(P) > 0$ for all points Q in the neighbourhood of P. Similarly, we shall say that P is a local maximum of f if $f(Q) - f(P) < 0$ for all points Q in the neighbourhood of P.

We offer no further discussion of these matters aside from their application to the special problem of polynomial fitting by *least squares.* This is the name given to the process whereby a polynomial of given degree m

$$Y(x) = c_0 + c_1 x + c_2 x^2 + \ldots + c_m x^m$$

with unknown coefficients $c_0, c_1, \ldots, c_m$ is fitted to n pairs of points $(x_1, y_1), (x_2, y_2), \ldots, (x_n, y_n)$ with $n \geq$ m. The fitting is carried out in such a manner that the sum of the squares of the differences $\sum_{r=1}^{n} (Y(s_r) - y_r)^2$ is minimized.

In graphical terms this amounts to obtaining the best fit in the least-squares sense of a polynomial curve of degree m to a set of n points which are connected by an unknown functional relationship. This process is of importance in statistics when the points usually represent the result of the measurement of determinate quantities which have random errors associated with them.

Our task is to minimize the sum $E(c_0, c_1, \ldots, c_m)$ of the squares of the errors at the known points, where

$$E(c_0, c_1, \ldots, c_m) = \sum_{r=1}^{n} (c_0 + c_1 x_r + c_2 x_r^2 + \ldots + c_m x_r^m - y_r)^2.$$

The square error $E(c_0, c_1, \ldots, c_m)$ is a differentiable function of the unknown quantities $c_0, c_1, \ldots, c_m$, which we shall now regard in the role of variables.We must determine them so that $E(c_0, c_1, \ldots, c_m)$ is minimized. From our earlier remarks we see that $E(c_0, c_1, \ldots, c_m)$ will have a stationary value if

$$\frac{\partial E}{\partial c_0} = \frac{\partial E}{\partial c_1} = \ldots = \frac{\partial E}{\partial c_m} = 0.$$

We must thus solve these $m+1$ simultaneous equations for $c_0, c_1, \ldots, c_m$. Performing the indicated differentiation in the general case, we find

$$\frac{\partial E}{\partial c_p} = \sum_{r=1}^{n} 2(c_0 + c_1 x_r + c_2 x_r^2 + \ldots + c_m x_r^m - y_r) x_r^p,$$

for $p = 0, 1, \ldots, m$. Hence the numbers $c_0, c_1, \ldots, c_m$ must be obtained by solving the $m+1$ simultaneous equations

$$\sum_{r=1}^{n} (c_0 + c_1 x_r + c_2 x_r^2 + \ldots + c_m x_r^m - y_r) x_r^p = 0, \qquad (12.38)$$

for $p = 0, 1, \ldots, m$. When matters are well behaved there is only one solution to this set of equations, and as $E(c_0, c_1, \ldots, c_m)$ is essentially positive it is not difficult to verify that the corresponding solution $Y(x)$ minimizes $E(c_0, c_1, \ldots, c_m)$.

A specially simple and useful case arises when a straight line $Y = c_0 + c_1 x$ is to be fitted by means of least squares to n points. Results (12.38) simplify and the general calculation may then be organized as follows.

To fit by means of least squares the straight line

$$Y = c_0 + c_1 x$$

to the n data points $(x_1, y_1), (x_2, y_2), \ldots, (x_n, y_n)$, it is necessary to solve the equations

$$\left(\sum_{i=1}^{n} x_i\right) c_0 + \left(\sum_{i=1}^{n} x_i^2\right) c_1 = \sum_{i=1}^{n} x_i y_i$$

and

$$nc_0 + \left(\sum_{i=1}^{n} x_i\right) c_1 = \sum_{i=1}^{n} y_i$$

for c_0 and c_1.

Example 12.28 Use least squares to fit the straight line $Y = c_0 + c_1 x$ to the four data points (0, 0.2), (1, 1.1), (2, 1.8), (3, 3.2).

Solution It is best to carry out the required calculations in tabular form as follows:

i	x_i	y_i	x_i^2	$x_i y_i$
1	0	0.2	0	0
2	1	1.1	1	1.1
3	2	1.8	4	3.6
4	3	3.2	9	9.6
Σ	6	6.3	14	14.3

Thus $n = 4$, $\Sigma x_i = 6$, $\Sigma y_i = 6.3$, $\Sigma x_i^2 = 14$, $\Sigma x_i y_i = 14.3$, so we must solve

$$6c_0 + 14c_1 = 14.3$$

$$4c_0 + 6c_1 = 6.3.$$

We find $c_0 = 0.12$ and $c_1 = 0.97$, so the least-squares straight line is $Y = 0.12 + 0.97x$. ■

12.5.4 Derivation of the Laplace finite difference approximation

In Section 10.10.6 a finite difference approximation was used without proof when describing a numerical approach to the solution of the Laplace equation

$$\frac{\partial^2 u}{\partial x^2} + \frac{\partial^2 u}{\partial y^2} = 0 \qquad (12.39)$$

in some region D of the (x, y)-plane. It is the purpose of this section to show how the difference equation is derived. For simplicity, suppose the region D is the rectangular one shown in Fig. 12.6(a) with its sides of length a parallel to the y-axes and its sides of length b parallel to the x-

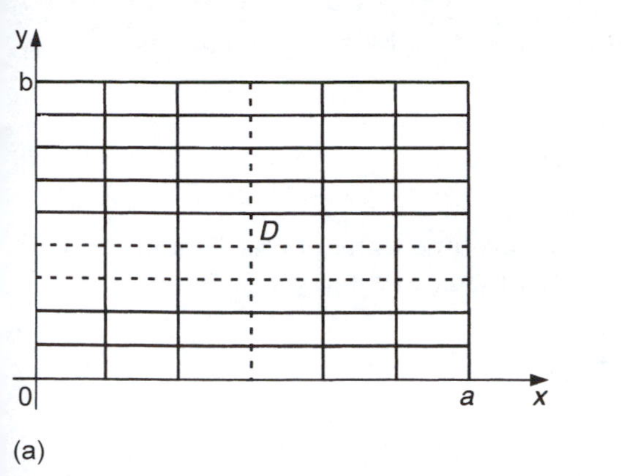

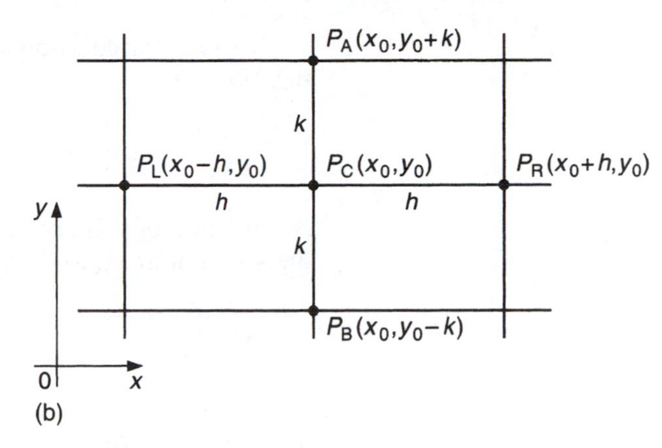

Fig. 12.6 The rectangular grid used in the approximation of partial derivatives.

axis. Then, as in section 10.10.6, a network of equally spaced grid lines parallel to the x and y-axes is superimposed on D causing it to be subdivided into a number of small rectangles, each of the same size. If there are m equi-spaced intervals of length h in the x-direction and n of length k in the y-direction, it follows that

$$h = a/m \quad \text{and} \quad k = b/n. \tag{12.40}$$

Each intersection of a pair of grid lines at the corner of a rectangle forms a mesh point at which the solution of Laplace's equation (or the Poisson equation), is to be approximated. Consider the gridlines through a typical mesh point P_c in Fig. 12.6(b) and denote the mesh points to the immediate left and right of it by P_L and P_R, and those immediately above and below it by P_A and P_B, respectively. The next step is to approximate the solution $u(x, y)$ of the Laplace equation (12.39) at points P_L and P_R in 12.6(b) by using the first three terms of a Taylor series expansion as follows (Theorem 12.11):

$$\text{Point } P_L\text{:} \quad u(x_0 - h, y_0) = u(x_0, y_0) - h\left(\frac{\partial u}{\partial x}\right)_{(x_0, y_0)} + \frac{h^2}{2!}\left(\frac{\partial^2 u}{\partial x^2}\right)_{(x_0, y_0)} + O(h^3),$$

$$\text{Point } P_R\text{:} \quad u(x_0 + h, y_0) = u(x_0, y_0) + h\left(\frac{\partial u}{\partial x}\right)_{(x_0, y_0)} + \frac{h^2}{2!}\left(\frac{\partial^2 u}{\partial x^2}\right)_{(x_0, y_0)} + O(h^3).$$

Adding these results, rearranging terms and neglecting the $O(h^3)$ terms gives

$$\left(\frac{\partial^2 u}{\partial x^2}\right)_{(x_0, y_0)} \approx \frac{1}{h^2}[u(x_0 - h, y_0) + u(x_0 + h, y_0) - 2u(x_0, y_0)]$$

or, denoting the solution u at P_c, P_L and P_R by u_c, u_L and u_R respectively, this becomes

$$\left(\frac{\partial^2 u}{\partial x^2}\right)_{P_c} \approx \frac{1}{h^2}[u_L - 2u_c + u_R]. \tag{12.41}$$

A repetition of this argument, but this time using points P_B and P_A, with the solution at P_A and P_B denoted by u_A and u_B, gives

$$\left(\frac{\partial^2 u}{\partial y^2}\right)_{P_c} \approx \frac{1}{k^2}[u_B - 2u_C + u_A]. \tag{12.42}$$

Combining results (12.41) and (12.42) shows that the expression $\partial^2 u/\partial x^2 + \partial^2 u/\partial y^2$at P_C, called the *Laplacian* at P_C, may be approximated by the finite difference scheme

$$\left(\frac{\partial^2 u}{\partial x^2} + \frac{\partial^2 u}{\partial y^2}\right)_{P_c} \approx \frac{1}{h^2}[u_L - 2u_C + u_R] + \frac{1}{k^2}[u_B - 2u_C + u_A]. \tag{12.43}$$

On account of this, Laplace's equation may be approximated at P_C by the finite difference scheme

$$\frac{1}{h^2}[u_L - 2u_C + u_R] + \frac{1}{k^2}[u_B - 2u_C + u_A] = 0. \tag{12.44}$$

A result analogous to (12.44) applies to each internal mesh point of D.

When this finite difference approximation is applied to each of the mesh points in D it leads to $(m-1)(n-1)$ linear simultaneous equations for the $(m-1)(n-1)$ unknown approximations to u, with values at the mesh points adjacent to the boundary incorporating the known values of u on the boundary. The solution is found by solving this inhomogeneous set of equations. When the rectangles formed by the grid are square, so that $h = k$, finite difference scheme (12.44) simplifies to

$$u_L + u_R + u_B + u_A - 4u_C = 0, \tag{12.45}$$

which is the finite difference scheme used in section 10.10.6.

Problems

Section 12.1

12.1 Write down the general term in each of the following series:

(a) $\frac{3}{4} + \frac{5}{4^2} + \frac{7}{4^3} + \ldots;$

(b) $\frac{2}{5} + \frac{4}{8} + \frac{6}{11} + \frac{8}{14} + \ldots;$

(c) $1 + \frac{1.3}{1.4} + \frac{1.3.5}{1.4.7} + \frac{1.3.5.7}{1.4.7.10} + \ldots;$

(d) $\frac{1}{2}+\frac{1}{3}+\frac{1}{2^2}+\frac{1}{3^2}+\frac{1}{2^3}+\frac{1}{3^3}+\ldots;$

(e) $\frac{1}{3}+\frac{1}{8}+\frac{1}{15}+\frac{1}{24}+\ldots.$

12.2 The series $a+ar+ar^2+\ldots+ar^n+\ldots$ is called either the *geometric progression* or the *geometric series* with initial term a and common ratio r. Denote by S_n the sum of its first n terms so that

$$S_n=\sum_{m=0}^{n-1} ar^m.$$

By considering the difference $S_n - rS_r$ prove that

$$S_n=a\left(\frac{1-r^n}{1-r}\right).$$

If $r<1$ deduce that

$$\sum_{m=0}^{\infty} ar^m=\frac{a}{1-r}.$$

What is the remainder R_n of the series after n terms?

12.3 Sum the following infinite series and find their remainders after n terms:

(a) $2+1+\frac{2}{4}+\frac{1}{5}+\frac{2}{4^2}+\frac{1}{5^2}+\frac{2}{4^3}+\frac{1}{5^3}+\ldots;$

(b) $2-1+\frac{2}{4}-\frac{1}{5}+\frac{2}{4^2}-\frac{1}{5^2}+\frac{2}{4^3}-\frac{1}{5^3}+\ldots.$

12.4 State which of the following series is divergent by Theorem 12.2:

(a) $\left(1-\frac{1}{2}\right)+\left(1+\frac{1}{2^2}\right)+\left(1-\frac{1}{2^3}\right)+\ldots+\left(1+\frac{(-1)^n}{2^n}\right)+\ldots;$

(b) $\frac{1}{3}+\frac{4}{9}+\frac{9}{19}+\ldots+\frac{n^2}{2n^2+1}+\ldots;$

(c) $\frac{1}{1}+\frac{2^2}{2^2}+\frac{3^2}{6^2}+\ldots+\frac{n^2}{(n!)^2}+\ldots;$

(d) $1+r+r^2+\ldots+r^n+\ldots;$

(e) $1+\frac{1}{\sqrt{2}}+\frac{1}{\sqrt{3}}+\ldots+\frac{1}{\sqrt{n}}+\ldots.$

12.5 Prove the divergence of the harmonic series by obtaining simple underestimates for the sums of each of the indicated groupings of its terms and showing that they themselves form a series which is obviously divergent.

$$1+\tfrac{1}{2}+\underbrace{\tfrac{1}{3}+\tfrac{1}{4}}_{2\text{ terms}}+\underbrace{\tfrac{1}{5}+\tfrac{1}{6}+\tfrac{1}{7}+\tfrac{1}{8}}_{4\text{ terms}}+\underbrace{\tfrac{1}{9}+\tfrac{1}{10}+\ldots+\tfrac{1}{16}}_{8\text{ terms}}+\ldots.$$

12.6 Use the comparison test to classify the following series as convergent or divergent:

(a) $1+(\tfrac{2}{3})+(\tfrac{2}{3})^4+(\tfrac{2}{3})^9+\ldots+(\tfrac{2}{3})^{n^2}+\ldots;$

(b) $\tfrac{3}{4}+\tfrac{4}{7}+\tfrac{5}{12}+\ldots+\left(\dfrac{n^2+2}{n^2+3}\right)+\ldots;$

(c) $1+\dfrac{1}{3^2}+\dfrac{1}{5^3}+\dfrac{1}{7^4}+\ldots+\dfrac{1}{(2n-1)^n}+\ldots;$

(d) $\dfrac{1}{3-1}+\dfrac{1}{3^2-2}+\dfrac{1}{3^3-3}+\ldots+\dfrac{1}{3^n-n}+\ldots.$

12.7 Use the integral test to determine the convergence or divergence of the following series:

(a) $\dfrac{1}{2\ln 2}+\dfrac{1}{3\ln 3}+\dfrac{1}{4\ln 4}+\ldots+\dfrac{1}{n\ln n}+\ldots;$

(b) $\dfrac{1}{2(\ln 2)^2}+\dfrac{1}{3(\ln 3)^2}+\dfrac{1}{4(\ln 4)^2}+\ldots+\dfrac{1}{n(\ln n)^2}+\ldots.$

Where appropriate, estimate the remainder after six terms.

12.8 Classify the following convergent series as conditionally convergent or absolutely convergent:

(a) $1-\dfrac{1}{\sqrt{2}}+\dfrac{1}{\sqrt{3}}-\ldots+\dfrac{(-1)^{n+1}}{\sqrt{n}}+\ldots;$

(b) $\dfrac{-3}{5}+\left(\dfrac{5}{9}\right)^2-\left(\dfrac{7}{13}\right)^3+\ldots+(-1)^n\left(\dfrac{2n+1}{4n+1}\right)^n+\ldots;$

(c) $1-\dfrac{1}{3^2}+\dfrac{1}{5^2}-\dfrac{1}{7^2}+\ldots+(-1)^{n+1}\dfrac{1}{(2n-1)^2}+\ldots.$

12.9 Test the following series for convergence by the ratio test:

(a) $\dfrac{2}{1}+\dfrac{2.5}{1.5}+\dfrac{2.5.8}{1.5.9}+\ldots+\dfrac{2.5.8\ldots(3n-1)}{1.5.9\ldots(4n-3)}+\ldots;$

(b) $\dfrac{1}{\sqrt{3}}+\dfrac{3}{3}+\dfrac{5}{(\sqrt{3})^3}+\ldots+\dfrac{2n-1}{(\sqrt{3})^n}+\ldots;$

(c) $\dfrac{2}{5}+\dfrac{5}{14}+\dfrac{10}{29}+\ldots+\dfrac{n^2+1}{3n^2+2}+\ldots;$

(d) $1+\dfrac{2^{30}}{2!}+\dfrac{3^{30}}{3!}+\ldots+\dfrac{n^{30}}{n!}+\ldots;$

(e) $\dfrac{1}{10}+\dfrac{2!}{10^2}+\dfrac{3!}{10^3}+\ldots+\dfrac{n!}{10^n}+\ldots.$

12.10 Test the following series for convergence by the nth root test:

(a) $1-\left(\dfrac{4}{5}\right)^2-\left(\dfrac{6}{8}\right)^3+\left(\dfrac{8}{11}\right)^4+\ldots+(-1)^{n(n-1)/2}\left(\dfrac{2n}{3n-1}\right)^n+\ldots;$

(b) $\dfrac{2}{3}+\left(\dfrac{3}{7}\right)^3+\left(\dfrac{4}{11}\right)^5+\ldots+\left(\dfrac{n+1}{4n-1}\right)^{2n-1}+\ldots;$

(c) $\dfrac{1}{1+a}+\left(\dfrac{2}{2+a}\right)^{2^2}+\left(\dfrac{3}{3+a}\right)^{3^2}+\ldots+\left(\dfrac{n}{n+a}\right)^{n^2}+\ldots.$

12.11 Test the following alternating series for convergence:

(a) $\dfrac{1}{2^2}-\dfrac{1}{4^2}+\dfrac{1}{6^2}-\dfrac{1}{8^2}+\ldots+(-1)^{n+1}\dfrac{1}{(2n)^2}+\ldots;$

(b) $\dfrac{1}{\sqrt{5}}-\dfrac{1}{\sqrt[3]{5}}+\dfrac{1}{\sqrt[4]{5}}-\ldots+(-1)^{n+1}\dfrac{1}{\sqrt[n+1]{5}}+\ldots;$

(c) $\dfrac{1}{3}-\dfrac{1}{2}\left(\dfrac{1}{3}\right)^2+\dfrac{1}{3}\left(\dfrac{1}{3}\right)^3-\ldots+(-1)^{n+1}\dfrac{1}{n}\left(\dfrac{1}{3}\right)^n+\ldots.$

Where appropriate, estimate the remainder after ten terms.

12.12 Indicate the fault in the following contradictory argument. Let

$$S_n=\sum_{r=1}^{n}\frac{1}{r(r+2)},$$

then

$$S_n=\sum_{r=1}^{n}\left(\frac{1}{2r}-\frac{1}{2(r+2)}\right)$$

and for $n=2m+1$ (n odd)

$$S_{2m+1} = \frac{3}{4} - \frac{1}{2(m+2)},$$

while for $n = 2m (n \text{ even})$

$$S_{2m} = \frac{3}{4} - \frac{1}{2(m+1)} - \frac{1}{2(m+2)}.$$

Hence for all n it is true that

$$S = \lim_{n\to\infty} S_n = \tfrac{3}{4}.$$

However, alternatively,

$$S = \sum_{n=1}^{\infty}\left(\frac{n+2}{4n} - \frac{n+4}{4(n+2)}\right) = \tfrac{3}{4} + \tfrac{1}{2} + \sum_{m=3}^{\infty}\frac{m+2}{3m} - \sum_{m=3}^{\infty}\frac{m+2}{3m} = \frac{5}{4}.$$

Section 12.2

12.13 Find the interval of convergence of each of the following power series:

(a) $x + \frac{2!}{2^2}x^2 + \frac{3!}{3^3}x^3 + \ldots + \frac{n!}{n^n}x^n + \ldots;$

(b) $x + \frac{(2!)^2}{4!}x^2 + \frac{(3!)^2}{6!}x^3 + \ldots + \frac{(n!)^2}{(2n)!}x^n + \ldots;$

(c) $1 + 5x + \frac{5^2x^2}{2!} + \frac{5^3x^3}{3!} + \ldots + \frac{5^nx^n}{n!} + \ldots;$

(d) $1 + \frac{x}{1} + \frac{x^2}{2^2} + \frac{x^3}{3^3} + \ldots + \frac{x^n}{n^n} + \ldots;$

(e) $\frac{x+7}{9} + \frac{(x+7)^3}{2.9^2} + \frac{(x+7)^5}{3.9^3} + \ldots + \frac{(x+7)^{2n+1}}{n.9^n} + \ldots.$

12.14 Find the radius of convergence of each of the following power series and verify that the differentiated series has the same radius of convergence:

(a) $\sum_{n=1}^{\infty}\frac{(x-3)^n}{(2n+1)2^n};$

(b) $\sum_{n=0}^{\infty}\frac{(-1)^n(x-2)^n}{(2n+3)\sqrt{(n+1)}};$

(c) $\displaystyle\sum_{n=1}^{\infty} \frac{(x+9)^n}{(n+1)^2}$.

12.15 Find the power series representation of $\arcsin x$ by considering the integral

$$\arcsin x = \int_0^x \frac{\mathrm{d}t}{\sqrt{(1-t^2)}},$$

and find its radius of convergence. Does the power series converge at the end points of its interval of convergence?

12.16 Find a series representation for the *elliptic integral*

$$\int_0^{\pi/2} \sqrt{(1-k^2\sin^2\phi)}\,\mathrm{d}\phi.$$

For what values of k is the resulting series convergent?

12.17 By using the power series representation for $\sin x$, find a series representation for the definite integral

$$\int_0^c \frac{\sin x}{x}\,\mathrm{d}x.$$

For what values of c is the resulting series convergent?

12.18 By using the power series representation for $\ln(1-x)$, find a power series representation for the definite integral

$$\int_0^x \frac{\ln(1-t)}{t}\,\mathrm{d}t.$$

For what values of x is the resulting power series convergent?

Section 12.3

12.19 Derive the Maclaurin series expansion of each of the following functions, together with its interval of convergence:

(a) $x\mathrm{e}^{-2x}$;

(b) $\cosh(x^2/2)$;

(c) $(1+\mathrm{e}^x)^3$;

(d) $\dfrac{1}{1+3x+2x^2}$; (*Hint*: Use partial fractions.)

(e) $\ln[x+\sqrt{(1+x^2)}]$.

12.20 Show that $f(x) = \exp(\arccos x)$ satisfies the relation

$$(1-x^2)f'' - xf' - f = 0.$$

Use Leibniz's theorem to obtain the result

$$(1-x^2)f^{(n+2)} - (2n+1)xf^{(n+1)} - (n^2+1)f^{(n)} = 0.$$

Hence write down the first three terms of Maclaurin series for $\exp(\arccos x)$.

12.21 Show that $f(x) = \sin(k \arcsin x)$ satisfies the relation

$$(1-x)f^{(n+2)} - (2n+1)xf^{(n+1)} - (n^2-k^2)f^{(n)} = 0.$$

Hence write down the first four terms of the Maclaurin series for $\sin(k \arcsin x)$ and show that it reduces to the single term x when $k=1$.

12.22 Taking $n=1$, use Taylor's theorem with a remainder to determine whether the following functions increase or decrease with x for $x>0$:
(a) $x - \tanh x$; (b) $\arctan x - x$; (c) $\ln(1+x) - x$.

12.23 By writing $\sin 4x = \sin(3x+x)$ and $\sin 2x = \sin(3x-x)$, and expanding the results, find an expression for $\sin x \cos 3x$ in terms of $\sin 2x$ and $\sin 4x$. Use this expression and the method of Example 12.19(b) to find the Maclaurin series for $\sin x \cos 3x$.

12.24 How many terms need to be taken in the series

$$\ln(1+x) = x - \frac{x^2}{2} + \frac{x^3}{3} - \ldots,$$

in order to determine ln 2 to within an error of
(a) 0.01
(b) 0.001.

12.25 Determine the value of the integral

$$\int_0^{0.8} \frac{\sin x}{x}\,dx$$

accurate to within an error of 0.0001.

12.26 Expand $f(x,y) = x^2 + 3xy^2 + y$ about the point (1, 1).

12.27 Write down the first three terms of the Taylor series expansion of $f(x,y) = e^x \sin y$ about the origin.

12.28 Write down the first three terms of the Taylor series expansion of $f(x,y) = e^{x+y}$ about the point (1, −1).

Section 12.4

12.29 Evaluate the limit

$$\lim_{x\to 2} \frac{\sin^2 \pi x}{2e^{x/2} - xe}.$$

12.30 Given that $f(x, y) = e^{xy} - 1 - y(e^x - 1)$, and that b is neither equal to 1 nor 0, evaluate

$$\lim_{x\to 0} \frac{f(x, a)}{f(x, b)}.$$

12.31 Evaluate the limit

$$\lim_{x\to 0} \frac{(4^x - 2^x) - x(\log 4 - \log 2)}{x^2}.$$

12.32 Evaluate the limit

$$\lim_{x\to 0}\left(\frac{\cot x - 1/x}{\coth x - 1/x}\right).$$

12.33 Evaluate the limit

$$\lim_{x\to \pi/4} \frac{\sec^2 x - 2\tan x}{1 + \cos 4x}.$$

12.34 Locate and identify the nature of the stationary points of the function

$$f(x, y) = x^3 + 3xy^2 - 15x - 12y.$$

Find the functional value of any maxima or minima.

12.35 Locate and identify the stationary points of the function

$$f(x, y) = 2x^4 - 3x^2y^2 + y^4 + 8x^2 + 3y^2.$$

12.36 Locate and identify the stationary point of the function

$$f(x, y) = x^3y^2(6 - x - y)$$

which lies in the first quadrant $x > 0$, $y > 0$.

12.37 Locate and identify the stationary points of the function

$$f(x, y) = x^3 + y^3 - 3xy.$$

12.38 By considering the proof of Theorem 12.14 show that the conditions stated in (c) are equivalent to, and may be replaced by, $f_{yy}(x_0, y_0) < 0$ for a local maximum, and $f_{yy}(x_0, y_0) > 0$ for a local minimum.

12.39 If $2s$ denotes the perimeter of a triangle with sides of length a, b, c, then its area A is given by the formula

$$A = \sqrt{[s(s - a)(s - b)(s - c)]}.$$

Show that for a given perimeter, the equilateral triangle has maximum area. (*Hint:* Consider the function $F(a,b,s)=s(s-a)(s-b)(a+b-s)$.)

12.40 Locate and identify the extrema of $f(x,y)=6-4x-3y$ subject to the constraint $x^2+y^2=1$. Interpret the problem and your result in geometrical terms.

12.41 Locate and identify the stationary point of $f(x,y)=x^2+y^2$ subject to the constraint $(x/2)+(y/3)=1$.

12.42 Locate and identify the stationary points of $f(x,y)=\cos^2 x+\cos^2 y$ subject to the constraint $y-x=\frac{1}{4}\pi$.

12.43 Locate and find the minimum value of $f(x,y,z)=x^2+y^2+z^2+1$ subject to the constraint $2x+y-z=3$.

12.44 Locate and find the minimum value of $f(x,y,z)=4x^2+y^2+5z^2+3$ subject to the constraint $2x+3y+4z=12$.

12.45 Locate and find the maximum value of $f(x,y,z)=3x^2y^2z^2$ subject to the constraint $x^2+y^2+z^2=1$.

12.46 Find the point on the curve of intersection of the ellipsoid $x^2+y^2+2z^2=1$ and the plane $x+y+z=1$ that is closest to the origin.

12.47 Find the two points on the curve of intersection of the cone $z^2=x^2+y^2$ and the plane $x+y-z+2=0$ which are nearest to and farthest from the origin.

12.48 Find the largest volume of a cylinder of length L and radius r if the sum of its length and the circumference of its cross-section equals C.

12.49 An airline transports freight in rectangular boxes subject to the requirement that the sum of the length of the box and the perimeter of its cross-section normal to the length is 120 inches. Find the dimensions of the box with maximum volume.

12.50 Find the values of three positive numbers if their product is to be maximized subject to the requirement that their sum equals a given positive number C.

12.51 Find the extrema of $4xy$ given that $x^2/9+y^2/16=1$.

12.52 Use the method of least squares to fit the straight line $Y=a+bx$ to the four points (1, 2.5), (2.5, 2.5), (3, 4), (4, 5).

12.53 Use the method of least squares to fit the straight line $Y=a+bx$ to the four points (0, 0.1), (1, 1.1), (2, 1.6), (3, 3.3).

13 Differential equations and geometry

Differential equations model many physical situations, and the form of a solution depends on the structure of the differential equation involved. This chapter shows how a general solution to a differential equation must contain arbitrary constants of integration equal in number to the order of the equation – that is, to the order of the highest derivative involved.

It is shown that to obtain a solution to a specific problem it is necessary to specify auxilliary conditions that lead to the determination of the values to be assigned to the arbitrary constants in the general solution. The notions of direction fields and isoclines for first-order equations are introduced and used to provide qualitative information about the set of all solutions to a differential equation once its form has been specified. The notion of a direction field is used later to deduce a simple numerical method for the solution of an initial value problem.

13.1 Introductory ideas

Special examples of differential equations have already been encountered; for example, the one that gave rise to the exponential function

$$\frac{dy}{dx} - y = 0,$$

and the one that gave rise to the sine and cosine functions

$$\frac{d^2y}{dx^2} + y = 0.$$

It is now appropriate to make a systematic study of certain differential equations that are both useful and of frequent occurrence. We shall begin by examining a number of simple examples to illustrate the basic ideas.

Any equation involving one or more derivatives of a differentiable function of a single independent variable is called an *ordinary differential equation*. The following related equations taken from elementary dynamics are familiar examples:

$$\frac{d^2x}{dt^2} = g \qquad \text{(acceleration equation)}$$

$$\frac{dx}{dt} = u + gt \qquad \text{(velocity equation).}$$

They describe, respectively, the acceleration and velocity of a particle falling freely under the action of gravity. Here g is the acceleration due to gravity, x is the distance of the particle from a fixed origin in its line of motion at the elapsed time t, and u is the initial velocity of the particle. In these simple equations the dependent variable is represented by the displacement x and the independent variable by the time t. The integration of these equations is straightforward. The velocity equation is in fact the integral of the acceleration equation with the arbitrary constant of integration set equal to the initial velocity u, since the velocity equation must describe the velocity at the start of the motion (when $t = 0$).

The first step in a systematic study of commonly occurring ordinary differential equations, aimed at producing general methods of solution whenever possible, is a straightforward classification of the equations. This we achieve by associating two numbers with each equation which we shall refer to as its *order* and its *degree*. We define the order of an ordinary differential equation to be the order of the highest derivative appearing in the equation, and the degree to be the exponent to which this highest derivative is raised when fractions and radicals involving y or its derivatives have been removed from the equation. The notion of degree is only applicable when a differential equation has a simple algebraic structure allowing such a classification to be made. Thus both the simple dynamical equations just described are of degree 1, but the acceleration equation is of second order whereas the velocity equation is of first order. These are, in fact, examples of a specially important class of equations known as *linear differential equations*.

All linear differential equations are characterized by the fact that the dependent variable and its derivatives only occur with degree 1, while the coefficients multiplying them are either constants or functions of the independent variable. Thus of the following three second-order differential equations, only the first two are linear, since the last involves the nonlinear product $y(\mathrm{d}y/\mathrm{d}x)$. In general, differential equations that are not linear are termed *nonlinear*.

$$\frac{\mathrm{d}^2y}{\mathrm{d}x^2} + 3\frac{\mathrm{d}y}{\mathrm{d}x} + 2y = 0, \qquad \text{(linear)}$$

$$x^2\frac{\mathrm{d}^2y}{\mathrm{d}x^2} + x\frac{\mathrm{d}y}{\mathrm{d}x} + (x^2 - n^2)y = 0, \qquad \text{(linear)}$$

$$\frac{\mathrm{d}^2y}{\mathrm{d}x^2} + y\frac{\mathrm{d}y}{\mathrm{d}x} - y = 0. \qquad \text{(non-linear)}$$

The classification of a more complicated differential equation is illustrated by the following example, involving both fractions and radicals, in which k is a constant:

$$\frac{(y'')^{3/2}}{y + (y'')^2} = k.$$

Clearing the fractions and radicals gives rise to the ordinary differential equation

$$k^2y''^4 - y''^3 + 2k^2yy''^2 + k^2y^2 = 0,$$

showing that the order is 2 and the degree is 4.

If y', y'', ..., $y^{(n)}$ respectively denote successive derivatives, up to order n, of a differentiable function $y(x)$ with independent variable x, then a general nth-order ordinary differential equation has the form

$$F(x, y, y', \ldots, y^{(n)}) = 0, \tag{13.1}$$

where F is an arbitrary function of the variables involved.

Definition 13.1 (solution of a differential equation) A solution of the ordinary differential Eqn (13.1) is a function $y = \phi(x)$ that is differentiable a suitable number of times in some interval I containing the independent variable x, and which has the property that

$$F(x, \phi(x), \phi'(x), \ldots, \phi^{(n)}(x)) = 0 \tag{13.2}$$

for all x belonging to I.

Notice that it is important to define the interval I since the differential equation does not necessarily describe the solution for unrestricted values of the argument x.

Thus a solution of the velocity equation just used as an example would be a differentiable function $x = \phi(t)$ defined for some interval I of time t with the property that

$$\phi'(t) - gt - u = 0, \tag{13.3}$$

for all t in the interval I. In this case I would be of finite size since the particle could not fall for an unlimited time without being arrested by contact with the ground, after which the ordinary differential equation giving rise to solution $x = \phi(t)$ would no longer be valid.

The prefix *ordinary* is used to describe differential equations involving only one dependent and one independent variable, in contrast with *partial differential equations*, as encountered in section 10.6 in the context of the Laplace equation which involve partial derivatives, and so have at least two independent variables and may also contain more than one dependent variable. Normally, when the type of differential equation being discussed is clear from the context, the adjectives ‘ordinary’ and ‘partial’ are omitted.

It is possible to develop the theory of differential equations in considerable generality, but our approach, as mentioned before, will be to examine a number of useful special forms of equation. We shall, however, first examine a few of the ways in which important forms of ordinary differential equation may arise.

13.2 Possible physical origin of some equations

At this stage it will be useful to illustrate some typical forms of differential equation, showing their manner of derivation from physical situations. We shall consider a number of essentially different physical problems and in each case take the discussion as far as the derivation of the governing differential equation.

Example 13.1 Experiment has shown that certain objects falling freely in air from a great height experience an air resistance that is proportional to the square of the velocity of the body. Let us determine the differential equation that describes this motion, and for convenience take our origin for the time t at the start of the motion. We shall assume that the body has a constant mass m and that at time t the velocity of fall is v, so that the air resistance at time t becomes λv^2 units of force, where λ is a constant of proportionality.

Now by definition, the acceleration a is the rate of change of velocity, so that $a = \mathrm{d}v/\mathrm{d}t$ and, since the body has constant mass m, it immediately follows from Newton's second law that the force accelerating the body is $m(\mathrm{d}v/\mathrm{d}t)$. To obtain the equation of motion this force must now be equated to the other forces acting vertically downwards which are, taking account of the sign, the weight mg and the resistance $-\lambda v^2$, where the negative sign is necessary because the resistance opposes the downward motion. The equation of motion is thus

$$m\frac{\mathrm{d}v}{\mathrm{d}t} = mg - \lambda v^2,$$

or, dividing throughout by the constant m,

$$\frac{\mathrm{d}v}{\mathrm{d}t} = g - \frac{\lambda}{m}v^2, \quad \blacksquare$$

The equation in Example 13.1 is a special case of a differential equation in which the variables are separable. A general differential equation of this form involving the independent variable x and the dependent variable y can be written in either of the two general forms:

$$\frac{\mathrm{d}y}{\mathrm{d}x} = M(x).N(y) \tag{13.4}$$

or

$$P(x)Q(y)\mathrm{d}x + R(x)S(y)\mathrm{d}y = 0. \tag{13.5}$$

Example 13.2 In many simple chemical reactions the conversion of a raw material to the desired product proceeds under constant conditions of temperature and pressure at a rate directly proportional to the mass of raw material remaining at any time. If the initial mass of the raw material is Q, and the mass of the product chemical at time t is q, then the unconverted mass remaining at time t is $Q - q$. Then, if $-k(k > 0)$ denotes the proportionality factor governing the rate of the reaction, the reaction conversion rate

$\mathrm{d}(Q-q)/\mathrm{d}t$ must be equal to $-k$ times the unconverted mass $Q-q$. The desired reaction rate equation thus has the form

$$\frac{\mathrm{d}}{\mathrm{d}t}(Q-q) = -k(Q-q),$$

where the minus sign has been introduced into the definition of k to allow for the fact that $Q-q$ decreases as t increases. ■

Example 13.3 A simple closed electrical circuit contains an inductance L and a resistance R in series, and a current i is caused to flow by the application of a voltage $V_0 \sin \omega t$ across two terminals located between the resistance and inductance. The equation governing this current i may be obtained by a simple application of Kirchhoff's second law, which tells us that the algebraic sum of the drops in potential around the circuit must be zero. Thus, since the driving potential is $V_0 \sin \omega t$ and the changes in potential across the inductance and resistance are in the opposite sense to i and so are, respectively, $-L(\mathrm{d}i/\mathrm{d}t)$ and $-Ri$, it follows that

$$V_0 \sin \omega t - L\frac{\mathrm{d}i}{\mathrm{d}t} - Ri = 0,$$

or

$$\frac{\mathrm{d}i}{\mathrm{d}t} + \frac{R}{L}i = \frac{V_0}{L}\sin \omega t. \quad ■$$

The final equations of Examples 13.2 and 13.3 are both specially simple cases of linear first-order differential equations. If the dependent variable is denoted by y and the independent variable by x, then all linear first-order differential equations have the general form

$$\frac{\mathrm{d}y}{\mathrm{d}x} + P(x)y = Q(x). \tag{13.6}$$

Example 13.4 Mechanical vibrations occur frequently in physics and engineering, and they are usually controlled by the introduction of some suitable dissipative force. A typical situation might involve a mass m on which acts a restoring force proportional to the displacement x of the mass from an equilibrium position, and a resistance to motion that is proportional to the velocity of the mass. Such a system, which to a first approximation could represent a vehicle suspension involving a spring and shock absorber (damper), is often tested by subjecting it to a periodic external force $F \cos \omega t$ in order to simulate varying road conditions. In this situation the displacement x would represent the movement of the centre of gravity of the vehicle about an equilibrium position as a result of passage of the vehicle along a road with a sinusoid profile. If the resisting force F_{d} has a proportionality constant k, and the restoring force F_{r} has a proportionality constant λ, then $F_{\mathrm{d}} = k(\mathrm{d}x/\mathrm{d}t)$ and $F_{\mathrm{r}} = \lambda x$. Applying Newton's second law, as in Example 13.1, and equating forces acting on the system, we obtain the equation of motion

$$m\frac{d^2x}{dt^2} = F\cos\omega t - k\frac{dx}{dt} - \lambda x,$$

or

$$\frac{d^2x}{dt^2} + \frac{k}{m}\frac{dx}{dt} + \frac{\lambda}{m}x = \frac{F}{m}\cos\omega t. \quad ■$$

The equation in Example 13.4 is a particular case of a linear constant-coefficient second-order differential equation; equations of this type have the general form

$$\frac{d^2y}{dx^2} + a\frac{dy}{dx} + by = f(x), \tag{13.7}$$

where x is the independent variable, y the independent variable, and a and b are constants. Equations (13.6) and (13.7) are said to be *inhomogeneous* when they contain a term involving only the independent variable; otherwise they are said to be *homogeneous*. The differential equation of Example 13.2 is thus homogeneous of order 1 with dependent variable $(Q-q)$, while that of Example 13.4 is inhomogeneous of order 2; both are linear and involve constant coefficients. If in Example 13.2 the temperature of the reaction were allowed to vary with time, then in general the velocity constant k of the reaction would become a function of the time t, and the equation would assume the homogeneous form of Eqn (13.6) with a variable coefficient.

The special importance of the types of differential equations singled out here lies in their frequent occurrence throughout the physical sciences. We shall later proceed with a systematic study of solution methods for these standard forms, together with other common cases of interest.

13.3 Arbitrary constants and initial conditions

If we consider the simple differential equation

$$\frac{d^2x}{dt^2} = g, \tag{13.8}$$

then a single integration with respect to time gives $dx/dt = gt$ as a possible first integral. This is certainly a solution of Eqn (13.8) in the sense defined in Eqn (13.2), but it is not the most general solution since

$$\frac{dx}{dt} = c_1 + gt, \tag{13.9}$$

where c_1 is an arbitrary constant, is also a solution. This specific example illustrates the general result that in order to obtain the most complete form of solution, each integral involved in the solution of a differential equation must be interpreted as an indefinite integral. When maximum generality is sought, the result containing arbitrary constants of integration is termed the *general solution* of the differential equation. It is, therefore, important that when obtaining the general solution of a differential equation, an arbitrary constant should be introduced immediately after each integration. Thus the general solution of Eqn (13.8), which is obtained after two integrations, is

$$x = c_2 + c_1 t + \tfrac{1}{2} g t^2, \tag{13.10}$$

where c_2 is another arbitrary constant.

These arbitrary constants may be given definite values, and a *particular solution* obtained, if the solution is required to satisfy a set of conditions, at some starting time $t = t_0$, equal in number to the order of the differential equation. If, for example, Eqn (13.8) describes the acceleration of a body falling under the influence of gravity, and air resistance may be neglected, then Eqn (13.10) is the general solution of the problem of the position of the body at time t. In the event that the body started to fall with an initial velocity u at time $t = 0$, it follows from Eqn (13.9) that $c_1 = u$. Similarly, if the body was at position $x = x_0$ at time $t = 0$, it follows from Eqn (13.10) that $c_2 = x_0$, and so the particular solution corresponding to the conditions $x = x_0$, $\mathrm{d}x/\mathrm{d}t = u$ at $t = 0$ is

$$x = x_0 + ut + \tfrac{1}{2} g t^3. \tag{13.11}$$

General starting conditions of this type are known as *initial conditions* by analogy with time-dependent problems such as this in which the solution evolves away from some known initial state. On occasion it is convenient to write initial conditions in an abbreviated form which we illustrate by repeating the initial conditions that gave rise to solution Eqn (13.11):

$$x\bigg|_{t=0} = x_0, \qquad \frac{\mathrm{d}x}{\mathrm{d}t}\bigg|_{t=0} = u.$$

Another way of indicating initial conditions employs the notation for functions. Thus,

$$y(\alpha) = \beta$$

means $y = \beta$ when $x = \alpha$ and, similarly,

$$y'(\alpha) = \beta$$

means $\mathrm{d}y/\mathrm{d}x = \beta$ when $x = \alpha$. More generally still,

$$y^{(n)}(\alpha) = \beta$$

means $\mathrm{d}y^n/\mathrm{d}x^n = \beta$ when $x = \alpha$.

Something more of the role of arbitrary constants may be appreciated if they are eliminated by differentiation from a general expression describing a family of curves in order that the differential equation describing the family may be obtained. Suppose, for example, that a general two-parameter family of curves is defined by the expression

$$y(x) = A\cosh 2x + B\sinh 2x,$$

where A and B are arbitrary constants (which we now regard as parameters). Then differentiation shows that

$$y' = 2(A\sinh 2x + B\cosh 2x) \quad \text{and} \quad y'' = 4(A\cosh 2x + B\sinh 2x),$$

from which it follows that elimination of A and B gives the differential equation

$$y'' = 4y.$$

This is the differential equation that has the two-parameter family of curves $y(x)$ as its general solution.

We should now see whether, having found a particular solution of a differential equation with given initial conditions, this is indeed the only possible solution. This is called the *uniqueness problem* for the solution and is obviously important in physical applications. To answer uniqueness questions for general classes of differential equations is difficult, but in the case of the dynamical problem just discussed a simple argument will suffice.

Let $v = \mathrm{d}x/\mathrm{d}t$ denote the velocity of the body so that Eqn (13.8) takes the form $\mathrm{d}v/\mathrm{d}t = g$. Now suppose that some other function w is also a solution of Eqn (13.8), satisfying the same initial conditions, so that $\mathrm{d}w/\mathrm{d}t = g$ and $v = w = u$ at $t = 0$. Then, setting $V = v - w$, it is easily established by subtraction of the two linear differential equations that the differential equation satisfied by the difference between the two postulated solutions is $\mathrm{d}V/\mathrm{d}t = 0$, thereby showing that $V = \text{constant}$. However, as the initial conditions require that $V = 0$, it follows at once that $w \equiv v$. The velocity is thus uniquely determined by the differential equation and the initial condition. To complete the proof that the position of the body is also uniquely determined it is only necessary to apply the foregoing argument to the velocity equation $\mathrm{d}x/\mathrm{d}t = u + gt$ obtained by direct integration of $\mathrm{d}v/\mathrm{d}t = g$.

It is not always necessary, or indeed possible, to prescribe only initial conditions for a differential equation, as we now illustrate by reformulating the previous example.

We have seen that the velocity v of the body is uniquely determined by Eqn (13.9) once it has been specified at some given instant of time. Similarly, when the velocity is known, the position is uniquely determined by Eqn (13.10) once it has been specified at some given instant of time. Velocity and position were specified at the same instant of time in the initial value problem just discussed; we now illustrate an alternative problem that could equally well have been considered.

As the solution (13.10) implies the result (13.9), it would be quite permissible to determine a particular solution by requiring the body to be at the positions x_0 and x_1 at the respective times t_0 and t_1. These conditions would enable the determination of the arbitrary constants c_1 and c_2 in order that this is possible and would, of course, completely determine the velocity. Conditions such as these that are imposed on the solution of a differential equation at two different values of the independent variable are called *two-point boundary conditions*. This name is derived from the fact that in many important applications the conditions to be imposed are prescribed at two physical boundaries associated with the problem.

In the simple initial value problem discussed here the question of the existence of a solution was never in doubt since we were able to find the general solution by direct integration. This is not usually the case; with more complicated differential equations the question to be asked is whether a solution exists and, if so, whether it is unique.

To illustrate this, let us again consider Eqn (13.8), but this time with different two-point boundary conditions. At first sight it might appear reasonable to specify the velocity rather than the position at two different times, but a moment's reflection shows that this is not possible. This arises because Eqn (13.9) determines the velocity, and unless the two preassigned velocities were in agreement with this equation there could obviously be no solution satisfying such two-point boundary conditions.

Having made this point we shall not pursue it further in this first course on differential equations.

13.4 First-order equations – direction fields and isoclines

We now describe two geometrical methods by which the *general behaviour* of solutions of the first-order equation

$$\frac{dy}{dx} = f(x, y) \tag{13.12}$$

may be determined. Arising out of these will come a simple numerical method for finding the solution of Eqn (13.12) subject to an initial condition $y(a) = b$; that is, the requirement that the solution curve is such that $y = b$ when $x = a$. Because the solution of a differential equation involves integration, the solution curves are also called *integral curves*.

Recalling the interpretation of the derivative dy/dx as the gradient of the curve $y = y(x)$, we see that at a point (α, β) on the solution $y = y(x)$ of Eqn (13.12) the gradient is $f(\alpha, \beta)$. Expressed differently, this says that a line with slope $f(\alpha, \beta)$ passing through the point (α, β) must be *tangent* to the solution curve at that point. The line will make an angle $\theta = \arctan f(\alpha, \beta)$ with the positive x-axis.

This process associates with each point in the (x, y)-plane a line that is tangent to the solution curve passing through that point. As a result a *direction field* of tangent lines is defined throughout the (x, y)-plane. To construct such a direction field in practice a grid of points is taken in the region $a \leq x \leq b$ and $c \leq y \leq d$, and through each grid point a small line segment of the associated tangent line is drawn, to which is added an arrow to show the direction in which y changes as x increases. The examples that follow illustrate typical direction fields and also representative solution curves for the purpose of comparison.

Example 13.5 Construct the direction field for the equation

$$\frac{dy}{dx} = y + \tfrac{1}{2}x + e^{-x}$$

in the region $-1.5 \leq x \leq 2.5$ and $-2.5 \leq y \leq 1.5$.

Solution The direction field is given in Fig. 13.1(a), while representative solution curves are shown in Fig. 13.1(b). ■

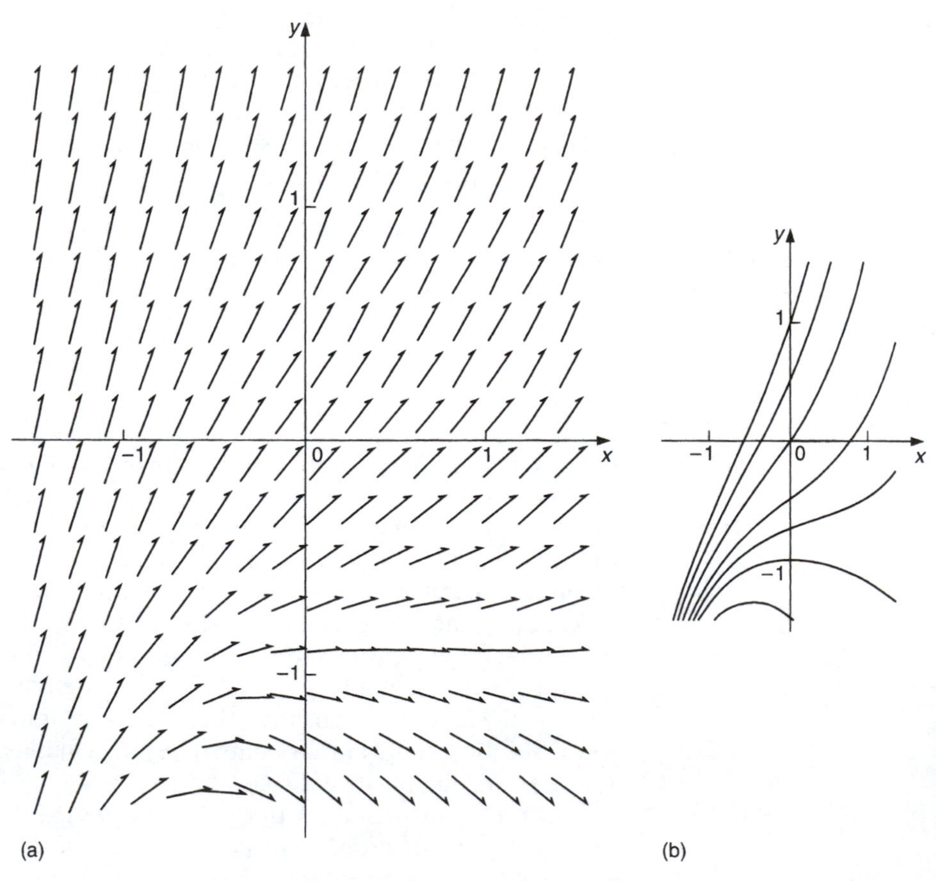

Fig. 13.1 (a) Direction field for $y' = y + \frac{1}{2}x + e^{-x}$; (b) representative solution curves.

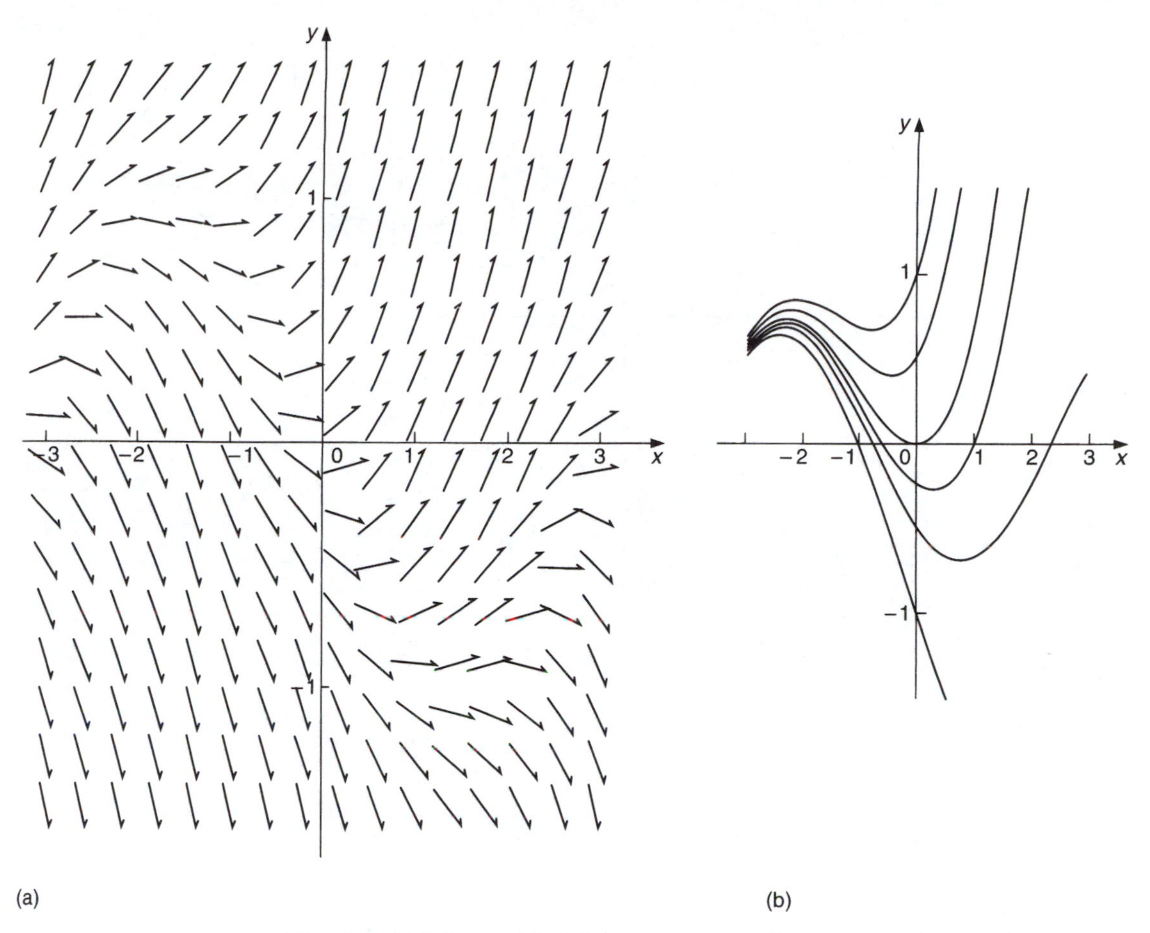

Fig. 13.2 (a) Direction field for $y' = y + \sin x$; (b) representative solution curves.

Example 13.6 Construct the direction field for the equation

$$\frac{dy}{dx} = y + \sin x$$

in the region $-3 \leq x \leq 3$ and $-1.5 \leq y \leq 1.5$.

Solution The direction field is given in Fig. 13.2(a), while representative solution curves are shown in Fig. 13.2(b). ∎

The two previous examples involved linear differential equations, so for the last example of a direction field we consider a nonlinear differential equation.

Example 13.7 Plot the direction field for the nonlinear differential equation

$$\frac{dy}{dx} = x - y^2$$

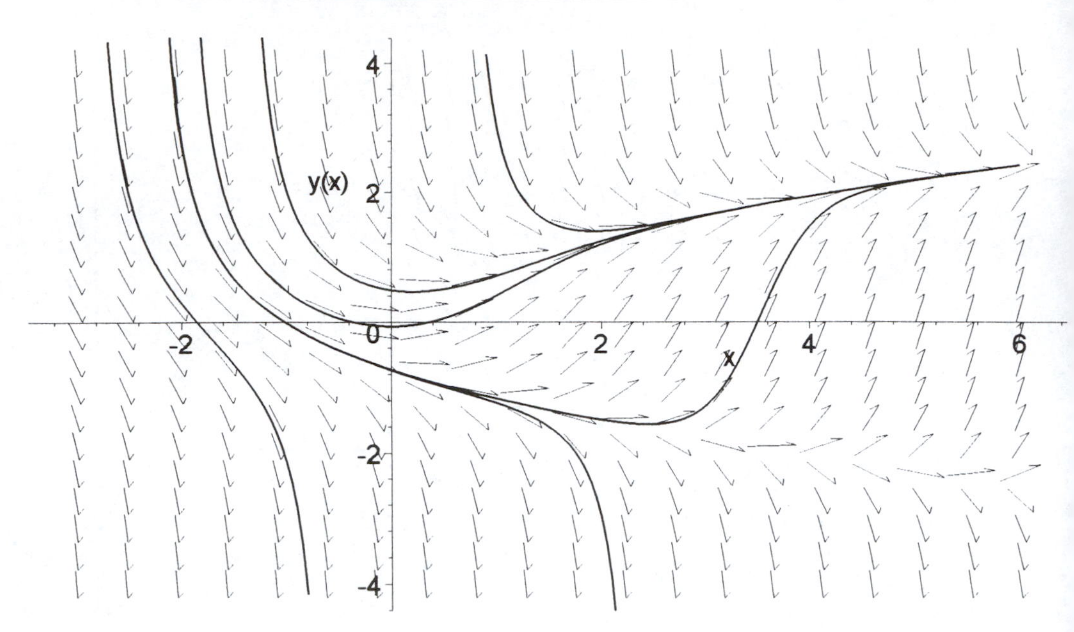

Fig. 13.3 The direction field and solution curves for $y' = x - y^2$.

over the region $-3 \leq x \leq 6$, $-4 \leq y \leq 4$, and superimpose on it some representative solution curves.

Solution The nonlinearity in the differential equation is due to the term y^2 on the right. The required computer plot is shown in Fig. 13.3, together with six representative solution curves. Inspection of the direction field shows why in the first quadrant one set of solution curves all become asymptotic to one another as x increases, while in the other set all solutions decrease without bound, without ever becoming asymptotic to one another. It is this type of general qualitative information about solutions provided by direction fields which often proves useful in applications. ■

The task of constructing a direction field is considerable unless a computer package is available for the purpose. However, the idea underlying a direction field can be put to use to provide a simple method for solving initial value problems for Eqn (13.12). The method, illustrated in Fig. 13.4, simply involves using a tangent line approximation to the true solution curve in order to advance from an initial point P_0 at (x_0, y_0) to a nearby point P_1 located at (x_1, y_1). The new point (x_1, y_1) is then taken as an approximation to the next point on the solution curve, and thereafter the process is repeated as often as necessary. We have the simple result

$$y_1 = y_0 + \Delta,$$

where

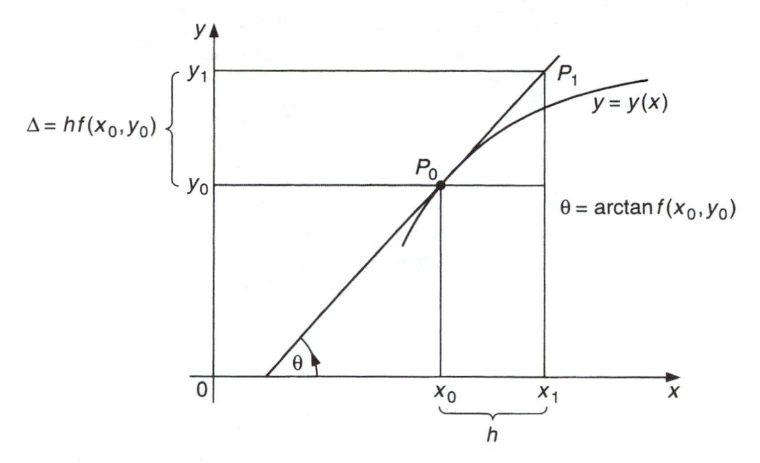

Fig. 13.4 Numerical solution of $y' = f(x, y)$, $y(x_0) = y_0$.

$$\Delta = h \tan \theta = hf(x_0, y_0), \tag{13.13}$$

and so

$$y_1 = y_0 + hf(x_0, y_0). \tag{13.14}$$

Setting $x_1 = x_0 + h$, we see that the next point on the approximate solution curve is (x_1, y_1). To advance the calculation further, (x_1, y_1) is taken as the next 'initial condition' and the calculation is repeated. When this process is carried out n times with step length b, the solution will be advanced from (x_0, y_0) to (x_n, y_n), with $x_n = x_0 + nh$. There is no necessity for all steps to be of equal length, though this is often taken to be the case by using a uniform *step length* h.

The algorithm for computing the approximate solution to the initial value problem

$$\frac{dy}{dx} = f(x, y) \quad \text{with initial condition } y(x_0) = y_0, \tag{13.15}$$

and n equal steps of length h, known as *Euler's method*, is

$$y_{i+1} = y_i + hf(x_i, y_i), \quad \text{with } i = 0, 1, 2, \ldots, n-1. \tag{13.16}$$

The polygonal approximation to the solution curve that results is called the *Cauchy polygon* approximation. The following example illustrates the process.

Example 13.8 Find the approximate value of y at $x = 0.5$, given that $y' = xy$ and $y(0) = 1$.

Solution If we take five equal sub-intervals, so that $h = 0.1$, the results of the calculations will be as follows:

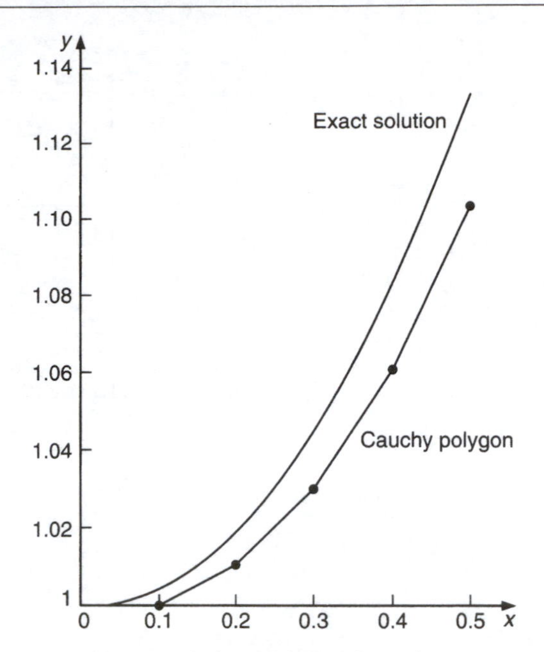

Fig. 13.5 Comparison of exact solution and Cauchy polygon.

i	x_i	y_i	$hf(x_i, y_i)$	Exact solution $e^{(1/2)x^2}$
0	0	1	0	1
1	0.1	1	0.01	1.0050
2	0.2	1.01	0.0202	1.0202
3	0.3	1.0302	0.0309	1.0460
4	0.4	1.0611	0.0424	1.0833
5	0.5	1.1035		1.1331

A comparison of the third column with the final column, which tabulates the exact solution $y = e^{x^2/2}$, demonstrates the relatively poor accuracy obtainable by this simple approach. The approximate value $y(0.5) = 1.1035$ obtained by Euler's method is seen to be already 2.6% low, and attempts to determine y for values of $x > 0.5$ would result in a very rapid growth of error. The Cauchy polygon is compared with the exact solution in Fig. 13.5. In a later chapter we shall show how a simple modification to this method will produce a considerable improvement by taking some account of the curvature of the solution curve. ■

An alternative approach to the determination of the general behaviour of solution curves is related to the direction field approach and requires much less computation, but it is only easy to interpret in simple cases.

The approach is as follows. If in the differential equation

$$\frac{dy}{dx} = f(x, y) \tag{13.17}$$

we set $\mathrm{d}y/\mathrm{d}x = k$, with k a constant of our choosing, then the curve

$$f(x, y) = k \tag{13.18}$$

in the (x, y)-plane has the property that wherever it is interesected by a solution curve the slope of the curve will be k, and so will be at an angle $\theta = \arctan k$ to the positive x-axis. Such a curve is called an *isocline*; it is not a solution curve.

Different values of k define different isoclines, and it is quite possible that no isocline exists for some values of k. If a representative set of isoclines is drawn, added to each of which are small line segments inclined at the angle $\theta = \arctan k$ to the positive x-axis, a solution curve through a given initial point can be sketched by drawing a curve with the appropriate tangent at each point of intersection with an isocline. This approach is illustrated in the following examples.

Example 13.9 Consider the simple differential equation $y' = x + 1$, which is easily seen to have the general solution $y = \frac{1}{2}x^2 + x + C$. Setting $y' = k$ then shows that the isoclines of this differential equation are the lines $x = k - 1$. Representative isoclines are illustrated in Fig. 13.6 as the full vertical lines. Short inclined lines have been added to these isoclines to indicate the direction of the tangents to the integral curves that intersect them.

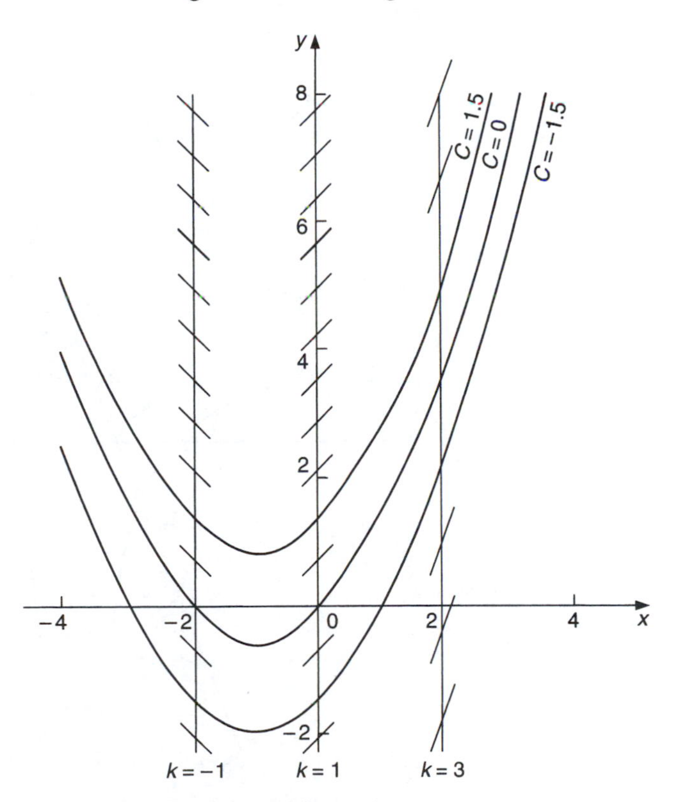

Fig. 13.6 Isoclines, direction field and integral curves.

Three solution curves, represented by curved full lines, have been drawn to show the relationship between isoclines, the tangents or gradients associated with isoclines and the integral curves themselves. The pattern of these tangents associated with the isoclines shows the direction taken by integral curves.

Figure 13.6 also serves to illustrate the geometrical analogue of Euler's method – namely, to use a map of the isoclines, each marked with their associated tangents indicating the direction field of the integral curves, in order to trace a solution that starts from a given point and always intersects each isocline at an angle equal to the gradient associated with it. ■

It is easily seen that with the simple equation $y' = x + 1$ there are no points in the finite (x, y)-plane at which the gradient is either infinite or ambiguous. The next two examples show more complicated situations involving characteristic behaviour of direction fields and solution curves at special points.

Example 13.10 In the case of the equation $y' = (1 - y)/(1 + x)$, the general solution is given by $y = 1 + C/(1 + x)$. As always, the isoclines are determined by setting $y' = k$ in the differential equation, thereby giving rise to the

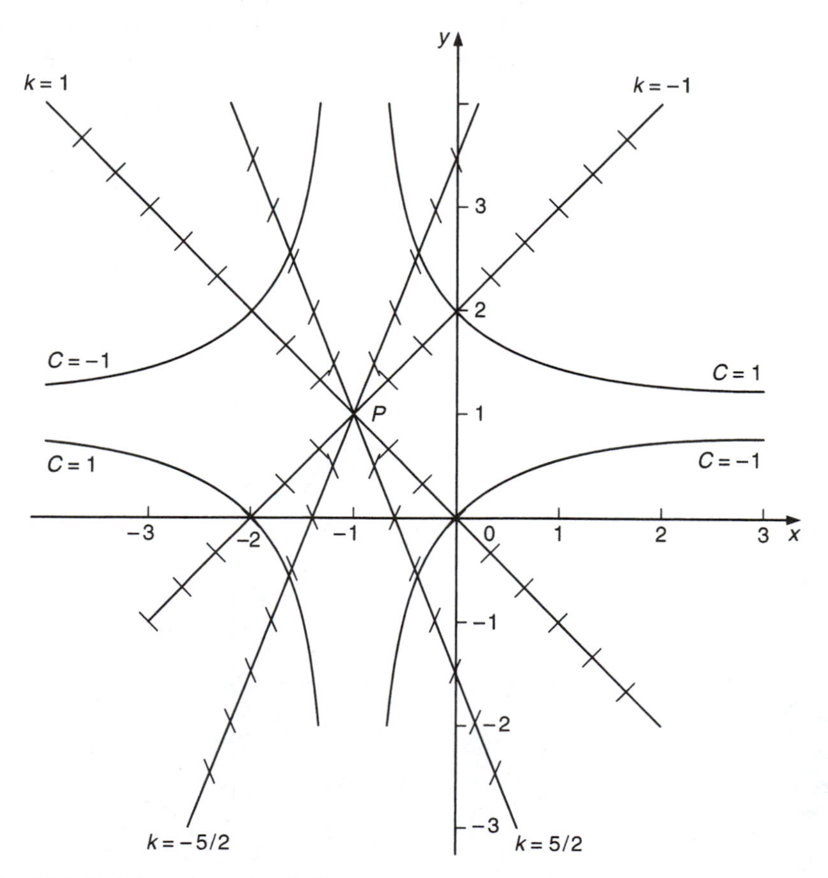

Fig. 13.7 Saddle point at $(-1, 1)$.

equation $1 - y = k(1 + x)$. This is simply a family of straight lines through the point P in the (x, y)-plane with the coordinates $(-1, 1)$. Solution curves for $C = \pm 1$ are shown in Fig. 13.7, together with representative isoclines. It is easily seen from the differential equation that the line $y = 1$ is both a degenerate solution curve and an isocline corresponding to $k = 0$. As all the isoclines pass through the point P, it is obviously a special point in the direction field. We shall call such a point P at which the derivative y' is indeterminate a *singular point* of the direction field of the differential equation in question. The hyperbola-like pattern of the integral curves in the vicinity of P is characteristic of a certain important form of behaviour, and any family of solution curves having this property is said to have a *saddle point* at P. ■

Example 13.11 A direction field of a different kind is provided by the differential equation $y' = 2y/x$, which has the lines $y = \frac{1}{2}kx$ as its isoclines and the curves $y = Cx^2$ as its solution curves. Their interrelationship is illustrated in Fig. 13.8, which also shows quite clearly that the singular point at the origin is of an essentially different kind than that of the previous example. Again the isoclines all pass through this point but, whereas in Example 13.8 there was only one degenerate solution curve through the point P, in the present case every solution curve passes through the singular point. The parabola-like behaviour of the integral curves in the vicinity of the origin is characteristic of a different form of singularity, and solution curves with this general property are said to have a *node* at the origin. ■

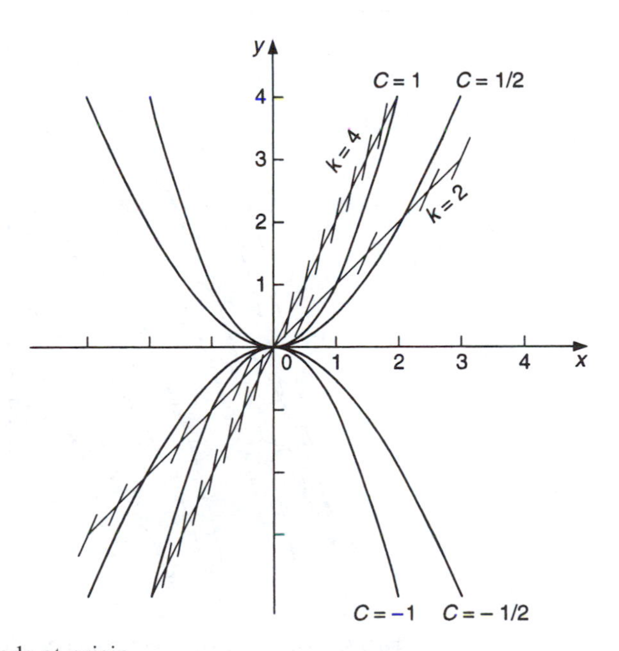

Fig. 13.8 Node at origin.

The last two examples also serves to illustrate that initial conditions to differential equations may not always be prescribed arbitrarily without reference to the equation in question, since there may either be no solution or an infinity of solutions satisfying a differential equation and arbitrarily prescribed initial conditions. For example, no integral curve passes through the point $(-1, 2)$ in Fig. 13.7, whereas every integral curve passes through the point $(0, 0)$ in Fig. 13.8. Since, in the first case, solutions have infinities along the line $x = -1$, and in the second case, the direction field is indeterminate at $(0, 0)$, this suggests that for a unique solution to exist, corresponding to initial conditions at a point in the

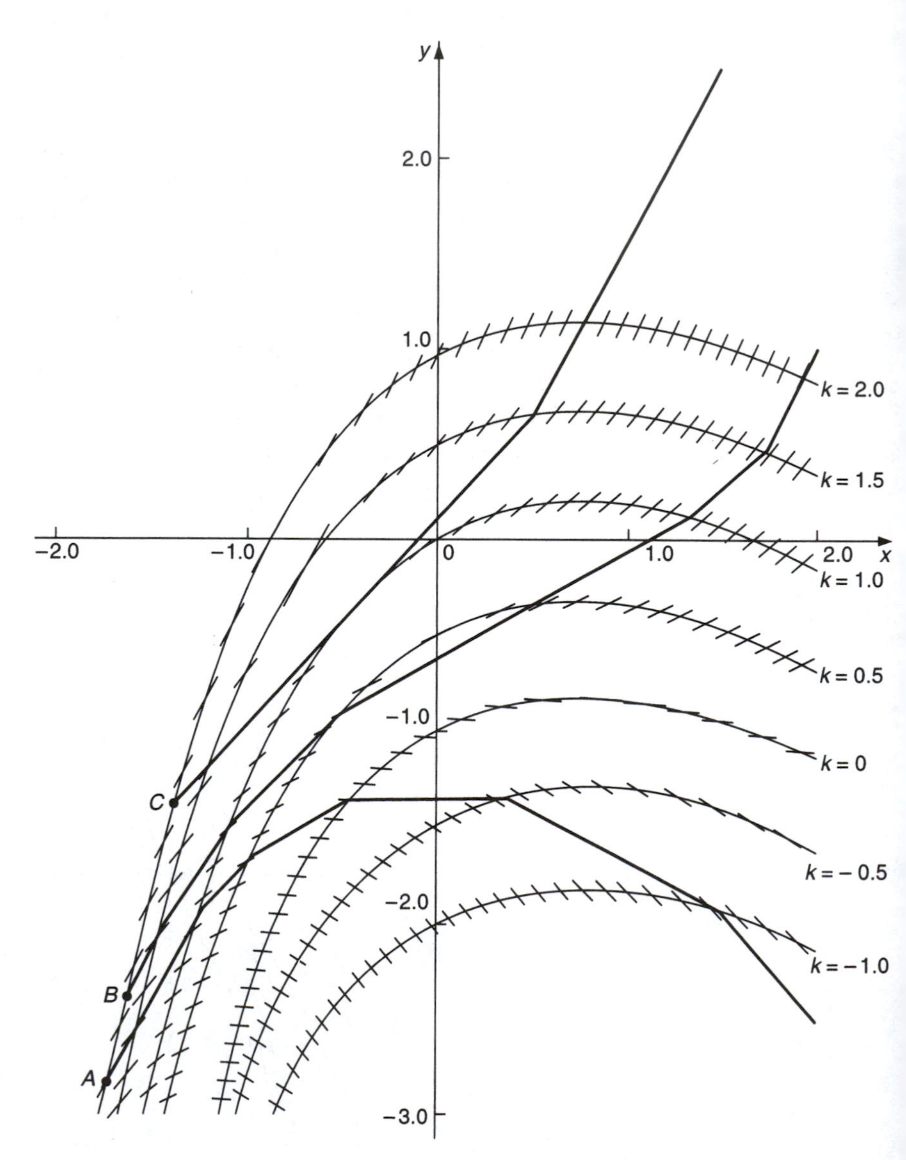

Fig. 13.9 Isoclines and approximate solutions to $y' = y + \frac{1}{2}x + e^{-x}$.

(x, y)-plane, that point must not be such that an infinite slope occurs there or that the slope is indeterminate.

Example 13.12 Draw isoclines for the equation

$$\frac{dy}{dx} = y + \tfrac{1}{2}x + e^{-x}$$

whose direction field and representative solution curves were given in Fig. 13.1. Choose an initial point on an isocline, and by producing the tangent line associated with it until it reaches the next isocline, and then repeating the process from isocline to isocline, construct a polygonal approximation to the exact solution with the same initial point. Using two different initial points, construct two other polygonal approximations to exact solutions.

Solution The isoclines are determined by the equation

$$y = k - \tfrac{1}{2}x - e^{-x}.$$

Isoclines for $k = -1.0, -0.5, 0, 0.5, 1.0, 1.5$ and 2.0 are shown in Fig. 13.9 in which a polygonal solution approximating the exact solution (Cauchy polygon approximation) has been constructed for each of the initial points A, B and C. ■

13.5 Orthogonal trajectories

The geometrical argument that led to the notion of an isocline helps to provide a simple solution to the problem of the determination of trajectories *orthogonal* to a given family of curves. Expressed another way, this is asking how, if a family of curves determined by a parameter α is specified in the form

$$F(x, y, \alpha) = 0, \tag{13.19}$$

may another family of curves

$$G(x, y, \beta) = 0 \tag{13.20}$$

with parameter β be determined so that each curve of the family G intersects all curves of the family F at right angles.

Questions of this nature are common in many branches of science and engineering. Similar questions occur in magnetism and heat conduction. We shall now solve the general problem we have formulated for plane *one-parameter families* of curves, as systems of the type (13.19) and (13.20) are usually termed.

In section 13.3 we have already seen how, by differentiation, an arbitrary constant α may be eliminated from a one-parameter family of curves of the form (13.19), thereby giving rise to the differential equation that characterizes all the curves of the family. In general, this will have the form

$$y' = f(x, y) \tag{13.21}$$

which, as we have just seen, then defines the direction field of all members of the family of integral curves represented by Eqn (13.19). Elementary coordinate geometry tells us that the product of the gradients of orthogonal straight lines must equal -1 and so, at every point of intersection of curves from the orthogonal families F and G, the product of the gradients of the tangents to these curves must also equal -1. Consequently, the differential equation of the trajectories of family G that are orthogonal to those of F is

$$\frac{\mathrm{d}y}{\mathrm{d}x} = \frac{-1}{f(x,\ y)}. \tag{13.22}$$

Integration of this equation then yields the required family of orthogonal curves.

Example 13.13 Let us determine the trajectories orthogonal to the family of parabolas $y^2 = ax$, in which a is an arbitrary parameter. In the notation of Eqn (13.19) this is equivalent to $F(x, y, a) = 0$, with $F(x, y, a) \equiv y^2 - ax$ and the parameter α set equal to a. First, we must obtain the differential equation characterizing this family of curves, by differentiation and elimination of a. Differentiating $F(x, y, a) = 0$ with respect to x gives $2yy' = a$, which on elimination of a by use of the original equation gives the differential equation $y' = y/2x$. The next step is to use this differential equation of the family of parabolas to determine the differential equation of the family of curves forming the orthogonal trajectories. As the gradient of the parabola through the general point (x, y) is $y/2x$, we see by Eqn (13.22) that the gradient of the orthogonal trajectory through the same point must be $-2x/y$. Thus from Eqn (13.22) the differential equation of the trajectories orthogonal to the parabolas is seen to be $y' = -2x/y$. This equation is one of the type already mentioned in which the variables are separable, and the final step in the determination of the actual family of orthogonal trajectories is the integration of this equation. We shall postpone discussing the actual method to be used until the next chapter. Nevertheless, it is easily verified by differentiation that the solution is the family of ellipses $x^2 + \frac{1}{2}y^2 = C^2$, where C^2 is a positive parameter. ■

Problems

Section 13.1

13.1 Determine the order and degree of each of the following equations:

(a) $x^2y''' + y'^2 + y = 0$;

(b) $y'^2 + 2xy = 0$;

(c) $\dfrac{(2y'' + x)^2}{(xy + 1)} = 3$;

(d) $\left(\dfrac{y'' + xy}{y'' + 3}\right)^{3/2} = (2y'' + xy' + y)$;

(e) $y''' + 2y''^2 + 6xy = e^x$.

Section 13.2

13.2 Determine the differential equation of the curve that has the property that the ratio of the length of the interval of the x-axis contained between the intercepts of the tangent and ordinate to a general point on the curve has a constant value k.

13.3 Obtain the differential equation governing the motion of a particle of mass m that is projected vertically upwards in a medium in which the resistance is λ times the square of the particle velocity v.

13.4 Derive the differential equation which describes the rate of cooling of a body at a temperature T on the assumption that the rate of cooling is k times the excess of the body temperature above the ambient temperature T_0 of the surrounding air. This is known as *Newton's law of cooling*, and it is a good approximation for small temperature differences.

Section 13.3

13.5 Eliminate the arbitrary constants in the following expressions to determine the differential equations for which they are the general solutions:

(a) $\frac{1}{2}x^2 + y^2 = C^2$; (b) $y = Cx + C^3$;

(c) $x^3 = C(x + y)^2$; (d) $\log\left(\dfrac{1 + x}{y}\right) = Cy$;

(e) $y = Ae^x + Be^{2x}$; (f) $y = (C + Dx)e^{2x}$.

13.6 Determine whether the following expressions satisfy the associated differential equations for all real x:

(a) $y = x^2 - 2x$; $xy'' + y = x^2$;

(b) $y = \dfrac{1}{x}$; $y' = y^2 + \dfrac{2}{x^2}$;

(c) $y = \sin 3x + \cos 3x$; $y'' + 9y = 0$;

(d) $y = e^{-x}(A\cos 2x + B\sin 2x)$; $y'' + 2y' + 5y = 0$;

(e) $y = 2x(e^x + C)$; $xy' - y = x^2e^x$;

(f) $y = A\cos x + B\sin x - \frac{1}{2}x\cos x$; $y'' + y = \sin x$.

Section 13.4

13.7 Determine whether the following differential equations have the associated functions as their solution over the stated intervals:

(a) $y' = x$, $y(x) = \begin{cases} \frac{1}{2}x^2 + 1, & x \leq 1 \\ \frac{1}{2}x^2 + 2, & 1 < x \end{cases}$ $(-1 \leq x \leq 1)$;

(b) $y'' - 9y = 0$, $y(x) = A\cosh 3x + B\sin 3x \quad (-\infty < x < \infty)$;

(c) $y' = 4$, $y(x) = \begin{cases} 4x+2, & x<0 \\ 4x-2, & x>0 \end{cases} \quad (-2 \le x \le -\frac{1}{2})$;

(d) $y' + y^2 = 0$, $y(x) = \dfrac{1}{1+x} \quad (-2 \le x \le 0)$;

(e) $x^4 y'' + y = 0$, $y(x) = x\sin\dfrac{1}{x} \quad (0 \le x \le 1)$;

(f) $y' - 3x^2 y = 0$, $y(x) = e^{x^3} \quad (-\infty < x \le 10)$.

13.8 Taking interval $\Delta x = 0.2$, use Euler's method to determine $y(1)$, given that $y' + y = 0$ and $y(0) = 1$. Compare your results with the exact solution $y = e^{-x}$. Construct the Cauchy polygon.

13.9 Taking intervals $\Delta x = 0.1$, use Euler's method to determine $y(1)$, given that $dy/dx = (x^2 + y)/x$ and $y(0.5) = 0.5$. Compare your results with the exact solution $y = \frac{1}{2}x + x^2$. Construct the Cauchy polygon.

13.10 Taking intervals $\Delta x = 0.2$, use Euler's method to determine $y(1)$, given that $y' = y + e^{-x}$ and $y(0) = 0$. Compare your results with the exact solution $y = \sinh x$. Construct the Cauchy polygon.

13.11 Repeat Problem 13.10 taking $\Delta x = 0.1$, and determine the improvement in accuracy.

13.12 Draw the isoclines, and sketch the direction fields for the differential equations in Problems 13.8 to 13.10.

13.13 Use the isoclines and direction field for the differential equation $dy/dx = (x^2 + y)/x$ to deduce the behaviour of the integral curves close to the origin. What form of singularity occurs at the origin?

13.14 Using isoclines and the associated direction field, determine the approximate value of $y(1)$, given that $y' = y + e^{-x}$ and $y(0) = -1$. Compare your result with the exact solution $y = -\cosh x$.

13.15 It can be shown that if the functions $f(x, y)$ and $f_y(x, y)$ are continuous in a rectangle R of the (x, y)-plane containing the point (x_0, y_0), then, for some sufficiently small positive number h, there exists a unique solution $y = y(x)$ of the differential equation $y' = f(x, y)$ that is defined on the interval $x_0 - h \le x \le x_0 + h$ and is such that $y(x_0) = y_0$. Use this result to determine whether the following differential equations have a unique solution passing through all points (x_0, y_0) of the given regions:

(a) $\dfrac{dy}{dx} = \left(\dfrac{x^2 + y}{x}\right) \quad (-1 \le x \le 3;\ -10 \le y \le 10)$;

(b) $\dfrac{dy}{dx} = \left(\dfrac{3y + 1}{2x}\right) \quad (\frac{1}{2} \le x \le 1;\ 2 \le y \le 5)$;

(c) $\dfrac{dy}{dx} = \left(\dfrac{2y}{x+y}\right)$ $(-3 \leq x \leq 3;\ y \leq -\frac{1}{2}x)$;

(d) $y' = y + e^{-x}$ (all finite points of the (x, y)-plane).

Section 13.5

13.16 Sketch the orthogonal trajectories in Example 13.12.

13.17 Derive the differential equation describing the trajectories that are orthogonal to the family of curves $y = ax$. Show, by differentiation, that the family of circles $x^2 + y^2 = C^2$, with C^2 a positive parameter, is the general solution describing the orthogonal trajectories.

13.18 Derive the differential equations of the families of trajectories that are orthogonal to the following differential equations:
(a) $x^2y' - y = y^2$;
(b) $y' \sin x + y + 1 = 0$;
(c) $e^x y' + e^{-x}(x + y) = 0$.

Supplementary computer problems

The following problems require the use of a computer direction field package to plot the direction field and some representative solution curves for the given differential equations.

13.23 $y' = x - \frac{1}{2}y^2$, for $-3 \leq x \leq 4$, $-3 \leq y \leq 3$.

13.24 $y' = x(1 + x)y$, for $-4 \leq x \leq 2$, $-3 \leq y \leq 3$.

13.25 $y' = 2x + \sin y - \frac{1}{3}y^2$, for $-3 \leq x \leq 3$, $-3 \leq y \leq 3$.

13.26 $y' = 2 + x^2 - y^2$, for $-3 \leq x \leq 3$, $-3 \leq y \leq 3$.

14 First-order differential equations

This chapter is concerned with the development of methods for the solution of the simplest classes of first-order differential equations for which solutions may always be obtained in an analytical form. The equations involved for which methods of solution are developed are those with separable variables, homogeneous equations, exact equations and first-order linear differential equations. Of these the general solution of the first-order linear differential equation is especially important because of its many applications. The Bernoulli equation is also considered because, although it is a nonlinear differential equation, a simple change of variable reduces it to a first-order linear differential equation so that its general solution can always be found.

The form of general solutions is, of course, dependent on the integrals involved being capable of evaluation in terms of elementary functions. The last section of the chapter indicates by means of a simple example how direct deductions can often be made from a differential equation without the necessity to solve it completely.

14.1 Equations with separable variables

The class of differential equations in which the variables are separable was identified in section 13.2, where it was remarked that either of the two forms

$$\frac{dy}{dx} = M(x)N(y) \tag{14.1}$$

and

$$P(x)Q(y)dx + R(x)S(y)dy = 0 \tag{14.2}$$

may arise. Here, Eqn (14.2) must, of course, be interpreted in the sense of differentials already defined elsewhere.

As the name implies, such equations are solved by rewriting so that functions of the variable x, together with its differential dx, and functions of the variable y, together with its differential dy, occur on opposite sides of the equation. The general solution may then be obtained by direct integration and the introduction of an arbitrary constant.

Written in symbolic form the solutions of Eqns (14.1) and (14.2) may be expressed as

$$\int \frac{\mathrm{d}y}{N(y)} = \int M(x)\mathrm{d}x + C, \tag{14.3}$$

provided $N(y)$ is non-vanishing, and

$$\int \frac{S(y)}{Q(y)}\mathrm{d}y = -\int \frac{P(x)}{R(x)}\mathrm{d}x + C, \tag{14.4}$$

provided $Q(y)$ and $R(x)$ are non-vanishing; C, of course, is an arbitrary constant.

Example 14.1 Solve the differential equation

$$\frac{\mathrm{d}y}{\mathrm{d}x} = \frac{1 - y}{1 + x}$$

which occurred in Example 13.7.

Solution Rewrite the differential equation as

$$\frac{\mathrm{d}y}{1 - y} = \frac{\mathrm{d}x}{1 + x}.$$

On integrating this we obtain

$$\int \frac{\mathrm{d}y}{1 - y} = \int \frac{\mathrm{d}x}{1 + x}$$

so

$$-\ln|1 - y| + \ln C = \ln|1 + x|,$$

where, for convenience, we write the arbitrary constant in the form $\ln C$. The general solution is thus $|(1 - y)(1 + x)| = C$ where, by virtue of the form of this expression, C is, of course, an essentially positive constant. Alternatively, the modulus signs may be removed and the general solution written as

$$y = 1 \pm \frac{C}{1 + x}.$$

In arriving at this solution we divided by the factor $(1 - y)$, which was assumed to be non-zero. To complete the solution we must now recognize that if this factor vanishes then the method of solution just outlined will fail. Clearly, when this happens we must also enquire whether the vanishing of the factor itself will give rise to a solution. Now the factor $(1 - y)$ vanishes when $y = 1$, and it is simple to substitute $y = 1$ into the original differential equation to verify that it is in fact a degenerate solution. ■

Example 14.2 Solve the following equation, already expressed in differential form:

$$x \cos y \, \mathrm{d}x - \mathrm{e}^{-x} \sec y \, \mathrm{d}y = 0.$$

Solution This is of the form shown in Eqn (14.2) and, as implied by Eqn (14.4), after division by $\mathrm{e}^{-x} \cos y$ the solution may be written

$$\int \sec^2 y \, \mathrm{d}y = \int x\mathrm{e}^x \, \mathrm{d}x.$$

Performing the indicated integrations then gives

$$\tan y = \mathrm{e}^x(x-1) + C \quad \text{or} \quad y = \arctan[\mathrm{e}^x(x-1) + C].$$

Again we must check to see if the vanishing of the divisor gives rise to another solution. Here the divisor was $\mathrm{e}^{-x} \cos y$, which only has zeros when $y = (2n+1)\frac{1}{2}\pi$ with n an integer. Substitution of these values of y into the original differential equation shows that they are not solutions. ■

Example 14.3 Find the general solution of

$$\frac{\mathrm{d}y}{\mathrm{d}x} = ky(L-y),$$

with $k > 0$ and $L > 0$, and hence find the solution in the interval $0 < y < L$ subject to the initial condition $y(0) = y_0$.

Solution Separating the variables we find, as in Eqn (14.3), that

$$\int \frac{\mathrm{d}y}{y(L-y)} = \int k \, \mathrm{d}x.$$

Now by means of partial fractions we have

$$\frac{1}{y(L-y)} = \frac{1}{L}\left(\frac{1}{y} + \frac{1}{L-y}\right),$$

so the integrals may be rewritten to give

$$\frac{1}{L}\int \frac{\mathrm{d}y}{y} + \frac{1}{L}\int \frac{\mathrm{d}y}{L-y} = \int k \, \mathrm{d}x.$$

Performing the integrations we find

$$\ln|y| - \ln|L-y| = L(kx + C),$$

which is equivalent to

$$\ln\left|\frac{y}{L-y}\right| = L(kx + C),$$

or, taking the exponential of both sides, to

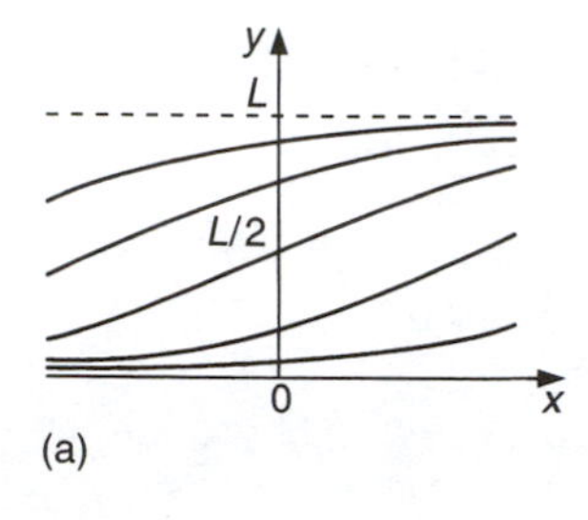

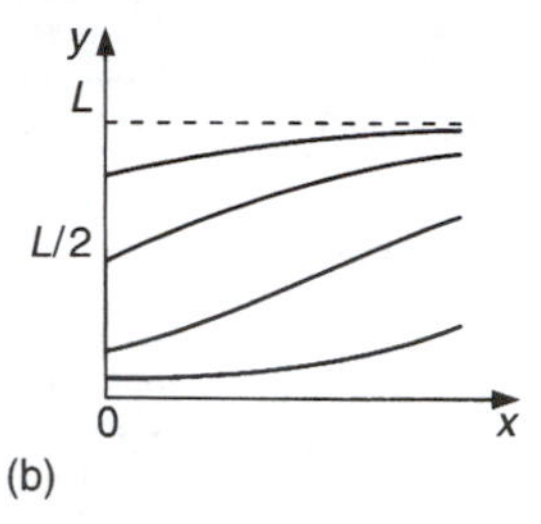

Fig. 14.1 (a) General solution curves; (b) logistic curves.

$$\left|\frac{y}{L-y}\right| = Ae^{Lkx},$$

where $A = e^{LC}$ is an arbitrary positive constant. The constant is positive because the exponential function is positive for all x, and it is arbitrary because, although L is a given parameter, C is an arbitrary constant causing the product LC to be arbitrary.

Since an absolute value is involved in this solution, the form of the solution will depend on the sign of $y/(L-y)$. We shall confine attention to the solution for which $0 < y < L$, for then both y and $L-y$ are positive, so $|y/(L-y)| = y/(L-y)$. This case has many important physical applications, a few of which will be mentioned later. Solving for y gives

$$y = \frac{ALe^{LKx}}{1 + Ae^{Lkx}},$$

which is in the form of the general solution for $k > 0$, $L > 0$ and $0 < y < L$.

Some typical solution curves are shown in Fig. 14.1(a), and they are seen to be asymptotic to the lines $y = 0$ and $y = L$, and to increase monotonically with x. Each solution curve has a point of inflection at $y = L/2$. This fact may be deduced from the general solution, but it is simplest to deduce it from the differential equation itself. To see this, first differentiate the differential equation with respect to x to obtain

$$\frac{d^2y}{dx^2} = k\frac{dy}{dx}(L-y) - ky\frac{dy}{dx}.$$

Then, setting $d^2y/dx^2 = 0$ to locate the points of inflection gives

$$k(L-2y)\frac{dy}{dx} = 0.$$

Now dy/dx only vanishes at $x = \pm\infty$, so it follows at once that there is only one point of inflection on each solution curve, which must lie at $y = L/2$. Our result is proved.

Let us now consider the initial value problem for this equation with $y(0) = y_0$: that is, we shall find the solution such that $y = y_0$ when $x = 0$. Setting $x = 0$ in the general solution shows $y_0 = y(0) = AL/(1 + A)$, from which the arbitrary constant A may be determined in order that the solution agrees with the initial condition. Using this value of A in the general solution and simplifying the result, we find

$$y = \frac{y_0 L}{y_0 + (L - y_0)e^{-Lkx}}, \qquad \text{for} \quad x \geq 0.$$

This, then, is the required solution of the initial value problem for the differential equation. Typical solution curves are shown in Fig. 14.1(b), where the intercept on the y-axis represents the initial condition appropriate to the solution curve which passes through that point. These curves defined for $x \geq 0$ are called *logistic curves*.

The importance of this equation is that it arises in population growth studies, product marketing, ecological problems, the spread of epidemics, the theory of learning and in certain nuclear problems. In qualitative terms the equation asserts that in a system in which the quantity y is limited to the maximum value L, the rate of growth dy/dx is proportional to the present quantity y and also to the quantity $L - y$ which remains. This means that the growth of y is slow at the start when y is small, and again when y approaches the limiting value L, when $L - y$ is small. The differential equation is known as the *logistic equation*, though sometimes it is called the *limited growth equation*. ■

14.2 Homogeneous equations

A function $F(x, y)$ is said to be *homogeneous of degree n* in the algebraic sense if $F(kx, ky) = k^n F(x, y)$ for every real number k. Thus the function $F(x, y) = ax^2 + bxy + cy^2$ is homogeneous of degree 2, since $F(kx, ky) = k^2 F(x, y)$ If a change of variable $y = sx$ is made then any homogeneous function $F(x, y)$ of degree n may be written in the form $F(x, y) = F(x \times 1, x \times s) = x^n F(1, s)$, since here x is now cast in the role of k. This variable change has resulted in $F(x, y)$ being expressed as the product of x^n, which is only a function of x, and $F(1, s)$, which is only a function of s.

This facilitates the integration of a differential equation of the form

$$M(x, y)dx + N(x, y)dy = 0, \tag{14.5}$$

in which the functions $M(x, y)$ and $N(x, y)$ are both homogeneous functions of the same degree. Such equations are called *homogeneous equations*. 'Homogeneous' here refers to the algebraic property shared by M and N and is not to be confused with the same term introduced in section 13.2.

Differentiating $y = sx$ gives

$$\frac{dy}{dx} = x\frac{ds}{dx} + s, \qquad \text{or} \quad dy = x\,ds + s\,dx,$$

and so if M and N are of degree n, then $M(x, y) = x^n M(1, s)$ and $N(x, y) = x^n N(1, s)$.

If, for simplification, we write $P(s) = M(1, s)$ and $Q(s) = N(1, s)$, Eqn (14.5) becomes

$$x^n P(s)dx + x^n Q(s)(s\,dx + x\,ds) = 0 \tag{14.6}$$

or, on cancelling x^n and rearranging,

$$[P(s) + sQ(s)]dx + xQ(s)ds = 0.$$

This equation is one in which the variables are separable, it has the general solution

$$\int \frac{Q(s)\mathrm{d}s}{P(s) + sQ(s)} = \ln\left|\frac{C}{x}\right|, \tag{14.7}$$

where C is an arbitrary constant.

The final solution in terms of x and y may be obtained by using the relation $s = y/x$. The vanishing of the divisor x^n also implies a possible solution $x = 0$ which must be tested for validity against the original differential equation.

Example 14.4 Integrate the following homogeneous differential equation (a) directly, and (b) by an application of Eqn (14.7):

$$(y - x)\mathrm{d}x - x\,\mathrm{d}y = 0.$$

Solution (a) The functions multiplying dx and dy are both of degree 1, so the equation is homogeneous. Setting $y = sx$ produces the equation

$$x(s - 1)\mathrm{d}x - x(s\,\mathrm{d}x + x\,\mathrm{d}s) = 0.$$

Cancelling the factor x and rearranging, we see that

$$\int \mathrm{d}s = -\int \frac{\mathrm{d}x}{x},$$

whence

$$s = C - \ln|x| \quad \text{or} \quad y = x(C - \ln|x|).$$

The original equation shows that the vanishing of the cancelled factor x gives rise to an additional solution $x = 0$.

(b) Here $M(x, y) = y - x$ and $N(x, y) = -x$ so that $M(1, s) = s - 1$ and $N(1, s) = -1$. Hence $P(s) = s - 1$ and $Q(s) = -1$.

$$\int \mathrm{d}s = \ln\left|\frac{C}{x}\right| \quad \text{or} \quad s = \ln\left|\frac{C}{x}\right|.$$

Again using $s = y/x$ we find that

$$y = x(\ln|C| - \ln|x|),$$

which agrees with the previous answer apart from the form of the arbitrary constant. ■

A slight variant of the homogeneous equation is the so-called *near-homogeneous equation*, which has the general form

$$\frac{\mathrm{d}y}{\mathrm{d}x} = \frac{ax + by + p}{cx + dy + q} \quad \text{with } ad - bc \neq 0,$$

and which is not homogeneous as it stands. However, by setting $x = X + \alpha$, $y = Y + \beta$, where α, β are constants, it follows that $\mathrm{d}y/\mathrm{d}x = \mathrm{d}Y/\mathrm{d}X$; and hence that

$$\frac{dY}{dX} = \frac{aX + bY + a\alpha + b\beta + p}{cX + dY + c\alpha + d\beta + q}.$$

Now if the constants α, β are chosen such that $a\alpha+b\beta+p=0$ and $c\alpha+d\beta+q=0$, which is always possible if $ad-bc\neq 0$, then the transformed equation becomes homogeneous in X, Y. It may then be solved as indicated above and the solution transformed back in terms of x, y by setting $X=x-\alpha$, $Y=y-\beta$.

For example, if

$$\frac{dy}{dx} = \frac{2x + 3y + 3}{x + y + 2},$$

then α, β would be determined by solving $2\alpha+3\beta+3=0$ and $\alpha+\beta+2=0$, whence $\alpha=-3$, $\beta=1$, so that $X=x+3$, $Y=y-1$. The details of the solution of the resulting homogeneous equation

$$\frac{dY}{dX} = \frac{2X + 3Y}{X + Y}$$

are left to the reader.

If $ad-bc=0$, then the equation may be solved by setting $u=ax+by+p$ and separating the variables.

14.3 Exact equations

Definition 14.1 (exact differential equation) A differential equation of the form

$$M(x, y)dx + N(x, y)dy = 0, \tag{14.8}$$

with the property that $M(x, y)$ and $N(x, y)$ are related to a differentiable function $F(x, y)$ by the equations

$$M(x, y) = \frac{\partial F}{\partial x}, \quad N(x, y) = \frac{\partial F}{\partial y}, \tag{14.9}$$

is said to be *exact*.

Let us examine Eqn (14.8) in order to derive a test for exactness. To do so we now consider the equation

$$F(x, y) = C \quad (C = \text{constant})$$

where for the moment $F(x, y)$ is an arbitrary differentiable function with continuous first- and second-order partial derivatives. Next we form the total derivative of $F(x, y) = C$

$$\frac{\partial F}{\partial x}\mathrm{d}x + \frac{\partial F}{\partial y}\mathrm{d}y = 0, \tag{14.10}$$

and compare Eqns (14.8) and (14.10).

We then see that by setting

$$M(x, y) = \frac{\partial F}{\partial x} \quad \text{and} \quad N(x, y) = \frac{\partial F}{\partial y}, \tag{14.11}$$

it follows that $F(x, y) = C$ is a solution of Eqn (14.8). However, any function $F(x, y)$ with continuous first- and second-order partial derivatives must be such that its mixed partial derivatives are equal, so that $\partial^2 F/\partial y \partial x = \partial^2 F/\partial x \partial y$. From Eqns (14.11) this condition is seen to be equivalent to $\partial M/\partial y = \partial N/\partial x$, which is the condition we are seeking that ensures Eqn (14.8) is exact. We have established the following theorem.

Theorem 14.1 If the functions $M(x, y)$ and $N(x, y)$ and their derivatives $\partial M/\partial y$ and $\partial N/\partial x$ are continuous and, furthermore, $\partial M/\partial y = \partial N/\partial x$ everywhere, then the differential equation

$$M(x, y)\mathrm{d}x + N(x, y)\mathrm{d}y = 0$$

is exact. ■

The proof of the test for exactness forming Theorem 14.1 also shows how to solve Eqn (14.8). To see this we integrate the first of Eqn (14.11) partially with respect to x (regarding y as a constant) to give

$$F(x, y) = \int M(x, y)\mathrm{d}x + A(y), \tag{14.12}$$

where for the moment $A(y)$ is an unknown function. In Eqn (14.12) the function $A(y)$ takes the place of the ordinary constant of integration that would have arisen were M to be only a function of x. To check that this is indeed correct we only need to differentiate Eqn (14.12) partially with respect to x to see that we recover the result $\partial F/\partial x = M(x, y)$. If, now, Eqn (14.12) is differentiated partially with respect to y, we find that

$$\frac{\mathrm{d}A}{\mathrm{d}y} = \frac{\partial F}{dy} - \frac{\partial}{\partial y}\int M(x, y)\mathrm{d}x,$$

but from the second of Eqn (14.11) $\partial F/\partial y = N(x, y)$, and so

$$\frac{\mathrm{d}A}{\mathrm{d}y} = N(x, y) - \frac{\partial}{\partial y}\int M(x, y)\mathrm{d}x. \tag{14.13}$$

The function $A(y)$ now follows by integrating Eqn (14.13) partially with respect to y to get

$$A(y) = \int\left[N(x, y) - \frac{\partial}{\partial y}\int M(x, y)\mathrm{d}x\right]\mathrm{d}y + \text{const.} \tag{14.14}$$

Thus we see that the solution of Eqn (14.18) is given by Eqn (14.12) in which we substitute for $A(y)$ from Eqn (14.14).

Example 14.5 Solve the differential equation

$$(2x + 3\cos y)\mathrm{d}x + (2y - 3x\sin y)\mathrm{d}y = 0.$$

The equation is exact since, as $M(x, y) = 2x + 3\cos y$ and $N(x, y) = 2y - 3x\sin y$, it follows that $\partial M/\partial y = \partial N/\partial x = -3\sin y$, satisfying the conditions of Theorem 14.1. Now integrating $\partial F/\partial x = 2x + 3\cos y$ partially with respect to x gives

$$F(x, y) = \int(2x + 3\cos y)\mathrm{d}x + A(y),$$

or

$$F(x, y) = x^2 + 3x\cos y + A(y).$$

Differentiating partially with respect to y, we find that

$$\frac{\partial F}{\partial y} = -3x\sin y + \frac{\mathrm{d}A}{\mathrm{d}y}$$

which, since $\partial F/\partial y = N(x, y) = 2y - 3x\sin y$, reduces the equation to $\mathrm{d}A/\mathrm{d}y = 2y$ and hence $A(y) = y^2 + C'$. The function $F(x, y)$ is thus

$$F(x, y) = x^2 + 3x\cos y + y^2 + \text{const.},$$

and so the general solution is

$$x^2 + 3x\cos y + y^2 = C.$$

If a particular solution is required, the value of the constant C must be determined by requiring the integral curve to pass through a specified point. For example, suppose that our equation is required to have the initial value $y(0) = \frac{1}{2}\pi$; then clearly $C = \frac{1}{4}\pi^2$ and the integral curve representing the solution is

$$x^2 + 3x\cos y + y^2 = \tfrac{1}{4}\pi^2. \quad \blacksquare$$

Not all differential equations of the form (14.8) are exact, but it can be proved that they can always be made exact by multiplying by a suitable function $\mu(x, y)$, called an *integrating factor*. So, if $\mu(x, y)$ is an integrating factor for Eqn (14.8),

$$\mu(x, y)M(x, y)\mathrm{d}x + \mu(x, y)N(x, y)\mathrm{d}y = 0 \tag{14.15}$$

is exact. However, from Theorem 14.1, we know that the integrating factor must, be such that

$$\frac{\partial}{\partial y}(\mu M) = \frac{\partial}{\partial x}(\mu N). \tag{14.16}$$

Although this is a condition that must be satisfied by an integrating factor μ, it is unfortunate that there is no general method by which it may be found.

Two special forms of differential equation in which an integrating factor may always be found have already been encountered in Eqns (14.2) and (14.5). The integrating factor for Eqn (14.2), in which the variables are separable, is $\mu(x, y) = 1/[R(x)Q(y)]$ and leads to the result (14.4). Similarly, the integrating factor for Eqn (14.5), with homogeneous coefficients, is $\mu(x, s) = 1/[P(s) + sQ(s)]$, where $s = y/x$, and gives rise to the solution shown in Eqn (14.7).

When the integrating factor is believed to be of a known simple form in which only certain constants need to be determined, direct substitution in Eqn (14.16) would confirm this, and also indicate conditions by which these constants may be determined. The arguments used in this trial and error method will now be applied to a search for an integrating factor having the form $\mu(x, y) = x^m y^n$ in the following simple example.

Example 14.6 Solve the differential equation

$$(2xy + y^2)\mathrm{d}x + (2x^2 + 3xy)\mathrm{d}y = 0.$$

Solution First we notice that $\partial M/\partial y \neq \partial N/\partial x$, so the equation is not exact and an integrating factor μ is required. As M and N are simple algebraic functions we shall try to find an integrating factor of the form $\mu(x, y) = x^m y^n$, in which the constants m and n must be determined so that

$$x^m y^n(2xy + y^2)\mathrm{d}x + x^m y^n(2x^2 + 3xy)\mathrm{d}y = 0$$

is exact. By condition (14.16) this implies that if $\mu(x, y) = x^m y^n$ is in fact an integrating factor, then m and n must be chosen so that

$$\frac{\partial}{\partial y}[x^m y^n(2xy + y^2)] = \frac{\partial}{\partial x}[x^m y^n(2x^2 + 3xy)].$$

This condition gives rise to the equation

$$\begin{aligned} nx^m y^{n-1}(2xy + y^2) + x^m y^n(2x + 2y) = mx^{m-1}y^n(2x^2 + 3xy)& \\ + x^m y^n(4x + 3y),& \end{aligned}$$

from which we must determine m and n if the chosen form of integrating factor is correct. Since this expression must be an identity, we now equate coefficients of terms of equal degree in x and y and, if possible, select m and n such that all conditions are satisfied. In this case only two conditions arise:

(a) terms involving $x^m y^{n+1}$:

$$n + 2 = 3m + 3;$$

(b) terms involving $x^{m+1}y^n$:

$$2n + 2 = 2m + 4.$$

These conditions are satisfied if $m = 0$, $n = 1$, so an integrating factor of the type assumed does exist, and in this case $\mu = y$. The exact differential equation is thus

$$(2xy^2 + y^3)\mathrm{d}x + (2x^2y + 3xy^2)\mathrm{d}y = 0,$$

which is easily seen to have the general solution

$$x^2y^2 + xy^3 = C.$$

Had values of m and n not been found that produced an identity from condition (14.16), then the integrating factor would not have been of the form $\mu(x, y) = x^m y^n$. ■

14.4 The linear equation of first order

The general linear equation of first order already encountered in Eqn (13.6) has the form

$$\frac{\mathrm{d}y}{\mathrm{d}x} + P(x)y = Q(x). \qquad (14.17)$$

To solve it, let us seek an integrating factor μ that will make $\mu(\mathrm{d}y/\mathrm{d}x) + \mu Py$ a derivative – namely, the derivative $(\mathrm{d}/\mathrm{d}x)(\mu y)$. Since the equation is linear, μ must be independent of y, so we need only consider μ to be of the form $\mu = \mu(x)$. Thus the integrating factor μ is required to be a solution of the equation $(\mathrm{d}/\mathrm{d}x)(\mu y) = \mu(\mathrm{d}y/\mathrm{d}x) + \mu Py$. Expanding the left-hand side and simplifying then leads to the simple differential equation $y[(\mathrm{d}\mu/\mathrm{d}x) - P\mu] = 0$. As the solution of y in Eqn (14.17) is not identically zero, it follows that μ must be the solution of

$$\frac{\mathrm{d}\mu}{\mathrm{d}x} = P(x)\mu. \qquad (14.18)$$

The variables x and μ are separable, giving

$$\frac{\mathrm{d}\mu}{\mu} = P(x)\mathrm{d}x,$$

showing that $\ln|\mu| = \int P(x)\mathrm{d}x + C'$, where C' is an arbitrary constant. Taking exponentials we find that the most general integrating factor is $\mu = \exp(C')\exp(\int P(x)\mathrm{d}x)$.

However, as the arbitrary factor $\exp(C')$ is always non-zero, and so may be cancelled when this expression is used as a multiplier in Eqn (14.17), we may always take as the integrating factor the expression

$$\mu = \exp(\int P(x)\mathrm{d}x). \qquad (14.19)$$

Multiplying eqn (14.17) by μ, and using its properties then gives

$$(y \exp(\int P(x)\mathrm{d}x)) = Q(x) \exp(\int P(x)\mathrm{d}x).$$

After a final integration and simplification, we obtain the general solution

$$y = \exp(-\int P(x)\mathrm{d}x)[C + \int Q(x) \exp(\int P(x)\mathrm{d}x)\mathrm{d}x], \qquad (14.20)$$

where C is the arbitrary constant of the final integration and must be retained.

Although this general solution is useful, in applications it is usually better to use the fact that expression (14.19) is an integrating factor and to proceed directly from that point without recourse to Eqn (14.20).

An important general point illustrated by Eqn (14.20), which, as we shall see later, characterizes all linear equations is that the general solution comprises the sum of two parts. The first part, $C \exp(-\int P(x)\mathrm{d}x)$, is the solution of the homogeneous equation (corresponding to $Q(x) \equiv 0$) and the second part, $\exp(-\int P(x)\mathrm{d}x)\int Q(x) \exp(\int P(x)\mathrm{d}x)\mathrm{d}x$, is particular to the form of the inhomogeneous term $Q(x)$. This second part is called the *particular integral*, while the first part is the *complementary function*. Notice that the two parts of the solution are additive and that the arbitrary constant is associated with the complementary function. These observations characterize linear differential equations of any order and will be encountered again in the next chapter.

Example 14.7 Solve

$$\frac{\mathrm{d}y}{\mathrm{d}x} + ky = a \sin mx$$

subject to the initial condition $y = 1$ when $x = 0$.

Solution In this case $P(x) = k$, so that the integrating factor $\mu = \mathrm{e}^{kx}$. Hence

$$\frac{\mathrm{d}}{\mathrm{d}x}(y\mathrm{e}^{kx}) = a\mathrm{e}^{kx} \sin mx,$$

giving rise to

$$y\mathrm{e}^{kx} = a\int \mathrm{e}^{kx} \sin mx \, \mathrm{d}x + C,$$

or

$$y = \mathrm{e}^{-kx}(C + a\int \mathrm{e}^{kx} \sin mx \, \mathrm{d}x).$$

Performing the indicated integration gives the general solution

$$y = Ce^{-kx} + \frac{a}{k^2 + m^2}(k \sin mx - m \cos mx),$$

the first term being the complementary function and the second term the particular integral.

To determine the constant C, we now utilize the initial conditions $y(0)=1$ by writing

$$1 = C - \frac{am}{k^2 + m^2}.$$

The particular solution is thus

$$y = \left(\frac{k^2 + m^2 + am}{k^2 + m^2}\right)e^{-kx} + \frac{a}{k^2 + m^2}(k\sin mx - m\cos mx).$$

If we make the identifications $y \equiv i$, $x \equiv t$, $k \equiv R/L$, $a \equiv V_0/L$ and $m \equiv \omega$, we discover that we have just obtained the solution of Example 13.3 (concerning the electric circuit), subject to the initial condition that a unit current flows when the circuit is closed. ■

Example 14.8 Find the general solution of the linear equation

$$\frac{dy}{dx} + xy = x.$$

Solution The integrating factor $\mu = \exp(\int x\,dx)$ so that $\mu = e^{x^2/2}$. Hence

$$\frac{d}{dx}(ye^{x^2/2}) = xe^{x^2/2}$$

or

$$ye^{x^2/2} = C + \int xe^{x^2/2}\,dx.$$

Since the indefinite integral on the right-hand side is $e^{x^2/2}$, it follows that

$$y = Ce^{-x^2/2} + 1.$$

The particular integral in this case is simply the constant unity. ■

The *Bernoulli equation*

$$\frac{dy}{dx} + R(x)y = S(x)y^n, \tag{14.21}$$

is nonlinear due to the inhomogeneous term y^n, though it may be transformed into an equivalent linear equation and solved by the method of this section if the new dependent variable $u = y^{1-n}$ is introduced. To see this, notice that $du/dx = [(1-n)/y^n](dy/dx)$, so that substituting in Eqn (14.21) for dy/dx and dividing throughout by a factor y^n gives the equivalent linear equation

$$\frac{\mathrm{d}u}{\mathrm{d}x} + (1-n)R(x)u = (1-n)S(x). \tag{14.22}$$

This is an equation of the form (14.17), with $P(x) = (1-n)R(x)$ and $Q(x) = (1-n)S(x)$. The solution of the Bernoulli equation may be recovered once Eqn (14.22) has been solved by setting $u = y^{1-n}$.

Example 14.9 Solve the Bernoulli equation

$$\frac{\mathrm{d}y}{\mathrm{d}x} + y = xy^3.$$

Solution In the notation of Eqn (14.21), $R(x) = 1$, $S(x) = x$ and $n = 3$. Thus either using these results in Eqn (14.22), or working from first principles and making the substitution $u = y^{-2}$ in the Bernoulli equation, we arrive at the linear equation

$$\frac{\mathrm{d}u}{\mathrm{d}x} - 2u = -2x.$$

This equation has the integrating factor $\mu = \exp(\int -2\mathrm{d}x) = \mathrm{e}^{-2x}$, so it may be rewritten as

$$\frac{\mathrm{d}}{\mathrm{d}x}(\mathrm{e}^{-2x}u) = -2x\mathrm{e}^{-2x}.$$

Integrating this result gives

$$\mathrm{e}^{-2x}u = -2\int x\mathrm{e}^{-2x}\,\mathrm{d}x + C,$$

or

$$\mathrm{e}^{-2x}u = (x + \tfrac{1}{2})\mathrm{e}^{-2x} + C,$$

so that

$$u = x + \tfrac{1}{2} + C\mathrm{e}^{2x}.$$

Finally, substituting $u = y^{-2}$ shows the required solution to be

$$y = \frac{1}{\sqrt{(x + \frac{1}{2} + C\mathrm{e}^{2x})}}. \quad \blacksquare$$

A special class of differential equations that may also be solved by means of the method described in this section is the class of linear second-order equations that do not depend explicitly on the dependent variable y. They have the general form

$$\frac{\mathrm{d}^2y}{\mathrm{d}x^2} + P(x)\frac{\mathrm{d}y}{\mathrm{d}x} = Q(x), \tag{14.23}$$

and may be reduced to the form of Eqn (14.17) by writing $z = \mathrm{d}y/\mathrm{d}x$. The solution then amounts to the determination of z from the first-order differential equation

$$\frac{\mathrm{d}z}{\mathrm{d}x} + P(x)z = Q(x), \tag{14.24}$$

together with the subsequent integration of $\mathrm{d}y/\mathrm{d}x = z(x)$ to give the general solution of Eqn (14.23)

$$y = \int z(x)\,\mathrm{d}x + C. \tag{14.25}$$

Notice that two constants of integration are involved, one arising in the determination of z and a second appearing as the integration constant in Eqn (14.25), as is to be expected in the solution of a second-order equation.

Example 14.10 Find the general solution of

$$\frac{\mathrm{d}^2y}{\mathrm{d}x^2} - \frac{1}{x}\frac{\mathrm{d}y}{\mathrm{d}x} = x\mathrm{e}^{2x}.$$

Solution Setting $z = \mathrm{d}y/\mathrm{d}x$, we obtain

$$\frac{\mathrm{d}z}{\mathrm{d}x} - \frac{z}{x} = x\mathrm{e}^{2x},$$

which has the integrating factor $1/x$ and so

$$\frac{\mathrm{d}}{\mathrm{d}x}\left(\frac{z}{x}\right) = \mathrm{e}^{2x}.$$

The solution z is thus $z = x(C + \frac{1}{2}\mathrm{e}^{2x})$. Finally, since $z = \mathrm{d}y/\mathrm{d}x$, we have

$$y = \int x(C + \tfrac{1}{2}\mathrm{e}^{2x})\mathrm{d}x + D,$$

or

$$y = \tfrac{1}{4}\mathrm{e}^{2x}\int(x - \tfrac{1}{2}) + \tfrac{1}{2}Cx^2 + D.$$

Because C is an arbitrary constant there is no necessity to retain the factor $\frac{1}{2}$ in the second term. ■

It is appropriate to remind the reader that arithmetic is not usually performed on arbitrary constants. Thus $3C$ and $-4C$ are usually written C, when C is an arbitrary constant. Similarly, $A+3B$ is just another arbitrary constant when A and B are arbitrary so it could, for example, either be written A or B.

Obviously, this general method of reduction of order applies to any differential equation not explicitly involving y, though the simplification is most striking in the situation just discussed.

14.5 Direct deductions

So far we have discussed a variety of special methods for the exact solution of first-order differential equations together with several possible numerical techniques of varying applicability and accuracy. We close this chapter by indicating how these techniques may often be usefully supplemented by a direct examination of the differential equation itself.

On occasion it is helpful to deduce properties of solutions of a differential equation directly from the equation itself without first obtaining the general solution. Indeed, in many practical situations the general solution is not known, as is the case with most nonlinear equations. In these circumstances any property of the solutions that may be deduced directly is likely to be of considerable value.

This idea is best illustrated by a simple example, and for this purpose we shall use Example 13.1. The equation in question is

$$\frac{dv}{dt} = g - \frac{\lambda}{m}v^2, \tag{14.26}$$

in which v denotes the velocity of a falling particle of mass m at time t in a resisting medium with a quadratic velocity resistance law having proportionality constant λ.

Suppose it is required to find the limiting constant velocity of fall c of the particle, often called the *terminal velocity*. Then we know that under these conditions the acceleration dv/dt must be identically zero and hence Eqn (14.26) reduces to

$$0 = g - \frac{\lambda}{m}c^2,$$

giving

$$c = (mg/\lambda)^{1/2}. \tag{14.27}$$

This result has been obtained directly without integration of Eqn (14.26) and is typical of many situations in which an examination of the differential equation with the physical problem clearly in mind will yield direct and useful information. Result (14.27) may, of course, be obtained from the solution of Eqn (14.26), but with far greater effort. If, for example, the case of a particle falling from rest at time $t = 0$ were to be considered, then separation of the variables and integrating would give for the solution

$$v = c\tanh\frac{\lambda ct}{m}. \tag{14.28}$$

Thus, as $\lim_{x\to\infty} \tanh x = 1$, it follows directly from Eqn (14.28) that $\lim_{t\to\infty} v = c$.

Problems

Section 14.1

14.1 Find the general solution of the following problems by separating the variables:

(a) $\tan x \sin^2 y \, dx + \cos^2 x \cot y \, dy = 0$;

(b) $xy' - y = y^3$;

(c) $xyy' = 1 - x^2$;

(d) $y - xy' = a(1 + x^2 y')$;

(e) $3e^x \tan y \, dx + (1 - e^x)\sec^2 y \, dy = 0$;

(f) $y' \tan x = y$.

Section 14.2

14.2 Find the solutions of the following homogeneous and near-homogeneous equations:

(a) $(x - y)y \, dx - x^2 \, dy = 0$, if initially $x = 1$, $y = 2$;

(b) $(x + y - 3)dx - (x - y - 1)dy = 0$;

(c) $(x^2 + y^2)dx - 2xy \, dy = 0$;

(d) $(x^2 - 3y^2)dx + 2xy \, dy = 0$;

(e) $zxy \, dx - (x^2 - y^2) \, dy = 0$.

Section 14.3

14.3 Show that the stated functions μ are integrating factors for their associated differential equations and find the general solutions:

(a) $\left(\dfrac{2x + y + 3}{x}\right)dx + \left(\dfrac{x + 2y + 3}{y}\right)dy = 0 \quad (\mu = xy)$;

(b) $y \, dx + x \, dy + \left(\dfrac{2x}{1 + \cos xy}\right)dx = 0 \quad (\mu = 1 + \cos xy)$;

(c) $(ye^{-y/2} + e^{y/2})dx + 2x \cosh \tfrac{1}{2}y \, dy \quad (\mu = e^{y/2})$.

14.4 Determine if $\mu = x^m y^n$ is an integrating factor for the following differential equations and, if so, deduce appropriate values for m and n. Use your results to determine the solution in each case:

(a) $(9y^2 + 4xy^2 + 3y)dx + (6xy + 2x^2 y + x)dy = 0$;

(b) $y(2x + \cosh x)\mathrm{d}x + 2(x^2 + \sinh x)\mathrm{d}y = 0$;

(c) $(x + y)\mathrm{d}x + x(x + 3y)\mathrm{d}y = 0$;

(d) $2e^y\,\mathrm{d}x + \left(xe^y + \frac{2y}{x}\right)\mathrm{d}y = 0$, if initially $x = 1$, $y = 0$;

(e) $y^2\,\mathrm{d}x + (x + \cos y)\mathrm{d}y = 0$;

(f) $y(2 + 3xy^2)\mathrm{d}x + 2x(1 + 2xy^2)\mathrm{d}y = 0$.

Section 14.4

14.5 Find the general solutions of the following differential equations, and when initial conditions are given determine the particular solution appropriate to them:

(a) $\frac{\mathrm{d}y}{\mathrm{d}x} - \frac{y}{x} = 2x$;

(b) $\frac{\mathrm{d}y}{\mathrm{d}x} + \frac{2y}{x} = x^3 \quad (y = 0,\ x = 1)$;

(c) $\frac{\mathrm{d}y}{\mathrm{d}x} - \frac{2}{x + 1}y = 3(x + 1)^3$;

(d) $xy' + y - e^x = 0 \quad (y = 1,\ x = 2)$;

(e) $y' - y\tan x = \sec x \quad (y = 1,\ x = 0)$;

(f) $y\,\mathrm{d}x - [y^2 + (y + 1)x]\mathrm{d}y = 0$;

(g) $y' + xy = xy^2$;

(h) $y' + \frac{y}{x} = xy^2 \sin x$.

14.6 Show that the solution of the homogeneous form of the differential equation

$$\frac{\mathrm{d}y}{\mathrm{d}x} + P(x)y = Q(x)$$

is

$$y = A\exp(-\int P(x)\mathrm{d}x).$$

If A is now regarded as a function of x show, by substituting back into the inhomogeneous equation, that $A(x)$ is the solution of the equation

$$\frac{\mathrm{d}A}{\mathrm{d}x} = Q(x)\exp(\int P(x)\mathrm{d}x),$$

in which the variables are separable. Integrate this result and find the general solution of the original differential equation.

14.7 By applying the previous method to the differential equation

$$\frac{dy}{dx}+\frac{y}{x}=e^{x},$$

show that $A(x)=C+e^{x}(x-1)$ and hence find the solution that satisfies the initial condition $y=1$ when $x=1$.

14.8 Find the general solution of

$$\frac{d^2y}{dx^2}+\frac{1}{x}\frac{dy}{dx}=1.$$

14.9 Find the solution of

$$\frac{d^2y}{dx^2}+\frac{dy}{dx}=e^{x},$$

such that $y=1$ and $dy/dx=0$ when $x=0$.

Section 14.5

14.10 A particle moves in a resisting medium with velocity v at time t governed by the differential equation

$$\frac{dv}{dt}=k-\lambda(1+e^{-t})v^{\alpha},$$

with k, λ and α positive constants. Deduce the terminal velocity c of the particle.

14.11 The differential equation

$$\frac{dy}{dt}=\alpha y-\sin y,$$

in which α is a parameter, y is a displacement and t is the time describes the response of part of a nonlinear control system. Obtain the conditions on the parameter α in order that there should only be four steady-state modes of operation characterized by non-zero values of y. Use graphical methods to determine the approximate values assumed by y in these steady-state modes of operation.

14.12 The reaction rate m in a chemical engineering process is governed by the differential equation

$$\frac{dm}{dt}=a+bm+km^2,$$

where a, b and k are constants. Find the possible steady-state reaction rates and deduce the conditions on the constants in order that these rates should be positive.

Supplementary computer problems

Problems in this section require the use of a graphics package.

14.13 In Example 14.2 the general solution of the differential equation

$$x \cos y \, dx - e^{-x} \sec y \, dy = 0$$

was shown to be

$$y = \arctan[e^{x}(x-1) + C].$$

Use this result to show that the solution subject to the initial condition $y(1) = 0$ is

$$y = \arctan[e^{x}(x-1)],$$

and display the result for $1 \leq x \leq 3$.

14.14 In Example 14.3 the general solution of the logistic equation

$$\frac{dy}{dx} = ky(L-y),$$

subject to the initial condition $y(0) = y_0$, was shown to be

$$y = \frac{y_0 L}{y_0 + (L - y_0)e^{-Lkx}} \quad \text{for} \quad x \geq 0.$$

Set $L = k = 1$ and display the solution for $y_0 = 0.25, 0.5, 0.75$ in the interval $0 \leq x \leq 4$.

14.15 In Example 14.4 the general solution of the differential equation

$$(y - x)dx - x\,dy = 0$$

was shown to be

$$y = x(C - \ln|x|).$$

Use the result to show that the solution subject to the initial condition $y(1) = 1$ is

$$y = x(1 - \ln|x|),$$

and display the solution in the interval $1 \leq x \leq 10$.

14.16 The differential equation

$$(1 + x^2)\frac{dy}{dx} + 3xy = 2x$$

subject to the initial condition $y(1) = 0$ has the solution

$$y = \frac{2}{3} - \frac{4}{3} \frac{\sqrt{2}}{(1 + x^2)^{3/2}}.$$

Display the solution for $1 \le x \le 5$.

14.17 The differential equation

$$x \frac{dy}{dx} + 3y = 3x^2$$

subject the initial condition $y(1) = 2$ has the solutions

$$y = \frac{3}{5} x^2 + \frac{7}{5} \frac{1}{x^3}.$$

Display the solution for $1 \le x \le 3$.

14.18 The differential equation

$$x \frac{dy}{dx} - 3y = x^4 \cos x$$

subject to the initial condition $y(\pi/2) = 0$ has the solution

$$y = x^3(\sin x - 1).$$

Display the solution for $\pi/2 \le x \le 3$.

14.19 The Bernoulli equation

$$\frac{dy}{dx} + y = 2xy^3$$

subject to the initial condition $y(0) = 1$ has the solution

$$y = \frac{1}{\sqrt{1 + 2x}}.$$

Display the solution for $0 \le x \le 10$.

15 Higher-order linear differential equations

Linear higher-order constant-coefficient differential equations are capable of modelling many fundamental physical processes, so they represent an important and useful class of differential equations. The chapter starts by showing that their general solution comprises the sum of two distinct parts, called the complementary function and the particular integral, each of which is also clearly identifiable in the general solution of the linear first-order equation that was solved in Chapter 14. The complementary function is the solution of the homogeneous form of the differential equation, and so contains all of the arbitrary constants that are always present in a general solution. The particular integral is determined by the form of the non-homogeneous term, and it contains no arbitrary constants.

The complementary function is shown to be determined as a result of finding the roots of a polynomial derived from the homogeneous form of the differential equation. Its form depends on whether the roots of this polynomial, called the characteristic polynomial, are real or complex. The complementary function will contain (i) an exponential function corresponding to each real root, (ii) a sine and a cosine function corresponding to each pair of purely imaginary complex conjugate roots, and (iii) the product of an exponential function and a linear combination of a sine and a cosine function corresponding to each pair of complex conjugate roots with a non-zero real part. When dealing with problems involving stable oscillations, each term of the complementary function corresponds to a possible natural mode of oscillation of the system described by the differential equation. In such problems the particular integral describes the behaviour of the transient solution at the start of the oscillations, and its precise form is determined by the initial conditions. When stable oscillations are involved, the complementary function is shown to be the solution that does not vanish after the particular integral has decayed to zero, and so it describes the long-time solution, or so-called steady-state solution.

The most elementary but nevertheless useful way of determining a particular integral when the non-homogeneous term is simple in form is by means of the method of undetermined coefficients. This involves deducing the functional form of the particular integral, but leaving as undetermined constants the numerical multipliers of the functions involved. The values of these constants are found by substituting the functional form into the differential equation and equating coefficients of corresponding terms on each side of the equation to make the result an identity. When complicated non-homogeneous terms are involved the method fails, and it becomes necessary to use the more powerful method of variation of parameters that is also described in order to find the particular integral.

The study of stable coupled oscillations described by systems of equations leads naturally to the use of eigenvalues and eigenvectors. The eigenvalues determine the natural frequencies of oscillation, while the eigenvectors describe the associated normal modes of oscillation of the system modelled by the differential equations.

Boundary value problems are more complicated and, depending on the differential equation involved and the associated boundary conditions, it is shown that they may have a unique solution, a non-unique solution or no solution at all. Problems of this type have many applications, and even the case in which the solution is not unique (there may be many solutions) has important physical implications.

An alternative method of approach to the solution of initial value problems for linear constant-coefficient differential equations of any order is provided by making use of the Laplace transform that is developed in the last part of the chapter. The Laplace transform $F(s)$ of a function $f(t)$ is defined as

$$F(s) = \int_0^\infty e^{-st} f(t) dt,$$

provided the integral exists. It is shown how functions, together with their derivatives, can be transformed in routine fashion to arrive at a table of Laplace transform pairs which lists the function, or derivative, and against it the corresponding Laplace transform. Then, with the aid of such a table of transform pairs, an initial value problem for a linear differential equation, complete with its initial conditions, can be transformed into an equivalent algebraic problem from which the Laplace transform of the solution can be found. By simplifying the transformed solution using partial fractions, and applying some simple theorems that are proved in the chapter, the table of Laplace transform pairs may be used in reverse to arrive at the required solution. Thus when applying the Laplace transform approach, instead of using integration followed by the imposition of the initial conditions to determine the values of the arbitrary constants and thus arriving at the solution, this process is replaced by a transformation and algebraic manipulation, coupled with the use of a table of transform pairs in reverse to arrive at the solution.

The Laplace transform is a powerful method for solving initial value problems, and its use can be extended to encompass the solution of linear variable-coefficient differential equations, and also equations of quite different types.

15.1 Linear equations with constant coefficients – homogeneous case

Thus far our encounter with differential equations of order greater than unity has been essentially confined to the second-order equation which served to introduce the sine and cosine functions in Chapter 6. We now take up the solution of higher-order differential equations in more general terms and, among other matters, extend systematically the notion of a complementary function and a particular integral first introduced in connection with a linear first-order differential equation in Chapter 14.

Differential equations of the form

$$\frac{d^n y}{dx^n} + a_1 \frac{d^{n-1} y}{dx^{n-1}} + \ldots + a_n y = f(x), \tag{15.1}$$

in which $a_1, a_2, \ldots, a_n$ are constants (usually real), are called *linear constant-coefficient equations*. We begin by studying the homogeneous form of the equation (that is, $f(x) = 0$) which in this context is usually called the *reduced equation*:

$$\frac{d^n y}{dx^n} + \frac{d^{n-1} y}{dx^{n-1}} + \ldots + a_n y = 0. \tag{15.2}$$

Equation (15.2) may be solved by using the fact that $y = Ce^{\lambda x}$ is obviously a solution for arbitrary constant C, provided λ satisfies the equation that results when this expression is substituted into Eqn (15.2) giving

$$(\lambda^n + a_1 \lambda^{n-1} + \ldots + a_n)e^{\lambda x} = 0. \tag{15.3}$$

The substitution $y = Ce^{\lambda x}$ has thus associated a *characteristic polynomial* in λ,

$$P(\lambda) \equiv \lambda^n + a_1 \lambda^{n-1} + \ldots + a_n, \tag{15.4}$$

with the differential Eqn (15.1) and, as $e^{\lambda n} \neq 0$, it follows that the permissible values of λ in the solutions $y = Ce^{\lambda x}$ must be the roots of $P(\lambda) = 0$. The values of λ are thus determined by solving the *characteristic equation*

$$P(\lambda) = \lambda^n + a_1 \lambda^{n-1} + \ldots + a_n = 0. \tag{15.5}$$

This equation is also known as the *indicial* or *auxiliary* equation.

Since $P(\lambda)$ is a polynomial of degree n, it follows that $P(\lambda) = 0$ has precisely n roots $\lambda = \lambda_i$; furthermore, we know that when the coefficients $a_1, a_2, \ldots, a_n$ are real, these roots must therefore either be real or must occur in complex conjugate pairs. Each expression $y_i = C_i e^{\lambda_i x}$, $i = 1, 2, \ldots, n$ is a solution of Eqn (15.2) for arbitrary C_i, and direct substitution into Eqn (15.2) verifies that

$$y = C_1 e^{\lambda_1 x} + C_2 e^{\lambda_2} x + \ldots + C_n e^{\lambda_n x} \tag{15.6}$$

is also a solution. The additive property of solutions of higher-order linear differential equations expressed by this result is often referred to as the *linear superposition* of solutions.

This solution of the reduced Eqn (15.2) is called the *complementary function*, and if the λ_i are distinct, the n individual solutions y_i are linearly independent. We shall now prove this assertion of linear independence indirectly, by assuming the result to be invalid and producing a contradiction, thereby showing that the assumption of linear dependence is incorrect. This contradiction proves the result.

Consider the case in which the roots λ_i are all distinct, but assume that the y_i are linearly dependent. Then there must exist some constants C_1, C_2, ..., C_n, not all zero, such that

$$C_1 e^{\lambda_1 x} + C_2 e^{\lambda_2 x} + \ldots + C_n e^{\lambda_n x} = 0. \tag{15.7}$$

Successive differentiation of Eqn (15.7) shows that also

$$\left.\begin{aligned} C_1\lambda_1 e^{\lambda_1 x} + C_2\lambda_2 e^{\lambda_2 x} + \ldots + C_n\lambda_n{}^{\lambda_n x} = 0,\\ \cdots\\ C_1\lambda_1^{n-1} e^{\lambda_1 x} + C_2\lambda_2^{n-1} e^{\lambda_2 x} + \ldots + C_n e\lambda_n^{n-1} e^{\lambda_n x} = 0. \end{aligned}\right\} \tag{15.8}$$

Now if Eqns (15.7) and (15.8) are to be true for a non-trivial set of constants C_i, the determinant $|W|$ of the coefficients of the C_i must vanish for all x, thereby giving rise to the condition

$$|W| \equiv \begin{vmatrix} e^{\lambda_1 x} & e^{\lambda_2 x} & \cdots & e^{\lambda_n x} \\ \lambda_1 e^{\lambda_1 x} & \lambda_2 e^{\lambda_2 x} & \cdots & \lambda_n e^{\lambda_n x} \\ \cdots & \cdots & \cdots & \cdots \\ \lambda_1^{n-1} e^{\lambda_1 x} & \lambda_2^{n-1} e^{\lambda_2 x} & \cdots & \lambda_n^{n-1} e^{\lambda_n x} \end{vmatrix} = 0. \tag{15.9}$$

The determinant $|W|$ formed in this manner is known as the *Wronskian* of the n solutions y_i, and plays an important role in more general studies of differential equations. In this case $|W|$ has a simple form and, as the common exponential factor in each column is non-zero, it follows that these may be removed as factors of $|W|$, showing that Eqn (15.9) implies the vanishing of an *alternant determinant* $|A|$, where

$$|A| \equiv \begin{vmatrix} 1 & 1 & \cdots & 1 \\ \lambda_1 & \lambda_2 & \cdots & \lambda_n \\ \cdots & \cdots & \cdots & \cdots \\ \lambda_1^{n-1} & \lambda_2^{n-1} & \cdots & \lambda_n^{n-1} \end{vmatrix} = 0. \tag{15.10}$$

Now we know from the theory of determinants that the value of $|A|$ is simply the product of all possible factors of the form $(\lambda_i - \lambda_j)$ with a suitable sign appended, so that if the roots λ_i are all distinct $|A|$ cannot vanish. This contradiction thus proves that Eqn (15.7) can be true only if all the C_i are zero, and hence the n solutions y_i must be linearly independent. We have thus proved:

Theorem 15.1 (linear independence of solutions – single roots) A differential equation

$$y^{(n)} + a_1 y^{(n-1)} + \ldots + a_n y = 0,$$

which has n distinct roots λ_i of its characteristic equation $P(\lambda) = 0$, has

n linearly independent solutions $y_i = C_i e^{\lambda i x}$. Its general solution, the complementary function, is of the form

$$y = C_1 e^{\lambda_1 x} + C_2 e^{\lambda_2 x} + \ldots + C_n e^{\lambda_n x}. \quad \blacksquare$$

Example 15.1 Suppose that $y'' + 3y' + 2y = 0$, then $P(\lambda) \equiv \lambda^2 + 3\lambda + 2$ and the roots of $P(\lambda) = 0$ are $\lambda = -1$, $\lambda = -2$. The linearly independent solutions are $y_1 = C_1 e^{-x}$ and $y_2 = C_2 e^{-2x}$, and the general solution or complementary function is $y = C_1 e^{-x} + C_2 e^{-2x}$. A simple calculation shows that the Wronskian $|W| = -e^{-3x}$. ■

When r of the roots of the characteristic polynomial $P(\lambda)$ coincide and equal λ^*, say, then $\lambda = \lambda^*$ is said to be a root of *multiplicity* r. The form of the general solution Eqn (15.6) is then inapplicable because r of its terms are linearly dependent. In this situation an additional $r - 1$ linearly independent solutions need to be determined to complete the general solution.

Theorem 15.2 (linear independence of solutions – repeated roots) When $\lambda = \lambda_1$ is a root of multiplicity r of the characteristic equation $P(\lambda) = 0$ belonging to

$$y^{(n)} + a_1 y^{(n-1)} + \ldots + a_n y = 0,$$

then $e^{\lambda_1 x}$, $xe^{\lambda_1 x}$, $x^2 e^{\lambda_1 x}$, $\ldots$, $x^{r-1} e^{\lambda_1 x}$ are linearly independent solutions of the differential equation corresponding to the r-fold root λ_1.

Proof Because the stated form of the linearly independent solutions may be established more easily by a different technique, which we shall discuss later, we only prove the result for a root of multiplicity 2 (a double root). The assertion of linear independence, however, will be proved for any value of r.

When $\lambda = \lambda_1$ is a root of multiplicity 2 the characteristic polynomial $P(\lambda)$ may be written

$$P(\lambda) \equiv \lambda^n + a_1 \lambda^{n-1} + \ldots + a_{n-1}\lambda + a_n = (\lambda - \lambda_1)^2 Q(\lambda), \qquad (15.11)$$

where $Q(\lambda)$ is a polynomial of degree $(n-2)$. Clearly, by definition, $P(\lambda_1) = 0$, and by differentiation of $P(\lambda) = 0$ with respect to λ it also follows that

$$n\lambda_1^{n-1} + (n-1)a_1 \lambda_1^{n-2} + \ldots + a_{n-1} = 0, \qquad (15.12)$$

which is a result that will be needed shortly.

If now we write $y = xe^{\lambda_1 x}$, then $y^{(r)} = r\lambda_1^{r-1} e^{\lambda_1 x} + \lambda_1^r x e^{\lambda_1 x}$ and, as $P(\lambda_1) = 0$, substitution into $y^{(n)} + a_1 y^{(n-1)} + \ldots\ a_n y$ gives $[n\lambda_1^{n-1} + (n-1)a_1\lambda_1^{n-2} + \ldots + a_{n-1}]e^{\lambda_1 x}$. This vanishes by virtue of Eqn (15.12), so we have shown that $y = xe^{\lambda_1 x}$ is actually a solution of Eqn (15.2) when $\lambda = \lambda_1$ is a double root. As $y = e^{\lambda_1 x}$ is obviously also a solution, we have established that when $\lambda = \lambda_1$ is a double root, the general solution must take the form

$$y = C_1 e^{\lambda_1 x} + C_2 x e^{\lambda_1 x} + C_3 e^{\lambda_3 x} + \ldots + C_n e^{\lambda_n x}, \qquad (15.13)$$

where the remaining $(n-2)$ roots $\lambda_3, \lambda_4, \ldots, \lambda_n$ are assumed to be distinct. ■

Whether or not the λ_i are multiple roots, n initial conditions must be specified in order to construct a particular solution from the appropriate form of the general solution. Used in conjunction with the general solution, they enable n simultaneous equations to be formed for the determination of the n arbitrary constants $C_1, C_2, \ldots, C_n$. The usual initial conditions for an nth-order differential equation are the specification of the values of $y, y^{(1)}, y^{(2)}, \ldots, y^{(n-1)}$ at some initial point $x = x_0$.

Now in connection with the Wronskian, we have already seen that a set of exponential functions is linearly independent provided the exponents are distinct. Hence, to show that the functions in Eqn (15.13) are linearly independent, it will be sufficient to show that $e^{\lambda_1 x}$ are $xe^{\lambda_1 x}$ are linearly independent. This result is self-evident, because removal of the common factor $e^{\lambda_1 x}$ leaves the functions 1 and x, which are obviously linearly independent.

For completeness, and for application to roots of multiplicity greater than 2, we prove the more general result that the functions $1, x, x^2, \ldots, x^m$ are linearly independent.

Assuming first that this is not true and that these functions are linearly dependent, it follows that there must exist a non-trivial set of constants $C_0, C_1, \ldots, C_m$ such that, for all x,

$$C_0 + C_1 x + C_2 x^2 + \ldots + C_x x^m = 0.$$

However, we know that as this is an algebraic equation of degree m, there can at most be only m distinct values of x for which it can be true. The expression cannot thus be true for all x, and so the assumption of linear dependence is false. This establishes the result.

Example 15.2 Find the general solution of

$$y''' + 4y'' + 5y' + 2y = 0,$$

and determine the particular solution satisfying the initial conditions

$$y = 1, \; y' = 0 \qquad \text{and} \quad y'' = 1 \text{ when } x = 0.$$

Solution The characteristic polynomial is

$$P(\lambda) = \lambda^3 + 4\lambda^2 + 5\lambda + 2 = (\lambda + 1)^2(\lambda + 2).$$

Thus the roots of $P(\lambda) = 0$ are

$$\lambda = -1 \text{ (multiplicity 2)} \qquad \text{and} \quad \lambda = -2.$$

The linearly independent solutions corresponding to the double root $\lambda = -1$ are $C_1 e^{-x}$ and $C_2 x e^{-x}$, while the solution corresponding to the single root $\lambda = -2$ is $C_3 e^{-2x}$. Thus the general solution is

$$y = C_1 e^{-x} + C_2 x e^{-x} + C_3 e^{-2x}.$$

To determine the particular solution satisfying the initial conditions

$$y(0) = 1, \quad y'(0) = 0 \quad \text{and} \quad y''(0) = 1$$

we must match the constants C_1, C_2 and C_3 in the general solution to the initial conditions. When this is done we obtain

$$\begin{aligned} &(y(0) = 1) && 1 = C_1 + C_3 \\ &(y'(0) = 0) && 0 = -C_1 + C_2 - 2C_3 \\ &(y''(0) = 1) && 1 = C_1 - 2C_2 + 4C_3. \end{aligned}$$

These have the solutions $C_1 = -1$, $C_2 = 3$ and $C_3 = 2$, so the required particular solution is

$$y = -e^{-x} + 3xe^{-x} + 2e^{-2x}$$

or

$$y = e^{-x}(3x - 1) + 2e^{-2x}. \quad \blacksquare$$

Example 15.3 Find the general solution of

$$y^{(5)} + 3y^{(4)} - y^{(3)} - 7y^{(1)} + 4y = 0.$$

Solution The characteristic polynomial is

$$P(\lambda) = \lambda^5 + 3\lambda^4 - \lambda^3 7\lambda + 4 = (\lambda - 1)^2(\lambda + 2)^2(\lambda + 1).$$

Thus the roots of $P(\lambda) = 0$ are

$$\lambda = 1 \text{ (multiplicity 2)}, \quad \lambda = -2 \text{ (multiplicity 2)} \quad \text{and} \quad \lambda = -1.$$

The linearly independent solutions corresponding to the double root $\lambda = 1$ are $C_1 e^x$ and $C_2 x e^x$, while those corresponding to the double root $\lambda = -2$ are $C_3 e^{-2x}$ and $C_4 x e^{-2x}$. The single root $\lambda = -1$ corresponds to the term in the solution $C_5 e^{-x}$. Thus the general solution is

$$y = C_1 e^x + C_2 x e^x + C_3 e^{-2x} + C_4 x e^{-2x} + C_5 e^{-x}. \quad \blacksquare$$

Example 15.4 Find the general solution of

$$y^{(5)} - 3y^{(4)} + 3y^{(3)} - y^{(2)} = 0.$$

Solution The characteristic polynomial is

$$P(\lambda) = \lambda^5 - 3\lambda^4 + 3\lambda^3 - \lambda^2 = \lambda^2(\lambda - 1)^3.$$

Thus the roots of $P(\lambda) = 0$ are

$$\lambda = 0 \text{ (multiplicity 2)} \quad \text{and} \quad \lambda = 1 \text{ (multiplicity 3)}.$$

To the double root $\lambda = 0$ there correspond the terms in the solution C_1 and C_2x, while to the triple root $\lambda = 1$ there correspond the terms C_3e^x, C_4xe^x and $C_5x^2e^x$. Thus the general solution is

$$y = C_1 + C_2x + C_3e^x + C_4xe^x + C_5x^2e^x. \quad \blacksquare$$

Finally we must give consideration to the situation in which the roots λ_i occur in complex conjugate pairs. Suppose that $\lambda_s = \mu + iv$ and its complex conjugate $\bar{\lambda}_s = \mu - iv$ are roots of the characteristic polynomial $P(\lambda) = 0$ of Eqn (15.2). Then, by analogy with the case of real roots, $y_s = \exp[(\mu + iv)x]$ and $y_s^* = \exp[(\mu - iv)x]$ must be solutions of Eqn (15.2). Linear combinations of y_s and y_s^* will also be solutions of Eqn (15.2) and so, in particular, $u = \frac{1}{2}(y_s + y_s^*)$ and $v = (1/2i)(y_s - y_s^*)$ will be solutions. A simple calculation then shows that $u = e^{\mu x} \cos vx$ and $v = e^{\mu x} \sin vx$. Hence the combination of terms corresponding to the complex root λ_s and its complex conjugate $\bar{\lambda}_s$ in the general solution gives rise to the solution

$$e^{\mu x}(A \cos vx + B \sin vx),$$

where A and B are arbitrary real constants. We have established the following general result:

Theorem 15.3 (form of solution – complex roots)

When $\lambda = \mu + iv$ and its complex conjugate $\bar{\lambda}$ are single roots of the characteristic equation $P(\lambda) = 0$ of the differential equation

$$y^{(n)} + a_1 y^{(n-1)} + \ldots + a_n y = 0,$$

and the remaining roots $\lambda_3, \lambda_4, \ldots, \lambda_n$ are real and distinct, the general solution has the form

$$y = e^{\mu x}(C_1 \cos vx + C_2 \sin vx) + C_3 e^{\lambda_3 x} + C_4 e^{\lambda_4 x} + \ldots + C_n e^{\lambda_n x}. \quad \blacksquare$$

The extension of Theorem 15.3 when other complex conjugate pairs of roots occur in the characteristic polynomial, or when there are multiple real roots, is obvious and immediate. When a complex root has multiplicity greater than 1, the results of Theorem 15.2 may be incorporated into Theorem 15.3 to modify the constants C_1 and C_2. Thus, for example, if complex root λ_s had multiplicity 2, the general solution stated in Theorem 15.3 would take the form

$$\begin{aligned} y = {} & e^{\mu x}[(C_1 + C_2x)\cos vx + (C_3 + C_4x)\sin vx] + C_5 e^{\lambda_5 x} \\ & + C_6 e^{\lambda_6 x} + \ldots + C_n e^{\lambda_n x}. \end{aligned}$$

The results of Theorem 15.3 may be extended to the case in which pairs of complex conjugate roots occur with multiplicity s by employing the same form of argument as that used in Theorem 15.2. Taken together, the results we have just established may be combined to form the following rule for deciding the form of the general solution (complementary function).

Rule 15.1 (Form taken by the general solution of a homogeneous equation)

It is required to find the general solution of the nth-order homogeneous equation

$$\frac{d^n y}{dx^n} + a_1 \frac{d^{n-1} y}{dx^{n-1}} + \ldots + a_n y = 0,$$

with real constant coefficients $a_1, a_2, \ldots, a_n$. The solution is called the *complementary function* and its form is determined as follows.

1. Find the roots of the characteristic equation

   $$\lambda^n + a_1 \lambda^{n-1} + \ldots + a_n = 0.$$

2. For every simple root $\lambda = \alpha$ (multiplicity 1) include in the solution the term

   $$Ae^{\alpha x},$$

   where A is an arbitrary real constant.
3. For every real root $\lambda = \beta$ of multiplicity r include in the solution the terms

   $$B_1 e^{\beta x} + B_2 x e^{\beta x} + \ldots + B_r x^{r-1} e^{\beta x},$$

   where $B_1, B_2, \ldots, B_r$ are arbitrary real constants.
4. For every pair of complex conjugate roots

   $$x = \mu + iv \quad \text{and} \quad x = \mu - iv$$

   include in the solution the terms

   $$e^{\mu x}(C \cos vx + D \sin vx),$$

   where C and D are arbitrary real constants.
5. For every pair of complex conjugate roots $x = \gamma + i\omega$ and $x = \gamma - i\omega$, each with multiplicity s, include in the solution the terms

   $$e^{\gamma x}(E_1 \cos \omega x + F_1 \sin \omega x + E_2 x \cos \omega x + F_2 x \sin \omega x + \ldots + E_s x^{s-1} \cos \omega x + F_s x^{s-1} \sin \omega x)$$

   where $E_1, F_1, \ldots, E_s, F_s$ are arbitrary real constants.
6. The general solution is then the sum of all the terms generated in steps 2 to 5.

Example 15.5 The differential equation $y''+4y'+13y=0$ has the characteristic polynomial $P(\lambda)\equiv\lambda^2+4\lambda+13=(\lambda+2+3i)(\lambda+2-3i)$ and the roots of $P(\lambda)=0$ are $\lambda=-2-3i$ and $\lambda=-2+3i$. The general solution is $y=e^{-2x}(C_1\cos 3x+C_2\sin 3x)$. ■

Example 15.6 The differential equation $y^{(5)}+3y^{(4)}+10y^{(3)}+6y^{(2)}+5y^{(1)}-25y=0$ has the characteristic polynomial $P(\lambda)\equiv\lambda^5+3\lambda^4+10\lambda^3+6\lambda^2+5\lambda-25=(\lambda-1)(\lambda+1-2i)^2(\lambda+1+2i)^2$. The complex roots $\lambda=-1-2i$ and $\lambda=-1+2i$ of $P(\lambda)=0$ are double roots, and the single root $\lambda=1$ is the only real root. The general solution is $y=e^{-x}[(C_1+C_2x)\cos 2x+(C_3+C_4x)\sin 2x]+C_5e^x$. ■

15.2 Linear equations with constant coefficients – inhomogeneous case

We now examine methods of solution of the inhomogeneous differential Eqn (15.1). Our approach will be to progress from a semi-intuitive method known as the method of undetermined coefficients to a systematic approach called the method of variation of parameters. To complete the chapter, a brief introduction is given to the solution of linear differential equations by means of the Laplace transform.

15.2.1 The structure of a general solution

It is an easily verified fact that $y=C_1\cos x+C_2\sin x+\frac{1}{2}e^{-x}$ is a solution of the inhomogeneous equation $y''+y=e^{-x}$. The first two terms of this solution obviously comprise the complementary function of the reduced equation $y''+y=0$, while the last term is a function which, when substituted into the differential equation, gives rise to the inhomogeneous term. There thus appear to be two distinct parts to this solution, the first being the general solution to the reduced equation and the second, which is additive, being a solution particular to the form of the inhomogeneous term. We now prove a theorem that establishes that this is in fact the pattern of solution that applies to all inhomogeneous linear equations. The sum of the two parts is termed the *general solution* or the *complete primitive* of the inhomogeneous equation.

To simplify manipulation it will be convenient to introduce a concise notation for the left-hand side of differential Eqn (15.1), and we achieve this be defining $L[y]\equiv y^{(n)}+a_1y^{(n-1)}+\ldots+a_ny$. In terms of this notation, in which $a_1, a_2, \ldots, a_n$ are understood to be constants, we now state:

Theorem 15.4 (form of general solution of linear inhomogeneous equations) The general solution of the inhomogeneous equation $L[y]=f(x)$ is of the form $y(x)=y_c(x)+y_p(x)$, where $y_c(x)$ is the general solution or complementary function of the reduced equation $L[y]=0$, and $y_p(x)$ is a particular solution of $L[y]=f(x)$.

Proof The proof is straightforward. Firstly, $y(x)=y_c(x)+y_p(x)$ does satisfy the equation since $L[y_c(x)]=0$ and $L[y_p(x)]=f(x)$, and as $(d^r/dx^r)(y_c+y_p)=(d^ry_c/dx^r)+(d^ry_p/dx^r)$ for $r=1, 2, \ldots, n$, it follows that

$$L(y) = L[y_c(x) + y_p(x)] = L[y_c(x) + L[y_p(x)]$$
$$= 0 + f(x) = f(x).$$

As $y_c(x)$ contains n arbitrary constants, we choose to write it in the form $y_c(x;\, C_1, C_2, \ldots, C_n)$ to make this explicit. Then, clearly, adding two complementary functions with differing constants C_i and C_i' gives

$$y_c(x;\; C_1,\; C_2,\; \ldots,\; C_n) + y_c(x;\; C_1',\; C_2',\; \ldots,\; C_n')$$
$$= y_c(x;\; C_1 + C_1',\; C_2 + C_2',\; \ldots,\; C_n + C_n'),$$

which is simply the same form of complementary function but with modified constants. Suppose next that $y_{1p}(x)$ and $y_{2p}(x)$ are two particular solutions of $L[y] = f(x)$. Then $L[y_{1p}(x) - y_{2p}(x)] = L[y_{1p}(x)] - L[y_{2p}(x)] = f(x) - f(x) = 0$ and so $y_{1p}(x) - y_{2p}(x)$ is a solution of $L[y] = 0$. Hence $y_{1p}(x) - y_{2p}(x) = y_c(x;\, C_1^0, C_2^0, \ldots, C_4^0)$, which is again the same form of complementary function but with some other set of constants.

Now

$$\begin{aligned} y_c(x; C_1, C_2,\; \ldots,\; C_n) + y_{1p}(x) &= y_c(x; C_1, C_2, \ldots, C_n) + y_{1p}(x) \\ &\quad - y_{2p}(x) + y_{2p}(x) \\ &= y_c(x; C_1, C_2, \ldots, C_n) \\ &\quad + y_c(x; C_1^0, C_2^0, \ldots, C_n^0) \\ &\quad + y_{2p}(x) \\ &= y_c(x; C_1 + C_1^0, C_2 + C_2^0, \ldots, \\ &\quad C_n + C_n^0) + y_{2p}(x), \end{aligned}$$

and so we have shown that any two particular solutions give rise to the same form of general solution. The arbitrary constants in this general solution must be determined by applying the initial conditions to the complete primitive $y = y_c(x) + y_p(x)$. ■

We have already seen the simplest example of this theorem in connection with Eqn (14.20), which clearly displays the two parts of the general solution of an inhomogeneous linear equation of first order. As in that section, we shall call the particular solution $y_p(x)$ of the inhomogeneous equation a *particular integral* of the differential equation.

15.2.2 The method of undetermined coefficients

The determination of simple particular integrals by means of the *method of undetermined coefficients* is best illustrated by example. In essence, the method is based on the fact that simple forms of the inhomogeneous term $f(x)$ in Eqn (15.1) can only arise as the result of differentiation of obvious functions in which the values of certain constants are the only things that need determination. A solution is achieved by substitution of a trial function into the inhomogeneous equation and subsequent comparison of coefficients of corresponding terms. There are four cases to consider.

(a) $f(x)$ *a polynomial in* x. Suppose $y'' + y = x^2$, then by inspection we see that the particular integral y_p can only be a polynomial in x and, furthermore, that it cannot be of degree higher than 2 since the equation contains an undifferentiated term y. Let us set $y_p = ax^2 + bx + c$; then $y_p'' = 2a$ and substitution into the original equation gives $2a + ax^2 + bx + c = x^2$. Equating the coefficients of corresponding powers of x shows that $a = 1$, $b = 0$, $2a + c = 0$, so that $c = -2$, and the required particular integral must be $y_p = x^2 - 2$. The general solution, or complete primitive, is $y = C_1 \cos x + C_2 \sin x + x^2 - 2$. To determine the solution appropriate to the initial conditions $y = -2$, $y' = 0$, say, at $x = \frac{1}{2}\pi$ we notice that the condition on y gives $-2 = C_1 \cos\frac{1}{2}\pi + C_2 \sin\frac{1}{2}\pi + \frac{1}{4}\pi^2 - 2$, or $C_2 = -\frac{1}{4}\pi^2$, while the condition on y' gives $0 = -C_1 \sin\frac{1}{2}\pi + C_2 \cos\frac{1}{2}\pi + \pi$, or $C_1 = \pi$. Hence the solution appropriate to these initial conditions is $y = \pi(\cos x - \frac{1}{4}\pi \sin x) + x^2 - 2$.

The method extends directly to linear constant coefficient equations of any order and to polynomials $f(x)$ of any degree. For example, using the same argument to determine the particular integral of $y'' + 3y' + y = 1 + x^3$, we would try a particular integral of the form $y_p = ax^3 + bx^2 + cx + d$. Substitution in the equation and comparison of the coefficients of corresponding powers of x would then show that $a = 1$, $9a + b = 0$, $6a + 6b + c = 0$ and $2b + 3c + d = 1$. These equations have the solution $a = 1$, $b = -9$, $c = 48$ and $d = -125$. The particular integral must be $y_p = x^3 - 9x^2 + 48x - 125$.

(b) $f(x)$ *an exponential function*. Suppose now that $y'' + 3y' + 2y = 3e^{2x}$. As e^{2x} does not appear in the complementary function $y_c = C_1 e^{-x} + C_2 e^{-2x}$ (in which case it would be a solution of the homogeneous equation), it can only arise in the inhomogeneous term as a result of differentiation of a function of the form $y_p = ke^{2x}$. Substituting this in the equation and cancelling the common factor e^{2x} shown that $4k + 6k + 2k = 3$ or $k = \frac{1}{4}$. The required particular integral must thus be $y_p = \frac{1}{4}e^{2x}$. A sum of exponentials occurring in the inhomogeneous term would be treated analogously, the constant multiplier of each being determined separately by the above method.

A complication arises if an exponential in the inhomogeneous term also occurs in the complementary function, as is the case with $y'' + y' - 2y = 2e^x$, which has the complementary function $y_c = C_1 e^x + C_2 e^{-2x}$. Attempting to find a solution by substituting $y_p = ke^x$ would fail here since e^x is a solution of $y'' + y' - 2y = 0$. A moment's reflection and consideration along lines similar to those concerning Theorem 15.2 shows that in this case we must try $y_p = kx\,e^x$. Then $y_p' = k(1 + x)e^x$

and $y_p'' = k(2+x)e^x$, so that substitution into the differential equation and cancellation of the common factor e^x gives the condition $3k = 2$. In this case the required particular integral is $y_p = \frac{2}{3}xe^x$. By an obvious extension, if an exponential term $e^{\lambda x}$ appears in the inhomogeneous term $f(x)$, and also occurs in the complementary function as a result of a root of the characteristic equation which has multiplicity r, then the particular integral will be of the form $y_p = kx^r e^{\lambda x}$. Suppose, for example, that $y'' - 2y' + y = 2e^x$; then the complementary function $y_c = (C_1 + C_2 x)e^x$ arises as a result of the double root $\lambda = 1$ of the characteristic equation $P(\lambda) \equiv \lambda^2 - 2\lambda + 1 = 0$. Hence we must seek a particular integral of the form $y_p = kx^2 e^x$. Substituting this in the equation and cancelling the common factor e^x shows that $k = 1$; hence the particular integral is $y_p = x^2 e^x$. The general solution is $y = (C_1 + C_2 x)e^x + x^2 e^x$.

(c) $f(x)$ *a trigonometric function*. The same method may be applied to an inhomogeneous trigonometric term involving $\sin mx$ or $\cos mx$. In this case, provided neither function occurs solely as a term in the complementary function, a particular integral of the form $y_p = a\cos mx + b\sin mx$ must be sought. Let us consider $y'' - 2y' + 2y = \sin x$, which has the complementary function $y_c = e^x(C_1 \cos x + C_2 \sin x)$. Substituting $y_p = a\cos x + b\sin x$ gives the result $-a\cos x - b\sin x + 2a\sin x - 2b\cos x + 2a\cos x + 2b\sin x = \sin x$. For this to be an identity we must equate the coefficients of $\sin x$ and $\cos x$ on each side of the equation. For the term $\sin x$ this gives rise to the equation $2a + b = 1$, and for the term $\cos x$, the equation $a - 2b = 0$. Hence $a = 2/5$, $b = 1/5$, so that $y_p = \frac{1}{5}(2\cos x + \sin x)$. The general solution is of course $y = e^x(C_1 \cos x + C_2 \sin x) + \frac{1}{5}(2\cos x + \sin x)$.

If, however, the trigonometric function in the inhomogeneous term occurs as part of the complementary function then considerations similar to those in the latter part of case (b) will apply. We must then try to find a particular integral of the form $y_p = x^r(a\cos mx + b\sin mx)$, where r is the multiplicity of the root of the characteristic polynomial giving rise to the term $\cos mx$ or $\sin mx$ in the complementary function.

Suppose, for example, that $y'' + y = \cos x$; then the characteristic polynomial is $P(\lambda) \equiv \lambda^2 + 1 = (\lambda - i)(\lambda + i)$, showing that terms in the complementary function $y_c = C_1 \cos x + C_2 \sin x$ arise from roots of $P(\lambda) = 0$ having multiplicity 1. We must thus attempt a particular solution of the form $y_p = ax\cos x + bx\sin x$. Substitution into the equation gives $-2a\sin x + 2b\cos x - ax\cos x - bx\sin x + ax\cos x + bx\sin x = \cos x$. Equating coefficients of $\cos x$ and $\sin x$ as before shows that $a = 0$, $b = \frac{1}{2}$, and hence $y_p = \frac{1}{2}x\sin x$.

(d) $f(x)$ *a product of exponential and trigonometric functions*. The previous methods extend to allow the determination of a particular integral when the inhomogeneous term is of the form $e^{kx}\cos mx$ or $e^{kx}\sin mx$. In this case a solution of the form $y_p = e^{kx}(a\cos mx + b\sin mx)$ must be sought. In the case of the equation $y'' - 3y' + 2y = e^{-x}\sin x$, as the complementary function $y_c = C_1 e^x + C_2 e^{2x}$ does not contain the inhomogeneous term, we seek a solution of the form $y_p = e^{-x}(a \cos x + b \sin x)$. Substituting and proceeding as before it is easily established that $a = b = 1/10$, from which we deduce that $y_p = \frac{1}{10} e^{-x}(\cos x + \sin x)$.

The treatment of the foregoing cases of commonly occuring inhomogeneous terms may be summarized in the form of a rule as follows.

Rule 15.2 (Determination of the form of a particular integral by the method of undetermined coefficients)

It is required to find the form of a particular integral y_p of

$$L[y] = f(x),$$

where

$$L[y] = \frac{d^n y}{dx^n} + a_1 \frac{d^{n-1} y}{dx^{n-1}} + \ldots + a_n y,$$

and $a_1, a_2, \ldots, a_n$ are constants.

1. $f(x) = \text{constant}$.
   Include in y_p the constant term K.
2. $f(x)$ is a polynomial in x of degree r.

   (a) If $L[y]$ contains an undifferentiated term y, include in y_p terms of the form

   $$A_0 x^r + A_1 x^{r-1} + \ldots + A_r.$$

   (b) If $L[y]$ does not contain an undifferentiated y, and $y^{(s)}$ is its lowest-order derivative, include in y_p terms of the form

   $$A_0 x^{r+s} + A_1 x^{r+s-1} + \ldots + A_r x^s.$$

3. $f(x) = e^{ax}$.

   (a) If e^{ax} is not contained in the complementary function, include in y_p the term

   $$B e^{ax}.$$

   (b) If the complementary function contains the terms $e^{ax}, xe^{ax}, \ldots, x^m e^{ax}$, include in y_p the term

   $$B x^{m+1} e^{ax}.$$

4. $f(x)$ contains terms in $\cos qx$ and/or $\sin qx$.

   (a) If $\cos qx$ and/or $\sin qx$ are not contained in the complementary function, include in y_p terms of the form

   $$C \cos qx + D \sin qx.$$

   (b) If the complementary function contains the terms $x^s \cos qx$ and/or $x^s \sin qx$ with $s = 0, 1, 2, \ldots, m$, include in y_p terms of the form

$$x^{m+1}(C\cos qx + D\sin qx).$$

5. $f(x)$ contains terms in $e^{px}\cos qx$ and/or $e^{px}\sin qx$.
   (a) If $e^{px}\cos qx$ and/or $e^{px}\sin qx$ are not contained in the complementary function, include in y_p terms of the form
   $$e^{px}(E\cos qx + F\sin qx).$$
   (b) If the complementary function contains $x^s e^{px}\cos qx$ and/or $x^s e^{px}\sin qx$ with $s = 0, 1, \ldots, m$, include in y_p terms of the form
   $$x^{m+1}e^{px}(E\cos qx + F\sin qx).$$
6. The form of the particular integral y_p is the sum of all the terms generated as a result of identifying each term in $f(x)$ with one of the special forms of $f(x)$ listed above in 1 to 5.
7. The numerical values of the undetermined constants K, A_0, A_1,..., E and F are determined by substituting the form of particular integral y_p obtained in step 6 into
   $$L[y_p] = f(x),$$
   and then equating the coefficients of corresponding terms on each side of the equation to make the result an identity.

The following examples illustrate the application of Rule 15.2.

Example 15.7 Find the general solution of

$$y'' + 4y = 5 + 3\cos 2x + 4\sin 2x,$$

and hence find the solution satisfying the initial conditions $y(0) = 0$ and $y'(0) = 1$.

Solution The general solution is the sum of the complementary function y_c and a particular integral y_p. The reduced equation is

$$y'' + 4y = 0,$$

so the characteristic polynomial is

$$P(\lambda) = \lambda^2 + 4 = (\lambda - 2i)(\lambda + 2i).$$

Thus from Rule 15.1, step 4, we see that

$$y_c = A\cos 2x + B\sin 2x.$$

To construct the particular solution we now make use of Rule 15.2. To allow for the presence of the constant 5 in the inhomogeneous term, it follows from Rule 15.2, step 1, that we must include a constant term K.

As the remaining inhomogeneous terms $3\cos 2x + 4\sin 2x$ occur in the complementary function, it follows from Rule 15.2, step 4(b), that we must include the terms $Cx\cos 2x + Dx\sin 2x$. Thus for the particular integral we must adopt a function of the form

$$y_p = K + Cx\cos 2x + Dx\sin 2x.$$

Substituting $y = y_p$ in

$$y'' + 4y = 5 + 3\cos 2x + 4\sin 2x,$$

we obtain the result

$$4K - 4C\sin 2x + 4D\cos 2x = 5 + 3\cos 2x + 4\sin 2x.$$

Equating the coefficients of corresponding terms on either side of this identity gives

$$4K = 5, \quad \text{so} \quad K = \tfrac{5}{4}$$

$$-4C = 4, \quad \text{so} \quad C = -1$$

$$4D = 3, \quad \text{so} \quad D = \tfrac{3}{4}.$$

Thus substituting these results back into the expression for y_p gives

$$y_p = \tfrac{5}{4} - x\cos 2x + \tfrac{3}{4}x\sin 2x.$$

This is the required particular integral. Hence the general solution is

$$y = y_c + y_p = A\cos 2x + B\sin 2x + \tfrac{5}{4} - x\cos 2x + \tfrac{3}{4}x\sin 2x.$$

To find the solution satisfying the initial conditions $y(0) = 0$, $y'(0) = 1$, we must match the arbitrary constants A and B to these conditions. Using the condition $y(0) = 0$ we obtain

$$0 = A + \tfrac{5}{4}, \quad \text{so} \quad A = -\tfrac{5}{4}.$$

Now

$$y' = -2A\sin 2x + 2B\cos 2x - \cos 2x + 2x\sin 2x$$
$$+ \tfrac{3}{4}\sin 2x + \tfrac{3}{2}x\cos 2x,$$

so the condition $y'(0) = 1$ gives

$$1 = 2B - 1, \quad \text{or} \quad B = 1.$$

Thus the solution satisfying the initial value problem is

$$y = \tfrac{5}{4} - \tfrac{5}{4}\cos 2x + \sin 2x - x\cos 2x + \tfrac{3}{4}x\sin 2x. \quad \blacksquare$$

Example 15.8 Find the general solution of

$$y'' + 3y' + 2y = 1 + x^2 + e^{3x}.$$

Solution The complementary function has already been found in Example 15.1, where it was shown to be

$$y_c = Ae^{-x} + Be^{-2x}.$$

Inspection of the inhomogeneous term in conjunction with Rule 15.2 shows that because of the polynomial $1+x^2$, we must include the terms $C+Dx+Ex^2$ (step 2(a)), and because of the exponential e^{3x} which is not contained in the complementary function we must include the term Fe^{3x} (step 3(a)). Thus for the particular integral we must adopt a function of the form

$$y_p = C + Dx + Ex^2 + Fe^{3x}.$$

Subsituting $y = y_p$ into

$$y'' + 3y' + 2y = 1 + x^2 + e^{3x}$$

gives

$$2E + 9Fe^{3x} + 3D + 6Ex + 9Fe^{3x} + 2C + 2Dx + 2Ex^2 + 2Fe^{3x} = 1 + x^2 + e^{3x}.$$

For this expression to be an identity it is necessary that the coefficients of corresponding terms on either side of the equation should be the same.

(constant terms) $2E+3D+2C=1$
(terms in x) $6E+2D=0$
(terms in x^2) $2E=1$
(terms in e^{3x}) $20F=1$.

Thus

$$C=\tfrac{9}{4},\quad D=-\tfrac{3}{2},\quad E=\tfrac{1}{2}\quad\text{and}\quad F=\tfrac{1}{20},$$

and so the particular integral is

$$y_p = \tfrac{9}{4} - \tfrac{3}{2}x + \tfrac{1}{2}x^2 + \tfrac{1}{20}e^{3x}.$$

Thus the general solution is

$$y = y_c + y_p = Ae^{-x} + Be^{-2x} + \tfrac{9}{4} - \tfrac{3}{2}x + \tfrac{1}{2}x^2 + \tfrac{1}{20}e^{3x}. \quad \blacksquare$$

Example 15.9 Find the general solution of

$$y'' - 4y' + 13y = 3 + 2e^{2x}\sin 3x.$$

Solution The reduced equation is

$$y'' - 4y' + 13y = 0,$$

so the characteristic polynomial is

$$P(\lambda) = \lambda^2 - 4\lambda + 13 = (\lambda - 2 - 3i)(\lambda - 2 + 3i).$$

As the roots of $P(\lambda)=0$ are $\lambda = 2 \pm 3i$, it follows from Rule 15.1, step 4, that the complementary function is

$$y_c = e^{2x}(A\cos 3x + B\sin 3x).$$

When constructing the particular integral it follows from Rule 15.2, step 1, that we must include a term K to take account of the constant 3 in the inhomogeneous term.

The remainder of the inhomogeneous term $2e^{2x}\sin 3x$ is contained in y_p, so from Rule 15.2, step 5(b), we must include in y_p the terms

$$xe^{2x}(C\cos 3x + D\sin 3x).$$

Thus the form of particular integral we must adopt is

$$y_p = K + xe^{2x}(C\cos 3x + D\sin 3x).$$

Substituting $y = y_p$ into

$$y'' - 4y' + 13y = 3 + 2e^{2x}\sin 3x$$

gives

$$13K - 6Ce^{2x}\sin 3x + 6De^{2x}\cos 3x = 3 + 2e^{2x}\sin 3x.$$

Equating coefficients, we obtain

$$13K = 3, \quad \text{so} \quad K = 3/13,$$

$$-6C = 2, \quad \text{so} \quad C = -1/3$$

$$\text{and} \quad 6D = 0, \quad \text{so} \quad D = 0.$$

Thus the particular integral is

$$y_p = \tfrac{3}{13} - \tfrac{1}{3}xe^{2x}\cos 3x,$$

and so the general solution is

$$y = e^{2x}(A\cos 3x + B\sin 3x) + \tfrac{3}{13} - \tfrac{1}{3}xe^{2x}\cos 3x. \quad ■$$

Example 15.10 Find the general solution of

$$y''' + y'' + 3y' - 5y = 2 + x + \sin 2x.$$

Solution The reduced equation is

$$y''' + y'' + 3y' - 5y = 0,$$

so the characteristic polynomial is

$$P(\lambda) = \lambda^3 + \lambda^2 + 3\lambda - 5 = (\lambda - 1)(\lambda + 1 - 2i)(\lambda + 1 + 2i).$$

Thus the roots of $P(\lambda) = 0$ are $\lambda = 1$ and $\lambda = -1 \pm 2i$. It then follows from Rule 15.1 that to the real root $\lambda = 1$ there corresponds a term Ae^x, while to the pair of complex conjugate roots there correspond the terms

$$e^{-x}(B\cos 2x + C\sin 2x).$$

Thus the complementary function is

$$y_c = Ae^x + e^{-x}(B\cos 2x + C\sin 2x).$$

As the inhomogeneous term $2+x+\sin 2x$ is not contained in the complementary function it follows from Rule 15.2 that the particular integral y_p must be of the form

$$y_p = D + Ex + F \sin 2x.$$

Here we made use of Rule 15.2, step 2(a), to arrive at the terms $D+Ex$, and Rule 15.2, step 4(a), to arrive at the term $F \sin 2x$.

Substituting $y = y_p$ into

$$y''' + y'' + 3y' - 5y = 2 + x + \sin 2x$$

gives

$$-8F \cos 2x - 4F \sin 2x + 3E + 6F \cos 2x - 5D - 5Ex - 5F \sin 2x$$
$$= 2 + x + \sin 2x.$$

Equating coefficients gives

$$3E - 5D = 2, \quad -5E = 1 \quad \text{and} \quad -9F = 1,$$

so $\quad D = -13/25, \quad E = -1/5 \quad \text{and} \quad F = -1/9.$

Thus the particular integral is

$$y_p = -\tfrac{13}{25} - \tfrac{1}{5}x - \tfrac{1}{9}\sin 2x,$$

so the general solution is

$$y = y_c + y_p = Ae^x + e^{-x}(B \cos 2x + C \sin 2x) - \tfrac{13}{25} - \tfrac{1}{5}x - \tfrac{1}{9}\sin 2x. \quad ■$$

Example 15.11 Find the general solution of

$$y''' + 3y'' = x.$$

Solution The reduced equation is

$$y''' + 3y'' = 0,$$

so the characteristic polynomial is

$$P(\lambda) = \lambda^3 + 3\lambda^2 = \lambda^2(\lambda + 3).$$

Thus the roots of $P(\lambda) = 0$ are $\lambda = 0$ (twice) and $\lambda = -3$. It then follows from Rule 15.1, step 2, that for the single root $\lambda = -3$ we must include the term Ae^{-3x}, while for the double root, Rule 15.1, step 3, shows that we must include the terms $B+Cx$. Hence the complementary function

$$y_c = Ae^{-3x} + B + Cx.$$

Now as the term x in the inhomogeneous term is included in the complementary function, and y is absent from the differential equation, and the lowest-order derivative to appear is 2, it follows from Rule 15.2, step 2(b), that we must include the terms

$$Dx^2 \quad \text{and} \quad Ex^3.$$

Thus the form of the particular integral to be adopted is

$$y_p = Dx^2 + Ex^3.$$

Substituting $y = y_p$ in

$$y''' + 3y'' = x$$

gives

$$6E + 6D + 18Ex = x.$$

Equating coefficients, we obtain

$$6(E + D) = 0, \quad \text{so} \quad D = -E$$

and

$$18E = 1, \quad \text{so} \quad E = 1/18.$$

Thus, $D = -1/18$, $E = 1/18$ and

$$y_p = -\tfrac{1}{18}x^2 + \tfrac{1}{18}x^3.$$

Hence the general solution is

$$y = y_c + y_p = Ae^{-3x} + B + Cx - \tfrac{1}{18}x^2 + \tfrac{1}{18}x^3. \quad \blacksquare$$

15.3 Variation of parameters

An alternative approach to the determination of the general solution of an inhomogeneous equation when the complementary function is known is provided by the method of variation of parameters. The importance of this method lies in the fact that it offers a systematic approach to the problem of the determination of a particular integral. This approach relies for its success on routine integration, and not on guessing the form of a particular integral and then matching constants, as in the method of undetermined coefficients.

We shall see that this method deals automatically with all the special cases considered in section 15.2.2 and, in addition, that it enables us to find particular integrals whose form is unlikely to be guessed. Although we shall only discuss in detail the application of the method to a second-order equation, we will outline later how it may be extended to equations of any order.

Consider the linear constant-coefficient inhomogeneous second-order equation

$$y'' + ay' + by = f(x). \tag{15.14}$$

Then if y_1 and y_2 are two nearly independent solutions of the reduced equation

$$y'' + ay' + by = 0, \tag{15.15}$$

the complementary function is

$$y_c(x) = C_1 y_1(x) + C_2 y_2(x), \tag{15.16}$$

with C_1 and C_2 arbitrary constants. Let us now attempt to find a solution of inhomogeneous Eqn (15.14) by setting

$$y(x) = C_1(x)y_1(x) + C_2(x)y_2(x), \tag{15.17}$$

where $C_1(x)$ and $C_2(x)$ are functions of x which must be determined.

If Eqn (15.17) is to be a solution of Eqn (15.14) then, denoting differentiation with respect to x by a prime, it follows that we may write

$$y'(x) = C_1(x)y_1'(x) + C_2(x)y_2'(x), \tag{15.18}$$

if we impose the special condition that

$$C_1'(x)y_1(x) + C_2'(x)y_2(x) = 0. \tag{15.19}$$

Differentiating Eqn (15.18) with respect to x, we also have

$$y''(x) = C_1(x)y_1''(x) + C_2(x)y_2''(x) + C_1'(x)y_1'(x) + C_2'(x)y_2'(x). \tag{15.20}$$

Substituting Eqns (15.17), (15.18) and (15.19) into Eqn (15.14) and grouping terms gives

$$\begin{aligned} &C_1(x)[y_1''(x) + ay_1'(x) + by_1(x)] + C_2(x)[y_2''(x) + ay_2'(x) + by_2(x)] \\ &\quad + C_1'(x)y_1'(x) + C_2'(x)y_2'(x) = f(x). \end{aligned} \tag{15.21}$$

Now as $y_1(x)$ and $y_2(x)$ are solutions of the reduced equation it follows that the multipliers of $C_1(x)$ and $C_2(x)$ in Eqn (15.21) are zero, so it reduces to

$$C_1'(x)y_1'(x) + C_2'(x)y_2'(x) = f(x). \tag{15.22}$$

Solving Eqns (15.19) and (15.22) for the functions $C_1'(x)$ and $C_2'(x)$, and then integrating, enables us to find $C_1(x)$ and $C_2(x)$. Using these functions in Eqn (15.17) gives the required solution. To arrive at the solution in this manner, we first solve for $C_1'(x)$ and $C_2'(x)$; we find that

$$C_1'(x) = -\frac{y_2(x)f(x)}{|W|}, \tag{15.23}$$

and

$$C_2'(x) = \frac{y_1(x)f(x)}{|W|}, \tag{15.24}$$

where

$$|W| = y_1(x)y_2'(x) - y_1'(x)y_2(x). \tag{15.25}$$

It will be recognized that $|W|$ is the Wronskian of the functions $y_1(x)$ and $y_2(x)$, and it will not vanish because these functions are linearly independent. Integrating Eqns (15.23) and (15.24) gives

$$C_1(x) = C_1 - \int \frac{y_2(x)f(x)}{|W|}\,dx \tag{15.26}$$

and

$$C_2(x) = C_2 + \int \frac{y_1(x)f(x)}{|W|}\,\mathrm{d}x, \tag{15.27}$$

where C_1 and C_2 are arbitrary constants.

Thus, incorporating Eqns (15.26) and (15.27) into Eqn (15.17), we arrive at the general solution of Eqn (15.14) in the form

$$y(x) = C_1 y_1(x) + C_2 y_2(x) - y_1(x)\left\{\int \frac{y_2(x)f(x)}{|W|}\,\mathrm{d}x\right\} + y_2(x)\left\{\int \frac{y_1(x)f(x)}{|W|}\,\mathrm{d}x\right\}. \tag{15.28}$$

The first two terms in Eqn (15.28) are seen to comprise the complementary function of Eqn (15.14):

$$y_c(x) = C_1 y_1(x) + C_2 y_2(x).$$

The remaining two terms in Eqn (15.28) represent the particular integral

$$y_p(x) = -y_1(x)\left\{\int \frac{y_2(x)f(x)}{|W|}\,\mathrm{d}x\right\} + y_2(x)\left\{\int \frac{y_1(x)f(x)}{|W|}\,\mathrm{d}x\right\}. \tag{15.29}$$

Thus we see that by omitting the arbitrary constants C_1 and C_2 in Eqn (15.28) the method of variation of parameters generates the particular integral $y_p(x)$.

Example 15.12 Find the general solution of

$$y'' + 5y' + 6y = x\mathrm{e}^{-2x}.$$

Solution The complementary function is easily seen to be

$$y_c(x) = C_1\mathrm{e}^{-3x} + C_2\mathrm{e}^{-2x},$$

so we shall set

$$y_1(x) = \mathrm{e}^{-3x} \quad \text{and} \quad y_2(x) = \mathrm{e}^{-2x}.$$

Then the Wronskian

$$|W| = y_1 y_2' - y_1' y_2 = \mathrm{e}^{-5x},$$

and as $f(x) = x\mathrm{e}^{-2x}$ it follows from Eqn (15.28) that the general solution is

$$y(x) = C_1\mathrm{e}^{-3x} + C_2\mathrm{e}^{-2x} - \mathrm{e}^{-3x}\int x\mathrm{e}^{x}\,\mathrm{d}x + \mathrm{e}^{-2x}\int x\,\mathrm{d}x.$$

Thus the required general solution is

$$y(x) = C_1\mathrm{e}^{-3x} + C_2\mathrm{e}^{-2x} + (1-x)\mathrm{e}^{-2x} + \tfrac{1}{2}x^2\mathrm{e}^{-2x}.$$

The particular integral in this case is

$$y_p(x) = (1 - x)e^{-2x} + \tfrac{1}{2}x^2 e^{-2x}. \quad \blacksquare$$

Example 15.13 Find the general solution of

$$y'' - 2y' + y = 3xe^x.$$

Solution The complementary function is seen to be

$$y_c(x) = C_1 e^x + C_2 x e^x,$$

so we shall set

$$y_1(x) = e^x \quad \text{and} \quad y_2(x) = xe^x.$$

The Wronskian becomes

$$|W| = e^{2x},$$

so setting $f(x) = 3xe^x$ and substituting into Eqn (15.28) gives

$$y(x) = C_1 e^x + C_2 x e^x - e^x \int 3x^2 \, dx + x\, e^x \int 3x \, dx,$$

and so the general solution is

$$y(x) = C_1 e^x + C_2 x e^x + \tfrac{1}{2}x^3 e^x.$$

Notice that although the inhomogeneous term is included in the complementary function the method still generates the general solution without added complication. ■

Example 15.14 Find the general solution of

$$y'' + y = \tan x \quad \text{for} \quad -\pi/2 < x < \pi/2.$$

Solution The complementary function is

$$y_c(x) = C_1 \cos x + C_2 \sin x,$$

so we shall set

$$y_1(x) = \cos x \quad \text{and} \quad y_2(x) = \sin x.$$

The Wronskian becomes

$$|W| = \cos^2 x + \sin^2 x \equiv 1.$$

Substituting these results into Eqn (15.28), together with $f(x) = \tan x$, we obtain the general solution in the form

$$y(x) = C_1 \cos x + C_2 \sin x - \cos x \int \sin x \tan x \, dx + \sin x \int \cos x \tan x \, dx.$$

Now

$$\sin x \tan x = \frac{\sin^2 x}{\cos x} = \frac{1 - \cos^2 x}{\cos x} = \sec x - \cos x,$$

and

$$\cos x \tan x = \sin x,$$

so that

$$y(x) = C_1 \cos x + C_2 \sin x - \cos x \int (\sec x - \cos x)\,\mathrm{d}x - \sin x \cos x.$$

Routine integration shows that

$$\int (\sec x - \cos x)\mathrm{d}x = \ln|\sec x + \tan x| - \sin x,$$

so using this in the above result gives

$$y(x) = C_1 \cos x + C_2 \sin x - \cos x \ln|\sec x - \tan x|.$$

The particular integral for this equation is seen to be

$$y_{\mathrm{p}}(x) = -\cos x \ln|\sec x + \tan x|,$$

which could not have been derived by using the method of undetermined coefficients. ■

In conclusion, we will outline how the method of variation of parameters may be applied to the nth-order inhomogeneous equation

$$y^{(n)} + a_1 y^{(n-1)} + \ldots + a_n y = f(x). \tag{15.30}$$

The reduced equation associated with Eqn (15.30) is

$$y^{(n)} + a_1 y^{(n-1)} + \ldots + a_n y = 0, \tag{15.31}$$

and this will have n linearly independent solutions

$$y_1(x),\ y_2(x),\ \ldots,\ y_n(x). \tag{15.32}$$

We now seek a general solution of Eqn (15.30) of the form

$$y(x) = C_1(x)y_1(x) + C_2(x)y_2(x) + \ldots + C_n(x)y_n(x), \tag{15.33}$$

where the functions $C_1(x)$, $C_2(x)$, ..., $C_n(x)$ are to be determined.

Proceeding as in the second-order case, it follows that

$$y'(x) = C_1(x)y_1'(x) + C_2(x)y_2'(x) + \ldots + C_n(x)y_n'(x), \tag{15.34}$$

if we impose the condition

$$C_1'(x)y_1(x) + C_2'(x)y_2(x) + \ldots + C_n'(x)y_n(x) = 0. \tag{15.35}$$

Repeated differentiation of this last result, together with the imposition of further conditions similar to Eqn (15.35) and use of the fact that $y_1(x)$, $y_2(x)$, ..., $y_n(x)$ are solutions of the homogeneous Eqn (15.31), leads to the following set of n equations from which $C_1'(x)$, $C_2'(x)$, ..., $C_n'(x)$, and hence $C_1(x)$, $C_2(x)$, ..., $C_n(x)$, are to be determined:

$$\begin{aligned}
&C_1'(x)y_1(x) + C_2'(x)y_2(x) + \ldots + C_n'(x)y_n(x) = 0\\
&C_1'(x)y_1'(x) + C_2'(x)y_2'(x) + \ldots + C_n'(x)y_n'(x) = 0\\
&\ldots\\
&C_1'(x)y_1^{(n-2)}(x) + C_2'(x)y_2^{(n-2)}(x) + \ldots + C_n'(x)y_n^{(n-2)}(x) = 0\\
&C_1'(x)y_1^{(n-1)}(x) + C_2'(x)y_2^{(n-1)}(x) + \ldots + C_n'(x)y_n^{(n-1)}(x) = f(x).
\end{aligned} \tag{15.36}$$

Substitution of the $C_1(x), C_2(x), \ldots, C_n(x)$ so determined into Eqn (15.33) yields the required general solution of Eqn (15.30). We remark in passing that, as before, the Wronskian $|W|$ of $y_1(x), y_2(x), \ldots, y_n(x)$ arises quite naturally when determining the functions $C_1(x), C_2(x), \ldots, C_n(x)$, but this time in the form

$$|W| = \begin{vmatrix} y_1 & y_2 & \cdots & y_n \\ y_1' & y_2' & \cdots & y_n' \\ \cdots & \cdots & \cdots & \cdots \\ y_1^{(n-1)} & y_2^{(n-2)} & \cdots & y_n^{(n-1)} \end{vmatrix}. \tag{15.37}$$

15.4 Oscillatory solutions

Although we have already discussed the general method of solution of second-order differential equations with constant coefficients, their importance in practical applications merits special mention when the inhomogeneous term is periodic. The second-order differential equation

$$a\frac{d^2y}{dt^2} + b\frac{dy}{dt} + cy = f(t) \tag{15.38}$$

characterizes many important physical situations. For example, when a represents a mass, b a damping force proportional to velocity, and c a restoring force, Eqn (15.38) could represent a mechanical vibration damper. Alternatively, if a represents an inductance, b a resistance and c a capacitance, Eqn (15.38) would describe an R–L–C circuit.

By analogy with a mechanical system in which $f(t)$ represents the input driving the system, the inhomogeneous term is sometimes called the *forcing function*. It is the inhomogeneous term that gives rise to the particular integral, and we again remark that part of the general solution is attributable solely to the function $f(t)$.

We shall confine attention to the following particular form of Eqn (15.38)

$$y'' + 2\zeta y' + \Omega^2 y = a \sin \omega t, \tag{15.39}$$

where a is called the *amplitude* of the forcing function $\sin \omega t$, which has *frequency* ω rad/s and period $2\pi/\omega$. The number ζ is usually called the *damping* of the system described by Eqn (15.39), and Ω is then called the *natural frequency* of the system. Several cases must be distinguished, and first we assume that $\zeta \neq 0$ with $\zeta^2 < \Omega^2$. Then, setting $\omega_0^2 = \Omega^2 - \zeta^2$, the roots of the characteristic equation $P(\lambda) \equiv \lambda^2 + 2\zeta\lambda + \Omega^2 = 0$ become $\lambda = -\zeta \pm i\omega_0$. In terms of this new notation, the complementary function y_c can be written

$$y_c = Ae^{-\zeta t}\sin(\omega_0 t + \varepsilon), \tag{15.40}$$

where A and ε are arbitrary constants, with ε called the *phase angle*. There is no loss of generality in adopting this form for y_c because by expanding $\sin(\omega_0 t+\varepsilon)$ it can be seen that $y_c = e^{-\zeta t}(C\sin\omega_0 t + D\cos\omega_0 t)$, where $C = A\cos\varepsilon$ and $D = A\sin\varepsilon$.

A routine calculation shows that the particular integral y_p is

$$y_p = P\sin\omega t + Q\cos\omega t, \tag{15.41}$$

where

$$P = \frac{a(\Omega^2 - \omega^2)}{(\Omega^2 - \omega^2)^2 + 4\zeta^2\omega^2} \quad \text{and} \quad Q = \frac{-2\zeta a\omega}{(\Omega^2 - \omega^2)^2 + 4\zeta^2\omega^2}.$$

By writing

$$y_p = (P^2 + Q^2)^{1/2}\left[\frac{P}{(P^2 + Q^2)^{1/2}}\sin\omega t + \frac{Q}{(P^2 + Q^2)^{1/2}}\cos\omega t\right],$$

setting

$$\cos\delta = P/(P^2 + Q^2)^{1/2}, \quad \sin\delta = Q/(P^2 + Q^2)^{1/2}$$

and using the expansion of $\sin(\omega t+\delta)$, it follows that y_p may be written

$$y_p = \frac{a}{[(\Omega^2 - \omega^2)^2 + 4\zeta^2\omega^2]^{1/2}}\sin(\omega t + \delta), \tag{15.42}$$

where $\delta = \arctan(Q/P)$. The complete solution can then be written

$$y = Ae^{-\zeta t}\sin(\omega_0 t + \varepsilon) + \frac{a}{[(\Omega^2 - \omega^2)^2 + 4\zeta^2\omega^2]^{1/2}}\sin(\omega t + \delta). \tag{15.43}$$

If $\zeta > 0$, then as time increases the influence of the complementary function on the complete solution will diminish. In these circumstances, after a suitable lapse of time, only the particular integral will remain and will describe what is often called the *steady-state behaviour*. This is to be interpreted in the sense that the complementary function, which essentially describes how the solution started, has ceased to influence the solution. It is for this reason that the complementary function is often said to describe the *transient behaviour* of the solution.

If we agree to call a solution stable when it is bounded in magnitude for all time, it can be seen from the form of y_p in Eqn (15.42) and our discussion of y_c, that the solution Eqn (15.43) is stable provided $\zeta > 0$.

Examining the steady-state solution Eqn (15.42) for a stable equation, we notice that the sine function has an amplitude $A(\omega)$ which is frequency-dependent:

$$A(\omega) = \frac{a}{[(\Omega^2 - \omega^2)^2 + 4\zeta^2\omega^2]^{1/2}}. \tag{15.44}$$

It is readily established that the denominator of $A(\omega)$ has a minimum when $\omega = \omega_c$, where $\omega_c^2 = \Omega^2 - 2\zeta^2$. Hence the maximum amplitude A_{max} attained by the steady-state solution must occur when $\omega = \omega_c$, and it has the value:

$$A_{max} = \frac{a}{2\zeta\Omega\left(1 - \frac{\zeta^2}{\Omega^2}\right)^{1/2}}. \tag{15.45}$$

The frequency ω_c at which A_{max} occurs is called the *resonant frequency*, and it can be seen that when there is zero damping ($\zeta = 0$), the original Eqn (15.38) describes simple harmonic motion for which $A_{max} \to \infty$ as $\omega \to \Omega$, which is then the natural frequency of the system. That is to say, Ω is the frequency of oscillations when the forcing function is removed.

If $0 < \zeta < \Omega$ the complementary function is oscillatory, and physical systems having a damping ζ in this range are said to be *normally damped*. If, however, $\zeta > \Omega$ the complementary function or transient solution becomes

$$y_c = C_1 e^{k_1 t} + C_1 e^{k_2 t},$$

where $k_1 = -\zeta + (\zeta^2 - \Omega^2)^{1/2}$ and $k_2 = -\zeta - (\zeta^2 - \Omega^2)^{1/2}$, and is no longer oscillatory. The associated physical system is then said to be *over-damped*.

A critical case occurs when $\zeta = \Omega$, for which the complementary function becomes

$$y_c = (C_1 + C_2 t)e^{-\Omega t}. \tag{15.46}$$

In these circumstances the associated physical system is said to be *critically damped*.

The amplitude $A(\omega)$ is essentially an amplification factor for the forcing function input $a \sin \omega t$ and the effect of changing the forcing frequency ω and the damping ζ is best seen by rewriting (15.44) in the non-dimensional form

$$\frac{A(\omega)\Omega^2}{a} = \left[\left(1 - \left(\frac{\omega}{\Omega}\right)^2\right)^2 + 4\left(\frac{\zeta}{\Omega}\right)^2\left(\frac{\omega}{\Omega}\right)^2\right]^{-1/2}.$$

Fig. 15.1 shows a plot of the non-dimensional amplitude $A(\omega)\Omega^2/a$ against the non-dimensional frequency ω/Ω for different values of the non-dimensional damping ζ/a.

The reason for the infinite amplification factor at $\omega = \omega_c$ in the case of zero damping may be readily appreciated by solving the equation

$$y'' + \Omega^2 y = a \sin \Omega t.$$

The complete solution here is

$$y = A \sin(\Omega t + \varepsilon) - \frac{a}{2\Omega} t \cos \Omega t, \tag{15.47}$$

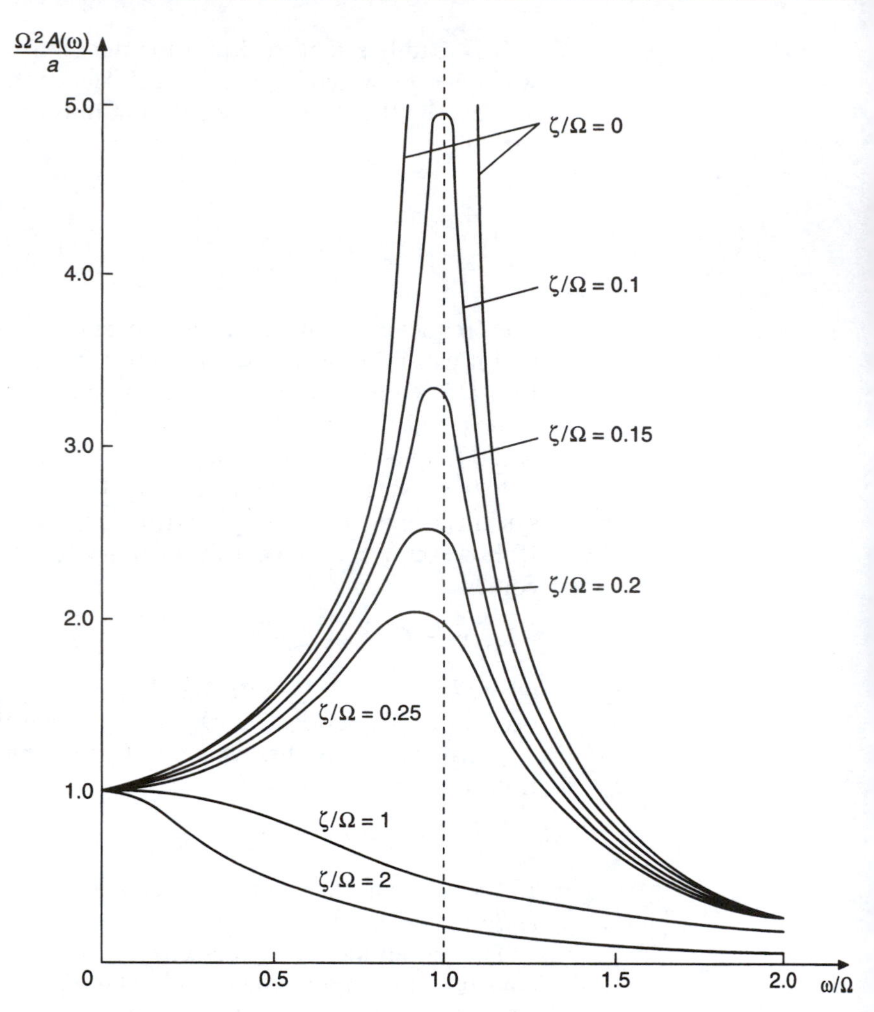

Fig. 15.1 Non-dimensional amplification factor $\Omega^2 A(\omega)/a$ as a function of normalized damping ζ/Ω and non-dimensional frequency ω/Ω.

and although the complementary function is bounded for all time, the particular integral is not as its amplitude increases linearly with time t. A differential equation of this form could, for example, describe the motion of a simple pendulum excited by a periodic disturbance at exactly its natural frequency. The disturbing force would always be in phase with the motion and so would continually reinforce it, thereby causing the amplitude to increase without bound.

15.5 Coupled oscillations and normal modes

A great many physical situations can be described approximately in terms of coupled oscillatory systems, each having properties of the type discussed in section 15.4. Such is the case in electrical circuits containing inductance, in many mechanical oscillation problems, and in certain forms of interacting control system.

A systematic examination of these problems is not appropriate here so, instead, attention will be confined to a typical but simple form of the problem containing neither damping nor inhomogeneous terms in the equations. Expressed in more physical terms, we shall confine attention to coupled simple harmonic type equations involving no forcing function.

The following is a typical mechanical vibration problem. We suppose that a light elastic string stretched between two fixed points A and B has masses $3m$ and $2m$ attached to it at points P and Q, where $AP = l$, $PQ = l$, $QB = l$. The tension in the string is kml, where k is the elastic constant of the string. Our task will be to determine whether there are preferred frequencies and, if so, the manner of vibration of the system when only small displacements are to be considered. We shall also determine the subsequent motion of the system if initially only the mass $3m$ is given a small lateral displacement d and is then released from rest.

The small lateral displacements of masses $3m$ and $2m$ at time t will be denoted by x and y, respectively (see Fig. 15.2).

Neglecting gravity and using the fact that the system is non-dissipative, energy considerations lead easily to the equations of motion

$$3m\frac{d^2x}{dt^2} + km(2x - y) = 0$$

and

$$2m\frac{d^2y}{dt^2} + km(2y - x) = 0.$$

Thus we must consider the solution of the simultaneous differential equations

$$3\frac{d^2x}{dt^2} + 2kx - ky = 0$$

and

$$2\frac{d^2y}{dt^2} + 2ky - kx = 0.$$

Now although the use of matrices can easily be avoided when solving second-order systems of equations of this kind, it will be more instructive to utilize them. Accordingly, defining the matrices **X**, **M** and **A** to be

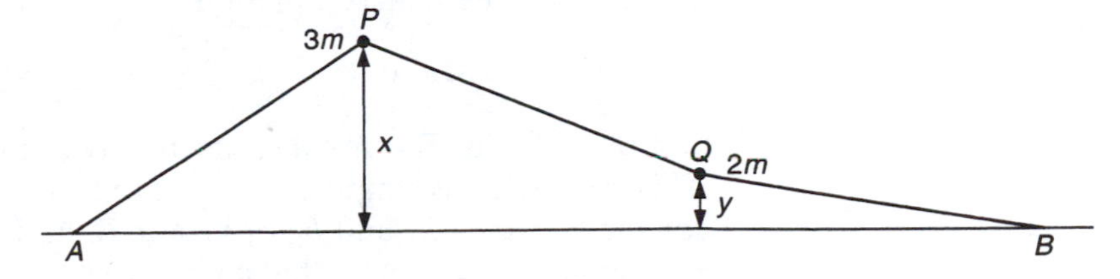

Fig. 15.2 Elastic string and mass system.

$$\mathbf{X} = \begin{bmatrix} x \\ y \end{bmatrix}, \quad \mathbf{M} = \begin{bmatrix} 3 & 0 \\ 0 & 2 \end{bmatrix}, \quad \mathbf{A} = \begin{bmatrix} 2k & -k \\ -k & 2k \end{bmatrix},$$

and defining $\mathrm{d}^2\mathbf{X}/\mathrm{d}t^2$ by the expression

$$\frac{\mathrm{d}^2\mathbf{X}}{\mathrm{d}t^2} = \begin{bmatrix} \dfrac{\mathrm{d}^2x}{\mathrm{d}t^2} \\ \dfrac{\mathrm{d}^2y}{\mathrm{d}t^2} \end{bmatrix},$$

we see that we must solve the matrix differential equation

$$\mathbf{M}\frac{\mathrm{d}^2\mathbf{X}}{\mathrm{d}t^2} + \mathbf{AX} = 0.$$

This now bears a striking resemblance to the familiar simple harmonic equation encountered when dealing with simple pendulum problems. Indeed, the resemblance becomes even closer if we notice that as $\det \mathbf{M} = 6 \neq 0$, we may pre-multiply the matrix differential equation by $\mathbf{M}^{-1}$ to obtain

$$\frac{\mathrm{d}^2\mathbf{X}}{\mathrm{d}t^2} + \mathbf{KX} = 0, \quad \text{with} \quad \mathbf{K} = \mathbf{M}^{-1}\mathbf{A}.$$

To find if there are preferred frequencies and periodic solutions let us set $\mathbf{X} = \mathbf{B}\sin(\omega t + \varepsilon)$, where ω is a frequency, ε is a phase, and $\mathbf{B}$ is the constant column vector

$$\mathbf{B} = \begin{bmatrix} b_1 \\ b_2 \end{bmatrix}.$$

It is now necessary to find relationships that exist between ω, b_1 and b_2. Using the equation $\mathbf{X} = \mathbf{B}\sin(\omega t + \varepsilon)$, it follows by differentiation that $\mathrm{d}^2\mathbf{X}/\mathrm{d}t^2 = -\omega^2\mathbf{B}\sin(\omega t + \varepsilon)$, so that substitution into the matrix differential equation gives the result

$$(\mathbf{KB} - \omega^2\mathbf{B})\sin(\omega t + \varepsilon) = \mathbf{0}.$$

Now the scalar multiplier $\sin(\omega t + \varepsilon)$ is not identically zero, so we may cancel it and, after taking out the constant post-multiplier $\mathbf{B}$ as a factor, we arrive at the matrix equation

$$(\mathbf{K} - \omega^2\mathbf{I})\mathbf{B} = \mathbf{0}.$$

Notice that as $\mathbf{B}$ is not a scalar it may not be cancelled from the result. This expression is simply a pair of homogeneous simultaneous equations for the elements b_1 and b_2 of vector $\mathbf{B}$, and from our previous study of such equations we know that a non-trivial solution will only be possible if the determinant of the coefficient matrix vanishes. That is, b_1 and b_2 may be determined, not both zero, provided that

$$|\mathbf{K} - \omega^2 \mathbf{I}| = 0.$$

This is usually called the *characteristic determinant* of the system.

We see from section 10.7 that we have arrived at an eigenvalue problem in which the eigenvalues of $\mathbf{K}$ determine the two possible values of ω^2, and hence of ω. The two eigenvectors of $\mathbf{K}$ corresponding to these eigenvalues then determine the ratio of b_1 to b_2 in each case.

Returning to the data of the problem, we see that

$$\mathbf{M}^{-1} = \begin{bmatrix} \frac{1}{3} & 0 \\ 0 & \frac{1}{2} \end{bmatrix},$$

showing that

$$\mathbf{K} = \mathbf{M}^{-1}\mathbf{A} = \begin{bmatrix} \frac{2}{3}k & -\frac{1}{3}k \\ -\frac{1}{2}k & k \end{bmatrix},$$

and hence the characteristic determinant of the system is

$$\begin{vmatrix} \frac{2}{3}k - \omega^2 & -\frac{1}{3}k \\ -\frac{1}{2}k & k - \omega^2 \end{vmatrix} = 0.$$

This is just an equation for ω^2. Expanding the determinant, we arrive at the characteristic equation of the system:

$$\omega^4 - \tfrac{5}{3}k\omega^2 + \tfrac{1}{2}k^2 = 0.$$

Solving this for ω^2, we find that the characteristic determinant will vanish, so that the system will give rise to values b_1 and b_2, not both identically zero, only when $\omega^2 = \omega_1^2 = k(5 - \sqrt{7})/6$ or $\omega^2 = \omega_2^2 = k(5 + \sqrt{7})/6$. ω_1 and ω_2 are called the *natural frequencies* of the system as they describe the only purely sinusoidal oscillations that occur naturally in the system.

To find the values of b_1 and b_2 corresponding to these natural frequencies, we return to the equation

$$(\mathbf{K} - \omega^2 \mathbf{I})\mathbf{B} = \mathbf{0}$$

and solve for $\mathbf{B}$, first with $\omega^2 = \omega_1^2$ and then with $\omega^2 = \omega_2^2$.

We begin by setting $\omega^2 = \omega_1^2$ to obtain the matrix equation

$$\begin{bmatrix} \left(\dfrac{-1 + \sqrt{7}}{6}\right)k & -\frac{1}{3}k \\ -\frac{1}{2}k & \left(\dfrac{1 + \sqrt{7}}{6}\right)k \end{bmatrix} \begin{bmatrix} b_1 \\ b_2 \end{bmatrix} = \mathbf{0},$$

which, as $k \neq 0$, reduces to the two scalar equations

$$\left(\frac{-1 + \sqrt{7}}{6}\right)b_1 - \tfrac{1}{3}b_2 = 0$$

and

$$-\tfrac{1}{2}b_1 + \left(\frac{1+\sqrt{7}}{6}\right)b_2 = 0.$$

Solving either of these homogeneous equations, which because of the manner of determination of ω^2 are of course compatible, we find that $b_2^{(1)} = 3b_1^{(1)}/(1+\sqrt{7})$; the superscript 1 indicates that these are the values assumed by b_1 and b_2 when $\omega^2 = \omega_1^2$.

As the equations are homogeneous they only determine the ratio $b_1^{(1)} : b_2^{(1)}$, and the value of either $b_1^{(1)}$ or $b_2^{(1)}$ may thus be assigned arbitrarily. Accordingly we shall choose to make $b_1^{(1)} = 1$, so that

$$\mathbf{B}^{(1)} = \begin{bmatrix} 1 \\ \dfrac{3}{1+\sqrt{7}} \end{bmatrix}$$

and, consequently,

$$\mathbf{X}(t)^{(1)} = \begin{bmatrix} 1 \\ \dfrac{3}{1+\sqrt{7}} \end{bmatrix} \sin(\omega_1 t + \varepsilon_1),$$

where the superscript 1 indicates that these are the forms assumed by $\mathbf{B}$ and $\mathbf{X}(t)$ when $\omega^2 = \omega_1^2$.

A similar argument in the case $\omega^2 = \omega_2^2$ shows that

$$\mathbf{B}^{(2)} = \begin{bmatrix} 1 \\ \dfrac{3}{1-\sqrt{7}} \end{bmatrix} \quad \text{and} \quad \mathbf{X}(t)^{(2)} = \begin{bmatrix} 1 \\ \dfrac{3}{1-\sqrt{7}} \end{bmatrix} \sin(\omega_2 t + \varepsilon_2).$$

Thus $\mathbf{X}(t)^{(1)}$ and $\mathbf{X}(t)^{(2)}$ describe the purely sinusoidal forms of disturbance that are possible when $\omega^2 = \omega_1^2$ and $\omega^2 = \omega_2^2$, respectively. Both of these are possible solutions to the original system of differential equations and, as the differential equations are linear, the general solution $\mathbf{X}(t)$ must be of the form

$$\mathbf{X}(t) = \alpha\mathbf{X}(t)^{(1)} + \beta\mathbf{X}(t)^{(2)},$$

where α, β are arbitrary constants. In more advanced works the solutions $\mathbf{X}(t)^{(1)}$ and $\mathbf{X}(t)^{(2)}$ are called *eigensolutions*, while the numbers ω_1^2 and ω_2^2 are, of course, the eigenvalues of $\mathbf{K}$.

The solution $\mathbf{X}(t)$ is the matrix equivalent of the complementary function encountered at the start of this chapter. To find the solution satisfying any given initial conditions it now only remains to determine the constants α and β and the arbitrary phase angles ε_1 and ε_2.

To complete the problem in question, we now make use of the fact that the system starts from rest at time $t = 0$ with $x = d$, $y = 0$. In terms of $\mathbf{X}(t)$ this yields the initial conditions

$$\left.\frac{d\mathbf{X}}{dt}\right|_{t=0} = \mathbf{0} \quad \text{and} \quad \mathbf{X}(0) = \begin{bmatrix} d \\ 0 \end{bmatrix}.$$

Applying the first of these conditions to $\mathbf{X}(t)$, we obtain

$$\left.\frac{d\mathbf{X}}{dt}\right|_{t=0} = \mathbf{0} = \alpha\omega_1^2\mathbf{B}^{(1)}\cos\varepsilon_1 + \beta\omega_2^2\mathbf{B}^{(2)}\cos\varepsilon_2,$$

showing that $\varepsilon_1 = \varepsilon_2 = \frac{1}{2}\pi$. The second condition gives

$$\mathbf{X}(0) = \begin{bmatrix} d \\ 0 \end{bmatrix} = \alpha \begin{bmatrix} 1 \\ \dfrac{3}{1+\sqrt{7}} \end{bmatrix} + \beta \begin{bmatrix} 1 \\ \dfrac{3}{1-\sqrt{7}} \end{bmatrix}.$$

Hence,

$$d = \alpha + \beta$$

$$0 = \frac{3\alpha}{1+\sqrt{7}} + \frac{3\beta}{1-\sqrt{7}},$$

and so

$$\alpha = \frac{(1+\sqrt{7})d}{2}, \quad \beta = \frac{-6d}{(1+\sqrt{7})^2 - 6}.$$

In terms of these constants α, β, the solution to the explicit initial value problem posed at the start of the section is

$$\mathbf{X}(t) = \alpha\mathbf{B}^{(1)}\sin(\omega_1 t + \tfrac{1}{2}\pi) + \beta\mathbf{B}^{(2)}\sin(\omega_2 t + \tfrac{1}{2}\pi).$$

The roles played by the phase angles ε_1 and ε_2 are most important since they serve to adjust the time origins of the eigensolutions $\mathbf{X}(t)^{(1)}$ and $\mathbf{X}(t)^{(2)}$ at the start of the prescribed motion. The constants α and β are just scale factors.

The four constants α, β, ε_1 and ε_2 are the four arbitrary constants that our previous work has led us to expect to be associated with two simultaneous second-order equations, though the manner of their appearance here is perhaps slightly unfamiliar.

In vibration problems it is common to refer to the fundamental eigensolutions $\mathbf{X}(t)^{(1)}$ and $\mathbf{X}(t)^{(2)}$ as the *normal modes* associated with the problem. This arises on account of the fact that each solution of this kind is a pure sinusoid disturbance describing a specially simple and characteristic mode of vibration. Thus, for example, in the first mode $\mathbf{X}(t)^{(1)}$, vibrations will be of the form

$$x = A\sin(\omega_1 t + \tfrac{1}{2}\pi), \quad y = \left(\frac{3A}{1+\sqrt{7}}\right)\sin(\omega_1 t + \tfrac{1}{2}\pi),$$

with A an arbitrary constant.

The apparent choice of sign that is possible for both ω_1 and ω_2 is immaterial, since it may be absorbed into the determination of α, β, ε_1 and ε_2. Compare the arguments used here with the discussion of eigenvalues and eigenvectors given in section 9.7.

15.6 Systems of first-order equations

In this section we introduce the simplest method by which the solution of a system of linear first-order constant-coefficient differential equations may be found. A more general method will be given when we discuss the Laplace transform. The systems to which the method of this section may be applied can involve two or more dependent variables and one independent variable, usually the time t. In general, there must be the same number of equations as there are dependent variables.

The approach is most easily illustrated by means of the following example which is typical and involves the two dependent variables $x(t)$ and $y(t)$.

Example 15.15 Solve

$$2\frac{dx}{dt}+\frac{dy}{dt}-2x-2y=5e^{t},$$

$$\frac{dx}{dt}+\frac{dy}{dt}+4x+2y=5e^{-t},$$

subject to the initial conditions $x(0)=2$, $y(0)=0$.

Solution The method of the solution is first to arrive at a second-order differential equation for a dependent variable by means of differentiation of the differential equations themselves with respect to t, followed by elimination. Then, when this equation has been solved, the result is used to determine the other dependent variable. This system is said to be *inhomogeneous* because the right-hand side contains terms depending on the independent variable t. If such terms are absent, a system is said to be *homogeneous*.

Let us arrive at a differential equation for x. Subtraction of the differential equations shows

$$\frac{dx}{dt}-6x-4y=5(e^{t}-e^{-t}), \tag{A}$$

and differentiation with respect to t then gives

$$\frac{d^2x}{dt^2}-6\frac{dx}{dt}-4\frac{dy}{dt}=5(e^{t}+e^{-t}). \tag{B}$$

Addition of the original differential equations followed by rearrangement gives

$$2\frac{dy}{dt} = -3\frac{dx}{dt} - 2x + 5(e^t + e^{-t}). \qquad \text{(C)}$$

Elimination of dy/dt between (B) and (C) gives the following second-order differential equation for x:

$$\frac{d^2x}{dt^2} + 4x = 15(e^t + e^{-t}).$$

This has the complementary function

$$x_c = A\cos 2t + B\sin 2t,$$

involving the arbitrary constants A and B. Using the method of undetermined coefficients, it is easily shown that the particular integral is

$$x_p = 3(e^t + e^{-t}).$$

Thus the general solution $x = x_c + x_p$ is

$$x = A\cos 2t + B\sin 2t + 3(e^t + e^{-t}). \qquad \text{(D)}$$

Using this result in (A) to eliminate x and dx/dt gives

$$y = \tfrac{1}{2}(B - 3A)\cos 2t - \tfrac{1}{2}(A + 3B)\sin 2t - 5e^t - 4e^{-t}. \qquad \text{(E)}$$

Results (D) and (E) thus represent the general solution to the system, as they contain the two arbitrary constants A and B. To find the particular solution corresponding to the initial conditions $x(0) = 2$, $y(0) = 0$, we use these conditions in (D) and (E) to obtain

$$2 = A + 6, \quad 0 = \tfrac{1}{2}(B - 3A) - 9,$$

showing $A = -4$, $B = 6$. The required solution is thus

$$x = -4\cos 2t + 6\sin 2t + 3(e^t + e^{-t}),$$

$$y = 9\cos 2t - 7\sin 2t - 5e^t - 4e^{-t}.$$

Notice that in this approach the two arbitrary constants which would be expected enter into the general solution for x, and a different combination of these same two arbitrary constants appears in the general solution for y. ■

15.7 Two-point boundary value problems

In addition to initial value problems, higher-order differential equations also give rise to what are called *two-point boundary value problems*. These are problems in which instead of all the conditions determining a particular solution being given at a single point (initial value problems), they are prescribed at two different points, say $x = a$ and $x = b$.

The term *boundary value problem* derives from the fact that the points $x = a$ and $x = b$ at which conditions on the dependent variable are specified usually coincide with some physical boundaries in the problem. For example, in the theory of the strength of materials, a *simply supported beam* of length L is a flexible beam supported at each end in such a way that the points of support are on the same horizontal level. Since the beam is flexible, its weight causes it to sag. The situation is illustrated in Fig. 15.3, in which the beam is supported at $x = 0$ and $x = L$, and the sag is y at a distance x from 0.

It is shown in the elementary theory of the strength of materials that if the beam is homogeneous with mass M per unit length, the moment of inertia of its cross-section is I, and Young's modulus for the material is E, then provided the deflection y is small it satisfies the differential equation

$$\frac{d^2y}{dx^2} = \frac{M}{2EI}(x^2 - Lx). \tag{15.48a}$$

The determination of the deflection y is a two-point boundary value problem for Eqn (15.48a), because y must satisfy the two boundary conditions

$$y(0) = 0 \quad \text{and} \quad y(L) = 0. \tag{15.48b}$$

These conditions require the end points to experience no deflection, as the beam is rigidly supported at its ends.

It is trivial to integrate Eqn (15.48a) and to use the boundary conditions (15.48b) to show that the deflection is given by

$$y = \frac{M}{24EI}(x^4 - 2Lx^3 + L^3x), \tag{15.49}$$

for $0 \leq x \leq L$.

Depending on the nature of the two-point boundary value problem, a unique solution may be determined (as with the simply supported beam), a set of solutions may be determined, or there may be no solution at all. In this first account of the subject we can do no more than illustrate these three different cases by example. The first example, which is the simpler of the two, gives rise to a unique solution.

Example 15.16 Solve the two-point boundary value problem

$$\frac{d^2y}{dx^2} + 3\frac{dy}{dx} + 2y = 0$$

with $y(0) = 0$ and $y(1) = 2$.

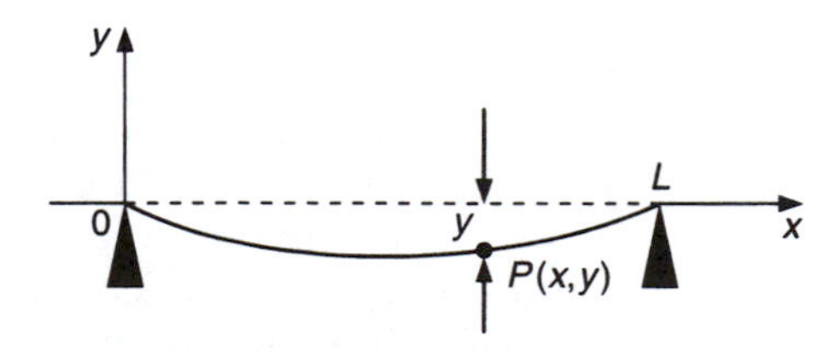

Fig. 15.3 Simply supported beam.

Solution The differential equation has the characteristic equation

$$\lambda^2 + 3\lambda + 2 = 0,$$

with the roots $\lambda = -1$ and $\lambda = -2$, so the general solution is

$$y = Ae^{-x} + Be^{-2x}.$$

To determine the arbitrary constants A and B, we must use the boundary conditions. The condition $y(0) = 0$ gives

$$0 = A + B,$$

and the condition $y(1) = 2$ gives

$$2 = Ae^{-1} + Be^{-2},$$

so that

$$A = -B = \frac{2e^2}{e - 1}.$$

Thus the unique solution is

$$y = \frac{2e^2}{e - 1}(e^{-x} - e^{-2x}),$$

for $0 \leq x \leq 1$. ■

The nature of the solution in the following example will be seen to depend on the choice of the constants in the problem.

Example 15.17 Solve the two-point boundary value problem

$$\frac{d^2y}{dx^2} + y = 0$$

when $y\left(\frac{\pi}{2}\right) = 0$ and $y(a) = K$, when a and K are arbitrary.

Solution The general solution is seen to be

$$y = A\cos x + B\sin x,$$

so, using the boundary condition $y\left(\frac{\pi}{2}\right) = 0$, we conclude that $B = 0$. However the remaining boundary condition $y(a) = K$ shows that

$$K = A\cos a. \tag{F}$$

If $K \neq 0$ we may solve this for A provided $\cos a \neq 0$; that is, if $a \neq n\pi/2$ with n an odd integer. We then find the unique solution

$$y = \left(\frac{K}{\cos a}\right)\cos x, \quad \pi/2 \le x \le a.$$

However, if $K = 0$ and $a = n\pi/2$, with n an odd integer, $\cos a = 0$, and (F) is then true for any constant A. Thus we then arrive at the non-unique solution

$$y = A\cos x,$$

for arbitrary A and $\pi/2 \le x \le a$.

If $K = 0$ and $a \ne n\pi/2$, with n an odd integer, we find (F) has the solution $A = 0$, leading to the solution

$$y = 0, \quad \pi/2 \le x \le a.$$

This is called the *trivial solution* of the problem.

Finally, if $K \ne 0$ but $a = n\pi/2$, with n an odd integer, then no value of A satisfies (F), so that no solution y then exists for this two-point boundary value problem. ■

In conclusion, we remark that if a boundary value problem arises in connection with an inhomogeneous differential equation, then the arbitrary constants must be determined by applying the boundary conditions to the complete solution. That is, if the complementary function is y_c and the particular integral is y_p, the boundary conditions must be applied to the complete solution $y = y_c + y_p$. It is left as an exercise for the reader to check that this was done when integrating (15.48a) and using (15.48b) to arrive at the expression in (15.49) for the deflection y of the beam at point x.

15.8 The Laplace transform

The *Laplace transform* is a powerful method for solving initial value problems for linear differential equations. Unlike the methods discussed so far, which first require the general solution to be found, and then the arbitrary constants to be chosen to suit the initial conditions, the Laplace transform determines the solution to the initial value problem directly.

In the simple but important cases which will be considered here we shall see that the usual method of solution is replaced by simple algebraic operations followed by the use of a table of Laplace transform pairs. When the equations are more complicated, the Laplace transform method requires the use of functions of a complex variable, though such problems will not be discussed here.

Definition 15.4 (Laplace transform) Let $f(t)$ be a piecewise continuous function defined for $t \ge 0$. Then, when the integral exists, the Laplace transform $F(s)$ of $f(t)$ is defined as

$$F(s) = \int_0^\infty e^{-st} f(t)\,dt.$$

Whenever it is necessary to emphasise that the function being transformed is $f(t)$, the Laplace transform is denoted by $\mathscr{L}\{f(t)\}$, so that $F(s) = \mathscr{L}\{f(t)\}$. It is conventional to denote the function and its Laplace transform by the lower- and upper-case forms of the same symbol, so that $F(s)$ is the Laplace transform of $f(t)$ and $Y(s)$ is the Laplace transform of $y(t)$.

We shall now determine the Laplace transform $\mathscr{L}\{f(t)\}$ of some simple functions $f(t)$.

If $f(t) = a$, with $a = \text{const.}$, we have

$$\mathscr{L}\{a\} = F(s) = \int_0^\infty a e^{-st}\,dt = \left.\frac{-ae^{-st}}{s}\right|_0^\infty = \lim_{t\to+\infty}\left(\frac{-ae^{-st}}{s}\right) + \frac{a}{s}.$$

This limit on the right-hand side is zero for all $s > 0$, and infinity for all $s < 0$. Thus the improper integral defining $\mathscr{L}\{a\}$ is only convergent when $s > 0$, so we have shown that

$$\mathscr{L}\{a\} = F(s) = \frac{a}{s}, \quad \text{for} \quad s > 0.$$

If $f(t) = t$, then

$$\mathscr{L}\{t\} = F(s) = \int_0^\infty te^{-st}dt = \left.\frac{-te^{-st}}{s}\right|_0^\infty + \frac{1}{s}\int_0^\infty e^{-st}dt$$

$$= \lim_{t\to+\infty}\left(\frac{-te^{-st}}{s}\right) - \lim_{t\to+\infty}\left(\frac{e^{-st}}{s^2}\right) + \frac{1}{s^2}.$$

Here again, the limits are both finite only when $s > 0$, when Theorem 6.3 shows they are both zero. Thus we have shown that

$$\mathscr{L}\{t\} = F(s) = \frac{1}{s^2}, \quad \text{for} \quad s > 0.$$

If $f(t) = e^{at}$, then

$$\mathscr{L}\{e^{at}\} = F(s) = \int_0^\infty e^{(a-s)t}dt = \left.\frac{e^{(a-s)t}}{a-s}\right|_0^\infty$$

$$= \lim_{t\to+\infty}\left(\frac{e^{(a-s)t}}{a-s}\right) + \frac{1}{s-a}.$$

The limit on the right-hand side will only be finite if $s > a$, when it becomes zero. Thus we have shown that

$$\mathscr{L}\{e^{at}\} = F(s) = \frac{1}{s-a}, \quad \text{for} \quad s > a.$$

These typical results show that a restriction must be placed on the permissible values of s in order that the improper integral defining the Laplace transform is convergent.

However, not every function has a Laplace transform even if s is restricted, since the function may grow so rapidly as $t \to +\infty$ that the defining integral is divergent for all s. An example of such a function is $f(t) = e^{t^2}$.

Using arguments similar to those used to establish Theorem 7.4, it is not difficult to prove that a piecewise continuous function $f(t)$ will have a Laplace transform $\mathscr{L}\{f(t)\} = F(s)$ for $s > a$ whenever, for sufficiently large t,

$$|f(t)| < M\mathrm{e}^{at}, \tag{15.50}$$

for some suitable choice of constants M and a. Any function satisfying (15.50) is said to be of *exponential order*. In effect, condition (15.50) says that $f(t)$ will have a Laplace transform if, when t is large, $|f(t)|$ grows no faster than $M\mathrm{e}^{at}$. Thus $f(t) = a$, $f(t) = t$ and $f(t) = \mathrm{e}^{at}$ are all of exponential order, and so have Laplace transforms, but $f(t) = e^{t^2}$ is not of exponential order and so has no Laplace transform.

It is possible to prove that if $f(t)$ and $g(t)$ are of exponential order and such that $\mathscr{L}\{f(t)\} = \mathscr{L}\{g(t)\}$, then $f(t) = g(t)$ whenever both functions are continuous. Thus for all such functions a given Laplace transform $\mathscr{L}\{f(t)\}$ corresponds to a unique function $f(t)$, and conversely.

If $\mathscr{L}\{f(t)\} = F(s)$, we shall write $f(t) = \mathscr{L}^{-1}\{F(s)\}$ to signify the function $f(t)$ whose Laplace transform is $F(s)$. Often this notation is simplified and written $f = \mathscr{L}^{-1}\{F\}$. Here $f(t)$ is called the *inverse Laplace transform*, or just the *inverse*, of $F(s)$. Thus $\mathscr{L}$ signifies the Laplace transform operation and $\mathscr{L}^{-1}$ the inverse Laplace transform operation.

The functions $f(t)$ and $\mathscr{L}\{f(t)\}$ are called *Laplace transform pairs*, and a short list of such pairs is given in Table 15.1. All entries in the table may be found by means of routine integration by parts coupled with use of the limiting properties of the exponential function.

Entries in tables of Laplace transforms may be extended by use of Theorem 7.8 whenever differentiation of an integral with respect to a parameter is permissible. Thus from entry 6 in Table 15.1 we have

$$\frac{a}{s^2 + a^2} = \int_0^\infty \mathrm{e}^{-st} \sin at \, \mathrm{d}t,$$

so, differentiating with respect to s, we find

$$\begin{aligned} \frac{2as}{(s^2 + a^2)^2} &= \int_0^\infty \mathrm{e}^{-st} t \sin at \, \mathrm{d}t \\ &= \mathscr{L}\{t \sin at\}, \end{aligned}$$

which is entry 8 in the table.

15.8.1 Linearity of the Laplace transform

It follows directly from the definition of integration that if $f(t)$ and $g(t)$ are both of exponential order, then

Table 15.1 Laplace transform pairs

$f(t) = \mathscr{L}^{-1}\{F(s)\}$	$F(s) = \mathscr{L}\{f(t)\}$	
1. a	$\dfrac{a}{s}$,	$s > 0$
2. t	$\dfrac{1}{s^2}$,	$s > 0$
3. t^n, n a positive integer	$\dfrac{n!}{s^{n+1}}$,	$s > 0$
4. e^{at}	$\dfrac{1}{s-a}$,	$s > a$
5. $t^n e^{at}$, n a positive integer	$\dfrac{n!}{(s-a)^{n+1}}$,	$s > a$
6. $\sin at$	$\dfrac{a}{s^2+a^2}$,	$s > 0$
7. $\cos at$	$\dfrac{s}{s^2+a^2}$,	$s > 0$
8. $t \sin at$	$\dfrac{2as}{(s^2+a^2)^2}$,	$s > 0$
9. $t \cos at$	$\dfrac{s^2-a^2}{(s^2+a^2)^2}$,	$s > 0$
10. $e^{at} \sin bt$	$\dfrac{b}{(s-a)^2+b^2}$,	$s > a$
11. $e^{at} \cos bt$	$\dfrac{s-a}{(s-a)^2+b^2}$,	$s > a$
12. $\sinh at$	$\dfrac{a}{s^2-a^2}$,	$s > \lvert a \rvert$
13. $\cosh at$	$\dfrac{s}{s^2-a^2}$,	$s > \lvert a \rvert$
14. $f(t) = u_a(t) = \begin{cases} 0 & \text{for} \quad t < a \\ 1 & \text{for} \quad t > a \end{cases}$	$\dfrac{e^{-as}}{s}$,	$s > 0$
15. $\delta(t-a)$	e^{-as},	$s > 0$, $a \geq 0$

$$\mathscr{L}\{af(t) + bg(t)\} = a\mathscr{L}\{f(t)\} + b\mathscr{L}\{g(t)\}. \tag{15.51}$$

This result shows the *linearity* of the Laplace transform operation and simplifies the task of finding the Laplace transform for sums of functions. For example, if $f(t) = t$ and $g(t) = \sin t$, it follows from result (15.51) and entries 2 and 6 of Table 15.1 that

$$\mathscr{L}\{4t + 7\sin t\} = \frac{4}{s^2} + \frac{7}{s^2+1}, \quad \text{for} \quad s > 0.$$

Conversely, taking the inverse Laplace transform of this result, we find

$$4t + 7\sin t = \mathscr{L}^{-1}\left\{\frac{4}{s^2} + \frac{7}{s^2+1}\right\} = 4\mathscr{L}^{-1}\left\{\frac{1}{s^2}\right\} + 7\mathscr{L}^{-1}\left\{\frac{1}{s^2+1}\right\}.$$

In order to solve initial value problems it will be necessary to transform derivatives and to interpret shifts in the variables t and s. So before discussing the solution of initial value problems we shall first examine the transformation of derivatives and then prove two shift theorems.

15.8.2 Transformation of derivatives

Let us determine the Laplace transform of $\mathrm{d}y/\mathrm{d}t$ when $y(t)$ is of exponential order. From Definition 15.4 we have

$$\mathscr{L}\{y'\} = \int_0^{\infty} \mathrm{e}^{-st} y' \,\mathrm{d}t$$

$$= y\mathrm{e}^{-st}\Big|_0^{\infty} + s\int_0^{\infty} \mathrm{e}^{-st} y \,\mathrm{d}t = \lim_{t\to+\infty} (y(t)\mathrm{e}^{-st}) - y(0) + s\mathscr{L}\{y\}.$$

Since $y(t)$ is assumed to be of exponential order the first term vanishes, and we obtain the result

$$\mathscr{L}\{y'\} = s\mathscr{L}\{y\} - y(0). \tag{15.52}$$

This shows that $\mathscr{L}\{y'\}$ depends not only on $\mathscr{L}\{y\}$, but also on $y(0)$ – that is, on the initial value of y.

Similar calculations establish that

$$\mathscr{L}\{y''\} = s^2\mathscr{L}\{y\} - y'(0) - sy(0), \tag{15.53}$$

and

$$\mathscr{L}\{y'''\} = s^3\mathscr{L}\{y\} - y''(0) - sy'(0) - s^2y(0). \tag{15.54}$$

By using mathematical induction it is not difficult to prove the general result

$$\mathscr{L}\{y^{(n)}\} = s^n\mathscr{L}\{y\} - y^{(n-1)}(0) - sy^{(n-2)}(0) - \ldots - s^{n-1}y(0). \tag{15.55}$$

Example 15.18 Find

(a) $\mathscr{L}\{y''\}$, given that $y(0) = 3$ and $y'(0) = -2$;
(b) $\mathscr{L}\{y'''\}$, given that $y(t) = \sin 2t$;
(c) $y(t)$ given that

$$\mathscr{L}\{y\} = \frac{7s^2 + 18}{s(s^2+9)}.$$

Solution

(a) $\mathscr{L}\{y''\} = s\mathscr{L}\{y\} - y'(0) - sy(0)$

$= s\mathscr{L}\{y\} + 2 - 3s.$

(b) $y' = 2\cos 2t$ and $y'' = -4\sin 2t$,

so $y(0) = 0$, $y'(0) = 2$ and $y''(0) = 0$.

Thus

$$\begin{aligned}\mathscr{L}\{y'''\} &= s^3\mathscr{L}\{y\} - y''(0) - sy'(0) - s^2y(0)\\ &= s^3\mathscr{L}\{\sin 2t\} - 2s.\end{aligned}$$

From entry 6 in Table 15.1 we have

$$\mathscr{L}\{\sin 2t\} = \frac{2}{s^2+4}, \quad \text{for} \quad s > 0,$$

so that

$$\mathscr{L}\{y'''\} = \frac{2s^3}{s^2+4} - 2s = \frac{-8s}{s^2+4}, \quad \text{for} \quad s > 0.$$

This result may be checked by using entry 7 in Table 15.1, because $y''' = -8\cos 2t$.

(c) This problem requires us to find the inverse Laplace transform of the given function of s. Using partial fractions we first re-express $\mathscr{L}\{y\}$ as

$$\mathscr{L}\{y\} = \frac{2}{s} + \frac{5s}{s^2+9}.$$

Then, taking the inverse Laplace transform gives

$$\begin{aligned}y(t) &= \mathscr{L}^{-1}\left\{\frac{2}{s} + \frac{5s}{s^2+9}\right\}\\ &= 2\mathscr{L}^{-1}\left\{\frac{1}{s}\right\} + 5\mathscr{L}^{-1}\left\{\frac{s}{s^2-9}\right\}.\end{aligned}$$

Finally, using entries 1 and 7 of Table 15.1, we find

$$y(t) = 2 + 5\cos 3t. \quad \blacksquare$$

15.8.3 Shift theorems for the Laplace transform

There are two shift theorems which simplify the task of working with the Laplace transform. The first involves a shift of the variable s to $s - a$, and the second a shift of the variable t to $t - a$, where $a > 0$ is an arbitrary constant.

Theorem 15.6 (first shift theorem) Let $\mathscr{L}\{f(t)\} = F(s)$ for $s > \gamma$. Then $\mathscr{L}\{e^{at}f(t)\} = F(s-a)$ for $s - a > \gamma$.

Proof We have

$$F(s) = \int_0^\infty e^{-st}f(t)\,dt,$$

so

$$\mathscr{L}\{e^{at}f(t)\} = \int_0^\infty e^{-(s-a)t}f(t)\,dt = F(s-a). \quad \blacksquare$$

If $F(s)$ exists for $s > \gamma$, then $F(s-a)$ exists for $s - a > \gamma$. The name 'shift theorem' is derived from the fact that multiplication of $f(t)$ by e^{at} shifts the variable s in the Laplace transform to $s-a$.

Example 15.19 Find

$$\mathscr{L}\{e^{2t}(t^3 + 5t\cos 4t)\}.$$

Solution We start from the result that

$$\mathscr{L}\{t^3 + 5t\cos 4t\} = \mathscr{L}\{t^3\} + 5\mathscr{L}\{t\cos 4t\}$$
$$= \frac{6}{s^4} + \frac{5(s^2-16)}{(s^2+16)^2}, \quad \text{for } s > 0,$$

which follows from entries 3 and 9 of Table 15.1. Then it follows from Theorem 15.6 that the required Laplace transform is obtained from this result by replacing s by $s-2$, since in this case $e^{at} = e^{2t}$, showing that $a = 2$. We have established that

$$\mathscr{L}\{e^{2t}(t^3 + 5t\cos 4t)\} = \frac{6}{(s-2)^4} + \frac{5((s-2)^2-16)}{((s-2)^2+16)^2}. \quad \blacksquare$$

Before proceeding to the second shift theorem we must define the function $u_a(t)$ which appears in entry 14 of Table 15.1. The function $u_a(t)$, called the *Heaviside unit step function*, is defined as

$$u_a(t) = \begin{cases} 0 & \text{for } t < a \\ 1 & \text{for } t > a. \end{cases}$$

Its graph is shown in Fig. 15.4, and the function is so called because of the shape of the graph. The notation $H(t-a)$ is often used instead of $u_a(t)$.

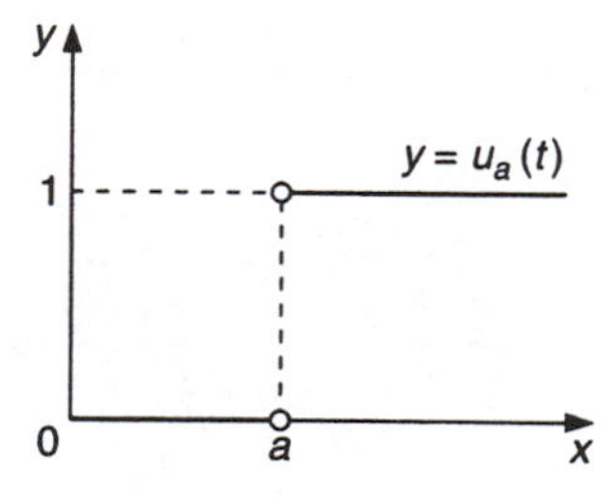

Fig. 15.4 Heaviside unit step function.

This function is useful because many inputs to physical systems are switched on and off, and the Heaviside function represents such a switching process.

If $y = f(t)$, then $y = u_a(t)f(t)$ is zero for $0 < t < a$ and becomes the function $f(t)$ for $t > a$, as shown in Fig. 15.5. Thus $f(t)$ is 'switched on' at $t = a$.

The Heaviside function is also useful when representing functions which are shifted in time. Thus, if $y = f(t)$ then $y = u_a(t)f(t-a)$ is zero for $0 < t < a$ and becomes the function $f(t)$ translated by an amount a for $t > a$. This is shown in Fig. 15.6; the function in Fig. 15.6(b) is obtained

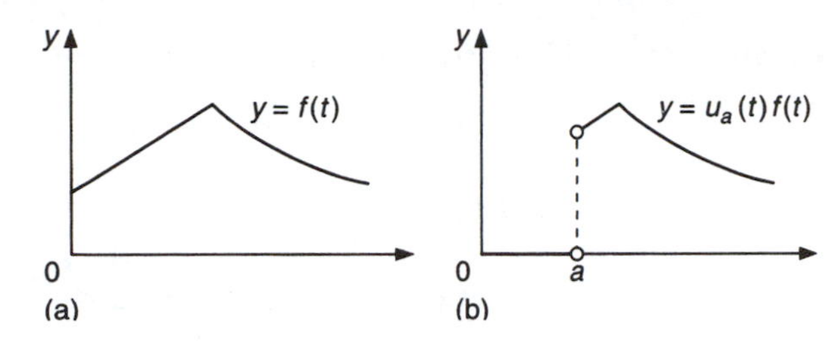

Fig. 15.5 Switching in $f(t)$ at $t=a$.

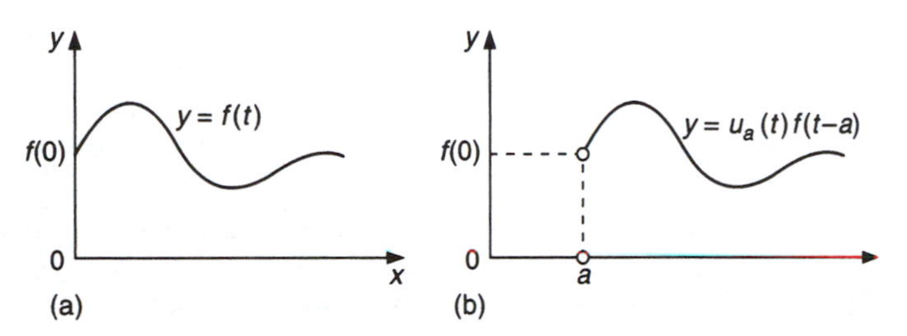

Fig. 15.6 Shifting a function $y=f(t)$.

by shifting to the right by an amount a the function in Fig. 15.6(a), while setting the shifted function equal to zero for $0<t<a$.

An important form of input to a system often occurs in the form of a rectangular pulse. This may be represented as a combination of two Heaviside step functions, one of which 'switches on' the pulse, and the other of which 'switches off' the pulse. Thus the function

$$y = u_a(t) - u_b(t), \ (b > a)$$

represents a rectangular pulse of unit height which is 'switched on' at $t = a$ and 'switched off' at $t=b$. This is shown in Fig. 15.7.

The Laplace transform of $u_a(t)$ is given by

$$\mathscr{L}\{u_a(t)\} = \int_a^{\infty} e^{-st}u_a(t)dt = \int_a^{\infty} e^{-st}dt = \frac{e^{-as}}{s},$$

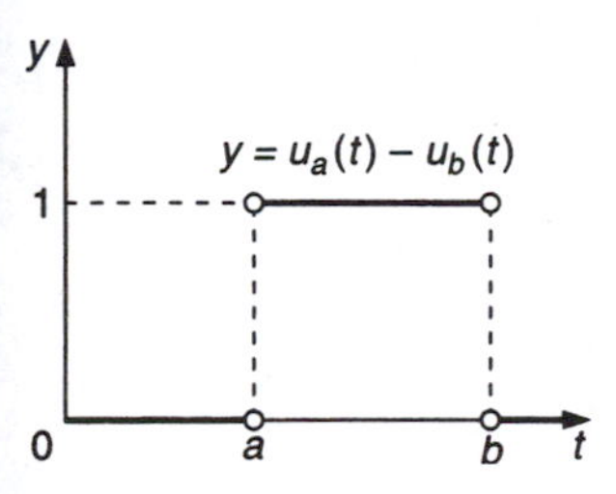

Fig. 15.7 Pulse of unit height $y=u_a(t)-u_b(t)$.

which is entry 14 in Table 15.1. It follows from this that if we consider the pulse shown in Fig. 15.7, we have

$$\mathscr{L}\{u_a(t) - u_b(t)\} = \frac{1}{s}(e^{-as} - e^{-bs}).$$

Thus the Laplace transform of a rectangular pulse has a simple form.

We are now ready to prove the second shift theorem, after which we shall proceed with the solution of initial value problems.

Theorem 15.7 (second shift theorem) If $\mathscr{L}\{f(t)\} = F(s)$, then $\mathscr{L}\{u_a(t)f(t-a)\} = e^{-as}F(s)$ or, equivalently,

$$\mathscr{L}^{-1}\{e^{-as}F(s)\} = u_a(t)f(t-a).$$

Proof By definition,

$$\mathscr{L}\{u_a(t)f(t-a)\} = \int_a^{\infty} e^{-as}f(t-a)\,dt.$$

Changing the variable in the integral to $\tau = t - a$, this becomes

$$\begin{aligned}\mathscr{L}\{u_a(t)f(t-a)\} &= \int_0^{\infty} e^{-s(a+\tau)}f(\tau)\,d\tau \\ &= e^{-as}\int_0^{\infty} e^{-st}f(\tau)\,d\tau \\ &= e^{-as}F(s),\end{aligned}$$

and the result is proved. ■

Entry 14 in Table 15.1 is now seen to be a trivial consequence of the second shift theorem.

Example 15.20 Find

(a) The Laplace transform of $u_{\pi/4}(t)f\left(t-\frac{\pi}{4}\right)$, when $f(t) = t\sin 2t$;

(b) the function $f(t)$ whose Laplace transform is

$$\frac{12se^{-4s}}{(s^2+9)^2};$$

(c) the function $f(t)$ whose Laplace transform is

$$\frac{2e^{-s}}{s^3} + \frac{6e^{-4s}}{s^2-9}.$$

Solution (a) From entry 8 of Table 15.1, we have

$$\mathscr{L}\{f(t)\} = \mathscr{L}\{t\sin 2t\} = \frac{4s}{(s^2+4)^2}.$$

Now the function to be transformed is $u_{\pi/4}(t)f\left(t-\frac{\pi}{4}\right)$, so we must set $a = \pi/4$ in Theorem 15.7, showing that

$$\mathscr{L}\left\{u_{\pi/4}(t)f\left(t-\frac{\pi}{4}\right)\right\} = \frac{4se^{-\pi s/4}}{(s^2+4)^2}.$$

(b) We are required to find the inverse Laplace transform of

$$\frac{12se^{-4s}}{(s^2+9)^2},$$

and as e^{-4s} appears as a factor it follows from Theorem 15.7 that a shift $a=4$ is involved. Now from entry 8 of Table 15.1 we see that

$$\mathscr{L}^{-1}\left\{\frac{12s}{(s^2+9)^2}\right\}=2t\sin 3t,$$

so incorporating the shift and using Theorem 15.7 we find

$$\mathscr{L}^{-1}\left\{\frac{12se^{-4s}}{(s^2+9)^2}\right\}=u_4(t)2(t-4)\sin[3(t-4)].$$

(c) Two separate shifts are involved when finding this inverse Laplace transform; namely, a shift of 1 in the first term and a shift of 4 in the second term. From entries 3 and 12 of Table 15.1 we see that

$$\mathscr{L}^{-1}\left\{\frac{2}{s^3}\right\}=t^2 \quad \text{and} \quad \mathscr{L}^{-1}\left\{\frac{6}{s^2-9}\right\}=2\sinh 3t.$$

So, incorporating the shifts and using Theorem 15.7, we find

$$\mathscr{L}^{-1}\left\{\frac{2e^{-s}}{s^3}+\frac{6e^{-4s}}{s^2-9}\right\}=u_1(t)(t-1)^2+2u_4(t)\sinh[3(t-4)]. \quad \blacksquare$$

Table 15.2 gathers together all the essential results of a general nature which have been established so far. Tables 15.1 and 15.2 together form the basic reference material necessary when an initial value problem is to be solved by means of the Laplace transform.

Table 15.2 Transformation of derivatives and shift theorems

Transformation of derivatives

$\mathscr{L}\{y'\}=s\mathscr{L}\{y\}-y(0)$

$\mathscr{L}\{y''\}=s^2\mathscr{L}\{y\}-y'(0)-sy(0)$

$\mathscr{L}\{y^{(n)}\}=s^n\mathscr{L}\{y\}-y^{(n-1)}(0)-sy^{(n-1)}(0)-\ldots-s^{n-1}y(0).$

First shift theorem

If $\mathscr{L}\{f(t)\}=F(s)$, then $\mathscr{L}\{e^{at}f(t)\}=F(s-a)$.

Second shift theorem

If $\mathscr{L}\{f(t)\}=F(s)$, then $\mathscr{L}\{u_a(t)f(t-a)\}=e^{-as}F(s)$
or equivalently,

$$\mathscr{L}^{-1}\{e^{-as}F(s)\}=u_a(t)f(t-a) \quad \text{for} \quad a>0.$$

15.8.4 Solution of initial value problems

The method of solution of initial value problems for linear constant-coefficient differential equations is best illustrated by means of example.

Example 15.21 Solve

$$y'' - 3y' + 2y = e^{-t},$$

given that $y(0) = 1$, $y'(0) = 0$.

Solution We take the Laplace transform of both sides of the equation to obtain

$$\mathscr{L}\{y''\} - 3\mathscr{L}\{y'\} + 2\mathscr{L}\{y\} = \mathscr{L}\{e^{-t}\}.$$

Transforming the derivatives by means of Table 15.2, and transforming e^{-t} by means of entry 4 of Table 15.1, we obtain

$$s^2 Y - y'(0) - sy(0) - 3[sY - y(0)] + 2Y = \frac{1}{s+1},$$

where $Y = \mathscr{L}\{y\}$.

Using the initial conditions and simplifying then gives

$$(s^2 - 3s + 2)Y = \frac{s^2 - 2s - 2}{s+1}.$$

Notice that the factor multiplying Y has the same form as the characteristic polynomial, though with s replacing λ. Solving for the Laplace transform Y of the required solution y, we find

$$\mathscr{L}\{y\} = Y(s) = \frac{s^2 - 2s - 2}{(s+1)(s^2 - 3s + 2)}.$$

Using partial fractions, this becomes

$$Y(s) = \frac{3}{2}\left(\frac{1}{s-1}\right) - \frac{2}{3}\left(\frac{1}{s-2}\right) + \frac{1}{6}\left(\frac{1}{s+1}\right).$$

Taking the inverse Laplace transform and using entry 4 of Table 15.1, we arrive at the complete solution to the initial value problem

$$y(t) = \frac{3}{2}e^{t} - \frac{2}{3}e^{2t} + \frac{1}{6}e^{-t}. \quad \blacksquare$$

Example 15.22 Solve

$$y'' - 2y' + y = e^{t},$$

given that $y(0) = 0, y'(0) = 1$.

Solution Proceeding as before, we find

$$s^2Y - y'(0) - sy(0) - 2[sY - y(0)] + Y = \frac{1}{s-1}.$$

Using the initial conditions and simplifying, we obtain

$$Y(s) = \frac{s}{(s-1)^3}.$$

Simplifying the right-hand side by means of partial fractions, we find

$$Y(s) = \frac{1}{(s-1)^2} + \frac{1}{(s-1)^3},$$

showing

$$y(t) = \mathscr{L}^{-1}\left\{\frac{1}{(s-1)^2}\right\} + \mathscr{L}^{-1}\left\{\frac{1}{(s-1)^3}\right\}.$$

Identifying the expressions on the right-hand side with entry 5 of Table 15.1, we arrive at the complete solution to the initial value problem

$$y(t) = te^t + \tfrac{1}{2}t^2e^t.$$

This same result could also have been obtained by use of the first shift theorem. This follows because $\mathscr{L}\{t\} = 1/s^2$, and $\frac{1}{2}\mathscr{L}\{t^2\} = 1/s^3$, showing that $\mathscr{L}\{te^t\} = 1/(s-1)^2$ and $\frac{1}{2}\mathscr{L}\{t^2e^t\} = 1/(s-1)^3$. Notice that although in this case the complementary function contained the inhomogeneous term, this presented no special difficulties when solving the problem by means of the Laplace transform. ■

Example 15.23 Solve the initial value problem

$$y'' - 3y' + 2y = e^{2t}$$

with $y(0) = 0$ and $y'(0) = 2$.

Solution Transforming the equation, we obtain

$$s^2Y - y'(0) - sy(0) - 3sY + 3y(0) + 2Y = \frac{1}{s-2},$$

but as $y(0) = 0$ and $y'(0) = 2$. this reduces to

$$Y = \frac{2s-3}{(s-1)(s-2)^2}.$$

The appropriate partial fraction expansion for this function is

$$\frac{2s-3}{(s-1)(s-2)^2} = \frac{A}{s-1} + \frac{B}{s-2} + \frac{C}{(s-2)^2},$$

and so

$$A(s-2)^2 + B(s-1)(s-2) + C(s-1) = 2s-3.$$

Equating the coefficients of corresponding powers of s on either side of this equation in the usual way, we find that

$$A = -1, \quad B = 1 \quad \text{and} \quad C = 1.$$

Thus

$$Y(s) = \frac{-1}{s-1} + \frac{1}{s-2} + \frac{1}{(s-2)^2},$$

and inverting these transforms with the help of Table 15.1 gives the solution

$$y(t) = -e^t + e^{2t} + te^{2t}. \quad \blacksquare$$

Example 15.24 Solve

$$y'' + 4y = u_0(t) - u_{\pi/2}(t),$$

given that $y(0) = 1, y'(0) = 0$.

Solution This equation has a pulse of unit amplitude and duration $\pi/2$ as its inhomogeneous term. The pulse is switched on at $t = 0$ and off at $t = \pi/2$. Taking the Laplace transform, we find

$$s^2 Y - y'(0) - sy(0) + 4Y = \frac{1}{s} - \frac{e^{-\pi s/2}}{s}.$$

Using the initial conditions and simplifying, this becomes

$$(s^2 + 4)Y = s + \frac{1}{s} - \frac{e^{-\pi s/2}}{s}.$$

so that

$$Y(s) = \frac{s}{s^2+4} + \frac{1}{s(s^2+4)}(1 - e^{-\pi s/2}).$$

Now, using partial fractions, we have

$$\frac{1}{s(s^2+4)} = \frac{1}{4}\left(\frac{1}{s}\right) - \frac{1}{4}\left(\frac{s}{s^2+4}\right),$$

so that finally we arrive at the result

$$Y(s) = \frac{3}{4}\left(\frac{s}{s^2+4}\right) + \frac{1}{4s} - \frac{1}{4}\left(\frac{e^{-\pi s/2}}{s}\right) + \frac{1}{4}\left(\frac{se^{-\pi s/2}}{s^2+4}\right).$$

The inverse Laplace transforms of the first three terms may be determined directly from Table 15.1, entries 7, 6 and 1, respectively, while the inverse Laplace transforms of the remaining two terms follow from Table 15.1 entries 1 and 7, respectively, together with the second shift theorem. Proceeding in this manner, we find that the complete solution is

$$y(t) = \frac{3}{4}\cos 2t + \frac{1}{4} - \frac{1}{4}u_{\pi/2}(t)\left\{1 - \cos\left[2\left(t - \frac{\pi}{2}\right)\right]\right\}. \quad \blacksquare$$

Example 15.25 Solve the initial value problem

$$y'' + y = f(t),$$

given that

$$f(t) = \begin{cases} 1-t, & 0 \leq t < 1 \\ 0, & t \leq 1, \end{cases}$$

when initially $y(0) = y'(0) = 0$.

Solution Before it is possible to take the Laplace transform of this differential equation, it is first necessary to find $\mathscr{L}\{f(t)\}$. Using the Heaviside unit step function, we can write

$$f(t) = (1-t) - (1-t)u_1(t) = (1-t) + (t-1)u_1(t),$$

so

$$\mathscr{L}\{f(t)\} = \mathscr{L}\{(1-t)\} + \mathscr{L}\{(t-1)u_1(t)\}.$$

The transform of the first expression follows directly from Table 15.1, while the transform of the second may be obtained by using the second shift theorem (Theorem 15.7). Proceeding in this manner brings us to the result

$$\mathscr{L}\{f(t)\} = \frac{1}{s} - \frac{1}{s^2} + \frac{e^{-s}}{s^2}.$$

Thus, using this result when transforming the differential equation, we obtain

$$s^2 Y - sy(0) - y'(0) + Y = \frac{1}{s} - \frac{1}{s^2} + \frac{e^{-s}}{s^2}.$$

However, $y(0) = y'(0) = 0$, so it follows that

$$Y(s) = \frac{s-1}{s^2(s^2+1)} + \frac{e^{-s}}{s^2(s^2+1)}.$$

We must now invert these Laplace transforms to find the required solution. Making the partial fraction expansion

$$\frac{s-1}{s^2(s^2+1)} = \frac{A}{s} + \frac{B}{s^2} + \frac{Cs+D}{s^2+1},$$

the usual arguments show that

$$A = 1, \quad B = -1, \quad C = -1 \quad \text{and} \quad D = 1,$$

and so

$$\frac{s-1}{s^2(s^2+1)} = \frac{1}{s} - \frac{1}{s^2} - \frac{s}{s^2+1} + \frac{1}{s^2+1}.$$

Consequently, as

$$\frac{1}{s^2(s^2+1)} = \frac{1}{s^2} - \frac{1}{s^2+1},$$

it follows that the Laplace transform of the solution is

$$Y(s) = \frac{1}{s} - \frac{1}{s^2} - \frac{s}{s^2+1} + \frac{1}{s^2+1} + \mathrm{e}^{-s}\left(\frac{1}{s^2} - \frac{1}{s^2+1}\right).$$

Inverting the first four terms of this transform with the aid of Table 15.1, we obtain

$$y(t) = 1 - t - \cos t + \sin t + \mathscr{L}^{-1}\left\{\mathrm{e}^{-s}\left(\frac{1}{s^2} - \frac{1}{s^2+1}\right)\right\}.$$

However, we see from the second shift theorem that

$$\mathscr{L}^{-1}\left\{\mathrm{e}^{-s}\left(\frac{1}{s^2} - \frac{1}{s^2+1}\right)\right\} = h(t-1)u_1(t),$$

where $h(t) = t - \sin t$. Thus the solution to the initial value problem is

$$y(t) = 1 - t - \cos t + \sin t + \{(t-1) - \sin(t-1)\}u_1(t). \quad \blacksquare$$

Example 15.26 Solve the initial value problem

$$y'' + \Omega^2 y = a \sin \Omega t,$$

subject to the arbitrary initial conditions $y(0) = \alpha$ and $y'(0) = \beta$.

Solution Transforming the equation gives

$$s^2 Y - y(0) - sy'(0) + \Omega^2 Y = \frac{a\Omega}{s^2 + \Omega^2},$$

and using $y(0) = \alpha$, $y'(0) = \beta$, we find that

$$Y(s) = \frac{\alpha}{s^2 + \Omega^2} + \frac{\beta s}{s^2 + \Omega^2} + \frac{a\Omega}{(s^2 + \Omega^2)^2}.$$

Now we shall rewrite the last term as follows:

$$\frac{a\Omega}{(s^2+\Omega^2)^2} = (a\Omega)\left(\frac{1}{2\Omega^2}\right)\left(\frac{(s^2+\Omega^2)-(s^2-\Omega^2)}{(s^2+\Omega^2)^2}\right)$$

$$= \frac{a}{2\Omega^2}\left(\frac{\Omega}{s^2+\Omega^2}\right) - \frac{a}{2\Omega}\left(\frac{s^2-\Omega^2}{(s^2+\Omega^2)^2}\right).$$

Thus using this result in $Y(s)$ and rearranging terms gives

$$Y(s) = \left(\frac{\alpha}{\Omega}+\frac{a}{2\Omega^2}\right)\left(\frac{\Omega}{s^2+\Omega^2}\right) + \beta\left(\frac{s}{s^2+\Omega^2}\right) - \left(\frac{a}{2\Omega}\right)\left(\frac{s^2-\Omega^2}{(s^2+\Omega^2)^2}\right).$$

Using Table 15.1 to determine the inverse transform, we arrive at the solution

$$y(t) = \left(\frac{\alpha}{\Omega}+\frac{a}{2\Omega^2}\right)\sin\Omega t + \beta\cos\Omega t - \frac{at}{2\Omega}\cos\Omega t.$$

Setting

$$C = \frac{\alpha}{\Omega}+\frac{a}{2\Omega^2} \quad \text{and} \quad D = \beta$$

enables the first two terms of $y(t)$ to be written as

$$\left(\frac{\alpha}{\Omega}+\frac{a}{2\Omega^2}\right)\sin\Omega t + \beta\cos\Omega t = A\sin(\Omega t+\varepsilon),$$

where

$$A = (C^2+D^2)^{1/2}, \quad \cos\varepsilon = C/A \text{ and } \sin\varepsilon = D/A.$$

Thus the solution may be expressed in the more concise form

$$y(t) = A\sin(\Omega t+\varepsilon) - \frac{at}{2\Omega}\cos\Omega t.$$

This is the resonance solution of an undamped oscillation already given in Eqn (15.47). Due to the absence of damping, the magnitude of the oscillations of this solution grows unboundedly with increasing time. ■

As a final example we return to the question of solving initial value problems for systems. It will be seen that the Laplace transform provides a very convenient method for the solution of such problems.

Example 15.27 Solve

$$y' = y + 5z, \quad z' = -(y+3z),$$

given that $y(0) = 1$, $z(0) = 0$.

Solution Two variables are now involved, so we introduce the Laplace transforms $Y(s)=\mathscr{L}\{y(t)\}$, $Z(s)=\mathscr{L}\{z(t)\}$. Then, taking the Laplace transform of each equation, we have

$$sY - y(0) = Y + 5Z,$$

and

$$sZ - z(0) = -Y - 3Z.$$

Using the initial conditions and simplifying brings us to

$$(s-1)Y - 5Z = 1,$$

$$Y + (s+3)Z = 0,$$

which when solved gives

$$Y(s) = \frac{s+3}{s^2+2s+2}, \quad Z(s) = \frac{-1}{s^2+2s+2}.$$

Now to determine the inverse Laplace transforms of Y and Z it is necessary to manipulate the algebraic expressions into forms which can be identified with entries in Table 15.1. The starting point is the denominator of both Y and Z, each of which may, by means of completing the square, be replaced by $(s+1)^2+1$.

We have

$$Y(s) = \frac{(s+1)}{(s+1)^2+1} + \frac{2}{(s+1)^2+1},$$

and

$$Z(s) = \frac{-1}{(s+1)^2+1};$$

entries 10 and 11 of Table 15.1 now show the complete solution to be

$$y(t) = e^{-t}(\cos t + 2\sin t), \quad z(t) = -e^{-t}\sin t. \quad \blacksquare$$

15.9 The Delta function

A useful generalization of a function is the *delta function* denoted by $\delta(t)$, which in mechanics can be used to describe an impulsive force, or in electric circuit theory the application of a very large voltage for a very short time. The delta function is not a function in the ordinary sense, but the result of a limiting process. The delta function $\delta(t-a)$ located at $t=a$ can be considered to be the limit as $h\to\infty$ of a rectangular pulse of very large amplitude h at $t=a$ and very small width $1/h$, so that however large h becomes, the area of the pulse remains constant at $h.(1/h)=1$.

A possible representation of the delta function at time $t=a$ can be taken as the limit as $h\to\infty$ of the difference between the two step functions of amplitude h, with one located at $t=a$ and the other located at $t=a+1/h$, as shown in Fig. 15.8.

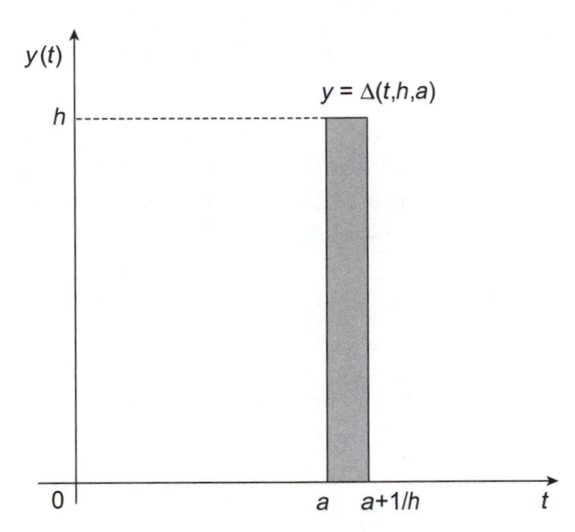

Fig. 15.8 The delta function $\delta(t-a)$ as the limit of the difference between two step functions.

Denote the rectangular pulse function in Fig. 15.8 by $\Delta(t, h, a)$, where

$$\Delta(t,\ h,\ a) = \begin{cases} 0, & 0 \le t < a \\ h, & a < t < a + 1/h \\ 0, & a + 1/h < t. \end{cases}$$

Then we have already seen from the reasoning following Fig. 15.7 that

$$\mathscr{L}\{\Delta(t,h,a)\} = \frac{h}{s}(\mathrm{e}^{-as} - \mathrm{e}^{-(a+1/h)s}) = \frac{h}{s}\mathrm{e}^{-as}(1 - \mathrm{e}^{-s/h}) \quad \text{for } a > 0,\ s > 0.$$

Proceeding to the limit as $h \to \infty$ it then follows that

$$\mathscr{L}\{\delta(t-a)\} = \mathrm{e}^{-as} \quad \text{for } a > 0,\ s > 0,$$

and in particular

$$\mathscr{L}\{\delta(t)\} = \mathrm{e}^{-s} \quad \text{for } s > 0.$$

An interesting and useful property of the delta function is that if a function $f(t)$ is continuous at $t = a$ then

$$\int_0^\infty f(t)\delta(t-a)\,\mathrm{d}t = f(a).$$

This is called the *filtering property* of the delta function, and the result can be established as follows. Consider the integral

$$\int_0^\infty \Delta(t,h,a)f(t)\,\mathrm{d}t = \int_a^{a+1/h} hf(t)\,\mathrm{d}t,$$

then from the mean value theorem for integrals (Theorem 7.5) it follows that

$$\int_0^\infty \Delta(t, h, a)f(t)\,dt = \frac{1}{h}hf(\xi) \quad \text{with } a < \xi < a + 1/h.$$

The result then follows by proceeding to the limit as $h \to \infty$, because then the limit of the expression on the left becomes $\int_0^\infty f(t)\delta(t-a)\,dt$ while ξ is squeezed between a and $a+1/h$ so that in the limit the expression on the right becomes $f(a)$.

Example 15.28 Solve the initial value problem

$$y'' + 4y = t + 2\delta(t-1)$$

with the initial conditions $y(0) = 1,\ y'(0) = 0$.

Solution Taking the Laplace transform of the equation and using the fact that $\mathscr{L}\{2\delta(t-1)\} = 2e^{-s}$, we find that

$$s^2Y - y'(0) - sy(0) + 4Y = 1/s^2 + 2e^{-s},$$

so as $y(0) = 1$ and $y'(0) = 0$, after simplification this becomes

$$Y(s) = \frac{s}{s^2+4} + \frac{1}{s^2(s^2+4)} + \frac{2e^{-s}}{s^2+4}.$$

Taking the inverse Laplace transform in the usual manner gives

$$y(t) = \cos 2t + \tfrac{1}{4}t - \tfrac{1}{8}\sin 2t + u_1(t)\sin(2t-2).$$

An examination of the solution shows, as would be expected, that until $t = 1$ the solution develops like the solution of $y'' + 4y = t$ with the initial conditions $y(0) = 1,\ y'(0) = 0$, but from $t = 1$ it is modified by the effect of the impulse $2\delta(t-1)$, which produces the last term in the solution. ■

15.10 Applications of the Laplace transform

The purpose of this section is to illustrate a few of the diverse applications of the Laplace transform which arise in engineering. The examples chosen are from mechanical, aeronautical, electrical, control and chemical engineering. Different though these topics are, they have as a unifying theme the fact that they all give rise to physical situations which are capable of description in terms of initial value problems for linear differential equations. This being the case, they may all be solved analytically by means of the Laplace transform using the methods outlined in section 15.8. Indeed, in some of the rather simple examples described here, it is the same second-order differential equation in each case. In general, different applications will lead to different differential equations, though their method of solution will always be the same.

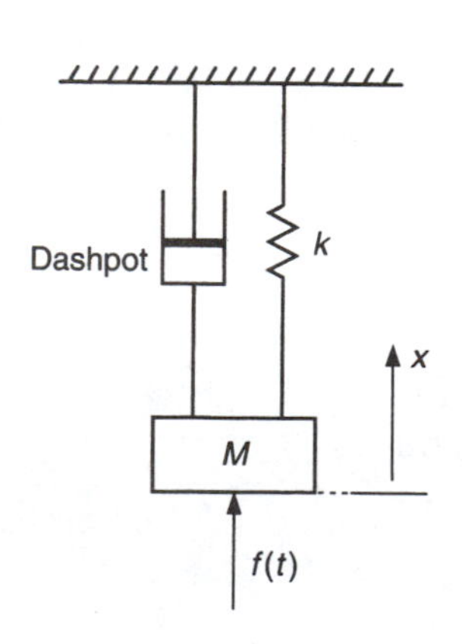

Fig. 15.9 Vibration damper.

15.10.1 Vibration damper

A typical one-dimensional damping mechanism for mechanical vibrations is shown in Fig. 15.9. This contains the mass M caused to vibrate by an applied force $f(t)$, a spring which exerts a force equal to k times its displacement, and a dashpot (piston in a viscous oil-filled cylinder). In its simplest form a dashpot produces a force-opposing motion which is proportional to velocity, so the actual force is equal to $\mu \mathrm{d}x/\mathrm{d}t$, where x is the displacement of the mass from its equilibrium position, μ is a constant and t is the time.

Applying Newton's law of motion to the system, we find that the equation of motion for the mass M is

$$M\frac{\mathrm{d}^2x}{\mathrm{d}t^2} + \mu\frac{\mathrm{d}x}{\mathrm{d}t} + kx = f(t),$$

or

$$\frac{\mathrm{d}^2x}{\mathrm{d}t^2} + 2\zeta\Omega\frac{\mathrm{d}x}{\mathrm{d}t} + \Omega^2 x = \frac{1}{M}f(t), \tag{15.57}$$

where

$$\Omega = (k/M)^{1/2}, \quad \zeta = \frac{\mu}{2(kM)^{1/2}}. \tag{15.58}$$

If the force $f(t)$ is applied to the mass M at time $t = 0$ when it is at rest, the corresponding initial conditions are

$$x(0) = 0, \quad x'(0) = 0. \tag{15.59}$$

Taking the Laplace transform of Eqn. (15.57) and using the initial conditions (15.59), we obtain

$$s^2X + 2\zeta\Omega sX + \Omega^2 X = \frac{1}{M}F(s),$$

where $X(\mathrm{s}) = \mathscr{L}\{x(t)\}$ and $F(s) = \mathscr{L}\{f(t)\}$. Thus the Laplace transform of the displacement $x(t)$ is

$$X(s) = \left(\frac{1}{M}\right)\frac{1}{s^2 + 2\zeta\Omega + \Omega^2}F(s). \tag{15.60}$$

Let us now suppose force $f(t)$ is suddenly applied at time $t = 0$ and maintained at a constant value F_0, so that

$$f(t) = F_0 u_0(t).$$

Then $\mathscr{L}\{f(t)\} = F(s) = F_0/\mathrm{s}$ and Eqn (15.60) becomes

$$X = \left(\frac{F_0}{M}\right)\frac{1}{s(s^2 + 2\zeta\Omega s + \Omega^2)}. \tag{15.61}$$

Completing the square shows

$$s^2 + 2\zeta\Omega s + \Omega^2 \equiv (s + \zeta\Omega)^2 + (1 - \zeta^2)\Omega^2,$$

so this quadratic will have complex conjugate zeros if $\zeta^2 < 1$ and real zeros if $\zeta^2 \geq 1$. Since it is the most interesting case, we shall determine the solution when $\zeta^2 < 1$.

Using partial fractions, Eqn (15.61) becomes

$$X = \frac{F_0}{k}\left(\frac{1}{s} - \frac{s + 2\zeta\Omega}{(s + \zeta\Omega)^2 + (1 - \zeta^2)\Omega^2}\right). \tag{15.62}$$

To find the inverse Laplace transform of Eqn. (15.62) by means of entries 1, 10 and 11 of Table 15.1, we first rewrite it in the form

$$X = \frac{F_0}{k}\left(\frac{1}{s} - \frac{s + \zeta\Omega}{(s + \zeta\Omega)^2 + (1 - \zeta^2)\Omega^2} - \frac{\zeta}{(1 - \zeta^2)^{1/2}}\frac{(1 + \zeta^2)^{1/2}\Omega}{(s + \zeta\Omega)^2 + (1 - \zeta^2)\Omega^2}\right),$$

from which the solution is easily seen to be

$$x(t) = \frac{F_0}{k}\left(1 - e^{-\zeta\Omega t}\cos[\Omega(1 - \zeta^2)^{1/2}t] - \frac{\zeta}{(1 - \zeta^2)^{1/2}}e^{-\zeta\Omega t}\sin[\Omega(1 - \zeta^2)^{1/2}t]\right), \tag{15.63}$$

for $t > 0$.

Comparison of this solution with the work of section 15.4 shows Ω to be the natural frequency of the system in Fig. 15.9, corresponding to the natural oscillations of the system without frictional resistance ($\mu = 0$) and with no applied external force ($f(t) = 0$). The parameter ζ is the damping of the system. When $\zeta^2 < 1$ the system is normally damped, when $\zeta^2 > 1$ it is over-damped, and when $\zeta = 1$ it is critically damped.

Inspection of Eqn (15.63) reveals that as $t \to +\infty$, so $x \to F_0/k$, the static equilibrium position. In the normally damped case discussed here, oscillations occur about this static equilibrium position and decay to zero exponentially as $e^{-\zeta\Omega t}$. We leave as an exercise the task of showing that when $\zeta^2 \geq 1$ the solution takes the form

$$x(t) = \frac{F_0}{k}\left\{1 - \frac{1}{2(\zeta^2 - 1)^{1/2}}\left(\frac{e^{\lambda_1 t}}{\zeta - (\zeta^2 - 1)^{1/2}} - \frac{e^{\lambda_2 t}}{\zeta + (\zeta^2 - 1)^{1/2}}\right)\right\} \tag{15.64}$$

where

$$\lambda_1 = \Omega[-\zeta + (\zeta^2 - 1)^{1/2}] \quad \text{and} \quad \lambda_2 = \Omega[-\zeta - (\zeta^2 - 1)^{1/2}].$$

The general behaviour of the solution $x(t)$ for different values of ζ is shown in Fig. 15.10. It is seen that to cause x to approach the static equilibrium position quickly it is necessary to have ζ small, but then persistent oscillations occur. If, instead, ζ is large, then no oscillations occur but the approach to the static equilibrium position is extremely

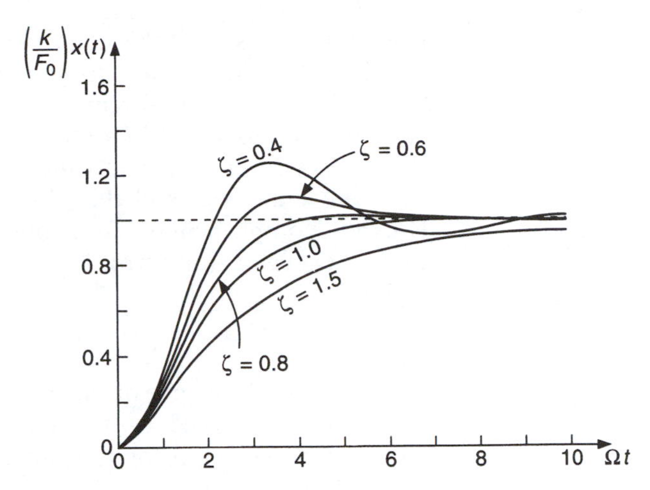

Fig. 15.10 Response of system to a step function input.

slow. In practice a compromise is usually made, and ζ is chosen within the interval $0.4 < \zeta < 1$.

The sudden application of a load to a system in the manner considered here is one of the standard ways of testing its overall response. Others involve applying the load for a finite time, so that $f(t)$ becomes a finite-width pulse, subjecting the system to a sinusoidal input whose frequency may be altered, or subjecting the system to an impulse. For our purposes an *impulse* may be considered to be an extremely narrow pulse of very large amplitude, with the product of the pulse width and its amplitude measuring the magnitude of the impulse. The response to a sinusoidal input has, in fact, already been considered in section 15.4. In this form of analysis the amplitude and phase of the output are determined as a function of the frequency of the input whose amplitude is kept constant.

At this point it is useful to generalize matters a little, and to consider the situation in which an arbitrary nth-order constant-coefficient differential equation is involved, so that the equation of motion may be written

$$G(x) = f(t), \tag{15.65}$$

where $G(x) \equiv a_0 x^{(n)} + a_1 x^{(n-1)} + \ldots + a_n x$, with $x^{(r)} = \mathrm{d}^r x/\mathrm{d}t^r$. Then, transforming this equation by the Laplace transform, we obtain

$$G(s)X = N(s) + F(s), \tag{15.66}$$

where $G(s) = a_0 s^n + a_1 s^{n-1} + \ldots + a_n$, $X = \mathscr{L}\{x(t)\}$, $N(s)$ represents the terms derived from the transformation of $G(x)$ due to the initial conditions, and $F(s) = \mathscr{L}\{f(t)\}$. Thus for the system in Fig. 15.9, $G(s) = Ms^2 + \mu s + k$, while $N(s) \equiv 0$, because $x(0) = x'(0) = 0$.

Hence the Laplace transformation of the solution to the initial value problem may be written

$$X(s) = \frac{N(s)}{G(s)} + \frac{F(s)}{G(s)}. \tag{15.67}$$

The right-hand side of this equation is seen to contain two distinct parts, the first, $N(s)/G(s)$, due to the initial conditions, and the second, $F(s)/G(s)$, due to the forcing function $f(t)$. Comparing this situation with the one discussed in section 15.2, it is appropriate, in the context of initial value problems, to call the solution due to $N(s)/G(s)$ the *complementary function*, and that due to $F(s)/G(s)$ the *particular integral*. Thus the complementary function vanishes when the initial conditions are homogeneous (all zero).

Since

$$\mathscr{L}\{x^{(n)}\} = s^n X - x^{(n-1)}(0) - sx^{(n-2)}(0) - \ldots - s^{n-1}x(0),$$

it follows that the degree of the polynomial $N(s)$ formed by transforming $G(x)$ can never exceed $n-1$, whereas the degree of the polynomial $G(s)$ itself is n. The nature of the function $F(s)$ depends on the choice of forcing function $f(t)$.

15.10.2 Rolling motion of an aircraft

Let the moment of inertia of an aircraft about its longitudinal axis be I, the roll angle be θ, and the rates of change of angular momentum due to aerodynamic damping and stabilizing effects be $k\,\mathrm{d}\theta/\mathrm{d}t$ and $a\theta$, respectively. Then if the applied torque causing roll is $f(t)$, the equation of motion of the aircraft becomes

$$I\frac{\mathrm{d}^2\theta}{\mathrm{d}t^2} + k\frac{\mathrm{d}\theta}{\mathrm{d}t} + a\theta = f(t). \tag{15.68}$$

Setting

$$\Omega = (a/I)^{1/2}, \quad \zeta = \frac{k}{2(aI)^{1/2}}, \tag{15.69}$$

the equation of motion becomes

$$\frac{\mathrm{d}^2\theta}{\mathrm{d}t^2} + 2\zeta\Omega\frac{\mathrm{d}\theta}{\mathrm{d}t} + \Omega^2\theta = \frac{1}{I}f(t), \tag{15.70}$$

which has the same form as Eqn (15.56). If, for example, a constant torque H_0 is applied at time $t=0$ when the aircraft is in equilibrium, so $\theta(0)=0$, $\theta'(0)=0$, then $f(t)=H_0u_0(t)$.

With the obvious modifications, solutions (15.63) and (15.64) again apply, depending on the value of the damping ζ. If, for example, $f(t)=H_0u_0(t)$, the aircraft will eventually settle down to its equilibrium orientation with roll angle $\theta = H_0/I$.

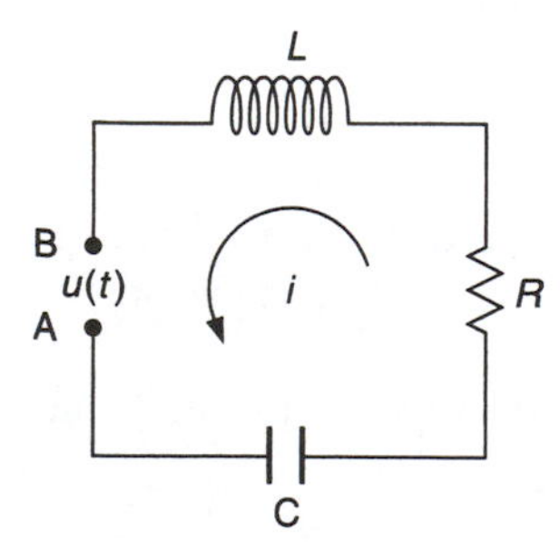

Fig. 15.11 The $R-L-C$ circuit.

15.10.3 The $R-L-C$ circuit

A simple electric circuit is shown in Fig. 15.11 containing a resistor R, an inductance L and a capacitor C. The current i flows due to an e.m.f. $v(t)$

which is suddenly applied across the terminals A and B at $t=0$, and then held constant at the value V_0, so that initially the charge q on the plate of the condenser into which current will flow is zero, as is $\mathrm{d}q/\mathrm{d}t$.

Applying Kirchhoff's law to the circuit brings us to the result

$$L\frac{\mathrm{d}i}{\mathrm{d}t}+Ri+\frac{1}{C}q=v(t).$$

Since the current $i=\mathrm{d}q/\mathrm{d}t$, the build up of the charge on the condenser plate is governed by the equation

$$L\frac{\mathrm{d}^2q}{\mathrm{d}t^2}+R\frac{\mathrm{d}q}{\mathrm{d}t}+\frac{1}{C}q=v(t). \tag{15.71}$$

Setting

$$\Omega=(LC)^{-1/2},\quad \zeta=\frac{1}{2}R(L/C)^{-1/2}, \tag{15.72}$$

Eqn (15.71) becomes

$$\frac{\mathrm{d}^2q}{\mathrm{d}t^2}+2\zeta\Omega q+\Omega^2q=\frac{1}{L}v(t).$$

Since $v(t)=V_0u_0(t)$, $q(0)=0$ and $q'(0)=0$, the build up of the current may be found by making the obvious identifications in Eqns. (15.63) or (15.64), depending on the value of ζ.

15.10.4 Control system for the position of an antenna

A *control system*, or *servomechanism*, is an automatic device designed to relate the response (output) of a system to an arbitrary input in a particular manner. It operates by comparing the desired response of the system with the actual one, and then altering the input so the difference is reduced to zero.

A symbolic representation of a simple device for controlling the bearing of an antenna is shown in Fig. 15.12. The input to the system is the required bearing θ_I, and the output is the actual bearing θ_O. The misalignment error $\varepsilon=\theta_I-\theta_O$ is used to provide a voltage proportional to ε by which to operate a servomotor. This is a device whose angle of rotation

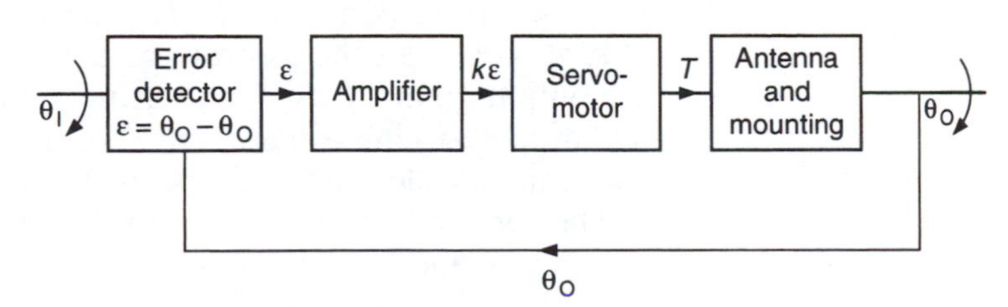

Fig. 15.12 Control mechanism for antenna bearing.

is proportional to the actuating voltage ε, producing a rotation in one sense if $\varepsilon > 0$ and in the other if $\varepsilon < 0$.

Let the combined moments of inertia of the servomotor and antenna system about their axis of rotation be I, the resisting torque in the antenna be $\mu\,\mathrm{d}\theta_{\mathrm{O}}/\mathrm{d}t$ (resistance assumed to be proportional to the rate of rotation), and the applied torque $T = k(\theta_{\mathrm{I}} - \theta_{\mathrm{O}})$, with k a constant and t the time. Then the equation governing the rotation of the system may be written

$$I\frac{\mathrm{d}^2\theta_{\mathrm{O}}}{\mathrm{d}t^2} + \mu\frac{\mathrm{d}\theta_{\mathrm{O}}}{\mathrm{d}t} = k(\theta_{\mathrm{I}} - \theta_{\mathrm{O}}),$$

or

$$\left(\frac{I}{k}\right)\frac{\mathrm{d}^2\theta_{\mathrm{O}}}{\mathrm{d}t^2} + \left(\frac{\mu}{k}\right)\frac{\mathrm{d}\theta_{\mathrm{O}}}{\mathrm{d}t} + \theta_{\mathrm{O}} = \theta_{\mathrm{I}}. \tag{15.74}$$

For the sake of simplicity, let us assume the system starts from rest with $\theta_{\mathrm{O}}(0) = 0$, $\theta'_{\mathrm{O}}(0) = 0$. Then taking the Laplace transform of Eqn (15.74), using the initial conditions, setting $\Theta_{\mathrm{O}} = \mathscr{L}\{\theta_{\mathrm{O}}\}$, $\Theta_{\mathrm{I}} = \mathscr{L}\{\theta_{\mathrm{I}}\}$ and solving for Θ_{O} we find

$$\Theta_{\mathrm{O}} = \frac{1}{(I/k)s^2 + (\mu/k)s + 1}\Theta_{\mathrm{I}}. \tag{15.75}$$

In control theory the function $[(I/k)s^2 + (\mu/k)s + 1]^{-1}$ in Eqn (15.75) is called the *transfer function* of the system, because it transfers the effect of the input to the system to the output.

Once the input $\theta_{\mathrm{I}}(t)$ is specified, its Laplace transform $\Theta_{\mathrm{I}} = \mathscr{L}\{\theta_{\mathrm{I}}\}$ is known. Then taking the inverse Laplace transform of Eqn (15.75) determines the output $\theta_{\mathrm{O}}(t)$. Suppose, for example, that θ_{I}, is a step function, corresponding to a sudden change from a zero bearing to some other fixed bearing $\theta_{\mathrm{I}} = \beta$. Then $\theta_{\mathrm{I}}(t) = \beta u_0(t)$, and with the obvious change of notation the solution to Eqn (15.75) may be found from either (15.63) or (15.64), depending on the value of $\zeta = \frac{1}{2}\mu(kI)^{1/2}$.

Notice that the quantities μ and I are characteristics of the mechanical design, and so are difficult to alter. However, the amplification factor (gain) k is part of the control system, and so may be selected at will by the designer. Usually k would be chosen to provide a compromise between a value of ζ which minimizes oscillation and one which gives a suitable frequency of response $\Omega = (k/I)^{1/2}$ for the controlled system.

In this example the control was provided by feeding back to the input the output quantity itself, and using a direct comparison between input and output to actuate the system. For this reason the quantity fed back to the input is called the *feedback* in the system.

The performance of the system may be modified still further, without altering the mechanical design, if the feedback quantity is a function of θ_{O} rather than θ_{O} itself. We conclude by examining this more general approach to control.

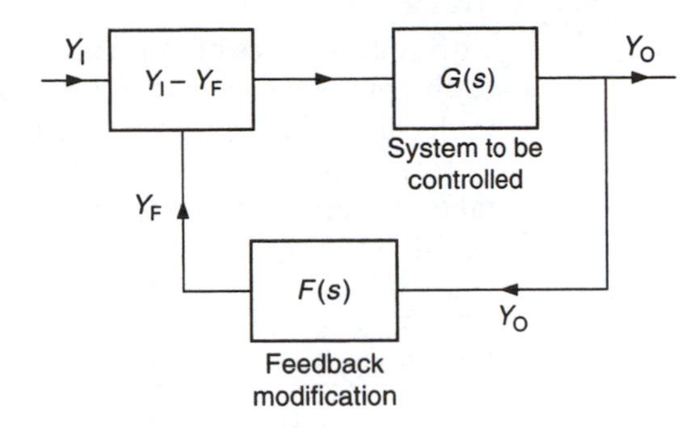

Fig. 15.13 General feedback control system.

Let us agree to call the transfer function $G(s)$ of an element in a control system the function of s which relates the Laplace transform $Y_O(s)$ of the output $y_O(t)$ to the Laplace transform $Y_I(s)$ of the input $y_I(t)$, when initially $y_I(t)$ and all its derivatives are zero, so

$$Y_O(s) = G(s)Y_I(s). \tag{15.76}$$

Working in terms of the Laplace transform, we now consider the control system shown diagramatically in Fig. 15.13, in which the feedback quantity is Y_F. The governing equations are

$$Y_O(s) = G(s)[Y_I(s) - Y_F(s)] \quad \text{and} \quad Y_F(s) = F(s)Y_O(s) \tag{15.77}$$

showing

$$Y_O(s) = \frac{G(s)}{1 + F(s)G(s)}Y_I(s). \tag{15.78}$$

The transfer function of the system is Fig. 15.13 is thus seen to be $G(s)[1+F(s)G(s)]^{-1}$, in which the feedback transfer function $F(s)$ is a function at the disposal of the designer. By means of a suitable choice of $F(s)$ the response of the system may now be modified within very wide limits, without the need for redesign of the original system whose transfer function is $G(s)$.

The solution of Eqn (15.78) by means of the Laplace transform follows as before once $F(s)$, $G(s)$ and $Y_I(s)$ have been specified. This constitutes the basis of control theory, an important aspect of which is ensuring that the choice of feedback does not give rise to *instability* in the system.

15.10.5 Irreversible chemical reactions

Many chemical reactions involve intermediate products, so that when in a reaction a molecule of chemical A is changed into a molecule of chemical D, this may take place via molecules of chemicals B and C. Let us determine the equations governing such a process if it is irreversible, the reaction rates $A \to B$, $B \to C$ and $C \to D$ are k_1, k_2 and k_3, respectively, and the number of molecules of A, B, C and D present at time t are x, y, z and w respectively. We shall assume that the process starts at time $t = 0$

with N molecules of A present and none of B, C or D. The irreversibility of the process means reactions can only proceed in one direction, so that molecules of B cannot revert to molecules of A, etc.

To take account of the overall progress of the reaction it is necessary to take account of each separate reaction. This is most easily accomplished as follows.

Reaction	Reaction rate of appearance	Reaction rate of appearance
$A \to B$	$\left(\frac{dx}{dt}\right)_{A\to B} = -k_1 x,$	$\left(\frac{dy}{dt}\right)_{A\to B} = k_1 x$
$B \to C$	$\left(\frac{dy}{dt}\right)_{B\to C} = -k_2 y,$	$\left(\frac{dz}{dt}\right)_{B\to C} = k_2 y$
$C \to D$	$\left(\frac{dz}{dt}\right)_{C\to D} = -k_3 z,$	$\left(\frac{dw}{dt}\right)_{C\to D} = k_3 z.$

The total reaction rates then follow by algebraic addition of the rates for each reaction, giving

$$\frac{dx}{dt} = \left(\frac{dx}{dt}\right)_{A\to B} = -k_1 x,$$

$$\frac{dy}{dt} = \left(\frac{dy}{dt}\right)_{A\to B} + \left(\frac{dy}{dt}\right)_{B\to C} = k_1 x - k_2 y,$$

$$\frac{dz}{dt} = \left(\frac{dz}{dt}\right)_{B\to C} + \left(\frac{dz}{dt}\right)_{C\to D} = k_2 y - k_3 z.$$

Thus the numbers of molecules x, y and z of chemicals A, B and C present at time t may be found by solving the simultaneous differential equations

$$\begin{aligned} &\frac{dx}{dt} + k_1 x = 0 \\ &\frac{dy}{dt} - k_1 x + k_2 y = 0, \\ &\frac{dz}{dt} - k_2 y + k_3 z = 0, \end{aligned} \tag{15.79}$$

subject to the initial conditions $x(0) = N$, $y(0) = 0$ and $z(0) = 0$. Once x, y and z have been determined from Eqns (15.79) the number of molecules w of chemical D present at time t follows from the *conservation of matter* which means that

$$x + y + z + w = N. \tag{15.80}$$

Taking the Laplace transform of Eqns (15.79) and using the initial conditions gives

$$sX + k_1 X = N,$$

$$sY - k_1 X + k_2 Y = 0,$$

$$sZ - k_2 Y + k_3 Z = 0.$$

Solving these sequentially, we find that

$$X = \frac{N}{s + k_1},$$

$$Y = \frac{k_1 N}{(s + k_1)(s + k_2)},$$

$$Z = \frac{k_1 k_2 N}{(s + k_1)(s + k_2)(s + k_3)}.$$

Using partial fractions together with entry 4 of Table 15.1, we finally arrive at the solution

$$x = N\mathrm{e}^{-k_1 t},$$

$$y = \frac{k_1 N}{k_1 - k_2}(\mathrm{e}^{-k_1 t} - \mathrm{e}^{-k_2 t}),$$

$$z = k_1 k_2 N\left(\frac{1}{(k_2 - k_1)(k_3 - k_1)}\mathrm{e}^{-k_1 t} + \frac{1}{(k_1 - k_2)(k_3 - k_2)}\mathrm{e}^{-k_2 t} + \frac{1}{(k_1 - k_3)(k_2 - k_3)}\mathrm{e}^{-k_3 t}\right), \quad (15.81)$$

so that from (15.80)

$$w = N - x - y - z.$$

15.10.6 Xenon transient after shut-down of a nuclear reactor

In a nuclear reactor, fission of the uranium fuel U^{235} to produce neutrons also gives rise to the radioactive isotope of iodine I^{135}. This has a moderately short half-life of 6.68 hours and decays to a radioactive isotope of xenon Xe^{135} which is a strong absorber of neutrons. The xenon in turn decays to the isotope of caesium Cs^{135} with a half-life of 9.13 hours, which is not a strong absorber of neutrons. When the reactor is operating at power, the I^{135} and Xe^{135} are destroyed by the flux of neutrons, and have relatively little effect on the operation of the reactor.

However, if the reactor is shut down, the neutron flux falls almost to zero, but the iodine continues to decay producing xenon, so there is a build-up of absorbing material (poison) within the reactor core. The analysis of this situation is important, because if the build-up of xenon becomes too large the reactor cannot be restarted until the xenon has decayed to a suitable level.

To study this process let m_{Xe} and m_I, denote the amounts of xenon and iodine present at time t; then the iodine production obeys the equation

$$\frac{dm_I}{dt} = -\lambda_I m_I, \tag{15.82}$$

and the xenon production obeys the equation

$$\frac{dm_{Xe}}{dt} = -\lambda_{Xe} m_{Xe} + \lambda_I m_I, \tag{15.83}$$

where λ_I and λ_{Xe} are the half-lives of the iodine and xenon isotopes.

Suppose the initial quantities of I^{135} and Xe^{135} present in the reactor are $(m_I)_0$ and $(m_{Xe})_0$, respectively. Then taking the Laplace transform of (15.82) we find

$$sM_I - (m_I)_0 = -\lambda_I M_I,$$

where $M_I = \mathscr{L}\{m_I\}$. Hence

$$M_I = \frac{(m_I)_0}{s + \lambda_I},$$

and taking the inverse Laplace transform gives

$$m_I = (m_I)_0 e^{-\lambda_I t}. \tag{15.84}$$

Using this result in Eqn (15.83), we find that the xenon obeys the equation

$$\frac{dm_{Xe}}{dt} = -\lambda_{Xe} m_{Xe} + \lambda_I (m_I)_0 e^{-\lambda_I t}. \tag{15.85}$$

Taking the Laplace transform of this result and setting $M_{Xe} = \mathscr{L}\{m_{Xe}\}$ gives

$$sM_{Xe} - (m_{Xe})_0 = -\lambda_{Xe} M_{Xe} + \frac{\lambda_I (m_I)_0}{s + \lambda_I},$$

from which, after use of partial fractions, we obtain

$$M_{Xe} = \frac{(m_{Xe})_0}{s + \lambda_{Xe}} + \frac{\lambda_I (m_I)_0}{(\lambda_I - \lambda_{Xe})} \left(\frac{1}{s + \lambda_{Xe}} - \frac{1}{s + \lambda_I} \right). \tag{15.86}$$

Taking the inverse Laplace transform of this result then brings us to the equation governing the build-up and subsequent decay of the xenon poison in the reactor core,

$$m_{Xe} = (m_{Xe})_0 e^{-\lambda_{Xe} t} + \frac{\lambda_I (m_I)_0}{(\lambda_I - \lambda_{Xe})} (e^{-\lambda_{Xe} t} - e^{-\lambda_I t}). \tag{15.87}$$

This function attains a maximum at some time $t_m > 0$, after which it decays to zero. The time t_m, and the maximum fractional increase of xenon $(m_{Xe}(t_m) - m_{Xe}(0))/m_{Xe}(0)$ depend on $(m_I)_0$ and $(m_{Xe})_0$.

Problems Section 15.1

15.1 Find the characteristic polynomials and complementary functions of the following differential equations and, where initial conditions are given, find the appropriate particular solution:

(a) $y'' + 5y' - 14y = 0$;

(b) $y'' - y = 0; \quad y = 1, \quad y' = 0 \quad \text{at} \quad x = 0$;

(c) $y'' + 4y' + 3y = 0$;

(d) $y''' + 5y'' + 2y' - 8y = 0$;

(e) $y''' + 7y'' + 12y' = 0; \quad y = 0, \quad y' = 9, \quad y'' = -39 \quad \text{at} \quad x = 0$;

(f) $y''' - y'' - y' + y = 0$;

(g) $y^{(4)} - 2y''' - 3y'' + 4y' + 4y = 0$.

15.2 By using the definition of linear dependence, state whether the following sets of functions are linearly dependent:

(a) x^2, x^4, x^6; (b) $\cos x, \ -3\cos x, \ 9\cos x$;

(c) $\cosh 2x, \ \sinh 2x, 1$; (d) $\cosh^2 3x, \ \sinh^2 3x, \ 2$;

(e) $x + 1, x + 2, x + 3$; (f) $x + 1, x + 2$.

15.3 Obtain the general solution of $y''' - 6y'' + 11y' - 6y = 0$, and, by finding the Wronskian of its three constituent functions, prove that they are linearly independent.

15.4 By forming the general solution and eliminating the arbitrary constants by differentiation, determine the differential equations that have the following sets of functions as linearly independent solutions:

(a) $e^{2x}, \ e^{-3x}$; (b) $e^x, \ xe^x, \ x^2e^x$; (c) $1, x, \ e^{2x}$.

15.5 Find the general solutions of the following differential equations:

(a) $y''' - y'' + y' - y = 0$;

(b) $y'' + y' + y = 0$;

(c) $y''' - 3ay'' + 3a^2y' - a^3y = 0$;

(d) $y^{(4)} + 2y'' + y = 0$;

(e) $y^{(4)} + 2y'' + 9y = 0$;

Section 15.2

15.6 Determine the general solutions of the following differential equations using the method of undetermined coefficients.

(a) $y'' + 2y' - 3y = x^2 + x + 1$;

(b) $y''' - 3y'' + 3y' - y = 6$;

(c) $y'' + 2y' + y = e^{3x}$;

(d) $y'' + 4y' + 5y = 6 \sin x$;

(e) $y'' - y = 2e^x$;

(f) $y'' + 4y = \cos 2x$;

(g) $y'' + 9y = \sinh x$;

(h) $y'' + 2y' + 5y = e^x(1 + 2e^x)$;

(i) $y^{(4)} + 3y'' + 2y = 1$;

(j) $y'' - y' - 6y = e^x + \sin x$;

(k) $y'' + 4y' = x + e^x$.

Section 15.3

15.7 Obtain the general solutions of the following differential equations by using the method of variation of parameters:

(a) $y'' - y = xe^x$;

(b) $y'' - 2y' + y = x \sin x$;

(c) $y'' - 2y' + 2y = 4e^x \sin x$;

(d) $y''' + y'' + y' + y = xe^x$;

(e) $y'' + y = \tan x$;

(f) $y'' + y = \cot x$;

(g) $y'' - 6y' + 13y = e^x$.

Section 15.4

15.8 The equation of motion of a forced oscillation is

$$\ddot{y} + 2\dot{y} + 5y = 10 \sin \omega t.$$

Find the complete solution, indicating the difference between the transient and steady-state terms. Find also the maximum value of the amplitude of the steady-state oscillation that may be obtained by varying ω.

15.9 Sketch the variation of the phase angle δ of the particular integral occurring in Eqn (15.42) as a function of the normalized excitation frequency ω/Ω, for the cases $\zeta = \frac{1}{2}$, $\zeta = 1$ and $\zeta = 2$.

15.10 Derive an expression for x in the case of a critically damped oscillator for which

$$\ddot{x} + 2n\dot{x} + n^2 x = 0,$$

where $\dot{x} = u$ and $x = s$ at time $t = 0$. Show that if this equation describes the motion of a particle, then it will come to rest when $x = u/ne$ if $s = 0$.

15.11 When $\Omega^2 = \zeta^2$, the general solution of the damped harmonic motion described by

$$\ddot{x} + 2\zeta\dot{x} + \Omega^2 x = 0$$

is

$$x = e^{-\zeta t}(A\cos\omega_0 t + B\sin\omega_0 t), \tag{G}$$

where $\omega_0^2 = \Omega^2 - \zeta^2$. Deduce that the extrema of x occur when

$$\tan\omega_0 t = (B\omega_0 - \zeta A)/(A\omega_0 + \zeta B). \tag{H}$$

Denote the positive solutions of this equation by

$$\omega_0 t = \delta_0 + r\pi,$$

where $r = 0, 1, 2, \ldots$, and δ_0 is the smallest positive angle satisfying (H). Thus, defining the sequence of times $\{t_r\}$ by

$$t_r = (\delta_0 + r\pi)/\omega_0, \quad r = 0, 1, 2, \ldots,$$

and the corresponding sequence of displacements $\{x_r\}$ by setting $t = t_r$ in (G), prove that

$$x_{r+1}/x_r = \exp(-\zeta\pi/\omega_0).$$

This establishes that the ratio of the amplitude of successive oscillations decreases by the constant factor $\exp(-\zeta\pi/\omega_0)$. The constant $\zeta\pi/\omega_0$ is called the *logarithmic decrement* of the oscillations.

Section 15.5

15.12 Repeat the solution of the vibration problem in section 15.5 without the use of matrices starting from the assumption that

$$x = \alpha\sin(\omega t + \varepsilon_1) \quad \text{and} \quad y = \beta\sin(\omega t + \varepsilon_2).$$

15.13 A thin light elastic string is stretched between two fixed points A and B and unit masses are attached to it at points P and Q, where $AP = PQ = QB$. The equations of motion determining small lateral displacements x and y of the masses at points P and Q are

$$\ddot{x} + 2x - 2y = 0 \quad \text{and} \quad \ddot{y} + 2y - 2x = 0.$$

Determine the subsequent motion of the system if it is initially released from rest at time $t = 0$ with $x = a$, $y = \frac{1}{2}a$.

15.14 Repeat Problem 15.13 subject to the initial conditions that the system is released from rest at time $t = 0$ with $x = a$, $y = -a$.

15.15 In a certain vibration problem, displacements x, y and z are described by the system of equations:

$$m\frac{d^2x}{dt^2}+(a+b)x-by=0,$$

$$m\frac{d^2y}{dt^2}+(a+2b)y-bx-bz=0,$$

$$m\frac{d^2z}{dt^2}+(a+b)z-by=0,$$

in which a, b and m are constants. Express these differential equations in matrix form, and by writing $\mathbf{X}(t)=\mathbf{B}\sin(\omega t+\varepsilon)$, with $\mathbf{B}$ an arbitrary three-element column vector, show that the system has three natural frequencies ω_1^2, ω_2^2 and ω_3^2 and find their values. Use your results to deduce the form of the three normal modes.

Section 15.6

15.16 Find the general solution of the homogeneous system

$$\frac{dx}{dt}+2y+3x=0,$$

$$\frac{dy}{dt}+3x-2y=0.$$

15.17 Find the solution of the homogeneous system

$$\frac{dx}{dt}-6x+y=0,$$

$$\frac{dy}{dt}-3x-2y=0,$$

subject to the initial conditions $x(0)=4$, $y(0)=3$.

15.18 Find the solution of the homogeneous system

$$\frac{dx}{dt}-6x+3y=0,$$

$$\frac{dy}{dt}-2x-y=0,$$

subject to the initial conditions $x(0)=-1$, $y(0)=2$.

15.19 Find the solution of the inhomogeneous system

$$\frac{dx}{dt} + 4x + 3y = 0,$$

$$\frac{dy}{dt} + 3x + 4y = 2,$$

subject to the initial conditions $x(0) = 0$, $y(0) = 0$.

15.20 Find the solution of the inhomogeneous system

$$\frac{dx}{dt} - 2x - 3y = t - 5,$$

$$\frac{dy}{dt} + 3x - 2y = 5t - 5,$$

subject to the initial conditions $x(0) = 1$, $y(0) = 2$.

Section 15.7

Solve the following two-point boundary value problems.

15.21 $y'' + y' - 6y = 0$, with $y(0) = 0, y'(1) = 1$.

15.22 $y'' + 2y' - 3y = 0$, with $y(1) = 0$, $y'(2) = 1$.

15.23 $y'' + 9y = 0$, with $y(0) = 0$, $y'(\pi/4) = 1$.

15.24 $y'' + 6y' + 9y = 2e^{3x}$, with $y(0) = 1, y(1) = 3$.

15.25 $y'' + y = \cos x$, with $y(0) = 0, y(\pi/2) = 1$.

Find the non-trivial solution, or show that no solution exists, in each of the following problems.

15.26 $y'' + y = 0$, with $y(0) = 0$, $y'(\pi/2) = 0$.

15.27 $y'' + y = 0$, with $y(0) = 0$, $y(\pi/2) = 3$.

15.28 $y'' + y' = 0$, with $y(0) = 0, y(1) = 1$.

15.29 $y'' + y = 0$, with $y(0) = 0$, $y'(\pi/2) = 1$.

Section 15.8

Derive the Laplace transform of each of the following functions, and verify the result by means of Table 15.1.

15.30 $\sin at$ **15.31** $\cos at$ **15.32** $t \cos at$

15.33 $e^{at} \cos bt$ **15.34** $\sinh at$

Use differentiation of an integral with respect to a parameter to find the

Laplace transform of each of the following functions.

15.35 $te^{at}\sin bt$ **15.36** $t^2\cos at$ **15.37** $f(t) = \begin{cases} 0 & \text{for } 0 < t < a \\ t & t > a \end{cases}$

15.38 $t\cosh at$

Find the Laplace transform of each of the following functions.

15.39 $2e^{3t} + t\cos 3t$ **15.40** $\sinh at + \cosh at$

15.41 $1 + 3\sin^2 at$ **15.42** $\cosh^2 t$

15.43 Verify Eqn (15.53).

15.44 Verify Eqn (15.54).

15.45 Find $\mathscr{L}\{y'''\}$, given that $y(0) = 1$, $y'(0) = 2$ and $y''(0) = -4$.

15.46 Find $\mathscr{L}\{y^{(4)}\}$, given that $y(0) = 2$, $y'(0) = -1$, $y''(0) = 3$ and

$y'''(0) = -7$.

15.47 Find $\mathscr{L}\{y'''\}$, given that $y(t) = e^{-2t}$.

15.48 Find $\mathscr{L}\{y'''\}$, given that $y(t) = \sinh 3t$.

Find the inverse Laplace transform of each of the following functions.

15.49 $\dfrac{6}{s^5}$ **15.50** $\dfrac{2s^3 - 3s^2 + 18s - 12}{(s^2 + 4)(s^2 + 9)}$

15.51 $\dfrac{1}{s(s^2 + 1)}$ **15.52** $\dfrac{1}{s^2 + 6s + 13}$

15.53 $\dfrac{s - 2}{s^2 - 6s + 10}$ **15.54** $\dfrac{1}{s^3 + 4s^2 + 3s}$

15.55 Find $\mathscr{L}\{e^{3t}(\sin 2t + \cos 5t)\}$.

15.56 Find $\mathscr{L}\{e^{2t}t\sin 5t\}$.

15.57 Find $\mathscr{L}\{e^{-2t}\sinh^2 t\}$.

15.58 Find $\mathscr{L}\{e^{-3t}\cosh 2t\}$.

15.59 Represent the following function $f(t)$ in terms of the Heaviside unit step function:

$$f(t) = \begin{cases} 0, & 0 < t < 1 \\ 2, & 1 < t < 3 \\ \frac{1}{2}, & 3 < t < 5 \\ 0, & x > 5. \end{cases}$$

15.60 Represent the following function $f(t)$ in terms of the Heaviside unit step function:

$$f(t) = \begin{cases} \sin t, & 0 < t < \pi \\ 0, & \pi < t < 2\pi \\ \sin t, & 2\pi < t < 3\pi \\ 0, & t > 3\pi. \end{cases}$$

15.61 Sketch the function $f(t) = u_{\pi/2}(t)\sin\left(t - \frac{\pi}{2}\right)$.

15.62 Sketch the function $f(t) = u_1(t)(t-1)^2$.

Find the Laplace transform of each of the following functions.

15.63 $u_1(t)\,f(t-1)$, when $f(t) = t^5 e^{2t}$.

15.64 $u_4(t)\,f(t-4)$, when $f(t) = e^{3t}\cos 2t$.

Find the inverse Laplace transform of each of the following functions.

15.65 $\dfrac{(s-2)e^{-3s}}{s^2-4s+13}$

15.66 $\dfrac{(s^2-16)e^{-s}}{(s^2+16)^2}$

15.67 $\dfrac{(4s^3-2s^2-16s+18)e^{-s}}{(s^2-9)(s^2-4)}$

15.68 $\dfrac{se^{-3s}}{s^2+4} + \dfrac{6se^{-4s}}{(s^2+9)^2}$

Solve the following initial value problems by means of the Laplace transform.

15.69 $y' - 2y = e^{5t}$, given that $y(0) = 3$.

15.70 $y'' - 5y' + 6y = 0$, given $y(0) = 3$, $y'(0) = 8$.

15.71 $y' - y = \sin t$, given $y(0) = 0$.

15.72 $y'' + 2y' + 5y = e^t$, given $y(0) = 0$, $y'(0) = 0$.

15.73 $y''' + 3y'' + 3y' + y = t$, given $y(0) = 0$, $y'(0) = 1$, $y''(0) = 0$.

15.74 $y'' + 2y' + y = t^2$, given $y(0) = 0$, $y'(0) = 0$.

15.75 $y'' + y = \sin t$, given $y(0) = 0$, $y'(0) = 0$.

15.76 $y'' + 4y' + 4y = 4\cos 2t$, given $y(0) = 1$, $y'(0) = 2$.

15.77 $y'' + 2y' + y = [1 - u_1(t)]$, given $y(0) = 0$, $y'(0) = 0$.

15.78 $y'' + 2y' + y = (1 - \delta(t - \pi/2))\sin t$, $y(0) = 1$, $y'(0) = 1$.

15.79 $4y'' + 4y' + 17y = t\delta(t-1)$, $y(0) = 1$, $y'(0) = 0$.

15.80 $y'' + y' - 6y = u_2(t)$, given $y(0) = 0$, $y'(0) = 0$.

15.81 $y' + 3y + z = 0$ and $z' - y + z = 0$ given $y(0) = 0$, $z(0) = 1$.

15.82 $y' - y + 4z = 0$ and $z' - y - z = 0$, given $y(0) = 2$, $z(0) = 0$.

15.83 $y' - 6y + 3z = 8e^t$, $z' - 2y - z = 4e^t$, given $y(0) = -1$, $z(0) = 0$.

Supplementary computer problems

The problems in this section require the use of graphics package.

15.84 The differential equation

$$y''' + 4y'' + 5y' + 2y = 0,$$

subject to the initial conditions $y(0) = 1,\ y'(0) = 0,\ y''(0) = 1$, was shown in Example 15.2 to have the solution

$$y = e^{-x}(3x - 1) + 2e^{-2x}.$$

Display the solution for $0 \le x \le 2$.

15.85 The differential equation

$$y'' + 4y' + 13y = 0$$

was shown in Example 15.5 to have the general solution

$$y = e^{-2x}(A\cos 3x + B\sin 3x).$$

Use this result to show that the solution subject to the initial conditions $y(0) = 1,\ y'(0) = 1$ is

$$y = e^{-2x}(\sin 3x + \cos 3x).$$

Use this result to display the solution for $0 \le x \le 3$.

15.86 The differential equation

$$y'' + 4y = 5 + 3\cos 2x + 4\sin 2x,$$

subject to the initial conditions $y(0) = 0,\ y'(0) = 1$, was shown in Example 15.7 to have the solution

$$y = \tfrac{5}{4} - \tfrac{5}{4}\cos 2x + \sin 2x - x\cos 2x + \tfrac{3}{4}x\sin 2x.$$

Display the solution for $0 \le x \le 4$.

15.87 The differential equation

$$y'' + 3y' + 2y = 1 + x^2 + e^{3x}$$

was shown in Example 15.8 to have the general solution

$$y = Ae^{-x} + Be^{-2x} + \tfrac{9}{4} - \tfrac{3}{2}x + \tfrac{1}{2}x^2 + \tfrac{1}{20}e^{3x}.$$

Use this result to show that the solution subject to the initial conditions $y(0) = 1,\ y'(0) = 0$ is

$$y = \tfrac{1}{2}x^2 - \tfrac{3}{2}x + \tfrac{9}{4} + \tfrac{1}{20}e^{3x} - \tfrac{1}{20}e^{-2x} - \tfrac{5}{4}e^{-x},$$

and display the solution for $0 \le x \le 1$.

15.88 The differential equation

$$y'' - 4y' + 13y = 3 + 2e^{2x}\sin 3x$$

was shown in Example 15.9 to have the general solution

$$y = e^{2x}(A\cos 3x + B\sin 3x) + \tfrac{3}{13} - \tfrac{1}{3}xe^{2x}\cos 3x.$$

Use this result to show that the solution subject to the initial conditions $y(0) = 0,\ y'(0) = 0$ is

$$y = \tfrac{3}{13} + \tfrac{31}{117}e^{2x}\sin 3x - \tfrac{3}{13}e^{2x}\cos 3x - \tfrac{1}{3}xe^{2x}\cos 3x,$$

and display the solution for $0 \le x \le 1$.

15.89 The differential equation

$$y''' + 3y'' = x$$

was shown in Example 15.11 to have the general solution

$$y = Ae^{-3x} + B + Cx - \tfrac{1}{18}x^2 + \tfrac{1}{18}x^3.$$

Use this result to show that the solution subject to the initial conditions $y(0) = 1,\ y'(0) = 0,\ y''(0) = 0$ is

$$y = \tfrac{1}{18}x^3 - \tfrac{1}{18}x^2 + \tfrac{1}{27}x + \tfrac{80}{81} + \tfrac{1}{81}e^{-3x},$$

and display the solution for $0 \le x \le 2$.

15.90 The differential equation

$$y'' + 5y' + 6y = xe^{-2x}$$

was shown in Example 15.12 to have the general solution

$$y = Ae^{-3x} + Be^{-2x} + (1 - x)e^{-2x} + \tfrac{1}{2}x^2e^{-2x}.$$

Use this result to show that the solution subject to the initial conditions $y(0) = 0,\ y'(0) = 3$ is

$$y = -4e^{-3x} + \tfrac{1}{2}x^2e^{-2x} - xe^{-2x} + 4e^{-2x},$$

and display the solution for $0 \le x \le 2$.

15.91 The simultaneous differential equations

$$2\frac{dx}{dt} + \frac{dy}{dt} - 2x - 2y = 5e^t$$

$$\frac{dx}{dt} + \frac{dy}{dt} + 4x + 2y = 5e^{-t},$$

subject to the initial conditions $x(0) = 2,\ y(0) = 0$, were shown in Example 15.15 to have the solution

$$x = -4\cos 2t + 6\sin 2t + 3(e^t + e^{-t})$$

$$y = 9\cos 2t - 7\sin 2t - 5e^t - 4e^{-t}.$$

Display these solutions for $0 \le x \le 3$.

16 Fourier series

The purpose of this chapter is to develop the basic theory of Fourier series. The idea underlying a Fourier series is to represent an arbitrary function $f(x)$, defined for $-\pi \leq x \leq \pi$, as a trigonometric series of the form

$$f(x) = \tfrac{1}{2}a_0 + \sum_{n=1}^{\infty}(a_n \cos nx + b_n \sin nx).$$

Thus a Fourier series represents a decomposition of the function $f(x)$ into the sum of its different frequency components, called its harmonic components. This is useful in many different contexts, but it proves to be particularly valuable when problems involving oscillations are involved. The coefficients $a_0, a_1, a_2, \ldots, b_1, b_2, b_3, \ldots$, called the Fourier coefficients of $f(x)$, are shown to be determined from the given function by means of elementary integrals involving the product of $f(x)$ with either a cosine or a sine function.

The periodicity of the sine and cosine functions in a Fourier series representation of $f(x)$ for $-\pi \leq x \leq \pi$ causes the representation itself to be periodic with period 2π. The interval $-\pi \leq x \leq \pi$ is called the fundamental interval of the Fourier series, and each periodic repetition of the functional behaviour in this interval is called a periodic extension of the function. A Fourier series representation of a function differs from a power series representation in a number of important ways. One significant difference is that whereas a Taylor series representation of $f(x)$ expanded about the point z_0 requires the function to be differentiable infinitely many times at x_0, a Fourier series representation makes no such demand and only requires the product of $f(x)$ and a sine or cosine function to be integrable over the interval $-\pi \leq x \leq \pi$. A direct consequence of this is that a Fourier series can be used to represent a discontinuous function, which is impossible when a Taylor series is involved.

A Fourier series converges to the function $f(x)$ it represents at all points where $f(x)$ is continuous, and to the mid-point of the jump at points where $f(x)$ has a finite jump discontinuity. These results are sometimes useful when seeking the sum of simple infinite numerical series, and it is shown how such series can be summed by relating the series to a Fourier series in which the substitution of a particular value of the argument x gives rise to the required numerical series.

Different forms of Fourier series are described, depending on whether $f(x)$ is defined on the interval $-\pi \leq x \leq \pi$ or on $a \leq x \leq b$, or if $f(x)$ is an even or an odd function, or whether a complex form of the series is required.

As with power series, it is sometimes desirable to differentiate or integrate Fourier series term by term, and an explanation of the conditions under which these operations may be performed forms the last part of the chapter.

16.1 Introductory ideas

The fundamental idea underlying Fourier series is that all functions $f(x)$ of practical importance which are defined on the interval $-\pi \leq x \leq \pi$ can be expressed in terms of a convergent *trigonometric series* of the form

$$f(x) = \frac{a_0}{2} + \sum_{n=1}^{\infty}(a_n \cos nx + b_n \sin nx), \tag{16.1}$$

in which the constant coefficients a_n, b_n are related to $f(x)$ in a special way. The apparent restriction of $f(x)$ to the interval $[-\pi, \pi]$ is unimportant, since an elementary change of variable will always reduce an arbitrary interval $[a, b]$ to $[-\pi, \pi]$.

Notice here that because of the periodicity properties of the sine and cosine functions, the right-hand side of (16.1) must of necessity be periodic with period 2π. This implies that the best we can expect of such a representation is that, at each point of the interval $[-\pi, \pi]$, the trigonometric series has for its sum the function $f(x)$. Naturally, although the trigonometric series will assign functional values to $f(x)$ for all real x, it does not follow that these need agree with the actual functional values of $f(x)$ outside the *fundamental interval* $[-\pi, \pi]$. In fact, the series will provide a *periodic extension* of the functional behaviour of $f(x)$ over the fundamental interval $[-\pi, \pi]$ to every interval of the form $[(2r-1)\pi, (2r+1)\pi]$, in the sense that $f(x) = f(x+2r\pi)$ for $r = 0, \pm 1, \pm 2, \ldots$.

To deduce the relationship between $f(x)$ and the coefficients a_n, b_n, let us first reinterpret results (8.34) to (8.36) of Chapter 8 in terms of definite integrals taken over the interval $[-\pi, \pi]$. We find at once that for any integers $m, n = 0, 1, \ldots$,

$$\int_{-\pi}^{\pi} \sin mx \cos nx \, dx = 0 \quad \text{for all } m, n, \tag{16.2}$$

$$\int_{-\pi}^{\pi} \sin mx \sin nx \, dx = \begin{cases} 0 & \text{for } m \neq n \\ \pi & \text{for } m = n \neq 0, \end{cases} \tag{16.3}$$

$$\int_{-\pi}^{\pi} \cos mx \cos nx \, dx = \begin{cases} 0 & \text{for } m \neq n \\ \pi & \text{for } m = n \neq 0, \\ 2\pi & \text{for } m = n = 0. \end{cases} \tag{16.4}$$

In mathematical terms, the facts expressed by Eqn (16.2) and by the first results of Eqns (16.3) and (16.4) are described by saying that the functions belonging to the system

$$1,\ \cos x,\ \sin x,\ \cos 2x,\ \sin 2x,\ \ldots,\ \cos nx,\ \sin nx,\ \ldots, \tag{16.5}$$

are *orthogonal* over the interval $[-\pi, \pi]$. In words, these equations say that the product of any two different functions of this sequence when integrated over the interval $[-\pi, \pi]$ will yield zero.

The significance of the orthogonality property of system (16.5) is seen when Eqn (16.1) is multiplied by $\cos mx$, and the result is then integrated over the interval $[-\pi, \pi]$. We find that

$$\int_{-\pi}^{\pi} f(x)\cos mx\,\mathrm{d}x = \frac{a_0}{2}\int_{-\pi}^{\pi}\cos mx\,\mathrm{d}x + \sum_{n=1}^{\infty}\left(a_n\int_{-\pi}^{\pi}\cos mx\cos nx\,\mathrm{d}x + b_n\int_{-\pi}^{\pi}\cos mx\sin nx\,\mathrm{d}x\right),$$

which on account of the above results immediately reduces to

$$\int_{-\pi}^{\pi} f(x)\cos mx\,\mathrm{d}x = \pi a_m \qquad \text{for } m = 0, 1, \ldots. \tag{16.6}$$

Had Eqn (16.1) been multiplied by $\sin mx$ and the result been integrated over the interval $[-\pi, \pi]$, an exactly similar argument would have yielded the result

$$\int_{-\pi}^{\pi} f(x)\sin mx\,\mathrm{d}x = \pi b_m \quad \text{for } m = 1, 2, \ldots. \tag{16.7}$$

Thus we have found that for $f(x)$ to have the trigonometric series representation (16.1), we must define the constant coefficients a_n, b_n by the following relationships, called the *Euler formulas*,

$$a_n = \frac{1}{\pi}\int_{-\pi}^{\pi} f(x)\cos nx\,\mathrm{d}x \qquad \text{for } n = 0, 1, \ldots, \tag{16.8}$$

and

$$b_n = \frac{1}{\pi}\int_{-\pi}^{\pi} f(x)\sin nx\,\mathrm{d}x \qquad \text{for } n = 1, 2, \ldots. \tag{16.9}$$

The coefficients a_n, b_n so defined are called the *Fourier coefficients* of $f(x)$, and the corresponding right-hand side of (16.1) is then called the *Fourier series* of $f(x)$. In our simple derivation of the form of the Fourier series of $f(x)$ we have presupposed that $f(x)$ is integrable, that termwise integration of (16.1) after multiplication by $\sin mx$ or $\cos mx$ is permissible and that the sum of the Fourier series for $-\pi \le x \le \pi$ is the function $f(x)$ itself. These are major assumptions, and perhaps the best way to indicate that they need questioning is by considering some typical examples of Fourier series. However, before proceeding with this plan, let us first make some general comparisons between Taylor series and Fourier series.

The idea that a function may be represented by its Taylor series has already been discussed at some length, and it was seen in Chapter 12 that for a function $f(x)$ to be so expressed it needed to be infinitely differentiable. This is a severe restriction on a function and is one which most functions do not satisfy. Even when Taylor's theorem with a remainder is employed the function still needs to be differentiable a finite number of times and this, like infinite differentiability, certainly implies that the function must be continuous. Nevertheless, many functions used to describe important physical phenomena are discontinuous and so cannot be represented by a Taylor series. For example, the function used to describe the voltage behaviour with time in a circuit in which a switch is suddenly operated is discontinuous, as is the variation of the gas pressure across a shock in which a sudden jump in pressure occurs.

In principle, at least, Fourier series would appear to offer the possibility of representation of discontinuous as well as continuous functions, because whereas for a Taylor series expansion a function needs to be differentiable, for a Fourier series expansion it would appear that it only needs to be integrable. This assertion follows because in Eqn (7.19) we have already seen that the integration of piecewise continuous functions presents no difficulty, and so the Fourier coefficients a_n, b_n can even be computed when $f(x)$ is piecewise continuous. Naturally, we must examine the functional value which a Fourier series attributes to a point of discontinuity of the function which it represents, since at such points it is reasonable to expect the behaviour of the series to differ from that of the function itself.

Another important feature of a Fourier series is that it offers a method of synthesis of a function in terms of simple harmonic components having periodicities which are sub-multiples of 2π. This is particularly valuable when an oscillatory problem is being studied since, in effect, it describes the function involved in terms of the simple harmonic oscillatory modes which occur naturally in the problem. At this point it is appropriate to comment on the use of the term 'orthogonal' in connection with the system of functions (16.5). This term is used deliberately on account of the similarity that exists between the resolution of an ordinary vector into three orthogonal components and the decomposition of a function into an infinite number of Fourier components which are orthogonal in the sense appropriate to system (16.5). There are important similarities and dissimilarities between a vector space of three dimensions and one with an infinite number of dimensions, but these will not be pursued here.

For our first encounter with Fourier series we now choose to consider the following examples.

Example 16.1 Determine the Fourier series expansion of the function

$$f(x) = \pi^2 - x^2 \qquad \text{for} \quad -\pi \leq x \leq \pi.$$

Solution As $f(x) = \pi^2 - x^2$, we see from Eqn (16.8) that the Fourier coefficients a_n are determined by the integral

$$a_n = \frac{1}{\pi}\int_{-\pi}^{\pi}(\pi^2 - x^2)\cos nx\,dx,$$

where $n = 0, 1, \ldots$ When $n = 0$ this yields

$$a_0 = \frac{1}{\pi}\int_{-\pi}^{\pi}(\pi^2 - x^2)dx = \frac{4}{3}\pi^2.$$

For the case $n \neq 0$ we have

$$\begin{aligned}a_n &= \pi\int_{-\pi}^{\pi}\cos nx\,dx - \frac{1}{\pi}\int_{-\pi}^{\pi}x^2\cos nx\,dx,\\ &= \pi\left(\frac{1}{n}\sin nx\right)\bigg|_{-\pi}^{\pi} - \frac{1}{\pi}\left(\frac{2x\cos nx}{n^2} + \frac{(n^2x^2 - 2)\sin nx}{n^3}\right)\bigg|_{-\pi}^{\pi}\\ &= (-1)^{n+1}\frac{4}{n^2},\end{aligned}$$

where we have used the fact that $\sin n\pi = 0$ and $\cos n\pi = (-1)^n$.

To determine the Fourier coefficients b_n we must use Eqn (16.9), which gives

$$b_n = \frac{1}{\pi}\int_{-\pi}^{\pi}(\pi^2 - x^2)\sin nx\,dx,$$

where $n = 1, 2, \ldots$. Instead of evaluating this integral directly, let us divide the interval of integration and rewrite the result as the sum of two integrals. First we write

$$b_n = \frac{1}{\pi}\int_{-\pi}^{0}(\pi^2 - x^2)\sin nx\,dx + \frac{1}{\pi}\int_{0}^{\pi}(\pi^2 - x^2)\sin nx\,dx,$$

and then, setting $x = -z$ in the first integral, this becomes

$$b_n = -\frac{1}{\pi}\int_{\pi}^{0}(\pi^2 - z^2)\sin(-nz)dz + \frac{1}{\pi}\int_{0}^{\pi}(\pi^2 - x^2)\sin nx\,dx.$$

However, $\sin(-nz) = -\sin nz$, and the minus sign in front of the first integral may be utilized to reverse the order of the limits of integration, so that finally we arrive at

$$b_n = -\frac{1}{\pi}\int_{0}^{\pi}(\pi^2 - z^2)\sin nz\,dx + \frac{1}{\pi}\int_{0}^{\pi}(\pi^2 - x^2)\sin nx\,dx.$$

for $n = 1, 2, \ldots$. As the variable in a definite integral is only a dummy variable, we may replace z by x in the first integral to deduce that

$$b_n = 0 \quad \text{for all } n.$$

This result could, of course, have been obtained by direct evaluation of the definite integral for b_n using integration by parts, though the argument would have been more tedious.

Inserting the Fourier coefficients a_n, b_n into Eqn (16.1) then gives the Fourier series of $f(x) = \pi^2 - x^2$ for $-\pi \le x \le \pi$. The result obtained is

$$f(x) = \frac{2}{3}\pi^2 + 4\left(\cos x - \frac{1}{2^2}\cos 2x + \frac{1}{3^2}\cos 3x - \ldots \right.$$
$$\left. + \frac{(-1)^{n+1}}{n^2}\cos nx + \ldots\right).$$

The relationship between the Fourier series representation of $f(x) = \pi^2 - x^2$ in the fundamental interval $[-\pi, \pi]$, the periodic extension it assigns to $f(x)$ outside the fundamental interval, and the actual behaviour of $f(x)$ both inside and outside the fundamental interval are illustrated in Fig. 16.1(a). The full curve denotes both the functional behaviour of $f(x)$ and that of its Fourier series in the fundamental interval, the dotted curve denotes the periodic extension of $f(x)$ and the chaindotted curve denotes the actual behaviour of $f(x)$ outside the fundamental interval. Of course, it still remains for us to justify our assertion that the Fourier series converges to $f(x)$ in $[-\pi, \pi]$, but it certainly does so when $x = 0$ and $x = \pm\pi$. This follows by employing the standard results

$$\frac{\pi^2}{12} = 1 - \frac{1}{2^2} + \frac{1}{3^2} - \frac{1}{4^2} + \ldots,$$

and

$$\frac{\pi^2}{6} = 1 + \frac{1}{2^2} + \frac{1}{3^2} + \frac{1}{4^2} + \ldots,$$

to evaluate the Fourier series at those points.

Let us define the mth partial sum $S_m(x)$ of this Fourier series to be

$$S_m(x) = \tfrac{2}{3}\pi^2 + 4\sum_{n=1}^{m-1} \frac{(-1)^{n-1}}{n^2}\cos nx.$$

Then, when working numerically with the Fourier series, the function will need to be approximated by a partial sum. The behaviour of the second, third and fourth partial sums

$$S_2(x) = \tfrac{2}{3}\pi^2 + 4\cos x,$$

$$S_3(x) = \tfrac{2}{3}\pi^2 + 4\cos x - \cos 2x,$$

$$S_4(x) = \tfrac{2}{3}\pi^2 + 4\cos x - \cos 2x + \tfrac{4}{9}\cos 3x,$$

is shown in Fig. 16.1(b), and they certainly suggest the convergence of $S_n(x)$ to $f(x)$ over $[-\pi, \pi]$. ■

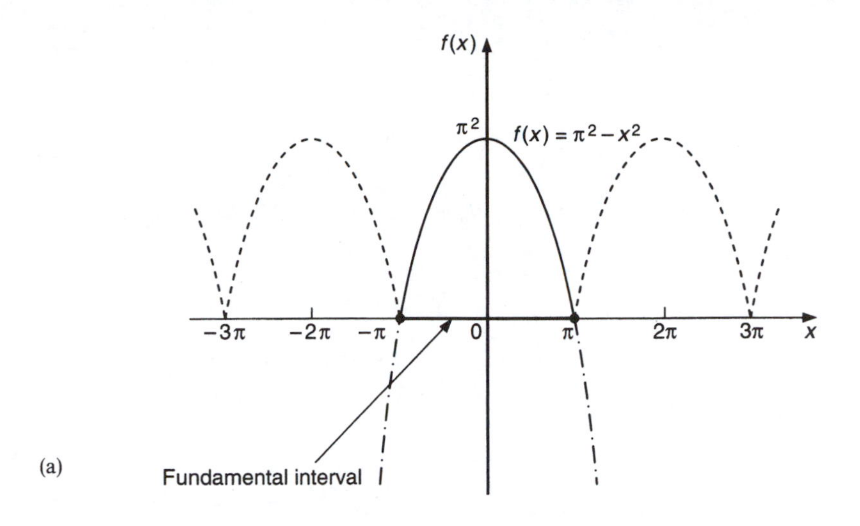

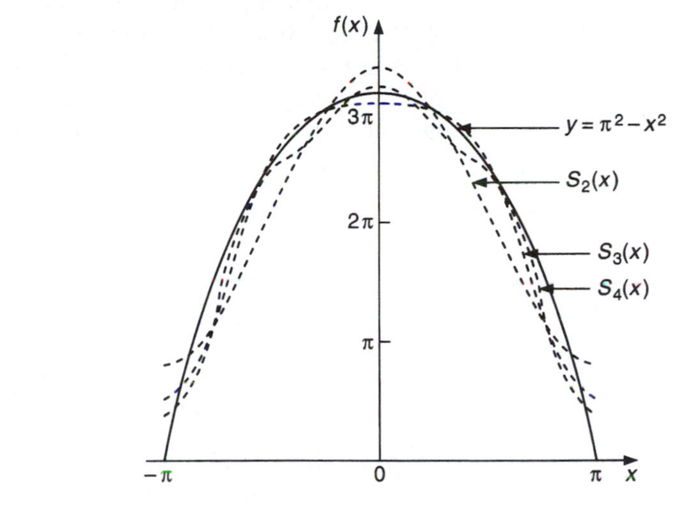

Fig. 16.1 (a) Fourier series representation of $f(x) = \pi^2 - x^2$; (b) approximation of $\pi^2 - x^2$ by partial sums.

Example 16.2 Determine the Fourier series expansion of the function

$$f(x) = |x| \qquad \text{for} \quad -\pi \le x \le \pi.$$

Solution As before, we have from Eqn (16.8) that

$$a_n = \frac{1}{\pi}\int_{-\pi}^{\pi} |x| \cos nx \, dx,$$

where $n = 0, 1, \ldots$. When $n = 0$ this yields

$$a_0 = \frac{1}{\pi}\int_{-\pi}^{0} (-x)dx + \frac{1}{\pi}\int_{0}^{\pi} x \, dx$$

$$= \frac{2}{\pi}\int_0^{\pi} x\,dx = \pi.$$

When $n \neq 0$, we have

$$a_n = \frac{1}{\pi}\int_{-\pi}^{0} (-x)\cos nx\,dx + \frac{1}{\pi}\int_0^{\pi} x\cos nx\,dx$$

$$= \frac{2}{\pi}\int_0^{\pi} x\cos nx\,dx$$

$$= \frac{2}{\pi}\left(\frac{x\sin nx}{n} + \frac{\cos nx}{n^2}\right)\Bigg|_0^{\pi}$$

$$= \frac{-4}{\pi(2n+1)^2}.$$

A moment's reflection shows that the form of argument used to establish that $b_n = 0$ for all n in the previous example succeeded because the function $f(x)$ involved was an *even* function, while $\sin nx$ is an *odd* function. Here, as $f(x) = |x|$, we are again dealing with an even function so that once again we may conclude that

$$b_n = 0 \qquad \text{for} \quad \text{all } n.$$

Hence the Fourier series of $f(x) = |x|$ in $-\pi \leq x \leq \pi$ has the form

$$f(x) = \frac{1}{2}\pi - \frac{4}{\pi}\left(\cos x + \frac{1}{3^2}\cos 3x + \frac{1}{5^2}\cos 5x + \ldots + \frac{1}{(2n-1)^2}\cos(2n-1)x + \ldots\right).$$

Again, this certainly converges to $f(x)$ when $x = 0$ and $x = \pm\pi$, as may be seen by assigning to x the appropriate values and employing the standard result

$$\frac{1}{8}\pi^2 = 1 + \frac{1}{3^2} + \frac{1}{5^2} + \frac{1}{7^2} + \ldots$$

to evaluate the Fourier series at those points.

In Fig. 16.2(a) the full line denotes the behaviour of the function $f(x) = |x|$ and its Fourier series in the fundamental interval $[-\pi, \pi]$, the dotted line denotes the periodic extension of $f(x)$ outside $[-\pi, \pi]$, and the chain-dotted line denotes the actual behaviour of $f(x)$ outside $[-\pi, \pi]$. The behaviour of the third partial sum

$$S_3(x) = \frac{1}{2}\pi - \frac{4}{\pi}\left(\cos x + \frac{1}{3^2}\cos 3x\right)$$

is shown in Fig. 16.2(b). Again, it appears reasonable to suppose that in the limit of large n, $S_n(x)$ will converge to $f(x)$ for all $x \in [-\pi, \pi]$. ■

Example 16.3 Determine the Fourier series expansion of the function

$$f(x) = \begin{cases} a & \text{for } -\pi \le x < 0 \\ b & \text{for } 0 \le x \le \pi. \end{cases}$$

Solution Proceeding as before and using the notation of Eqn (7.19),

$$a_n = \frac{1}{\pi}\int_{-\pi}^{0-} a\cos nx \, dx + \frac{1}{\pi}\int_{0+}^{\pi} b\cos nx \, dx$$

for $n = 0, 1, 2, \ldots$, and

$$b_n = \frac{1}{\pi}\int_{-\pi}^{0-} a\sin nx \, dx + \frac{1}{\pi}\int_{0+}^{\pi} b\sin nx \, dx$$

for $n = 1, 2, \ldots$.

A simple calculation shows that

$$a_0 = a + b, \quad a_n = 0 \quad \text{for } n = 1, 2, \ldots,$$

and

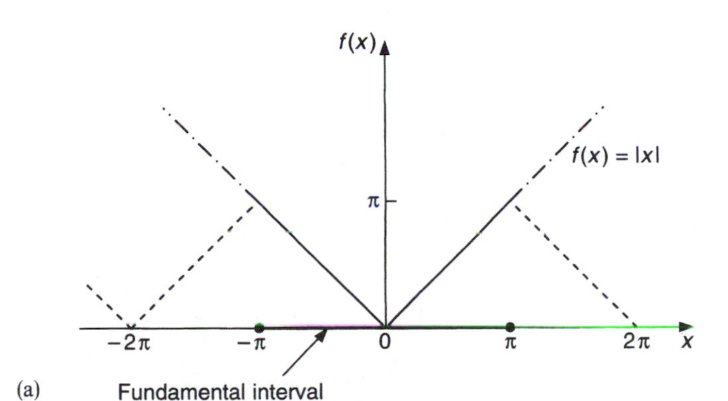

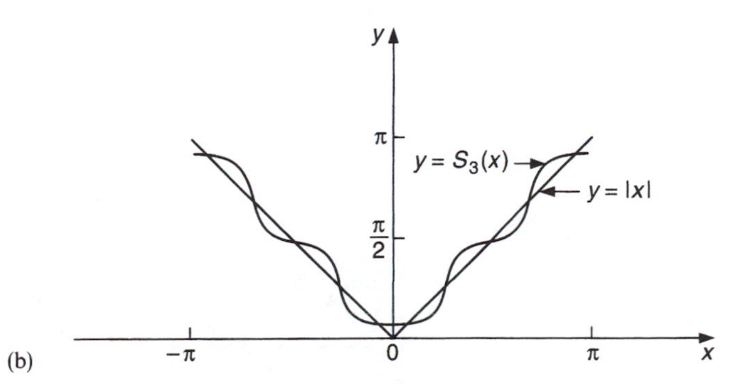

Fig. 16.2 (a) Fourier series representation of $f(x) = |x|$ for $-\pi \le x \le \pi$; (b) approximation of $|x|$ by the partial sum $S_3(x)$.

$$b_{2n} = 0, \quad b_{2n+1} = \frac{2(b-a)}{(2n+1)\pi} \quad \text{for } n = 1, 2, \ldots.$$

Substitution of these Fourier coefficients into Eqn (16.1) then shows that the Fourier series of $f(x)$ in $[-\pi, \pi]$ is

$$f(x) = \left(\frac{a+b}{2}\right) + \frac{2}{\pi}(b-a)(\sin x + \tfrac{1}{3}\sin 3x$$

$$+ \tfrac{1}{5}\sin 5x + \ldots + \frac{1}{(2n+1)}\sin(2n+1)x + \ldots).$$

This Fourier series certainly converges to the function $f(x)$ when $x = \pm\frac{1}{2}\pi$, as can be seen by assigning these values to x and employing the final result of Example 12.12 to sum the series. Observe, though, that in this case the Fourier series assumed the value $(a+b)/2$ for $x = 0$ and $x = \pm\pi$ which is not in agreement with the actual functional values at those points. It is, in fact, the average of the functional values to the immediate left and right of the discontinuity at $x = 0$. This result is not coincidental, and it can be shown to be true of all Fourier series at jump discontinuities. The approximation of $f(x)$ by the third partial sum

$$S_3(x) = \left(\frac{a+b}{2}\right) + \frac{2}{\pi}(b-a)(\sin x + \tfrac{1}{3}\sin 3x)$$

in the case $a = 0$, $b = 1$ is shown in Fig. 16.3, in which a circle denotes an end point not included and a dot an end point that is included. ■

Example 16.4 Find the Fourier series expansion for the function

$$f(x) = \begin{cases} 0, & -\pi < x < 0 \\ \sin x, & 0 \le x \le \pi, \end{cases}$$

and hence deduce a series for π.

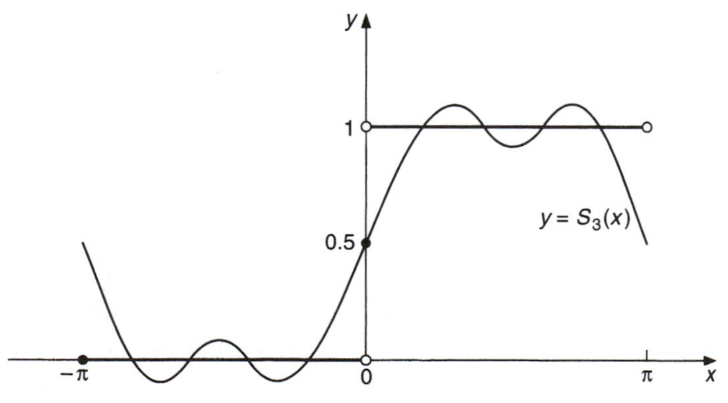

Fig. 16.3 Approximation of $f(x) = 0$ for $-\pi \le x < 0$ and $f(x) = 1$ for $0 \le x \le \pi$ by the partial sum $S_3(x)$.

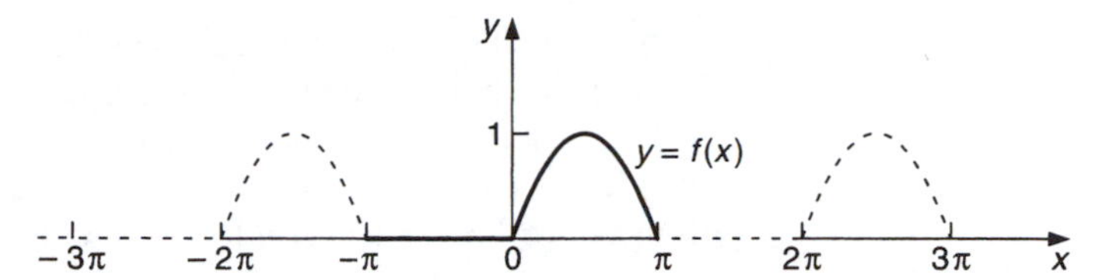

Fig. 16.4 The function $f(x)$ with periodic extensions.

Solution The function, together with its periodic extension to the intervals $-3\pi < x \leq -\pi$ and $\pi < x \leq 3\pi$, is shown in Fig. 16.4.

As usual, to derive the Fourier series representation, we start from Eqns (16.8) and (16.9). However, this example illustrates the need for care when evaluating Fourier coefficients. By dividing the interval of integration into the two parts $-\pi \leq x \leq 0$ and $0 \leq x \leq \pi$, on each of which the function is defined differently, we see from Eqn (16.8) that

$$a_0 = \frac{1}{\pi}\int_{-\pi}^{\pi} f(x)\mathrm{d}x = \frac{1}{\pi}\int_{-\pi}^{0} f(x)\mathrm{d}x + \frac{1}{\pi}\int_{0}^{\pi} f(x)\mathrm{d}x$$
$$= \frac{1}{\pi}\int_{0}^{\pi} \sin x\,\mathrm{d}x = \frac{2}{\pi}.$$

Arguing in similar fashion, it follows that

$$a_n = \frac{1}{\pi}\int_{0}^{\pi} \sin x \cos nx\,\mathrm{d}x, \quad \text{for } n = 1, 2, \ldots.$$

Using the trigonometric identity.

$$\sin x \cos nx = \tfrac{1}{2}\{\sin[(1+n)x] + \sin[(1-n)x\},$$

we find that

$$a_1 = \frac{1}{2\pi}\int_{0}^{\pi} \sin 2x\,\mathrm{d}x = 0,$$

and

$$a_n = \frac{1}{2\pi}\int_{0}^{\pi} \{\sin[(1+n)x] + \sin[(1-n)x]\}\mathrm{d}x$$
$$= \frac{-1}{2\pi}\left\{\frac{\cos[(1+n)x]}{1+n} + \frac{\cos[(1-n)x}{1-n}\right\}\Bigg|_0^{\pi}$$
$$= \frac{1+(-1)^n}{\pi(1-n^2)}, \quad \text{for } n = 2, 3, \ldots,$$

where use has been made of the fact that $\cos[(1+n)\pi] = \cos[(1-n)\pi] = (-1)^{n+1}$.

Notice the need to treat the case $n = 1$ separately. This is because the divisor $1-n$ introduced when $\sin[(1-n)x]$ is integrated vanishes when $n=1$, thereby causing the result to be invalid for this one value of n. This type of behaviour is often encountered in Fourier series expansions.

The numerator $1+(-1)^n$ vanishes when n is odd and equals 2 when n is even. Thus as $2m$ is always even and $2m+1$ is always odd, when m is integral, we have succeeded in showing that

$$a_0 = \frac{2}{\pi}, \quad a_1 = 0,$$

$$a_{2m} = \frac{-2}{(2m-1)(2m+1)} \text{ and } a_{2m+1} = 0, \qquad \text{for } m = 1, 2, \ldots.$$

Now from Eqn (16.9) it follows by means of similar arguments that

$$b_n = \frac{1}{\pi}\int_0^{\pi} \sin x \sin nx \, dx, \qquad \text{for } n = 1, 2, \ldots,$$

so as

$$\sin x \sin nx = \tfrac{1}{2}\{\cos[(1-n)x] - \cos[(1+n)]\},$$

we have

$$b_1 = \frac{1}{2\pi}\int_0^{\pi}(1-\cos 2x)dx = \frac{1}{2},$$

while

$$b_n = \frac{1}{2\pi}\int_0^{\pi}\{\cos[(1-n)x] - \cos[(1+n)x]\}dx = 0, \qquad \text{for } n = 2, 3, \ldots.$$

Notice that here again it was necessary to treat the case $n=1$ separately for essentially the same reason as before.

Thus the required Fourier series expansion of $f(x)$ is

$$f(x) = \frac{1}{\pi} + \frac{1}{2}\sin x - \frac{2}{\pi}\sum_{m=1}^{\infty} \frac{\cos 2mx}{(2m-1)(2m+1)}.$$

To determine a series for π we set $x=\pi/2$ in this Fourier series and assume that, as $f(x)$ is continuous of $x=\pi/2$, the Fourier series at this point will converge to the function itself which has the value $f(\pi/2) = \sin(\pi/2) = 1$. As a result, we obtain the series

$$1 = \frac{1}{\pi} + \frac{1}{2} + \frac{2}{\pi}\sum_{m=1}^{\infty} \frac{(-1)^{m+1}}{(2m-1)(2m+1)},$$

which is equivalent to

$$\pi = 2 + \frac{4}{1.3} - \frac{4}{3.5} + \frac{4}{5.7} - \ldots.$$

This is the required series for π. The reason for choosing to set $x = \pi/2$ was because this produces a series of especially simple form from which the constant π can be determined. Had we set $x = \pi/4$ it is easily verified that we would have obtained the different series

$$\pi = 2^{3/2}\left[1 + \sum_{n=1}^{\infty} \frac{(-1)^{n+1}}{16n^2 - 1}\right].$$

Before leaving this problem, it will be instructive to use our knowledge of alternating series to determine where this series must be truncated if it is to be used to determine π to within some prescribed accuracy.

If an alternating series is truncated so that the first term to be omitted is of magnitude $|a_N|$, then it follows from Corollary 12.7 that the absolute error ε which is involved is such that $\varepsilon < |a_N|$. To use this result, suppose, for example, that π is to be determined to within an accuracy of 0.001. Then as

$$a_N = \frac{(-1)^{N+1}4}{(2N - 1)(2N + 1)} = \frac{(-1)^{N+1}4}{4N^2 - 1},$$

it follows that the suffix N must be such that it is the smallest integer for which

$$\frac{4}{4N^2 - 1} < 0.001.$$

A routine calculation shows that $N = 32$, so the last term to be retained in the series in order to achieve the required accuracy must be

$$\frac{4}{61.63}.$$

Thus, counting the initial term 2 in the series, it will be necessary to sum 32 terms of this series to determine π to within an accuracy of 0.001. ■

Two results that can save much unnecessary effort when calculating Fourier coefficients are the following:

(i) The Fourier series of an even function $f(x)$ defined on $[-\pi, \pi]$ contains only cosine terms, so $b_n = 0$ for $n = 1, 2, 3, \ldots$, while

$$a_n = \frac{2}{\pi}\int_0^{\pi} f(x)\cos nx\,dx, \qquad \text{for } n = 0, 1, 2, \ldots. \tag{16.10}$$

(ii) The Fourier series of an odd function $f(x)$ defined on $[-\pi, \pi]$ contains only sine terms, so $a_n = 0$ for $n = 0, 1, 2, \ldots$, while

$$b_n = \frac{2}{\pi}\int_0^{\pi} f(x)\sin nx\,dx, \qquad \text{for } n = 1, 2, 3, \ldots. \tag{16.11}$$

The proof of these results is simple. To establish (i) notice that if $f(x)$ is even, the product $f(x)\cos nx$ is even but the product $f(x)\sin nx$ is odd. The integral of an odd function over an interval $-a \leq x \leq a$ vanishes because the areas above and below the x-axis representing the integral must be equal, but they carry opposite signs and so cancel. However, because of symmetry about the y-axis, the areas of an even function to the left and right of the y-axis are equal so the integral over $-a \leq x \leq a$ is twice the integral over the interval $0 \leq x \leq a$ Thus, in particular, if $f(x)$ is even we arrive at the results in (i).

The proof of (ii) is similar, but now if $f(x)$ is odd then $f(x) \sin nx$ is even while $f(x) \cos nx$ is odd. Thereafter the proof proceeds as before, and we arrive at the results in (ii).

In concluding this section, let us emphasize how general the class of functions may be for which Fourier series can be found. Observe that in Example 16.1 the function was continuous and it has a continuous derivative throughout $[-\pi, \pi]$; in Example 16.2 the function was continuous everywhere throughout $[-\pi, \pi]$ but its derivative was not defined at the origin; in Example 16.3 the function was discontinuous at the origin; while in Example 16.4 the function was continuous throughout the interval $-\pi \leq x \leq \pi$, but identically zero for $-\pi \leq x \leq 0$.

Having considered some typical examples of Fourier series we now summarize the steps which must be followed when arriving at this form of representation of a function. We present this summary in the form of a set of rules and, at the same time, record in step 6 the important convergence properties of Fourier series. The justification for these properties will be given in the next section, where it forms the content of Theorem 16.1. The derivation of Theorem 16.1 may be omitted by the student who merely wishes to use Fourier series.

Rules for the determination of the Fourier series representation of a function $f(x)$ defined for $-\pi \leq x \leq \pi$

Let the function $f(x)$ be defined on the interval $-\pi \leq x \leq \pi$ on which it is either continuous or, if discontinuous, has at most a finite number of points at which bounded jump discontinuities occur. Then the Fourier series representation of $f(x)$ for $-\pi \leq x \leq \pi$ is determined in the following manner, and has the convergence properties described in step 6.

1. Determine the coefficient a_0 from the expression

$$a_0 = \frac{1}{\pi} \int_{-\pi}^{\pi} f(x)\, dx.$$

2. Determine the coefficients a_n by evaluating the integral

$$a_n = \frac{1}{\pi} \int_{-\pi}^{\pi} f(x) \cos nx\, dx, \qquad \text{for } n = 1, 2, \ldots.$$

Check that the a_n so defined are valid for all n. If a_n is not defined for $n = N$, say, due to the occurrence of a zero factor in the denominator of a_N, determine the value of a_N directly from the expression

$$a_N = \frac{1}{\pi}\int_{-\pi}^{\pi} f(x)\cos Nx\,dx.$$

This calculation will need to be repeated if a zero factor occurs in the denominator of a_n for more than one value of n.

3. Determine the coefficients b_n by evaluating the integral

$$b_n = \frac{1}{\pi}\int_{-\pi}^{\pi} f(x)\sin nx\,dx, \qquad \text{for } n = 1, 2, \ldots.$$

Check that the b_n so defined are valid for all n. If b_n is not defined for $n = N$, say, due to the occurrence of a zero factor in the denominator of b_N, determine the value of b_N directly from the expression

$$b_N = \frac{1}{\pi}\int_{-\pi}^{\pi} f(x)\sin Nx\,dx.$$

This calculation will need to be repeated if a zero factor occurs in the denominator of b_n for more than one value of n.

4. (i) The Fourier series of an even function $f(x)$ defined on $[-\pi, \pi]$ contains only cosine terms, so $b_n = 0$ for $n = 1, 2, 3, \ldots$, while

$$a_n = \frac{2}{\pi}\int_{0}^{\pi} f(x)\cos nx\,dx, \qquad \text{for } n = 0, 1, 2, \ldots.$$

(ii) The Fourier series of an odd function $f(x)$ defined on $[-\pi, \pi]$ contains only sine terms, so $a_n = 0$ for $n = 0, 1, 2, \ldots$, while

$$b_n = \frac{2}{\pi}\int_{0}^{\pi} f(x)\sin n\,dx, \qquad \text{for } n = 1, 2, 3, \ldots.$$

5. Substitute the a_n and b_n determined in steps 1 to 3 into the expression

$$\frac{1}{2}a_0 + \sum_{n=1}^{\infty}(a_n\cos nx + b_n\sin nx),$$

which is the required Fourier series representation of $f(x)$ for $-\pi \leq x \leq \pi$.

6. (i) Wherever $f(x)$ is continuous, its Fourier series representation will converge to the function $f(x)$.

(ii) Wherever $f(x)$ has a finite jump discontinuity its Fourier series representation will converge to the value corresponding to the mid-point of the jump. Thus if a bounded jump discontinuity occurs at x_0, to the immediate left of which $f(x)$ has the value L_- and to the immediate right the value L_+, then at x_0 the Fourier series of $f(x)$ will converge to the value $\frac{1}{2}(L_- + L_+)$.

(iii) The properties exhibited by the Fourier series representation of $f(x)$ in the inteval $-\pi \leq x \leq \pi$ will be repeated in all its periodic extensions.
Instead of the relationship between a function $f(x)$ and its Fourier series in step 5 above being indicated by an equals sign,

$$f(x) \sim \tfrac{1}{2}a_0 + \sum_{n=1}^{\infty}(a_n \cos nx + b_n \sin nx),$$

with the understanding that the symbol $\sim$ is to be interpreted as an equality at points of continuity of $f(x)$, and that it means that the Fourier series converges to the mid-point of a jump at infinite jump discontinuity.

16.2 Convergence of Fourier series

The examples studied in the previous section suffice to indicate not only that Fourier series can be associated with widely differing types of function, but also that the convergence properties of the resulting series require careful attention. In fact, in the examples considered, the series could only be summed and shown to converge to the functional value of $f(x)$ at special points in $[-\pi, \pi]$ and, although the behaviour of the partial sums was suggestive, nothing rigorous could be inferred from this about the general convergence properties of Fourier series at any other points in $[-\pi, \pi]$.

In this section, for the sake of completeness, a formal definition of Fourier series is given and the fundamental Fourier expansion theorem (Theorem 16.1) is stated without proof.

Notice that, in line with more advanced accounts of Fourier series, in the formal Definition 16.1 of a Fourier series, the symbol $\sim$ has been used between the function $f(x)$ and its Fourier series representation on the right. This is to take account of the fact that the symbol is to be interpreted as an equality sign when $f(x)$ is continuous, and as the average of the values of $f(x)$ to the immediate left and right of a point $x = a$, say, across which $f(x)$ is discontinuous.

However, in what follows, instead of using the symbol $\sim$ we will always use the equality sign and leave to the reader the task of interpreting its meaning at isolated points where $f(x)$ is discontinuous.

Definition 16.1 (Fourier series) The *Fourier series* of an integrable function $f(x)$ defined on the interval $[-\pi, \pi]$ is the series

$$f(x) \sim \frac{a_0}{2} + \sum_{n=1}^{\infty}(a_n \cos nx + b_n \sin nx),$$

in which the *Fourier coefficients* a_n, b_n are given by

$$a_n = \frac{1}{\pi}\int_{-\pi}^{\pi} f(x)\cos nx \, \mathrm{d}x \qquad \text{for } n = 0, 1, \ldots,$$

and

$$b_n = \frac{1}{\pi}\int_{-\pi}^{\pi} f(x)\sin nx\,dx \qquad \text{for } n = 1, 2, \ldots,$$

Theorem 16.1 (Fourier's theorem) Let $f(x)$ be a piecewise continuous function defined arbitrarily on the interval $[-\pi, \pi]$, and by periodic extension outside it. Then, if $[f(x)]^2$ is integrable over this interval, and $f(x)$ has finite left-hand and right-hand derivatives at its points of jump discontinuity, it follows that

(a) when $x = x_0$ is a point of continuity of $f(x)$ then

$$\lim_{n\to\infty} S_n(x_0) = f(x_0),$$

(b) when $x = x_0$ is a point of discontinuity of $f(x)$, then

$$\lim_{n\to\infty} S_n(x) = \tfrac{1}{2}[f(x_{0-}) + f(x_{0+})]. \quad \blacksquare$$

This theorem justifies our intuitive approach to Fourier series, and leads to series expansions for mathematical constants. It also leads to the *Parseval relation*

$$\frac{1}{2}a_n^2 + \sum_{n=1}^{\infty}(a_n^2 + b_n^2) = \frac{1}{\pi}[f(x)]^2\,dx. \tag{16.12}$$

Example 16.5 (a) Deduce a series expansion for π^2 using the Fourier series expansion for $f(x) = \pi^2 - x^2$ in the interval $[-\pi^2, \pi]$

(b) Given that the function

$$f(x) = \begin{cases} 0 & \text{for } -\pi \le x \le 0 \\ x & \text{for } 0 \le x \le \pi \end{cases}$$

has the Fourier series

$$f(x) = \frac{\pi}{4} - \frac{2}{\pi}\sum_{n=1}^{\infty}\frac{\cos[(2n-1)x]}{(2n-1)^2} - \sum_{n=1}^{\infty}(-1)^n\frac{\sin nx}{n},$$

deduce a series expansion for π^2 and compare it with result (a).

Solution (a) The function $f(x)$ is continuous in $[-\pi, \pi]$ so that by Theorem 16.1 its Fourier series converges to $f(x)$ at every point of that interval. Hence using the Fourier series derived in Example 16.1 with $x = \frac{1}{2}\pi$ yields

$$\pi^2 - \tfrac{1}{4}\pi^2 = \tfrac{2}{3}\pi^2 + 4\left(\cos\frac{1}{2}\pi - \frac{1}{2^2}\cos\pi + \frac{1}{3^2}\cos\frac{3}{2}\pi - \ldots\right),$$

which, after simplificaion, gives the result

$$\pi^2 = 48\left(\frac{1}{2^2} - \frac{1}{4^2} + \frac{1}{6^2} - \frac{1}{8^2} + \dots\right).$$

This is an alternating series with nth term $a_n = (-1)^{n+1}(12/n^2)$, so that by Corollary 12.7 the remainder R_n after n terms is such that $0 < |R_n| < 12/(n+1)^2$. Thus, summing to nine terms, the absolute value of the remainder cannot exceed 0.12. In fact the ninth partial sum yields the approximation $\pi \approx 3.1528$. Notice that this series is more convenient for the determination of π than the one encountered in Example 16.4 because fewer terms are required for a given accuracy.

(b) The function $f(x)$ is continuous everywhere in $[-\pi, \pi]$ except at $x = \pi$ where, by Theorem 16.1 and the periodic extension of $f(x)$ beyond that interval, the Fourier series converges to $\frac{1}{2}[f(\pi-) + f(\pi+)] = \frac{1}{2}(\pi + 0) = \frac{1}{2}\pi$. Hence, setting $x = \pi$ and $f(\pi) = \frac{1}{2}\pi$ in the given Fourier series, we obtain

$$\frac{1}{2}\pi = \frac{1}{4}\pi - \frac{2}{\pi}\sum_{n=1}^{\infty}\frac{\cos[(2n-1)\pi]}{(2n-1)^2} - \sum_{n=1}^{\infty}(-1)^n\frac{\sin n\pi}{n},$$

whence

$$\pi^2 = 8\left(1 + \frac{1}{3^2} + \frac{1}{5^2} + \dots\right).$$

This is, in fact, the series quoted in Example 16.2. Both the series derived for π^2 converge at roughly the same rate, but the series in (a) has the advantage that the error due to truncation of the series is easy to estimate because it is an alternating series. ■

16.3 Different forms of Fourier series

A number of special forms of Fourier series occur depending on whether or not the fundamental interval is $[-\pi, \pi]$, or the function $f(x)$ is even or odd. These are of sufficient importance to merit recording them formally.

16.3.1 Change of interval

If $f(x)$ is defined arbitrarily in the fundamental interval $[-\pi, \pi]$, and by periodic extension outside it, then the integrands of the integrals defining the Fourier coefficients a_n, b_n in Definition 16.1 will be periodic with period 2π.

Consider, for example, the integral

$$\int_{\alpha-\pi}^{\alpha+\pi} f(x)\cos nx\,dx.$$

Then from the property of the definite integral we may write

$$\int_{\alpha-\pi}^{\alpha+\pi} f(x)\cos nx\,dx = \int_{\alpha-\pi}^{-\pi} f(x)\cos nx\,dx$$

$$+\int_{-\pi}^{\pi} f(x)\cos nx\,\mathrm{d}x + \int_{\pi}^{\alpha+\pi} f(x)\cos nx\,\mathrm{d}x.$$

If, in the first integral on the right-hand side, the variable change $u = x + 2\pi$ is made then, using the fact that reversing the limits of a definite integral changes its sign, and the periodicity of f implies $f(u-2\pi) = f(u)$, we see that

$$\int_{\alpha+\pi}^{\pi} f(u-2\pi)\cos nu\,\mathrm{d}u = -\int_{\pi}^{\alpha+\pi} f(u)\cos nu\,\mathrm{d}u.$$

Combining the above results, and changing the dummy variable u to x, gives

$$\int_{\alpha-\pi}^{\alpha+\pi} f(x)\cos nx\,\mathrm{d}x = -\int_{\pi}^{\alpha+\pi} f(x)\cos nx\,\mathrm{d}x + \int_{-\pi}^{\pi} f(x)\cos nx\mathrm{d}x$$
$$+\int_{\pi}^{\alpha+\pi} f(x)\cos nx\mathrm{d}x = \int_{-\pi}^{\pi} f(x)\cos nx\,\mathrm{d}x.$$

The implication of this result is that for any α, the Fourier coefficient a_n is given by

$$a_n = \frac{1}{\pi}\int_{\alpha-\pi}^{\alpha+\pi} f(x)\cos nx\,\mathrm{d}x. \tag{16.13}$$

A similar argument establishes an equivalent result for b_n, and we have thus proved the next theorem.

Theorem 16.2 (Shift of fundamental interval) If $f(x)$ is defined arbitrarily in the fundamental interval $[-\pi, \pi]$ and by periodic extension outside it, then, for any α, the Fourier coefficients a_n, b_n of $f(x)$ are given by

$$a_n = \frac{1}{\pi}\int_{\alpha-\pi}^{\alpha+\pi} f(x)\cos nx\,\mathrm{d}x \qquad \text{for } n = 0, 1, \ldots,$$

$$b_n = \frac{1}{\pi}\int_{\alpha-\pi}^{\alpha+\pi} f(x)\sin nx\,\mathrm{d}x \qquad \text{for } n = 1, 2, \ldots.$$

The Fourier series of $f(x)$ then has the form

$$f(x) = \frac{a_0}{2} + \sum_{n=1}^{\infty}(a_n \sin nx + b_n \sin nx) \qquad \text{for } \alpha - \pi \le x \le \alpha + \pi. \quad \blacksquare$$

It often happens that the fundamental interval to be used is $[-L, L]$ instead of $[-\pi, \pi]$. When this occurs, the simple variable change $t = \pi x/L$ maps the interval $-L \le x \le L$ onto the interval $-\pi \le t \le \pi$ for which we already have a Fourier expansion theorem. So, if $f(x)$ is defined on the fundamental interval $-L \le x \le L$ we see that the function $f(x) =$

$f\left(\frac{Lt}{\pi}\right) = g(t)$, say, is defined, on the interval $-\pi \leq t \leq \pi$. Hence, using our ordinary Fourier expansion on $-\pi \leq t \leq \pi$ we can write

$$g(t) = \frac{a_0}{2} + \sum_{n=1}^{\infty} (a_n \cos nt + b_n \sin nt)$$

with

$$a_n = \frac{1}{\pi} \int_{-\pi}^{\pi} g(t) \cos nt \, \mathrm{d}t \qquad \text{for } n = 0, 1, \ldots,$$

and

$$b_n = \frac{1}{\pi} \int_{-\pi}^{\pi} g(t) \sin nt \, \mathrm{d}t \qquad \text{for n} = 1, 2, \ldots .$$

To complete the argument, transforming back to the variable x, we arrive at the following theorem.

Theorem 16.3 (change of interval length)

The Fourier series of an integrable function $f(x)$ defined on the interval $[-L, L]$ is the series

$$f(x) = \frac{a_0}{2} + \sum_{n=1}^{\infty} \left(a_n \cos \frac{n\pi x}{L} + b_n \sin \frac{n\pi x}{L} \right)$$

with

$$a_n = \frac{1}{L} \int_{-L}^{L} f(x) \cos \frac{n\pi x}{L} \mathrm{d}x \qquad \text{for } n = 0, 1, \ldots,$$

and

$$b_n = \frac{1}{L} \int_{-L}^{L} f(x) \sin \frac{n\pi x}{L} \mathrm{d}x \qquad \text{for } n = 1, 2, \ldots .$$

Example 16.6 (a) Deduce the Fourier series of the function

$$f(x) = \begin{cases} -x & \text{for } -\frac{1}{2}\pi \leq x < 0 \\ x & \text{for } 0 \leq x < \pi \\ 2\pi - x & \text{for } \pi \leq x \leq \frac{3}{2}\pi. \end{cases}$$

(b) Deduce the Fourier series of the function

$$f(x) = x^3 \qquad \text{for } -1 \leq x \leq 1.$$

Solution (a) Defining $f(x)$ by periodic extension outside the fundamental interval $-\frac{1}{2}\pi \leq x \leq \frac{3}{2}\pi$, it is easily seen that the graph of $f(x)$ is part of a sawtooth function (Fig. 16.5). Comparison of Fig. 16.5 and Fig. 16.2(a) then shows that the function $f(x)$ is in fact just that part of the function $f(x) = |x|$ and its periodic extension outside $[-\pi, \pi]$ that lies in the interval $[-\frac{1}{2}\pi, \frac{3}{2}\pi]$. Hence, from Theorem 16.2, the Fourier series deduced in Example 16.2 and the one required here must be the same. Consequently, without further work, we may write

$$f(x) = \tfrac{1}{2}\pi - \frac{4}{\pi}\left(\cos x + \frac{1}{3^2}\cos 3x + \frac{1}{5^2}\cos 5x + \ldots\right)$$

$$\text{for } \tfrac{1}{2}\pi \leq x \leq \tfrac{3}{2}\pi.$$

Notice that had the result of Example 16.2 not been available, rather than integrating the three-part functional form given in the present example over the interval $[-\frac{1}{2}\pi, \frac{3}{2}\pi]$, it would have been easier to use Theorem 16.2 to justify integrating the simpler function $f(x) = |x|$ over the more convenient interval $[-\pi, \pi]$.

(b) In this case the fundamental interval is $[-1, 1]$ so that setting $L = 1$ in Theorem 16.3 gives for the Fourier coefficients of $f(x) = x^3$ the values

$$a_n = \int_{-1}^{1} x^3 \cos n\pi x \, dx \qquad \text{and} \qquad b_n = \int_{-1}^{1} x^3 \sin n\pi x \, dx.$$

Routine integration then shows $a_n = 0$ for $n = 0, 1, \ldots$, and

$$b_n = (-1)^n \frac{2(6 - n^2\pi^2)}{n^3\pi^3} \qquad \text{for } n = 1, 2, \ldots.$$

Hence the Fourier series of $f(x) = x^3$ in the fundamental interval $-1 \leq x \leq 1$ is

$$f(x) = \frac{2}{\pi^3}\sum_{n=1}^{\infty}(-1)^n \frac{(6 - n^2\pi^2)}{n^3} \sin n\pi x.$$

Since the periodic extension of $f(x)$ is discontinuous at $x = \pm 1$, the Fourier series at these points will converge to the value $\frac{1}{2}[f(1-) + f(1+)]$. In this case this value is zero, since $f(1-) = 1$ and $f(1+) = -1$. ■

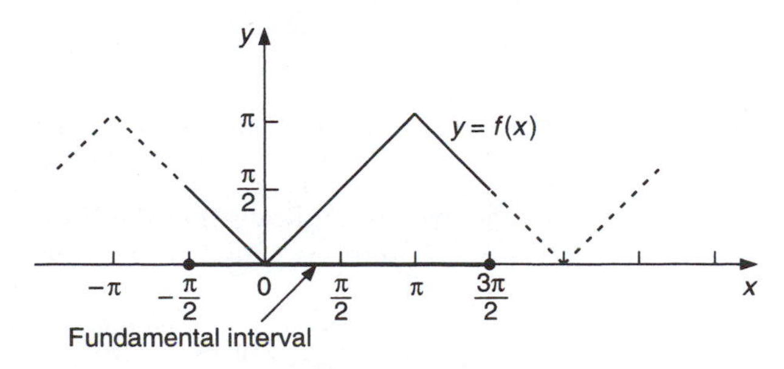

Fig. 16.5 Fourier series with fundamental interval $[-\pi/2, 3\pi/2]$.

16.3.2 Fourier sine and cosine series

When $f(x)$ is an even function defined over the interval $[-\pi, \pi]$, then $f(-x) = f(x)$. It thus follows directly that $f(x) \cos nx$ is an even function, because $\cos nx$ is even, and $f(x) \sin nx$ is an odd function, because $\sin nx$ is odd. Consider the Fourier coefficient a_n of an even function $f(x)$ which we choose to write in the form

$$a_n = \frac{1}{\pi}\int_{-\pi}^{0} f(x) \cos nx \, dx + \int_0^{\pi} f(x) \cos nx \, dx.$$

Then, changing the variable in the first integrand by writing $u = -x$, employing the even nature of the integrand to replace $f(-u) \cos[n(-u)]$ by $f(u) \cos nu$ and changing the sign of the integral by reversing the limits, we find

$$a_n = \frac{1}{\pi}\int_0^{\pi} f(u) \cos nu \, du + \frac{1}{\pi}\int_0^{\pi} f(x) \cos nx \, dx.$$

The variable u is only a dummy variable so it may be replaced by x to give the result

$$a_n = \frac{2}{\pi}\int_0^{\pi} f(x) \cos nx \, dx \qquad \text{for } n = 0, 1, \ldots. \tag{16.14}$$

This same argument applied to the coefficient b_n shows that

$$b_n = 0 \qquad \text{for } n = 1, 2, \ldots. \tag{16.15}$$

Consequently, when $f(x)$ is an even function in $[-\pi, \pi]$, its Fourier series contains only cosine functions and is of the form

$$f(x) = \frac{a_0}{2} + \sum_{n=1}^{\infty} a_n \cos nx \qquad \text{for } -\pi \leq x \leq \pi. \tag{16.16}$$

This is called the *Fourier cosine expansion* of the even function $f(x)$ in $[-\pi, \pi]$.

When $f(x)$ is an odd function defined over the interval $[-\pi, \pi]$, then $f(-x) = f(x)$. The form of argument just used then establishes that

$$a_n = 0 \qquad \text{for } n = 0, 1, \ldots \tag{16.17}$$

and

$$b_n = \frac{2}{\pi}\int_0^{\pi} f(x) \sin nx \, dx \qquad \text{for } n = 1, 2, \ldots, \tag{16.18}$$

from which it follows that the Fourier series of an odd function defined in $[-\pi, \pi]$ contains only sine function and is of the form

$$f(x) = \sum_{n=1}^{\infty} b_n \sin nx \qquad \text{for } -\pi \leq x \leq \pi. \tag{16.19}$$

This is called the *Fourier sine expansion* of the odd function $f(x)$ in $[-\pi, \pi]$.

These results can be usefully interpreted in terms of any arbitrary function $f(x)$ which is to be expanded in the half-interval $[0, \pi]$. Defining a new function $g(x)$ by the rule

$$g(x) = \begin{cases} f(-x) & \text{for } -\pi \le x \le 0 \\ f(x) & \text{for } 0 \le x \le \pi, \end{cases} \tag{16.20}$$

we see that $g(x)$ is an even function which is equal to $f(x)$ in the required interval $[0, \pi]$. Thus, as a Fourier cosine expansion of $g(x)$ only requires knowledge of $g(x)$ in the half-interval $[0, \pi]$ in which $g(x) = f(x)$, it follows that Eqn (16.16) provides the desired expansion of $f(x)$ when x is restricted to the interval $0 \le x \le \pi$.

Alternatively, we may expand the same function $f(x)$ in the half-interval $[0, \pi]$ in a Fourier sine expansion as follows. Define a new function $h(x)$ by the rule

$$h(x) = \begin{cases} -f(-x) & \text{for } -\pi \le x < 0 \\ f(x) & \text{for } 0 \le x \le \pi. \end{cases} \tag{16.21}$$

Then $h(x)$ is an odd function which is equal to $f(x)$ in the required interval $[0, \pi]$. The Fourier sine expansion of $h(x)$ only requires knowledge of $h(x)$, in the half-interval $[0, \pi]$ where $h(x) = f(x)$, so that Eqn (16.19) provides the desired expansion if x is restricted to the interval $0 \le x \le \pi$. For obvious reasons, these expansions are often called the *half-range expansions* of $f(x)$. We have now proved the next theorem.

Theorem 16.4 (Fourier sine and cosine series) If $f(x)$ is an arbitrary function defined and integrable on $[0, \pi]$, then it may either be expanded as a Fourier cosine series

$$f(x) = \frac{a_0}{2} + \sum_{n=1}^{\infty} a_n \cos nx \qquad \text{for } 0 \le x \le \pi,$$

in which

$$a_n = \frac{2}{\pi}\int_0^{\pi} f(x)\cos nx \, dx \qquad \text{for } n = 0, 1, \ldots,$$

or as a Fourier sine series

$$f(x) = \sum_{n=1}^{\infty} b_n \sin nx \qquad \text{for } 0 \le x \le \pi,$$

in which

$$b_n = \frac{2}{\pi}\int_0^{\pi} f(x)\sin nx \, dx \qquad \text{for } n = 1, 2, \ldots. \quad ■$$

Example 16.7 Deduce the Fourier cosine and sine series of the function $f(x) = x$ in $[0, \pi]$.

Solution From the first part of Theorem 16.4 we have

$$a_n = \frac{2}{\pi}\int_0^{\pi} \cos nx \, dx \qquad \text{for } n = 0, 1, \ldots .$$

A simple integration then shows

$$a_0 = \pi, \quad a_{2n-1} = \frac{-4}{\pi(2n-1)^2}, \quad a_{2n} = 0.$$

The Fourier cosine series of $f(x) = x$ thus has the form

$$f(x) = \frac{\pi}{2} - \frac{4}{\pi}\sum_{n=1}^{\infty} \frac{\cos[(2n-1)x]}{(2n-1)^2} \qquad \text{for } 0 \leq x \leq \pi.$$

From the second part of Theorem 16.4 we find

$$b_n = \frac{2}{\pi}\int_0^{\pi} x \sin nx \, dx \qquad \text{for } n = 1, 2, \ldots .$$

Evaluating this integral gives

$$b_n = (-1)^{n+1}\frac{2}{n} \qquad \text{for } n = 1, 2, \ldots,$$

so that the Fourier sine series of $f(x) = x$ is seen to have the form

$$f(x) = 2\sum_{n=1}^{\infty} (-1)^{n+1}\frac{\sin nx}{n}.$$

From the practical point of view the cosine series is preferable to the sine series in this instance because it converges more rapidly. ■

16.3.3 Complex form of Fourier series

Suppose the arbitrary function $f(x)$ is defined on the interval $[-L, L]$. Then, as in Theorem 16.3, its Fourier series will contain sine and cosine functions with argument $n\pi x/L$. If these are expressed in terms of the complex exponential function $\exp\left(i\frac{n\pi x}{L}\right)$, it transpires that the series may be written in a convenient complex form. To achieve this, observe first that a simple integration gives

$$\int_{-L}^{L} \exp\left(i\frac{m\pi x}{L}\right)\exp\left(i\frac{n\pi x}{L}\right)dx = \begin{cases} 0 & \text{for } m \neq -n \\ 2L & \text{for } m = -n. \end{cases} \tag{16.22}$$

First we assume that $f(x)$ is continuous and satisfies the conditions of Theorem 16.1, so that its Fourier series always converges to $f(x)$. Then, in place of Definition 16.1, it is natural to write

$$f(x) = \sum_{m=-\infty}^{\infty} A_m \exp\left(i\frac{m\pi x}{L}\right), \tag{16.23}$$

where the complex coefficients A_m are to be determined, and the summation indicates that $m = 0, \pm 1, \pm 2, \ldots$. Multiplication of Eqn (16.23) by $\exp\left(-i\frac{n\pi x}{L}\right)$ and integration over $[-L, L]$ then lead to the result

$$A_n = \frac{1}{2L}\int_{-L}^{L} f(x)\exp\left(-i\frac{n\pi x}{L}\right)\mathrm{d}x \quad \text{for } n = 0, \pm 1, \pm 2, \ldots. \tag{16.24}$$

Using these coefficients in Eqn (16.23) gives the complex Fourier series of $f(x)$ which, when simplified, naturally reduces to the same series that would have been obtained had Definition 16.1 been employed. If the symbol $\sim$ is used in place of the equality in (16.23) with the sense that was explained after Theorem 16.1, we immediately arrive at the following alternative definition of a Fourier series.

Definition 16.2 (complex Fourier series)

The *complex Fourier series* of an integrable function $f(x)$ defined on the interval $[-L, L]$ is the series

$$f(x) \sim \sum_{n=-\infty}^{\infty} A_n \exp\left(i\frac{n\pi x}{L}\right),$$

in which the *complex Fourier coefficients* A_n are given by

$$A_n = \frac{1}{2L}\int_{-L}^{L} f(x)\exp\left(-i\frac{n\pi x}{L}\right)\mathrm{d}x \qquad \text{for } n = 0, \pm 1, \pm 2, \ldots .$$

Example 16.8 Find and simplify the complex Fourier series of the function

$$f(x) = \begin{cases} a & \text{for } -\pi \le x < 0 \\ b & \text{for } 0 \le x \le \pi. \end{cases}$$

Solution Using Definition 16.2 with $L = \pi$, we can at once write

$$A_0 = \frac{1}{2\pi}\int_{-\pi}^{0} a\mathrm{d}x + \frac{1}{2\pi}\int_{0}^{\pi} b\mathrm{d}x = \frac{a+b}{2},$$

$$A_n = \frac{a}{2\pi}\int_{-\pi}^{0} \exp(-inx)\,\mathrm{d}x + \frac{b}{2\pi}\int_{0}^{\pi}\exp(-inx)\,\mathrm{d}x$$

$$= \frac{i(a-b)}{2\pi n}[1-(-1)^n]$$

and

$$A_{-n} = \frac{a}{2\pi}\int_{-\pi}^{0} \exp(inx)\,dx + \frac{b}{2\pi}\int_{0}^{\pi} \exp(inx)\,dx$$
$$= \frac{-i(a-b)}{2\pi n}[1-(-1)^n].$$

Hence the complex Fourier series is

$$f(x) = \frac{a+b}{2} + \sum_{n=-\infty}^{\infty} A_n \exp(inx),$$

in which A_n, A_{-n} have the values just computed.

Combining the general terms for $\pm n$ in the complex Fourier series, we obtain

$$A_n \exp(inx) + A_{-n}\exp(-inx) = \begin{cases} 0 & \text{for even } n \\ \dfrac{(b-a)}{\pi n}[1-(-1)^n]\sin nx & \text{for odd } n. \end{cases}$$

Hence the complex Fourier series of $f(x)$ for $-\pi \leq x \leq \pi$ reduces to the real series

$$f(x) = \left(\frac{a+b}{2}\right) + \frac{2}{\pi}(b-a)\sum_{m=1}^{\infty} \frac{\sin[(2m-1)x]}{2m-1},$$

which, as would be expected, is exactly the result obtained in Example 16.3. ■

16.4 Differentiation and integration

In this section we complete our discussion of Fourier series by stating without proof the theorem concerning the differentiation and integration of Fourier series.

Theorem 16.5 (integration of Fourier series)

If $f(x)$ is an integrable piecewise continuous function in the interval $[-\pi, \pi]$ with the Fourier series

$$f(x) = \frac{a_0}{2} + \sum_{n=1}^{\infty} (a_n \cos nx + b_n \sin nx),$$

then

$$\int_{x_1}^{x} f(t)\mathrm{d}t = \frac{a_0(x-x_1)}{2} + \sum_{n=1}^{\infty} \int_{x_1}^{x} (a_n \cos nt + b_n \sin nt)\mathrm{d}t. \quad ■$$

Example 16.9 Discuss the result of termwise integration of the Fourier series of the function

$$f(x) = \begin{cases} -1 & \text{for } -\pi < x < 0 \\ 1 & \text{for } 0 \leq x \leq \pi \end{cases}$$

over the interval $[-\pi, x]$ with $-\pi < x \leq \pi$.

Solution Setting $a = -1$, $b = 1$ in the result of Example 16.3 shows that the Fourier series of $f(x)$ is

$$f(x) = \frac{4}{\pi}\left(\sin x + \frac{1}{3}\sin 3x + \frac{1}{5}\sin 5x + \ldots\right). \tag{A}$$

We now need to apply Theorem 16.5 with $x_1 = -\pi$ and $x_2 = x$ and, in doing so, because of the discontinuity in $f(x)$, to consider the two cases $-\pi < x < 0$ and $0 \leq x \leq \pi$.

(*i*) $-\pi < x < 0$. From Theorem 16.5 we have

$$\int_{-\pi}^{x}(-1)\mathrm{d}t = \frac{4}{\pi}\sum_{n=1}^{\infty}\int_{-\pi}^{x}\frac{\sin[(2n-1)t]}{(2n-1)}\mathrm{d}t$$

or

$$-(x+\pi) = -\frac{4}{\pi}\left\{\sum_{n=1}^{\infty}\frac{\cos[(2n-1)t]}{(2n-1)^2}\right\}\Bigg|_{-\pi}^{x}$$

$$= -\frac{4}{\pi}\left\{\left(\cos x + \frac{\cos 3x}{3^2} + \frac{\cos 5x}{5^2} + \ldots\right) + \left(1 + \frac{1}{3^2} + \frac{1}{5^2} + \ldots\right)\right\}.$$

The numerical series has the sum $\pi^2/8$, so that this result simplifies to

$$-x = \frac{1}{2}\pi - \frac{4}{\pi}\left(\cos x + \frac{\cos 3x}{3^2} + \frac{\cos 5x}{5^2} + \ldots\right) \quad \text{for } -\pi < x < 0. \tag{B}$$

(*ii*) $0 \leq x \leq \pi$. Here, as a result of an application of Theorem 16.5, we have

$$\int_{-\pi}^{0}(-1)\,\mathrm{d}t + \int_{0}^{x}1\,\mathrm{d}t = \frac{4}{\pi}\sum_{n=1}^{\infty}\int_{-\pi}^{x}\frac{\sin[(2n-1)t]}{(2n-1)}\mathrm{d}t,$$

which after simplification as before gives

$$x = \frac{1}{2}\pi - \frac{4}{\pi}\left(\cos x + \frac{\cos 3x}{3^2} + \frac{\cos 5x}{5^2} + \ldots\right) \quad \text{for } 0 \leq x \leq \pi. \tag{C}$$

Results (B) and (C), taken together, comprise the series representing the function

$$f(x) = \begin{cases} -x & \text{for } -\pi < x < 0 \\ x & \text{for } 0 \le x \le \pi. \end{cases}$$

This is simply the function $f(x) = |x|$ examined in Example 16.2.

Theorem 16.6 (differentiation of Fourier series) Let $f(x)$ be a continuous function with a piecewise continuous derivative in the interval $[-\pi, \pi]$, and let $f(\pi) = f(-\pi)$. Then, if the Fourier series of $f(x)$ is

$$f(x) = \frac{a_0}{2} + \sum_{n=1}^{\infty} (a_n \cos nx + b_n \sin nx),$$

termwise differentiation of this series is permissible and

$$f'(x) = \sum_{n=1}^{\infty} \frac{\mathrm{d}}{\mathrm{d}x}(a_n \cos nx + b_n \sin nx)$$

for $-\pi < x < \pi$, except at points where $f'(x)$ is discontinuous. ■

Example 16.10 Deduce the Fourier series of $f(x) = x$ in $[-\pi, \pi]$ from the Fourier series of $f(x) = \pi^2 - x^2$. Give an example where termwise differentiation of a Fourier series is not permissible.

Solution From Example 16.1 we see that the Fourier series of $f(x) = \pi^2 - x^2$ in $[-\pi, \pi]$ is

$$\pi^2 - x^2 = \frac{2}{3}\pi^2 + 4\left(\cos x - \frac{1}{2^2}\cos 2x + \frac{1}{3^2}\cos 3x - \ldots\right).$$

As $f(\pi) = 0 = f(-\pi)$ and $f(x)$ and $f'(x)$ are continuous, Theorem 16.6 is applicable so that differentiation of the above result yields

$$-2x = -4\left(\sin x - \tfrac{1}{2}\sin 2x + \tfrac{1}{3}\sin 3x - \ldots\right),$$

whence

$$x = 2\left(\sin x - \tfrac{1}{2}\sin 2x + \tfrac{1}{3}\sin 3x - \ldots\right)$$

for $-\pi < x < \pi$. A direct calculation easily verifies that this is the correct Fourier expansion of the function $f(x) = x$ for $-\pi \le x \le \pi$. ■

For our example of a Fourier series where termwise differentiation is not permissible we need look no further than Example 16.3, for there $f(x)$ is not continuous in $[-\pi, \pi]$ and also $f(\pi) \neq f(-\pi)$. If the resulting Fourier series in that example is differentiated termwise it yields

$$\frac{2}{\pi}(b - a)(\cos x + \cos 3x + \cos 5x + \ldots)$$

which is divergent for all x by virtue of the test in Theorem 12.2, because the nth term $\cos nx$ does not tend to zero.

Problems

Section 16.1

Find the Fourier series of each of the following functions.

16.1 $f(x) = x$ for $-\pi \le x \le \pi$.

16.2 $f(x) = x^2$ for $-\pi \le x \le \pi$.

16.3 $f(x) = \begin{cases} 0 & \text{for } -\pi \le x \le 0 \\ -x & \text{for } 0 \le x \le \pi. \end{cases}$

16.4 $f(x) = |\cos x|$ for $-\pi \le x \le \pi$.

16.5 $f(x) = |\sin x|$ for $-\pi \le x \le \pi$.

16.6 $f(x) = x^2 - 2x$ for $-\pi \le x \le \pi$.

16.7 $f(x) = \cos \alpha x$ for $-\pi \le x \le \pi$ with $\alpha \ne 0, \pm 1, \pm 2, \ldots$.

16.8 $f(x) = \sin \alpha x$ for $-\pi \le x \le \pi$ with $\alpha \ne 0, \pm 1, \pm 2, \ldots$.

16.9 $f(x) = \sin^2 x$ for $-\pi \le x \le \pi$.

16.10 $f(x) = \begin{cases} -1 & \text{for } -\pi < x < -\frac{1}{2}\pi \\ 1 & \text{for } -\frac{1}{2}\pi \le x \le \frac{1}{2}\pi \\ -1 & \text{for } \frac{1}{2}\pi < x \le \frac{1}{2}\pi. \end{cases}$

16.11 $f(x) = \begin{cases} 2 & \text{for } -\pi \le x < -\frac{1}{2}\pi \\ 1 & \text{for } -\frac{1}{2}\pi \le x \le \frac{1}{2}\pi \\ 2 & \text{for } \frac{1}{2}\pi < x \le \pi. \end{cases}$

Section 16.2

16.12 Apply Parseval's relation to the Fourier series of the function $f(x) = \pi^2 - x^2$ for $-\pi \le x \le \pi$ to deduce a series for $\pi^4/90$.

16.13 Apply Parseval's relation to the Fourier series of the function $f(x) = x^2$ for $-\pi \le x \le \pi$ to deduce a series for $\pi^4/90$.

16.14 Deduce an expansion for π by using the Fourier series for the function

$$f(x) = \begin{cases} 0 & \text{for } -\pi \le x < 0 \\ \sin x & \text{for } 0 \le x \le \pi. \end{cases}$$

16.15 Using the Fourier series expansion of the function

$$f(x) = \begin{cases} -\frac{1}{4}\pi & \text{for} \quad -\pi \le x < 0 \\ \frac{1}{4}\pi & \text{for} \quad 0 \le x < \pi, \end{cases}$$

deduce that

$$\frac{\pi}{4} = 1 - \frac{1}{3} + \frac{1}{5} - \frac{1}{7} + \frac{1}{9} - \cdots.$$

16.16 Express the function $f(x) = e^x$ as a Fourier series in the interval $-\pi \le x \le \pi$. Use the resulting series to find a series representation for e.

16.17 Use the Fourier series expansion of the function $f(x) = x + x^2$ in $-\pi < x \le \pi$ to deduce a series for $\pi^2/12$ by setting $x = 0$.

Section 16.3

16.18 Find the Fourier series of the function $f(x) = x$ for $0 \le x \le 2\pi$.

16.19 Find the Fourier series of the function

$$f(x) = \begin{cases} x^2 & \text{for} \quad -\frac{1}{2}\pi \le x \le \pi \\ (2\pi - x)^2 & \text{for} \quad \pi \le x \le \frac{3}{2}\pi. \end{cases}$$

16.20 Find the Fourier series of the function

$$f(x) = \begin{cases} 0 & \text{for} \quad -\frac{1}{2}\pi \le x < 0 \\ \sin x & \text{for} \quad 0 \le x \le \pi \\ 0 & \text{for} \quad \pi < x \le \frac{3}{2}\pi. \end{cases}$$

16.21 Find the Fourier series of the function $f(x) = x^2$ for $-\pi \le x \le 3\pi$.

16.22 Find the Fourier series of the function $f(x) = x$ for $-1 \le x \le 1$.

16.23 Find the Fourier series of the function

$$f(x) = |x| \text{ for } -3 \le x \le 3.$$

16.24 Find the Fourier series of the function

$$f(x) = \sin x \text{ for } 0 \le x \le \pi.$$

16.25 Find the Fourier sine series for the function

$$f(x) = \cos x \text{ in } 0 \le x \le \pi.$$

16.26 Find the Fourier cosine series for the function

$$f(x) = x^3 \text{ for } 0 \le x \le \pi.$$

16.27 Find the Fourier cosine series for the function

$$f(x) = e^x \text{ for } 0 \leq x \leq \pi.$$

16.28 Find the complex Fourier series for the function $f(x) = e^x$ for $-1 \leq x \leq 1$. Use the result to deduce the ordinary Fourier series for this function.

Section 16.4

16.29 Given that the Fourier series of the function $f(x) = 2x+1$ for $-\pi \leq x \leq \pi$ is

$$f(x) = 1 - 4\sum_{n=1}^{\infty}(-1)^n \frac{\sin nx}{n},$$

deduce a series expansion for the function $g(x) = x^2+x$. Is this new series a Fourier series?

16.30 Apply termwise differentiation to the Fourier series of the function

$$f(x) = \begin{cases} 0 & \text{for} \quad -\pi \leq x < 0 \\ \sin x & \text{for} \quad 0 \leq x < \pi, \end{cases}$$

to obtain the Fourier series of the function

$$g(x) = \begin{cases} 0 & \text{for} \quad -\pi \leq x \leq 0 \\ \cos x & \text{for} \quad 0 \leq x \leq \pi. \end{cases}$$

Is the Fourier series of $g(x)$ termwise differentiable for any x in $-\pi \leq x \leq \pi$? Examine the convergence properties of the differentiated series for $g(x)$.

16.31 Theorem 16.6 is stated for a continuous function $f(x)$ which has a derivative which is only piecewise continuous, though the proof was only given on the assumption that the derivative was continuous. Modify the process of integration by parts to allow for the functions $f(x)$ which have derivatives which are only piecewise continuous and so complete the proof of the theorem as stated.

General problems

16.32 Find the Fourier series of the function $f(x) = \sinh \alpha x$ for $-\pi \leq x \leq \pi$. What is the value assumed by the Fourier series at $x = \pm\pi$?

16.33 Find the Fourier series of the function $f(x) = \cosh \alpha x$ for $-\pi \leq x \leq \pi$.

16.34 Find the Fourier series of the function $f(x) = x \sin x$ for $-\pi \leq x \leq \pi$.

16.35 Find the Fourier sine series of the function

$$f(x)=\begin{cases} x & \text{for } 0\le x\le 1\\ 2-x & \text{for } 1<x\le 2.\end{cases}$$

16.36 Find the Fourier cosine series of the function

$$f(x)=x(\pi-x)\quad \text{for}\quad 0\le x\le \pi.$$

16.37 It can be proved that if $f(x)$, $g(x)$ are any two integrable functions defined on the interval $[-\pi,\pi]$ and their Fourier coefficients are a_n, b_n and a_n', b_n' respectively, then

$$\frac{1}{\pi}\int_{-\pi}^{\pi} f(x)g(x)\,\mathrm{d}x=\frac{a_0\,a_0'}{2}+\sum_{n=1}^{\infty}(a_na_n'+b_nb_n').$$

Apply this result to the case $f(x)=|x|$, $g(x)=x^2$ for $-\pi\le x\le\pi$ and hence deduce a series representation for $\pi^4/96$.

16.38 Let

$$T_n(x)=\frac{\alpha_0}{2}+\sum_{r=1}^{n}(\alpha_r\cos rx+\beta_r\sin rx)$$

for $-\pi\le x\le\pi$ where α_r, β_r are arbitrary coefficients. Then, if $f(x)$ is an arbitrary integrable function for $-\pi\le x\le\pi$, use the orthogonality property of the sine and cosine functions to prove that the integral of the square of the error between $f(x)$ and $T_n(x)$

$$\Delta_n=\int_{-\pi}^{\pi}[f(x)-T_n(x)]^2\mathrm{d}x$$

is a minimum when the coefficients α_r, β_r assume the values of the Fourier coefficients of $f(x)$. This is called the *least-squares property* of Fourier series. (*Hint*: Regard Δ_n as a function of the variables α_r, β_r and choose their values so that Δ_n is a minimum.)

Supplementary computer problems

16.39 The Fourier series representation of $f(x)=\pi^2-x^2$ for $-\pi\le x\le\pi$ was shown in Example 16.1 to be given by

$$f(x)=\tfrac{2}{3}\pi^2+4\sum_{r=1}^{\infty}(-1)^{r+1}\frac{\cos rx}{r^2}.$$

Let

$$S_n(x)=\tfrac{2}{3}\pi^2+4\sum_{r=1}^{n}(-1)^{r+1}\frac{\cos rx}{r_1^2}$$

and display $S_n(x)$ for $n=4$, 8, 12 in the interval $-\pi\le x\le\pi$.

16.40 The Fourier series representation of $f(x)=|x|$ for $-\pi\le x\le\pi$ was shown in Example 16.2 to be given by

$$f(x) = \tfrac{1}{2}\pi - \frac{4}{\pi}\sum_{r=1}^{\infty} \frac{\cos[(2r-1)x]}{(2r-1)^2}.$$

Let

$$S_n(x) = \tfrac{1}{2}\pi - \frac{4}{\pi}\sum_{r=1}^{n} \frac{\cos[(2r-1)x]}{(2r-1)^2},$$

and display $S_n(x)$ for $n = 4$, 8, 12 in the interval $-\pi \le x \le \pi$.

16.41 The Fourier series representation of

$$f(x) = \begin{cases} 0 & \text{for} \quad -\pi < x < 0 \\ 1 & \text{for} \quad 0 < x < \pi \end{cases}$$

follows from Example 16.3 in the form

$$f(x) = \frac{1}{2} + \frac{2}{\pi}\sum_{r=0}^{\infty} \frac{\sin[(2r+1)x]}{2r+1}.$$

Let

$$S_n(x) = \frac{1}{2} + \frac{2}{\pi}\sum_{r=0}^{n} \frac{\sin[(2r+1)x]}{2r+1},$$

and display $S_n(x)$ for $n = 4$, 8, 12 in the interval $-\pi \le x \le \pi$.

16.42 The Fourier series representation of

$$f(x) = \begin{cases} 0 & \text{for} \quad -\pi < x < 0 \\ \sin x & \text{for} \quad 0 \le x \le \pi \end{cases}$$

was shown in Example 16.4 to be given by

$$f(x) = \frac{1}{\pi} + \frac{1}{2}\sin x - \frac{2}{\pi}\sum_{r=1}^{\infty} \frac{\cos 2rx}{(2r-1)(2r+1)}.$$

Let

$$S_n(x) = \frac{1}{\pi} + \frac{1}{2}\sin x - \frac{2}{\pi}\sum_{r=1}^{n} \frac{\cos 2rx}{(2r-1)(2r+1)},$$

and display $S_n(x)$ for $n = 4$, 8, 12 in the interval $-\pi \le x \le \pi$.

16.43 In Example 16.7 the half-range sine and cosine series for $f(x) = x$ in the interval $0 \le x \le \pi$ were shown to be given by

$$f(x) = 2\sum_{n=1}^{\infty} (-1)^{r+1} \frac{\sin rx}{r} \qquad \text{(sine series)}$$

$$f(x) = \frac{\pi}{2} - \frac{4}{\pi}\sum_{\pi=1}^{\infty} \frac{\cos[(2r-1)x]}{(2r-1)^2}. \quad \text{(cosine series)}.$$

Set

$$S_n(x) = 2\sum_{r=1}^{n} (-1)^{r+1} \frac{\sin rx}{r},$$

$$C_n(x) = \frac{\pi}{2} - \frac{4}{\pi}\sum_{r=1}^{n} \frac{\cos[(2r-1)x]}{(2r-1)^2},$$

and display $S_n(x)$ and $C_n(x)$ for $n = 4, 8, 12$ in the interval $0 \leq x \leq \pi$. Compare the corresponding sine and cosine approximations.

17 Numerical analysis

The purpose of this chapter is to provide a first account of the elements of numerical analysis. The practical application of mathematics to physical problems leads to numerical computation whenever explicit analytical results are not available, or their form makes it impractical for them to be used. When employing numerical methods it is important to distinguish between, and understand the nature of, errors caused by round-off and by approximations used in the method of calculation. These are matters that are examined in section 17.1.

Some of the most commonly used numerical methods have been developed for the solution of linear systems of equations. Section 17.2 describes the Gaussian elimination method, the iterative method due to Jacobi and the iterative Gauss–Seidel scheme. This section is followed by accounts of elementary interpolation methods and techniques for numerical integration, also known as numerical quadrature.

Many applications of mathematics require the roots of an equation to be determined. Accordingly, section 17.5 is devoted to the determination of the roots of polynomial and transcendental equations by various techniques. These include a graphical method, the secant method, and Newton's method, also known as the Newton–Raphson method. Section 17.3 on interpolation includes an introductory account of spline function approximation that now plays such an important role in computer-aided design techniques.

The numerical integration of differential equations is discussed in Section 17.6; the methods included range from the simple Euler and modified Euler methods to a predictor–corrector method and the accurate fourth-order Runge–Kutta method.

Because of the importance of the application of the eigenvalues and eigenvectors of a matrix, section 17.7 describes a simple method for their determination by numerical means. The method described is an iterative technique that finds the eigenvalue with the largest magnitude, together with its associated eigenvector. It is then shown how this eigenvalue may be used in conjunction with a modification of the same iterative technique to arrive at the eigenvalue of next largest magnitude, together with its associated eigenvector. A repetition of this process leads to the determination of all the eigenvalues and eigenvectors of the matrix. In general, the accurate determination of eigenvalues and eigenvectors is difficult, and numerical packages and computer algebra packages employ far more sophisticated methods than those described here.

17.1 Errors and efficient methods of calculation

Whenever numerical results have to be obtained from mathematical expressions the possibility of error arises. Numerical computation and its attendent error is generally inescapable when seeking to solve any moderately complicated problem. On account of this, errors must be understood and, when unavoidable, they must always be controlled.

Four types of error can occur:

(a) round-off errors;
(b) error due to approximations in the method of calculation;
(c) errors in the data;
(d) human error.

Only human error may be avoided. Round-off errors and errors in data derived from observation will always occur. The errors introduced by the method of calculation may sometimes be reduced by improvement of the method, but they can seldom be eliminated. For example, whenever an irrational number is involved in a calculation it must of necessity be approximated by a rational number, and when two n-digit decimal numbers are multiplied together their product contains $2n$ decimal digits, not all of which are usually retained.

17.1.1 Round-off errors

A number is often approximated by setting all digits to the right of the nth equal to zero, and modifying the nth digit to take account of the discarded part of the number. This process is called *rounding off* and it introduces an error. The convention for rounding off is that the nth digit should not be changed if the digit following it is 0, 1, 2, 3 or 4. If the digit following the nth digit is 5, 6, 7, 8 or 9 then the nth digit should be increased by one. In the event that the first discarded digit is 5, and the subsequent digits are all zeros, the last digit to be retained should only be increased by one if it is odd. A number is said to be rounded off to n *significant figures* when the rounding off has taken place after n digits from the first non-zero digit to occur in its representation. Sometimes the rounding off is expressed in terms of the number of decimal places that are retained.

Thus 2.8284 is 3 when rounded off to one significant figure and 2.828 when rounded off to four significant figures. This last result could also be said to be rounded off to three decimal places. Similarly 0.0340450 is 0.0340 when rounded off to three significant figures. It is, however, 0.03405 when rounded to five decimal places using the convention for a number terminating in a 5 followed by a zero.

17.1.2 Absolute and relative error

When x is an approximation to the exact number X we call the number $\varepsilon = x - X$ the *absolute error*, and the number $r = |x - X|/|X| = |\varepsilon|/|X|$ the *relative error*. We only work with relative errors. As X is normally unknown the relative error r is approximated by the result $r \approx |\varepsilon|/|x|$, where the sign $\approx$ is to be read 'is approximately equal to'.

17.1.3 Development of errors during arithmetic operations

Addition and subtraction

It is a direct consequence of the properties of the absolute value (Theorem 1.1) that if $x_3 = x_1 \pm x_2$ is an approximation to $X_3 = X_1 \pm X_2$, then $|\varepsilon_3| \leq |\varepsilon_1| + |\varepsilon_2|$. Thus, if we consider the sum $39.35 + 38.65 \approx 78.00$, in which these numbers have been rounded off to two decimal places, we know that $|\varepsilon_1| \leq 0.005$ and $|\varepsilon_2| \leq 0.005$, and so $|\varepsilon_3| \leq 0.01$. Consequently we may write the sum as 78.00 ± 0.01. The relative error is then 0.01/78. However, when considering the difference $39.35 - 38.65$ we still have $|\varepsilon_3| \leq 0.01$, so that the difference is 0.70 ± 0.01, but the relative error is now 0.01/0.7. The increase in the relative error has come about because we have subtracted two nearly equal numbers.

Multiplication

If $x_3 = x_1 x_2$ is an approximation to $X_3 = X_1 X_2$, then $x_3 - X_3 = x_1 x_2 - X_1 X_2$. In terms of ε_1, ε_2 and ε_3 this becomes $\varepsilon_3 = x_1 x_2 - (x_1 - \varepsilon_1)(x_2 - \varepsilon_2)$, which after division by $x_3 = x_1 x_2$ can be written

$$\frac{\varepsilon_3}{x_3} = \frac{\varepsilon_1}{x_1} + \frac{\varepsilon_2}{x_2} - \frac{\varepsilon_1 \varepsilon_2}{x_1 x_2}.$$

Taking absolute values we arrive at the approximate result

$$r_3 \approx r_1 + r_2$$

involving the relative errors.

So in the product $14.11 \times 0.3472 \approx 4.8990$ involving numbers rounded off to two and four decimal places, respectively, we have $|\varepsilon_1| = 0.005$ and $|\varepsilon_2| = 0.00005$. Thus $r_1 \leq 0.005/14.11$ and $r_2 \leq 0.00005/0.3472$, showing that $r_3 \leq 0.50 \times 10^{-3}$. Setting $x_3 \approx 4.8990$ we have $\varepsilon_3 \approx 0.50 \times 10^{-3} \times 4.8990 = 0.0024$, so that when working to four decimal places the error will be ± 0.0024. The product is thus 4.8990 ± 0.0024.

Division

If $x_3 = x_1/x_2$ is an approximation to $X_3 = X_1/X_2$, it follows that

$$x_3 - X_3 = \frac{x_1}{x_2} - \frac{x_1}{x_2}\left(1 - \frac{\varepsilon_1}{x_1}\right) \Big/ \left(1 - \frac{\varepsilon_2}{x_2}\right).$$

Expanding the last term by the binomial theorem and dividing by x_3 gives

$$\frac{\varepsilon_3}{x_3} \approx \frac{\varepsilon_1}{x_1} - \frac{\varepsilon_2}{\varepsilon_2}.$$

Taking the absolute value and using Theorem 1.1 then shows that the relative errors satisfy the result

$$r_3 \approx r_1 + r_2,$$

where again the relative errors r_1 and r_2 have been added.

Functional relationships

If $x_2 = f(x_1)$ is an approximation to $X_2 = f(X_1)$, then

$$x_2 - X_2 = f(x_1) - f(x_1 - \varepsilon_1),$$

so that from the mean value theorem we may write

$$x_2 - X_2 \approx \varepsilon_1 f'(x_1).$$

Dividing by x_2 and taking the absolute value shows that the relative errors r_1 and r_2 are related by

$$r_2 \approx \left| \frac{x_1 f'(x_1)}{f(x_1)} \right| r_1.$$

17.1.4 Precision and accuracy

These brief considerations about errors should make clear the difference between the precision and accuracy of a calculation. The *precision* is the number of digits used in the calculation, which since it is only approximate may not all be correct. The *accuracy* is the number of digits to which the result is correct.

The minimization of errors by the reorganization of calculations

An appropriate reorganization of a calculation can often lead to a considerable reduction in the effect of rounding off and the use of fixed precision. This is illustrated by considering the determination of the roots x_1 and x_2 of the quadratic equation

$$ax^2 + bx + c = 0$$

given by the formula

$$x_1 = \frac{-b + \sqrt{b^2 - 4ac}}{2a} \quad \text{and} \quad x_2 = \frac{-b - \sqrt{b^2 - 4ac}}{2a}.$$

Consider the equation

$$x^2 + 73.21x + 1 = 0$$

with the approximate roots

$$x_1 = -0.0136619 \qquad \text{and} \quad x_2 = -73.1963381.$$

Errors can arise when working with an inappropriate fixed precision because $-b$ and $\surd(b^2 - 4ac)$ only differ by a small quantity, so the determination of the smaller of the two roots x_1 will involve large errors if the precision is too low. If, for example, the calculations are rounded off to four digits, we find that

$$x_1 = \frac{-73.21 + \sqrt{(73.21)^2 - 4}}{2} = \frac{-73.21 + \sqrt{5630 - 4}}{2}$$

$$= \frac{-73.21 + 73.18}{2} = -0.0150,$$

and

$$x_2 = -73.20.$$

Thus the relative error r in x_1 is seen to be almost 10%, because

$$r = \frac{|-0.0150 - (-0.0136619)|}{|-0.0136619|} = 0.098.$$

In this case the accuracy can be improved by rationalizing the expression for x_1 as follows:

$$x_1 = \frac{-b + \sqrt{b^2 - 4ac}}{2a} = \left(\frac{-b + \sqrt{b^2 - 4ac}}{2a}\right)\left(\frac{-b - \sqrt{b^2 - 4ac}}{-b - \sqrt{b^2 - 4ac}}\right)$$

$$= \frac{-2c}{b + \sqrt{b^2 - 4ac}}.$$

Again rounding off to four digits, we now find that

$$x_1 = -0.0137.$$

Nested multiplication

The efficient and accurate evaluation of polynomials is best accomplished by means of *nested multiplication*. We illustrate this by showing how the following polynomial $P(x)$ may be expressed in nested form, and then by using the result to evaluate $P(2.17)$.

The polynomial

$$P(x) = 2x^4 + 1.7x^3 + 3.21x^2 - 1.3x + 3.21$$

may be written in *nested form* by proceeding as follows:

$$\begin{aligned} P(x) &= (2x^3 + 1.7x^2 + 3.21x - 1.3)x + 3.21 \\ &= ((2x^2 + 1.7x + 3.21)x - 1.3)x + 3.21 \\ &= (((2x^2 + 1.7)x + 3.21)x - 1.3)x + 3.21 \end{aligned}$$

and so, in nested form,

$$P(x) = (((2x + 1.7)x + 3.21)x - 1.3)x + 3.21.$$

Setting $x = 2.17$ in the above result gives

$$P(2.17) = (((2\times 2.17 + 1.7)2.17 + 3.21)2.17 - 1.3)2.17 + 3.21$$

$$= ((6.04\times 2.17 + 3.21)2.17 - 1.3)2.17 + 3.21$$

$$= (16.3168\times 2.17 - 1.3)2.17 + 3.21$$

$$= 34.1074\,56\times 2.17 + 3.21$$

$$= 77.2\,23\,17\,952.$$

The nested multiplication procedure is also known as *Horner's method* or *synthetic division*. The process reduces the number of multiplications needed for the evaluation of a polynomial and, when working to a fixed precision, it reduces the errors involved.

17.2 Solution of linear equations

We shall consider a system of n linear inhomogeneous equations in n variables $x_1, x_2, \ldots, x_n$ of the general form

$$\begin{aligned} a_{11}x_1 + a_{12}x_2 + \ldots + a_{1n}x_nx_n &= k_1 \\ a_{21}x_1 + a_{22}x_2 + \ldots + a_{2n}x_nx_n &= k_2 \\ &\ldots \\ a_{n1}x_1 + a_{n2}x_2 + \ldots + a_{nn}x_n &= k_n. \end{aligned} \qquad (17.1)$$

When the number of variables involved becomes large enough to make Gaussian elimination impracticable, then approximate methods of solution must be used. The idea underlying the most important of these, which are known as *iterative* methods, is simply that given an approximation to the solution, a numerical rule or *algorithm* is developed which when starting from this approximation gives rise to a better one. If such an algorithm is applied repetitively, then each successive stage in the calculation is called an *iteration*. The successive approximations themselves are called *iterates*. Of the many possible iterative methods we mention only the *Jacobi* and the *Gauss–Seidel* schemes. As an alternative to the direct solution of Eqns (17.1) by Gaussian elimination we also outline the **LU** *factorization method*.

The Jacobi and Gauss–Seidel schemes apply to systems of equations having $n \times n$ coefficient matrices $\mathbf{A} = [a_{ij}]$ with the property that for each i the coefficient a_{ii} in the ith row of $\mathbf{A}$ has magnitude $|a_i|$ greater than the sum of the magnitudes of the other coefficients in that same row. Such matrices are called *diagonally dominant* because in iterative schemes the elements a_{ii} in the leading diagonal dominate the calculations.

This means, for example, that the matrix

$$\mathbf{A} = \begin{bmatrix} a_{11} & a_{12} & a_{13} \\ a_{21} & a_{22} & a_{23} \\ a_{31} & a_{32} & a_{33} \end{bmatrix}$$

will be diagonally dominant if

$$|a_{11}| > |a_{12}| + |a_{13}|, \ |a_{22}| > |a_{21}| + |a_{23}| \ \text{and} \ |a_{33}| > |a_{31}| + |a_{32}|.$$

If any one of these conditions is violated then **A** will not be diagonally dominant. Sometimes when a system has a coefficient matrix that is not diagonally dominant it can be made so by changing the order of the equations. This is the case with the system

$$3x_1 + 6x_2 - x_3 = 1$$

$$2x_1 - 4x_2 - 7x_3 = 4$$

$$5x_1 + x_2 - 2x_3 = -3,$$

whose coefficient matrix is not diagonally dominant because in the first equation $|3| \not> |6| + |-1|$. This is sufficient to ensure the non-diagonal dominance of the system, though the condition is also violated by the second and third equations, because $|-4| \not> |2| + |-7|$ and $|-2| \not> |5| + |1|$. However, the system becomes diagonally dominant if the order of the equations is changed to

$$5x_1 + x_2 - 2x_3 = -3$$

$$3x_1 + 6x_2 - x_3 = 1$$

$$2x_1 - 4x_2 - 7x_3 = 4,$$

because now in the first equation $|5| > |1| + |-2|$, while in the second and third equations $|6| > |3| + |-1|$ and $|-7| > |2| + |-4|$.

Diagonal dominance is important because it is a condition that ensures the convergence of these numerical schemes. It is a *sufficient* condition for convergence, but not a necessary one.

Diagonally dominant matrices occur frequently in connection with certain numerical methods for the solution of differential equations where the matrices **A** that arise are both large and sparse (they contain numerous zero elements).

Both of the methods we now describe have essentially the same starting point. We shall work with a system of linear simultaneous equations of the form shown in (17.1). On the assumption that the system is diagonally dominant, we begin by rewriting it in the form

$$\begin{aligned}
x_1 &= (k_1 - a_{12}x_2 - a_{13}x_3 - \ldots - a_{1n}x_n)/a_{11},\\
x_2 &= (k_2 - a_{21}x_1 - a_{23}x_3 - \ldots - a_{2n}x_n)/a_{22},\\
&\ldots\\
x_n &= (k_n - a_{n1}x_1 - a_{n2}x_2 - \ldots - a_{nn-1}x_{n-1})/a_{nn}.
\end{aligned} \tag{17.2}$$

Now suppose the $(r-1)$th iteration gives rise to the approximations $x_1^{(r-1)}, x_2^{(r-1)}, \ldots, x_n^{(r-1)}$ to the required solutions $x_1, x_2, \ldots, x_n$. In the *Jacobi iterative method* the rth iterates are determined from the $(r-1)$th iterates by means of the algorithm

$$\begin{aligned}
x_1^{(r)} &= (k_1 - a_{12}x_2^{(r-1)} - a_{13}x_3^{(r-1)} - \ldots - a_{1n}x_n^{(r-1)})/a_{11},\\
x_2^{(r)} &= (k_2 - a_{21}x_1^{(r-1)} - a_{23}x_3^{(r-1)} - \ldots - a_{2n}x_n^{(r-1)})/a_{22},\\
x_3^{(r)} &= (k_3 - a_{31}x_1^{(r-1)} - a_{32}x_2^{(r-1)} - \ldots - a_{3n}x_n^{(r-1)})/a_{33},\\
&\ldots\\
x_n^{(r)} &= (k_n - a_{n1}x_1^{(r-1)} - a_{n2}x_2^{(r-1)} - \ldots - a_{nn-1}x_{n-1}^{(r-1)})/a_{nn}.
\end{aligned}$$

This has been obtained from the rewritten system (17.2) by simply employing the $(r-1)$th iterates in the right-hand side to determine the rth iterates. To start the iteration procedure any values may be assumed for $x_1^{(0)}, x_2^{(0)}, \ldots, x_n^{(0)}$. Unless some approximate solution is known, it is customary to start either by setting each $x_i^{(0)} = 1$ or, alternatively, each $x_i^{(0)} = 0$. The iteration process is terminated once the desired accuracy has been attained. This is achieved by stopping the calculation when the n numbers $|x_i^{(r)} - x_i^{(r-1)}|$, for $i = 1, 2, \ldots, n$, are all less than some predetermined small number $\varepsilon > 0$.

The *Gauss – Seidel iterative method* differs from the Jacobi method in that it uses each approximation $x_i^{(r)}$ obtained in the rth iteration *as soon* as it is available. The algorithm for the Gauss–Seidel method is thus as follows:

$$\begin{aligned}
x_1^{(r)} &= (k_1 - a_{12}x_2^{(r-1)} - a_{13}x_3^{(r-1)} - \ldots - a_{1n}x_n^{(r-1)})/a_{11},\\
x_2^{(r)} &= (k_2 - a_{21}x_1^{(r)} - a_{23}x_3^{(r-1)} - \ldots - a_{2n}x_n^{(r-1)})/a_{22},\\
x_3^{(r)} &= (k_3 - a_{31}x_1^{(r)} - a_{32}x_2^{(r)} - \ldots - a_{3_x}x_n^{(r-1)})/a_{33},\\
&\ldots\\
x_n^{(r)} &= (k_n - a_{n1}x_1^{(r)} - a_{n2}x_2^{(r)} - \ldots - a_{nn-1}x_{n-1}^{(r)})/a_{nn}.
\end{aligned}$$

The Gauss–Seidel iterative method is started and terminated in exactly the same fashion as the Jacobi method, though its rate of convergence is roughly double that of the Jacobi method. If the condition of diagonal dominance is not satisfied the successive iterations may diverge.

Example 17.1 Solve by the Jacobi and the Gauss–Seidel iterative methods the system of equations

$$\begin{aligned}
5x_1 + x_2 - x_3 &= 4\\
x_1 + 4x_2 + 2x_3 &= 15\\
x_1 - 2x_2 + 5x_3 &= 12,
\end{aligned}$$

which has the exact solution $x_1 = 1$, $x_2 = 2$ and $x_3 = 3$.

Solution As $|5| > |1| + |-1|$, $|4| > |1| + |2|$ and $|5| > |1| + |-2|$ the system is diagonally dominant and so each of the two iterative methods may be used. Both calculations start from the equations derived by rewriting the system in the form

$$x_1 = 0.8 - 0.2x_2 + 0.2x_3$$

$$x_2 = 3.75 - 0.25x_1 - 0.5x_3$$

$$x_3 = 2.4 - 0.2x_1 + 0.4x_2,$$

leading to:

Jacobi scheme

$$x_1^{(r)} = 0.8 - 0.2x_2^{(r-1)} + 0.2x_3^{(r-1)}$$
$$x_2^{(r)} = 3.75 - 0.25x_1^{(r-1)} - 0.5x_3^{(r-1)}$$
$$x_3^{(r)} = 2.4 - 0.2x_1^{(r-1)} + 0.4x_2^{(r-1)}$$

Gauss – Seidel scheme

$$x_1^{(r)} = 0.8 - 0.2x_2^{(r-1)} + 0.2x_3^{(r-1)}$$
$$x_2^{(r)} = 3.75 - 0.25x_1^{(r)} - 0.5x_3^{(r-1)}$$
$$x_3^{(r)} = 2.4 - 0.2x_1^{(r)} + 0.4x_2^{(r)}.$$

Starting by setting $x_i^{(0)} = 1$, for $i = 1, 2, 3$, the result of iteration by each method is as follows and clearly shows the faster convergence of the Gauss–Seidel method.

Iteration number	*Jacobi*			*Gauss–Seidel*		
r	$x_1^{(r)}$	$x_2^{(r)}$	$x_3^{(r)}$	$x_1^{(r)}$	$x_2^{(r)}$	$x_3^{(r)}$
0	1.0	1.0	1.0	1.0	1.0	1.0
1	0.8	3.0	2.6	0.8	3.05	3.46
2	0.72	2.25	3.44	0.882	1.7995	2.934
3	1.038	1.85	3.632	1.0288	2.0211	3.0026
4	1.1564	1.6745	2.9324	0.9963	1.9996	3.0005
5	1.0516	1.9947	2.8385	1.0002	1.9997	2.9999
6	0.9688	2.0681	2.9876	1.0	2.0	3.0
7	0.9839	2.0140	3.0334			
8	1.0039	1.9873	3.0088			

■

On occasion the solution of a system of equations becomes very difficult because it is unusually sensitive to the values of the coefficients involved, and hence to round-off errors. In general this is associated with systems for which the determinant of the coefficient matrix is close to zero. Such systems are said to be *ill-conditioned*, and it is necessary to work to a very high accuracy if a reliable solution is to be obtained. A simple example is provided by the system

$$x_1 + x_2 = 4$$

$$x_1 + 1.0001\,x_2 = 1,$$

for which the determinant of the coefficient matrix is equal to 0.0001, while the exact solution is $x_1 = 30\,004$ and $x_2 = -30\,000$. Changing the system very slightly to

$$x_1 + x_2 = 4$$

$$x_1 + 1.00015\, x_2 = 1,$$

produces a totally different exact solution $x_1 = 20\,004$ and $x_2 = -20\,000$. Thus, in either case, a slight numerical error due, say, to round-off, could give rise to an extremely inaccurate result. In graphical terms the reason for this difficulty is clear. The solution of each system may be represented by the point of intersection of two straight lines; these lines are almost parallel, so their point of intersection is difficult to determine accurately.

This shows very clearly that not only does the solution of such a system need to be computed to an appropriate accuracy, but also if the coefficients involved are themselves obtained by computation, then the accuracy required for their determination is of crucial importance. When computations of this type are undertaken on a large computer, *double precision arithmetic* is usually used. That is to say, instead of working to eight significant figures, 16 will be used. Sometimes even greater precision is needed.

An example of a class of $n \times n$ matrices whose determinant becomes vanishingly small as n increases is provided by the *Hilbert matrix* $\mathbf{H}_n$ in which the element h_{ij} is given by

$$h_{ij} = \frac{1}{i + j - 1} \qquad \text{for } i, j = 1, 2, \ldots, n.$$

Thus

$$\mathbf{H}_2 = \begin{bmatrix} 1 & \frac{1}{2} \\ \frac{1}{2} & \frac{1}{3} \end{bmatrix}, \quad \mathbf{H}_3 = \begin{bmatrix} 1 & \frac{1}{2} & \frac{1}{3} \\ \frac{1}{2} & \frac{1}{3} & \frac{1}{4} \\ \frac{1}{3} & \frac{1}{4} & \frac{1}{5} \end{bmatrix} \quad \text{and} \quad \mathbf{H}_4 = \begin{bmatrix} 1 & \frac{1}{2} & \frac{1}{3} & \frac{1}{4} \\ \frac{1}{2} & \frac{1}{3} & \frac{1}{4} & \frac{1}{5} \\ \frac{1}{3} & \frac{1}{4} & \frac{1}{5} & \frac{1}{6} \\ \frac{1}{4} & \frac{1}{5} & \frac{1}{6} & \frac{1}{7} \end{bmatrix}.$$

The values of the determinants of these three matrices are

$$|\mathbf{H}_2| = 1/12 = 0.8333333\ldots, \quad |\mathbf{H}_3| = 4.62962963 \times 10^{-4} \quad \text{and}$$

$$|\mathbf{H}_4| = 1.65343915 \times 10^{-7}.$$

Thus for any matrix such as this, attempting to find its inverse even when n is comparatively small will give rise to extremely inaccurate results unless very high precision is used. The same conclusion is true, of course, if Gaussian elimination is applied to a system of equations with a coefficient matrix of this type that has a determinant close to zero.

To conclude this section, we discuss a method that is often used for the solution of large systems of equations

$$\mathbf{AX} = \mathbf{B};$$

this is the so-called **LU** *factorization method*. The idea underlying this method is to adapt the Gaussian elimination process to arrive at a factorization of the $n \times n$ matrix $\mathbf{A}$ of the form

$$\mathbf{A} = \mathbf{LU}$$

where $\mathbf{L}$ is a *lower-triangular matrix* with 1s on its leading diagonal, and $\mathbf{U}$ is an *upper-triangular matrix*. The system of equations

$$\mathbf{AX} = \mathbf{B}$$

is solved by first finding the column vector $\mathbf{Y}$ that is the solution of

$$\mathbf{LY} = \mathbf{B},$$

and then with this $\mathbf{Y}$ solving the system of equations

$$\mathbf{UX} = \mathbf{Y}.$$

Although this approach might seem cumbersome it is, in fact, very efficient. This is because once $\mathbf{L}$ and $\mathbf{U}$ have been found the elements of $\mathbf{Y}$ follow by use of *forward substitution*, while the elements of $\mathbf{X}$, the required solution vector, follow by *backward substitution*. The method also has the advantage that once $\mathbf{L}$ and $\mathbf{U}$ are known, the vector $\mathbf{B}$ can be changed without the need to redetermine $\mathbf{L}$ and $\mathbf{U}$.

In the following algorithm we describe, without justification, the simplest case in which none of the multipliers m_{ij} vanish.

LU factorization

1. Apply Gaussian elimination to $\mathbf{A}$, and at the completion of the process the matrix obtained is the required upper-triangular matrix $\mathbf{U}$.
2. Record the multipliers m_{ij} used in step 1 to reduce $\mathbf{A}$ to $\mathbf{U}$, and record them as follows to form the required lower-triangular matrix $\mathbf{L}$.

$$\mathbf{L} = \begin{bmatrix} 1 & 0 & 0 & \dots & 0 \\ m_{21} & 1 & 0 & \dots & 0 \\ m_{31} & m_{32} & 1 & \dots & 0 \\ \dots & \dots & \dots & \dots & \dots \\ m_{n1} & m_{n2} & m_{n3} & \dots & 1 \end{bmatrix}.$$

3. The required factorization of $\mathbf{A}$ is then

$$\mathbf{A} = \mathbf{LU}.$$

Example 17.2 Obtain the LU factorization of

$$\mathbf{A} = \begin{bmatrix} 2 & 2 & 1 & 3 \\ 1 & 2 & 2 & 1 \\ 4 & 1 & 0 & 2 \\ 2 & 0 & 2 & 1 \end{bmatrix},$$

and hence solve the system of equations

$$\begin{aligned}2x_1 + 2x_2 + x_3 \quad + 3x_4 &= 1\\ x_1 \quad + 2x_2 + 2x_3 + x_4 &= 4\\ 4x_1 + x_2 \qquad\qquad + 2x_4 &= -1\\ 2x_1 \qquad\quad + 2x_3 + x_4 &= 0.\end{aligned}$$

Solution Using the notation for now operations introduced in Chapter 9, we have

$$\begin{bmatrix}2&2&1&3\\1&2&2&1\\4&1&0&2\\2&0&2&1\end{bmatrix}\quad \begin{matrix}R_1^1 = R_1(m_{11}=1)\\ R_2^1 = R_2 - \frac{1}{2}R_1(m_{21}=\frac{1}{2})\\ R_3^1 = R_3 - 2R_1(m_{31}=2)\\ R_4^1 = R_4 - R_1(m_{41}=1)\end{matrix}\quad \begin{bmatrix}2&2&1&3\\0&1&\frac{3}{2}&-\frac{1}{2}\\0&-3&-2&-4\\0&-2&1&-2\end{bmatrix}$$

$$\begin{matrix}R_1^1 = R_1\\ R_2^1 = R_2(m_{22}=1)\\ R_3^1 = R_3 + 3R_2(m_{32}=-3)\\ R_4^1 = R_4^1 + 2R_2(m_{42}=-2)\end{matrix}\quad \begin{bmatrix}2&2&1&3\\0&1&\frac{3}{2}&-\frac{1}{2}\\0&0&\frac{5}{2}&-\frac{11}{2}\\0&0&4&-3\end{bmatrix}$$

$$\begin{matrix}R_1^1 = R_1\\ R_2^1 = R_2\\ R_3^1 = R_3(m_{33}=1)\\ R_4^1 = R_4 - \frac{8}{5}R_3(m_{43}=\frac{8}{5})\end{matrix}\quad \begin{bmatrix}2&2&1&3\\0&1&\frac{3}{2}&-\frac{1}{2}\\0&0&\frac{5}{2}&-\frac{11}{2}\\0&0&0&\frac{29}{5}\end{bmatrix}.$$

Thus

$$\mathbf{L} = \begin{bmatrix}1&0&0&0\\\frac{1}{2}&1&0&0\\2&-3&1&0\\1&-2&\frac{8}{5}&1\end{bmatrix} \quad\text{and}\quad \mathbf{U} = \begin{bmatrix}2&2&1&3\\0&1&\frac{3}{2}&-\frac{1}{2}\\0&0&\frac{5}{2}&-\frac{11}{12}\\0&0&0&\frac{29}{5}\end{bmatrix}.$$

As

$$\mathbf{X} = \begin{bmatrix}x_1\\x_2\\x_3\\x_4\end{bmatrix}, \quad \mathbf{B} = \begin{bmatrix}1\\4\\-1\\0\end{bmatrix},$$

by setting

$$\mathbf{Y} = \begin{bmatrix}y_1\\y_2\\y_3\\y_4\end{bmatrix}$$

we see that $\mathbf{LY} = \mathbf{B}$ becomes

$$\begin{aligned}
y_1 \quad\quad\quad\quad &= 1\\
\tfrac{1}{2}y_1 + y_2 \quad &= 4\\
2y_1 - 3y_2 + y_3 &= -1\\
y_1 - 2y_2 + \tfrac{8}{5}y_3 + y_4 &= 0.
\end{aligned}$$

Using forward substitution we find that

$$y_1 = 1,\ y_2 = \tfrac{7}{2},\ y_3 = \tfrac{15}{2} \quad \text{and} \quad y_4 = -6.$$

Thus the system $\mathbf{UX} = \mathbf{Y}$ becomes

$$\begin{aligned}
2x_1 + 2x_2 + x_3 + 3x_4 &= 1\\
x_2 + \tfrac{3}{2}x_3 - \tfrac{1}{2}x_4 &= \tfrac{7}{2}\\
\tfrac{5}{2}x_3 - \tfrac{11}{2}x_4 &= \tfrac{15}{2}\\
\tfrac{29}{5}x_4 &= -6.
\end{aligned}$$

Using backward substitution, we arrive at the solution

$$x_4 = -\tfrac{30}{29},\quad x_3 = \tfrac{21}{29},\quad x_2 = \tfrac{55}{29} \quad \text{and} \quad x_1 = \tfrac{-6}{29}. \quad \blacksquare$$

17.3 Interpolation

The result of the numerical computation of a function $y = f(x)$ for x equal to one of the discrete set of values $x_1, x_2, \ldots x_n$ is a set of numbers y_1, $y_2, \ldots, y_n$, where $y_r = f(x_r)$, for $r = 1, 2, \ldots, n$. If a value of y is required corresponding to the number $x = \xi$ which is intermediate between any two of the numbers $x_1, x_2, \ldots, x_n$ at which y is known, then either the computation must be repeated for $x = \xi$ or the value $f(\xi)$ must be deduced from the known values $y_1, y_2, \ldots, y_n$. Any method by which such an intermediate value may be deduced from the known values is called *interpolation*. Frequently interpolation is essential since the determination of the appropriate functional value might be impossible. This happens, for example, when the numbers y_r corresponding to x_r are only available in tabular form, possibly as a result of experimental observations.

17.3.1 Linear interpolation

The simplest interpolation method is called *linear interpolation*. This involves approximating the graph of $y = f(x)$ by a straight line between each successive pair of points at which y is known. Consider the two points $P(x_0, y_0)$ and $Q(x_1, y_1)$ shown in Fig. 17.1.

Linear interpolation between points P and Q involves approximating the true value of y at S by the value y_R at R on the chord joining P and Q. The equation of the line PQ is easily seen to be

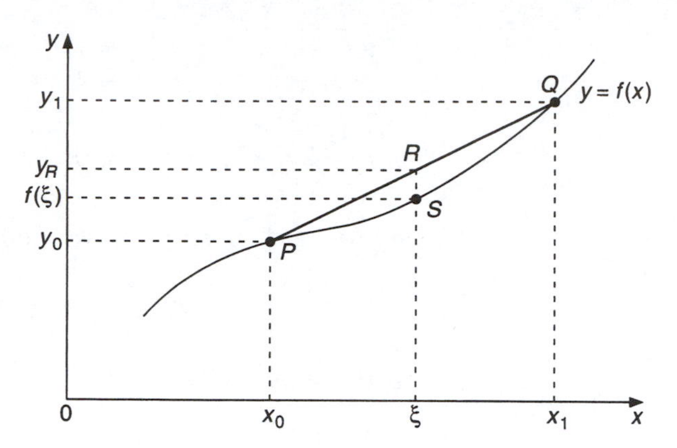

Fig. 17.1 Linear interpolation.

$$y = y_0 + \left(\frac{y_1 - y_0}{x_1 - x_0}\right)(x - x_0),$$

so that the approximation y_R to $f(\xi)$ is just

$$f(\xi) \approx y_0 + \left(\frac{y_1 - y_0}{x_1 - x_0}\right)(\xi - x_0). \tag{17.3}$$

The error involved has magnitude equal to the length RS.

Example 17.3 By using linear interpolation find the value of y corresponding to $x = 0.14$, given that $y(0.1) = 1.1052$ and $y(0.2) = 1.2214$.

Solution Setting $x_0 = 0.1$, $x_1 = 0.2$, $\xi = 0.14$, $y_0 = 1.1052$ and $y_1 = 1.2214$ in Eqn (17.3) gives $y(0.14) \approx 1.1517$ when rounded off to four decimal places. In actual fact the function $y = f(x)$ used here was $y = e^x$, and the correct value for $y(0.14) = 1.1503$. The precision used in this calculation is four decimal places, but the accuracy of the result is only two decimal places. ■

17.3.2 Lagrange interpolation

One of the more general forms of interpolation that uses n points at which the function $y = f(x)$ is known is *Lagrange interpolation*. We only give its form for $n = 3$, but its generalization to larger n is obvious. We assume that $f(x)$ is a continuous function in the interval $x_0 \leq x \leq x_2$ and that $f(x_0) = y_0$, $f(x_1) = y_1$ and $f(x_2) = y_2$, with $x_0 < x_1 < x_2$. Then the approximate value of $f(x)$ for $x_0 \leq x \leq x_2$ is given by

$$f(x) \approx \frac{(x-x_1)(x-x_2)}{(x_0-x_1)(x_0-x_2)}y_0 + \frac{(x-x_0)(x-x_2)}{(x_1-x_0)(x_1-x_2)}y_1 + \frac{(x-x_0)(x-x_1)}{(x_2-x_0)(x_2-x_1)}y_2. \tag{17.4}$$

This result is known as the *three-point Lagrange interpolation* formula. Like linear interpolation, this method has the property that $f(x)$ assumes precisely the given values of y at the given values of x. It is also useful because the three points x_0, x_1 and x_2 need not be equally spaced.

Example 17.4 Use the three point Lagrangian interpolation formula to find $y(0.1424)$, given that $y(0.1)=1.1052$, $y(0.2)=1.2214$ and $y(0.3)=1.3499$.

Solution Substituting into (17.4) gives the result

$$y(0.1424) \approx 0.4539 \times 1.1052 + 0.6682 \times 1.2214 - 0.1221 - 1.3499 = 1.1530.$$

Here again the function $y=e^x$ was used, so that this time when rounded off the result is correct to four decimal places. This is an example where both the precision and the accuracy are the same. ■

It is inadvisable to use Lagrange interpolation for more than three or four points since this can lead to large errors. This comes about because the method fits a high-order polynomial to the points in question, and the effect of forcing the interpolating curve through the known points can be to introduce large oscillations in between them.

Unlike other more advanced methods of interpolation, there is no error estimate available for the Lagrange interpolation formula. However, its simplicity, coupled with the flexibility that results from the fact that the points need not be equally spaced, makes it very useful when the function that is to be interpolated is known to be smoothly varying. The Lagrange interpolation formula for n intervals ($n+1$ points) is often denoted by $L_n(x)$, so that the right-hand side of result (17.4) may be written $L_2(x)$.

17.3.3 Extrapolation

When the function $y=f(x)$ is known for the discrete set of values $y_0, y_1, \ldots, y_n$, with $y_r=f(x_r)$ for $r=0, 1, \ldots, n$, the deduction of the value of y for $x=\xi$ when ξ lies outside the interval $x_0 \le x \le x_n$ is called *extrapolation*. It is a process that is liable to considerable error and should only be attempted when absolutely necessary. Either of the interpolation methods already discussed may be used for extrapolation provided $x=\xi$ does not lie too far outside the interval in question. It is probably best to use the three-point Lagrange interpolation formula $L_2(x)$ for this purpose, since it fits a parabola through the three points and so takes some account of the curvature of the graph of $y=f(x)$.

17.3.4 Approximation by spline functions

The rapid development of computer-aided design (CAD) has given rise to the need for a new form of interpolation that allows complicated curves extending over large intervals, and often exhibiting rapid changes of curvature, to be approximated with a high degree of accuracy. The method that is used is called *spline function approximation*. The idea underlying this method is to allow a curve whose form is usually only known graphically, and so is specified in tabular form as a set of ordinates at points in an interval $a \leq x \leq b$, to be approximated in a piecewise manner by using a set of low-degree polynomials $S_0(x)$, $S_1(x), \ldots, S_{n-1}(x)$ in the adjacent sub-intervals $a \leq x \leq t_1$, $t_1 \leq x \leq t_2, \ldots, t_{n-1} \leq x \leq b$.

The special feature of a spline function approximation is that not only is it continuous at the points $t_1, t_2, \ldots, t_{n-1}$ where adjacent approximations meet, so that $S_i(t_i) = S_{i+1}(t_i)$ for $i = 1, 2, \ldots, n-1$, but also one or more of the derivatives of the functions $S_i(x)$ and $S_{i+1}(x)$ are equal at these same points. Thus a spline function allows a complicated curve to be approximated in a smooth manner by a composite function $S(x)$ defined in a piecewise fashion over the interval $a \leq x \leq b$ in such a way that it is continuous and at least once differentiable throughout the interval.

Spline functions have the advantage of allowing a curve of complicated shape that is specified at many points to be approximated with a high degree of accuracy, while avoiding the excessive oscillations that would be introduced if the approximation were to be attempted by using high-degree polynomials.

In this section it is only possible to introduce the basic idea of spline function approximation so, for simplicity, attention will be confined to a special type of quadratic spline in which the ith segment of the interval $a \leq x \leq b$ is approximated by a function $S_i(x)$ of the form $S_i(x) = p_i x^2 + a_i x + r_i$ for $i = 0, 1, 2, \ldots, n-1$.

We now introduce some of the terminology of spline functions. The points t_i at which adjacent approximations meet are called the *knots* of the spline function. As a quadratic spline is involved, the individual functions $S_i(x)$ in the composite function $S(x)$ must be constructed so that $S(x)$ is continuous and has a continuous first-order derivative at each knot. The points on the x-axis at which the ordinates $y_0, y_1, \ldots, y_{n+1}$ of the curve to be approximated are specified are called the *nodes* of the spline function.

In the special case considered here the location of the knots must be specified first, and once this has been done the positions of the nodes can be determined. The knots should be placed along the curve that is to be approximated so that there is one at each end of the curve, with the others positioned so that each segment of the curve between a pair of knots can be well approximated by a quadratic curve. A knot should always be located at any point of inflection of the curve that is to be represented by the spline function approximation. The nodes at which the ordinates of the curve are to be recorded are placed one at each of the curve and one at the mid-point between each pair of knots. Thus if there are $n+1$ knots with the x-coordinates $a=t_0, t_1, \ldots, t_n=b$, the x-coordinates $\tau_0, \tau_1, \ldots, t_{n+1}$ of the nodes will be given by

$$\tau_0 = t_0 = a, \quad \tau_{n+1} = t_n = b$$

and

$$\tau_i = \tfrac{1}{2}(t_{i-1} + t_i) \quad \text{for} \quad i = 1, 2, \ldots, n.$$

The input data for the calculation are thus

(i) the x-coordinates $t_0, t_1, \ldots, t_n$ of the $n+1$ knots;
(ii) a table listing the x-coordinates τ_i of each of the $n+1$ nodes together with the corresponding ordinates y_i of the curve to be approximated:

x	τ_0	$\tau_1 \ldots \tau_n + 1$
y	y_0	$y_1 \ldots y_n + 1$

The subsequent calculation will also need the lengths $h_0, h_1, \ldots, h_{n-1}$ of the intervals between the knots given by

$$h_i = (t_{i+1} - t_i) \quad \text{for} \quad i = 1, 2, \ldots, n-1.$$

Figure 17.2 illustrates the location of the knots (K) and the nodes (N) for the case of five knots, corresponding to $n=4$.

We now seek a piecewise continuous once differentiable function $S(x)$ of the form

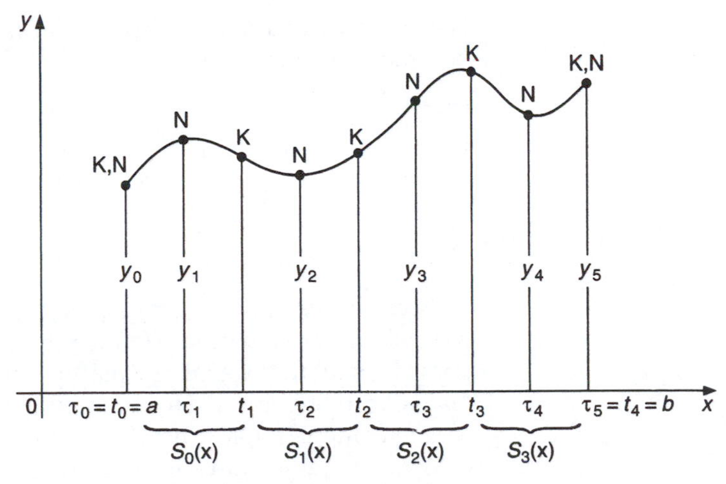

Fig. 17.2 Location of knots (K) and nodes (N) for $n=4$.

$$S(x) = \begin{cases} S_0(x), & t_0 \le x \le t_1 \\ S_1(x), & t_1 \le x \le t_2 \\ \vdots & \\ S_{n-1}(x), & t_{n-1} \le x \le t_n, \end{cases}$$

in which

$$S_{i-1}(t_i) = S_i(t_i) \quad \text{for } i = 1, 2, \ldots, n-1 \quad \text{(continuity at knot)}$$

$$\left[\frac{\mathrm{d}}{\mathrm{d}x} S_{i-1}(x)\right]_{x=t_i} = \left[\frac{\mathrm{d}}{\mathrm{d}x} S_i(x)\right]_{x=t_i} = z_i(\text{say}), \quad \text{for} i = 1,\ 2,\ \ldots,\ n-1$$

(equality of gradient z_i to left and right of knot)

with

$$S_0(t_0) = y_0 \qquad \text{and} \qquad S_{n-1}(t_n) = y_n, \qquad \text{(end conditions)}$$

where the gradients $z_0,\ z_1,\ \ldots,\ z_{n-1}$ of the approximation at the knots have yet to be determined.

When the mathematical consequences of these conditions are followed through, they show that the quadratic interpolation polynomial $S_i(x)$ can be written

$$S_i(x) = y_{i+1} + \tfrac{1}{2}(z_i + z_{i+1})(x - \tau_{i+1}) + \frac{1}{2h_i}(z_{i+1} - z_i)(x - \tau_{i+1}),$$

for $t_i \le x \le t_{i+1}$. The gradients $z_0,\ z_1,\ \ldots,\ z_{n-1}$ of $S(x)$ at the $n+1$ knots that enter into the interpolation functions $S_i(x)$ are found by solving the system of $n+1$ equations

$$\begin{aligned} &3h_0 z_0 + h_0 z_1 = 8(y_1 - y_0) \\ &h_{i-1} z_{i-1} + 3(h_{i-1} + h_i) z_i + h_i z_{i+1} = 8(y_{i+1} - y_i) \ \text{ for } i = 1,\ 2, \ldots, n-1 \\ &h_{n-1} z_{n-1} + 3h_{n-1} z_n = 8(y_{n+1} - y_n). \end{aligned}$$

When these equations are written in matrix form their simple structure becomes obvious:

$$\begin{bmatrix} 3h_0 & h_0 & 0 & 0 & \ldots & 0 & 0 & 0 \\ h_0 & 3(h_0 + h_1) & h_1 & 0 & \ldots & 0 & 0 & 0 \\ 0 & h_1 & 3(h_1 + h_2) & h_2 & \ldots & 0 & 0 & 0 \\ 0 & 0 & h_2 & 3(h_2 + h_3) & \ldots & 0 & 0 & 0 \\ \ldots & \ldots & \ldots & \ldots & & \ldots & \ldots & \ldots \\ 0 & 0 & 0 & 0 & \ldots & h_{n-2} & 3(h_{n-2} + h_{n-1}) & h_{n-1} \\ 0 & 0 & 0 & 0 & \ldots & 0 & h_{n-1} & 3h_{n-1} \end{bmatrix} \begin{bmatrix} z_0 \\ z_1 \\ z_2 \\ z_3 \\ \ldots \\ z_{n-1} \\ z_n \end{bmatrix} \begin{bmatrix} 8(y - y_0) \\ 8(y_2 - y_1) \\ 8(y_3 - y_2) \\ 8(y_4 - y_3) \\ \ldots \\ 8(y_n - y_{n-1}) \\ 8(y_{n+1} - y_n) \end{bmatrix}$$

Once the values of $z_0,\ z_1,\ \ldots,\ z_{n-1}$ have been found, the composite piecewise approximation $S(x)$ can be constructed. When the number of knots is small the system of equations determining the z_i can be solved by Gaussian elimination. However, if the number of knots is very large, special computer algorithms are available that minimize the computation involved by taking advantage of the fact that the matrix of coefficients is tridiagonal in form and so contains many zeros.

In practice quadratic spline approximations are not usually used, and instead the more flexible cubic spline based on a cubic approximation takes its place. This is because a cubic spline allows continuity of the function and its first *two* derivatives at knots, thereby improving the representation of the curvature of the function that is being approximated. In addition, unlike the special quadratic spline just described, cubic splines allow a choice of end condition at $x = a$ and $x = b$ which is desirable in many applications.

To illustrate the application of this quadratic spline function, we shall approximate the curve shown in Fig. 17.3 on which five knots have been selected and the corresponding nodes marked.

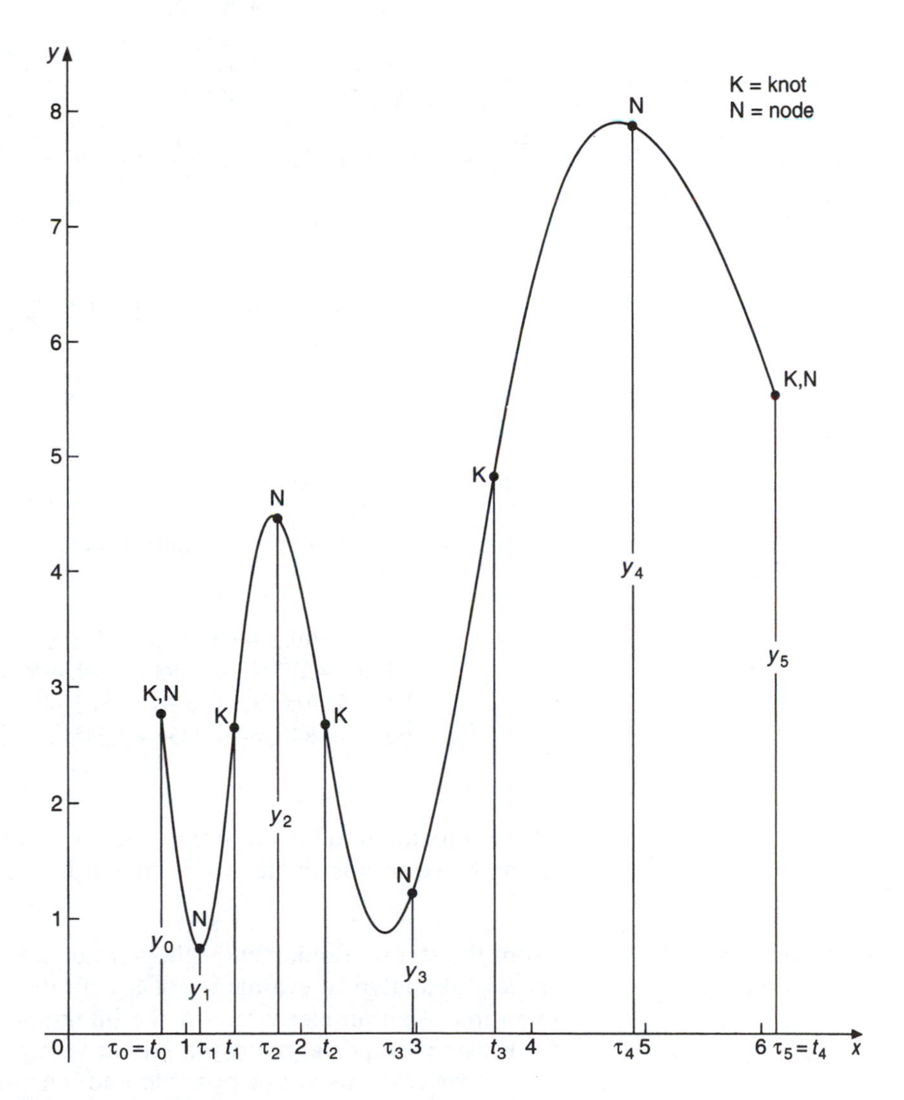

Fig. 17.3 The function to be approximated.

The input data for the calculation to fit a spline function with four segments to this curve are as follows:

(i) the positions of the knots: $t_0 = 0.833$, $t_1 = 1.433$, $t_2 = 2.233$, $t_3 = 3.7$, $t_4 = 6.167$;
(ii) the positions of the nodes and the corresponding ordinates of the curve:

i	τ_i	y_i
0	0.833	2.667
1	1.133	0.7
2	1.833	4.467
3	2.967	1.133
4	4.933	7.9
5	6.167	5.533

$h_0 = 0.6$, $\quad h_1 = 0.8$

$h_2 = 1.467$, $\quad h_3 = 2.467$

The gradients z_i follow by solving the system of equations

$$\begin{bmatrix} 1.8 & 0.6 & 0 & 0 & 0 \\ 0.6 & 4.2 & 0.8 & 0 & 0 \\ 0 & 0.8 & 6.801 & 1.467 & 0 \\ 0 & 0 & 1.467 & 1.802 & 2.467 \\ 0 & 0 & 0 & 2.467 & 7.401 \end{bmatrix} \begin{bmatrix} z_0 \\ z_1 \\ z_2 \\ z_3 \\ z_4 \end{bmatrix} = \begin{bmatrix} -15.736 \\ 30.136 \\ -26.672 \\ 54.136 \\ -18.936 \end{bmatrix}$$

which has the solution $z_0 = -12.123$, $z_1 = 10.143$, $z_2 = -6.489$, $z_3 = 6.373$, $z_4 = -4.683$.

The spline function approximation is thus

$$S(x) = \begin{cases} 0.7 - 0.99(x - 1.833) + 18.555(x - 1.833)^2, & 0.833 \leq x \leq 1.433 \\ 4.467 + 1.827(x - 1.833) - 10.395(x - 1.833)^2, & 1.433 \leq x \leq 2.233 \\ 1.133 - 0.058(x - 2.967) + 4.384(x - 2.967)^2, & 2.233 \leq x \leq 3.7 \\ 7.9 + 0.845(x - 4.933) - 2.241(x - 4.933)^2, & 3.7 \leq x \leq 6.167. \end{cases}$$

The approximation to the curve shown in Fig. 17.3 produced by this spline function approximation is shown in Fig. 17.4.

17.4 Numerical integration

From the second fundamental theorem of calculus we have seen that the successful analytical evaluation of a definite integral involves the determination of an antiderivative of the integrand. Although in many practical cases of importance an antiderivative can be found, the fact remains that in general this is not possible and Theorem 7.7 is therefore of no avail. Such, for example, is the case with an integral as simple as

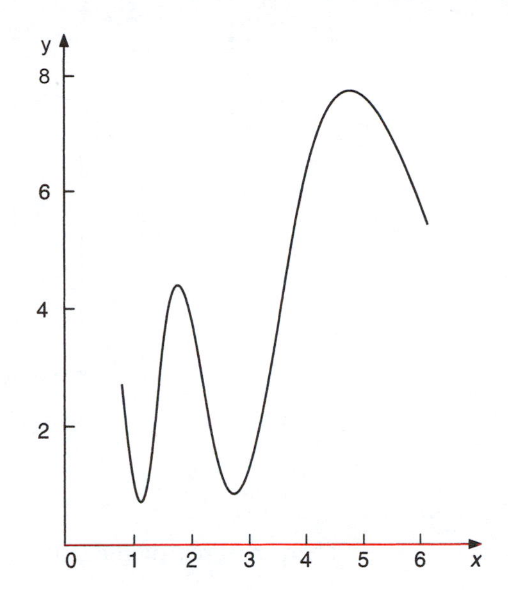

Fig. 17.4 The quadratic spline function approximation.

$$\int_1^3 e^{-x^2} dx,$$

for although an antiderivative of e^{-x^2} certainly exists on theoretical grounds, it is not expressible in terms of elementary functions.

Of the many possible methods whereby a numerical estimate of the value of a definite integral may be made, we choose to mention only the very simplest ones here. The general process of evaluating a definite integral by numerical means will be referred to as *numerical integration*, though the original term *numerical quadrature* is still often employed for such a process. The matter of the accuracy of these methods will be taken up in section 17.4.4.

17.4.1 Trapezoidal rule

Although a strictly analytical derivation of the so-called *trapezoidal rule* for integration may be given we shall not use this approach, and instead make appeal to the area representation of a definite integral. Consider Fig. 17.5, and let us estimate the shaded area below the curve $y = f(x)$ which we know has the value

$$\int_a^b f(x) dx.$$

Let us begin by taking any set of $n+1$ points $a = x_0 < x_1 < \ldots < x_n = b$, and on each interval $[x_{i-1}, x_i]$, approximate the true area above it by the trapezium obtained by replacing the arc of the curve through the points $(x_{i-1}, f(x_{i-1}))$, $(x_i, f(x_i))$ by the chord joining these two points.

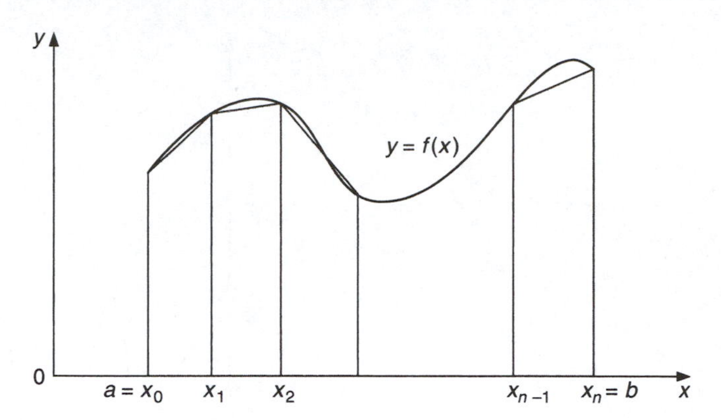

Fig. 17.5 Trapezoidal approximation of area.

Then the area of the trapezium on the interval $[x_{i-1}, x_i]$ is

$$\tfrac{1}{2}(f(x_{i-1}) + f(x_i))\Delta x_i,$$

where $\Delta x_i = x_i - x_{i-1}$.

Thus, adding the n contributions of this type, we arrive at the *general trapezoidal rule*

$$\int_a^b f(x)\mathrm{d}x \approx \tfrac{1}{2}(f(x_0) + f(x_1))\Delta x_1 + \tfrac{1}{2}(f(x_1) + f(x_2))\Delta x_2 + \ldots$$

$$+ \tfrac{1}{2}(f(x_{n-1}) + f(x_n))\Delta x_n. \qquad (17.5)$$

If the interval $[a, b]$ is divided into n equal parts of length $h = (b - a)/n$, then (17.5) becomes the *trapezoidal rule for equal intervals*

$$\int_a^b f(x)\mathrm{d}x = h[\tfrac{1}{2}(f(x_0) + f(x_1) + f(x_2) + \ldots + f(x_{n-1})$$

$$+ \tfrac{1}{2}f(x_n)] + E(h), \qquad (17.6)$$

where an equality sign has now been used because we have included the *error term* $E(h)$, which recognizes that the error is, in part, dependent on the magnitude of h. The multipliers $\frac{1}{2}$, 1, 1, ..., $\frac{1}{2}$ of the numbers $f(x_0)$, $f(x_1), f(x_2), \ldots, f(x_{n-1}), f(x_n)$ are called the *weights* for the trapezoidal rule.

The evaluation of a definite integral by means of the trapezoidal rule is best accomplished by means of the steps set out in the following rule. For the sake of completeness, step 5 provides an error estimate for the trapezoidal rule, the derivation of which is outlined in Problem 17.12.

Rule 17.1 Integration by the trapezoidal rule

To determine

$$\int_a^b f(x)\mathrm{d}x$$

by means of the trapezoidal rule it is necessary to proceed as follows.

1. Choose the number n of equal-length intervals into which the interval of integration $a \le x \le b$ is to be divided.
2. Calculate $h = (b - a)/n$.
3. Setting $x_r = a + rh$, calculate the $n+1$ numbers

   $$f(x_0),\ f(x_1),\ f(x_2),\ \ldots,\ f(x_n)$$

   where, of course, $f(x_0) = f(a)$ and $f(x_n) = f(b)$.
4. Calculate the sum

   $$S = \tfrac{1}{2}f(x_0) + f(x_1) + f(x_2) + \ldots + f(x_{n-1}) + \tfrac{1}{2}f(x_n).$$

   Then the estimate of the definite integral provided by the trapezoidal rule is

   $$\int_a^b f(x)\mathrm{d}x \approx hS.$$

5. The magnitude of the error $E(h)$ involved when the trapezoidal rule is used is such that

   $$\frac{nh^3 m}{12} < E(h) < \frac{nh^3 M}{12},$$

   where $m = \min|f''(x)|$ and $M = \max|f''(x)|$ for $a \le x \le b$.

When the maximum and minimum values of $|f''(x)|$on $a \le x \le b$ are not easily determined, the error in the trapezoidal rule may be approximated by

$$E(h) \approx \frac{nh^3 k}{12},$$

where

$$k = \frac{1}{3}\left[|f''(a)| + \left|f''\left(\frac{a+b}{2}\right)\right| + |f''(b)|\right].$$

This replaces m and M by the average of the values of $|f''(x)|$ at the end points and mid-point of the interval of integration.

Example 17.5 Calculate the definite integral

$$I = \int_1^6 \frac{\mathrm{d}x}{1 + x^2}$$

by means of the trapezoidal rule using five equal-length intervals, and compare the result with the exact analytical value. Estimate the error by using the result of step 5 of Rule 17.1 and show that it is compatible with the actual error.

Solution We start by tabulating the functional values of the integrand at intervals of $h = 1$. Then in the column adjacent to each functional value we enter the appropriate trapezoidal weight w_i. The last column contains the products of the functional values and their weights, and the sum of these entries is the quantity

$$S = \tfrac{1}{2}f(1) + f(2) + f(3) + f(4) + f(5) + \tfrac{1}{2}f(6).$$

The estimate of the integral obtained by the trapezoidal rule is the product hS, where in this case $h = 1$.

i	x_i	$f(x_i) = 1/(1 + x_i^2)$	Weight w_i	$w_i f(x_i)$
0	1.0	0.5	0.5	0.25
1	2.0	0.2	1	0.2
2	3.0	0.1	1	0.1
3	4.0	0.05 882	1	0.05 882
4	5.0	0.03 846	1	0.03 846
5	6.0	0.02 703	0.5	0.01 352

Thus

$$S = \sum_{i=0}^{5} w_i f(x_i) = 0.66\,080.$$

Hence the estimate of the definite integral provided by the trapezoidal rule is

$$\int_1^6 \frac{\mathrm{d}x}{1 + x^2} \approx hS = 0.66\,080,$$

because in this case $h = 1$.

The exact result is, of course, $\arctan 6 - \arctan 1 = 0.62025$, so the magnitude of the actual error is $E(h) = |0.62025 - 0.66080| = 0.04055$.

To estimate the error using step 5 of Rule 1 we begin by computing $f''(x)$, which is given by

$$f''(x) = \frac{6x^2 - 2}{(1 + x^2)^3}.$$

Inspection shows that in the interval of integration $1 \leq x \leq 6$, $|f''(x)|$ is greatest when $x = 1$ and least when $x = 6$, so $m = \min |f''(x)| = |f''(6)| = 0.00422$ and $M = \max|f''(x)| = |f''(1)| = 0.5$. Thus from step 5 of Rule 17.1, as $n = 5$ and $h = 1$, the magnitude of the error must be such that

$$\frac{5m}{12} < E(h) < \frac{5M}{12},$$

and thus

$$0.00176 < E(h) < 0.20833.$$

The actual magnitude of the error was seen to be $E(h) = 0.04055$ which agrees with the above inequality. ■

17.4.2 Simpson's rule

A different approach involves dividing $[a, b]$ into an even number of sub-intervals of equal length h, and then approximating the function over consecutive pairs of sub-intervals by a quadratic polynomial. That is to say fitting a parabola to the three points $(a, f(a))$, $(a+h, f(a+h))$, $(a+2h, f(a+2h))$ comprising the first two sub-intervals, and therefore repeating the process until the whole of the interval $[a, b]$ has been covered. The value of the definite integral can then be estimated by integrating the successive quadratic approximations over their respective intervals of length $2h$ and adding the results. This simple idea leads to *Simpson's rule* for numerical integration which we now formulate in analytical terms.

Consider the first interval $[a, a+2h]$, and represent the function $y = f(x)$ in this interval by the quadratic

$$y = c_0 + c_1x + c_2x^2. \tag{17.7}$$

Then the approximation to the desired integral taken over this interval is

$$\int_a^{a+2h} f(x)\mathrm{d}x \approx \int_a^{a+2h} (c_0 + c_1x + c_2x^2)\mathrm{d}x$$

$$= \left(c_0x + \frac{c_1x^2}{2} + \frac{c_2x^3}{3} \right)\Bigg|_a^{a+2h}. \tag{17.8}$$

To determine the coefficients c_0, c_1 and c_2 in order that the quadratic should pass through the three points $(a, f(a))$, $(a+h, f(a+h))$, $(a+2h, f(a+2h))$ we must solve the three simultaneous equations

$$\begin{aligned} f(a) &= c_0 + c_1a + c_2a^2, \\ f(a+h) &= c_0 + c_1(a+h) + c_2(a+h)^2, \\ f(a+2h) &= c_0 + c_1(a+2h) + c_2(a+2h)^2. \end{aligned} \tag{17.9}$$

When this is done and the results are substituted into Eqn (17.8) we arrive at the desired result

$$\int_a^{a+2h} f(x)\mathrm{d}x = \frac{h}{3}[f(a) + 4f(a+h) + f(a+2h)] + E(h), \qquad (17.10)$$

where again we have included the error term $E(h)$. In its simplest form Eqn (17.10), together with its error term, is called Simpson's rule.

Now let the interval of integration $a = x \le b$ be divided into $2n$ intervals (an even number) each of length $h = (b-a)/2n$. Then the integral taken over the complete interval $a \le x \le b$ may be estimated by Simpson's rule by applying result (17.10) to the successive intervals $[a, a+2h]$, $[a+2h, a+4h]$, ..., $[b-2h, b]$ and then adding the results. When this is done we arrive at Simpson's rule in its more general form for $2n$ intervals of length $h = (b-a)/2n$:

$$\int_a^b f(x)\mathrm{d}x = \frac{h}{3}[f(x_0) + 4f(x_1) + 2f(x_2) + 4f(x_3) + \ldots$$

$$+ 2f(x_{2n-2}) + 4f(x_{2n-1}) + f(x_{2n})] + E(h), \qquad (17.11)$$

where $x_r = a + rh$, with $r = 0, 1, 2, \ldots, 2n$, and $E(h)$ represents the error term. The multipliers 1, 4, 2, 4, ..., 2, 4, 1 of the numbers $f(x_0)$, $f(x_1)$, $f(x_2), f(x_3), \ldots, f(x_{2n-2}), f(x_{2n-1}), f(x_{2n})$ are called the *weights* for Simpson's rule.

When Simpson's rule with $2n+1$ points, that is $2n$ intervals, is used to integrate $f(x)$ over the interval $[a, b]$ the error estimate becomes

$$\frac{nh^5 m}{90} \le |E(h)| \le \frac{nh^5 M}{90},$$

where $m = \min|f^{(4)}(x)|$ and $M = \max|f^{(4)}(x)|$, for $a \le x \le b$. An alternative form of this estimate is

$$\frac{(b-a)h^4 m}{180} \le |E(h)| \le \frac{(b-a)h^4 M}{180}.$$

We now summarize the above results in the form of a rule by which to calculate a definite integral using Simpson's rule, and then to estimate the error involved.

Rule 17.2 Integration by Simpson's rule

To determine

$$\int_a^b f(x)\mathrm{d}x$$

by Simpson's rule it is necessary to proceed as follows:

1. Choose the even number $2n$ of intervals into which the interval of integration $a \leq x \leq b$ is to be divided.
2. Calculate $h = (b - a)/2n$.
3. Setting $x_r = a + rh$, calculate the $(2n+1)$ numbers $f(x_0)$, $f(x_1)$, $f(x_2), \ldots, f(x_{2n})$ where, of course, $f(x_0) = f(a)$ and $f(x_{2n}) = f(b)$.
4. Calculate the sum

$$S = f(x_0) + 4f(x_1) + 2f(x_2) + 4f(x_3) + \ldots + 4f(x_{2n-1}) + f(x_{2n}).$$

5. The estimate of the definite integral provided by Simpson's rule is

$$\int_a^b f(x)\mathrm{d}x \approx \frac{1}{3}hS.$$

6. The magnitude $E(h)$ of the error involved when Simpson's rule is used may be estimated from the inequality

$$\frac{nh^5m}{90} \leq E(h) \leq \frac{nh^5M}{90},$$

where $m = \min |f^{(4)}(x)|$ and $M = \max |f^{(4)}(x)|$ for $a \leq x \leq b$.

When the maximum and minimum values of $|f^{(4)}(x)|$ on $a \leq x \leq b$ are not easily determined, the error in Simpson's rule may be approximated by

$$E(h) \approx \frac{nh^5k}{90},$$

where

$$k = \frac{1}{3}\left[|f^{(4)}(a)| + \left|f^{(4)}\left(\frac{a+b}{2}\right)\right| + |f^{(4)}(b)|\right].$$

This replaces m and M by the average of the values of $|f^{(4)}(x)|$ at the end points and mid-point of the interval of integration.

Example 17.6 Calculate the definite integral

$$\int_1^2 x \ln x \, \mathrm{d}x$$

by Simpson's rule, using four intervals, and compare the result with the exact result. Use Rule 17.2, step 6, to estimate the error and show that the actual error obeys this estimate.

Solution In this case $2n = 4$, so that $n = 2$ and $h = (2 - 1)/4 = 1/4$. We start by tabulating the functional values of $f(x) = x \ln x$ at the points $x_0 = 1$, $x_1 = 1.25$, $x_2 = 1.5$, $x_3 = 1.75$ and $x_4 = 2.0$. Then adjacent to the column of functional values we enter the weights w_i for Simpson's rule. The last column contains the products $w_i f(x_i)$, and the sum of its entries is S. Then the estimate of the definite integral provided by Simpson's rule is

$$\int_a^b x \ln x \, dx \approx \frac{1}{3} hS = \frac{1}{12} S.$$

i	x_i	$f(x_i) = x_i \ln x_i$	Weight w_i	$w_i f(x_i)$
0	1.0	0	1	0
1	1.25	0.27 893	4	1.11 572
2	1.5	0.60 820	2	1.21 640
3	1.75	0.97 933	4	3.91 732
4	2.0	1.38 629	1	1.38 629

Thus

$$S = \sum_{i=0}^{4} w_i f(x_i) = 7.63\ 573$$

and so as $h = \frac{1}{4}$ the estimate of the definite integral provided by Simpson's rule is

$$\int_1^2 x \ln x \, dx = \frac{1}{3} hS = \frac{1}{12}(7.63\ 573) = 0.63\ 631.$$

The actual value is

$$\int_1^2 x \ln x \, dx = \left(\frac{x^2}{2} \ln x - \frac{x^2}{4}\right)\Bigg|_1^2 = 0.63\ 629,$$

so the magnitude of the true error is $E(h) = |0.63\,629 - 0.63\,631| = 2 \times 10^{-5}$.

To estimate the magnitude of the error using the inequality of Rule 17.2, step 6, it is first necessary to calculate $f^{(4)}(x)$. We find that

$$f^{(4)}(x) = \frac{2}{x^3},$$

which attains its maximum value when $x = 1$ and its minimum value when $x = 2$. Thus for $1 \le x \le 2$ we have $m = \min |f^{(4)}(x)| = |f^{(4)}(2)| = 0.25$ and $M = \max |f^{(4)}(x)| = |f^{(4)}(1)| = 2$.

As $n = 2$ and $h = \frac{1}{4}$, substitution into the inequality of step 6 gives

$$0.43 \times 10^{-6} \le E(h) \le 4.34 \times 10^{-5},$$

which is in agreement with the true magnitude of the error 2×10^{-5}. ■

17.5 Solution of polynomial and transcendental equations

A problem that occurs frequently in mathematics, and which is of fundamental importance, is the numerical computation of the zeros of an nth-degree polynomial $P(x)$, where

$$P(x) = a_0x^n + a_1x^{n-1} + \ldots + a_n. \tag{17.12}$$

The so-called *zeros* of a polynomial $P(x)$ are those values of x which make $P(x) = 0$ and so, expressed differently, they are the *roots* of the equation $P(x) = 0$. A similar problem that is also of importance is the numerical computation of the roots of an equation that is not algebraic but involves trigonometric, hyperbolic, logarithmic and other mathematical functions. These equations are called *transcendental equations*, and a typical example is

$$\sin x - \cosh x + 1 = 0.$$

17.5.1 Graphical methods

The simplest method of determining the zeros of a function $f(x)$, that may either be a polynomial or a transcendental function, is to draw its graph $y = f(x)$ and to use it to find those values of x for which $f(x) = 0$. These will, of course, be the real roots of the equation. As some of the zeros of a polynomial with real coefficients may occur in complex conjugate pairs, this method cannot always be used to locate all the zeros of a polynomial.

On occasion it is helpful to obtain a graphical solution of this type in a slightly different manner by writing $f(x)$ in the form $f(x) = g(x) - h(x)$, when the zeros of $f(x)$ will correspond to the equation $g(x) = h(x)$. To see how this may help, suppose that we seek the roots of the transcendental equation

$$\sin x - \cosh x + 1 = 0,$$

where $f(x) = \sin x - \cosh x + 1$. Then by writing $f(x) = g(x) - h(x)$, with $g(x) = \sin x$ and $h(x) = \cosh x - 1$, the zeros of $f(x) = 0$ will correspond to those values of x for which the graphs of $y = g(x)$ and $y = h(x)$ intersect, as shown in Fig. 17.6. Familiarity with the two graphs involved makes it easier to sketch them and to appreciate how many zeros there are likely to be. The two zeros are seen to be $x = 0$ (exact) and $x \approx 1.3$.

17.5.2 Method of false position or the secant method

This method, which in other books is often encountered under its Latin name *regula falsi*, offers the simplest approach to the numerical computation of the real zeros of polynomial and transcendental functions $f(x)$. It depends for its success on the intermediate value theorem (Theorem 5.9) and linear interpolation. First, by trial and error, two fairly close values $x = x_0$ and $x = x_1$ are found for which $f(x_0)$ and $f(x_1)$ have opposite

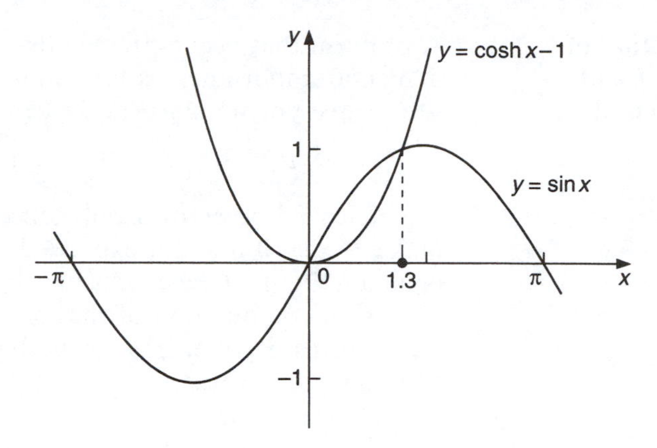

Fig. 17.6 Graphical determination of zeros.

signs. Then by the intermediate value theorem at least one zero $x = \xi$ of $f(x)$ must lie between x_0 and x_1. A straight line is drawn between the points $(x_0, f(x_0))$ and $(x_1, f(x_1))$, and its point of intersection $x = \xi_p$ with the x-axis is taken as an approximation to the true zero $x = \xi$, as shown in Fig. 17.7.

If the points involved are $(x_0, f(x_0))$ and $(x_1, f(x_1))$, then substitution into the linear interpolation formula (17.3), followed by setting $f(\xi) = 0$ and solving for ξ, gives

$$\xi_p = \frac{x_0 f(x_1) - x_1 f(x_0)}{f(x_1) - f(x_0)}. \tag{17.13}$$

By taking $x = \xi_p$, and whichever of the other two points $x = x_0$ and $x = x_1$ makes $f(x)$ have a sign opposite to that of $f(\xi_p)$, the process may be repeated and a better approximation found to the true zero $x = \xi$. Setting

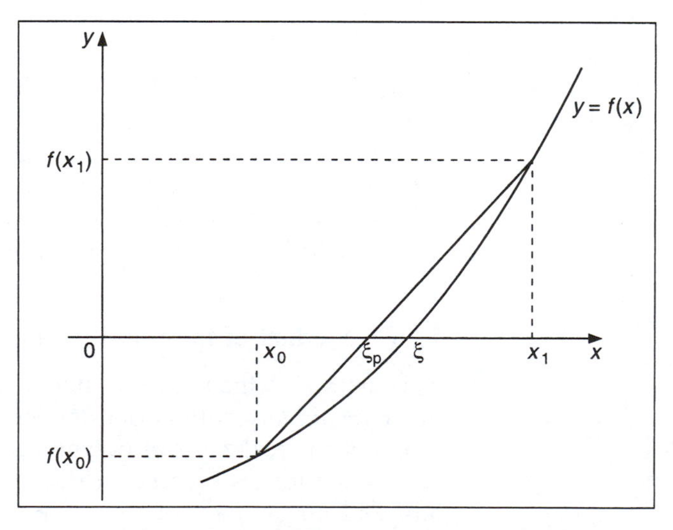

Fig. 17.7 Method of false position or the secant method.

$f(x) = \sin x - \cosh x + 1$, and using the fact that $f(1.25) = 0.06056$ and $f(1.30) = -0.00736$, we find from (17.17) that $\xi_p = 1.2946$. As $f(1.2946) = 0.00033$, the next pair of points to take if a more accurate result is required would be $f(1.2946) = 0.00033$ and $f(1.30) = -0.00736$. These lead to the improved estimate $\xi_p = 1.2948$, but if greater accuracy is required for ξ_p it must correspondingly be reflected in the calculation of $f(x_0)$ and $f(x_1)$ to more significant figures.

17.5.3 Newton's method

Newton's method is a simple and powerful method for the accurate determination of the roots of an equation $f(x) = 0$, and is based on Taylor's theorem with the Lagrange remainder $R_2(x)$.

Suppose x_0 is an approximate root of $f(x) = 0$ and h is such that $x = x_0 + h$ is an exact root. Then by Taylor's theorem

$$f(x_0 + h) = f(x_0) + hf'(x_0) + \frac{h^2}{2} f''(\xi),$$

where $x_0 < \xi < x_0 + h$.

As, by supposition, $f(x_0 + h) = 0$, we find

$$0 = f(x_0) + hf'(x_0) + \frac{h^2}{2} f''(\xi).$$

Now ξ is not known, but on the assumption that h is small we may define a first approximation h_1 to h by neglecting the third term and writing

$$h_1 = -\frac{f(x_0)}{f'(x_0)}.$$

The next approximation to the root itself must be $x_1 = x_0 + h_1$, whence, by the same argument, the approximation h_2 to the correction needed to make x_1 an exact root is

$$h_2 = -\frac{f(x_0 + h_1)}{f'(x_0 + h_1)}.$$

Proceeding in this manner we find that the nth approximation x_n to the exact root of $f(x) = 0$ is, in terms of the $(n-1)$th approximation x_{n-1},

$$x_n = x_{n-1} - \frac{f(x_{n-1})}{f'(x_{n-1})}. \tag{17.14}$$

The successive calculation of improved approximations in this manner is an *iterative process*, in which x_n is the nth *iterate*.

If the sequence $\{x_n\}$ tends to a limit x^*, it follows that this limit must be the desired root, for then the numerator of the correction term vanishes. The choice of an approximate root x_0 with which to start the

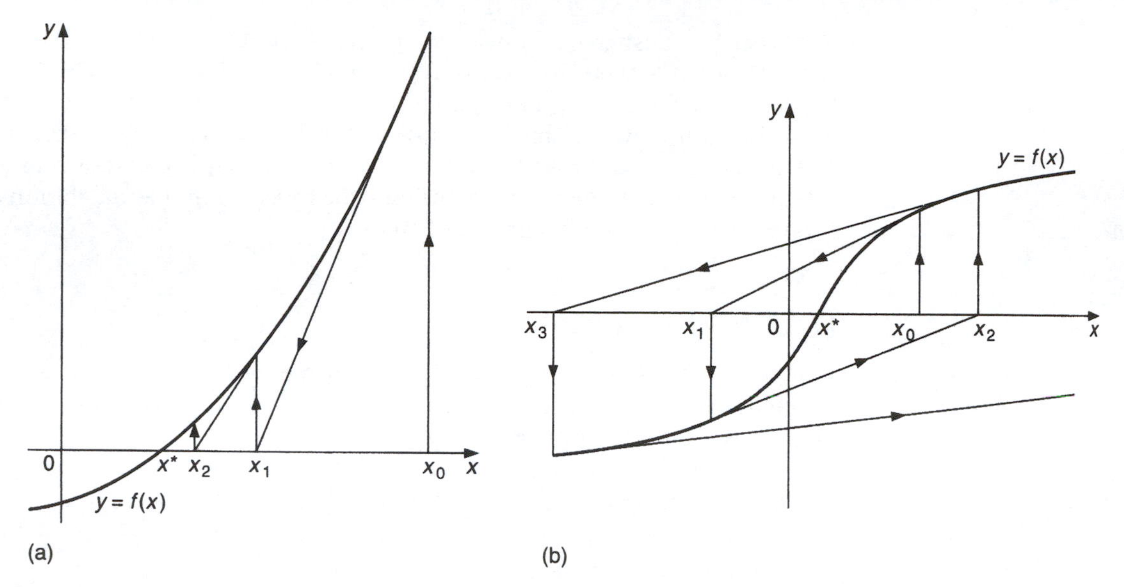

Fig. 17.8 (a) Convergent Newton iteration process; (b) divergent Newton iteration

process may be made in any convenient manner. The most usual method is to seek to show that the root lies between two fairly close values $x = a$, $x = b$ and then to take for x_0 any value that is intermediate between them. The numbers a, b are usually found by direct calculation, which is used to prove that $f(a)$ and $f(b)$ are of opposite sign, so that by the intermediate value theorem a zero of $y = f(x)$ must occur in the interval $a < x < b$.

The reasons for both the success and failure of Newton's method are best appreciated in geometrical terms. The calculation of x_n from x_{n+1} amounts to tracing back the tangent to the curve $y = f(x)$ at x_{n-1} until it intersects the x-axis at the point x_n. If x_n lies between x_{n-1} and x^* for all n then the process converges; otherwise it diverges. Fig. 17.8(a) illustrates a convergent iteration and Fig. 17.8(b) a divergent one.

To avoid the possibility of divergence when working with Newton's method, a suitable starting approximation is often found by carrying out a few iterations of the secant method described in section 17.5.1. Alternatively, a starting approximation may be found by graphing $y = f(x)$ and identifying the approximate value of x for which $f(x) = 0$.

Example 17.7 Locate the real root of the cubic

$$x^3 + x^2 + 2x + 1 = 0.$$

Use the result to find the remaining roots.

Solution Setting $f(x) = x^3 + x^2 + 2x + 1$ we see that $f(0) = 1 > 0$ and $f(-1) = -1 < 0$, so that by the intermediate value theorem a root of the equation $f(x) = 0$ must lie in the interval $-1 < x < 0$. Take $x_0 = -0.5$, since this lies within the desired interval.

Now $f'(x) = 3x^2 + 2x + 2$ so that Newton's method requires us to employ the relation

$$x_n = x_{n-1} = \frac{x_{n-1}^3 + x_{n-1}^2 + 2x_{n-1} + 1}{3x_{n-1}^2 + 2x_{n-1} + 2}$$

starting with $x_0 = -0.5$.

A straightforward calculation shows that to four decimal places $x_1 = -0.5714$, $x_2 = -0.5698$ and $x_3 = -0.5698$. The iteration process has thus converged to within the required accuracy after only three iterations. The real root is $x^* = -0.5698$, and the remaining two roots can now be found by dividing $f(x) = 0$ by the factor $(x + 0.5698)$ and then solving the remaining quadratic in the usual manner. If this is done, long division gives

$$\frac{x^3 + x^2 + 2x + 1}{x + 0.5698} = x^2 + 0.4302x + 1.7549,$$

from which we find the other two roots are

$$x = -0.2151 + i1.3071 \quad \text{and} \quad x = -0.2151 - i1.3071. \quad \blacksquare$$

Equation (17.18) shows that the absolute error involved in Newton's method is $h^2|f''(\xi)|/2 \approx h^2|f''(x_n)|/2$. Thus the error converges as h^2, so that this is a very efficient method. On account of this convergence in terms of h the method is said to have a *second-order rate of convergence*, as opposed to that of the method of false position which uses linear interpolation and so has only a *first-order rate of convergence*.

In many books the name of Raphson is coupled with that of Newton, and the method is known as the Newton – Raphson method. The method works equally well for locating the zeros of both polynomial and transcendental functions.

Newtons's method works equally well when the function $f(x)$ is more complicated than a polynomial, as can be seen in the next example.

Example 17.8 Find the root of the equation

$$2e^x + 3x + 4 = 0,$$

and show the effect of the iterations graphically.

Solution Setting $f(x) = 2e^x + 3x + 4$, it follows that $f'(x) = 2e^x + 3$, so Newton's iterative scheme becomes

$$x_n = x_{n-1} - (2\exp(x_{n-1}) + 3)/(2\exp(x_{n-1}) + 3x_{n-1} + 4),$$

for $n = 0, 1, \ldots$, with a suitable choice for x_0. To illustrate the steps in the iterations, we set $x_0 = 3$ and calculate successive iterates to six decimal places. This gives the following results

$$x_0 = 3,\ x_1 = 1.768363,\ x_2 = 0.340101,\ x_3 = -1.007617,$$
$$x_4 = -1.465323,\ x_5 = -1.484398,\ x_6 = -1.484422,$$
$$x_7 = -1.484422.$$

Thus working, to six decimal places, the iterative scheme has converged to the root $x^* = -1.484422$ after only six iterations. To find the error involved, if x^* was an exact root then substitution of $x = x^*$ into $f(x)$ should yield zero, whereas it gives $f(x^*) = 0.6 \times 10^{-6}$, showing an error of approximately 1 in the sixth decimal place. To obtain a more accurate value of x^* it would be necessary to work to more decimal places, and possibly to perform more iterations. To show the effect of this, working to nine decimal places the result still converges after six iterations yielding $x^* = -1.484422172$ when the error is found to have reduced to $f(x^*) = 0.2 \times 10^{-9}$.

The evolution of the iterations is shown in Fig. 17.9 where the graph of $f(x)$ is shown together with the tangents to the curve at each step. Only the effect of the first four iterations can be seen because of the scale of the diagram. ■

Newton's method can also be used to find the zeros of functions defined in a more complicated way than that of Example 17.8. An example

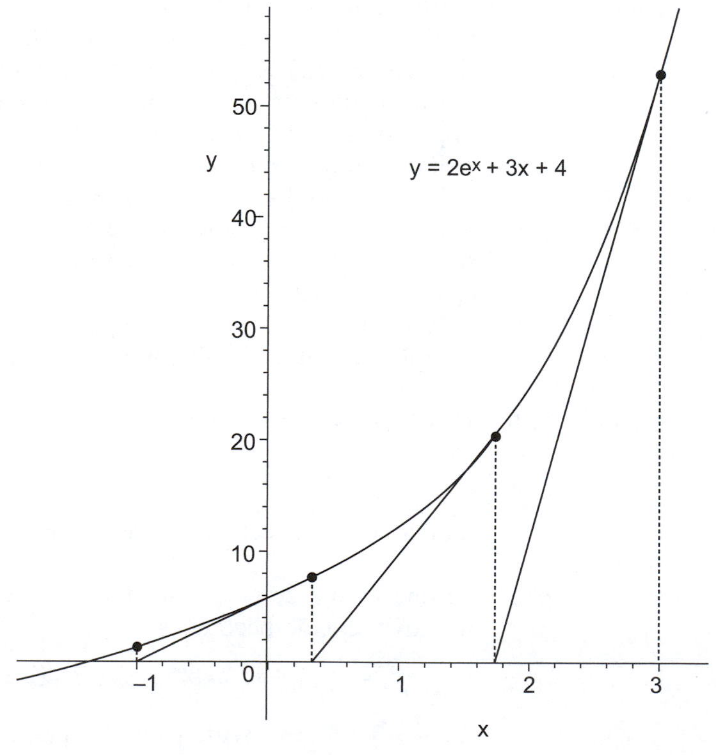

Fig. 17.9 The first four stages in Newton's iterative scheme for the root of $2e^x + 3x + 4 = 0$, starting with $x_0 = 3$.

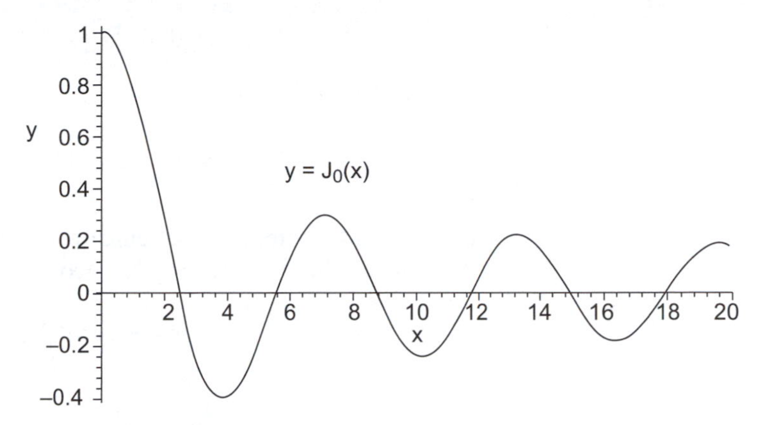

Fig. 17.10 A plot of the Bessel function $J_0(x)$ for $0 \le x \le 20$, showing its oscillatory nature and its first few positive zeros.

of this type involves a solution of the linear second order variable coefficient differential equation

$$x^2\frac{d^2y}{dx^2} + x\frac{dy}{dx} + (x^2 - n^2)y = 0,$$

called Bessel's equation of order n where $n = 0, 1, 2, \ldots$, which can be expressed in terms of a definite integral involving x as a parameter. Although we will not derive the result, when $n = 0$ one of the two linearly independent solutions of this equation, denoted by $J_0(x)$ and called the Bessel function of the first kind of order zero, takes the form

$$J_0(x) = \frac{2}{\pi}\int_0^{\pi/2} \cos(x \cos\theta)d\theta,$$

in which x appears as a parameter in the integrand. In this notation the term of *order zero* refers to the fact that the parameter $n = 0$ in Bessel's equation.

The solution, scaled so that $J_0(0) = 1$, is oscillatory and looks like a decaying cosine function. It has infinitely many zeros (places where its graph crosses the x-axis), the first few of which can be seen from its graph in Fig. 17.10.

The example that follows shows one way in which a zero of $J_0(x)$ can be found. A quite different way of finding the zeros of Bessel functions is used in practice, but the purpose of this example is to demonstrate something of the versatility of Newton's method.

Example 17.9 Use Newton's method and numerical integration with the integral definition of $J_0(x)$ to find, accurate to six decimal places, the first zero of $J_0(x)$.

Solution We are required to find the first value of x, say $x = x^*$, for which $J_0(x^*) = 0$. From Eqn (17.4), if x_n is the nth approximation to a zero x^* of $J_0(x)$, the iterative scheme to find x^* becomes

$$x_n = x_{n-1} - J_0(x_{n-1})/\mathrm{d}/\mathrm{d}x[J_0(x_{n-1})], \text{ where } x_0 \text{ is given.}$$

From Theorem 7.6 we have

$$\frac{\mathrm{d}}{\mathrm{d}x}(J_0(x)) = \frac{2}{\pi}\int_0^{\pi/2} \cos(x \cos\theta)\mathrm{d}\theta = -\frac{2}{\pi}\int_0^{\pi/2} \sin(x\cos\theta)\cos\theta\,\mathrm{d}\theta.$$

So the Newton iterative scheme to compute x^*, starting with a given initial approximation x_0 to x^*, becomes

$$x_n = x_{n+1} + \int_0^{\pi/2} \cos(x\cos\theta)\mathrm{d}\theta \Big/ \int_0^{\pi/2} \sin(x\cos\theta)\cos\theta\,\mathrm{d}\theta,$$

for $n = 1, 2, \ldots$. These definite integrals must be evaluated numerically, so to reduce the effort involved a computer numerical integration routine was used to obtain the following results.

Examination of Fig. 17.10 shows that the first positive zero x^* lies between 1 and 4, so by way of illustration we choose as the initial approximation to x^* the value $x_0 = 2$, when we find that:

$$x_1 = 2 + 0.388211 = 2.388211,$$

$$x_2 = 2.388211 + 0.016559 = 2.404770,$$

$$x_3 = 2.404770 + 0.000056 = 2.404826,$$

$$x_4 = 2.404826 + 0.696 \times 10^{-9} = 2.404826.$$

The approximations to x^* have converged to six decimal places after only four iterations, showing that the first positive zero x^* of $J_0(x)$ is $x^* = 2.404826$. The actual value, accurate to eight decimal places, is $x^* = 2.40482555$.

It is necessary to remark here that this approach is not suitable for the determination of large zeros of $J_0(x)$, because when x is large the integrand $\cos(x\cos\theta)$ will oscillate rapidly, so to obtain an accurate result in the numerical integration the number of steps used must be increased significantly to allow for the oscillations and, furthermore, the method is likely to diverge unless the initial approximation x_0 is close to the required zero x^*. ■

17.6 Numerical solutions of differential equations

17.6.1 Modified Euler method

The Euler method for the numerical solution of a first-order differential equation provides a means of determining the solution of an initial value problem but, as we have already seen in an example and several problems, the accuracy is poor. We now show that attention to the geometrical implications of the method can greatly improve its accuracy. In Fig. 13.3 the gradient appropriate to point P_0 was used to determine the change Δ in the functional value of y over the entire interval of length h. This is obviously only a first approximation to the true situation, and a better approximation to the increment in y consequent upon a step of

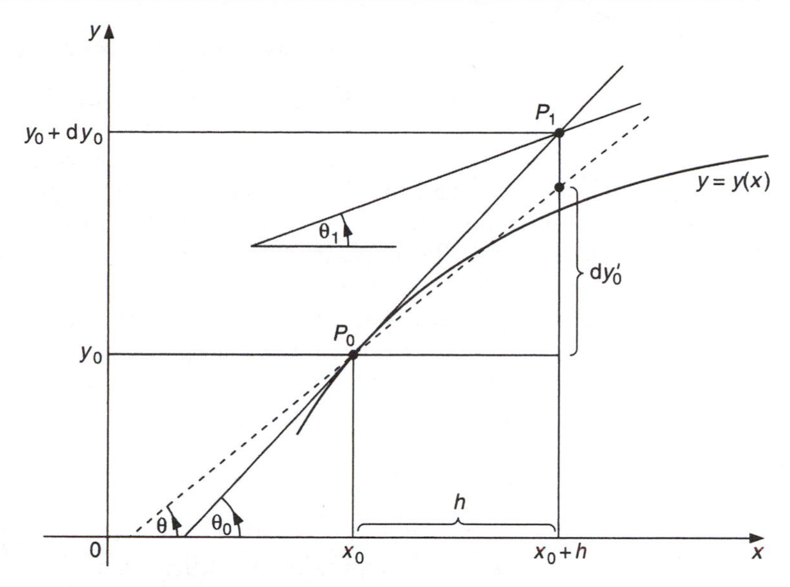

Fig. 17.11 Euler's modified method.

length h in x would be provided by using the average of the gradients at P_0 and P_1 in place of $f(x_0, y_0)$ in the Euler method. This simple refinement applied to the previous argument is known as the *modified Euler method* in which the error at each step is of the order of h^3.

The proposed modification is shown diagrammatically in Fig. 17.11, in which the full straight lines passing through points P_0 and P_1 have respective gradients $m_0 = f(x_0, y_0)$ and $m_1 = f(x_0+h, y_0+dy_0)$. Then, if the dotted line through P_0 has gradient $m_0' = \frac{1}{2}(m_0 + m_1)$, the improved approximation dy_0' to the increment in y is simply $dy_0' = m_0'h$. In terms of the angles θ, θ_0 and θ_1 defined in the figure, $\tan\theta = \frac{1}{2}\{\tan\theta_0 + \tan\theta_1\}$.

The improved accuracy is best illustrated by repeating the numerical Example 13.7 to determine the value of y at $x = 0.5$, given that $y' = xy$ and $y(0) = 1$. To simplify the headings on the tabulation we set $m_i = f(x_i, y_i)$ and $m_{i+1} = f(x_i+h, y_i+dy_i)$ and, as before, use increments $h = 0.1$. The exact result is shown in the last column.

i	x_i	y_i	m_i	dy_i	m_{i+1}	$m_i = \frac{1}{2}(m_i + m_{i+1})$	dy_i	$e^{(1/2)x^2}$
0	0.0	1.0	0.0	0.0	0.1	0.05	0.005	1.0
1	0.0	1.0050	0.1005	0.0101	0.2030	0.1517	0.0152	1.0050
2	0.2	1.0202	0.2040	0.0204	0.3122	0.2581	0.0258	1.0202
3	0.3	1.0460	0.3138	0.0314	0.4310	0.3724	0.0372	1.0460
4	0.4	1.0832	0.4333	0.0433	0.5633	0.4983	0.0498	1.0833
5	0.5	1.1330						1.1331

The approximate value $y(0.5) = 1.1330$ shown in the third column is now only 0.0001 less than the true value, demonstrating the superiority of the modified Euler method over its predecessor.

17.6.2 A simple predictor–corrector method

Despite the improvement in accuracy brought about by the modification of Euler's method, it is nevertheless quite apparent that it can only be used with any degree of accuracy close to the initial point. Later we shall be describing the Runge–Kutta method, which overcomes many of these limitations, but in the meantime it will be useful to give a brief outline of an alternative method, using a predictor and a corrector formula.

The method we describe is perhaps the simplest of its kind but it has, nevertheless, an accuracy of the order of h^5 for an integration step of length h. This time our approach will differ in that it will be based on direct integration of a differential equation of the form

$$\frac{dy}{dx} = f(x, y).$$

We shall suppose that by some means, possibly by the modified Euler method, approximate values y_0, y_1, y_2 and y_3 of the solution $y(x)$ are known at points x_0, x_1, x_2 and x_3, equally spaced with step interval h. Now bearing in mind the definition of a solution $y(x)$ as expressed by Eqn (13.2), we may rewrite our differential equation in the form

$$\frac{dy}{dx} = f[x, y(x)]. \tag{17.15}$$

If we integrate this result over the interval $[x_0, x_4]$ we obtain

$$\int_{x_0}^{x_4} \left(\frac{dy}{dx}\right) dx = \int_{x_0}^{x_4} f[x, y(x)]dx,$$

or

$$y(x_4) = y(x_0) + \int_{x_0}^{x_4} f[x, y(x)]dx, \tag{17.16}$$

where, in general, $x_m = x_0 + mh$, with m an integer. Thus, if the integral in Eqn (17.16) be estimated using the information available to us at the starting points x_2 to x_3, then the equation can be used to predict $y(x_4)$. Since an error will be involved in approximate methods of integration we shall write $y_p(x_4)$ for the predicted value of $y(x_4)$. Now in Problem 17.10 (D) we already have a formula for evaluating the integral in Eqn (17.16) that only uses the values of y' at points x_1, x_2 and x_3. Expressing the result in terms of the points x_0 to x_4 we thus have

$$y_p(x_4) = y(x_0) + \frac{4h}{3}[2y'[x_3) - y'(x_2) + 2y'(x_1)], \tag{17.17}$$

where $y'(x_m)$ signifies the value of dy/dx at $x = x_m$. However, from the four given starting values we may use Eqn (17.15) to calculate approximate values of $y'(x_m)$, so that Eqn (17.17) becomes

$$y_p(x_4) = y(x_0) + \frac{4h}{3}[2f(x_3, x_3) - f(x_2, y_2) + 2f(x_1, y_1)]. \tag{17.18}$$

Expressed in terms of any five consecutive points x_{n-3} to x_{n+1}, this result, which is called a *predictor formula*, has the general form

$$y_p(x_{n+1}) = y(x_{n-3}) + \frac{4h}{3}[2f(x_n, x_n) - f(x_{n-1}, y_{n-1}) + 2f(x_{n-2}, y_{n-2})]. \tag{17.19}$$

Returning to the determination of $y_p(x_4)$ by Eqn (17.18), we may now use this value, together with Eqn (17.15), to find, as an approximation to $y'(x_4)$, the value $f[x_4, y_p(x_4)]$. Using this estimate of $y'(x_4)$ we now correct the estimate $y_p(x_4)$ by use of Simpson's rule. We first write down the obvious result

$$y(x_4) - y(x_2) = \int_{x_2}^{x_4} \left(\frac{dy}{dx}\right) dx, \tag{17.20}$$

and then use Simpson's rule to evaluate the right-hand side in terms of the known values of $y'(x)$ at the three points x_2, x_3 and x_4. This result will then express the value of $y(x_4)$ in terms of known quantities, and we shall take this value as the corrected value for $y(x_4)$. Since an error will again be involved in the numerical integration we shall write $y_c(x_4)$ to denote the corrected estimate of $y(x_4)$, and Eqn (17.20) becomes

$$y_c(x_4) = y(x_2) + \frac{h}{3}\{f[x_4, y_p(x_4)] + 4f(x_3, y_3) + f(x_2, y_2)\}. \tag{17.21}$$

The improvement in the accuracy of $y_c(x_4)$ so determined arises from the fact that the error term in the predictor formula has been shown to have magnitude $|28h^5y^{(5)}/90|$, whereas the magnitude of the error in the corrector formula has been shown to be only $|h^5y^{(5)}/90|$. Here $y^{(5)} = \left(\frac{d^5y}{dx^5}\right)_{x=\xi}$, with ξ an interior point of the interval of integration.

Using this value $y_c(x_4)$ we again use Eqn (17.16) to recalculate $y'(x_4)$, obtaining the corrected value $f[x_4, y_c(x_4)]$. This completes the calculation since we now know $y_c(x_4)$ and $f[x_4, y_c(x_4)]$, which we take as the true values of $y(x_4)$ and $y'(x_4)$, respectively.

Again, expressed in terms of any three consecutive points x_{n-1}, x_n and x_{n+1}, result (17.21) which is called a *corrector formula*, has the general form

$$y_c(x_{n+1}) = y(x_{n-1}) + \frac{h}{3}[f(x_{n+1}, y_{n+1}) + 4f(x_n, y_n) + f(x_{n-1}, y_{n-1})]. \tag{17.22}$$

Writing $y_4 = y(x_4)$, we then use the known values y_1, y_2, y_3 and y_4 at points x_1, x_2, x_3 and x_4, and repeat the process to determine $y_5 = y(x_5)$. Thereafter, repetition of the method will advance the solution in increments h in x as far as is desired. This manner of solution is known as *Milne's method*. The modified Euler method can be used to obtain starting values for the predictor–corrector approach.

Example 17.10 Given that $dy/dx = xy$ and $y(0) = 1$, $y(0.2) = 1.02020$, $y(0.4) = 1.08329$, $y(0.6) = 1.19722$, use the predictor–corrector method to compute $y(0.8)$ and $y(1.0)$.

Solution

n	x	y	y'
0	0.0	1.00000	0.0
1	0.2	1.02020	0.20404
2	0.4	1.08329	0.43332
3	0.6	1.19722	0.71833

Here $h = 0.2$ so that from Eqn (17.29) we have

$$y_p(x_4) = 1.0000 + \frac{0.8}{3}(2 \times 0.71833 - 0.4332 + 2 \times 0.20404),$$

giving $y_p(x_4) = 1.37638$. Our first predicted value $(y_4')_p$ of $y'(x_4)$ is thus

$$y_p'(x_4) = 0.8 \times 1.37638 = 1.10110.$$

Using this value to calculate $y_c(x_4)$ from Eqn (17.32), we have

$$y_c(x_4) = 1.08329 + \frac{0.2}{3}[1.10110 + 4 \times 0.71833 + 0.43332],$$

giving $y_c(x_4) = 1.37714$. The corrected value of $y'(x_4)$ is then

$$y_c'(x_4) = 0.8 \times 1.37714 = 1.10171.$$

This completes the determination of $y(x_4)$ and $y'(x_4)$, since we set $y(x_4) = y_c(x_4) = 1.37714$ and $y'(x_4) = y_c'(x_u) = 1.10171$.

To determine $y(x_5) = y(1.0)$ we use as starting values the entries in the following table:

n	x	y	y'
1	0.2	1.02 020	0.20 404
2	0.4	1.08 329	0.43 332
3	0.6	1.19 722	0.71 833
4	0.8	1.37 714	1.10 171

Then, as before, but now using Eqn (17.30), with $n = 4$, we find

$$y_p(x_5) = 1.02020 + \frac{0.8}{3}(2 \times 1.10171 - 0.71833 + 2 \times 0.43332),$$

giving $y_p(x_5) = 1.64733$. This then gives $y'_p(x_5) = 1.0 \times 1.64733 = 1.64733$. Computing $y_c(x_5)$ from Eqn (17.33), with $n = 4$, we obtain

$$y_c(x_5) = 1.19722 + \frac{0.2}{3}[1.64733 + 4 \times 1.10171 + 0.71833],$$

or $y_c(x_5) = 1.64872$. The value $y'_c(x_5)$ for a further integration step, should it be desired, is $y'_c(x_5) = 1 \times 1.64872 = 1.64872$.

The correction to $y_p(x_4)$ was $+0.00076$ and the correction to $y_p(x_5)$ was -0.00139.

Comparison of $y_c(x_4)$ and $y_c(x_5)$ with the actual values obtained from the exact solution $y = e^{x^2/2}$ with $x_4 = 0.8$ and $x_5 = 1.0$, shows that to five places of decimals, the error in $y_c(x_4)$ was -0.00001, whereas $y_c(x_5)$ was exact. ■

If at the completion of an integration it is desired to change the integration step length from h to h', then this may be accomplished by means of interpolation. Using the available tabular entries of y, an interpolation formula must be used to deduce appropriate functional values at four new values of x equally spaced with the new interval h'. Thereafter the method proceeds as before, using these new starting values and the step length h'.

17.6.3 Runge–Kutta method

Although the predictor–corrector method just described has much better accuracy than the modified Euler method, it also has two disadvantages. These are, first, that it requires four accurate starting values, which may not always be available, and second, that change of interval length is not a straightforward matter.

The very useful and flexible numerical method that we now describe was first introduced by C. Runge around 1900, and subsequently modified and improved by W. Kutta. It is essentially a generalization of Simpson's rule, and it can be shown that the error involved when integrating over a step of length h is of the order of h^5. The method is simple to use and, unlike the predictor–corrector method outlined in section 17.6.2, allows adjustment of the length of the integration step from point to point without modification of the method.

We suppose that x and y assume the values x_n, y_n after the nth integration step in the numerical integration of

$$\frac{dy}{dx} = f(x, y). \qquad (17.23)$$

Then the value y_{n+1} of the dependent variable y that is to be associated with argument $x_{n+1} = x_n + h$ is computed as follows.

Use an integration step of length h and, at the $(n+1)$th stage of the calculation, let

$$\begin{aligned}
k_{1n} &= f(x_n\ y_n).h \\
k_{2n} &= f(x_n + \tfrac{1}{2}h,\ y_n + \tfrac{1}{2}k_{1n}).h \\
k_{3n} &= f(x_n + \tfrac{1}{2}h,\ y_n + \tfrac{1}{2}k_{2n}).h \\
k_{4n} &= f(x_n + h,\ y_n + k_{3n}).h \\
\Delta y_n &= \tfrac{1}{6}(k_1 + 2k_{2n} + 2k_3 + k_{4n}).
\end{aligned} \tag{17.24}$$

Then the value y_{n+1} of y corresponding to $x = x_n + h$ is determined by

$$y_{n+1} = y_n + \Delta y_n. \tag{17.25}$$

Rule 17.3 Runge–Kutta algorithm.

Let the differential equation to be integrated be

$$\frac{dy}{dx} = f(x,\ y).$$

subject to the initial condition

$$y(x_0) = y_0.$$

Set $x_n = x_0 + nh$ and $y_n = y(x_n)$, where h is the step length. The algorithm determines the value y_{n+1} at x_{n+1} from y_n at x_n by means of the following steps:

1. Compute

$$\begin{aligned}
k_{1n} &= hf(x_n,\ y_n) \\
k_{2n} &= hf(x_n + \tfrac{1}{2}h,\ y_n + \tfrac{1}{2}k_{1n}) \\
k_{3n} &= hf(x_n + \tfrac{1}{2}h,\ y_n + \tfrac{1}{2}k_{2n}) \\
k_{4n} &= hf(x_n + h,\ y_n + k_{3n}).
\end{aligned}$$

2. Compute

$$\Delta y_n = \tfrac{1}{6}(k_{1n} + 2k_{2n} + 2k_{3n} + k_{4n}).$$

3. The numerical estimate of y_{n+1} at x_{n+1} is given by

$$y_{n+1} = y_n + \Delta y_n.$$

Example 17.11 Let us again determine the value $y(0.5)$ given that $y' = xy$, with $y(0) = 1$ and $h = 0.1$. In this simple example, already used to illustrate Euler's method and its modification, we have $f(x, y) = xy$. As we must anticipate an error of the order of $(0.1)^5$ we shall work to five decimal places so that we may compare our solution with the exact result $y = e^{x^2/2}$.

Solution

n	x_n	y_n	$f(x_n, y_n)$	k_{1n}	k_{2n}	k_{3n}	k_{4n}	y_n+_1	$e^{x^2/2}$
0	0.0	1.0	0.0	0.0	0.0050	0.00 501	0.010 05	1.00 501	1.0
1	0.1	1.00 501	0.10 050	0.01 005	0.01 515	0.01 519	0.02 040	1.02 020	1.00 501
2	0.2	1.02 020	0.20 404	0.02 040	0.02 576	0.02 583	0.03 138	1.04 603	1.02 020
3	0.3	1.04 603	0.31 381	0.03 138	0.03 716	0.03 726	0.04 333	1.08 329	1.04 603
4	0.4	1.08 329	0.43 332	0.04 332	0.04 972	0.04 987	0.05 666	1.13 315	1.08 329
5	0.5	1.13 315							

Comparison of the results of column 3 with the analytical solution $y = e^{x^2/2}$ shows that it is in fact accurate to five decimal places, so that in this case our rough error estimate was too severe. ■

The superiority of the Runge–Kutta method over the Euler and modified Euler methods is clearly demonstrated if the Ruge–Kutta solution is compared with the previous solutions. This improvement is uniformly true and not just in this instance, since it may be shown that the errors involved in the Euler and modified Euler methods are, respectively, of the order h^2 and h^3. No discussion will be offered here of the more subtle finite difference methods that may be used to provide integration formulae having extremely high accuracy.

The Runge–Kutta method readily extends to allow the numerical solution of simultaneous and higher-order equations. Suppose the equations involved are

$$\frac{dy}{dx} = f(x, y, z)$$

$$\frac{dz}{dx} = g(x, y, z) \tag{17.26}$$

subject to the initial conditions $y = y_0$, $z = z_0$ at $x = x_0$.

Then, with a step of length h in x, define

$$\begin{aligned} k_{1n} &= f(x_n, y_n, z_n).h \\ k_{2n} &= f(x_n + \tfrac{1}{2}h, y_n + \tfrac{1}{2}k_{1n}, z_n + \tfrac{1}{2}K_{1n}).h \\ k_{3n} &= f(x_n + \tfrac{1}{2}h, y_n + \tfrac{1}{2}k_{2n}, z_n + \tfrac{1}{2}K_{2n}).h \\ k_{4n} &= f(x_n + h, y_n + k_{3n}, z_n + K_{3n}).h \end{aligned} \tag{17.27}$$

and

$$\begin{aligned} K_{1n} &= g(x_n, y_n, z_n).h \\ K_{2n} &= g(x_n + \tfrac{1}{2}h, y_n + \tfrac{1}{2}k_{1n}, z_n + \tfrac{1}{2}K_{1n}).h \\ k_{3n} &= g(x_n + \tfrac{1}{2}h, y_n + \tfrac{1}{2}k_{2n}, z_n + \tfrac{1}{2}K_{2n}).h \\ K_{4n} &= g(x_n + h, y_n + k_{3n}, z_n + K_{3n}).h \end{aligned} \tag{17.28}$$

The following formulae are then used to compute the increments Δy_n and Δz_n in y and z:

$$\Delta y_n = \tfrac{1}{6}(k_{1n} + 2k_{2n} + 2k_{3n} + k_{4n})$$

and

$$\Delta z_n = \tfrac{1}{6}(K_{1n} + 2K_{2n} + 2K_{3n} + K_{4n}). \tag{17.29}$$

The values of y and z at the $(n+1)$th step of integration are $y_{n+1} = y_n + \Delta y$, $z_{n+1} = zn + \Delta z_n$.

These results may also be used to integrate a second-order equation by introducing the first derivative as a new dependent variable. Suppose $y'' - 2y' + 2y = 0$ with $y(0) = y'(0) = 1$. Then setting $y' = z$, the second-order equation is seen to be equivalent to the two first-order simultaneous equations $y' = z$ and $z' = 2(z - y)$, with $y(0) = 1$ and $z(0) = 1$. Applying Eqns (17.38) to (17.40) with $h = 0.2$, $f \equiv z$ and $g \equiv 2(y - z)$ in order to determine $y(0.2)$, we find

$$\begin{aligned} k_{11} &= 0.2, & K_{11} &= 0 \\ k_{21} &= 0.2, & K_{21} &= -0.04 \\ k_{31} &= 0.196, & K_{31} &= -0.048 \\ k_{41} &= 0.1904, & K_{41} &= -0.0976, \end{aligned}$$

so that $\Delta y_t = 0.19706$ and $\Delta z_1 = \Delta(y') = -0.04560$. Hence $y(0.2) = 1.9706$ and $y'(0.2) = 0.95440$, which are in complete agreement with the analytical solution $y = e^x \cos x$.

17.7 Determination of eigenvalues and eigenvectors

In section 10.7 the eigenvalues $\lambda_1, \lambda_2, \ldots, \lambda_n$ of an nth-order square matrix $\mathbf{A}$ were defined to be the roots of the characteristic determinant

$$|\mathbf{A} - \lambda \mathbf{I}| = 0. \tag{17.30}$$

When expanded, this determinant gives rise to a polynomial equation of degree n in λ called the *characteristic equation*, which may always be written

$$\lambda^n + \alpha_1 \lambda^{n-1} + \alpha_2 \lambda^{n-2} + \ldots + \alpha_n = 0. \tag{17.31}$$

Thus, to find the eigenvalues of $\mathbf{A}$, it is necessary to locate the n roots of Eqn (17.31).

The vector $\mathbf{X}_i$ that is a solution to the homogeneous system

$$(\mathbf{A} - \lambda_i \mathbf{I})\mathbf{X}_i = \mathbf{0}, \tag{17.32}$$

which may also be written in the form

$$\mathbf{A}\mathbf{X}_i = \lambda_i \mathbf{X}_i, \tag{17.33}$$

is the eigenvector of $\mathbf{A}$ associated with λ_i. There will be n such eigenvectors corresponding to the n eigenvalues λ_i, with $i = 1, 2, \ldots, n$, and they will all be different provided the λ_i are all distinct (i.e., different one from the other).

The homogeneity of system (17.32) means that if $\mathbf{X}_i$ is a solution, then so also is $\beta\mathbf{X}_i$, for any constant $\beta \neq 0$. It is useful to use this fact to normalize the eigenvectors $\mathbf{X}_i$ by making a convenient choice for β. Hereafter we shall set $\beta_i = 1/a_i$, where a_i is the element of vector $\mathbf{X}_i$ with the greatest absolute value. This normalization will make the largest element in $\mathbf{X}_i$ become unity. Not only will this keep the numbers involved scaled to lie within a manageable range, but we shall see that it will also make our task of finding the eigenvalues easier. When an eigenvector $\mathbf{X}_i$ is normalized in this manner it will be denoted by $\tilde{\mathbf{X}}_i$. It then follows that $\mathbf{X}_i = \beta_i\tilde{\mathbf{X}}_i$. Other normalizations are also possible, though less convenient for our purpose. The most common one is to set $\beta = 1/(x_{i1}^2 + x_{i2}^2 + \ldots + x_{in}^2)^{1/2}$, where $x_{i1}, x_{i2}, \ldots, x_{in}$ are the elements of $\mathbf{X}_i$. This was the normalization used in section 10.7.

The method of calculation of eigenvalues and eigenvectors used in section 10.7 involved first solving for the roots of Eqn (17.31), and then using them in (17.32) to find the eigenvectors. Although this works satisfactorily for matrices $\mathbf{A}$ of very low order, it begins to become difficult even when the order of $\mathbf{A}$ is as low as 4. This is because, as we have already seen, the location of the zeros of a fourth-degree polynomial (a quartic) is not a trivial matter, and the process becomes progressively more difficult as the degree increases. In addition to this, errors in the determination of the eigenvalues are likely to influence the accuracy of the eigenvectors, particularly when some eigenvalues are close together and the system has a tendency to become ill-conditioned.

To avoid these troubles it is necessary to adopt a different approach and to determine both an eigenvalue and its associated eigenvector directly and simultaneously from the matrix $\mathbf{A}$ by means of an iterative numerical method. There are many such methods, but in this section we mention only one of the simplest which is suitable for hand calculation.

The method starts from the general result that any arbitrary n-element column vector $\mathbf{u}_0$ may always be expressed in the form

$$\mathbf{u}_0 = c_1\mathbf{X}_1 + c_2\mathbf{X}_2 + c_3\mathbf{X}_3 + \ldots + c_n\mathbf{X}_n, \tag{17.34}$$

where $\mathbf{X}_i$ are the eigenvectors of $\mathbf{A}$ and the c_i are suitable constants. In this representation of an arbitrary column vector in terms of the eigenvectors $\mathbf{X}_1, \mathbf{X}_2, \ldots, \mathbf{X}_n$, the role of the eigenvectors is directly analogous to that of the triad of unit vectors $\mathbf{i}$, $\mathbf{j}$ and $\mathbf{k}$ in geometrical vectors.

On the understanding that we consider only the case in which all the eigenvalues are real and distinct, we shall assume them to be ordered so that

$$|\lambda_1| > |\lambda_2| > \ldots > |\lambda_n|. \tag{17.35}$$

The eigenvalue λ_1 with the greatest absolute value is called the *dominant eigenvalue of* $\mathbf{A}$.

Let us now examine the effect of repeatedly pre-multiplying $\mathbf{u}_0$ in (17.34) by the matrix $\mathbf{A}$. Define $\mathbf{u}_r = \mathbf{A}^r\mathbf{u}_0$, and use Eqn (17.34) and Eqns (17.33) for $i = 1, 2, \ldots, n$, to obtain

$$\begin{aligned}\mathbf{u}_r &= \mathbf{A}^r(c_1\mathbf{X}_1 + \mathbf{c}_2\mathbf{X}_2 + c_3\mathbf{X}_3 + \ldots + c_n\mathbf{X}_n) \\ &= c_1\lambda_1^r\mathbf{X}_1 + c_2\lambda_2^r\mathbf{X}_2 + c_3\lambda_3^r\mathbf{X}_3 + \ldots + c_n\lambda_n^r\mathbf{X}_n \\ &= \lambda_1^r\{c_1\mathbf{X}_1 + c_2(\lambda_2/\lambda_1)^r\mathbf{X}_2 + c_3(\lambda_3/\lambda_1)^r\mathbf{X}_3 + \ldots + c_n(\lambda_n/\lambda_1)^r\mathbf{X}_n.\end{aligned} \tag{17.36}$$

Since $1 > |\lambda_2/\lambda_1| > |\lambda_3/\lambda_1| > \ldots > |\lambda_n/\lambda_1|$, we see that as r increases so the factors $(\lambda_s/\lambda_1)^r$, $s = 2, 3, \ldots, n$, tend to zero. Hence, for large r,

$$\mathbf{u}_r \to \lambda_1^r c_1\mathbf{X}_1. \tag{17.37}$$

This shows that as r increases, so the vector $\mathbf{u}_r$ tends to become proportional to the eigenvector $\mathbf{X}_1$ associated with the dominant eigenvalues λ_1. As $\mathbf{u}_r = \mathbf{A}^r\mathbf{u}_0 = \mathbf{A}(\mathbf{A}^{r-1}\mathbf{u}_0) = \mathbf{A}\mathbf{u}_{r-1}$, we see that the ratio of corresponding elements in the vectors $\mathbf{u}_r$ and $\mathbf{u}_{r-1}$ tends to the dominant eigenvalue λ_1. This process will fail if, by chance, the arbitrary vector $\mathbf{u}_0$ has been chosen so that $c_1 = 0$. All that is necessary to overcome this problem is to start with a different arbitrary vector $\mathbf{u}_0$. For convenience, $\mathbf{u}_0$ is usually taken to be the n-element column vector which has each of its elements equal to unity.

In point of fact, when the numerical calculation is actually performed, this process is slightly modified to prevent the elements of $\mathbf{u}_r$ from becoming unreasonably large. This is accomplished by modifying the definition of $\mathbf{u}_r$, by setting $\mathbf{u}_r = \mathbf{A}\tilde{\mathbf{u}}_{r-1}$, where $\tilde{\mathbf{u}}_{r-1}$ is the normalized vector defined here as $\mathbf{u}_{r-1}$ divided by the largest element in $\mathbf{u}_{r-1}$. Then, if β_r is the element of $\mathbf{u}_r$ with the greatest absolute value, the equation $\mathbf{u}_r = \mathbf{A}\tilde{\mathbf{u}}_{r-1}$ becomes

$$\mathbf{A}\mathbf{u}_{r-1} = \beta_r\tilde{\mathbf{u}}_r. \tag{17.38}$$

Since $\tilde{\mathbf{u}}_{r-1} \to \tilde{\mathbf{u}}_r$ as r increases, and the largest element in the normalized eigenvector $\tilde{\mathbf{u}}_r$ is unity, it follows at once that $\beta_r \to \lambda_1$, the dominant eigenvalue of $\mathbf{A}$.

The remaining eigenvalues, sometimes called the *sub-dominant eigenvalues* of $\mathbf{A}$, may be found by the same iterative process after a simple modification has been made to matrix $\mathbf{A}$. To see how this may be accomplished, let k be a constant, and define

$$\mathbf{B} = \mathbf{A} - k\mathbf{I}.$$

Then, as

$$\mathbf{A}\mathbf{X}_i = \lambda_i\mathbf{X}_i,$$

we have

$$\begin{aligned}\mathbf{B}\mathbf{X}_i &= \mathbf{A}\mathbf{X}_i - k\mathbf{X}_i \\ &= (\lambda_i - k)\mathbf{X}_i.\end{aligned} \tag{17.39}$$

So the eigenvectors of $\mathbf{A}$ and $\mathbf{B}$ are identical, but the eigenvalues of $\mathbf{B}$ are those of $\mathbf{A}$ reduced by the constant k. When λ_1 is known, by setting $k = \lambda_1$, a further eigenvalue λ_2 and eigenvector $\mathbf{X}_2$ may be found iteratively, by working with the matrix $\mathbf{B}$ in place of $\mathbf{A}$, and so on. When $n-1$ of the n eigenvalues have been determined the remaining one may be found by using the result that the sum of the eigenvalues of $\mathbf{A}$ is equal to the sum $a_{11}+a_{22}+\ldots+a_{nn}$ of elements of the leading diagonal of $\mathbf{A}$. As mentioned also in section 10.7, this sum is called the *trace* of $\mathbf{A}$ and is usually written tr($\mathbf{A}$). The proof of this follows by expanding $|\mathbf{A}-\lambda\mathbf{I}|$ in terms of elements of the first column by means of Theorem 10.5, and it is left as an exercise for the reader.

The rate at which the iterations converge is determined in part by the initial vector $\mathbf{u}_0$ that is used, but in the main by the separation of the eigenvalues. If some eigenvalues are close together it is inevitable that the convergence will be slow when determining them.

Example 17.12 Find the eigenvalues and normalized eigenvectors of the matrix

$$\mathbf{A}=\begin{bmatrix}-1 & 1 & 2\\ 0 & 3 & -1\\ 0 & 0 & 5\end{bmatrix}.$$

Solution We start the iteration by setting $\mathbf{u}_0=\begin{bmatrix}1\\1\\1\end{bmatrix}$, so that $\mathbf{u}_1=\mathbf{A}\mathbf{u}_0\begin{bmatrix}2\\2\\5\end{bmatrix}=5\begin{bmatrix}0.4\\0.4\\1\end{bmatrix}$,

and so $\beta_1=5$ and $\tilde{\mathbf{u}}_1=\begin{bmatrix}0.4\\0.4\\1\end{bmatrix}$. Similarly, $\mathbf{u}_2=\mathbf{A}\tilde{\mathbf{u}}_1=\begin{bmatrix}2\\0.2\\5\end{bmatrix}=5\begin{bmatrix}0.4\\0.04\\1\end{bmatrix}$,

and so $\beta_2=5$ and $\tilde{\mathbf{u}}_2=\begin{bmatrix}0.4\\0.04\\1\end{bmatrix}$. The outcome of the full calculation is set out in the table below in which $\mathbf{u}_1^{(r)}$, $\mathbf{u}_2^{(r)}$ and $\mathbf{u}_3^{(r)}$ are the three elements of the vector $\mathbf{u}_r$, and $\tilde{\mathbf{u}}_1^{(r)}$, $\tilde{\mathbf{u}}_2^{(r)}$ and $\tilde{\mathbf{u}}_3^{(r)}$ are the corresponding elements of the normalized vector $\tilde{\mathbf{u}}_r$.

Iteration $\mathbf{u}_r=\mathbf{A}\tilde{\mathbf{u}}_{r-1}$

Iteration	0	1	2	3	4	5	13	14
$\mathbf{u}_1^{(r)}$	1	2	2	1.54	1.515	1.391	1.252	1.251
$\mathbf{u}_2^{(r)}$	1	2	0.2	−0.88	−1.528	−1.917	−2.490	−2.494
$\mathbf{u}_3^{(r)}$	1	5	5	5	5	5	5	5
β_r	1	5	5	5	5	5	5	5
$\tilde{\mathbf{u}}_1^{(r)}$	1	0.4	0.4	0.308	0.303	0.278	0.251	0.25
$\tilde{\mathbf{u}}_2^{(r)}$	1	0.4	0.04	−0.176	−0.306	−0.383	−0.498	−0.5
$\tilde{\mathbf{u}}_3^{(r)}$	1	1	1	1	1	1	1	1

In this case the normalization factor β_r converged to the value 5 after only one iteration, but the vector $\mathbf{u}_r$, required 14 iterations before it converged to the eigenvector to an accuracy of three decimal places.

The dominant eigenvalue is thus $\lambda_1 = 5$, and the associated exact normalized eigenvector is

$$\mathbf{X}_1 = \begin{bmatrix} \tilde{\mathbf{u}}_1^{(14)} \\ \tilde{\mathbf{u}}_2^{(14)} \\ \tilde{\mathbf{u}}_3^{(14)} \end{bmatrix} = \begin{bmatrix} 0.25 \\ -0.5 \\ 1 \end{bmatrix}.$$

To find another eigenvalue of $\mathbf{A}$ we now set $k = \lambda_1 = 5$, and consider the matrix

$$\mathbf{B} = \mathbf{A} - k\mathbf{I} = \begin{bmatrix} -6 & 1 & 2 \\ 0 & -2 & -1 \\ 0 & 0 & 0 \end{bmatrix}.$$

Iterating the equation $\mathbf{u}_r = \mathbf{B}\tilde{\mathbf{u}}_{r-1}$ using the same initial vector $\mathbf{u}_0$ produces the following results.

Iteration $\mathbf{u}_r = \mathbf{B}\tilde{\mathbf{u}}_{r-1}$

Iteration	0	1	2	3	4	5	6	7	8	9	10
$\mathbf{u}_1^{(r)}$	1	-3	-5	-5.6	-5.857	-5.951	-5.984	-5.995	-5.998	-5.999	$-$
$\mathbf{u}_2^{(r)}$	1	-3	-2	-0.8	-0.286	-0.098	-0.031	-0.010	-0.003	-0.001	0
$\mathbf{u}_3^{(r)}$	1	0	0	0	0	0	0	0	0	0	0
β_r	1	-3	-5	-5.6	-5.857	-5.951	-5.984	-5.995	-5.998	-5.999	$-$
$\tilde{\mathbf{u}}_1^{(r)}$	1	1	1	1	1	1	1	1	1	1	1
$\tilde{\mathbf{u}}_2^{(r)}$	1	1	0.4	0.143	0.049	0.016	0.005	0.002	0.001	0	0
$\tilde{\mathbf{u}}_3^{(r)}$	1	0	0	0	0	0	0	0	0	0	0

The dominant eigenvalue of $\mathbf{B}$ is thus -6, but as the eigenvalues λ of $\mathbf{B}$ are those of $\mathbf{A}$ reduced by $k = 5$ we have $-6 = \lambda - 5$, or $\lambda = -1$. However the eigenvectors of $\mathbf{A}$ and $\mathbf{B}$ are the same, and so matrix $\mathbf{A}$ has the eigenvalue $\lambda = -1$ with the corresponding normalized eigenvector

$$\tilde{\mathbf{X}} = \begin{bmatrix} \tilde{\mathbf{u}}_1^{(10)} \\ \tilde{\mathbf{u}}_2^{(10)} \\ \tilde{\mathbf{u}}_3^{(10)} \end{bmatrix} = \begin{bmatrix} 1 \\ 0 \\ 0 \end{bmatrix}.$$

Since the sum of two eigenvalues of $\mathbf{A}$ is $5 - 1 = 4$, and $\mathrm{tr}(\mathbf{A}) = -1 + 3 + 5 = 7$, the final eigenvalue of $\mathbf{A}$ is $\lambda = 7 - 4 = 3$. A direct calculation shows that the corresponding exact normalized eigenvector is

$$\tilde{\mathbf{X}} = \begin{bmatrix} 0.25 \\ 1 \\ 0 \end{bmatrix}.$$

The eigenvalues and corresponding normalized eigenvectors of $\mathbf{A}$ are thus

$$\lambda_1 = 5,\ \tilde{\mathbf{X}}_1 = \begin{bmatrix} 0.25 \\ -0.5 \\ 1 \end{bmatrix}; \quad \lambda_2 = 3,\ \tilde{\mathbf{X}}_2 = \begin{bmatrix} 0.25 \\ 1 \\ 0 \end{bmatrix};$$

$$\lambda_3 = -1,\ \tilde{\mathbf{X}}_3 = \begin{bmatrix} 1 \\ 0 \\ 0 \end{bmatrix}. \quad ■$$

Problems

Section 17.1

17.1 (a) Calculate the greatest value of the absolute error modulus of $0.146 - 2.3122 + 4.72311$, in which the values are rounded, and hence round the answer off to a meaningful number of figures.

(b) Calculate 0.146×4.72311, in which the numbers are rounded off, and by estimating the relative error, give the answer to a meaningful number of figures.

(c) If 0.706 is the result of rounding a number of three significant figures, estimate the absolute error modulus of exp(0.706), and hence give the result to a meaningful number of figures.

17.2 If $X_2 \approx X_1^p$, show from first principles that with the usual notation the relative errors r_1 and r_2 are related by $r_2 \approx pr_1$.

Section 17.2

17.3 Set $x_i^{(0)} = 1$, for $i = 1, 2, 3$, and complete four iterations of the Jacobi and Gauss–Seidal methods for the system of equations

$$7x_1 - x_2 + x_3 = 7.3$$

$$2x_1 - 8x_2 - x_3 = -6.4$$

$$x_1 + 2x_2 + 9x_3 = 13.6.$$

Compare the results of the fourth iteration with the exact solution $x_1 = 1$, $x_2 = 0.9$ and $x_3 = 1.2$.

17.4 Rearrange the following equations to form a diagonally dominant system and perform the first four iterations of the Gauss–Seidel method:

$$x_1 + 5x_2 - x_3 = 8$$

$$-9x_1 + 3x_2 + 2x_3 = 3$$

$$x_1 + 2x_2 + 7x_3 = 26.$$

Compare the results of the fourth iteration with the exact solution $x_1 = 1$, $x_2 = 2$ and $x_3 = 3$.

Section 17.3

17.5 Use the three-point Lagrange interpolation formula to find $y(2.17)$, given that $y(2.1) = 8.1662$, $y(2.2) = 9.0250$ and $y(2.3) = 9.9742$. Compare the result with $\exp(2.17) = 8.7583$, which is the exact result rounded off to four decimal places.

17.6 Use the three-point Lagrange interpolation formula to find $y(0.65)$, given that $y(0.6109) = 0.7002$, $y(0.6981) = 0.8391$ and $y(0.8727) = 1.1918$. Compare the result with $\tan 0.65 = 0.7602$, which is the exact result rounded off to four decimal places.

Section 17.4

17.7 Evaluate the definite integral

$$\int_1^2 (x^3 + 2x + 1)\mathrm{d}x$$

by the trapezoidal rule using four intervals of equal length, and then by Simpson's rule for the same intervals. Compare the results with that obtained by direct integration. Infer from your comparison that Simpson's rule is exact for cubic equations, despite the fact that it is based on a parabolic fitting of the function.

17.8 Evaluate the definite integral

$$\int_0^{1.5} \mathrm{e}^{-x^2/2}\mathrm{d}x$$

by means of Simpson's rule, using four and eight intervals of equal length, respectively. Compare the results with the exact value 1.0859.

17.9 The gamma function $\Gamma(n)$ is defined by the improper integral

$$\Gamma(n) = \int_0^{\infty} t^{n-1}\mathrm{e}^{2t}\mathrm{d}t.$$

When n is a positive integer it follows from integration by parts that $\Gamma(n) = (n-1)!$. Prove this and then, for the case $n = 4$, approximate the integral by

$$\Gamma(4) \approx \int_0^{10} t^3\mathrm{e}^{-t}\mathrm{d}t$$

and evaluate it by means of Simpson's rule using 20 intervals of equal length. Compare your result with $\Gamma(4) = 3! = 6$. Justify the integral inequality

$$\int_{10}^{\infty} t^3\mathrm{e}^{-t}\mathrm{d}t > 10^3\int_{10}^{\infty}\mathrm{e}^{-t}\mathrm{d}t,$$

and hence estimate the error that can result from truncating the infinite interval of integration.

17.10 This problem is concerned with the derivation of a numerical integration formula using five equally spaced ordinates in which the functional value is specified at the first and last point, and its derivative is specified at the three intermediate points. To be precise, it establishes that

$$\int_{-2h}^{2h} f'(x)\mathrm{d}x = \frac{4h}{3}[2f'(h) - f'(0) + zf'(-h)] + E(h), \qquad \text{(A)}$$

where the error $E(h)$ is such that

$$\frac{28h^5 m}{90} \leq E(h) \leq \frac{28h^5}{90} M,$$

with $m = \min|f^{(5)}(x)|$ and $M = \max|f^{(5)}(x)|$ for $-2h \leq x \leq 2h$. Expand $f'(x)$ in a Maclaurin series with a remainder term of the form $x^4 f^{(5)}(\zeta x)/4!$, where $0 < \zeta < 1$, and show that

$$\int_{-2h}^{2h} f'(x)\mathrm{d}x = 4hf'(0) + \frac{8h^3}{3} f^{(3)}(0) + \int_{-2h}^{2h} \frac{x^4}{4!} f^{(5)}(\zeta x)\mathrm{d}x. \qquad \text{(B)}$$

Show from the Maclaurin series with a remainder that

$$h^2 f^{(3)}(0) = f'(h) - 2f'(0) + f'(-h) - \frac{h^4}{4!}[f^{(5)}(\zeta x) + f^{(5)}(\eta x)], \qquad \text{(C)}$$

where the term $f^{(5)}(\eta x)$ arises from the remainder term in the expansion of $f'(-x)$, and $0 < \eta < 1$.
Deduce result (A) from (B) and (C), and show that it may also be written in the form

$$f(2h) = f(-2h) + \frac{4h}{3}[2f'(h) - f'(0) + 2f'(-h)] + E(h). \qquad \text{(D)}$$

17.11 Use the method of the above problem to show that

$$\int_{-h}^{h} f'(x)\mathrm{d}x = \frac{h}{3}[f'(h) + 4f'(0) + f'(-h)] + E(h),$$

where $(h^5 m)/90 \leq E(h) \leq (h^5 M)/90$, $m = \min|f^{(5)}(x)|$ and $M = \max|f^{(5)}(x)|$ for $-h \leq x \leq h$. Deduce that it may also be written in the form

$$f(h) = f(-h) + \frac{h}{3}[f'(h) + 4f'(0) + f'(-h)] + E(h).$$

This result is, of course, Simpson's rule applied to the derivative $f'(x)$ and could have been deduced directly from the result of section 17.4; conversely, replacing $f'(x)$ by $f(x)$ and $f^{(5)}(x)$ by $f^{(4)}(x)$, this provides an alternative derivation of the error term in Simpson's rule.

17.12 Using the method of Problem 17.10, derive the trapezoidal rule, together with its error estimate. Namely, show that

$$\int_0^h f(x)\mathrm{d}x = \frac{h}{2}[f(0) + f(h)] + E(h),$$

where

$$\frac{h^3}{12}m \le E(h) \le \frac{h^3}{12}M,$$

with $m = \min|f^{(2)}(x)|$ and $M = \max|f^{(2)}(x)|$ for $0 \le x \le h$.

Section 17.5

17.13. Find by the method of false position the smallest positive root of the equation

$$\tan x + \tanh x = 0,$$

rounded off accurately to three places of decimals.

17.14 Find by the method of false position the root of

$$2x - \mathrm{e}^{-x} = 0,$$

rounded off accurately to three places of decimals.

17.15 Use Newton's method to calculate $\sqrt{21}$ accurately to four decimal places by seeking the zero of the function $f(x) = 21 - x^2$. Start your iteration with $x_0 = 4$.

17.16 Use Newton's method to find to an accuracy of three decimal places the positive root of

$$\tan x = \tanh x.$$

Start your iteration with $x_0 = 3.9$. How many roots will this equation have?

17.17 Show by Newton's method that the positive root of the equation

$$x^4 - 3x^3 + 2x^2 + 2x - 7 = 0$$

is $x = 2.3267$.

17.18 Use Newton's method to find the positive root of

$$\sinh x + 1 - 2x - 2x^2 = 0.$$

17.19 Use Newton's method to find the negative root of

$$x^4 + 5x^3 + 3x^2 - 5x - 9 = 0.$$

Section 17.6

17.20 Taking intervals $h = 0.2$, use Euler's modified method to determine $y(1)$, given that $y' + y = 0$ and $y(0) = 1$. Compare your results with the exact solution $y = e^{-x}$. Construct the Cauchy polygon.

17.21 Taking intervals $h = 0.1$, use Euler's modified method to determine $y(1)$, given that $y' = (x^2 + y)/x$ and $y(0.5) = 0.5$. Compare your results with the exact solution $y = \frac{1}{2}x + x^2$. Construct the Cauchy polygon.

17.22 Given that $y(0) = 1$, use the predictor-corrector method to integrate $y' = xy$ as far as $x = 0.3$, taking increments $h = 0.05$ and obtaining the starting values by means of the modified Euler method. Compare the results with the exact solution $y = \exp(\frac{1}{2}x^2)$.

17.23 Given that $y(0) = 1$, use the predictor-correcter method to integrate $y' = y + \sin x$ as far as $x = 0.3$, taking increments $h = 0.05$ and obtaining the starting values by means of the modified Euler method. Compare the results with the exact solution $y = \frac{1}{2}(3e^x - \sin x - \cos x)$.

17.24 Using the Runge–Kutta method with $h = 0.1$, and working to four decimal places, determine $y(1)$ given that $y' = (x^2 + y)/x$ with $y(0.5) = 0.5$. Compare the result with the exact solution $y = \frac{1}{2}x + x^2$.

17.25 Using the Runge–Kutta method with $h = 0.2$, and working to four decimal places, determine $y(1)$ given that $y' = y + e^{-x}$ and $y(0) = 0$. Compare the result with the exact solution $y = \sinh x$.

17.26 Using the Runge–Kutta method with $h = 0.1$, and working to four decimal places, determine $y(0.3)$ given that $y'' - 3y' + 2y = 0$ with $y(0) = 1$ and $y'(0) = 0$. Compare the result with the exact solution $y = 2e^x - e^{2x}$.

Section 17.7

17.27 Find by iteration the eigenvalues and normalized eigenvectors of the matrix

$$\mathbf{A} = \begin{bmatrix} 1 & 1 & 3 \\ 0 & 6 & -2 \\ 0 & 0 & 11 \end{bmatrix}.$$

17.28 Find by iteration the eigenvalues and normalized eigenvectors of the matrix

$$\mathbf{A} = \begin{bmatrix} 8 & -12 & 5 \\ 15 & -25 & 11 \\ 24 & -42 & 19 \end{bmatrix}.$$

Supplementary computer problems

17.29 Use a computer matrix package and the Jacobi method to solve the diagonally dominant system of equations

$$\begin{aligned} 6.3x_1 + 2.87x_2 - 1.62x_3 + 1.08x_4 &= 3.62 \\ -1.54x_1 + 8.19x_2 + 2.23x_3 + 1.92x_4 &= -4.26 \\ -3.16x_1 + 1.71x_2 + 6.81x_3 - 0.24x_4 &= 6.84 \\ -1.24x_1 + 2.13x_2 - 1.16x_3 - 7.31x_4 &= -1.61, \end{aligned}$$

starting with any choice of initial values $x_1^{(0)}, x_2^{(0)}, x_3^{(0)}$ and $x_4^{(0)}$ and iterating until the difference between two successive approximations to each of x_1 to x_4 first differ by less than 0.0001. Compare your result with the actual solution $x_1 = 1.40578503$, $x_2 = -0.63111071$, $x_3 = 1.79801703$ and $x_4 = -0.48743351$. Alter the order of the equations and perform five iterations to demonstrate the divergence of the calculations when the system is no longer diagonally dominant.

17.30 Repeat the calculations of Problem 17.29, but this time using the Gauss–Seidel scheme, and compare the rate of convergence with that of the Jacobi scheme.

17.31 Use the trapezoidal rule with 20 equal integration steps to find

$$\int_0^{2\pi} e^{-x/2} \sin mx \, dx,$$

for $m = 1$, 2 and 3. Compare the results with the analytical solutions

$$\int_0^{2\pi} e^{-x/2} \sin x \, dx = \tfrac{4}{5}(1 - e^{-\pi}) = 0.76542887$$

$$\int_0^{2\pi} e^{-x/2} \sin 2x \, dx = \tfrac{8}{17}(1 - e^{-\pi}) = 0.45025227$$

$$\int_0^{2\pi} e^{-x/2} \sin 3x \, dx = \tfrac{12}{37}(1 - e^{-\pi}) = 0.31030900.$$

Notice that the accuracy decreases as m increases due to the increased oscillations of the integrand that is not matched by an increase in the number of integration steps necessary to approximate the integrand.

17.32 Repeat the calculations in Problem 17.31, but this time using Simpson's rule. Compare the accuracy obtained using Simpson's rule with that obtained in Problem 17.31 using the trapezoidal rule.

17.33 Use the secant method to find the three roots of the polynomial equation

$$x^3 + 2x^2 - 3x - 2 = 0,$$

stopping the iterations when two successive iterates first differ by 0.001. Use a graph of the function to find appropriate pairs of starting values x_1 and x_2 or each of the three roots.

17.34 Use the secant method to find the three roots of the polynomial equation

$$x^3 + 3x^2 - 4x - 3 = 0,$$

stopping the iterations when two successive iterates first differ by 0.001. Examine the way the method converges to a root if more than one root lies between the starting values x_1 and x_2. Use a graph of the function to find appropriate pairs of starting values for each of the three roots.

17.35 Use Newton' method to find the three real zeros of $f(x) = x^5 + 3x^4 - 3x^2 - x - 4$. You will find it useful to plot the graph of $f(x)$ in order to find starting values for Newton's method.

17.36 Use Newton's method to find the two roots of

$$4 + \sin x - 2\sqrt{1 + x^2} = 0.$$

17.37 Use Newton's method to find the two roots of

$$e^x \sin x - x^2 + 4x - 3 = 0.$$

17.38 Use Newton's method to find the first three positive roots of

$$\tan x - \tfrac{1}{3} x = 0.$$

17.39 Use Newton's method with a symbolic algebra integration package to find the first three positive zeros of

$$f(x) = \int_0^x \cos(\tfrac{1}{2}t^2\pi)\mathrm{d}t - \tfrac{1}{2}.$$

The integral in $f(x)$ is called the *Fresnel cosine integral*, and it is denoted by $C(x)$, and it first arose in connection the study of the diffraction of light. In this case, because of the rapid oscillation of the integrand, the zero of $f(x)$ to which Newton's method converges is very sensitive to the choice of starting value, as will be seen if the value $x = 1.7$ is used. Fig. 17.12 will be useful when choosing starting values and confirming that the zeros found are the ones that are required.

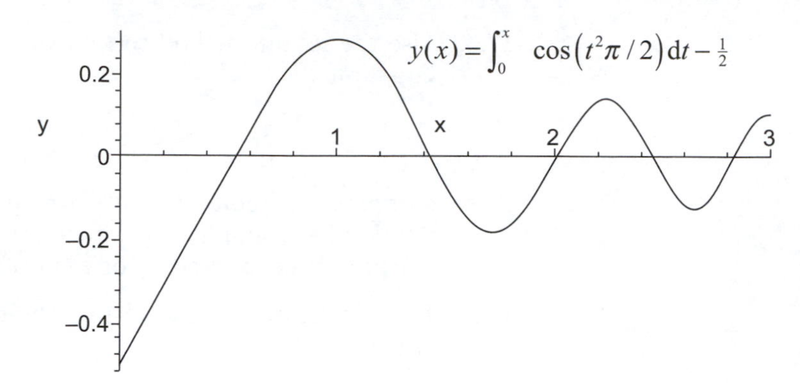

Fig. 17.12 A plot of $y(x) = \int_0^x \cos(\frac{1}{2}t^2\pi)\mathrm{d}t - \frac{1}{2}$. (Hint: When constructing the Newton iterative scheme you will need to use the second result in Theorem 7.6.)

17.40 Use the method of Example 17.9 to find the first three positive zeros of the Bessel function $J_1(x)$ defined as

$$J_1(x) = \frac{2}{\pi}\int_0^{\pi/2} \sin(x \cos\theta)\cos\theta \, \mathrm{d}\theta.$$

Fig. 17.13 will be useful when choosing starting values and confirming that the zeros found are the ones required.

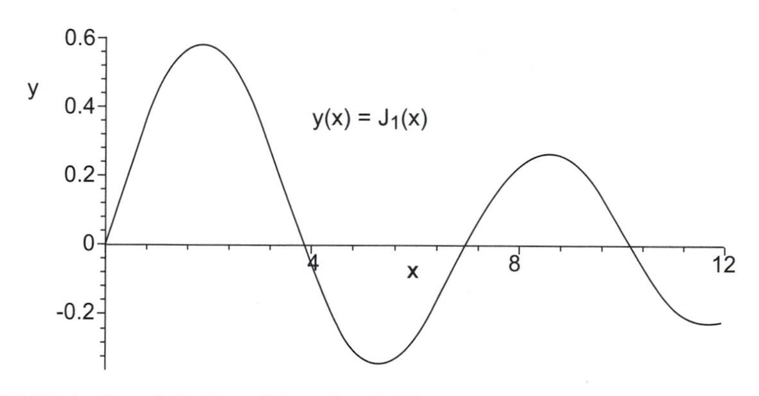

Fig. 17.13 A plot of the Bessel function $J_1(x)$.

17.41 Use Euler's method to find the numerical solution of

$$\frac{\mathrm{d}y}{\mathrm{d}x} = \frac{3 + x^2}{1 + x^2} \quad \text{with } y(0) = 1$$

in the interval $0 \leq x \leq 1$, using 10 steps each with $h = 0.1$. Compare the result with the analytical solution $y\ (x) = 1 + x + 2\arctan x$.

17.42 Use Euler's method to find the numerical solution of

$$\frac{dy}{dx} = \frac{2 + 3x^3}{1 + 2x} \quad \text{with } y(0) = 1$$

in the interval $0 \le x \le 1$, using 10 steps each with $h = 0.1$. Compare the result with the analytical solution $y(x) = 1 - \frac{11}{4}\ln 2 + \frac{11}{8}\ln(4 + 8x) - \frac{3}{4}x + \frac{3}{4}x^2$.

17.43 Use the Runge–Kutta method to find the numerical solution of

$$\frac{dy}{dx} = \frac{1 + \sin x}{1 + x^2} \quad \text{with } y(1) = 1$$

in the interval $0 \le x \le 1$, using 10 steps each with $h = 0.1$.

17.44 Use the Runge–Kutta method to find the numerical solution of

$$\frac{dy}{dx} = \frac{y\sin x}{1 + \sinh x} \quad \text{with } y(1.5) = 2.5$$

in the interval $1.5 \le x \le 2.5$, using 10 steps each with $h = 0.1$.

17.45 Use the Runge–Kutta method to find the numerical solution of

$$\frac{dy}{dx} = \frac{y^2}{1 + 2x^2} \quad \text{with } y(0) = 1$$

in the interval $0 \le x \le 1$, using 10 steps each with $h = 0.1$. Compare the results with the analytical solution

$$y(x) = \frac{2}{2 - \sqrt{2}\arctan(x\sqrt{2})}.$$

17.46 Use the Runge–Kutta method to find the numerical solution of

$$\frac{dy}{dx} = \frac{1 + 2y^2}{2 + x^2} \quad \text{with } y(0) = 0$$

in the interval $0 \le x \le 1$, using 10 steps each with $h = 0.1$. Compare the results with the analytical solution $y(x) = \frac{1}{2}x$.

18 Probability and statistics

This chapter overview summarizes the topics that are to be found in this chapter together with some of the reasons for their inclusion. Many physical situations requiring analysis do not depend on data or events that are determinate, in the sense that an outcome can be predicted exactly. Thus the number of cars waiting each time a traffic light turns red cannot be specified in advance. Variables of this type are called *random variables*. The laws governing the behaviour of random variables are the laws of probability, and they apply subject to certain hypotheses being satisfied. The probability p of the occurrence of an event is a number in the interval $0 \leq p \leq 1$, with $p = 0$ corresponding to an event that never occurs and $p = 1$ to an event that is a certainty.

Random variables can be discrete in nature, like the number of telephone calls to an office during a given period, which must be an integral number. Conversely, random variables can also be continuous, as occurs when the actual resistance of a nominally 1000 ohm resistor produced by a production line is measured. Thus both discrete and continuous random variables must be considered. This in turn leads to the introduction of discrete and continuous probability distributions, and to the study of their consequences when related to randomly occurring events. The most commonly occurring discrete probability distributions are the binomial and Poisson distributions, while the most important continuous distribution is certainly the normal or Gaussian distribution, all of which are considered in this chapter.

Whereas probability theory studies the theoretical behaviour to be expected as a result of infinitely many repetitions of a random event, statistics analyses the actual behaviour based on a finite number of events, called a *sample*, and recorded in a series of actual observations. The determination of unbiased estimates of the mean and standard deviation of a sample of observations, and the confidence with which the mean can be predicted, are some of the simple but widely applicable and useful results that emerge from the study of statistics described in this chapter. Here the confidence with which a variable X can be predicted is specified by asserting that X lies in the confidence interval $a \leq X \leq b$ with probability p.

The fitting of a straight line to experimental data by using the method of least squares is known in statistics as *linear regression*. The theory of confidence intervals can be applied to the linear regression process to determine a confidence interval for the gradient of the regression line. Such a result is of considerable value in experimental work, because when a confidence interval for the gradient is small it is reasonable to conclude that the corresponding straight line will provide a good fit to the data. Conversely, if the confidence interval is large, it is unlikely that the straight-line representation of the data will be adequate.

18.1 The elements of set theory for use in probability and statistics

In probability and statistics it is necessary to study the relationships that exist between sets of events, such as sequences of 'heads' and 'tails' when different coins are tossed, or the number of defective components in batches of a hundred produced by different but supposedly similar production lines. The notation of set theory provides a natural way of describing relationships that exist between sets of events, and so in what follows it will be used to introduce some of the elementary ideas involved in probability theory.

The simplest situation that can occur is that from a set of events A, a new set of events B is formed, such that all the events in set B are also events in set A. Such a set of events B is called a *subset* of A, and we write

$$B \subseteq A,$$

which is to be read as 'B is a subset of A'. If, for example, the set A of events is the record of 'heads' and 'tails' obtained as a result of tossing a coin 100 times, the set B of 'heads' would be a subset of A.

If x is an event in set A, so that we may write $x \in A$, then either $x \in B$, or $x \notin B$. When there are some events $x' \in A$ that are not to be found in set B, so that $x' \notin B$, then B is called a *proper subset* of A, the result being written

$$B \subset A.$$

The definition of a subset B of A does not preclude the possibility that for every element $x \in A$ it is also true that $x \in B$. When this occurs the sets of events A and B contain the same events and are said to be equal, the result being written

$$A = B.$$

It is clear from the definition of equality of sets that when $A = B$ both the statements $A \subset B$ and $B \subset A$ must be true. These last two statements are often useful as an alternative definition of the equality of two sets of events.

A more general situation arises when two sets of events A and B are involved, each of which contains events that are not contained in the other so that neither $A \subset B$ nor $B \subset A$ is true. The set of events C that is common to the two sets A and B is called the *intersection* of the two sets, and we write

$$C = A \cap B.$$

This statement is usually read as 'A cap B'. A typical example occurs when two coins are each tossed 100 times, and the record of the sequence

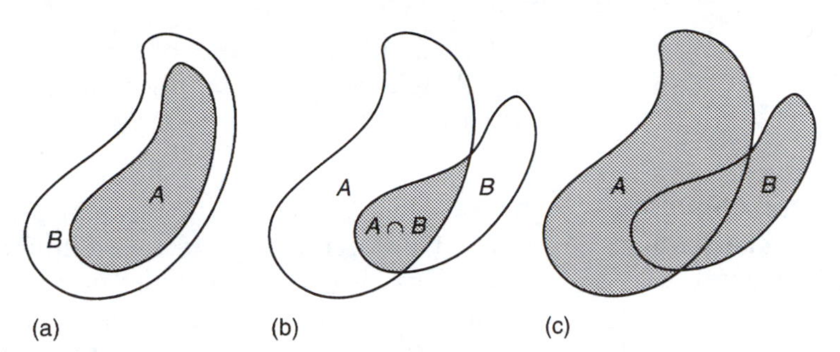

Fig. 18.1 Symbolic representation of set operations: (a) proper subset, $A \subset B$; (b) intersection, $A \cap B$; (c) union, $A \cup B$.

of 'heads' and 'tails' for the first coin is taken as set A while that for the second is taken as set B. An example of the intersection of sets A and B is provided by the set C comprising the occasions on which a 'head' occurs after the same number of tosses in each record.

When it happens that there are *no* events common to sets A and B we write

$$A \cap B = \varnothing,$$

with the understanding that $\varnothing$ is the *null* set that contains no events.

Another important set of events that is related to sets of events A and B is the set C containing all of the events that belong to A, to B or to both A and B. This is called the *union of sets* A and B, and it is written

$$C = A \cup B,$$

and read as 'A cup B'.

These seemingly abstract ideas can be illustrated symbolically by means of a convenient diagram called the *Venn diagram*. The Venn diagram uses a pictorial representation for the sets in question. Sets are represented by the interiors of closed curves, usually of arbitrary shape, and their relationships are then illustrated by the relationships that exist between these curves. Thus, when as in Fig. 18.1(a) a curve A representing a set of events A lies within a curve B representing a set of events B, we have the situation that A is a proper subset of B, so that $A \subset B$. Figures 18.1(b), (c) illustrate, respectively, the intersection $A \cap B$ and the union $A \cup B$ of sets A and B, which are shown as shaded areas on those figures.

In general this representation is only symbolic, but when the events in sets A and B can be unambiguously represented by points in the plane, the Venn diagrams become true representations, as we shall see later in Fig. 18.3.

A final concept we now introduce in connection with sets of events A and B is the *complement* of set B relative to set A, which we shall write as $A \backslash B$. This is a generalization of the notion of subtraction and comprises the set of events in A that do not belong to set B. The notation used here, namely $A \backslash B$, usually read as 'A minus B' or as 'the set of events in A but not in B', is illustrated in Fig. 18.2.

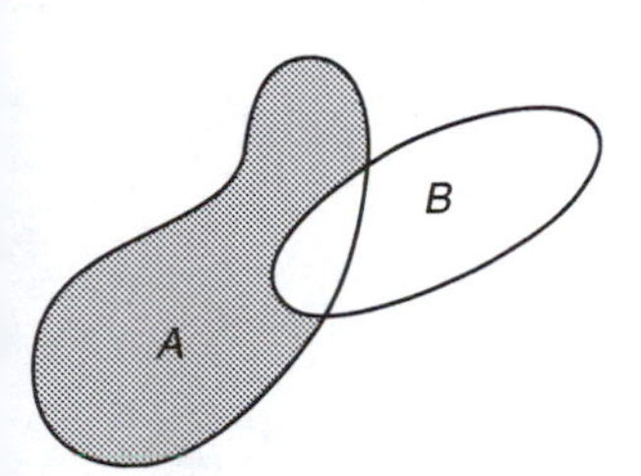

Fig. 18.2 Symbolic representation of complement of B relative to A. The area shaded is $A \backslash B$.

The complement of a set of events A relative to a set of events to which A belongs, but which is otherwise unspecified, is denoted by $\bar{A}$, read either as 'A bar' or, more precisely, as 'the set of points not in A'.

18.2 Probability, discrete distributions and moments

As indicated in section 18.1, one of the most direct applications of the elements of set theory is to be found in a formal introduction to probability theory. Because the notion of a probability is fundamental to many branches of engineering and science we choose to introduce some basic ideas and definitions now, making full use of the notions of set theory.

In some situations the outcome of an experiment is not determinate, so one of several possible events may occur. Following statistical practice, we shall refer to an individual event of this kind as the result or outcome of a *trial*, whereas an agreed number of trials, say N, will be said to constitute an *experiment.* If an experiment comprises throwing a die N times, then a trial would involve throwing it once and the outcome of a trial would be the score that was recorded as a result of the throw. The experiment would involve recording the outcome of each of the N trials.

In general, if a trial has m outcomes we shall denote them by E_1, E_2, ..., E_m and refer to each as a *simple event.* Hence a trial involving tossing a coin would have only two simple events as outcomes: namely 'heads', which could be labelled E_1, and 'tails', which would then be labelled E_2. In this instance an experiment would be a record of the outcomes from a given number of such trials. A typical record of an experiment involving tossing a coin eight times would be E_1, E_2, E_1, E_1, E_1, E_2, E_2, E_1. With such a simple experiment the E_1, E_2 notation has no apparent advantage over writing H in place of E_1 and T in place of E_2 to obtain the equivalent record H, T, H, H, H, T, T, H. The advantage of the E_i notation accrues from the fact that the subscripts attached to the E may be ordered numerically, thereby enabling easier manipulation of the outcomes during analysis.

Events such as the result of tossing a coin or throwing a die are called chance or *random* events, since they are indeterminate and are supposedly the consequence of unbiased chance effects. Experience suggests that the relative frequency of occurrence of each such event averaged over a series of similar experiments tends to a definite value as the number of experiments increases.

The *relative frequency of occurrence* of the simple event E_i in a series of N trials is thus given by the expression

$$\frac{\text{number of occurrences of event } E_i}{N}.$$

By virtue of its definition, this ratio must be either positive and less than unity, or zero. For any given N, this ratio provides an estimate of the theoretical ratio that would have been obtained were N to have been made arbitrarily large. This theoretical ratio will be called the *probability* of occurrence of event E_i and will be written $P(E_i)$. In many simple situations its value may be arrived at by making reasonable postulates concerning the mechanisms involved in a trial. Thus when fairly tossing an unbiased coin it would be reasonable to suppose that over a large number of trials the number of 'heads' would closely approximate the number of 'tails' so that $P(H) = P(T) = \frac{1}{2}$. Here, of course, $P(H)$ signifies the probability of occurrence of a 'head', and $P(T)$ signifies the probability of occurrence of a 'tail'.

If there are m outcomes $E_1, E_2, \ldots, E_m$ of a trial, and they occur with the respective frequencies $n_1, n_2, \ldots, n_m$ in a series of N trials, then we have the obvious identity

$$\frac{n_1 + n_2 + \ldots + n_m}{N} = 1.$$

When N becomes arbitrarily large we may interpret each of the relative frequency ratios $n_i/N(i = 1, 2, \ldots, m)$ occurring on the left-hand side as the probability of occurrence $P(E_i)$ of event E_i, thereby giving rise to the general result

$$P(E_1) + P(E_2) + \ldots + P(E_m) = 1. \tag{18.1}$$

By this time a careful reader will have noticed that the definition of probability adopted here has a logical difficulty associated with it, namely, the question whether a relative frequency ratio such as n_i/N can be said to approach a definite number as N becomes arbitrarily large. We shall not attempt to discuss this philosophical point more fully but rather be content that our simple definition in terms of the relative frequency ratio is in accord with everyday experience.

An examination of Eqn (18.1) and its associated relative frequency ratios is instructive. It shows the obvious results that:

(a) if event E_i never occurs, then $n_i = 0$ and $P(E_i) = 0$;
(b) if event E_i is certain to occur, then $n_i = N$ and $P(E_i) = 1$;
(c) if event E_i occurs less frequently than event E_j, then $n_i < n_j$ and $P(E_i) < P(E_j)$;
(d) if the m possible events $E_1, E_2, \ldots, E_m$ occur with equal frequency, then $n_1 = n_2 = \ldots = n_m = N/m$ and $P(E_1) = P(E_2) = \ldots = P(E_m) = 1/m$.

The relationship between sets and probability begins to emerge once it is appreciated that a trial having m different outcomes is simply a rule by which an event may be classified unambiguously as belonging to one of m different sets. Often a geometrical analogy may be used to advantage when representing the different outcomes of a particular trial.

A convenient example is provided by the simple experiment which involves throwing two dice and recording their individual scores. There will be in all 36 possible outcomes which may be recorded as the ordered number pairs (1, 1), (1, 2), (1, 3), ..., (2, 1), (2, 2), ..., (6, 5), (6, 6). Here

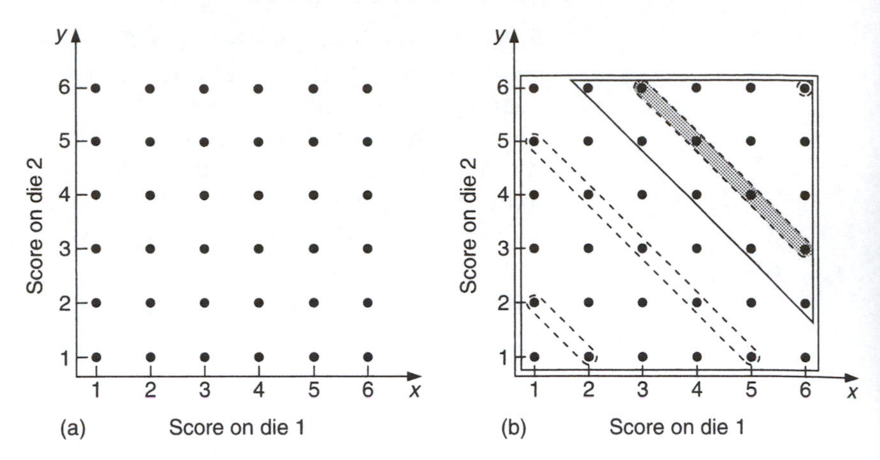

Fig. 18.3 Sample space for two dice: (a) complete sample space; (b) sample space for specific outcome.

the first integer in the ordered number pair represents the score on die 1 and the second the score on die 2. These may be plotted as 36 points with integer coordinates as shown in Fig. 18.3(a).

Because each of the indicated points in Fig. 18.3(a) lies in a two-dimensional geometrical space (that is, they are specific points in a plane), and in their totality they describe all possible outcomes, the representation is usually called the *sample space* of events. The probability of occurrence of an event chracterized by a point in the sample space is, of course, the probability of occurrence of the simple event it represents.

As a sample space will require a 'dimension' for each of its variables it is immediately apparent that only in simple cases can it be represented graphically.

The points in the sample space may be regarded as defining points in a set D so that specific requirements as to the outcome of a trial will define a subset A of D, at each point of which the required event will occur. Typical of this situation would be the case in which a simple event is the throw of two dice, and the requirement defining the subset is that the combined score after throwing the two dice equals or exceeds 8. Here the set D would be the 36 points within the square in Fig. 18.3(b) and the set A the points within the triangle. Using set notation, we may write $A \subset D$.

The sample-space representation becomes particularly valuable when trials are considered whose outcome depends on the combination of events belonging to two different subsets A and B of the sample space. Thus, again using our previous example and taking for A the points within the triangle in Fig. 18.3(b), the points in B might be determined by the requirement that the combined score be divisible by the integer 3. The set of points B is then those contained within the triangle and the dotted curves of Fig. 18.3(b).

A new set C may be derived from two sets A and B in two essentially different ways according to whether

(a) C contains points in A or B or in both A and B;
(b) C contains points in A and B.

For convenience, in statistics the situation in (a) is abbreviated to 'C contains points in A or B,' while as the one in (b) is unambiguous it is left as 'C contains points in A and B'.

If desired, these statements about sets may be rewritten as statements about events. This is so because there is an unambiguous relationship between an event and the set of points S in the sample space at which that event occurs. Thus, for example, we may paraphrase the first statement by saying 'the event corresponding to points in C denotes the occurrence of the events corresponding to points in A or B, or both'. Because of this relationship it is often convenient to regard an event and the subset of points it defines in the sample space as being synonymous.

The statements provide yet another connection with set theory, since in (a) we may obviously write $C = A \cup B$, whereas in (b) we must write $C = A \cap B$. In terms of the sets A and B defined in connection with Fig. 18.3(b), the set $C = A \cup B$ contains the points in the triangle together with those within the two dotted curves exterior to the triangle. The set $C = A \cap B$ contains only the five points within the two dotted curves lying inside the triangle.

Here it should be remarked that the statistician often prefers to avoid the set theory symbols $\cup$ and $\cap$, writing instead either $A+B$ or often 'A or B' in place of $A \cup B$, and AB or often 'A and B' in place of $A \cap B$. This largely arises because of the duality we have already mentioned that exists between an event and the set of points it defines, the statistican naturally preferring to think in terms of events rather than sets. However, to emphasize the connection with set theory we shall preserve the set theory notation.

Using this duality, we now denote by $P(A)$ the probability that an event corresponding to a point in the sample space lies within subset A, and define its value to be as follows.

Definition 18.1 $P(A)$ is the sum of the probabilities associated with every point belonging to the subset A.

A very useful general result concerning $P(A)$ follows from the definition of the complement of a set introduced in section 18.1. To derive it we proceed as follows. Let A be a subset of all the points in some set S. Then the sets A and $S\backslash A$, the complement of A relative to S, are such that

$$A \cup (S\backslash A) = S.$$

Expressed in words, this merely asserts the obvious result that the points in A together with the points in S which are *not* in A comprise the set S itself. If now we use the usual abbreviation and write $\bar{A} = S\backslash A$, this becomes

$$A \cup \bar{A} = S.$$

Now let S be the set of all points in a sample space, and A be the set of points belonging to S at which an event A occurs with probability $P(A)$. Then, in terms of probabilities, the statement equivalent to $A \cup \bar{A} = S$ becomes

$$P(A) + P(\bar{A}) = P(S),$$

but $P(S) = 1$, so

$$P(\bar{A}) = 1 - P(A).$$

An application of this result will be found in Example 18.3(iv).

In Fig. 18.3(b) the set A contains the 15 points within the triangle and, since for unbiased dice each point in the sample space is equally probable, it follows at once that the probability $1/36$ is to be associated with each of these points. Hence from our definition we see that in this case, $P(A) = 15 \times (1/36) = 5/12$. Similarly, for the set B comprising the 12 points contained within the dotted curves we have $P(B) = 12 \times (1/36) = 1/3$.

We can now introduce the idea of a conditional probability through the following definition.

Definition 18.2 $P(A|B)$ is the *conditional probability* that an event known to be associated with set B is also associated with set A.

Clearly we are only interested in the relationship that exists between A and B, with B now playing the part of a sample space. Because in Definition 18.2 B plays the part of a sample space, but is itself only a subset of the complete sample space, it is sometimes called the *reduced sample space*.

In terms of set theory, Definition 18.2 is easily seen to be equivalent to

$$P(A|B) = \frac{P(A \cap B)}{P(B)} \quad \text{and, equivalently,} \quad P(B|A) = \frac{P(A \cap B)}{P(A)}, \tag{18.2}$$

which immediately shows us how $P(A|B)$ may be computed. Namely, $P(A|B)$ is obtained by dividing the sum of the probabilities at points belonging to the intersection $A \cap B$ of sets A and B by the sum of the probabilities at points belonging to B. This ensures that $P(B|B) = 1$, as would be expected.

Alternative statements of (18.2) employing the different notations just mentioned are

$$P(A|B) = \frac{P(A \text{ and } B)}{P(B)} \qquad \text{and, equivalently,}$$

$$P(B|A) = \frac{P(A \text{ and } B)}{P(A)}$$

or

$$P(A|B) = \frac{P(AB)}{P(B)} \quad \text{and, equivalently,} \quad P(B|A) = \frac{P(AB)}{P(A)}.$$

We can illustrate this by again appealing to the sets A and B defined in connection with Fig. 18.3(b). It has already been established that $P(B) = 1/3$, and since there are only five points in $A \cap B$, each with a probability $1/36$, it follows that $P(A \cap B) = 5/36$. Hence $P(A \mid B) = (5/36)/(l/3) = 5/12$. This result, expressed in words, states that when two dice are thrown and their score is divisible by the integer 3, then the probability that it also equals or exceeds 8 is $5/12$.

A direct consequence of Eqn (18.2) is the so-called *probability multiplication rule*.

Theorem 18.1 If two events define subsets A and B of a sample space, then

$$P(A \cap B) = P(B)P(A|B) = P(A)P(B|\mathrm{A}). \quad \blacksquare$$

Sometimes, when it is given that the event corresponding to points in subset B occurs, it is also true that $P(A|B)$ depends only on A, so that $P(A|B) = P(A)$. The events giving rise to subsets A and B will then be said to be *independent*. The probability multiplication rule then simplifies in an obvious manner which we express as follows:

Corollary 18.1 If the events giving rise to subsets A and B of a sample space are independent, then

$$P(A \cap B) = P(A)P(B). \quad \blacksquare$$

Equivalent statements using the alternative notations are

$$P(A \text{ and } B) \quad \text{and} \quad P(A)P(B) \quad \text{and} \quad P(AB) = P(A)P(B).$$

Consideration of the interpretation of $P(A \cup B)$ leads to another important result known as the *probability addition rule*.

Theorem 18.2 If two events define subsets A and B of a sample space, then

$$P(A \cup B) = P(A) + P(B) - P(A \cap B). \quad \blacksquare$$

Equivalent statements using the alternative notations are

$$P(A \text{ or } B) = P(A) + P(B) - P(A \text{ and } B),$$

and

$$P(A + B) = P(A) + P(B) - P(AB).$$

The proof of this theorem is self-evident once it is remarked that when computing $P(A)$ and $P(B)$ from subsets A and B and then forming the expression $P(A)+P(B)$, the sum of probabilities at points in the intersection $A \cap B$ is counted twice. Hence $P(A)+P(B)$ exceeds $P(A \cup B)$ by an amount $P(A \cap B)$.

The probability addition rule also has an important special case when sets A and B are disjoint so that $A \cap B = \varnothing$. When this occurs the events corresponding to sets A and B are said to be *mutually exclusive*, and we express the result as follows:

Corollary 18.2 If the events giving rise to subsets A and B of a sample space are mutually exclusive, then

$$P(A \cup B) = P(A) + P(B). \quad \blacksquare$$

Equivalent statements using the alternative notations are

$$P(A \text{ or } B) = P(A) + P(B) \quad \text{and} \quad P(AB) = P(A) + P(B).$$

As a simple illustration of Theorem 18.2 we again use the sets A and B defined in connection with Fig. 18.3(b) to compute $P(A \cup B)$. The result is immediate, for we have already obtained the results $P(A) = 5/12$, $P(B) = 1/3$ and $P(A \cap B) = 5/36$, so from Theorem 18.2 follows the result

$$P(A \cup B) = 5/12 + 1/3 - 5/36 = 11/18.$$

The applications of these theorems and their corollaries are well illustrated by the following simple examples.

Example 18.1 A bag contains a very large number of red and black balls in the ratio one red ball to four black. If two balls are drawn successively from the bag at random, what is the probability of selecting

(a) two red balls,
(b) two black balls,
(c) one red and one black ball?

Solution Let A_1 denote the selection of a red ball first (and either colour second), and A_2 the selection of a red ball second (and either colour first). Then $A_1 \cap A_2$ is the selection of two red balls and, similarly, $B_1 \cap B_2$ is the selection of two black balls. As the balls occur in the ratio one red to four black it follows that their relative frequency ratios are $1/5$ for a red ball and 4/5 for a black ball, so $P(A_1) = 1/5$ and $P(B_1) = 4/5$.

The fact that the bag contains a *large* number of balls implies that the drawing of one or more does not materially alter the relative frequency ratio that existed at the start, so $P(A_i) = 1/5$ and $P(B_i) = 4/5$. This, together with the fact that the balls are drawn at random, implies that the drawing of each ball is an *independent* event. The independence of events A and B then alows the use of Corollary 18.1 to determine the required solutions to (a) and (b). We find that

(a) $P(A_1 \cap A_2) = (1/5).(1/5) = 1/25$,
(b) $P(B_1 \cap B_2) = (4/5).(4/5) = 16/25$.

Now to answer (c) we notice that there are two mutually exclusive orders in which a red and a black ball may be selected: namely as the event $C \cup D$ where $C = A_1 \cap B_2$ (red then black) and $D = B_1 \cap A_2$ (black then red). From Corollary 18.2 we then have that $P(C \cup D) = P(C) + P(D)$, where $P(C)$ and $P(D)$ are determined by Corollary 18.1. This shows that $P(C) = P(A_1)P(B_2)$ and $P(D) = P(B_1)P(A_2)$, so that $P(C) = P(D) = (1/5).(4/5) = 4/25$. This solution to (c) becomes

$$P(C \cup D) = 4/25 + 4/25 = 8/25.$$

The three forms of selection (a), (b) and (c) are themselves mutually exclusive, and it must follow that $P(A_1 \cap A_2) + P(B_1 \cap B_2) + P(C \cup D) = 1$, as is readily checked. Indeed, this result could have been used directly to calculate $P(C \cup D)$ from $P(A_1 \cap A_2)$ and $P(B_1 \cap B_2)$ in place of the above argument using Corollary 18.2. ■

The previous situation becomes slightly more complicated if only a limited number of balls are contained in the bag.

Example 18.2 A bag contains 50 balls of which 10 are red and the remander black. If two balls are drawn successively from the bag at random, what is the probability of selecting

(a) two red balls,
(b) two black balls,
(c) one red and one black ball?

Solution This time the approach must be slightly different because, unlike in Example 18.1, the removal of a ball from the bag now materially alters the probabilities involved when the next ball is drawn. In fact this is a problem involving conditional probabilities.

Here we shall define A to be the event that the first ball selected is red, and B to be the event that the second ball selected is red. The probability we must now evaluate is the probability of occurrence of event B given that event A has occurred. Expressed in set notation, we have to find $P(A \cap B)$, the probability of occurrence of the event associated with

$(A \cap B)$. This is a conditional probability with the set associated with event A playing the role of the reduced sample space. Utilizing this observation, we now make use of Theorem (18.2) to write

$$P(A \cap B) = P(A)P(B|A).$$

Now the relative frequency of occurrence of a red ball at the first draw is $10/(10+40) = 1/5$, so that $P(A) = 1/5$. (Not till later will we use the fact that the relative frequency of occurrence of a black ball is $40/(10+40) = 4/5$.)

Given that a red ball has been drawn, 9 red balls and 40 black balls remain in the bag. If the next ball to be drawn is red then its probability of occurrence is the conditional probability $P(B \mid A) = 9/(9+40) = 9/49$. Hence it follows that the solution to (a) is

$$P(A \cap B) = (1/5).(9/49) = 9/245.$$

It is interesting to compare this with the value 1/25 that was obtained in Example 18.1. on the assumption that there was virtually an infinite number of balls in the bag.

If C is defined to be the event that the first ball drawn is black and D the event that the second ball drawn is black, then to answer (b) we must compute $P(C \cap D)$. Obviously, $P(C) = 4/5$, and by using an argument analogous to that above it follows that $P(D \mid C) = 39/(10+39) = 39/49$. Hence the solution to question (b) is

$$P(C \cap D) = (4/5).(39/49) = 156/245.$$

Again this should be compared with the value 16/25 obtained in Example 18.1.

The simplest way to answer (c) is to use the fact that events (a), (b) and (c) describe the only possibilities and so are mutually exclusive. Hence the sum of the three probabilities must equal unity. Denoting the probability of event (c) by P, we have

$$P = 1 - P(A \cap B) - P(C \cap D),$$

showing that $P = 1 - 9/245 - 156/245 = 16/49$. ■

Example 18.3 Given that events A and B occur with probabilities $P(A) = 1/5$ and $P(B) = 1/3$, and that $P(A \cup B) = 2/5$, find (a) $P(A|B)$, (b) $P(B|A)$, (c) $P(A \cap \bar{B})$ and (d) $P(A|\bar{B})$.

Solution To calculate results (a) and (b) we need to know $P(A \cap B)$, which may be found from Theorem 18.2 when it is written in the form

$$P(A \cap B) = P(A) + P(B) - P(A \cup B).$$

(a) We have

$$P(A|B) = \frac{P(A \cap B)}{P(B)} = \frac{P(A) + P(B) - P(A \cup B)}{P(B)}$$
$$= \frac{(1/5) + (1/3) - (2/5)}{1/3} = \frac{2}{5},$$

so

$$P(A|B) = 2/5.$$

(b) We have

$$P(B|A) = \frac{P(A \cap B)}{P(A)} = \frac{P(A) + P(B) - P(A \cup B)}{P(A)}$$
$$= \frac{(1/5) + (1/3) - (2/5)}{1/5} = \frac{2}{3},$$

so

$$P(B|A) = 2/3.$$

(c) We recall from section 18.1 that if the points of A belong to a set S, say, then the set of points belonging to S but not to A is denoted by $S \backslash A$ or, more simply, by $\bar{A}$. Then if A and B are any two sets (both is S), $A \cap B$ is the set of points in both A and B, while $A \cap \bar{B}$ is the set of points in A but not in B. Consequently the union of these two sets is simply A, so we have proved that

$$A = (A \cap B) \cup (A \cap \bar{B}).$$

Now if A and B are events with probabilities $P(A)$ and $P(B)$, the equivalent statement in terms of probabilities is

$$P(A) = P(A \cap B) + P(A \cap \bar{B}),$$

so that

$$P(A \cap \bar{B}) = P(A) - P(A \cap B).$$

This result is sufficiently useful for it to be worth restating using the alternative notations often used by statisticians. It is equivalent either to

$$P(A \text{ and } \bar{B}) = P(A) - P(A \text{ and } B),$$

or to

$$P(A\bar{B}) = P(A) - P(AB).$$

Returning to the problem, and using the fact that $P(A) = 1/5$ and

$$P(A \cap B) = P(A) + P(B) - P(A \cup B) = 1/5 + 1/3 - 2/5 = 2/15,$$

it follows at once that

$$P(A\cap \bar{B}) = 1/5 - 2/15 = 1/15.$$

(d) $$P(A|\bar{B}) = \frac{P(A\cap \bar{B})}{P(\bar{B})} = \frac{P(A\cap \bar{B})}{1 - P(B)} = \frac{1/15}{1-(1/3)} = \frac{1}{10},$$

so

$$P(A|\bar{B}) = 1/10. \quad \blacksquare$$

Example 18.4 The power supply system illustrated in Fig. 18.4 contains four circuit breakers A, B, C and D. The performance of each circuit breaker is independent of the others, and their respective probabilities of cutting off the power supply to their segment of the transmission line are p_A, p_B, p_c, p_C. Find the probability that the transmission line will transmit power.

Solution If q_T and q_F are, respectively, the probabilities that the line transmits and fails to transmit power, then as these are mutually exclusive events

$$q_T + q_F = 1,$$

so

$$q_T = 1 - q_F.$$

To determine q_F, which is a little simpler to find than q_T, we notice first that

$$q_F = \text{probability of failure of } (A \text{ or } (B \text{ and } C) \text{ or } D).$$

In terms of the usual probability notation this becomes

$$q_F = P(A\cup (B\cap C)\cup D).$$

Setting $A\cup (B\cap C) = E$, it follows from Theorem 19.2 that

$$q_F = P(E) + P(D) - P(E\cap D).$$

Now

$$P(E) = P(A\cup (B\cap C)),$$

so again applying Theorem 18.2 we find that

$$P(E) = P(A) + P(B\cap C) - P(A\cap (B\cap C)).$$

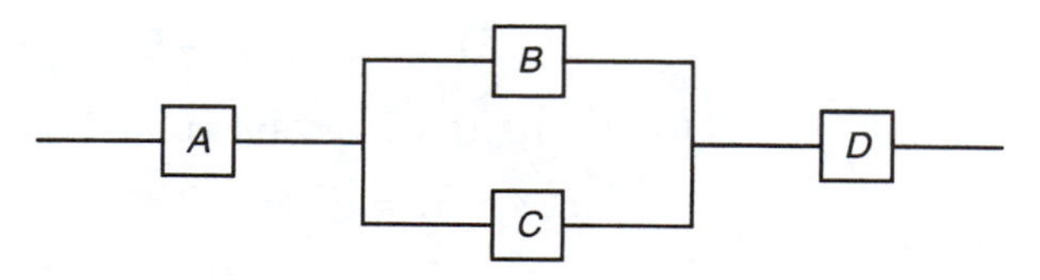

Fig. 18.4 The circuit breaker configuration.

As the failures are independent events,

$$P(B \cap C) = P(B)P(C) \quad \text{and} \quad P(A \cap (B \cap C)) = P(A)P(B)P(C),$$

and thus

$$P(E) = P(A) + P(B)P(C) - P(A)P(B)P(C).$$

Once again, as the failures are independent events, we must have

$$P(E \cap D) = P(E)P(D),$$

so it follows that

$$q_{\mathrm{F}} = P(E) + P(D) - P(E)P(D).$$

Finally, setting $P(A) = p_A$, $P(B) = p_B$, $P(C) = p_C$ and $P(D) = p_D$ and substituting for $P(E)$, we arrive at the probability of failure

$$q_F = p_A + p_B p_C - p_A p_B p_C + p_D - p_A p_D - p_B p_C p_D + p_A p_B p_C p_D.$$

The probability q_T that the line transmits power is thus

$$q_T = 1 - p_A - p_B p_C + p_A p_B p_C - p_D + p_A p_D + p_B p_C p_D \\ - p_A p_B p_C p_D. \quad \blacksquare$$

On occasion a diagram helps when determining the probability of occurrence of events. To illustrate matters, suppose out of a batch of 70 transistors 9 are in fact defective. Then to determine the probability that there will be one defective transistor in a random sample of three when replacement is not involved, we may proceed as follows. In Fig. 18.5, called a tree branch diagram, the symbol G represents a

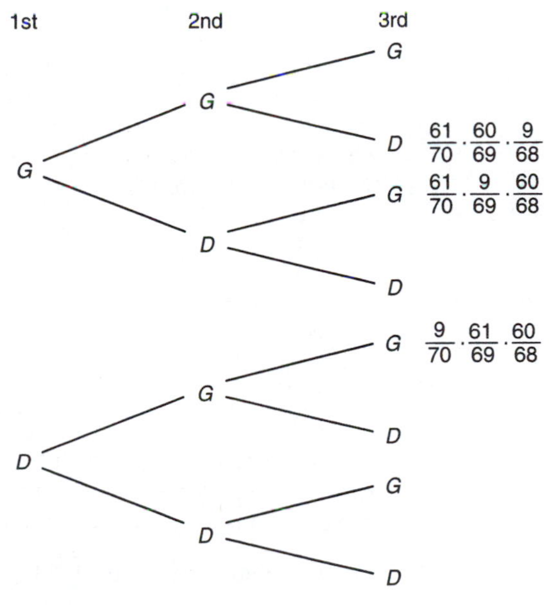

Fig. 18.5 Tree branch diagram.

good transistor, D a defective one, and the columns numbered 1st, 2nd and 3rd show the result of the first, second and third random sample without replacement and its possible outcome. Each factor in the products at the right represents the probability of the successive outcomes leading to one defective transistor in the sample of three. Since each product of factors equals 0.1003, the probability of there being one defective transistor in a random sample of three without replacement is 0.3009.

To close this section with a brief examination of repeated trials, the ideas of permutation and combination must be utilized. The student will already be familiar with these concepts from elementary combinatorial algebra, and so we shall only record two definitions.

Definition 18.3 A *permutation* of a set of n mutually distinguishable objects r at a time is an arrangement or enumeration of the objects in which their order of appearance counts.

Thus of the five letters a, b, c, d, e the arrangements a, b, c and a, c, b represent two different permutations of three of the five letters. These are described as permutations of five letters taken three at a time. Other permutations of this kind may be obtained by further rearrangement of the letters a, b, c and by the replacement of any of them by either or both of the remaining two letters d and e.

The total number of different permutations of n objects r at a time will be denoted by ${}^nP_r$, and it is left to the reader to prove as an exercise that

$${}^nP_r = \frac{n!}{(n-r)!}, \tag{18.3}$$

where $n!$ (factorial n) $= n(n-1)(n-2)\ldots 3.2.1$, and, as usual, we adopt the convention that $0! = 1$.

Definition 18.4 A *combination* of a set of n mutually distinguishable objects r at a time is a selection of r objects from the n without regard to their order of arrangement.

It follows from the definition of a permutation that a set of r objects may be arranged in $r!$ different ways so that the number of different combinations of n objects r at a time must be $n!/r!(n-r)!$, which is simply the binomial coefficient $\binom{n}{r}$. We have thus found the result that the number of different combinations of n objects r at a time is

$$\binom{n}{r} = \frac{n!}{r!(n-r)!}. \tag{18.4}$$

In many books it will often be found that the expression ${}^nC_r$ is written in place of $\binom{n}{r}$. It will be recalled that the numbers $\binom{n}{r}$ are called *binomial coefficients* because of their occurrence in the binomial expansion

$$(p+q)^n = \sum_{r=0}^{n} \binom{n}{r} p^r q^{n-r}, \quad \text{with } n \text{ a positive integer.} \tag{18.5}$$

Example 18.5 Given the eight digits 1, 2, 3, 4, 5, 6, 7, 8, find

(a) how many different eight-digit even numbers can be formed,
(b) how many different eight-digit even numbers can be formed which start with an odd number,
(c) how many different four-digit numbers can be formed in which no digit is repeated.

Solution (a) An even number is, by definition, exactly divisible by 2, so each number formed must terminate in an even digit. Consequently the required number of eight-digit numbers must be

$$7! \times 4 = 20\,160.$$

This is so because the first seven digits can be arranged in 7! different ways while the last digit, which must be even, can only be chosen in 4 different ways.

(b) The required number of eight-digit numbers must be

$$4 \times 6! \times 4 = 11\,520$$

because the first and last digits can each be chosen in 4 different ways, while the six digits in between can be arranged in 6! different ways.

(c) The result here is simply

$${}^8P_4 = \frac{8!}{(8-4)!} = 1\,680. \quad \blacksquare$$

Now consider an experiment involving a series of independent trials in each of which only one of two events A or B may occur. Then if the probabilities of occurrence of events A and B are p and q, respectively, we must obviously have $p+q=1$. If n such trials constitute an experiment, we might wish to know with what probability the experiment may be expected to yield r events of type A. The statistician will call such a situation *repeated independent trials*.

An experiment will be deemed to be successful if r events of type A and $n-r$ events of type B occur, irrespective of their order of occurrence.

Clearly this can happen in $\binom{n}{r}$ different ways, and by Corollary 18.2, since the trials are independent, the probability of occurrence of any one of these events will be $p^r(1-p)^{n-r}$. Hence, as the results of trials are also mutually exclusive, it follows from Corollary 18.3 that the required probability $P(r)$ of occurrence of r events of A each with probability of occurrence p in n independent trials is

$$P(r) = \binom{n}{r} p^r(1-p)^{n-r}. \tag{18.6}$$

Identifying the p and q of Eqn (18.5) with the probabilities of occurrence of the events A and B just discussed, we see that $q = 1-p$, so that Eqn (18.5) takes the form

$$1 = \sum_{r=0}^{n} \binom{n}{r} p^r(1-p)^{n-r}. \tag{18.7}$$

Each term on the right-hand side of Eqn (18.7) then represents the probability of occurrence of an event of the form just discussed. For example, the first term

$$P(0) = \binom{n}{0}(1-p)^n$$

is the probability that event A will never occur in a series of n independent trials, while the third term

$$P(2) = \binom{n}{2} p^2(1-p)^{n-2}$$

is the probability that event A will occur exactly twice in a series of n independent trials.

The $n+1$ numbers $P(r)$, $r = 0, 1, \ldots, n$ have, by definition, the property that

$$P(0) + P(1) + \ldots + P(n) = 1, \tag{18.8}$$

and they are said to define a *discrete probability distribution*. It is conventional to plot them in histogram fashion when they illustrate the probabilities to be associated with the $n+1$ possible outcomes of an experiment involving n trials. Fig. 18.6(a) illustrates the case in which $n = 4$ and $p = \frac{1}{4}$, so that $P(0) = \binom{4}{0}\left(\frac{1}{4}\right)^0\left(\frac{3}{4}\right)^4 = \frac{81}{256}$, $P(1) = \binom{4}{1}\left(\frac{1}{4}\right)^1\left(\frac{3}{4}\right)^3 = \frac{27}{64}$, and, similarly, $P(2) = 54/256$, $P(3) = 3/64$, and $P(4) = 1/256$. Because of the origin of this distribution, Eqn (18.6) is said to define the *binomial distribution*. This distribution is historically associated with Jakob Bernoulli (1654–1705) and experiments of the type

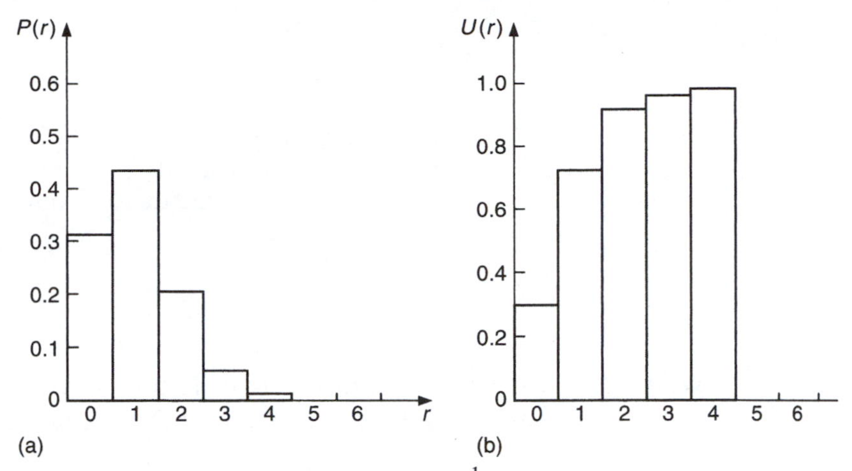

Fig. 18.6 Binomial distribution with $n = 4, p = \frac{1}{4}$: (a) probability density function; (b) cumulative distribution function.

just examined are sometimes referred to as *Bernoulli trials*. When the cumulative total

$$U(r) = \sum_{t=0}^{r} P(t), \tag{18.9}$$

is plotted in histogram fashion against r the result is called the *cumulative distribution function*. The cumulative distribution function corresponding to Fig. 18.6(a) is shown in Fig. 18.6(b). It is conventional to refer to $P(r)$ as the probability density function or the frequency function since it describes the proportion of observations appropriate to the value of r.

Example 18.6 If an unbiased coin is tossed six times, what is the probability that only two 'heads' will occur in the sequence of results?

Solution As the coin is unbiased $p = q = \frac{1}{2}$ and so

$$P(2) = \binom{6}{2}\left(\frac{1}{2}\right)^2\left(\frac{1}{2}\right)^4 = \frac{15}{64}. \quad ■$$

It is an immediate consequence of Eqn (18.6) that

(a) if A occurs with probability p in independent trials then the probability that it will occur *at least* r times in n trials is

$$\sum_{s=r}^{n} \binom{n}{s} p^s(1-p)^{n-s};$$

(b) if A occurs with probability p in independent trials then the probability that it will occur *at most* r times in n trials is

$$\sum_{s=0}^{r}\binom{n}{s}p^s(1-p)^{n-s};$$

and to this we may add Eqn (18.6) in this form;

(c) if A occurs with probability p in independent trials then the probability that it will occur *exactly* r times in n trials is

$$\binom{n}{r}p^r(1-p)^{n-r}.$$

Example 18.7 What is the probability of hitting a target when three shells are fired, assuming each to have a probability $\frac{1}{2}$ of making a hit?

Solution Obviously here $p=\frac{1}{2}$, and we will have satisfied the conditions of the question if *at least* one shell finds the target. Accordingly, using (a) above, the result is

$$\sum_{s=1}^{3}\binom{3}{s}\left(\frac{1}{2}\right)^s(1-\tfrac{1}{2})^{3-s}.$$

Hence the required probability is $\frac{3}{8}+\frac{3}{8}+\frac{1}{8}=\frac{7}{8}$. ∎

The binomial distribution (18.6) is characterized by the two parameters n, p and its form varies considerably with the values they assume. One important case occurs when p is very small and n is large, in such a way that the product np is not negligible. The binomial distribution can then be approximated by a new and simpler distribution characterized by the single parameter $\mu=np$. To see this, observe first that Eqn (18.6) can be written in the form

$$P(r)=Q(r,\ n)\frac{\mu^r}{r!}\left(1-\frac{\mu}{n}\right)^n, \tag{18.10}$$

where

$$Q(r,\ n)=\left(1-\frac{1}{n}\right)\left(1-\frac{2}{n}\right)\dots\left(1-\frac{r-1}{n}\right)\left(1-\frac{\mu}{n}\right)^{-r}.$$

However, for any fixed r we have

$$\lim_{n\to\infty}Q(r,\ n)=1,$$

while, from section 3.3,

$$\lim_{n\to\infty}\left(1-\frac{\mu}{n}\right)^n=\mathrm{e}^{-\mu},$$

so that in the limit as $n\to\infty$,

$$P(r) = \frac{\mu^r e^{-\mu}}{r!}. \tag{18.11}$$

This is a probability density function in its own right. We are able to assert this because $P(r) \geq 0$ for all $r \geq 0$ and the probability that events of all frequencies occur, namely

$$\sum_{r=0}^{\infty} P(r) = \sum_{r=0}^{\infty} \frac{\mu^r e^{-\mu}}{r!} = e^{-\mu} \sum_{r=0}^{\infty} \frac{\mu^r}{r!} = e^{-\mu} e^{\mu} = 1,$$

as is required of a probability density function.

The probability density function in Eqn (18.11) is called the *Poisson distribution* and differs from the binomial distribution, for which we have seen it is a limiting approximation, in that it only involves the single parameter μ. It finds useful application in a variety of seemingly different circumstances. Thus, for example, it can be used to describe the particle count rate in a radioactive process, the rate at which telephone calls arrive at an exchange and the rate of arrival of people in a queue.

The Poisson probability distribution is illustrated in Fig. 18.7(a) for the choice of parameter $\mu = 3$. This could, for instance, be taken to be an approximation to a binomial distribution in which $n = 150$ and $p = 0.02$. When the equivalent histogram for a binomial distribution with these values of n and p is shown for comparison on the same diagram, it is virtually indistinguishable from the Poisson distribution. The Poisson distribution in Fig. 18.7(a) would not, however, be a very good approximation for a binomial distribution with $n = 30$, $p = 0.1$. As a general rule, the approximation of the binomial distribution by the Poisson distribution is satisfactory when $n \geq 100$ and $p < 0.05$. The

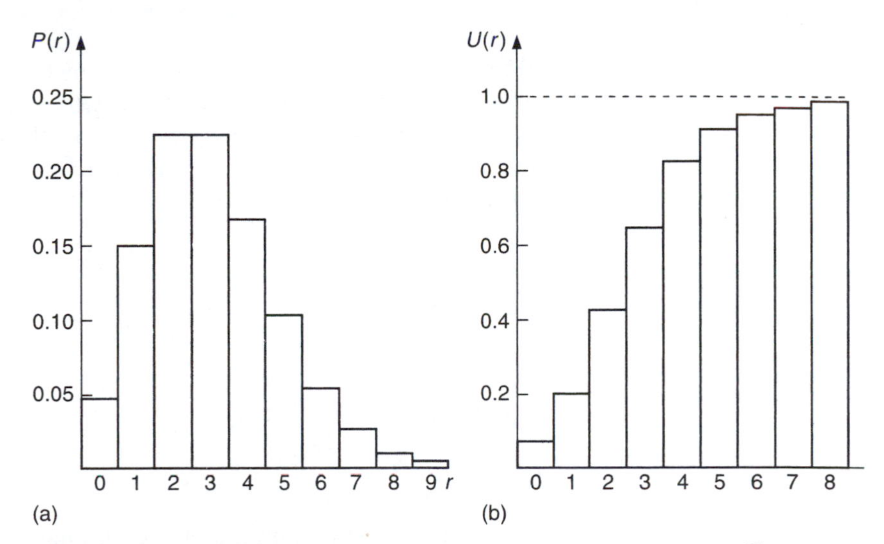

Fig. 18.7 Poisson distribution with $\mu = 3$ and standard deviation $\sigma = \sqrt{3}$ (a) probability density function; (b) cumulative distribution function.

Poisson cumulative distribution function corresponding to Fig. 18.7(a) is shown in Fig. 18.7(b).

Whereas $P(r)$ is the probability that exactly r events will occur, the cumulative probability $U(r)$ is the probability that up to and including r events will occur. Thus $U(3)$ is the probability that 0, 1, 2 or 3 events will occur.

Example 18.8 An ammunition manufacturer knows from experience that 5 out of every 1000 shells manufactured can be expected to be defective. What is the probability that out of a batch of 600, (a) exactly 4 will be defective, (b) no more than 4 will be defective.

Solution Here $p = 5/1000 = 1/200$ and $n = 600$ so that $\mu = np = 3$. The Poisson distribution is thus an appropriate approximation to the binomial distribution which really describes this situation, and it has the probability density function

$$P(r) = 3^r e^{-3}/r!.$$

Now as $P(r)$ is the probability that exactly r shells are defective, the answer to (a) is $P(4) = 3^4 e^{-3}/4! = 0.1680$. The answer to (b) is just $U(4)$, where

$$U(4) = P(0) + P(1) + P(2) + P(3) + P(4) = 0.8152.$$

The exact result of (a) is relatively more difficult to calculate because it is given by $(600!/596!4!)(1/200)^4(199/200)^{596} = 0.1685$, while the exact result of (b) involves a prohibitively large calculation. ■

By the same form of argument, had we needed to know the probability that at least m but no more than s shells were defective in a batch, then this would have been given by the sum $P(m)+P(m+1)+P(m+2)+\ldots+P(s)$.

When a discrete probability density function is known, it is often useful to give a simple description of the distribution in terms of two numbers known as the first and second *moments* of the distribution. If the density function is $P(r)$, then when the sums are convergent it is conventional to denote the *first moment* or the *mean* by μ, and the *second moment* or the *variance* by σ^2, where

$$\mu = \sum_{r=0}^{\infty} rP(r), \tag{18.12}$$

and

$$\sigma^2 = \sum_{r=0}^{\infty} (r - \mu)^2 P(r). \tag{18.13}$$

Expanding the expression under the summation sign in Eqn (18.13), using Eqn (18.12) and the fact that $\sum_{r=0}^{\infty} P(r) = 1$, it is a simple matter to show that an alternative, and often more convenient, expression for the variance s^2 is the following:

$$\sigma^2 = \sum_{r=0}^{\infty} r^2 P(r) - \mu^2. \tag{18.14}$$

Using a mechanical analogy, the mean μ can be thought of as the perpendicular distance of the centre of gravity of the area under the distribution from the $P(r)$ axis. That is, the turning effect about the $P(r)$ axis of a unit mass at a perpendicular distance μ from that axis will be the same as the sum of the individual turning effects of masses $P(r)$ at perpendicular distances r from the axis. It is, indeed, this analogy that gives rise to the term 'moment' of a distribution already mentioned. The square root σ of the variance is called the *standard deviation*, and provides a measure of the spread of the distribution about the mean. The advantage of using the standard deviation is that it makes it unnecessary to work with squared units of the measured quantity.

We leave to the reader as exercises in manipulation the task of proving the following results:

Binomial distribution

probability density $P(r) = \binom{n}{r} p^r (1-p)^{n-r}, \ r = 0, 1, \ldots, n$

mean $\mu = np$, variance $\sigma^2 = np(1-p)$.

Poisson distribution

probability density $P(r) = \dfrac{\mu^r e^{-\mu}}{r!}, \quad r = 0, 1, 2, \ldots$

mean $\mu = np$, variance $\sigma^2 = \mu$.

18.3 Continuous distributions and the normal distribution

18.3.1 Continuous distributions

So far, the sample spaces we have used have involved discrete points, and it is for this reason that the term 'discrete' has been used in conjunction with the definition of the binomial and Poisson distributions. In other words, in discrete distributions, no meaning is to be attributed to points that are intermediate between the discrete sample-space points.

When different situations involving probability are examined it is often necessary to consider sample spaces in which all points are allowed. This happens, for instance, when considering the actual measured lengths of manufactured items having a specific design length, or when considering the actual measured capacity of capacitors, all having the same nominal capacity.

The probability density for situations like these can be thought of as being arrived at as the limit of an arbitrarily large number of discrete observations in the following sense. Suppose first that all the measured quantities x lie within the interval $a \leq x \leq b$. Next, consider the division of $[a, b]$ into n equal intervals of length $\Delta = (b-a)/n$ and number them sequentially from $i=1$ to n starting from the left-hand side of the interval $[a, b]$. Thus the ith interval corresponds to the inequality $a+(i-1)\Delta < x < a+i\Delta$. Now in the set or *sample* of N measurements taken from the *population* or totality of all possible measurements, it will be observed that f_i of these lie within the ith interval. This number f_i is the observed *frequency of occurrence* of a measurement in the ith interval, and it is obvious that $\sum_{i=1}^{n} f_i = N$. If a histogram were to be constructed at this stage by plotting the numbers f_i/N against i then it would appear as in Fig. 18.8(a).

As the number of observations N becomes arbitrarily large, so as ultimately to include all the population, so each number f_i/N will tend to a limiting value. Then, by allowing n to become arbitrarily large, it is possible to associate a number $P(x)$ with every value of x in $a \le x \le b$. The type of smooth limiting curve that results is illustrated in Fig. 18.8(b). Because x is indeterminate within the interval $a \le x \le b$ associated with $P(x)$ it will be said to be a *random variable* having the distribution $P(x)$. It is important at this stage to distinguish between the general random variable x occurring in a continuous probability distribution of this type, which is a theoretical distribution, and its realization in terms of an actual measurement carried out as a result of an experiment. Later we look briefly at the problems of inferring properties of a governing probability distribution from measurements performed on a sample of the population.

The intuitive approach to continuous probability distributions must suffice for our purposes, and we shall not look deeper into the important matters of interpretation that arise in connection with the sense in which the limit f_i/N is to be understood. It is, nevertheless, clear from these arguments that a continuous probability distribution $P(x)$, defined for some interval $a \le x \le b$, is one with the following properties:

(i) $P(x) \ge 0$ for $a \le x \le b$.

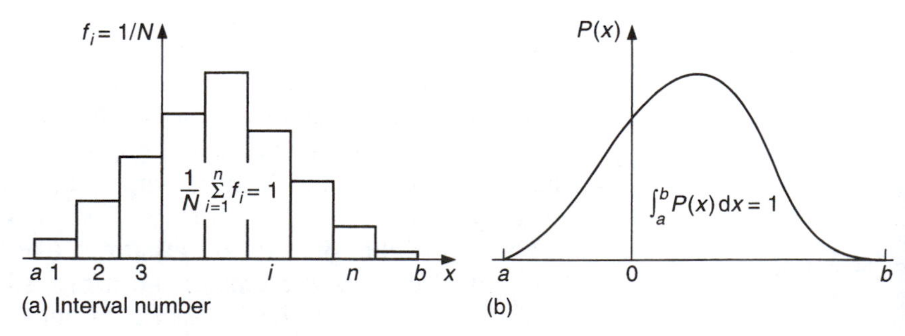

Fig. 18.8 (a) Relative frequency polygon or histogram; (b) continuous probability distribution obtained as the limit of the frequency polygon as $n, N \to \infty$.

(ii)

$$\int_a^b P(x)\mathrm{d}x = 1.$$

(iii) The cumulative distribution function $U(x)$ is defined by

$$U(x) = \int_a^x P(t)\mathrm{d}t.$$

$U(a)$ has the property that it is the probability that the random variable x will be such that $x \leq a$. Thus the probability that $c < x < d$, usually written $P\{c < x < d\}$, is given by

$$P\{c < x < d\} = \int_c^d P(t)\mathrm{d}t = U(d) - U(c).$$

(iv) The connection between the probability density function $P(x)$ and the cumulative distribution function $U(x)$ is provided by the result (Theorem 7.6)

$$P(x) = \frac{\mathrm{d}}{\mathrm{d}x}\int_a^x U(t)\mathrm{d}t.$$

When the integrals are convergent, a continuous distribution has associated with it a mean

$$\mu = \int_a^b xP(x)\mathrm{d}x, \tag{18.15}$$

and a variance

$$\sigma^2 = \int_a^b (x-\mu)^2 P(x)\mathrm{d}x. \tag{18.16}$$

The simplest example of a continuous distribution is the *uniform* or *rectangular distribution* defined by

$$P(x) = \begin{cases} 1/(b-a) & \text{for } a \leq x \leq b \\ 0 & \text{elsewhere.} \end{cases}$$

This characterizes an event associated with a value of x that is equally likely to occur for any x in the interval $a \leq x \leq b$. From Eqns (18.15) and (18.16) it follows that for the uniform distribution

$$\mu = \frac{a+b}{2} \quad \text{and} \quad \sigma x^2 = \frac{(b-a)^2}{12}.$$

An example of a continuous distribution for which only the mean μ is defined is provided by the Cauchy distribution,

$$P(x) = \frac{1}{\pi(1+x^2)} \quad \text{for} \quad -\infty < x < \infty.$$

This has zero mean, because

$$\mu = \frac{1}{\pi}\int_{-\infty}^{\infty} \frac{x\mathrm{d}x}{1 + x^2} = 0,$$

but infinite variance σ^2, because

$$\sigma^2 = \frac{1}{\pi}\int_{-\infty}^{\infty} \frac{x^2\mathrm{d}x}{1 + x^2},$$

and the integral is divergent.

Example 18.9 A target moves along a straight path of length l joining points A and B. The probability density that a single shot fired at it will score a hit is described by the uniform distribution over AB. If a hit is scored at a point C distant x from A, what is the probability that the ratio AC/CB lies between the two positive numbers k_1 and $k_2 (k_1 < k_2)$.

Solution As C is distant x from A we know that $AC/CB = x/(l—x)$, so that

$$k_1 \leq \frac{x}{l - x} \leq k_2 \qquad \text{or, equivalently,} \qquad \frac{k_1 l}{1 + k_1} \leq x \leq \frac{k_2 l}{1 + k_2}.$$

Now the uniform probability density $P(x)$ over AB is simply $P(x) = 1/l$ for $0 \leq x \leq l$. Thus the desired probability P must be

$$P = \int_{k_1 l/(1+k_1)}^{k_2 l/(1+k_2)} \frac{1}{l}\mathrm{d}x = \frac{k_2 - k_1}{(1 + k_1)(1 + k_2)}. \quad \blacksquare$$

18.3.2 Normal distribution

By using more advanced methods than those we have at our disposal in this book it can be shown that

$$1 = \frac{1}{(2\pi)^{1/2}}\int_{-\infty}^{\infty} \exp\left(-\frac{X^2}{2}\right)\mathrm{d}X, \tag{18.17}$$

and

$$1 = \frac{1}{(2\pi)^{1/2}}\int_{-\infty}^{\infty} X^2 \exp\left(-\frac{X^2}{2}\right)\mathrm{d}x. \tag{18.18}$$

The fact that the integrand of Eqn (18.17) is essentially positive shows that the function

$$P(X) = \frac{1}{(2\pi)^{1/2}} \exp\left(-\frac{X^2}{2}\right) \quad \text{for} \quad -\infty < X < \infty \tag{18.19}$$

can be regarded as a probability density function. Now $XP(X)$ is an odd function and tends rapidly to zero as $|X| \to \infty$, and so it follows immediately that

$$0 = \frac{1}{(2\pi)^{1/2}} \int_{-\infty}^{\infty} X \exp\left(-\frac{X^2}{2}\right) \mathrm{d}X. \tag{18.20}$$

We have thus established that $P(X)$ as defined in Eqn (18.19) has a mean $\mu = 0$, and hence from Eqn (18.18) that $P(X)$ has a variance $\sigma^2 = 1$. When the random variable X satisfies the probability density (18.19), it is said to have a *standardized normal distribution*. The normal distribution is the most important of all the continuous probability distributions and for historical reasons it is also known as the *Gaussian distribution*.

Making the change of variable $x = \sigma X + \mu$, it is a simple matter to transform integral (18.17) and to deduce that the corresponding unstandarized normal distribution obeyed by x has the probability density function

$$P(x) = \frac{1}{(2\pi\sigma^2)^{1/2}} \exp\left(-\frac{(x-\mu)^2}{2\sigma^2}\right) \quad \text{for} \quad -\infty < x < \infty. \tag{18.21}$$

By appeal to Eqns (18.18) and (18.20) it is then easily established that (18.21) describes a *normal distribution* with mean μ and variance σ^2. It is customary to represent the fact that the random variable x obeys a normal distribution with mean μ and variance σ^2 by saying that x obeys the distribution $N(\mu, \sigma^2)$. In this notation the random variable X of (18.19) obeys the normal distribution $N(0, 1)$.

The cumulative standardized normal distribution is usually denoted by a special symbol, and we choose to use $\Phi(x)$, where

$$\Phi(x) = \frac{1}{(2\pi)^{1/2}} \int_{-\infty}^{x} \exp\left(-\frac{t^2}{2}\right) \mathrm{d}t. \tag{18.22}$$

As before, $\Phi(x)$ is the probability that the random variable X is such that $X < x$.

The $N(0, 1)$ normal probability density (18.19) and the corresponding cumulative normal distribution function (18.22) are readily available in tabulated form in any statistical tables. An abbreviated table is given in Table 18.1 for $x \geq 0$, and the corresponding graphs are illustrated in Fig. 18.9. To deduce values of $P(x)$ and $\Phi(x)$ for negative x from this table it is only necessary to use the fact that $P(x)$ is an even function, so that $P(-x) = P(x)$, and it then follows that $\Phi(-x) = 1 - \Phi(x)$.

Table 18.1 $N(0, 1)$ normal and cumulative distribution

x	$P(x)$	$\Phi(x)$	x	$P(x)$	$\Phi(x)$	x	$P(x)$	$\Phi(x)$
0	0.39894	0.50000	1.2	0.19419	0.88493	2.5	0.01753	0.99379
0.2	0.39104	0.57926	1.4	0.14973	0.91924	2.75	0.00909	0.99702
0.4	0.36827	0.65542	1.6	0.11092	0.94520	3.0	0.00443	0.99865
0.6	0.33322	0.72575	1.8	0.07895	0.96407	3.25	0.00203	0.99942
0.8	0.28969	0.78814	2.0	0.05399	0.97725	3.5	0.00087	0.99977
1.0	0.24197	0.84134	2.25	0.03174	0.98778	4.0	0.00013	0.99997

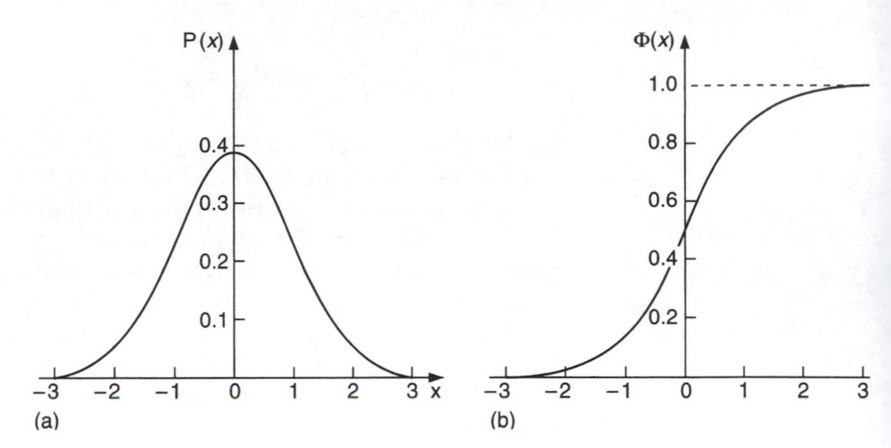

Fig. 18.9 Normal distribution $N(0, 1)$: (a) probability density function; (b) cumulative distribution function.

Examination of Table 18.1 shows that when the random variable X is distributed as $N(0, 1)$, the probability that X lies within the interval $-1 \leq X \leq 1$ is $\Phi(1) - \Phi(-1) = 2\Phi(1) - 1 = 0.6827$. Similarly, the probability that $-2 \leq X \leq 2$ is 0.9545 and the probability that $-3 \leq X \leq 3$ is 0.9973. Recalling that when X is $N(0, 1)$ the mean $\mu = 0$ and the standard deviation $\sigma = 1$, and interpreting the results in terms of $N(\mu, \sigma^2)$, we may deduce that when x belongs to $N(\mu, \sigma^2)$ there is approximately

(i) a probability $\frac{2}{3}$ that $|x - \mu| \leq \sigma$,
(ii) a probability 0.95 that $|x - \mu| \leq 2\sigma$,
(iii) a probability 0.997 that $|x - \mu| \leq 3\sigma$.

The distribution $N(\mu, 1)$ is easily seen to be the distribution obtained from Fig. 18.7(a) by shifting the peak to $x = \mu$. It follows directly from Eqn (18.21) that, if $\sigma_1 > \sigma_2$, the distribution $N(\mu, \sigma_1)$ and $N(\mu, \sigma_2)$ will both have the same symmetrical bell shape with their peaks located at $x = \mu$, but the $N(\mu, \sigma_1)$ distribution will have a bell-shaped curve which is lower and broader than that of the $N(\mu, \sigma_2)$ distribution. That is, the smaller σ, the more sharply peaked is the distribution, and the larger σ, the flatter and broader is the distribution.

Example 18.10 Let the random variable x obey the standardized normal distribution $N(0, 1)$. Then, denoting by $P\{a < x < b\}$ the probability that $a < x < b$, find (a) $P\{0 < x < 1.2\}$, (b) $P\{-0.6 < x < 0\}$ and $P\{-0.4 < x < 0.4\}$.

Solution (a) By definition

$$P\{0 < x < 1.2\} = \int_0^{1.2} P(t)\mathrm{d}t = \int_{-\infty}^{1.2} P(t)\mathrm{d}t - \int_{-\infty}^{0} P(t)\mathrm{d}t$$
$$= \Phi(1.2) - \Phi(0).$$

However, from Table 18.1, we find $\Phi(0) = 0.5$ and $\Phi(1.2) = 0.88493$, so

$$P\{0 < x < 1.2\} = 0.88493 - 0.5 = 0.38493.$$

(b) As $P(x)$ is an even function (it is symmetric about the line $x = 0$), it follows that

$$P\{-0.6 < x < 0\} = P\{0 < x < 0.6\}.$$

Thus

$$P\{-0.6 < x < 0\} = \int_{0}^{0.6} P(t)\mathrm{d}t = \int_{-\infty}^{0.6} P(t)\mathrm{d}t - \int_{-\infty}^{0} P(t)\mathrm{d}t$$
$$= \Phi(0.6) - \Phi(0).$$

A further appeal to Table 18.1 shows that $\Phi(0.6) = 0.72575$, so

$$P\{-0.6 < x < 0\} = 0.72575 - 0.5 = 0.22575.$$

(c) Again, because $P(x)$ is an even function, it follows directly that

$$P\{-0.4 < x < 0.4\} = 2P\{0 < x < 0.4\}.$$

Thus

$$P\{-0.4 < x < 0.4\} = 2\left\{\int_{-\infty}^{0.4} P(t)\mathrm{d}t - \int_{-\infty}^{0} P(t)\mathrm{d}t\right\}$$
$$= 2\{\Phi(0.4) - \Phi(0)\}$$
$$= 2\Phi(0.4) - 1,$$

but as $\Phi(0.4) = 0.65542$ it follows that

$$P\{-0.4 < x < 0.4\} = 0.31084. \quad \blacksquare$$

Most random variables x which belong to a normal distribution do not have a zero mean and a unit variance, but a mean $\mu \neq 0$ and a variance $\sigma \neq 1$ so that they belong to the $N(\mu, \sigma)$ distribution. To determine the probability that $x < a$ or, say, that $a < x < b$, it is necessary to relate the $N(\mu, \sigma)$ distribution to which x belongs to the standardized normal distribution $N(0, 1)$ which is available in tabulated form in statistical tables.

This is easily accomplished by considering the arguments leading to Eqn (18.22). These showed that the change of variable

$$x = \sigma X + \mu$$

transforms the variable x in the $N(\mu, \sigma)$ distribution into the variable X in the standardized normal distribution $N(0, 1)$. Thus as

$$X = \frac{x - \mu}{\sigma},$$

this means that

$$P\{x < a\} = P\left\{X < \frac{a-\mu}{\sigma}\right\}$$

and, correspondingly, that

$$P\{a < x < b\} = P\left\{\frac{a-\mu}{\sigma} < X < \frac{b-\mu}{\sigma}\right\}.$$

Example 18.11 Given that the random variable x obeys the $N(1.5, 2)$ distribution, find $P\{2.3 < x < 5.1\}$.

Solution In this case $\mu = 1.5$ and $\sigma = 2$, so

$$P\{2.3 < x < 5.1\} = P\left\{\frac{2.3-1.5}{2}\right\} < X < \left\{\frac{5.1-1.5}{2}\right\}$$
$$= P\{0.4 < X < 1.8\}$$

where X obey the $N(0, 1)$ distribution. Thus

$$P\{2.3 < x < 5.1\} = \int_{-\infty}^{1.8} P(t)\mathrm{d}t - \int_{-\infty}^{0.4} P(t)\mathrm{d}t = \Phi(1.8) - \Phi(0.4)$$
$$= 0.96470 - 0.65542 = 0.30928. \quad \blacksquare$$

18.4 Mean and variance of a sum of random variables

It is necessary to return briefly to the idea of the independence of events encountered first in connection with conditional probability in Theorem 18.1. When several random variables are of interest and the selection of an event corresponding to any one of these has no effect on the selection of the others then these are called *independent random variables* corresponding to mutually independent events; when this is not the case they are said to exhibit *dependence*.

These ideas are of fundamental importance when it comes to examining the mean and variance of sums or linear combinations of n random variables $x_1, x_2, \ldots, x_n$. Denoting the mean of the random variable x_i by μ_i and the variance by σ_i^2, it is possible in a more advanced account of statistics to establish various important theorems for linear combinations of random variables. The main results of interest to us are expressed in the next theorem and its corollary.

Theorem 18.3 (mean and variance of a sum of random variables) Let $x_1, x_2, \ldots, x_n$ be n random variables with respective means μ_i and variances σ_i^2, and let $a_1, a_2, \ldots, a_n, b$ be real constants. Then,

(i) irrespective of whether or not the x_i are independent random variables,

mean of $(a_1x_1 + a_2x_2 + \ldots + a_nx_n + b)$

$$= a_1\mu_1 + a_2\mu_2 + \ldots + a_nx_n + b;$$

(ii) when the x_i are independent random variables,

$$\text{variance of } (a_1x_1 + a_2x_2 + \ldots + a_nx_n + b) = a_1\sigma_1^2 + a_1\sigma_2^2 + \ldots + a_ns_n^2.$$

Corollary 18.3 If n independent random variables x_i are all drawn from distributions having the same mean μ and variance σ^2, then the random variable $(x_1+x_2+\ldots+x_n)/n$ has the same mean μ but variance σ^2/n.

18.5 Statistics – inference drawn from observations

Thus far, we have considered the ideal situation involving random variables as characterized by probability theory. The situation confronting the user of statistics is somewhat different. He or she is usually required to draw some *inference* concerning the theoretical probability distribution governing some observations of interest from the imperfect data provided by only a sample taken from the possible population. Here, by *population*, we mean simply the totality of possible observational outcomes in the situation of interest.

For example, when attempting to assess the quality of a production line for 1000-ohm resistors it is customary to do so by a periodic check on a random sample of a given size. All resistors could be measured, but the cost might be prohibitive. The situation is rather different with ammunition, for all shells could be tested by firing, but then there would be no useful output from the production line. These are, respectively, typical examples of non-destructive and destructive testing.

Let us first outline the simplest way in which data or *observations* may be categorized. In the first instance data may be grouped into categories, each with a value of a random variable associated with it, and a histogram constructed similar to that in Fig. 18.7(a), irrespective of whether or not the observations are governed by a continuous distribution. Thus we might take 5-ohm intervals about the desired value of a resistor and record the frequency of occurrence f_i, with which the resistance of a resistor drawn from a given random sample of N falls in the ith such interval. The observed relative frequency of occurrence in that interval is then f_i/N and the results could be recorded in the form shown in Table 18.2.

By associating a nominal resistance x_i with the ith interval, equal, say, to the resistance at the mid-point of that interval, it is then possible to define the *sample mean* $\bar{x}$ in a natural manner by means of the expression

$$\bar{x} = \sum_{i=1}^{n} f_i x_i / N \quad \text{with} \quad N = \sum_{i=1}^{n} f_i. \tag{18.23}$$

It would then also seem reasonable to define the *sample variance* by the following expression

Table 18.2 Grouping observation into categories

Interval i	1	2	3	...	i	...	n
Frequency	f_1	f_2	f_3	...	f_i	...	f_n
Relative frequency	f_1/N	f_2/N	f_3/N	...	f_i/N	...	f_n/N

$$\sum_{i=1}^{n}(x_i - \bar{x})^2 f_i/N,$$

since this is in accord with (18.16). However, it can be shown that this definition does not take fully into account the fact that a finite sample size biases the estimate of the population variance. In advanced books on statistics it is established that this bias in the estimate is corrected if the sample variance s^2 is computed from the modified expression

$$s^2 = \frac{1}{N-1}\sum_{i=1}^{n}(x_i - \bar{x})^2 f_i, \tag{18.24}$$

which is the one we shall use throughout this chapter. It is clear that this definition only differs significantly from the previous one when N is small, which is as would be expected, for then the bias is greatest. Computationally, it is usually best to combine Eqns (18.23) and (18.24), and to determine the sample variance from the result

$$s^2 = \frac{1}{N-1}\left(\sum_{i=1}^{n} f_i x_i^2 - N\bar{x}^2\right), \tag{18.25}$$

which is equivalent to (18.24). It can be established that the estimate of the mean provided by (18.22) is free from bias due to sample size. As before, s is known as the *sample standard deviation*.

When only single observations are involved, so that all the frequencies $f_i = 1$ and $n = N$, the expression for the sample mean $\bar{x}$ of the N observations $x_1, x_2, \ldots, x_N$ corresponding to (18.23) becomes

$$\bar{x} = \frac{1}{N}\sum_{i=1}^{N} x_i, \tag{18.23$'$}$$

and the expression for the sample variance s^2 corresponding to (18.25) becomes

$$s^2 = \frac{1}{N-1}\left(\sum_{i=1}^{N} x_i^2 - N\bar{x}^2\right). \tag{18.26}$$

Other descriptive parameters associated with the histogram that are sometimes used are the following:

(a) The *range* of the observations, which is the difference between the largest and smallest observations in the sample under consideration.

(b) The *mode*, which is equal to the random variable x_i corresponding to the most frequently occurring observation when this exists. It is not defined if more than one observation x_i has the maximum frequency of occurrence. This provides the simplest measure of the location of the bulk of the observations.

(c) The *median*, which is the value of the random variable x_i below which exactly half of the observations lie. When there is no grouping and the number N of observations involved in the sample is even, the median is defined to be equal to the mean of the two adjacent central observations. In the event that observations are grouped, it usually becomes necessary to introduce a fictitious median by linear interpolation. This can also be found by inspection of a graph of the histogram in which adjacent points are joined by straight lines. The median provides a better measure of location for the observations than does the mode, and is usually perferable to the mean when the relative frequency polygon is badly asymmetric and the distribution is not known.

(d) The first, second and third *quartiles* Q_1, Q_2 and Q_3 which are an immediate extension of the idea underlying the median. They represent the values of the random variable x_i below which one-quarter, one-half and three-quarters of the observations lie. The second quartile Q_2 is identical with the median.

(e) The *mean deviation* which, like the standard derivation, provides a measure of the spread or dispersion of the observations about their mean $\bar{x}$. It is defined as the number

$$\sum_{i=1}^{n} |x_i - \bar{x}| f_i / N. \tag{18.27}$$

Example 18.12 Find the range, mode, mean, median, variance, standard deviation and mean deviation of the observations x_i, occurring with frequencies f_i that are contained in the following table.

i	1	2	3	4	5	6	7	8
x_i	0.2	0.4	0.6	0.8	1.0	1.2	1.4	1.6
f_i	2	6	10	9	7	8	4	2
Σf_i	2	8	18	27	34	42	46	48

Solution The range is $1.6 - 0.2 = 1.4$, the mode is 0.6 and the mean

$$\begin{aligned}\bar{x} &= (0.2 \times 2 + 0.4 \times 6 + 0.6 \times 10 + 0.8 \times 9 + 1.0 \times 7 + 1.2 \times 8 \\ &\quad + 1.4 \times 4 + 1.6 \times 2)/48 \\ &= 0.8625.\end{aligned}$$

As there are 48 observations, inspection of the cumulative frequency in the bottom row of the table shows that a fictitious median is required corresponding to 24 observations. It must lie between 0.6 and 0.8, and the linearly interpolated value of the median is

$$0.6 + \left(\frac{24-18}{27-18}\right)(0.8-0.6) = 0.733.$$

The sample variance

$$s^2 = \frac{1}{N-1}\left(\sum_{i=1}^{n} f_i x_i^2 - N\bar{x}^2\right) = \tfrac{1}{47}(41.88 - 48 \times 0.8625^2) = 0.1313,$$

and the sample standard deviation $s = \sqrt{0.1313} = 0.362$.

The mean deviation is computed as follows:

$$\{2|0.2-0.863| + 6|0.4-0.863| + 10|0.6-0.863| \\ +9|0.8-0.863| + 7|1-0.863| + 8|1.2-0.863| \\ +4|1.4-0.863| + 2|1.6-0.863|\}/48 = 0.304. \quad \blacksquare$$

When the sample mean $\bar{x}$ and sample variance s^2 need to be calculated and the observations involve large numbers there is often a loss of accuracy due to round-off, and possibly even to overflow. This can usually be avoided by first subtracting from each observation a suitably large number N_0, chosen to reduce the set of observations to more convenient values. The modified observations will have the same sample variance s^2 as the original set of observations, and a sample mean $\bar{x}_1 = \bar{x} - N_0$. Thus the addition of N_0 to the sample mean $\bar{x}_1$ of the modified observations will yield the required sample mean $\bar{x}_1$ of the original data. This is a simple example of what a statistician calls *coding* the data. More generally, coding involves both scaling observations and adjusting their mean, and then determining the sample mean and sample variance of the original observations from the sample mean and sample variance of the coded data.

Suppose, for example, we need to find the sample mean $\bar{x}$ and sample variance of s^2 of the ten observations

$$995, 1002, 1019, 998, 1011, 1004, 991, 1021, 993, 1003.$$

Setting $N_0 = 990$ and subtracting N_0 from each observation gives the more convenient modified set of observations

$$5, 12, 29, 8, 21, 14, 1, 31, 3, 13.$$

These have the sample mean $\bar{x}_1 = 13.7$, so as the sample mean $\bar{x}$ of the original observations is $\bar{x} = \bar{x}_1 + N_0$ it follows that the sample mean $\bar{x} = 1003.7$. The sample variance s^2 of both the original and the modified set of observations is

$$s^2 = \tfrac{1}{9}[(5-13.7)^2 + (12-13.7)^2 + (29-13.7)^2$$

$$+(8-13.7)^2 + (21-13.7)^2 + (14-13.7)^2$$

$$+(1-13.7)^2 + (31-13.7)^2 + (3-13.7)^2 + (13-13.7)^2],$$

so the sample variance

$$s^2 = 108.23,$$

and the sample standard deviation

$$s = 10.40.$$

The inferences drawn so far from the data of Table 18.2 are rather general and, for example, take no account of the fact that different samples drawn from the same population would vary, and so affect our conclusions. In what follows we examine a few basic statistical tests which take account of this variability, and look also at their possible application. We make no rigorous attempt to derive these tests since to do so would require aspects of probability theory it has been necessary to omit for reasons of mathematical simplicity.

In considering these tests we shall have in mind a random sample of given size N drawn from the possible population in such a way that each member of the population has an equal chance of inclusion in the sample. Usually some simple *statistical hypothesis* will be in mind which involves some property of the underlying probability distribution that is involved. We shall make some hypothesis or assumption about this distribution and then test to see if it is in agreement with the sample data. In statistical language such a hypothesis is called the *null hypothesis*, and it is usually abbreviated by the notation H_0. Thus H_0 might, for example, be the hypothesis that the population mean is 3.7. The test would then indicate with what probability this was in agreement with the calculated sample mean $\bar{x}$. Using statistical language, we would *accept* H_0 if the probability of agreement equalled or exceeded a predetermined probability; otherwise H_0 would be *rejected*. That is to say, when H_0 is rejected, we would conclude that the sample was not drawn from a population with mean 3.7.

Inherent in such tests is the notion of a confidence interval which has already been encountered at the end of section 18.3, though this term was not then used. There we saw, for example, that with probability approximately 0.95 the random variable x drawn from the distribution $N(\mu, \sigma^2)$ will be contained in the interval $\mu - 2\sigma \leq x \leq \mu + 2\sigma$. For a given distribution the width of the confidence interval depends on the probability, or confidence, with which the location of the random variable x is required to be known. Thus for this same distribution we saw that to locate x with a probability of about 0.997 it is necessary to widen the interval to $\mu - 3\sigma \leq x \leq \mu + 3\sigma$.

Suppose that an arbitrary probability density function $P(x)$ for the random variable x is involved, that an estimate x_1 of x is obtained from such a sample and that a number r is given $(0 \leq r \leq 100)$. Then, corresponding to a probability of occurrence $1 - r/100$ of x in the interval, the $(100 - r)$ per cent *confidence interval* for random variable x

$$x_1 - d \leq x \leq x_1 + d \tag{18.28}$$

is obtained by determining d from the equation

$$P(|x - x_1| \leq d) = 1 - \frac{r}{100}$$

or, equivalently,

$$P(x_1 - d \leq x \leq x_1 + d) = 1 - \frac{r}{100}. \tag{18.29}$$

The number $1 - r/100$ is called the *confidence coefficient* for this $(100 - r)$ per cent confidence interval for x.

Let us suppose now that we have a random sample of n independent observations $x_1, x_2, \ldots, x_n$ drawn from a normal distribution $N(\mu, \sigma^2)$ in which σ^2 is known. Then their mean $\bar{x}$ is also a random variable and the more advanced theory that gave rise to Theorem 18.3 also asserts that $\bar{x}$ has a normal distribution. The distribution of $\bar{x}$ is known as the *sampling distribution of the mean*. Now from Corollary 18.3 we know that the population mean of $(x_1+x_2+\ldots+x_n)/n$ is μ, but that its variance is σ^2/n, and we are assuming σ^2 to be known. So, as the variance of the sample means is smaller by a factor $1/n$ than the variance of the x_i, we can take our sample mean $\bar{x}$ as a good estimate of the population mean μ if n is large $(n \geq 50)$.

Thus $\bar{x}$ belongs to the distribution $N(\bar{x}, \sigma^2/n)$ and so we are immediately able to determine a confidence interval for μ. Following result (ii) given before example 18.10 at the end of section 18.3 we know that with approximately 95 per cent confidence μ lies within the interval $\bar{x} - 2\sigma/\sqrt{n} \leq \mu \leq \bar{x} + 2\sigma/\sqrt{n}$. Alternatively, expressed in terms of Eqns (18.27) and (18.28), we have found that the sampling distribution of the mean $P(\bar{x})$ is $N(\bar{x}, \sigma^2/n)$ and that corresponding to $r = 5$, $x_1 = \bar{x}$, d is approximately $2\sigma/\sqrt{n}$. By interpolation in Table 18.1 the approximate results at the end of section 18.3 may be made exact and turned into the following statistical rule.

Rule 18.1 Confidence interval for the population mean μ of a normal distribution with known variance

First select a $(100 - r)$ per cent confidence level and use this to determine α_r from the following table.

r	20	10	5	1	0.1
Confidence level	80%	90%	95%	99%	99.9%
α_r	1.282	1.645	1.960	2.576	3.291

Using this value of α_r, and denoting the computed sample mean by $\bar{x}$ and the known variance by σ^2, the $(100-r)$ per cent confidence interval for the population mean μ is

$$\bar{x} - \alpha_r\sigma/\sqrt{n} \leq \mu \leq \bar{x} + \alpha_r\sigma/\sqrt{n}.$$

Example 18.13 A measuring microscope used for the determination of the length of precision components has a standard deviation $\sigma = 0.057$ cm. Find the 90 per cent and 99 per cent confidence intervals for the mean μ given that a sample mean $\bar{x} = 3.214$ cm was obtained as the result of 64 observations.

Solution Here $n = 64$, $\sigma = 0.057$ cm and for 90 per cent confidence interval $\alpha_{10} = 1.645$, while for the 99 per cent confidence interval $\alpha_1 = 2.576$. Thus $\alpha_{10}\sigma/\sqrt{n} = 0.0117$ cm and $\alpha_1\sigma/\sqrt{n} = 0.0184$ cm, and so the 90 per cent confidence interval for μ is $3.202 \text{ cm} \leq \mu \leq 3.226$ cm and the 99 per cent confidence interval is $3.196 \text{ cm} \leq \mu \leq 3.232$ cm ■

The form of Rule 18.1 can be adapted to the problem of determining the confidence interval for the parameter p in the binomial distribution when n is large and $0.1 < p < 0.5$. It follows from this that as the mean and variance of the binomial distribution are simple functions of p, confidence intervals may also be deduced for them. The result we now discuss comes about because when n and p satisfy these criteria, the binomial distribution is closely approximated by the normal distribution with population mean $\mu = np$ and population variance $\sigma^2 = np(1-p)$. That is, we may take the distribution $N(np, np(1-p))$ as an approximation to the binomial distribution.

Because of this we can define a new variable $X = (x - np)/\sqrt{(np(1-p))}$ which obeys the normal distribution $N(0, 1)$. The $(100-r)$ per cent confidence interval $-X_r \leq X \leq X_r$ for X is then given by solving

$$P(-X_r \leq X \leq X_r) = 1 - \frac{r}{100}.$$

As X belongs to the distribution $N(0, 1)$, the number X_r is precisely the number α_r defined in the table given in Rule 18.1. The $(100-r)$ per cent confidence interval for X is thus given by

$$-\alpha_r \leq (x - np)/\sqrt{(np(1-p))} \leq \alpha_r$$

or, equivalently, by

$$\frac{(x - np)^2}{np(1-p)} \leq \alpha_r^2.$$

So, we arrive at an inequality determining p in terms of x of the form

$$(n + \alpha_r^2)p^2 - (2x + \alpha_r^2)p + \frac{x^2}{n} \leq 0. \tag{18.30}$$

Now if n is large and the frequency of occurrence of an event of interest is m, where both m and $n - m$ are large, we can reasonably approximate the random variable x by m. In terms of our criterion for p, this is equivalent to requiring $n < 10m < 5n$. Inequality (18.30) then determines an interval $p_1 \leq p \leq p_2$ in which p must lie for (18.30) to be true. This is the $(100 - r)$ per cent confidence interval for p and the result gives us our next rule.

Rule 18.2 Confidence interval for p in the binomial distribution

This approximate rule applies to a series of n independent trials governed by the binomial distribution in which n is large and the frequency of occurrence m of an event of interest is such that $n < 10m < 5n$. Then the $(100 - r)$ per cent confidence interval for the parameter p is

$$p_1 \leq p \leq p_2,$$

where p_1 and p_2 are the roots of the equation

$$(n + \alpha_r^2)p^2 - (2m + \alpha_r^2)p + \frac{m^2}{n} = 0,$$

and α_r is determined from the table given in Rule 19.1.

Example 18.14 In a test, 240 microswitches were subjected to 25 000 switching cycles. Given that 28 failed to operate after this test, what is the 95 per cent confidence interval for the mean number of failures to be expected in similar batches of the same design of microswitch?

Solution The situation is governed by a binomial distribution since a microswitch will either pass or fail the test. Here $n = 240$ is large, and m is such that $n < 10m < 5n$, so that for 95 per cent confidence Rule 18.2 applies with $\alpha_5 = 1.960$. Using these results, the 95 per cent confidence interval for p is found to be $0.082 \leq p \leq 0.163$. As the mean $\mu = np$, the corresponding 95 per cent confidence interval for the mean number of failures in further batches of 240 microswitches is $20 < \mu < 39$. ■

Often it is necessary to determine the confidence interval for the mean of a normal distribution when small samples ($n < 50$) are involved and the variance is not known. If in Rule 18.1 the population variance σ^2 were to be replaced by the sample variance s^2 given in Eqn (18.24) then a serious error could result. This is because of the dependence of that test on the fact that the population parameter σ^2 is assumed to be known. The difficulty can be overcome, however, by replacing the numbers contained in the table given in Rule 18.1 by a different set derived from another distribution known as the *t-distribution* with $n-1$ degrees of freedom. Because the originator of the t-distribution, W. S. Gosset, published his results under the pseudonym 'Student' and denoted the random variable in his distribution by t this is also known as *Student's t-distribution*. This distribution does not depend on known population parameters and takes account of the fact that the number n of observations in the sample is not necessarily large.

The t-distribution is symmetric, but a little flatter and broader than the normal distribution, though it tends rapidly to the normal distribution as n becomes large. The number of *degrees of freedom* involved in this test can be thought of as the number of observations x_i whose values may be assigned arbitrarily when a value of $\bar{x}$ is specified. Clearly, in this case, as n observations are involved, the number of degrees of freedom $\nu = n - 1$. The modified rule is as follows.

Rule 18.3 Confidence interval for the population mean μ of a normal distribution with unknown variance

Given a sample of n observations, first select a $(100 - r)$ per cent confidence level and use this to determine α_r from the following table, looking under the nearest entry appropriate to the number of degrees of freedom $\nu = n - 1$ that is involved.

	a_{10}	a_5	a_1		a_{10}	a_5	a_1
	$(100-r)\%$				$(100-r)\%$		
ν	90%	95%	99%	ν	90%	95%	99%
2	2.920	4.303	9.925	20	1.725	2.086	2.845
4	2.132	2.776	4.604	30	1.697	2.042	2.750
6	1.943	2.447	3.707	40	1.684	2.021	2.704
8	1.860	2.306	3.355	50	1.678	2.011	2.682
10	1.812	2.228	3.169	∞	1.645	1.960	2.576

Then, using the computed sample mean $\bar{x}$ and variance s^2 determined by Eqn (18.25), the $(100 - r)$ per cent confidence interval for the population mean μ is

$$\bar{x} - \alpha_r s/\sqrt{n} \leq \mu \leq \bar{x} + \alpha_r s/\sqrt{n}.$$

Example 18.15 In a series of nine length measurements of components drawn at random from a production run, the sample mean was found to be $\bar{x} = 11.972$ cm and the sample standard deviation $s = 0.036$ cm. If it is known that the errors are normally distributed, find the 99 per cent confidence limits for the population mean μ.

Solution As $n=9$, the number of degrees of fredom involved is $v=9-1=8$. At the 99 per cent confidence level, the appropriate α_r from the table in Rule 18.2 is $\alpha_1 = 3.355$. Now $\alpha_1 s/\sqrt{n} = 0.043cm$, so that the 99 per cent confidence interval for the mean μ is $11.929 \text{ cm} \leq \mu \leq 12.015 \text{ cm}$. ■

We conclude this section by describing a rule and a statistical test which have frequent application. They employ the t-distribution and concern the difference between two sample means $\bar{x}_1, \bar{x}_2$ computed from two different independent samples of sizes n_1 and n_2, each of which is assumed to be drawn from a normal distribution with the same unknown variance. In the case of the test, the null hypothesis H_0 will be that the two unknown population means μ_1 and μ_2 are equal.

This situation can arise, for example, when two similar machines are used to produce the same components and their adjustment is in question. They will each be assumed to give rise to normally distributed errors, say in the length of the component, which have the same variance, though the length settings which affect the means may differ. The statistical problem is essentially one of deducing a $(100-r)$ per cent confidence interval for the difference $\mu_1 - \mu_2$ of the two population means from n_1, n_2, $\bar{x}_1$, $\bar{x}_2$, s_1 and s_2. The test then follows by finding whether or not the confidence interval contains the point 0. If the point 0 is contained in the confidence interval, then the null hypothesis H_0 is possible and H_0 is accepted; that is, in our illustration, there is no reason to dispute the assumption that the means are equal, so that the machines need not be adjusted. However, if the point 0 lies outside the confidence interval then H_0 is rejected; that is, we conclude that $\mu_1 \neq \mu_2$, and we would conclude that the setting of the machines does require adjustment.

Rule 18.4 Confidence interval for the difference $\mu_1 - \mu_2$ between two means of normal distributions with the same variance

Given two samples comprising n_1 and n_2 observations, compute their sample means $\bar{x}_1$, $\bar{x}_2$ and their sample variances s_1^2 and s_2^2. Select a $(100-r)$ per cent confidence level and use this to determine α_r from the table given in Rule 18.3, looking under the nearest entry appropriate to the number of degrees of freedom $v = n_1 + n_2 - 2$.

The $(100-r)$ per cent confidence interval for the difference of the means $\mu_1 - \mu_2$ has for its two end points the numbers

$$\bar{x}_1 - \bar{x}_2 \pm \alpha_r d,$$

where

$$d = \left[\frac{(n_1 + n_2)\{(n_1 - 1)s_1^2 + (n_2 - 1)s_2^2\}}{n_1 n_2 (n_1 + n_2 - 2)}\right]^{1/2}.$$

The test follows directly from Rule 18.4 and provides an illustration of how the idea of a confidence interval may be turned into a statistical test of significance. The test takes the following form, and we remark here that the number r used in Rule 18.4 is called the *significance level* of the test. The meaning of the significance level is that there is a probability $r/100$ that the null hypothesis H_0 will be accepted by mistake.

Rule 18.5 Test of the null hypothesis H_0 that $\mu_1 = \mu_2$ against the alternative H_1 that $\mu_1 \neq \mu_2$ for the two means of normal distributions with the same variance

Given the information in Rule 18.4, the null hypothesis H_0 is to be

(i) *accepted* at the r per cent significance level

$$\text{if } |(\bar{x}_1 - \bar{x}_2)|/d < \alpha_r,$$

and

(ii) rejected at the r per cent significance level

$$\text{if } |(\bar{x}_1 - \bar{x}_2)|/d < \alpha_r.$$

Example 18.16 Two similar machines produce a certain component. A sample of five components from machine 1 has sample mean length $\bar{x}_1 = 11.866$ cm and sample standard deviation $s_1 = 0.071$ cm, while the corresponding quantities based on a sample of seven components from machine 2 are $\bar{x}_2 = 11.943$ cm and $s_2 = 0.063$ cm. On the assumption that the samples are drawn from normal distributions with the same unknown variance, compute the 95 per cent confidence interval for the difference of the population means $\mu_1 - \mu_2$. Test at the 5 per cent significance level the hypothesis $\mu_1 = \mu_2$ against $\mu_1 \neq \mu_2$.

Solution Here $n_1 = 5$, $n_2 = 7$, $\bar{x}_1 = 11.866$ cm, $\bar{x}_2 = 11.943$ cm, $s_1 = 0.071$ cm and $s_2 = 0.063$ cm. The number of degrees of freedom $\nu = 5 + 7 - 2 = 10$ and the value of α_r from the table in Rule 18.3 corresponding to $r = 5$, the 95 per cent confidence level, is $\alpha_5 = 2.228$. Using Rule 18.4, we find $d = 0.03883$, so that the 95 per cent confidence interval for $\mu_1 - \mu_2$ has for its end points the lengths

$$(11.866 - 11.943) \pm 2.228 \times 0.03883 \text{ cm}.$$

Thus we know that with 95 per cent confidence

$$-0.164 \text{ cm} \leq \mu_1 - \mu_2 \leq 0.010 \text{ cm}.$$

The null hypothesis H_0 is to be accepted at the 5 per cent significance level, because

$$|(\bar{x}_1 - \bar{x}_2)/d| = |-0.077/0.03883| = 1.983 \leq 2.228.$$

Acceptance of H_0 at the 5 per cent significance level is, of course, implied by the 95 per cent confidence interval since this contains the point 0. Had the significance level been raised to 10 per cent, then the corresponding $\alpha_{10} = 1.812$ and so that at this level H_0 must be rejected because $1.983 > 1.812$. This is reasonable, because the 95 per cent confidence interval is obviously wider than the 90 per cent confidence interval. Expressed another way, the larger the permitted probability of accepting H_0 in error, the closer together must the limits of the confidence interval be taken. ■

18.6 Linear regression

A very frequently occurring situation in engineering and science is one in which pairs of observations are recorded and then a straight line is fitted to the data. This is not always an easy task, because experimental errors tend to scatter the points so that a variety of different straight lines can often seem to provide an equally good visual fit to the data. *Linear regression* using least squares provides a way of fitting such data to a straight line, while with the aid of the t-distribution a confidence interval may be found for the gradient of the line. The gradient is usually a quantity of considerable physical significance in an experiment.

A typical physical example is provided by the experimental determination of the adiabatic exponent γ occurring in the law $pv^{\gamma} = \text{constant}$, relating the pressure p and volume v of a fixed mass of gas. Setting $Y = \ln p$ and $X = \ln v$, the law takes the form $Y = -\gamma X + \text{constant}$, showing that $-\gamma$ is the gradient of the straight line involved. By obtaining n pairs of experimental observations $(p_1, v_1), (p_2, v_2) \ldots, (p_n, v_n)$, and plotting their logarithms as indicated, we at once arrive at a linear regression problem. Knowledge of the confidence interval for the gradient then provides information about the accuracy with which γ can be determined.

Let us suppose that n pairs of observations $(x_1, y_1), (x_2, y_2), \ldots, (x_n, y_n)$ are given and that we wish to fit a linear relationship of the form

$$Y = aX + b. \tag{18.31}$$

Our approach using the method of least squares will be to fit this straight line to the n points (x_i, y_i) in such a manner that we minimize the sum of the squares of the n quantities $(Y(x_i) - y_i)^2$. That is, we minimize the sum of the squares of the deviations of these points from the straight line, the deviations being measured in the Y-direction. To achieve this we first define S to be the sum of the squares of the errors

$$S = \sum_{i=1}^{n} (ax_i + b - y_i)^2, \tag{18.32}$$

and observe that S is a function of the two unknown parameters a, b of Eqn (18.31). From the work of section 12.5, we know that the conditions for S as a function of the variables a, b to attain an extremum are $\partial S/\partial a = 0$ and $\partial S/\partial b = 0$. In terms of Eqn (18.32) these equations take the form

$$\sum_{i=1}^{n} x_i(ax_i + b - y_i) = 0 \tag{18.33}$$

and

$$\sum_{i=1}^{n}(ax_i + b - y_i) = 0. \qquad (18.34)$$

Solving these inhomogeneous linear equations for the unknown parameters a, b then leads to the results

$$a = \frac{\sum_{i=1}^{n} x_i y_i - n\bar{x}\bar{y}}{\sum_{i=1}^{n} x_i^2 - n\bar{x}^2} \quad \text{and} \quad b = \bar{y} - a\bar{x}, \qquad (18.35)$$

where $\bar{x}, \bar{y}$ denote the means of the x_i and y_i, respectively. The quantity S will be minimized by this choice of a, b because it is essentially positive, so that its smallest possible value will be zero which it will only attain when all the n points $(x_i,\ y_i)$ lie on a straight line. It follows directly from (18.35) that Eqn (18.31) can be written in the form

$$Y - \bar{y} = a(X - \bar{x}), \qquad (18.36)$$

which shows that the straight line that has been fitted passes through the point $(\bar{x},\ \bar{y})$. In statistics, the line with Eqn (18.36), in which X is regarded as the independent variable, is called the *regresson line* of Y on X.

If the errors in the y_i are all normally distributed with the same variance, and the n pairs of observations are independent, the t-distribution may be used to determine a $(100 - r)$ per cent confidence interval for the true gradient $\tilde{a}$ of the regression line of Y on X, of which a is the sample estimate. The rule by which this may be achieved is as follows.

Rule 18.6 Confidence interval for the gradient $\tilde{a}$ of the regression line of Y on X

Given n pairs of observations $(x_1,\ y_1), (x_2,\ y_2), \ldots, (x_n,\ y_n)$ select a $(100 - r)$ per cent confidence level and determine α_r from the table given in Rule 3 appropriate to $v = n - 2$ degrees of freedom. Then the $(100 - r)$ per cent confidence interval for the true gradient $\tilde{a}$ of the regression line of Y on X, of which a in Eqn (19.34) is the sample estimate, is

$$a - \delta \le \tilde{a} \le a + \delta,$$

where

$$\delta = \alpha_r \left\{ \frac{\sum_{i=1}^{n} [a(x_i - \bar{x}) - (y_i - \bar{y})]^2}{(n-2)\left[\sum_{i=1}^{n} x_i^2 - n\bar{x}^2\right]} \right\}^{1/2}.$$

Example 18.17 Find the regression line of Y on X for the following six pairs of observations.

i	1	2	3	4	5	6
x_i	1	3	4	5	7	8
y_i	2	8	9	10	14	19

Find the 95 per cent confidence interval for the gradient of this line.

Solution It follows from the data that $n = 6$, $\bar{x} = 4.667$, $\bar{y} = 10.333$, $\Sigma\, x_i^2 = 164$, $\Sigma\, x_i y_i = 362$. Then from (19.34) we obtain $a = 2.18$, $b = 0.16$ and $\Sigma[a(x_i - \bar{x}) - (y_i - \bar{y})]^2 = 8.317$. As $v = 6 - 2 = 4$, we find from the table in Rule 19.3 corresponding to the 95 per cent confidence level that $a_5 = 2.776$. Thus $d = 0.69$, so that the 95 per cent confidence interval for the true gradient $\tilde{a}$ of this regression line is $2.18 - 0.69 \leq \tilde{a} \leq 2.18 + 0.69$, or $1.49 \leq \tilde{a} \leq 2.87$. ■

Problems **Section 18.2**

18.1 Toss a coin 50 times and plot the relative frequency of 'heads'.

18.2 Suggest a graphical representation for the sample space characterizing the score recorded in a trial involving the tossing of a die together with a coin which has faces numbered 1 and 2. Give example of:
(a) two disjoint subsets of the sample spaces;
(b) two intersecting subsets of the sample space, indicating the points in their intersection.

18.3 A bag contains 30 balls of which five are red and the remainder are black. A trial comprises drawing a ball from the bag at random, recording the result and then replacing the ball and shaking the bag. This process is called *sampling with replacement*. If this process is carried out twice, what is the probability of selecting
(a) two red balls;
(b) two black balls;
(c) one red and one black ball?

The next four problems concern transmission lines in which the probability of failure of components A, B, C, D and E is p_A, p_B, p_C, p_D and p_E, respectively. If these probabilities of failure are independent, find the probability that the transmission line will stop transmitting power.

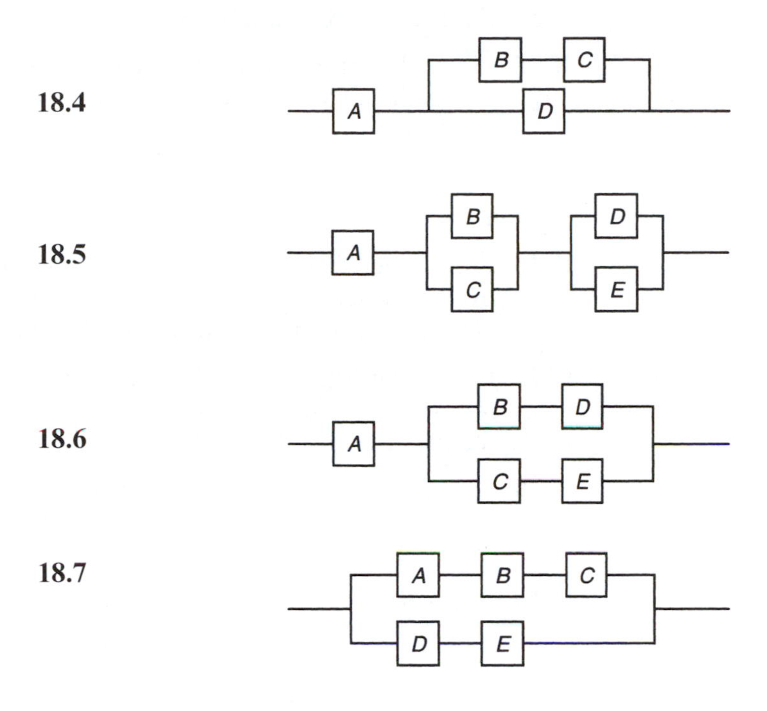

18.8 By considering arrangements of the five letters A, B, C, D, E, verify that

$$^{5}P_2 = 20 \text{ and } \binom{5}{2} = 10.$$

18.9 On the assumption that a participant in a raffle will buy either two or four numbered tickets, how many different sets of tickets may he choose from a book of 20 tickets?

18.10 A coin is biased so that the probability of 'heads' is 0.52. What is the probability that:
(a) three heads will occur in six throws;
(b) three or more heads will occur in six throws?

18.11 Shells fired from a gun have a probability $\frac{1}{2}$ of hitting the target. What is the probability of missing the target if four shells are fired?

18.12 Draw the probability density function for the binomial distribution in which $p = \frac{1}{2}$ and $n = 6$. Use your result to draw the corresponding cumulative distribution function.

18.13 By considering Fig. 18.3(a), deduce and draw the probability density function describing the sum of the scores of the two dice.

18.14 A light bulb manufacturer knows that 0.3% of every batch of 2500 bulbs will fail before attaining their stated lifetime. What is the probability that out of a batch of 500 (a) no more than six will fail, (b) exactly five will fail?

18.15 Using the information provided in Problem 18.14, find the probability that at least two but no more than five bulbs will fail out of a batch of 500 bulbs.

Section 18.3

18.16 Prove that $P(x) = 3(b^2 - x^2)/4b^3$ defines a continuous distribution over the interval $-b \leq x \leq b$. Find its mean μ and variance σ^2, and determine the cumulative distribution $U(x)$.

18.17 Prove that $P(x) = a\mathrm{e}^{-ax}$ is a continuous distribution over the interval $0 \leq x \leq \infty$. Find the probability that $x \geq \mu$, where μ is the mean of this distribution.

18.18 By considering Fig. 18.3, construct the probability density function describing the sum of the scores of the two dice and hence find the mean μ and the variance σ^2.

18.19 A target moves along a straight path of length l joining points A and B. The probability that a single shot fired at it will score a hit at a point distant x from A is $P(x) = 2x/l^2$. If a hit is scored at a point C what is the probability P that the ratio CB/AC exceeds the positive number k?

18.20 Given that the random variable x belongs to the $N(0, 1)$ distribution, find
(a) $P(x > -0.4)$, (b) $P(x < -0.2)$ and (c) $P(-1.8 < x < -0.6)$.

18.21 Given that the random variable x belongs to the $N(0, 1)$ distribution, find
(a) $P(x > 2.25)$, (b) $P(x < 0.8)$ and (c) $P(-0.2 < x < 1.2)$.

18.22 Given that the random variable x belongs to the $N(2, 2.5)$ distribution find (i) $P(x > 7)$ and (ii) $P(0.5 < x < 6)$.

18.23 Given that the random variable x belongs to the $N(1.2, 1.8)$ distribution, find (i) $P(x < -0.24)$ and (ii) $P(x > 4.44)$.

Section 18.4

18.24 Find the range, mode, mean, median and standard deviation of the observations x_i occurring with frequency f_i that are contained in the following table.

x_i	0.3	0.5	0.7	0.9	1.1
f_i	3	5	11	10	9

Section 18.5

18.25 A sample containing 112 observations has a variance $\sigma = 0.079$ cm and a sample mean $\bar{x} = 4.612$ cm. Find the 99.9% confidence interval for the population mean μ.

18.26 In a mechanical shock test on 320 insulators 39 were found to have failed. What is the 99% confidence interval for the mean number of failures to be expected in similar-sized batches of insulators?

18.27 In a series of eleven diameter measurements taken from a random sample of disk blanks, the sample mean diameter was found to be $\bar{x} = 2.536$ cm. Given that the sample standard deviation $s = 0.027$ cm, find the 95% confidence limits for the population mean μ.

18.28 Repeat the calculation in Problem 18.27 using the same values of $\bar{x}$ and s, but this time assuming that they had been obtained from a sample of four measurements. Use linear interpolation to determine α_r.

18.29 Repeat the calculations in Example 18.11 using the same data, with the exception that this time it is assumed that a sample of seven components was taken from machine 1 and a sample of five components from machine 2.

Section 18.6

18.30 Find the regression line of Y on X for the following data.

x_i	0	2	5	7	9
y_i	-2	3	8	14	16

Find the 90% confidence interval for the gradient $\tilde{a}$ of this line, and sketch the two possible extreme regression lines of Y on X that pass through $(\bar{x}, \bar{y})$.

18.31 Find the regression line of Y on X for the following data.

x_i	1	2	5	6	8	9
y_i	3	5	9	8	10	12

Find the 99% confidence interval for the gradient $\tilde{a}$ of this line, and sketch the two possible extreme regression lines of Y on X that pass through $(\bar{x}, \bar{y})$.

19 Symbolic algebraic manipulation by computer software

In recent years several types of computer software have become available that allow algebra and calculus to be performed symbolically on a computer. In general, this is called *computer algebra software*, though its capabilities extend far beyond purely algebraic operations. Such software can plot graphs in two and three-dimensions to allow the geometrical properties of functions to be explored, it can be used to understand limiting operations in calculus, to perform differentiation and integration in symbolic form, as would be done with pencil and paper, and it can solve differential equations. By removing the possible introduction of errors in problems where a solution using pencil and paper would involve tiresome and prohibitively lengthy calculations, computer algebra software also makes it possible to solve realistic problems in the physical sciences very early on in their study.

To understand the capabilities of this type of software, it is necessary to know what is meant by software that can perform *symbolic operations*. This means software that cannot only perform purely numerical operations on functions, but also software with the ability to perform algebraic and calculus operations on functions like $\sqrt{1+x^2}$ and $\cosh x$ when x is treated as a symbol, and not as a specific number. If, for example, the function $f(x) = x \sin x$ is entered into a computer with the instruction to differentiate $f(x)$ once with respect to x, after displaying $f(x)$ on the screen, symbolic algebraic software will then show on the screen the derivative of $f(x)$ as the function of x that would be found by a hand calculation, namely, $\sin x + x \cos x$.

A purely algebraic example of the use of symbolic software, involves entering the algebraic expression $(x+3)/(x^2+3x+2)$ into the computer with the instruction to express it in partial fraction form. The result shown on the computer screen by symbolic software will then be

$$-\frac{1}{x+2} + 2\frac{1}{x+1}.$$

Symbolic algebra software is also capable of acting as an ordinary calculator, because the numerical value of a symbolic expression $f(x)$ can be found once a specific numerical value has been given to its argument x. So if the value of the derivative of $f(x) = x \sin x$ is required when

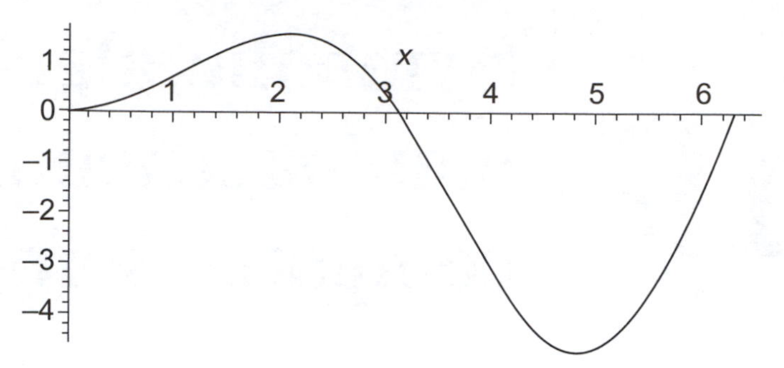

Fig. 19.1 A plot of $f(x) = x \sin x$.

$x = 0.6$ (remember that in mathematics when working with trigonometric functions the argument x is in *radians*), once the software has found the derivative $f'(x) = \sin x + x \cos x$, the result of entering $x = 0.6$ will be to produce on the screen the result 1.059843842.

Symbolic software can graph functions of one or two independent variables, and the instruction to plot the graph of $f(x) = x \sin x$ for $0 \leq x \leq 2\pi$ will produce on the screen the plot shown in Fig. 19.1

Similarly, the instruction to produce a three-dimensional plot of the function $f(x, y) = x^2/4 - y^2/9$ for $-2 \leq x \leq 2$, $-2 \leq y \leq 2$ displaying the appropriate Cartesian axes will produce on the screen a plot like the one in Fig. 19.2, from which it is clear that the function describes a surface with a saddle point at the origin $x = 0$, $y = 0$.

Symbolic algebra software uses what is called *exact arithmetic* when working with numbers, both in symbolic calculations and when functioning as an ordinary calculator. This means that, provided a number is not irrational, it is always represented in the computer as a rational number.

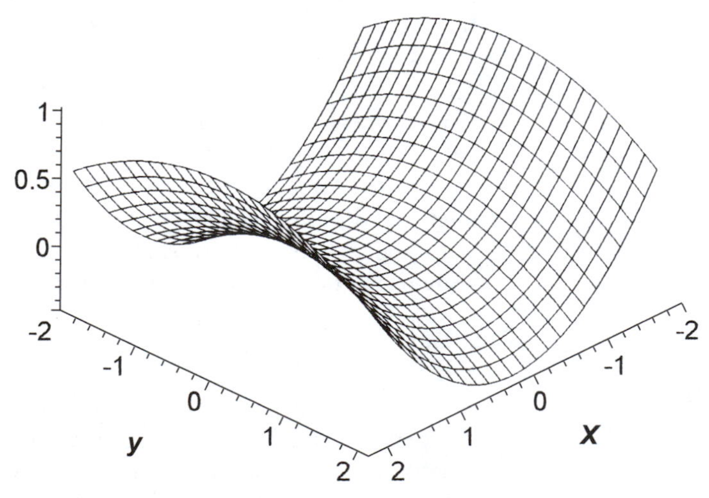

Fig. 19.2 A three-dimensional plot of $f(x, y) = x^2/4 - y^2/9$.

That is, numbers are always treated by the software as fractions, that may be either proper or improper. Consider the result of an internal computer symbolic operation leading to, say, the sum of the numbers $2\frac{17}{342}+3\frac{21}{47}$. The software will treat this sum as the sum of the rational numbers (improper fractions) $\frac{701}{342}+\frac{162}{47}$, and the result given will be the rational number (improper fraction) $\frac{88351}{16074}$. If the decimal approximation of this sum is required, the software can be instructed to approximate this rational number to a given number n of floating-point digits, so if $n=10$ it will return the result 5.496516113.

By working with rational numbers in this way, the accuracy of calculations is preserved until the time when a decimal approximation of a calculation is required. Again, to preserve accuracy, irrational numbers like π, $\sqrt{2}$ and $\sqrt{3}$ are left in this form by the software until a decimal approximation of a final result involving them is required. Thus an internal calculation producing a number like $(3\frac{47}{91})\pi$ will be treated by the software as $\frac{320}{91}\pi$, and if displayed on the screen it will appear in this form. Should the decimal approximation of this result be needed to ten floating-point digits, the result shown on the screen will be 11.04735878.

Symbolic algebra software can integrate a wide range of functions. This is accomplished inside the computer by transforming the integrand into a standard form that can be checked against an internal reference list of known results, after which the result is transformed back to the original variable. Naturally there will be integrals for which this method fails, either because the software is not powerful enough to find the result, or because no integral exists involving elementary functions. This ability of the software to integrate also enables it to find analytical solutions for a wide variety of differential equations, both with and without initial conditions.

For the software to function in the ways just described, it is necessary that the information specifying the functions involved, and the symbolic operations to be performed, is given to the computer in a very special manner. The mathematical operational instructions for different types of symbolic software differ, depending on the internal workings of the software, but these are all described in detail in the literature accompanying the software. Many excellent books are available that describe the different types of symbolic algebraic software now in use. All of these books explain the instructions and operations in considerable detail, and they also illustrate how to use them by making typical applications.

Because of the significant differences in the way the instructions for different types of symbolic algebraic software are entered into a computer, this brief introduction can only mention two particular types of such software. The first is called MAPLE, and it is the registered trademark of Waterloo Maple Inc., 57 Erb Street West, Waterloo, Ontario, Canada N2L 5J2. The other, called MATLAB, is the registered trademark of The MathWorks Inc., 24 Prime Park Way, Natick, MA 01760-1500,USA.

MAPLE MAPLE was used in the preparation of this sixth edition of *Mathematics for Engineers and Scientists*, but because of its many functions, this short introduction can only outline a few of the most important ones that are relevant to this book. In what follows, the version described is called MAPLE 8, because it is the eighth issue of the software to have been released. Technically, MAPLE 8 is a *Symbolic Algebraic Computation System* whose main function is to manipulate information symbolically according to the usual rules of algebra and calculus. For conciseness, in what follows MAPLE 8 will simply be referred to as MAPLE.

Once a symbolic computation has been completed, numerical values can be assigned to the symbolic quantities involved, and MAPLE will then evaluate the numerical expressions and, if required, plot graphs in two or three-dimensions. Various coordinate systems are supported by MAPLE, including the Cartesian coordinates and plane polar coordinates used throughout this book.

If input information is given in purely numerical form MAPLE acts like an ordinary calculator, but if the information is entered in symbolic form MAPLE manipulates it symbolically. For example, MAPLE can find either the antiderivative (indefinite integral) or definite integral of $x \ln x$, though when finding antiderivatives the arbitrary additive integration constant associated with indefinite integration is always omitted. So the integral of $x \ln x$ is given by MAPLE as $\frac{1}{2}x^2 \ln x - \frac{1}{4}x^2$, without the usual arbitrary additive constant C.

Before introducing some of MAPLE's many procedures, it is necessary to know how to start MAPLE, and how to exit from it. MAPLE is started by placing the mouse pointer arrow on the MAPLE icon seen on the screen when the computer is first switched on, and double clicking with the left mouse button. This causes the name MAPLE with its logo to appear briefly on the screen, followed a few moments later by a prompt at the top left of the screen in the form of the arrow head symbol $>$. This prompt is the signal to start to input information. To exit from MAPLE, the 'File' icon at the top left of the screen must be clicked once with the left mouse button. An instruction will then appear in the middle of the screen asking if the work should be saved, not saved, or if the exit command should be cancelled. Until some familiarity with MAPLE has been developed it is advisable to click the 'No' icon, when the screen will return to the form seen when the computer was first switched on.

Where possible MAPLE uses a notation similar to standard mathematical notation. For example, to find $2/3 \times (2/7 + 3/2)$ we would proceed as follows, using the fact that in MAPLE notation, addition and subtraction are denoted by the usual + and − signs, multiplication by an asterisk *, and division by a solidus /, so the result of this calculation when entered in Maple, which uses brackets (...) in the usual way, will appear as

```
>    2/3*(2/7+3/2);
```

$$\frac{25}{21}.$$

Notice the presence of the semicolon ; at the end of the first line. This is the instruction to MAPLE to perform the calculation and to display the result in the line below. If the semicolon is omitted nothing will happen.

MAPLE uses exact arithmetic and cancels common factors in fractions, so the result of a numerical calculation appears either in the form of an integer or rational number, but if irrational numbers like $\sqrt{2}$ or π occur in the calculation, these are displayed in this same form in the result. The number $\sqrt{2}$ is entered in Maple as sqrt(2), while the constant π is entered as Pi. If a decimal result is required the instruction for the calculation must be prefixed by the instruction 'evalf', after which the details that follow must be enclosed in parentheses, as shown below:

```
>     evalf(2/3*(2/7 + 3/2));
      1.190476191.
```

The default number of floating-point digits (that is the number used unless MAPLE is instructed to use a different number) is ten, but this can be controlled by prefixing the calculation at the prompt > by the instruction 'Digits: = n;' where n is the number of digits required. Notice that when defining n the symbol : = must follow the word Digits, and then after entering n either a semicolon ; or a colon : must be used to instruct Maple to round calculations to this number of floating-point digits in all subsequent calculations. When a semicolon is used Maple prints out the instruction *Digits*: = 6 as a reminder of what is to follow, and then uses this number of digits in all subsequent calculations, but when a colon is used Maple omits the print out, but still uses six floating-point digits in all subsequent calculations. To illustrate matters we will repeat the previous calculation using only six floating-point digits, and omit the print out of the instruction Digits: = 6 by using a colon. We type

```
>     Digits: = 6:
>     evalf(2/3*(2/7 + 3/2));
      1.19048
```

which is the previous result now rounded to six floating-point digits. When necessary, MAPLE can work to an arbitrarily large number of decimal digits, so if π is required to 30 decimal digits, this is obtained by typing

```
>     Digits: = 30;
      Digits: = 30
>     evalf(Pi);
      3.14159265358979323846264338328.
```

To illustrate the way MAPLE operates with irrational numbers like $\sqrt{2}$ and π, consider the calculation $\sqrt{2}(3 + 2\pi)$, that when entered into MAPLE after the prompt yields

```
>       sqrt(2)*(3 + 2*Pi);
```

$\sqrt{2}(3+2\pi)$.

Here MAPLE has returned the result unchanged, apart from displaying it in mathematical notation, because the nature of the irrational numbers $\sqrt{2}$ and π means that MAPLE cannot represent either exactly by a rational number. If a numerical result is required for the last calculation, say to 8 floating-point digits, we make use of the instruction 'evalf', and at the prompt write

```
>       Digits: = 8:
>       evalf(sqrt(2)*(3 + 2*Pi));
```

13.128407.

MAPLE can manipulate and perform arithmetic on complex numbers by representing the imaginary unit i by the symbol I together with the multiplication symbol * that must be used if a multiple of i occurs, so the complex number $3-2i$ is entered as 3 − 2*I or, equally well, as 3 − I*2. The quotient $(2-3I)/(4+2I)$ is found by typing

```
>       (2 − 3*I)/(4 + 2*I);
```

$$\frac{1}{10}-\frac{4}{5}I.$$

This result can be expressed in modulus/argument form by using the instruction 'convert', as follows

```
>       convert(1/10 − 4*I/5, polar);
```

$$\text{polar}\left(\frac{\sqrt{65}}{10}, -\arctan(8)\right).$$

In this result the term *polar* refers to the use of plane polar coordinates in the complex plane, with the first entry $\sqrt{65}/10$ representing the modulus of the complex number, and the second entry $-\arctan(8)$ its argument. In radian measure, the value of $-\arctan(8)$ is found by typing

```
>       evalf(−arctan(8));
```

−1.446441332.

The instruction *convert* is also useful when seeking a partial fraction representation of the quotient of two polynomials. For example, the partial fraction representation of $(1+2x+3x^2-x^3)/(1+2x+x^2)$ is found as follows, where in MAPLE notation the symbol ^ is used to indicate a power so, for example, 5^3 is entered as 5 ^ 3. For convenience, the label eq1 will be added to the original quotient so it can called again by this label (name) if required.

```
>       eq1: = (1 + 2*x + 3*x ^ 2 − x ^ 3)/(1 + 2*x + x ^ 2);
```

$$eq1 := \frac{(1 + 2x + 3x^2 - x^3)}{(1 + 2x + x^2)}$$

```
>    convert(eq1, parfrac, x);
```

$$-x + 5 + \frac{1}{x+1} + \frac{3}{(x+1)^2}.$$

The term 'parfrac, x' is the instruction to convert the expression eq1, which is rational a function of x, into partial fractions in terms of x. To see how the label eq1 can be used again, let us subtract $1/(x+1)^2$ from the partial fraction representation of eq1 by writing

```
>    eq1 − 1/(x + 1)^2;
```

$$-x + 5 + \frac{1}{x+1} + \frac{2}{(x+1)^2}$$

where now the required subtraction is seen to have been performed.

Functions like sine x, cosine x, tangent x, $\sinh x$, $\cosh x$, $\tanh x$, e^x and the natural logarithm $\ln x$ are entered into MAPLE as sin(x), cos(x), tan(x), sinh(x), cosh(x), tanh(x), exp(x) and ln(x). Inverse functions like $\arcsin x$, $\arctan x$ and arcsinh x, are entered as arcsin(x), arctan(x) and arcsinh(x). These, and all other functions used by MAPLE, can be plotted by using the instruction 'plot', once the interval over which the plot is required has been specified.

Denoting the function $\ln(1+x)+\sin 2x$ by eq2, at the prompt we type

```
>    eq2: = ln(1 + x) + sin(2*x);
```

$$eq2 := \ln(1 + x) + \sin(2x).$$

To plot this over the interval $0 \leq x \leq 2\pi$ using a thick line, at the prompt we enter

```
>    plot(eq2, x = 0..2*Pi, thickness = 3);
```

when the result seen on the screen is shown in Fig. 19.3. Here, we have used the ability to adjust the line thickness of the plot by adding at the end of the plot the command the instruction 'thickness = n' where n is typically an integer 1, 2 or 3, with the line corresponding to the choice n = 3 being much thicker than the line corresponding to n = 1.

Notice that when specifying the interval in x over which the plot is to be constructed, two full stops (called periods in American literature) must be entered between the value of the argument used to start the plot and the value of the argument where the plot is to finish. The plot can be resized by first clicking with the left mouse button at any point inside the plot. A box will appear around the plot with a small black rectangle at the mid-point of each side. If the arrow in the plot window is positioned on one of the rectangles at the side, a double sided arrow like ↔ will appear. Keeping the left mouse button depressed, the plot can be expanded or contracted laterally in the direction shown by the double headed arrow,

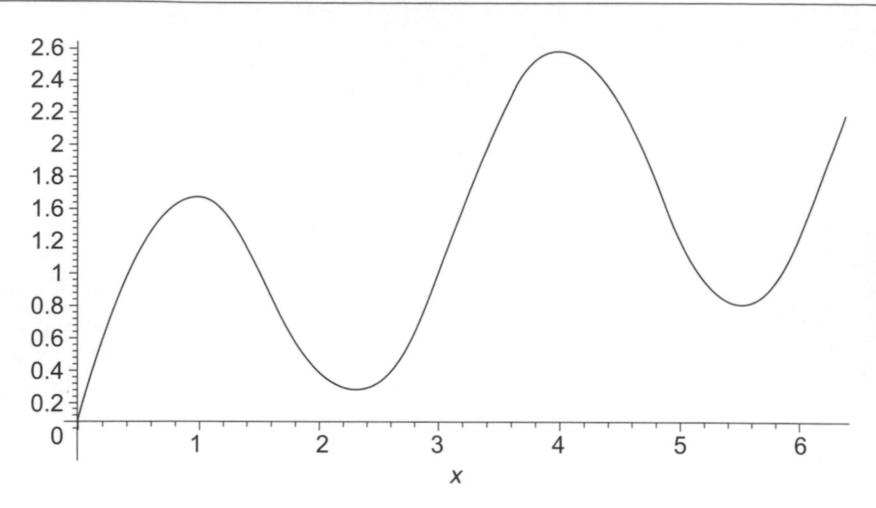

Fig. 19.3

by moving the double headed arrow either out of or into the plot. A similar procedure will expand or contract the plot in the vertical direction if a black rectangle at the top or bottom of the box is used.

The colour used in the screen plot is assigned by MAPLE, but it can be changed by adding at the end of the plot command the instruction 'color = blue' if, for example, the plot of eq2 is to be represented in blue. Many plot colours are available, but the most useful are black, red, blue and green, Notice that in this plot command the American spelling for 'color' *must* be used. MAPLE generates this plot by computing the value of eq2 at 50 points over the interval $0 \le x \le 2\pi$, and then joining the points by straight line segments.

MAPLE chooses the horizontal and vertical scales for a plot depending on the total variation of the function in the horizontal and vertical directions, but sometimes it needs help when choosing an appropriate vertical scale if the functional variation in the vertical direction is very large. For example, issuing the plot command 'plot(tan(x),x = 0..2);' will produce the result shown in Fig. 19.4, where the unbounded behaviour of $\tan x$ close to $x = \pi/2$ has biased the entire plot. This can be corrected by adding after the specification of the interval over which the function is to be plotted, an instruction of the form 'a.. b', where 'a' is the smallest value of the function to be plotted and 'b' is the largest value. If, in the case of Fig. 19.4, it is required that the plot of $\tan x$ is to vary from –10 to 10, the result of typing

```
>      plot(tan(x), x = 0..2, –10..10);
```

will be to produce the more useful result shown in Fig. 19.5.

If required, Maple can superimpose two or more plots, provided their arguments are over the same interval. The following instruction shows how plots of e^{-x} and $\sin x$ can be superimposed for $0 \le x \le 4$, but to accomplish this it is now necessary to enclose the two functions e^{-x} and

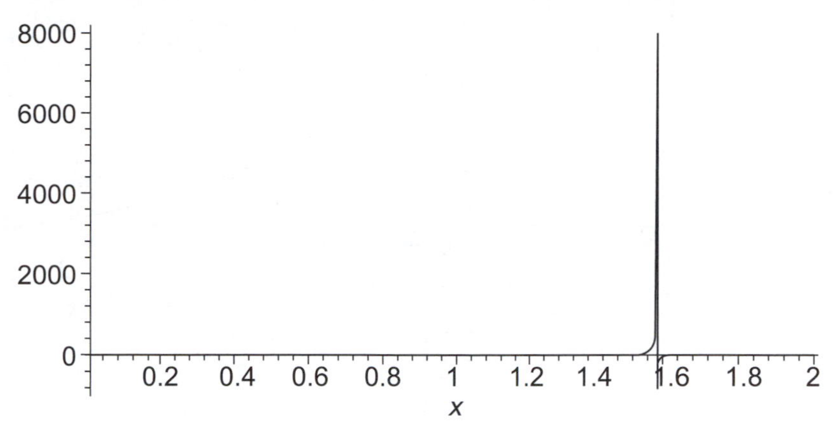

Fig. 19.4

$\sin x$ in parentheses $\{\ldots\}$, with a comma between each of the functions.

> plot({exp(−x), sin(x)}, x = 0..4, color = black).

The result is shown in Fig. 19.6, where each plot is black. The result of using a printer capable only of black printing when reproducing a line shown in a light colour on the computer screen is to produce a grey printed line. It was for this reason that the instruction 'color = black' was added to the print command.

This plot could be used to find the approximate values of x for which $e^{-x} = \sin x$ in the interval $0 \le x \le 4$, by finding the values of x in the interval where the two plots intersect. However if, say, this information is required to find starting values for Newton's method, it would be more sensible to plot the difference $e^{-x} - \sin x$, and to find where the plot intersects the x-axis. The result obtained in this manner by using the plot command

> plot(e^{-x} − sin(x), x = 0..4);

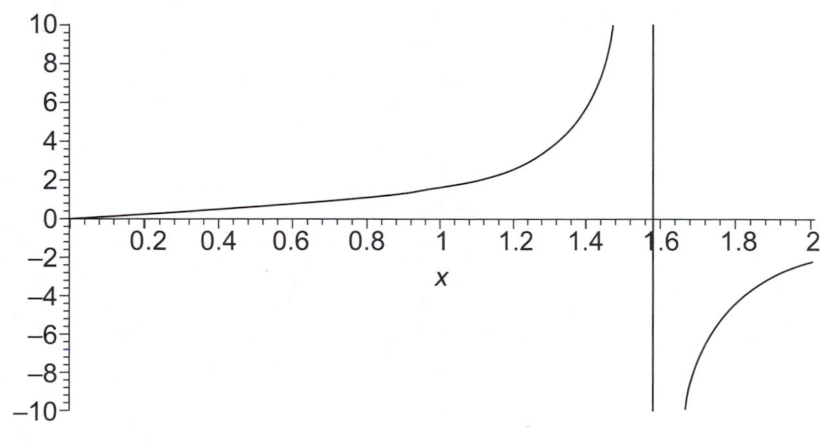

Fig. 19.5

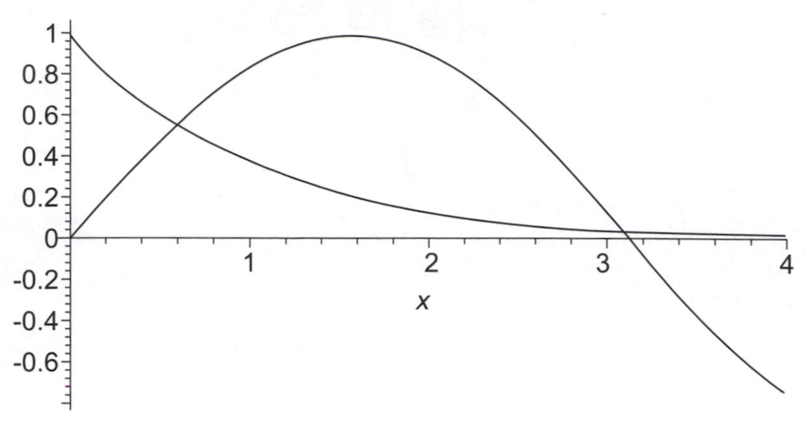

Fig. 19.6

is shown in Fig. 19.7, from which it can be seen quite clearly that the approximate zeros of the function $e^{-x} - \sin x$ in the interval $0 \leq x \leq 4$ occur at $x \approx 0.6$ and $x \approx 3.1$.

MAPLE can also make parametric plots in terms of a parameter t when the x-coordinate is specified as $x = f(t)$ and the y-coordinate as $y = g(t)$, say, with $0 \leq t \leq T$. The parametric plot of a curve called an *astroid*, with the parametric equations $x = \cos^3 t$ and $y = \sin^3 t$ with $0 \leq t \leq 2\pi$ shown in Fig. 19.8, was obtained by entering at the prompt

```
>       plot([cos(t)^3, sin(t)^3, t = 0..2*Pi]).
```

Notice that to specify a parametric plot it is necessary to use the square brackets [...], and to enclose inside them first the x-coordinate then, separated by a comma, the y-coordinate and finally, separated by another comma, the interval for the parameter t specified in the usual way.

Plots of functions with two independent variables can also be obtained by using a plot command of the form 'plot3d(eq, x = a..b, y = c..d);' where eq is the function of the two independent variables x and y to

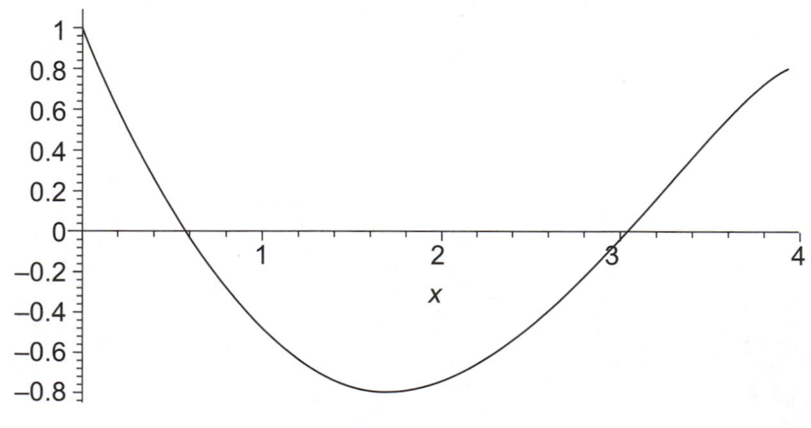

Fig. 19.7

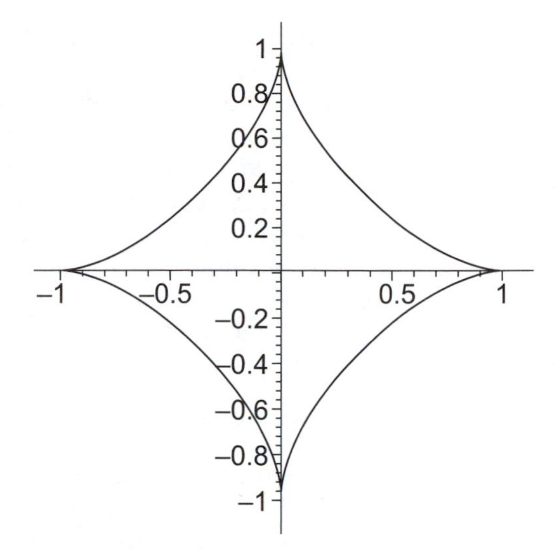

Fig. 19.8

be plotted in the region $a \le x \le b$, $c \le y \le d$. Fig. 19.9 is a typical MAPLE 3d plot of the function $\cos(x^2+y^2)/(1+x^2+y^2)$ in which the axes are shown. The necessary MAPLE instructions to generate Fig. 19.9 are

```
>    eq3: = cos(x^2 + y^2)/(1 + x^2 + y^2);
```

$$eq3{:} = \cos(x^2 + y^2)/(1 + x^2 + y^2)$$

```
>    plot3d(eq3, x = -4..4, y = -4..4).
```

MAPLE will produce a coloured plot on the screen unless the pointer arrow is clicked inside the plot to cause a bar of icons to appear at the top of the screen, when at the same time the plot will become enclosed in a rectangle. Clicking with the left button on the icon marked 'color', and then clicking on the entry 'no coloring' will produce the black and white

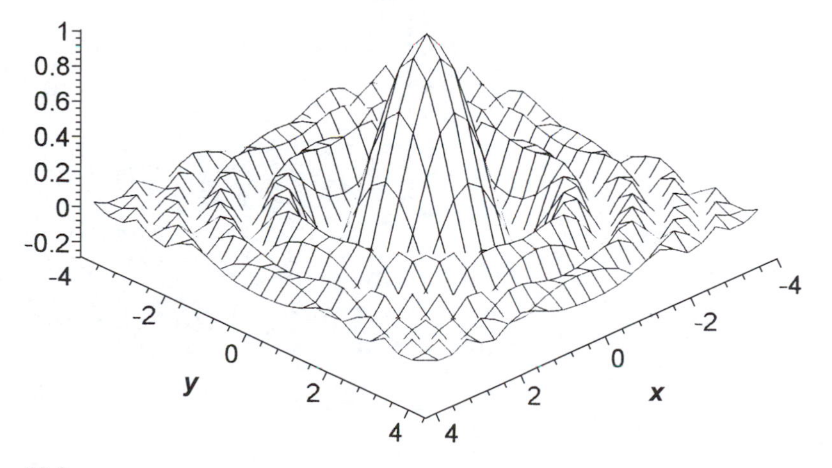

Fig. 19.9

wire-mesh type plot shown in Fig. 19.9. The rough appearance of the plot in Fig. 19.9 is due to MAPLE having computed function values using its default grid over the interval $-4 \le x \le 4$, and $-4 \le y \le 4$ (the grid used automatically if no other one is specified), which in this case happens to have its points spaced too far apart to provide a smooth representation of $f(x, y)$ when the computed points are joined by straight line segments.

A smoother plot can be obtained by instructing MAPLE to use a finer grid by including in the plot command the statement 'grid = [m,n]', where m is the number of equi-spaced points used in the x-direction and n is the corresponding number of equi-spaced points in the y-direction. Entering the plot command

```
>     plot3d(eq3, x = -4..4, y = -4..4, grid = [50, 50]);
```

produces the smoother plot shown in Fig. 19.10.

A 3dplot can be rotated to view it from any direction by first clicking inside the plot to cause the plot to become enclosed in a box, and then with the left mouse button depressed, dragging the plot into the desired orientation. A typical plot obtained in this way from the one in Fig. 19.10 is shown in Fig. 19.11.

Maple can also make plots using the plane polar coordinates (r, θ) by first calling a special plot facility by typing 'with(plots):' and then, at the prompt, typing 'polarplot(eq, $\theta = a..\ b$)'. Here, eq is the expression for the radius r in terms of the polar angle θ, and the plot starts with $\theta = a$ and finishes with $\theta = b$. For example, the instructions

```
>     with(plots):
>     polarplot(1 + 3*sin(5*theta), theta = 0..2*Pi);
```

produces the flower-like plot in Fig. 19.12. Notice that in MAPLE, lower case Greek characters like θ and α are entered by typing 'theta'

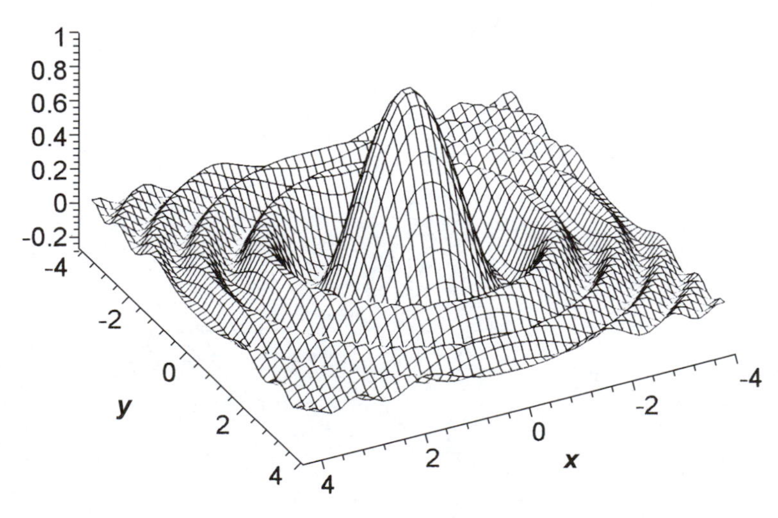

Fig. 19.10

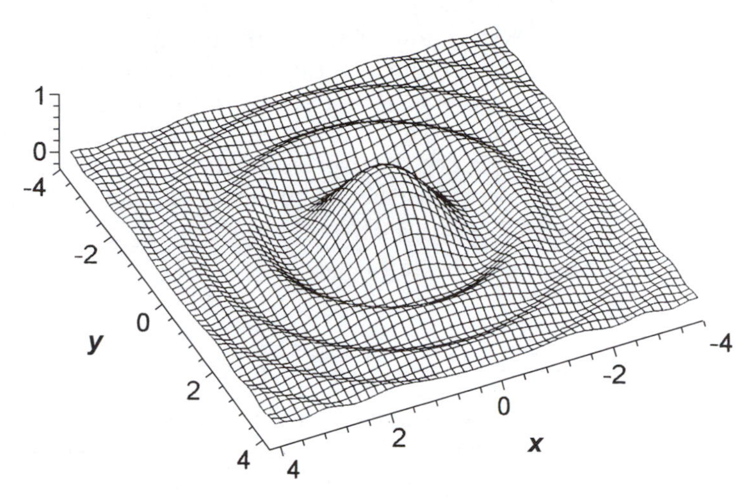

Fig. 19.11

and 'alpha', while an upper case character like Θ is entered by typing 'Theta'.

When reducing the screen plot to a size suitable for inclusion in this text, an arbitrary point inside the screen plot was clicked once to enclose the plot in a rectangular box. Then, in the bar above the plot, the button 1:1 was clicked. This rendered the scaling in both the horizontal and vertical directions the same, so when the plot was reduced by dragging inward one of the black rectangles on a side of the box, the horizontal and vertical dimensions both reduced at the same rate, thereby preserving the proportions of the plot.

Something of the symbolic manipulation capabilities of MAPLE can be appreciated by using the instruction 'int' signifying integrate. An in-

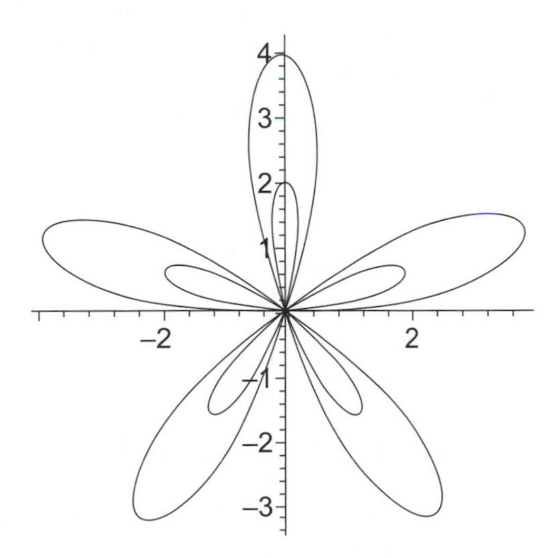

Fig. 19.12

struction of the form 'int(eq,x);' entered at the prompt will return an indefinite integral (antiderivative) of the expression eq as a function of the variable x, but always without the added arbitrary constant of integration, while the instruction 'int(eq,x, = a..b);' will return the definite integral of the expression eq from $x = a$ to $x = b$. As a typical example, the next instruction requiring the definite integral of e^{2x} between $x = 1$ and $x = 2$ yields

```
>    int(exp(2*x), x = 1..2);
```

$$\frac{1}{2}e^4 - \frac{1}{2}e^2.$$

A useful facility in MAPLE is provided by typing the symbol %, because this is interpreted by MAPLE to mean 'the previous result'. So typing 'evalf(%)' after this last result will produce its value to ten floating-point digits:

```
>    int(exp(2*x), x = 1..2);
```

$$\frac{1}{2}e^4 - \frac{1}{2}e^2$$

```
>    evalf(%);
```

23.60454697.

Depending on the integrand involved, MAPLE uses one of several different ways to evaluate an integral, but if it is unable to find an analytical result, either because one does not exist, or because the MAPLE routines are not powerful enough to find a result, this is indicated by returning the original instruction. The next instruction asks Maple to evaluate a definite integral of $\sin(x + x^3)$ from $x = 1$ to $x = \pi/2$ for which there is no known result in terms of elementary functions, so in this case MAPLE returns the original instruction in ordinary mathematical notation

```
>    int(sin(x + x^3), x = 1..π/2);
```

$$\int_1^{\pi/2} \sin(x + x^3)\mathrm{d}x.$$

When such integrals are well defined they can always be evaluated numerically by using the evalf instruction as follows:

```
>    evalf(int(sin(x + x^3), x = 1..π/2));
```

–0.1019805955.

This result was obtained automatically by MAPLE using a very accurate built in numerical integration routine.

The result of integrating a function can often be expressed in several different but equivalent ways, and the form returned by MAPLE will depend on the method it uses. Consequently the result of an integration performed by MAPLE is not always given on the screen in its most convenient form, so it may need to be manipulated to bring it into a more suitable form. The transformations used to accomplish such simplifications can all be performed within MAPLE, but to do so requires knowledge of the appropriate instructions to be issued, and an understanding of the mathematics involved in order to decide how best to simplify the result.

For example, if MAPLE is instructed to integrate $\sin^2 x \cos^2 x$ by using the instruction 'int(sin(x)^2 *cos(x)^2, x);' it produces the result

$$\frac{x}{8} + \frac{1}{8}\cos x \sin x - \frac{1}{4}\sin x \cos^3 x$$

(remember the arbitrary additive integration constant is always omitted). However, as $\sin 2x = 2\sin x \cos x$, this can be re-expressed as

$$\frac{x}{8} + \frac{1}{16}(1 - 2\cos^2 x)\sin x,$$

or equally, because $\cos^2 x = 1 – \sin^2 x$, as

$$\frac{x}{8} + \frac{1}{16}(2\sin^2 x - 1)\sin x.$$

For another example of integration let us consider the following indefinite integral, for which the result given by MAPLE is

$$\int e^x \sinh x \, dx = \tfrac{1}{2}\sinh x \cosh x - \tfrac{1}{2}x + \tfrac{1}{2}(\cosh x)^2.$$

If, however, we replace $\sinh x$ by $\frac{1}{2}(e^x - e^{-x})$ in the integrand and instruct MAPLE to integrate the equivalent result we obtain

$$\int e^x \tfrac{1}{2}(e^x - e^{-x}) dx = \tfrac{1}{4}e^{2x} - \tfrac{1}{2}x.$$

These two results are *not* the same, because when the first is expressed in terms of exponentials it reduces to

$$\int e^x \sinh x \, dx = \tfrac{1}{4}e^{2x} - \tfrac{1}{2}x + \tfrac{1}{4},$$

showing that the results differ by $\frac{1}{4}$. This apparent paradox is resolved once it is remembered that MAPLE does *not* include additive arbitrary integration constants in its results, so the constant $\frac{1}{4}$ occurring in the second result can be absorbed into the additive arbitrary constant that has been omitted, showing that in this sense the two results are, in fact, identical.

Finally, to illustrate another way in which a symbolic algebra program does not always give a result in its most convenient form, consider the following result obtained by using MAPLE

$$\int \sqrt{1+x^2}\,\mathrm{d}x = \tfrac{1}{2}\left[x\sqrt{1+x^2} + \operatorname{arcsinh} x\right],$$

where as usual the additive arbitrary constant is not shown. The function $\operatorname{arcsinh} x$ is not an elementary function that is convenient to use, but it can be transformed into a more useful form by using the result $\operatorname{arcsinh} x = \ln[x + \sqrt{1+x^2}]$ (see Example 8.5(a)), when the equivalent but more convenient result

$$\int \sqrt{1+x^2}\,\mathrm{d}x = \tfrac{1}{2}\left\{x\sqrt{1+x^2} + \ln\left[x + \sqrt{1+x^2}\right]\right\}$$

is obtained, where again the additive arbitrary constant has been omitted.

These examples illustrate a common difficulty that can arise when employing symbolic manipulation, usually because the first result obtained may not be in the form expected. This fact must also be remembered when reconciling two calculations obtained in different ways, as when considering the integral $\int e^x \sinh x\,\mathrm{d}x$ above, because although two results may appear different, provided no error has been made they must be equivalent.

MAPLE can perform all of the usual operations in calculus, and the one to be considered next involves partial differentiation. The function of two variables $f(x, y) = (x+2y)/(1+x^2+y)$ can be differentiated partially with respect to x and y as follows. Giving $f(x, y)$ the label eq4, the function is entered into MAPLE by typing

```
>    eq4: = (x + 2*y)/(1 + x^2 + y);
```

$$eq4{:} = \frac{x + 2y}{1 + x^2 + y}.$$

The partial derivative $\partial f/\partial x$ is then found by typing

```
>    eq5: = diff(eq4, x);
```

$$eq5{:} = \frac{1}{1 + x^2 + y} - \frac{2(x + 2y)}{(1 + x^2 + y)^2}.$$

Similarly, the partial derivative $\dfrac{\partial^3 f}{\partial y^2 \partial x}$ is found by typing

```
>    eq6: = diff(eq5, y$2);
```

$$eq6{:} = -\frac{4}{(1 + x^2 + y)} + \frac{2(x + 2y)}{(1 + x^2 + y)^3}.$$

Here 'diff' signifies differentiation, so 'diff(eq4, x);' is the instruction to compute $\partial f/\partial x$ and 'diff(eq5,y$2);' is the instruction to differentiate eq5 (that is $\partial f/\partial x$) *twice* with respect to y, hence the notation y$2 in which the dollar sign $ followed by an integer tells MAPLE how many times differentiation with respect to y is to be performed on $\partial f/\partial x$.

MAPLE can also determine indeterminate forms by using the instruction 'limit'. For example, to find the limit of $\tan 2x/\sin 3x$ as $x \to 0$, we first give the function the label eq7, and then use the limit instruction. Typing

```
>   eq7: = tan(2*x)/(sin(3*x);
```

$$eq7 := \frac{\tan(2x)}{\sin(3x)}$$

defines the function whose limit is required. Then, typing the limit instruction generates the required limit, so that

```
>   limit(eq7, x = 0);
```

$$\frac{2}{3}.$$

Here, the instruction x = 0 requires MAPLE to find the limit as $x \to 0$ in *either* direction.

If the indeterminate form is undefined MAPLE will say so, as happens when attempting to find

$$\lim_{x \to 1} \frac{\tan(1 - x^2)}{|\cos(\pi/(2x))|}.$$

This limit is undefined in the ordinary sense, because while the numerator changes sign across $x = 1$, the denominator does not, so that this function has a *different* limit depending on whether x decreases or increases to $x = 1$. Using the fact that MAPLE denotes the absolute value by 'abs', to see what happens if in the above case we attempt to find the ordinary limit as $x \to 1$, we type

```
>   limit(tan(1 - x^2)/(abs(cos(Pi/(2*x))), x = 1);
```

undefined

showing that MAPLE cannot find the limit.

This situation can be rectified by using the fact that MAPLE can compute directional limits, by adding the instructions 'left' or 'right' to show if the limit at $x = 1$ is to be determined by allowing x to increases or decrease to 1.

```
>   limit(tan(1 - x^2)/(abs(cos(Pi/(2*x))), x = 1, left);
```

$$\frac{4}{\pi}$$

and

```
>   limit(tan(1 - x^2)/(abs(cos(Pi/(2*x))), x = -1, right);
```

$$-\frac{4}{\pi}.$$

Limits can also be determined when $x \to \infty$, by entering the symbol ∞ as the word 'infinity'. To find $\lim_{x \to \infty} x^{1/x}$ we type

```
>    limit(x^(1/x), x = infinity);
     1.
```

So this infinite limit exists and is equal to 1.

MAPLE performs all matrix operations, but before these can be used it is necessary to type at the prompt 'with(linalg):' as this makes available all of the MAPLE linear algebra routines necessary for matrix operations. Once 'linalg' has been called, it remains available for all subsequent matrix calculations until cancelled. For example, the matrices

$$A = \begin{bmatrix} 1 & -1 & 2 \\ -1 & 4 & 1 \\ 2 & 1 & 2 \end{bmatrix} \quad \text{and} \quad B = \begin{bmatrix} -2 & 3 \\ 1 & -1 \\ 2 & 2 \end{bmatrix}$$

are entered by typing

```
>    with(linalg):
>    A: = array([[1, -1, 2], [-1, 4, 1], [2, 1, 2]]);
```

$$A{:} = \begin{bmatrix} 1 & -1 & 2 \\ -1 & 4 & 1 \\ 2 & 1 & 2 \end{bmatrix}$$

and

```
     B: = array([[-2, 3], [1, -1], [2, 2]]);
```

$$B{:} = \begin{bmatrix} -2 & 3 \\ 1 & -1 \\ 2 & 2 \end{bmatrix}.$$

Notice that each row of elements in a matrix is entered between square brackets [...], with a comma between each element, and a comma between each set of brackets [...] representing a row. Finally, the set of rows are themselves contained between square brackets [...].

The product AB is obtained by typing the instruction

```
>    multiply(A, B);
```

$$\begin{bmatrix} 1 & 8 \\ 8 & -5 \\ 1 & 9 \end{bmatrix}$$

and the transpose of A is obtained by typing 'transpose A'. After typing 'with(linalg)' to make the linear algebra routines available, when required an $n \times n$ diagonal matrix with elements $a_1, a_2, \ldots, a_n$ on its leading diagonal and zeros elsewhere is entered by typing

$$\text{diag}(a_1, a_2, \ldots, n)$$

so as a special case the 3×3 unit matrix is entered by typing

diag(1, 1, 1).

If a matrix product is not defined, like the product BA involving matrices A and B above, MAPLE will return an error message saying the two matrices are not conformable for such a product.

The determinant of A is found by the typing

```
>    det(A);
     -15.
```

The inverse of A, to which we will give the label A1, is found by typing the instruction

```
>    A1: = inverse(A);
```

$$A1: = \begin{bmatrix} \frac{-7}{15} & \frac{-4}{15} & \frac{3}{5} \\ \frac{-4}{15} & \frac{2}{15} & \frac{1}{5} \\ \frac{3}{15} & \frac{1}{5} & \frac{-1}{5} \end{bmatrix}.$$

The simultaneous equations $AX = U$ with $X = [x, y, z]^T$ and $U = [1, -2, 4]^T$ can be solved in terms of the matrix $A1$ inverse to A as $X = A1U$ as follows

```
>    X: = array([[x], [y], [z]]);
```

$$X: = \begin{bmatrix} x \\ y \\ z \end{bmatrix}$$

```
>    multiply(A1, U);
```

$$\begin{bmatrix} \frac{37}{15} \\ \frac{4}{15} \\ \frac{-3}{5} \end{bmatrix}$$

so equating corresponding elements gives $x = 31/15$, $y = 4/15$ and $z = -3/5$. MAPLE can also find the characteristic polynomial, the eigenvalues and the eigenvectors of an $n \times n$ matrix A, but for $n > 3$ the eigenvalues and eigenvectors are best found in decimal form by typing at the prompt 'evalf(eigenvects(A))'.

This brief introduction to MAPLE now closes with a few remarks about its ability to solve variables separable and constant coefficient differential equations. A first derivative of a function $y(x)$ of x is denoted by 'diff(y(x),x)', while second and third order derivatives of $y(x)$ are denoted in MAPLE by 'diff(y(x),x$2)' and 'diff(y(x),x$3)', respectively. The initial value $y(a) = k_1$, where k_1 is a given number, is specified by typing 'y(a) = k_1', a first derivative $y'(a) = k_2$, where k_2 is a given number, is specified by typing 'D(y)(a) = k_2', and a second order derivative $y''(a) = k_3$, where k_3 is a given number, is specified by typing 'D(D(y))(a) = k_3', with obvious extensions of this notation for higher order derivatives.

The instruction to MAPLE to find the general solution of a first order differential equation with the label eq, where eq is a function of x, is given by typing 'dsolve(eq, y(x))'. If MAPLE can find a solution it will give the result with the single arbitrary constant shown as _C1, otherwise no result will be returned. In the general solution of a second order equation there will be two arbitrary constants, and these will be denoted by _C1 and _C2, and so on for still higher order equations. Let us find the general solution of the variables separable equation

$$\frac{dy}{dx} = \frac{\sqrt{1 + y^2}}{xy}.$$

To do this we type

```
>      dsolve(diff(y(x), x)) = sqrt(1 + y(x)^2)/(x*y(x), y(x));
```

$$\ln x - \sqrt{1 + y(x)^2} + \_C1 = 0.$$

Notice that the dependent variable must be typed $y(x)$, and not simply y so that MAPLE knows it is a function of x, and not simply a constant. In this case, as usually happens with variables separable equations, the result returned by MAPLE gives $y(x)$ in an *implicit* form. This expression must be manipulated if an explicit expression for $y(x)$ in terms of x is required. Solutions of variable separable equations usually require manipulation if explicit solutions $y(x)$ are to be found, though of course it is not always possible to find an explicit form for $y(x)$.

To find the general solution of the second order constant coefficient equation

$$\frac{d^2y}{dx^2} + 2\frac{dy}{dx} - 3y = \sin x$$

we type

```
>      dsolve(diff(y(x), x$2) + 2*diff(y(x), x) - 3*y(x) = sin(x), y(x));
```

$$y(x) = e^{(-3x)}\_C_2 + e^x\_C1 - \frac{1}{10}\cos(x) - \frac{1}{5}\sin(x).$$

The positioning and numbering of the two arbitrary constants $\_C1$ and $\_C2$ in the general solution is determined by MAPLE, and were this result to have been obtained by hand it would probably have been written

$$y(x) = C_1 e^{-3x} + C_2 e^x - \frac{1}{10}\cos x - \frac{1}{5}\sin x.$$

Initial conditions for a second order equation can be imposed at $x = a$ by specifying the numerical values of $y(a)$ and $y'(a)$. For example, to solve the previous equation subject to the initial conditions $y(0) = 1$ and $y'(0) = 0$, we type

```
>    dsolve({diff(y(x), x$2) + 2*diff(y(x), x) – 3*y(x) = sin(x), y(0) = 1,
     D(y)(0) = 0}, y(x));
```

$$y(x) = \frac{9}{40}e^{(-3x)} + \frac{7}{8}e^x - \frac{1}{10}\cos(x) - \frac{1}{5}\sin(x).$$

Notice that this time the equation and the initial conditions must all be enclosed in parentheses $\{\dots\}$.

MAPLE can use Laplace transform methods to solve initial value problems at $t = 0$ for differential equations involving discontinuous functions like the Heaviside unit step function and the Dirac delta function. The Heaviside unit step function, denoted in this book by $u_T(t)$, is entered into MAPLE as 'Heaviside(T – t)' while the Dirac delta function, denoted in this book by $\delta(t - T)$, is entered into MAPLE as 'Dirac(t – T)'. Here we have chosen to use the independent variable t in place of x, because problems involving these functions usually depend on the time denoted by t. Let us solve the equation

$$\frac{d^2y}{dt^2} + y = \delta(t - 1)$$

subject to the initial conditions $y(0) = 1$ and $y'(0) = 0$, by instructing MAPLE to use the Laplace transform method. (You will recall that the Laplace transform method of solution *always* requires initial conditions to be specified at $t = 0$.) We have

```
>    dsolve({diff(y(t), t$2) + y(t) = Dirac(t – 1), y(0) = 1, D(y)(0) = 0},
     y(t), method = laplace);
```

$$y(t) = \cos(t) + \text{Heaviside}(t - 1)\sin(t - 1).$$

In terms of the notation used in this book this solution reads

$$y(t) = \cos t + u_1(t)\sin(t - 1).$$

We mention in passing that MAPLE can find many Laplace transforms and inverse transforms. To accomplish this it is first necessary to type at the prompt 'with(inttrans)' to make available a whole suite of integral transform and inverse integral transform routines. The Laplace transform $F(s)$ of a function $f(t)$ is found by typing at the prompt 'laplace(f(t),t,s)', where the word 'laplace' and the symbols 't, s' instruct MAPLE to find the Laplace transform of $f(t)$, as a function of t, in terms of the transform variable s. Conversely, the inverse Laplace transform of the function $F(s)$, as a function of s, is found by typing at the prompt 'invlaplace(F(s),s,t)', where the word 'invlaplace' and the symbols 's, t' instruct MAPLE to find the inverse Laplace transform of $F(s)$ as a function of s, in terms of the variable t. To find the Laplace transform of $\sin 2t$, we type at the prompt

```
>    with(inttrans):
     laplace(sin(2*t), t, s);
```

$$2\frac{1}{s^2+4}.$$

Similarly, to find the inverse Laplace transform of $e^{-s}/(s^2+4)$ as a function of t, at the prompt we type

```
>    with(inttrans):
     invlaplace(e^-s/(s^2 + 4), s, t);
```

$$\tfrac{1}{2}\text{Heaviside}(t-1)\sin(2t-2)$$

which in the notation of this book becomes

$$\tfrac{1}{2}u_1(t)\sin(2t-2).$$

MAPLE software is very well documented, and a full description of all its commands, together with examples of their use, can be found by clicking the Maple 'Help' button and typing in the name of the command.

Having read this brief introduction to MAPLE, it is important that a reader should not be left with the impression that using MAPLE will make it unnecessary to understand most of the detailed mathematics presented in this book. It is essential that the mathematics described here is studied with care and fully understood if it is to be applied correctly and effectively in engineering and science, and if physical problems are to be modelled properly in terms of mathematics. A thorough understanding of these mathematics is also necessary if MAPLE is to be used as a valuable analytical and computational tool, and if its capabilities are to be used to full advantage. MAPLE will certainly remove the necessity to perform tedious calculations by hand, it will prevent the introduction of errors in manipulation, it will plot excellent graphs, and it will permit analytical and numerical calculations to be performed that if carried out by hand might otherwise be prohibitively long, but MAPLE will *not* remove the necessity to think, to understand what MAPLE is being asked to do, or to tell it how best to do it. MAPLE is a valuable aid when carrying out mathematical operations and recording results in graphical form, but it is only really useful when the underlying mathematics is properly understood.

REFERENCES

The following are two useful and easily understood accounts of MAPLE that describe and illustrate its most important commands.

1. An Introduction to MapleV: Jack-Michel Cornill and Philippe Testud, Springer, 2001
2. The Maple Book: Frank Garvan, Chapman & Hall/CRC, 2002

MATLAB MATLAB is quite different from MAPLE, because all numerical calculations in MATLAB are based on a matrix type input. When it was first introduced, MATLAB was designed to perform state of the art numerical computing with a high quality graphical output, but without any symbolic computational capability. It is now available with many different options one of which, through a special additional package called the *Symbolic Math Toolbox*, allows it to perform symbolic algebraic operations. This *Toolbox* uses an essential core of material from MAPLE, and when the *Symbolic Math Toolbox* is installed on a computer already running MATLAB, it enables the computer to perform both high performance technical computing, and also to function as a *Symbolic Algebraic Computation System*.

MATLAB is started by double clicking the MATLAB icon, and it is closed by typing ‘quit’. The MATLAB prompt is the double arrowhead symbol >>, and some of the operations in MATLAB and MAPLE are denoted by same symbols. For example, addition and subtraction are denoted by + and −, multiplication by *, division by /, and raising to a power by ^, so 7^5 is entered as 7^5. Other similar symbols for elementary functions like $\sin x$ and $\arccos x$, are entered by typing sin(x) and arccos(x), etc.

However some other symbols used in MATLAB have quite different meanings from those in MAPLE, with $\sqrt{-1}$ denoted either by i or j, the mathematical constant π by pi, infinity by Inf and the natural logarithm by log. MATLAB also has many built in special operations and functions that have no counterpart in MAPLE, mainly due to the way MATLAB uses a matrix-type input. Two simple but important differences between MAPLE and MATLAB involve the way MATLAB uses the semicolon ; and the colon : as instructions. In MATLAB the semicolon ; performs the same task as the colon : in MAPLE, because it allows MATLAB to perform an operation after which it suppresses the subsequent print out of the result. The function of the colon operator in MATLAB is quite different from MAPLE, because the function it performs depends on how it is used. If the colon operator is used between two digits, say by typing 1: 6, the result in MATLAB will be to generate the row vector

1 2 3 4 5 6

containing the integers 1 to 6, while typing t = pi/4:pi/50:pi will produce a sequence of numbers represented by t that starts with $\pi/4$ and then steps up with increments of $\pi/50$ until reaching the value π. This type of instruction can be used when plotting a graph, because it determines how close the successive arguments t of a function $f(t)$ are to lie on the graph. For example, typing

```
t = 0:pi/100:2*pi;
y = cos(t);
plot(t, y)
```

will produce a 2dplot of the function cos t from $t = 0$ to 2π, that appears smooth because the computed points on the graph all lie close together at intervals in t of $\pi/100$.

MATLAB can also produce three-dimensional plots of functions, and the 3dplot in Fig. 19.10 can be reproduced using MATLAB by typing

```
[X, Y] = meshgrid(–4:0.2:4);
R = X.^2 + Y.^2;
Z = cos(R)./(1 + R);
mesh(X, Y, Z)
```

the result of which is shown in Fig. 19.13.

MATLAB is capable of sophisticated forms of graphical output. However these will not be described here, because they are numerous and the details of their function and how they are to be used are all listed in the

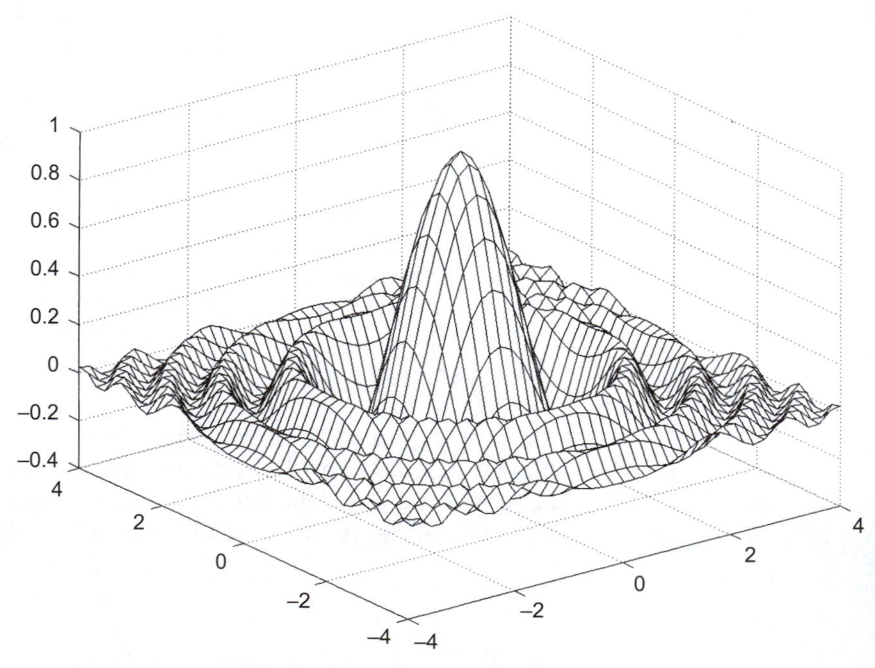

Fig. 19.13

MATLAB literature distributed with the software, and described in the references given at the end of this introduction to MATLAB.

Entering a purely numerical matrix is very simple, because the matrix A used at the start of the description of MAPLE is produced by typing

A = [1, −1, 2; −1, 4, 1; 2, 1, 2].

The result shown on the screen is

```
A =

 1  -1  2
-1   4  1
 2   1  2
```

from which it will be seen that MATLAB displays a purely numerical matrix *without* the enclosing square brackets [...] used in MAPLE. The transpose of A is entered as A′, its inverse is found by typing inv(A), and matrix multiplication is indicated by the symbol *, so that typing A*inv(A) will yield the 3×3 unit matrix.

Typing 'digits(n)' will return results rounded to n significant digits, where the default value of digits is 32. So, using the above matrix A, its inverse rounded to four decimal digits is found by typing

```
digits(4);

inv(A)

-0.4667  -0.2667   0.6
-0.2667   0.1333   0.2000
 0.6000   0.2000  -0.2000
```

The number of decimal digits displayed in MATLAB can also be controlled by using the 'format' instruction. Typing 'format short' will result in only four decimal digits being displayed, while typing 'format long' will result in fourteen decimal digits being displayed. The format command only influences the display, and not the accuracy of internal MATLAB calculations, which are always performed to the default number of digits. The result of typing A*inv(A) will, of course, be the 3×3 unit matrix, shown by MATLAB when using the 'format short' command as

```
1.0000  0.0000  0.0000
0.0000  1.0000  0.0000
0.0000  0.0000  1.0000.
```

If an $n \times n$ unit matrix is required in MATLAB it is entered by typing eye(n).

Matrix addition and subtraction is indicated, respectively, by typing + and − between matrices, and multiplication of matrix by the number 3, say, is indicated by typing 3*A. For example, using matrices A and inv(A), the matrix $2A - 3A^{-1}$ is found by typing

2*A − 3inv(A)

```
 3.4000  -1.2000  2.2000
-1.2000   7.6000  1.4000
 2.2000   1.4000  4.6000.
```

MATLAB can also find the eigenvalues and eigenvectors of $n \times n$ matrices, it can solve systems of linear algebraic equations, and it can transform rows and columns of matrices in many different ways.

Unlike MAPLE, the symbolic capabilities of MATLAB are only available when in addition to running MATLAB, the *Symbolic Math Toolbox* has been installed. To enter symbolic quantities it is necessary to identify these as such in MATLAB by using the instruction 'syms'. To enter the matrix

$$A = \begin{bmatrix} a & b & c \\ c & b & a \\ b & a & c \end{bmatrix},$$

where a, b and c are symbolic quantities, we identify these quantities as such by typing at the prompt

```
>>     syms a b c
```

with spaces as shown. A symbolic matrix involving these quantities can then be entered by typing

```
A = [a b c;  c b a;  b, a, c]

A =
[a, b, c]
[c, b, a]
[b, a, c]
```

where now MATLAB encloses each row in square brackets.

The matrix 2A, in which each element is multiplied by 2 is obtained by typing at the prompt

```
2*A

ans =
[2*a, 2*b, 2*c]
[2*c, 2*b, 2*a]
[2*b, 2*a, 2*c]
```

where MATLAB has given the matrix product the label (name) 'ans'. The result of subtracting matrix A from matrix 2A to return to matrix A if found by typing at the prompt

```
2*A – A

ans =
[a, b, c]
[c, b, a]
[b, a, c].
```

Definite integrals can be calculated numerically by MATLAB, and the numerical solution of initial value problems for both linear and nonlinear ordinary differential equations can found with great accuracy. To integrate symbolically it is necessary to use the instruction 'int', but first the symbolic variable must be identified. To find the antiderivative of $f(x) = 1+2\sin 2x$ we first identify x as a symbolic variable by typing

```
syms x.
```

Next we define the function $f(x)$ by typing

```
f = 1 + 2*sin(2*x)

f

= 1 + 2*sin(2*x).
```

To find the antiderivative of $f(x)$, where again the additive arbitrary constant is omitted, we type

```
int(f)

ans =

x - cos(2*x).
```

The definite integral of $f(x)$ over the interval $0 \leq x \leq \pi/2$ is obtained by typing

```
int(f, 0, pi/2)

ans =

1/2*pi + 2.
```

As an example of the way the truncated Taylor series expansion of the function $f(x)$ about $x = 0$ can be found, truncated after the term in x^5, and then plotted, we proceed as follows:

```
T = taylor(f, 6)

T =

1 + 4*x - 8/3*x^3 + 8/15*x^5.
```

This can be plotted for $0 \leq x \leq 2$ by using the simple and very convenient plot command 'ezplot' as follows:

```
ezplot(T, [0, 2]).
```

The truncation of the Taylor series after the term in x^5 means this approximation is only valid for small x. The validity of the approximation can be seen by comparing Fig. 19.14 with the plot of $f(x) = 1+2\sin 2x$ shown in Fig. 19.15, and obtained by typing

```
syms x

f = 1 + 2*sin(2*x)

ezplot(f, [0, 2]).
```

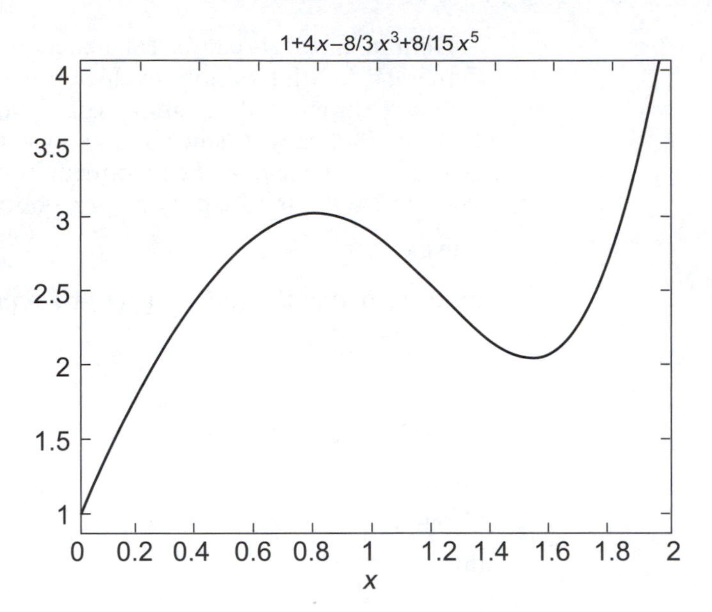

Fig. 19.14

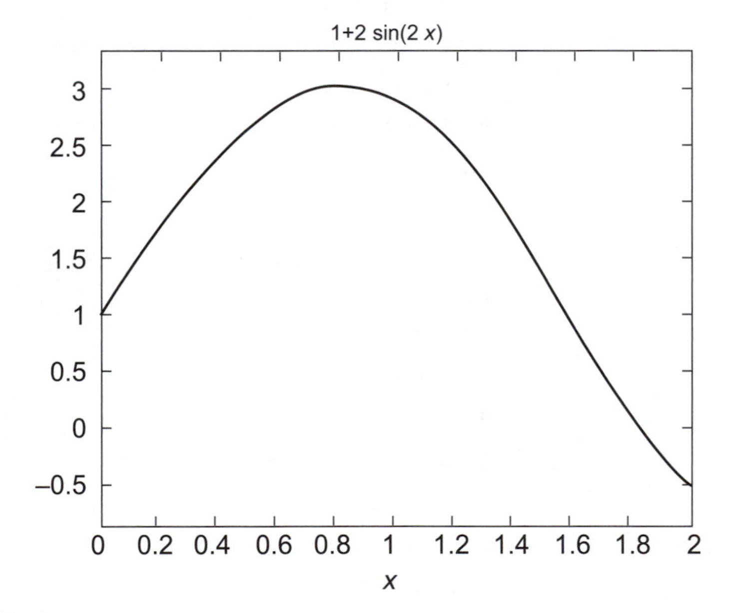

Fig. 19.15

The first order differential equation

$$\frac{dy}{dt} = y(t) + \sin(t)$$

is solved symbolically by typing

dsolve('Dy = y + sin(t)')

when the answer is returned by typing 'ans = ', giving

ans =

$\mathrm{C1e^{1}} - \frac{1}{2}\cos(t) - \frac{1}{2}\sin(t)$.

Notice that here the entry in the dsolve command is enclosed between primes '...' to indicate that symbolic quantities are involved, and that the symbol Dy represents the first derivative of y with respect to the independent variable, which here will be t, because it occurs in the nonhomogeneous term sin t. An initial value problem for this equation with initial condition $y(0) = 1$ is solved by typing

y = dsolve('Dy = y + sin(t)', 'y(0) = 1')

where now each individual entry in the dsolve command is enclosed between primes. The solution is returned as

$\mathrm{y} = \frac{3}{2}\mathrm{e^{t}} - \frac{1}{2}\cos(t) - \frac{1}{2}\sin(t)$.

Similarly, the initial value problem

$$\frac{d^2y}{dx^2} + 2\frac{dy}{dx} + y = x, \quad \text{with} \quad y(0) = 1, \quad y'(0) = 1$$

is solved by typing

y = dsolve('D2y + 2*Dy + y = 1', 'y(0) = 1', 'Dy(0) = 1', 'x')

where typing 'x' at the end of the dsolve instruction tells MATLAB that now the independent variable is x. The result of this operation is to return the solution

y =

−2 + x + 3*exp(−x)*x.

Notice that here D2y represents the result of differentiating the function $y(x)$ twice with respect to x. Similarly, if required, a third order derivative of $y(x)$ would be entered as D3y. For full details of the way the *Symbolic Math Toolbox* works, with various applications, we refer the reader to the documentation mentioned below supplied by with this Toolbox and to the other references for general information about numerical calculations performed by MATLAB.

The closing remarks made at the end of the description of MAPLE apply equally to MATLAB and to any other symbolic computation software. All such software provides valuable help when performing mathematics, but it does not remove the need to understand and think about the mathematics involved.

REFERENCES

1. MATLAB 5 for Engineering: Adrian Biran and Moshe Breiner, Addison Wesley, 1999

2. Symbolic Math Toolbox: User's Guide, Version 2, The Math Works Inc. 1997
3. The MATLAB 5 Handbook: Eva Part-Enander and Anders Sjoberg, Addison Wesley, 1999
4. Basics of MATLAB and Beyond: Andrew Knight, Chapman & Hall/CRC, 2000

The following is a useful book when seeking to develop an understanding of differential equations and their use in modelling, because it employs the symbolic capabilities of MAPLE to model and solve a variety of interesting case studies. The book is equally useful when working with MAPLE or MATLAB, because although directed towards MAPLE, it applies equally well to MATLAB as its symbolic MATLAB toolbox uses a core of MAPLE symbolic operations.

Mathematical Modelling with Case Studies: A Differential Equation Approach using MAPLE: Belinda Barens and Glenn R. Fulford, Taylor and Francis, 2002.

Answers

Chapter 1

1.1 (a) $-3, 3, -4, 4, -5, 5, -6, 6$
(b) 64, 125, 216
(c) $(0, 4), (0, -4), (4, 0), (-4, 0)$
(d) (0, 7, 7), (7, 0, 7), (0, 8, 8), (8, 0, 8), (1, 7, 8), (7, 1, 8),
(e) 1.

1.3 (a) For example, the ages of all people at a dance; a proper subset would then be the set of ages of all men at the dance.
(b) For example, the set of all people in telephone directory whose surname begins with the letter G; a proper subset would then be the set of all women in that section of the directory.

1.5 Method (i) could be the numbers $x_r = 1 + \left(\dfrac{1}{r^2 + 1}\right)$, with $r = 1, 2, \ldots, n$.

Method (ii) could be the numbers $x_r = 1 + \left(\dfrac{r^2}{r^2 + 1}\right)$, with $r = 1, 2, \ldots, n$.

A generalization of (i) is the set of numbers $x_r - a + \dfrac{(b - a)}{r^2 + 1}$, with $r = 1, 2, \ldots, n$.

1.7 We prove only the first result because the others follow in similar fashion. As $\sqrt{2}$ is irrational it cannot be represented in the form p/q with p and q integers without a common factor. However α is rational, so adding this rational number to $\sqrt{2}$ cannot yield another rational number, because the only way another rational number can be created by addition to α is by adding another rational number.

1.13 $\max(x + 2y) = 10$.

1.21 General solution $u_n = A(-3)^n + B2^n$. Particular solution

$u_n = -\frac{1}{15}(-3)^n + \dfrac{2}{5}2^n$.

1.23 $a^7 + 7a^6b + 21a^5b^2 + 35a^4b^3 + 35a^3b^4 + 21a^2b^5 + 7ab^6 + b^7$.

1.25 Coefficient of a^3b^5 is 56. Coefficient of a^6b^2 is 28.

1.31 (a) (i) 0.4712 rad, (ii) 5.7596 rad, (iii) 3.3161 rad, (iv) 4.7822 rad;
(b) (i) 41.253°, (ii) 77.922°, (iii) 104.278°, (iv) 164.439°.

1.37 (a) $y = y_1 + \{(y_2 - y_1)(x - x_1)/(x_2 - x_1)\}$; (b) $y = \frac{1}{4}(3x - 11)$.

1.39 $(x-1)^2 + (y-2)^2 = 5$; $y = 2 - \sqrt{5}$.

1.43 Hyperbola. Clockwise rotation through $\pi/6$ reduces it to $X^2 - Y^2 = 1$.

1.45 Parabola. Clockwise rotation through $\pi/3$ reduces it to $Y^2 = \frac{3}{4}(X - 1)$.

1.47 (i) Right circular cone with axis coincident with z-axis and vertex at $(1, 2, 0)$.
(ii) Elliptic cone with axis coincident with z-axis and vertex at $(-3, 3, 0)$.
(iii) Elliptic cone with axis parallel to z-axis and vertex at $(-2, 4, 3)$.
(iv) Right circular cone with axis coincident with x-axis and vertex at $(0, 1, 6)$.

1.59

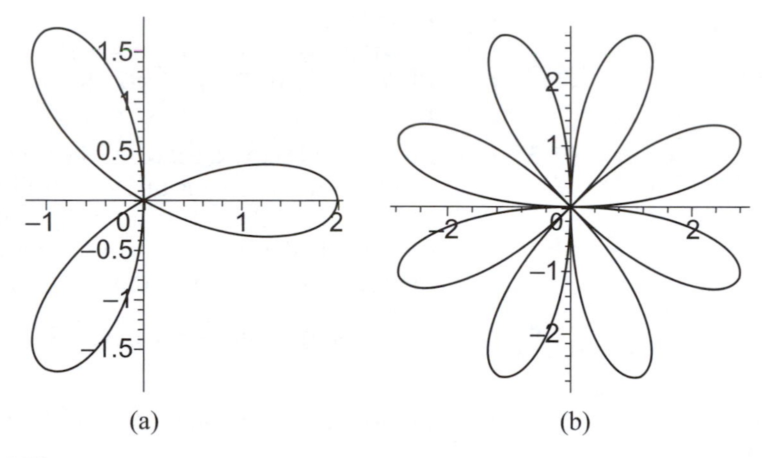

Fig. A1.59

Chapter 2

2.3 (a) [1, 2]: many–one; (b) (5, 17): one–one; (c) [1, 17]: many–one; (d) [1, 10]: many–one.

2.5 $f(n) = 3$ for $n = 7, 8, 9, 10, 11, 12, 13$.

2.9 (a) Neither even nor odd; (b) odd; (c) neither even nor odd; (d) odd; (e) even; (f) odd; (g) neither even nor odd. In (h) to (j) the first group of terms is even and the second is odd: (h) $f(x) = (1 + x\sin x) + x^3$ with the interval $[-2\pi, 2\pi]$; (i) $f(x) = 1 + (x + |x|\sin x)$ with the interval $[-3\pi, 3\pi]$; (j) $f(x) = (1 + 2x^2) - (x - 4x^3)$ with the interval $[-3, 3]$.

2.11 (a) 0 is simply a lower bound, 11 is a strict upper bound;
(b) 0 is a strict lower bound, 2 is simply an upper bound;
(c) neither upper bound 2 nor lower bound 1/6 is strict;
(d) 0 is simply a lower bound, 1 is a strict upper bound.

2.13

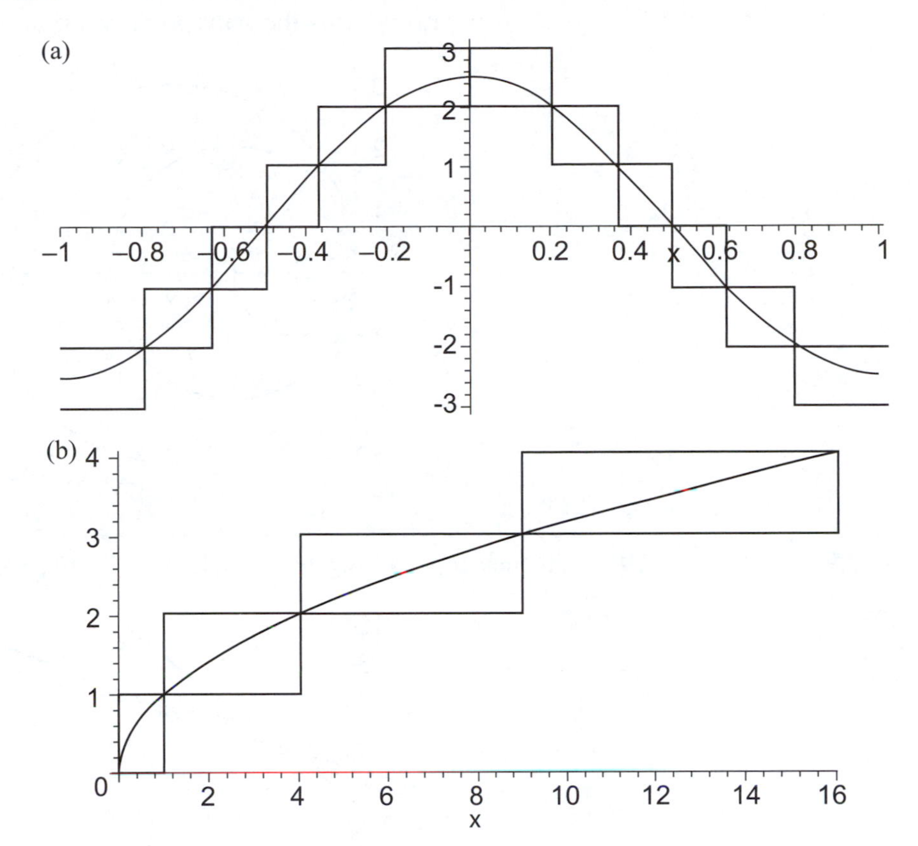

Fig. A2.13

2.15 The envelope is tangent to each curve in Fig. A2.15.

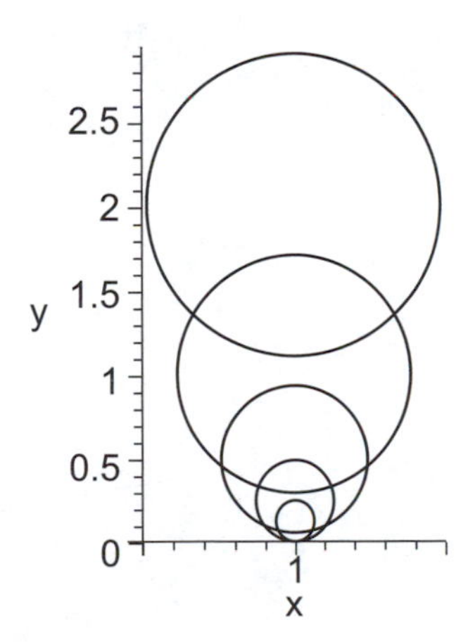

Fig. A2.15

2.17 The envelope has degenerated to the single point at the origin in Fig. A2.17, because that is the only point common to each circle.

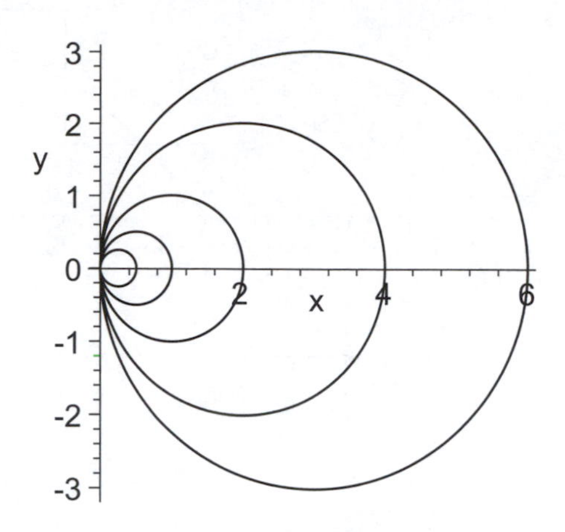

Fig. A2.17

2.19 The envelope is tangent to each curve in Fig. A2.19.

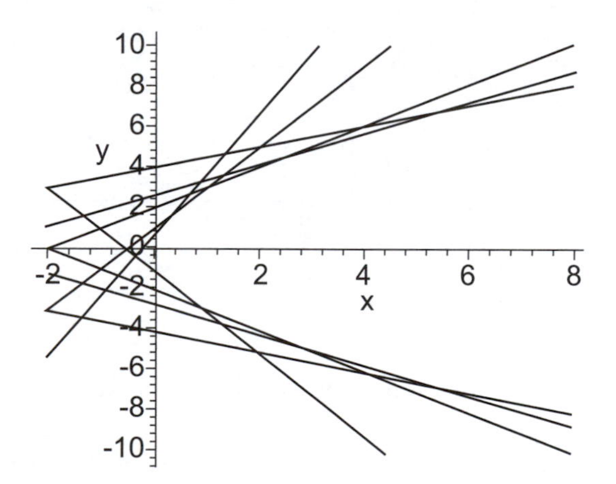

Fig. A2.19

2.21

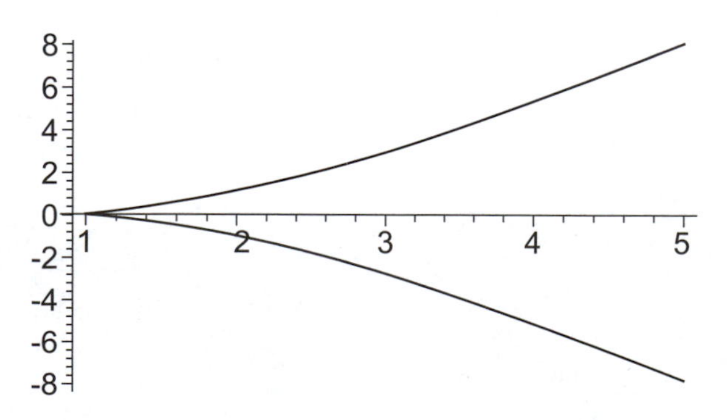

Fig. A2.21

2.23 Cross-sections by planes $x = \text{constant}$ are straight lines and cross-sections by planes $y = \text{constant}$ are parabolas.

2.25 Domain of definition is an annulus centred on (1, 2) with inner radius $\sqrt{2}$ and outer radius 3. Level curves are concentric circles centred on (1, 2).

2.27 Elliptic cylinders with axis coincident with the z-axis.

2.31 Step function: $y(x) = 2$ for $-4 \leq x < 2$, $y(x) = 6$ for $2 < x \leq 4$.

2.37 Envelope is y-axis.

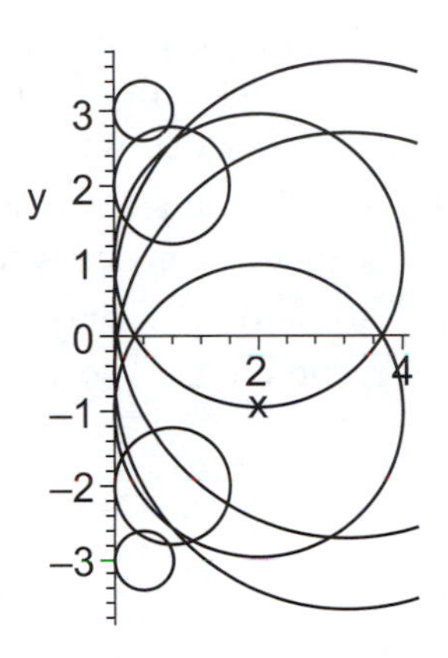

Fig. A2.37

Chapter 3

3.3 Limits points are:
0, corresponding to a sub-sequence with n odd;
1, corresponding to a sub-sequence with n a member of the sequence $\{4, 8, 12, \ldots\}$;
-1, corresponding to a sub-sequence with n a member of the sequence $\{2, 6, 10, \ldots\}$.

3.5 Limit points are 1, 0, and they are not members of the sequence.

3.8 Starting with $a_5 = 10\tan(\pi/5)$ and $b_5 = 10\sin(\pi/5)$ gives $a_{10} = 6.498394$, $b_{10} = 6.180340$, $a_{20} = 6.335378$, $b_{20} = 6.257379, \ldots, a_{2560} = 6.283188$, $b_{2560} = 6.283184$, so $3.141592 < \pi < 3.141594$.

3.11 (a) $u_1 = 1,\ u_2 = 3,\ u_3 = 1.7778,\ u_4 = 1.6799,\ u_5 = 1.7748$
(b) $u_1 = 2,\ u_2 = 1.625,\ u_3 = 1.7592,\ u_4 = 1.6874;$
$u_5 = 1.7217;\ \sqrt[3]{5} = 1.7100.$

3.15 (a) 1 (b) $\frac{1}{2}$ (c) $\dfrac{\pi^2}{4} + \sin\dfrac{\pi^2}{4}$ (d) 0 (e) infinite.

3.17 (a) all x (b) $(-\infty, -1), (-1, 1), (1, \infty)$ (c) all x
(d) $(-\infty, -4), (-4, \infty)$ (e) everywhere apart from the infinity of points $x = n\pi/2$.

3.21 (a) No neighbourhood exists because of infinites along $x = -1$ and $y = 2$.

(b) Interior of circle $x^2 + y^2 = 1$. Defined at P but undefined everywhere or the boundary of this circular neighbourhood.
(c) The entire (x, y)-plane. Defined everywhere in neighbourhood, including the point P.

3.23 (a) $\sqrt{2}$, (b) 0, (c) $\cos x$, (d) 1/32, (e) $\lim_{x\to 0-} |\sin x|/x = -1$, $\lim_{x\to 0+} |\sin x|/x = 1$.

3.25 (a) $x = \pm 1$; (b) $y = 4x - 8$ as $x \to +\infty$, $y = -2x + 4$ as $x \to -\infty$ and also $x = -1$; (c) $x = \pm 1$ and $y = \pm x$.

3.27 No, the sequence has the two limit points 1 and -1. By setting $n = 2r + 1$ only the behaviour at the limit point -1 is seen. To see the other one set $n = 2r$. This sequence does not converge.

3.29 Yes. The sine term is bounded by 1 and $1/n \to 0$ as $n \to \infty$, so the sequence converges to 0.

3.33 A computer works to a fixed precision and when n is large $1 + 1/n$ is recorded as 1, so $(1 + \frac{1}{n})^n$ becomes $1^n = 1$.

Chapter 4

4.1 (a) $1, -1, i\sqrt{2}; -i\sqrt{2}$; (b) $i\sqrt{2}, -i\sqrt{2}, i\sqrt{3}, -i\sqrt{3}$; (c) $\sqrt{2}, -\sqrt{2}, \sqrt{3}, -\sqrt{3}$.

4.3 (a) $z = -11 + i$; (b) $z = -3i$; (c) $z = 0$; (d) $z = 4 + 17i$.

4.5 (a) $z_1 + z_2 = 7 + 6i$; (b) $z_1 + z_2 = -i$; (c) $z_1 + z_2 = 2$; (d) $z_1 + z_2 = 2$.

4.7 (a) $z_1 z_2 = -1 + 5i$; (b) $z_1 z_2 = 34$; (c) $z_1 z_2 = 3 + 4i$; (d) $z_1 z_2 = 18 - 2i$.

4.9 $-\frac{1}{2}$.

4.11 (a) $z_1/z_2 = (1 + 5i)/2$; (b) $z_1/z_2 = 3$; (c) $z_1/z_2 = 2i$.

4.13 Roots are $z = 2 + 3i$, $z = 2 - 3i$, $z = 1$, $z = -1$; $P(z) = (z^2 - 4z + 13)(z - 1)(z + 1)$.

4.15 (a) $z_1 + z_2 = 1 + 5i$, $z_1 - z_2 = 3 + i$; (b) $z_1 + z_2 = 7 - i$, $z_1 - z_2 = -1 + i$; (c) $z_1 + z_2 = 3$, $z_1 - z_2 = -3 + 8i$; (d) $z_1 + z_2 = -2$, $z_1 - z_2 = -4i$.

4.19 (a) $|z| = 5$, $\arg z = 2.214$ rad; (b) $|z| = 5$, $\arg z = 4.067$ rad; (c) $|z| = 3\sqrt{2}$, $\arg z = 3\pi/4$; (d) $|z| = 4$, $\arg z = 5.761$ rad.

4.21 (a) $|z_1 z_2| = 3/2$, $\arg z_1 z_2 = \pi/2$: $|z_1/z_2| = 6$, $\arg(z_1/z_2) = -\pi/6$; (b) $|z_1 z_2| = 8$, $\arg z_1 z_2 = -\pi/12$: $|z_1/z_2| = 2$, $\arg(z_1/z_2) = -7\pi/12$; (c) $|z_1 z_2| = 2$, $\arg z_1 z_2 = -7\pi/4$: $|z_1/z_2| = 1/18$, $\arg(z_1/z_2) = 5\pi/4$.

4.23 $\sin 7\theta = 7\sin\theta - 56\sin^3\theta + 112\sin^5\theta - 64\sin^7\theta$
$\cos 7\theta = 64\cos^7\theta - 112\cos^5\theta + 56\cos^3\theta - 7\cos\theta$.

4.25 $\omega_k = \cos\dfrac{2k\pi}{7} + i\sin\dfrac{2k\pi}{7}$; $k = 0, 1, \ldots, 6$.

4.27 $\omega_k = 2^{1/4}\left[\cos\left(\dfrac{1 + 6k}{12}\right)\pi + i\sin\left(\dfrac{1 + 6k}{12}\right)\pi\right]$; $k = 0, 1, 2, 3$.

4.29 (a) $|OP| = \sqrt{6},\ l = 2/\sqrt{6},\ m = -1/\sqrt{6},\ n = -1/\sqrt{6}$: $\theta_1 = 0.625$ rad, $\theta_2 = 1.992$ rad, $\theta_3 = 1.992$ rad;
(b) $|OP| = 2\sqrt{5},\ l = 2/\sqrt{5},\ m = 0,\ n = 1/\sqrt{5}$: $\theta_1 = 0.465$ rad, $\theta_2 = \frac{1}{2}\pi,\ \theta_3 = 1.108$ rad;
(c) $|OP| = \sqrt{6},\ l = -1/\sqrt{6},\ m = 2/\sqrt{6},\ n = 1/\sqrt{6}$: $\theta_1 = 1.992$ rad, $\theta_2 = 0.625$ rad, $\theta_3 = 1.150$ rad.

4.31 (a) $\theta_1 = \frac{1}{6}\pi,\ \theta_2 = \frac{1}{2}\pi,\ \theta_3 = \frac{1}{3}\pi$; (b) $\theta_1 = \theta_2 = \theta_3 = 0.956$ rad;
(c) $\theta_1 = 1.231$ rad, $\theta_2 = 1.911$ rad, $\theta_3 = 0.490$ rad.

4.33 (a) $OP = \mathbf{i} + \mathbf{j} + \mathbf{k}$; (b) $OP = -2\mathbf{i} + 3\mathbf{j} + 7\mathbf{k}$;
(c) $OP = 3\mathbf{i} - \mathbf{j} + 11\mathbf{k}$; (d) $OP = \mathbf{j}$.

4.35 (a) $\mathbf{a} + \mathbf{b} = 2\mathbf{i} - 4\mathbf{j} + 4\mathbf{k},\ \mathbf{a} - \mathbf{b} = 4\mathbf{i} - 2\mathbf{k}$;
(b) $\mathbf{a} + \mathbf{b} = \mathbf{i} - 2\mathbf{j} + \mathbf{k},\ \mathbf{a} - \mathbf{b} = 3\mathbf{i} + 6\mathbf{j} - 3\mathbf{k}$;
(c) $\mathbf{a} + \mathbf{b} = 2\mathbf{i} + \mathbf{j} - 2\mathbf{k},\ \mathbf{a} - \mathbf{b} = -2\mathbf{i} + 3\mathbf{j} - 4\mathbf{k}$.

4.37 (a) $\mathbf{a} = \sqrt{14}\left(\frac{2}{\sqrt{14}}\mathbf{i} - \frac{1}{\sqrt{14}}\mathbf{j} + \frac{3}{\sqrt{14}}\mathbf{k}\right)$;
(b) $\mathbf{a} = \sqrt{19}\left(\frac{3}{\sqrt{19}}\mathbf{i} - \frac{3}{\sqrt{19}}\mathbf{j} + \frac{1}{\sqrt{19}}\mathbf{k}\right)$;
(c) $\mathbf{a} = -\frac{\sqrt{71}}{9}\mathbf{i} + \frac{1}{3}\mathbf{j} - \frac{1}{9}\mathbf{k}$.

4.39 (a) $\mathbf{a}.\mathbf{b} = -11,\ \theta = \arccos\left(\frac{-11}{3\sqrt{59}}\right)$;
(b) $\mathbf{a}.\mathbf{b} = 4,\ \theta = \arccos\left(\frac{4}{3\sqrt{34}}\right)$;
(c) $\mathbf{a}.\mathbf{b} = -28,\ \theta = \pi$.

4.43 (a) 8; (b) 56; (c) 32; (d) 0.

4.45 (a) 1; (b) -10; (c) 0.

4.47 (a) $\left(\frac{7\mathbf{i} - 5\mathbf{j} - 2\mathbf{k}}{\sqrt{78}}\right)$; (b) $\left(\frac{-\mathbf{i} - 6\mathbf{j} - 2\mathbf{k}}{\sqrt{41}}\right)$
Results are unique apart from sign since if $\mathbf{n}$ is a unit normal, then so also is $-\mathbf{n}$.

4.53 $\mathbf{r} = (2\mathbf{i} + \mathbf{j} - \mathbf{k}) + \lambda(-3\mathbf{i} + 3\mathbf{k})$; $l = -3/\sqrt{18},\ m = 0,\ n = 3/\sqrt{18}$.

4.57 $p = \sqrt{(24/11)}$.

4.59 $\arccos\left(\frac{19}{7\sqrt{14}}\right)$.

4.61 $(x-2)^2 + (y-3)^2 + (z-1)^2 = 9$.

4.63 $(\sqrt{6}x - \sqrt{6} - 4)^2 + (\sqrt{6}y - \sqrt{6} + 2)^2 + (\sqrt{6}z - 2\sqrt{6} - 2)^2 = 24$.

4.65 Resultant $= \frac{1}{\sqrt{2}}(5\mathbf{i} + 4\mathbf{j} + 3\mathbf{k})$; 5 units; $l = \frac{1}{\sqrt{2}},\ m = \frac{4}{5\sqrt{2}},\ n = \frac{3}{5\sqrt{2}}$.

4.69 $305/\sqrt{26}$ Nm.

4.71 $\mathbf{M} = -11\mathbf{i} + 7\mathbf{j} - 15\mathbf{k}$.

4.79 $P(3-2i) = \dfrac{783 - 2844i}{4 - 14i}$; $\mathrm{Re}\{P(z)\} = \dfrac{46\,863}{277}$; $\mathrm{Im}\{P(z)\} = \dfrac{-14\,634}{2777}$.

Chapter 5

5.3 $|x|$ is continuous at the origin but its left-hand derivative there is -1 and its right-hand derivative is $+1$.

5.5 (a) Left-hand derivative is -3, right-hand derivative is 4.
(b) Left-hand derivative is 2, right-hand derivative is 2.

5.7 $a = 1$, $b = (6\sqrt{3} - 2\pi)/12$. There is a unique tangent at $x = \pi/6$.

5.9 (a) $y' = \frac{1}{3}x^{-2/3}\sin x + x^{1/3}\cos x$;
(b) $y' = (2x+3)(1+\cos 2x) - 2(x^2+3x+1)\sin 2x$;
(c) $y' = 6\cos 6x \cos 2x - 2\sin 6x \sin 2x$;
(d) $y' = (3x^2+2)\cos 3x - 3(x^3+2x-1)\sin 3x$.

5.11 (a) $y' = 3(x+1)(x^2+2x+1)^{1/2}$;
(b) $y' = bx^2(a+bx^3)^{-2/3}$;
(c) $y' = 30\cos 2x(2+3\sin 2x)^4$;
(d) $y' = 6x^2\cos(1+2x^3)$;
(e) $y' = 2x\cos(1+x^2)\cos[\sin(1+x^2)]$;
(f) $y' = -2x^3(1+x^4)^{-1/2}\sin(1+x^4)^{1/2}$.

5.13 (a) $y' = \dfrac{-6\sin x}{(1-3\cos x)^3}$;
(b) $y' = \dfrac{1}{a^2(b^2+x^2)^{1/2}} - \dfrac{x^2}{a^2(b^2+x^2)^{3/2}}$;
(c) $y' = \dfrac{2x(1+2x^2)\sec^2(1+x^2+x^4)}{\sin(1+x^2)} - \dfrac{2x\cos(1+x^2)\tan(1+x^2+x^4)}{\sin^2(1+x^2)}$;
(d) $y' = -6\csc^2(1+3x)\cot(1+3x)$;
(e) $y' = -4/(\sin x - 2\cos x)^2$;
(f) $y' = \dfrac{(3x^2-2x-12)}{(x^2+4)^2}\csc^2\left(\dfrac{3x-1}{x^2+4}\right)$.

5.15 (a) $f(x)$ is discontinuous at $|x| = 1$, so that the intermediate value theorem will not apply to any interval containing this point. In particular, this is true of the interval [0, 6].
(b) $f(x)$ is continuous in $[-11, -2]$ so that the theorem applies. $f(x) = -0.5$ in $[-11, -2]$ when $x = -3$.

5.17 Critical points at $\xi_1 = (1+\sqrt{13})/3$ and $\xi_2 = (1-\sqrt{13})/3$. The point ξ_1 corresponds to a minimum and the point ξ_2 to a maximum.

5.19 Critical points at $\xi_1 = 0$, $\xi_2 = 3$ and $\xi_3 = 3/2$; ξ_1 and ξ_2 are minima; ξ_3 is a maximum.

5.21 (a) One at $x = 0$.
(b) Five.
(c) None. Derivative is discontinuous at $x = 0$ so theorem does not apply.
(d) None. Derivative is discontinuous at $x = 0$ so theorem does not apply.

5.25 $2^{3/2} + 3\sqrt{2}(x-1) < (1+x^2)^{3/2} < 2^{3/2} + 6\sqrt{5}(x-1)$ for $1 \le x \le 2$.

5.29 (a) $\pi^2/2$; (b) 5; (c) 3; (d) -1.

5.31 (a) 0; (b) 1/5; (c) 0; (d) 3; (e) $2/\pi$; (f) -1.

5.33 $\mathrm{d}V = 3\pi[\alpha R_0^2 H_0(1+\alpha t)^2 - \beta r_0 h_0(1+\beta t)^2)^2]\mathrm{d}t$.

5.35 $\lim_{x\to 1-} f'(x) = \lim_{x\to 1+} f'(x) = 5;\quad \lim_{x\to 1-} f''(x) = 14,\quad \lim_{x\to 1+} f''(x) = 10.$

5.37 (a) $x = 1$ is minimum; $x = -2$ is maximum; $x = -\frac{1}{2}$ corresponds to a point of inflection with gradient $-27/2$.
(b) Neither maxima nor minima; point of inflection with zero gradient at $x = 0$.
(c) Minimum at $x = 0$; maximum at $x = 6$; minimum at $x = 12$; points of inflection at $x_1 = 6 + \sqrt{12}$, $x_2 = 6 - \sqrt{12}$ with corresponding gradients $2x_i(x_i - 12)(2x_i - 12)$ for $i = 1, 2$.

5.39 (a) $f_x = 2x/y,\quad f_y = -x^2/y^2$;
(b) $f_x = 6xy + (x+y)^2 + 2x(x+y),\quad f_y = 3x^2 + 2x(x+y)$;
(c) $f_x = 2x\cos(x^2+y^2),\quad f_y = 2y\cos(x^2+y^2)$;
(d) $f_x = \cos(1+x^2y^2) - 2x^2y^2\sin(1+x^2y^2),\ f_y = -2x^3y\sin(1+x^2y^2)$.

5.41 (a) $f_x = 2xyz - \dfrac{1}{x^2yz^2},\quad f_y = x^2z - \dfrac{1}{xy^2z^2},\quad f_z = x^2y - \dfrac{2}{xyz^3}$;
(b) $f_x = \cos yz - yz\sin xz - yz\sin xy$,
$f_y = -xz\sin yz + \cos xz - xz\sin xy$,
$f_z = -xy\sin yz - xy\sin xz + \cos xy$;
(c) $f_x = -(2x+y)\sin(x^2+xy+yz)$,
$f_y = -(x+z)\sin(x^2+xy+yz)$,
$f_z = -y\sin(x^2+xy+yz)$.

5.43 (a) $\mathrm{d}u = \left(\dfrac{-2}{x^3yz} + yz\right)\mathrm{d}x + \left(\dfrac{-1}{x^2y^2z} + xz\right)\mathrm{d}y + \left(\dfrac{-1}{x^2yz^2} + xy\right)\mathrm{d}z$;
(b) $\mathrm{d}u = \sin(y^2+z^2)\,\mathrm{d}x + 2xy\cos(y^2+z^2)\,\mathrm{d}y + 2xz\cos(y^2+z^2)\,\mathrm{d}z$;
(c) $\mathrm{d}u = -3(1-x^2-y^2-z^2)^{1/2}(x\,\mathrm{d}x + y\,\mathrm{d}y + z\,\mathrm{d}z)$.

5.45 (a) $\dfrac{\partial z}{\partial x} = -\dfrac{x}{z},\quad \dfrac{\partial z}{\partial y} = -\dfrac{y}{z}$;
(b) $\dfrac{\partial z}{\partial x} = \dfrac{-yz - z^2\cos xz^2}{(xy + 2xz\cos xz^2)},\quad \dfrac{\partial z}{\partial y} = \dfrac{-z}{(y + 2z\cos xz^2)}$;
(c) $\dfrac{\partial z}{\partial x} = \dfrac{2x}{y - 6z},\quad \dfrac{\partial z}{\partial y} = \dfrac{1 - 4y - z}{y - 6z}$;
(d) $\dfrac{\partial z}{\partial x} = \dfrac{z\sin x - \cos y}{\cos x - y\sin z},\quad \dfrac{\partial z}{\partial y} = \dfrac{x\sin y - \cos z}{\cos x - y\sin z}$.

5.47 $y^2 = 6x$.

5.49 lines $x = 0$ and $y = 0$.

5.51 (a) $\dfrac{du}{dt} = z\sqrt{1+t^2} + \dfrac{2t^2}{\sqrt{1+t^2}} + 10\,t\cos(5t^2+1)$;

(b) $\dfrac{du}{dt} = 3(t + 3t^2 + 2t^3 + 3t^5)\sqrt{1 + t^2 + 2t^3 + 2t^4 + t^6}$;

(c) $\dfrac{du}{dt} = \dfrac{2t}{3}$.

5.55 (a) $\dfrac{du}{dx} = 3(x^2 + y) + \dfrac{3(x + y^2)(y\sin x - \cos y)}{(\cos x - x\sin y)}$;

(b) $\dfrac{du}{dx} = (2xy^2 + y\cos xy) + (2x^2y + x\cos xy)\dfrac{x}{2y}$.

5.57 $\dfrac{\partial u}{\partial x} = \dfrac{-(u+v)}{u^2+v^2}$, $\dfrac{\partial u}{\partial y} = \dfrac{2v+3u}{2(u^2+v^2)}$,

$\dfrac{\partial v}{\partial x} = \dfrac{v-u}{u^2+v^2}$, $\dfrac{\partial v}{\partial y} = \dfrac{2u-3v}{2(u^2+v^2)}$.

5.61 (a) $f'(x) = 2x\,\text{arcsec}\left(\dfrac{x}{a}\right) + \dfrac{ax^2}{|x|\sqrt{(x^2-a^2)}}$;

(b) $f'(x) = \dfrac{2x+1}{\arcsin(x^2-2)} - \dfrac{2x(x^2+x+1)}{[\arcsin(x^2-2)]^2\sqrt{[1-(x^2-2)^2]}}$;

(c) $f'(x) = \frac{3}{2}(1 + x + \arccos 2x)^{1/2}\left(\dfrac{\sqrt{(1-4x^2)}-2}{\sqrt{(1-4x^2)}}\right)$.

5.63 At $t = \frac{1}{2}\pi$, $\dfrac{dy}{dx} = 2$, $\dfrac{d^2y}{dx^2} = -2$.

5.65 $f_{xx}(1, 1) = 384$, $f_{xy}(1, 1) = 384$, $f_{yy}(1, 1) = 192$. $f_{xy} = f_{yx}$ everywhere because of Theorem 5.24.

Chapter 6

6.3 (a) $\frac{2}{3}$; (b) $\frac{1}{4}$; (c) $\frac{1}{6}$.

6.5 (a) $2e^{2x}/\sqrt{(1-e^{4x})}$; (b) $(e^x + xe^x + 1)/[2\sqrt{(xe^x + x)}]$;

(c) $(1+x)e^x\cos(xe^x + 2)$; (d) $2e^x/(e^x+1)^2$.

6.9 $\dfrac{\partial f}{\partial x} = -\dfrac{y}{x^2}\,e^{\sin(y/x)}\cos(y/x)$, $\dfrac{\partial f}{\partial x} = \dfrac{1}{x}\,e^{\sin(y/x)}\cos(y/x)$.

6.11 (a) 0; (b) $\frac{1}{4}\ln 2$; (c) $\ln 3$; (d) $\ln(3/2)$; (e) 1.

6.13 (a) $(1 + \ln x)x^x$;

(b) $(\ln\sin 2x + 2x\cot 2x)(\sin 2x)^x$;

(c) $\left(\cos x\ln x + \dfrac{\sin x}{x}\right)x^{\sin x}$;

(d) $(\cot x\ln 10)10^{\ln(\sin x)}$.

6.15 $\dfrac{\partial u}{\partial x} = yz(xy)^{z-1}$, $\dfrac{\partial u}{\partial y} = xz(xy)^{z-1}$, $\dfrac{\partial u}{\partial z} = (xy)^z\ln xy$.

6.19 (a) $2\cosh x(\cosh 2x \cosh x + \sinh 2x \sinh x)$;
(b) $3\sinh 3x \exp(1 + \cosh 3x)$;
(c) $\operatorname{sech}^2 x \coth x$;
(d) $-8x/[(2x^2+1)\sqrt{3-4x^2-4x^4}]$, $0 < x^2 < \frac{1}{2}$;
(e) $2\cos 2x \sinh(\sin 2x)$.

6.23 (a) $\sqrt{2}e^{i(1/4)\pi}$; (b) $\sqrt{2}e^{-i\pi/4}$; (c) $16e^{-\pi i/3}$; (d) $8e^{\pi i/2}$;
(e) $\sqrt{13}e^{i\alpha}$ with $\alpha = \arctan(3/2)$.

6.25 $\cos^2\theta \sin^3\theta = \frac{1}{8}\sin\theta + \frac{1}{16}\sin 3\theta - \frac{1}{16}\sin 5\theta$.

6.27 $z = (2n+1)\frac{1}{2}\pi$.

6.43 (a) all z; (b) for $z \neq 2$; (c) all z; (d) all z because $\sinh z = \sinh(x+iy) = \sinh x \cos y + i\cosh x \sin y$.

6.45 (a) $7+4i$; (b) $(2+4i)/5$; (c) $5(1-i)$.

6.47 (a) yes; (b) yes; (c) no; (d) yes; (e) no; (f) yes.

Chapter 7

7.1 $\underline{S}_{P_n} = \lambda(b-a)\left[a + \frac{(n-1)(b-a)}{2n}\right]$, $\overline{S}_{P_n} = \lambda(b-a)\left[a + \frac{(n-1)(b-a)}{2n}\right]$;

$\lim \underline{S}_{P_n} = \lim \overline{S}_{P_n} = \frac{\lambda}{2}(b^2 - a^2)$.

7.3 $S_{P_n} = (e^{\lambda a} - e^{\lambda b})\left(\frac{\Delta}{1 - e^{\lambda\Delta}}\right)$, where $\Delta = (b-a)/n$.
Since, by L'Hospital's rule, $\lim_{\Delta \to 0}\{\Delta/(1 - e^{\lambda\Delta})\} = -1/\lambda$,

we have $\int_a^b e^{\lambda x}dx = \lim_{n\to\infty} S_{P_n} = \frac{1}{\lambda}(e^{\lambda b} - e^{\lambda a})$.

7.9 $\mathbf{I} = 7$.

7.11 Required area is a triangle $+\int_1^\infty \frac{dx}{x^2} = \frac{1}{2} + 1 = 3/2$.

7.13 $\xi = \sqrt{\frac{7}{3}}$ is unique in first case. In second case ξ is not unique, for $\xi = \pm 2\sqrt{3}$.

7.15 Possible examples are:

$$f(x) = \begin{cases} 1 \text{ for } 1 \le x \le 2; & \text{for this then requires an } \xi \text{ in } [1, 4] \text{ for which} \\ 2 \text{ for } 2 < x \le 4; & f(\xi) = 5/3. \text{ Such a number } f(\xi) \text{ does not exist.} \end{cases}$$

$$f(x) = \begin{cases} 2 \text{ for } 1 \le x \le 2; & \text{for this then requires an } \xi \text{ in } [1, 4] \text{ for which} \\ 3 \text{ for } 2 < x \le 3; & f(\xi) = 3. \text{ This is true for any } \xi \text{ in the interval} \\ 4 \text{ for } 3 < x \le 4: & 2 < \xi \le 3. \end{cases}$$

7.17 $4 + \pi + \frac{\pi^4}{4}$.

7.21 $S = 2\pi \int_{\phi(a)}^{\phi(b)} \phi(y)\sqrt{(1 + [\phi'(y)]^2)}\,dy$.

7.23 $V=\pi\int_{\phi(a)}^{\phi(b)}[\phi(y)]^2\mathrm{d}y$.

7.31 1.099698.

7.33 $3+2\pi=9.283185$.

7.35 length $=\sqrt{17}-\frac{1}{4}\ln(\sqrt{17}-4)=4.646784$.

7.37 volume $=\pi(2+\frac{1}{4}\pi^2)=14.034754$.

Chapter 8

8.1 (a) $\frac{3}{16}\ln\left(\frac{x-2}{x+2}\right)+C,\ x>2$; (b) $-\frac{1}{3}\cos 3x+C$;

(c) $\frac{1}{6}\ln\left(\frac{x+3}{x-3}\right)+C,\ x>3$; (d) $\frac{1}{2}\arctan\frac{x}{2}+C$;

(e) $\frac{1}{12}\sin 4x+C$; (f) $3^x/\ln 3+C$.

8.5 (a) $\frac{x^3}{3}-3\cos x+x+C$; (b) $4^x/\ln 4+\sin 2x+C$;

(c) $4\cosh x-\cos x+C$; (d) $\frac{1}{a}\mathrm{e}^{ax}+3x+C$.

8.7 $\frac{1}{2}\operatorname{arccos}\left(\frac{2}{x}\right)+C$ for $x>2$, or $-\frac{1}{2}\arctan\left(\frac{2}{\sqrt{x^2-4}}\right)+C$ for $x>2$.

8.9 $\arctan\sqrt{(\cosh x-1)}+C$.

8.11 $\frac{1}{36}(3x^2+1)^6+C$.

8.13 $\sqrt{(x^2-1)}-\operatorname{arccos}(1/x)+C$, or $\sqrt{x^2-1}+\arctan\left(\frac{1}{\sqrt{x^2-1}}\right)+C$.

8.15 $(22\sqrt{2})/105-8/105$.

8.17 $\ln(121/25)$.

8.19 $\mathrm{e}^{ax}\left(\frac{x}{a}-\frac{1}{a^2}\right)+C$.

8.21 $\frac{1}{2}(\sin x\cosh x-\cos x\sinh x)+C$.

8.23 $x\ln^2 x-2x\ln x+2x+C$.

8.27 $\frac{13}{15}-\frac{\pi}{4}$.

8.29 $x+\ln\left|\frac{x-3}{x-2}\right|^3+C$.

8.31 $\frac{8}{49(x-5)}-\frac{27}{49(x+2)}+\frac{30}{343}\ln\left|\frac{x-5}{x+2}\right|+C$.

8.33 $\ln\left|\frac{\tan(x/2)-5}{\tan(x/2)-3}\right|+C$.

8.35 $\frac{1}{\sqrt{2}}\arcsin\left(\frac{4x-3}{5}\right)+C$.

8.37 $\ln\left|\frac{x}{1+\sqrt{(1-x^2)}}\right|+C$.

8.39 $\frac{\sin x}{2}+\frac{\sin 5x}{20}+\frac{\sin 7x}{28}+C$.

8.41 $1/(1-\lambda)$ if $\lambda < 1$; divergent if $\lambda \geq 1$.

8.43 π.

Chapter 9

9.1 $\int_0^2 \mathrm{d}x \int_1^3 (x + y^2)\,\mathrm{d}y = \frac{64}{3}$.

9.3 $\int_0^2 \mathrm{d}x \int_1^3 x^2 \sin 2y\,\mathrm{d}y = \frac{4}{3}(\cos 2 - \cos 6)$.

9.5 $\int_0^{\pi/4} \mathrm{d}x \int_0^{\pi/3} \sin(2x - 3y)\,\mathrm{d}y = \int_0^{\pi/4} (-\frac{2}{3}\cos 2x)\,\mathrm{d}x = -\frac{1}{3}$.

9.7 $\int_0^3 \mathrm{d}y \int_{y^2}^9 \mathrm{d}x = \int_0^9 \mathrm{d}x \int_0^{\sqrt{x}} \mathrm{d}y = 18$.

9.9 $\int_0^2 \mathrm{d}x \int_0^{\sqrt{9-y^2}} 1/\sqrt{9-y^2}\,\mathrm{d}y = 2$.

9.11 $\int_0^3 \mathrm{d}x \int_x^{2x^2} xy^2\,\mathrm{d}y = \frac{10854}{40}$.

9.13 $\int_0^2 \mathrm{d}x \int_0^{\sqrt{4-x^2}} x^2 y\,\mathrm{d}y = \frac{32}{15}$.

9.15 $f(x) = \int_0^{\sqrt{1-x^2}} \sqrt{1 - x^2 - y^2}\,\mathrm{d}y$

$$= \lim_{a \to \sqrt{1-x^1}} \left[\frac{1}{2} a\sqrt{1 - x^2 - a^2} + \frac{1}{2}\arctan\left(\frac{a}{\sqrt{1 - x^2 - a^2}}\right)(1 - x^2)\right]$$

$$= \frac{1}{4}\pi(1 - x^2) \int_0^1 f(x)\,\mathrm{d}x = \frac{1}{6}\pi.$$

9.17 9/4.

9.19 $\int_0^{\pi/2} \mathrm{d}\theta \int_0^2 r^4 \cos^2\theta \sin\theta\,\mathrm{d}r = (r^5/5)_{r=0}^2 (-\frac{1}{3}\cos 3\theta)_{\theta=0}^{\pi/2} = \frac{32}{15}$.

9.21 $\int_0^{\pi} \mathrm{d}\theta \int_1^4 r\sin(r\theta)\,\mathrm{d}r = 3$.

9.23 The volume is above the plane $z = 0$, so the volume V_1 below the upper plane and above $z = 0$ is

$$V_i = \int_0^{\pi/4} \mathrm{d}\theta \int_0^1 (2 - r\cos\theta - \tfrac{1}{2} r\sin\theta) r\,\mathrm{d}r = \int_0^{\pi/4} (1 - \tfrac{1}{3}\cos\theta - \tfrac{1}{3}\sin\theta)\,\mathrm{d}\theta$$

$$= \tfrac{1}{4}\pi - \tfrac{1}{6} - \tfrac{\sqrt{2}}{12}.$$

Similarly, the volume V_2 below the lower plane and above $z = 0$ is

$$V_2 = \int_0^{\pi/4} d\theta \int_0^1 (1 - r\cos\theta - \tfrac{1}{2} r\sin\theta) r\,dr = \int_0^{\pi/4} (\tfrac{1}{2} - \tfrac{1}{3}\cos\theta - \tfrac{1}{3}\sin\theta)\,d\theta$$
$$= \tfrac{1}{8}\pi - \tfrac{1}{6} - \tfrac{\sqrt{2}}{12}.$$

Thus the volume between the two planes is $V = V_1 - V_2 = \pi/8$.

9.25 Area $= \int_0^{\pi} d\theta \int_0^2 (r^2)^{3/2} r\,dr = \frac{32}{5}\pi$.

9.27 $\int_0^1 dr \int_0^{\pi} re^{-r^2}\,d\theta = \frac{1}{2}\pi(1 - 1/e)$.

9.29 $m = \int_0^{\pi/2} d\theta \int_1^{1+\cos\theta} (\rho/r) r\,dr = \rho$.

9.31 $I_0 = \int_0^{2\pi} d\theta \int_0^{2(1+\cos\theta)} (\rho r^2) r\,dr = 35\pi\rho$.

9.33 12/5.

Chapter 10

10.3 (a) 9; (b) 0; (c) 15.

10.7 (a) Equal if $a = 1, b = 2, c = 4$; (b) cannot be made equal because no solution to the two equations $a = 1$, $a^2 = 4$; (c) equal if $a = 1$, $b = 3$, $c = -1$.

10.9 $\mathbf{AB} = \begin{bmatrix} 4 & 3 & 3 & 0 \\ 7 & 5 & 5 & 0 \\ 3 & 4 & 2 & 2 \end{bmatrix}, \quad \mathbf{CD} = \begin{bmatrix} 20 \\ 29 \end{bmatrix}$.

10.11 $\mathbf{BAX} = \mathbf{BK} \Rightarrow \mathbf{IX} = \mathbf{BK} \Rightarrow \mathbf{X} = \mathbf{BK}$ so $x_1 = 15/8$, $x_2 = -3/8$, $x_3 = 1/4$.

10.13 $\mathbf{A}^2 = \begin{bmatrix} \cosh 2x & \sinh 2x \\ \sinh 2x & \cosh 2x \end{bmatrix}, \quad \mathbf{A}^3 = \begin{bmatrix} \cosh 3x & \sinh 3x \\ \sinh 3x & \cosh 3x \end{bmatrix}$

and $\mathbf{A}^n = \begin{bmatrix} \cosh nx & \sinh nx \\ \sinh nx & \cosh nx \end{bmatrix}$.

10.17 (a) -1; (b) 3; (c) 0:

10.19 (a) Row 1–10 row 3, $|\mathbf{A}| = -31$; (b) Remove factor 3 from row 1 and a factor 2 from row 2, $|\mathbf{A}| = -18$; (c) column 3 = column 1 + 3 column 2, $|\mathbf{A}| = 0$.

10.25 (a) Linearly independent; (b) linearly independent; (c) linearly dependent because row 4 = row 1 + row 2 + 2 row 3.

10.27 (a) $\begin{bmatrix} 6 & 1 & -5 \\ -2 & -5 & 4 \\ -3 & 3 & -1 \end{bmatrix}$; (b) $\begin{bmatrix} -7 & 6 & -1 \\ 1 & 0 & -1 \\ 1 & -2 & 1 \end{bmatrix}$; (c) $\begin{bmatrix} d & -b \\ -c & a \end{bmatrix}$.

10.29 $\mathbf{A}^{-1} = \begin{bmatrix} 1 & 2 & 3 \\ 2 & 5 & 7 \\ -2 & -4 & -5 \end{bmatrix}$.

10.31 $\dfrac{d\mathbf{A}}{dt} = \begin{bmatrix} 6t^2 & \sec^2 t & -\sin t \\ 0 & -2t & 1 \end{bmatrix}$ for $-\frac{1}{2}\pi < t < \frac{1}{2}\pi$.

10.35 $x_1 = 3,\ x_2 = 2,\ x_3 = 2.$

10.37 $x_1 = 1,\ x_2 = 2,\ x_3 = 1,\ x_4 = -1.$

10.39 Rank 2: $x_1 = (22 + x_3)/19, x_2 = (5 + 8x_3)/19.$

10.41 (a) Unique solution; (b) inconsistent; (c) consistent; (d) infinity of solutions.

10.43 $\lambda_1 = 2,\ \lambda_2 = -1;\quad \mathbf{X}_1 = \mu\begin{bmatrix} 1 \\ -1 \end{bmatrix},\quad \mathbf{X}_2 = \mu\begin{bmatrix} 1 \\ 2 \end{bmatrix}$ for arbitrary scalar μ.

10.45 $\lambda_1 = 1,\ \lambda_2 = 3,\ \lambda_3 = 4;\quad \mathbf{X}_1 = \begin{bmatrix} 0 \\ 0 \\ 1 \end{bmatrix},\quad \mathbf{X}_2 = \begin{bmatrix} 2 \\ -10 \\ -27 \end{bmatrix},\quad \mathbf{X}_3 = \begin{bmatrix} 0 \\ 1 \\ 2 \end{bmatrix};$

$$\hat{\mathbf{X}}_1 = \begin{bmatrix} 0 \\ 0 \\ 1 \end{bmatrix},\quad \hat{\mathbf{X}}_2 = \frac{1}{\sqrt{833}}\begin{bmatrix} 2 \\ -10 \\ -27 \end{bmatrix},\quad \hat{\mathbf{X}}_3 = \frac{1}{\sqrt{5}}\begin{bmatrix} 0 \\ 1 \\ 2 \end{bmatrix}.$$

10.53 $\mathbf{A}^{-1} = \begin{bmatrix} -\frac{1}{15} & -\frac{1}{5} & \frac{2}{5} & \frac{7}{15} \\ -\frac{1}{5} & -\frac{8}{5} & \frac{11}{5} & \frac{7}{5} \\ -\frac{1}{15} & \frac{4}{5} & -\frac{3}{5} & -\frac{8}{15} \\ \frac{7}{15} & \frac{7}{5} & -\frac{9}{5} & -\frac{19}{15} \end{bmatrix}$

10.55 $x_1 = 1.362442,\ x_2 = 0.021394,\ x_3 = 0.258751,\ x_4 = -0.292331$

10.57 $x_1 = -0.326900,\ x_2 = 1.059921,\ x_3 = 1.171250,\ x_4 = 0.730223$

10.69 The characteristic equation is $\lambda^4 - 4\lambda^3 - \lambda^2 + 16\lambda - 12 = 0$, and a routine calculation confirms that $\mathbf{A}$ satisfies the corresponding Cayley-Hamilton theorem. $\det\mathbf{A} = -12$ so $\mathbf{A}^{-1}$ exists. Pre-multiplying the corresponding version of the Cayley-Hamilton theorem gives $\mathbf{A}^3 - 4\mathbf{A}^2 - \mathbf{A} - \mathbf{I} + 16\mathbf{I} - 12\mathbf{A}^{-1} = \mathbf{0}$.
Rearrangement of terms followed by division by 12 then gives
$$\mathbf{A} = \begin{bmatrix} \frac{1}{2} & \frac{1}{2} & 0 & 0 \\ \frac{1}{6} & -\frac{1}{3} & \frac{1}{6} & 0 \\ -\frac{1}{3} & \frac{1}{6} & \frac{1}{6} & 0 \\ -\frac{1}{6} & \frac{4}{3} & -\frac{1}{6} & 1 \end{bmatrix}.$$

10.71 $u_1 = 0.0423,\ u_2 = 0.0729,\ u_3 = 0.0786,\ u_4 = 0.0517,\ u_5 = 0.0962,$
$u_6 = 0.1708,\ u_7 = 0.1898,\ u_8 = 0.1280,\ u_9 = 0.1717,\ u_{10} = 0.3243,$
$u_{11} = 0.3817,\ u_{12} = 0.2707,\ u_{13} = 0.2662,\ u_{14} = 0.5731,$
$u_{15} = 0.7420,\ u_{16} = 0.5732.$

10.73 $u_1 = 0.2924,\ u_2 = 0.1413,\ u_3 = 0.5285,\ u_4 = 0.2727,$
$u_5 = 0.6828,\ u_6 = 0.4209,\ u_7 = 0.7819,\ u_8 = 0.7282,$
$u_9 = 0.5890,\ u_{10} = 0.4555,\ u_{11} = 0.2534,\ u_{12} = 0.8507,$
$u_{13} = 1.1208,\ u_{14} = 1.1723,\ u_{15} = 0.9795,\ u_{16} = 0.5582.$

Chapter 11

11.1 (a) Constant pitch helix with elliptic cross-section; (b) variable pitch helix with elliptic cross-section lying entirely above (x, y)-plane; (c) curve with parabolic projection on (x, y)-plane and cubic projection on (y, z)-plane.

11.3 (a) $\dfrac{d\mathbf{u}}{dt} = -2a\pi(\sin 2\pi t)\mathbf{i} + 2b\pi(\cos 2\pi t)\mathbf{j} + \mathbf{k}$,

$\dfrac{d^2\mathbf{u}}{dt^2} = -4a\pi^2(\cos 2\pi t)\mathbf{i} - 4b\pi^2(\sin 2\pi t)\mathbf{j}$;

(c) $\dfrac{d\mathbf{u}}{dt} = \mathbf{i}+2t\mathbf{j} + 3t^2\mathbf{k}$,

$\dfrac{d^2\mathbf{u}}{dt^2} = 2\mathbf{j} + 6t\mathbf{k}$

and

$$\frac{d\mathbf{u}}{dt} = \frac{1}{r^2}\frac{d\mathbf{r}}{dt} - \frac{2}{r^3}\left(\frac{dr}{dt}\right)\mathbf{r} + \left(\mathbf{a}\cdot\frac{d\mathbf{r}}{dt}\right)\mathbf{b} + \mathbf{a}\times\frac{d^3\mathbf{r}}{dt^3}.$$

11.5 $\mathbf{T}(0) = \mathbf{i}, \quad \mathbf{T}(1) = \dfrac{1}{\sqrt{14}}(\mathbf{i} + 2\mathbf{j} + 3\mathbf{k})$.

11.7 (a) $\frac{1}{2}(\sinh 2t)\mathbf{i} + (\log t)\mathbf{j} + \dfrac{t^4}{4}\mathbf{k} + \mathbf{C}$;

(b) $[(2 - t^2)\cos t + 2t\sin t]\mathbf{i} + e^t\mathbf{j} + t(\ln t - 1)\mathbf{k} + \mathbf{C}$.

11.9 $\left(\dfrac{d\mathbf{r}}{dt}\right)^2 = -\mathbf{\Omega}^2\mathbf{r}^2 + \text{const.}$

11.13 29/28.

11.15 (a) 8; (b) 8; (c) 8.

11.17 transverse velocity $= \dfrac{u\cos\theta}{\cos\theta - \sin\theta}$;

radial acceleration $= \dfrac{-u^2\cos\theta}{ae^{\theta}(\cos\theta - \sin\theta)^2}$;

transverse acceleration $= \dfrac{2u^2(\cos\theta - \sin\theta) + 2u^2\sin\theta\cos\theta}{ae^{\theta}(\cos\theta - \sin\theta)^2}$.

11.19 $J_1 = -1, \; J_2 = -2, \; J_3 = -3$.

11.21 (a) Yes; (b) Yes; (c) No.

11.23 (a) $\frac{1}{3}(6x + y^2) + \frac{2}{3}(2xy + z) - \frac{2}{3}y$;

(b) $\frac{2}{3}xyz + \frac{2}{3}(x^2z - \sin y) - \frac{2}{3}x^2y$;

(c) $-\dfrac{1}{3}\dfrac{1}{x^2yz} - \dfrac{2}{3}\dfrac{1}{xy^2z} + \dfrac{2}{3}\dfrac{1}{xyz^2}$.

11.27 $\mathbf{n} = \dfrac{1}{3\sqrt{2}}(\mathbf{i} + 4\mathbf{j} - \mathbf{k})$; $x + 4y - z = 8$.

11.31 $e + 2\pi + 4$.

11.33 $2(1+x)$.

11.37 $2x(y-1)\mathbf{i} + y(1-y)\mathbf{j} + z\mathbf{k}$.

11.43 $\phi = (x+y)\cos z + xz + C$.

11.45 $\phi = x^2yz^3 + C$.

11.47 $a = 2;\ \ b = 6,\ \ \phi = 2x^2y + 3y^2z^2 + C$.

Chapter 12

12.1 (a) $\dfrac{2n+1}{4^n}$; (b) $\dfrac{2n}{3n+2}$; (c) $\dfrac{1.3.5\ldots(2n-1)}{1.4.7\ldots(3n-2)}$;
(d) $a_{2m+1} = 1/2^{m+1},\ a_{2m} = 1/3^m$; (e) $a_n = 1/n(n+2)$.

12.3 (a) 47/12, remainder $\dfrac{8}{3}\left(\dfrac{1}{4}\right)^n + \dfrac{5}{4}\left(\dfrac{1}{5}\right)^n$;
(b) 17/12, remainder $\dfrac{8}{3}\left(\dfrac{1}{4}\right)^n - \dfrac{5}{4}\left(\dfrac{1}{5}\right)^n$.

12.7 (a) divergent; (b) convergent $R_6 < 1/\log 6$.

12.9 (a) convergent; (b) convergent; (c) divergent;
(d) convergent; (e) divergent.

12.11 (a) convergent, $0 < |R_{10}| < 1/22^2$; (b) divergent;
(c) convergent, $0 < |R_{10}| < \dfrac{1}{11}\left(\dfrac{1}{3}\right)^{11}$.

12.13 (a) $-1 < x < 1$; (b) $-4 < x < 4$; (c) $-\infty < x < \infty$;
(d) $-\infty < x < \infty$; (e) $-16 < x < 2$.

12.15 $\arcsin x = x + \dfrac{1}{2}\cdot\dfrac{x^3}{3} + \dfrac{1.3}{2.4}\dfrac{x^5}{5} + \dfrac{1.3.5}{2.4.6}\dfrac{x^7}{7} + \ldots$, radius of convergence $r = 1$, interval of convergence $-1 < x < 1$, divergent at $x = \pm 1$.

12.17 $C - \dfrac{C^3}{3.3!} + \dfrac{C^5}{5.5!} - \dfrac{C^7}{7.7!} + \ldots$, convergent for all C.

12.19 (a) $x + \sum_{n=2}^{\infty} \dfrac{(-1)^{n-1}2^{n-1}x^n}{(n-1)!}, \quad -\infty < x < \infty$;
(b) $\sum_{n=0}^{\infty} \dfrac{x^{4n}}{2^{2n}(2n)!}, \quad -\infty < x < \infty$;
(c) $8 + 3\sum_{n=1}^{\infty}\left(\dfrac{1 + 2^n + 3^{n-1}}{n!}\right)x^n, \quad -\infty < x < \infty$;
(d) $\sum_{n=0}^{\infty}(-1)^n(2^{n+1}-1)x^n, \quad -\dfrac{1}{2} < x < \dfrac{1}{2}$;
(e) $x - \dfrac{1}{2}\dfrac{x^3}{3} + \dfrac{1.3}{2.4}\dfrac{x^5}{5} - \dfrac{1.3.5}{2.4.6}\dfrac{x^7}{7} + \ldots, \quad -1 \le x \le 1$; this is the series for $\arcsin x$.

12.23 $\sin 4x = \sin(3x + x) = \sin 3x\cos x + \cos 3x \sin x$,
$\sin 2x = \sin(3x - x) = \sin 3x \cos x - \cos 3x \sin x$.

So $\sin 4x - \sin 2x = 2 \sin x \cos 3x$, giving
$\sin x \cos 3x = \frac{1}{2}(\cos 4x - \cos 2x)$.

Thus $\sin x \cos 3x = \frac{1}{2}\sum_{n-0}^{\infty}(-1)^n \frac{(4^{2n+1} - 2^{2n+1})}{(2n+1)!} x^{2n+1}$.

12.25 0.7721.

12.27 $f(x, y) = y + xy + (3x^2y - y^3)/3!$.

12.29 $4\pi^2/\mathrm{e}$.

12.31 $3(\ln 2)^2/2$.

12.33 1/2.

12.35 min at (0, 0), saddle points at (5, 6), (−5, 5), (5, −6), (−5, −6).

12.37 (0, 0) saddle point; (1, 1) minimum.

12.41 minimum at (18/13, 12/13).

12.43 $x = 1,\ y = \frac{1}{2},\ z = -\frac{1}{2}$.

12.45 1/9 at $\left(\frac{1}{\sqrt{3}}, \frac{1}{\sqrt{3}}, \frac{1}{\sqrt{3}}\right)$ and $\left(\frac{-1}{\sqrt{3}}, \frac{-1}{\sqrt{3}}, \frac{-1}{\sqrt{3}}\right)$.

12.47 (i) $x = -2 - \sqrt{2},\ y = -2 - \sqrt{2},\ z = -2 - 2\sqrt{2}$;
(ii) $x = -2 + \sqrt{2},\ y = -2 + \sqrt{2},\ z = -2 + 2\sqrt{2}$.

12.49 length 60″, width = height = 30″.

12.51 Identical extrema at $(3/\sqrt{2}, 4/\sqrt{2})$ and $(-3/\sqrt{2}, -4/\sqrt{2})$.

12.53 $Y = 0.01 + 1.01x$.

Chapter 13

13.1 (a) order 3, degree 1; (b) order 1, degree 2;
(c) order 2, degree 2; (d) order 2, degree 3;
(e) order 3, degree 1.

13.3 $m\,\mathrm{d}v/\mathrm{d}t = -mg - \lambda v^2$.

13.5 (a) $x + 2y\frac{\mathrm{d}y}{\mathrm{d}x} = 0$; (b) $y'(x + y'^2) - y = 0$;
(c) $2xy' - 3y - x = 0$; (d) $(1 + x)y'\left[1 + \ln\left(\frac{1 + x}{y}\right)\right] - y = 0$;
(e) $y'' - 3y' + 2y = 0$; (f) $y'' - 4y' + 4y = 0$.

13.7 (a) No: solution is discontinuous when $x = 1$;
(b) No: $\sin 3x$ is not a solution;
(c) Yes; (d) No: solution is infinite when $x = -1$.
(e) Yes; (f) Yes.

13.9 0.5, 0.65, 0.8183, 1.0052, 1.2109, 1.4354.

13.11 0, 0.2005, 0.4067, 0.6265, 0.8681, 1.1405.

13.13 Node.

13.15 (a) No: y' is infinite when $x = 0$;
(b) Yes;

(c) No: y' is infinite when $y = -x$;

(d) Yes.

13.17 $xy' - y = 0$; so orthogonal trajectories satisfy $y' = -x/y$. Integration then shows the orthogonal trajectories are the circles $x^2 + y^2 = c^2$.

13.19

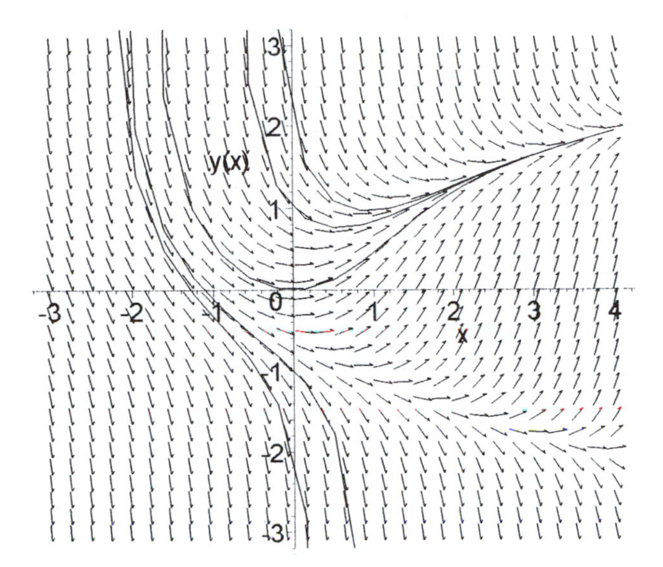

Fig. A13.19

13.21

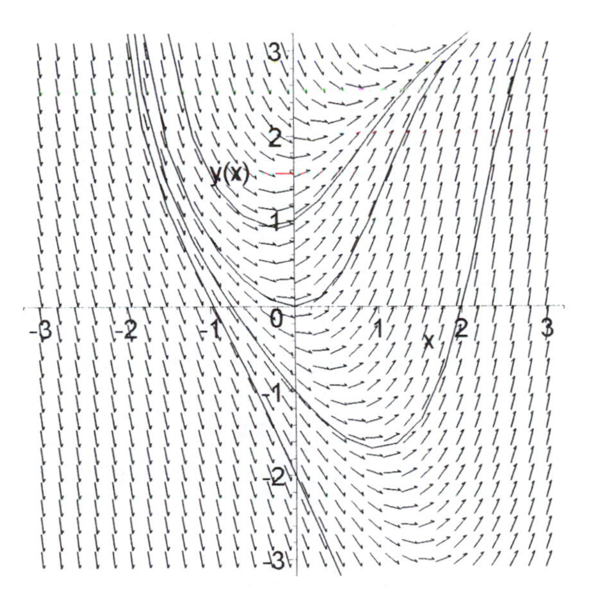

Fig. A13.21

Chapter 14

14.1 (a) $\operatorname{cosec}^2 y = \sec^2 x + C$;
(b) $y^2(x^2 - C)^2 = x^2(C - x^2)$;
(c) $x^2 + y^2 = \ln|Cx^2|$;
(d) $y = (a + Cx)/(1 + ax)$;
(e) $\tan y = C(1 - e^x)^3$; $x \neq 0$.
(f) $y = C \sin x$.

14.3 (a) $x^2 y + xy^2 + 3xy = C$;
(b) $xy + \sin xy + x^2 = C$;
(c) $x(y + e^y) = C$.

14.5 (a) $y = Cx + 2x^2$;
(b) $y = \frac{1}{6}x^4 - \frac{1}{6x^2}$;
(c) $y = \frac{3(x+1)^4}{2} + C(x+1)^2$;
(d) $y = (2 - e^2 + e^x)/x$;
(e) $y = \frac{1+x}{\cos x}$;
(f) $x = y(Ce^y - 1)$;
(g) $y = \frac{1}{1 + Ce^{x^2/2}}$;
(h) $y = 1/(x \cos x + Cx)$.

14.7 $y = \frac{1}{x}[1 + \mathrm{e}^x(x - 1)]$.

14.9 $y = \cosh x$.

14.11 In the steady state mode $y' = 0$, so then $\alpha = \sin y/y$.
If $\alpha < 0$, line $z = \alpha y$ must intersect $z = \sin y$ between π and 2π for positive y.
If $\alpha > 0$, line $z = \alpha y$ must be tangent to $z = \sin y$ at point ξ in $2\pi < \xi < \frac{5\pi}{2}$.

Chapter 15

15.1 (a) $P(\lambda) = \lambda^2 + 5\lambda - 14$; $y = C_1\mathrm{e}^{2x} + C_2\mathrm{e}^{-7x}$;
(b) $P(\lambda) = \lambda^2 - 1$; $y = \cosh x$;
(c) $P(\lambda) = \lambda^2 + 4\lambda + 3$; $y = C_1\mathrm{e}^{-3x} + C_2\mathrm{e}^{-x}$;
(d) $P(\lambda) = \lambda^3 + 5\lambda^2 + 2\lambda - 8$; $y = C_1\mathrm{e}^{x} + C_2\mathrm{e}^{-2x} + C_3\mathrm{e}^{-4x}$;
(e) $P(\lambda) = \lambda^3 + 7\lambda^2 + 12\lambda$; $y = C_1 + C_2\mathrm{e}^{-3x} + C_3\mathrm{e}^{-4x}$;
$y = 2 + \mathrm{e}^{-3x} - 3\mathrm{e}^{-4x}$;
(f) $P(\lambda) = \lambda^3 - \lambda^2 - \lambda + 1$; $y = C_1\mathrm{e}^{x} + C_2\mathrm{e}^{-x} + C_3 x\mathrm{e}^{x}$;
(g) $P(\lambda) = \lambda^4 - 2\lambda^3 - 3\lambda^2 + 4\lambda + 4$;
$y = C_1\mathrm{e}^{-x} + C_2 x\mathrm{e}^{-x} + C_3\mathrm{e}^{2x} + C_4 x\mathrm{e}^{2x}$.

15.3 The general solution is $y = C_1 e^x + C_2 e^{2x} + C_3 e^{3x}$.
The Wronskian $|W| = 2e^{6x}$ is non-vanishing for finite x proving that the functions e^x, e^{2x} and e^{3x} are linearly independent for finite x.

15.5 (a) $y = C_1 e^x + C_2 \cos x + C_3 \sin x$;
(c) $y = (C_1 + C_2 x + C_3 x^2)e^{ax}$;
(e) $y = (C_1 \cos\sqrt{(2x)} + C_2 \sin\sqrt{(2x)})e^{-x} + (C_3 \cos\sqrt{(2x)} + C_4 \sin\sqrt{(2x)})e^x$.

15.7 (a) $y = C_1 e^x + C_2 e^{-x} + \frac{1}{4}x(x-1)e^x + \frac{1}{4}\sinh x$;
(b) $y = C_1 e^x + C_2 x e^x$;
(c) $y = e^x(C_1 \cos x + C_2 \sin x + 2\sin x - 2x\cos x)$;
(d) $y = C_1 \sin x + C_2 \cos x + C_3 e^{-x} - \frac{3}{8}e^x + \frac{1}{4}xe^x$;
(e) $y = C_1 \sin x + C_2 \cos x - \cos x(\ln \sec x + \tan x|)$;
(f) $y = \sin x \ln[\sin x/(1 + \cos x)]$;
(g) $y = (C_1 \cos 2x + C_2 \sin 2x)e^{3x} + \dfrac{1}{65}(7\cos x - 4\sin x)e^x$.

15.13 $x(t) = \dfrac{3a}{4} + \dfrac{a}{4}\cos 2t, \quad y(t) = \dfrac{3a}{4} - \dfrac{a}{4}\cos 2t$.

15.15 This problem arises from the case in which three identical pendulums swinging from a common line of support are coupled one to another by identical springs between their masses.

$\omega_1^2 = a/m, \quad \omega_2^2 = (a+b)/m, \quad \omega_3^2 = (a+3b)/m$;

$$\mathbf{X}(t)^{(1)} = \begin{bmatrix} 1 \\ 1 \\ 1 \end{bmatrix} \sin(\omega_1 t + \varepsilon_1), \quad \mathbf{X}(t)^{(2)} = \begin{bmatrix} 1 \\ 0 \\ -1 \end{bmatrix} \sin(\omega_2 t + \varepsilon_2),$$

$$\mathbf{X}(t)^{(3)} = \begin{bmatrix} 1 \\ -2 \\ 1 \end{bmatrix} \sin(\omega_3 t + \varepsilon_3).$$

15.17 $x = e^{4t}(4\cos(t\sqrt{2}) + \sqrt{2}\sin(t\sqrt{2})), \quad y = 3e^{4t}(\sqrt{2}\sin(t\sqrt{2}) + \cos(t\sqrt{2}))$.

15.19 $x = e^{-t} - \frac{1}{7}e^{-7t} - \frac{6}{7}, \quad y = -e^{-t} - \frac{1}{7}e^{-7t} + \frac{8}{7}$.

15.21 $y = \left(\dfrac{e^3}{e^5 + 3}\right)(e^{2x} - e^{-3x})$.

15.23 $y = -(\dfrac{\sqrt{2}}{3})\sin 3x$.

15.25 $y = \frac{3}{2}(x + \cos x \sin x)\sin x - \frac{3}{2}\sin^2 x \cos x + (1 - 3\pi/4)\sin x$.

15.27 $y = 3\sin x$.

15.29 No solution.

15.35 $\dfrac{2b(s-a)}{((s-a)^2 + b^2)^2}$.

15.37 $\dfrac{e^{-as}(1 + as)}{s^2}$.

15.39 $\dfrac{2}{s-3} + \dfrac{s^2 - 9}{(s^2+9)^2}$.

15.41 $\dfrac{5}{2s} - \dfrac{3s}{2(s^2+4)}$.

15.45 $s^3 Y = 4 + 2s - s^2$.

15.47 $\dfrac{s^3}{s+2} - 4 + 2s - s^2 = \dfrac{-8}{s+2}$.

15.49 $\dfrac{1}{4}t^4$.

15.51 $1 - \cos t = 2\sin^2(t/2)$.

15.53 $e^{3t}(\cos t + \sin t)$.

15.55 $\dfrac{2}{(s-3)^2+4} + \dfrac{(s-3)}{(s-3)^2+25}$.

15.57 $\dfrac{2}{(s+2)((s+2)^2-4)} = \dfrac{1}{4(s+4)} + \dfrac{1}{4s} - \dfrac{1}{2(s+2)}$.

15.59 $f(t) = 2u_1(t) - \frac{3}{2}u_3(t) - \frac{1}{2}u_5(t)$.

15.61 $f(t) = 0$ for $0 < t < \pi/2$ and $f(t) = \sin\left(t - \dfrac{\pi}{2}\right)$ for $t > \pi/2$.

15.63 $\dfrac{5!e^{-s}}{(x-2)^6}$.

15.65 $u_3(t)e^{2(t-3)}\cos[3(t-3)]$.

15.67 $u_1(t)\{4\cosh[3(t-1)] - \sinh[2(t-1)]\}$.

15.69 $y = \frac{8}{3}e^{2t} + \frac{1}{3}e^{5t}$.

15.71 $y = \frac{1}{2}(e^t - \sin t - \cos t)$.

15.73 $y = (3 + 3t + \frac{3}{2}t^2)e^{-t} + t - 3$.

15.75 $y = (3 + 3t + \frac{3}{2}t^2)e^{-t} + t - 3$.

15.77 $y = 1 - (1+t)e^{-t} + u_1(t)(te^{1-t} - 1)$.

15.79 $y = e^{-t/2}(\cos 2t + \frac{1}{4}\sin 2t) + \frac{1}{8}u_1(t)e^{(1-t)/2}\sin(2t-2)$.

15.81 $y = -te^{-2t}, \quad z = (1+t)e^{-2t}$.

15.83 $y = 2e^t + e^{4t}, \quad z = \frac{2}{3}(e^{-4t} - e^t)$.

Chapter 16

16.1 $f(x) = 2\displaystyle\sum_{n=1}^{\infty}(-1)^{n+1}\frac{\sin nx}{n}$.

16.3 $f(x) = -\frac{1}{4}\pi + \displaystyle\sum_{n=1}^{\infty}\left[(-1)^n\frac{\sin nx}{n} + \frac{1-(-1)^n}{\pi n^2}\cos nx\right]$.

16.5 $f(x) = \dfrac{2}{\pi} - \dfrac{4}{\pi}\displaystyle\sum_{n=1}^{\infty}\frac{\cos 2nx}{4n^2-1}$.

16.7 $f(x) = \dfrac{2}{\pi}(\sin\alpha\pi)\left(\dfrac{1}{2\alpha} + \alpha\displaystyle\sum_{n=1}^{\infty}(-1)^{n+1}\frac{\cos nx}{n^2-\alpha^2}\right)$.

16.9 $f(x) = \frac{1}{2} - \frac{1}{2}\cos 2x$.

16.11 $f(x) = \dfrac{3}{2} - \dfrac{2}{\pi}\displaystyle\sum_{n=1}^{\infty}\frac{\sin(n\pi/2)}{n}\cos nx$.

16.13 $\dfrac{\pi^2}{90} = \sum_{n=1}^{\infty} \dfrac{1}{n^4}$.

16.15 $f(x) = \sum_{n=1}^{\infty} \dfrac{[1-(-1)^n]}{2n} \sin nx$; set $x = \pi/2$ when $f(\pi/2) = \pi/4$.

16.17 $f(x) = \dfrac{\pi^2}{3} + 4\sum_{n=1}^{\infty}(-1)^{n+1}\left(\dfrac{\sin nx}{2n} - \dfrac{\cos nx}{n^2}\right)$; $\dfrac{\pi^2}{12} = \sum_{n=1}^{\infty}(-1)^{n+1/n^2}$.

16.19 $f(x) = \dfrac{\pi^2}{3} + 4\sum_{n=1}^{\infty}(-1)^n \dfrac{\cos nx}{n^2}$.

16.21 $f(x) = \dfrac{13\pi^2}{2} + 4\sum_{n=1}^{\infty}(-1)^n\left[\dfrac{\cos nx}{n^2} - \dfrac{2\pi \sin nx}{n}\right]$.

16.23 $f(x) = \dfrac{3}{2} + \dfrac{6}{\pi^2}\sum_{n=1}^{\infty}\dfrac{[(-1)^n - 1]}{n^2}\cos\left(\dfrac{n\pi x}{3}\right)$.

16.25 $f(x) = \dfrac{8}{\pi}\sum_{n=1}^{\infty}\dfrac{n \sin 2nx}{4n^2 - 1}$.

16.27 $f(x) = \dfrac{e^{\pi} - 1}{\pi} - \dfrac{2}{\pi}\sum_{n=1}^{\infty}[1 - (-1)^n e^{\pi}]\dfrac{\cos nx}{n^2 + 1}$.

16.29 Integrate $f(t) = 2t + 1$ over [0, 1] to obtain

$$g(x) = x + \frac{\pi^2}{3} + 4\sum_{n=1}^{\infty}(-1)^n \frac{\cos nx}{n^2}.$$

The series for $g(x)$ is not a Fourier series because of the presence of the term x in the Fourier series for $g(x)$.

16.33 $f(x) = \dfrac{2\sinh \alpha\pi}{\pi}\left[\dfrac{1}{2\alpha} + \sum_{n=1}^{\infty}(-1)^n\dfrac{\alpha \cos nx}{n^2 + \alpha^2}\right]$ with $f(\pm\alpha) = \cosh \alpha\pi$.

16.35 $f(x) = \dfrac{8}{\pi^2}\sum_{n=1}^{\infty}\dfrac{\sin(n\pi/2)}{n^2}\sin\left(\dfrac{n\pi x}{2}\right)$.

16.37 The Fourier coefficients for $f(x)$ are:

$$a_0 = \pi,\ a_{2n} = 0,\ a_{2n-1} = \frac{-4}{\pi}\cdot\frac{1}{(2n-1)^2} \text{ and } b_n = 0 \text{ for all } n.$$

The Fourier coefficients for $g(x)$ are:

$$a_0' = \frac{2\pi^2}{3},\ a_n' = (-1)^n\frac{4}{n^2} \text{ and } b_n = 0 \text{ for all } n.$$

Hence $\dfrac{\pi^4}{96} = \sum_{n=1}^{\infty}\dfrac{1}{(2n-1)^4}$.

Chapter 17

17.1 (a) Absolute error $< 0.56 \times 10^{-3}$. Answer 2.56 to three significant figures.

(b) Relative error $< 3.4 \times 10^{-3}$. Absolute error $< 2.8 \times 10^{-3}$. Answer 0.6896 ± 0.0028, or 0.7 to one significant figure.

(c) Absolute error $< 4 \times 10^{-3}$. Answer 4.10 to three significant figures.

17.5 $y(2.17) = 8.7583$.

17.9 $\Gamma(4) \approx 5.4364$. Max error due to truncation is 0.0454. Remaining error is due to inaccuracy in determining integral over the interval [0, 10] caused by taking too few points.

17.13 2.365.

17.15 4.5826.

17.19 -4.101859.

17.27 $\lambda_1 = 11,\ \tilde{\mathbf{X}}_1 = \begin{bmatrix} 0.26 \\ -0.4 \\ 1 \end{bmatrix};\quad \lambda_2 = 6,\ \tilde{\mathbf{X}}_2 = \begin{bmatrix} 0.2 \\ 0 \\ 1 \end{bmatrix};\quad \lambda_3 = 1,\ \tilde{\mathbf{X}}_3 = \begin{bmatrix} 1 \\ 0 \\ 0 \end{bmatrix}.$

17.31 $(m = 1)$ 0.757556, $(m = 2)$ 0.434429, $(m = 3)$ 0.286375.

17.33 -2.813606, -0.529317, 1.342923.

17.35 -2.464399, -1.714047, 1.233398.

17.37 0.587074, 3.129528.

17.39 Newton's iterative scheme is $x_{n+1} = x_n - \left(\int_0^{x_n} \cos(\tfrac{1}{2}t^2\pi)\,\mathrm{d}t - \tfrac{1}{2}\right)/\cos(\tfrac{1}{2}x_n^2\pi)$: Zeros are 0.508309, 1.443186, 2.011757.

17.41 $y(x)$ at increments of 0.1 starting at $x = 0$: 1.0, 1.3, 1.598020, 1.890327, 2.173814, 2.446228, 2.706228, 2.953286, 3.187515, 3.409466, 3.619963.

17.43 $y(x)$ at increments of 0.1 starting at $x = 1$: 1.0, 1.088821, 1.171186, 1.247254, 1.317265, 1.381507, 1.440295, 1.493951, 1.542801, 1.587162, 1.627340.

17.45 $y(x)$ at increments of 0.1 starting at $x = 0$: 1.0, 1.110298, 1.242094, 1.396107, 1.572376, 1.770557, 1.990221, 2.231067, 2.493041, 2.776397, 3.081709.

Chapter 18

18.3 (a) 1/36; (b) 25/36; (c) 5/18.

18.4 Probability of failure of A or ((B or C) and D)
$= P(A \cup ((B \cup C) \cap D))$
$= p_A + (p_B + p_C - p_B p_C)p_D - p_A p_D(p_B + p_C - p_B p_C)$.

18.5 Probability of failure of A or ((B and C) or (D and E))
$= P(A \cup ((B \cap C) \cup (D \cap E))$
$= p_A + p_B p_C + p_D p_E - p_B p_C p_D p_E - p_A(p_B p_C + p_D p_E - p_B p_C p_D p_E)$.

18.7 Probability of failure of (A or B or C) and (D or E))
$= P((A \cup B) \cup C) \cap P(D \cup E))$
$= (p_A + p_B + p_C - p_A p_B - p_C(p_A + p_B - p_A p_B))(p_D + p_E - p_E p_D)$.

18.9 $\binom{20}{2} + \binom{20}{4}$.

18.11 16/81.

18.13

Score	2	3	4	5	6	7	8	9	10	11	12
$P(x)$	1/36	1/18	1/12	1/9	5/36	1/6	5/36	1/9	1/12	1/18	1/36

18.15 0.4349.

18.17 $\int_0^\infty P(x)\,\mathrm{d}x = 1$; $\mu = 1/a$, $P(x \le \mu) = 1/\mathrm{e}$.

18.19 $P = (1+k)^{-2}$.

18.21 (i) $P(x > 2.25) = 1 - \Phi(2.25) = 0.01222$.
(ii) $P(x > 0.8) = \Phi(0.8) = 0.78814$.
(iii) $P(-0.2 < x < 1.2) = \Phi(1.2) - \Phi(-0.2) = \Phi(1.2) - (1 - \Phi(0.2))$
$= 0.46419$.

18.23 (i) $P(x < -0.24) = P(X < -0.8) = 1 - \Phi(0.8) = 0.21186$.
(ii) $P(x > 4.44) = P(X > 1.8) = 1 - \Phi(1.8) = 0.03593$.

18.25 $4.587\,\mathrm{cm} \le \mu \le 4.637\,\mathrm{cm}$.

18.27 $2.518\,\mathrm{cm} \le \mu \le 2.554\,\mathrm{cm}$.

18.29 $-0.166\ \mathrm{cm} \le \mu_1 - \mu_2 \le 0.012\,\mathrm{cm}$: accept H_0.

18.31 $a = 1.00$, $b = 2.63$, $Y = X + 2.63$; $0.41 \le \tilde{a} \le 1.59$.

Reference list 1

Useful identities and constants

Trignometric identities

$$\sin^2 x + \cos^2 x = 1$$
$$\sec^2 x = 1 + \tan^2 x$$
$$\csc^2 x = 1 + \cot^2 x$$
$$\sin 2x = 2 \sin x \cos x$$
$$\begin{aligned}\cos 2x &= \cos^2 x - \sin^2 x \\ &= 1 - 2\sin^2 x \\ &= 2\cos^2 x - 1\end{aligned}$$
$$\sin^2 x = \tfrac{1}{2}(1 - \cos 2x)$$
$$\cos^2 x = \tfrac{1}{2}(1 + \cos 2x)$$
$$\sin(x + y) = \sin x \cos y + \cos x \sin y$$
$$\sin(x - y) = \sin x \cos y - \cos x \sin y$$
$$\cos(x + y) = \cos x \cos y - \sin x \sin y$$
$$\cos(x - y) = \cos x \cos y + \sin x \sin y$$
$$\tan(x + y) = \frac{\tan x + \tan y}{1 - \tan x \tan y}$$
$$\tan(x - y) = \frac{\tan x - \tan y}{1 + \tan x \tan y}$$

Hyperbolic identities

$$\cosh^2 x - \sinh^2 x = 1$$
$$\operatorname{sech}^2 x = 1 - \tanh^2 x$$
$$\operatorname{csch}^2 x = \coth^2 x - 1$$
$$\sinh 2x = 2 \sinh x \cosh x$$
$$\begin{aligned}\cosh 2x &= \cosh^2 x + \sinh^2 x \\ &= 1 + 2\sinh^2 x \\ &= 2\cosh^2 x - 1\end{aligned}$$
$$\sinh^2 x = \tfrac{1}{2}(\cosh 2x - 1)$$
$$\cosh^2 x = \tfrac{1}{2}(\cosh 2x + 1)$$
$$\sinh(x + y) = \sinh x \cosh y + \cosh x \sinh y$$
$$\sinh(x - y) = \sinh x \cosh y - \cosh x \sinh y$$
$$\cosh(x + y) = \cosh x \cosh y + \sinh x \sinh y$$
$$\cosh(x - y) = \cosh x \cosh y - \sinh x \sinh y$$

$$\tanh(x+y) = \frac{\tanh x + \tanh y}{1 + \tanh x \tanh y}$$

$$\tanh(x-y) = \frac{\tanh x - \tanh y}{1 - \tanh x \tanh y}$$

Inverse hyperbolic functions

$$\operatorname{arcsinh} x = \ln(x + \sqrt{x^2+1}), \quad -\infty < x < \infty$$

$$\operatorname{arccosh} x = \ln(x + \sqrt{x^2-1}), \quad 1 \le x < \infty$$

$$\operatorname{arctanh} x = \tfrac{1}{2}\ln\left(\frac{1+x}{1-x}\right), \quad -1 < x < 1$$

Complex relationships

$$e^{ix} = \cos x + i \sin x$$

$$\sinh x = \frac{e^x - e^{-x}}{2}$$

$$\sin x = \frac{e^{ix} - e^{-ix}}{2i}$$

$$\sin ix = i \sinh x$$

$$\sinh ix = i \sin x$$

$$(\cos x + i \sin x)^n = \cos nx + i \sin nx$$

$$\cosh x = \frac{e^x + e^{-x}}{2}$$

$$\cos x = \frac{e^{ix} + e^{-ix}}{2}$$

$$\cos ix = \cosh x$$

$$\cosh ix = \cos x$$

Constants

$$e = 2.7182\,81828\,45904$$

$$\pi = 3.1415\,92653\,58979$$

$$\pi^2 = 9.8696\,04401\,08935$$

$$(2\pi)^{-2} = 0.3989\,42280\,40143$$

$$\ln 10 = 2.3025\,85092\,99404$$

$$\log_{10} e = 0.4342\,94481\,90325$$

Reference list 2

Basic derivatives and rules

Basic derivatives

	$f(x)$	$f'(x)$		$f(x)$	$f'(x)$
1	x^n	nx^{n-1}	13	$\sinh ax$	$a\cosh ax$
2	e^{ax}	$a e^{ax}$	14	$\cosh ax$	$a\sinh ax$
3	$\ln x$	$1/x$	15	$\tanh ax$	$a\,\mathrm{sech}^2\,ax$
4	$\sin ax$	$a\cos ax$	16	$\mathrm{csch}\,ax$	$-a\,\mathrm{csch}\,ax\,\coth ax$
5	$\cos ax$	$-a\sin ax$	17	$\mathrm{sech}\,ax$	$-a\,\mathrm{sech}\,ax\,\tanh ax$
6	$\tan ax$	$a\sec^2 ax$	18	$\coth ax$	$-a\,\mathrm{csch}^2\,ax$
7	$\csc ax$	$-a\csc ax\cot ax$			
8	$\sec ax$	$a\sec ax\tan ax$	19	$\mathrm{arcsinh}\,\frac{x}{a}$	$1/\sqrt{x^2+a^2}$
9	$\cot ax$	$-a\csc^2 ax$			
10	$\arcsin\frac{x}{a}$	$1/\sqrt{a^2-x^2}$	20	$\mathrm{arccosh}\,\frac{x}{a}$	$1/\sqrt{x^2-a^2}, \quad x/a>1$
11	$\arccos\frac{x}{a}$	$-1/\sqrt{a^2-x^2}$			
12	$\arctan\frac{x}{a}$	$a/(a^2+x^2)$	21	$\mathrm{arctanh}\,\frac{x}{a}$	$a/(a^2-x^2), \quad \lvert x/a\rvert<1$

Rule of differentiation and integration

1 $$\frac{d}{dx}(u+v)=\frac{du}{dx}+\frac{dv}{dv} \quad \text{(sum)}$$

2 $$\frac{d}{dx}(uv)=u\frac{dv}{dx}+v\frac{du}{dx} \quad \text{(product)}$$

3 $$\frac{d}{dx}\left(\frac{u}{v}\right)=\left(v\frac{du}{dx}-u\frac{dv}{dx}\right)\Big/v^2 \quad \text{for } v\neq 0 \quad \text{(quotient)}$$

4 $$\frac{d}{dx}[f\{g(x)\}]=f'\{g(x)\}\frac{dg}{dx} \quad \text{(function of a function)}$$

5 $\int (u+v)\mathrm{d}x = \int u\,\mathrm{d}x + \int v\,\mathrm{d}x$ (sum)

6 $\int u\,\mathrm{d}v = uv - \int v\,\mathrm{d}u$ (integration by parts)

7 $$\frac{\mathrm{d}}{\mathrm{d}x}\int_{\phi(\alpha)}^{\psi(\alpha)} f(x,\alpha)\mathrm{d}x = \left(\frac{\mathrm{d}\psi}{\mathrm{d}\alpha}\right) f(\psi,\alpha) - \left(\frac{\mathrm{d}\phi}{\mathrm{d}\alpha}\right) f(\phi,\alpha) + \int_{\phi(\alpha)}^{\psi(\alpha)} \frac{\partial f}{\partial \alpha}\,\mathrm{d}x$$

(differentiation of an integral containing a parameter)

Reference list 3

Laplace transform pairs

	$f(t) = \mathscr{L}^{-1}(F(s))$	$F(s) = \mathscr{L}(f(t))$
1.	a	$\dfrac{a}{s}, \quad s > 0$
2.	t	$\dfrac{1}{s^2}, \quad s > 0$
3.	t^n, n a positive integer	$\dfrac{n!}{s^{n+1}}, \quad s > 0$
4.	e^{at}	$\dfrac{1}{s-a}, \quad s > a$
5.	$t^n e^{at}$, n a positive integer	$\dfrac{n!}{(s-a)^{n+1}}, \quad s > a$
6.	$\sin at$	$\dfrac{a}{s^2+a^2}, \quad s > 0$
7.	$\cos at$	$\dfrac{s}{s^2+a^2}, \quad s > 0$
8.	$t \sin at$	$\dfrac{2as}{(s^2+a^2)^2}, \quad s > 0$
9.	$t \cos at$	$\dfrac{s^2-a^2}{(s^2+a^2)^2}, \quad s > 0$
10.	$e^{at} \sin bt$	$\dfrac{b}{(s-a)^2+b^2}, \quad s > a$

11.	$e^{at}\cos bt$	$\dfrac{s-a}{(s-a)^2+b^2}, \quad s>a$
12.	$\sinh at$	$\dfrac{a}{s^2-a^2}, \quad s>\lvert a\rvert$
13.	$\cosh at$	$\dfrac{s}{s^2-a^2}, \quad s>\lvert a\rvert$
14.	$f(t)=u_a(t)=\begin{cases}0 & \text{for } t<a\\ 1 & \text{for } t>a\end{cases}$	$\dfrac{e^{-as}}{s}, \quad s>0$
15.	$\delta(t-a)$	$e^{-as}, \quad s>0, \quad a\geq 0$

Reference list 4

Short table of integrals

Common standard forms

1. $\int x^n \, dx = \frac{1}{n+1} x^{n+1} + C, \quad n \neq -1$

2. $\int \frac{1}{x} \, dx = \ln|x| + C = \begin{cases} \ln x + C, & x > 0 \\ \ln(-x) + C, & x < 0 \end{cases}$

3. $\int e^{ax} \, dx = \frac{1}{a} e^{ax} + C$

4. $\int a^x \, dx = \frac{a^x}{\ln a} + C \quad a \neq 1, \ a > 0$

5. $\int \ln x \, dx = x \ln x - x + C$

6. $\int \sin ax \, dx = -\frac{1}{a} \cos ax + C$

7. $\int \cos ax \, dx = \frac{1}{a} \sin ax + C$

8. $\int \tan ax \, dx = -\frac{1}{a} \ln|\cos ax| + C$

9. $\int \sinh ax \, dx = \frac{1}{a} \cosh ax + C$

10. $\int \cosh ax \, dx = \frac{1}{a} \sinh ax + C$

11. $\int \tanh ax \, dx = \frac{1}{a} \ln \cosh ax + C$

12. $\int \frac{1}{\sqrt{a^2 - x^2}} \, dx = \arcsin \frac{x}{a} + C, \quad x^2 < a^2$

13. $\int \frac{1}{\sqrt{x^2 - a^2}} \, dx = \operatorname{arccosh} \frac{x}{a} + C, \quad a^2 < x^2$

14. $$\int \frac{1}{\sqrt{a^2+x^2}}\,\mathrm{d}x = \begin{cases} \operatorname{arcsinh}\dfrac{x}{a} + C \\ \ln(x+\sqrt{a^2+x^2}) + C \end{cases}$$

15. $$\int \frac{1}{x^2+a^2}\,\mathrm{d}x = \frac{1}{a}\arctan\frac{x}{a} + C$$

16. $$\int \frac{1}{x^2-a^2}\,\mathrm{d}x = \frac{1}{2a}\ln\left|\frac{x-a}{x+a}\right| + C, \quad a^2 \le x^2$$

Algebraic forms

17. $$\int (a+bx)^n\mathrm{d}x = \frac{(a+bx)^{n+1}}{b(n+1)} + C, \quad n \neq -1$$

18. $$\int \frac{1}{a+bx}\,\mathrm{d}x = \frac{1}{b}\ln|a+bx| + C$$

19. $$\int x(a+bx)^n\mathrm{d}x = \frac{(a+bx)^{n+1}}{b^2}\left[\frac{a+bx}{n+2} - \frac{a}{n+1}\right] + C, \quad n \neq -1, -2$$

20. $$\int \frac{x}{a+bx}\,\mathrm{d}x = \frac{x}{b} - \frac{a}{b^2}\ln|a+bx| + C$$

21. $$\int \frac{x^2}{a+bx}\,\mathrm{d}x = \frac{1}{b^3}\left(\frac{1}{2}b^2x^2 - abx + a^2\ln|a+bx|\right) + C$$

22. $$\int \frac{x}{(a+bx)^2}\,\mathrm{d}x = \frac{1}{b^2}\left[\frac{a}{a+bx} + \ln|a+bx|\right] + C$$

23. $$\int \frac{x^2}{(a+bx)^2}\,\mathrm{d}x = \frac{1}{b^3}\left[bx - \frac{a^2}{a+bx} - 2a\ln|a+bx|\right] + C$$

24. $$\int \frac{1}{x(a+bx)}\,\mathrm{d}x = \frac{1}{a}\ln\left|\frac{x}{a+bx}\right| + C$$

25. $$\int \frac{1}{x^2(a+bx)}\,\mathrm{d}x = -\frac{1}{ax} + \frac{b}{a^2}\ln\left|\frac{a+bx}{x}\right| + C$$

26. $$\int \frac{1}{x(a+bx)^2}\,\mathrm{d}x = \frac{1}{a(a+bx)} + \frac{1}{a^2}\ln\left|\frac{x}{a+bx}\right| + C$$

27. $$\int \frac{1}{x\sqrt{a+bx}}\,\mathrm{d}x = \frac{1}{\sqrt{a}}\ln\left|\frac{\sqrt{a+bx}-\sqrt{a}}{\sqrt{a+bx}+\sqrt{a}}\right| + C \quad \text{if } a>0,\ a+bx>0$$

28. $$\int \frac{1}{x\sqrt{a+bx}}\,\mathrm{d}x = \frac{2}{\sqrt{-a}}\arctan\sqrt{\frac{a+bx}{-a}} + C \quad \text{if } a<0$$

29. $$\int \frac{1}{x^n\sqrt{a+bx}}\,\mathrm{d}x = -\frac{\sqrt{a+bx}}{a(n-1)x^{n-1}} - \frac{b(2n-3)}{2a(n-1)}\int \frac{1}{x^{n-1}\sqrt{a+bx}}\,\mathrm{d}x$$

$$n \neq 1, \quad a \neq 0$$

30. $$\int \frac{x}{\sqrt{a+bx}}\,\mathrm{d}x = \frac{2}{3b^2}(bx-2a)\sqrt{a+bx}+C$$

31. $$\int \frac{x^2}{\sqrt{a+bx}}\,\mathrm{d}x = \frac{2}{15b^3}(8a^2+3b^2x^2-4abx)\sqrt{a+bx}+C$$

32. $$\int \frac{x^n}{\sqrt{a+bx}}\,\mathrm{d}x = \frac{2x^n\sqrt{a+bx}}{b(2n+1)} - \frac{2an}{b(2n+1)}\int \frac{x^{n-1}}{\sqrt{a+bx}}\,dx$$

$$n \neq -\tfrac{1}{2}, \quad b \neq 0$$

33. $$\int (\sqrt{a+bx})^n \mathrm{d}x = \frac{2}{b}\frac{(\sqrt{a+bx})^{n+2}}{n+2}+C, \quad n \neq -2$$

34. $$\int \frac{\sqrt{a+bx}}{x}\,\mathrm{d}x = 2\sqrt{a+bx} + a\int \frac{1}{x\sqrt{a+bx}}\,\mathrm{d}x$$

35. $$\int \frac{\sqrt{a+bx}}{x^n}\,\mathrm{d}x = -\frac{(a+bx)^{3/2}}{a(n-1)x^{n-1}} - \frac{b(2n-5)}{2a(n-1)}\int \frac{\sqrt{a+bx}}{x^{n-1}}\,\mathrm{d}x \quad n \in \mathbf{N}$$

36. $$\int x\sqrt{a+bx}\,\mathrm{d}x = \frac{2}{15b^2}(3bx-2a)(a+bx)^{3/2}+C$$

37. $$\int \sqrt{a^2+x^2}\,\mathrm{d}x = \frac{x}{2}\sqrt{a^2+x^2} + \frac{a^2}{2}\ln(x+\sqrt{a^2+x^2})+C$$

38. $$\int x^2\sqrt{a^2+x^2}\,\mathrm{d}x = \frac{x}{8}(a^2+2x^2)\sqrt{a^2+x^2} - \frac{a^4}{8}\ln(x+\sqrt{a^2+x^2})+C$$

39. $$\int \frac{\sqrt{a^2+x^2}}{x}\,\mathrm{d}x = \sqrt{a^2+x^2} - a\,\mathrm{arctanh}(a/\sqrt{a^2+x^2})+C$$

40. $$\int \frac{\sqrt{a^2+x^2}}{x^2}\,\mathrm{d}x = \ln(x+\sqrt{a^2+x^2}) - \frac{\sqrt{a^2+x^2}}{x}+C$$

41. $$\int \frac{1}{x\sqrt{a^2+x^2}}\,\mathrm{d}x = \frac{1}{a}\ln\left|\frac{x}{a+\sqrt{a^2+x^2}}\right|+C$$

42. $$\int \frac{1}{x^2\sqrt{a^2+x^2}}\,\mathrm{d}x = -\frac{\sqrt{a^2+x^2}}{a^2x}+C$$

43. $$\int \sqrt{a^2-x^2}\,\mathrm{d}x = \frac{x}{2}\sqrt{a^2-x^2} + \frac{a^2}{2}\arcsin\frac{x}{a}+C, \quad x^2 \leq a^2$$

44. $$\int \frac{1}{x\sqrt{a^2-x^2}}\,\mathrm{d}x = \frac{1}{a}\ln\left|\frac{x}{a+\sqrt{a^2-x^2}}\right|+C, \quad x^2 \leq a^2$$

45. $$\int \sqrt{x^2-a^2}\,\mathrm{d}x = \frac{x}{2}\sqrt{x^2-a^2} - \frac{a^2}{2}\ln(x+\sqrt{x^2-a^2})+C, \quad a^2 \leq x^2$$

46. $$\int \frac{\sqrt{x^2-a^2}}{x}\,\mathrm{d}x = \sqrt{x^2-a^2} - a\,\mathrm{arcsec}\left|\frac{x}{a}\right|+C, \quad a^2 \leq x^2$$

47. $\displaystyle\int \frac{1}{x\sqrt{x^2 - a^2}}\,dx = \frac{1}{a}\,\mathrm{arcsec}\left|\frac{x}{a}\right| + C = \frac{1}{a}\arccos\left|\frac{a}{x}\right| + C, \quad a^2 \le x^2$

48. $\displaystyle\int \frac{1}{(a^2 + x^2)^2}\,dx = \frac{x}{2a^2(a^2 + x^2)} + \frac{1}{2a^3}\arctan\frac{x}{a} + C$

49. $\displaystyle\int \frac{1}{(a^2 - x^2)^2}\,dx = \frac{x}{2a^2(a^2 - x^2)} + \frac{1}{4a^3}\ln\left|\frac{a + x}{a - x}\right| + C$

Trigonometric forms

50. $\displaystyle\int \sin ax\,dx = -\frac{1}{a}\cos ax + C$

51. $\displaystyle\int \sin^2 ax\,dx = \frac{x}{2} - \frac{\sin 2ax}{4a} + C$

52. $\displaystyle\int \sin^n ax\,dx = \frac{-\sin^{n-1} ax\cos ax}{na} + \frac{n - 1}{n}\int \sin^{n-2} ax\,dx$

53. $\displaystyle\int \cos ax\,dx = \frac{1}{a}\sin ax + C$

54. $\displaystyle\int \cos^2 ax\,dx = \frac{x}{2} + \frac{\sin 2ax}{4a} + C$

55. $\displaystyle\int \cos^n ax\,dx = \frac{\cos^{n-1} ax\sin ax}{na} + \frac{n - 1}{n}\int \cos^{n-2} ax\,dx$

56. $\displaystyle\int_0^{\pi/2} \sin^n a\,dx = \int_0^{\pi/2} \cos^n x\,dx = \begin{cases} \dfrac{1.3.5\ldots(n - 1)}{2.4.6\ldots n}\cdot\dfrac{\pi}{2}, & \text{if } n \text{ is an even integer } \ge 2, \\ \dfrac{2.4.6\ldots(n - 1)}{3.5.7\ldots n}, & \text{if } n \text{ is an odd integer } \ge 3 \end{cases}$

57. $\displaystyle\int \sin ax\sin bx\,dx = \frac{\sin(a - b)x}{2(a - b)} - \frac{\sin(a + b)x}{2(a + b)}, \quad a^2 \ne b^2$

58. $\displaystyle\int \cos ax\cos bx\,dx = \frac{\sin(a - b)x}{2(a - b)} + \frac{\sin(a + b)x}{2(a + b)}, \quad a^2 \ne b^2$

59. $\displaystyle\int \sin ax\cos bx\,dx = -\frac{\cos(a + b)x}{2(a + b)} - \frac{\cos(a - b)x}{2(a - b)} + C, \quad a^2 \ne b^2$

60. $\displaystyle\int \sin ax\cos ax\,dx = -\frac{\cos 2ax}{4a} + C = \frac{\sin^2 ax}{2a} + C$

61. $\displaystyle\int \sin^n ax\cos ax\,dx = \frac{\sin^{n+1} ax}{(n + 1)a} + C, \quad n \ne -1$

62. $\displaystyle\int \cos^n ax\sin ax\,dx = -\frac{\cos^{n+1} ax}{(n + 1)a} + C, \quad n \ne -1$

63. $$\int \sin^n ax \cos^m ax\,\mathrm{d}x = \frac{\sin^{n+1} ax \cos^{m-1} ax}{a(m+n)} + \frac{m-1}{m+n}\int \sin^n ax \cos^{m-2} ax\,\mathrm{d}x, \quad m \neq -n$$

64. $$\int x \sin x\,\mathrm{d}x = -x\cos x + \sin x + C$$

65. $$\int x^2 \sin x\,\mathrm{d}x = -x^2\cos x + 2x\sin x + 2\cos x + C$$

66. $$\int x^n \sin x\,\mathrm{d}x = -x^n\cos x + n\int x^{n-1}\cos x\,\mathrm{d}x + C$$

67. $$\int x \cos x\,\mathrm{d}x = x\sin x + \cos x + C$$

68. $$\int x^2 \cos x\,\mathrm{d}x = x^2\sin x + 2x\cos x - 2\sin x + C$$

69. $$\int x^n \cos x\,\mathrm{d}x = x^n\sin x - n\int x^{n-1}\sin x\,\mathrm{d}x$$

70. $$\int \mathrm{e}^{ax}\sin bx\,\mathrm{d}x = \frac{\mathrm{e}^{ax}}{a^2+b^2}(a\sin bx - b\cos bx) + C$$

71. $$\int \mathrm{e}^{ax}\cos bx\,\mathrm{d}x = \frac{\mathrm{e}^{ax}}{a^2+b^2}(a\cos bx + b\sin bx) + C$$

72. $$\int \frac{\mathrm{d}x}{a+b\cos x} = \frac{2}{\sqrt{a^2-b^2}}\arctan\left[\frac{(a-b)\tan(x/2)}{\sqrt{a^2-b^2}}\right] + C, \quad a^2 > b^2$$

73. $$\int \frac{\mathrm{d}x}{a+b\cos x} = \frac{1}{\sqrt{b^2-a^2}}\ln\left|\frac{(b-a)\tan(x/2)+\sqrt{b^2-a^2}}{(b-a)\tan(x/2)-\sqrt{b^2-a^2}}\right| + C, \quad b^2 > a^2$$

74. $$\int \frac{\mathrm{d}x}{a+b\cos x} = \frac{2}{\sqrt{b^2-a^2}}\operatorname{arctanh}\left[\frac{(b-a)\tan(x/2)}{\sqrt{b^2-a^2}}\right] + C, \quad b^2 > a^2, \quad |(b-a)\tan(x/2)| < \sqrt{b^2-a^2}$$

75. $$\int \frac{\mathrm{d}x}{a+b\cos x} = \frac{2}{\sqrt{b^2-a^2}}\operatorname{arccot}\left[\frac{(b-a)\tan(x/2)}{\sqrt{b^2-a^2}}\right] + C, \quad b^2 > a^2, \quad |(b-a)\tan(x/2)| > \sqrt{b^2-a^2}$$

76. $$\int \sec ax\,\mathrm{d}x = \frac{1}{a}\ln|\sec ax + \tan ax| + C$$

77. $$\int \csc ax\,\mathrm{d}x = -\frac{1}{a}\ln|\csc ax + \cot ax| + C$$

78. $$\int \cot ax\,\mathrm{d}x = \frac{1}{a}\ln|\sin ax| + C$$

79. $$\int \tan^2 ax\,\mathrm{d}x = \frac{1}{a}(\tan ax - ax) + C$$

80. $$\int \tan^n ax\,dx = \frac{\tan^{n-1} ax}{a(n-1)} - \int \tan^{n-2} ax\,dx, \quad n \neq 1$$

81. $$\int \sec^2 ax\,dx = \frac{1}{a}\tan ax + C$$

82. $$\int \csc^2 ax\,dx = -\frac{1}{a}\cot ax + C$$

83. $$\int \cot^2 ax\,dx = -\frac{1}{a}\cot ax - x + C$$

84. $$\int \sec^n ax\,dx = \frac{\sec^{n-2} ax \tan ax}{a(n-1)} + \frac{n-2}{n-1}\int \sec^{n-2} ax\,dx, \quad n \neq 1$$

85. $$\int \csc^n ax\,dx = -\frac{\csc^{n-2} ax \cot ax}{a(n-1)} + \frac{n-2}{n-1}\int \csc^{n-2} ax\,dx, \quad n \neq 1$$

86. $$\int \cot^n ax\,dx = -\frac{\cot^{n-1} ax}{a(n-1)} - \int \cot^{n-2} ax\,dx, \quad n \neq 1$$

87. $$\int \sec^n ax \tan ax\,dx = \frac{\sec^n ax}{na} + C, \quad n \neq 0$$

88. $$\int \csc^n ax \cot ax\,dx = -\frac{\csc^n ax}{na} + C, \quad n \neq 0$$

Inverse trigonometric forms

89. $$\int \arcsin ax\,dx = x \arcsin ax + \frac{1}{a}\sqrt{1 - a^2x^2} + C, \quad a^2x^2 \leq 1$$

90. $$\int \arccos ax\,dx = x \arccos ax - \frac{1}{a}\sqrt{1 - a^2x^2} + C, \quad a^2x^2 \leq 1$$

91. $$\int \arctan ax\,dx = x \arctan ax - \frac{1}{2a}\ln(1 + a^2x^2) + C$$

Exponential and logarithmic forms

92. $$\int e^{ax}\,dx = \frac{1}{a}e^{ax} + C$$

93. $$\int b^{ax}\,dx = \frac{1}{a}\frac{b^{ax}}{\ln b} + C, \quad b > 0, \quad b \neq 1$$

94. $$\int xe^{ax}\,dx = \frac{e^{ax}}{a^2}(ax - 1) + C$$

95. $$\int x^n e^{ax}\,dx = \frac{1}{a}x^n e^{ax} - \frac{n}{a}\int x^{n-1}e^{ax}\,dx$$

96. $$\int x^n b^{ax}\,dx = \frac{x^n b^{ax}}{a \ln b} - \frac{n}{a \ln b}\int x^{n-1}b^{ax}\,dx, \quad b > 0, \quad b \neq 1$$

97. $$\int \ln ax\,dx = x \ln ax - x + C$$

98. $\int x^n \ln ax\,\mathrm{d}x = \frac{x^{n+1}}{n+1}\ln ax - \frac{x^{n+1}}{(n+1)^2} + C, \quad n \neq -1$

99. $\int \frac{\ln ax}{x}\,\mathrm{d}x = \frac{1}{2}(\ln ax)^2 + C$

100. $\int \frac{1}{x\ln ax}\,\mathrm{d}x = \ln|\ln ax| + C$

Hyperbolic forms

101. $\int \sinh ax\,\mathrm{d}x = \frac{1}{a}\cosh ax + C$

102. $\int \sinh^2 ax\,\mathrm{d}x = \frac{\sinh 2ax}{4a} - \frac{x}{2} + C$

103. $\int \sinh^n ax\,\mathrm{d}x = \frac{\sinh^{n-1}ax\cosh ax}{na} - \frac{n-1}{n}\int \sinh^{n-2}ax\,\mathrm{d}x, \quad n \neq 0$

104. $\int x\sinh ax\,\mathrm{d}x = \frac{x}{a}\cosh ax - \frac{1}{a^2}\sinh ax + C$

105. $\int x^n \sinh ax\,\mathrm{d}x = \frac{x^n}{a}\cosh ax - \frac{n}{a}\int x^{n-1}\cosh ax\,\mathrm{d}x$

106. $\int \cosh ax\,\mathrm{d}x = \frac{1}{a}\sinh ax + C$

107. $\int \cosh^2 ax\,\mathrm{d}x = \frac{\sinh 2ax}{4a} + \frac{x}{2} + C$

108. $\int \cosh^n ax\,\mathrm{d}x = \frac{\cosh^{n-1}ax\sinh ax}{na} + \frac{n-1}{n}\int \cosh^{n-2}ax\,\mathrm{d}x, \quad n \neq 0$

109. $\int x\cosh ax\,\mathrm{d}x = \frac{x}{a}\sinh ax - \frac{1}{a^2}\cosh ax + C$

110. $\int x^n \cosh ax\,\mathrm{d}x = \frac{x^n}{a}\sinh ax - \frac{n}{a}\int x^{n-1}\sinh ax\,\mathrm{d}x$

111. $\int \mathrm{e}^{ax}\sinh bx\,\mathrm{d}x = \frac{\mathrm{e}^{ax}}{2}\left[\frac{\mathrm{e}^{bx}}{a+b} - \frac{\mathrm{e}^{-bx}}{a-b}\right] + C, \quad a^2 \neq b^2$

112. $\int \mathrm{e}^{ax}\cosh bx\,\mathrm{d}x = \frac{\mathrm{e}^{ax}}{2}\left[\frac{\mathrm{e}^{bx}}{a+b} + \frac{\mathrm{e}^{-bx}}{a-b}\right] + C, \quad a^2 \neq b^2$

113. $\int \tanh ax\,\mathrm{d}x = \frac{1}{a}\ln(\cosh ax) + C$

114. $\int \tanh^2 ax\,\mathrm{d}x = x - \frac{1}{a}\tanh ax + C$

115. $\int \tanh^n ax\,\mathrm{d}x = -\frac{\tanh^{n-1}ax}{(n-1)a} + \int \tanh^{n-2}ax\,\mathrm{d}x, \quad n \neq 1$

116. $\int \coth ax \, dx = \frac{1}{a} \ln |\sinh ax| + C$

117. $\int \coth^2 ax \, dx = x - \frac{1}{a} \coth ax + C$

118. $\int \coth^n ax \, dx = -\frac{\coth^{n-1} ax}{(n-1)a} + \int \coth^{n-2} ax \, dx, \quad n \neq 1$

119. $\int \operatorname{sech} ax \, dx = \frac{1}{a} \arcsin(\tanh ax) + C$

120. $\int \operatorname{sech}^2 ax \, dx = \frac{1}{a} \tanh ax + C$

121. $\int \operatorname{csch} ax \, dx = \frac{1}{a} \ln \left| \tanh \frac{ax}{2} \right| + C$

122. $\int \operatorname{csch}^2 ax \, dx = -\frac{1}{a} \coth ax + C$

Index